South-Central Section
of the
Geological Society of America

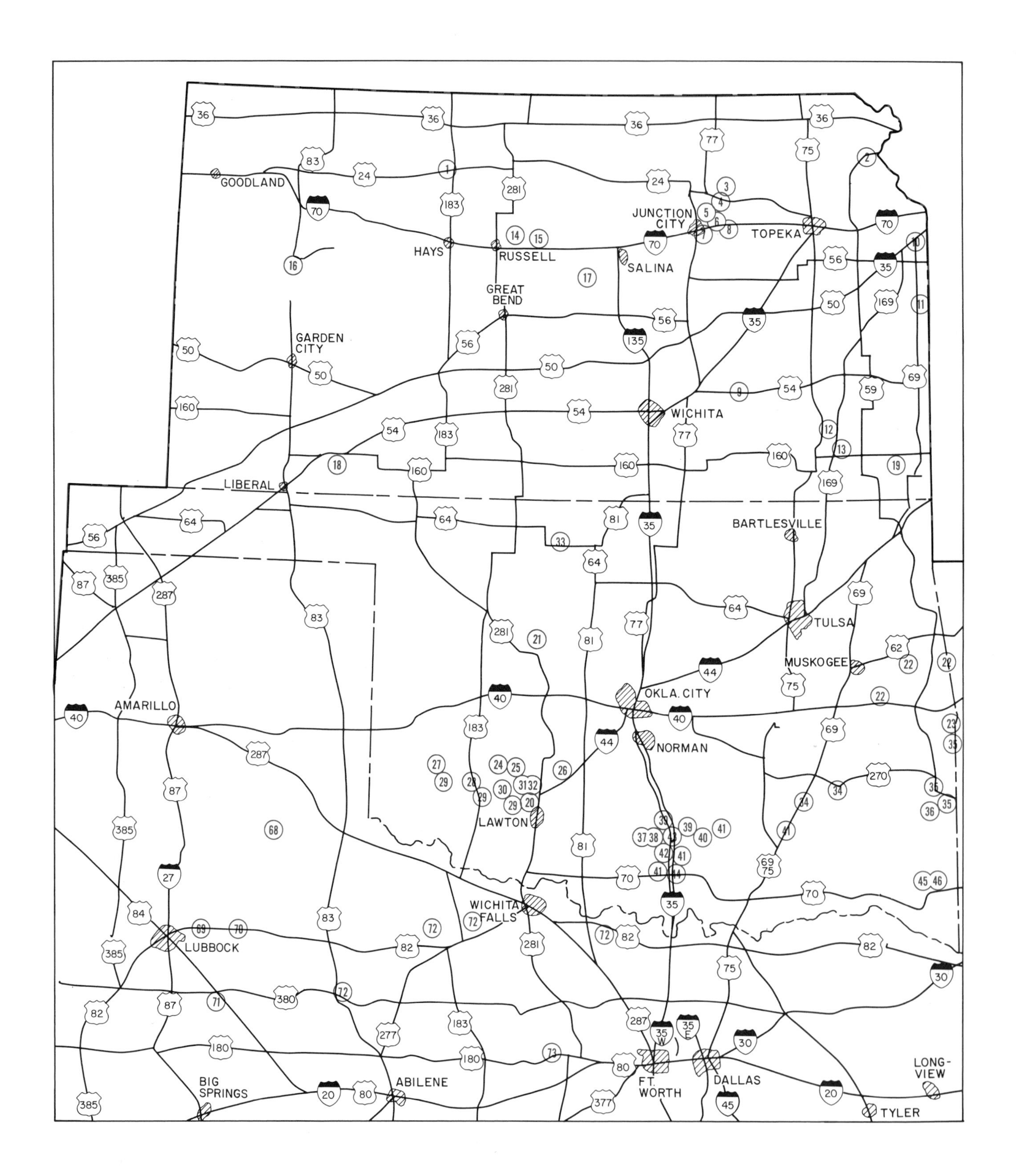
GOODLAND
HAYS
RUSSELL
JUNCTION CITY
TOPEKA
SALINA
GREAT BEND
GARDEN CITY
WICHITA
LIBERAL
BARTLESVILLE
TULSA
MUSKOGEE
OKLA. CITY
NORMAN
AMARILLO
LAWTON
WICHITA FALLS
LUBBOCK
BIG SPRINGS
ABILENE
FT. WORTH
DALLAS
LONG-VIEW
TYLER

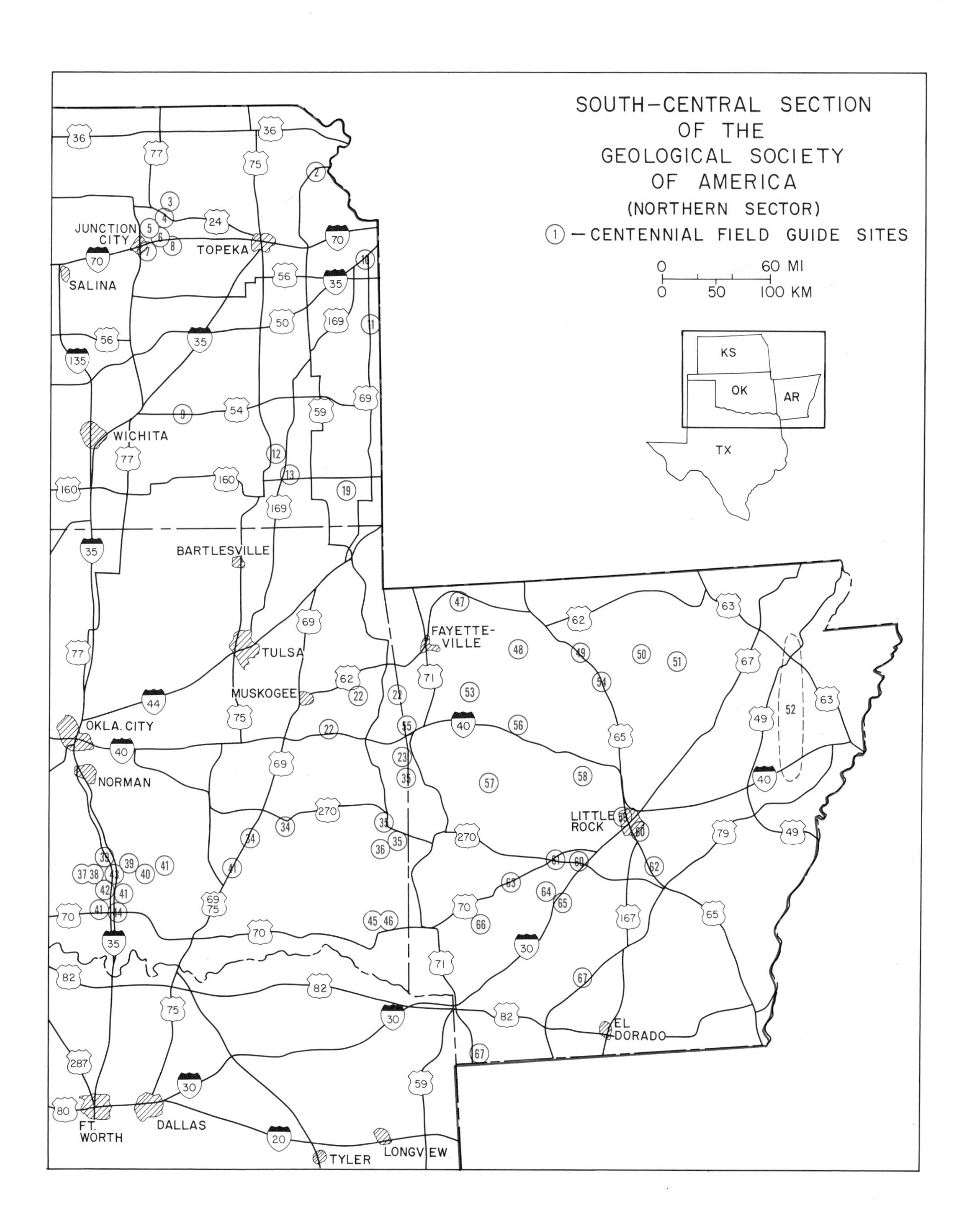

SOUTH–CENTRAL SECTION
OF THE
GEOLOGICAL SOCIETY
OF AMERICA
(NORTHERN SECTOR)
1 — CENTENNIAL FIELD GUIDE SITES
0 60 MI
0 50 100 KM
KS
OK
AR
TX
JUNCTION CITY
TOPEKA
SALINA
WICHITA
BARTLESVILLE
TULSA
MUSKOGEE
OKLA. CITY
NORMAN
FAYETTE-VILLE
LITTLE ROCK
EL DORADO
FT. WORTH
DALLAS
TYLER
LONGVIEW

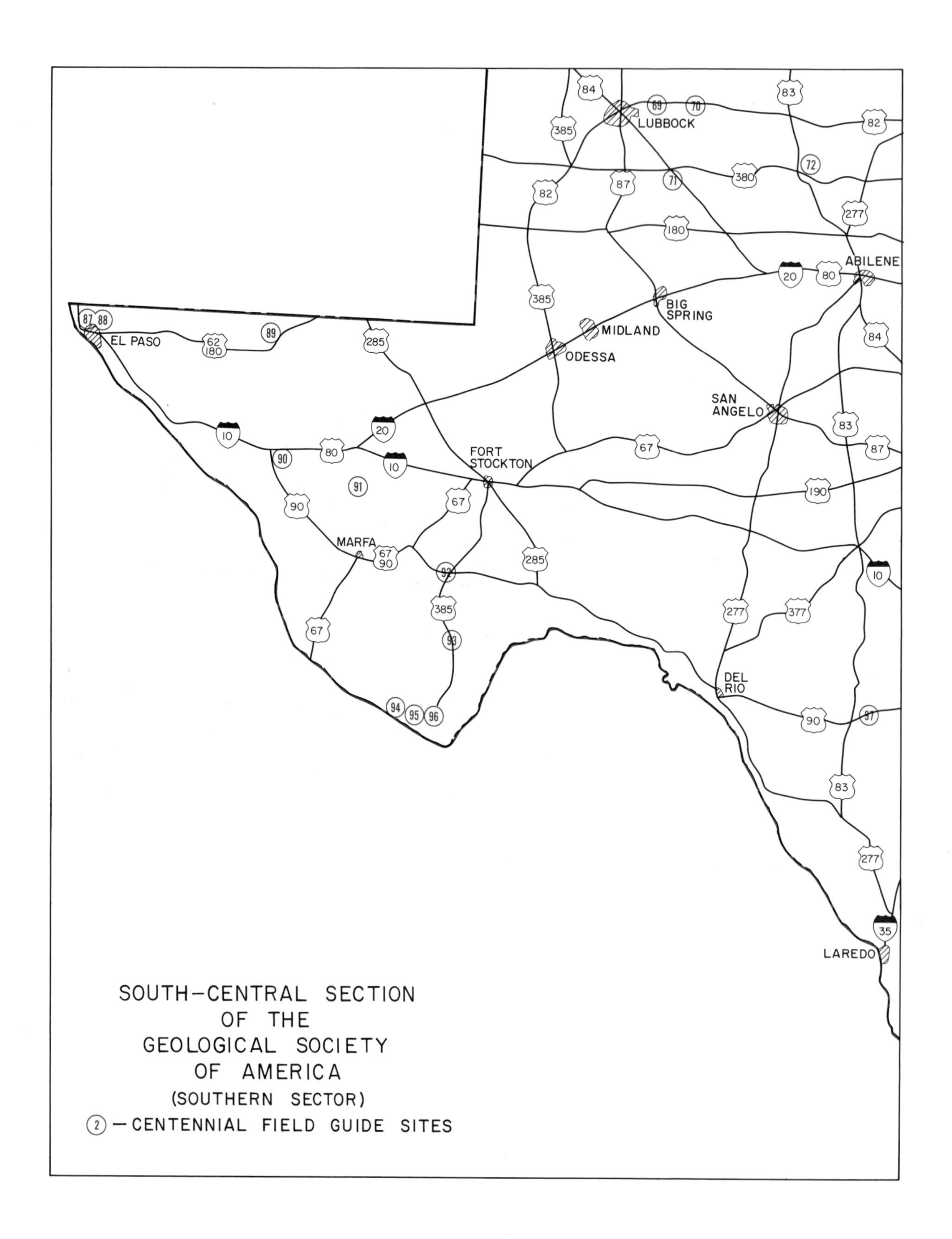
LUBBOCK
ABILENE
BIG SPRING
MIDLAND
ODESSA
EL PASO
SAN ANGELO
FORT STOCKTON
MARFA
DEL RIO
LAREDO
SOUTH–CENTRAL SECTION
OF THE
GEOLOGICAL SOCIETY
OF AMERICA
(SOUTHERN SECTOR)
(2) — CENTENNIAL FIELD GUIDE SITES

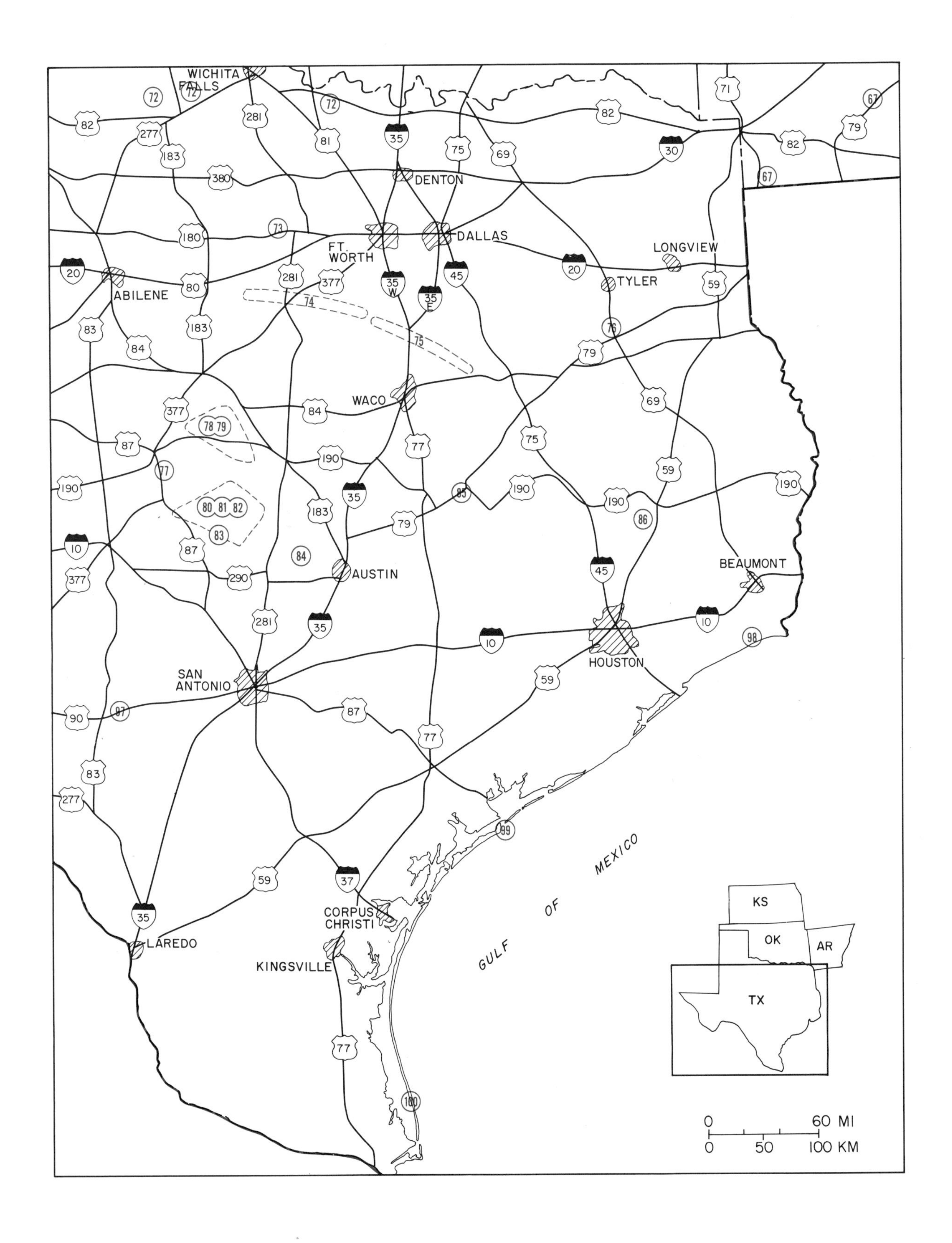
WICHITA FALLS
DENTON
DALLAS
FT. WORTH
LONGVIEW
TYLER
ABILENE
WACO
AUSTIN
BEAUMONT
HOUSTON
SAN ANTONIO
CORPUS CHRISTI
KINGSVILLE
LAREDO
GULF OF MEXICO
KS
OK
AR
TX
0 60 MI
0 50 100 KM

Centennial Field Guide Volume 4

South-Central Section of the Geological Society of America

Edited by

O. T. Hayward
Geology Department
Baylor University
Waco, Texas 76798

1988

Acknowledgment

Publication of this volume, one of the Centennial Field Guide Volumes of *The Decade of North American Geology Project* series, has been made possible by members and friends of the Geological Society of America, corporations, and government agencies through contributions to the Decade of North American Geology fund of the Geological Society of America Foundation.

Following is a list of individuals, corporations, and government agencies giving and/or pledging more than $50,000 in support of the DNAG Project:

Amoco Production Company
ARCO Exploration Company
Chevron Corporation
Diamond Shamrock Exploration Corporation
Exxon Production Research Company
Getty Oil Company
Gulf Oil Exploration and Production Company
Paul V. Hoovler
Kennecott Minerals Company
Kerr McGee Corporation
Marathon Oil Company
McMoRan Oil and Gas Company
Mobil Oil Corporation
Pennzoil Exploration and Production Company
Phillips Petroleum Company
Shell Oil Company
Caswell Silver
Standard Oil Production Company
Sun Exploration and Production Company
Superior Oil Company
Tenneco Oil Company
Texaco, Inc.
Union Oil Company of California
Union Pacific Corporation and its operating companies:
 Union Pacific Resources Company
 Union Pacific Railroad Company
 Upland Industries Corporation
U.S. Department of Energy

Published by The Geological Society of America, Inc.
3300 Penrose Place, P.O. Box 9140, Boulder, Colorado 80301

Printed in U.S.A.

Front Cover: Lampasas Cut Plain, central Texas. The landscape, developed on rocks of the Fredericksburg Group (Comanchean, Cretaceous), is typical of landscapes developed on a variety of Late Paleozoic and Mesozoic rocks over broad areas of Texas, Oklahoma, and Kansas. Photo by O. T. Hayward.

Library of Congress Cataloging-in-Publication Data
(Revised for vol. 4)

Centennial field guide.

"Prepared under the auspices of the regional Sections of the Geological Society of America as a part of the Decade of North American Geology (DNAG) Project"—Vol. 6, pref.
Vols. 1–   : maps on lining papers.
Includes bibliographies and indexes.
Contents: v. 1. Cordilleran Section of the Geological Society of America / edited by Mason L. Hill—v. 2. Rocky Mountain Section of the Geological Society of America / edited by Stanley S. Beus— — v. 4. South-central Section of the Geological Society of America / edited by O. T. Hayward.
1. Geology—United States—Guide-books. 2. Geology—Canada—Guide-books. 3. United States—Description and travel—1981–   —Guide-books. 4. Canada—Description and travel—1981–   Guide-books. 5. Decade of North American Geology Project. I. Geological Society of America.
QE77.C46 557.3 86-11986
ISBN 0-8137-5401-1 (v. 1)

Contents

Preface xvii

Foreword xix

1. *Exposures of the Fort Hays Limestone Member, Niobrara Chalk (Upper Cretaceous) near Stockton, Kansas* 1
   Alan P. Laferriere

2. *West Atchison Drift Section* 5
   James S. Aber

3. *Spillway fault system, Tuttle Creek reservoir, Pottawatomie County, northeastern Kansas* 11
   James R. Underwood, Jr., and Allyn Polson

4, 5. *Minor T-R units in the Lower Permian Chase Group, northeast Kansas* 17
   Richard M. Busch

6. *Barneston Limestone (Lower Permian) of eastern Kansas* 25
   Page C. Twiss and James R. Underwood, Jr.

7. *The Crouse Limestone in north-central Kansas; Lateral changes and paleoecological analysis* 29
   R. R. West and P. C. Twiss

8, 9. *Beattie Limestone (Lower Permian) of eastern Kansas* 35
   Page C. Twiss

10, 11, 12, 13. *Classic "Kansas" cyclothems* 43
   Philip H. Heckel

14. *Bunker Hill section of Upper Cretaceous rocks in Kansas* 57
   Donald E. Hattin and Page C. Twiss

15. *Rocktown channel in Dakota Formation of central Kansas* 61
   Page C. Twiss

16. ***Ogallala Formation (Miocene), western Kansas*** 63
Edwin D. Gutentag

17. ***Lower Cretaceous Kiowa Formation; The eastern margin of the Western Interior Seaway*** 67
Robert W. Scott

18. ***Plio-Pleistocene rocks, Borchers Badlands, Meade County, southwestern Kansas*** 69
Richard J. Zakrzewski

19. ***Mine 19 geologic section, Pittsburg and Midway Coal Mining Company, Cherokee County, Kansas*** 75
Lawrence L. Brady

20. ***The Meers fault scarp, southwestern Oklahoma*** 79
R. N. Donovan

21. ***Evaporites and red beds in Roman Nose State Park, northwest Oklahoma*** 83
Kenneth S. Johnson

22. ***Carbonate platform facies of the Morrowan Series (Lower Pennsylvanian), northeastern Oklahoma and northwestern Arkansas*** 85
Patrick K. Sutherland and Walter L. Manger

23. ***Deltaic facies of the Hartshorne Sandstone in the Arkoma Basin, Arkansas-Oklahoma border*** 91
David W. Houseknecht

24. ***Carlton rhyolite and lower Paleozoic sedimentary rocks at Bally Mountain in the Slick Hills of southwestern Oklahoma*** 93
R. N. Donovan, D. Ragland, K. Cloyd, S. Bridges, and R. E. Denison

25. ***Some aspects of the geology of Zodletone Mountain, southwestern Oklahoma*** 99
R. N. Donovan, P. Younger, and C. Ditzell

26. ***Hydrocarbon-induced diagenetic aureole at Cement-Chickasha Anticline, Oklahoma*** 103
Zuhair Al-Shaieb

27, 28, 29, 30. ***Igneous geology of the Wichita Mountains, southwestern Oklahoma*** 109
M. C. Gilbert and B. N. Powell

31, 32. ***Pennsylvanian deformation and Cambro-Ordovician sedimentation in the Blue Creek Canyon, Slick Hills, southwestern Oklahoma*** 127
R. N. Donovan, D. Ragland, M. Rafalowski, D. McConnell, W. Beauchamp, W. R. Marcini, and D. J. Sanderson

33. ***Great Salt Plains and hourglass selenite crystals, Salt Fork of the Arkansas River, northwest Oklahoma*** ... 135
Kenneth S. Johnson

34. ***Shelf to slope facies of the Wapanucka Formation (lower-Middle Pennsylvanian), frontal Ouachita Mountains, Oklahoma*** ... 139
Robert C. Grayson, Jr., and Patrick K. Sutherland

35. ***Depositional and deformational characteristics of the Atoka Formation, Arkoma Basin, and Ouachita frontal thrust belt, Oklahoma*** ... 145
David W. Houseknecht and Michael B. Underwood

36. ***Carboniferous flysch, Ouachita Mountains, southeastern Oklahoma; Big Cedar–Kiamichi Mountain section*** ... 149
R. J. Moiola, G. Briggs, and G. Shanmugam

37, 38. ***Turner Falls Park; Pleistocene tufa and travertine and Ordovician platform carbonates, Arbuckle Mountains, southern Oklahoma*** ... 153
R. N. Donovan, D. A. Ragland, and D. Schaefer

39. ***Pennsylvanian conglomerates in the Arbuckle Mountains, southern Oklahoma*** ... 159
R. N. Donovan and W. D. Heinlen

40. ***Basal sandstone of the Oil Creek Formation in the quarry of the Pennsylvania Glass Sand Corporation, Johnson County, Oklahoma*** ... 165
J. G. McPherson, R. E. Denison, D. W. Kirkland, and D. M. Summers

41. ***Middle Ordovician strata of the Arbuckle and Ouachita Mountains, Oklahoma; Contrasting lithofacies and biofacies deposited in southern Oklahoma Aulacogen and Ouachita Geosyncline*** ... 171
Stanley C. Finney

42. ***Structural styles in the Arbuckle Mountains, southern Oklahoma*** ... 177
J. Bryan Tapp

43. ***I-35 Roadcuts; Geology of Paleozoic strata in the Arbuckle Mountains of southern Oklahoma*** ... 183
Robert O. Fay

44. ***Lithostratigraphy and depositional environments of the Springer and lower Golf Course Formations, Ardmore Basin, Oklahoma*** ... 189
Frederick B. Meek, R. Douglas Elmore, and Patrick K. Sutherland

45, 46. ***Beavers Bend State Park, Broken Bow Uplift, Oklahoma*** ... 195
K. C. Nielsen

47. ***Beaver Dam, northwestern Arkansas*** 203
Walter L. Manger

48. ***The Paleozoic rocks of the Ponca region, Buffalo National River, Arkansas*** 207
John David McFarland III

49. ***Geology of the Buffalo River Valley in the vicinity of U.S. 65, Arkansas Ozarks*** 211
William W. Craig

50. ***Post-St. Peter Ordovician strata in the vicinity of Allison, Stone County, Arkansas*** 215
William W. Craig and Michael J. Deliz

51. ***Silurian rocks of northern Arkansas*** 221
O. A. Wise and W. M. Caplan

52. ***Crowley's Ridge, Arkansas*** 225
M. J. Guccione, W. L. Prior, and E. M. Rutledge

53. ***Atokan stratigraphy of the Cherry Bend area, northwestern Arkansas*** 231
Gary D. Harris, Dean A. Ramsey, and Doy L. Zachry

54. ***Peyton Creek Road Cut, northern Arkansas*** 235
Walter L. Manger

55. ***Atoka Formation in the Lee Creek area, Arkansas*** 239
J. D. McFarland III

56. ***Horsehead Lake spillway, Arkansas*** 241
J. D. McFarland III

57. ***Blue Mountain Dam and Magazine Mountain, Arkansas*** 243
Richard R. Cohoon and Victor K. Vere

58. ***Late Paleozoic rocks, eastern Arkoma basin, central Arkansas Valley Province*** 249
P. L. Kehler

59. ***I-430 bypass, Little Rock, Arkansas*** 255
J. D. McFarland III

60. ***Igneous rocks at Granite Mountain and Magnet Cove, Arkansas*** 259
J. Michael Howard and Kenneth F. Steele

61. ***Geological features at Hot Springs, Arkansas*** 263
J. D. McFarland III

62. ***The Eocene Jackson Group at White Bluff, Arkansas*** 265
George W. Colton and William V. Bush

63. ***Lower Stanley Shale and Arkansas Novaculite, western Mazarn Basin and Caddo Gap, Ouachita Mountains, Arkansas*** 267
Jay Zimmerman and James Timothy Ford

64. ***Turbidite exposures near DeGray Lake, southwestern Arkansas*** 273
J. D. McFarland III

65. ***The upper Jackfork Section, Mile Post 81, I-30, Arkadelphia, Arkansas*** 277
Denise M. Stone, David N. Lumsden, and Charles G. Stone

66. ***The DeQueen Formation at the old Highland Quarry, Arkansas*** 281
J. D. McFarland III

67. ***The Claiborne Group in southwest Arkansas*** 283
William Lee Prior and Quin Baber

68. ***The Lingos formation, western Rolling Plains of Texas*** 287
S. Christopher Caran and Robert W. Baumgardner, Jr.

69. ***Late Pleistocene and Holocene stratigraphy, Southern High Plains of Texas*** 293
Vance T. Holliday

70. ***Ogallala and post-Ogallala sediments of the Southern High Plains, Blanco Canyon and Mt. Blanco, Texas*** 299
Paul N. Dolliver and Vance T. Holliday

71. ***The Triassic section of the West Texas High Plains*** 305
Ted Gawloski

72. ***Permian strata of North-Central Texas*** 309
James O. Jones and Tucker F. Hentz

73. ***Middle and late Pennsylvanian rocks, North-Central Texas*** 317
E. L. "Jack" Trice and Robert C. Grayson, Jr.

74. ***The Comanchean Section of the Trinity shelf, central Texas*** 323
O. T. Hayward

75. ***Gulfian rocks, western margin of the East Texas Basin*** 329
O. T. Hayward

76. ***The Claiborne Group of East Texas*** 335
Patricia S. Sharp and Austin A. Sartin

77. ***Cambrian algal reefs of the upper Wilberns Formation, central Texas, The Camp San Saba locality*** 339
Wayne M. Ahr

78, 79. ***Middle and Upper Pennsylvanian (Atokan-Missourian) strata in the Colorado River Valley of Central Texas*** 343
Robert C. Grayson, Jr., and E. L. "Jack" Trice

80, 81. ***Paleozoic strata of the Llano region, Central Texas*** 351
Robert S. Kier

82. ***The Precambrian of Central Texas*** 361
Virgil E. Barnes

83. ***Enchanted Rock dome, Llano and Gillespie counties, Texas*** 369
Robert M. Hutchinson

84. ***The middle Comanchean section of Central Texas*** 373
David L. Amsbury

85. ***The lower Tertiary of the Texas Gulf Coast*** 377
Thomas E. Yancey and Elizabeth S. Yancey

86. ***The Catahoula Formation; A volcaniclastic unit in east Texas*** 383
Ernest B. Ledger

87, 88. ***Precambrian and Paleozoic stratigraphy; Franklin Mountains, west Texas*** 387
David V. LeMone

89. ***The southern Guadalupe Mountains, Texas; Permian stratigraphy and Great Plains/Basin and Range structural transition*** 395
Cleavy L. McKnight

90. ***Lower Cretaceous of western Trans-Pecos Texas*** 401
David L. Amsbury and Donald F. Reaser

91. ***The Davis Mountains volcanic field, west Texas*** 407
Don F. Parker

92. ***Geology of the Marathon Uplift, west Texas*** 411
Earle F. McBride

93. ***Persimmon Gap in Big Bend National Park, Texas; Ouachita facies and Cretaceous cover deformed in a Laramide overthrust*** 417
Peter R. Tauvers and William R. Muehlberger

94. ***Basal Gulfian and Comanchean Section, Anguila Fault Zone and Santa Elena Canyon, Big Bend National Park, Trans-Pecos Texas, and Chihuahua, Mexico*** 423
J. B. Stevens

95. ***Mid-Tertiary and Pleistocene sections, Sotol Vista to Cerro Castellan, Big Bend National Park, southwestern Brewster County, Trans-Pecos Texas*** 429
James B. Stevens

96. ***Dikes in Big Bend National Park; Petrologic and tectonic significance*** 435
Jonathan G. Price and Christopher D. Henry

97. ***The Anacacho Limestone of southwest Texas*** 441
R. W. Rodgers

98, 99, 100. ***Late Quaternary geology of the Texas coastal plain*** 445
Robert A. Morton

Index 459

TOPICAL CROSS-REFERENCES FOR FIELD GUIDE SITES

Topic			Arkansas	Kansas	Oklahoma	Texas
Geomorphology			52, 57, 58		24, 29, 30, 31, 33	74, 75, 95
Igneous Geology			60		24, 27, 28, 29, 30, 31	82, 83, 87, 91, 95, 96
Metamorphic Geology						82, 87
Sedimentary Geology	Diagenesis				26	
Sedimentary Geology	Stratigraphy	Quaternary	52, 61	2, 18	24, 37	68, 69, 98, 99, 100
		Tertiary	52, 62, 67	16		70, 76, 85, 86
		Cretaceous/ Tertiary Boundary				85
		Cretaceous	66	1, 14, 15, 17		74, 75, 84, 90, 94, 97
		Triassic				71
		Permian		3, 4, 5, 6, 7, 8, 9	21, 30	72, 89
		Pennsylvanian	22, 23, 48, 53, 55, 56, 57, 58, 59, 64, 65	10, 11, 12, 13, 19	22, 34, 35, 36, 39, 43, 44	73, 78, 79, 81, 88, 92
		Mississippian/ Pennsylvanian Boundary	54			
		Mississippian	47, 48, 49, 59, 61, 63		43, 45	81, 88
		Devonian	47		43	88, 92
		Silurian	49, 51		43	88
		Ordovician	47, 48, 49, 50, 59, 61		24, 32, 38, 40, 41, 43, 45	80, 88, 92
		Cambrian			24, 25, 32, 43	77, 80
Structural Geology			63	3	20, 31, 35, 42, 46	92, 93, 95

Preface

This volume is one of a six-volume set of Centennial Field Guides prepared under the auspices of the regional Sections of the Society as a part of the Decade of North American Geology (DNAG) Project. The intent of this volume is to highlight, for the geologic traveler and for students and professional geologists interested in major geologic features of regional significance, 100 of the best and most accessible geologic localities in the area of the South-Central Section. The leadership provided by the editor, O. T. Hayward, and the support provided to him by the South-Central Section of the Geological Society of America and the Department of Geology of Baylor University are greatly appreciated.

Drafting services were offered by the DNAG Project to those authors of field guide texts who did not have access to drafting facilities. Particular thanks are given here to Ms. Karen Canfield of Louisville, Colorado, who prepared final drafted copy of many figures from materials provided by the authors.

In addition to Centennial Field Guides, the DNAG Project includes a 29 volume set of syntheses that constitute *The Geology of North America,* and 8 wall maps at a scale of 1:5,000,000 that summarize the geology, tectonics, magnetic and gravity anomaly patterns, regional stress fields, thermal aspects, seismicity, and neotectonics of North America and its surroundings. Together, the synthesis volumes and maps are the first coordinated effort to integrate all available knowledge about the geology and geophysics of a crustal plate on a regional scale. They are supplemented, as a part of the DNAG project, by 23 Continent–Ocean Transects providing strip maps and both geologic and tectonic cross-sections strategically sited around the margins of the continent, and by several related topical volumes.

The products of the DNAG Project have been prepared as a part of the celebration of the Centennial of the Geological Society of America. They present the state of knowledge of the geology and geophysics of North America in the 1980s, and they point the way toward work to be done in the decades ahead.

Allison R. Palmer
Centennial Science Program Coordinator

Foreword

Miracles happen!

This Field Guide is finished! While it has been a long time coming, it has been worth the wait. Even now it is not exactly what we had wanted, but it comes very close to being the "Field Guide to the South-Central Section"—Kansas, Oklahoma, Arkansas, Texas—that we had visualized.

The original guidelines from the DNAG Committee specified that our volume was to consist of 100 of the "best" outcrop sites in our four-state area. It left the definition of "best" up to our local committee. Each site was to be limited to four text pages of description, with the provision for a limited number of multiple sites for those geological features which could not effectively be limited to four pages. It recommended that in our selection of localities we acknowledge the great breadth of geologic interest. Finally, it recognized that a major problem for our treatment of the South-Central Section as a region is that it is divided on political boundaries, not geological ones. It includes pieces of geological provinces which transect political boundaries into other Sections, so it is often difficult to describe major geologic subdivisions of the continent while confined to the political boundaries of the Section.

To these guidelines from the DNAG Committee, the South-Central Field Guide Committee added its own. We wanted the sites to describe the regional geology of the South-Central Section to the degree that 10 sites could describe the geology of 500,000 square miles (1,300,000 square kilometers). We defined "best" (as in "best localities") to mean the most descriptive and representative sites to illustrate major geologic subdivisions of the region. Where thick shale sections are indeed the characteristic of a geologic province, then the "best" site may be remarkably uninspiring, and yet it does represent the province.

In many cases, we elected to present major geologic sections, consisting of numbers of outcrops, as single numbered sites. Only in this way could we convey the magnitude and complexity of the geological provinces which make up the South-Central Section. To facilitate access to chosen sites, and to protect both landowners and geologists from the mutual problems of trespass, we attempted to select localities on public property wherever possible. To a significant degree we managed to do all this. We were not completely successful (as the distribution of sites indicates) but we came close. This is a "Field Guide to the South-Central Section." Authors did a remarkable job of staying within the guidelines while at the same time accomplishing the major purposes of the volume. It goes beyond description, tying what we know locally to what we interpret regionally, even on the grand scale.

However it is probably almost as interesting for what it tells us about ourselves as for what it says about the region. As is evident from the distribution of sites, geologists from the South-Central Section still tend to rush to the hills and breaks to do their geology. We apparently need to see those rocks in near vertical faces, and near vertical faces are highly localized in the South-Central Section. As the evidence of site treatment suggests, most of us think of geology as stratigraphy; a very few recognize the element of structure; and it is a rare individual who sees the landscape. Yet the most conspicuous of field geological elements will always be landscape. Stratigraphy, as represented by outcrops with near vertical faces, will always be the least represented in that landscape.

To a major degree we have become specialists, though we yet recognize broad regional problems. Most of us (even the most specialized) have broad regional problems sequestered in our minds, and these tend to be those about which we can become almost poetic. All these tendencies appear to be the products of changing lifestyles; the need to publish (big problems take BIG time); the availability of funding (research tends to be directed by money); time (field work has become an afternoon and weekend activity—the long field seasons of our first half-century are largely gone now); and there is even a different perspective about what constitutes acceptable field accommodations (a fly-blown garbage-dish dinner shared with cows on a 100° night at a West Texas water hole lacks the appeal it once had).

Because of constraints on our time we tend to return to familiar outcrops. We need to know that when we arrive, those rocks will be there. We have less time to hunt. However, as the blank spaces on the index map indicate, there are very large areas with very large problems, mostly ignored in this overall effort. For those of us now looking for that grand plan that will occupy the better part of a professional lifetime, it is these blank spaces which offer the greatest potential. Many of these areas are better known in the subsurface than on the surface, but true to form, subsurface geology rarely rises above the bottom of the surface casing. Field geology, and all it implies, is still there, waiting.

In those blank spaces the landscapes are typically subdued; structure is most notable for its simplicity and scale; the problems are inconspicuous and monumental. In effect, our omissions from this volume have handed us keys to lifetimes of satisfying effort. Despite the coverage and quality of the sites described in this book, one of its principal contributions to the future may be what it does **not** tell us about the field geology of the South-Central Section. In planning for the second-century revision of this volume, it is our intention to concentrate on the blank spaces so conspicuous on the index map. If you wish to be included, you need now to start setting aside those prairie areas of red muds which you plan to submit as sites for the 2088 edition.

Finally, we are proud of this book. Even with the omissions it is remarkably informative. While normally it is customary for the editors to accept responsibility for errors in their publications, in this case this will not be necessary.

As we have indicated, we have exceptional authors. They can take responsibility for their own work. Call them.

The Editorial Committee
O. T. Hayward, Chairman
James O. Jones
Kenneth S. Johnson
John D. McFarland
Page C. Twiss

Exposures of the Fort Hays Limestone Member, Niobrara Chalk (Upper Cretaceous) near Stockton, Kansas

***Alan P. Laferriere**, Department of Geology, Indiana University, Bloomington, Indiana 47405*

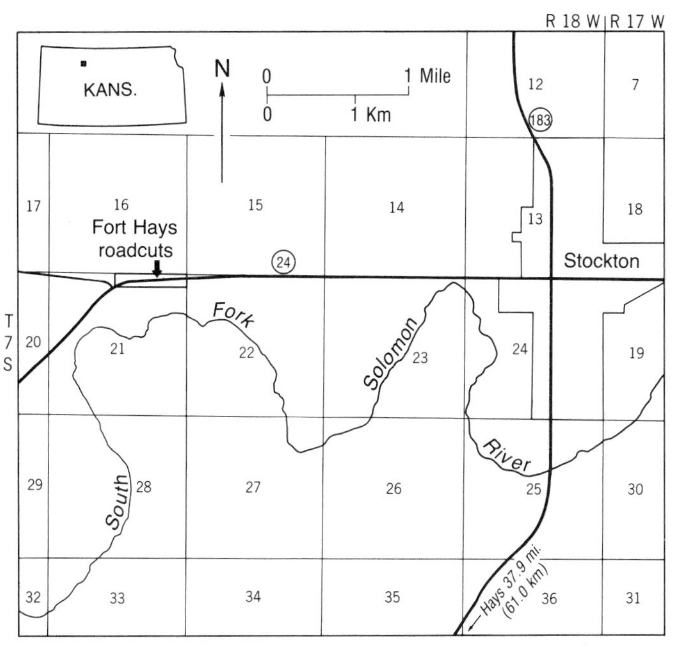

Figure 1. Location of Fort Hays outcrops west of Stockton, Kansas.

Figure 2. Exposure of Fort Hays Member west of Stockton, Kansas, showing rhythmically interbedded limestone and calcareous shale.

LOCATION

A complete section of the Fort Hays Limestone Member occurs within roadcuts along Kansas 24, 2.4 mi (3.8 km) west of Stockton (Stockton 7½-minute Quadrangle) in NE¼Sec.21, T.7S.,R.18W., Rooks County, Kansas (Fig. 1). At this locality, a series of three roadcuts occurs within a 0.25-mi (0.4 km) stretch of highway. Stratigraphically higher beds of the Fort Hays crop out at the east end of the exposures. By utilizing all three outcrops, most beds of the unit can be examined near road level with a minimum of climbing.

SIGNIFICANCE

The Fort Hays Limestone Member provides an excellent example of epicontinental pelagic carbonate deposition that occurred during one of the greatest Cretaceous marine transgressions in North America. Geographically widespread, rhythmically interbedded shales and limestones within the Fort Hays yield valuable information regarding Cretaceous climate, tectonics, and sea-level changes in the U.S. Western Interior region.

DESCRIPTION

Williston (1897) defined the Upper Turonian–Coniacian Fort Hays Limestone Member as the basal member of the Niobrara Formation. It is named for outcrops along streams and within badlands near Hays, Kansas. The Fort Hays Limestone member comprises a sequence of massive light olive-gray to yellowish-gray limestone beds that alternate cyclically with beds of olive gray calcareous shale or marlstone (Figs. 2, 3). Limestone beds are predominantly biomicrites and biomicrosparites that contain about 5 percent bivalve remains and 10 percent pelagic foraminiferal test fragments in a coccolith-rich matrix. Macrofossils found in outcrop consist mainly of a low-diversity assemblage of bivalves of the genus *Inoceramus,* which are commonly encrusted by small oysters. Trace fossils are abundant in limestone beds and include *Planolites, Chondrites, Teichichnus, Thalassinoides,* and *Trichichnus.* During limestone deposition, progressive early lithification of relatively pure carbonate mud resulted in increased substrate firmness and induced a succession of burrow structures, culminating in open dwelling structures (Hattin, 1986). These dwelling structures are now preserved as mineral-filled burrows. A thorough description of trace and body fossils of the Fort Hays in Kansas is provided by Frey (1970, 1972).

Limestones are thoroughly homogenized by bioturbation and display few physical sedimentary structures. Conversely, laminations and thin bentonite seams are common in many shale beds where bioturbation was less pervasive. In Kansas, most shaly intervals are thin, and many occur only as partings between massive limestone beds (Fig. 3). Shale and marlstone beds

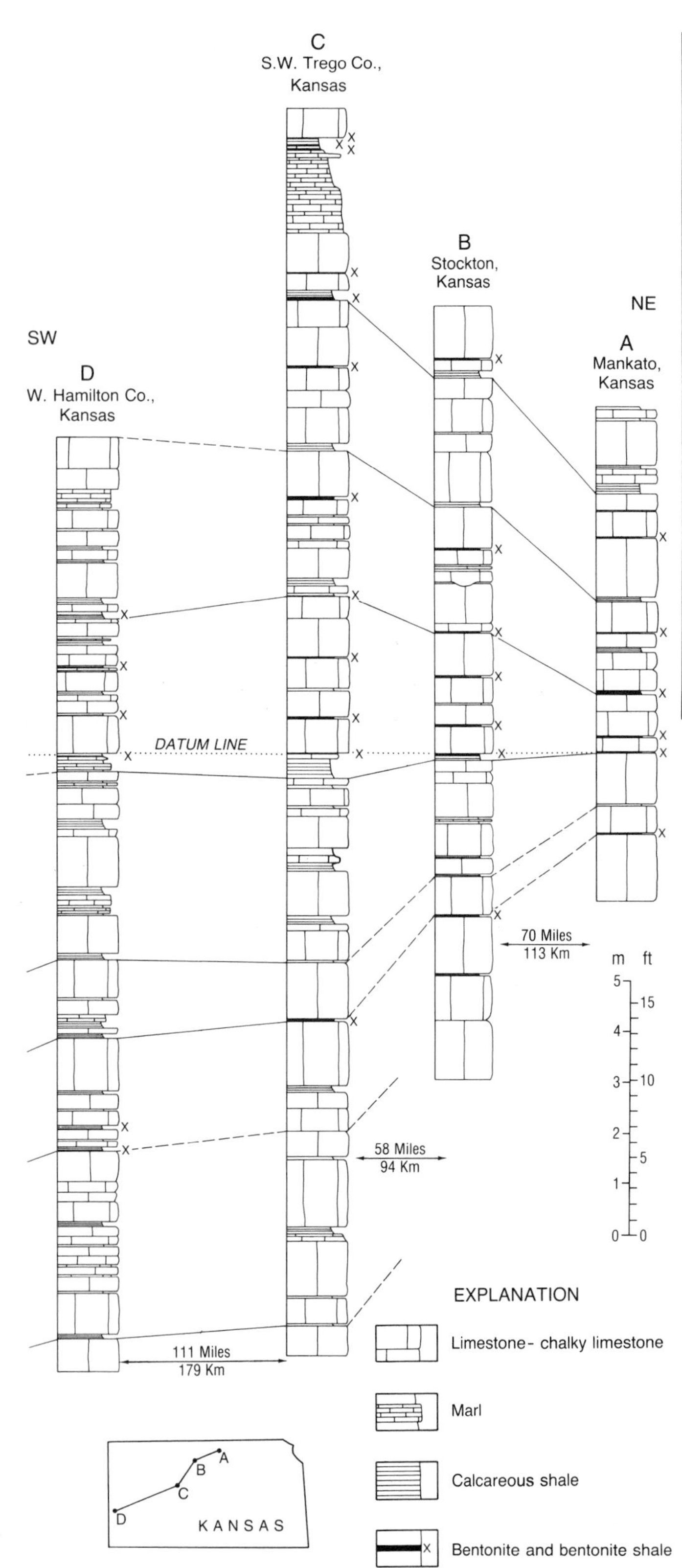

Figure 3. Graphic sections of Fort Hays Member along transect from north-central to western Kansas. Localities: A, SE¼Sec.16,T.3S.,R.7W., Jewell County. B, NE¼Sec.21,T.7S.,R.18W., Rooks County (Stockton locality). C, SW¼Sec.24,NW¼Sec.25,T.14S.,R.25W., Trego County. D, NW¼Sec.25,T.22S.,R.42W., Hamilton County.

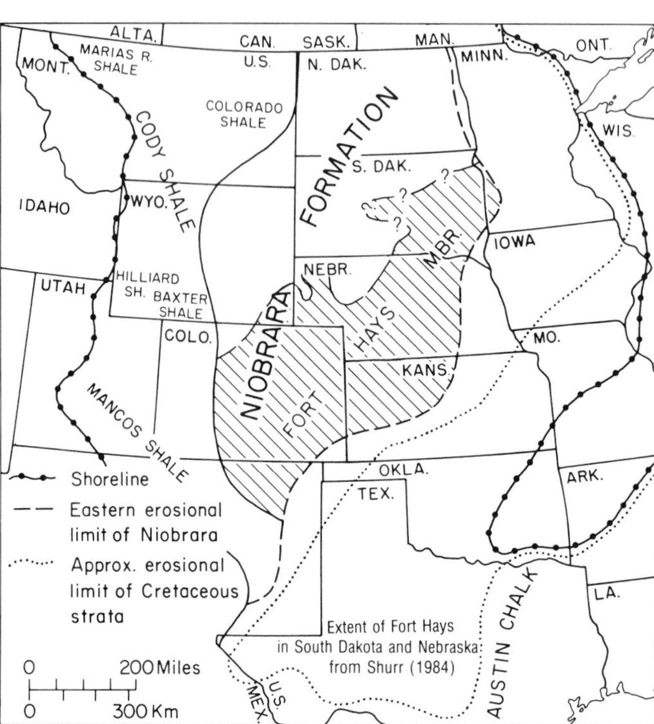

Figure 4. Distribution of Niobrara Formation and Fort Hays Member. Extent of Fort Hays in South Dakota and Nebraska from Shurr (1984).

thicken in a westward direction, closer to the major source of siliciclastic sediments. Fort Hays deposition occurred over a large geographic area (Fig. 4), and individual shale-limestone couplets can be correlated more than 500 mi (800 km), throughout much of Kansas (Fig. 3) and Colorado (Laferriere and others, 1987). Evidence from widespread bentonite seams demonstrates that Fort Hays depositional cycles were regionally synchronous events and individual beds are time-parallel.

In Kansas, the Fort Hays unconformably overlies the Codell Sandstone Member of the Carlile Shale (Hattin, 1975). However, at the Stockton locality, the Codell is poorly developed and represented only by 0.4 ft (12.0 cm) of fine-grained sandstone underlain by olive-gray siltstone and dark gray shale of undetermined thickness. This unconformity, which involves a northeastwardly widening lacuna, denotes the boundary between the Greenhorn and Niobrara cyclothems, two major transgressive-regressive depositional events that occurred in the U.S. Western Interior during Cretaceous time (Weimer, 1960; Kauffman, 1977). The Fort Hays is conformably overlain by the Smoky Hill Chalk Member of the Niobrara Chalk (Hattin, 1982).

REGIONAL SETTING

Fort Hays sediments were deposited in the Western Interior Sea during the transgressive phase of the Niobrara marine cycle.

During this transgression, rising sea level widened the seaway, influx of siliciclastic sediments to the center of the basin was greatly reduced, and coccolith-rich pelagic ooze became the predominant sediment type in the eastern and central parts of the seaway (Fig. 4). West of the sea, uplifts of the Sevier orogeny furnished most of the terrigenous detritus that reached the basin (Armstrong, 1968). The eastern part of the basin was a relatively stable shelf area that received little detrital material (Reeside, 1957; Hattin, 1975).

Cyclical alternations of shale and marl beds with pelagic limestone beds, such as those observed in the Fort Hays, have been attributed on the basis of geochemical and paleontologic evidence to alternating wet and dry climatic periods (Pratt, 1984). During wet periods, increased influx of siliciclastics diluted pelagic carbonates, and shale or marl deposition occurred. During dry periods, coccolith-rich lime mud accumulated in a well-circulated seaway that received little runoff from terrigenous sediment source areas. Fischer (1980) and Fischer and others (1985) suggested that the climatic oscillations that generated Fort Hays cycles were controlled by Milankovitch-type orbital variations. Regional analysis of bedding cycles in the Fort Hays over much of Kansas and Colorado revealed that these climatically induced bedding patterns were complicated by tectonic effects of the Sevier orogeny, as well as by erosional events associated with sea level changes (Laferriere and others, 1987).

REFERENCES CITED

Armstrong, R. L., 1968, Sevier orogenic belt in Nevada and Utah: Geological Society of America Bulletin, v. 79, p. 429–458.

Fischer, A. G., 1980, Gilbert bedding rhythms and geochronology, *in* Yochelson, E. L., ed., The scientific ideas of G. K., Gilbert: Geological Society of America Special Paper 183, p. 93–104.

Fischer, A. G., Herbert, T., and Premoli Silva, I., 1985, Carbonate bedding cycles in Cretaceous pelagic and hemipelagic sediments, *in* Pratt, L. M., and others, eds., Fine-grained deposits and biofacies of the Cretaceous Western Interior Seaway; Evidence of cyclic sedimentary processes: Society of Economic Paleontologists and Mineralogists Guidebook 4, p. 1–10.

Frey, R. W., 1970, Trace fossils of the Fort Hays Limestone Member of Niobrara Chalk (Upper Cretaceous), west-central Kansas: University of Kansas Paleontological Contributions, art. 53, 41 p.

—— , 1972, Paleoecology and depositional environment of Fort Hays Limestone Member, Niobrara Chalk (Upper Cretaceous), west-central Kansas: University of Kansas Paleontological Contributions, art. 58, 72 p.

Hattin, D. E., 1975, Stratigraphic study of the Carlile-Niobrara (Upper Cretaceous) unconformity in Kansas and northeastern Nebraska, *in* Caldwell, W.G.E., ed., The Cretaceous system in the Western Interior of North America: Geological Association of Canada Special Paper 13, p. 195–210.

—— , 1982, Stratigraphy and depositional environment of Smoky Hill Chalk Member, Niobrara Chalk (Upper Cretaceous) of the type area, western Kansas: Kansas Geological Survey Bulletin 225, 108 p.

—— , 1986, Carbonate substrates of the Late Cretaceous sea, central Great Plains and southern Rocky Mountains: Palaios, v. 1, p. 347–367.

Kauffman, E. G., 1977, Geological and biological overview; Western Interior Cretaceous Basin, *in* Kauffman, E. G., ed., Cretaceous facies, faunas, and paleoenvironments across the Western Interior Basin: Mountain Geologist, v. 14, no. 3, 4, p. 75–99.

Laferriere, A. P., Hattin, D. E., and Archer, A. W., 1987, Effects of climate, tectonics, and sea-level changes on rhythmic bedding patterns in the Niobrara Formation (Upper Cretaceous), U.S. Western Interior: Geology v. 15, p. 233–236.

Pratt, L. M., 1984, Influence of paleoenvironmental factors on preservation of organic matter in middle Cretaceous Greenhorn Formation, Pueblo, Colorado: American Association of Petroleum Geologists Bulletin, v. 68, p. 1146–1159.

Reeside, J. B., Jr., 1957, Paleoecology of the Cretaceous seas of the Western Interior of the United States, *in* Ladd, H. S., ed., Treatise on marine ecology and paleoecology; Volume 2, Paleoecology: Geological Society of America Memoir 67, p. 505–541.

Shurr, G. W., 1984, Regional setting of Niobrara Formation in northern Great Plains: American Association of Petroleum Geologists Bulletin, v. 68, no. 5, p. 598–609.

Weimer, R. J., 1960, Upper Cretaceous stratigraphy, Rocky Mountain area: American Association of Petroleum Geologists Bulletin, v. 44, p. 1–20.

Williston, S. W., 1897, Kansas Niobrara Cretaceous: Kansas Geological Survey, v. 2, p. 235–246.

West Atchison Drift Section

James S. Aber, Department of Earth Science, Emporia State University, Emporia, Kansas 66801

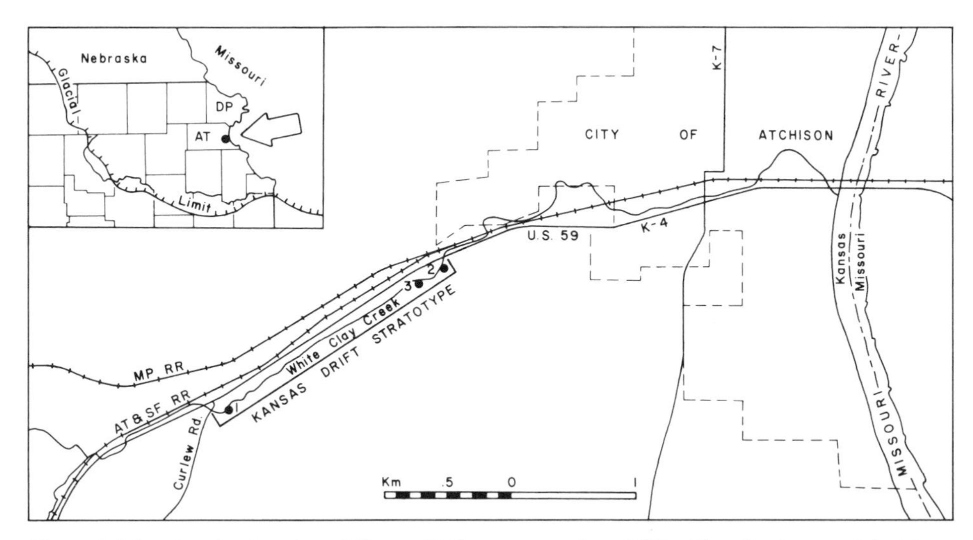

Figure 1. Map showing location of Kansas Drift stratotype along White Clay Creek west of Atchison, Kansas. Numbered locations are: 1, Atchison No. 1; 2, Atchison No. 2; and 3, type section of Atchison Formation. Inset map shows counties and glacial limit in northeastern Kansas; AT, Atchison County, DP, Doniphan County. Base map from Atchison West and Atchison East, Kansas, 7½-minute Quadrangles.

LOCATION

A series of long-studied exposures of glacial sediments is found immediately west of the city of Atchison, Atchison County, Kansas, along the south side of White Clay Creek in Sec. 2,10,11,T.6S.,R.20E. (Fig. 1). The stream cuts are readily accessible by foot from U.S. 59, which parallels the north side of White Clay Creek. It is advisable to ask local residents for permission to cross private property to reach the stream cuts.

SIGNIFICANCE

The West Atchison section has traditionally been recognized as the type section for Kansan glacial sediments, and the Kansan Stage is regarded as an important episode of Early Pleistocene continental glaciation. The Kansan glaciation represented, in fact, the greatest ice coverage to ever take place on the plains of central North America, and the only Pleistocene glaciation to enter the region covered in this field guide. Recent chronometric studies (Easterbrook and Boellstorff, 1981) combined with new field observations (Aber, 1985) have demonstrated the complex regional stratigraphy for the Kansan glacial deposits.

PREVIOUS STUDIES

These exposures were first described by Schoewe (1938, p. 227), who stated that, "... two distinct tills are present. The lower till is a typical unaltered or fresh, dark gray to blue, compact boulder clay, as much as 20 ft (6 m) thick. Its upper surface is irregular. The upper till, from 10 to 15 ft (3 to 5 cm) thick, is separated from the lower one by 50 ft (15 m) or more of stratified sand. This till in contrast to the lower one is brown in color, and is very stony . . . Whether this first drift sheet is to be correlated with the Nebraskan glacier, or whether it represents the first of two advances of the Kansan ice sheet, must remain unsettled until further studies are made."

A few years later, Frye (1941) decided that the lower gray till was Nebraskan, the middle sand Aftonian, and the upper brown till Kansan. He traced the lower gray till from southern Atchison County to southern Doniphan County. However, Frye and Leonard (1952) later included all three formations in a reference section for the Kansas Till, and designated the middle sand as the Atchison Formation. The Kansas Till (formation rank) was placed stratigraphically above the Atchison Formation. Following recognition of multiple Kansan and Nebraskan tills in Nebraska (Reed and Dreeszen, 1965), the two tills in these exposures were both considered to be Kansan. The upper and lower tills were correlated with the Cedar Bluffs and Nickerson Tills, respectively, of Nebraska (Ward, 1973). Recently, Aber (1985) has designated this site as the Kansas Drift stratotype, a lithostratigraphic unit of group rank containing three formations: Upper Kansas Till, Atchison Formation, and Lower Kansas Till.

Two sections are presently well exposed. Atchison No. 1 (AT-1) is located at the southwestern end of the site in NW¼, Sec. 10, and Atchison No. 2 (AT-2) is found at the northeastern end of the site on the center line between the SW¼ and SE¼, Sec. 2.

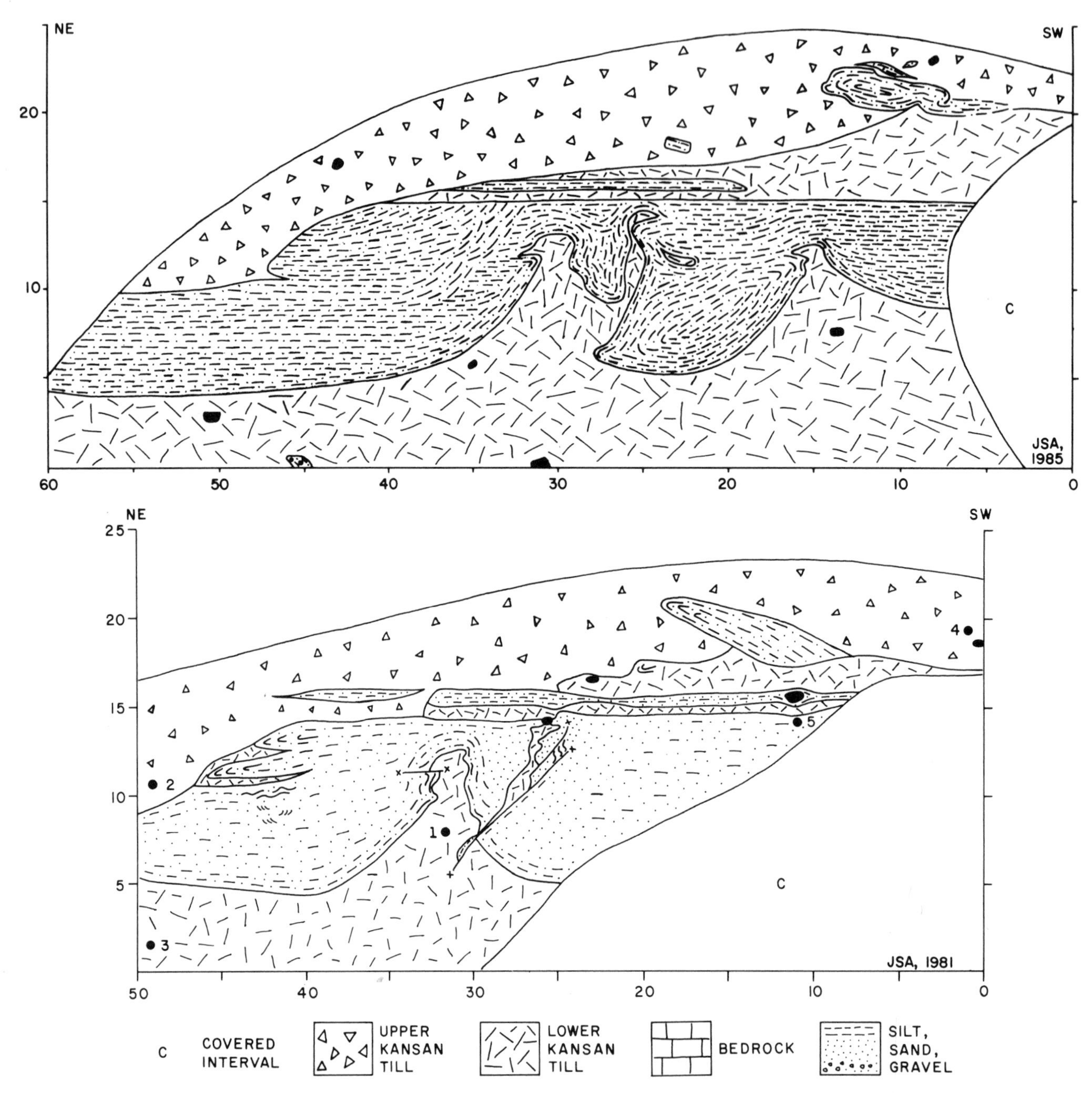

Figure 2. Atchison No. 1 sections showing two tills separated by sand and disturbed by several ice-pushed structures. Upper section measured in 1985; lower section measured in 1981 (Aber, 1985, Fig. 4). Scale given in meters; sample sites shown by solid dots.

DESCRIPTION OF SECTIONS

Atchison No. 1

Continuing stream erosion during the last half-century has enlarged this impressive exposure, which is the Kansas Till reference section of Frye and Leonard (1952). Actually, two pebbly clay tills separated by sand and displaying several kinds of ice-pushed structures are revealed in a cut more than 160 ft (50 m) long and nearly 80 ft (25 m) high (Fig. 2). The tills differ mainly in color: the upper is moderate to dark-yellowish-brown (10 YR 5/4, 4/2) and the lower is medium-dark-gray (N4), but they are otherwise similar. Intervening sand of the Atchison Formation is mainly a fine-rippled sand containing scattered lenses of pebbly coarse sand.

The lower gray till is dislocated in a pair of large diapirs that

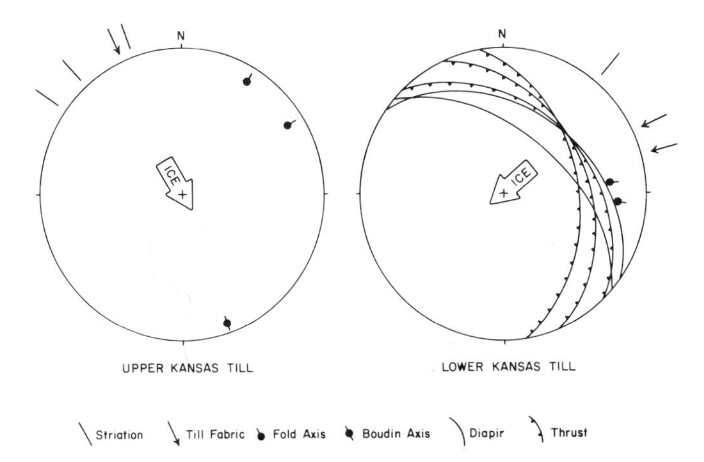

Figure 3. Summary equal-area stereonets of structural and directional features at Atchison No. 1 section. Lower Kansas Till was disturbed by northeasterly ice movement, whereas northwesterly ice advance was responsible for the Upper Kansas Till. (Data from Dellwig and Baldwin, 1965, and Aber, 1985.)

intrude up into the overlying sand. Prior to 1983, only the main, or larger, of these diapirs was exposed (lower section, Fig. 2). The main diapir is, overall, a wedge-shaped body trending northwest-southeast and pushed from the northeast. The nose of the diapir displays several minor folds that trend roughly east-west, and the left side of the diapir is cut by a small underthrust. The enclosing sand wraps smoothly around the diapir nose, and the boundary between sand and till is quite sharp with no indication of mixing or till apophyses. The right side of the diapir is cut by a prominent thrust fault, which can be traced for about 30 ft (10 m). Several smaller thrusts parallel the main fault, and a narrow finger of contorted gray till is displaced above the main fault. This is presumably the same thrust described by Dellwig and Baldwin (1965). They likewise concluded that the thrust diapir had been deformed by ice movement from an easterly or northeasterly direction.

Stream erosion during and since 1983 has enlarged the section. The main diapir has changed little, but some new features are now exposed, including a second smaller, but nearly identical, diapir (upper section, Fig. 2). The remarkable similarity in form and orientation between the two diapirs is most striking. Both have enlarged heads with extensions toward the northeast; both terminate upward at about the same level, and both strike northwest-southeast. The right side of the main diapir, which previously displayed a thrust, now branches upward into an irregular till dike, and isolated blobs of gray till are present in the sand between the two diapirs.

The intrusive character of the diapirs indicates deformation in a thawed state, and orientations of both diapirs and thrusts correspond to ice pushing from approximately N50°E (Fig. 3). This general direction of ice movement is also confirmed by two till fabric measurements and by striations on a bevelled boulder top.

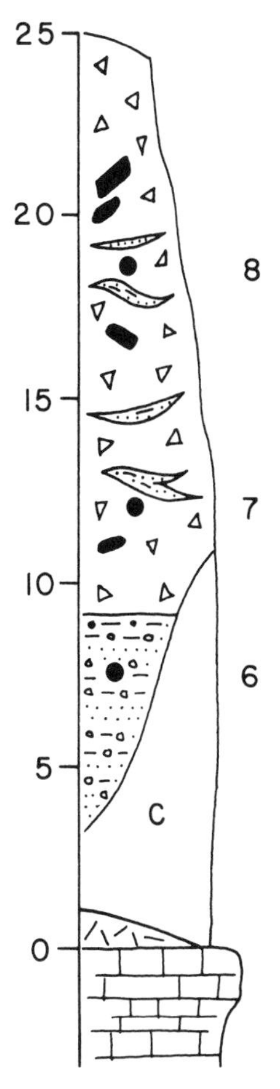

Figure 4. Atchison No. 2–measured section showing thick Upper Kansas Till overlying pebbly sand of the Atchison Formation. Lower Kansas Till rests on bedrock at base of section. Symbols same as in Figure 2.

The lower gray till and middle sand are discordantly covered by an upper brown till, which contains several sand bodies, gray till masses, and limestone boulders. Ice movement, which laid down the upper till, also produced a pair of recumbent folds in the underlying sand near the northeastern end of the exposure. The axes of these two folds plunge slightly toward the northeast (Fig. 3). This is clearly related to ice movement from about N30 W, and corresponds to northwest-southeast–trending striations observed on the bevelled tops of three in situ boulders and to a till fabric sample with a northwesterly maximum.

Atchison No. 2

This exposure is immediately east of the original type section for the Atchison Formation (Fig. 1), which is now covered. The present exposure reveals 50 ft (15 m) of moderate brown to dark-yellowish-brown Upper Kansas Till containing numerous large limestone boulders and masses of sand (Fig. 4). The till becomes grayish toward its base; this transitional color change is certainly due to weathering, as the brown oxidation color is seen penetrating downward along fractures, which are usually filled with calcite veins.

The lower portion of the stream cut is largely covered; however, at the northeastern end of the exposure, the upper till rests on at least 5 m of sandy pebble gravel of the Atchison Formation. The gravel contains many erratic pebbles and is heavily iron stained and cemented. Approximately 3 ft (1 m) of dark-gray Lower Kansas Till rests directly on limestone bedrock in the stream channel below the cut.

At the Atchison Formation type section, Frye and Leonard (1952) reported 8 ft (2.5 m) of bouldery till overlying 70 ft (21 m) of sand, noting that the till appeared to thicken markedly to the southwest. It appears the till also thickens rapidly to the east, as shown by the present exposure. The type Atchison Formation consisted mainly of fine to very fine cross-bedded sand; however, the basal 8 ft (2.5 m) was described as sand and gravel, locally cemented. This corresponds to the gravel seen at the present exposure. On this basis, it seems the Atchison Formation must undergo rapid lateral variations in thickness and texture.

DRIFT COMPOSITION

More than 25 drift samples have been analyzed from Atchison and Doniphan Counties, including several samples from the West Atchison sections (Aber and others, 1982). The size grade of 4 to 8 mm was used for pebble counts, and roughly 300 pebbles were counted for each sample into the following categories: quartzite, quartz, felsic crystalline, mafic crystalline, limestone, chert, sandstone plus shale, limonite plus pyrite, and other. The first four categories are regarded as erratics. The "other" category includes such unusual finds as an Upper Cretaceous shark tooth and gypsum.

Pebble composition seems to vary widely, but this is at least partly the result of weathering. Samples leached of all limestone have more than 40 percent erratic pebbles, whereas unleached samples contain less than 20 percent erratics. Felsic crystallines make up roughly half or more of total erratics in nearly all samples, and quartzite, mainly pink or white colored, is present in most samples, although rarely in excess of 1 to 2 percent. Kansan drift samples show similar and overlapping ranges of pebble compositions.

According to Frye and Leonard (1952, p. 71): " . . . the Atchison Formation was deposited as outwash and lacustrine deposits in pro-glacial lakes produced by the advancing Kansan glacier." Size distribution analysis of Atchison Formation sand supports this interpretation (Fig. 5). Good sorting is evident in both samples, with about 50 percent of each sample contained within one phi-size grade. Statistical measures, such as sorting coefficient and skewness, are typical of fluvial or deltaic sands.

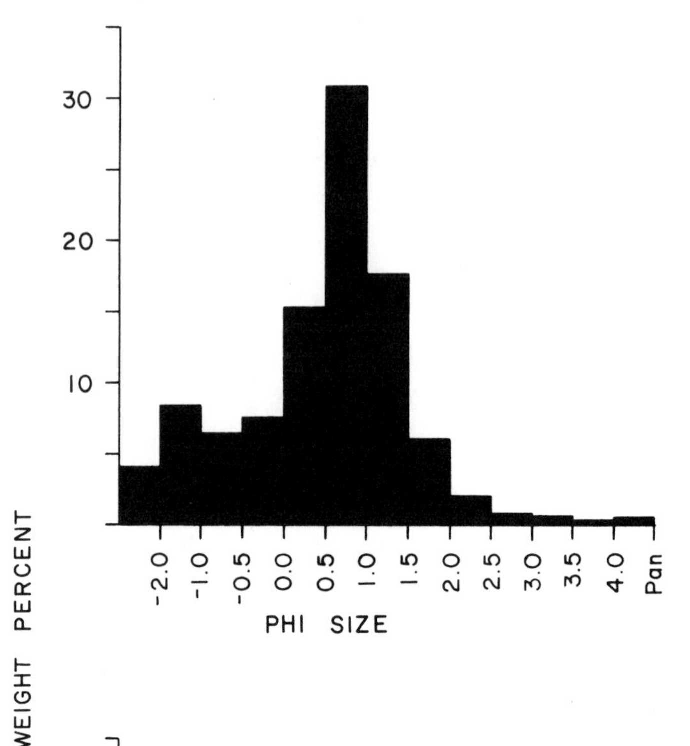

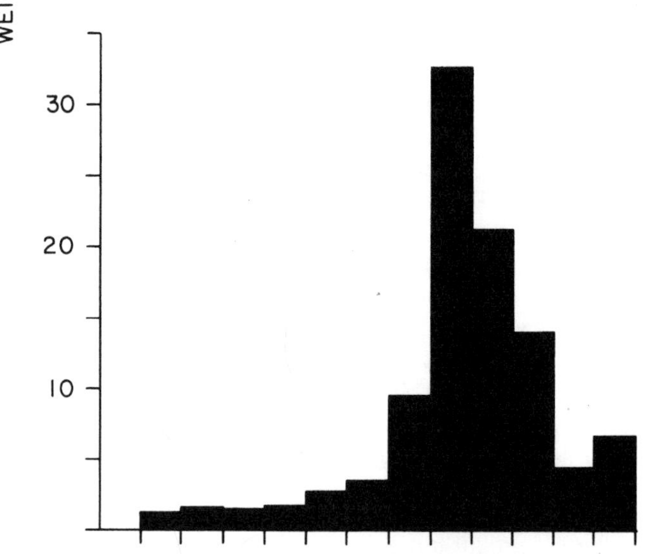

Figure 5. Histograms showing grain-size analyses of Atchison Formation sand (AT-1): coarse pebbly sand above, fine rippled sand below. Each histogram represents cumulative weight percentages of three subsamples. Good sorting is shown in both cases.

Till clay mineralogy is dominated by montmorillonite, illite, and kaolinite, which are present in roughly equal amounts in most samples (Fig. 6). Upper and Lower Kansas Till samples plot near the center of the ternary diagram with no clear distinction, whereas the sub-Kansan till sample (from Doniphan County) is marked as the only one with less than 20 percent montmorillonite. Among the heavy minerals in the fine sand fraction, opaques, mainly magnetite and pyrite, dominate in all samples, along with small percentages of epidote, amphibole, zircon, garnet, tourmaline, and others. Preliminary copper analyses show that the Lower Kansas Till contains 10 to 50 ppm of copper, whereas the Upper Kansas Till has little or no detectable copper. The significance of this is not known at present.

The Lower Kansas Till at AT-1 and in Doniphan County

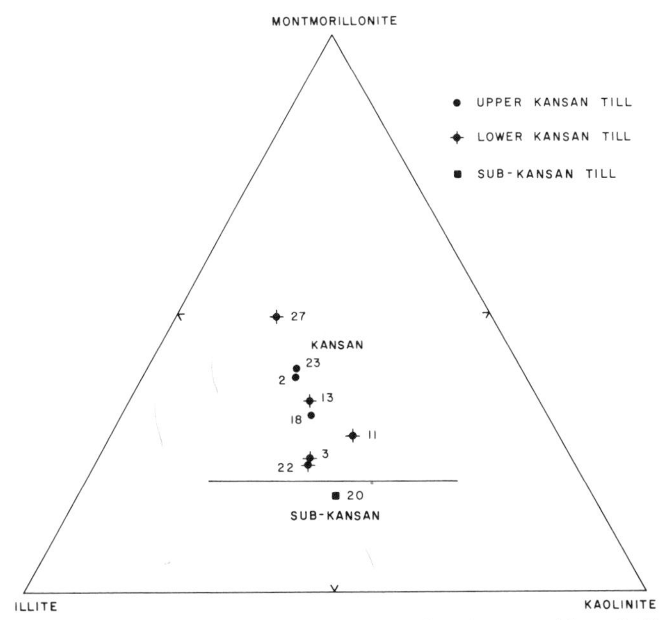

Figure 6. Ternary diagram showing clay mineral composition of till samples from Atchison and Doniphan Counties. Samples 2 and 3 from AT-1 (Fig. 2). Percentages of montmorillonite, illite, kaolinite normalized to 100 percent. Horizontal line represents a 20 percent value for montmorillonite.

contains abundant well-preserved fragments of wood up to 1 ft (30 cm) long. The wood has been identified as spruce by Hedstrom (1986). The wood fragments appear freshly broken, not abraded or weathered, and were probably not transported far by the ice advance. A boreal spruce forest was evidently growing in northeastern Kansas in front of the advancing Kansan ice sheet.

GEOLOGIC HISTORY AND REGIONAL CORRELATION

The structural and fabric evidence is consistent with ice movement from the northeast during deposition and subsequent displacement of the Lower Kansas Till and Atchison Formation sand. Multiple northeasterly advances are probable. The initial advance laid down the lower till, and the sand was then conformably deposited over the till during a brief ice recession. The sand deposits probably accumulated in ice-marginal lakes, indicating the ice margin had retreated only a short distance, perhaps to near the present Missouri River valley. Renewed northeasterly advances, possibly due to ice surging, then dislocated till and sand with thrusting and diapiric intrusions. After retreat of the northeastern ice, a northwesterly advance deformed the Atchison Formation and deposited the Upper Kansas Till.

The stratigraphy and the structures displayed at the Kansas Drift stratotype demonstrate the development of two ice lobes during the Kansan glaciation (Aber, 1982). The Minnesota ice lobe advanced southward through Iowa, into Missouri, and entered Kansas from the northeast. Conversely, the Dakota ice lobe came from the Dakotas, across Nebraska, and moved into Kansas from the northwest. These two lobes followed bedrock troughs on either side of the Coteau des Prairies upland and Sioux Quartzite bedrock ridge in southwestern Minnesota. At times the two lobes coallesced over the Coteau forming a broad Kansan ice-fan to the south.

The lower, gray, wood-bearing till at West Atchison is equivalent to the lower McCredie Formation till in northern Missouri (Guccione, 1983) and to the lowest gray till at Fremont cliffs, Nebraska (Wayne, 1987). This widespread till is everywhere dark gray or olive gray in color and contains abundant buried wood. This till has been variously referred to as Nebraskan, upper Nebraskan, lower Kansan, or pre-Illinoian. It is probably identical to the type Nebraskan till of Shimek (1909), but it also occurs in sections long regarded as Kansan. The upper brown till at West Atchison is correlated with the upper McCredie Formation till in northern Missouri and with the Nickerson Till at Fremont cliffs, Nebraska. The Upper Kansas Till stretches south of the Kansas River valley in northeastern Kansas, whereas the Lower Kansas Till does not extend south of Atchison County.

In view of the confusing stratigraphy, Aber (1986) proposed to demote the Nebraskan to the first substage of the Kansan Stage, and thus the Lower Kansas Till and Atchison Formation belong to the Nebraskan Substage. The term Nebraskan and its relative stratigraphic position are thereby retained within a more complex stratigraphy. The Aftonian Substage, an interval of ice retreat, follows the Nebraskan Substage. The Upper Kansas Till along with still-younger tills to the north (Cedar Bluffs Till) fall into the last substage of the Kansan. The term Potawatomian, from an Indian reservation in northeastern Kansas, has been used for this substage.

The available time control from ash dating and paleomagnetic polarity (Easterbrook and Boellstorff, 1981) indicates that the Kansan Stage correlates with marine oxygen-isotope stages 16 through 22: Potawatomian Substage = stages 16 through 18 (about 0.6 to 0.69 Ma); Aftonian Substage = stages 19 through 21 (about 0.69 to 0.78 Ma); Nebraskan Substage = stage 22 (about 0.78 to 0.85 Ma). Stage 16/Potawatomian Substage represents the greatest continental glaciation during the Pleistocene Epoch.

REFERENCES CITED

Aber, J. S., 1982, Two-ice-lobe model for Kansan glaciation: Transactions of the Nebraska Academy of Sciences, v. 10, p. 25–29.

—— , 1985, Definition and model for Kansan glaciation: Tertiary-Quaternary Symposium Series, v. 1, p. 53–60.

—— , 1986, Stratigraphy of Kansan and Nebraskan tills west of the Missouri River: Kansas Academy of Science, Abstract, v. 5, p. 1.

Aber, J. S., Gundersen, J. N., Medina, K. J., Schroeder, D. C., and Rose, R. K., 1982, Preliminary report on drift composition in northeastern Kansas:

Kansas Academy of Science Abstract, v. 1, p. 1.
Dellwig, L. F., and Baldwin, A. D., 1965, Ice-push deformation in northeastern Kansas: Kansas Geological Survey Bulletin 175, pt. 2, 16 p.
Easterbrook, D. J., and Boellstorff, J., 1981, Paleomagnetic chronology of "Nebraskan-Kansan" tills in midwestern U.S., *in* Sibrava, V., and Shotton, F. W., eds., Quaternary Glaciations in the Northern Hemisphere: IUGS-UNESCO International Correlation Program, Project 73-1-24, Report No. 6, Ostrava, Czechoslovakia (1979), p. 72–82.
Frye, J. C., 1941, Reconnaissance of ground-water resources in Atchison County, Kansas: Kansas Geological Survey, Bulletin 38, pt. 9, p. 237–260.
Frye, J. C., and Leonard, A. B., 1952, Pleistocene Geology of Kansas: Kansas Geological Survey, Bulletin 99, 230 p.
Guccione, M. J., 1983, Quaternary sediments and their weathering history in northcentral Missouri: Boreas, v. 12, p. 217–226.
Hedstrom, B. L., 1986, Identification of glacial wood from Kansas and Nebraska: Kansas Academy of Science, Abstract v. 5, p. 20.
Reed, E. C., and Dreeszen, V. H., 1965, Revision of the classification of the Pleistocene deposits of Nebraska: Nebraska Geological Survey, Bulletin 23.
Schoewe, W. H., 1938, The west Atchison glacial section: Transactions of the Kansas Academy of Science, v. 41, p. 227.
Shimek, B., 1909, Aftonian sands and gravels in western Iowa: Geological Society of America Bulletin, v. 20, p. 399–408.
Ward, J. R., 1973, Geohydrology of Atchison County, northeastern Kansas: U.S. Geological Survey Hydrological Investigations Atlas no. HA-467.
Wayne, W. J., 1987, The Platte River and Todd Valley near Fremont, Nebraska, *in* Biggs, D. L., ed., North-central Section of the Geological Society of America: Boulder, Colorado, Geological Society of America, Centennial Field Guide, v. 3, p. 19–22.

Spillway fault system, Tuttle Creek reservoir, Pottawatomie County, northeastern Kansas

James R. Underwood, Jr., and Allyn Polson, Department of Geology, Kansas State University, Manhattan, Kansas 66506

LOCATION AND ACCESS

The best exposure of the Spillway fault system is in the northwest wall of the northwest-trending spillway of Tuttle Creek reservoir (SW¼,SE¼SW¼,Sec.18,T.9S.,R.8E.) in westernmost Pottawatomie County, immediately northwest of the point where Kansas 13 crosses the spillway (Fig. 1).

From Manhattan, Kansas, drive north–northwest on Tuttle Creek Boulevard (also US 24, Kansas 13 and 177) 3.6 mi (5.7 km) from its intersection with Kimball Avenue to the point where Kansas 13 turns right (northeast). Drive 1.8 mi (2.9 km) across the dam to the parking areas on both sides of the road that overlook the spillway to the northeast.

SIGNIFICANCE

Good exposures of faults in Kansas are rare; thus of special interest are the Spillway fault and associated minor faults. (Table 1). The spillway section also provides a good exposure of approximately 140 ft (43 m) of the Council Grove Group (Gearyan Stage) of the Permian System (Fig. 2), a classic stratigraphic sequence representing cyclic deposition on a broad, continental shelf.

STRATIGRAPHIC SECTION

The alternating hard and soft strata of limestone and mudstone exposed at the spillway (Fig. 2) are typical of the rocks that underlie the gentle, subdued ledge-and-slope topography of the Flint Hills and that express the cyclic deposition that characterized the midcontinent during the late Paleozoic.

SPILLWAY FAULT SYSTEM

The gently arcuate, almost east–west-trending trace of the Spillway fault can be followed for a total distance of about 11 mi (18 km) (Scott and others, 1959; Kirkwood, 1970). The fault is believed to die out both to the east and to the west near where the trace disappears beneath surficial deposits. Kirkwood (1970) determined the throw at 10 localities along the trace; the range is 5.6 to 28.4 ft (1.7 to 8.6 m). The fault (Figs. 1, 3) does not lie beneath any part of the dam or the control gates of the spillway; the closest approach of the fault to the spillway gates is the point where the fault crosses Kansas 13, 377 ft (115 m) northeast of the spillway.

DISCUSSION

The U.S. Army Corps of Engineers, aware of the fault from the earliest stage of planning and design of the Tuttle Creek project, mapped the fault, studied drill cores taken along it, and concluded that the fault had not been active in geologically recent time and posed no significant danger to the dam and spillway. Monitoring of microseismic activity in north–central Kansas since late 1977 by the Kansas Geological Survey has identified some 100 events (about one a month); none has been related to the Spillway fault (D. W. Steeples, personal communication, 1986).

Seven possible origins of the Spillway fault are discussed below:

1. The parallelism of the major structural features of the region—the Abilene and Nemaha anticlines and the Irving syncline—suggests that they had a common origin. They may have formed in response to a single compressional stress field, or they may be the result of reactivation of faults in the basement that may have formed originally in response to a common stress field. Based on the orientation of these features (Fig. 1), the greatest principal compressive stress would have been oriented N65°W–S65°E. The strike of the Spillway fault, N78°E, is close to one of two possible conjugate shear directions, N85°E. The spillway fault, then, may have resulted from reactivation, with mostly dip-slip movement, of an old strike-slip fault formed during a late Paleozoic phase of compression.

2. The reactivation of very ancient faults related to the 1.2-Ga nearby midcontinent rift system has been described recently by Van Schmus and Hinze (1985), Steeples (1982), and Yarger (1981). A complex system of faults striking both parallel and transverse to the axis of the midcontinent rift system has been identified along the rift (Serpa and others, 1984; Morey and Green, 1982). Similar faults associated with the Midcontinent rift system in north-central Kansas may persist, with periodic movement, through time and through overlying younger rock. All of the faults recognized in the region, including the Spillway fault, may be the result of periodic reactivation of very ancient faults related to the Proterozoic episode of rifting.

3. Still older faults, those in the 1.6-Ga terrain that was disrupted by the incipient rifting at 1.2 Ga (Van Schmus and Hinze, 1985), also may have been active periodically through time. Very little is known of the structural characteristics of these ancient rocks.

4. Chelikowsky (1972) suggested that right-lateral strike-slip faulting could have produced the deflection in the regional joint pattern (Fig. 1). Movement along such a fault parallel to, and approximately coincident with, the Abilene anticline would also be parallel to the Midcontinent rift and possibly related. Visualizing the orientation of the strain and stress ellipsoids asso-

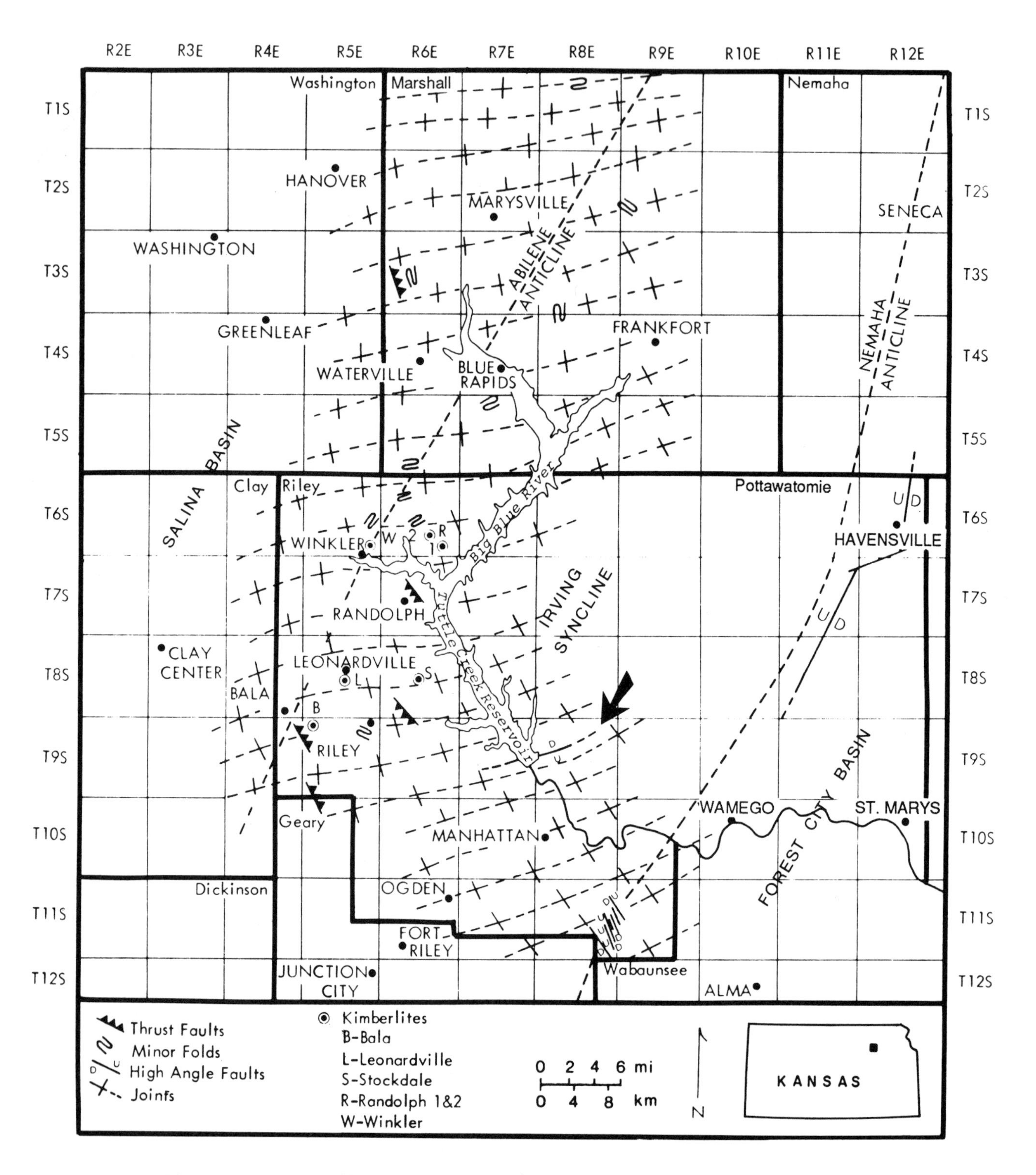

Figure 1. Location map of Tuttle Creek reservoir; arrow indicates Spillway fault (full length of trace not shown). Map shows trends of regional joints and locations of kimberlite pipes, faults, and folds. From Chelikowsky (1972), who suggested that regional joints were an important controlling factor in determining the trend of high-angle faults and emplacement of kimberlite pipes.

Stratigraphic Section—Spillway, Tuttle Creek Dam

FORMATION	MEMBER	SECTION	DESCRIPTION	TK. (m)
Speiser Sh			**Speiser Shale** Mudstone; varicolored, indistinctly bedded to laminated, moderately calcareous; present only on downthrown block north-northwest of fault; upper part eroded.	3.0 ft (0.92 m)
Funston Ls			**Funston Limestone** Limestone (skeletal calcilutite); grayish yellow, weathers yellowish gray to grayish orange pink; sparsely fossiliferous; scattered chert nodules; 1.19 m thin bedded, contains 0.30 m laminated shaly mudstone midway; overlies 0.86 m thick-bedded resistant calcilutite underlain by 0.15 m mudstone, light gray, weathers yellowish gray; 0.43 m thin-bedded calcilutite at base.	8.7 ft (2.63 m)
Blue Rapids Shale			**Blue Rapids Shale** Mudstone; upper part yellowish gray, laminated; lower part varicolored (dusky yellow green, dark greenish gray, dusky red); indistinctly bedded; iron stained; contains two thin discontinuous calcilutite seams.	19.0 ft (5.79 m)
Crouse Limestone			**Crouse Limestone** Limestone (skeletal calcilutite); light olive gray, weathers pale to dark yellowish orange; upper part very platy, argillaceous, thin-bedded to laminated; limonite-replaced gastropods common; middle, very thin-bedded argillaceous calcilutite interbedded with pale yellowish orange very calcareous mudstone; lower third slabby to massive, thick-bedded, honeycombed calcilutite.	9.3 ft (2.85 m)
Easly Creek Shale			**Easly Creek Shale** Mudstone; upper part light olive brown, laminated, calcareous; overlies 0.51 m grayish yellow skeletal calcilutite with brachiopod fragments, internal worm molds; lower mudstone varicolored (moderate yellowish green, dusky red), mica-rich, coarsely laminated with 5-cm thick greenish black mudstone 1.07 m above base.	18.8 ft (5.74 m)
Bader Limestone	Middleburg Ls		**Middleburg Limestone** Limestone (skeletal calcilutite); some brachipod, mullusk fragments; upper part light brown, argillaceous, thin bedded, honeycombed; overlies 15-cm thick light gray, laminated and shaly mudstone; lower calcilutite 0.76-m thick, yellowish gray, weathers grayish yellow; blocky, thick bedded.	4.25 ft (1.30 m)
	Hooser Sh		**Hooser Shale** Mudstone; upper part light gray, coarsly laminated, calcareous, contains persistent 10–18 cm calcilutite seam; underlying, two non-persistent thin-bedded calcilutite beds separated by 30-cm thick pale olive laminated mudstone; basal mudstone varicolored (dark reddish brown, pale olive), very thin bedded to laminated.	9.1 ft (2.77 m)
	Eiss Ls		**Eiss Limestone** Limestone (skeletal calcilutite); upper skeletal calcilutite 1.27-m thick, grayish yellow, massive; honeycombed, secondary mineralization in voids; sparse brachiopods; overlies 40- to 45-cm thick mudstone, medium light gray, laminated, calcareous, containing brachiopod and bryozoan fragments, and crinoid columnals; basal limestone skeletal and clayey calcilutite 35- to 45-cm thick, light olive gray, very thin bedded to coarsely laminated; contains gastropods, brachiopods, crinoid columnals.	6.0 ft (2.08 m)
Stearns Shale			**Stearns Shale** Mudstone; light gray to pale olive; partly dusky red in lower part; two calcareous seams in upper half; 1.83 m above base, 30-cm thick olive black laminated shale with iron oxide-coated fracture surfaces.	13.9 ft (4.24 m)
Beattie Limestone	Morrill Ls		**Morrill Limestone** Limestone (calcilutite); yellowish gray, weathers grayish yellow; argillaceous; some chert; honeycombed with voids filled by secondary calcite crystals.	1.8 ft (0.56 m)
	Florena Sh		**Florena Shale** Mudstone; light gray to grayish yellow; laminated, calcareous; very thin-bedded calcareous seam midway; abundant fragments bryozoans, brachiopods, crinoids, some trilobites.	8.25 ft (2.52 m)
	Cottonwood Ls		**Cottonwood Limestone** Limestone (skeletal, cherty calcilutite); light gray to grayish orange, weathers light gray; thick-bedded upper half intensely pitted by weathering out of fusulinids; some chert; lower half cherty with some fusulinids.	5.9 ft (1.80 m)
Eskridge Shale			**Eskridge Shale** Mudstone; upper part medium light gray to light olive gray, laminated, calcareous, with thin highly calcareous seam; lower part varicolored, very thin-bedded to laminated, containing two thin clayey calcilutite seams with brachiopods, abundant mollusks.	26.0 ft (7.93 m)
Grenola Ls	Neva Ls		**Neva Limestone** Limestone (skeletal calcilutite); 23-cm thick thin-bedded light gray unfossiliferous calcilutite; some secondary mineralization; overlies light gray, thick-bedded fossiliferous skeletal calcilutite with algae, fusulinids, and fragments of echinoids and brachipods.	5.8 ft (1.78 m)

Figure 2. Stratigraphic section at spillway of Tuttle Creek reservoir, measured directly northwest of the point where Kansas 13 crosses the northeast wall of the spillway.

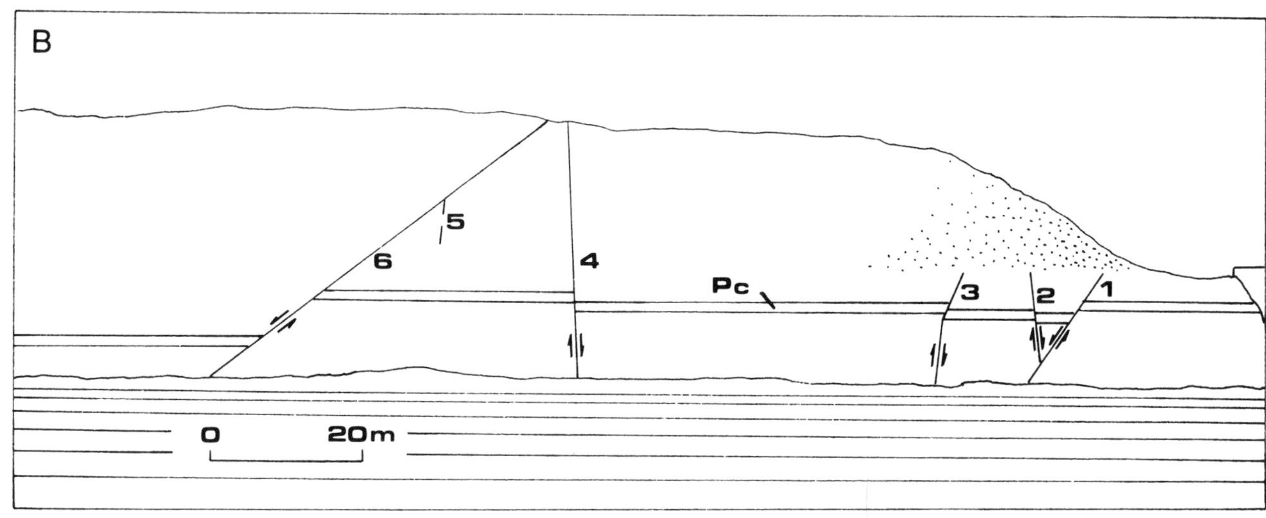

Figure 3. A. Northeast view of the exposure of the Spillway fault system photographed from the parking area overlooking the spillway from southwest. Spillway wall slopes 40° toward camera and trends N57°W. B. Spillway fault (Fault 6) strikes N78°E, dips 81°NW; in this view, fault trace is inclined 37° to horizontal. Symbol P_c identifies Cottonwood Limestone Member of the Beattie Limestone. Scale is approximate for a position midway between top and bottom of slope.

ciated with right-lateral strike-slip movement along the fault results in the following observations: a. The Spillway fault, a tensional feature, is oriented perpendicular to the axis of greatest strain and parallel to the axis of the greatest principal stress. b. The small thrust faults striking north–northeast to northeast (Fig. 1) are perpendicular to the least strain axis, i.e., to the greatest principal stress axis. c. The fractures perpendicular to the axis of the Abilene anticline and into which the kimberlite pipes intruded (Brookins, 1970) may be conjugate shear fractures oriented east–northeast and west–southwest. These observations support the hypothesis that the Spillway fault could be related to the postulated strike-slip fault suggested by Chelikowsky (1972).

5. The Spillway fault could have resulted from the stress field developed by a right-lateral shear couple produced by a right-lateral strike-slip fault parallel to and approximately coincident with the Abilene anticline and a second right-lateral strike-slip fault to the east, parallel to and approximately coincident with the Nemaha uplift. The orientation of the stress and strain ellipsoids and resulting structural features can be visualized as in item 4, above. The strike-slip faults hypothesized above (items 4 and 5) may have been very ancient faults that experienced reactivation. The diverse orientation of the axes of a number of small folds between the communities of Winkler and Randolph (Fig. 1) do not support this hypothesis, but these folds may be related to local

TABLE 1. FAULTS OF SPILLWAY FAULT SYSTEM EXPOSED IN NORTHEAST WALL OF SPILLWAY

Fault No. (Fig. 3)	Type	Strike	Dip	Vertical Separation ft	(m)	Distance* ft	(m)	Comments
1	Normal	N80°E	85°NW	5.3	(1.63)	65	(19.7)	Slickensides, rake angle = 68°SW; fault breccia.
2	Normal	N62°E	83°SE	1.6	(0.48)	89	(25.6)	Antithetic to F1.
3	Reverse	N60°E N40°E	82°NW	3.8	(1.17)	118	(36.1)	Strike changes across spillway wall; displacement decreases downward.
4	Normal	?	88°SE (App. dip)	4.3	(1.32)	278	(84.9)	
5	Normal	?	85°NW (App. dip)	1.2	(0.36)	---	---	Originates in Easly Creek; extends downward approx. 9m from fault plane of F6. Does not affect Cottonwood.
6	Normal	N78°E	81°NW	23.0	(7.01)	390	(129)	Breccia; secondary mineralization; fault drag; slickensides, rake angle = 79°SW; small horse.
7	Normal	N35°E	86°SE	2.0	(0.61)	1030	(314)	Not shown on Fig. 3; slickensides, rake angle = 84°SW.

Note: Limestone strata broken by joints into blocks that may have tilted as spillway was constructed. Bewtwen Faults 1 and 3, joint-bounded blocks of upper part of Cottonwood appear to have been removed from outcropping ledge. Displacement of Faults 1, 2, and 3 measured at base of Cottonwood.

*Indicates distance along Cottonwood Limestone outcrop in spillway wall northwest of Kansas 13.

stresses associated with the emplacement of known kimberlite pipes in the vicinity and, perhaps, to emplacement of pipes not yet discovered.

6. Relatively local, differential compaction of Phanerozoic sedimentary rocks, influenced perhaps by inhomogeneities within them, by relief on the basement surface, and by tidal stresses repeatedly flexing the crust could have produced the Spillway fault. Body forces such as those created by the gravitational attraction of relatively nearby planetary bodies in the solar system would be expected to produce numerous, pervasive fractures or joints. Joints produced by this process and controlled by other structural features of the region, such as deep-seated faults, could have influenced the development of the Spillway fault. The apparent dying out both east and west of the Spillway fault suggests that the fracture may be viewed most logically as having originated by compaction.

7. Some combination of mechanisms, items 1–6, also is possible.

Coveney and others (1987) have suggested that deep-seated processes related to the Midcontinent rift system still may be affecting the area. They suggested that the origin of the recently discovered hydrogen-rich natural gas, some 30 mi (50 km) southwest in Morris and Geary counties, may be related to serpentinization or deserpentinization of kimberlite or layered ultramafites of the old rift system.

No disruption of Quaternary surficial deposits nor of stream terraces has been observed along the Spillway fault. The abundant secondary calcite that cements fault gouge and breccia is intact where visible, and cores through the fault zone showed it to be "well sealed" with mudstone (U.S. Army Corps of Engineers, n.d.). No evidence is known that suggests that the Spillway fault has been active in geologically recent time.

The intriguing possibility exists that the Midcontinent rift system, the Riley County kimberlite pipes, the hydrogen-rich natural gas of Morris and Geary counties, and the Spillway fault system all may be interrelated. Recently drilled and proposed petroleum exploration wells, targeted to explore sedimentary basins associated with the Midcontinent rift system, should provide much-needed additional subsurface information about this very interesting region of the midcontinent.

REFERENCES CITED

Brookins, D. G., 1970, The kimberlites of Riley County, Kansas: Kansas Geological Survey Bulletin 200, 32 p.

Chelikowsky, J. R., 1972, Structural geology of the Manhattan, Kansas, area: Kansas Geological Survey Bulletin 204, pt. 4, 13 p.

Coveney, R. M., Goebel, E. D., Zeller, E. J., Dreschoff, G.A.M., and Angino, E. E., 1987, Hydrogen in well gases from Kansas may have originated from redox processes, such as serpentinization: American Association of Petroleum Geologists Bulletin, V. 71, p. 39–48.

Kirkwood, S. G., 1970, A study of the Tuttle Creek fault, Riley and Pottawatomie counties, Kansas: Unpublished problem report, Department of Geology, Kansas State University, 4 p.

Morey, G. B., and Green, J. C., 1982, Status of the Keweenawan as a stratigraphic unit in the Lake Superior region, *in* Wold, R. J., and Hinze, W. J., eds., Geology and tectonics of the Lake Superior basin: Geological Society of America Memoir 156, p. 15–26.

Scott, G. R., Foster, F. W., and Crumpton, C. F., 1959, Geology and construction material resources of Pottawatomie County, Kansas: U.S. Geological Survey Bulletin 1060-C, 178 p.

Serpa, L., and 6 others, 1984, Structure of the southern Keweenawan rift from COCORP surveys across the Midcontinent Geophysical Anomaly in northwestern Kansas: Tectonics, v. 3, p. 367–384.

Steeples, D. W., 1982, Structure of the Salina–Forest City interbasin boundary from seismic studies, *in* Proctor, P. D. and Koenig, J. W., eds., Selected structural basins of the Midcontinent, U.S.A.: V. H. McNutt Geology—Geophysics Colloquium Series no. 3, Rolla, University of Missouri, p. 55–81.

U. S. Army Corps of Engineers, n.d., Relation of Tuttle Creek dam to geologic faults: Report in files of Corps of Engineers, Tuttle Creek Project office, 3 p.

Van Schmus, W. R., and Hinze, W. J., 1985, The Midcontinent rift system: Annual Review of Earth and Planetary Sciences, v. 13, p. 345–383.

Yarger, H. L., 1981, Aeromagnetic survey of Kansas: EOS, v. 62, p. 173–178.

Minor T-R units in the Lower Permian Chase Group, northeast Kansas

Richard M. Busch, *Department of Geology, Thompson Hall, Kansas State University, Manhattan, Kansas 66506*

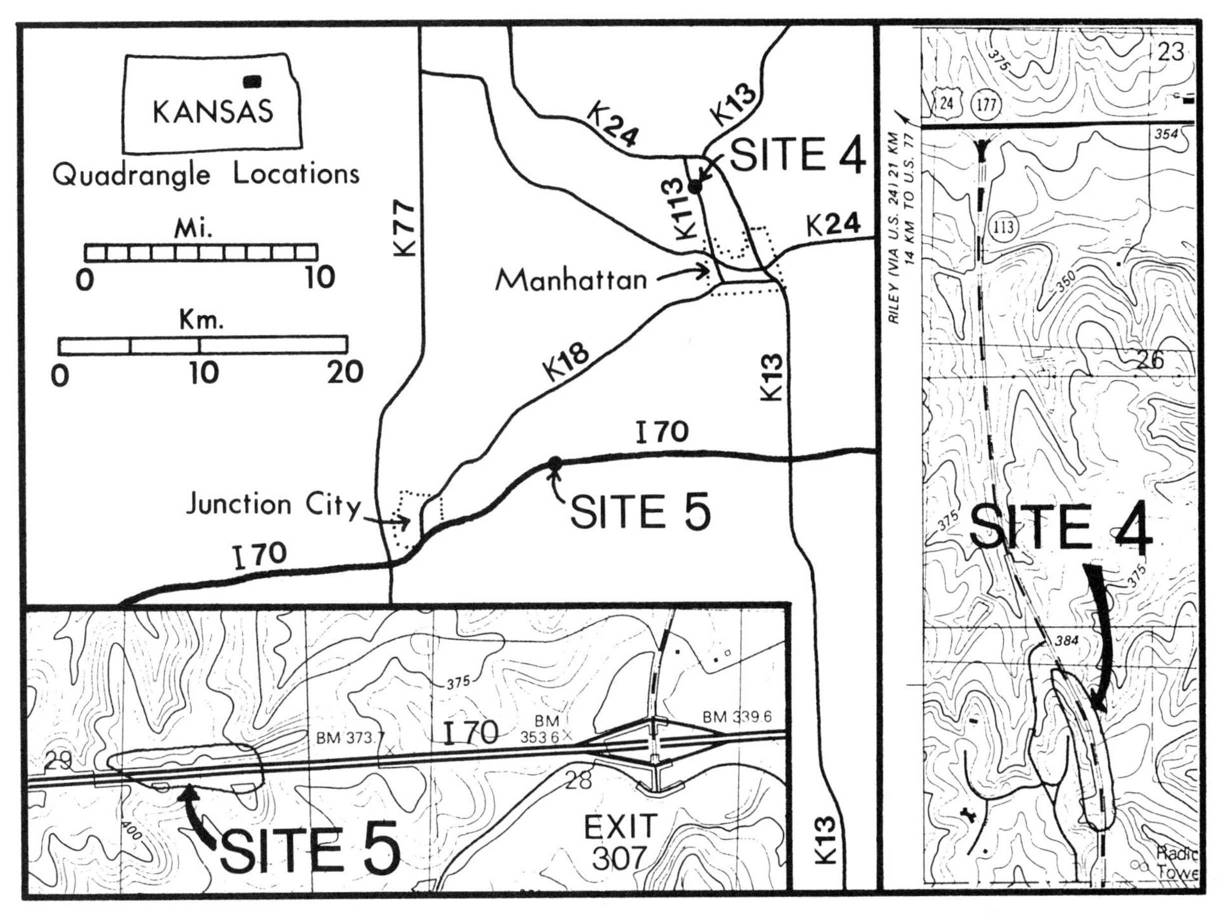

Figure 1. Highway map showing location of sites 4 and 5 in northeast Kansas and topographic map inserts of both sites.

LOCATION

Site 4: Roadcuts between mile markers 4.4 and 4.6 along east and west sides of Kansas 113 (Seth Childs Road), from 1.25 to 1.45 mi (2 to 2.3 km) south of its intersection with Kansas 24, at the northwest edge of Manhattan, Riley County, Kansas, NW¼,Sec.35,T.9S.,R.7E.

Site 5: Roadcuts along the north side of I-70, between mile markers 306.6 and 307, starting 0.85 mi (1.4 km) west of the overpass for Exit 307 to McDowell Creek Road, Geary County, Kansas, NW¼,Sec.28,T.11S.,R.7E.

SIGNIFICANCE

Permo-Carboniferous rocks of Kansas are tyically described and interpreted relative to idealized units composed of rhythmic or cyclic alternations of certain specific lithofacies. These lithostratigraphic units have been termed cyclothems (after Wanless and Weller, 1932) by the North American Commission on Stratigraphic Nomenclature (1983). The same rocks can also be described and interpreted relative to their constituent transgressive-regressive (T-R) (i.e., deepening-shallowing) units by considering the total range of facies and facies contacts present in a stratigraphic sequence.

This chapter examines two exposures of the basal part of the Lower Permian Chase Group (Gearyan-Wolfcampian age) in northeast Kansas (Fig. 1). The presence of cyclothems is noted and discussed relative to the definition and correlation of fifth-order and sixth-order T-R units (Busch and Rollins, 1984).

SETTING

Lower Permian rocks of eastern Kansas consist largely of alternating limestones and terrigenous mudstones that crop out in

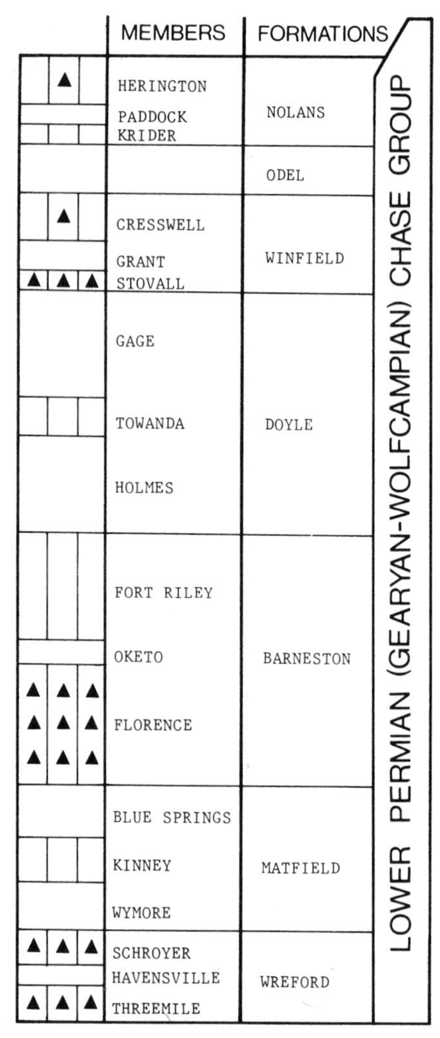

Figure 2. Schematic illustration of Lower Permian Chase Group of northeast Kansas, showing position of named members and formations. Unshaded intervals indicate shale; triangles, chert nodules; vertically lined intervals, limestone.

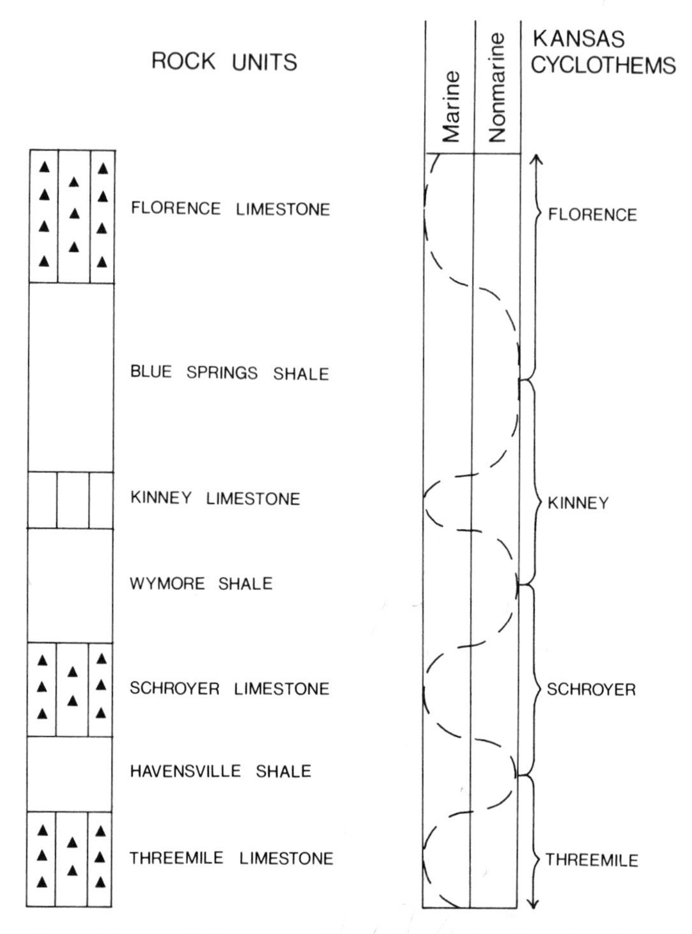

Figure 3. Schematic illustration of base of Lower Permian Chase Group, including study interval from Schroyer Limestone to Florence Limestone. Rock units can be arranged into limestone-mudstone or mudstone-limestone rhythmic cyclothems. They can also be arranged into Kansas cyclothems, as noted, which are fifth-order T-R units (see text). Triangles indicate chert nodules.

a north-south belt, along the western margin of the Osage Plains Physiographic Province. These rocks dip very slightly (i.e., about 15 to 20 ft/mi; 3 to 4 m/km) to the west-northwest, and the limestones form north-northeast–trending escarpments that face eastward. Because some of these limestones contain resistant chert (flint) nodules, they form particularly prominent escarpments, and the region is called the "Flint Hills."

The basal part of the Lower Permian Chase Group (Fig. 2) in the Flint Hills contains three limestones having abundant chert nodules (O'Connor and others, 1968). They are the Threemile Limestone (5 to 30 ft; 1.5 to 9 m thick), marking the base of the group, and the superjacent Schroyer Limestone (5 to 15 ft; 1.5 to 4.5 m thick) and Florence Limestone (15 to 45 ft; 4.5 to 14 m thick). The prominence and lateral extent of these three cherty limestones makes them excellent marker beds, two of which (Schroyer and Florence) are utilized in this chapter.

CYCLOTHEMS

The interval examined at Sites 4 and 5 extends from the Schroyer cherty limestone marker bed to the superjacent Florence cherty limestone marker bed. It also contains the Kinney noncherty limestone marker bed, which is separated from the Schroyer Limestone by the Wymore Shale and from the Florence Limestone by the Blue Springs Shale (shown schematically in Fig. 3). Two types of cyclothems can be defined within this study interval: one type is rhythmic, or of the form A-B, A-B, A-B; a second type is cyclic, or of the form A-B-A, A-B-A, A-B-A.

The Schroyer, Kinney, and Florence Limestones are prominent marine carbonates, whereas the Wymore and Blue Springs Shales are predominantly nonmarine terrigenous mudstones. Therefore, each rhythm of limestone-mudstone or mudstone-limestone is a cyclothem.

Alternatively, cyclic cyclothems can be defined relative to Heckel's (1977) model for "Kansas cyclothems." According to that model, Kansas cyclothems consist of five members (facies associations). In ascending order they are the Outside Shale (nonmarine), Transgressive Limestone (shallow marine), Core Shale (offshore or deepest marine), Regressive Limestone (shallow marine), and Outside Shale (nonmarine) Members. The Wymore and Blue Springs Shales are Outside Shale Members according to this model. The limestones, however, do not contain internal "core shales" and are essentially a combination of both transgressive and regressive facies. A Kansas Cyclothem thus defined at Sites 4 and 5 would be a cycle of the form: Outside Shale—Transgressive/Regressive Limestone—Outside Shale. The Threemile, Schroyer, Kinney, and Florence cyclothems illustrated in Figure 3 have been defined here using such a concept.

T-R UNITS

Heckel (1980) has noted that Kansas cyclothems have genetic significance, because they are T-R units with an average periodicity of about 400,000 years. Kansas cyclothems are thus fifth-order T-R units according to a hierarchical classification of six scales of Permo-Carboniferous T-R units (based on average periodicity) compiled by Busch and others (1985) after Busch and Rollins (1984). Busch and others (1985) have also shown that Permo-Carboniferous sequences contain smaller sixth-order T-R units that are best delineated by utilizing a punctuated aggradational cycle (PAC) approach to outcrop/core analysis (Goodwin and Anderson, 1985).

According to the hypothesis of punctuated aggradational cycles (Goodwin and Anderson, 1985), most sedimentary sequences are composed of T-R units about 3.3 to 16.5 ft (1 to 5 m) thick that generally shallow upward and are bounded by marine transgressive surfaces. Transgressive surfaces are easiest to discern as contacts between a marine facies and a subjacent nonmarine facies that was transgressed. They may also, however, be more cryptic surfaces that separate one marine facies from a second, subjacent marine facies that is relatively more shallow.

PAC-scale (sixth-order) T-R units seem to be mainly the depositional record of cyclic, minor sea-level oscillations caused by climate changes (Busch and Rollins, 1984; Goodwin and Anderson, 1985). Therefore, in nonmarine sequences it is possible to define "climate change surfaces" as PAC boundaries (Busch, 1985). Climate change surfaces are defined as contacts between nonmarine facies presumed to have formed under subaerial conditions (e.g., paleosols or calcretes) and superjacent nonmarine facies presumed to have formed under more humid (or even nonmarine subaqueous) conditions (e.g., coal, lacustrine, facies, or alluvium).

By defining climate change surfaces, transgressive surfaces, and detailed facies changes relative to the Schroyer, Kinney, and Florence Limestone marker beds, the stratigraphic intervals at Sites 4 (Figs. 4, 5) and 5 (Figs. 6, 7) have been subdivided into eight correlative sixth-order T-R units. The numbered sixth-order T-R units can also be arranged into larger fifth-order T-R units that are equivalent in scale to Kansas cyclothems.

DESCRIPTION OF SITE 4

Site 4 is evaluated in terms of Busch and others (1985) (Figs. 4 and 5). Five transgressive surfaces, four climate change surfaces, and 38 numbered facies serve to delineate the eight numbered sixth-order T-R units.

Sixth-order T-R unit number 8 contains the Schroyer Limestone (facies 3 through 8). There is a marginal marine shale facies (facies 2) at the base of this unit, on an obvious transgressive surface. Facies 2 is transgressive, because facies 3 contains a more normal marine fauna and was undoubtedly deposited at maximum transgression within the unit. Facies 4 through 8 are regressive, as diversity of marine taxa is less than it is in facies 3. Regression is also marked by the superjacent facies 9 through 13, which are nonmarine siliciclastic mudstones.

A climate change surface marks the base of sixth-order T-R unit 7, which contains a basal supratidal or lacustrine dolostone (facies 14) and superjacent, regressive, nonmarine siliciclastic mudstones (facies 15 through 17).

Sixth-order T-R unit 6 begins basally at a climate change surface (between facies 17 and 18), marking the start of a transgressive phase that proceeded through the time of deposition of facies 18 and 19. Facies 18 is a shale containing plant fossils that formed above facies 17 (which contains halite crystal molds formed under more arid conditions) during earliest transgression. Facies 19 is a marginal marine facies, containing bivalves, indicating that transgression of marine facies had proceeded to Site xx. The maximum marine transgression is represented by the Kinney Limestone (facies 20), containing a more normal marine fauna. The Kinney is overlain by a nonmarine shale (facies 21), indicating regression only in the top of T-R unit 6.

Sixth-order T-R units 5 and 4 both consist of basal marine carbonates marking transgression and maximum transgression (facies 22 and 27) and superjacent nonmarine siliciclastic mudstones (facies 23 through 26 and 28 through 30), indicating regression. The top of sixth-order T-R unit 5 contains gypsiferous facies (facies 25 through 26), indicating arid conditions, and the top of sixth-order T-R unit 4 (facies 30) is root-mottled, indicating an early stage of soil development (fluvent type).

The red color (oxidized) and root bioturbation of unit 30 seems to indicate paleosol development under subaerial conditions, whereas facies 31 is a bedded olive (reduced?) shale presumed to have formed under subaqueous conditions. Therefore, T-R unit 3 is bounded at its base and top by climate change surfaces.

The climate change surface at the base of sixth-order T-R unit 2 is succeeded by silty bedded mudstones (alluvium?) of facies 33 and 34, another root mottled zone (facies 35), and a blocky mudstone (facies 36). Facies 35 and 36, therefore, probably represent another time of paleosol development.

Sixth-order T-R unit 1 contains at least part of the Florence Limestone. There is a basal marine carbonate facies (facies 37) containing an open marine fauna that represents maximum transgression. It is overlain by facies 38, which contains a less diverse

State: Kansas County: Riley Quadrangle: Manhattan

Locality Description:

Roadcuts along east (units 1 through 8) and west (units 3 through 38) sides of State Route 113, from 1.25 to 1.45 miles south of its intersection with State Route 24, northwest of Manhattan, Kansas: NW1/4 section 35, T9S R7E. This locality is known locally as "Top of the World".

Interval measured May, 1985.

5th order T-R units/boundaries	6th order T-R units/boundaries	UNIT DESCRIPTIONS (Transgressive Surface —— ----- Climate Change Surface)	Unit Thicknesses ft	in	m
FLORENCE	1	38. Florence Limestone: medium to thick bedded, tan to very pale orange, calcilutite (wackestone) with dark gray, contorted, medium beds of chert.	2	8	0.81
		37. Florence Limestone: medium bedded, tan to very pale orange calcilutite (wackestone) with dark gray, contorted, medium beds of chert; fusulinids, crinoids, brachiopods, and bryozoa common.	13	0	3.97
KINNEY	2	36. Blue Springs Shale: dark olive, indurated, blocky claystone.	1	2	0.35
		35. Blue Springs Shale: mottled pale olive to green-gray, indurated, silty mudstone with root mottling in upper foot.	3	3	0.98
		34. Blue Springs Shale: dark brown-red, massive, indurated, silty mudstone.	0	8	0.20
		33. Blue Springs Shale: pale green, flaggy, silty shale.	0	9	0.23
	3	32. Blue Springs Shale: red-brown, thin bedded siltstone and very fine-grained sandstone; massive weathering; green root mottling common throughout unit.	2	0	0.60
		31. Blue Springs Shale: olive, thinly laminated, fissile shale.	0	4	0.10
		----?----------?----------?----------?----------?----------?----------?---------			
	4	30. Blue Springs Shale: same as unit 32.	1	5	0.43
		29. Blue Springs Shale: same as unit 31.	0	7	0.18
		28. Blue Springs Shale: red-brown to olive, silty, flaggy shale.	6	9	2.06
		27. Blue Springs Shale: tan, dolomitic calcilutite (wackestone), skeletal, and fenestral near top; ostracodes, gastropods, bivalves, and brachiopods.	1	11	0.58
	5	26. Blue Springs Shale: dark green-gray, silty shale with gypsum crystals.	1	8	0.51
		25. Blue Springs Shale: dark green-gray, nonbedded, massive, blocky mudstone with gypsum crystals.	0	9	0.23
		24. Blue Springs Shale: dark green-gray to olive clay shale with olive, silty interlaminae; platy to fissile.	2	5	0.73
		23. Blue Springs Shale: dark gray, platy to flaggy clay shale; weathers medium to light gray.	8	2	2.49
		22. Blue Springs Shale: massive, tan, skeletal dolostone.	0	3	0.08
	6	21. Blue Springs Shale: yellow-gray to pale gray, platy to flaggy clay shale.	1	6	0.45
		20. Kinney Limestone: tan, argillaceous calcilutite (wackestone); thin to medium bedded; echinoids, crinoids, bivalves, and high-spired gastropods.	3	9	1.13
		19. Wymore Shale: yellow-gray to tan, laminated to thin bedded, platy to flaggy, silty shale; plant fragments and marine bivalves including Volsellina, Aviculopectin, and Permophorus.	5	9	1.76
		18. Wymore Shale: yellow-gray shale as unit 19; plant stems and geodes.	3	0	0.91
SCHROYER	7	17. Wymore Shale: tan, calcareous siltstone with angular, light green shale lithoclasts and halite crystal molds.	0	2	0.05
		16. Wymore Shale: indurated claystone (basally) becomes more silty and more indurated upwards; almost a siltstone in upper six inches.	4	10	1.65
		15. Wymore Shale: red-brown to dusky red, massive, blocky mudstone.	0	9	0.23
		14. Wymore Shale: yellow-gray to green-gray, silty, argillaceous dolostone with rootlets and invertebrate burrows -- lacustrine limestone?	0	11	0.28
	8	13. Wymore Shale: variegated grayish-red, red-brown, and olive, silty, blocky mudstone; nonbedded.	2	0	0.61
		12. Wymore Shale: green-gray to olive, silty, blocky, nonbedded mudstone.	1	9	0.53
		11. Wymore Shale: green-gray, platy to fissile, clay shale.	0	9	0.23
		10. Wymore Shale: lens to 4 inches thick of platy to fissile, green-gray to black shale with common plants including Neuropteris.	0	4	0.10
		9. Wymore Shale: yellow-gray to pale olive, silty shale.	0	3	0.08
		8. Schroyer Limestone: very pale yellow-orange, skeletal, calcilutite to fine calcarenite; top 3 inches with pebble-sized, rounded intraclasts.	1	10	0.55
		7. Schroyer Limestone: dark gray to black chert; weathers gray to white; fossil fragments of invertebrates and burrows.	0	7	0.18
		6. Schroyer Limestone: very pale orange to tan, skeletal calcilutite.	0	4	0.10
		5. Schroyer Limestone: dusky yellow, mottled grayish-olive, silty, flaggy shale; weathers tan to yellow-gray; crinoids, echinoids, Derbya, and Productids.	1	4	0.40
		4. Schroyer Limestone: dark gray to black, medium to thick bedded chert with fossil fragments and interbedded with thin to medium beds of skeletal calcilutite; crinoids, Derbya, small brachiopods, and Productids.	2	4	0.62
		3. Schroyer Limestone: very pale orange to yellow-gray, skeletal calcilutite (wackestone), massive, with chert nodules, Productids, Composita, crinoids, rugosans, bryozoa, echinoids, and other brachiopods.	1	7	0.48
		2. Havensville Shale: dark gray, platy to flaggy, clay shale with nuculoid bivalves, shark's teeth, and Lingulids plus silty interlaminae.	1	7	0.48
		1. Havensville Shale: dark gray to black, nonindurated clay shale.	1	0	0.30

Figure 4. Description of facies, contacts, and T-R units at site 4.

Figure 5. Photograph of part of site 4, showing exact locations of transgressive surface at the base of sixth-order T-R unit 1 (T_1), climate change surfaces at the base of sixth-order T-R units 2 (C_2) and 3 (C_3), and transgressive surface at the base of sixth-order T-R unit 4 (T_4).

fauna that probably lived under more restricted or shallow conditions than facies 37. The top of unit 1 is eroded.

Fifth-order T-R units can also be defined at Site 4 (Fig. 4) by considering the relative extent of marine transgression within the sixth-order T-R units. For example, marine facies are well developed in sixth-order T-R unit 7. Thus net shallowing is indicated at the fifth-order scale from sixth-order T-R unit 8 to 7 for the Schroyer fifth-order T-R unit.

The Kinney fifth-order T-R unit has well-developed marine facies basally in sixth-order T-R unit 6, but there is relatively poor development of marine facies in superjacent sixth-order T-R units 5 and 4. Finally, sixth-order T-R units 3 and 2 have no marine facies. Thus, the Kinney fifth-order T-R unit is a transgressive deposit from facies 18 through 20 of sixth-order T-R unit 6 (the latter representing maximum transgression). It is a regressive deposit from facies 21 through facies 36 (top of sixth-order T-R unit 6 through sixth-order T-R unit 2). Most sixth-order T-R units (PACs) of the Kinney fifth-order T-R unit are highly asymmetric in favor of thin or absent transgressive facies and thick regressive facies, as predicted by the hypothesis of punctuated aggradational cycles (Goodwin and Anderson, 1985). A notable exception is sixth-order T-R unit 6 at this site. The Kinney fifth-order T-R unit at Site 4 is also a good example of a shallowing sequence of PACs.

Sixth-order T-R unit 1 contains well-developed marine facies, especially at its base, compared to sixth-order T-R unit 2. Therefore, it is the basal sixth-order T-R unit of the Florence fifth-order T-R unit.

DESCRIPTION OF SITE 5

Site 5 is also evaluated in terms of Busch and others (1985) (Figs. 6, 7). Five transgressive surfaces, two climate change surfaces, and 29 numbered facies serve to delineate the eight numbered sixth-order T-R units.

Sixth-order T-R unit 8 is incomplete but contains at least the top of the Schroyer Limestone (facies 1). Nonmarine facies 2 and 3 above the Schroyer are largely concealed but represent shallowing to the top of the unit.

Sixth-order T-R unit 7 is transgressive, and reached maximum transgression, in facies 4. Shallowing occurred through the superjacent nonmarine shale at the top of the unit (facies 5).

Sixth-order T-R unit 6 was initiated as the transgressive to maximum transgressive facies 6 (Kinney Limestone) developed at its base. Shallowing through the top of the unit is indicated by marine claystones of facies 7 (containing mainly bivalves and lingulids) and superjacent nonmarine claystones of facies 8.

Sixth-order T-R unit 5 contains a basal grainstone with

State: Kansas **County:** Geary **Quadrangle:** Ogden

Locality Description:

Roadcuts along the north side of Interstate Route 70, between mile markers 306.6 and 307, starting 0.85 mile west of the overpass for Exit 307 to McDowell Creek Road: NW1/4 section 28, T11S R7E.

Crouse Limestone through superjacent, cherty Threemile Limestone is exposed adjacent to the western end of the entrance ramp for westbound I-70 from McDowell Creek Road

Interval measured May, 1985.

5th order T-R units/boundaries	6th order T-R units/boundaries	UNIT DESCRIPTIONS (Transgressive Surface ——— ----- Climate Change Surface)	Unit Thicknesses ft	in	m
FLORENCE	1	29. Florence Limestone: interbedded tan, skeletal calcilutite and gray chert; common marine fossils including brachiopods, fusulinids, bryozoa, crinoids.	12+	0	3.66
		28. Florence Limestone: pale yellow-orange dolostone, thin to medium bedded, with argillaceous partings; Derbya, Productids, bryozoa, gastropods, crinoids, Chonetids, echinoids, microgastropods.	4	8	1.60
KINNEY	2	27. Blue Springs Shale: olive to dark gray, blocky to flaggy, clay shale.	1	1	0.32
		26. Blue Springs Shale: olive, weathers pale olive, fissile to platy clay shale.	5	9	1.76
	3	25. Blue Springs Shale: red-brown siltstone to very fine-grained sandstone, massive, with green root mottling.	1	6	0.45
		24. Blue Springs Shale: same as unit 25.	0	5	0.13
		23. Blue Springs Shale: pale green, poorly indurated, silty shale.	0	7	0.18
		----?----------?----------?----------?----------?----------?----------?---------			
	4	22. Blue Springs Shale: same as unit 25.	1	2	0.36
		21. Blue Springs Shale: same as unit 23.	0	6	0.15
		20. Blue Springs Shale: intraclastic limestone lens (caliche?)	0	1	0.02
		19. Blue Springs Shale: brown-red, silty, platy to fissile, clay shale.	2	11	0.88
		18. Blue Springs Shale: olive to dark gray, indurated, blocky mudstone.	0	11	0.28
		17. Blue Springs Shale: pale gray-green, platy to fissile, clay shale.	0	9	0.23
		16. Blue Springs Shale: pale gray-green, platy to fissile, clay shale.	0	10	0.25
		15. Blue Springs Shale: pale gray-green, silty, flaggy shale with common fossil fragments of invertebrate taxa.	0	8	0.20
		14. Blue Springs Shale: pale gray-green, weathers tan to red-brown, skeletal calcilutite to fine-grained calcarenite (wackestone to packstone); ostracodes, gastropods, and brachiopods.	3	1	0.94
	5	13. Blue Springs Shale: very pale olive, platy to fissile, nonindurated shale with gypsum veins and crystals.	2	5	0.73
		12. Blue Springs Shale: brown-red to magenta, fissile, nonindurated clay shale with gypsum veins and crystals.	1	0	0.30
		11. Blue Springs Shale: gray to olive, platy shale with rare plant fragments.	3	2	0.96
		10. Blue Springs Shale: flaggy to blocky, calcareous claystone; pale yellow-orange to tan; common fragments of marine fossils.	1	9	0.53
		9. Blue Springs Shale: skeletal calcarenite (grainstone) with an abundant and diverse marine faunal association.	0	4	0.10
	6	8. Blue Springs Shale: flaggy to blocky, calcareous claystone; pale yellow-orange to tan.	10	0	3.05
		7. Blue Springs Shale: pale yellow-orange to tan, platy to flaggy, clay shale with microfossils, bivalves, and Lingulids.	2	0	0.61
		6. Kinney Limestone: gray, very crinoidal, intraclastic wackestone to packstone with brachiopods; weathers tan and massive.	1	2	0.36
SCHROYER	7	5. Wymore Shale: tan to light brown, papery to flaggy clay shale.	1	8	0.51
		4. Wymore Shale: tan to light brown, flaggy, calcareous, clay shale with marine bivalves.	0	6	0.15
	8	3. Wymore Shale: fissile to flaggy, tan clay shale.	6	0	1.83
		2. Wymore Shale: concealed interval.	16	0	4.88
		1. Schroyer Limestone: top of unit consisting of thinly interbedded, tan, skeletal calcilutite and gray chert; common marine fossils.	-	-	-

Figure 6. Description of facies, contacts, and T-R units at site 5.

good marine faunal diversity (facies 9) that probably developed as a transgressive lag or at maximum transgression. Superjacent facies 10 contains only fragments of marine taxa, and facies 11 through 13 are nonmarine. This indicates shallowing from facies 9 through facies 13. Gypsiferous facies 12 and 13 at the top of unit 5 indicate an arid paleoenvironment.

Sixth-order T-R unit 4 also contains a basal skeletal carbonate facies with marine taxa (facies 14) that is overlain by a flaggy shale having only fragments of marine taxa (facies 15). Superjacent facies 16 through 22 are nonmarine, including a rooted zone (poorly developed paleosol) at its top (facies 22). Therefore, T-R unit 4 exhibits its maximum transgressive facies basally (facies 14), as seen in sixth-order T-R unit 5.

As seen at Site 4, the oxidized red color and root bioturbation of facies 22, 24, and 25 seem to indicate paleosol development under subaerial conditions, whereas facies 23 and 26 are bedded green (reduced?) mudstones that are probably floodplain alluvium. Thus, climate change surfaces bounding sixth-order T-R unit 3 (Fig. 7) are present between facies 22 and 23 and between facies 25 and 26.

The climate change surface between facies 25 and 26 also marks the base of sixth-order T-R unit 2. The top of facies 27 at the top of T-R unit 2 is blocky in part and may represent a poorly developed paleosol (Fig. 7).

The base of T-R unit 1 is marked by a very well developed transgressive surface between facies 27 and 28 (Fig. 7). Facies 28 is a fine-grained dolostone containing a variety of marine taxa. It is a transgressive facies, as facies 29 is more skeletal, contains fusulinids, and probably represents more open or deeper conditions than facies 28. Facies 29 probably represents maximum transgression within sixth-order T-R unit 1, though the unit is incomplete.

Fifth-order T-R units can be defined for Site 5 (Fig. 6), as was done at Site 4 (Fig. 4), by considering the relative extent of marine transgression within the sixth-order T-R units. For example, marine facies are once again better developed in sixth-order T-R unit 8 than they are in sixth-order T-R unit 7. This indicates net shallowing in the represented upper portion of the Schroyer fifth-order T-R unit.

The Kinney fifth-order T-R unit contains the prominent marine facies of the Kinney Limestone at the base of sixth-order T-R unit 6, indicating transgression and the development of maximum transgression at the fifth-order scale with deposition of facies 6. Shallowing through the remainder of this fifth-order T-R unit is indicated by the less well developed marine facies in sixth-order T-R units 5 and 4, plus the lack of marine facies in sixth-order T-R units 3 and 2. As at Site 4, the sixth-order T-R units (PACs) of the Kinney fifth-order T-R unit have thin or absent transgressive facies overlain by thicker regressive (aggradational) facies, as predicted by the hypothesis of punctuated aggradational cycles (Goodwin and Anderson, 1985). This includes unit 6 at this site. The Kinney fifth-order T-R unit at Site 5 is also a good example of a shallowing sequence of PACs.

The base of the Florence fifth-order T-R unit is marked by prominent marine facies in sixth-order T-R unit 1, as compared to sixth-order T-R unit 2, which has no marine facies.

Figure 7. Photograph of part of site 5, showing exact locations of transgressive surface at base of sixth-order T-R unit 1 (T_1), climate change surfaces at base of sixth-order T-R units 2 (C_2) and 3 (C_3), and transgressive surface at base of sixth-order T-R unit 4 (T_4).

CORRELATION OF T-R UNITS

Correlation of the fifth-order T-R units was achieved, because at both sites the Schroyer fifth-order T-R unit contains the Schroyer cherty limestone marker bed, the Kinney fifth-order T-R unit contains the Kinney limestone marker bed, and the Florence fifth-order T-R unit contains the Florence cherty limestone marker bed. A similar rationale can be used to confirm the correlation of sixth-order T-R units. For example, sixth-order T-R unit 8 contains the Schroyer marker bed at both Sites, sixth-order T-R unit 6 contains the Kinney marker bed at both Sites, and sixth-order T-R unit 1 contains the Florence marker bed at both Sites.

Correlation of sixth-order T-R units 5, 3, and 2 can be confirmed using less conspicuous marker beds that were revealed after the above correlation of sixth-order T-R units 8, 6, and 1 was achieved using the obvious marker beds. For example, there are gypsiferous facies at the top of sixth-order T-R unit 5 at both sites, and red paleosols are developed at the top of sixth-order T-R units 4 and 3 at both sites.

SUMMARY

Cyclothem methods of stratigraphic analysis used at Sites 4 and 5 revealed facies rhythms and cycles at the fifth-order scale of observation. On the other hand, a hierarchical T-R unit method of stratigraphic analysis, that incorporated the PAC-approach, revealed more detailed information at the fifth- and sixth-order scales of observation. For example, there are unnamed marine zones developed at the base of sixth-order T-R units 5 and 4 within the Blue Springs Shale (a nonmarine Outside Shale Member based on the Kansas cyclothem concept). As the entire Kinney fifth-order T-R unit probably represents a depositional period of about 400,000 years, sixth-order T-R units 6 through 2 within it have an average periodicity of about 80,000 years (i.e., are PAC scale).

Interpretation of these two sites is based mainly upon outcrop analysis and is a modification of Busch and others (1985), who recognized only seven sixth-order T-R units in the same interval. More detailed laboratory investigation may reveal even more cryptic sixth-order T-R units within this interval.

REFERENCES CITED

Busch, R. M., 1985, Stratigraphic Analysis of Pennsylvanian Rocks Using a Hierarchy of Transgressive-Regressive Units [Ph.D. thesis]: Pittsburgh, Pennsylvania, University of Pittsburgh; Ann Arbor, Michigan, University Microfilms International, 427 p.

Busch, R. M., and Rollins, H. B., 1984, Correlation of Carboniferous strata using a hierarchy of transgressive-regression units: Geology, v. 12, p. 471–474.

Busch, R. M., West, R. R., Barrett, F. J., and Barrett, T. R., 1985, Cyclothems versus a hierarchy of transgressive-regressive units, *in* Watney, W. L., Kaesler, R. L., and Newell, K. D., eds., Recent Interpretations of Late Paleozoic Cyclothems: Conference Symposium, Society of Economic Paleontologists and Mineralogists, Mid-Continent Section, p. 141–153.

Goodwin, P. W., and Anderson, E. J., 1985, Punctuated aggradational cycles; A general hypothesis of episodic stratigraphic accumulation: Journal of Geology, v. 93, p. 515–533.

Heckel, P. H., 1977, Origin of phosphatic black shale facies in Pennsylvanian cyclothems of Midcontinent North America: American Association of Petroleum Geologists Bulletin, v. 61, p. 1045–1068.

——, 1980, Paleogeography of eustatic model for deposition of Midcontinent Upper Pennsylvanian cyclothems, *in* Fouch, T. D., and Magathan, E. R., eds., Paleozoic Paleogeography of the west-central United States: Paleogeography Symposium 1, Society of Economic Paleontologists and Mineralogists, Rocky Mountain Section, p. 197–215.

North American Commission on Stratigraphic Nomenclature, 1983, North American stratigraphic code: American Association of Petroleum Geologists Bulletin, v. 67, p. 841–875.

O'Connor, H. G., Zeller, D. E., Bayne, C. K., Jewett, J. M., and Swineford, A., 1968, Permian System, *in* Zeller, D. E., ed., The stratigraphic succession in Kansas: State Geological Survey of Kansas Bulletin, v. 189, p. 43–53.

Wanless, H. R., and Weller, J. M., 1932, Correlation and extent of Pennsylvanian cyclothems: Geological Society of America Bulletin, v. 47, p. 1177–1206.

Barneston Limestone (Lower Permian) of eastern Kansas

Page C. Twiss and James R. Underwood, Jr., Department of Geology, Kansas State University, Manhattan, Kansas 66506

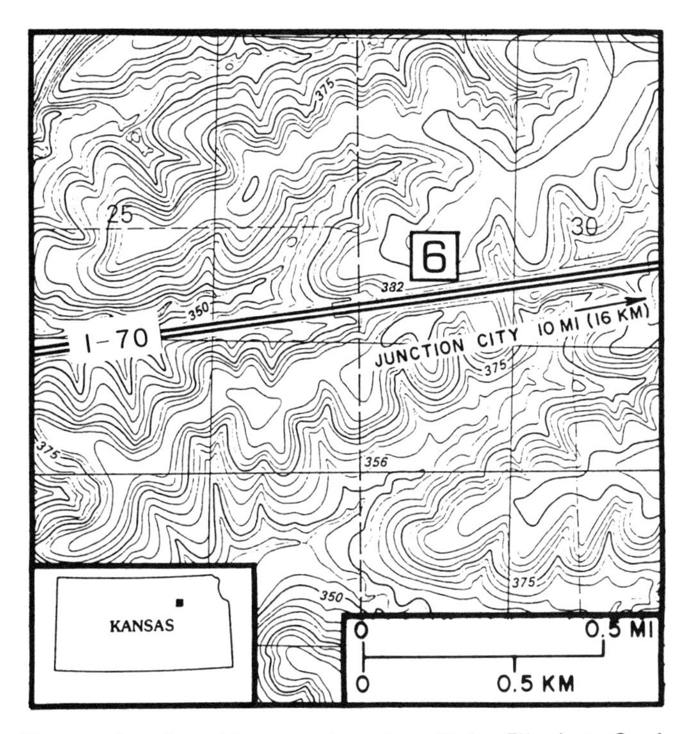

Figure 1. Location of Barneston Limestone, Ogden 7½-minute Quadrangle, Geary County, Kansas. Site number is at milepost 305.

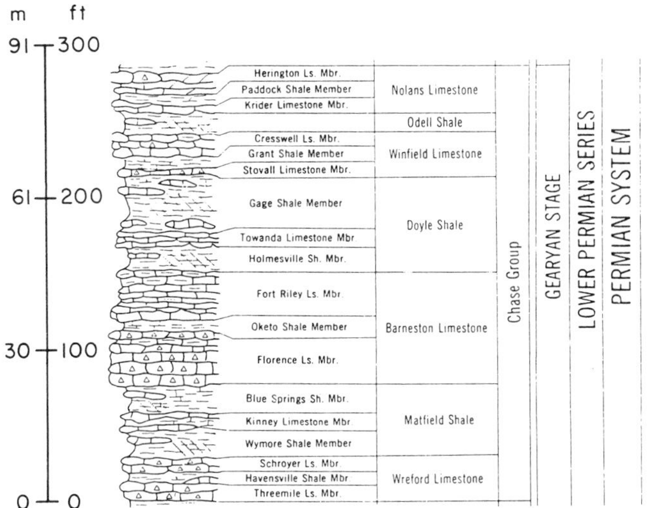

Figure 2. Stratigraphic section of Chase Group, Gearyan Stage, Lower Permian Series, Permian System in Kansas (adapted from Zeller, 1968).

LOCATION

The Barneston Limestone is well exposed on the north side of the road cut at milepost 305 on I-70 in NW¼SW¼Sec.30, T.11S.,R.6E., Geary County, Kansas (Fig. 1), about 10 mi (16 km) east of Junction City. Vehicles traveling east should continue to exit 307 (McDowell Creek Road), pass under I-70, and return westward to milepost 305. Because Kansas law prohibits pedestrians from crossing a divided highway, vehicles should be parked on the north side, off the pavement and passengers should exit on the right-hand side. Keep far away from the traffic lanes while examining the rocks. The upper two-thirds of this section can be examined closely by walking up the hill on the west side of the road cut.

SIGNIFICANCE

The Florence Limestone Member of the Barneston Limestone is the thickest of the chert-bearing limestone beds in the Lower Permian rocks of Kansas. With the other limestone units of the Chase Group it forms the "backbone" of the Flint Hills Uplands, a physiographic feature that trends northward through east-central Kansas (Frye and Leonard, 1959). Twenty-eight horizontal layers of chert nodules that are enclosed in skeletal calcilutite occur in this road cut. This site offers the opportunity to make observations on the origin of chert and depositional environments of the limestone, siltstone, and mudstone.

SITE INFORMATION

The Chase Group (Fig. 2) is about 280 ft (85 m) thick and consists of alternating limestone and shale couplets. The Barneston Limestone consists of three members, in ascending order: Florence Limestone Member, Oketo Shale Member, and Fort Riley Limestone Member. Exposed at milepost 305 are the upper part of the Blue Springs Shale Member of the Matfield Shale and the lower approximately 19 ft (6 m) of the overlying Barneston Limestone. In addition to the 28 nodular chert beds in the Florence, the most arresting aspect of the outcrop and the adjacent slopes is the prominent "rimrock" limestone of the lower part of the Fort Riley. Figure 3 is a view of the road cut at this site, and Figure 4 is a measured section of the outcrop. Details of this section are in Table 2.

Units 1, 2, and 3 of the Blue Springs Shale Member (Fig. 4) have gradational boundaries, lack macrofossils, and probably represent phase 2 of the idealized Permian cyclothem (Table 2) of Elias (1937). The contact between the Blue Springs Shale Member and the Florence Limestone Member is sharp and undulating (Fig. 5A). Unit 4, the basal unit of the Florence Limestone Member, contains the lowest occurrence of platy algal fragments; these are overlain by echinoid spines and brachiopod fragments. This unit probably represents phase 5 or 6 of Elias's (1937) idealized cyclothem. Unit 6 marks the lowest of the chert-bearing horizons that extend upward through Unit 12 (Fig. 5B); the chert occurs in skeletal calcilutite. The chert nodules contain silicified fragments of algae, brachiopods, echinoids, and bryozoans in the same proportion as calcitic ones in the limestone; this interval includes phase 6 of Elias's (1937) cyclothem model.

The Oketo Shale Member, Unit 15, is a medium gray mud-

Figure 3. Barneston Limestone, general view northeast, showing thick Florence Limestone Member, thin Oketo Shale Member along line of bushes in upper quarter, and Fort Riley Limestone Member, with massive "rimrock" near top.

TABLE 1. IDEALIZED PERMIAN CYCLOTHEM OF ELIAS (1937), SHOWING PROGRESSIVE (TRANSGRESSIVE) AND REGRESSIVE HEMICYCLES, PHASES OF PALEOBIOTOPES, AND CORRESPONDING LITHOLOGIES

	No.	Phases established chiefly on paleontologic evidence	Corresponding typical lithology
Regressive hemicycle	1.	Red shale	Clayey to fine sandy shale, rarely consolidated.
	2r.	Green shale	
	3r.	Lingula phase	Sandy, often varved (?), rarely clayey shale.
	4r.	Molluscan phase	Clayey shale, mudstone to bedded limestone.
	5r.	Mixed phase	Massive mudstone, shaly limestone.
	6r.	Brachiopod phase	Limestone, flint, calcareous shale.
	7.	Fusulinid phase	
Progressive hemicycle	6p.	Brachiopod phase	
	5p.	Mixed phase	Massive mudstone, shaly limestone.
	4p.	Molluscan phase	Clayey shale, mudstone to bedded limestone.
	3p.	Lingula phase	Sandy, often varved (?), rarely clayey shale.
	2p	Green shale	Clayey to fine sandy shale, rarely consolidated.
	1.	Red shale	

Stage / Group / Formation	Member	Unit	Thick. m	Thick. ft	Elias' Phases (1937)
Gearyan Stage, Chase Group		S			
Barneston Limestone	Fort Riley Ls. Mbr.	18	2.2	6.6	6. Brachiopod phase
		17	1.3	4.3	
		16	2.4	8.0	
	Oketo Sh. Mbr.	15	0.4	1.5	
		14	0.5	1.7	
	Florence Limestone Member	13	0.8	2.5	
		12	0.8	2.5	
		11	0.6	2.0	
		10	1.6	5.1	
		9	1.6	5.3	
		8	1.8	6.0	
		7	1.1	3.6	
		6	0.1	0.3	
		5	0.4	1.4	
		4	0.5	1.5	
Matfield Shale	Blue Springs Shale Mbr.	3	0.06	0.2	2. Green shale phase
		2	0.7	2.3	
		1	1.3	4.2	

Figure 4. Measured section of Upper Matfield Shale and Barneston Limestone, at this site. Measured by P. C. Twiss and J. R. Underwood, Jr., 1985.

stone that is about 2 ft (0.6 m) thick; it has not been examined for microfossils. The contact with the overlying Fort Riley Limestone Member is abrupt.

Unit 17 is the massive Fort Riley "rimrock" that forms a prominent bench on grass-covered slopes and is a good marker bed on aerial photographs. It is interstratified with clayey skeletal calcilutite of Units 16 and 18. The Fort Riley Limestone lacks chert but contains the same skeletal debris as the Florence and probably belongs to phase 6 of Elias (1937). At other localities in Riley and Geary counties, Hattin (1957) and Grossnickle (1961) reported a molluscan fauna in the Fort Riley Limestone Member.

Jewett (1933) recognized that the Lower Permian rocks of Kansas, including those of the Chase Group, consist of from eight to ten cycles of nonmarine and marine sedimentation. Moore (1936) designated these cycles as Permian cyclothems and named each for the limestone unit that represented the maximum marine transgression. The Permian cyclothems contain beds of red and green siltstone and mudstone and evaporites, but lack the sandstone, underclay, coal, and "core shale" of the Pennsylvanian cyclothems of Kansas (Heckel, 1977).

Elias (1937) also recognized the vertical cyclicity of the terrestrial and marine rocks in the Lower Permian and proposed seven transgressive-regressive phases of fossil assemblages and corresponding lithologies (Table 1). He noted that one cycle extended from red mudstone of the upper Blue Springs Shale, through the Barneston Limestone, and up to the red mudstone of the middle of the overlying Holmsville Shale. This cyclothem marks the uppermost known occurrence of fusulinids (phase 7) in

TABLE 2. ROCK DESCRIPTIONS FOR FIGURE 4:

Unit S. Soil and regolith—not measured

Unit 18. Fort Riley Limestone Member, Barneston Limestone. Calcareous mudstone and clayey calcilutite; upper few ft (1 m) covered; forms a shalelike slope; skeletal fragments of brachiopods, crinoids, and fenestrate bryozoans, many encrusted with algae, occur on weathered slope; grayish yellow (5Y7/4) to grayish orange (10YR7/4), weathers light brownish gray (5YR6/1) to pale yellowish orange (10YR7/6).

Unit 17. Skeletal calcilutite; porous; algal-encrusted crinoid fragments, crinoid columnals, and trace of echinoid spines; grayish yellow (5Y8/4), weathers brownish gray (5YR7/1); massive ridge former; principal joint set N65°E, secondary joint set N20°W, joints widened 2 to 4 in (5 to 10 cm) by solution; upper surface is rounded and pitted by solution; well-developed miniature karst surface, conical solution pits 0.25 in (0.6 cm), most are vertical or near vertical; some 2 to 10 in (5 to 25 cm) solution pans contain soil and support vegetation. Fort Riley "rimrock" forms prominent outcrop on grass-covered slopes; good marker bed on aerial photos.

Unit 16. Clayey calcilutite; poorly exposed except in vertical face; thin-bedded, lacks chert; grayish yellow (5Y8/4), weathers yellowish brown (10YR7/6); sparsely fossiliferous containing echinoid spines, algal-coated skeletal debris of brachiopods and bryozoans.

Unit 15. Oketo Shale Member, Barneston Limestone. Shale; medium gray (N5), weathers easily; very thin bedded; no fossils observed.

Unit 14. Florence Limestone Member, Barneston Limestone. Skeletal calcilutite; massive; yellowish gray (5Y7/2), weathers yellowish gray (5Y7/1) to medium gray (N5); silicified skeletal fragments abundant; no chert.

Unit 13. Clayey calcilutite or marl; largely covered; forms shalelike slope.

Unit 12. Chert-bearing skeletal calcilutite; massive; colors same as unit 7; two chert horizons in upper 1 ft (30 cm); upper chert horizon contains nodules 2 to 5 in (5 to 25 cm) in diameter; lower-chert horizon, 6 to 12 in (15 to 30 cm) below upper chert, 1 to 5 in (2.5 to 12.5 cm) thick and continuous; chert with silicified echinoid spines, brachiopod fragments, and fenestrate bryozoans enclosed in skeletal calcilutite; forms vertical cliff.

Unit 11. Clayey calcilutite or marl; shalelike slope; less resistant to weathering than lower units; color as in unit 7; largely covered.

Unit 10. Chert-bearing skeletal calcilutite; contains 5 chert zones evenly spaced vertically; lower 4, 2.5 in (6.5 cm) thick; uppermost, 6 in (15 cm) thick and more continuous; capped by skeletal calcilutite 0.5 to 2 in (1.5 to 5 cm) thick, sparsely fossiliferous with algal-coated grains; some brachiopod and fenestrate bryozoan fragments; not as fossiliferous as lower units; color as in 7; forms near-vertical cliff.

Unit 9. Chert-bearing skeletal calcilutite; massive, forms near-vertical cliff; color as in unit 7; six horizons of irregular chert nodules. Capped by 5-in (12.5-cm) thick skeletal calcilutite with some chert nodules, 0.5 to 1 in (1 to 2.5 cm) diameter, more rounded and rodlike; calcilutite contains brachiopod fragments, whole valves of *Derbyia* sp., echinoid spines, and fenestrate bryozoan fragments. 5th chert horizon approximately 6 in (15 cm) thick; nodules in skeletal calcilutite. 4th chert horizon 20 in (50 cm) thick and in contact with third horizon; irregular tabular chert masses 2 to 4 in (5 to 10 cm) long and 2 to 3 in (5 to 7.5 cm) thick, smaller than those below. 3rd chert horizon 7 in (18 cm) above 2nd; lower half of horizon contains horizontal and discontinuous chert zones. 2nd chert horizon; base 12 in (30 cm) above 1st chert horizon; ranges from 2 to 6 in (5 to 15 cm) thick; intervening skeletal calcilutite. Skeletal calcilutite; very irregular chert nodules 0 to 3 in (0 to 8 cm) thick, elongated horizontally; not as numerous as in unit 8.

Unit 8. Chert-bearing calcilutite; massive, forms vertical cliff; weathers to rubble of chert masses; color as in unit 7; 9 horizons of discontinuous, horizontally elongated chert masses enclosed in skeletal calcilutite; contains algal-coated skeletal grains of fenestrate bryozoans and algae, traces of brachiopod and echinoid fragments and spines; upper surface is microkarst with silicified fenestrate bryozoan fragments. Skeletal calcilutite cap, 4 in (10 cm) thick. 9th chert horizon: 1.5 to 3 in (4 to 7.5 cm) thick; 4 to 5 in (10 to 12 cm) above 8th. 8th chert horizon: 2 to 5 in (5 to 12 cm) thick; 7 in (18 cm) above 7th. 7th chert horizon: 0 to 5 in (0 to 12.5 cm) thick; 5 in (12.5 cm) above 6th. 6th chert horizon: 2 to 5 in (5 to 12.5 cm) thick; 12 in (30 cm) above 5th. 5th chert horizon: 2 to 7 in (5 to 18 cm) thick; 3 in (7.5 cm) above 4th. 4th chert horizon: 1 to 5 in (2.5 to 12.5 cm) thick; 2 in (5 cm) above 3rd. 3rd chert horizon: 2 to 4 in (5 to 10 cm) thick; 1 to 3 in (2.5 to 7.5 cm) above 2nd. 2nd chert horizon: 2 to 5 in (5 to 12.5 cm) thick; 2 to 4 in (5 to 10 cm) above 1st. 1st chert horizon: 2 to 3 in (5 to 7.5 cm) thick; 5 in (12.5 cm) above top of unit 7.

Unit 7. Chert-bearing calcilutite; massive, forms vertical cliff; contains algal-coated fenestrate bryozoans, echinoid spines, and miscellaneous skeletal debris; yellowish gray (5Y7/2), weathers yellowish gray (5Y7/1) to medium gray (N5); contains 5 chert horizons. 5th chert horizon: 2 in (5 cm) thick; 6 in (15 cm) above 4th. 4th chert horizon: 3 in (7.5 cm) thick; 6 in (15 cm) above 3rd. 3rd chert horizon: 4 to 5 in (10 to 12.5 cm) thick; horizontal diameter up to 14 in (35 cm); 8 in (20 cm) above 2nd. 2nd chert horizon: 4 to 5 in (10 to 12.5 cm) thick; horizontal diameter up to 14 in (35 cm); 2 in (5 cm) above 1st. 1st chert horizon: about 2 in (5 cm) thick; horizontal diameter up to 8 in (20 cm). Chert nodules range from very light gray (N8) to medium gray (N5), weather medium gray (N5) or dark yellowish orange (10YR6/6); nodules contain algal-coated, mostly silicified skeletal debris; chert nodules irregular, discontinuous, and elongate parallel to bedding.

Unit 6. Clayey chert-bearing calcilutite; more calcareous and more resistant than unit 5; contains some widely spaced chert masses 2 in (5 cm) thick and 2 to 4 in (5 to 10 cm) across horizontally; color as in 5.

Unit 5. Clayey calcilutite; platy and lenticular; yellowish gray (5Y7/1) mottled to moderate yellowish brown (10YR6/4); contains crinoid columnals and fragments of brachiopods and platy algae.

Unit 4. Clayey calcilutite; massive, forms steep slope; light gray (N7), weathers very pale orange (10YR7/2), horizontal zones of grayish orange (10YR7/4); contains platy algal fragments near base that decrease upward; echinoid spines and brachiopod fragments in upper part; possibly bioturbated near top; basal contact sharp and undulating.

Unit 3. Blue Springs Shale Member, Matfield Shale. Mudstone; weathered; dark yellowish brown (10YR4/2); tabular bedding; nonfossiliferous; transitional with unit 2 at base, but forms sharp undulating contact with overlying Florence Limestone.

Unit 2. Clayey calcilutite; blocky structure; yellowish gray (5Y7/2), weathers pale yellowish orange (10YR7/2) with moderate yellowish brown (10YR5/4) iron-oxide splotches; nonfossiliferous; forms steep slope.

Unit 1. Mudstone; greenish gray (5GY5/1), weathers light greenish gray (5GY7/1); nonfissile, tabular to blocky; nonfossiliferous; forms shalelike slope.

the Permian rocks of Kansas; they occur near the base of the Florence Limestone Member. Elias recognized all seven phases of paleobiotopes in the lower part of the sequence, which according to his diagram (1937, p. 406) represented a rapid marine transgression within about 15 ft (4.5 m) of vertical section.

Moore (1964) described 23 different paleoecosystems that occur in the Pennsylvanian and Permian cyclothems of Kansas. He showed the "Tarkio-Type (*Triticites*) Assemblage" in north-central Kansas to be about 5 ft (1.5 m) above the base of the Florence Limestone Member. No fusulinids were observed in Section 9, but crinoid columnals and skeletal fragments of brachiopods, echinoderms, and bryozoans, many of which are coated with algae, occur throughout the limestone.

The origin of the chert in the Barneston Limestone is unknown. Hattin (1957), in his study of the Wreford Megacyclothem (Fig. 2), recorded the stratigraphic and lithologic relationships of the nodular and bedded cherts in the limestone and noted the similarities with the overlying Barneston Limestone. He described compact, noncalcareous chert that he thought to be a primary inorganic precipitate in relatively deep seawater of normal salinity. Support for this origin is based on the following: (1) many chert beds are laterally persistent over long distances without change in thickness or in position; (2) most chert beds occur between, and are parallel to, bedding planes; (3) isolated chert nodules tend to be elliptical with long axes parallel to bedding; and (4) larger skeletal fragments are concentrated on the surface of the nodules.

These characteristics, except for (4), are typical of the Florence Limestone Member at Section 9. There, the chert nodules contain silicified skeletal debris of brachiopods, echinoderms, and bryozoans in the same proportion as the calcitic skeletal debris that occurs in the enclosing calcilutite. No concentration of skeletal debris on the surface of the chert nodules was observed. The areal and stratigraphic distribution of the chert has not been determined.

Although gypsum occurs in the subsurface in many beds of red mudstone and siltstone of the Chase Group, none was observed in the Florence Limestone. The diversity and number of fossils, both in the limestone and in the chert, are too high to support the hypothesis that chert has replaced original gypsum in limestone.

Detailed stratigraphic, petrographic, and geochemical studies of these chert horizons and the enclosing limestone are lacking, but examination of this section may provide some hypotheses for the origin of the chert.

A

B

Figure 5. Barneston Limestone: A: Undulating and sharp contact between base of Florence Limestone Member (Unit 4) and top of Blue Springs Shale Member (Unit 3). Jacob staff 5.3 ft (1.6 m) long. B: Unit 7, Florence Limestone, showing lighter gray chert nodules at the lower third, middle, and upper quarter of Jacob staff. Visible height of staff 4.9 ft (1.5 m).

REFERENCES

Elias, M. K., 1937, Depth of deposition of the Big Blue (Late Paleozoic) sediments in Kansas: Geological Society of America Bulletin, v. 48, p. 403–432.

Frye, J. C., and Leonard, A. B., 1959, Flint Hills physiography: Wichita, Kansas Geological Society 24th Field Conference Guidebook, p. 79–85.

Grossnickle, W. E., 1961, Petrology of the Fort Riley Limestone from four Kansas quarries [M.S. thesis]: Manhattan, Kansas State University, 82 p.

Hattin, D. E., 1957, Depositional environment of the Wreford megacyclothem (Lower Permian) of Kansas: Kansas Geological Survey Bulletin 124, 150 p.

Heckel, P. H., 1977, Origin of phosphatic black shale facies in Pennsylvanian cyclothems of Mid-Continent North America: American Association of Petroleum Geologists Bulletin, v. 61, p. 1045–1068.

Jewett, J. M., 1933, Evidence of cyclic sedimentation in Kansas during the Permian Period: Kansas Academy of Science Transactions, v. 36, p. 137–140.

Moore, R. C., 1936, Stratigraphic classification of the Pennsylvanian rocks of Kansas: Kansas Geological Survey Bulletin 22, 256 p.

——, 1964, Paleoecological aspects of Kansas Pennsylvanian and Permian cyclothems: Kansas Geological Survey Bulletin 169, v. 1, 287–380.

Zeller, D. E., 1968, The stratigraphic succession in Kansas: Kansas Geological Survey Bulletin 189, 81 p.

The Crouse Limestone in north-central Kansas; Lateral changes and paleoecological analysis

R. R. West and P. C. Twiss, Department of Geology, Thompson Hall, Kansas State University, Manhattan, Kansas 66506

LOCATION

Exposures of the Crouse Limestone along I-70 south and east of Manhattan, Kansas (Fig. 1), illustrate the stratigraphy, structural geology, palaeoecology, and cyclic sedimentation in the Permian rocks of eastern Kansas. This text provides details of four exposures (between mileposts 319 and 305) that emphasize the interrelationships among these four areas of geology.

GENERAL SETTING

The Crouse Limestone is one of twelve limestone formations that compose the Gearyan Stage of the Lower Permian in eastern Kansas. These formations crop out in a band that strikes northeast-southwest and dips, regionally, at a low angle, to the northwest. The Crouse is underlain by the Easly Creek Shale and overlain by the Blue Rapids Shale (Fig. 2).

Twenty-three stratigraphic sections of the Crouse Limestone have been studied in detail (Fig. 1). These detailed studies began with the work of Huber (1965) and continued with the work of West and others (1972) and Voran (1977).

Upper and lower limestone members of the Crouse Limestone are separated by a calcareous, and in places fossiliferous, mudstone (Fig. 3). Generally, the upper limestone is a thin to platy-bedded, yellowish brown to gray limestone that often weathers like a platy shale. This upper unit contains ostracodes and foraminiferids and has the planar laminations of a stromatolitic limestone. The basal limestone of the Crouse is nearly constant in thickness, is typified by molluscs (especially bivalves), and ranges from a thin and platy-bedded to a single massive bed of clayey limestone.

West and others (1972) reported that the Crouse reaches a maximum thickness of 18 ft (5.5 m) in central Kansas. It thins to less than 8 ft (2.4 m) near the Kansas-Nebraska border and the thinnest (4.5 ft; 1.4 m) Crouse is reported about 30 mi (50 km) northeast of this fieldguide site in Jackson County, Kansas (Walters, 1953), where the middle mudstone unit is absent.

Structurally, all sections (except the one in Wabaunsee County) are between the Nemaha anticline on the east and the Abilene anticline on the west (Fig. 1). They are along the eastern margin of the Irving syncline, which is an asymmetrical structure with its axis near the axis of the Abilene syncline (Condra and Upp, 1931, p. 9). This syncline plunges southwest and, assuming that it was a low area during Crouse time, could have created an embayed area on an otherwise open, shallow, marine shelf. Northward thinning of the Crouse and detailed studies by Huber (1965), West and others (1972), and Voran (1977) support the embayed shelf interpretation.

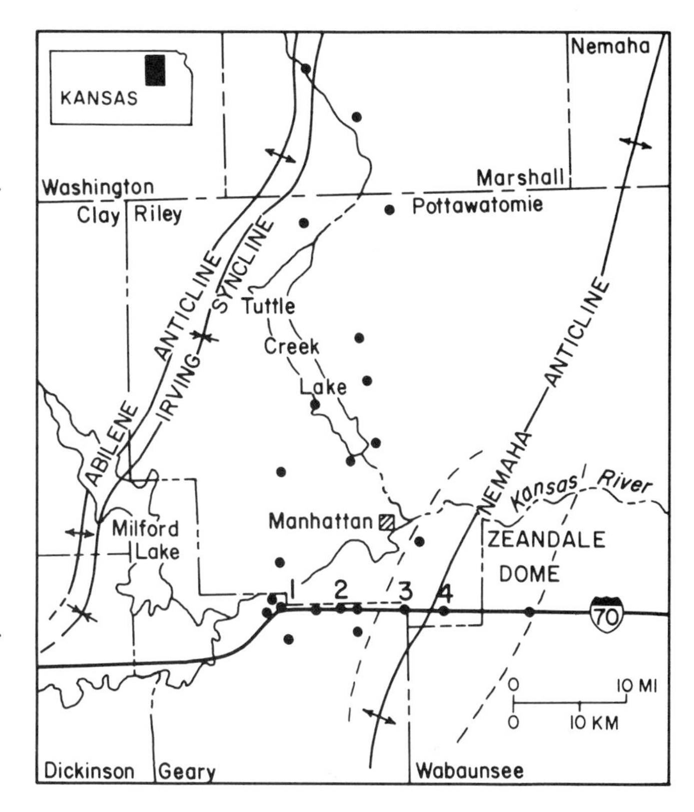

Figure 1. Map of north-central Kansas showing the structural setting, the location of the twenty-three stratigraphic sections including sections 1 to 4 of Crouse Limestone that have been studied in detail.

Immediately north and east of section 2, and in the area of sections 3 and 4, is the Zeandale Dome (Fig. 1). This domal feature represents a slight east-west broadening of the north-trending Nemaha anticline. The effect of this positive feature is shown by the four sections in this text which illustrate the importance of, and interrelationships among, stratigraphy, structural geology, paleoecology, and sedimentology in understanding depositional environments and earth history. Section 4 is on the axis of the Nemaha anticline, and sections 3, 2, and 1 extend westward toward the axis of the Irving syncline. After considering the general characteristics of each section, beginning with 4 and proceeding west to 1, we will look in detail at the biotic and lithologic features at 2, and then summarize the interpretations and significance of our observations.

Section 4 is near the center of the Zeandale Dome in the SE¼,NE¼,SE¼,Sec.30,T.11S.,R.9E., Riley County, Kansas. It is in a roadcut on the north side of I-70, 0.1 mi (0.16 km) west of

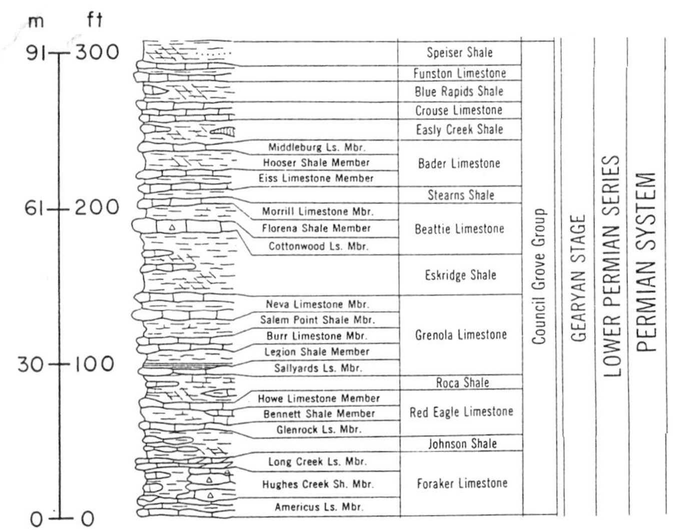

Figure 2. Stratigraphic section of Council Grove Group (Gearyan Stage, Lower Permian Series, Permian System; adapted from Zeller, 1968).

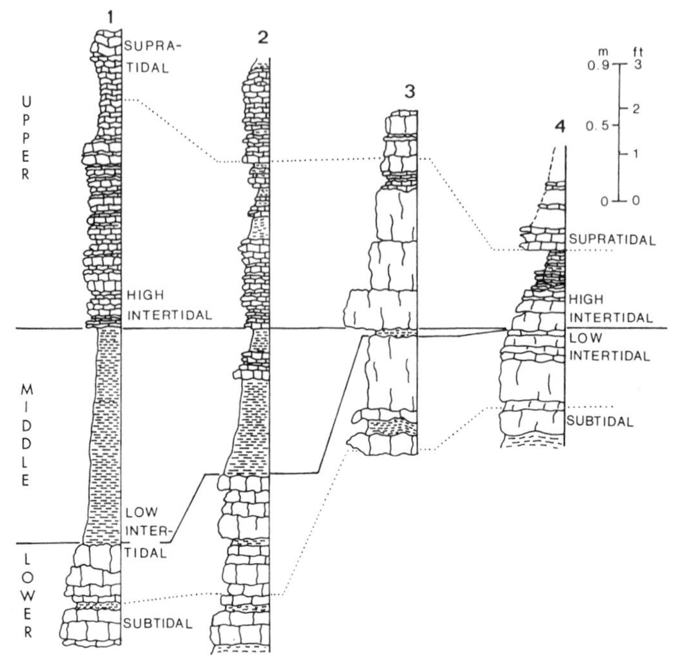

Figure 3. Lateral changes of Crouse Limestone at sections 1 to 4.

mile post 319 (Fig. 4B). The Crouse is at the top of the outcrop; its upper beds are partially covered, and its top is eroded, thus the total thickness is unknown. This exposure is a little over 5 ft (1.5 m) thick and the middle calcareous mudstone is absent (Fig. 3). The easily recognized thin- to platy-bedded upper limestone rests, with a sharp undulating contact, on the molluscan limestone beds of the lower Crouse Limestone. Planar laminations characteristic of a stromatolitic limestone are readily visible, whereas the ostracodes and foraminiferids are easily seen only with some magnification. Beds of the lower Crouse limestone are conspicuously fossiliferous with the bedding surfaces, particularly of the lower beds, revealing the internal molds of bivalves, such as *Permophorus, Aviculopecten,* and *Septimyalina. Permophorus* is the dominant bivalve in the lower Crouse limestone in this area (sections 3 and 4). Pyramidellid gastropods and ostracodes are also preserved and can be seen with a hand lens; indeed, ostracodes make up over 80 percent of the fossil assemblage of beds in this interval in the next section to the east (West and McMahon, 1972). Biotic, lithologic, and detailed petrographic study suggests that the Crouse at this locality, from the base to the top, reflects subtidal to supratidal conditions (Fig. 3). Detailed documentation of these environments of deposition is given by West and others (1972).

Section 3 is on the west side of the Zeandale Dome, near the center of the north line, SE¼,Sec.27,T.11S.,R.8E., Geary County, Kansas. It is a roadcut on the north side of I-70, 0.2 mi (0.3 km) west of mile post 316 (Fig. 4B). At this locality the underlying Easly Creek Shale and overlying Blue Rapids Shale are also well exposed. Immediately above the 7.7 ft (2.3 m) of Crouse Limestone is a yellowish brown platy to blocky claystone followed by a yellowish brown limestone with thin, contorted bedding and boxwork. Both units are part of the Blue Rapids Shale and are noted here to orient the reader to the overlying Crouse Limestone.

The upper Crouse at section 3 is composed of thick to thin beds of regular to mildly undulatory laminated carbonate mudstone. Because of the laminations, this unit weathers like a platy shale. It has the typical biotic and lithologic characteristics of the upper Crouse (West, 1972, p. 34–37). The middle calcareous mudstone is only 0.1 ft (0.03 m) thick (Fig. 3). Underlying this blocky calcareous mudstone are the limestone beds and thin claystone to shale beds of the lower Crouse. These lower Crouse limestones are very similar to those at section 4, and the vugs characteristic of the upper beds of the lower Crouse are well developed. Some obscure planar stromatolitic laminations occur in some of the vuggy beds. Bivalves are the conspicuous invertebrate fossils, but ostracodes are dominant in terms of density. The lowest bed in the Crouse at this locality contains areas of fine micritic calcarenite (packstone; West, 1972).

These rocks, like those at section 4, are interpreted as representing a regressive sequence from subtidal at the base to supratidal at the top (Fig. 3). Details for this interpretation are given in West and others (1972).

Section 2, which is approximately 6 mi (9.7 km) west of section 3, is in the NE¼,SW¼,NE¼,Sec.27,T.11S.,R.7E., Geary County, Kansas. It is a roadcut on the north side of I-70, 0.4 mi (0.64 km) west of milepost 309 (Fig. 4A). The complete Crouse Limestone (12.4 ft; 3.8 m thick) is exposed here, as is the upper

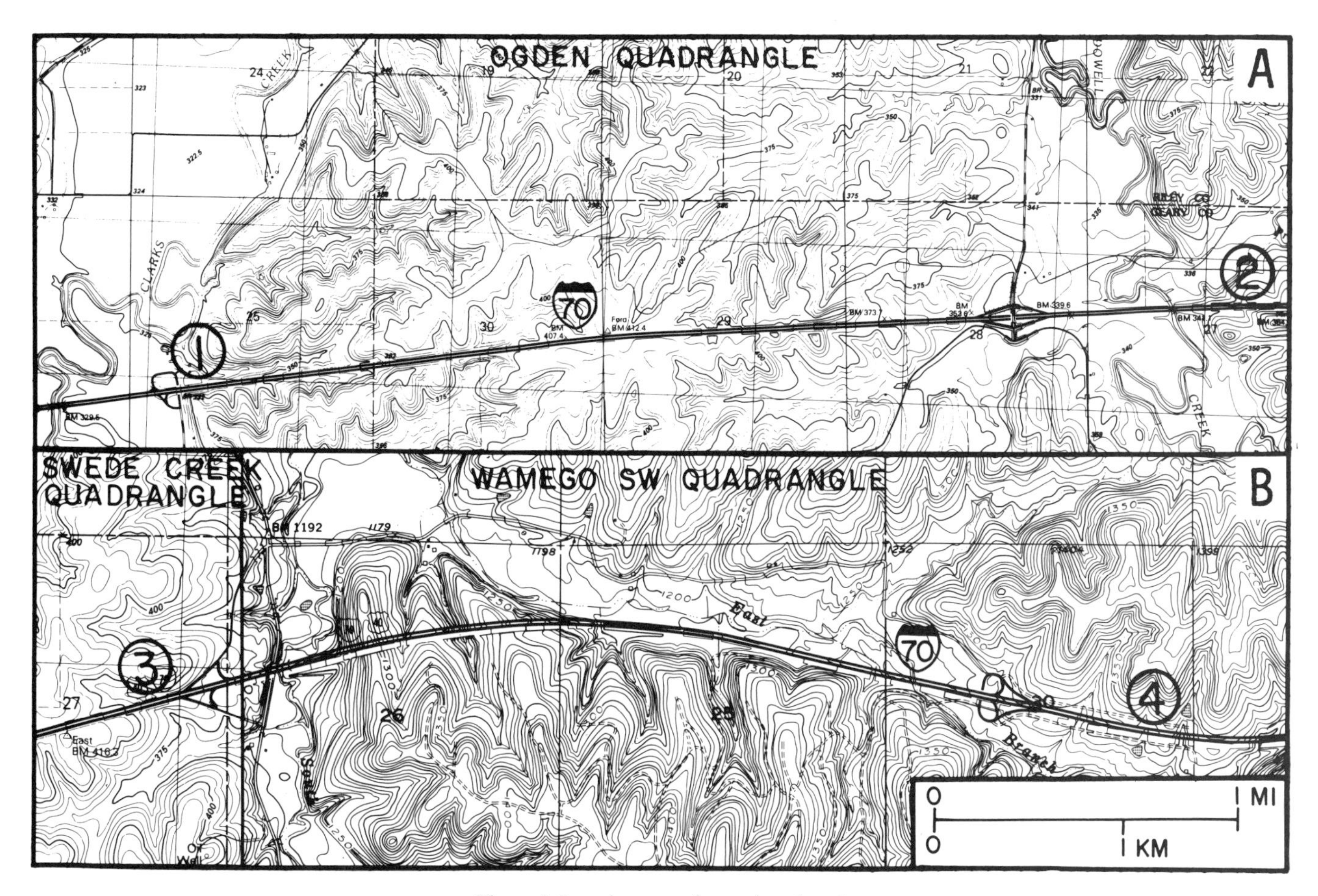

Figure 4. Location maps for sections 1 to 4.

part of the underlying Easly Creek Shale (Fig. 3). The overlying Blue Rapids Shale is mostly covered at this locality.

The most conspicuous feature of section 2 is the middle mudstone; it is 3.2 ft (1 m) thick and contains a 0.6 ft (0.18 m) unit of thin, wavy bedded, laminated, ostracodal carbonate mudstone. Mudcracks have been observed on some of the bedding planes in this carbonate (West, 1972). The mudstone units above and below this carbonate are platy, laminated, foraminiferal to ostracodal, calcareous mudstone that contain very fine sand to silt-sized quartz.

Beds of grayish orange to yellowish gray, thin, wavy, laminated, ostracodal carbonate mudstone compose the upper Crouse. Beds near the top of the Crouse appear to be a planar stromatolitic boundstone, and the undulations at the top of the Crouse suggest megaripples. Mudcracks, gypsum crystal molds, and some possible root-traces, although typical of the upper Crouse in north-central Kansas, are particularly conspicuous at this locality.

Carbonate mudstones, the dominant lithology of the lower Crouse at section 2 are typically bioturbated and vuggy. Bioturbation is especially obvious in the middle bed and less so in the lower and upper beds. Their orange and yellow colors reflect the abundance of iron oxides, and all beds are clayey. Molluscs (gastropods and bivalves) and ostracodes are the major invertebrate components. In the lower beds, bivalves (aviculopectinids, myalinids, carditids, and pholadomyoids) compose 100 percent of the assemblages preserved on bedding surfaces, but near the middle of the lower limestone, pyamidellid snails and ostracodes are the dominant components and bivalves are lacking (West and McMahon, 1972). These fossil assemblages, in the middle beds of the lower Crouse, are similar to those noted in the lower Crouse limestone at section 4.

Inferred environments of deposition are in Figure 3. Data supporting these inferences are given in West and others (1972), and will be discussed further after section 1.

Section 1 is one of the westernmost exposures of the Crouse Limestone in the Irving syncline. It is in the SW¼,NE¼,SW¼, Sec.25,T.11S.,R.6E., Geary County, Kansas, in a roadcut on the north side of I-70, 0.9 mi (1.5 km) west of mile post 305 (Fig. 4A) not quite 5 mi (8 km) west of section 2. This exposure is very much like that at section 2: both underlying and overlying units are present, though the overlying one is largely covered. Total thickness is 13.5 ft (4.1 m) with only 2.2 ft (0.7 m) of lower Crouse (thinner than at most Crouse exposures). Both the middle calcareous mudstone and the upper platy limestone are thicker here than at section 2, whereas the lower limestone is as thin here as it is at Section 4. These relationships are shown in Figure 3.

No "shaly" beds occur in the upper Crouse, and no limestone beds occur in the middle Crouse mudstone as they do in section 3 (Fig. 3). Mudcracks, birdseye structures, gypsum crystal molds, and planar laminations—sedimentary features indicative of supratidal deposition—are well exposed in the upper beds. The gray, calcareous, thin to platy-bedded middle Crouse mudstone is nearly homogeneous; the only fossils observed are ostracodes in the lower part.

Lithologically, the lower Crouse is as noted for section 2; however, unlike section 2, bioturbation is less conspicuous. Most beds are fossiliferous carbonate mudstone. Fossils in the lower beds of the lower Crouse appear to be algae-coated and include brachiopods and bivalves. Higher in the lower Crouse, assemblages are dominated by pyramidellid snails whereas ostracodes and bivalves are less conspicuous (West and McMahon, 1972).

A regressive sequence, like those at sections 2, 3, and 4 is inferred for the Crouse at section 1. Indeed lateral changes in the lower Crouse limestone from west to east, as shown by bedding plane studies (West and McMahon, 1972), suggest a pattern similar to that shown by vertical changes in these exposures.

ANALYSIS OF SECTION 2

Because the Crouse limestone at section 2 is typical of the Crouse throughout this area, we have chosen to present the detailed lithologic, petrographic, biotic, and sedimentologic data available for this exposure. These data are in Figure 5, which shows the lower, middle, and upper Crouse Limestone. The data have been gathered for the Crouse Limestone at all localities shown on the locality map (Fig. 1). Four basic patterns are used to indicate the carbonate rock type: (1) vertical lines = recrystallized, clayey, molluscan to ostracodal carbonate mudstone; (2) horizontal lines = recrystallized, ostracodal to molluscan carbonate mudstone; (3) horizontal lines with dots = recrystallized, clayey, bioturbated, molluscan to ostracodal carbonate mudstone; and (4) vertical lines with dots = recrystallized, clayey, laminated, ostracodal carbonate mudstone. Columns for insoluble residues, sedimentary structures, and color are self-explanatory. Data on biotic elements are from detailed thin-section studies and careful examination of washed residues (West, 1972). Lithologic symbols are standard: blocky symbol denotes limestones and the short dashes designate shale, mudstone, or claystone.

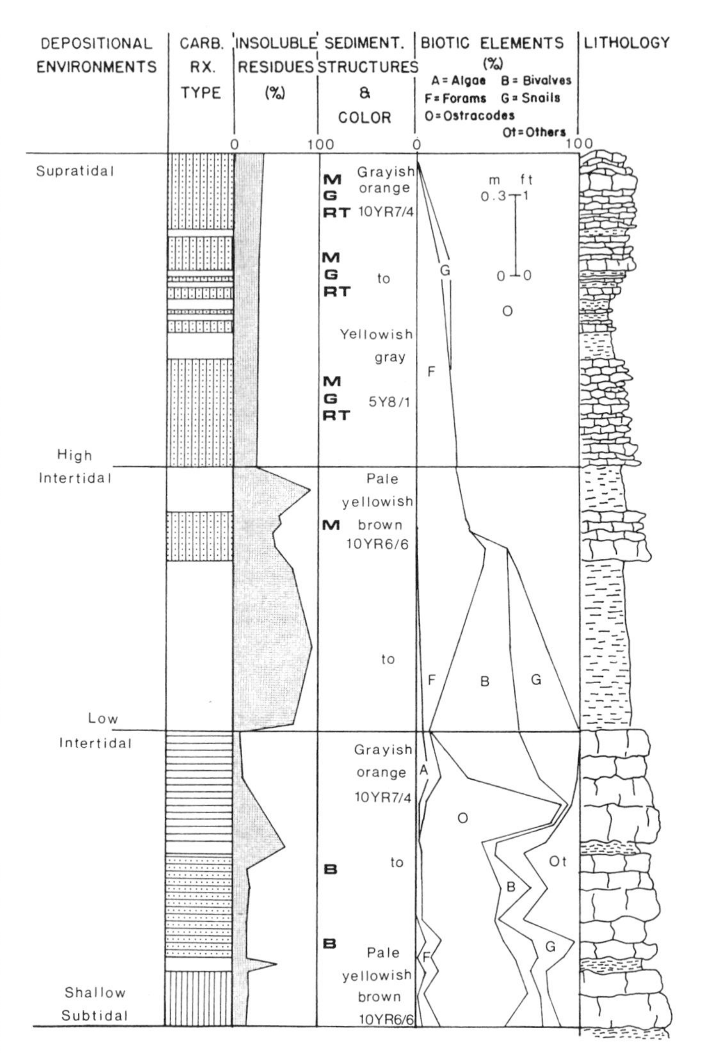

Figure 5. Lithologic, petrographic, biotic, and sedimentologic data and inferred depositional environments for the Crouse Limestone at section 2. Symbols for sedimentary structures are: M, mudcracks; G, crystal molds of gypsum; RT, possible root traces; and B, bioturbation.

Lower Crouse Limestone. Recrystallized, clayey carbonate mudstone dominates this part of the Crouse with bioturbation an obvious feature in the middle bed (Fig. 5). The percent of insoluble residue is low except for the thin shaly beds that separate the major limestone units. The color, grayish orange to pale yellowish brown, is a reflection of different natural compounds of iron and oxygen, such as goethite, limonite, and hematite. Diversity decreases upward with ostracodes a conspicuous component except near the top. What appear to be calcareous worm tubes, compose part of the "other" category; the size, geometry, and microstructure supports this identification. Bivalves and gastropods (snails) occur throughout this member and are particularly abundant in the upper beds. Some bedding surfaces in the lower limestone contain only bivalves (Fig. 5). One square decimenter of one bedding surface contained 56 bivalves, including 24 aviculopectinids, 16 myalinids, 9 carditids, 3 pholadomyoids, and 4 unidentifiable internal molds. Only 3 of the 56 were broken, none were articulated, the majority were oriented in a hydrodynamically stable position, and nearly half (47.2 percent) were right valves (West and McMahon, 1972).

The lower beds represent shallow subtidal conditions, whereas the middle and upper beds reflect low intertidal carbonate sedimentation. These beds are similar to those interpreted as subtidal and low intertidal by Walker and Laporte (1970) from the Ordovician Devonian of New York.

Gebelein (1971), discussing the Cape Sable area of Florida, indicated that high species diversity and abundance, many macroscopic invertebrates, and a wide size range of individuals are characteristics of subtidal carbonate areas. When compared to the overlying Crouse, this criteria fit, with the possible exception of the last one. There seems to be a wide size range of ostracode individuals and a moderate size range of molluscan individuals. Gebelein (1971) indicated that the sediment of subtidal carbonate areas is fine grained (less than 4 microns) and lacks burrows. The basal bed of the Crouse is a carbonate mudstone lacking bioturbation.

The upper beds of the lower Crouse are bioturbated, contain ostracodes, pyramidellid snails, and possible calcareous worm tubes, some of which contain a few pellets. These characteristics are analogous to the low diversity, high abundance of worms, mainly microscopic-sized individuals, and bioturbated sediment with some pellets that Gebelein (1971) reported for low intertidal carbonate areas of the Holocene.

Middle Crouse Limestone. Regardless of the lithology, whether limestone or mudstone, this unit is laminated, which may reflect activity of cyanobacteria. The limestone near the top is dominated by ostracodes and foraminiferids; bivalves and gastropods disappeared in the lower part of the limestone (Fig. 5). Because macrofossils are absent, this unit seems barren, but washed residues and microscopic study of bedding planes reveal an abundance of microfossils. Ostracodes include bairdiids, hollinids, healdids and the genus *Geisina.* Ammovertellids, *Ammodiscus* sp., *Bathysiphon* sp., *Globovalvulina* sp., and *Eggerella*-like foraminiferids compose the foraminiferal fraction. The bivalve, *Permophorus* sp. can be found in a "butterflied" position on some bedding surfaces, indicating burial in a quiet environment. A few mudcracks are also preserved with these bivalves. Diversity continues to decrease upward. As expected, insolubles are high in the mudstones.

Using criteria presented by Walker and Laporte (1970) for some Ordovician and Devonian carbonate rocks in New York and those presented by Gebelein (1971) for Holocene sediments of Cape Sable, Florida, we feel this middle "shaly" unit represents continued regression into the middle and upper intertidal zones. Overall diversity continues to decrease, and the stromatolitic (laminated) limestone and associated mudstones point to a higher intertidal environment.

Studies by Lane (1964) on the paleoecology of microfossil assemblages in the Council Grove Group provide additional evidence. He defined five different assemblages: (1 and 2) two different neritic assemblages, (3) a marginal assemblage, and (4 and 5) two different fresh water assemblages. Components of the neritic assemblages in the middle Crouse are hollinid, bairdiid and healdid ostracodes, *Globovalvulina* sp., *Tetrataxis* sp., ammovertellids, and holothurian sclerites. *Ammodiscus* sp. is in the middle Crouse and is one of two genera in Lane's marginal assemblage. The ostracode genus *Geisina* and fish debris, components of one of the fresh water assemblages, also occur in the middle Crouse. Thus, there is a mixing of marine and fresh water components, which seems plausible in an intertidal environment where the gradient is low and the subtidal areas shallow.

Upper Crouse Limestone. Laminations of this recrystallized, clayey, ostracode carbonate mudstone account for its platy bedding. Indeed, Huber (1965) pointed out that these "plates" often cover the upper and middle Crouse, making field differentiation difficult. As in the middle Crouse, these laminations are thought to be the result of cyanobacteria that produce beds of stromatolitic limestone and mudstone. Mudcracks, gypsum crystal molds, and possible root traces are common in this unit. Diversity continues to decrease until only ostracodes occur in the uppermost beds (Fig. 5).

We believe that the upper "platy" Crouse was formed in a high intertidal to supratidal environment. Biotic diversity is low, and molluscs (absent at section 2, but present elsewhere in the upper Crouse) are relatively small individuals that are probably a single species. This corresponds to the biotic aspects that indicate supratidal conditions (Gebelein, 1971).

Sedimentary features listed by Walker and Laporte (1970) as criteria for high intertidal to supratidal conditions are "birdseye" structures, mudcracks, early dolomitization, intraclasts, massive laminated bedding, algal laminae, and vertical burrows. Of these, mudcracks, algal laminae, possible vertical burrows, and "birdseye" structures occur in the upper unit of the Crouse Limestone. There is no direct evidence of dolomitization, but there are areas of microscopic rhomb-shaped voids, which may suggest dedolomitization. Huber (1965) reported rhombs filled with granular calcite enclosed in the clayey carbonate mudstone.

Thus it seems reasonable to us that the upper Crouse reflects a continuance of conditions typical of a regressive sequence in an area of carbonate sedimentation.

SUMMARY

The details described in section 2 are typical of all exposures of the Crouse Limestone studied to date in north-central Kansas. This area is interpreted as representing the eastern side of an embayed portion of a shallow open marine shelf. This embayed area was bounded on the east and west by the Nemaha and Abilene anticlines respectively, and occupied the Irving syncline. The anticlines were not major dry land areas, but rather slight topographic highs on the sea floor that might have extended to mean high tide level or slightly above. Thinning to zero of the middle mudstone unit of the Crouse and east and west dips off the axis of the Nemaha structure (West and others, 1972) indicate that the Nemaha, at least, affected sedimentation during Crouse deposition. Drying of this embayed area during Crouse time started in the north and northeast and along the edges of the structural "highs" where the water was shallowest and progressively extended south and southwest toward the deeper parts of the Irving syncline. Such conditions would record a regressive sequence. Information on thickness of lithologic units, petrography, sedimentary structures, and fossil assemblages all support a regressive interpretation for the Crouse, starting with subtidal conditions at the base and ending with supratidal conditions at the top. This sequence appears to be part of a punctuated aggradational cycle as defined by Goodwin and Anderson (1985). Additionally, lateral changes in the fossil assemblages preserved on bedding surfaces in the lower Crouse reflect a similar pattern, that is subtidal to intertidal, from west to east (West and McMahon, 1972).

The Crouse Limestone, or any interval of earth history, is best understood by looking in detail at the "total picture": stratigraphy, structural geology, palaeoecology, and sedimentology.

REFERENCES CITED

Condra, G. E. and Upp, J. E., 1931, Correlation of the Big Blue series in Nebraska: Nebraska Geological Survey, 2nd series, Bulletin 6, 74 p.

Gebelein, C. D., 1971, Sedimentology and ecology of Holocene carbonate facies mosaic, Cape Sable, Florida: American Association of Petroleum Geologists Bulletin, v. 55, p. 339–340.

Goodwin, P. W., and Anderson, E. J., 1985: Punctuated aggradational cycles; A general hypothesis of episodic stratigraphic accumulation: Journal of Geology, v. 93, p. 515–533.

Huber, D. D., 1965, Petrology of the Crouse Limestone in the vicinity of Manhattan, Kansas [M.S. thesis]: Manhattan, Kansas State University, 133 p.

Lane, N. G., 1964, Paleoecology of the Council Grove Group (Lower Permian) in Kansas, based upon microfossils: Kansas Geological Survey Bulletin 170, pt. 5, 23 p.

Voran, R. L., 1977, Fossil assemblages, stratigraphy, and depositional environments of the Crouse Limestone (Lower Permian) in north-central Kansas [M.S. thesis]: Manhattan, Kansas State University, 208 p.

Walker, K. R., and Laporte, L. F., 1970, Congruent fossil communities from Ordovician and Devonian carbonates of New York: Journal of Paleontology, v. 44, p. 298–944.

Walters, K. L., 1953, Geology and ground-water resources of Jackson County, Kansas: Kansas Geological Survey Bulletin 101, 91 p.

West, R. R., 1972, ed., Stratigraphy and depositional environments of the Crouse Limestone (Permian) in north-central Kansas: Guidebook for 6th Annual Meeting of South-Central Section of Geological Society of America, 110 p.

West, R. R., and McMahon, E. C., 1972, Significance of lower Crouse fossil assemblages: Guidebook for 6th Annual Meeting of South-Central Section of Geological Society of America, p. 91–95.

West, R. R., Jeppeson, J. A., Pearce, R. W., and Twiss, P. C., 1972, Crouse Limestone; Its stratigraphy and depositional environment: Guidebook for 6th Annual Meeting of South-Central Section of Geological Society of America, p. 81–89.

Zeller, D. E., ed., 1968, The stratigraphic succession in Kansas: Kansas Geological Survey Bulletin 189, 81 p.

Beattie Limestone (Lower Permian) of eastern Kansas

Page C. Twiss, *Department of Geology, Kansas State University, Manhattan, Kansas 66506*

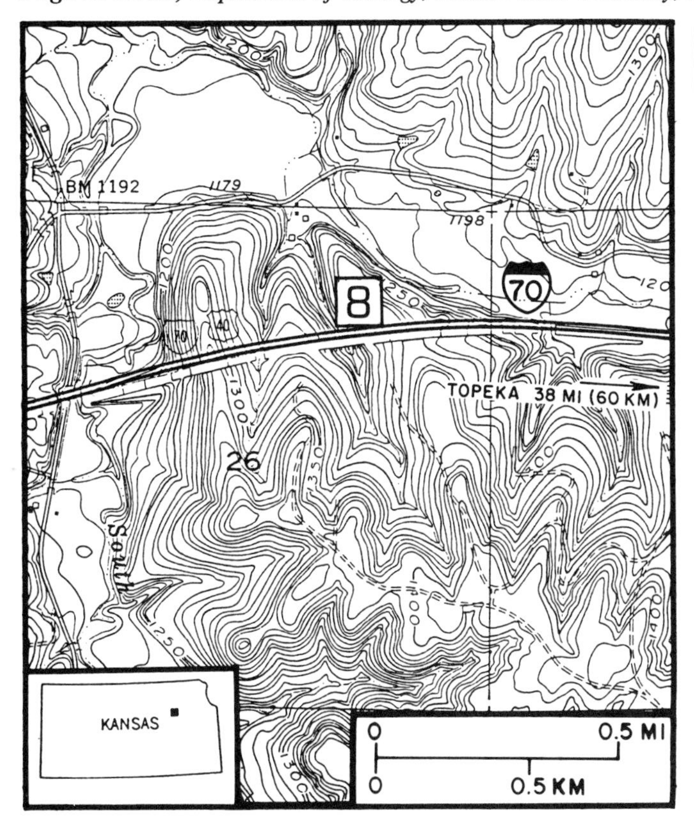

Figure 1. Location of Beattie Cyclothem, Site 8 (Wamego SW 7½-minute Quadrangle, 1953, photorevised 1978, Riley County, Kansas).

Figure 2. Location of platy algal facies of Cottonwood Limestone Member of Beattie Limestone, Site 9 (Rosalia, 7½-minute Quadrangle, 1961, Greenwood County, Kansas).

LOCALITIES

The Beattie cyclothem (site 8) is exposed on the north side of a road cut about 325 ft (100 m) west of milepost 317 on I-70 in the center of NE¼,Sec.26,T.11S.,R.8E., Riley County, Kansas (Fig. 1). If traveling eastward, continue to exit 318, pass under I-70, and return to the road cut (Kansas law prohibits pedestrians from crossing divided highways). Vehicles should be parked off the pavement, and passengers should exit vehicles on the north side. Keep as far as possible from the traffic lane while examining the rocks.

The Cottonwood Limestone Member of the Beattie Limestone (site 9) is exposed on the north side of a road cut at milepost 272 on U.S. 54 about 0.3 mi (0.5 km) east of the Butler County line. The outcrop is near the north edge of SE¼,SW¼,Sec.3, T.26S.,R.8E., Greenwood County, Kansas (Fig. 2). Vehicles should be parked on the south side of the highway where the old road enters the present highway.

SIGNIFICANCE

The Beattie Limestone of the Council Grove Group, Gearyan Stage, Lower Permian Series (Fig. 3), crops out in a northeast-trending, narrow belt from southern Nebraska to northern Oklahoma. The formation ranges from about 10 to 25 ft (3 to 8 m) thick and contains three members, in ascending order: Cottonwood Limestone Member, Florena Shale Member, and the Morrill Limestone Member (Zeller, 1968). This formation is typical of Lower Permian cyclothems (Moore and others, 1934) that lack beds of sandstone, underclay, coal, and "core shale" of the Pennsylvanian cyclothems of Kansas (Moore, 1936; Heckel, 1977). Further, Imbrie and others (1964) have identified ten facies along the narrow outcrop belt (Fig. 4) and have documented some departures from the ideal cyclothem (Imbrie, 1955; Laporte, 1962; Laporte and Imbrie, 1964; Imbrie and others, 1964).

PERMIAN CYCLOTHEMS

The Lower Permian rocks of Kansas consist of cyclic sequences of red mudstone and siltstone, green to gray mudstone,

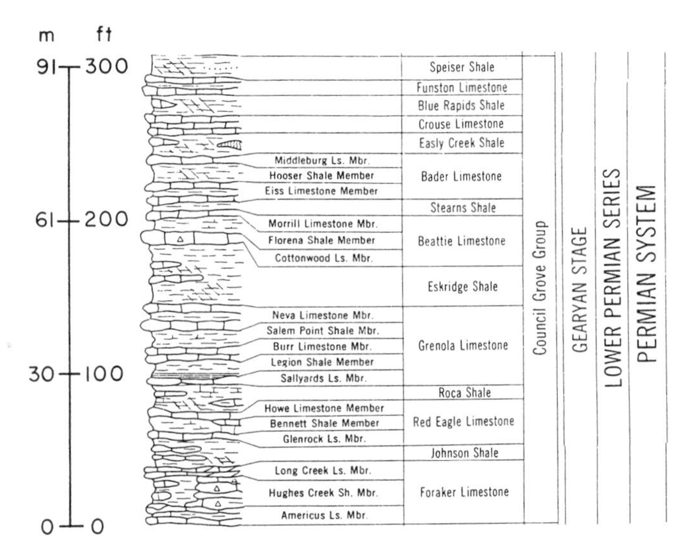

Figure 3. Stratigraphic section of Council Grove Group, Gearyan Stage, Lower Permian Series, Permian System in Kansas (adapted from Zeller, 1968).

skeletal calcilutite, and yellowish-gray mudstone. The lack of sandstone and conglomerate and only sporadic traces of coal led Moore and others (1934) to propose an "ideal Permian cyclothem" of transgressive and regressive rock types and Elias (1937) to describe associated "phases" of assemblages of fossils, or paleobiotopes (Fig. 5). The Permian cyclothem is bounded by continental (nonmarine) red siltstone and mudstone and is named for the limestone that represents maximum marine transgression.

In their study of the uppermost Pennsylvanian and lowermost Permian rocks of Kansas, Mudge and Yochelson (1962, p. 5) tabulated the phases for each member of each formation. Nowhere in these 35 rock units do all phases of the "ideal" cyclothem occur, not even in the Beattie Cyclothem. The implication that these rocks record orderly cycles of marine transgression and regression is not justified. Many contacts between beds are abrupt, and successive phases are absent. Any depositional model must accommodate these abrupt contacts.

Busch and Rollins (1984), Busch and others (1985), and Goodwin and Anderson (1985) have proposed a hierarchy of transgressive-regressive (T-R) units in which PACs (punctuated aggradational cycles) are designated as the basic sixth-order (T-R) units. PACs are about 3 to 16 ft (1 to 5 m) thick and are estimated to last from 80,000 to 225,000 years. They are asymmetrical T-R units that contain thin or absent transgressive bases followed by thicker, shallowing-upward sequences (Busch and

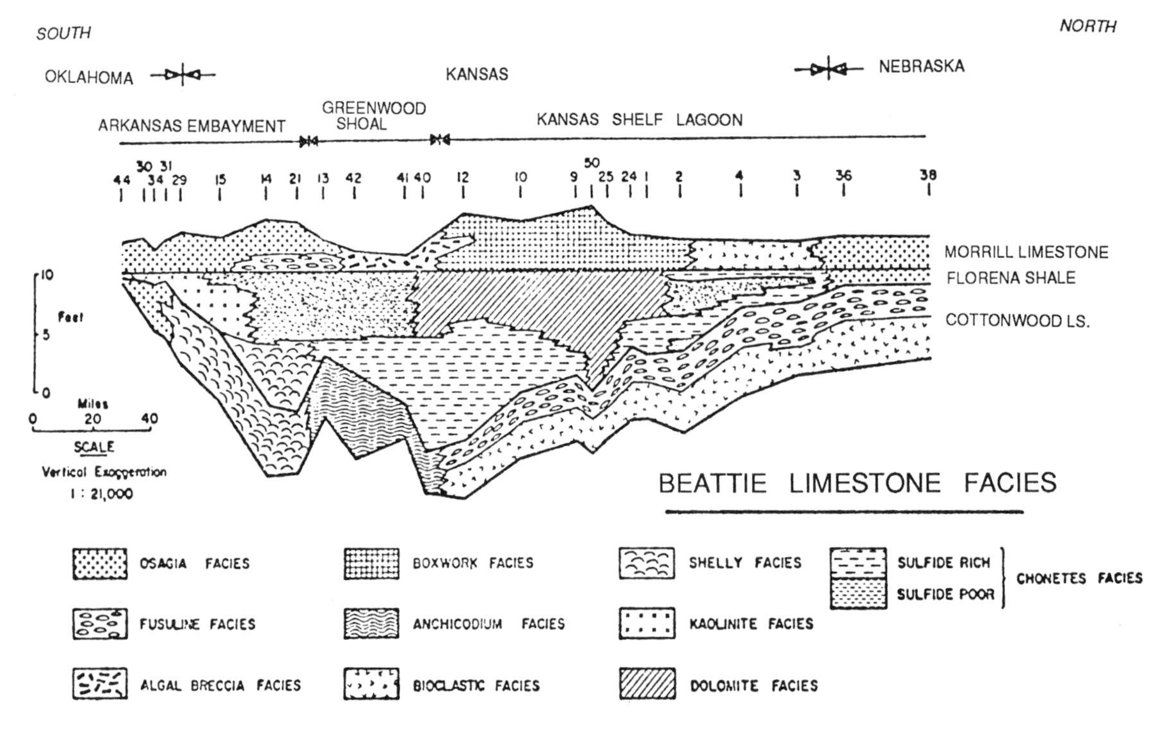

Figure 4. Restored stratigraphic cross section showing facies distribution in Beattie Limestone along line of outcrop from Nemaha County, Nebraska, to Osage County, Oklahoma. Datum is top of Florena Shale. Facies boundaries determined from data obtained in foot-by-foot sampling at each of 24 measured sections (Imbrie and others, 1959, p. 73).

others, 1985). Rapid, basin wide, minor rises in sea level are followed by longer periods of sea-level stasis during which aggradation and progradation occur slowly. Boundaries between PACs are either transgressive surfaces or climatic change surfaces.

Busch and Rollins (1984) designated the fifth-order T-R units as cyclothemic PAC sequences; these are equivalent to the Kansas cyclothems that are estimated to last from 400,000 to 500,000 years (Heckel and others, 1979; Heckel, 1980). Cyclothemic PAC sequences differ from cyclothems in that the latter are lithostratigraphic units that consist of cyclic repetitions of specific lithofacies that may lack genetic boundaries or lateral persistence. Because many Kansas cyclothems have sharp contacts between units and have certain phases missing from the idealized sequence of phases (paleobiotopes) of Elias's (1937) model, Kansas provides a good testing ground for the PAC hypothesis of Goodwin and Anderson (1985).

Transgressive surfaces are recognized by sharp contacts between rocks containing marine fossils and subjacent nonmarine rocks that were transgressed. Climatic change surfaces are recognized by contacts between underlying nonmarine facies capped by paleosols or calcretes and overlying nonmarine coal or lacustrine limestone (Busch and others, 1985; Goodwin and Anderson, 1985).

SITE 8: BEATTIE CYCLOTHEM

The road cut at this site exposes most of the Beattie Cyclothem, beginning with the upper 11 ft (2 m) of the Eskridge Shale and continuing through the Beattie Limestone, Stearns Shale, and the base of the Eiss Limestone Member of the Bader Limestone at the top of the hill. This exposure illustrates the shallow marine facies of the Beattie Limestone in the northern half of the outcrop belt and is near site 24 of Imbrie and others (1964). It includes, in ascending order, the bioclastic facies and fusulinid facies of the Cottonwood Limestone Member, the chonetid facies and dolomite facies of the Florena Shale Member, and the boxwork facies of the Morrill Limestone Member.

This measured section (Fig. 6) illustrates sharp and transitional boundaries between rock units (Fig. 7). The Cottonwood Limestone Member is a massive ridge former that has a sharp lower contact and transitional upper contact. The Florena Shale Member is predominately carbonate mudstone, dolomitic mudstone, and mudstone with both sharp and transitional contacts. The Morrill Limestone Member has a sharp base and a transitional top.

Units 1 and 2 of the Eskridge Shale correspond to phases 1 and 2 of Elias's (1937) cyclothem model and are separated by a sharp contact. Both are nonfossiliferous and probably record continental environments. An abrupt contact occurs at the top of Unit 2, and Unit 3 marks the beginning of the Cottonwood Limestone Member. Units 3 and 4 compose the bioclastic facies of Laporte (1962) and phase 6 of Elias (1937). This skeletal calcilutite consists of broken fragments of *Osagia* and algal coatings of fragments of bryozoans, echinoderms, and brachiopods.

IDEAL PERMIAN CYCLOTHEM & PHASES
FROM MOORE, ELIAS, AND NEWELL, 1934
MODIFIED BY MUDGE AND YOCHELSON, 1962

PHASES	ROCK TYPE	DESCRIPTION
1		RED SHALE — NO MARINE FOSSILS
2		GREEN SHALE — NO MARINE FOSSILS
3		*LINGULA* PHASE
4		PELECYPOD PHASE
5		PELECYPOD *ALLORISMA, AVICULOPINNA, ORTHOMYALINA, MYALINA,* & BRACHIOPOD PHASE
6		BRACHIOPOD PHASE
7		FUSULINID PHASE
6		BRACHIOPOD PHASE
5		PELECYPOD *ALLORISMA, AVICULOPINNA, ORTHOMYALINA, MYALINA,* & BRACHIOPOD PHASE
4		PELECYPOD PHASE
3		*LINGULA* PHASE
2		GREEN SHALE — NO MARINE FOSSILS
1		RED SHALE — NO MARINE FOSSILS

Figure 5. Ideal Permian cyclothem and paleobiotopes (phases), adapted from Moore and others (1934), Elias (1937), and modified by Mudge and Yochelson (1962, p. 5, fig. 2). Note: *Allorisma* is no longer a valid generic name; this bivalve is probably *Wilkingia.*

The upper 8 in (20 cm) of Unit 4 contains some irregular lenses of fusulinids that may have accumulated in burrows or in scour depressions.

Units 5 and 6 constitute the fusuline facies of Laporte (1962) and phase 7 of Elias (1937). Although the fusulinid *Schwagerina* sp. is prominent in the outcrop, *Osagia*-coated skeletal fragments are also abundant. Notice that the chert nodules contain silicified fusulinids in the same abundance as calcitic ones in the limestone. The boundary between the fusulinid facies and the underlying bioclastic facies is abrupt and may represent an erosional surface that does not always coincide with the contact between Units 4 and 5.

The Florena Shale Member consists of four units. The lower contact between Unit 6 and Unit 7 is gradational, and fusulinids occur in the thin-bedded, clayey calcilutite (phase 7). Unit 8 is a nonfossiliferous, laminated, gray shale that has sharp upper and lower boundaries. The clayey calcilutite of Unit 9 contains horizons of unbroken and articulated valves of *Neochonetes granulif-*

Rock Descriptions For Figure 6, Site 8

Unit 13. Stearns Shale. Clayey calcilutite to carbonate mudstone; grades upward to medium light gray (N6) mudstone; nonfossiliferous; not measured.
Unit 12. Morrill Limestone Member, Beattie Limestone. Limestone (calcilutite); light olive gray (5YR7/4), weathers grayish orange (10YR7/4); massive, skeletal fragments of brachiopods and echinoids; boxwork structure at base; sharp basal contact, gradational upper contact; irregular thickness; USGS reference marker Deep No. 2 emplaced at top. Boxwork facies of Imbrie and others (1964).
Unit 11. Clayey dolomitic limestone; boxwork with calcite veins; abrupt upper contact. Boxwork facies of Imbrie and others (1964).
Unit 10. Florena Shale Member, Beattie Limestone. Clayey limestone to marl, slightly dolomitic; medium gray (N5.5), weathers yellowish gray (5Y5/2); thin-bedded; weathers blocky and tabular; nonfossiliferous. Dolomitic facies of Imbrie and others (1964).
Unit 9. Clayey Limestone (calcilutite); pale yellowish brown (10YR6/2), weathers yellowish gray (5Y8/2); contains horizons of *Neochonetes granuliferer* (unbroken and articulated valves) and compacted, disarticulated, and broken *Derbyia* sp.; a few thin 0.39–1.18 in (1–3 cm) skeletal calcilutite lenses. Chonetid facies of Imbrie and others (1964).
Unit 8. Shale; bluish gray (5B2.5/1), weathers medium gray (N5); laminated, weathers platy; nonfossiliferous.
Unit 7. Clayey limestone; pale yellowish brown (10YR6/2), weathers yellowish gray (5Y7/2); gradational lower contact, sharp upper contact; grades upward from clayey limestone to marl; some fusulinids near base; thin bedded, weathers blocky to platy. Fusulinid facies of Imbrie and others (1964).
Unit 6. Cottonwood Limestone Member, Beattie Limestone. Chert-bearing fusulinid limestone (biomicrosparite); massive; yellowish gray (5Y7/2), weathers pale grayish orange (10YR8/4) to moderate yellowish brown (10YR5/4); profusion of fusulinids that preferentially weather leaving a porous and chalky rock; two horizons of chert nodules, 9 and 14 in (23 and 36 cm) above base, elongated horizontally, circular (1 in or 2.5 cm) to platy (6 in or 15 cm); nodules contain silicified fusulinids in same abundance as calcareous ones in limestone. Fusulinid facies of Laporte (1962).
Unit 5. Clayey limestone (fusulinid wackestone); same as unit 6, but less resistant. Fusulinid facies of Laporte (1962).
Unit 4. Limestone (skeletal calcilutite); massive; yellowish gray (5Y7/2), weathers grayish orange (10YR8/4); some chert nodules (0.4 to 1 in or 1 to 3 cm) widely separated horizontally and 18 in (46 cm) above base, some horizontally elongated and widely separated chert nodules 9 in (23 cm) above the first horizon; skeletal fragments of bryozoans, crinoids, and brachiopods, many encrusted by *Osagia*; upper 8 in (20 cm) contains some irregular lenses (6 × 3 in or 15 × 7 cm) of fusulinids that may have accumulated in burrows or in scour depressions. Bioclastic facies of Laporte (1962).
Unit 3. Limestone (skeletal calcilutite); massive; yellowish gray (5Y7/2), weathers grayish orange (10YR8/4); many *Osagia*-encrusted skeletal fragments of bryozoans and crinoids; parting 6.5 in (16 cm) above base; sharp basal contact. Bioclastic facies of Laporte (1962).
Unit 2. Eskridge Shale. Calcareous mudstone and marl; yellowish gray (5Y6/2), weathers yellowish gray (5Y7/2); blocky; nonfossiliferous.
Unit 1. Mudstone; grayish red purple (5RP4/2), weathers grayish purple (5P4/2); blocky; nonfissile; nonfossiliferous; noncalcareous.

Stage	Group	Formation	Member	Graphic column	Unit	Thick. m	Thick. ft	Elias' phases (1937)
Gearyan Stage	Council Grove Group	Stearns Sh.			13	0.41	1.3	6. Brachiopod phase
		Beattie Limestone	Morrill L.S. Mbr.		12	0.49 to 0.53	1.5 to 1.7	
					11	0.25	0.8	
			Florena Shale Member		10	0.71	2.4	
					9	0.68	2.3	
					8	0.26	0.8	
					7	0.23	0.8	7. Fusulinid phase
			Cottonwood Limestone Member		6	0.82 to 0.87	2.7 to 2.8	
					5	0.09	0.3	
					4	0.56 to 0.63	1.8 to 2.1	6. Brachiopod phase
					3	0.33	1.1	
		Eskridge Shale			2	0.92	3.0	2. Green shale phase
					1	1.0	3.3	1. Red shale phase

Figure 6. Measured section of Beattie Cyclothem, Site 8. Measured by P. C. Twiss, 1985.

era and compacted, disarticulated, and broken values of *Derbyia* sp. This is the chonetid facies of Imbrie and others (1964) and phase 6 of Elias (1937). Continuing upward, Unit 10 becomes less fossiliferous and more dolomitic; it grades into the boxwork facies of Unit 11 of the Morrill Limestone Member.

The contact between Unit 11 and Unit 12 is abrupt. Unit 12 is a calcilutite that contains skeletal fragments of brachiopods and echinoids (phase 6); it grades upward into the Stearns Shale, which is mostly covered.

Summary

The Beattie Cyclothem, a fifth-order T-R unit, is bounded by red mudstone and siltstone and contains several abrupt and gradational contacts. The abrupt contacts and absence of phases suggest that several sixth-order T-R units occur within the Beattie Cyclothem. Preliminary work shows that these T-R units may be

A

B

Figure 7. Beattie Cyclothem at Site 8. A: Upper 3.3 ft (1 m) of Eskridge Shale, just above road level, is overlain by massive Cottonwood Limestone Member, Florena Shale Member, and Morrill Limestone Member (all of Beattie Limestone) in foreground. Stearns Shale (lower part covered) is capped by lower part of the Eiss Limestone Member of the Bader Limestone in background. B: Beattie Limestone with massive Cottonwood Limestone Member at base (Jacob Staff placed at the base is 5.3 ft (1.6 m) long); Florena Shale Member (partly covered) in the middle; capped by Morrill Limestone Member.

traced over the northern part of the outcrop belt of the Beattie Limestone. The PAC hypothesis of Goodwin and Anderson (1985) provides a model to apply to further study of this cyclothem in northern Kansas and southern Nebraska.

SITE 9: PLATY ALGAL FACIES OF COTTONWOOD LIMESTONE

This road cut, locality 42 of Laporte (1962), was designated as the platy algal facies that marks the center of the postulated Greenwood Shoal. The phylloid alga *Anchicodium* sp. occurs as detached "potato chip" thalli oriented parallel to contacts (Fig. 8), and in the upper part *Anchicodium* sp. occurs in growth position. The diversity of fossils is high and includes, in addition to the phylloid alga, *Osagia,* trilobites, bryozoans, gastropods, bivalves, ostracodes, echinoderms, brachiopods, and pelloids.

This measured section of the Cottonwood Limestone Member (Fig. 9) is the result of additional study by Tyrone Black, Jim Mathewson, Bob Sawin, Roxie Voran, Ron West, and Page Twiss. Information from polished sections, acetate peels, and thin sections is added to data recorded by Laporte (1962), Imbrie and others (1964), and West and Sawin (1984).

Unit 1 is a nonfossiliferous calcareous mudstone that corresponds to phase 2 of Elias (1937); it forms a sharp and undulating contact with unit 2. Unit 2 has a high diversity that corresponds to either phase 5 or 6 of Elias (1937); it contains *Osagia* and skeletal fragments of echinoderms, trilobites, brachiopods, and bryozoans. The upper boundary is sharp and undulatory in places and gradational in others.

Unit 3 is Laporte's (1962) platy algal facies, a massive coarse calcirudite containing abundant "potato chip" thalli of *Anchicodium* sp. and skeletal fragments of bryozoans, bivalves, echinoderms, ostracodes, brachiopods (especially *Crurithyris?* sp.), and gastropods. This assemblage approximates the mixed-fauna (molluscan and brachiopod) phase 5 of Elias (1937).

The very coarse calcirudite of unit 4 contains *Anchicodium* in growth position and the same diverse fossil assemblage as unit 3. The lower and upper contacts range from sharp to gradational. Units 5 and 6 are bounded by sharp contacts; the fossil diversity remains high and is phase 5 of Elias (1937). Laporte (1962) found a thin fusulinid-rich layer at the top of this location that is phase 7 of Elias (1937).

Imbrie and others (1964) proposed the Greenwood Shoal to account for the abrupt facies change of the Beattie Limestone at this location (Fig. 4). The Kansas Shelf Lagoon and the Nebraska Shelf include facies similar to those seen at the above Beattie cyclothem locality (Site 8). A few miles south of the algal facies the Cottonwood Limestone Member contains more mud and is difficult to recognize. Terrigenous sediment was poured into the shallow sea from the Oklahoma mountains as evidenced by the increase in kaolinite southward. Imbrie and others (1964) compared the platy algal facies of the Greenwood Shoal to the *Thallasia*-covered mudbanks of Florida Bay. Although they did not record any *Anchicodium* in growth position at this locality as noted by West and Sawin (1984), Imbrie and others (1964,

A B

Figure 8. Platy algal facies of Cottonwood Limestone Member at Site 9. A: Unit 3 showing massive algal calcirudite with horizontal lenticular cavities (Jacob staff placed at the base is 5.3 ft [1.6 m] long; the scale at top is 3.5 in [8.8 cm] long). B: Closeup of unit 3 showing horizontal "potato chip" fronds of *Anchicodium* sp. and large lenticular cavity.

Stage	Group	Formation	Member	Unit	Thick. m	Thick. ft	Elias' phases (1937)
Gearyan Stage	Council Grove Group			S			
		Beattie Limestone	Cottonwood Limestone Member	6	0.16	0.5	5. Mixed phase
				5	0.09	0.3	
				4	0.36	1.2	
				3	1.14	3.7	
				2	0.13	0.4	5 or 6. Mixed phase, or brachiopod phase
		Eskridge Shale		1	0.18	0.5	2. Green shale phase

Figure 9. Measured section of platy algal facies of Cottonwood Limestone Member, Site 9. Measured by Tyrone Black, Jim Mathewson, Bob Sawin, Roxie Voran, Ron West, and Page Twiss, 1976.

Rock Description For Figure 9, Site 9

Unit S. Soil and regolith—not measured.

Unit 6. Cottonwood Limestone Member, Beattie Limestone. Very coarse skeletal calcirudite; algal packstone; dark yellowish orange (10YR6/6), weathers grayish orange (10YR7/4); thin bedded to platy; irregular fracture; covered upper contact; skeletal fragments of *Anchicodium* sp. thalli, echinoderms, bryozoans, gastropods, bivalves, and pelloids; sharp and undulating lower contact, upper contact covered. Immature recrystallized biomicrudite. Phase 5 of Elias (1937).

Unit 5. Fine skeletal calcirudite; algal packstone; pale yellowish orange (10YR8/6), weathers grayish yellow orange (10YR7/6); thin bedded; irregular fracture; fair outcrop; sharp upper contact; skeletal fragments of echinoderms, bryozoans, trilobites, ostracodes, forams, and bivalves. Phase 5 of Elias (1937).

Unit 4. Very coarse calcirudite; immature recrystallized biomicrudite; very pale orange (10YR8/2), weathers pale yellowish orange (10YR8/6); undulated bedding over *Anchicodium* sp. in growth position; skeletal fragments of echinoderms, bryozoans, bivalves, ostracodes, forams, and *Anchicodium* sp. surrounding algal heads; irregular fracture; sharp upper contact; biolithite or algal boundstone. Phase 5 of Elias (1937).

Unit 3. Coarse skeletal calcirudite; immature recrystallized biomicrudite; very pale orange (10YR8/2), weathers to yellowish orange (10YR7/6); massive; irregular fracture; forms vertical outcrop; sharp upper contact; porous with cavities up to 6 in (15 cm) long and 2 in (5 cm) thick; skeletal debris of *Anchicodium* sp. thalli, bryozoans, bivalves, echinoderms, ostracodes, brachiopods (especially *Crurithyris*? sp.), and gastropods; differential weathering due to different cementation, "potato chip" texture due to algal thalli. Phase 5 of Elias (1937).

Unit 2. Medium skeletal calcarenite; dusty yellowish gray (5Y6.5/3), weathers yellowish gray (5Y7/2), thin bedded, high clay content at base decreasing upward; irregular fracture; sharp upper and lower contact; irregular iron stains; silty biomicrudite; about 50% skeletal fragments of *Osagia,* echinoderms, trilobites, brachiopods, and bryozoans. Phase 5 of Elias (1937).

Unit 1. Eskridge Shale. Calcareous mudstone; yellowish gray (5Y7/2); laminated; irregular fracture; sharp and undulating upper contact; forms shalelike slope; thin zone (0.2 in or 0.5 cm) of calcite concretions (up to 1.2 in or 3 cm in diameter) 5 in (13 cm) below top; irregular and laminar iron stains (10YR6/6).

p. 236) did refer to "a pinching and swelling of algal layers suggestive of in situ growth" at locality 46, which is about 2 mi (3.2 km) north-northeast.

An alternative explanation for the phylloid algae is that proposed by Heckel (1978) for the phylloid algal mounds of the regressive carbonate members of the Pennsylvanian cyclothems in southeast Kansas. Those mounds mark the boundary between the "open marine" carbonates of the north and the "terrigenous detrital" facies of Oklahoma. These conditions persisted into the Early Permian, and the phylloid algal carbonate facies of the Cottonwood Limestone Member marks the southern boundary of the "normal" marine facies of the north. South of this facies the underlying Eskridge Shale becomes more sandy, whereas the Cottonwood Limestone Member, if traceable at all, and the Florena Shale Member become more clayey with an increase of kaolinite (Twiss and Stindl, 1968; Imbrie and others, 1964). The overlying Morrill Limestone Member becomes an Osagite that has been thoroughly burrowed and was probably formed nearshore; in the extreme southern part of the exposure the upper few inches may have formed under subaerial conditions. Acceptance of this model does not require a Greenwood Shoal or an Arkansas Embayment.

SUMMARY

These two sections of the Beattie Limestone and associated stratigraphic units illustrate typical transgressive-regressive sequences in the Lower Permian rocks of eastern Kansas. The units are thin and widespread and contain several facies that can be traced laterally and vertically. Although the earlier studies of these rocks have described them accurately, the interpretation of the depositional history needs to be refined by considering the PAC approach of Goodwin and Anderson (1985). Much work remains to be done on these cyclic units.

REFERENCES

Busch, R. M., and Rollins, H. B., 1984, Correlation of Carboniferous strata using a hierarchy of transgressive-regressive units: Geology, v. 12, p. 471–474.

Busch, R. M., West, R. R., Barrett, F. J., and Barrett, T. R., 1985, Cyclothems versus a hierarchy of transgressive-regressive units, *in* Watney, W. L., Kaesler, R. L., and Newell, K. D., eds., Recent interpretations of Late Paleozoic cyclothems: Lawrence, Kansas Geological Survey, p. 141–153.

Elias, M. K., 1937, Depth of deposition of the Big Blue (Late Paleozoic) sediments in Kansas: Geological Society of America Bulletin, v. 48, p. 403–432.

Goodwin, P. W., and Anderson, E. J., 1985, Punctuated aggradational cycles; A general hypothesis of episodic stratigraphic accumulation: Journal of Geology, v. 93, p. 515–533.

Heckel, P. H., 1977, Origin of phosphatic black shale facies in Pennsylvanian cyclothems of Mid-Continent North America: American Association of Petroleum Geologists Bulletin, v. 61, p. 1045–1068.

—— , 1978, Field guide to Upper Pennsylvanian cyclothemic limestone facies in eastern Kansas: Kansas Geological Survey Guidebook Series 2, 79 p.

—— , 1980, Paleogeography of eustatic model for deposition of Mid-Continent Upper Pennsylvanian cyclothems, *in* Fouch, T. D., and Magathan, E. R., eds., Paleozoic paleogeography of West-central United States: Society of Economic Paleontologists and Mineralogists, Rocky Mountain Section Paleogeography Symposium I, p. 197–215.

Heckel, P. H., Brady, L. L., Ebanks, W. J., Jr., and Pabian, R. K., 1979, Pennsylvanian cyclic platform deposits of Kansas and Nebraska (Ninth International Congress of Carboniferous Stratigraphy and Geology, Field Trip No. 10): Kansas Geological Survey Guidebook Series 4, 79 p.

Imbrie, J., 1955, Quantitative lithofacies and biofacies study of Florena Shale (Permian) of Kansas: American Association of Petroleum Geologists Bulletin, v. 39, p. 649–670.

Imbrie, J., Laporte, L. F., and Merriam, D. F., 1959, Beattie Limestone facies and their bearing on cyclic sedimentation theory: Wichita, Kansas Geological Society 24th Field Conference Guidebook, p. 69–78.

—— , 1964, Beattie Limestone facies (Lower Permian) of the northern Midcontinent: Kansas Geological Survey Bulletin 169, v. 1, p. 219–238.

Lanning, F. C., and Twiss, P. C., 1965, Vertical variations in the chemical composition of the Cottonwood Limestone bed in Riley County, Kansas: Kansas Academy of Science Transactions, v. 68, p. 438–442.

Laporte, L. F., 1962, Paleoecology of the Cottonwood Limestone (Permian), northern Mid-Continent: Geological Society of America Bulletin, v. 73, p. 521–544.

Laporte, L. F., and Imbrie, J., 1964, Phases and facies in the interpretation of cyclic deposits: Kansas Geological Survey Bulletin 169, v. 1, p. 249–263.

Moore, R. C., 1936, Stratigraphic classification of the Pennsylvanian rocks of Kansas: Kansas Geological Survey Bulletin 22, 256 p.

Moore, R. C., Elias, M. K., and Newell, N. D., 1934, Stratigraphic sections of Pennsylvanian and Permian rocks of Kansas River Valley: Kansas Geological Survey chart, issued in December.

Mudge, M. R., and Yochelson, E. L., 1962, Stratigraphy and paleontology of the uppermost Pennsylvanian and lowermost Permian rocks in Kansas: U.S. Geological Survey Professional Paper 323, p. 1–213.

Twiss, P. C., and Stindl, H., 1968, Clay mineralogy of Cottonwood Limestone: Geological Society of America Program with Abstracts, p. 301.

West, R. R., and Sawin, R. S., 1984, *In situ* associations are essential for studies of community evolution; An example from the Lower Permian of North America: 27th International Geological Congress Abstracts, v. 9, pt. 2, p. 69.

Zeller, D. E., 1968, The stratigraphic succession in Kansas: Kansas Geological Survey Bulletin 189, 81 p.

Classic "Kansas" cyclothems

Philip H. Heckel, *Department of Geology, University of Iowa, Iowa City, Iowa 52242*

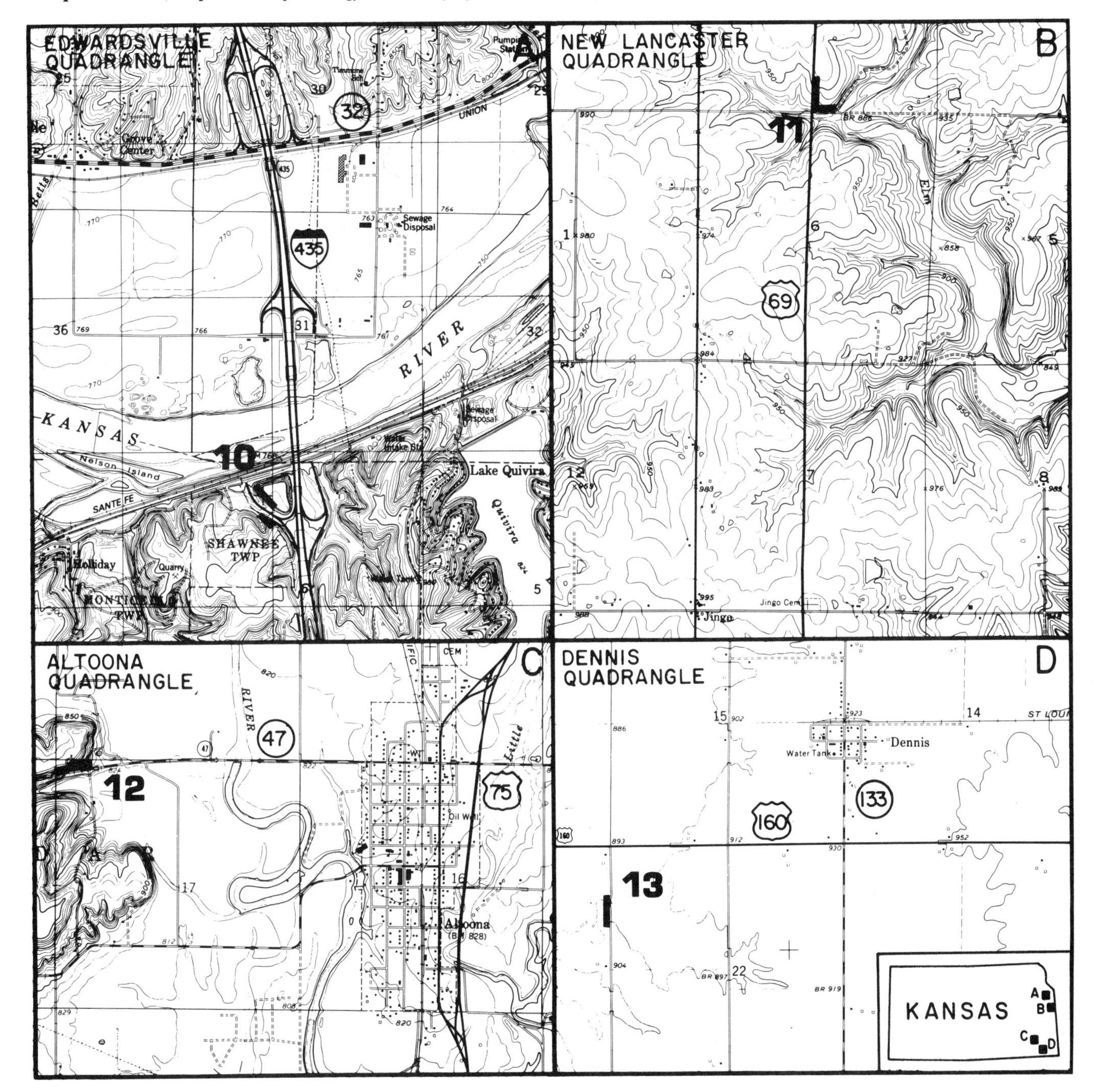

Figure 1. Index maps showing locations of sites 10 to 13.

LOCATIONS

This quadruple site (Fig. 1) illustrates typical examples of Kansas cyclothems of Pennsylvanian age.

Site 10: Iola and Wyandotte cyclothems (Figs. 2, 6). I-435 exit at Holliday Road, NE¼NW¼,Sec.6,T.12S.,R.24E., Edwardsville 7½-minute Quadrangle, Johnson County, Kansas.

The best way to get to this section is to proceed south on I-435 from Kansas 32, cross the Kansas River and take the Holliday Road exit around to the right, pulling off to park on the shoulder before meeting Holliday Road. The lower part of this section (up to the lower Argentine) is easily accessible on the east

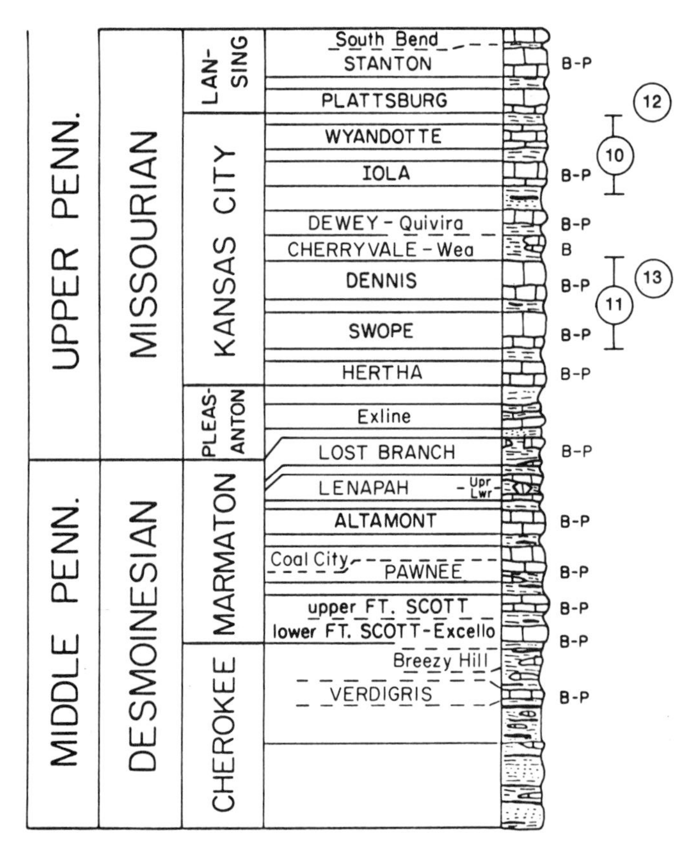

Figure 2. Part of Middle and Upper Pennsylvanian sequence along Midcontinent outcrop belt, showing stratigraphic positions of sites 10 to 13. Horizontal names are formations and members (mainly limestones) that are developed to various degrees as cyclothems. B-P denotes presence of black phosphatic shales that characterize classic cyclothems.

side of this ramp road; the upper part is accessible on the west side of this road, but requires more careful climbing.

Site 11: Swope and Dennis cyclothems (Figs. 2, 7). U.S. 69 roadcut by Jingo, SW¼SE¼,Sec.31,T.18S.,R.25E., New Lancaster 7½-minute Quadrangle, Miami County, Kansas.

This section is exposed continuously along the east side of U.S. 69 and the north side of a gravel section-line road that crosses U.S. 69, 13 mi (21 km) south of Kansas 68 (Louisburg) exit and 6 mi (9.6 km) north of Kansas 135 (LaCygne) exit. Park along the section-line road east of U.S. 69.

Site 12: Plattsburg cyclothem (Figs. 2, 8). Kansas 47 Altoona lower roadcut, near NW corner, Sec. 17,T.29S.,R.16E., Altoona 7½-minute Quadrangle, Wilson County, Kansas.

This section is a little more than 1 mi (1.6 km) west of Altoona, Kansas, about the middle of the long roadcut on Kansas 47 up the prominent escarpment along the west side of the Verdigris River. The best place to park is on the shoulder of the gravel road that extends north and west from Kansas 47, at the foot of the escarpment, because the heavily traveled highway has shoulders too narrow for safe parking.

Site 13: Dennis cyclothem (Figs. 2, 9). Roadcut southwest of Dennis, center of east line, NE¼,Sec.21,T.31S.,R.18E., Dennis 7½-minute Quadrangle, Labette County, Kansas.

This section is exposed along a north-south gravel road, 0.3 mi (0.5 km) south of U.S. 160, 1 mi (1.6 km) west of Dennis, Kansas (Kansas 133), and 6 mi (9.6 km) east of the junction with U.S. 169. Park on the shoulder of the gravel road.

SIGNIFICANCE

The cyclic alternation of limestone and shale formations that dominates the Middle and Upper Pennsylvanian sequence (Fig. 2) along the Midcontinent outcrop belt has intrigued geologists ever since Moore (1931) first described it, and Wanless and Weller (1932) applied the name cyclothem to the component unit of repeating rock types in Illinois. This term was soon adopted in the Midcontinent outcrop area by Moore (1936). Weller (1930) invoked a model of periodic tectonism to explain both the overall alternation and the individual cyclothem. In contrast, Wanless and Shepard (1936) related both these features to periodic eustatic changes in sea level brought about by waxing and waning of Gondwanan ice caps. More recently, autocyclic models of delta shifting have been applied to cyclic sequences in the Appalachians (Ferm, 1970) and Texas (e.g., Galloway and Brown, 1973), and these have been extended to Illinois (Merrill, 1975). In the meantime, Wanless (1964, 1967) suggested that the glacial eustatic model readily accommodates delta shifting as a mechanism to explain otherwise anomalous clastic wedges in many of the cyclothems. This view has recently been more fully developed by Heckel (1977, 1980, 1984a), who recognized the cyclothems as marine transgressive-regressive sequences, centered on the thin, nonsandy, black phosphatic ("core") shales, which represent maximum inundation of the shelf; most deltas formed during the succeeding regressive phase.

BASIC CYCLOTHEM

The term cyclothem has been applied to a number of different, but related and quite specific, repeating lithic successions in the Pennsylvanian (e.g., Moore, 1936, 1950; Weller, 1958; see review in Heckel, 1984b). Current work on the mid-Desmoinesian-to mid-Virgilian sequence in the Midcontinent has established the nature of the basic transgressive-regressive ("Kansas") cyclothem (Figs. 3, 4, and 5). This cyclothem resulted from a major rise and fall of sea level over the northern Midcontinent shelf, which extended from the Arkoma-Anadarko basinal region of central Oklahoma across the lower parts of the North American craton northward into Iowa and Nebraska. In ascending order, this cyclothem consists of the following.

Transgressive ("middle") limestone, deposited in deepening water. This is typically a thin marine skeletal calcilutite with a diverse open-marine biota, deposited below effective wave base later during transgression; however, it locally includes calcarenites at the base, deposited in shallower water earlier ring

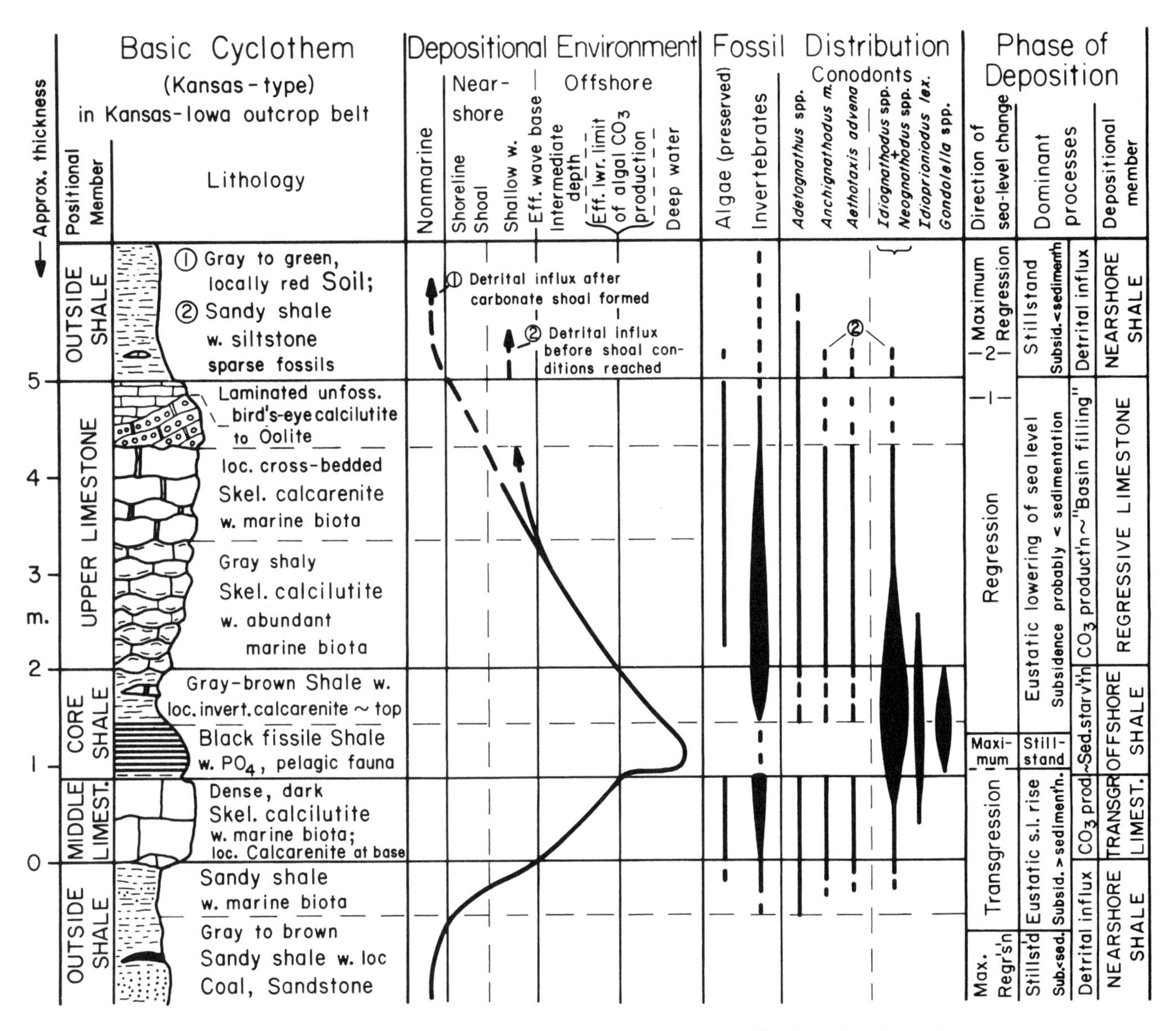

Figure 3. Basic "Kansas" cyclothem characterizing with minor modification all major and many intermediate marine cycles of deposition across northern Midcontinent shelf (Iowa to southern Kansas). Phases of deposition reflect ranges of sea-level stand. Conodont information is derived mainly from Heckel and Baesemann (1975) and Swade (1985). (From Heckel, 1984a.)

transgression. Transgressive limestones are typically dense, dark, nonpelleted calcilutites that contain neomorphosed aragonite grains, and overpacked calcarenites that generally lack evidence of early marine cementation or meteoric leaching and cementation. This is because the calcarenites remained in the marine phreatic environment of deposition until buried by higher marine strata of the cyclothem, which acted as a barrier to meteoric diagenesis (Heckel, 1983). Thus, they underwent slow compaction before cementation in a decreasingly oxygenated burial environment, in which much of the fine organic matter became preserved in the rock.

Offshore ("core") shale formed at maximum transgression. This is typically a thin, nonsandy, gray to black phosphatic shale deposited under conditions of near sediment starvation. In most cyclothems the water became deep enough for a thermocline to develop over much or all of the lower areas of the northern Midcontinent shelf. The thermocline reduced bottom-oxygen replenishment enough over shallower areas to produce the gray dysaerobic facies that includes only low oxygen–tolerant benthic invertebrates, such as certain brachiopods and crinoids. It eliminated bottom oxygen over the deeper areas, thus producing the black anoxic facies containing only pelagic

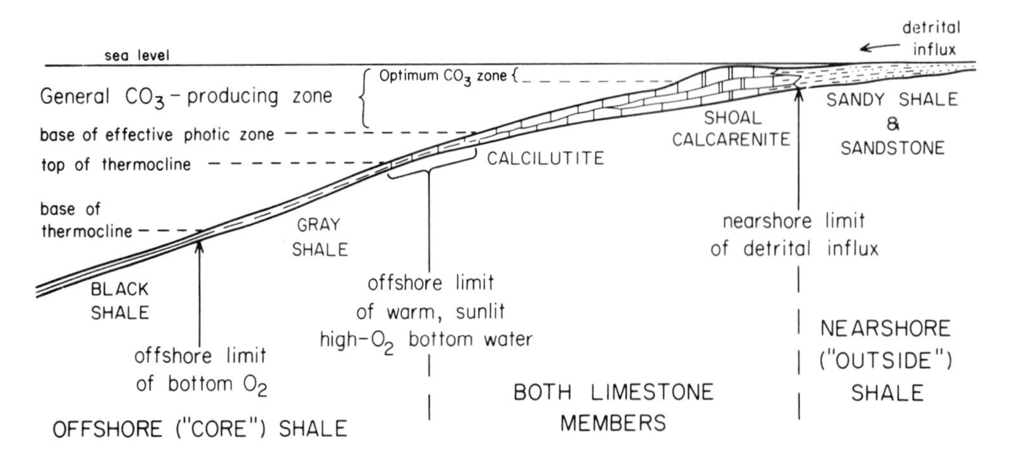

Figure 4. General model for limits of carbonate deposition on gently sloping tropical shelf, showing lateral distribution of rock types that became superposed by transgression and regression to form classic "Kansas" cyclothem. (From Heckel, 1984a.)

fossil remains such as certain conodonts, fish debris, radiolarians, and occasional ammonoids. Conodonts in particular are abundant in both gray and black facies of these offshore shales.

Sedimentation was so slow at this time in the northern Midcontinent that in the gray facies, aragonite fossils apparently were dissolved and calcite fossils were locally corroded (Malinky, 1984). Further evidence for this lies in the appearance of great numbers of originally aragonitic molluscs (snails, clams, and ammonoids; see Boardman and others, 1984) now preserved as siderite, pyrite, or phosphorite in thicker developments of these gray shales in southern Kansas and Oklahoma. Ammonoids are also preserved locally in early diagenetic carbonate nodules ("bullion") in the black facies in Kansas and Missouri. In the first case, rapid burial and early mineralization of the molluscs, and in the second case, early matrix mineralization around the ammonoids, prevented sea-floor dissolution of the fossils in the colder, undersaturated waters below the top of the thermocline.

Quasi-estuarine circulation and upwelling associated with the thermocline in this deeper phase of the shelf sea caused deposition, in both the gray and black shale facies, of nonskeletal phosphorite as granules, laminae, and nodules. These are analogous to modern phosphorite nodules forming under similar conditions of periodic upwelling associated with a thermocline in low-oxygen sediment on the offshore shelf along the coast of Peru (Kidder, 1985).

Regressive ("upper") limestone deposited in shallowing water. This is typically a thick marine skeletal calcilutite that has a diverse biota, and was deposited below effective wave base. It grades upward into skeletal calcarenite that has abraded grains, algae, and cross-bedding, all evidence for traction transport in shallow water.

In some cyclothems, a distinctly different calcarenite appears at the base, associated with the offshore shale. This type of calcarenite (seen at Site 10) contains only invertebrates (particularly crinoids, brachiopods, bryozoans, and encrusting foraminifers), often shows evidence of grain corrosion, and lacks any evidence of algae, grain abrasion, or cross-bedding. It therefore must have formed below effective wave base and probably below the effective photic base for the algae in this sea, and represents proliferation of invertebrates as the thermocline weakened and sufficient oxygen returned to the bottom. Its overcompacted nature resulted from relatively deep burial before cementation, as in the transgressive limestone (Heckel, 1983).

The tops of some regressive limestones display cross-bedded oolite formed in agitated, highly supersaturated waters. The tops of others, particularly northward (Fig. 5), display sparsely fossiliferous, laminated to birdseye-bearing lagoonal to peritidal carbonates that represent passage of the strandline during later regression. Examples of oolite and peritidal calcilutite are exposed at site 11.

Even without development of these shoreline facies, passage of the strandline is recorded in subaerial exposure surfaces on the tops or within the upper parts of some regressive limestones (seen at site 11). These exposure surfaces resulted from the effects of meteoric weathering and include cracking, in-place brecciation, pitting, and formation of "solution-tubes" from plant rooting and infiltration of fresh water and terrestrial organic matter. This often led to a mottled to rubbly to rusty, punky appearance when various amounts of overlying shale and iron oxides were carried down into the open spaces in the limestone. In local areas where the meteoric water on the surface of the limestone became saturated, both laminar and vaguely pisolitic caliche crusts formed. The recognition and lateral tracing of these widespread exposure surfaces on regressive limestones, which cannot be explained by a depositional model involving delta shifting alone, has corroborated the eustatic model for cyclothem formation in the northern Midcontinent (Watney, 1984).

Subaerial exposure and infiltration of oxygenating, under-

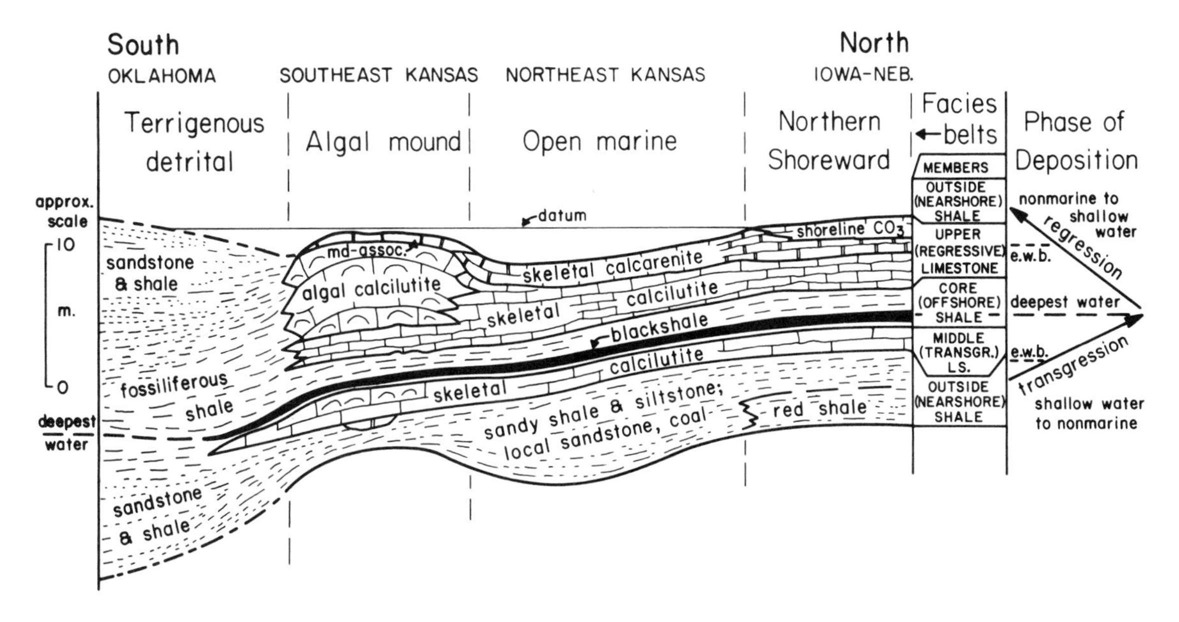

Figure 5. Generalized north-south cross section of classic Kansas cyclothem along Midcontinent outcrop belt showing lateral relations of members and facies to phases of transgression and regression. Datum is interpreted sea level at time when increased detrital influx terminated deposition of regressive limestone member. (From Heckel, 1980.)

saturated meteoric water farther down into the regressive limestone before much grain-to-grain compaction took place also oxidized much of the original organic matter in the sediment, leached most of the aragonitic grains, and eventually, as the water became saturated, precipitated blocky calcite in both the intergranular and moldic voids. This preserved original peloidal fabric, original depositional packing of grains, and, in the case of incomplete cementation, porosity. Thus the lighter-colored, more porous, and more conspicuously sparry, upper regressive limestones often stand in contrast with the darker, denser, overcompacted transgressive limestones and the lower, more offshore facies of regressive limestones (Heckel, 1983).

Nearshore ("outside") shales. These shales (which lie outside the limestone formations comprising the three previously described marine members of the cyclothem) encompass a variety of nearshore marine and terrestrial deposits on the shelf, all deposited at lower stands of sea level. They include thick, sparsely fossiliferous prodeltaic shales (sites 10, 12) that prograde out over regressive limestones in areas where deltas were active during later stages of sea-level fall. In places, these shales grade upward into delta-front and delta-plain sandstones and coals (site 13).

Outside shales also include thinner blocky mudstones (site 11) that range from gray to red in color, from inches to several feet in thickness, and typically overlie exposure surfaces on regressive limestones. Those that have been studied in detail show upward decrease in crystallinity of illite and upward increase in mixed-layer and kaolinite proportions relative to illite (e.g., Galesburg Shale at site 11; Schutter, 1983), which are characteristics of clay mineralogy expected in a soil profile. The blocky fabric of the paleosol mudstone is a result of disturbance of the normally flat-lying clay minerals (which account for the fissility of marine shales) by plant rooting, animal burrowing, and rain-water illuviation of clay minerals into holes formed by the organic agents and into cracks formed by periodic desiccation. This blocky fabric provides a strong suggestion of the origin of these mudstones as paleosols in cases where the clay minerals have not been analyzed. In addition, these mudstones often contain irregular carbonate nodules, some of which show internal clotted, mottled, and cracked fabric characteristic of formation in a soil (Watney, 1980).

The blocky mudstones at several horizons in the sequence are overlain by the most widespread and thickest coals in the Midcontinent Pennsylvanian. These coals apparently formed in response to the early stages of sea-level rise of the succeeding transgression, which ponded fresh-water runoff to form broad swamps on the surface of low relief. These swamps then migrated shelfward ahead of the transgression. A thin coal of this type is a short distance below the transgressive marine deposits of the Dennis cycle at site 13. Coals at this position are more characteristic of the underlying Desmoinesian Series, when the overall climate was apparently wetter than during the Missourian (Schutter and Heckel, 1985), which includes all the rocks seen at sites 10 to 13 (see Fig. 2).

OTHER CYCLES

Many units represent lesser inundations of the shelf that did not develop all members of a classic cyclothem; therefore, I have classified the irregular spectrum of transgressive-regressive cycles into three categories for the purpose of further analysis, based partly on the extent of transgression onto the shelf: (1) Major cycles are those inundations far and deep enough onto the shelf to form a conodont-rich shale to the northern limit of outcrop in Nebraska and Iowa, and to develop enough of the other facies to be recognized as classic cyclothems over much of their extent. (2) Intermediate cycles extend as marine horizons into Iowa and Nebraska, but have conodont-rich horizons only on the lower shelf; most have been recognized only in some places as cyclothems or parts of them. (3) Minor cycles typically extend as marine horizons only a short distance from the basinal region of Oklahoma into Kansas or Missouri, or represent a minor reversal within a more major cycle, and have not generally been recognized as cyclothems nor, in many cases, have they been named as separate units. The recognition of minor eustatic cycles requires correlation of the horizon over a reasonable geographic area, because in any one section or small geographic area autocyclic processes such as delta shifting can produce minor cycles of deposition.

ULTIMATE CONTROLS

Analysis of the probable lengths of time during which each of these types of marine cycles of transgression and regression took place is based on sets of assumptions explained elsewhere (Heckel, 1986). Heckel has estimated a range of lengths of 235,000 to about 400,000 yr for the major cyclothems, 120,000 to 220,000 yr for the intermediate cycles, and 44,000 to 120,000 yr for the minor cycles. The estimated ranges for all cycles fall within the range of periods of the Earth's orbital cycles that constitute the Milankovitch insolation theory for control of the Pleistocene ice ages. These cyclic orbital parameters are: eccentricity, with two dominant periods, one about 413,000 yr, and the other ranging from 95,000 to 136,000 yr and averaging about 100,000 yr; obliquity, with a dominant period near 41,000 yr; and precession, with two dominant periods averaging 19,000 and 23,000 yr (Imbrie and Imbrie, 1980).

The thin shallowing-upward depositional sequences observed within the regressive Winterset Limestone of the Dennis cycle at site 10 appear to be punctuated aggradational cycles (PAC), which were proposed by Goodwin and Anderson (1985) as basic, small-scale stratigraphic units present throughout the geologic column. They estimated the lengths of time for formation of individual PAC to be within the range of the shorter orbital periods.

SITE DESCRIPTIONS

Site 10 (Fig. 6). This stop shows two Kansas-type cyclothems as described by Heckel (1977, 1980). Each of these cyclothems resulted from a widespread glacial-eustatic transgression and regression of the sea across the northern Midcontinent shelf from the basinal area of Oklahoma. The well-developed Iola cyclothem (described by Mitchell, 1981) has a typical black phosphatic offshore ("core") shale, but the Wyandotte cyclothem (described by Crowley, 1969, before cyclothem concepts were updated), here developed on a delta-lobe high, has a gray offshore shale, and also displays minor cycles of deposition at the top.

The **Chanute Shale** here records an influx of prodeltaic clastics during the lowstand of sea level prior to the Iola transgression. Southward, the Chanute is dominated by thicker shale and included sandstones and coal, which together represent a delta-plain environment.

The **Paola Limestone** is a thin skeletal calcilutite containing a diverse biota recording a major advance of the sea to quiet water below effective wave base. Traced with little change from Iowa to Oklahoma, a distance of about 400 mi (640 km), it is a typical transgressive limestone at the base of the Iola cyclothem and was deposited in deepening water. Conodonts are used to supplement paleoenvironmental interpretations of lithic sequences in the Midcontinent Pennsylvanian (Fig. 3). In the Paola, conodonts number tens per kilogram and are dominated by *Idiognathodus,* typical of the open shelf; *Anchignathodus,* typical of shelf carbonates, is present throughout; *Adetognathus,* typical of nearshore environments, is present at the base; and *Idioprioniodus,* typical of more offshore environments, is present at the top (data from Mitchell, 1981).

The **Muncie Creek Shale** is the offshore ("core") shale of the Iola cyclothem. The black shale facies in the middle, which contains phosphorite nodules, records anoxic, phosphate-rich bottom conditions that developed below a thermocline with quasi-estuarine circulation and upwelling during maximum transgression (Heckel, 1977). The gray shales above and below reflect low bottom-oxygen (dysaerobic) conditions, which developed as the thermocline was forming and as it was breaking up. Conodont abundance in this shale member is the highest (100s/kg) in the entire cyclothem, due to slow sedimentation in the deep offshore-shelf environment. The conodont fauna is strongly dominated by *Idiognathodus*; also present is *Idioprioniodus,* and *Gondolella,* the most offshore genus, which was probably tolerant of low-oxygen water. A scattering of the other, shallower water genera is found in some samples (data from Heckel and Baesemann, 1975).

The **Raytown Limestone** is the regressive limestone of the Iola cyclothem, and was deposited in shallowing water. The basal overpacked invertebrate calcarenite lacks cross-bedding, grain abrasion, algal grains, micrite envelopes, and early marine or meteoric cement. It records accumulation of invertebrates in quiet water, below both effective wave base and the lower effective limit of algal activity, after the thermocline broke up and sufficient oxygen for abundant benthic life returned to the sea bottom. This bed was cemented very late with ferroan cements, after burial compaction eliminated much of the original intergranular space (Heckel, 1983). It has high conodont abundance; *Idiognathodus* dominates *Idioprioniodus,* as in the underlying core shale. The upper Raytown is skeletal calcilutite; conspicuous phylloid algae records deposition in water that had become shallow

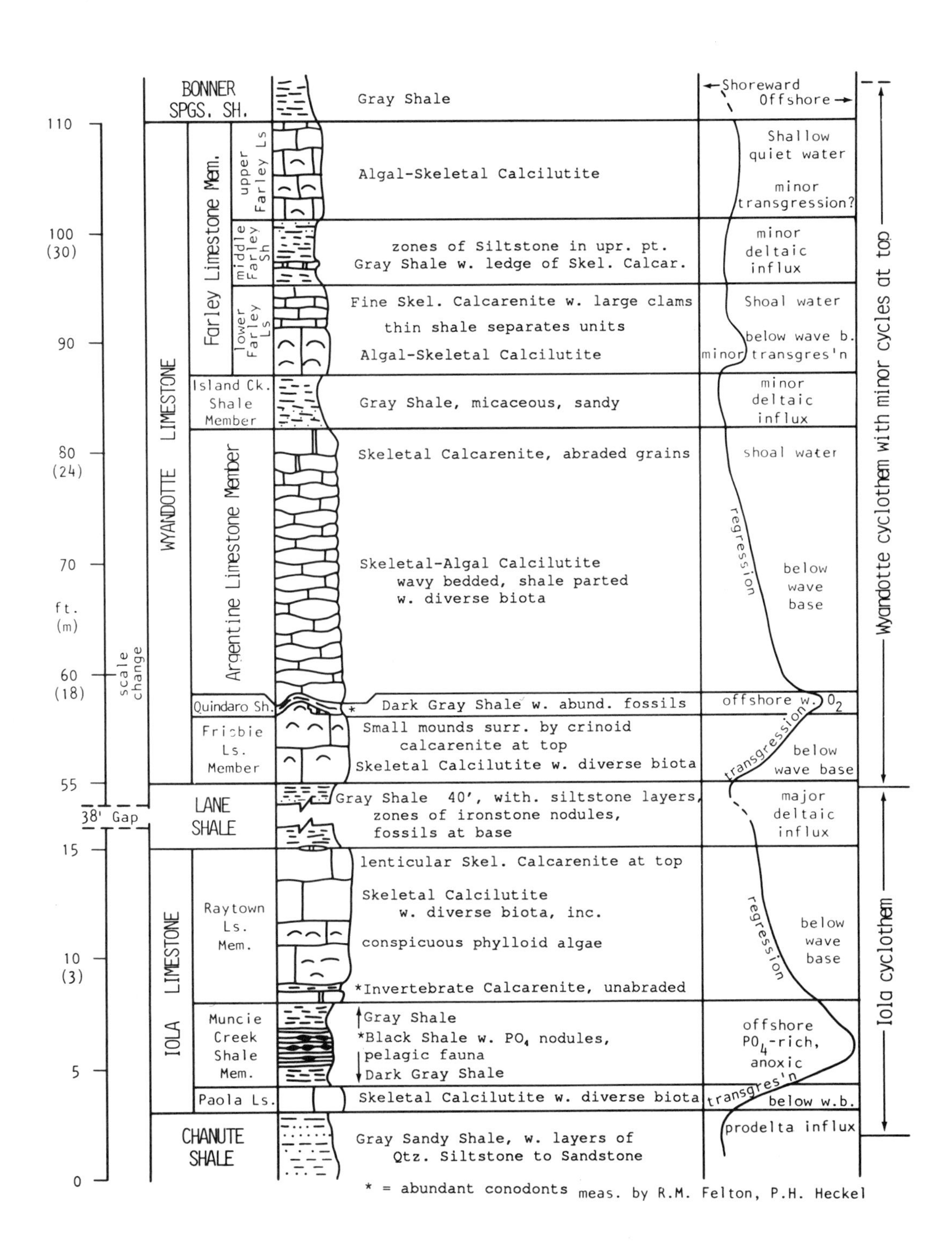

Figure 6. Stratigraphic column and environmental interpretations of section at site 10 (from Heckel and others, 1985, p. 29).

enough for algae, but was still below effective wave base. The thin lenticular skeletal calcarenite at the top may record a storm event. Conodont faunas are sparser, still generally dominated by *Idiognathodus*; *Idioprioniodus* is in the base, and *Anchignathodus* and *Adetognathus* appear in greater numbers toward the top. Thus the pattern of conodont distribution in the regressive Raytown is a mirror image of that in the transgressive Paola, symmetrical about the offshore Muncie Creek Shale.

The **Lane Shale** records a major advance of prodeltaic clastics from the northeast that smothered accumulation of Raytown carbonates while still in a quiet marine environment before regression reached the phase of shoal-water deposition. The fossiliferous zone in the base reflects the onset of terrigenous mud accumulation from the approaching delta as it began to smother the clear water, open marine, fossiliferous Raytown environment.

The **Frisbie Limestone** is the transgressive limestone of the Wyandotte cyclothem. It records a marine inundation strong enough both to swamp the detrital influx that formed the Lane delta and to push the shoreline at least 200 mi (320 km) northward, past the Iowa outcrop, where the Lane Shale is represented by a thin terrestrial deposit. At the top of the Frisbie in this exposure, there are several small mounds where phylloid algae locally built up the sea bottom in the deepening water while becoming surrounded by crinoid thickets. The Frisbie has a sparse *Idiognathodus*-dominated conodont fauna similar to that described for the homologous Paola (data from Heckel and Baesemann, 1975).

The **Quindaro Shale** is the offshore shale of the Wyandotte cyclothem. It is gray and fossiliferous because it lay above the thermocline (i.e., in oxygenated water), even at maximum transgression. The abundant *Idiognathodus*-dominated conodont fauna also includes *Idioprioniodus*.

The **Argentine Limestone** is the main regressive limestone of the Wyandotte cyclothem. It consists of skeletal calcilutite overlain by skeletal calcarenite, which contains abraded and sorted grains, and was deposited in shallowing water that attained shoal conditions at the top prior to detrital influx. This is in contrast with the homologous Raytown regressive limestone in the underlying Iola cyclothem, which was smothered by detrital influx before shoal conditions were attained. The Argentine forms the main part of the bedded phylloid-algal bank complex described in the Kansas City area by Crowley (1969). It has a sparse *Idiognathodus*-dominated conodont fauna, and *Anchignathodus* is relatively abundant throughout (data from Heckel and Baesemann, 1975).

The **Island Creek Shale** records a minor deltaic influx from the north (Crowley, 1969). It has a conodont fauna dominated by *Adetognathus,* the typical nearshore genus.

The **Farley Limestone** here comprises two carbonates (lower and upper) separated by the middle Farley shale. Conodont faunas throughout are sparse and dominated by *Adetognathus,* with *Anchignathodus,* but very few *Idiognathodus,* which suggests that the entire sequence was deposited in shallow nearshore environments. The lower Farley is a shallowing-upward sequence of algal-skeletal calcilutite deposited below effective wave base, overlain by fine skeletal calcarenite deposited in shoal water. The middle Farley is a minor, perhaps prodeltaic influx from the north (Crowley, 1969). The upper Farley is algal-skeletal calcilutite throughout, suggesting a shallow, but quiet off-shoal environment.

Site 11 (Fig. 7). The section here shows the well-developed Swope and Dennis cyclothems, both containing black phosphatic "core"shales. It also shows evidence for subaerial exposure at the top of the Swope, and for minor cycles, with subaerial exposure at the top of at least one, within the regressive phase of the Dennis.

The **Middle Creek Limestone** is exposed low in the ditch north of the section-line road below the bluff. It is the transgressive limestone of the Swope cyclothem. Traced from Iowa to southern Kansas, it is a dense skeletal calcilutite resembling the transgressive Paola Limestone seen at site 10. It also records a major transgression of the sea to quiet water below effective wave base, above the nearshore Ladore Shale (not exposed here).

The **Hushpuckney Shale** is poorly exposed in the ditch just above the Middle Creek Limestone. It is the offshore "core" shale of the Swope cyclothem with a black phosphatic facies like that seen in the Muncie Creek Shale of the Iola cyclothem at site 10. It also is traceable from Iowa to southern Kansas: its characteristic abundant conodont fauna is dominated by *Idiognathodus,* and includes *Idioprioniodus* and *Gondolella.*

The **Bethany Falls Limestone** is the regressive limestone of the Swope cyclothem. Only the top is well exposed in the bluff here. Although homologous with the regressive Raytown Limestone of the Iola cyclothem (site 10), and consisting of similar skeletal calcilutite in the lower two-thirds, the Bethany Falls differs in having a cap of oolite and pelleted calcilutite with faint birdseye fabric, which record late-regressive shoaling before any detrital influx reached this area. Vertical rusty tubelike structures in the upper beds strongly suggest pathways of intense meteoric leaching which were filled with ferroan dolomite during burial diagenesis. The exposure surface at the top is marked a few miles away by paleocaliche, including both laminar and vaguely pisolitic forms described by Heckel (1983, p. 753–754). Recognition of widespread exposure surfaces and meteoric diagenesis in regressive carbonates is an important argument for the significance of eustatic drops of sea level as a major factor in the formation of Midcontinent cyclothems.

The **Galesburg Shale** is exposed in the middle of the bluff along the north side of the section-line road. It is a gray blocky mudstone with a pattern of clay mineralogy that strongly suggests formation as a paleosol (Schutter, 1983). These features characterize the Galesburg from east-central Kansas to the northern erosional limit of outcrop in Iowa and Nebraska. Southward, the Galesburg displays fluvial facies in southeastern Kansas (seen at site 13), and a large deltaic complex in northeastern Oklahoma that extends tens of miles south of Tulsa. This shows the great distance of marine withdrawal between the two major Swope and Dennis marine cycles, at least to southeastern Kansas. Both

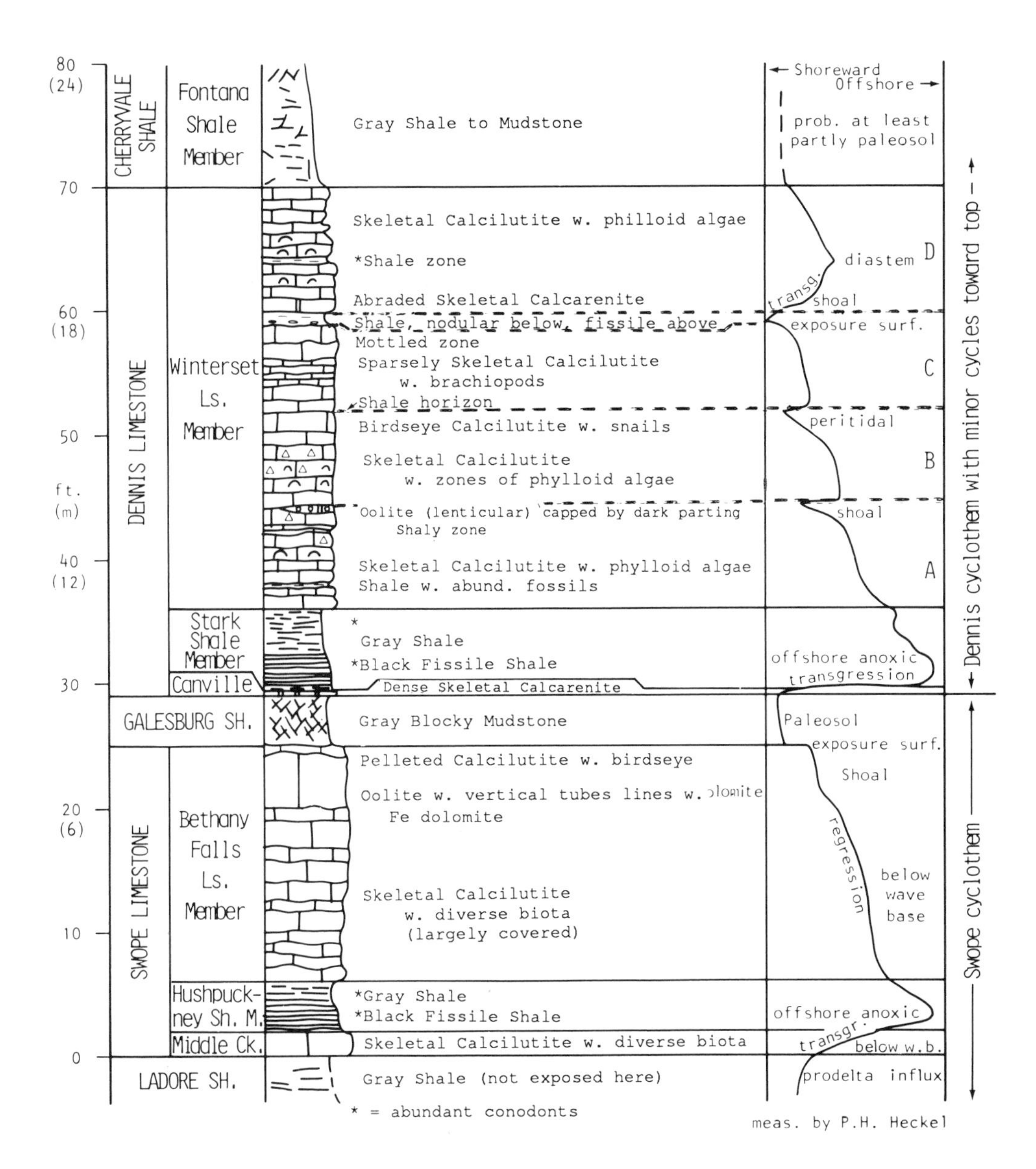

Figure 7. Stratigraphic column and environmental interpretations of section at site 11 (from Heckel and others, 1985, p. 47).

of these cycles retain their offshore black phosphatic shale facies to the northern outcrop limit north of Omaha, Nebraska, about 400 mi (644 km) from Tulsa, implying that the strandline of both transgressions went much farther to the north.

The **Canville Limestone** is the transgressive limestone of the Dennis cyclothem, and is here a thin, dense, overpacked skeletal calcarenite, which suggests compaction before late burial cementation such as in transgressive calcilutites and offshore calcarenites. But, unlike the offshore calcarenite in the basal Raytown at site 10, it has abraded grains, suggesting deposition in shallow agitated water early during transgression. The Canville is not as continuous northward, as are most transgressive limestones; this suggests a more rapid transgression with insufficient time for carbonate production, resulting in sediment starvation. The overlying offshore black shale rests on the Galesburg paleosol in most areas to the north.

The **Stark Shale** is the offshore shale of the Dennis cyclothem and displays its characteristic black fissile phosphatic facies in the upper part of the bluff. The Stark is traceable from Iowa into northeastern Oklahoma, and the black facies has an abundant conodont fauna, *Idiognathodus* dominating *Idioprioniodus* and *Gondolella,* just as in the homologous Hushpuckney and Muncie Creek shales in the Swope and Iola cyclothems.

The **Winterset Limestone** is the regressive limestone of the Dennis cyclothem. It forms the top of the bluff, and is best exposed northward along the east side of U.S. 69. Its regressive position has been corroborated by the evidence of infiltration of meteoric water from the top described by Railsback (1984). But, unlike the rather simple regression displayed by the homologous Bethany Falls Limestone of the underlying Swope cycle, the Winterset displays several conspicuous minor cycles of shallowing-upward lithology that are well exposed along U.S. 69. Because the transgressive surfaces are abrupt, with little or no deepening-upward deposits, three of these units appear to represent the punctuated aggradational cycles (PAC) of Goodwin and Anderson (1985). If correlatable over significant distances, these would be minor eustatic cycles, and thus would probably represent the climatic control of the shorter periods of the Earth's orbital parameters, according to the Milankovitch theory of climate change and glacial waxing and waning. In the probable subsurface equivalent of the Winterset Limestone, 400 mi (640 km) to the northwest, DuBois (1985) traced two minor shoaling-upward cycles in numerous cores across an entire county in southwestern Nebraska, which hints at the possibility of the widespread extent of some of these minor cycles.

The lowest minor cycle in the Winterset (A) is open marine skeletal calcilutite capped with lenticular oolite that records a minor shoal. The next cycle up (B) is similar skeletal calcilutite capped by birdseye-bearing calcilutite containing snails, recording a peritidal environment. The next cycle up (C) is sparsely fossiliferous calcilutite grading up to a mottled zone which has fitted clasts of limestone separated by shale-filled fissures containing small fragments of limestone and brownish-stained calcite that is probably a form of caliche. The mottled zone is capped by a shale, the lower one-third of which contains irregular limestone clasts but no fossils, and appears to be a paleosol containing erosional remnants of the underlying limestone. These lower three minor cycles together show progressive shallowing of the capping facies upward from shoal to tidal flat to soil, and are therefore easily accommodated within the regressive phase of the Dennis cyclothem. The sparse conodont faunas collected at selected horizons corroborate this sequence, in that open-shelf *Idiognathodus* was found in low numbers only in minor cycle A and the base of minor cycle B, whereas nearshore *Adetognathus* was found only stratigraphically higher.

In contrast to the PAC-like lower three minor cycles, the topmost minor cycle (D) displays a deepening-upward or transgressive lithic sequence in the lower half. The dark shale at the base that contains few conodonts probably represents a nearshore, perhaps lagoonal environment. It is overlain by an abraded-grain skeletal calcarenite suggestive of a shoal environment, and this is followed upward by more offshore skeletal-algal calcilutites deposited below effective wave base. This sequence culminates in a thin shale containing relatively abundant conodonts (over 100/kg) that probably represents a slight diastem and would be analogous to a core shale of a more major cycle. However, all of the conodonts picked from this shale are *Adetognathus,* the nearshore genus, which shows that this cycle involved only shallow nearshore-shelf environments. The upper part of minor cycle D is regressive up into the overlying Fontana Shale, which may represent in part another paleosol. The main difference between cycle D and the underlying PAC is that the transgression in cycle D was slow enough for a deepening-upward lithic sequence to be deposited. Cycle D is possibly the nearshore end of an intermediate cycle better represented southward in Oklahoma.

Site 12 (Fig. 8). The section here shows the offmound facies of the Plattsburg cyclothem, a marine cycle of intermediate extent that lacks black phosphatic facies in the offshore shale and contains shaly limestones and fossiliferous shales which yield abundant fossils.

The **Lane–Bonner Springs Shale** is about 165 ft (50 m) thick in this region. It records prodeltaic detrital influx during the sea-level lowstand prior to the Plattsburg transgression. The prodeltaic influx waned abruptly, perhaps through delta abandonment, sometime before the Plattsburg transgression, allowing the accumulation of enough shells to form a regional zone of shaly limestone lenses in the upper part. The assemblage of clams, snails, and crinoid fragments in these lenses probably represents a marine fauna dominated by organisms adapted to soft substrate and relatively turbid water. Above these lenses, a later shift of prodeltaic influx back into this region formed the upper 13 ft (4 m) of this shale. Waning of this influx, as the Plattsburg transgression pushed the shoreline northward, established first the molluscan fauna, then a clear-water fauna at the top, containing bryozoans and more brachiopods and crinoids than below.

The **Merriam Limestone** is the transgressive limestone of the Plattsburg cyclothem, deposited in clearer water when algae produced enough carbonate mud to form a shaly calcilutite. The Merriam is lenticular and appears to be developed on a mega-rippled surface, which suggests current activity strong enough to produce bedforms on the underlying shale surface and to prevent even distribution of lime mud, but not strong enough to winnow sufficient mud to form a calcarenite.

The **Hickory Creek Shale** is the core shale of the Plattsburg cyclothem. It was deposited far enough offshore for maintenance of stable open marine salinity, but not in water deep enough for establishment of the thermocline and quasi-estuarine circulation cell that led to loss of bottom oxygen and phosphorite deposition in other cyclothems. The suppression of carbonate production was probably due to its position near the base of the effective photic zone; any fine detrital influx that may have helped subdue algal growth in water this deep was nonetheless slow enough to allow establishment of a good filter-feeding fauna of most of the major invertebrate phyla, including conspicuous calcisponges.

The **Spring Hill Limestone** is the regressive limestone of the Plattsburg cyclothem. Its more pervasive shaliness than most other regressive limestones suggests a more persistent detrital

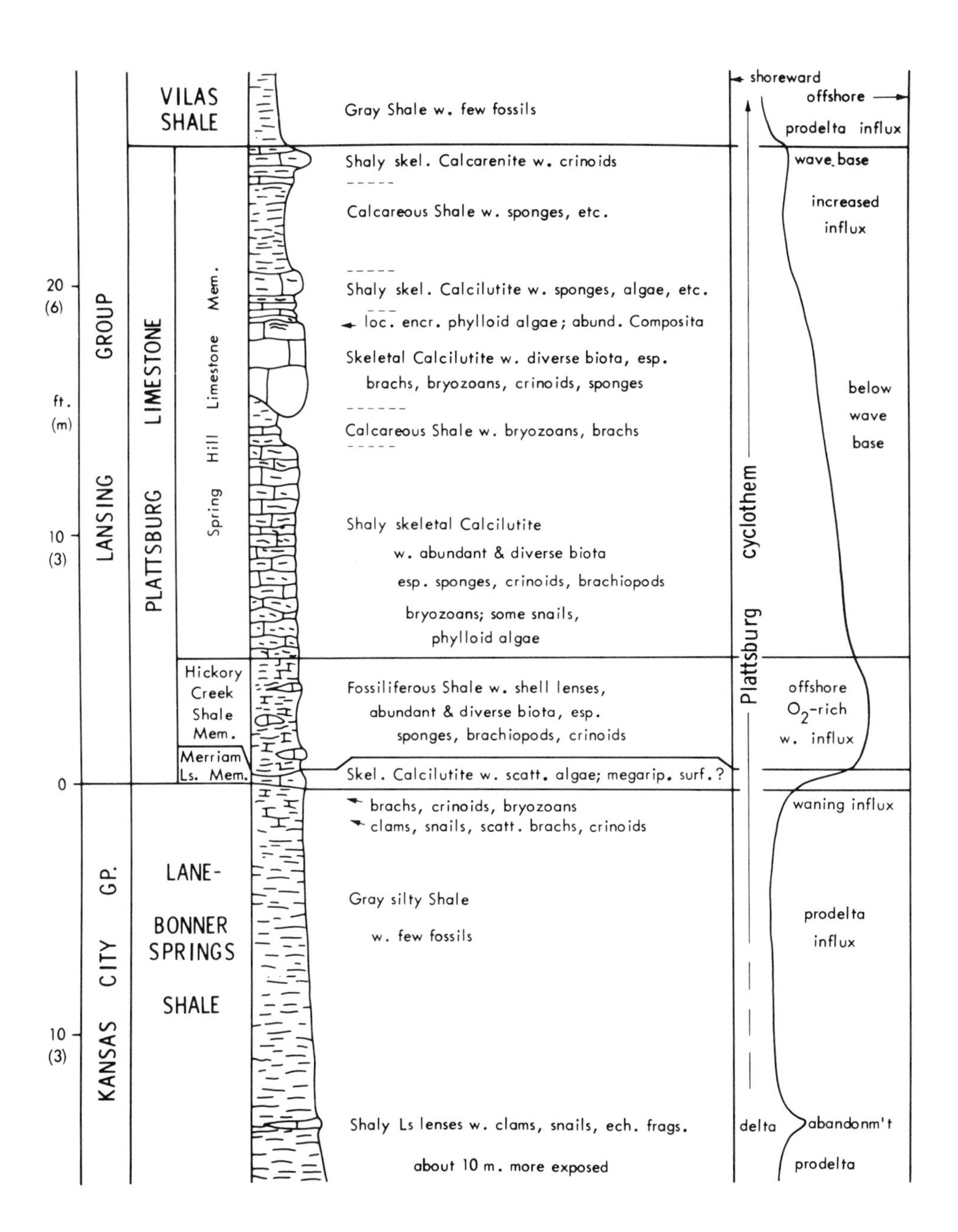

Figure 8. Stratigraphic column and environmental interpretations of section at site 12 (from Heckel and others, 1979, p. 32).

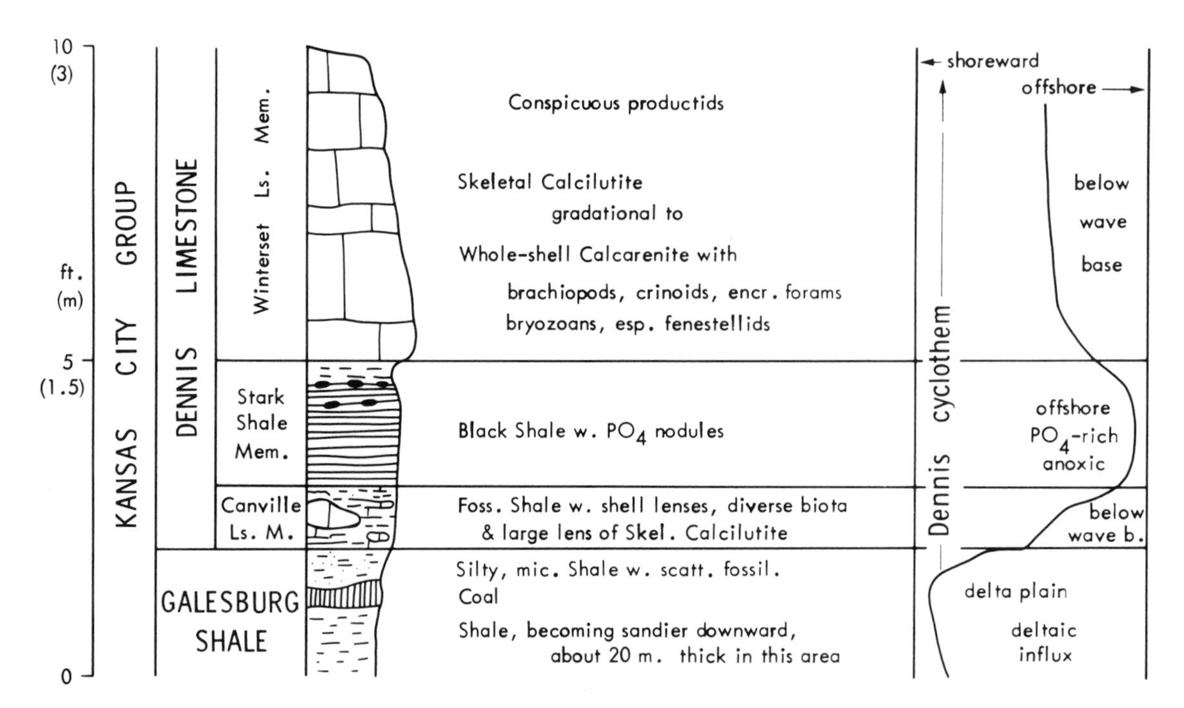

Figure 9. Stratigraphic column and environmental interpretations of section at site 13 (from Heckel and others, 1979, p. 34).

source, which apparently remained closer to the present outcrop because less water depth was attained during the Plattsburg transgression. Although the Plattsburg Limestone is known on the Iowa-Nebraska outcrop, there it is thin and shaly, and lacks the offshore shale development that characterizes major cyclothems such as the Swope and Dennis (seen at site 11), even on the Iowa-Nebraska outcrop. The local proliferation of encrusting phylloid algae in the upper part of the Spring Hill here is the only indication that the north end of one of the thickest (82 ft; 25 m) algal-mound complexes in the Kansas Pennsylvanian, equivalent to the upper half of the Spring Hill, is about 2 mi (3.2 km) to the southwest (Harbaugh, 1959). Following a further increase in detrital influx, the calcarenite at the top records shallowing through the zone of wave agitation during regression.

The **Vilas Shale** records another prodeltaic detrital influx that ended Plattsburg deposition here and filled in around the north end of the Plattsburg algal mound.

Site 13 (Fig. 9). The section here shows the widespread Dennis cyclothem, with its black phosphatic offshore shale, 90 mi (145 km) south of the exposure at site 11. Here the cyclothem displays a transition of characteristics between Kansas-type and Illinois-type cyclothems, the latter of which lack transgressive limestones. Another example of this cyclothem is described in the Centennial Field Guide for the North-Central Section (Heckel, 1987).

The **Galesburg Shale** records delta-plain aggradation and coal-swamp formation during the lowstand of sea level prior to the widespread Dennis transgression. The Galesburg displays thick, sandstone-rich deltaic facies in northeastern Oklahoma, and a thinner blocky mudstone facies with clay mineralogy characteristic of a soil profile (Schutter, 1983) from east-central Kansas northward to Iowa; this shows that the entire northern Midcontinent shelf was emergent during this glacial sea-level low stand.

The **Canville Limestone** records transgression with waning of detrital influx, increase in invertebrates to form small shell lenses, and localized proliferation of carbonate mud-producing algae to form calcilutite lenses. The lack of widespread algae such as produced the laterally continuous transgressive limestones in other cyclothems may relate to inhibition by (1) insufficient waning of detrital influx until water depth was too great; (2) proximity of buried peat to the substrate, which inhibited algal growth; or (3) transgression too rapid to allow enough time for sufficient algal carbonate production to be spread laterally to form a continuous bed of limestone.

The **Stark Shale** is a typical black phosphatic offshore ("core") shale that records anoxic bottom conditions below a thermocline with quasi-estuarine circulation at maximum transgression. The Stark and overlying Winterset Limestone can be traced from Iowa and Nebraska into northeastern Oklahoma, a distance of about 400 mi (645 km). This is only the minimum lateral extent of the Dennis transgression above the terrestrial Galesburg, because the Stark still displays the deep-water black facies at the northern limit of outcrop, and thus must have originally extended a substantial distance farther northward.

The **Winterset Limestone** records enough regression and

shallowing of water to destroy the thermocline and quasi-estuarine cell and reoxygenate the sea bottom for reestablishment of an abundant invertebrate fauna. Scarcity of algae suggests that greater water depths may have been maintained here for a longer time than in the algal-mound facies belt to the north (Heckel and Cocke, 1969), perhaps because of greater subsidence in this more southern area of the shelf, closer to the more basinal area of Oklahoma.

The vertical sequence through the Canville skeletal calcilutite lens is a typical Kansas-type cyclothem, as seen in the Iola Limestone at site 10, whereas the vertical sequence through the shaly Canville is more like an Illinois-type cyclothem, which typically has only a thin, shelly marine shale between the coal and the black shale. The lateral transition of these two types along a single exposure shows the genetic similarity between Kansas-type and Illinois-type cyclothems that contain phosphatic black shales; they are differentiated only by the presence or absence of a conspicuous transgressive limestone.

REFERENCES CITED

Boardman, D. R., and others, 1984, A new model for the depth-related allogenic community succession within North American Pennsylvanian cyclothems and implications on the black shale problem, *in* Hyne, N. J., ed., Limestones of the mid-continent: Tulsa Geological Society Special Publication 2, p. 141–182.

Crowley, D. J., 1969, Algal-bank complex in Wyandotte Limestone (Late Pennsylvanian) in eastern Kansas: Kansas Geological Survey Bulletin 198, 52 p.

DuBois, M. K., 1985, Application of cores in development of an exploration strategy for the Lansing-Kansas City "E" Zone, Hitchcock County, Nebraska: Kansas Geological Survey Subsurface Geology Series 6, p. 120–132.

Ferm, J. C., 1970, Allegheny deltaic deposits: Society of Economic Paleontologists and Mineralogists Special Publication 15, p. 246–255.

Galloway, W. E., and Brown, L. F., Jr., 1973, Depositional systems and shelf-slope relations on cratonic basin margin, uppermost Pennsylvanian of north-central Texas: American Association of Petroleum Geologists Bulletin, v. 57, p. 1185–1218.

Goodwin, P. W., and Anderson, E. J., 1985, Punctuated aggradational cycles; A general hypothesis of episodic stratigraphic accumulation: Journal of Geology, v. 93, p. 515–533.

Harbaugh, J. W., 1959, Marine bank development in Plattsburg Limestone (Pennsylvanian), Neodesha-Fredonia area, Kansas: Kansas Geological Survey Bulletin 134, pt. 8, p. 289–331.

Heckel, P. H., 1977, Origin of phosphatic black shale facies in Pennsylvanian cyclothems of Midcontinent North America: American Association of Petroleum Geologists Bulletin, v. 61, p. 1045–1068.

——, 1980, Paleogeography of eustatic model for deposition of Midcontinent Upper Pennsylvanian cyclothems, *in* Fouch, T. D., and Magathan, E. R., eds., Paleozoic paleogeography of west-central United States: Society of Economic Paleontologists and Mineralogists, Rocky Mountain Section, Paleogeography Symposium I, p. 197–215.

——, 1983, Diagenetic model for carbonate rocks in Midcontinent Pennsylvanian eustatic cyclothems: Journal of Sedimentary Petrology, v. 53, p. 733–759.

——, 1984a, Factors in Mid-Continent Pennsylvanian limestone deposition, *in* Hyne, N. J., ed., Limestones of the mid-continent: Tulsa Geological Society Special Publication 2, p. 25–50.

——, 1984b, Changing concepts of Midcontinent Pennsylvanian cyclothems: IX International Carboniferous Congress 1979, Compte Rendu, v. 3, p. 535–553.

——, 1986, Sea-level curve for Pennsylvanian eustatic marine transgressive-regressive depositional cycles along Midcontinent outcrop belt: Geology, v. 14, p. 330–334.

——, 1987, Pennsylvanian cyclothems near Winterset, Iowa, *in* Biggs, D. L., ed., North-Central Section of the Geological Society of America: Boulder, Colorado, Geological Society of America, Centennial Field Guide, v. 5 (in press).

Heckel, P. H., and Baesemann, J. F., 1975, Environmental interpretation of conodont distribution in Upper Pennsylvanian (Missourian) megacyclothems in eastern Kansas: American Association of Petroleum Geologists Bulletin, v. 59, p. 486–509.

Heckel, P. H., and Cocke, J. M., 1969, Phylloid algal-mound complexes in outcropping Upper Pennsylvanian rocks of Mid-Continent: American Association of Petroleum Geologists Bulletin, v. 53, p. 1058–1074.

Heckel, P. H., and others, 1979, Field guide to Pennsylvanian cyclic deposits of Kansas and Nebraska: Kansas Geological Survey Guidebook series 4, p. 4–60.

Heckel, P. H., and others, 1985, Guidebook for Midcontinent SEPM Field Trip, October 12, 1985, *in* Watney, W. L., and others, eds., Recent interpretations of late Paleozoic cyclothems: Proceedings, 3rd Annual Meeting and Field Conference, Society of Economic Paleontologists and Mineralogists, Midcontinent Section, Kansas Geological Survey, p. 23–69.

Imbrie, J., and Imbrie, J. Z., 1980, Modeling the climatic response to orbital variations: Science, v. 207, p. 943–953.

Kidder, D. L., 1985, Petrology and origin of phosphate nodules from the Midcontinent Pennsylvanian epicontinental sea: Journal of Sedimentary Petrology, v. 55, p. 809–816.

Malinky, J. M., 1984, Paleontology and paleoenvironment of "core" shales (Middle and Upper Pennsylvanian) Midcontinent North America [Ph.D. thesis]: Iowa City, University of Iowa, 327 p.

Merrill, G. K., 1975, Pennsylvanian conodont biostratigraphy and paleoecology of northwestern Illinois: Geological Society of America Microform Publication 3, 2 cards.

Mitchell, J. C., 1981, Stratigraphy, petrography, and depositional environments of the Iola Limestone, Mid-Continent U.S. [Ph.D. thesis]: Iowa City, University of Iowa, 364 p.

Moore, R. C., 1931, Pennsylvanian cycles in the northern Mid-Continent region: Illinois Geological Survey Bulletin 60, p. 247–257.

——, 1936, Stratigraphic classification of the Pennsylvanian rocks of Kansas: Kansas Geological Survey Bulletin 22, 256 p.

——, 1950, Late Paleozoic cyclic sedimentation in central United States: 18th International Geological Congress, Great Britain, 1948, pt. 4, p. 5–16.

Railsback, L. B., 1984, Carbonate diagenetic facies in the Upper Pennsylvanian Dennis Formation in Iowa, Missouri, and Kansas: Journal of Sedimentary Petrology, v. 54, p. 986–999.

Schutter, S. R., 1983, Petrology, clay mineralogy, paleontology, and depositional environments of four Missourian (Upper Pennsylvanian) shales of Mid-Continent and Illinois Basin [Ph.D. thesis]: Iowa City, University of Iowa, 1208 p.

Schutter, S. R., and Heckel, P. H., 1985, Missourian (early Late Pennsylvanian) climate in Midcontinent North America: International Journal of Coal Geology, v. 5, p. 111–140.

Swade, J. W., 1985, Conodont distribution, paleoecology, and preliminary biostratigraphy of the upper Cherokee and Marmaton Groups (upper Desmoinesian, Middle Pennsylvanian) from two cores in south-central Iowa: Iowa Geological Survey Technical Information Series 14, 71 p.

Wanless, H. R., 1964, Local and regional factors in Pennsylvanian cyclic sedimentation: Kansas Geological Survey Bulletin 169, p. 593–606.

——, 1967, Eustatic shifts in sea level during the deposition of late Paleozoic sediments in the central United States: West Texas Geological Society Publi-

cation 69-56, p. 41–54.
Wanless, H. R., and Shepard, F. P., 1936, Sea level and climatic changes related to late Paleozoic cycles: Geological Society of America Bulletin, v. 47, p. 1177–1206.
Wanless, H. R., and Weller, J. M., 1932, Correlation and extent of Pennsylvanian cyclothems: Geological Society of America Bulletin, v. 43, p. 1003–1016.
Watney, W. L., 1980, Cyclic sedimentation of the Lansing-Kansas City Groups in northwestern Kansas and southwestern Nebraska: Kansas Geological Survey Bulletin 220, 72 p.
—— , 1984, Recognition of favorable reservoir trends in Upper Pennsylvanian cyclic carbonates in western Kansas, *in* Hyne, N. J., Limestones of the Mid-Continent: Tulsa Geological Society Special Publication 2, p. 201–245.
Weller, J. M., 1930, Cyclic sedimentation of the Pennsylvanian period and its significance: Journal of Geology, v. 38, p. 97–135.
—— , 1958, Cyclothems and larger sedimentary cycles of the Pennsylvanian: Journal of Geology, v. 66, p. 195–207.

Bunker Hill section of Upper Cretaceous rocks in Kansas

Donald E. Hattin, Department of Geology, Indiana University, Bloomington, Indiana 47405
Page C. Twiss, Department of Geology, Kansas State University, Manhattan, Kansas 66506

LOCATION

The Bunker Hill section of the upper part of the Graneros Shale, Greenhorn Limestone, and of the lower part of the Carlile Shale is along the east side of Bunker Hill–Luray Road (Kansas FAS 47) about 4.5 mi (7.2 km) north of I-70, along the center of the west line of Sec.18,T.13S.,R.12W., Russell County, Kansas (Fig. 1). The top of the section is at the intersection with the unpaved side road to the east at an elevation of 1,790 ft (546 m); the base of the section is about 0.5 mi (0.8 km) north at an elevation of about 1,700 ft (518 m). Safest parking is on the side road near the top of the section (Fig. 1T).

Vehicles traveling on I-70 should exit at the Bunker Hill exit (exit 193) and travel 4.5 mi (7.2 km) north through Bunker Hill to Sec. 18. The Bunker Hill exit is about 9 mi (14 km) east of Russell.

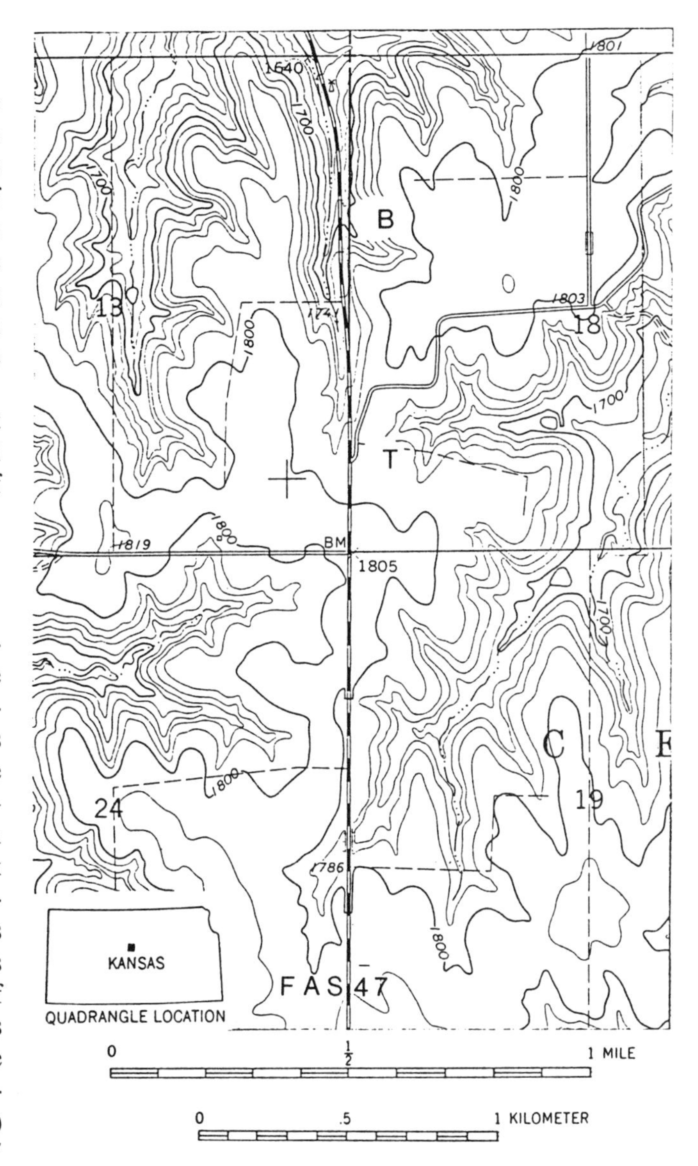

Figure 1. Location of Bunker Hill section of Upper Cretaceous rocks in Kansas (Dorrance NW, 7½-minute Quadrangle, 1967, Russell County, Kansas). B = base; T = section top.

SIGNIFICANCE

The Bunker Hill section was deposited during the Greenhorn third-order transgressive-regressive (T-R) cycle of marine deposition, which is one of the most widely recognizable tectono-eustatic events of the Late Cretaceous (Kauffman, 1985). This section illustrates the vertical transition from quartzose sandstone and dark gray shale in the Graneros Shale through dark olive-gray chalky shale, chalk, and chalky limestone of the Greenhorn Limestone, into the lower 6.5 ft (2 m) of the Fairport Chalk Member of the Carlile Shale (Fig. 2). Chalky strata of the Greenhorn Limestone were deposited during near-peak and peak stages of marine transgression (Fig. 3), and include pelagic rhythmites that have been attributed to Milankovitch-scale cyclic changes of climate (R.O.C.C. Group, 1986). Individual bedding couplets displayed in chalky parts of the section are isochronous and have been traced over thousands of kilometers (Hattin, 1971). Twenty-three seams of bentonite also occur within this 118-ft-(36 m) thick section, some of which have been traced laterally (surface/subsurface) for several hundred kilometers.

SITE INFORMATION

Only the upper 22 ft (6.7 m) of the Graneros Shale are exposed in the Bunker Hill section (Fig. 4); they are in a hillside that is just inside a private fenceline (access has not been a problem in the past). The Graneros Shale consists of dark gray silty shale, quartzose sandstone, and skeletal limestone, and includes the uppermost beds of the *Aphrodina lamarensis* Assemblage Zone and all of the *Ostrea beloiti* Assemblage Zone. Unit 14 is a bentonite seam that is 1 ft (0.3 m) thick. The Graneros forms grass-covered slopes and is exposed only in steep roadcuts and cutbanks of major streams (Hattin and others, 1978, p. 15).

The Greenhorn Limestone is 91 ft (27.8 m) thick and includes four members (Fig. 2) that consist predominantly of olive gray, generally laminated, chalky, or marly shale. These are differentiated on basis of character and abundance of contained limestones (Hattin, 1975). The contact between the Greenhorn Limestone and the underlying Graneros Shale is disconformable;

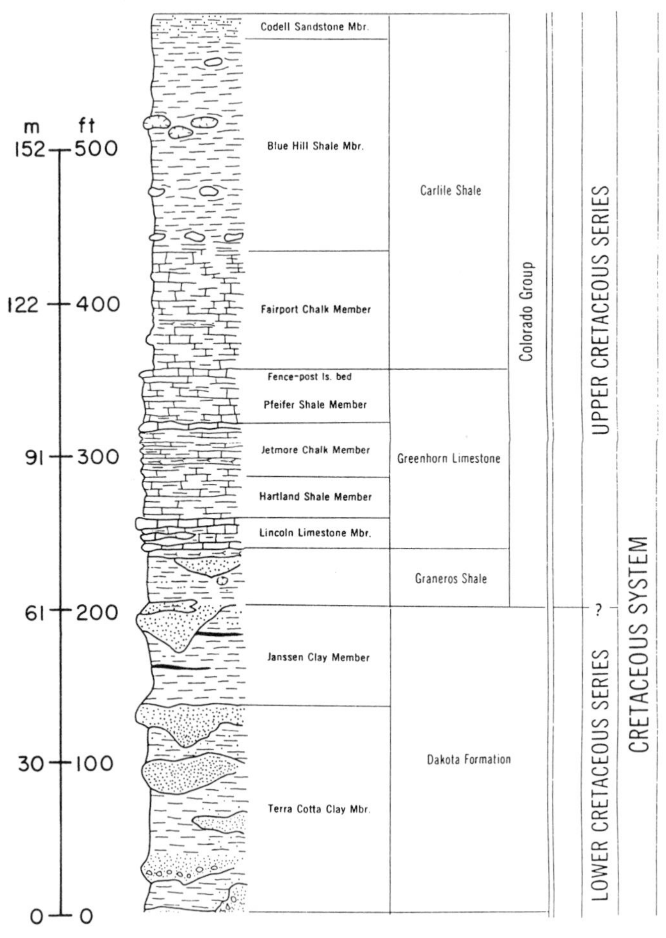

Figure 2. Stratigraphic section of Upper Cretaceous Series, Cretaceous System in Kansas (adapted from Zeller, 1968).

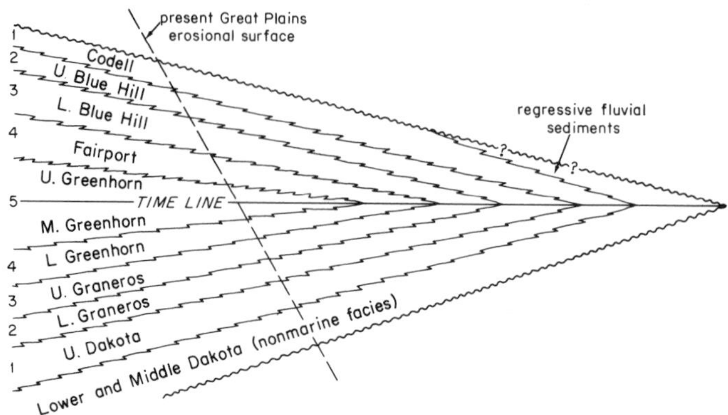

Figure 3. Schematic representation of wedgelike facies arrangement in lithologically symmetrical Greenhorn cyclothem in Kansas. Major erosion surfaces (wavy lines) bound deposits of cyclothem. Marine facies include: 1, marginal marine sandstone and shale; 2, sandy shale; 3, silty shale with septarian concretions; 4, chalky and marly shale with beds and lenses of skeletal limestone; 5, chalky limestone, chalky shale, and marly shale that manifest rhythmic bedding. Regressive alluvial deposits are not preserved on top of Codell sandstones because of incomplete withdrawal of sea combined with prolonged episode of submarine scour and sedimentary bypass, or because of prolonged episode of subaerial erosion, which stripped away regressive fluvial deposits that once lay above Codell. Great vertical exaggeration accounts for steepness of dashed line labeled "present Great Plains erosional surface."

it is sharp, and one fossil assemblage zone is missing (Hattin and others, 1978, p. 26).

The basal Lincoln Limestone Member is 21.3 ft (6.5 m) thick and is characterized by many thin lenses and wavy beds of skeletal limestone and numerous bentonite seams. Commonly, the basal bed is calcarenite (bed 16 of Fig. 4). The upper part forms gentle slopes and is poorly exposed. The lower part forms steeper slopes, cliffs, and cutbanks along streams (Hattin and others, 1978, p. 15).

The Hartland Shale Member is 28.9 ft (8.8 m) thick and includes the thickest bentonites of the Kansas Greenhorn Limestone. It also includes a few beds of bioturbated chalky limestone that are concentrated in the well-known *Sciponoceras gracile* Assemblage Zone (beds 35 to 37 in Fig. 4). Above the *S. gracile* zone, this member contains thin beds of soft bioturbated chalk that thicken westward and continue into Colorado and New Mexico as beds of hard, brittle limestone. The lower part of the Hartland Member forms poorly exposed gentle slopes, whereas the upper part forms steep slopes below cliffs and bluffs of the Jetmore Chalk Member (Hattin and others, 1978, p. 15).

The Jetmore Chalk Member is 21 ft (6.4 m) thick and is characterized by 13 subequally spaced beds of bioturbated limestone that range from chalky to hard and brittle. One or more layers of oblate spheroidal chalky limestone concretions lie between the twelfth and thirteenth limestone beds; a thin bentonite seam lies on the tenth bed. The Jetmore Member forms cliffs and steep bluffs along courses of larger streams (Hattin and others, 1978, p. 15).

Only the lower 6.2 ft (1.9 m) of the Fairport Chalk Member of the Carlile Shale are exposed at the Bunker Hill section, and consist of olive-gray, laminated, chalky shale with several limestone concretions; the section weathers yellowish-gray to grayish-orange (Hattin and others, 1978, p. 15).

Hattin (1964) described seven phases of transgression and regression in about 485 ft (148 m) of section that contains the upper Dakota Formation, Graneros Shale, Greenhorn Limestone, and Carlile Shale, which he designated as the Greenhorn cyclothem (third-order cyclothem of Kauffman, 1985). The Bunker Hill section contains phases 3 and 4 of the transgressive part of the Greenhorn third-order cyclothem.

The uppermost silty shale beds of the Graneros Shale (Fig. 4) were deposited under offshore shelf conditions, where influence of terrigenous detrital distribution systems masked any effects of pelagic carbonate sedimentation (Hattin, 1964; phase

3). Calcarenite-rich beds of the Greenhorn lie unconformably on the weakly to noncalcareous Graneros Shale. The unconformity and the overlying basal calcarenite bed (bed 16 of Fig. 4) were produced by sea-floor scour and subsequent lag deposition associated with transgression during the early part of Hattin's (1964) phase 4. Scattered sparry skeletal grainstones in the upper part of the Graneros Shale are earliest manifestations of phase 4 deposition. A vertical upward shift from predominantly siliciclastic Graneros sediments to predominantly pelagic carbonate sediments of the Greenhorn is a direct reflection of progressively greater distance from terrigenous source areas during culmination of the Cenomanian-Turonian third-order transgression.

Pelagic sediments deposited at the peak of the Greenhorn third-order transgression (Hartland, Jetmore, and Pfeifer Members), have marked rhythmicity involving alternations of chalk or chalky limestone beds with chalky or marly shale beds (Fig. 5). Sequence, fossils, and associated bentonite seams prove that each bed in the rhythmic sequence is isochronous. The most readily identifiable isochronous marker beds, which have been assigned alphanumeric codes (Fig. 4), can be traced across vast expanses of the U.S. Western Interior (Hattin, 1971, 1975, 1985). The rhythmic sequence of more or less pure chalky strata resulted from climatically linked dilution/redox cycles (Pratt, 1984; Hattin, 1986). These have average periodicities of 40,000 yr and were probably driven by Milankovitch cycles of obliquity (Pratt, 1985; R.O.C.C. Group, 1986).

In the Bunker Hill section, most beds of chalky limestone are pervasively bioturbated, reflecting bottom conditions favorable to a mobile endobenthos. Pratt (1985) inferred that beds of pelagic limestone were deposited during more arid conditions, when the entire water column was well mixed and the bottom waters were well oxygenated. Beds of marly or chalky shale are generally laminated, and denote oxygen depleted (but not usually anoxic) bottom conditions. Pratt (1985) suggested that these beds were deposited during pluvial intervals when the water column was poorly mixed and the bottom waters were oxygen-depleted; the clay was derived from discharging rivers.

Throughout the Bunker Hill section, inoceramid bivalves are the most abundant macroinvertebrate fossils; ostreid bivalves are second. Ammonites, invariably preserved as molds, are common throughout, with notable abundance in units 16 (zone of *Dunveganoceras pondi*), 35 (zone of *Sciponoceras gracile*), 41 (zone of *Vascoceras birchbyi*), 44 (zone of *Mammites nodosoides*), and 50 (zone of *Collignoniceras woollgari*). Common inoceramids of the Bunker Hill section include *Inoceramus rutherfordi* (upper Graneros), *I. prefragilis* and *I. ginterensis* (Lincoln), *I. pictus* (lower Hartland), *Mytiloides.* cf. *M. columbianus* (lower Jetmore), *M. mytiloides* (upper Jetmore), *I. cuvieri* and *M. subhercynicus* (upper Pfeifer). *C. woollgari, I. cuvieri,* and *M. subhercynicus* range upward into the Fairport Member, Carlile Shale, the lower few feet of which are exposed at the top of the Bunker Hill section.

The Graneros Shale at the Bunker Hill section contains four beds of bentonite, but the upper bed, Unit 14 of Figure 4, is a

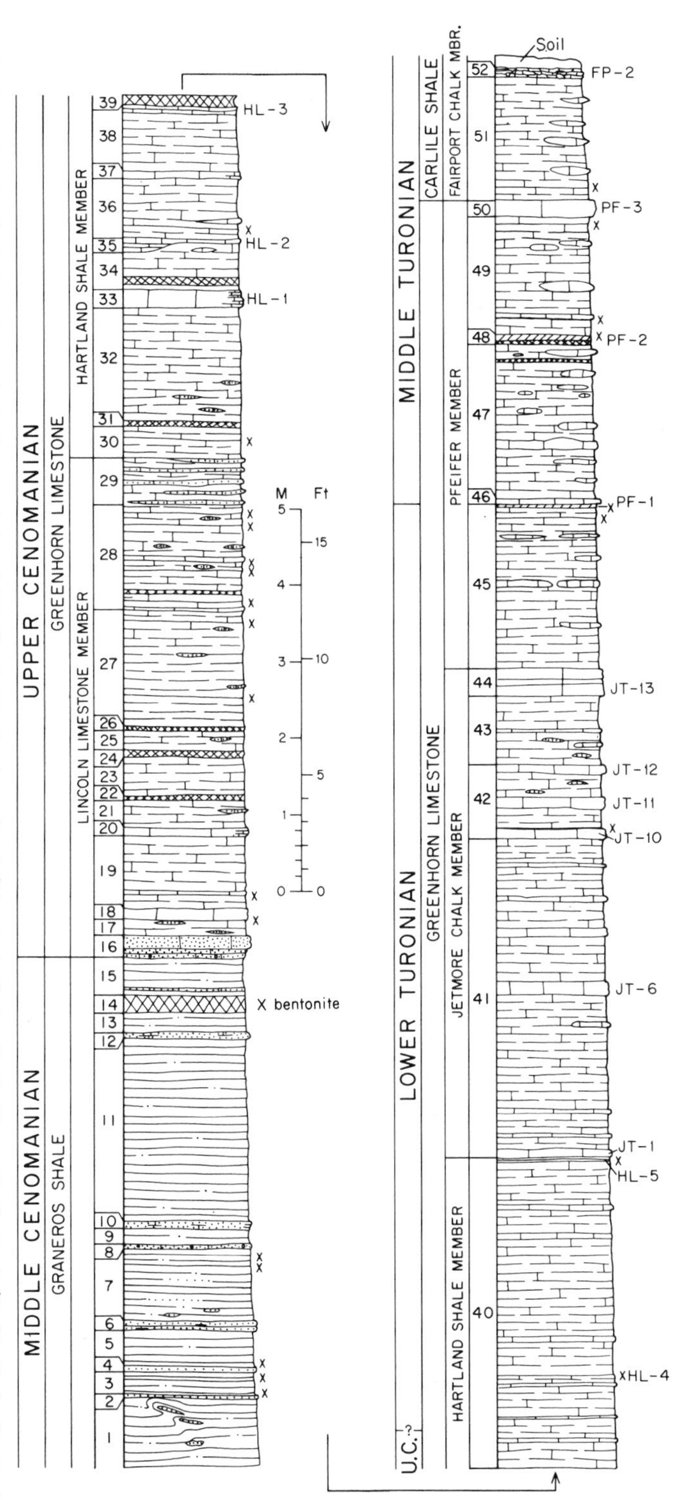

Figure 4. Graphic section of Cenomanian-Turonian strata exposed in Bunker Hill section. Bed numbers refer to section description published in Hattin (1965). Key beds used in precise correlation have alphanumeric designations.

Figure 5. Field photograph of middle part of Bunker Hill section, showing features of Hartland and Jetmore Members, Greenhorn Limestone.

widespread marker bed designated as the "X" bentonite. Hattin (1965, p. 38) reported that the "X" bentonite has been traced in the subsurface across Kansas (Merriam, 1957), throughout the Denver Basin (Haun, 1959, plates 1 and 2), and into Wyoming (McCrae, 1956). It has subsequently been traced as far south as New Mexico (e.g., Watkins, 1986) and at least as far north as Manitoba (McNeil and Caldwell, 1981).

Although nine beds of bentonite are shown in the Greenhorn Limestone (Fig. 4), Hattin (1975, p. 116–119) listed 21 in Key Section 2, which is a more detailed description of the section. Most seams range from 1 to 6 in (0.3 to 16 cm) thick, are light gray to dark yellowish-orange, and are mostly smectite (montmorillonite); some have been traced laterally for several hundreds of miles. The traceable bentonite seams provide isochronous time markers that are useful in reconstructing ancient paleogeographies.

SUMMARY

The Bunker Hill section illustrates the complexity of the marginal marine and marine depositional environments of part of the Western Interior Sea (Hattin and others, 1978) as recorded in the Graneros Shale, Greenhorn Limestone, and Carlile Shale. Suites of body fossils, sedimentary structures, bentonite seams, and vertical progression of rock types are well exposed. Planktonic coccoliths (Cepek and Hay, 1969; Watkins, 1986) and planktonic foraminifers (Eicher and Worstell, 1970) are abundant and well preserved throughout the entire Bunker Hill section.

REFERENCES CITED

Cepek, P., and Hay, W. W., 1969, Calcareous nannoplankton and biostratigraphic subdivision of the Upper Cretaceous: Transactions of the Gulf Coast Association of Geological Societies, v. 19, p. 323–336.

Eicher, D. L., and Worstell, P., 1970, Cenomanian and Turonian Foraminifera from the Great Plains, United States: Micropaleontology, v. 16, p. 269–324.

Hattin, D. E., 1964, Cyclic sedimentation in the Colorado Group of west-central Kansas: Kansas Geological Survey Bulletin 169, p. 205–217.

—— , 1965, Upper Cretaceous stratigraphy, paleontology, and paleoecology of western Kansas: Geological Society of America Field Conference Guidebook, Annual Meeting, Kansas City, Missouri, 69 p. (contains a 3-p. section on the Pierre Shale by W. A. Cobban).

—— , 1971, Widespread, synchronously deposited, burrow-mottled limestone beds in Greenhorn Limestone (Upper Cretaceous) of Kansas and southeastern Colorado: American Association of Petroleum Geologists Bulletin, v. 55, p. 412–431.

—— , 1975, Stratigraphy and depositional environment of Greenhorn Limestone (Upper Cretaceous) of Kansas: Kansas Geological Survey Bulletin 209, 128 p.

—— , 1985, Distribution and significance of widespread, time-parallel pelagic limestone beds in the Greenhorn Limestone (Upper Cretaceous) of the central Great Plains and Southern Rocky Mountains: Society of Economic Paleontologists and Mineralogists Field Trip Guidebook no. 4, 1985 Midyear Meeting, Golden, Colorado, p. 28–37.

—— , 1986, Interregional model for deposition of Upper Cretaceous pelagic rhythmites, U.S. Western Interior: Paleoceanography, v. 1, p. 483–494.

Hattin, D. E., Siemers, C. T., and Stewart, G. F., 1978, Upper Cretaceous stratigraphy and depositional environments of western Kansas: Kansas Geological Survey Guidebook Series 3, 102 p.

Haun, J. D., 1959, Lower Cretaceous stratigraphy of Colorado: Rocky Mountain Association of Geologists Guidebook, 11th Field Conference, p. 1–8.

Kauffman, E. G., 1985, Cretaceous evolution of the Western Interior Basin of the United States: Society of Economic Paleontologists and Mineralogists Field Trip Guidebook no. 4, 1985 Midyear Meeting, Golden, Colorado, p. IV-XIII.

McCrae, R. O., 1956, Subsurface stratigraphy of the pre-Niobrara formations in the Julesburg basin, southeastern Wyoming: Wyoming Geological Association, Subsurface Stratigraphy of the Pre-Niobrara Formations in Wyoming, pt. 1, p. 85–89.

McNeil, D. H., and Caldwell, W.G.E., 1981, Cretaceous rocks and their Foraminifera in the Manitoba Escarpment: Geological Association of Canada Special Paper 21, 439 p.

Merriam, D. F., 1957, Subsurface correlation and stratigraphic relation of rocks of Mesozoic age in Kansas: Kansas Geological Survey Oil and Gas Investigations 14, 25 p.

Pratt, L. M., 1984, Influence of paleoenvironmental factors on preservation of organic matter in Middle Cretaceous Greenhorn Formation, Pueblo, Colorado: American Association of Petroleum Geologists Bulletin, v. 68, p. 1146–1159.

Pratt, L. M., 1985, Isotopic studies of the organic matter and carbonate in rocks of the Greenhorn Marine Cycle: Society of Economic Paleontologists and Mineralogists Field Trip Guidebook no. 4, 1985 Midyear Meeting, Golden, Colorado, p. 38–48.

R.O.C.C. (Research on Cretaceous Cycles) Group, 1986, Rhythmic bedding in Upper Cretaceous pelagic carbonate sequences; Varying sedimentary response to climatic forcing: Geology, v. 14, p. 153–156.

Watkins, D. K., 1986, Calcareous nannofossil paleoceanography of the Cretaceous Greenhorn Sea: Geological Society of America Bulletin, v. 97, p. 1239–1249.

Zeller, D. E., ed., 1968, The stratigraphic succession in Kansas: Kansas Geological Survey Bulletin 189, 81 p.

Rocktown channel in Dakota Formation of central Kansas

Page C. Twiss, *Department of Geology, Kansas State University, Manhattan, Kansas 66506*

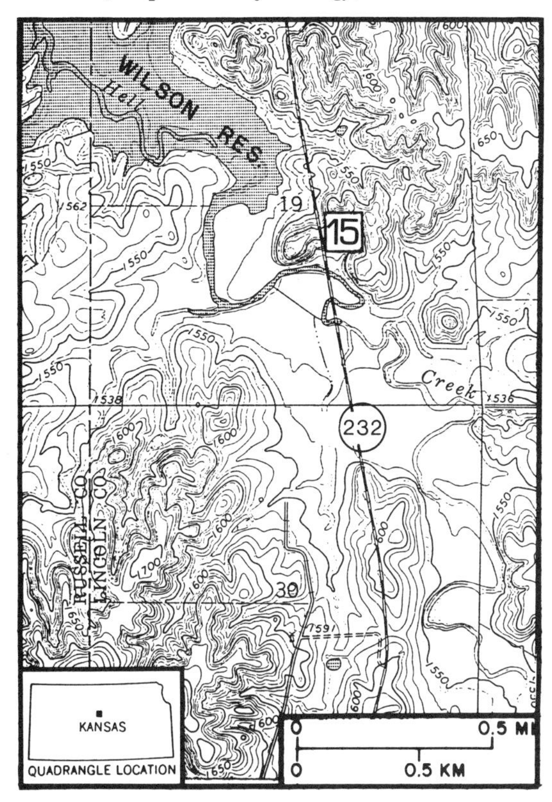

Figure 1. Location of Rocktown channel sandstone in upper Dakota Formation, Wilson NW, 7½-minute Quadrangle, 1964, photorevised 1979, Lincoln County, Kansas.

Figure 2. Stratigraphic section of Lower Cretaceous Series, Cretaceous System in Kansas (adapted from Zeller, 1968).

LOCATION

The Rocktown channel sandstone of the Upper Dakota Formation is exposed on both sides of a road cut on Kansas 232, about 4.1 mi (6.6 km) north of I-70 in the NW¼SE¼ Sec.19,T.12S.,R.10W., Lincoln County, Kansas (Fig. 1). Vehicles traveling on I-70 should exit at the Wilson and Wilson Reservoir exit (exit 206) and travel north on Kansas 232 toward Wilson Reservoir.

SIGNIFICANCE

The section exposed in the road cut shows the upper part of the Dakota Formation (Fig. 2) and its complex depositional history, represented by floodplain deposits, fluvial sandstone, and deltaic facies. The dominant feature is the Rocktown channel sandstone body that Siemers (1976, p. 121) interpreted to have been deposited in a diverse deltaic environment that was modified by the "eastward transgression of a Late Cretaceous sea across central Kansas." He recognized two depositional phases within this sandstone body (Siemers, 1976, p. 121): (1) a lower, sinuous, trough-shaped, cross-bedded sandstone body and its laterally associated floodplain facies that suggest a river system on a low relief coastal plain and perhaps an upper deltaic plain, and (2) an upper, elongate, tabular-shaped, horizontally laminated, and ripple-bedded sandstone body that suggests a transgressed delta-front distributary with a wider, estuarine-like channel. The entire depositional system grades vertically into the marine strata of the Graneros Shale.

SITE INFORMATION

This site is Stop 1 of Hattin and others (1978, p. 62–65). The section description (Table 1) is taken verbatim from their work, except that the descriptions of units have been inverted to place the youngest unit (Unit 15) at the top. For additional detail, visitors should refer to the 1978 guidebook by Hattin and others while examining this exposure.

The lithologic sketch (Fig. 3) of Hattin and others (1978, fig. 14, p. 63) is of the west side of the road cut. It shows the stratigraphic record of the eastward marine transgression of the Western Interior Sea over the lower, nonmarine, fluvial channel and floodplain facies and the upper, marginal-marine, deltaic facies in the upper part of the Dakota Formation. This site is more complex, but similar to other exposures of the upper Dakota Formation in central Kansas.

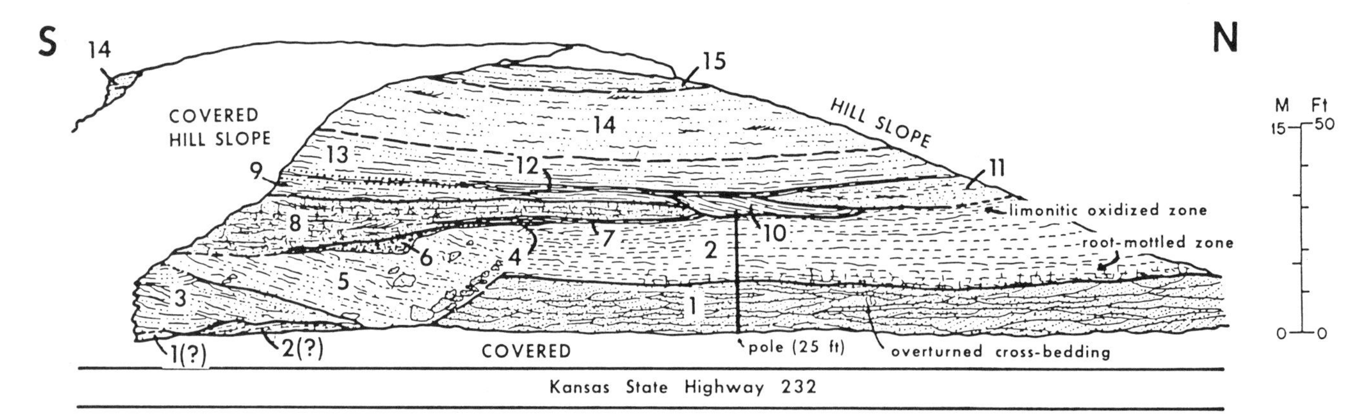

Figure 3. Lithologic sketch of Dakota strata exposed at this site. This exposure illustrates complex relationships of lithologic units in nonmarine fluvial channel and floodplain facies and marginal-marine facies in upper part of Dakota Formation. In general, unit 1 represents active channel; units 2–6 represent complex floodplain and partly abandoned channel-fill processes; units 7–12 probably represent complex fluvial to marginal-marine deposition in deltaic environment; units 13–15 may represent marginal-marine sand deposition across a deltaic surface during initial phase of Graneros transgression. Elsewhere Graneros Shale commonly overlies units such as 14 and 15, but has been stripped off here. Discordant surface, labeled 4, is a spectacular feature of this exposure and may represent faulting of unit 5 against units 1 and 2 prior to deposition of unit 6. This surface may, however, be an erosional scarp of some sort, as indicated by numerous large exotic siltstone blocks in close proximity to discordant surface of unit 5.

Table 1. Descriptions of Units in Figure 3:

Unit 15. Interbedded silty sandstone and clayey siltstone with silty shale, caps exposure; plant debris common, contains sparse *Planolites* burrows and a possible *Teredolithus* unit as much as 5 ft (1.5 m) thick.

Unit 14. Flat-bedded sandstone unit that caps hill; small-scale cross-bedding and wavy bedding present; possible small burrow structures sparsely represented; unit 16 ft (4.9 m) thick.

Unit 13. Thin-bedded and wavy-bedded sandstone with interlaminated shale; sharply overlies unit 12 and grades upward into unit 14; 10 to 11 ft (3.1 to 3.4 m) thick.

Unit 12. Black, carbonaceous shale layer; relatively continuous across cut, thins northward over unit 11.

Unit 11. Very friable, yellowish-gray sand containing carbonaceous wood fragments; 0 to 4.5 ft (0 to 1.4 m) thick.

Unit 10. Brownish-black, carbonaceous silty shale overlying units 2, 7, 8, and 9; truncated by units 11 and 12; as much as 4.5 ft (1.4 m) thick.

Unit 9. Interlaminated yellowish sand and dark gray shale; as much as 2 ft (0.6 m) thick; truncated by unit 10.

Unit 8. Fine-grained, well-sorted sandstone unit with abundant carbonized wood fragments and abundant root molds; unit thickens to about 10 ft (3.1 m) southward where covered and thins northward to about 3 ft (0.9 m) where it is truncated by unit 10.

Unit 7. Thin, discontinuous, black carbonaceous shale with abundant fine plant debris; wedges out southward and truncated by unit 10 to north; 0 to 2 ft (0 to 0.6 m) thick.

Unit 6. Discontinuous, highly ferruginous, sandy conglomeritic rubble layer with sandstone and mudstone clasts and abundant carbonized wood fragments; unit truncates discordant surface 4 below; as much as 2 ft (0.6 m) thick.

Unit 5. Interlaminated, carbonaceous, silty shale and clayey siltstone with large exotic blocks of dense siltstone; laminae steeply inclined (15–20°) northward and truncated against discordant surface 4; siltstone blocks display irregular bedding and contain pyrite nodules and carbonized plant debris; as much as 17 to 18 ft (5.2 to 5.5 m) exposed.

Surface 4. Prominent discordant surface separating units 1 and 2 from unit 5; in upper part surface brings dark gray shale of unit 5 against yellowish-gray mudstone of unit 2; the surface appears to be a fault with small drag folds in the shale; near top the surface has strike orientation of N 37°E and dip of approximately 60°SE; in lower part where surface separates units 1 and 4 it is lined with limonitic sandy rubble and has an orientation of N 45–65°E and dip of 23–28° SE; shale appears drag folded against sandstone.

Unit 3. Complex sedge-trough cross-beds sets of laminated sand and carbonaceous shale; unit as much as 9 ft (2.7 m) thick.

Unit 2. Blocky, variegated (yellowish- to reddish-gray) mudstone with abundant oxidized siderite (?) pellets, plant debris, and root mottling; main exposure as much as 12 ft (3.7 m) thick; also probable exposure of same unit near lower left of road cut.

Unit 1. Trough cross-bedded, medium- to fine-grained, yellowish-gray to grayish-orange, friable sandstone; plant and wood debris and root mottling at top; main exposure as much as 7 ft (2.1 m) thick; also possible small exposure of unit at lower left of cut.

REFERENCES

Franks, P. C., 1975, The transgressive-regressive sequence of the Cretaceous Cheyenne, Kiowa, and Dakota formations of Kansas, *in* Caldwell, W.G.E., ed., The Cretaceous System in the Western Interior of North America: Geological Association of Canada Special Paper 13, p. 459–521.

Franks, P. C., Coleman, G. L., Plummer, N., and Hamblin, K., 1959, Cross-stratification, Dakota Sandstone (Cretaceous), Ottawa County, Kansas: Kansas Geological Survey Bulletin 134, part 6, p. 223–238.

Hattin, D. E., Siemers, C. T., and Stewart, G. F., 1978, Guidebook Upper Cretaceous stratigraphy and depositional environments of western Kansas: Kansas Geological Survey Guidebook Series 3, 102 p.

Merriam, D. E., 1963, The geologic history of Kansas: Kansas Geological Survey Bulletin 162, 317 p.

Siemers, C. T., 1976, Sedimentology of the Rocktown channel sandstone, upper part of the Dakota Formation (Cretaceous), central Kansas: Journal of Sedimentary Petrology, v. 46, p. 97–123.

Zeller, D. E., 1968, The stratigraphic succession in Kansas: Kansas Geological Survey Bulletin 189, 81 p.

(Other basic references are in Hattin and others, 1978)

Ogallala Formation (Miocene), western Kansas

Edwin D. Gutentag, *Water Resources Division, Central Region, U.S. Geological Survey, Denver Federal Center, Lakewood, Colorado 80225*

LOCATION

The area in the vicinity of Lake Scott State Park, Scott County, Kansas (Fig. 1), contains excellent outcrops of the Ogallala Formation. One of the best vertically continuous exposures is in the west side of the roadcut at the ridge called Devils Backbone, which is located in the NW¼,NE¼,Sec.25,T.16S.,R.32W. This section is described in Table 1 and diagramatically sketched in Figure 2.

INTRODUCTION

Tertiary and Quaternary rocks underlie the High Plains physiographic region. The Ogallala Formation is the principal geologic unit of the High Plains and underlies 134,000 mi^2 (347,000 km^2) or 77 percent of the High Plains surface in parts of eight states—Colorado, Kansas, Nebraska, New Mexico, Oklahoma, South Dakota, Texas, and Wyoming.

The Ogallala Formation was first considered to be late Tertiary in age (Darton, 1899). In his classic report for Wallace County, Kansas, Elias (1931) considered the mammalian fossils of the Ogallala to represent Late Miocene and Early Pliocene time. Waite (1947) assigned the age of clastic material of the Ogallala to the Pliocene and indicated that, when present, the age of the lower clay facies of the Ogallala, consisting of brightly colored unctuous clays, would be either Miocene or Early Pliocene. The age of the Ogallala was considered as Neogene (Miocene and Pliocene) by Frye and others (1956). The only definitive age determination of the Ogallala Formation was by Boellstorff (1976) who used the fission-track method to determine the age of volcanic glass from the Ogallala type section to be 7.6 ± 0.7 Ma, which is late Miocene. Miocene time ranges from 24 to 5 Ma (Geologic Names Committee, 1984).

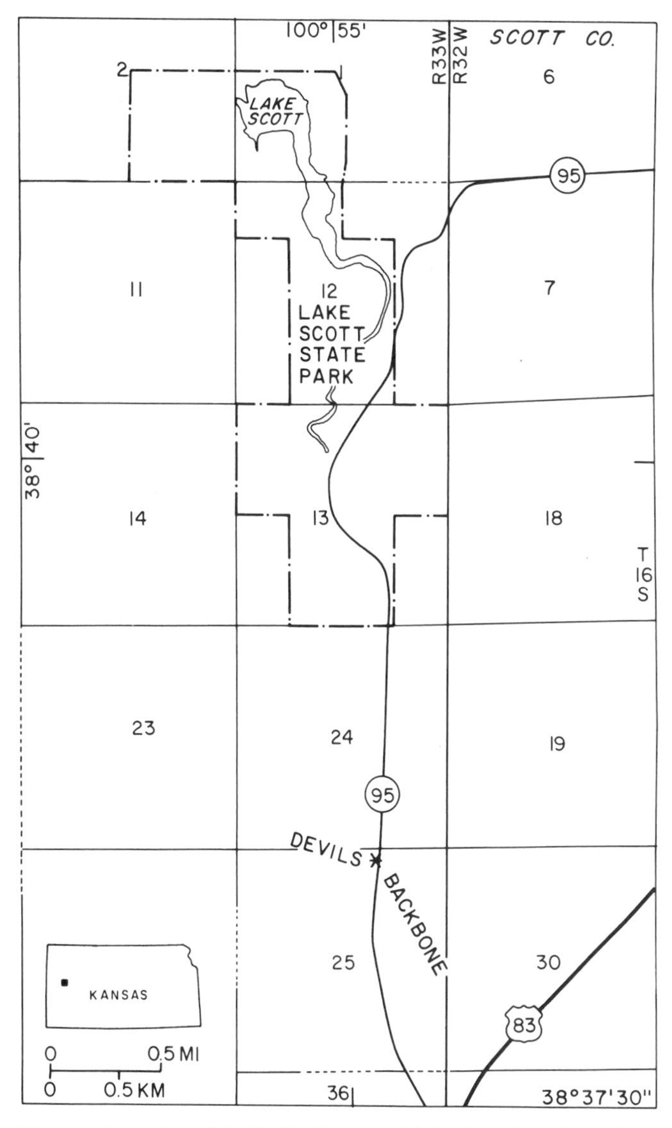

Figure 1. Location of Devils Backbone and Lake Scott State Park, Scott County, Kansas. Base from Lake Scott 7½-minute Quadrangle.

SITE DESCRIPTION

The Ogallala Formation in this road cut at Devils Backbone is typified by the amount of calcium-carbonate cement found throughout the section. Lithologies grade from one lithotype to another; for example, the contact between sand (unit 2) and caliche (unit 3) is not uniform, and the underlying sand interfingers with the lowermost layer of caliche (Fig. 3).

Units 3, 5, 6, 8, 10, 12, and 14 (Fig. 2) for the most part are ledge maker beds. The hard ledges are usually unevenly cemented and form weathered benches and cliffs that resemble mortar beds (Elias, 1931). These beds are clastic limestone consisting of silty sand cemented by calcium carbonate. In present geologic terminology these beds are calcretes. They are described in the measured lithologic section (Table 1) as caliche, which is the term used when similar material is found in rotary drill-hole cuttings from the High Plains. Although cemented beds predominate at the outcrop, test-hole drilling (Gutentag and Stullken, 1976) indicates that in the subsurface, where the Ogallala has never been exposed, there may be only one massive caliche bed at the very top of the Ogallala. Drilling test holes away from the outcrops, by the author, indicated that the material in the saturated zone is less indurated and less cemented than at the outcrop.

Smith (1940) found that in the fresh roadcuts, such as the bottom part of the Devils Backbone section, the caliche is less

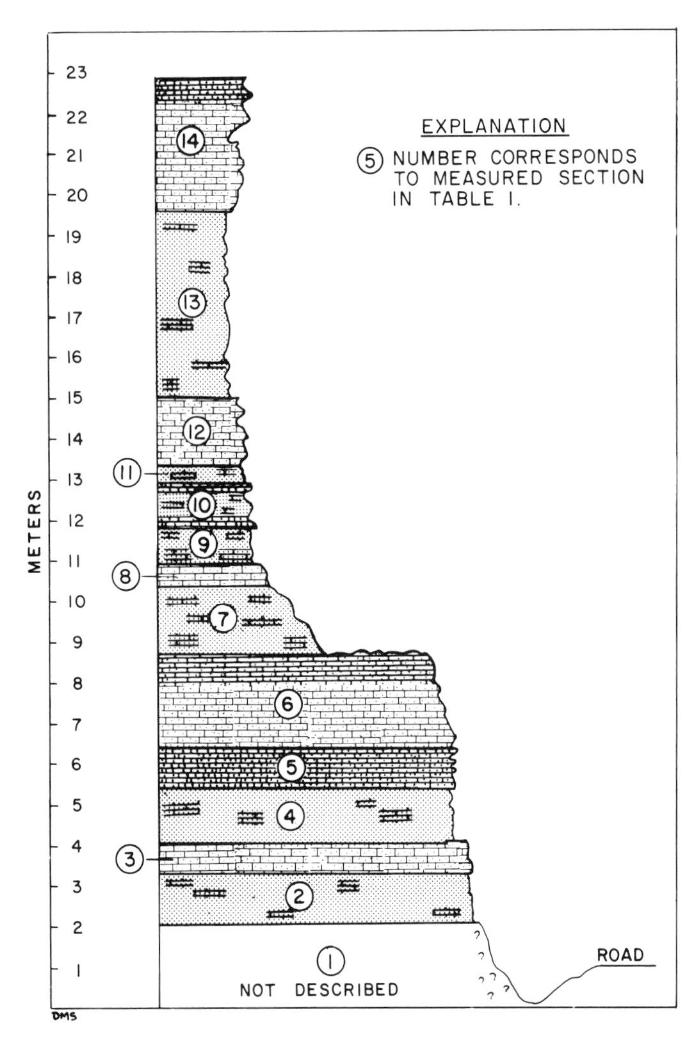

Figure 2. Measured section of Ogallala Formation on the west side of Devils Backbone, NW¼,NE¼,Sec.25,T.16S.,R.32W., Scott County, Kansas.

Figure 3. Contact between sand (unit 2) and caliche (unit 3) at the west side of Devils Backbone.

developed than on the upper long-exposed surface. Exposure to the atmosphere allows calcium carbonate to accumulate and be transported downward by percolation of precipitation; redeposition of this calcium carbonate forms caliche. At this outcrop, the most distinctive feature is the caliche honeycomb structure and root casts shown in Figure 4, which is an example of redeposition of calcium-carbonate cement. Although caliche honeycomb structure and root casts are found at both sides of the outcrop, both are best developed on the east side. The caliche honeycomb structure appears to be caused by adding more and more calcium carbonate to the sand at or near the surface of the outcrop, which expands the honeycomb structure (Fig. 5).

A distinctive bed in the Ogallala at the outcrop is the hard nodular caliche of unit 14 (Table 1), which contains the pisolitic limestone. The pisoliths found at the top of unit 14 generally are round and dark gray. Elias (1931) believed the pisolitic layer to be formed by algae and called the pisoliths and related concentric wavey structures "algal limestone." Frye and others (1956) thought that the algal limestone was formed on the surface of the plain of alluviation that marked the end of Ogallala deposition in the region. The plain of alluviation would have maintained surface stability for a long time; thus the algal limestone and associated pisoliths may be secondary solution effects produced by weathering on a relatively stable surface for a long time (Frye and others, 1956).

Visitors to the park may view other Ogallala outcrops that are described in Waite (1947). Many of the beds seen at the Devils Backbone can be traced into the other outcrops although the surfaces may be grass covered.

The Ogallala Formation is best known as the principal geologic unit in the High Plains aquifer. Most of the 17,800,000 acre-ft (22,000 hm^3) of water pumped for irrigation in the High Plains comes from the Ogallala Formation. Water from the Ogallala is available for drinking at Big Spring (Fig. 1). Big Spring issues from near the contact between the Ogallala and the underlying Niobrara Chalk of Cretaceous age.

TABLE 1. MEASURED SECTION OF OGALLALA FORMATION (MIOCENE)*

Unit	Description	Thickness ft	(m)
14	Caliche, white, very hard, nodular; contains layers of very hard pistolitic limestone (algal limestone of Elias, 1931) and lag gravel on top.	11.0	(3.4)
13	Sand, very fine to very coarse, silty, red; contains white caliche-cemented root casts, tubes, pipes, and nodules, that weather to a honeycomb surface.	15.0	(4.6)
12	Caliche, reddish white, nodular; consists of very fine to very coarse sand, very silty, with calcite cement; weathering produces tubes, pipes, and a caliche honeycomb.	5.5	(4.6)
11	Sand, very fine to very coarse, silty, red; contains hard white caliche nodules.	1.3	(0.4)
10	Caliche, white; consists of a nodular layer near the base, very sandy, silty, reddish-white, less cemented middle layer, and a very nodular caliche layer near the top; units grade laterally into very hard caliche or less cemented, very fine to coarse sand with many caliche-cemented root casts and tubes.	3.6	(1.1)
9	Sand, very fine to coarse, silty, red; contains nodules of very hard, white caliche.	3.0	(0.9)
8	Caliche, white, nodular; consists of fine to coarrse sand, very silty with calcite cement; nodules are hard and very well cemented.	1.3	(0.4)
7	Sand, very fine to very coarse, silty, very calcareous; reddish-tan; contains hard white caliche nodules.	5.6	(1.7)
6	Caliche, white; consists of very fine to very coarse sand, very silty, with calcite cement.	7.5	(2.3)
5	Caliche, white, very dense, hard; consists of very fine to coarse sand, very silty with calcite cement; contains many root casts and tubes.	3.3	(1.0)
4	Sand, very fine to coarse, silty, red; contains some root casts, tubes, and caliche honeycomb structures.	4.3	(1.3)
3	Caliche, white; consists of very fine to coarse sand, very silty, cemented with calcite and contains many tubes and root casts.	2.6	(0.8)
2	Sand, very fine to coarse, slightly silty, red; contzins some tubes, root casts, and caliche honeycomb.	4.3	(1.3)
1	Base of ditch to bottom of unit 2, rubble from section above (covered interval).	6.9	(2.1)
	Total	75.5	(23.0)

*NW1/4,NE1/4,Sec.25,T.16S.,R.32W., Scott County, Kansas (west side of road cut through Devils Backbone). Section measured by A. B. Gutentag and E. D. Gutentag, August 21,1985.

Figure 4. Contact between sand (unit 4) and caliche (unit 5) showing development of caliche honeycomb structure and root casts at the east side of Devils Backbone.

Figure 5. Closeup of caliche honeycomb structure and root casts in unit 4.

REFERENCES CITED

Boellstroff, J. D., 1976, The succession of Late Cenozoic volcanic ashes in the Great Plains; A progress report, *in* Bayne, C. K., ed., Guidebook 24th annual meeting Midwestern Friends of the Pleistocene: Kansas Geological Survey Guidebook Series 1, p. 37–71.

Darton, N. H., 1899, Preliminary report on the geology and water resources of Nebraska west of the one hundred and third meridian: U.S. Geological Survey 19th Annual Report, Hydrology, p. 719–785.

Elias, M. K., 1931, The geology of Wallace County, Kansas: Kansas Geological Survey Bulletin 18, 254 p.

Frye, J. C., Leonard, A. B., and Swineford, A., 1956, Stratigraphy of the Ogallala Formation (Neogene) of Northern Kansas: Kansas Geological Survey Bulletin 118, 92 p.

Geologic Names Committee, 1984, Geologic time chart, *in* Stratigraphic notes, 1983: U.S. Geological Survey Bulletin 1537-A, p. A1–A4.

Gutentag, E. D., and Stullken, L. E., 1976, Ground-water resources of Lane and Scott counties, western Kansas: Kansas Geological Survey Irrigation Series 1, 37 p.

Smith, H.T.U., 1940, Geologic studies in southwestern Kansas: Kansas Geological Survey Bulletin 34, 212 p.

Waite, H. A., 1947, Geology and ground-water resources of Scott County, Kansas: Kansas Geological Survey Bulletin 66, 216 p.

Lower Cretaceous Kiowa Formation; The eastern margin of the Western Interior Seaway

Robert W. Scott, *Amoco Production Company, Tulsa, Oklahoma 74102*

LOCATION

The Kiowa Formation is exposed discontinuously along the shores of Kanopolis Reservoir in central Kansas. This reservoir was formed by the damming of the Smoky River in Ellsworth County about 20 mi (32 km) southwest of Salina. One of the best exposures is about 660 ft (200 m) southwest of the dam control tower along the shore (NE¼,Sec.3,T.17S.,R.6W.; 97°58′W., 38°36′30″N.; Crawford 1964 USGS topographic sheet) (Fig. 1). The outcrop is accessible by a short hike west of the road leading across the dam. The condition of the exposure depends on the water level and varies from time to time. A similar section is exposed along the eastern bluff in the second bay west of the state campground. It is accessible by walking northwest a few hundred meters from the road (C.NE¼,Sec.4,T.17S.,R.6W.;97°59′W., 38°36′24″N.). The unfossiliferous lower part of the section is partly exposed on private land east of the dam (SW¼, Sec.1,T.17S.,R.6W.;97°56′36″W.,38°36′24″N.)

SIGNIFICANCE

The Albian Kiowa Formation is dominantly shale and secondarily sandstone that was deposited along the eastern shore of the Western Interior Seaway (Franks, 1975). This seaway transected North America in a north-south direction connecting the Gulf of Mexico with the Arctic Basin. Albian flooding was the first of the major Cretaceous transgressions into the Western Interior that produced economically important sediments. However, the Kiowa is not an important resource in Kansas, but it is time-equivalent with Albian source rocks and reservoirs in Colorado (Weimer, 1978).

STRATIGRAPHY

In central Kansas, the Kiowa Formation consists of shale and sandstone from 100 to more than 200 ft (30 to 60 m) thick. It is quite similar to the type section in Kiowa County in south-central Kansas (Cragin, 1895; Latta, 1946). The basal contact of the Kiowa is a sharp disconformity in contact with underlying greenish and reddish clays of the Permian Ninnescah Formation. The top of the Kiowa in the vicinity of Kanopolis Reservoir is a sharp contact between yellow-brown, fine-grained sandstone below and light-gray, red-mottled clay and siltstone of the overlying Dakota Formation (Bayne and others, 1971).

The Kiowa Formation is Upper Albian and is time-equivalent with much of the Washita Group in north-central Texas (Scott, 1970b). This age is based on several ammonites, inoceramids, and other bivalves and foraminifers (Loeblich and Tappan, 1950). A zonation based on megafossils was derived for the Kiowa in Kansas and equivalent strata in New Mexico and Colorado (Scott, 1970b). Correlation of these zones to Ellsworth County suggests that the lower part of the Kiowa is equivalent with the *Inoceramus comancheanus* zone, and the upper part correlates with the *Inoceramus bellvuensis* zone. The *I. comancheanus* zone is distinguished by the overlapping ranges of *Engonoceras belviderense* (Cragin), *Rastellum quadriplicata* (Shumard), and *Pteria salinensis* White, and by the top of *Texigryphaea corrugata* (Say) s.s. Imprints of *I. comancheanus* cragin are rare in the shale at the Kanopolis dam. The *I. bellvuensis* zone is characterized by the nomiate species with *Texigryphaea tucumcarii* (Marcou) and the top of *T. corrugata belviderensis* (Hill and Vaughan).

LITHOFACIES AND PALEOCOMMUNITIES

The Kiowa section near the dam consists of shale 20 ft (6 m) thick, grading up into sandstone 15 ft (4.5 m) thick, and represents shoreface progradation. The shale is dark gray to medium gray, fissile, and thinly laminated. Local laminae of silt to very fine-grained sandstone are uncommon. Molds and casts of fossils are not abundant; by splitting large volumes of shale, *Nuculana, Nucula,* and *Turritella* can be exposed. The shells are not abraded nor concentrated by currents and are thought to have accumulated where they died, a disturbed-neighborhood assemblage (Scott, 1970a). These species are part of the *Nucula-Nuculana* association characteristic of the muddy open bay or shelf (Scott, 1974).

An unusual lobster is a moderately common member of this community in the shale facies, *Huhatanka kiowana* (Scott, 1970a). Its large grasping appendages are very distinctive (Feldman and West, 1978).

This shale facies was deposited on a marine shelf below the normal wave base. The substrate was firm enough to support animal burrows. Bottom waters were oxygenated as indicated by the presence of a moderately diverse benthic assemblage. The dark color of the shales is probably a result of a mixture of ferrous iron and organic matter, which suggests that reducing conditions were present within the sediment column. Occasionally, storms deposited sand waves and shell beds that contain *Turritella* from the shelf and *Crassostrea* oysters transported from the nearby brackish bays.

The overlying sandstone consists of two shoaling-up cycles represented by thin-laminated sandstone, siltstone, and shale grading up to thin-bedded sandstone. The thin-laminated facies is medium gray to yellow-brown, and is very fine grained. Subparallel beds up to 6 in (15 cm) thick alternate from mudstone to siltstone to sandstone. These beds grade up into the thin-bedded sandstone by increasing bed thickness and grain size. The thin-bedded sandstone is grayish-orange and fine to very fine grained.

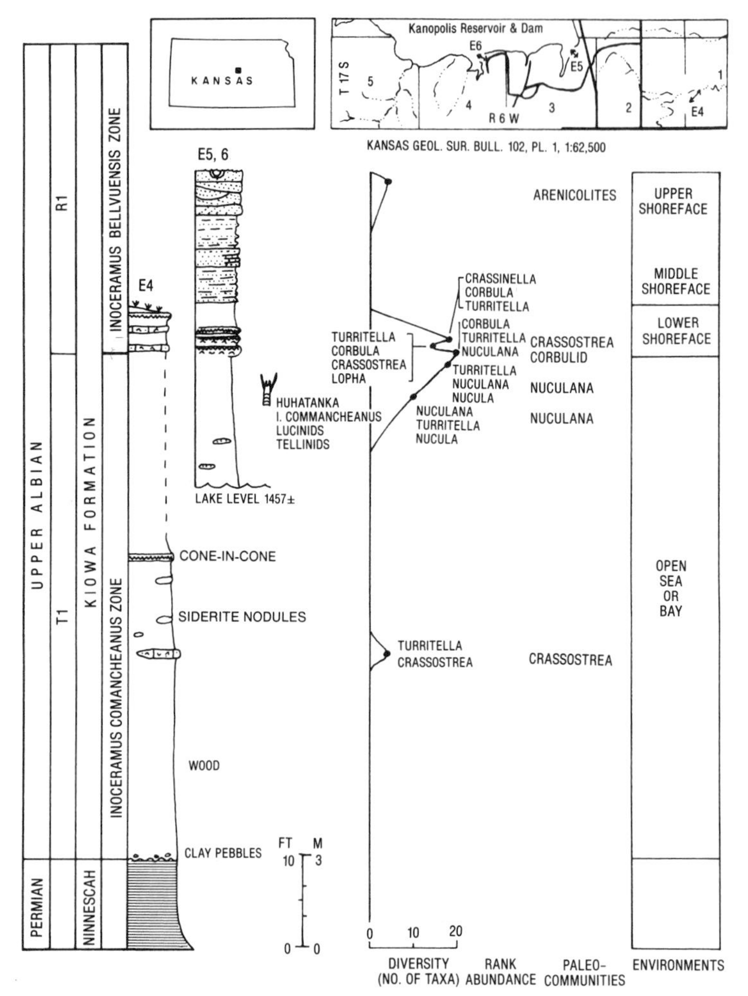

Figure 1. Kanopolis Reservoir sections, Ellsworth County, Kansas.

Tabular strata are parallel bedded or cross-bedded. Small- to medium-scale, tabular planar, and trough cross-strata are common. Trace fossils are uncommon in sandstone at this locality but *Arenicolites* occurs in the upper sandstone bed. At other localities, molds of bivalves are common and represent diverse assemblages of shoreface marine communities (Scott, 1974).

Another shale-sandstone cycle is developed above the sequence at the Kanopolis dam. It is exposed in bluffs at the upper end of the reservoir (Scott, 1970a, Fig. 11). Bluffs along the north shore expose thick lenticular sandstone that interfingers with this upper Kiowa cycle (Bayne and others, 1971a, p. 10). This may be either a channel or a bar in the paralic Kiowa facies. Deposition of the overlying Dakota Formation represents regression and deposition of nonmarine sediments, which completes the first transgressive-regressive megacycle (Kauffman, 1969).

REFERENCES CITED

Bayne, C. K., Franks, P. C., and Ives, W., Jr., 1971, Geology and ground-water resources of Ellsworth County, central Kansas: Kansas Geological Survey, Bulletin 201, 84 p.

Cragin, F. W., 1895, A study of the Belvidere beds: American Geologist, v. 16, p. 357–386.

Feldman, R. M., and West, R. R., 1978, *Huhatanka,* a new genus of lobster (Decapoda: Mecochiridae) from the Kiowa Formation (Cretaceous: Albian) of Kansas: Journal of Paleontology, v. 52, p. 1219–1226.

Franks, P. C., 1975, The transgressive-regressive sequence of the Cretaceous Cheyenne, Kiowa, and Dakota Formations of Kansas: Geological Association of Canada Special Paper 13, p. 469–521.

Kauffman, E. G., 1969, Cretaceous marine cycles of the Western Interior: The Mountain Geologist, v. 6, p. 227–245.

Latta, B. F., 1946, Cretaceous stratigraphy of the Belvidere area, Kiowa County, Kansas: Kansas Geological Survey, Bulletin 64, pt. 6, p. 217–260.

Loeblich, A. R., Jr., and Tappan, H., 1950, Foraminifera of the type Kiowa Shale, Lower Cretaceous of Kansas: University of Kansas Paleontological Contributions, Protozoa, Article 3, p. 1–15.

Scott, R. W., 1970a, Paleoecology and paleontology of the Lower Cretaceous Kiowa Formation, Kansas: University of Kansas Paleontological Contributions, Article 52 (Cretaceous 1), p. 1–94.

——, 1970b, Stratigraphy and sedimentary environments of Lower Cretaceous rocks, Southern Western Interior: American Association of Petroleum Geologists Bulletin, v. 54, p. 1225–1244.

——, 1974, Bay and shoreface benthic communities in the Lower Cretaceous: Lethaia, v. 7, p. 315–330.

Weimer, R. J., 1978, Influence of Transcontinental arch on Cretaceous marine sedimentation; A preliminary report, *in* Pruit, J. D., and Coffin, P. E., eds., Energy resources of the Denver basin: Rocky Mountain Association Geologists Symposium, p. 211–222.

Plio-Pleistocene rocks, Borchers Badlands, Meade County, southwestern Kansas

Richard J. Zakrzewski, *Department of Earth Sciences and Sternberg Memorial Museum, Fort Hays State University, Hays, Kansas 67601*

LOCATION

The Borchers Badlands, hereafter referred to as the badlands, are located in the south–central part of Meade County, Kansas, approximately 9 mi (14.4 km) south-southwest of Meade, the county seat (Fig. 1). This area is part of the transition zone between the High Plains and the Plains Border provinces. Meade is at the junction of U.S. 54 and 160 and Kansas 23. The latter highway bisects the badlands into a north half (Section 4) and a south half (Section 3). The badlands are located in Sec. 16; a small portion of the SE¼ of Sec. 17; the east half of Sec. 20; and the west half of Sec. 21,T33S.,R.28W. (Fig. 2), an area of approximately 2 mi^2 (5.2 km^2) (Lake Larabee and Irish Flats NE7½-minute Quadrangles). The western boundary is Crooked Creek, a fault-controlled stream. The eastern boundary is the undissected upland area.

SIGNIFICANCE

Meade County is perhaps the most geologically studied county in Kansas. For nearly 30 years the Plio-Pleistocene deposits in the county were the subject of intensive research by the late Claude W. Hibbard, his students, and colleagues. The county contains the best documented series of deposits and faunas spanning Plio-Pleistocene time in North America, if not the world (see Zakrzewski, 1975, and Bayne, 1976, for summaries). The badlands, a crucial area for understanding the relationships of the deposits and faunas in the county, contain the type section of one formation, two superposed volcanic ashes, and six fossil vertebrate sites. Study of the ashes and fossils suggests that the Plio-Pleistocene boundary lies within these sediments. The deposits in the badlands have been subjected to sedimentary (Kovach, 1979) and paleomagnetic (Lindsay and others, 1975) analyses; the ashes have been radiometrically dated and chemically analyzed (Boellstorff, 1976; Izett and Wilcox, 1982); and the fossil vertebrates have been described (Hibbard, 1941; Bayne, 1976; Eshelman and Hibbard, 1981; Zakrzewski, 1981; Izett and Wilcox, 1982). Therefore, all subsequent studies on the Plio-Pleistocene will need to use the badlands as a reference area.

GENERAL STATEMENT

The sediments in the badlands were deposited by streams flowing generally to the east. The units (Fig. 3) range in age from mid-Miocene to at least mid-Pleistocene. The original source for the clastics was the Rocky Mountains, although the units also contain reworked local material.

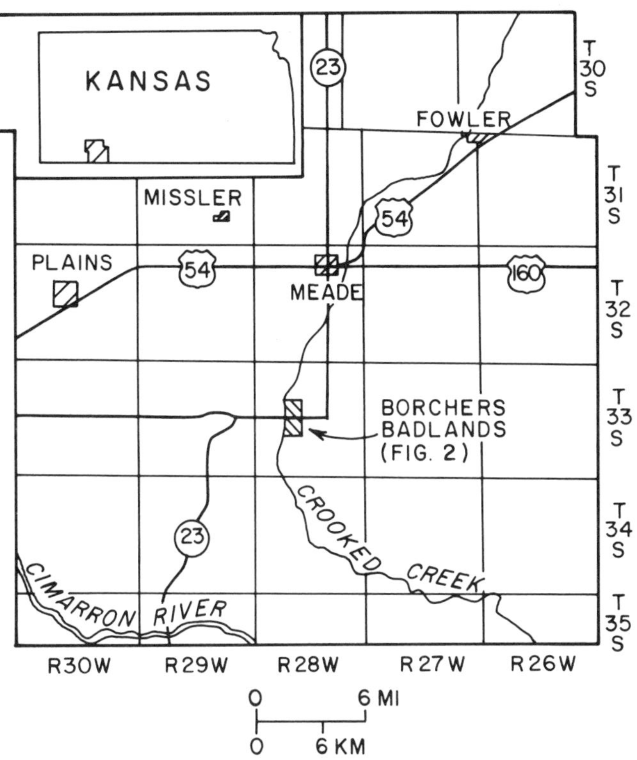

Figure 1. Map of Meade County, Kansas showing location of Borchers Badlands.

The thickness of the sediments was controlled by movement along the Crooked Creek fault, which runs approximately N–S at the western margin of the area. The original cause of the faulting is unknown, but solution and subsidence in the underlying Permian salt beds played a role in subsequent movements (Frye and Hibbard, 1941). As an example of sediment control by the fault, only 30 ft (9 m) of Rexroad Formation can be seen east of the fault; whereas, on the down-dropped block west of the fault the Rexroad ranges in thickness from 200 to 250 ft (60 to 75 m) (Frye and Hibbard, 1941).

Three of the formations, Ogallala, Ballard, and Crooked Creek, exhibit fining-upward sequences. The coarse gravels at the base of the latter two units may be a reflection of increased stream velocity produced by runoff from montane glaciation (Zakrzewski, 1975).

Originally the gravel members of the Ballard and Crooked Creek formations were correlated to the Nebraskan and Kansan glacial stages, respectively, and it was presumed that the volcanic ash in the county was the result of one fall and, therefore, a good stratigraphic marker. However, studies undertaken on the ash

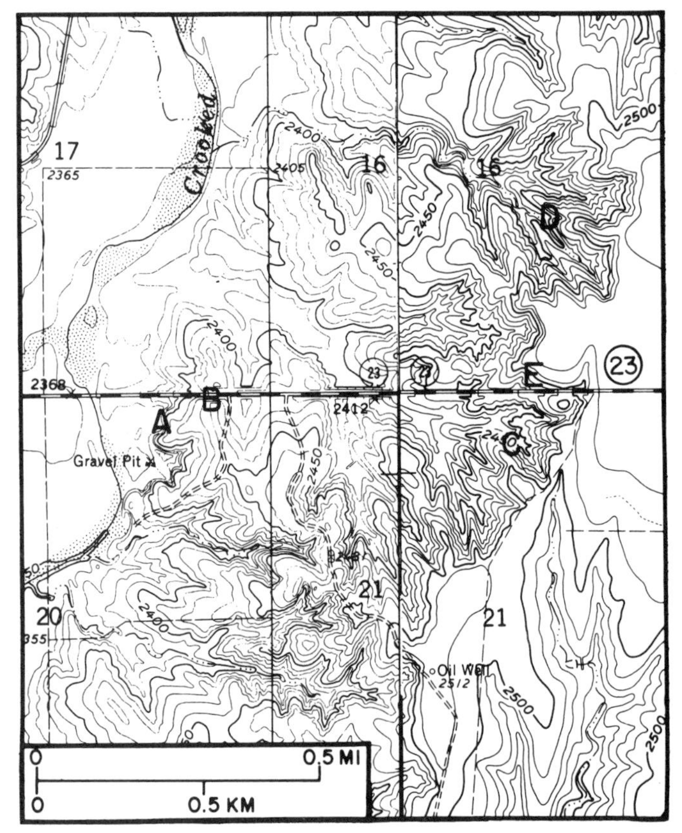

Figure 2. Topographic map of Borchers Badlands area showing locations of measured sections (from Lake Larabee and Irish Flats 7½-minute Quadrangles).

SERIES EPOCH	GROUP	FORMATION	ASH BEDS	PALEO MAG. DATA	LAND-MAMML. AGES	FAUNAS
PLEIS-TOCENE	SANBORN		UPPER B		IRVING-TONIAN	NASH = ARIES BORCHERS
PLIOCENE	MEADE	CROOKED CREEK	TYPE B	‖	BLANCAN	
		BALLARD		+		
		REXROAD		+		
MIOCENE		OGALLALA			HEMP-HILLIAN	
					CLAREN-DONIAN	UNNAMED

Figure 3. Stratigraphy and faunas in Borchers Badlands.

beds and the faunas by various workers have shown the situation to be more complex (Zakrzewski, 1975).

Sections south of Kansas 23. These deposits are located in Secs. 20 and 21 (Fig. 2). Composite sections have been measured and published in several papers (Kovach, 1979). The type localities of the Meade Formation of Frye and Hibbard (1941), the Crooked Creek Formation of Hibbard (1949), and the Type-B Ash of Naeser and others (1973) are in this area (Fig. 2, Site C). The Type-B Ash also is referred to as the Borchers Ash (Boellstorff, 1973) and the Huckleberry Ridge Ash (Izett and Wilcox, 1982). Also present are the type localities of the Borchers (Hibbard, 1941) and the Nash (Eshelman and Hibbard, 1981) local faunas (l.f.).

The base of the section is composed of the Ogallala Formation of Miocene age, which rests unconformably on the Permian Taloga Formation. In the badlands proper, the Ogallala is a silty sandstone, light brown at close range but orange when viewed from a distance. The Ogallala occurs near the bottom of the deeper ravines near the western margin of the badlands. The best exposure of the Ogallala can be seen in a gravel pit to the east of Crooked Creek in the NE¼NE¼Sec.20 (Fig. 2, Site A). A measured section is given in Bayne (1976), Kovach (1979), and Figure 4. Three facies can be seen in the outcrop (Kovach, 1979). The lower facies is composed of cemented sandstone and conglomerate containing clasts of Rocky Mountain origin, as well as locally derived clasts from Permian and Cretaceous rocks. Clasts of local origin dominate. The middle facies is a brown-to-reddish silty sand from which some vertebrate fossils (Bayne, 1976) have been obtained. The upper facies is a sandy caliche or mortar bed containing fossil nutlets.

Bayne (1976) incorrectly referred to these deposits as the Laverne Formation because teeth of the small three-toed horse, *Pseudhipparion gratum,* were collected from the middle facies. This horse is considered to be diagnostic for the Clarendonian land-mammal age. The Clarendonian ranges in age from about 12 to 8 Ma, which would make it mid-Miocene. However, prior to the revision of the Cenozoic time scale in the early 1970s, the Clarendonian was considered to be equivalent with the early Pliocene. The Hemphillian, the land-mammal age that follows the Clarendonian, was considered to be middle Pliocene. The Ogallala was thought to have been deposited during Hemphillian time, and the Laverne during Clarendonian time (Frye and Hibbard, 1941). Therefore, the name of the unit was changed to Laverne from Ogallala when the fossils suggested that it had been deposited during Clarendonian rather than Hemphillian time. However, a formation should be distinguished on lithologic grounds and not on the presence of some particular fossil. As these deposits can be distinguished from the type Laverne (Myers, 1959) they should be retained in the Ogallala Formation.

Overlying the Ogallala is the Rexroad Formation (Smith, 1940) of Pliocene age. The Rexroad is poorly exposed in the badlands. The best outcrop is in a roadcut just south of Kansas 23 (Fig. 2, Site B), approximately 0.25 mi (0.4 km) east of the bridge over Crooked Creek. At this point 30 ft (9 m) of the Rexroad are found unconformably overlying the caliche in the Ogallala (Fig. 5). The lithology is a brown, argillaceous, arkosic sandstone,

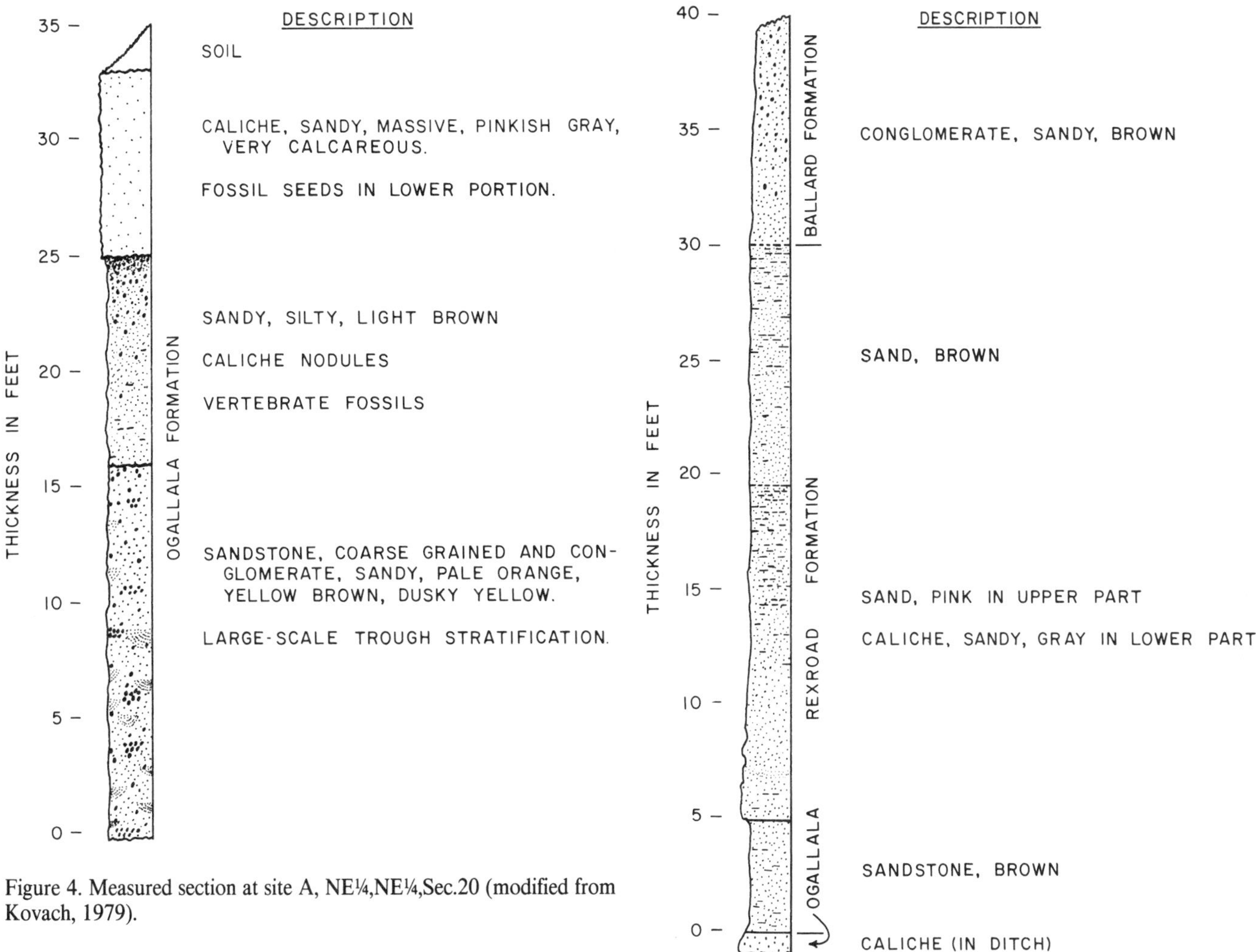

Figure 4. Measured section at site A, NE¼,NE¼,Sec.20 (modified from Kovach, 1979).

Figure 5. Measured section at site B, NE¼,NE¼,Sec.20 (modified from Kovach, 1979).

similar to the middle facies of the Ogallala (Kovach, 1979). In fact, if the caliche in the Ogallala were not present, these two lithofacies would be difficult to distinguish. Comparison with studies of the Rexroad west of the fault suggests that this facies represents the lower part of the unit. The Rexroad Formation is dominantly normally magnetized and, therefore, was deposited probably during the Gauss Epoch (Lindsay and others, 1975). No fossils have been found in the Rexroad in the badlands. However, west of the fault the Rexroad is very fossiliferous and contains one of the best-known faunas of the Blancan land-mammal age (Hibbard, 1970). The organisms suggest a warm, mild climate.

The Ballard Formation overlies the Rexroad. The Ballard consists of three lithofacies (Kovach, 1979). The lowest is a sandy, arkosic conglomerate designated the Angell Member (Hibbard, 1958). The clasts are mostly granitic and pegmatitic, suggesting an influx from the Rocky Mountains. This influx may be related to montane glaciation as one striated pebble has been reported from outside the badlands (Hibbard, 1944). The upper two lithofacies are designated the Missler Member (Hibbard, 1949). They are similar in being reddish, friable, argillaceous sandstones. The upper lithofacies is distinguished by being more silty and containing abundant caliche nodules and stringers. The Ballard, like the Rexroad, is normally magnetized and, therefore, also was deposited during the Gauss Epoch (Lindsay and others, 1975). Although no fossils have been collected from the Ballard in the badlands, vertebrate and molluscan fossils found elsewhere in the county are suggestive of a warm, mild climate.

The Crooked Creek Formation unconformably overlies the Ballard. It is a greenish friable sandstone in its lower part but grades into a siltstone higher in the section (Fig. 6). The type section of the Crooked Creek is located near the CNW¼NE¼Sec.21 (Fig. 2, Site C). The Crooked Creek exhibits a great deal of lateral variation, consisting of numerous interfingering lentils. Local channels are present also, but they are faint and difficult to trace.

Approximately 11 ft (3.3 m) above the Ballard–Crooked Creek contact is the Type-B Ash (=Borchers = Huckleberry Ridge). The ash is approximately 3.5 ft (1 m) thick, grayish white, with a blocky surface and abundant glass shards. The ash has been dated at 1.96 Ma by Naeser and others (1973). It is a member of the Pearlette family of ashes whose source is the Yellowstone region. Both the Crooked Creek and the Type-B Ash are reversely magnetized and, therefore, were deposited during the lower Matuyama Epoch (Lindsay and others, 1975).

Directly above the Type-B Ash is a gray clay that contains the Borchers l.f. About 7 to 10 ft (2 to 3 m) higher in the section; to the north and east of the Borchers l.f. are two localities of the Nash l.f. Fossils in the Borchers are indicative of a warm, mild climate. However, the Nash l.f. contains two organisms (see discussion below) that might indicate that the climate was becoming cooler.

Overlying the Crooked Creek Formation with a gradational contact are the sediments of the Sanborn Group. The Sanborn consists of the Kingsdown and Vanhem formations (Hibbard, 1958), but these cannot be differentiated in the badlands area. The Sanborn is a brown, friable, quartzose siltstone about 15 ft (4.5 m) thick. Caliche nodules and bands are common; gravel clasts are rare.

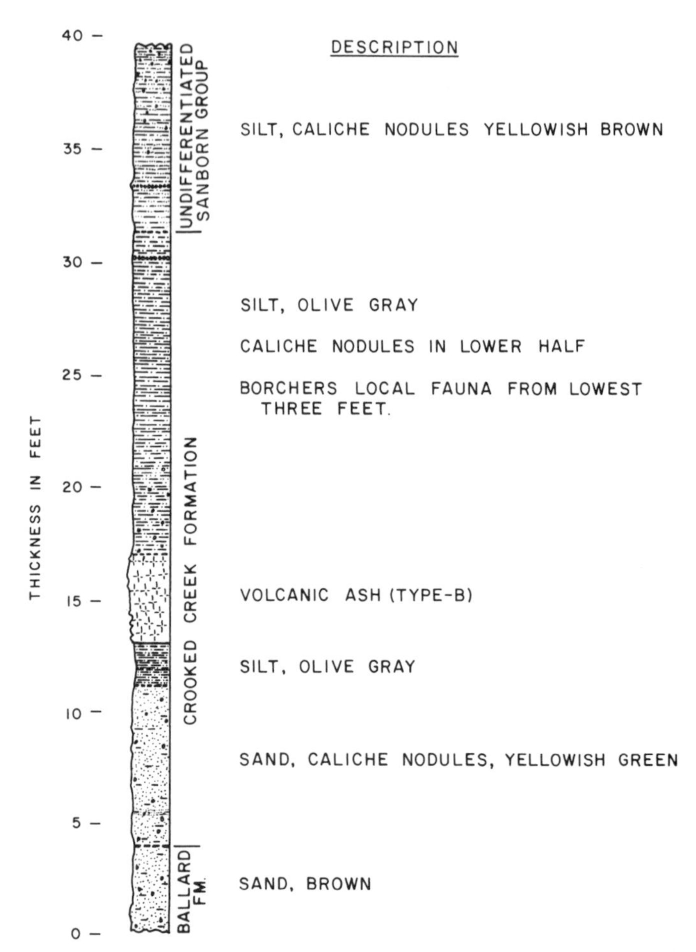

Figure 6. Measured section at site C, NW¼,NE¼,Sec.21 (modified from Kovach, 1979).

Discussion. When the sediments in this area were being studied for the first time and it was presumed there was only one volcanic ash, the Crooked Creek was correlated with the Kansan and Yarmouth stages of the classic glaciated area of the midwest, and the underlying Ballard was correlated with the Nebraskan and Aftonian stages. One reason for this correlation was that the fossils recovered from the Borchers l.f. (Hibbard, 1941) were indicative of a warm climate. As mentioned earlier, these fossils were taken from above a Pearlette-like ash. At the type section of the Pearlette Ash 6 mi (9.6 km) north of Meade, fossils (Cudahy l.f.) that indicated a cold climate were recovered below the ash (Hibbard, 1944; Paulson, 1961). The stage of evolution of the animals in the Cudahy indicated a mid-Pleistocene age for the fauna. Because of the cyclic nature of the sediments and the climatic implications of the faunas, the correlations mentioned above were made. Subsequently, when the radiometric dates (Naeser and others, 1973; Boellstorff, 1973) showed that the type Pearlette dated 0.6 Ma and the Type-B dated 1.96 Ma, Hibbard (Skinner and Hibbard, 1972) lowered the Crooked Creek to the earlier Nebraskan–Aftonian cycle and the Ballard to a cold-warm cycle in the Pleistocene prior to continental glaciation. The type Pearlette Ash and the Cudahy l.f. were retained in the Kansan–Yarmouth but the units in which they were found were unnamed.

In reviewing the development of the Pleistocene sequence in the Meade County area, Zakrzewski (1975) suggested that both the Ballard and the Crooked Creek were Pliocene in age and that the gravel members in both formations were the result of increased stream velocity produced by runoff from montane glaciation. One basis for this suggestion was that a restudy of the type Pleistocene in Europe (Bandy and Wilcoxin, 1970) provided a radiometric date of 1.8 Ma for the base of the Pleistocene. Therefore, the Type-B Ash would be in the Pliocene. Recently, Boellstorff (1978) demonstrated that continental glaciers reached Nebraska as early as 2.5 Ma and suggested that the classic concept of four continental glaciations in the Pleistocene be reevaluated. He also suggested that the base of the Pleistocene be set at 2.5 Ma. If this date becomes accepted, then the Ballard and Crooked Creek would revert to the Pleistocene.

Whether the date of 2.5 Ma eventually is accepted or the 1.8 Ma date is retained for the base of the Pleistocene, there is no faunal evidence for climatic deterioration (cooling) in the area that might be associated with glaciation until the time the animals of the Nash l.f. lived. This cooling is suggested by the presence of two arvicoline rodents, *Synaptomys (Mictomys) kansaensis* and *Pitymys ("Allophaiomys")* sp. (Eshelman and Hibbard, 1981). Arvicoline rodents primarily have a northern distribution. Today the subgenus *Mictomys* has a distribution that is limited essentially to Canada, and the *Pitymys* represents an early invasion of an advanced arvicoline stock from Asia. Whether the presence of these animals is the result of glaciation forcing populations

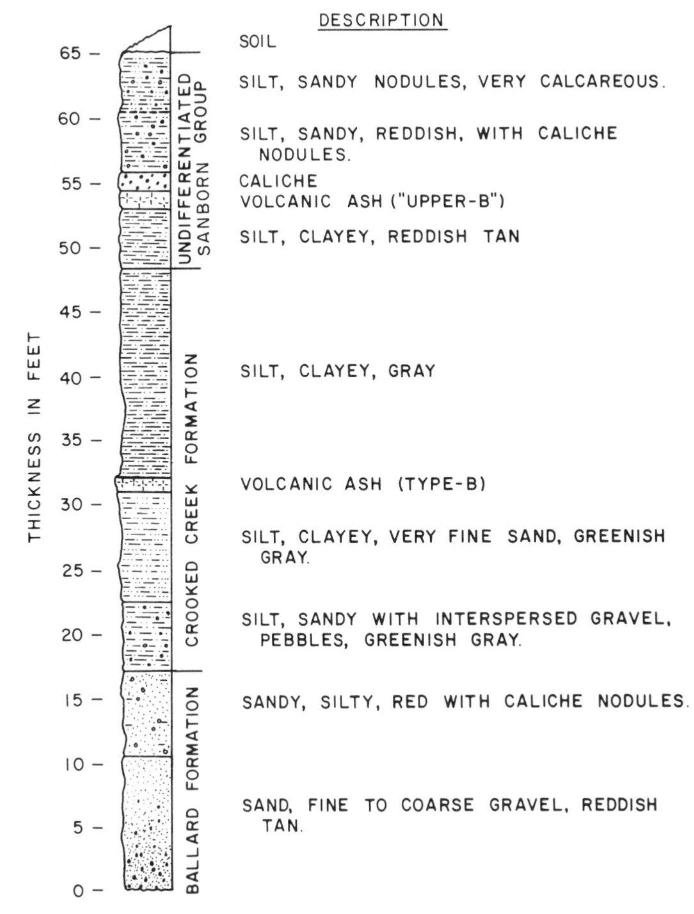

Figure 7. Measured section at site D, NW¼,SE¼,Sec.16 (modified from Bayne, 1976).

southward or the result of other factors cannot be determined at this time. Eshelman and Hibbard (1981) considered the Nash l.f. to be Pleistocene and placed it in the Aftonian interglacial stage because other elements in the fauna are not typical of those associated with glacial stages.

Section North of Kansas 23. The section on the north side of the road is similar to that on the south with two exceptions. The first is that no fauna equivalent to the Borchers has yet been found. The second is the presence of a younger volcanic ash (Fig. 7). The upper ash was dated at 1.2 Ma by Boellstorff (1976) and called the Upper Borchers. Izett and Wilcox (1982) referred to this ash as the Cerro Toledo, denoting that its source was in New Mexico rather than Yellowstone. The upper Borchers Ash occurs in deposits of the Sanborn Group about 33 ft (10 m) above the Type-B Ash in the NW¼SE¼ of Sec. 16 (Fig. 2, Site D). At this site, fossils comprising the Aries l.f. have been collected about 23.1 ft (7 m) above the Type-B Ash (Izett and Wilcox, 1982). Izett and Wilcox (1982) considered this l.f. to be transitional between the Blancan (Pliocene) and Irvingtonian (early Pleistocene) land-mammal ages, as a zebrine horse and a *Pitymys* at about the same stage of evolution as in the Nash l.f. have been found. *Synaptomys* has been reported as well, suggesting that the Nash and Aries l.f. may be temporal equivalents. The two ashes also can be seen in superposition in a small draw to the north of Kansas 23 (Fig. 2, Site E).

SUMMARY

Five formations, Ogallala (mid-Miocene), Rexroad and Ballard (Pliocene), Crooked Creek (Plio–Pleistocene), and Sanborn (Pleistocene), are present in the Borchers Badlands. They are fluvial in origin. The Rexroad and Ballard are normally magnetized (probably Gauss); the Crooked Creek is reversely magnetized (probably lower Matuyama). Coarse granitic and pegmatitic gravels at the base of the Ballard and Crooked Creek may be the result of montane glaciation in the Rockies. Two volcanic ashes, Borchers (1.96 Ma, Yellowstone source) and Upper Borchers (1.2 Ma, New Mexican source) are found within the Crooked Creek and Sanborn formations, respectively. The ashes are superposed at, at least, two sites on the north side of Kansas 23.

Three vertebrate local faunas have been collected. The Borchers, directly overlies the Borchers Ash and is indicative of a warm, mild climate. The Nash and Aries are found in sediments higher in the section, but bracketed by the two ashes, and may be temporal equivalents. They contain two arvicoline rodents, *Synaptomys* and *Pitymys,* which may be the first indication that climates are beginning to cool. The remainder of the taxa suggest a warm, mild climate.

If a date of 1.8 Ma is accepted for the base of the Pleistocene Epoch, then the Plio–Pleistocene boundary lies within the Crooked Creek Formation above the Borchers Ash and local fauna and below the Nash and Aries local faunas. If a date of 2.5 Ma is accepted, then the boundary lies at the base of the Ballard Formation.

REFERENCES CITED

Bandy, O. L., and Wilcoxin, J. A., 1970, The Plio–Pleistocene boundary, Italy and California: Geological Society America Bulletin, v. 81, p. 2939–2948.

Bayne, C. K., 1976, Early medial Pleistocene faunas of Meade County, Kansas: Kansas Geological Survey Guidebook Series 1, p. 1–36.

Boellstorff, J., 1973, Fission-track ages of Pleistocene volcanic ash deposits in the central plains, U.S.A.: Isochron/West, v. 8, p. 39–43.

—— , 1976, The succession of late Cenozoic volcanic ashes in the Great Plains; A progress report: Kansas Geological Survey Guidebook Series 1, p. 37–71.

—— , 1978, North American Pleistocene stages reconsidered in the light of probable Pliocene–Pleistocene continental glaciation: Science, v. 202, p. 305–307.

Eshelman, R. E., and Hibbard, C. W., 1981, Nash local fauna (Pleistocene: Aftonian) of Meade County, Kansas: University Michigan Museum Paleontology Contributions, v. 25, p. 317–326.

Frye, J. C., and Hibbard, C. W., 1941, Pliocene and Pleistocene stratigraphy and paleontology of the Meade Basin, southwestern Kansas: Kansas Geological

Survey Bulletin, v. 38, p. 389–424.
Hibbard, C. W., 1941, The Borchers fauna, a new Pleistocene interglacial fauna from Meade County, Kansas: Kansas Geological Survey Bulletin, v. 38, p. 197–220.
——, 1944, Stratigraphy and vertebrate paleontology of Pleistocene deposits of southwestern Kansas: Geological Society America Bulletin, v. 55, p. 707–754.
——, 1949, Pleistocene stratigraphy and paleontology of Meade County, Kansas: University Michigan Museum Paleontology Contributions, v. 7 p. 63–90.
——, 1958, New stratigraphic names for early Pleistocene deposits in southwestern Kansas: American Journal Science, v. 256, p. 54–59.
——, 1970, Pleistocene mammalian local faunas from the Great Plains and Central Lowland provinces of the United States, *in* Dort, W., and Jones, J. K., eds., Pleistocene and Recent environments of the central Great Plains: University of Kansas, Department of Geology Special Publication 3, p. 395–433.
Izett, G. A., and Wilcox, R. E., 1982, Map showing localities and inferred distributions of the Huckleberry Ridge, Mesa Falls, and Lava Creek ash beds (Pearlette Family ash beds) of Pliocene and Pleistocene age in the western United States and southern Canada: U.S. Geological Survey Miscellaneous Investigation Series Map, I-1325, scale 1:4,000,000.
Kovach, J. T., 1979, Geology of late Cenozoic sediments of the Borchers' Badlands, Meade County, Kansas [M.S. thesis]: Hays, Kansas, Fort Hays State University, 129 p.
Lindsay, E. H., Johnson, N. M., and Opdyke, N. D., 1975, Preliminary correlation of North American land mammal ages and geomagnetic chronology: University of Michigan Papers Paleontology, v. 12, p. 111–119.
Myers, A. J., 1959, Geology of Harper County, Oklahoma: Oklahoma Geological Survey Bulletin 80, p. 1–108.
Naeser, C. W., Izett, G. A., and Wilcox, R. E., 1973, Zircon fission-track ages of Pearlette Family ash beds in Meade County, Kansas: Geology, v. 1, p. 187–189.
Paulson, G. R., 1961, The mammals of the Cudahy fauna: Papers of the Michigan Academy of Science, v. 46, p. 127–153.
Skinner, M. F., and Hibbard, C. W., 1972, Early Pleistocene pre-glacial and glacial rocks and faunas of north-central Nebraska: American Museum Natural History Bulletin, v. 148, p. 1–148.
Smith, H.T.U., 1940, Geologic studies in southwestern Kansas: Kansas Geological Survey Bulletin, v. 34, p. 1–212.
Zakrzewski, R. J., 1975, Pleistocene stratigraphy and paleontology in western Kansas; The state of the art, 1974: University of Michigan Papers in Paleontology, v. 12, p. 121–128.
——, 1981, Kangaroo rats from the Borchers local fauna, Blancan, Meade County, Kansas: Transactions of the Kansas Academy of Sciences, v. 84, p. 78–88.

ACKNOWLEDGMENTS

I thank Ms. Gwenne Cash for drafting the columnar sections in Figures 4 through 7.

Mine 19 geologic section, Pittsburg and Midway Coal Mining Company, Cherokee County, Kansas

Lawrence L. Brady, Kansas Geological Survey, Lawrence, Kansas 66046

LOCATION

This site is an abandoned strip mine just south of West Mineral in northwest Cherokee County, in extreme southeastern Kansas. It can be reached most easily by traveling 5 mi (8 km) west from Kansas 7 on Kansas 102 to West Mineral and following local roads to the exposures shown in Figure 1.

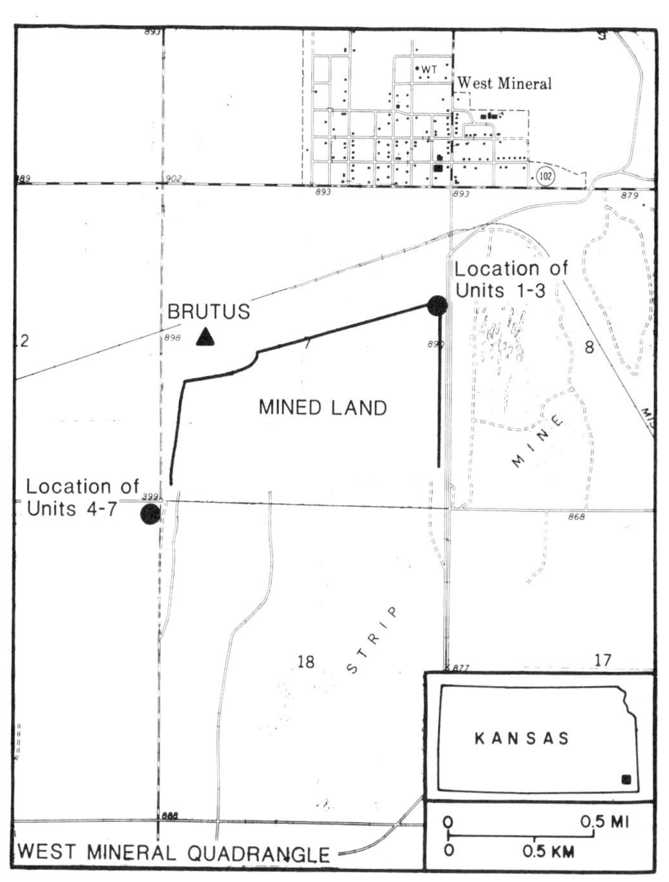

Figure 1. Location of the two mine high walls used for composite section description. The location of Brutus, the large mining shovel used to mine most of the land in this area, is also shown. This shovel forms the most prominent landmark in the area and can be used as a guide to this location.

SIGNIFICANCE

This locality consists of two high-wall exposures of limestone, shales, and coal that were uncovered while strip mining for coal (Fig. 2, A and B). These high-wall sections expose an important part of the Cherokee Group that has both widespread correlation potential and local economic importance.

GEOLOGY

The composite section (Fig. 3) consists of six geologic units (and a soil profile) that are located in the middle part of the Cabaniss Formation in the upper part of the Cherokee Group. Coals mined at these locations by the Pittsburg and Midway Coal Mining Company (Mine #19) include the Mineral, Fleming, and occasionally the Croweburg (where coal thickness was sufficient for economic recovery). A nearly complete core of the Cherokee Group (Fig. 4) was described by Harris (1984, p. 29, A-43, A-44). This core, drilled 4 mi (6.4 km) west of the west high-wall section, allows comparison of the Croweburg coal, the Verdigris Limestone, and adjacent shales and claystone at the measured section to the total section of the Cherokee Group.

The cyclicity of the Cherokee rocks in Kansas was recognized by Abernathy (1937, p. 19) as having 15 cyclothems that represent changes from nonmarine to marine conditions. Abernathy (1937) defined his *phases* of the normal Cherokee cyclothem in a depositional order as: (0.1) sandstone, (0.2) sandy shale, (0.3) underclay, (0.4) coal, (0.5) black shale, (0.6) gray shale, (0.7) limestone, and (0.8) calcareous shale. This sequence of nonmarine to marine rocks was close to the "ideal cyclothem" of Moore (1936, p. 24–25), except for the obvious lack of limestone, which is a feature more typical of cyclic deposits in the Upper Pennsylvanian rocks. In Figure 5, this normal Cherokee cyclothem of Abernathy is shown, along with the described composite geologic section. Different opinions exist on how this sequence of rock units fit this normal Cherokee cyclothem. Abernathy (1937) considered the underclay, coal, and black shale as part of the Croweburg cyclothem, and the shale below the limestone and the limestone as two phases of the Ardmore (Verdigris) cyclothem (Fig. 5).

In reviewing the Cherokee cyclothems, Moore (1949, p. 43, 45) considered Abernathy's Croweburg and Ardmore (Verdigris) cyclothems as a single cyclothem (Fig. 5). A gray shale overlying the Croweburg coal was not described by either Moore or Abernathy. Later, Moore and others (1951, p. 99, 101) referred to this sequence of rocks as the Ardmore cyclothem. The gray shale overlying the Croweburg was recognized at that time and it was considered by Moore and others (1951) that this shale, the Croweburg coal, underclay, and two underlying lithologic units might possibly constitute a distinct cycle.

Howe (1956) recognized 18 cyclothems in the Cherokee rocks. Five lithologic units were considered by Howe (1956, p. 21–26) to be common to most Cherokee cycles. These units

Figure 2. A: Location where units 1–3 were described—NE¼SE¼NE¼Sec.7,T.32S.,R.23E. Positive structure of the Croweburg coal (unit 2) shown on the north side of the high wall. This positive structure is also present on the east end of the pit just to the right of the photograph. B: Mine high wall showing prominent Verdigris Limestone (unit 6). Units 4–7 were described along the north side of this high wall, NE¼NE¼NE¼Sec.13,T.32S.,R.22E., Cherokee County, Kansas.

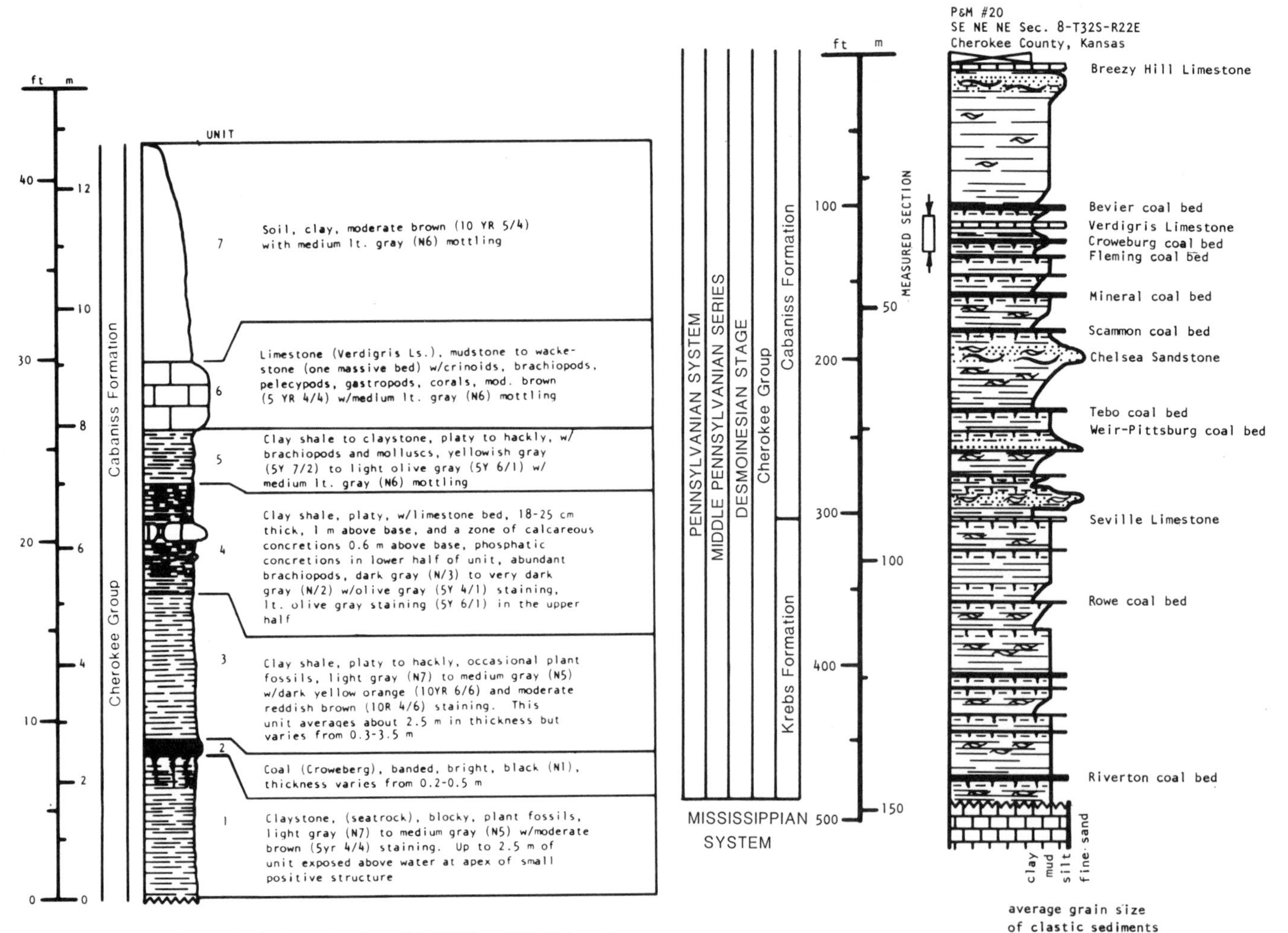

Figure 3. Composite geological section from P&M Mine #19 high walls. Actual mined section includes coal and shale down to the Mineral coal, approximately 26 ft (8 m) below the described section.

Figure 4. Cored section showing nearly complete Cherokee Group (less than 16 ft [5 m] missing). (Modified from Harris, 1984, p. 29.)

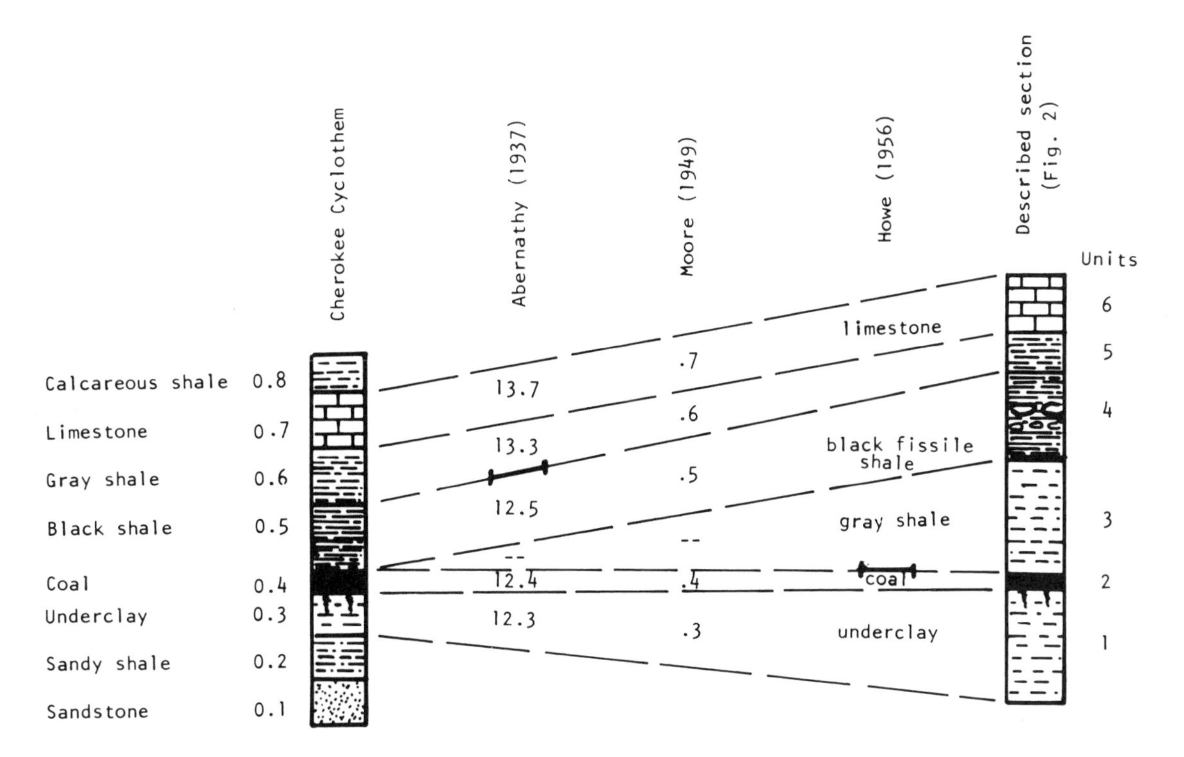

Figure 5. Generalized section comparing the described geologic section with the "normal" Cherokee cyclothem of Abernathy (1937) and interpretation of the rock sequence by various workers. Dashed lines between the geologic section and the "normal" Cherokee cyclothem are based on the interpretation of Moore (1949) and Moore and others (1951).

include (without the significance placed on the numbers as was the case by Abernathy [1937] and Moore [1949]), in depositional order of marine to nonmarine: (1) dark shale and dark irregular limestone, (2) gray shale, (3) underlimestone and sandstone, (4) underclay, and (5) coal. This sequence of lithologies was considered by Howe (1956, p. 23) to best fit the Cabaniss subgroup (Formation).

Use of Howe's Cherokee cycle to fit the described geological section indicates that parts of two of Howe's cycles are present—his Croweburg and Verdigris cycles (which he proposed as formations in his paper). The top of Howe's Croweburg cycle is the Croweburg coal.

Paleontological evidence was presented by Howe (1956, p. 72–74) and Williams (1937, p. 104–105) to indicate that at least part of the gray shale over the Croweburg coal might be of marine origin.

The Verdigris Limestone (unit 6) is the best-developed limestone in the Cherokee Group. The dark-gray shale (unit 4) has a high gamma-ray reading on geophysical logs, and both units are distinctive marker beds and have widespread distribution in the Midcontinent area.

Distribution of a large coal swamp—perhaps the most widespread in North America—resulted in the formation of the Croweburg coal in Kansas, Missouri, and Oklahoma, and its equivalents, the Colchester (No. 2) in Illinois, the Whitebrest coal in Iowa, the IIIa coal in Indiana, Shultztown coal in Kentucky, and the Lower Kittanning coal in the northern Appalachians (Wanless and others, 1969, p. 131–134; Wright, 1975, p. 78, Pl. 17). This widespread coal deposit was considered by Wright (1975) to have developed because sediments underlying the coal formed a broad platform on which the swamp could develop.

In Kansas, the marine dark-gray shale (unit 4) is unnamed. Its equivalent in Illinois is also unnamed, but in Indiana, the well-known Mecca Quarry Shale Member of the Linton Formation was considered its counterpart by Wright (1975, p. 79–180), who also noted that the physical characteristics of the shale are generally the same from Oklahoma to Indiana.

Distribution of the Verdigris Limestone (unit 6) and its equivalents extends from central Oklahoma to central Iowa (Wright, 1975, Pl. 17).

Additional information on the local stratigraphy of units observed at the high-wall section can be obtained from Pierce and Courtier (1937), and Howe (1956); paleontology is summarized by Williams (1937).

ECONOMIC GEOLOGY

Extensive mining for coal by surface and underground methods has been an important part of the economy of southeast

Kansas, especially in Cherokee and Crawford counties, where nearly 230 million metric tons of coal have been mined. The important coal beds mined other than the three mined at the P&M #19 Mine—the Mineral, Fleming, and Croweburg coals—are the Bevier and Weir-Pittsburg coal beds (see Fig. 4). Most of the underground mining (more than 50,000 acres [20,000 hectares]) of the Weir-Pittsburg coal was by the room-and-pillar system.

Coals from the Cherokee Group in southeast Kansas are high-volatile A bituminous in rank, and are currently used mainly for power generation and cement manufacturing.

A large amount of the observed strip-mined land in this area of Cherokee County was mined as part of P&M Mine #19. A significant amount of this mining was done by "Brutus," the large mining shovel located 0.5 mi (0.8 km) north of the west high-wall section. This Bucyrus-Erie 1850-B shovel has a bucket capacity of nearly 90 yd^3 (70 m^3) and was used by the Pittsburg & Midway Coal Mining Company to mine coal in this area from 1963 until 1974. At the time of its construction in 1962–63, this machine was the second-largest mining shovel in existence.

Coals mined at Mine #19 were processed at the company tipple near Hallowell, Kansas, a truck haul of 10 mi (16 km). The processed coal, consisting of blends of the Mineral, Fleming, and Croweburg coals, produced a product with 6.5% moisture, 12.5% ash, 3.3% sulfur, and a heat content of 12,300 Btu/lb (6,830 Kcal/kg).

REFERENCES CITED

Abernathy, G. E., 1937, The Cherokee Group of southeastern Kansas: Kansas Geological Society, Guidebook, 11th Annual Field Conference, p. 18–23.

Harris, J. W., 1984, Stratigraphy and depositional environments of the Krebs Formation-Lower Cherokee Group (Middle Pennsylvanian) in southeastern Kansas [M.S. thesis]: Lawrence, University of Kansas, 205 p. (Kansas Geological Survey Open-File report 84–9).

Howe, W. B., 1956, Stratigraphy of pre-Marmaton Desmoinesian (Cherokee) rocks in southeastern Kansas: Kansas Geological Survey Bulletin 123, 132 p.

Moore, R. C., 1936, Stratigraphic classification of the Pennsylvanian rocks of Kansas: Kansas Geological Survey Bulletin 22, 256 p.

——, 1949, Divisions of the Pennsylvanian System in Kansas: Kansas Geological Survey Bulletin 83, 203 p.

Moore, R. C., Frye, J. C., Jewett, J. M., Lee, W., and O'Connor, H. G., 1951, The Kansas rock column: Kansas Geological Survey Bulletin 89, 132 p.

Pierce, W. G., and Courtier, W. H., 1937, Geology and coal resources of the southeastern Kansas coal field: Kansas Geological Survey Bulletin 24, 122 p.

Wanless, H. R., Baroffio, J. R., and Trescott, P. C., 1969, Condition of deposition of Pennsylvanian coal beds, *in* Dapples, E. C., and Hopkins, M. E., eds., Environments of coal deposition: Geological Society of America Special Paper 114, p. 105–142.

Williams, J., 1937, Pennsylvanian invertebrate faunas of southeastern Kansas, *in* Pierce, W. A., and Courtier, W. H., eds., Geology and coal resources of the southeastern Kansas coal field: Kansas Geological Survey Bulletin 24, p. 92–122.

Wright, C. R., 1975, Environments within a typical Pennsylvanian cyclothem, *in* McKee and others, eds., Paleotectonic investigations of the Pennsylvanian System in the United States, Part II: Interpretive summary and special features of the Pennsylvanian System: U.S. Geological Survey Professional Paper 853, p. 73–84.

The Meers fault scarp, southwestern Oklahoma

***R. N. Donovan,** Geology Department, Texas Christian University, Fort Worth, Texas 76129*

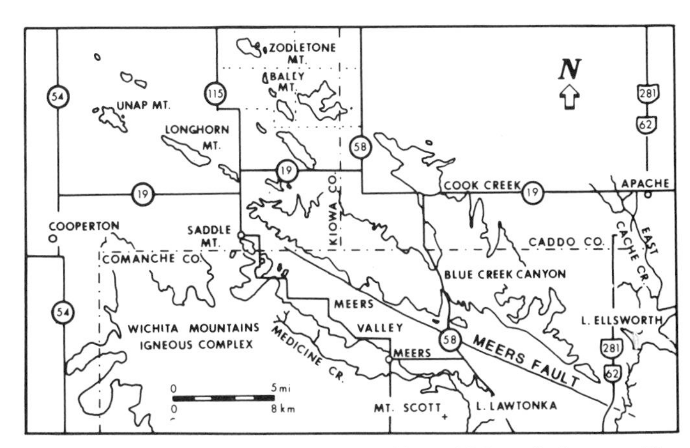

Figure 1. Location map illustrating position of Meers fault trace. The section described here is located about 1 mi west of Oklahoma 58.

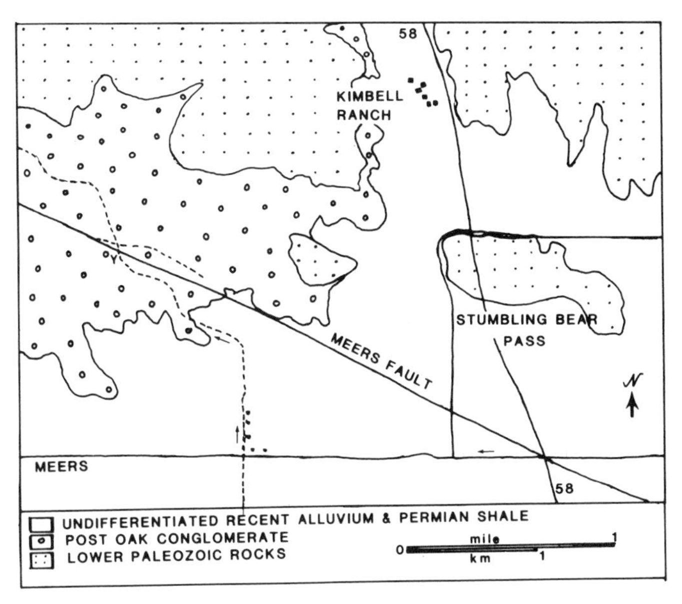

Figure 2. Map showing details of access to the "Oliver Section."

LOCATION

The Meers fault scarp trends ESE–WNW along the northern edge of the Meers Valley in Commanche County, southwestern Oklahoma (Fig. 1).The impressively straight scarp can be traced for 16 mi (26 km). The southeastern two-thirds of the scarp intends across land that is under relatively intensive agricultural use, and the northwestern third extends across hilly open range. This land usage reflects the underlying Permian bedrock: shales to the southeast and conglomerates to the northwest. The most visually impressive and instructive lengths of scarp are developed in the conglomerate terrain. Access procedure to a convenient length of the scarp in this terrain is as follows. First, contact Mr. Charlie Bob Oliver, manager of the Kimbell Ranch. The ranch house is located in Blue Creek Canyon on Oklahoma 58. Mr. Oliver has been closely involved in recent work on the fault and, if notified in advance, can make a key to the ranch tracks available. Oklahoma 58 must be followed south of Stumbling Bear Pass to a county road (Fig. 2). This road should then be taken for just over 1 mi (1.6 km) west to a group of houses north of the road. At this point a decision must be made with respect to the condition of a ranch track which trends north by the houses. This road is impassable for most vehicles after a rain (the alternative is a walk of about 1.5 mi; 2.4 km). The track is followed north for about 0.5 mi (0.8 km) to a large gate which marks the entrance to Kimbell Ranch. After opening (and closing) the gate, follow the track as it winds northwestward across hilly conglomerate terrain to the fault scarp. The point where the ranch track crosses the scarp is clearly marked—recent excavations by the Oklahoma Geological Survey have been left open for inspection. Park here and walk along the route suggested in Figure 3. This section of the fault scarp is informally known as the "Oliver Section" (Donovan, 1986). Noteworthy outcrops of Permian Post Oak Conglomerate add additional interest to this section. Beware of rattlesnakes.

SIGNIFICANCE

The Meers fault scarp has been described as the finest active fault scarp in the United States east of the Rocky Mountains (Slemmons, personal communication). The compellingly constant lineament has recently been interpreted (Donovan and others, 1983; Gilbert, 1983) as a record of modern rejuvenation of the Meers Fault, a distant echo of a time-honored past. The ancient fault is a major structure that was the southwestern boundary of the Frontal fault zone between the Wichita uplift and the Anadarko basin as these structures developed in Pennsylvanian time. The late Paleozoic displacement is difficult to quantify accurately, but probably involved at least 10,000 ft (3,000 m) of vertical separation, plus an unknown amount of left-lateral displacement. The modern movement is much more modest, involving a downthrow to the south that has a maximum displacement of about 20 ft (6 m) and a left-lateral displacement of similar dimensions.

The Holocene movement represents a reversal of the vertical sense of Paleozoic displacement. Furthermore, the younger fault plane is either vertical or dips to the north at angles of up to 70° (indicating a high-angle reverse structure). This dip is in the opposite sense to that suggested by COCORP deep subsurface data

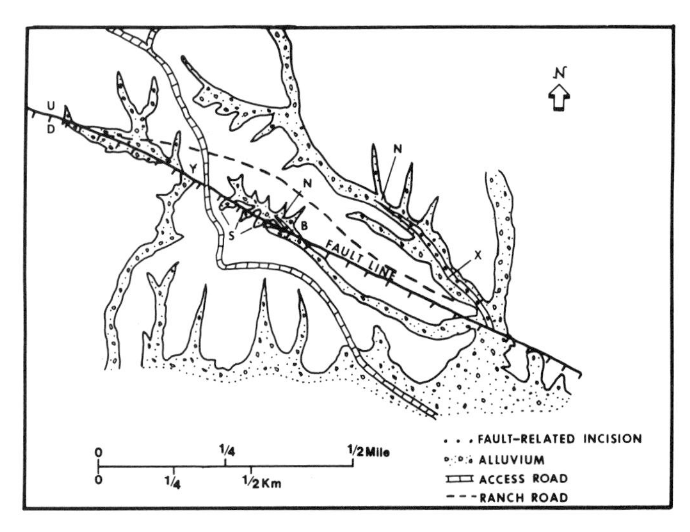

Figure 3. The "Oliver Section" of the Meers fault. S = sediment entrapment sites formed where the fault scarp has impeded drainage; N = fault-related knick points; B = recent gullying on fault trace; X = Post Oak Conglomerate exposures depicted in Figure 5; Y = excavation site on fault scarp (from Donovan, 1986).

(Brewer and others, 1983). It is worth noting that linkage of the COCORP structure to the surface trace is not easy.

Recent work (Ramelli and Slemmons, 1986; Madole, 1986; Crone and Luza, 1986) involving trenching, detailed geomorphic analysis, soil profiling, and radiocarbon dating has added much detail to the initial interpretation. Some of the important conclusions reached are: (1) The fault is unusual in that it is one of only a few known active faults in the midcontinent region (Ramelli and Slemmons, 1986); (2) Richter earthquake magnitudes of 7.5 to 8.0 are possible on the fault (Ramelli and Slemmons, 1986); (3) ^{14}C ages of soil humus and fluvial deposits suggest that a major movement occurred between 600 ± 50 yr and 1,740 ± 55 yr ago, most probably around 1,300 ± 100 yr (Madole, 1986); and (4) A slip rate on the fault of about 0.02 mm/yr is likely; an earthquake recurrence interval of tens of thousands of years has been suggested (Crone and Luza, 1986).

GEOMORPHIC EVIDENCES OF RECENT FAULT MOVEMENT

Quaternary rejuvenation of the Meers fault was initially suspected on geomorphic evidence (Fig 3). For example, the fault line is constant in its throw regardless of the resistance to weathering exhibited by the facies it cuts. In addition, recent drainage patterns have been interrupted, as evidenced by: (1) the sediment ponding of northward draining tributaries; (2) displacement of river terraces; (3) rejuvenation of streams crossing the fault scarp (with resultant knick points developed upstream); and (4) gullying developed along the fault scarp with resulting drainage piracy. All these features are admirably developed within the "Oliver Section" (Fig. 3). In addition, a constant amount of left-lateral offset of pre-fault topography is apparent (Fig. 4).

FRACTURE PATTERNS RELATED TO THE FAULT

The Oklahoma Geological Survey has recently trenched the fault close to the ranch road ("Y" in Fig. 1). Mr. Oliver has allowed this trench to remain open for inspection: it shows one principal and two or three minor lines of displacement that are essentially vertical and contain some silica-rich tan-colored infill. Near-horizontal slickensides are visible on some fracture surfaces. I have made 16 observations on separate slickenside striations along the fault trace. All indicate a variable, oblique, left-lateral sense of slip (from 8 to 82°).

Along its trace in the Post Oak terrain, the fault is associated with a large number of fractures. Major modes of distribution of these fractures have been interpreted as Riedel (R and R′) shear fractures developed in response to a left-lateral sense of shear on the main fault (Donovan and others, 1983). Pebbles showing both left-lateral (R) and right-lateral (R′) displacement can be found along the Oliver section. Few tensional fractures are present, presumably because deformation has been essentially transpressive.

THE POST OAK CONGLOMERATE

The Post Oak Conglomerate facies cut by the Meers fault in the Oliver section is a well-indurated limestone-pebble conglomerate consisting of subangular to subround (solution-modified) pebbles of the local Arbuckle Group rocks. The facies has been interpreted as the deposits of small alluvial fans debouching from the adjacent Slick Hills (Gilbert and Donovan, 1984). An interesting section is exposed in the fault-related rejuvenation gorge shown in Figure 3 (at "X"). At this point, a sharply defined channel is incised into earlier conglomerates.

By comparison to Permian sedimentation in the Anadarko basin, the rate of sediment entrapment in the Meers Valley was very low, beause the latter was not an area of pronounced subsidence. Some evidence in support of this hypothesis is the close spacing of calcareous pedocals (calcretes) in the Post Oak Conglomerate. Calcretes in the area have been described by Gilbert and Donovan (1982), Collins (1985), and Bridges (1985). These authors have noted that, in order to form, calcretes require a stable geomorphic setting that persists for thousands of years. The close spacing of calcretes in Post Oak sequences indicates slow episodic sedimentation punctuated by frequent periods of nondeposition and soil development.

The section illustrated in Figure 5 gives some idea of the episodic nature of sedimentation in the conglomerate (both pre- and post-channel incision). Red mudstones and siltstones demarcate pauses in sedimentation: calcretes formed when these pauses were of sufficient duration and/or the climate was favorable for their development (i.e., semi-arid). Episodic sedimentation on

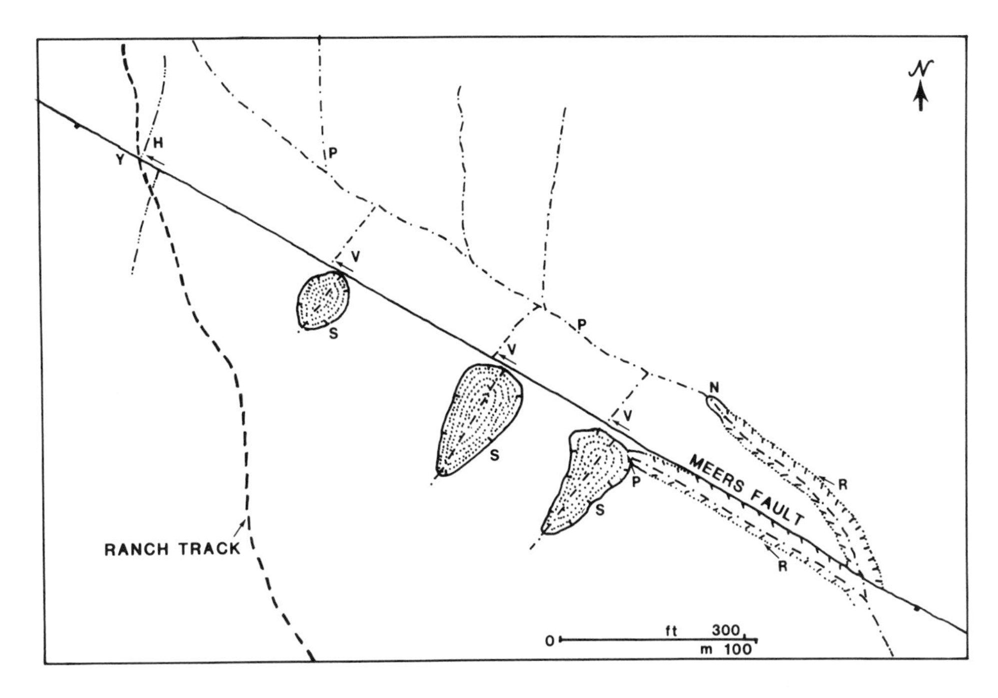

Figure 4. Detail of the "Oliver Section." Y = excavation site on fault scarp (see Fig. 3); H = displaced ridge axis; V = displaced valley axis; D = original (pre-fault) drainage; S = closed fault-related depression filled by soil; P = site of drainage piracy by fault-related rejuvenation; N = knick point; R = rejuvenation gorge.

alluvial fans may be of either climatic or tectonic origin. Both mechanisms may have been operative in the Meers Valley (Gilbert and Donovan, 1984). The channel cut represents an avulsion event. The stratified nature of the infill suggests that the channel was originally simply a sediment-supply corridor (located above the fan intersection point) that was subsequently clogged with sediment. It is pertinent that the upper slopes of this channel were calcreted before being overlain by sediment.

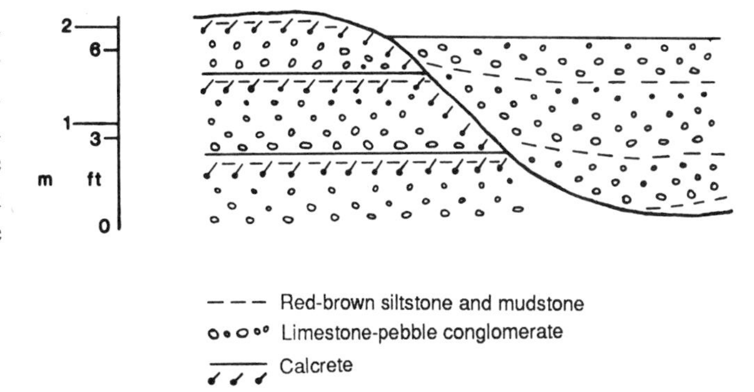

Figure 5. Diagrammatic sketch of a Post Oak Conglomerate exposure (Fig. 3; site X) near the "Oliver Section." Vegetation omitted.

REFERENCES

Brewer, J. A., Good, R., Oliver, J. E., Brown, L. D., and Kaufman, S., 1983, COCORP profiling across the southern Oklahoma aulacogen; Overthrusting of the Wichita Mountains: Geology, v. 11, p. 109–114.

Bridges, S. D., 1985, Mapping, stratigraphy, and tectonic implications of Lower Permian strata, eastern Wichita Mountains, Oklahoma [M.S. thesis]: Stillwater, Oklahoma State University, 129 p.

Collins, K., 1985, The evolution of the Meers Valley in the Wichita Mountains, Oklahoma [M.S. thesis]: Stillwater, Oklahoma State University, 128 p.

Crone, A. J., and Luza, K. V., 1986, Holocene deformation associated with the Meers fault, southwestern Oklahoma, *in* Donovan, R. N., ed., The Slick Hills of southwestern Oklahoma; Fragments of an aulocogen?: Oklahoma Geological Survey Guidebook no. 24, p. 68–74.

Donovan, R. N., 1986, Stop 5: The Meers fault: Modest finale for a hoary giant?, *in* Donovan, R. N., ed., The Slick Hills of southwestern Oklahoma; Fragments of an aulacogen?: Oklahoma Geological Survey Guidebook no. 24, p. 106–108.

Donovan, R. N., Gilbert, M. C., Luza, K. V., Marchini, D., and Sanderson, D. J., 1983, Possible Quaternary movement on the Meers fault, southwestern Oklahoma: Oklahoma Geology Notes, v. 43, p. 124–133.

Gilbert, M. C., 1983, The Meers fault of southwestern Oklahoma; Evidence for possible strong Quaternary seismicity in the midcontinent [abs.]: EOS American Geophysical Union Transactions, v. 64, p. 313.

Gilbert, M. C., and Donovan, R. N., editors, 1982, Geology of the eastern Wichita Mountains, southwestern Oklahoma: Oklahoma Geological Survey Guidebook 21, 160 p.

—— , 1984, Recent developments in the Wichita Mountains (Geological Society of America South-Central Section meeting guidebook, field trip 1): College Station, Texas A&M University, 101 p.

Madole, R. F., 1986, The Meers fault; Quaternary stratigraphy and evidence for late Holocene movement, *in* Donovan, R. N., ed., The Slick Hills of southwestern Oklahoma; Fragments of an aulacogen?: Oklahoma Geological Survey Guidebook no. 24, p. 55–67.

Ramelli, A. R., and Slemmons, D. B., 1986, Neotectonic activity of the Meers fault, *in* Donovan, R. N., ed., The Slick Hills of southwestern Oklahoma; Fragments of an aulacogen?: Oklahoma Geological Survey Guidebook no. 24, p. 45–54.

Evaporites and red beds in Roman Nose State Park, northwest Oklahoma

Kenneth S. Johnson, *Oklahoma Geological Survey, University of Oklahoma, Norman, Oklahoma 73019*

LOCATION

Roman Nose State Park comprises 540 acres (219 hectares) of public lands where layers of Permian gypsum, dolomite, and shale are well exposed and where fresh water flows from three natural springs. The site is in the central part of Blaine County, northwestern Oklahoma (Fig. 1), located about 6 mi (10 km) north of Watonga, in Secs. 23 and 24, T.17N., R.12W. The park allows free access at all times and provides picnic and camping facilities; the park also contains a resort lodge, meals, and cottages. For information and access to springs and other areas partially closed during winter months (October 1 till early April), contact the Park Superintendent, Roman Nose State Park, Route 1, Box 2-2, Watonga, OK 73772 (telephone: 405/623-4215). The area is on the Watonga Lake 7½-minute Quadrangle.

SIGNIFICANCE OF SITE

Permian evaporites and red beds underlie a vast area of the southwestern United States, and Roman Nose Park is an excellent, public-access site to examine cyclic sedimentation in an evaporite–red bed sequence. The site is located on the north flank of the Anadarko Basin, and is on the east side of the broad Permian Basin evaporite region that extends northward from west Texas through parts of New Mexico, Oklahoma, Kansas, and Colorado. Lithostratigraphic marker beds of the Blaine Formation that crop out here are correlative as discrete rock units over an area extending about 300 mi (480 km) east-west (western Oklahoma to eastern New Mexico) by 400 mi (640 km) north-south (Matador Arch to northwestern Kansas): this attests the epeirogenic crustal movements beneath the broad, epicontinental sea that covered the evaporite basin during Permian time.

The geology along the entire Blaine Escarpment (extending across northwest Oklahoma) is similar to that described here, and many other sites (particularly the gypsum quarries here in Blaine County) could be used for a similar examination. The description that follows is modified from Johnson (1972), which in turn is based upon a report by Fay (1959) on the history, geology, botany, and zoology of the park. Additional data are in reports by Fay (1964) and Fay and others (1962).

SITE INFORMATION

Outcropping rocks are the Blaine Formation and overlying Dog Creek Shale of Permian age (Figs. 2 and 3). Three resistant beds of white gypsum, each 6 to 12 ft (2 to 4 m) thick, are separated by red-brown and light gray shales 25 to 30 ft (8 to 10 m) thick; each gypsum bed is immediately underlain by a fossiliferous dolomite bed 1 ft (0.3 m) thick (see cross section). Fossils in the dolomites are molds of pelecypods, and belong to the genus

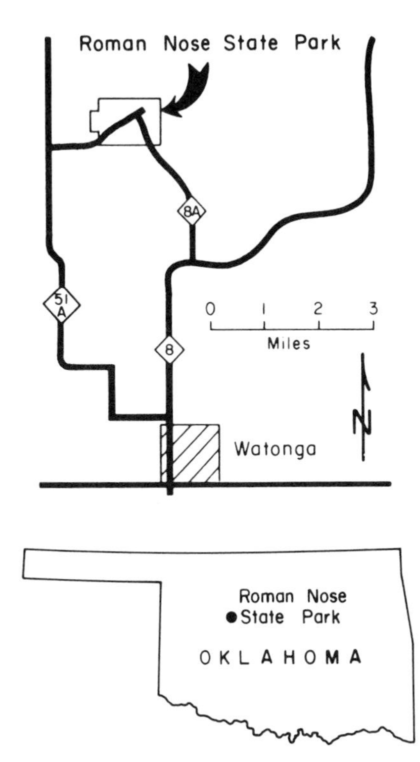

Figure 1. Location of Roman Nose State Park near Watonga in northwestern Oklahoma.

Permophorus (Fay and others, 1962, p. 39). Locally the Altona Dolomite contains ripple marks on its upper surface. Anhydrite, an anhydrous form of calcium sulfate, in layers or lenses 0.5 to 3 ft (0.15 to 1 m) thick, is present locally in the middle of both the Shimer and Nescatunga gypsum beds: the anhydrite is light gray and finely crystalline and is both harder and denser than the gypsum.

The succession or cycle of rock types making up an "evapoite" series is well exposed. Dolomite, arbitrarily considered the base of a cycle, is overlain by gypsum; this in turn is overlain by a thick unit of red-brown shale with thin interbeds of green-gray shale and gypsum; at the top is a thin layer of green-gray shale. Salt, which normally follows gypsum in the order of precipitation from sea water and should occur between the gypsum and red-brown shale, either was not deposited because of an interruption in evaporation of sea water or was deposited and has since been dissolved by ground water. The former explanation is more likely to be correct in this area. The cycle is repeated several times in the Blaine Formation, and a more incomplete cycle is seen in the Dog Creek Shale, where even the thick gypsum beds were not deposited.

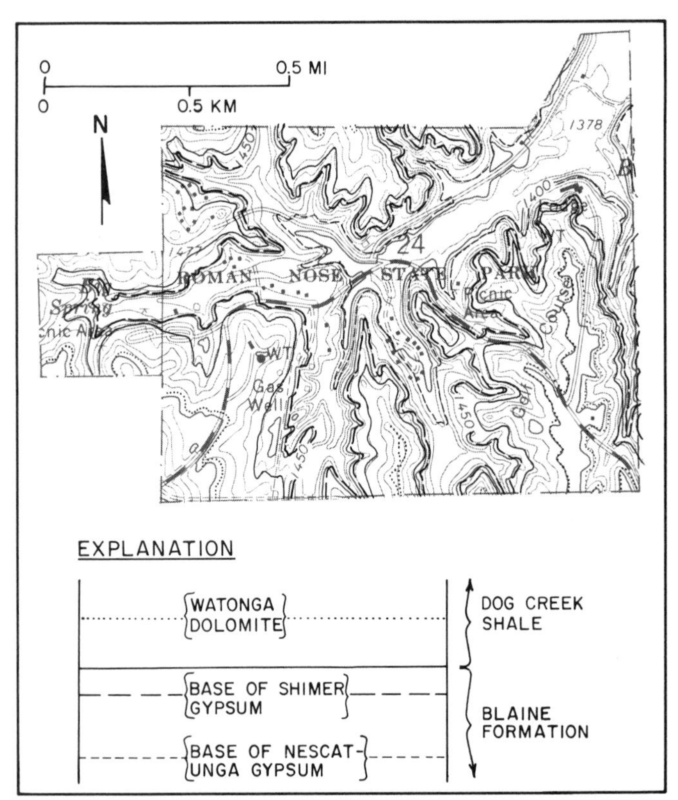

Figure 2. Geologic map of Roman Nose State Park.

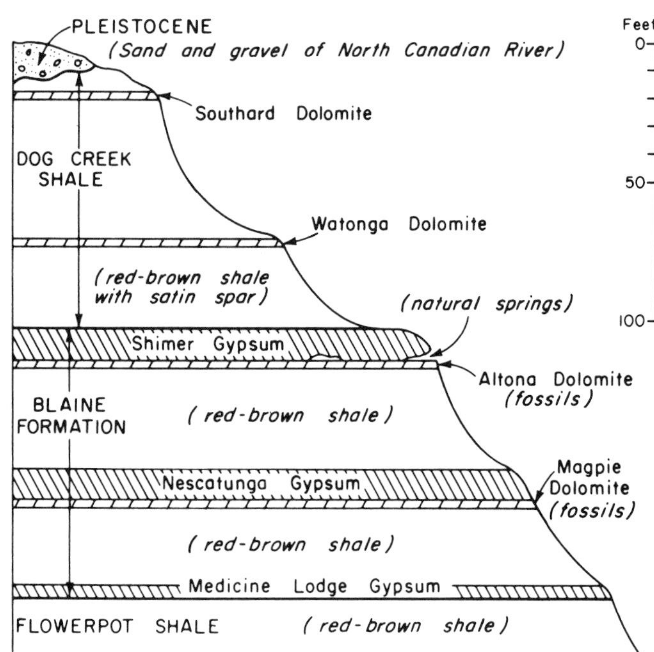

Figure 3. Stratigraphic column of rocks exposed in Roman Nose State Park and adjacent areas.

Two prominent ledges of thick white gypsum are easily seen in the eastern part of the park near the lake; the lower ledge (upon which the resort lodge is built) is the Nescatunga Gypsum, and the upper ledge (above the roof of the lodge) is the Shimer Gypsum. About 25 ft (8 m) above the Shimer Gypsum is 5 ft (1.5 m) of interbedded light gray dolomite and dark gray shale (Watonga Dolomite) that caps the highest hills in the park. The Shimer Gypsum and the Watonga Dolomite are easily recognized and can be identified throughout the park, whereas the Nescatunga Gypsum is conspicuous only in the east near the lake. The fact that these strata are essentially horizontal causes their outcrop to coincide with specific topographic contour lines throughout the park (Fig. 2). The youngest beds of the Dog Creek Shale are not exposed in the park: they were eroded by the North Canadian River when it flowed across the park area. Pleistocene sands and gravels deposited upon that eroded surface have been largely reworked by wind into a hummocky sand-dune surface to the south and west of the park: the dunes are now stabilized by vegetation.

Locally, large blocks of gypsum have fallen from the main ledges and rest on slopes made up of underlying shales. One large rockfall in December 1971 brought a block of Shimer Gypsum weighing about 150,000 pounds (68,000 kg) down to the road on the north side of the lake, 0.1 mi (0.16 km) northeast of the riding stables. The fall occurred after a heavy rain when the block was sufficiently undercut by shale erosion, and joints or slippage planes in the underlying shale were sufficiently "lubricated" by water.

Ground water is emitted at the surface through several natural springs near the picnic area in the west end of the park. Major springs, such as Big Spring, Middle Spring, and Little Spring, all issue from the base of the Shimer Gypsum. Rain water entering the ground farther south and west moves through karstic cavities and crevices in gypsum and dolomite until it emerges at the springs. The water is gypsiferous because it is actively dissolving the rock gypsum and enlarging underground caverns and caves. The combined flow from the three springs is generally 600 to 800 gallons (2,400 to 3,200 liters) per minute, and one set of measurements in 1958 recorded a flow of 80, 175, and 365 gpm (320, 700, and 1,460 lpm) from Little, Middle, and Big Springs, respectively.

REFERENCES CITED

Fay, R. O., 1959, Guide to Roman Nose State Park, Blaine County, Oklahoma: Oklahoma Geological Survey Guide Book 9, 31 p.

—— , 1964, The Blaine and related formations of northwestern Oklahoma and southern Kansas: Oklahoma Geological Survey Bulletin 98, 238 p.

Fay, R. O., Ham, W. E., Bado, J. T., and Jordan, L., 1962, Geology and mineral resources of Blaine County, Oklahoma: Oklahoma Geological Survey Bulletin 89, 252 p.

Johnson, K. S., 1972, Guidebook for geologic field trips in Oklahoma; Book II, Northwest Oklahoma: Oklahoma Geological Survey Educational Publication 3, 42 p.

Carbonate platform facies of the Morrowan Series (Lower Pennsylvanian), northeastern Oklahoma and northwestern Arkansas

Patrick K. Sutherland, School of Geology and Geophysics, University of Oklahoma, Norman, Oklahoma 73019
Walter L. Manger, Department of Geology, University of Arkansas, Fayetteville, Arkansas 72701

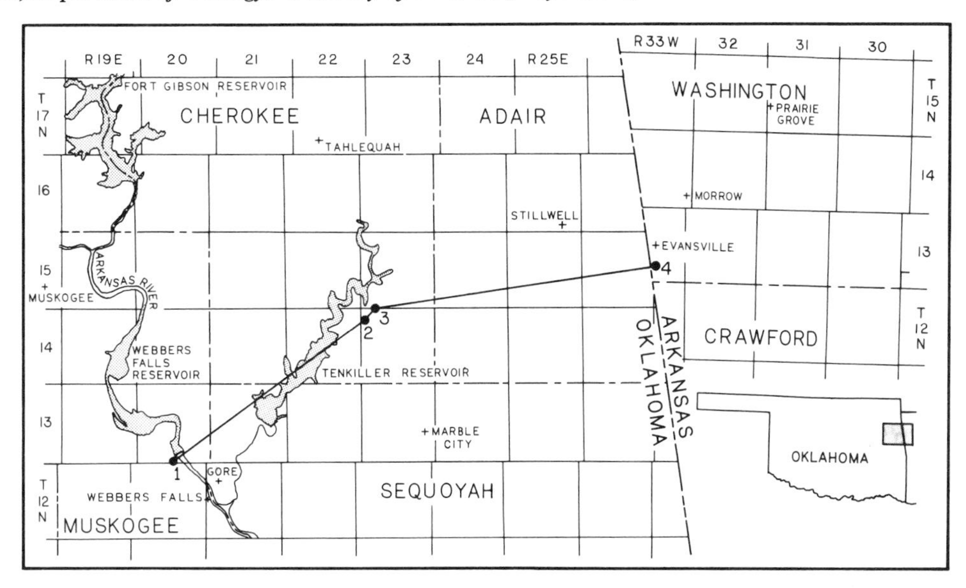

Figure 1. Index map. Numbered dots refer to localities shown in Figure 2. Township and Range squares are each 6 mi (9.6 km) across.

LOCATION

The localities described here form a transect from the Arkansas River, near Webbers Falls, Muskogee County, eastern Oklahoma, eastward to the Oklahoma/Arkansas boundary area near Evansville, Arkansas (Fig. 1). Detailed location data for each stop on the transect are given in the text.

SIGNIFICANCE

The Lower Pennsylvanian Morrowan Series in northeastern Oklahoma (Fig. 2) is represented by a well-developed shallow marine carbonate bank facies. It contains common macrofossils in many intervals. The sequence has been correlated in detail with the traditional type sequence of the Morrowan Series located 40 to 50 mi (64 to 80 km) to the northeast in Washington County, Arkansas. In that area, the Morrowan Series is composed predominantly of terrigenous shales and sandstones and contains only two well-developed limestones, the Brentwood and Kessler Limestone Members of the Bloyd Formation. Sutherland and Henry (1977) proposed that the type area of the Morrowan be extended westward to include the highly fossiliferous facies in Oklahoma, and they proposed new formation and member names for the Oklahoma sequence. Four localities provide opportunities for the examination of the carbonate bank facies in Oklahoma (Localities, 1, 2, and 3) and for a comparison of it with the traditional-type sequence of the Morrowan Series at Evansville Mountain (Locality 4), located in Arkansas 1.3 mi (2 km) east of the state line (Fig. 1). All localities are on public land, on or near paved roads, and are easily located.

Locality 1, Lock and Dam (Measured Section 1)

Location. The location is a large bluff on west side of the Arkansas River, at the west end of Webbers Falls Lock and Dam 16, 0.6 mi (1.0 km) east of an unnumbered paved road that leads to Lock and Dam Lookout, and 2.9 mi (4.7 km) north of the junction with U.S. 64. The lower parts of the section (Fig. 3), units 1 through 9, which are in the Sausbee Formation, were measured in the cuts on the bluff 50 to 400 ft (15 to 122 m) south of the west end of the dam, in SW¼,SW¼,SE¼,Sec.34,T.-13N,R.20E. The section was then offset northward on the base of the Brewer Bend Member, which forms the prominent bluff at the top of the cuts (Fig. 3), across a normal fault at the west end of the dam. The base of the Brewer Bend Member is brought down by faulting to the level of the top of the dam at this point. The Brewer Bend Member and the remaining units in the section were measured on the bluffs immediately north and above the end of

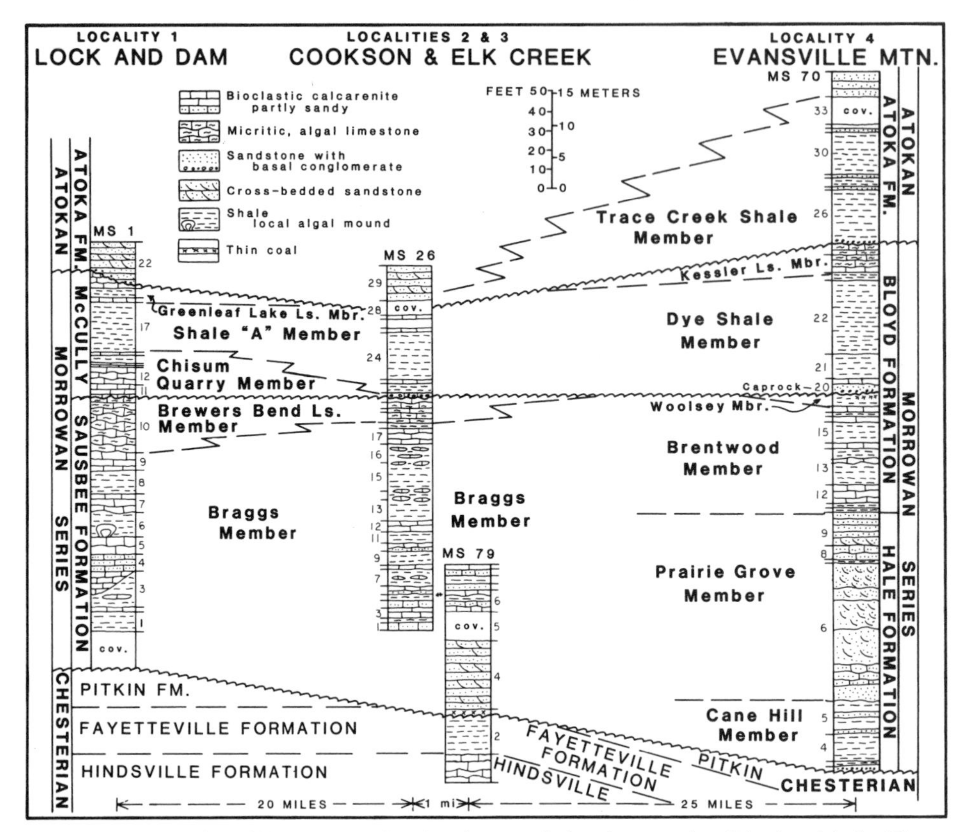

Figure 2. Cross section of Morrowan series, showing correlations between localities 1 and 4. See Figure 1 for location of each section.

the dam (Fig. 4). The section ends with the lowest exposure of the Atoka Formation in the bluff high above the west end of the dam, in SE¼,SE¼,SW¼,Sec.34,T.13N.,R.20E., Webbers Falls Quadrangle, Muskogee County, Oklahoma.

Description. Measured Section 1 is the most important Morrowan exposure in northeastern Oklahoma. It is the type sequence for the Sausbee and McCully Formations, along with the bluff exposures located on the west side of the Arkansas River 0.3 mi (0.5 km) to the northwest of the Lock and Dam (Sutherland and Henry, 1977).

The basal unconformable contact of the Sausbee Formation is not exposed at the Lock and Dam. The partial thickness of the Sausbee at this exposure averages 118.5 ft (36 m), compared with a total thickness for the formation of 132 ft (40 m) 0.3 mi (0.5 km) to the northwest where the Upper Mississippian Pitkin Limestone is exposed low on the Arkansas River Bluff. The covered interval below road level at the base of the Lock and Dam Section is therefore assumed to be no more than 20 ft (6 m). The thickness of the partial Morrowan section at this locality is 182.5 ft (55.6 m; Fig. 2).

The Sausbee Formation is subdivided into the Braggs Member and overlying Brewer Bend Limestone Member throughout Muskogee, Cherokee, and western Sequoyah Counties. The sequence in the Arkansas River area was deposited on the outer shelf with deeper water to the southwest and a terrigenous source to the northeast and east. The Braggs Member (Fig. 3) is characterized by interbedded limestones and shales, and these are partly sandy only in the lower part of the member. The broad bluff exposure at the Lock and Dam, about 1,500 ft in length, is exceptional in that it allows examination of the numerous rapid lateral changes in facies that are developed (Fig. 3). The limestones are mostly skeletal grainstones and packstones and some are channeloid. Units 6 and 8 are characterized by the occurrence of micritic, algal bryozoan bioherms in the Arkansas River area, but they are developed in the Lock and Dam section only in unit 6. They can mostly easily be examined, in both intervals, in the Chisum Quarry located across the Arkansas River to the east where the largest are about 6 ft (2 m) in height and 10 ft (3 m) across (see Sutherland, 1977, Stop 9, for description and directions; and Bonem, 1977, for a detailed description of the mounds).

The Sausbee Formation represents a shallowing-upward sequence with the Braggs gradational and interfingering with the overlying Brewer Bend Limestone Member. The Brewer Bend represents the development of a large, complex algal carbonate mudbank. When fully developed it extended over a distance at least 35 mi (56 km) across (Sutherland and Henry, 1977, Figs. 13, 14). It is characterized in the Arkansas River area by intensely burrowed algal wackestone and mudstone commonly interbedded with algal boundstone. The principal framebuilders of the

Figure 3. Exposure on bluff below Lock and Dam of Sausbee Formation: Braggs Member, below, forms most of cliff; Brewer Bend Limestone Member forms highest rounded cliff. Contact is marked by arrow. Exposure is 110 ft (33.5 m) high and includes units 1 through 10 of Figure 2.

Figure 4. Exposure at west end of Lock and Dam of Sausbee Formation: upper part of Brewer Bend Limestone Member (A). McCully Formation: Chisum Quarry Member (B), shale "A" member (C), Greenleaf Lake Limestone Member (D). Atoka Formation: (E). The arrow to left of E points to truncating unconformity at base of Atoka Formation. Undulating contact between A and B is unconformity between Sausbee and McCully Formations. Exposure includes units 10 through 22. Man at base provides scale.

boundstone are *Archaeolithophyllum* and *Cuneiphycus.* These structures are generally biostromal in nature, and small biohermal developments are extremely rare.

In the Arkansas River area the Brewer Bend was developed near the outer shelf margin; it has a maximum known thickness in this area of 41.5 ft (12.6 m). In the Lock and Dam section it is 28.5 ft (8.7 m) thick. In this area the Brewer Bend is highly fossiliferous, and the biotic diversity is great, particularly in the middle and upper parts. Algae followed by fenestrate bryozoans are the most common biotic elements, with corals and brachiopods also common. Scattered colonies of the colonial rugose coral *Petalaxis* up to 2 ft (0.6 m) in diameter occur in the upper parts. Locally developed, sharply defined, channel-shaped deposits, composed of skeletal packstone, typically 5 to 10 ft (1.5 to 3 m) across, cut the carbonate mudstone facies at scattered horizons and are interpreted to be tidal channels. Pelmatozoan debris (including much blastoid material) dominates these deposits, but fragments of algae, bryozoans, goniatites, brachiopods, and corals are also included. The goniatites occur locally as a coquina.

The Brewer Bend Limestone Member was measured and studied at the west end of the Lock and Dam (Fig. 4). At this locality the unconformity between the Sausbee and McCully Formations is well exposed in the cliff face and, as can be seen in Figure 4, it is a sharp and irregularly undulating contact with a relief of up to 3 ft (1 m). Small fractures in the upper surface of the Brewer Bend Member are commonly filled with material from the overlying sequence, and a thin, basal conglomerate, consisting of claystone clasts and scattered limestone pebbles, is commonly present at the base of the overlying formation. This is the only major regional break in deposition found within the Morrowan Series in the area. Its regional analysis is complicated, however, by lateral changes in facies of the units immediately below the erosional surface. It is a continuation of the regional, subaerially developed unconformity at the base of the Dye Shale Member of the Bloyd Shale in Arkansas (Fig. 2).

The McCully Formation in the Arkansas River area consists of about equal percentages of limestone and shale. At the Lock and Dam the thickness of the formation is 64 ft (19.5 m). Sutherland and Henry (1977) named a basal sequence of grainstones and thin interbedded shales the Chisum Quarry Member, an intermediate interval of noncalcareous shale the Shale A Member, and an upper interval of predominantly wackestone the Greenleaf Lake Limestone Member (Fig. 4). This subdivision of the McCully into three members is obscured about 6 mi (10 km) northeast of the Arkansas River by the irregular local development of lenses and layers of limestone within the interval equivalent to the shale A member.

The unconformity at the base of the Atoka Formation is

regionally truncating in northeastern Oklahoma west of Adair County (Fig. 1). The pre-Atokan surface in this area was tilted toward the south, because the higher Morrowan strata are progressively truncated northward. The post-McCully erosional surface is, in fact, distinctly irregular throughout all but the eastern part of the northeastern Oklahoma area, and at least one major pre-Atokan valley, with a relief of as much as 120 ft (37 m) within a lateral distance of less than 5 mi (8 km), is recorded by Sutherland and Henry (1977).

Biostratigraphic Correlations. Detailed regional correlations of the Morrowan sequence have been made between the Arkansas River area (Locality 1) and the type Morrowan area of Washington County, Arkansas, as exemplified particularly by Evansville Mountain (Locality 4) on the basis of conodonts (Lane, 1967; Henry, 1970), goniatites (Gordon, 1970; Saunders and others, 1977), foraminifers (Groves, 1983), and brachiopods (Henry and Sutherland, 1977).

In the Arkansas River area the lowermost fossiliferous strata contain the *Idiognathoides noduliferous* conodont zone, which in Arkansas occupies the lower half of the Prairie Grove Member of the Hale Formation. The middle and upper parts of the Braggs Member contain the *Neognathodus bassleri* conodont zone (unrestricted), which ranges in Arkansas from the middle part of the Prairie Grove Member upward into the Bloyd Shale. Also, the upper third of the Braggs Member and the whole of the overlying Brewer Bend Member in the Arkansas River area contain a variety of common goniatites characteristic of the *Branneroceras branneri* goniatite zone, as found in the Brentwood Limestone Member of the Bloyd Shale in Arkansas. This indicates a correlation of the upper Braggs with the lower part of the Brentwood and the Brewer Bend with the middle and upper part of the Brentwood, plus the Woolsey Member of the Bloyd Shale in Arkansas. Conodonts recovered from the mud-supported limestones of the Brewer Bend are extremely sparse and not diagnostic.

A significant biostratigraphic change in the composition of both the goniatite and brachiopod faunas occurs at the boundary between the Sausbee and McCully Formations. It forms the break between the *Branneroceras branneri* and *Axinolobus modulus* goniatite zones and the boundary between the *Plicochonetes? arkansanus* and *Linoproductus nodosus* brachiopod zones.

It is clear that the McCully Formation correlates biostratigraphically in a general way with the Dye Shale and Kessler Limestone Members of the Bloyd Formation in Arkansas, but none of the fossil groups provides a basis for subdivision of this interval. This is true in part because the section in Arkansas is composed mostly of the unfossiliferous Dye Shale. The *Axinolobus modulus* goniatite zone occurs in both the "caprock" at the base of the Dye Shale and in the overlying Kessler Limestone in Arkansas. In Oklahoma, specimens of *Axinolobus* have been recovered from the Chisum Quarry Member and from the lower part of the shale A member of the McCully Formation, but no goniatites have as yet been recovered from the Greenleaf Lake Limestone.

Locality 2, Cookson (Measured Section 26)

Location. This location comprises roadcuts on the east side of Oklahoma 82, 0.4 to 0.6 mi (0.6 to 1.0 km) north of the village of Cookson, Oklahoma; it begins in a ditch opposite the entrance to Fort Chickamauga and extends for about 0.25 mi (0.4 km) southwestward to the top of a hill: SE¼,Sec.1,T.14N.,R.22E., Cookson Quadrangle, Cherokee County, Oklahoma.

Locality 3, Elk Creek (Measured Section 79)

Location. This location is on the north side of Elk Creek near the steel-girder bridge on Oklahoma 82, 2.2 mi (3.5 km) south of the junction with Oklahoma 100 and 1.2 mi (2 km) north of Locality 2: NW¼,SE¼,Sec.31,T.15N.,R.23E., Cookson Quadrangle, Cherokee County, Oklahoma. The base of the section starts at top of Hindsville exposed on a bluff on the north side of Elk Creek about 900 ft (275 m) southeast of the bridge; the top of unit 4 is exposed in a roadcut on the southeast side of Oklahoma 82 just northeast of the bridge; unit 5 is covered by the highway; units 6 through 11 are exposed partly in roadcuts on the northwest side of Oklahoma 82 and up to 0.2 mi (0.3 km) to the northwest of the highway.

Description. The Elk Creek (Locality 3) and Cookson (Locality 2) sections, 1.2 mi (2 km) apart, together form an important reference section for the northern Lake Tenkiller Reservoir area (Fig. 2). They are 20 mi (32 km) northeast of the type section for the Sausbee and McCully Formations on the Arkansas River and 25 mi (40 km) west of Evansville Mountain, the primary reference section for the type Morrowan Series in Washington County, Arkansas (Fig. 2).

At Elk Creek the Pitkin Limestone is missing and the Sausbee Formation rests directly on the weathered surface of the Fayetteville Formation. There was obviously a depression on the pre-Morrowan erosional surface, which is indicated not only by the local removal of the Pitkin but also by the deposition in this local area of the thick cross-bedded sandstone of unit 4 (compare with Sutherland and Henry, 1977, Fig. 3). The correlation of the upper part of the Elk Creek with the lower part of the Cookson section is based partly on general lithologic similarity but, more importantly, on the lowest occurrence in both sequences of the important conodont form-species *Neognathodus symmetricus.* A composite thickness for the Sausbee and McCully Formations is 209 ft (64.7 m).

The Braggs Member of the Sausbee at Elk Creek and Cookson is equivalent to the Prairie Grove Member of the Hale and the lower part of the Brentwood Limestone Member of the Bloyd at Evansville (Fig. 2). This comparison brings out the striking lithologic differences between the Braggs at Cookson, in which shale is the most important lithic constituent, and the Prairie Grove at Evansville, where shales are lacking. The Prairie Grove Member in Washington County, Arkansas, is a prominant ridge-forming unit made up mostly of sandstone and calcareous sandstone. This member can be recognized westward into Oklahoma across most of Adair County (Fig. 1), but beginning in the western part of that county the sandstones are rapidly replaced westward by shales and thin limestones. The topographic expression of the Prairie Grove, which provides the primary mapping surface in Arkansas, is lost (see discussion at Locality 4).

At Cookson the Brewer Bend Limestone Member is only about 10 ft (3 m) thick in contrast to 30 ft (9 m) or more in the Arkansas River area. The unit is well exposed in the Cookson section as a bench on the hillside just east of Arkansas 82. The basic lithic type in the northern Tenkiller Reservoir area, which includes Cookson, is algal wackestone, but the percentage of algae is appreciably lower in this limestone than in that to the southwest. The member in the Cookson area is sparsely fossiliferous and has low biotic diversity, in marked contrast to the profusion and diversity of faunal elements in the Arkansas River area. Desiccation cracks are abundant. The unit is interpreted in the northern Tenkiller Reservoir area as a shoaling tidal flat area developed at or near sea level many miles away from shoreline. Circulation was relatively poor compared to that seen nearer to the shelf margin (Arkansas River area). One of the few fossils to occur fairly commonly is the colonial rugose coral *Petalaxis.* These colonies, locally more than 2 ft (0.6 m) in diameter, occur in clusters, in growth position on the algal limestone surfaces.

The Brewer Bend Limestone Member loses its distinctive character to the east and is replaced in central Adair County by marine shale at the top of the Brentwood. The same stratigraphic position is occupied in extreme eastern Adair County and in Washington County, Arkansas, by the nonmarine Woolsey Member of the Bloyd Shale that marks the land margin.

The thin basal conglomerate of the McCully Formation at Cookson has produced a specimen of the goniatite *Axinolobus,* which confirms its correlation with the "caprock" at the base of the Dye Shale at Evansville Mountain. Because of the thinness of the McCully Formation at Cookson, it seems doubtful that the limestone layers at the top of the roadside exposure (units 25 and 27) represent the Greenleaf Lake Limestone. More likely these are limestone layers within the shale A member, and the Greenleaf has been removed by erosion below the unconformity at the base of the overlying Atoka Formation.

Locality 4, Evansville (Measured Section 70)

Location. This location is composed of roadcuts on the west side of Arkansas 59, extending from 4.0 mi (6.4 km) (base of section at southern boundary of Washington County) to 2.7 mi (4.3 km) (in the gap near the crest of Evansville Mountain) south of Evansville, Arkansas: S½Sec.26 and Sec.35,T.13N.,R.33W., Evansville Quadrangle, Washington County, Arkansas. The section was measured as a series of offsets along a distance of 1.3 mi (2.1 km) as the road climbs up the side of Evansville Mountain.

Description. The Evansville Mountain section is included primarily as a basis for comparison with the Oklahoma sequences. The Morrowan Series in northwestern Arkansas is subdivided into the Hale and Bloyd Formations. This division resulted because the upper part of the Hale (Prairie Grove Member) is commonly a cliff-forming sequence and provides, in this area of normally poor exposures, the only readily recognizable and easily mappable feature within the Morrowan Series. The Hale and Bloyd Formations can be recognized in Oklahoma westward only as far as central to western Adair County (see the discussion under Localities 2 and 3).

The type localities for the Hale and Bloyd Formations and their members are scattered in Washington County, but the Evansville Mountain section constitutes, for all practical purposes, the primary overall reference section for the Morrowan Series. All of the Morrowan units are well exposed except for the middle part of the Dye Shale (Fig. 2).

The Hale Formation at Evansville Mountain, 136.5 ft (41.6 m) is subdivided into the Cane Hill and Prairie Grove Members. The Cane Hill, 37.5 ft (11.4 m) thick, consists of dark-gray, silty shale interlaminated locally with thin-bedded fine-grained sandstone and siltstone. The Cane Hill thins westward and strata of equivalent age are missing in Oklahoma except near the state line. The Prairie Grove Member, 99 ft (30.2 m) thick, consists of irregularly alternating, fine- to medium-grained sandstone, calcareous sandstone and sandy skeletal and/or oolitic grainstone. The massive beds of this member are commonly cross-laminated, intensely burrowed sandstone that has coarse pits on weathered surfaces.

The Bloyd Formation is 204.5 ft (62.3 m) thick at Evansville Mountain. The formation is subdivided, in ascending order, into the Brentwood Limestone, Woolsey, Dye Shale, and Kessler Limestone Members. The base of the Brentwood is gradational with the subjacent Prairie Grove Member and is marked by the change in lithology from cliff-forming sandstones to thin-bedded shales and limestones. A whole spectrum of limestone types is represented in the Brentwood, the most common being mixed skeletal grainstone and packstones, which may have an appreciable content of quartz sand and silt. The sequence is highly variable laterally in lithic character. The Brentwood is the most fossiliferous unit of the Morrowan Series in Washington County.

The Brentwood Limestone Member is overlain by the nonmarine Woolsey Member throughout much of Washington County. It includes the Baldwin Coal bed and is as much as 45 ft (13.7 m) thick in the central part of the county. It thins westward and is only 5.5 to 7 ft (1.7 to 2.1 m) thick at Evansville Mountain. Toward the southwest across western Washington County, the lower part of the nonmarine Woolsey Member is progressively replaced by the marine Brentwood facies. Locally, in some parts of the county, the Woolsey Member disconformably overlies the Brentwood with a basal conglomerate. The preserved western limit of the nonmarine facies is less than 1 mi (1.6 km) west of the Oklahoma state line. West of that area the entire sequence is marine.

The Dye Shale Member of the Bloyd is characterized by dark-gray shale. The base of this member, resting unconformably on the underlying Woolsey Member throughout Washington County, is marked by the "caprock of the Baldwin coal" (Henbest, 1953). This member is the basal calcareous sandstone and sandy grainstone unit of a transgressive marine sequence, and this regional unconformity is the same as that that separates the Sausbee and McCully Formations in northeastern Oklahoma (Fig. 2) (see discussion under Locality 1).

The Kessler Limestone Member is well exposed in the roadcuts at the crest of the gap in Evansville Mountain, 2.7 mi (4.4 km) south of Evansville, Arkansas. It is highly variable laterally in lithic character but is composed typically of algal and mixed skeletal wackestone and packstone, algal oncolith-bearing packstone and oolitic grainstone. The unit is 13 ft (4 m) thick at Evansville Mountain but varies in observed thickness from 9 to 32 ft (2.7 to 9.8 m) across Washington County.

The contact between the Kessler Limestone and the overlying Trace Creek Shale is well exposed in the section at Evansville Mountain. The contact is sharp and undulating, and small layers below the contact are locally truncated. This contact is believed to represent a regional unconformity. Sutherland and Grayson (1978) reported the occurrence of a conglomerate at the base of the Trace Creek at other localities in Washington and Adair Counties and the occurrence of Atokan conodonts within 3 ft (1 m) of the base of the Trace Creek elsewhere in Washington County. The Trace Creek Shale has traditionally been placed in the Morrowan Series in Washington County but Sutherland and others (1978) presented evidence that the regionally truncating unconformity at the base of the Atoka Formation in northeastern Oklahoma passes eastward into the regionally present disconformity at the base of the Trace Creek Shale, and that the Trace Creek is a lateral facies of the Atoka in Oklahoma (Fig. 2). They transferred the Trace Creek Shale from the Bloyd Formation to the Atoka Formation.

Regional Analysis. The strikingly different facies to be seen at Evansville Mountain, compared to those examined in Oklahoma, result primarily from two factors. First, the Evansville Mountain area was much closer to a terrigenous source, located to the northeast and east, as indicated by the development of the thick sandstones of the Prairie Grove Member of the Hale Formation in Arkansas. These change facies to shale and thin limestones west of Adair County in Oklahoma. A second factor is the fact that a regional regression during middle Morrowan time produced strikingly different effects in the two areas. In the Washington County area in Arkansas the shallow marine facies represented by the Brentwood Limestone Member gave way to the nonmarine facies and coal beds of the Woolsey Member of the Bloyd. However, the Arkansas River area in Oklahoma (Locality 1) lay much farther out on the shallow marine shelf, and the drop in sea level there resulted at first in shallowing but not in emergence. The open marine shallow shelf facies of the Braggs Member of the Sausbee Formation gradually changed to an intertidal algal bank facies represented by the Brewer Bend Member. As the regional regression continued, both regions became exposed and eroded, resulting in the only regional unconformity within the Morrowan Series (Sutherland and Henry, 1977).

The marine transgression above the unconformity distributed muds across the entire region (Dye Shale in Arkansas) but the Arkansas River region in Oklahoma, farther from the terrigenous source, continued with the development of a higher percentage of shallow water limestones than developed farther to the east. The depositional relationship of the Greenleaf Limestones in Oklahoma to that of the Kessler Limestone in Arkansas has not as yet been established. They may or may not be equivalent units.

REFERENCES CITED

Bonem, R. M., 1977, Stratigraphic setting, carbonate lithofacies, and development of Lower Pennsylvanian bioherms of northeastern Oklahoma, *in* Sutherland, P. K., and Manger, W. L., eds., Upper Chesterian-Morrowan Stratigraphy and the Mississippian Pennsylvanian Boundary in Northeastern Oklahoma and Northwestern Arkansas: Oklahoma Geological Survey Guidebook 18, p. 55–59.

Gordon, M., Jr., 1970, Carboniferous ammonoid zones of the south-central and western United States: Sheffield, Sixth International Congress of Carboniferous Stratigraphy and Geology, Compte Rendu, v. 2, p. 817–826.

Groves, J. R., 1983, Calcareous Foraminifers and Algae from the Type Morrowan (Lower Pennsylvanian) Region of Northeastern Oklahoma and Northwestern Arkansas: Oklahoma Geological Survey Bulletin 133, 65 p.

Henbest, L. G., 1953, Morrow group and lower Atoka formation of Arkansas: American Association of Petroleum Geologists Bulletin, v. 37, p. 1935–1953.

Henry, T. W., 1970, Conodont Biostratigraphy of the Morrow Formation (Lower Pennsylvanian) in Portions of Cherokee, Sequoyah, Muskogee, and Adair Counties, Northeastern Oklahoma [M.S. thesis]: Norman, University of Oklahoma, 170 p.

Henry, T. W., and Sutherland, P. K., 1977, Brachiopod biostratigraphy of Morrowan Series (Pennsylvanian) in northwestern Arkansas and northeastern Oklahoma, *in* Sutherland, P. K., and Manger, W. L., eds., Upper Chesterian–Morrowan Stratigraphy and the Mississippian–Pennsylvanian Boundary in Northeastern Oklahoma and Northwestern Arkansas: Oklahoma Geological Survey Guidebook 18, p. 107–116.

Lane, H. R., 1967, Uppermost Mississippian and Lower Pennsylvanian conodonts from the type Morrowan region, Arkansas: Journal of Paleontology, v. 41, p. 920–942.

Saunders, W. B., Manger, W. L., and Gordon, M., Jr., 1977, Upper Mississippian and Lower and Middle Pennsylvanian ammonoid biostratigraphy of northern Arkansas, *in* Sutherland, P. K., and Manger, W. L., eds., Upper Chesterian–Morrowan Stratigraphy and the Mississippian–Pennsylvanian Boundary in Northeastern Oklahoma and Northwestern Arkansas: Oklahoma Geological Survey Guidebook 18, p. 117–137.

Sutherland, P. K., 1977, Stop descriptions—second day, *in* Sutherland, P. K., and Manger, W. L., eds., Upper Chesterian–Morrowan Stratigraphy and the Mississippian–Pennsylvanian Boundary in Northeastern Oklahoma and Northwestern Arkansas: Oklahoma Geological Survey Guidebook 18, p. 117–137.

Sutherland, P. K., and Grayson, R. C., Jr., 1978, Redefinition of the Morrowan Series (Lower Pennsylvanian) in its type area, northwestern Arkansas: Geological Society of America Abstracts with Programs, v. 10, p. 501.

Sutherland, P. K., and Henry, T. W., 1977, Carbonate platform facies and new stratigraphic nomenclature of the Morrowan Series (Lower and Middle Pennsylvanian), northeastern Oklahoma: Geological Society of America Bulletin, v. 88, p. 425–440.

Sutherland, P. K., Grayson, R. C., Jr., and Zimbrick, G. D., 1978, Reevaluation of Morrowan Atokan Series boundary in northwestern Arkansas and northeastern Oklahoma: American Association of Petroleum Geologists Bulletin, v. 62, p. 566.

Deltaic facies of the Hartshorne Sandstone in the Arkoma Basin, Arkansas-Oklahoma border

David W. Houseknecht, Department of Geology, University of Missouri, Columbia, Missouri 65211

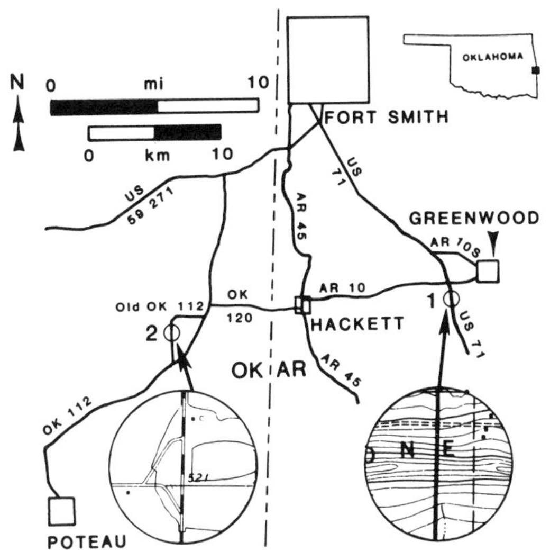

Figure 1. Highway map showing locations of two Hartshorne outcrops discussed in text. Note that Greenwood topographic map predates roadcut of locality 1.

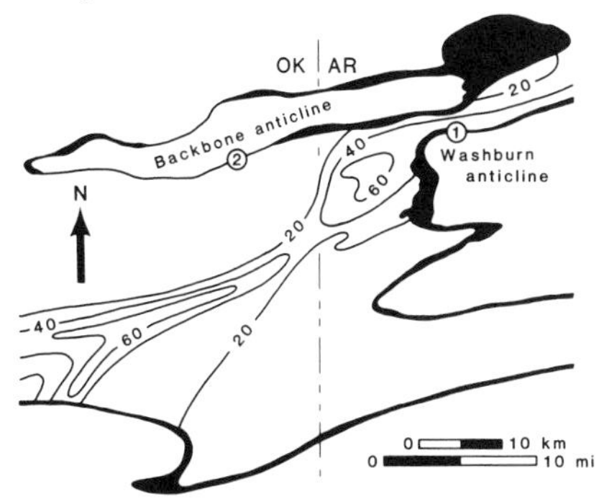

Figure 2. Sandstone isolith map of Hartshorne Sandstone illustrating field trip localities (numbered circles) relative to linear sand body. Black indicates Hartshorne outcrops; isolith contours are in meters (from Houseknecht and others, 1983).

LOCATION

Sedimentary facies deposited in channel and interchannel positions within an actively prograding, river-dominated delta are illustrated at two localities near the Arkansas-Oklahoma border south of Fort Smith, Arkansas.

Locality 1 (94°17′20″W.; 35°11′41″N.; Greenwood, Arkansas, Quadrangle) is a distinctive, terraced roadcut on U.S. 71, approximately 12 mi (20 km) south of Fort Smith (Fig. 1). The Hartshorne Sandstone dips steeply north (nearly vertical) along the north limb of the Washburn anticline (Fig. 2).

Locality 2 (94°31′20″W.; 35°10′30″N.; Spiro, Oklahoma, Quadrangle) is a roadcut on old Oklahoma 112, approximately 8 mi (12 km) west of Hackett, Arkansas (Fig. 1). The Hartshorne Sandstone dips 15° southward along the southern limb of the Backbone anticline (Fig. 2).

SIGNIFICANCE

From late Cambrian through early Pennsylvanian time, the area now occupied by the Arkoma basin was a passive continental margin dominated by carbonate sedimentation. During the Atokan, convergent tectonism along the Ouachita orogenic belt resulted in the formation of the Arkoma foreland basin, and detrital sedimentation prevailed in this basin through the Desmoinesian. Bordered by the Ozark uplift to the north and the Ouachita orogenic highlands to the south, the Arkoma basin evolved into a typical molasse environment; shallow marine and coal-bearing deltaic and fluvial facies dominate the Desmoinesian stratigraphic section (Houseknecht, 1986).

Within this foreland basin setting, the Hartshorne Sandstone was deposited in a river-dominated delta system, the seaward margin of which was modified primarily by tidal processes; this delta system migrated longitudinally westward through the basin (Houseknecht and others, 1983). The combination of fluvial domination and tidal reworking of the delta front resulted in the accumulation of elongate, linear sandstone bodies composed primarily of distributary channel facies. In addition, the development of marked lateral facies changes was dependent on proximity to the mouth of an active distributary channel.

Figure 2 shows the distribution of a Hartshorne channel deposit that trends southwestward and bifurcates near the western edge of the map. Paleocurrent data collected at both the eastern and western outcrops of that sandstone body indicate that paleoflow was westward (Houseknecht and others, 1983). The two outcrops spotted in Figure 2 and described below have been selected to illustrate the similarities and differences between facies deposited in channel (Locality 1) and interchannel (Locality 2) positions within this deltaic environment.

LOCATIONS

Locality 1

This roadcut exposes a 450-ft (137-m) section that represents a complete vertical facies sequence deposited in a channel position of a prograding delta (Fig. 3).

The lower 70 ft (20 m) (south end of outcrop) displays a generally coarsening upward sequence from shale to siltstone intercalated with shale and thin, rippled laminae of sandstone. This sequence is interpreted as prodelta to distal bar facies deposited on the aggradational slope in front of the prograding delta.

The overlying 70 ft (20 m) displays two interbedded subfacies. One comprises ripple-bedded sandstone beds that are 1 to 2 ft (30 to 60 cm) thick and intercalated with shale "drapes" that are 1 to 2 cm thick. Both asymmetrical and symmetrical ripples are present, although the former are more abundant; interference ripple-crest patterns are common on exposed bedding planes. The other subfacies comprises sandstone that displays trough cross-bedding in sets 1 to 3 ft (30 cm to 1 m) thick and 15 to 30 ft (5 to 10 m) wide, with unidirectional westward paleoflow indicated. This subfacies typically has an erosional base and commonly contains shale and siderite rip-up clasts. These two subfacies together represent distributary mouth bar deposits. The cross-bedded sand was probably deposited by fluvial currents during periods of high discharge, and the ripple-bedded sand with shale drapes was deposited by tidal and subordinate wave processes during periods of low discharge.

The thickest individual facies exposed in the roadcut comprises sandstone displaying ubiquitous trough cross-bedding. Beds range from 4 in to 6 ft (10 cm to 2 m) in thickness, and cross-beds indicate unidirectional westward paleoflow. At the top, this facies fines upward via a decrease in the thickness of sandstone beds and the appearance and upward increase in thickness of shale interbeds. These facies were deposited in a distributary channel within a prograding delta lobe, with the fining upward sequence representing channel abandonment. The channel geometries are the primary reason for the linear map pattern illustrated in Figure 2.

The upper 130 ft (40 m) of section consists predominantly of silty shale with locally abundant laminae of rippled sandstone, macerated plant debris, horizontal burrows, pelecypod fossils, and rooted horizons. Two coalbeds are also present. Together, these facies represent deposition in interdistributary bays, marshes, and swamps on the delta plain.

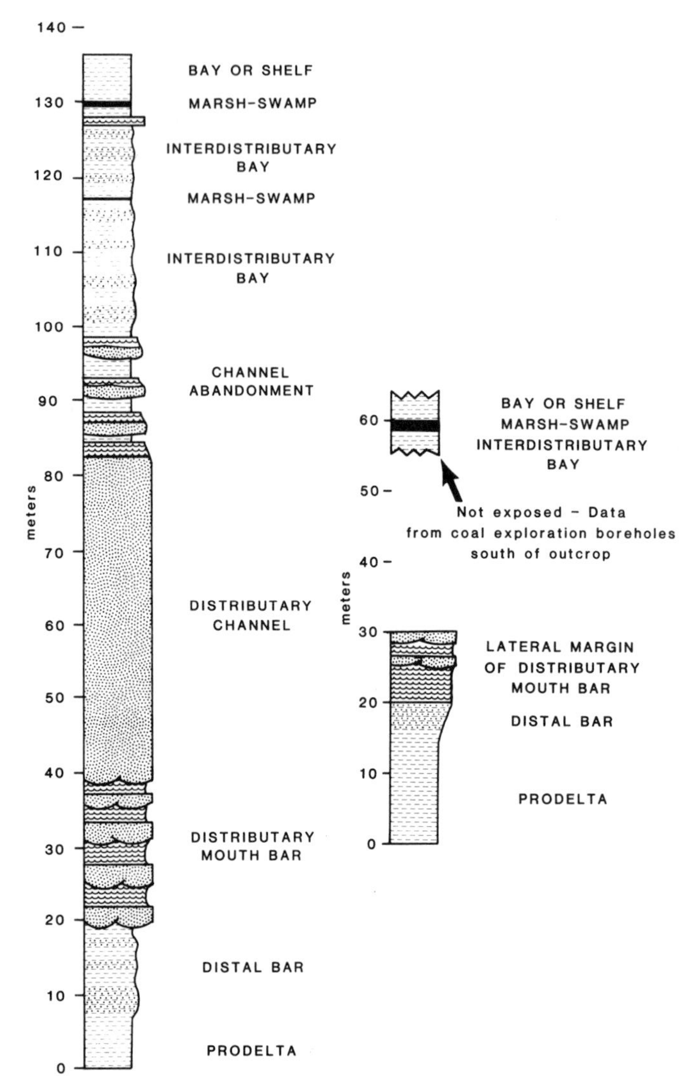

Figure 3. Measured sections for locality 1 (left) and locality 2 (right). Symbols explained in text.

Locality 2

This roadcut exposes 100 ft (30 m) of section deposited in an interchannel position of the same delta whose facies were examined at Locality 1 (Fig. 2).

The entire outcrop is a coarsening upward sequence, grading from shale into ripple-bedded siltstone into ripple-bedded sandstone (Fig. 3). Two sandstone beds near the top of the section display small-scale trough cross-bedding. This facies sequence was deposited in prodelta, distal bar, and distributary mouth bar subenvironments. The overall lack of sand compared to equivalent facies at Locality 1 is the result of deposition in an interchannel position that was well removed from active input of sand from a distributary channel (Fig. 2).

Even though shales and coalbeds of the delta plain are not exposed at this locality, data from coal exploration boreholes just south of this outcrop have been added to Figure 3 to facilitate comparison with Locality 1. The major coal bed is located about 92 ft (28 m) above the top of the coarsening upward sequence exposed in the roadcut, and most of the intervening section is shale. Thus, the total vertical facies sequence, from prodelta to delta-plain coal bed, is about 200 ft (60 m) in the interchannel position compared to 425 ft (130 m) in the channel position (Fig. 3).

REFERENCES CITED

Houseknecht, D. W., 1986, Evolution from passive margin to foreland basin; The Atoka Formation of the Arkoma basin, south-central U.S.A., *in* Allen, P. A., and Homewood, P. (eds.), Foreland Basins: International Association of Sedimentologists, Special Publication No. 8, p. 183–201.

Houseknecht, D. W., Zaengle, J. F., Steyaert, D. J., Matteo, Jr., A. P., and Kuhn, M. A., 1983, Facies and depositional environments of the Desmoinesian Hartshorne Sandstone, Arkoma basin, *in* Houseknecht, D. W., ed., Tectonic–Sedimentary Evolution of the Arkoma Basin: Columbia, Missouri, Mid-continent Section Society of Economic Paleontologists and Mineralogists, p. 53–82.

Carlton rhyolite and lower Paleozoic sedimentary rocks at Bally Mountain in the Slick Hills of southwestern Oklahoma

R. N. Donovan, *Geology Department, Texas Christian University, Fort Worth, Texas 76129*
D. Ragland, *Geology Department, Oklahoma State University, Stillwater, Oklahoma 74078*
K. Cloyd, *Geology Department, State University of New York, Binghamton, New York 13901*
S. Bridges, *Geology Department, Oklahoma State University, Stillwater, Oklahoma 74078*
R. E. Denison, *Dallas Research Lab, Mobil Research and Development Corporation, P.O. Box 819047, Dallas, Texas 734581-9047*

LOCATION

The Bally Mountain Range is located in eastern Kiowa County in southwestern Oklahoma (T.6N.,R.14W.). The closest important road is Oklahoma 58 (Fig. 1). This road can be accessed from the H. E. Bailey Turnpike by taking the Medicine Park exit a few miles north of Lawton. Alternatively, the route can be joined from the north by taking the Hydro exit from I-40. The easiest way of reaching the site is to proceed north on Oklahoma 58 for 3 mi (4.8 km) from its junction with Oklahoma 19, then turn west along a partly paved county road for 3 mi (4.8 km). Just before reaching the most prominent hill in view (which is Bally Mountain), turn right on an unpaved road and proceed north for a few hundred yards to the only inhabited house (on the west side of the road). This is the home of Mr. W. Hodges, the principal landowner in the area, where permission to visit the site should be sought. In common with several other landowners in the area, Mr. Hodges freely allows large parties of students access to his land on condition that they do not smoke or leave litter and that they close all gates, etc. His house is very conveniently situated; immediately to the west is the bulk of Bally Mountain, which is composed of Carlton Rhyolite. This hill is 1,900 ft (580 m) high and rises 350 ft (107 m) above the surrounding plains. To the north and east rises the remainder of the range, which is built mostly of limestones of the Arbuckle Group. The conspicuous serrated ridge ("Ragged Mountain") that runs to the east is formed by the Royer Dolomite.

A full visit to the site requires a good deal of walking with some (optional) scrambling and will take all day. Access to the area is difficult after heavy rain as the county roads become impassable. Beware of rattlesnakes, particularly in the Royer Dolomite terrain. (A student recently trapped 57 rattlesnakes there in three hours.) The quarry area (E2 on Fig. 3) is owned by Mr. D. Leatherbury, of Apache, who has given geologists full permission to visit the site.

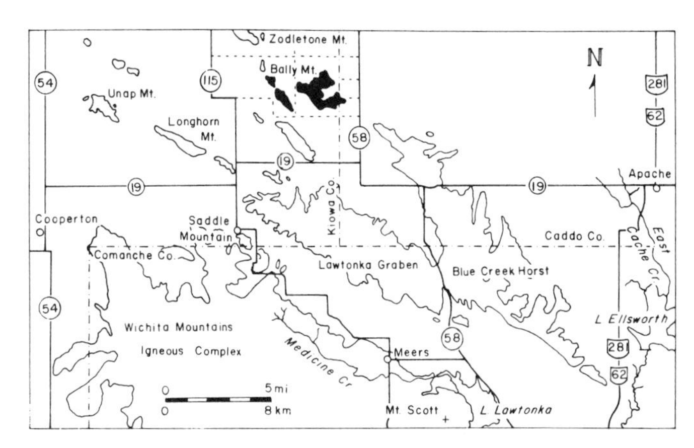

Figure 1. Map showing location of Bally Mountain Range.

SIGNIFICANCE

The area exposes Lower Paleozoic rocks deposited near the axis of a WNW-ESE-trending trough, which underwent relatively rapid subsidence during the Lower Paleozoic. Formerly referred to as the southern Oklahoma geosyncline, this trough is now most commonly explained as recording a period of thermally controlled subsidence during the development of the Southern Oklahoma aulacogen.

Bally Mountain itself is the type section of the Carlton Rhyolite; a thickness of 3,600 ft (1,100 m) can be inspected (Ham and others, 1964). The rhyolite lies unconformably beneath the sandstones and limestones of the Timbered Hills Group, which in turn underlies the Arbuckle Group. The West Springs Creek Formation tops the latter group and was eroded in Permian times; otherwise the area offers the most complete and easily accessible exposure of the group in Oklahoma. Counting the rhyolite, Bally Mountain displays an apparently complete section of c. 9,000 ft (2,750 m) of Lower Paleozoic rocks. The Bally Mountain Range is part of a Permian landscape that is being exhumed in this part of Oklahoma. Permian conglomerates and karst-fills in the limestones help to confirm that these are ancient hills formed of ancient rocks, not modern hills formed of ancient rocks.

SITE INFORMATION

The Carlton Rhyolite Group. The Carlton Rhyolite Group is part of an enormous acidic igneous complex of Cambrian (c. 525 Ma) age, which formed during the intiation of the southern Oklahoma aulacogen. The total volume of magma was enormous—perhaps 6,000 mi^3 (40,000 km^3) (Gilbert in Gilbert and Donovan, 1982). The rhyolites are the extrusive equivalent of the Wichita granites, the two rock groups having virtually identical chemistry. The maximum known thickness of rhyolite is at least 4,500 ft (1,370 m), and the volcanic rocks are believed to

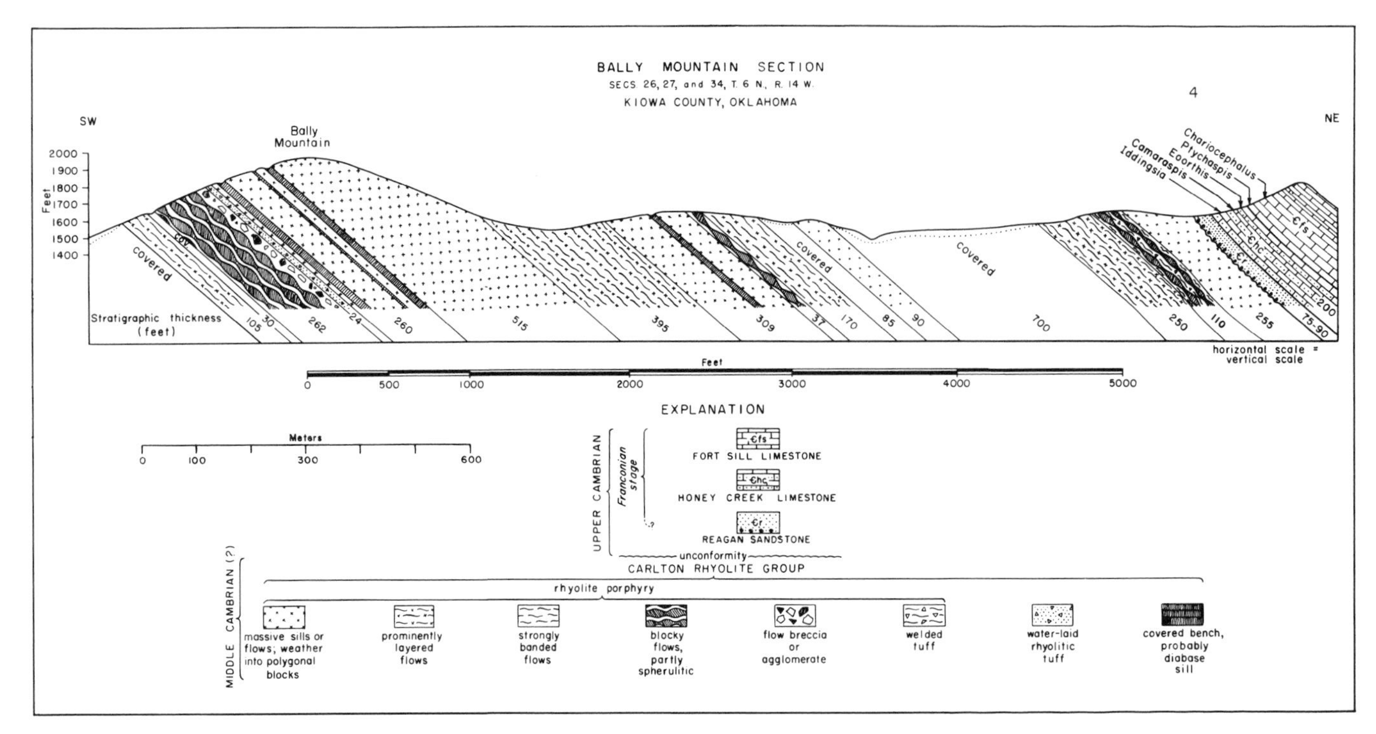

Figure 2. Measured section of Carlton Rhyolite Group at Bally Mountain (modified from Ham and others, 1964). Lower part of sedimentary-rock sequence is present at right side of section.

occupy an area of 17,000 mi^2 (44,000 km^2) in southern Oklahoma.

Ham and others (1964) located the type section of the Carlton Rhyolite Group on Bally Mountain. Here a mostly unbroken succession of 3,600 ft (c. 1,100 m) of flows, welded tuffs, water lain tuffs, and flow breccias is displayed in an easily accessible fashion (Fig. 2). Some flows are spherulitic, others are ignimbrites and show flow banding (e.g., Fig. 3, A1) while most are massive and essentially featureless.

The Unconformity Between the Carlton Rhyolite and Timbered Hills Groups. By Late Cambrian time (after an interval of perhaps 15 m.y. or less), the rhyolites had weathered to a range of low hills. This topography was gradually transgressed during Franconian times; the initial deposits were siliciclastics (the Reagan Formation); subsequently the limestones of the Honey Creek Formation formed. At this later period the higher parts of the Rhyolite topography formed an archipelago (Donovan, 1982; Donovan, in Gilbert and Donovan, 1984). A superb example of a Franconian "island" can be seen at B1 (Fig. 3); not only is the Reagan absent but a good part of the Honey Creek is missing too.

The unconformity is not profound, particularly when compared to that at the base of the Paleozoic transgression over much of the United States. The length of the unconformity is short, the topographic relief is modest, and there is only a mild angular unconformity (c. 5°) between igneous and overlying sedimentary beds.

The Timbered Hills Group. The Reagan Formation can conveniently be examined at C1 (Fig. 3) where it consists of a basal breccia of rhyolite clasts, a few feet of tan-colored lithic sandstones, and a topmost unit of about 30 ft (9 m) of dark green glauconitic sandstone (in places glauconite comprises 60% of the constituent grains). More than 200 ft (60 m) of the lower Reagan is not present (Tsegay, 1983); this is not surprising in view of the proximity of Franconian "land" (Fig. 3, B1). The upward passage from Reagan into Honey Creek is marked by orange-weathering stringers and thin beds of ankerite (Cloyd and Rafalowski, 1974). This orange-weathering zone is everywhere a distinctive feature at the contact between the two formations and appears to be a diagenetic product of the reaction between glauconite and carbonate-rich groundwaters related to the overlying limestone. The Honey Creek is an impure coarse-grained, cross-bedded, bioclastic limestone composed of vast amounts of broken pelmatozoans, subordinate robust-shelled trilobites, some brachiopods (including phosphatic inarticulates), and quartz and glauconite grains. It appears to have formed as a stacked series of tidally influenced sandbars (Rafalowski, 1984).

Point C1 is a convenient place to notice the "double unconformity" effect, which is found throughout the Slick Hills. Not only does the Reagan onlap the Carlton Rhyolite; it and the succeeding formations are onlapped by the Permian Post Oak conglomerate facies (Fig. 3).

The Arbuckle Group. About 5,000 ft (c. 1,500 m) of limestones assigned to the Arbuckle Group are well exposed in a

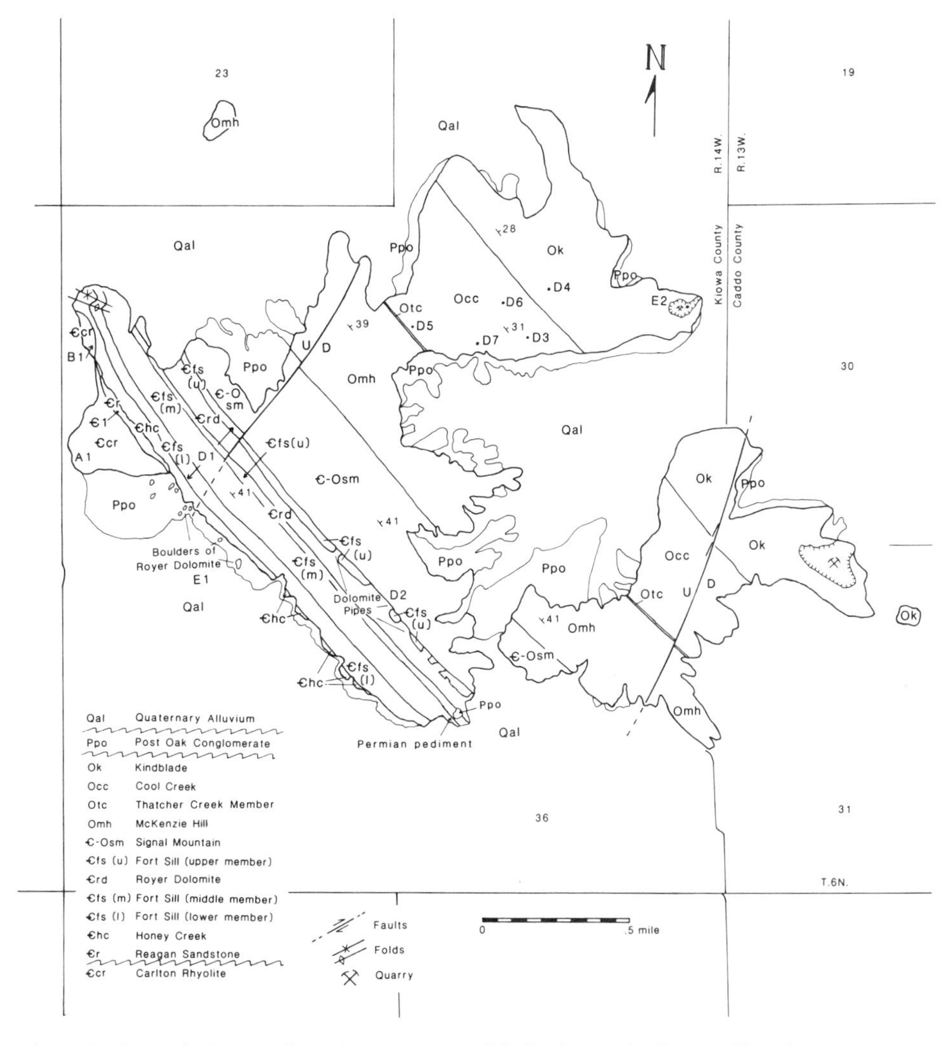

Figure 3. Geological map of northeastern part of Bally Mountain Range. Sites A1–E2 are areas highlighted in text.

continuous section that dips to the northeast at 25–45°. Various parts of this section have been described by workers from several areas in the Slick Hills (e.g., Barthelman, 1968; Brookby, 1969; Stitt, 1977; Donovan, 1982; Stitt, 1983; Ragland, 1983; Rafalowski, 1984; Ditzell, 1984). Fig. 4 is a compilation of data from these sources and our own work, which is applicable to the Slick Hills as a whole. Obviously it is impossible to describe this great section of platform carbonates in detail here. This account simply presents some of the highlights on Bally Mountain. The student of platform carbonates will undoubtedly spend at least a day on this section.

Throughout the Slick Hills, three informal members can be recognized in the Fort Sill Formation: (1) a basal thinly bedded limestone sequence dominated by mudstones but with important packstones and grainstones; (2) a middle sequence consisting of alternations of lime mudstones with very thinly bedded dolomitic siltstones and (3) a topmost sequence of more massive limestones containing abundant algal boundstones (D1, Fig. 3). On Bally Mountain and adjacent Zodletone Mountain (but not elsewhere in the Slick Hills), the central portion of the upper member has been comprehensively dolomitized to a coarsely crystalline texture. A similar dolomite, the Royer, occupies a close but not identical stratigraphic position in the Arbuckle Mountains. The same name has been applied here, although the two occurrences are not stratigraphically continuous.

For the most part, the Bally Mountain dolomite maintains a stratigraphic integrity, but toward the eastern end of the exposure (Fig. 3, D2), discrete "pipe"-like zones of complete dolomitization extend from the Royer to the top of the Fort Sill (but not into the overlying Signal Mountain Formation). The boundaries of the "pipes" appear to coincide with north-east-trending joints; the dolomite is particularly coarse grained and contains vugs lined by baroque crystals.

The overlying Signal Mountain is the most consistently

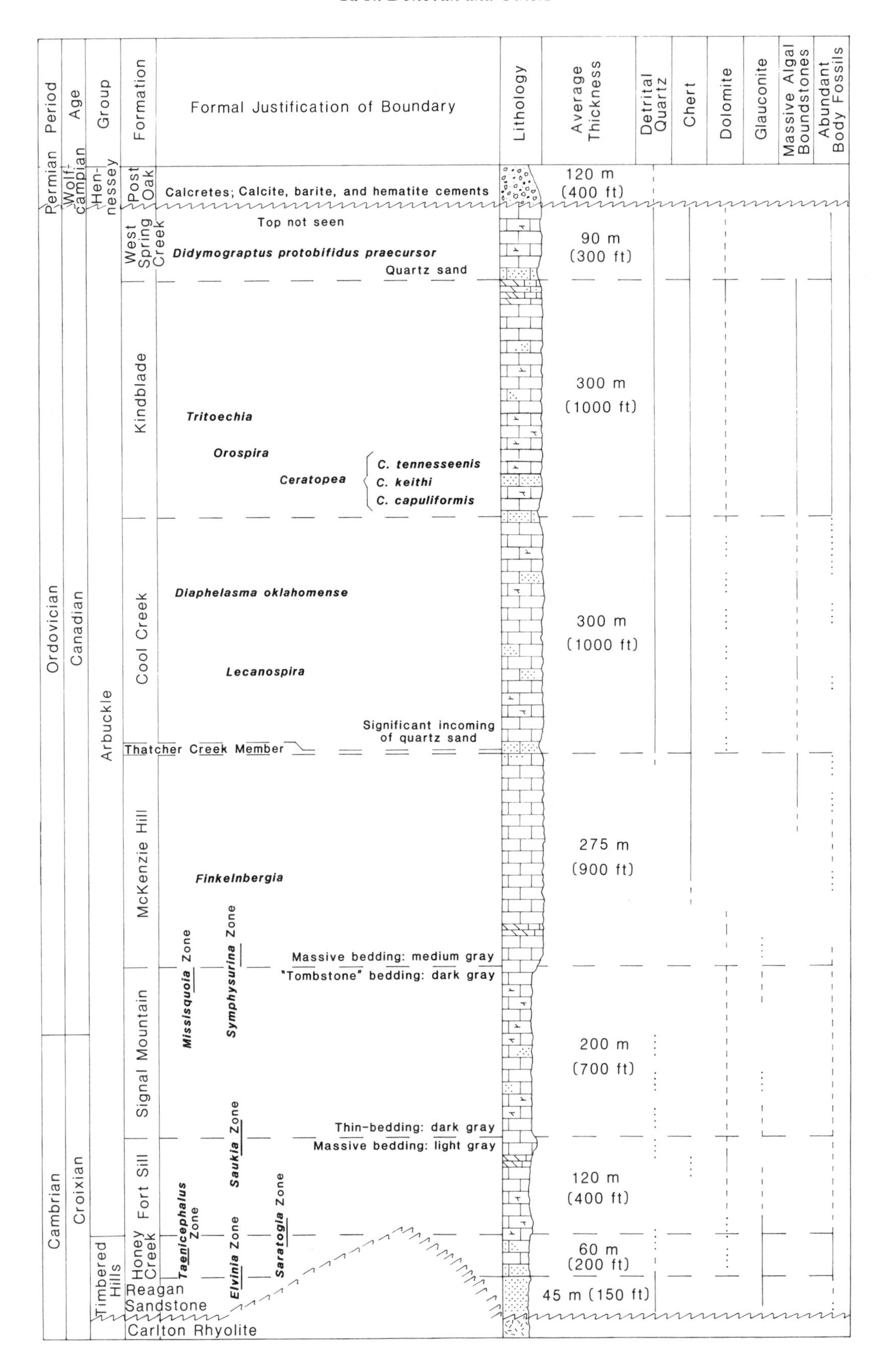

Figure 4. General log of Timbered Hills and Arbuckle groups as exposed in the Slick Hills.

thinly bedded formation in the Arbuckle Group. It is particularly fossiliferous and contains an abundant fauna of trilobites, gastropods, and brachiopods (Stitt, 1977, 1983). In fact the whole Arbuckle Group is highly fossiliferous, with the exception of the Cool Creek Formation where fossils are more scarce. In general the younger formations show a greater faunal diversity. However it should be noted that many bed boundaries throughout the group have been stylolitized and, as a practical result, fossils are not easy to spot. On the other hand, particularly in the Kindblade and upper Cool Creek Formation, some fossils are highly conspicuous because they have been silicified.

Throughout the Arbuckle Group the dominant lithologies are a textural spectrum of clastic limestones, from intraformational conglomerates (IFCs) to grainstones and packstones to wackestones and mudstones (many of the finer lithologies are bioturbated). IFCs are a very common lithology, those in the Signal Mountain and Cool Creek formations (Fig. 3, D3) are particularly well exposed. The textural immaturity of many of these limestones suggests a storm-affected environment. A constant interplay between fair-weather deposition, hard-ground cementation, and storm-induced erosion is suggested by the fact that lime mud intraclasts are the most common allochem, followed by peloids, fossil trash, and ooids. The latter are developed best (but not exclusively) in the Cool Creek and Kindblade formations, where they form the dominant constituent in cross-bedded grainstones (which presumably record migrating ooid shoals). Some herringbone cross-bedding (Fig. 3, D4) in these grainstones suggest tidal influence.

Detrital quartz is a component of the Fort Sill Formation but is largely absent from the Signal Mountain and McKenzie Hill Formations. It occurs conspicuously at the base of the Cool Creek Formation (the Thatcher Creek member of Ragland and Donovan, 1985; Fig. 3, D5) and at several higher horizons throughout the remainder of the Group. All of these quartz-rich units are similar in that they consist of cross-bedded mixed carbonate-quartz sandstones, which appear to have formed in small erosive-based channels.

The Arbuckle Group contains a spectacular assemblage of algal boundstones, particularly in the topmost member of the Fort Sill Formation, the upper part of the McKenzie Hill Formation, and throughout the Cool Creek and Kindblade Formations. No algal boundstones have been observed in the Signal Mountain Formation. The Kindblade boundstones include complex edifices that involve framework builders other than blue green algae (e.g., *Archaeoscyphia, Pulchrilamina;* Toomey and Nitecki, 1979). However, the greatest diversity of external form and internal construction in algal boundstones is seen in the Cool Creek Formation (Fig. 3, D6). External forms include encrustations, mats, mounds, and reefs; internal organization into varieties of stromatolite and thrombolite is apparent.

The boundstone varieties in the Cool Creek Formation may have been controlled by fluctuations in water depth. The same formation also contains evidence of emergence in the form of stratiform dolomite, which is intimately associated with cherts of the "cauliflower" type (Fig. 3, D7). The cherts contain inclusions of both anhydrite and celestite.

In addition to the "cauliflower" textures, cherts of several external forms (nodules of various forms and thinly laminated beds) are present from the upper part of the McKenzie Hill Formation to the top of the group. Some of the silicification is fabric selective (e.g., in burrows), and some is not, while part of the chert is a cement rather than a replacement. In addition, some of the cherts in the McKenzie Hill Formation are recognizable as siliceous sponges.

The Permian Imprint on Bally Mountain. As noted, Bally Mountain is an exhumed Permian Hill. It is surrounded by its detrital products (the "Post Oak Conglomerate"). Two aspects of the Permian imprint are particularly interesting. At the southwestern edge of the hill (Fig. 3, E1) house-size boulders of Royer Dolomite have clearly tumbled from the outcrop on the hill above. Some of these boulders may have fallen recently, but at least two of them are embedded in the calcrete that cemented the Post Oak Conglomerate. These huge boulders may represent sporadic rock falls; on the other hand, they may record a similar seismic effect to that documented in the nearby Permian of the Meers Valley (Donovan and others, 1985). At the southeastern edge of the "Royer" ridge, a small deposit of Post Oak sits on what appears to be a Permian pediment.

The second Permian highpoint is located in a quarry near the northeastern end of Bally Mountain (E2, Fig. 3). The rock quarried is the Kindblade Formation, but the geological interest in the quarry centers on the Permian karst fissures that follow fractures in the limestones. These fissures are securely dated by an abundant (although fragmented) vertebrate fauna documented by Simpson (1979). The vertebrates are mixed with laminated green clays and poorly sorted detritus (chert and limestone pebbles) derived from the Kindblade Formation. Associated karst deposits—which are earlier than the vertebrate-bearing detritus—include flowstones, cave pearls, and horizontally interlaminated beds of clay and travertine that show some soft-sediment deformation (Bridges, 1985). The most intriguing feature of the travertine is that the fibrous calcite crystals are interlaminated with hydrocarbon. This would appear to date a phase of hydrocarbon migration through these beds in Permian (specifically Leonardian) time.

SUMMARY

The geology of the Bally Mountain Range records in a single conveniently accessible exposure several aspects of the development of the southern Oklahoma aulacogen. These are:

1) The initial Cambrian igneous phase as manifested by the Carlton Rhyolite.

2) Deposition of platform carbonates in a Cambro-Ordovician subsidence phase associated with a decreasing geothermal gradient.

3) Pennsylvanian deformation, which in this area has produced gentle homoclinal tilting to the northeast.

4) Permian sculpting of the resulting topography.

REFERENCES CITED

Barthelman, W. B., 1968, Upper Arbuckle (Ordovician) outcrops in the Unap Mountain–Saddle Mountain area, northeastern Wichita Mountains, Oklahoma [M.S. thesis]: Norman, University of Oklahoma, 65 p.

Bridges, S. D., 1985, Mapping, stratigraphy, and tectonic implications of Lower Permian strata, eastern Wichita Mountains, Oklahoma [M.S. thesis]: Stillwater, Oklahoma State University, 129 p.

Brookby, H. E., 1986, Upper Arbuckle (Ordovician) outcrops in Richards Spur–Kindblade Ranch area, northeastern Wichita Mountains, Oklahoma [M.S. thesis]: Stillwater, Oklahoma State University, 165 p.

Ditzell, C., 1984, The sedimentary geology of the Cambro-Ordovician Signal Mountain Formation as exposed in the Wichita Mountains of southwestern Oklahoma [M.S. thesis]: Stillwater, Oklahoma State University, 165 p.

Donovan, R. N., 1982, Geology of Blue Creek Canyon, *in* Gilbert, M. C., and Donovan, R. N., eds., Geology of the eastern Wichita Mountains, southwestern Oklahoma: Oklahoma Geological Survey Guidebook, 21, p. 65–77.

—— , 1984, The geology of the Blue Creek Canyon area, *in* Gilbert, M. C., and Donovan, R. N., eds., Recent developments in the Wichita Mountains: Richardson, Texas, 18th Annual Meeting, South-Central Section Geological Society of America, Guidebook to Field Trip 1, p. 40–101.

Gilbert, M. C., 1984, Geological setting of the southern Oklahoma aulacogen, *in* Gilbert, M. C., and Donovan, R. N., eds., Recent developments in the Wichita Mountains: Richardson, Texas, 18th Annual Meeting, South-Central Section Geological Society of America, Guidebook to Field Trip 1, p. 1–40.

Ham, W. E., Denison, R. E., and Merrit, C. A., 1964, Basement rocks and structural evolution of southern Oklahoma: Oklahoma Geological Survey Bulletin 95, 302 p.

Rafalowski, M., The sedimentary geology of the Late Cambrian Honey Creek and Fort Sill formations as exposed in the Slick Hills of southwestern Oklahoma [M.S. thesis]: Stillwater, Oklahoma State University, 147 p.

Ragland, D. A., 1983, Sedimentary geology of the Ordovician Cool Creek Formation as it is exposed in the Wichita Mountains of southwestern Oklahoma [M.S. thesis]: Stillwater, Oklahoma State University, 170 p.

Ragland, D. A., and Donovan, R. N., 1985, The Thatcher Creek Member, basal unit of the Cool Creek Formation, in southern Oklahoma: Oklahoma Geology Notes, v. 45, no. 3, p. 83–91.

Simpson, L. C., 1979, Upper Gearyan and Lower Leonardian terrestrial vertebrate faunas of Oklahoma: Oklahoma Geology Notes, v. 39, no. 1, p. 3–19.

Stitt, J. H., 1977, Late Cambrian and earliest Ordovician trilobites, Wichita Mountains area, Oklahoma: Oklahoma Geological Survey Bulletin 124, 79 p.

—— , 1983, Trilobites, biostratigraphy, and lithostratigraphy of the McKenzie Hill Limestone (Lower Ordovician), Wichita and Arbuckle Mountains, Oklahoma: Oklahoma Geological Survey Bulletin 134, 54 p.

Toomey, D. F., and Nitecki, M. H., 1979, Organic buildups in the Lower Ordovician (Canadian) of Texas and Oklahoma: Fieldiana, Geology, n.s. 2, 181 p.

Tsegay, T., 1983, Sedimentary geology of the Reagan Formation (Upper Cambrian) of the Blue Creek Canyon, Slick Hills, southwest Oklahoma [M.S. thesis]: Stillwater, Oklahoma State University, 95 p.

Some aspects of the geology of Zodletone Mountain, southwestern Oklahoma

R. N. Donovan, *Geology Department, Texas Christian University, Fort Worth, Texas 76129*
P. Younger, C. Ditzell, *Geology Department, Oklahoma State University, Stillwater, Oklahoma 74078*

LOCATION

Zodletone Mountain, a 250-ft-high (76 m) eminence in Kiowa County, southern Oklahoma, constitutes the northernmost member of the Slick Hills Range and rises 1,649 ft (500 m) above sea level (SW¼SW¼ Sec.9,T.6N.,R.14W. Saddle Mountain 15-minute Quadrangle, Oklahoma). The best approach to the mountain is by following Oklahoma 58 south from Carnegie for 7 mi (11 km). At this point turn west along an unpaved section road for 3 mi (4.8 km), then south for 0.5 mi (800 m) to a white two-storied farmhouse, which is the home of Mr. Don Trobe (Fig. 1). Obtain permission to visit here. (Mr. Trobe, like everyone else hereabouts, has been very supportive of geologists. However, no smoking and no litter, please.) About 0.5 mi (0.8 km) south of Mr. Trobe's home, turn west along another section road and drive 1.7 mi (2.7 km) to a silver gate at SE¼SW¼Sec.16,T.6N.,R.14W. From here a fair-weather track runs north to the northeast side of Zodletone Mountain. Follow this track to its termination, park, and walk (but beware of rattlesnakes).

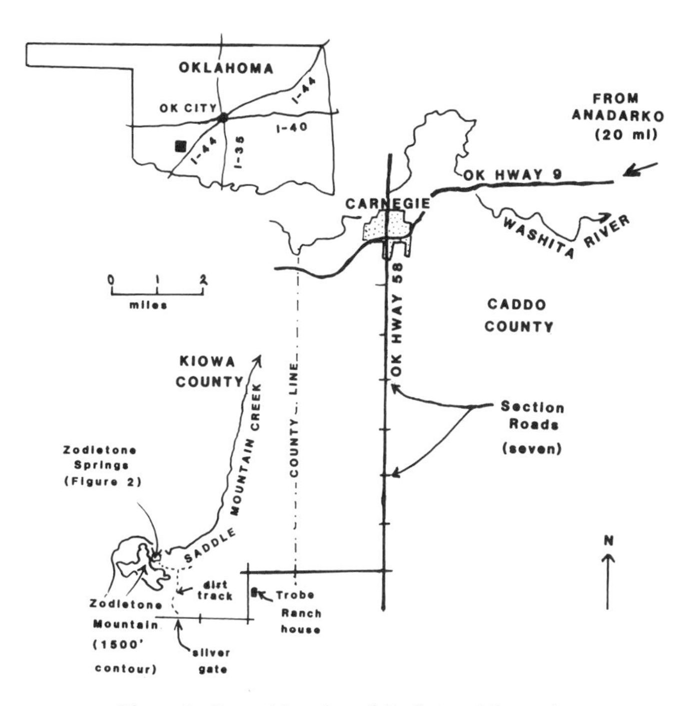

Figure 1. General location of Zodletone Mountain.

SIGNIFICANCE

Two geological themes are of unusual interest at Zodletone Mountain: the Cambrian sedimentary sequence and the modern barite-depositing spring on the northern edge of the mountain (Fig. 2). The barite travertine and tufa that has formed in the spring area is a most unusual deposit, known elsewhere in the world at only three sites: Hotchkiss, Colorado (Headden, 1905; Cadigan and others, 1976; Younger, 1986); Kokuto, Taiwan; and Akita, Japan (Ishizu, 1915).

The Cambrian sedimentary sequence includes the Timbered Hills Group, comprising the Reagan and Honeycreek Formations and the Cambrian part of the Arbuckle Group (the Fort Sill and lower Signal Mountain Formations) (Tsegay, 1983; Ditzell, 1984; Rafalowski, 1984). A generalized log for this succession is given with the account of the Bally Mountain site elsewhere in this guide (Donovan and others). At Zodletone the Timbered Hills Group is beautifully exposed; in particular the Reagan Formation is noteworthy for the geometry of its sandstones and their trace fossils.

The Modern Spring

The modern spring (Fig. 3) is enclosed by crumbling concrete walls that record an attempt to develop the area as a health spa. The surroundings are barren, resembling an industrial waste site, and the astringent smell of hydrogen sulphide pervades the air (Zodlestone is a corruption of the Indian term for Stinking Mountain). The spring water bubbles continually, and the spring's orifice is lined with purple sulphur bacteria (*Chromatiaceae*). The spring feeds a tiny stream which, after a short course, tumbles into and contaminates Saddle Mountain Creek. This stream is actively depositing calcite and barite as a tufa coating around vegetation.

The soil profile around the spring has been greatly altered; soil in contact with the water is black, passing upward and outward through a gray horizon that contains alunogen (a white sodium sulphate) and in places native sulphur and blades of selenitic gypsum. The latter are up to 1.25 in (3 cm) in length and may show chiasmatic twinning. The gray zone gradually passes into the normal (oxidized) red-brown soil profile.

Clearly at one time more springs were active in the area; several drainage depressions up to 35 ft (10 m) from the active spring record drainage into Saddle Mountain Creek from extinct orifices. The gray soil in some of these depressions is capped by tubules that are the unconsolidated remains of plant roots and stems and are coated by white barite.

Farther to the northwest (Fig. 3) are several exposures of

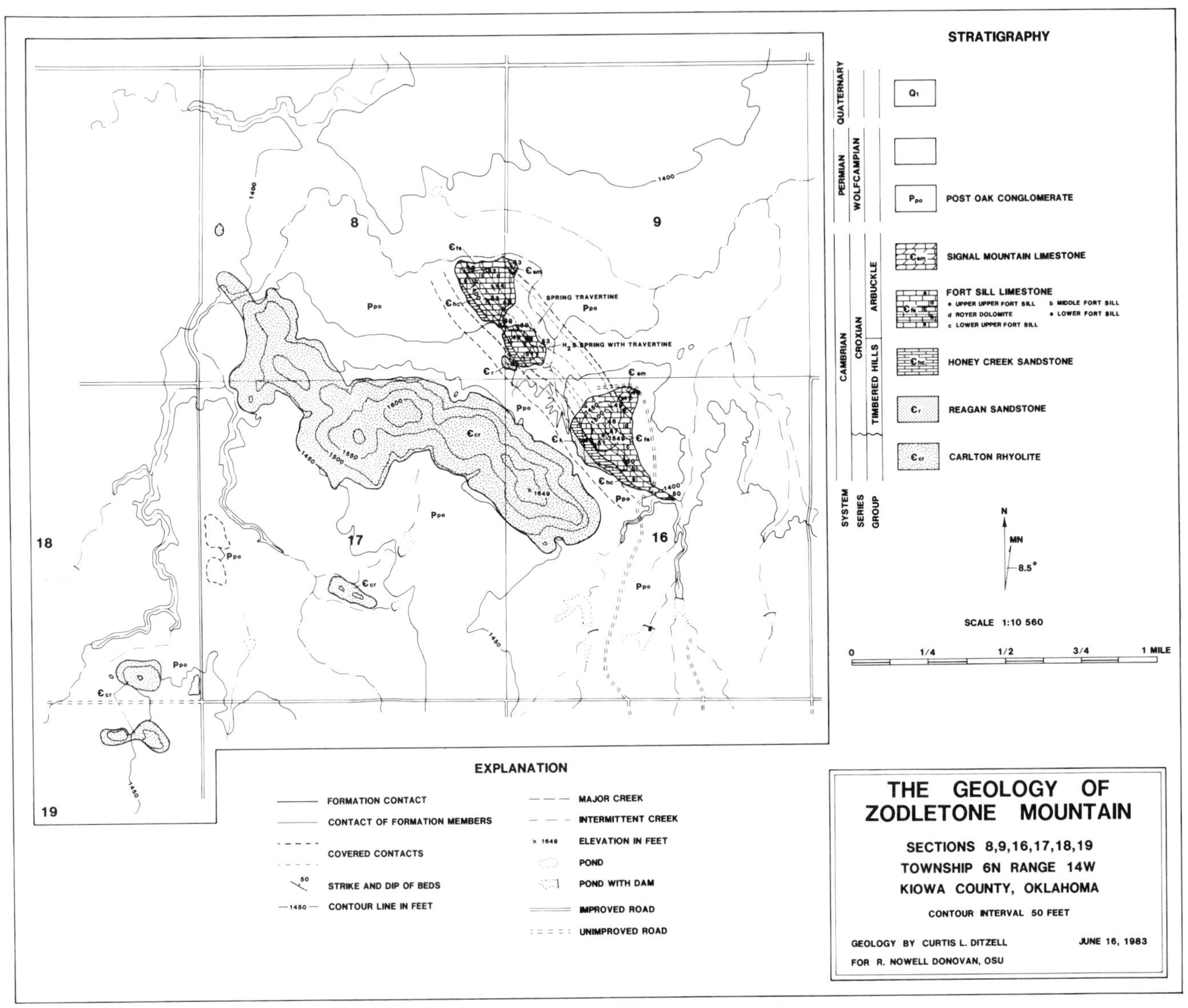

Figure 2. General geology of Zodletone Mountain.

indurated travertine, which mineralogically are either barite-cemented calcareous travertine or wholly barite travertine.

Thin section studies of these rocks reveal stunning mineralogical textures. In the unconsolidated travertine, barite occurs as isopachous coatings (typically 0.1 to 0.2 mm thick) on the inner and outer surfaces of tubules. In the more indurated deposits, various textures indicative of both vadose and phreatic conditions are found (Younger and others, 1985, and in preparation), including pendant and isopachous crystal growths (Fig. 4). Some cavities show drusy growth, and many crystals show growth lines delineated by inclusions of organic matter and micrite. Similar textures are also found in the calcareous tufa and travertine (which in general, though not exclusively, predates the barite).

The longest and finest barite crystals, up to 0.24 in (6 mm) across, are associated with the wholly barite travertine. Most of the rock is composed of barite euhedra arranged in well-defined laminae. Characteristic features include lamellar twinning, plumose habit, and a distinctive chevron zonation that records preferential episodic crystal growth of [110] faces at the expense of the [011] planes.

Most travertine suites are composed of either calcite or silica. As noted, barite deposits of this type are very rare. This is possibly because barite is practically insoluble at surface temperatures and pressures. However, the solubility of barite is much greater in warmer, more saline waters at higher pressures (Younger, 1986). Chemical analysis of the Zodletone spring water (Table 1) indicates that (1) it is unlike that of any other spring water in Oklahoma (Havens, 1983), (2) it cannot be derived from meteoric recharge through the adjacent Arbuckle Group, and (3) it is presently just at the saturation level for barite (presumably it exceeded this level in the recent past). Within the Sulin Oilfield Water Classification, the Zodletone water is a

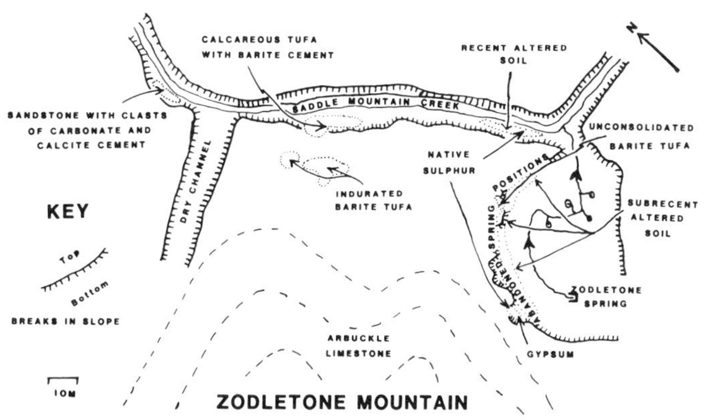

Figure 3. Detail of spring and associated deposits at Zodletone Mountain.

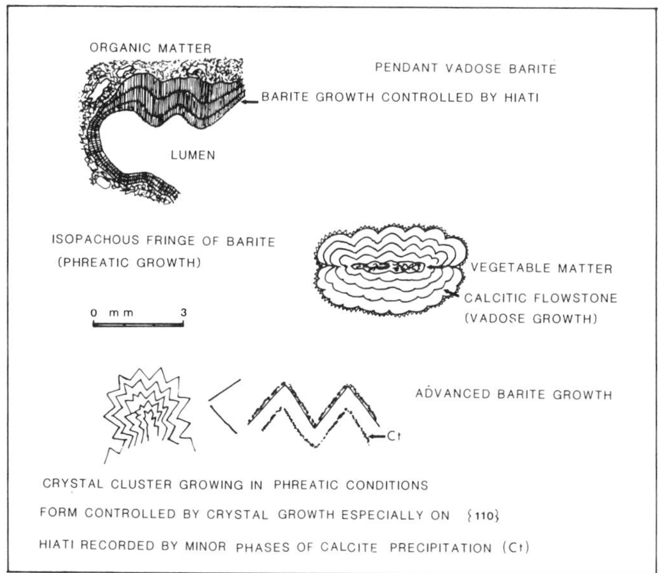

Figure 4. Textural varieties of barite tufa and travertine illustrating both phreatic and vadose form. Drawing from thin section photomicrograph.

Chloride-Calcium Type, Chloride Group, Sodium Sub-Group Class S1S2. Waters of this type are typical of deep stagnant subsurface brines most commonly associated with hydrocarbon occurrence (Collins, 1975).

This result is important, as it suggests that the Zodletone water is an oilfield brine (Zodletone Mountain is located spatially above the Anadarko basin, as suggested by COCORP data, Brewer, 1982). It seems likely that the underlying reverse faults and associated fractures provide an escape route whereby deep-seated brines confined at depth can rise under a potentiometric head to contaminate the surface aquifer.

If this is the case, then it would follow that sulphate-bearing water should not be used during secondary recovery in adjacent oilfields lest barite scale precipitation occur. This happens in other parts of the Anadarko basin (Cowan and Weintritt, 1976), suggesting that Zodletone water is a typical Anadarko brine.

The Lower Paleozoic Section

Zodletone Mountain is an exhumed Permian landform covered in places by a thin veneer of Permian detritus—the Post Oak Conglomerate facies. Some of this detritus obscures the unconformity between the Reagan Formation and the underlying Carlton Rhyolite, which forms the bulk of Zodletone Mountain (Fig. 2). Otherwise the Cambrian sediments are well exposed on two spurs that run northward from the main (rhyolite) hill mass. In many respects the sequence resembles that on adjacent Bally Mountain (including the "Royer" dolomite, for example). However, the Reagan is thicker and more cleanly exposed.

Both Reagan sections are dominated by quartz-rich sandstones of medium to coarse grain. The more basal sandstones contain numerous rhyolite grains and are cemented by quartz and hematite. The upper sandstones are glauconitic and cemented by quartz and calcite and contain abundant broken phosphatic brachiopod shells and (more rarely) phosphate nodules.

Sandstone geometries are mostly lenticular and dominated

TABLE 1. CHEMICAL ANALYSES OF THE ZODLETONE SPRING WATER*

Constituent	DMZW-1	ZT-1	Havens, 1983
Ba	43.5	35.0	---
Co	273.5	296.0	---
Fe	1.1	---	---
K	28.6	---	---
Li	1.1	---	---
Mg	149.0	144.0	---
Na	2500.0	2550.0	2900.0(Na+K)
SiO_2	10.9	---	---
Sr	12.4	---	---
Cl	4910.0	8800.0	5000.0
SO_4	70.4	55.0	90.0
HCO_3	---	481.0	270.0
CO_3	---	0.0	0.0
NO_3(N)	---	3.0	---
TDS	8001.0	9926.0	9160.0
Hardness (Ca, Mg)	---	---	1300.0
pH	7.4	7.7	8.2
Conductivity	>10000.0	15040.0	14800.0
Temperature °C	22	22	22

*All concentrations in mg/l; conductivities in microsiemens.

Notes: DMZW-1 was sampled by Mrs. C. Patterson of the Geology Department, Oklahoma State University, on July 12, 1984. ZT-1 was sampled by Paul Younger of the Geology Department, Oklahoma State University, and by the Agronomy Water and Soil Testing Laboraory of OSU, on April 30, 1985.

by trough cross-bedding in sets up to 5 ft (1.5 m) thick. The overall paleocurrent distribution is bipolar with opposed but unequal modes oriented NNW and SSE, suggesting a general tidal influence. However, the thicker lenticular sandstones appear to have migrated to the SSE, perhaps as a result of storm action. The most impressive of these sandstones is located on the eastern outcrop.

Apart from phosphate brachiopods, the Reagan has yielded few body fossils. However, it does display a fine array of trace fossils, particularly the vertical burrows of *Skolithos* ichnofacies, including *Areniciolites.* In general, trace fossils are more abundant in the tan-colored sandstones than in the highly glauconitic greensands, perhaps because the latter were less oxygenated. Such trace fossils as do occur in the glauconitic sandstones tend to have been formed by bottom-crawling organisms and are of a horizontal character.

The trace fossils of the western exposure are particularly fine and include one spectacular piece of cross-bedding where successive reactivation surfaces in the set are defined by burrows oriented normal to the cross set laminae (Fig. 5). This arrangement indicates a slow and episodic origin for this cross bed. When eventually stabilized, the unit was colonized by a group of *Skolithos* organisms that grew to a larger size than the earlier inhabitants of the migrating cross bed.

20 ins
50 cms

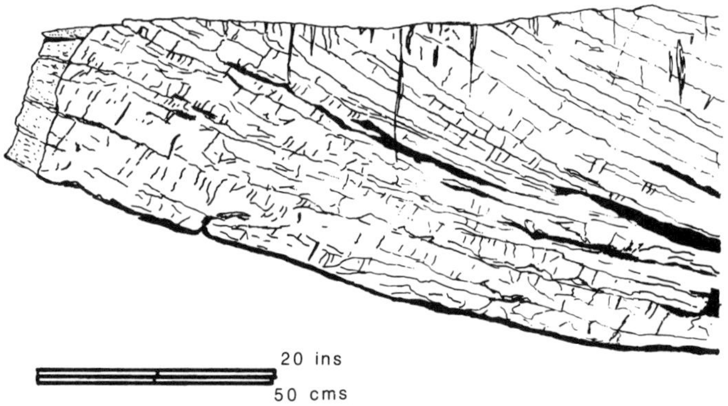

Figure 5. A single set of cross-bedding from the western outcrops of Reagan Formation. Note how *Skolithos* burrows are oriented normal to cross-bedding, indicating that successive surfaces were colonized by small individuals. This indicates that the cross bed is punctuated by numerous hiati and that in a sense each cross bed is a reactivation surface. Once completely constructed the top surface was then colonized by larger individuals.

The overlying Honey Creek Formation consists of a mixed quartz, glauconite, and pelmatazoan grainstone complex characterized by bipolar cross-bedding with similar orientations to those in the underlying Reagan. However, the cross beds are smaller (c. 6-in [15-cm] sets) and more uniform in size. They appear to have formed as small, tidally influenced lunate dunes rather than as the complex sand bars found in the Reagan.

REFERENCES

Brewer, T. A., 1982, Study of southern Oklahoma aulacogen, using COCORP deep seismic-reflection profiles, *in* Gilbert, M. C., and Donovan, R. N., eds., Geology of the eastern Wichita Mountains southwestern Oklahoma, Guidebook 21: Oklahoma Geological Survey, p. 31–39.

Cadigan, R. A., Felmlee, J. K., and Rosholt, J. N., 1976, Radioactive mineral springs in Delta County, Colorado: U.S. Geological Survey Open File Report No. 76-223.

Collins, A. G., 1975, The geochemistry of oilfield waters: Amsterdam, Elsevier, Developments in petroleum science, v. 1, 469 p.

Cowan, J. C., and Weintritt, D. J., 1976, Water-formed scale deposits: Houston, Texas, Gulf Publishing Company, 569 p.

Ditzell, C., 1984, The sedimentary geology of the Cambro-Ordovician Signal Mountain Formation as exposed in the Wichita Mountains of Southwestern Oklahoma [M.S. thesis]: Stillwater, Oklahoma State University, 165 p.

Havens, J. S., 1983, Reconnaissance of ground water in vicinity of Wichita Mountains, Southwestern Oklahoma: Oklahoma Geological Survey Circular 85, 13 p.

Headden, W. P., 1905, The Doughty Springs, a group of radium bearing springs on the North Fork of the Gunnison River, Delta County, Colorado: Proceedings Colorado Scientific Society, v. 8, p. 1–30.

Ishizu, R., 1915, The mineral springs of Japan (specially edited for the Panama-Pacific International Exposition, San Francisco): Tokyo, Imperial Hygienic Laboratory, pt. 1, 94 p.

Rafalowski, M. B., 1984, Sedimentary geology of the late Cambrian Honey Creek and Fort Sill Formations as exposed in the Slick Hills of southwestern Oklahoma [M.S. thesis]: Stillwater, Oklahoma State University, 147 p.

Tsegay, T., 1983, Sedimentary geology of the Reagan Formation (Upper Cambrian) of the Blue Creek Canyon, Slick Hills, southwest Oklahoma [M.S. thesis]: Stillwater, Oklahoma State University, 95 p.

Younger, P. L., 1986, Barite travertine in Southwestern Oklahoma and West-Central Colorado [M.S. thesis]: Stillwater, Oklahoma State University, 163 p.

Younger, P. L., Patterson, C., Donovan, R. N., and Hounslow, A. W., 1985, Barite tufa from Zodletone Mountain, southwestern Oklahoma: Geological Society of America Abstracts with Programs, v. 17, no. 3, p. 198.

Younger, P. L., Donovan, R. N., and Hounslow, A. W., in preparation, Barite tufa from Zodletone Mountain, Southwestern Oklahoma: Oklahoma Geological Survey Guidebook.

Hydrocarbon-induced diagenetic aureole at Cement-Chickasha Anticline, Oklahoma

Zuhair Al-Shaieb, *School of Geology, Oklahoma State University, Stillwater, Oklahoma 74078*

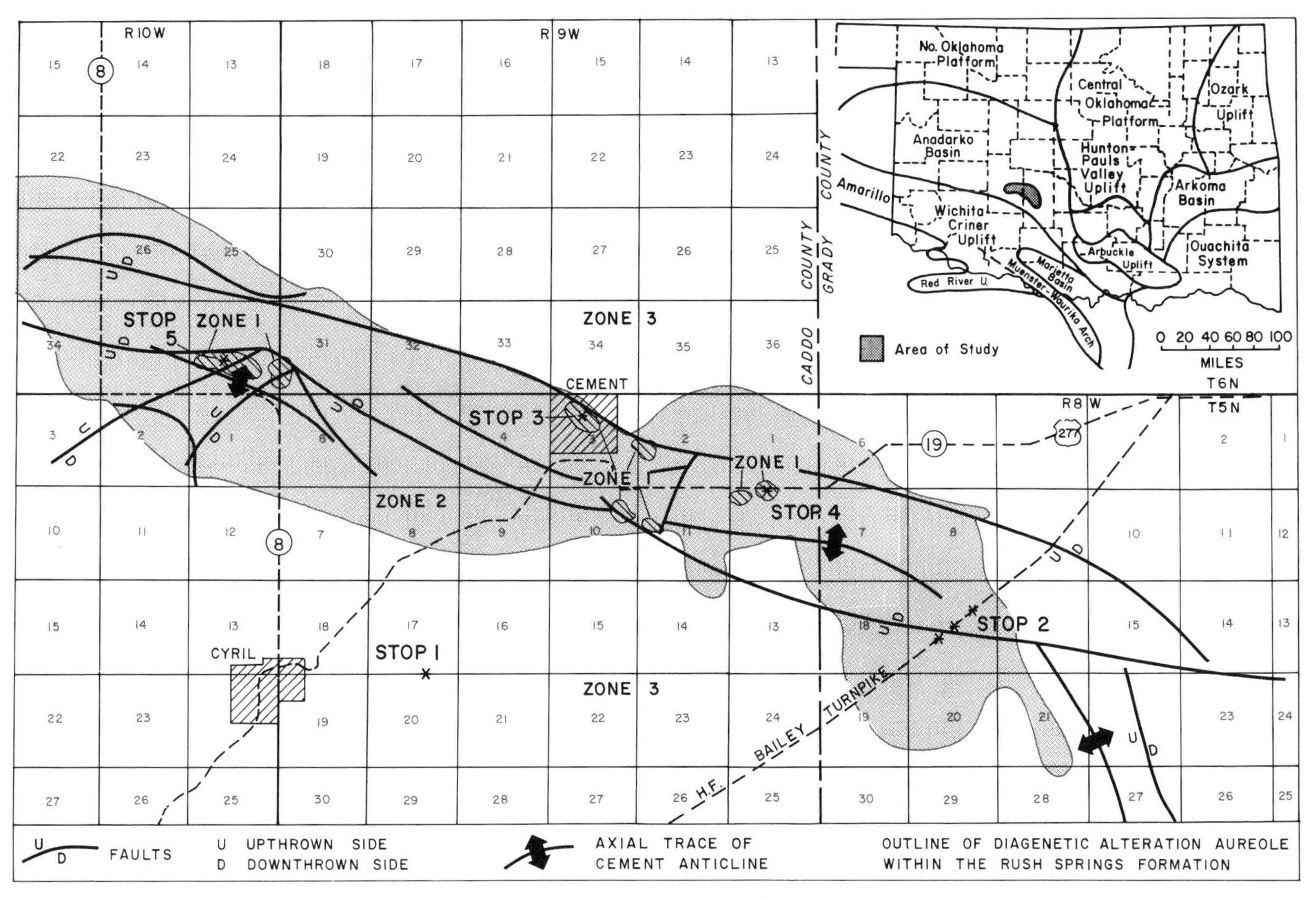

Figure 1. Map of pre-Permian subsurface structure and surface alteration zones, Cement-Chickasha Anticline. Inset; major geologic provinces in Oklahoma and location of this site (shaded).

LOCATION

The Cement-Chickasha Anticline includes parts of Caddo and Grady counties in southwestern Oklahoma (Fig. 1). The structure is elongated and trends in a northwesterly direction. The town of Cement is located in the central part of the structure. Three major highways intersect the site. The H. E. Bailey Turnpike extends diagonally across the southeastern portion of the site, while U.S. 277/Oklahoma 19 trends east–west through the southern half of the site and connects the towns of Cement and Cyril. Oklahoma 8 trends north–south through the western portion of the site and terminates at the town of Cyril.

SIGNIFICANCE

One of the most remarkable features of the Cement Chickasha Anticline area is alteration of the Permian red beds at the outcrops and in the subsurface. The chemical and mineralogical changes observed are directly related to seepage of hydrocarbon from underlying reservoirs. The vertical migration of hydrocarbon and other products was facilitated by a complex fault system.

SITE INFORMATION

Stratigraphy. Outcrops of upper Permian Rush Springs Formation 130 to 300 ft thick (41 to 90-m) are restricted to the Cement-Chickasha structure and the surrounding area. The Rush Springs Formation conformably overlies the 90 to 130 ft thick (27 to 40-m) Marlow Formation. The Rush Springs and Marlow Formations compose the Whitehorse Group of Guadalupian age (Fig. 2). Conformably overlying the Rush Springs Formation is

ERA	SYSTEM	SERIES	GROUP	FORMATION	OUTCROP COLUMNAR SECTION	LITHOLOGIC DESCRIPTION	THICKNESS meters	THICKNESS feet	DEPOSITIONAL ENVIRONMENT
CENOZOIC	QUATERNARY	HOLOCENE				Fluvial-terrace and flood-plain conglomerates, sands, and silts plus windblown sand and silt			
		PLEISTOCENE							
PALEOZOIC	PERMIAN	GUADALUPIAN		CLOUD CHIEF		Massive pink to white gypsum	10.7–26.0	35–85	Restricted Marine (Ham, 1960)
			WHITEHORSE	RUSH **		Dolomitic sandstones and siltstones	41.1–91.4	135–300	Near Shore (O'Brien, 1963)
				WEATHERFORD GYPSUM BED		Massive pink gypsum and some dolomite			
				SPRINGS *		Large-scale trough cross-laminated, very fine to fine grained, red sandstones			
				MARLOW		Red-brown, very fine grained sandstone; some silty shale; occasional thin gypsum and dolomite beds	32.0–41.1	105–135	Tidal Flat (MacLachlan, 1967)
			EL RENO	DOG CREEK SHALE		Red-brown silty shale	30.5–61.0	100–200	(North) Dog Creek Sh. Blaine Fm. Tidal Flat (Fay, 1964); (South) Chickasha Fm. Fluvial Deltaic (Fay, 1964)
				BLAINE		Interbedded gypsum, dolomite, and shale			
				FLOWERPOT SHALE		Red-brown and gray-green shale and gray-green silt interbeds	30.5–122.0	100–400	Fluvial Deltaic (Self, 1966)
				SAN ANGELO SANDSTONE		Interstratified sandstone, mudstone conglomerate, and shale			
		LEONARDIAN	HENNESSEY	HENNESSEY SHALE		Red-brown to gray shale; some tan sandstone	45.7–73.2	105–240	Tidal Flat (Stith, 1968)
				POST OAK CONGLOMERATE *		Conglomerates containing granitic or limestone detritus, and arkosic sandstones; red or gray-green shale and siltstone interbeds; some calcretes	152.4–>610.0	500–>2,000	(Southwest) Surface Post Oak Cong. Piedmont
			SUMNER	GARBER SANDSTONE *		Red-brown and gray-green fine-grained sands and silts; asphaltic sandstone or carbonate- and mudstone-pebble conglomerate at base	30.5–61.0	100–200	(Southwest) Subsurface Post Oak Cong. Piedmont; (Southwest) Tidal Flat and Supratidal (Flood, 1969); (East and Southwest) Fluvial Deltaic (Flood, 1969)
				WELLINGTON *		Maroon and red shales and red and gray silt and sand intercalations; gray-green and black sandstones at base	30.5–61.0	100–200	
	?PENNSYLVANIAN	WOLFCAMPIAN (GEARYAN)	OSCAR	*		Gray-brown and red shale, sandstone, and arkose; wedges of cherty and arkosic conglomerates; some thin sandy limestones near base	91.4–152.4	300–500	
				VIOLA					

*Favorable for uranium deposition
**Uranium ore deposits

Figure 2. Stratigraphic column with lithological description for the uppermost Pennsylvanian–Recent section overlying the pre-Permian Cement-Chickasha Anticline.

the Cloud Chief Formation, which is represented by approximately 30 ft (9 m) of the Moccasin Creek gypsum member (Haven, 1977).

Structure. The Cement-Chickasha structure is a west-northwest-trending anticline that is slightly overturned to the north (Fig. 1). A predominantly unfaulted, 2,500-ft-thick (760 m), Lower Wolfcampian to upper Guadalupian Permian succession unconformably overlies a faulted and tightly folded pre-Permian structure. A major south-dipping reverse fault intersects the pre-Permian unconformity along the north flank of the anticline (Herrmann, 1961). This fault parallels both the fold axis and a north-dipping normal-fault system, which has been truncated by the unconformity near the crest of the structure. The pre-Permian fold axis is also offset by several minor normal faults. Minor structural deformation in the post–Cloud Chief times produced a gentle, near-symmetric anticlinal fold in Permian time. On the top of the Rush Springs Formation, this anticline is approximately 11 mi (18 km) long and 2 mi (3 km) wide. Its crest is a topographic high dominated by two Domes, known as the East Cement and West Cement Domes, 4 mi (6 km) apart, and capped by the Moccasin Creek gypsum member of the Cloud Chief Formation.

Lithology. The Rush Springs Formation typically consists of medium to light red or (less commonly) orange-brown to light brown, very fine to medium-grained, predominantly medium- to large-scale trough cross-bedded, weakly indurated subarkosic sandstones. Very coarse, frosted, spherical quartz grains are common in the lower part of the formation. Locally, the Rush Springs Formation exhibits remarkable lithologic homogeneity; however, a silty shale phase is present in Grady County, and several sandstones within the formation grade laterally into gypsum beds (Tanaka and Davis, 1963) (Fig. 2). Massively thick gypsum (up to 60 ft; 18 m), the Weatherford gypsum bed, occurs in the upper part of the Rush Springs. It is separated from the overlying Moccasin Creek gypsum member of the Cloud Chief Formation by 10 to 15 ft (3 to 5 m) of dolomitic sandstones and siltstones.

Depositional Environments. The constituent sand facies of the Rush Springs Formation may represent shallow-marine offshore bars, plus shore-face, beach-barrier, and localized back-shore eolian dunes. Polydirectional paleocurrents are indicated by Rush Springs cross-bed orientations. Thin gypsum and dolomite beds may be coastal sabkha deposits; the thicker gypsums were probably precipitated after evaporative concentration from marine embayments of restricted circulation. Ham (1960) thought deposition of the Cloud Chief took place in a semi-enclosed arm of a sea that received periodic influxes of sulfate-rich waters during minor transgressions (Fig. 2).

Hydrocarbon Induced Diagenetic Aureole (HIDA). HIDA is a term coined here to include all chemical and mineralogical changes of the Permian red beds that overlie Cement-Chickasha and other oil fields in southwestern and south-central Oklahoma. The diagenetic minerals occur within a distinctly zoned aureole that delineates the position of the oil field. The geometry of Cement-Chickasha aureole strongly reflects the major structural elements that controlled emplacement of hydrocarbons in the underlying rocks. Calcite, ferroan calcite, manganese-rich calcite, dolomite, ankerite, pyrite, marcasite, and native sulfur are the major diagenetic minerals. The inner-most zone of the aureole is characterized by abundant carbonate cementation and generally coincides with major fault systems. Three diagenetic zones were mapped based on criteria determined primarily by petrographic and lithologic analysis. These are: Zone I—extensive carbonate-cemented sandstone and carbonate-replacing gypsum with or without pyrite; Zone II—altered (buff to yellow) red beds with minimal carbonate cement; and Zone III—original, unaltered red beds. The distribution of the various zones is shown in Figure 1. A well-exposed, unaltered Rush Springs sandstone (Zone III) and the overlying Moccasin Creek gypsum of the Cloud Chief Formation may be examined along the section road between Sec. 17 and 20,T.5N.,R.9W. (Stop 1; Figs. 1 and 3). The contact between unaltered Rush Springs sandstone (Zone III) and altered buff to yellow (Zone II) is encountered in Sec. 8,T.5N.,R.9W., along U.S. 277/Oklahoma 19. In addition, several exposures of Zone II altered sandstones are present in Sec. 17,T.5N.,R.8W., along H. E. Bailey Turnpike (Stop 2; Figs. 1 and 4). Intense cementation of the friable sandstones along the crest of the anticline resulted in preservation of topographically expressed structures known as East Cement and West Cement Domes. At the East Cement Dome, toward the north part of the city of Cement, Sec. 3,T.5N.,R.9W. (Stop 3, Figs. 1 and 5), an excellent example of the rock assemblage of Zone I is exposed. Cross-bedded and sparry calcite-cemented Rush Springs sandstone with typical buff color is underlying Moccasin Creek gypsum, which is completely replaced by calcite (Fig. 6). The pseudomorphous replacement of gypsum by calcite has resulted in a spectacular preservation of the original structure, such as the crinkly bedding of gypsum. Ferroan calcite and ankerite are also present in variable quantities. A complete replacement of the Rush Springs sandstone by carbonate minerals (mainly calcite, ferroan calcite, and dolomite to a lesser extent) can be examined at two different quarries. The first is an abandoned quarry located along U.S. 277/Oklahoma 19 and to the north of Sec. 12,T.5N.,R.9W. (Stop 4; Figs. 1 and 7). The second quarry is the property of Dolese Brothers Inc., located at the West Cement Dome, Sec. 36,T.6N.,R.10W. (Stop 5; Fig. 1). For all practical purposes this rock may be classified as diagenetic limestone. Pyrite in the Rush Springs sandstone and diagenetic limestone are present as cementing nodules and as disseminated cubes (Stop 3, 4, and 5; Figs. 1 and 8). Marcasite occurs as a minor constituent. Native sulfur occurs, filling small veinlets near sulfide zones. Reynolds (1985) reported the presence of pyrrhotite in the subsurface.

Lilburn and Al-Shaieb (1983) conducted an extensive iso-

Figure 3. Large-scale cross-bedding of the red to brown Rush Springs sandstone and the overlying Moccasin Creek gypsum of the Cloud Chief Formation (Stop 1: Zone III).

Figure 4. Altered yellow, trough cross-bedded, and friable Rush Springs sandstone of Zone II (Stop 2).

Figure 5. Carbonate-cemented Rush Springs sandstone and the overlying Moccasin Creek gypsum, which is completely replaced by calcite (Stop 3; Zone I).

Figure 6. Pseudomorphous replacement of the Moccasin Creek gypsum by calcite (Stop 3; Zone I). Note the preservation of the original gypsum structure.

Figure 7. Diagenetic limestone formed by complete replacement of the Rush Springs sandstone by carbonate minerals. This limestone is isotopically light with $\delta C^{13} = -30.7$ (Stop 4; Zone I).

Figure 8. Calcite-cemented Rush Springs sandstone with pyrite nodules (Stop 4; Zone I).

topic and geochemical study in the subsurface of the Cement-Chickasha Anticline area. Lithologic sections representing 39,000 ft (11,887 m) of cuttings and cores were constructed with emphasis on lithology, texture, color, and evidence of carbonate and pyrite mineralization. Figure 9 shows the distribution of pyrite in the subsurface of the Cement-Chickasha Anticline. The source of sulfur in pyrite is mainly H_2S gas associated with migrated hydrocarbons. Comparison of the average values of δS^{34} of 30 pyrite samples ($3.6^0/_{00}$) and of δS^{34} of 10 oil samples ($4.7^0/_{00}$) strongly suggests a direct genetic relationships. In addition, the distribution of sulfur isotope ratios in pyrite from Cement-Chickasha Oil Field is remarkably similar to the distribution of sulfur isotope ratios showing H_2S gas associated with petroleum reservoirs (Goldhaber and others, 1978).

Diagenetic carbonate samples were analyzed for carbon isotope ratios to determine the source of carbon in carbonates and their genetic relationship to hydrocarbon source. The mean δC^{13} value of the 23 samples collected and those reported by Donovan (1974) is $-19^0/_{00}$ PDB with a standard deviation of $\pm 7.02^0/_{00}$ and values ranging from -2 to $-30^0/_{00}$. This wide range reflects more than one carbon source or a mixture of sources.

Calcrete samples from the Wellington Formation show δC^{13} values between -4 and $-10^0/_{00}$, which suggests a fresh water carbonate origin. One sample showing fossil fragments yields a δC^{13} value of $-1.8^0/_{00}$, implying a marine origin. A group of values clusters between -10 to $-15^0/_{00}$, suggesting a hybrid carbon source from both hydrocarbon and fresh water.

The following model is proposed to explain the genesis for the isotopic composition of hydrocarbon-derived, fresh water, and hybrid-type carbonates. The average δC^{13} value for fresh water carbonate is $-4.9^0/_{00} \pm 2.8$ as reported by Keith and Weber (1964). The carbonate enriched in δC^{12} from hydrocarbon oxidation, as reported by this study and Donovan (1974), has a δC^{13} value of $-32^0/_{00}$. Therefore, the δC^{13} of each sample analyzed may be recalculated in terms of two end members as shown by the following equation:

$$-32 * X + -4.9 * (1-X) = Z$$

where X represents the fraction of hydrocarbon contribution and Z the measured δC^{13}. The graphic representation of this equation shown in Figure 10. The isotopic data plotted on Figure 10 appear to cluster in three regions: region A represents hydrocarbon-derived carbon; region B is fresh water-derived carbon; and region C represents a hybrid source of carbon. The hybrid source combines carbon made available from the oxidation of hydrocarbon and the dissolution of earlier calcrete deposits or any other fresh water carbonate species.

REFERENCES CITED

Al-Shaieb, Z., Thomas, R. G., and Stewart, G. F., 1982, Lawton Quadrangle, Oklahoma and Texas: National Uranium Resource Evaluation Program, Grand Junction, Colorado, Bendix Field Engineering Corporation Technical Library, Report no. GJQ-017(82), 48 p.

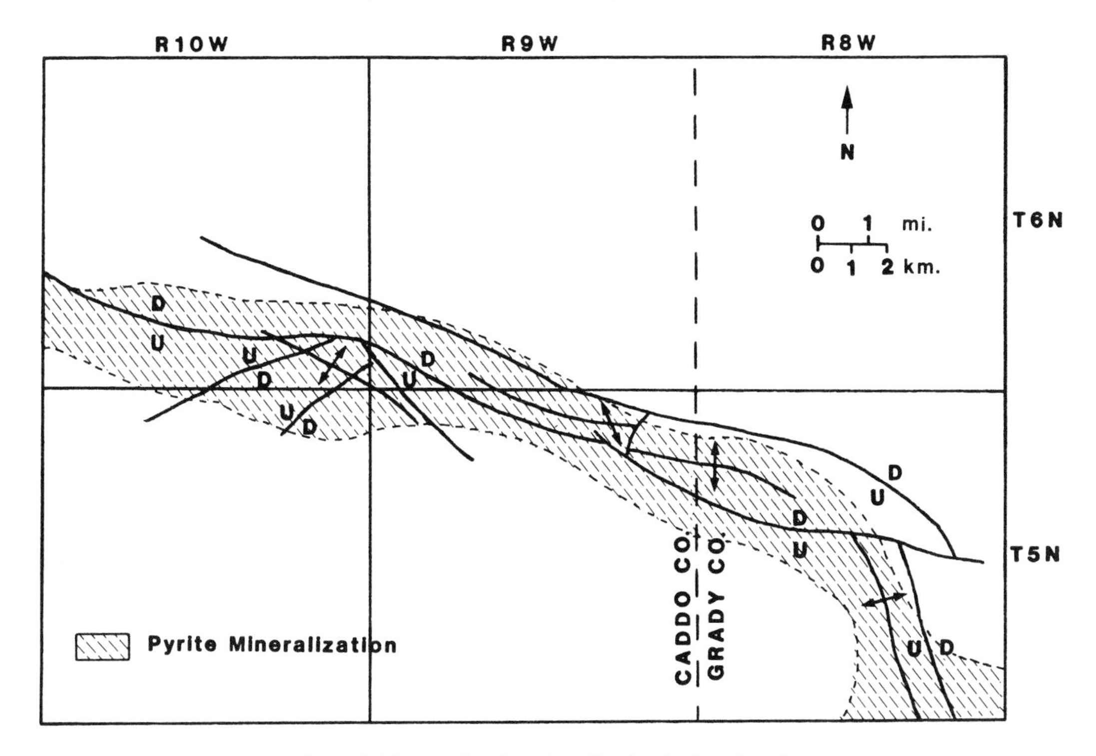

Figure 9. Extent of pyrite mineralization in the subsurface.

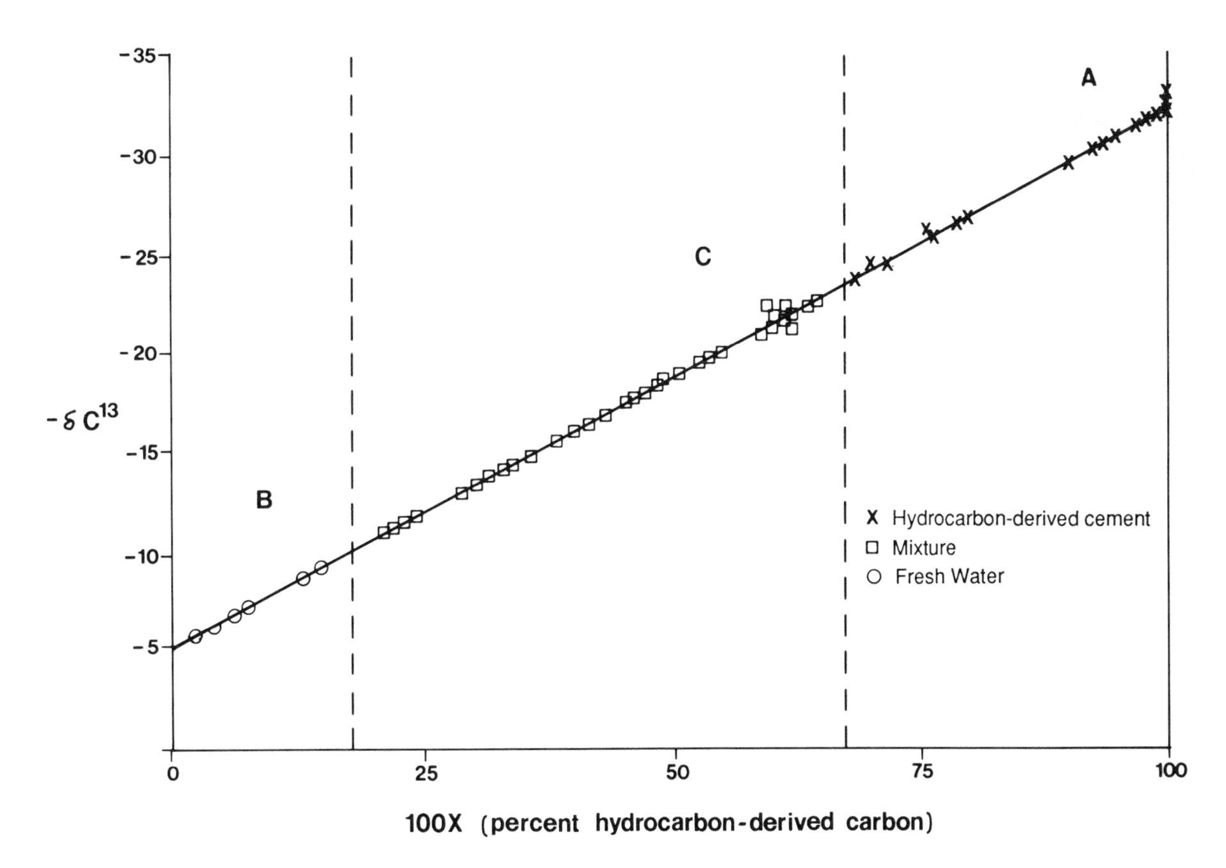

Figure 10. Percent of hydrocarbon-derived carbon in carbonate samples in terms of two end members –32 and –4.9%. (Data include δC^{13} values reported by Donovan, 1974).

Donovan, T. J., 1974, Petroleum microseepage at Cement, Oklahoma; Evidence and mechanism: American Association of Petroleum Geologists Bulletin, v. 58, no. 3, p. 429–446.

Goldhaber, M. B., Reynolds, R. L., and Rye, R. O., 1978, Origin of a south Texas roll-type uranium deposit; II. Sulfide petrology and sulfur isotopes studies: Economic Geology, v. 73, p. 1960–1705.

Ham, W. E., 1960, Middle Permian evaporites in southwestern Oklahoma: Copenhagen, International Geologic Congress, 21st, pt. 12, p. 138–151.

Havens, J. S., 1977, Reconnaissance of the water resources of the Lawton Quadrangle, southwestern Oklahoma: Norman, Oklahoma Geological Survey Hydrologic Atlas 6, 4 plates, scale 1:250,000.

Herrmann, L. A., 1961, Structural geology of Cement-Chickasha area, Caddo and Grady Counties, Oklahoma: American Association of Petroleum Geologists Bulletin, v. 45, p. 1971–1993.

Keith, M. L., and Weber, J. N., 1964, Carbon and oxygen isotopic composition of selected limestones and fossils: Geochimica et Cosmochimica Acta, v. 28, p. 1787–1816.

Lilburn, R. A., and Al-Shaieb, Z., 1983, Geochemistry and isotopic composition of hydrocarbon-induced diagenetic aureole (HIDA), Cement, Oklahoma: Shale Shaker, pt. I, v. 34, no. 4, p. 40–56.

——, 1984, Geochemistry and isotopic composition of hydrocarbon-induced diagenetic auerole (HIDA), Cement, Oklahoma: Shale Shaker, pt. II, v. 34, no. 5, p. 57–67.

Tanaka, H. H., and Davis, L. V., 1963, Ground water resources of the Rush Springs Sandstone in the Caddo County area, Oklahoma: Norman, Oklahoma Geological Survey Circular 61, 63 p.

Igneous geology of the Wichita Mountains, southwestern Oklahoma

*M. C. **Gilbert**, Department of Geology, Texas A&M University, College Station, Texas 77843*
*B. N. **Powell**, 237 GB, Phillips Petroleum Company, Bartlesville, Oklahoma 74004*

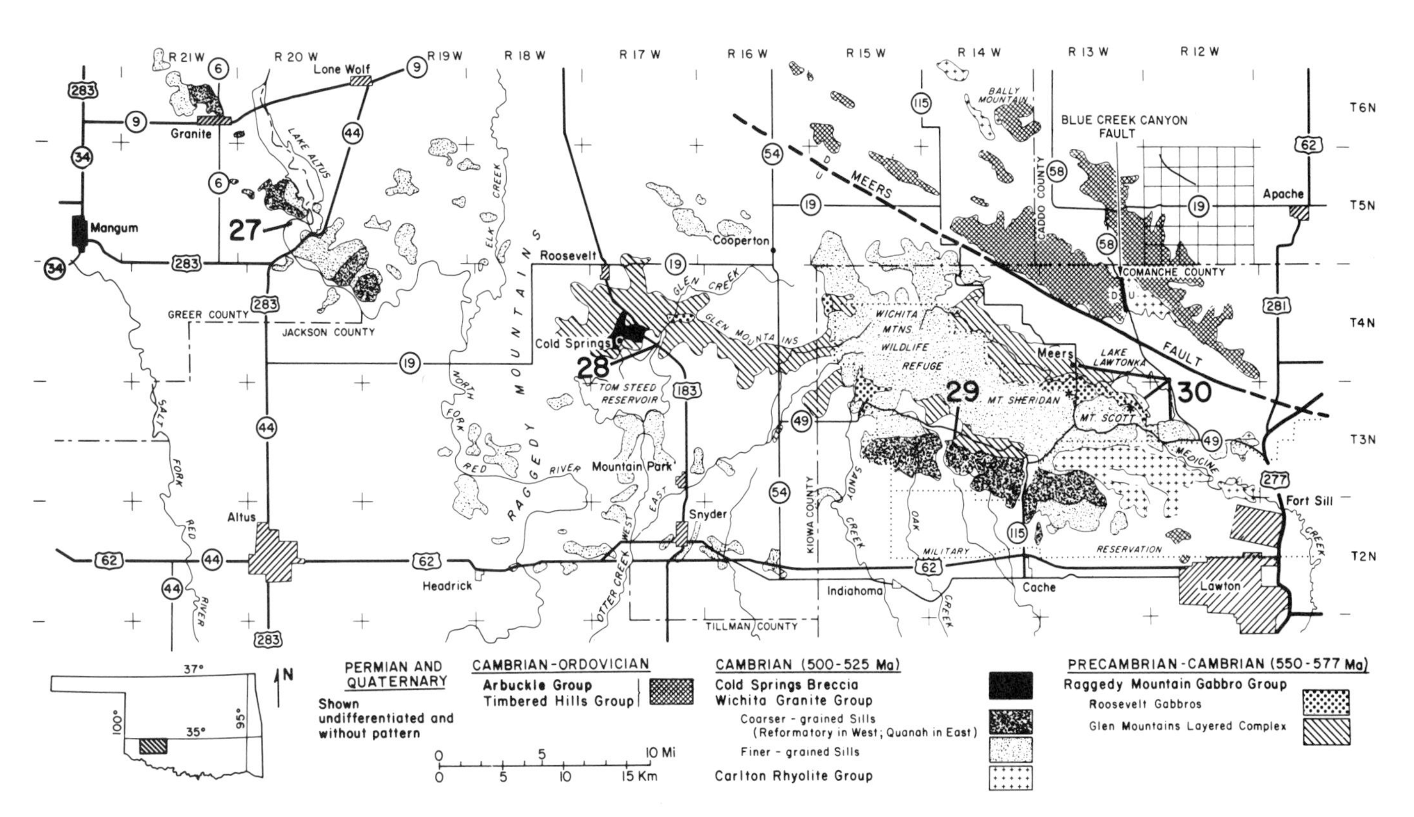

Figure 1. Simplified geologic map of the Wichita Mountains (after Powell and others, 1980) showing locations of sites 27 through 30, which highlight igneous relations in the region.

LOCATION

The Wichita Mountains extend for about 65 mi (105 km) northwest from Lawton, Oklahoma (Fig. 1). Four sites are described here: 27—Quartz Mountain State Park in the western part of the Wichita Mountains, near Oklahoma 44A; 28—The Tom Steed Reservoir area, along U.S. 183 in the central Wichita Mountains; 29—French Lake, south of Oklahoma 49; and 30—the Mt. Scott area, north of Oklahoma 49 in the eastern Wichita Mountains. Specific directions to each site are given with the site descriptions.

SIGNIFICANCE

The four sites described here display various aspects of several complexes of Cambrian igneous rocks that formed during the early stages of development of the southern Oklahoma aulacogen (Gilbert, 1983b). In addition, the sites provide excellent examples of exhumed Permian topography that developed on the igneous rocks.

27—QUARTZ MOUNTAIN STATE PARK

LOCATION

Quartz Mountain State Park is located 18 mi (29 km) north of Altus off of U.S. 283 and Oklahoma 44, on Oklahoma 44A (Fig. 1). Quartz Mountain and the park lie on the west side of Lake Altus on the North Fork of Red River, within the Lawton 2-degree Quadrangle and the Lake Altus 7½-minute Quadrangle.

ACCESSIBILITY

The state park contains only the east side of Quartz Mountain but to the south contains all or part of Williams Peak, King Mountain, and Mt. Lugert. The geology described here is found along the park road from the "Y" with Oklahoma 44A, north of the North Fork of Red River about 2 mi (3.2 km), to the Lodge.

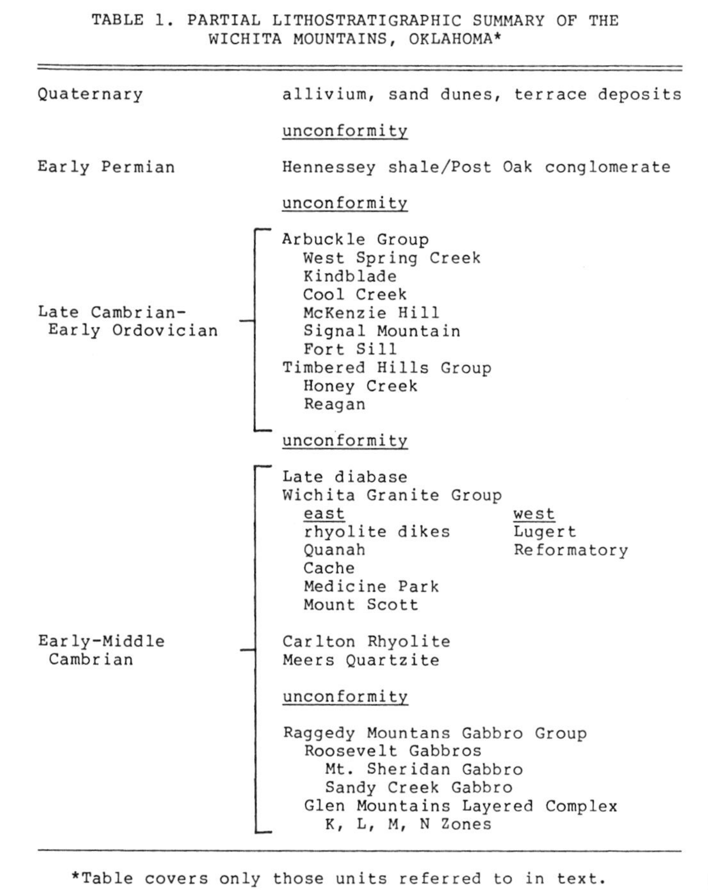

TABLE 1. PARTIAL LITHOSTRATIGRAPHIC SUMMARY OF THE WICHITA MOUNTAINS, OKLAHOMA*

Quaternary	allivium, sand dunes, terrace deposits
	unconformity
Early Permian	Hennessey shale/Post Oak conglomerate
	unconformity
Late Cambrian-Early Ordovician	Arbuckle Group West Spring Creek Kindblade Cool Creek McKenzie Hill Signal Mountain Fort Sill Timbered Hills Group Honey Creek Reagan
	unconformity
	Late diabase Wichita Granite Group east: rhyolite dikes, Quanah, Cache, Medicine Park, Mount Scott west: Lugert, Reformatory
Early-Middle Cambrian	Carlton Rhyolite Meers Quartzite
	unconformity
	Raggedy Mountans Gabbro Group Roosevelt Gabbros Mt. Sheridan Gabbro Sandy Creek Gabbro Glen Mountains Layered Complex K, L, M, N Zones

*Table covers only those units referred to in text.

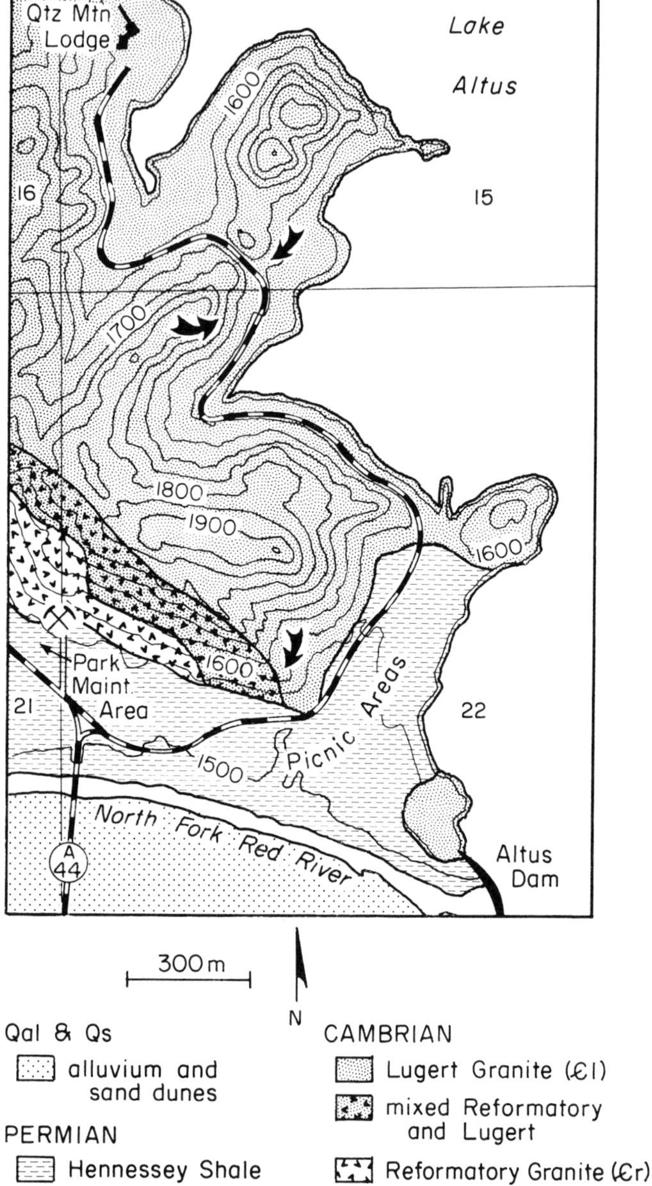

Figure 2. Geologic map of a part of the Quartz Mountain State Park. Arrows show features emphasized in the text.

GEOLOGIC SIGNIFICANCE

This area is one of the best exposed and most accessible for studying contact relations between shallow-seated granites: the Reformatory Granite and two phases of the Lugert Granite. It is near the type area for the Lugert Granite. Xenoliths are common in parts of the Lugert and Reformatory. Permian paleotopography and spheroidal weathering processes can be investigated.

DESCRIPTION

Figure 2 is a local geologic map showing some of the points of interest. A small quarry behind the park maintenance facility contains typical exposures of Reformatory Granite. Request permission to enter through the facility from the Nature Center about 0.5 mi (0.8 km) east on the road to the Lodge or climb around from the back. This granite and the Quanah Granite in the eastern Wichitas Mountains, with feldspars reaching only about 1 cm diameter, are the coarsest found in the mountains. The Reformatory Granite is hypersolvus and consists dominantly of alkali feldspar, now unmixed but originally about $Or_{45}Ab_{55}$, plus quartz plus amphibole, either hastingsitic or arfvedsonitic. Numerous inclusions are present. Most are metasedimentary but some are mafic, probably fragments of disrupted basaltic dikes. The inclusions show crude alignment reflecting flowage processes within the magma.

One-half mi (0.8 km) east of the river crossing are parking areas for the New Horizon Trail and Nature Center. Displayed on the south-facing slopes of Quartz Mountain from here (see southernmost arrow in Fig. 2) back toward the maintenance area are many beautiful examples of stoping and granite intruding granite. This is in the transition zone from Lugert Granite on the north and east to Reformatory Granite on the south and west. Several different textural varieties of granites are present, which make relations difficult to sort out without walking across the entire zone.

Essentially the sequence of events seems to be:

1) Reformatory Granite. The pre-existing matrix in this part of the Wichita Mountains may have been sedimentary (Meers(?), Tillman(?)), rather than Carlton rhyolite as thought to be the case in the east, because of the abundance of metasedimentary xenoliths. The Reformatory Granite fines to the east toward the younger Lugert Granite, suggesting this was near the original Reformatory Granite boundary.

2) Intrusion of the first phase of the Lugert Granite, a finer-grained, porphyritic, low C.I. granite with abundant granophyre. This texture developed on quenching of the Lugert Granite against the Reformatory Granite margins, indicating some significant time difference. Later quartz veins are prominent in some parts of this phase because, along with the quenching, contraction cracks would have developed localizing later solutions during final consolidation. Inclusions are rare in this phase.

3) Intrusion of the second and more typical phase of the Lugert Granite. This is somewhat coarser-grained, has little or no granophyre, and contains abundant inclusions similar in type to those found in the Reformatory Granite (about 10:1 metasedimentary to mafic) (Fig. 3). The metasedimentary ones commonly stand up in positive relief on low dipping rock surfaces. This phase of the Lugert Granite intrudes and stopes the finer Lugert Granite, breaking it into spectacular angular blocks. Both phases have the same assemblage—alkali feldspar + quartz + amphibole—and are also hypersolvus. On cursory examination, the second phase appears to have a higher C.I., but this is illusory due to the gray-colored cores of the feldspar and the abundance of xenoliths both adding to the somewhat darker color of the rock.

4) Formation of late pulses of fluid-saturated magma, aplitic and pegmatitic dikes and pods, and quartz veins. Although all indications are that the felsic magmas in the Wichita Granite Group/Carlton Rhyolite had relatively low water contents, the quite shallow depth of emplacement led to fluid saturation in the final stages of crystallization. Just north of the parking area, across the small stream gulley, and west, around the southernmost outcrop of Quartz Mountain, is a spectacular example of a late surge of fluid-saturated magma. The resulting rock is immediately recognizable because of its highly pock-marked character. The holes are miarolitic cavities that contained the exsolving vapor phase. Many of them are lined with small quartz and feldspar crystals precipitated from the vapor.

Many of the scattered aplitic dikes have associated pegmatitic cores, margins, or pods. Amphibole is the dominant femic phase, commonly found in elongated prismatic forms growing out from the walls, suggesting late growth after intrusion. Quartz veins seem to be last, representing subsolidus, hydrothermal activity of the exsolved vapor phase. These veins appear to be occupying cooling cracks.

Approximately 1 mi (1.6 km) past the Nature Center/New Horizon Trail parking areas, the road to the Lodge passes through the Twin Peaks cut (the northernmost arrow on Fig. 2). This road cut is wide enough to be a convenient place to pull off and see many interesting features: a) two phases of the Lugert Granite—one fine-grained, porphyritic, and granophyric, the other medium-grained, nongranophyric, and inclusion-rich; b) several types of inclusions—metasedimentary and mafic; and c) pervasively fractured granite with some local fault gouge: It is convenient to discuss features beginning from the northwest end of the cut on the northeast side. Here, a very obvious, steeply-dipping contact between the finer-phase and the coarser-phase of the Lugert may be seen striking at approximately the bearing of the cut (Fig. 4). Traversing southeastward, the face cut and the contact are nearly parallel, thus transecting each other many times. The inclusions within the coarser-phase are abundant but erratically distributed. No systematic study of their orientation has been done. The mafic ones are less abundant and range in size from 1 to 3 in (2 to 8 cm); the sedimentary ones range from 1 to 12 in (2 to 30 cm). Around the southeast end, toward the tip of the arrow, are well-exposed natural rock surfaces with many inclusions.

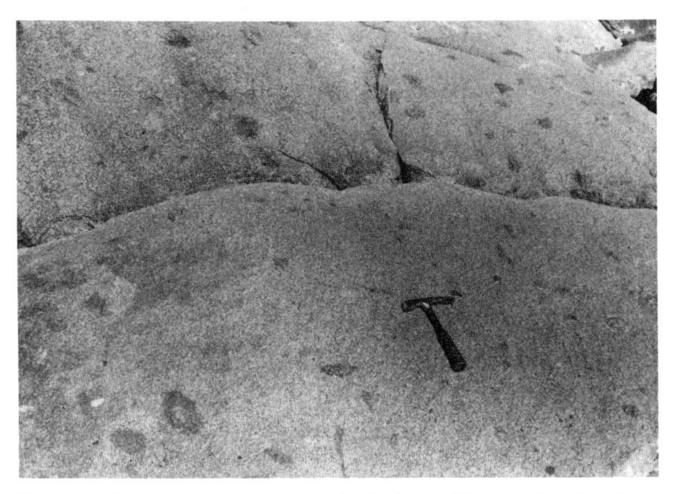

Figure 3. Photograph of coarser-grained phase of Lugert Granite showing scattered inclusions. Hammer is 33 cm long. Location is near arrow tip on most northerly arrow of Figure 1.

The dominant fracture system strikes 260° (80°)–270° and dips 85°S. However, the northeast face of the cut appears less fractured than the southwest face. Across the road, on the southwest face, there are isolated and local examples of consolidated, dark red aphanitic fault gouge along fractures (Fig. 5). Fracture spacing varies, and where more concentrated, perhaps represents "faulting," although total offset within any one set may not be large.

Near the southwest end of the west face, several clusters of inclusions are prominent. The mafic ones, in particular, seem to be undergoing disaggregation by the granite. Around the natural slope (toward the other arrow) are easily accessible illustrations

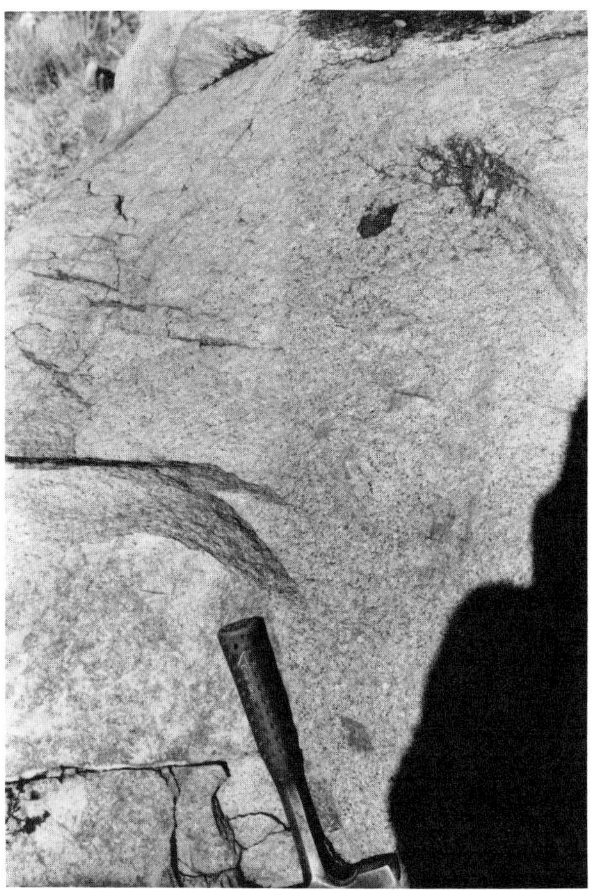

Figure 4. Photograph of nearly vertical contact between finer-grained, inclusion-free phase of Lugert Granite on the left, and coarser-grained, inclusion-rich phase on the right. Location is at the northeast side of the north end of the road cut to the Park Lodge.

Figure 5. Photograph of fractured Lugert Granite with dark red aphanitic material along some fractures. Material is either gouge or later clay infillings. Location is on the southwest wall of road cut to the Park Lodge.

of contact relations between finer and coarser phases of granite. Here it can be seen that angular blocks of the finer granite are surrounded by coarser. Within the finer, quartz veins occupy fractures which end against coarser. These fractures are genetically related to the intrusions through quenching of the finer granite and later filling upon solidification of the coarser. They are to be distinguished from the Pennsylvanian age tectonic fractures which are generally unfilled (or filled with later material) and which crosscut both finer and coarser phases. Aplitic dikes and pegmatitic pods with miarolitic cavities are in some cases positioned along the two granite contacts.

28—TOM STEED RESERVOIR AREA

LOCATION

This site is on the northeast side of Lake Tom Steed in the State Park Game Refuge, off U.S. 183, 8 mi (13 km) north of Mountain Park, 5 mi (8 km) south of Roosevelt, in Sec.27, T.4N.,R.17W. of the Glen Mountains 7½-minute Quadrangle and the Lawton 2-degree Quadrangle; lat. 34°47′30″; long. 98°57′ (Fig. 1).

ACCESSIBILITY

Turn south off of U.S. 183 (opposite a sign, "Mountain Park Wildlife Management Area," which points north) on to a gravel road that immediately crosses a railroad track (Fig. 6). Follow the gravel road 0.35 mi (0.5 km) to the parking area in front of a smelter foundation, once owned by the Gold Belle Mine and Milling Company. Around 1901, when the land was first opened for settlement, this area was involved in a gold rush for which there was no geological basis.

GEOLOGIC INTEREST

Outcrops of the L-Zone of the Glen Mountains Layered Complex (GMLC) and of the Cold Springs Breccia of the Wich-

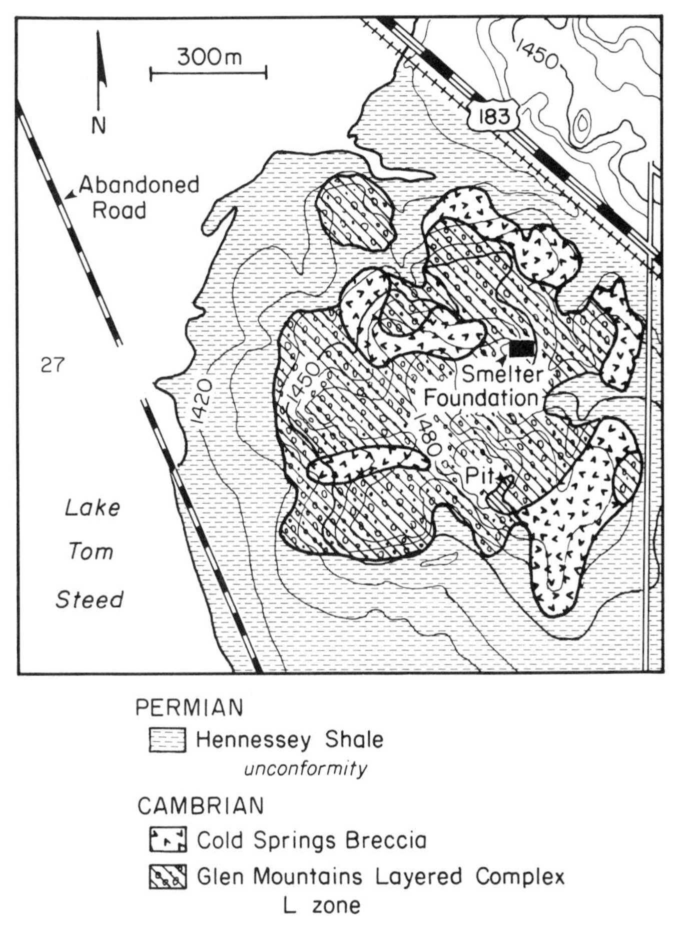

Figure 6. Geologic map of the area of the pre-Permian unconformity at Tom Steed Reservoir.

TABLE 2. GEOLOGIC PERSPECTIVE FROM TOP OF GOLD BELLE MILL FOUNDATION*

Azimuth	Feature
100°-0°-270°	Glen Mountains. Local range of the Raggedy Mountains underlain by the Glen Mountains Layered Complex.
355°	Highest peak of the Glen Mountains capped by M zone.
35°	"Window Rock" mountain, capped by M zone.
70°	Bear Mountain, on the distant horizon about 17 km away, underlain by Mt. Scott Granite (Wichita Granite Group), and one of the two highest in the Wichitas--2489 ft (737 m).
80°	Mt. Baker, at the northwest corner of the Wichita Mountain Wildlife Refuge. Cap composed of Mt. Scott Granite, but north side composed of Glen Mountains Layered Complex, and the west side, of Mt. Baker hornblende gabbro (one of the Roosevelt Gabbro bodies).
110°	Charon's Garden Mountain at southwest edge of Wildlife Refuge, composed of Quanah Granite.
175°	Mt. Radzminsky, composed mostly of Lugert Granite. Prominent large exfoliation surfaces facing south and well seen from the Snyder area.
190°	Mountains mostly composed of Mt. Scott Granite through which Otter Creek flows and was dammed to make Tom Steed Reservoir (and Great Plains State Park).
235°	Long Mountain, east-west elongated ridge composed of Long Mountain Granite.
250°	Twin Peak, mostly composed of Lugert(?) Granite but famous for an interesting outcrop of a mafic hybrid on its east side (See Powell and Fischer, 1976, Stop #10).
290°	Inactive quarry on west side of hill with water tower where many of the Cold Springs Breccia mixed rock relations are well-exposed (See Powell and Fischer, 1976, Stop #9).
300°	Hill capped with water tower is composed almost entirely of Cold Springs Breccia and is its largest outcrop.

*See Johnson and Denison, 1973, Stop #5; Powell and Fischer, 1976, Stop #1; and Gilbert and Donovan, 1982, Stop #1 for nearby geologic descriptions.

ita Mountains Igneous Complex of Lower to Middle Cambrian age (Table 1), and fossil topography and an unconformity of Permian age (Fig. 6) can be seen.

DESCRIPTION

This area is in the middle of the central region of the Wichita Mountains in a region called the Raggedy Mountains (Figs. 1, 7). The stop is near the crest of the Fort Sill South Anticline (Gilbert, in Gilbert and Donovan, 1982) (Fig. 8). From the top of the Gold Belle Mill foundation, the igneous core of the Wichitas can be seen in all directions (Table 2). This core was once covered by about 15,000 ft (4.5 km) of Upper Cambrian to Mississippian sedimentary units that were stripped off during Pennsylvanian uplift (see Figure 11 for a simplified regional cross-section). The physiography of the igneous knobs was formed in the Permian, just before they were buried and preserved by the Hennessy Shales.

All knobs readily visible to the south are composed of formations of the Wichita Granite Group. All immediate knobs to the east, north, and west are underlain by the Glen Mountains Layered Complex, locally intruded by Roosevelt Gabbros, Cold Springs Breccia, and/or late diabase (Table 2).

On the south side of the hill carrying the foundation are exposures of the Glen Mountains Layered Complex (GMLC) (Fig. 6). These outcrops exhibit typical L-Zone characteristics with some lithologies transitional to the M-Zone. Texturally the rocks are adcumulates and heteradcumulates; mineralogically they are gabbroic anorthosites and troctolitic anorthosites. Cumulus plagioclase (typically An_{73}) is the most abundant phase and imparts its gray color to the rock. The feldspar ranges in texture from sugary to well-laminated, and zoned rims can be seen on some crystals in thin section.

The second most abundant phase is ophitic (poikilitic) clinopyroxene, which occurs in two textural types. One type forms crystals 5–10 cm in diameter: the individual crystals can be readily seen when sunlight reflects off cleavage surfaces. This type—

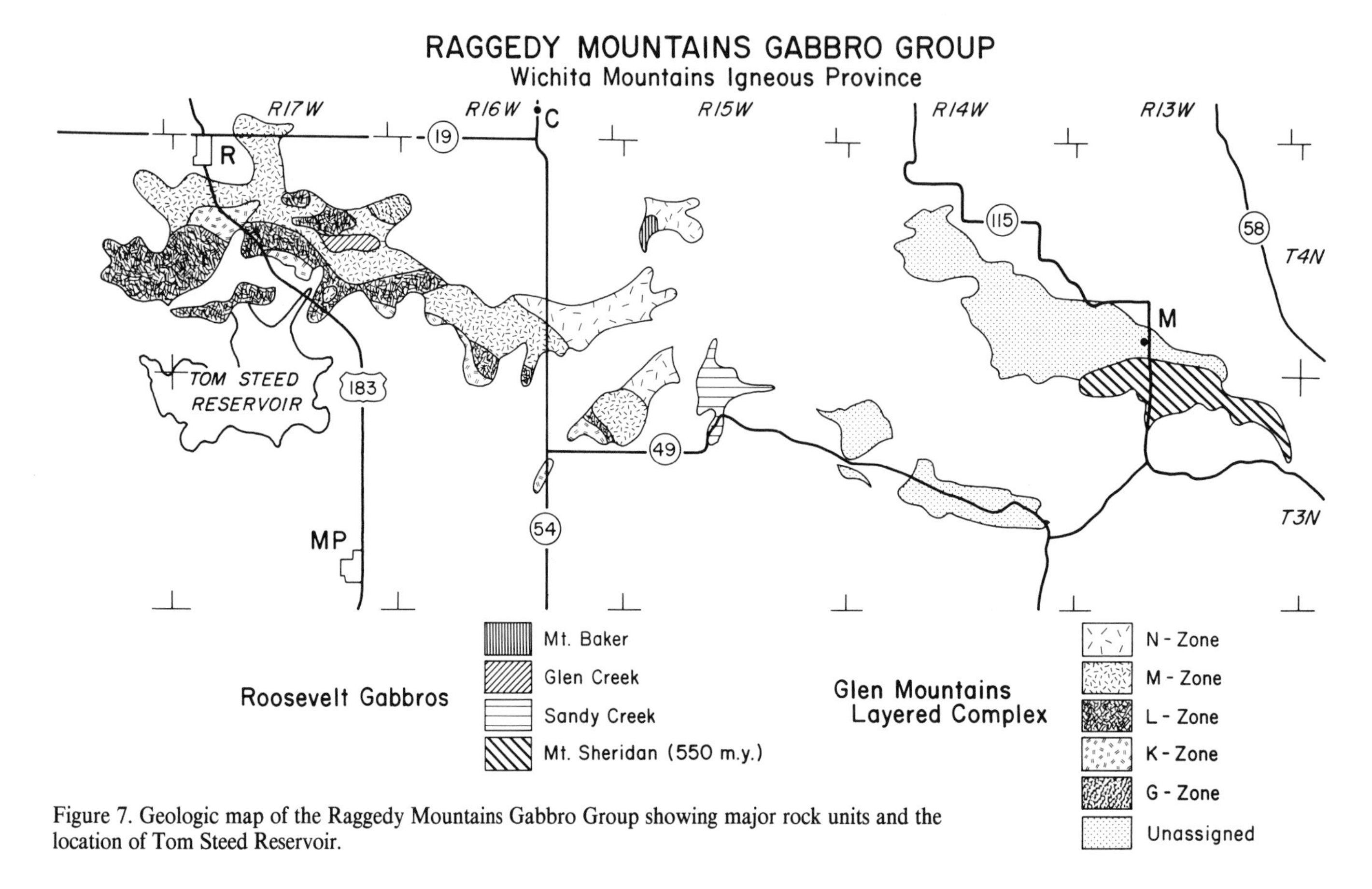

Figure 7. Geologic map of the Raggedy Mountains Gabbro Group showing major rock units and the location of Tom Steed Reservoir.

interpreted as the result of postcumulus nucleation and growth—is generally elongate in the plane of the plagioclase lamination, and the feldspar chadacrysts (inclusions) approximate in size and orientation the cumulus cores of free plagioclase. The other common textural type of clinopyroxene occurs in clusters of (or isolated) smaller (2–3 cm) crystals. These are also ophitic but contain smaller, randomly oriented plagioclase chadacrysts and are interpreted as cumulus (see Powell and others, 1980, for a more complete discussion of the ophitic pyroxenes in the GMLC). This type of clinopyroxene weathers to a lumpy positive relief on the outcrop.

The shiny black grains visible in the outcrop are "poikilitic" titaniferous magnetite. They weather to a positive relief. Some of these crystals interpenetrate with clinopyroxenes.

Olivine (typically Fo_{74}) occurs as a cumulus phase in irregular lenses. In these outcrops, well-defined horizons are not obvious but are common in the stratigraphically lower K-Zone and in many other outcrops of the nearby L- and M-Zones. Many of the cumulus cores show adcumulus growth so that the olivine rims become partially poikilitic. Peritectic rims of orthopyroxene as well as symplectic coronas of orthopyroxene + magnetite commonly are developed around olivine. These oxidize to a bright red and the symplectites form either pits, small pimples, or bumps on the weathered rock surface. Typical radiating expansion cracks around olivine formed during late-stage or postmagmatic hydrous alteration. Outcrops on the south and southwest sides of the hill are more olivine-rich than the rocks near the foundation.

Electron microprobe analyses reveal that plagioclase compositions and ratios of Fe/(Fe + Mg) in olivine (0.26), orthopyroxene (0.23), and augite (0.21) are generally consistent with values determined for the L Zone elsewhere in the Complex. (See Powell and Phelps, 1977 and Powell and others, 1980 for additional data.)

A substantial portion of the outcrop area at this stop is Cold Springs Breccia (Fig. 6). This unit consists of three parts: 1) Otter Creek Microdiorite (the mafic endmember), 2) the granite endmember, which is unnamed and whose relationship with the Wichita Granite Group is problematic, and 3) the mixed rock units, locally called the Cold Springs "granite," which are a suite of intermediate, mostly gray, granitoids ranging from mafic diorite to granodiorite and granite. Collectively these rocks have intruded the Glen Mountains Layered Complex as plugs, dikes, and quasi-sill-like bodies. Contact relations can be seen 1,000–1,300 ft (300–400 m) south of the parking area, just south of elevation 1,498 ft (456 m) in a shallow excavated pit. The rocks here are altered, but the structural relations are well-displayed.

The GMLC host rocks have been fractured and are cut by

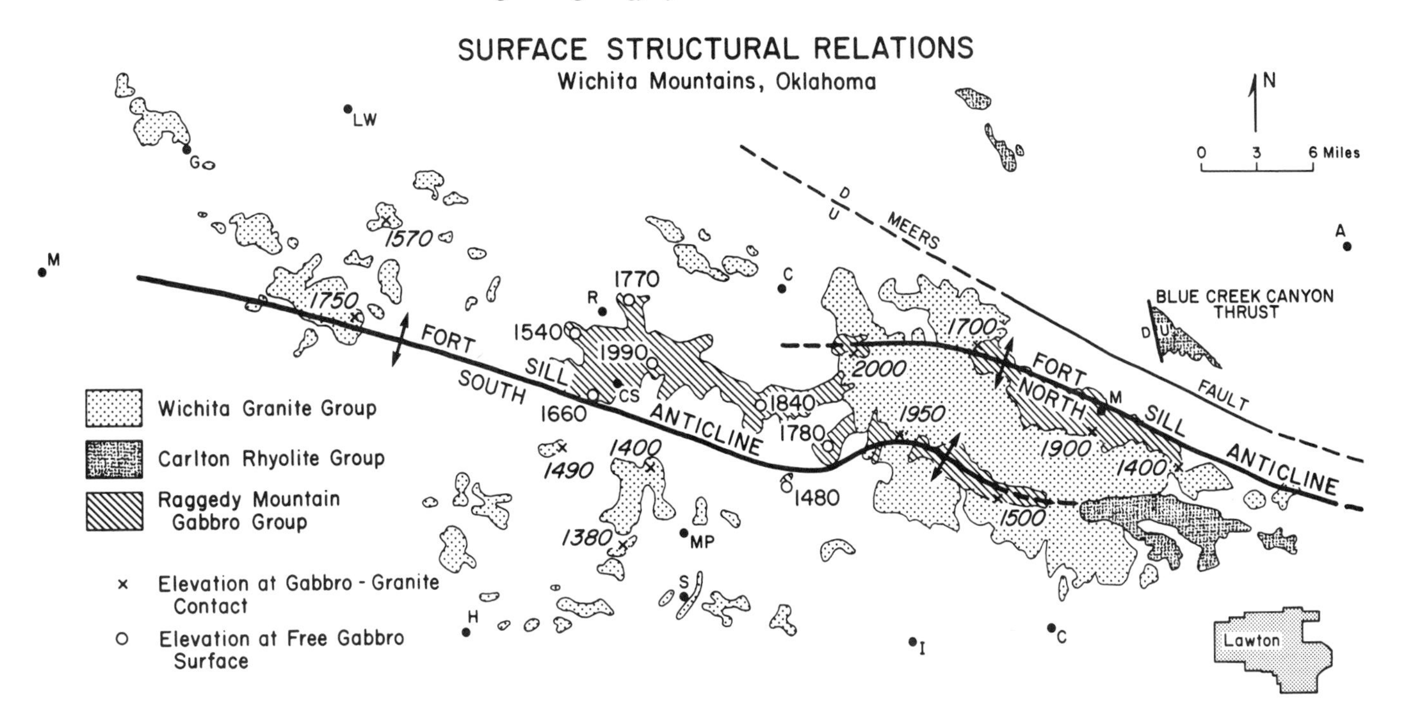

Figure 8. Generalized map of the Wichita Mountains showing major structural relations. This site is near CS in the center of the figure.

numerous small (thickness 1 mm–2 cm) veins, dikes, and sills. Presumably most of these are related to intrusion of the Cold Springs and/or the originally overlying Wichita Granite Group.

NEARBY STOPS

Table 2 includes references to stops in this general region that have been described in other guidebooks. Additionally, northwest of this stop two railroad cuts display many of the relations and contacts of the rocks described above. From the parking area, it is 2.35 mi (3.75 km) to the first cut and 3.45 mi (5.5 km) to the second. Take the gravel road to U.S. 183 0.35 mi (0.6 km), turn left (north) for 1.45 mi (2.3 km) on U.S. 183, turn left (west) off of U.S. 183 (at this turn are signs that point to Cold Springs and note leaving Mountain Park Wildlife Area). Continue westerly, turning left, then right, 0.55 mi (0.9 km) to a pile of large boulders on the south (left) side of the road, and park. Many of these blocks, excavated from the nearby railroad cut, illustrate well the mineralogy and textures of the L-Zone.

This cut (Cut #1), 0.3 mi (0.5 km) long and just south of the block pile and road, shows the relations between and among the Cold Springs Breccia with its associated Otter Creek Microdiorite and microgranite, and the Late Diabase dikes as they intrude the L-Zone of the Glen Mountains Layered Complex. A 130 to 165 ft-wide (40 to 50 m-) intrusion of Cold Springs is prominent at the east end of the cut. This intrusion has the common arrangement of a core of diabase with more granite on the margins. Farther west, a Late Diabase dike strikes 65° with primary en echelon offsets and is cut by a shear zone (Pennsylvanian?). Still farther west, several individual 3 to 12 ft-thick (1 to 4 m-) Cold Springs dikes occur. A distinct 4 in (10 cm) vertical microgranite dike strikes 24°. Nearer the west end of the cut, a Late Diabase dike clearly cuts Cold Springs. At the west end, a diorite-rich (Otter Creek) Cold Springs dike strikes 50° and dips 85° SE.

Driving farther west, 0.65 mi (1 km) (total distance from parking lot now 3.0 mi; 4.8 km) the road "Ts." North (right) about 0.5 mi (0.8 km) is the Cold Springs Quarry (see Stop #9 in Powell and Fischer, 1976). Turn left (south) and cross the railroad track as the road swings westerly, entering the Mountain Park Public Hunting and Fishing Area (at 3.1 mi; 5 km total). Proceed west and straight ahead up an incline to another "T" (3.25 mi; 5.2 km total), turning right past another pile of fresh boulders excavated from Cut #2 and park at about 3.45 mi (5.5 km) total. The second cut is just to the north.

The second cut shows the same relations as the first except that the sill-like nature of much of the Cold Springs is better displayed here in the east end of the cut. Ophitic clinopyroxenes of the L-Zone commonly range in size between 4 and 8 in (10 and 20 cm) and are distinct to the eye because of the strong reflectivity of their cleavage surfaces. The weathering profile on gabbroic anorthosites can be studied. Intensity of fracturing noticeably decreases from the surface into the fresh rock. A distinct 1.5 ft-thick (0.5 m) Late Diabase dike, striking 355° and dipping 85 to 90°E occurs at the west end of the cut.

29—FRENCH LAKE AREA

LOCATION

French Lake is wholly within the Wichita Mountain Wildlife Refuge. It is shown on the Quanah 7½-minute Quadrangle, the Lawton 1:100,000 Quadrangle, and the Lawton 2-degree Quadrangle. The localities described here, on the south and east sides of the lake, are all in the NW¼Sec.20T.3N.,R.14W. The area is south of Oklahoma 49 (extended), as it passes through the wildlife refuge, and east of the Indiahoma Road (Figs. 1, 9).

ACCESSIBILITY

Oklahoma 49, from Medicine Park, and Oklahoma 115, from Cache, lead to the principal east-west scenic highway (Oklahoma 49 extended). Oklahoma 49 can be followed west to the headquarters of the Wichita Mountains Wildlife Refuge. From there, proceed south on Indiahoma Road to the French Lake Parking area. Indiahoma Road can also be accessed from the west via Oklahoma 49 off of Oklahoma 54, or from the south. From Indiahoma, turn north off of U.S. 62 onto a county road that leads to the wildlife refuge's southwest entrance. Permits, obtainable from wildlife refuge headquarters, are required for collecting.

GEOLOGIC SIGNIFICANCE

This area documents some of the intrusive relations of the Quanah Granite into the Mount Scott Granite and the Glen Mountains Layered Complex (GMLC). A variety of xenoliths and dikes occur in the Quanah that are characteristic of the pluton's margin, and which allow inferences to be made about the nature of the granite contacts and the intrusive process. The Permian unconformity and overlying Post Oak conglomerate can be seen here. (Table 1 shows the stratigraphic column for the Wichita Mountains area.)

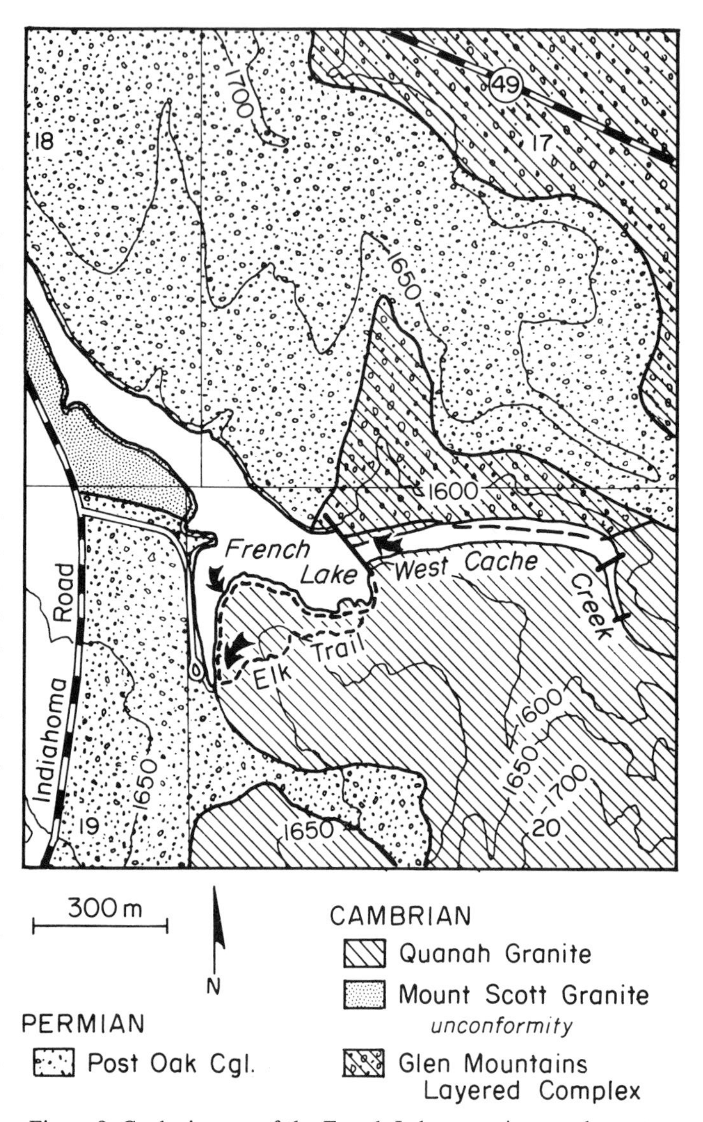

Figure 9. Geologic map of the French Lake area. Arrows show stops emphasized in the text.

DESCRIPTION

This area has previously been described as a field trip stop by Gilbert (1982a). It is one of the key places within the Central Lowland ("central area" and "interior lowland" of Hoffman, 1930) where Quanah Granite can be seen intruding the Glen Mountains Layered Complex (Fig. 9). This description is centered around the parking area, the shortest of the adjacent loop trails (Elk Trail), and the French Lake Dam area.

Post Oak Conglomerate outcrops on the west side of the parking area. Its age has been variously interpreted as Pleistocene (Hoffman, 1930) to Permian (Chase, 1954). Although there appear to be conglomerates/gravels of both ages within the Wichita Mountains area, these are taken as Permian. The conglomerate has prominent rounded clasts, which are evidently due to spheroidal weathering at the source region where they formed as corestones before being dislodged from the altered rock matrix by later erosion. All of the clasts are of Mount Scott Granite and thus derived from the north. The type section for the Post Oak, several kilometers south of here (see Chase, 1954), is now part of a restricted impact area within Fort Sill. A convenient stratotype has been selected near Camp Boulder and can be reached off of the main east-west scenic highway (Oklahoma 49 extended). The outcrop is along a small stream tributary to West Cache Creek on its south side in SW¼NW¼SW¼Sec.28,T.3N.,R.14W., just west of Camp Boulder.

Across the small bridge at the trail head on the east side of the parking lot are outcrops of Quanah Granite. The Quanah in this area is somewhat finer-grained and more varied compared to typical exposures elsewhere. Gilbert (1982a) argued that this indicates quenching, not only near the Quanah's northern contact with the Glen Mountains Layered Complex and Mount Scott Granite, but also near its upper contact, presumed to have been against Mount Scott Granite, now eroded away.

Upon crossing the bridge, turn left and proceed a few meters on the lakeside trail, then turn right off of the trail for a few meters more upslope to the east (near the arrow point on Fig. 9). A very shallowly-dipping 1-ft (30-cm) aplitic dike cuts the granite. Along its upper contact is developed an even later stage, possibly subsolidus, vuggy quartz vein representing fluid-excess conditions. This could be taken as a stratigraphic-up indicator, showing that no significant rotation of this crustal block has occurred. Aplitic dikes and quartz veins are common in this vicinity but are more abundant near the granite-anorthosite contact.

The area between here and French Lake Dam contains abundant angular inclusions of Mount Scott Granite, some several meters in diameter. Other inclusions appear to be metasedimentary, are similar to those seen at Quartz Mountain, and are more rounded. Less abundant are hornfelsed Carlton Rhyolite and mafic clots, which may be disrupted basaltic (diabasic) dikes. Many of these are hard to see at first and it is helpful to take the time to look carefully over several tens of square meters of outcrop to sensitize the eye.

The endpoint of this discussion is the outcrop just below French Lake Dam, shown as the eastern-most arrow on Figure 9. Two approaches are possible. The quickest way is to follow the lakeside trail, turn left after crossing the small bridge from the parking lot. The longer route follows the Elk Trail Loop.

To proceed on the Elk Trail Loop, cross the bridge, turn right (southerly) onto the trail system, and then turn left at the first opportunity to arrive on the Elk Trail segment. Immediately behind the trail signpost and between the trail segments are inclusions of Mount Scott Granite and metasediments; there are also aplites—some with pegmatitic margins. Farther up the Elk Trail, some of the pegmatitic dikes show quartz and alkali feldspars crystallizing perpendicular to the walls and into what was fluid-occupied space. There must have been low fluid velocities within the dikes at this stage. Finer-grained phases of the Quanah are ubiquitous. Rhyolite inclusions can be seen in the path. On the way to the crest of the ridge, there are good illustrations of corestones that result from spheroidal weathering and of corestones split by ice-wedging.

From the ridge crest, the view to the west shows two principal peaks; Mt. Lincoln to the southwest, with large exfoliation surfaces, and the east end of Elk Mountain. All but the north side of Elk Mountain is underlain by Quanah Granite. On that side is a transition zone where Quanah Granite has intruded Mount Scott Granite. Very large blocks of Mount Scott are engulfed and interfingered by Quanah. The "contact" between these granites thus requires some interpretation. Is it the northernmost extent of Quanah apophyses, or is it the southernmost extent of large Mount Scott xenoliths? The contact has been taken as the beginning of the zone where Mount Scott is the dominant granite in volume even though it is partially stoped. This is thought to have been the original "boundary" of the Mount Scott against its host, presumably the Carlton Rhyolite Group. Somewhat later (the actual time difference is unknown since the Mount Scott has never been successfully dated while the Quanah has: see Powell and others, 1980, for discussion of dating), it seems that the Quanah intruded against the side of and under, the Mount Scott Granite. This accounts for the inclusions of metasedimentary and metarhyolite xenoliths in the Quanah, since these are remnants of the original host rock.

The Quanah-Mount Scott contact is high-angle as it cuts across the north side of Elk Mountain. It is inferred to bend over and become shallowly dipping to the north in this immediate area. Our present position would have been under this contact, with the Mount Scott now eroded off. The Mount Scott would have overlain the Quanah, having originally been its intrusive roof, over much of the area east of here toward Quanah Parker Lake, 3 mi (5 km) away, and possibly as far south as Eagle Mountain. The present topography as shown on the Quanah Mountain quadrangle is suggestive of this former distribution of Mount Scott.

The view to the north and east from the ridge crest shows the central/interior lowland, which is floored by Glen Mountains Layered Complex and Post Oak Conglomerate. The mountains and ridges north of the lowland are all composed of Mount Scott Granite. Mt. Marcy is the prominent peak to the northeast. The lowland surface was the floor of the Mount Scott sill and the sill has been stripped back by erosion. The "lowland" was carved out in the Early Permian during the first stage of development of the tor topography. Subsequently, this paleovalley was filled with Post Oak Conglomerate and all the Permian topography was then buried in the Hennessey shale. To the north of the lake, the grassland area is underlain by Post Oak, sitting as a sheet on the old Permian valley floor, which is composed of Glen Mountains Layered Complex (GMLC). Just east of the dam, the grasslands are underlain by the GMLC where the Post Oak was removed during the Quaternary. On reflection, it is an incredibly rare situation that the GMLC surface of the lowland has been exposed three different times over geological time: once in the Cambrian when the upper parts of the GMLC were eroded off and rhyolite was deposited on it (after which the Mount Scott Granite intruded along that unconformity); once again in the Permian when the Mount Scott was eroded off just prior to deposition of the Post Oak; and now as the Permian layers are being stripped away.

Hoffman (1930) pointed out that the present drainage partly follows the Permian and is also superposed. For example, West Cache Creek drains the Central Lowland but exits south to the plains through newly-cut (Late Tertiary-Quaternary) gorges between Quanah and Eagle Mountains. A brief review of the gross stream network shown in Figure 1 indicates that regional drainage patterns appear to have been consistent from the Tertiary onward.

As one proceeds down the northeast side of the ridge, Carl-

Figure 10. Contact of the Glen Mountains Layered Complex (GMLC) and the Quanah Granite along West Cache Creek.

ton rhyolite inclusions are evident; the Quanah is relatively fine-grained, consistent with quenching against its presumed roof; spheroidally weathered boulders are abundant (formed, of course, during an earlier period of weathered rock cover while below the groundwater table); local, irregular fracture sets are obvious; there are some more examples of mafic inclusions that were being disaggregated; and a nice illustration of a polygonally fractured dike whose side wall is the face of the outcrop. The limestone gravel on the trail is the Cambro-Ordovician Arbuckle Group from the Slick Hills to the north. The trail passes by one of the hundreds of small prospect pits that were dug all over these mountains around the turn of the century. After the trail leaves the ridge slope and joins the main trail system, the next stop is the French Lake Dam area.

To reach the dam via the lakeside trail, turn left after crossing the bridge from the parking lot. About 600 ft (180 m) along the trail, indicated by an arrow on Figure 9, there is a small bend in the path around a small boulder. This is a convenient stop to observe a variety of inclusions. Please do not hammer on them. There are definite porphyritic granite inclusions with ovoid feldspars that are taken to be Mount Scott Granite. These may be seen everywhere in the vicinity, for example, as one moves eastward off the trail up the slope. The boulder around which the trail bends has an angular fragment of fractured rhyolite on its top. Were these fractures there before being engulfed in the Quanah, were they formed on immersion, or were they formed tectonically later in the Pennsylvanian?

On the bedrock surface immediately around the boulder's base, on the north side, are a cluster of inclusions. The metasedimentary inclusions are similar to those seen at Quartz Mountain. Several may be fragments of rhyolite and mafic rocks of unknown origin. Just at the west side of the exposed surface, highly quenched Quanah Granite with quartz pods is evident.

As the trail turns east, note that the north-facing rock slopes are almost completely covered with several varieties of lichens. This is characteristic of all north-facing slopes within the igneous parts of the Wichita Mountains. Consequently it is difficult to look for subtle rock variations on such slopes.

The dam site locality is shown by an arrow in Figure 9. It requires crossing the dam and its spillway. If water is flowing over the spillway, hold onto the cable, as the surface can be slippery. Alternatively, it is sometimes convenient to cross West Cache Creek 30 to 60 ft (10 to 20 m) downstream from the circular fish ladder. The outcrop to be studied extends from the base of the dam along the north side of the creek. Here the contact between the Glen Mountains Layered Complex (L-Zone?) and the Quanah Granite is displayed (Fig. 10). The state geologic map (Miser, 1954) and the 2-degree quadrangle map (Havens, 1977) show this as a fault. Gilbert (1982a, 1984) has concluded this is an igneous intrusive contact. The granite contact surface is more resistant to erosion and, locally, dips north under the GMLC. Several granite dikes are subparallel to the contact on the GMLC side, confirming the granite's intrusion. The anorthosite shows good in situ spheroidal weathering.

Most of the bare rock surface is Quanah Granite at its intrusive margin. It is full of scattered, angular to rounded inclusions. Most are metasedimentary, metarhyolite, or Mount Scott Granite, but there are a few deformed pancake-shaped mafic inclusions, which may be disrupted fragments of coeval basaltic (diabasic) dikes intruding into the partially molten Quanah sill. Some of the igneous inclusions themselves contain inclusions. Aplitic dikes abound, many subparallel to the contact. As the granite is not highly quenched here, substantial masses of magma must have already moved by this point before solidification.

30—MOUNT SCOTT AREA

LOCATION

The Mount Scott area, in Comanche County, Oklahoma, extends from Medicine Park on the east to Meers on the west (Fig. 1) Most of this area is within the Wichita Mountains Wildlife Refuge, U.S. Department of the Interior, but some is private land or belongs to the U.S. Army (Fort Sill), the City of Lawton

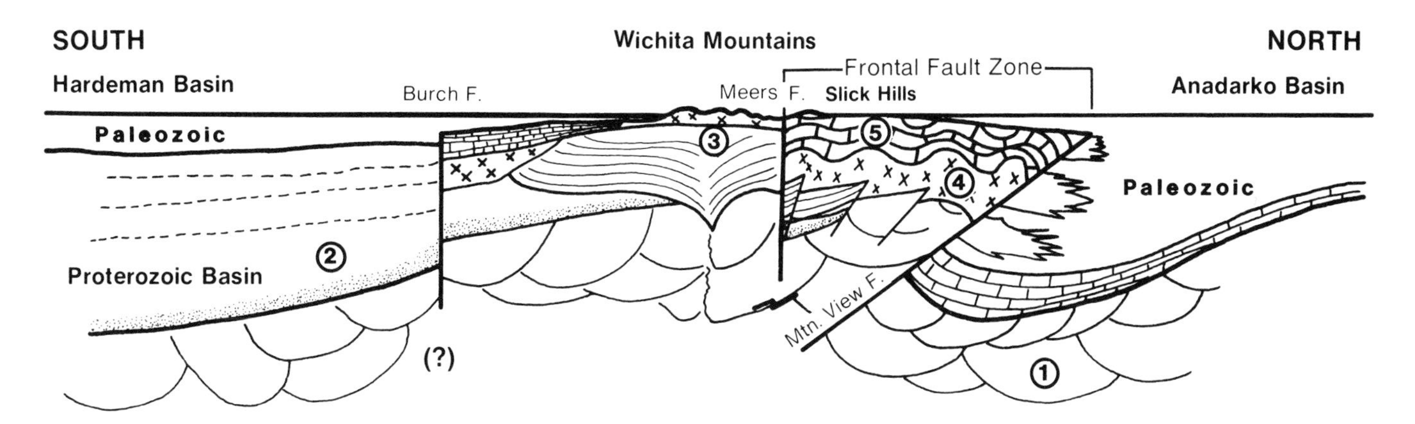

Figure 11. South-north schematic cross-section across the Wichita Mountains axis. The uplifted central block and the adjoining basins are part of the Southern Oklahoma Aulacogen. Explanation of numbered areas: 1) Proterozoic basement, presumably the 1.35–1.40 Ga granite-rhyolite terrane (Denison and others, 1984). How far south this extends is unknown. 2) Proterozoic basin inferred by Brewer and others, 1981. 3) An Early to Middle Cambrian large, layered, gabbroic anorthosite body akin to the Stillwater and Duluth complexes, called the Glen Mountains Layered Complex (Powell and others, 1980); also houses smaller, later gabbroic plutons called the Roosevelt Gabbros. 4) The Middle Cambrian silicic igneous cap to the Wichita igneous province: Carlton Rhyolite Group and Wichita Granite Group. 5) The Upper Cambrian to Mississippian pre-uplift sedimentary sequence, largely carbonates.

(Lake Lawtonka), or the Medicine Park State Fish Hatchery. The area is about 17 mi (27 km) north and west of Lawton, Oklahoma, and falls entirely within the Lawton 2-degree Quadrangle or the Lawton 1:100,000 Quadrangle. However, the following three 7½-minute quadrangles are most useful: Fort Sill (extreme NW corner), Mount Scott (north one-quarter), and Meers (south one-quarter). The immediate area falls mostly within T.3N., R.13W., with Mount Scott itself being located at about 34°44′45″ north latitude, 98°32′ west longitude.

ACCESSIBILITY

Oklahoma 49 (west of U.S. 277-281 and I-44) and Oklahoma 115 (north of U.S. 62), and their extensions through the Wildlife Refuge, provide easy access. Five stops are highlighted. Three of these are within the Refuge, which forbids hammering on outcrops or collecting samples without a permit. Permits can only be secured from Refuge headquarters, about 11 mi (17 km) west of Mount Scott.

GEOLOGIC SIGNIFICANCE

Mount Scott (elevation 2,464 ft [740 m], relief about 1,100 ft [330 m]) is one of the best known physiographic features of Oklahoma. It represents an exposed portion of anomalously young (Middle Cambrian) Midcontinent basement that is part of the Southern Oklahoma Aulacogen (Fig. 11). In the vicinity are parts of the type areas for the Mount Scott Granite, Quanah Granite, Medicine Park Granite, Cache Granite, Carlton Rhyolite, Mt. Sheridan Gabbro, Meers Quartzite, Fort Sill Formation, Signal Mountain Formation, McKenzie Hill Formation, and the Post Oak Conglomerate (Table 1). The area is strongly fractured. Several major faults are exposed. One of these, the Meers Fault, shows Holocene movement. Interesting fossil (Permian) geomorphic forms are being uncovered.

DESCRIPTION

The area in the vicinity of Mount Scott is described from five local stopping points, beginning on the east and moving westward: 1) the Medicine Park State Fish Hatchery, 2) the Mount Scott Picnic Area, 3) the parking area atop Mount Scott, 4) the Quetone Overlook-Eastern Lowland, and 5) Meers.

Medicine Park State Fish Hatchery

Location. Medicine Park, along Oklahoma 49, where 49 crosses the boundary between sections 19 and 20, T.3N.,R.12W.; Fort Sill 7½-minute Quadrangle (Fig. 12). Park at the small picnic site on the north side of the highway. The stop is the small bluff-roadcut immediately to the east.

Geology. Hornfelsed Carlton Rhyolite, Mount Scott Granite (facies B), rhyolite dike, Post Oak Conglomerate, Permian unconformity.

This outcrop was first noted by Ham, Denison, and Merritt (1964) as important in demonstrating the contemporaneity of granitic and rhyolitic units. Granite can easily be shown to have intruded rhyolite in certain outcrops on Fort Sill. Here, in the road cut on the north side of the highway, a rhyolitic dike cuts the granite. The contact has two interesting aspects: 1) At the eastern contact, rhyolite appears to have risen up underneath the granite and then intruded its way upward cutting through the granite. The contact is digitated, with rhyolite fingering into granite. Phenocrysts of alkali feldspar in the rhyolite show a rough alignment with the contact. All relations are consistent, with rhyolite intruding granite. 2) At the western contact, a few meters west of (1), granite extends to the road level. There the rhyolite reappears against granite along a smooth, high angle contact before being capped by granite. There is no direct evidence of faulting.

The rhyolite dike has extremely evolved chemistry (in oxide wt%: SiO_2, 76.62; TiO_2, 0.12; Al_2O_3, 11.85; total Fe as Fe_2O_3, 2.29; MnO, 0.01; MgO, 0.01, CaO, 0.02; Na_2O, 3.42; K_2O, 4.37, H_2O, 0.36, total 99.07 [XRF data from J. D. Myers and M. C. Gilbert]). CaO and MgO are very low. Rare earth element data show a distinct negative Eu anomaly. Phenocrysts are quartz, alkali feldspar, and an altered prismatic femic mineral, possibly hornblende. In thin section, some phenocrysts are broken into several segments, implying rapid and forceful injection. The groundmass was probably originally glassy but is now a very fine-grained granoblastic aggregate of quartz and feldspar.

The granite is tentatively assigned to Mount Scott, facies B, the basal part of the sill (Myers and others, 1981). It is fine-grained without the obvious ovoid feldspars characteristic of facies A.

The second important point about this dike, not hitherto discussed, is that texturally it is a rhyolite. It is a very rapidly quenched rock. The question arises: Why is it not an aplite or a micrograinte? This can only mean that 1) it intruded an already cooled granite and 2) there was little overburden above this stratigraphic level. Since the granite originally cooled slowly enough under sufficient cover to become texturally a "granite," significant overburden was removed before the time of rhyolite dike emplacement. Because Mount Scott Granite appears to be early in the volcanic-intrusive felsic sequence, and the volcanic pile built up over time after Mount Scott intrusion, the rhyolite dike must be very late in the sequence. Some Mid-Cambrian uplift must have occurred here to allow erosion to strip much of the Carlton Group and most of the Mount Scott away before the dike was emplaced.

The floor of the Mount Scott Granite can be seen here. Just to the northwest of the roadcut discussed above, a small knob of hornfelsed rhyolite (not the dike material) is exposed. This is also the older, presumably very nearly stratigraphically lowest, part of the Carlton Rhyolite Group. It is a fine-grained granoblastic mass of quartz and alkali feldspar with indistinct alkali feldspar phenocrysts. The recrystallization has blurred its original porphyritic character because the phenocryst margins are no longer sharp. This outcrop is not a small inclusion immersed in granite, as much more of the hornfelsed rhyolite is exposed immediately to the north and up the hill, and also to the south on the south edge of the fish ponds.

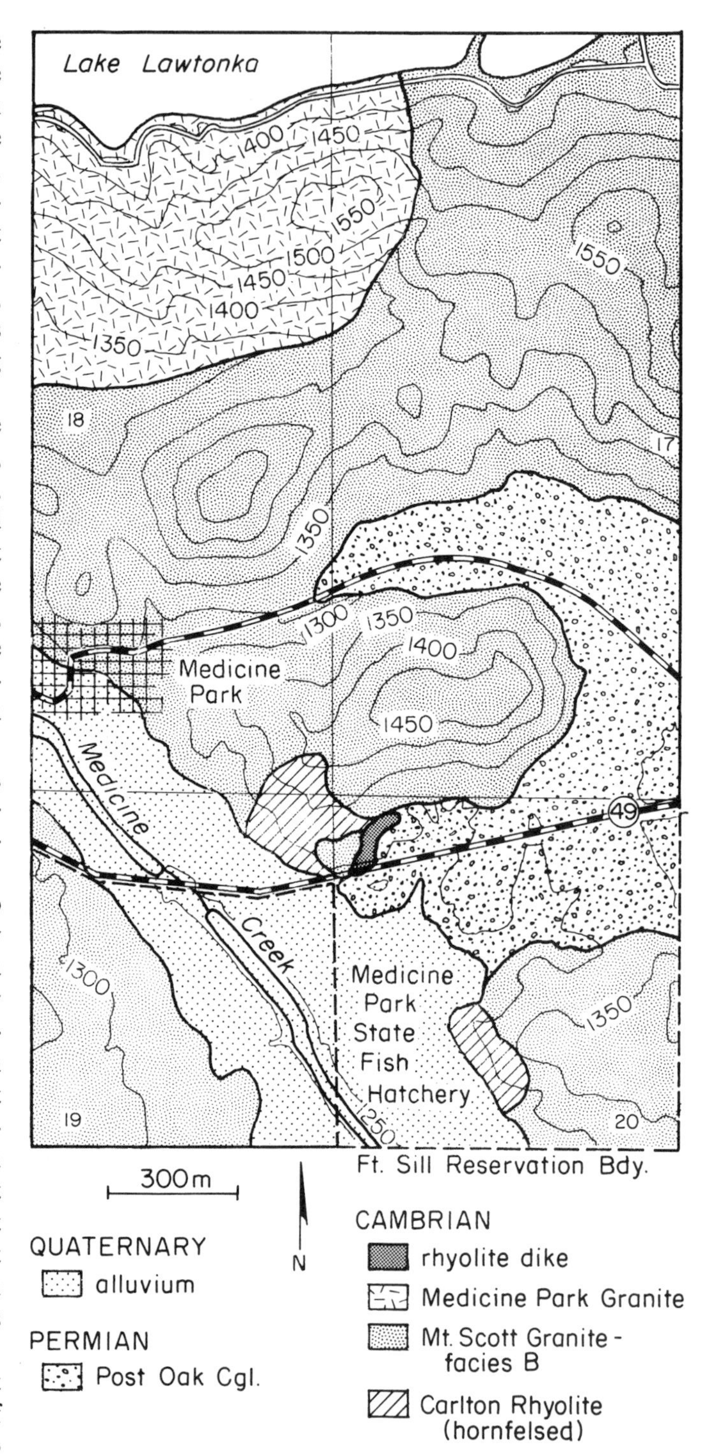

Figure 12. Geologic map of Medicine Park area.

Some feldspathic quartzite zones associated with the hornfelsed rhyolite have been reported as Meers Quartzite (e.g. Ham and others, 1964, p. 114, 115, 118). Gilbert, and Sides and Miller, have discussed aspects of this problem in Gilbert and Donovan (1982). This association may mean that early tuffaceous units and externally derived arenaceous units were intermingled in the beginning of the development of the Carlton Group. Any such quartzite would not be Tillman equivalent (Proterozoic) but would postdate the gabbro floor onto which all the felsic rhyolites and granites were situated. Thus there may be two kinds and ages of Meers Quartzite, or only one younger (Cambrian) kind. There probably cannot be only one older (Tillman-Proterozoic) kind.

The Post Oak Conglomerate, a poorly-cemented variety of the granite-rhyolite facies, is exposed on the east end of the cut. The boulders and cobbles are all derived from the Mount Scott Granite or the Carlton Rhyolite. The unconformity against the igneous units shows marked discoloration (whitening) due to ground water-induced alteration. Ground water flow seems to be concentrated along this surface, giving it a fairly distinctive appearance.

Mount Scott Picnic Area

Location. At the base of Mount Scott on the south side of Oklahoma 49 (extended) in the Wichita Mountains Wildlife Refuge NW¼Sec.14,T.3N.,R.13W., Mount Scott 7½-minute Quadrangle. The picnic area is closed at night.

Geology. Base of Mount Scott Granite sill, dirty facies of Meers Quartzite, Carlton Rhyolite, late diabase dike, late rhyolite plug, Post Oak Conglomerate, Permian unconformity.

This stop illustrates several key stratigraphic and structural relations among Mount Scott Granite, Carlton Rhyolite, and Meers Quartzite. The picnic area is mostly underlain by the "dirty" facies of the Meers. Interesting outcrops have been exposed in two broad, shallow, excavated pits immediately to the south of the east picnic area restrooms. These show the Meers to be highly fractured, greenish-gray and gritty in hand specimen and composed mostly of quartz and yellow-green to white micas. Several types of banding (layering) of unknown origin occur.

Near the west edge of the picnic area, the Meers is interpreted to dip under the Carlton Rhyolite. The contact may be located within several meters but is not exposed in detail. The Carlton is a salmon to dark-red-brown, fractured massive rhyolite. Phenocrysts of red-orange alkali feldspar (variable but commonly up to 10 modal %) and dark gray quartz (variable but commonly about 5 modal %) are prominent. The feldspars are typically aligned rather consistently over single outcrops. Layering so defined here has a strike of 50–80° and dip of 10–15° NW.

Immediately to the north of the picnic area rises Mount Scott, composed on this side entirely of Mount Scott Granite. The rock is a typical hypersolvus, fine-grained, A-type granite of sheet form, variably granophyric. Color index is 5–7%, consisting mostly of chloritized hornblende and magnetite. Commonly it contains ovoid feldspar phenocrysts. It is oxidized red, presumably during or shortly after the emplacement process. The basal "layer" of the granite is finer-grained and was called "facies B" by Myers and others (1981). This is the granite phase that crosses the highway at the small summit, just at and east of the picnic area entrance.

On the north side of Mount Scott, the Mount Scott Granite sill lies on top of the Mt. Sheridan Gabbro Member of the Roosevelt Gabbros (Raggedy Mountains Gabbro Group) (Table 1). On the south side, at the picnic area, the Granite lies on Meers and Carlton. Gabbro can be found at the far east end of Mount Scott, about 1 mi (1.6 km) east of the on-ramp to the summit road and 0.25 mi (0.4 km) north along a small access road, but the nature of the gabbro to rhyolite or quartzite contact in this vicinity has not been determined. Gilbert (1982b, Fig. 127) interpreted these relations to show that the gabbro dips more steeply to the south than the granite, allowing for an intervening wedge of quartzite and rhyolite. These two units should lie unconformably on the gabbro if previous regional interpretations are correct, but this has not been verified in the field.

Structurally, the granite sill appears to dip southward, whereas the rhyolite dips northward and the quartzite may be variably deformed. Continuing Cambrian tectonism is implied during development of these rock units.

Parking Area—Top of Mount Scott

Location. Northward up summit road 3 mi (5 km), which is entered off Oklahoma 49 extended, about 0.25 mi (0.4 km) west of the picnic area entrance, SE¼Sec.11T.3N.,R.13W. within the Mount Scott 7½-minute Quadrangle. The road and parking area are closed at night.

Geology. Traverse of Mount Scott Granite sill, type locality of the Granite; view of geologic, geomorphic, and geographic relations of the eastern Wichita Mountains and surrounding Permian-floored plains.

The 3-mi (5 km) drive to the Mount Scott summit parking area passes up through the largest well-exposed section of the Mount Scott Granite. This granite nomenclature was introduced by Merritt (1965) and has been accepted by Myers and others (1981). One feature characteristic of this granite is erratically distributed, generally small, mafic xenoliths. These are fine-grained and range in size from microscopic to cm-scale. The cuts along the road to the summit display them well, particularly from the half-way point on up to the summit, where they can be seen in the center outcrops. While the immediate substrate for most of the Mount Scott is medium- to coarse-grained gabbro and anorthosite, almost all of the mafic inclusions are fine-grained, indicating a source that was texturally a basalt. Whether this was Navajoe Mountain Group or some other liquid more directly associated with the Mount Scott Granite is unknown.

On the Mount Scott summit one is standing near the crest of the inverted block that once represented the floor of the Southern Oklahoma Aulacogen (See Fig. 11). This Middle Cambrian floor, after significant erosion, was buried by an estimated 15,000 ft (4.5 km) of sediment from Upper Cambrian through Mississippian age. The lower 10,000 ft (3 km) or so of this sequence is exposed directly to the north in the Slick (Limestone) Hills portion of the Wichita Mountains. Limited parts are also exposed to the southeast in Fort Sill and more complete sections, particularly of the mid and upper 15,000 ft (4.5 km) are prominent in the Arbuckle Mountains about 60 mi (100 km) southeast.

Only 9 mi (15 km) or so to the north, in the Anadarko Basin, this whole section is buried by about 23,000 ft (7 km) of Pennsylvanian and Permian syn- and post-uplift units. Thus, the basement, the equivalent rocks on which you are now standing, is perhaps 7.5 mi (12 km) deep only a short distance northward. This Pennsylvanian structural inversion is accomplished by a series of large throw, thrust (oblique-slip) faults through an interval called the Frontal Fault Zone (Harlton, 1963). The northernmost of the large faults, and the largest, is called the Mountain View (dipping 40° south) and is buried beneath the Permian. The southernmost of the large faults is called the Meers and passes 3 mi (5 km) north. Its major movement related to the inversion was Pennsylvanian. Permian movement with strike-slip component (Donovan and others, 1985) and Quaternary movement (Gilbert, 1983a; Donovan and others, 1983) have also been documented. (Donovan, Ragland and others, this volume).

After inversion was complete in the late Pennsylvanian, a low-relief plain above the level of the Mount Scott summit, was eroded on the uplifted block. The first phase of development of the presently-exposed tor topography then commenced: deep spheroidal weathering controlled by the joint, fracture, and fault pattern in the igneous rocks. This distinct cessation of tectonic activity was terminated in the Permian by a period of block faults with possible strong strike-slip components.

Stream erosion and mass wasting carved a landscape very similar to the present one. The rinds and corestones of the spheroidally weathered rock became the alluvial debris immediately surrounding the re-uplifted and denuded mountain blocks. This lithified debris of sandstone lenses, shaly interbeds, and conglomeratic wedges is the Post Oak facies of the Hennessey Group. Eventually, externally-derived muds and silts from the Ouachita belt to the east enveloped the Permian surface. Essentially, a paleotopography of Permian age is now being exhumed. So, technically, these are truly quite "old" mountains. The largest boulder streams in the Wichita Mountains are on Mount Scott and nearby peaks to the east. These streams are formed from toppled tors.

The views from the summit are instructive and will be described briefly. To the east and southeast, Lake Lawtonka, part of the water supply system of Lawton and Fort Sill, was formed by damming Medicine (Bluff) Creek. Medicine Creek is the principal drainage of the upper Meers Valley but it does not exit directly to Cache Creek. Note that it cuts through resistant granite and rhyolite hills, indicating it is superposed from Late Tertiary-Early Quarternary erosion surfaces. Medicine Park is the small community just below the Lawtonka dam. The higher of the wood-covered hills to the east-southeast are underlain by Mount Scott Granite, which can be traced for about 34 mi (55 km) from its farthest east to its farthest west point. Fort Sill and Lawton can be seen farther to the southeast, lying on the Permian Hennessey shales and Quarternary flood plain deposits of Cache Creek and its tributaries.

To the south, Lake Elmer Thomas was formed by damming Little Medicine Creek. The south half of the lake is within Fort Sill. Most of the smooth-surfaced and rolling hills are underlain by the Carlton Rhyolite on the Fort Sill Military Reservation. The differences in topography between all rhyolite- and some granite-floored hills, and other granite-floored hills results primarily from intensity and degree of fracturing. Since the bulk rock chemistry of all the silicic units is similar, differences in weathering, erosion, and vegetation must relate to the lower water retention of some of the more highly fractured units.

The two most noticeable, but low, east-west ridges on the south within Fort Sill are underlain by Carlton Rhyolite dipping uniformly south. A prominent cliff on Pratt Hill on the south side of Lake Elmer Thomas is a Late Tertiary-Quaternary cut on a Permian hill by the superposed Little Medicine Creek. Part of that cliff exposes the southernmost outcrop of Meers Quartzite, which is stratigraphically below rhyolite. Beyond the two east-west ridges is one of the principal impact areas of Fort Sill. This area and the Eastern Lowland are drained by Blue Beaver Creek almost directly to the south.

To the southwest, Mt. Sherman, 2,207 ft (673 m) high, Cross Mountain, 2,161 ft (659 m) high, Mt. McKinley, 2,035 ft (620 m) high, and Arapaho Point, 1,989 ft (606 m) high, form a prominent ridge flanking the south side of the Eastern Lowland. Mount Scott Granite forms the lower part of the mountain front, Quanah Granite and Cache Granite form the higher parts. The Eastern Lowland is floored mostly by rhyolite and is mostly within Fort Sill.

Extending west from the ridge, across Oklahoma 115, the lumpy-textured mountains are all composed of Quanah Granite and are within the Wildfire Refuge: the relatively low-lying Quanah Mountain, 1,700+ ft (518+ m) high, and associated knobs, 1,803 ft (549 m) high, and Eagle Peak, 1,800+ ft (548+m) high, and then far to the west, Mt. Lincoln, 2,200+ ft (690+ m) high, and flat-topped Elk Mountain, 2,280+ ft (695+ m) high. Elk Mountain has some of the largest corestones in the Wichitas, up to 30 to 65 ft (10 to 20 m) in diameter, on its south and west flanks. It is just southwest of the Refuge Headquarters. Quanah Mountain and Eagle Peak are south of the south side of the Central Lowland, which is itself floored by the Glen Mountains Layered Complex, but is not directly visible from here.

To the west, the main backbone of the eastern Wichita Mountains underlain by the Mount Scott Granite may be seen. To the far horizon, in that part of the Wichitas known as the Raggedy Mountains, because of the scattered, semi-isolated groups and clusters of peaks, which is west of the Wildlife Refuge, Lake Tom Steed can sometimes be seen about 25+ mi (40+ km) away. The main peak of the ridge immediately west of Mount Scott is Mount Wall 2,163 ft (659 m) high. West of Mount Wall is north-south Oklahoma 115 (Meers Road)—not visible—and west of that is Mt. Sheridan 2,450 ft (747 m) high, one of the most striking peaks of the Wichitas. The base of the Mount Scott Granite sheet can clearly be seen on Mt. Sheridan dipping south. The sheet there (and on the north face of Mount

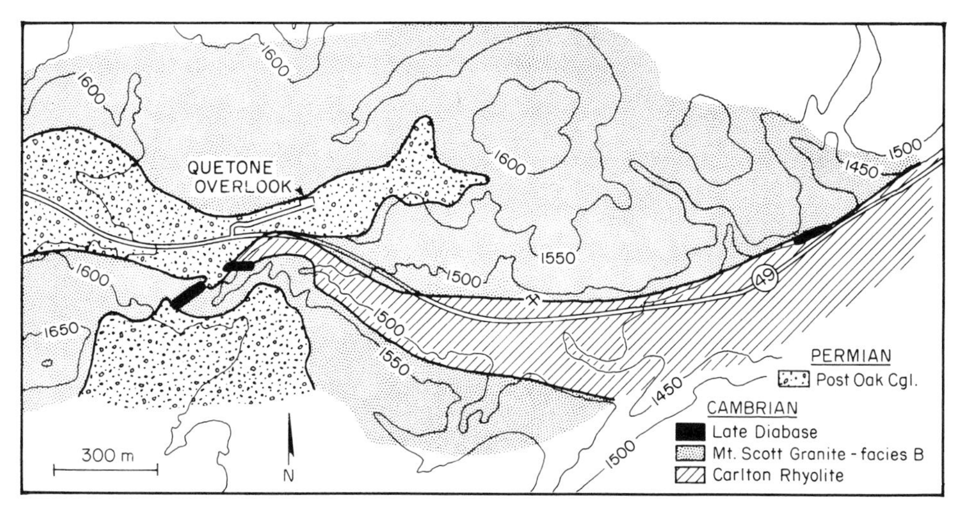

Figure 13. Geologic map of the Quetone Overlook area.

Scott and Mount Wall) is lying on the Mt. Sheridan Gabbro Member of the Roosevelt Gabbros. The heavily forested slopes are underlain by gabbro (and also mostly covered by granite talus). The granite-gabbro contact can be seen extending south from Mt. Sheridan to Mt. Roosevelt, 2,240+ ft (683 m) high, the haystack-shaped peak. The Holy City of the Wichitas is just south of Mt. Roosevelt. Mt. Sheridan and Mt. Roosevelt are both in the restricted area of the refuge and only accessible by permit.

To the northwest, looking along the WNW trend of the mountain system, past Mt. Sheridan, the north edge of the Mount Scott Granite forms the topographic mountain front marking the south side of the Meers Valley. Meers itself is located about 1 mi (1.5 km) north of Mt. Sheridan. The south part of the valley is floored by Mt. Sheridan Gabbro and its host, the Glen Mountains Layered Complex, and in places by a covering of Post Oak conglomerate, mostly of granitic facies. The north part of the valley is floored by the limestone facies of the Post Oak. The north side of the valley is formed by the Slick or Limestone Hills underlain by the Post Oak at the lower elevations and the Arbuckle Group carbonates at higher elevations. At the far end of the valley is a distinctive peak called Saddle Mountain, composed of Mount Scott Granite, as the granite sheet wraps around a gentle nose of the Fort Sill North Anticline.

The Meers Fault (striking about 300°) offsets the Post Oak-floored hills on the north side of the Valley. This fault may have played a key role in forming the Permian valley now being exhumed.

To the north, the main mass of the Slick Hills, whose elevations reach 2,000 to 2,100 ft (610 to 640 m), is underlain by Carlton Rhyolite on the east side of Blue Creek Canyon, and by the Timbered Hills and Arbuckle Groups everywhere else. Blue Creek Canyon (Stumbling Bear Pass on Oklahoma 58; also formerly known as Gore Gap) is the north-trending valley, the only break in this part of the Slick Hills, through which is exposed the Pennsylvanian Blue Creek Canyon thrust. As Upper Cambrian-Lower Ordovician sedimentary units once overlay Mount Scott (and the igneous basement), considerable downthrow to the north occurred along the Meers in the Pennsylvanian. However, Permian movement was up to the north (as well as left-lateral strike slip), and Holocene movement was also up to the north and left-lateral.

To the east, the Meers Valley merges with the modern drainage system of Cache Creek, which flows south to Red River. The bluff on the east side of Cache Creek is mapped as the Permian Garber Sandstone (lower part of the Hennessey Group) and has oil/asphalt seeps in places. The lake on Cache Creek is Lake Ellsworth also part of the Lawton-Fort Sill water supply. The large limestone quarry operation at the southeast end of the Slick Hills near Porter Hill is owned by the Dolese Company, which has removed several of the hills over the past 50 + years.

Quetone Overlook—Eastern Lowland

Location. Off Oklahoma 49: 1.8 mi (2.9 km) west of the intersection of Oklahoma 49 and the Mount Scott summit access road, 0.9 mi (1.4 km) E of Meers highway intersection, center E½,Sec.15,T.3N.,R.13W., Mount Scott 7½-minute Quadrangle (Fig. 13).

Geology. Post Oak conglomerate, Permian unconformity, late diabase, Mount Scott Granite, Carlton Rhyolite.

This stop is in a structurally complex region. All rock units below the Permian Post Oak are pervasively intensely fractured and locally host subvertical small breccia "dikes" or fins. Deformation was almost entirely brittle and presumably reflects not only a relatively low confining pressure but also an abrupt stress drop. Exact age of strain is somewhat problematic but the major event is here taken to be Pennsylvanian.

The overlook parking area is on Post Oak Conglomerate, granitic facies. The unconformity between the Post Oak and underlying igneous units is well displayed, both 1) on the outcrop scale in an adjacent excavated pit on the south side of the overlook and along the continguous highway road cut, and 2) on a more regional scale in the views to the south and west. The Post Oak is underlain here primarily by the Mount Scott Granite, facies B. Carlton Rhyolite is exposed locally along part of the eastern floor and lower sides of the small headwater canyon of Little Medicine Creek, paralleling the highway on the south. Most of the north-facing bluff on Little Medicine Creek is Mount Scott Granite.

In this immediate area, the main facies A of the Mount Scott Granite was eroded off in the Pennsylvanian-Permian. The hills and ridges to the north, from Mount Scott itself on the northeast to Mt. Sheridan and Mt. Roosevelt on the northwest are capped by facies A. In effect, one is looking into a partial vertical section of the Mount Scott sill in these hills. Facies B is the basal portion of the sill, which lies on Carlton Rhyolite in this part of the mountains.

The Post Oak at this stop is primarily a coarse conglomerate with locally interspersed arkosic sandy lenses. Within the confines of the excavated pit, clast size ranges up to 1 to 2 ft (0.3 to 0.5 m) but up to 3 ft (1 m) in nearby areas. All rock clasts are Mount Scott Granite, facies A, presumably derived from the immediately surrounding mountains. Because long axes of coarse clasts are aligned with bedding, transportation was by stream flow and not debris flow processes. Rounding of clasts was primarily due to spheroidal weathering before transportation to site of deposition. Lower parts of the unit are predominately gray, indicating an earlier "reducing" environment, while upper parts and some fractures are yellow-brown due to later oxidating groundwaters. The excavated pit floor is along the unconformity with the granite.

The underlying granite is severely broken. The timing of this fracturing and faulting is probably Pennsylvanian-related to the transpression of the Wichita Mountains block. The major deformation of this crustal block occurred at that time. Primary folds and faults of the Slick Hills, formation of the Frontal Fault Zone, and deepening of adjacent basins are all Pennsylvanian. Clastic wedges in the Anadarko Basin show several tectonic surges until unroofing of the igneous rocks occurred when "granite wash" became prominent. Topographic and structural highs had developed. Local rhyolite-granite relations at this stop show probable offsets in a wedge-like uplift consistent with the tectonic style reported by Donovan farther north. Thus, the Pennsylvanian is assumed to be the most likely time of occurrence.

However, a possible Permian event is not ruled out as some late fault movements are also recorded in the Slick Hills portion of the Wichitas during early Post Oak time (Donovan and others, 1985). These movements were sufficient to produce some local relief and record the end of a tectonically quiesent period when deep weathering occurred (Gilbert and Donovan, 1982). These faults would have cut these rock units at the near surface (low confining pressure situation). If Permian, the deformation would predate the Post Oak conglomerate at this stop because brecciation seems to stop at the unconformity.

The granite at this stop is fine-grained and hypersolvus, reflecting a rapidly quenched epizonal emplacement. Much of the southern midcontinent basement consists of epizonal granites and rhyolites (Denison and others, 1984). While the grain size of this granite is smaller than normal for Oklahoma basement granites, the mineralogy, chemistry, and mechanical properties would be similar. Therefore, the deformation seen here may be suggestive of that expected in faulted and deformed zones in basement rocks, such as along the Nemaha Ridge.

Rhyolite at this outcrop is somewhat hornfelsed and intensely oxidized. It is typical of other exposures of the "Fort Sill" section of the Carlton, having noticeable alkali feldspar and quartz phenocrysts. The rhyolite rises to a structural high just south of the highway, opposite and southwest of the excavated pit. The outcrop narrows westward from the valley floor, rising up the north flank of the canyon. The canyon is floored by granite on its westward end. Granite is on both sides of the rhyolite. The contacts between the granite and rhyolite here are probably at least partly faulted with rhyolite being uplifted as a wedge.

A peculiar late diabase dike 33 to 40 ft (10 to 12 m) wide, striking 30–35° and dipping southeastward at 55–75°, cuts the granite 1,300 to 1,500 ft (400 to 450 m) WSW of the excavated pit. It shows well where it crosses Little Medicine Creek. It is highly weathered but can be seen to contain apparent granite fragments 1 mm to 1 cm in size. This dike weathered down preferentially to the enclosing granite in the Permian because part of its outcrop is covered by Post Oak Conglomerate but its trace can still be followed due to the locally high-standing boundary walls of granite. The dike narrows to about 2 ft (0.5 m) and assumes a 90° strike separating granite and rhyolite on the east-facing slope of the valley, directly southwest of the excavated pit, after which it is lost locally. Another diabasic dike and dike fragment can be seen at the other (northern) rhyolite-granite contact just south of the concrete culvert under Oklahoma 49. (The culvert is about halfway up the highway as it climbs up from the valley). Analyses of the greenish material confirm its origin as a basaltic dike, although now it is highly altered and sheared out. Apparently the stress system took advantage of the altered diabase as a zone of weakness, wherever the diabase was close to the granite-rhyolite contact.

Meers

Location. Meers is at the intersection of Oklahoma 115 and the east-west county road west off of Oklahoma 58 along the head of Lake Lawtonka, SW¼Sec.28 and NW¼Sec.33 T.4N., R.13W., Meers 7½-minute Quadrangle (Fig. 1). One location and a short road log will be highlighted.

Geology. 1) Glen Mountains Layered Complex, late diabase, granitic dikes and plugs; and 2) Meers Quartzite—clean facies, Mt. Sheridan Gabbro, late diabase.

The location is the road cut in front of the Meers store on the south side of the county road just east of Meers. This cut is in the

Glen Mountains Layered Complex, M(?) Zone, which has been hydrothermally altered (Cambrian) and weathered (Permian).

The intrinsic layering is defined by plagioclase lamination, here rather subtle but elsewhere very pronounced, and strikes approximately 90° with a 10–15° N dip. Dark gray cumulus calcic plagioclase and post-cumulus poikilitic clinopyroxene dominate the primary mineralogy. Hydrothermal alteration is evident in the albitized and prehnitized rock, with its network of fine white veins, chlorite after clinopyroxene, and chalcopyrite blebs. This alteration may be associated with the intrusion of the hydrous Mt. Sheridan Gabbro, as seen in other outcrops near that contact, but here seems to be related to the granite dikes (discussed later). Finally, spheroidal weathering is prominent and partly accounts for the differing outcrop character along the cut.

The Layered Complex was intruded by a diabase dike of strike 340°, dip 75–80°E, near the east end of the cut. The dike varies in width from 1.5 in (4 cm) at a yoke to 12 to 15 in (30 to 40 cm). It is stepped to the left three times moving up the slope and once to the right near the top. These are intrusive steps and not later faults.

Intruding both the Complex and, indirectly, the diabase dike are two pink granite dikes of alkali feldspar + quartz + magnetite. The one near the center of the cut is sinuous; it pinches and swells in width from 1.5 to 30 in (4 to 80 cm). Strike is about 305°, dip varies from vertical to southward. There is a highly altered zone on both sides of the dike but the zone is most well-developed on its north side as well as around a small offshoot. The other main dike, near the west end of the cut, is 12 in (30 cm) in width, and strikes 295° with a vertical to SW dip. A prominent series of white veins, which strike 295–315° and dip vertical to 75°S, appear related to the granite dikes. On first impression, these white veins seem to be cut by the diabase. However, closer inspection shows that vein material occurs inside the dike but there follows fracture directions intrinsic to the dike. Thus the granite is judged younger than the diabase; this demonstrates mafic dike formation throughout the development of the whole igneous sequence, not just at its end.

The granite dikes are probably related to the Mount Scott Granite, which presumably overlay this location but was eroded off in the Permian.

Other Localities of Interest

From Meers South on Oklahoma 115. The following simplified short road log (crossing the axis of the Fort Sill North Anticline) highlights some other important relationships.

0–0.35 mi (0-0.5 km). Pass through Glen Mountains Layered Complex (GMLC), M(?) Zone, dipping northward.

0.35 mi (0.5 km). Cross the approximate east-west contact with the Mt. Sheridan Gabbro (MSG) member of the biotite-bearing Roosevelt Gabbros, intrusive into the GMLC. Contact is near base of small steep hill (part of a locally major east-west ridge) as the road passes by the Rowe Ranch headquarters. Northward, Mt. Sheridan Gabbro passes underneath the GMLC with an undulating contact and regionally dips north. Subtle cumulus plagioclase layering within the MSG, measured in a quarry about 0.6 mi (1 km) west, strikes 60 to 80° and dips 15 to 20° northwest. This is discordant with the more obvious primary cumulus layering of the GMLC and documents tectonism, block faulting, and rotation, between the times of intrusion of the GMLC and MSG.

0.55 mi (0.9 km). Cross a small sharp rise in the road, which is underlain by Meers Quartzite, clean facies, some of which is sillimanite-bearing, and is the type locality (Taylor, 1915; Hoffman, 1930; Merritt, 1948; Ham and others, 1964). The small east-west ridge about 650 ft (200 m) in length is formed by the Meers.

0.75 mi (1.2 km). New bridge (1985) over Medicine Creek with road cuts on the south side exposing collectable MSG and a major diabase dike.

1.3 mi (2.1 km). The view to the west of Mt. Sheridan clearly outlines the gently southward-dipping contact between Mount Scott Granite, which forms the cliffs and mountain top, and at the tree line, the underlying Mt. Sheridan Gabbro on which granite talus slopes are built. A recent rock fall (winter 1983–84) is evident. The strongly colonnaded granite, as exposed in the cliffs, is thought to be due to a combination of Cambrian-age columnar jointing formed on chilling of the base of the sill during intrusion and later fracturing during Pennsylvanian uplift.

1.55 mi (2.5 km). North (Meers) entrance to the Wichita Mountains Wildlife Refuge.

From Meers North on Oklahoma 115. **0.45 mi (0.7 km).** Contact with the Permian Post Oak Conglomerate, granitic facies. This conglomerate fills a previously cut Permian valley formed during left-lateral strike-slip movement on the Meers Fault (Donovan and others, 1985). The movement also has a dip-slip component, up on the north. The granite facies clasts are derived from the Mount Scott Granite on the south. This facies appears to be overlapped by the limestone facies shed from the Arbuckle Group outcrops of the higher Slick Hills on the north, and may be seen in the lower slopes of the Slick Hills ahead, and encountered in outcrop about 1 mi (1.6 km) ahead.

From Meers East. **0.35 mi (0.6 km).** Contact with the Post Oak Conglomerate. No other accessible exposures of gabbro east along this road.

REFERENCES

Al-Shaieb, Z., 1978, Guidebook to uranium mineralization in sedimentary and igneous rocks of Wichita Mountains region southwestern Oklahoma: Oklahoma City Geological Society, 73 p.

Brewer, J. A., Brown, L. D., Steiner, D., Oliver, J. E., Kaufaman, S., and Denison, R. E., 1981. Proterozoic basin in the southern Midcontinent of the United States revealed by COCORP deep seismic reflection profiling: Geology, v. 9, p. 569–575.

Chase, G. W., 1954, Permian conglomerate around Wichita Mountains, Oklahoma: American Association of Petroleum Geologists Bulletin, v. 38, p. 2028–2035.

Denison, R. E., Lidiak, E. G., Bickford, M. E., and Kisvarsanyi, E. G., 1984, Geology and geochronology of Precambrian rocks in the central interior region of the United States: U.S. Geological Survey Professional Paper 1241-C, 20 p.

Donovan, R. N., Gilbert, M. C., Luza, K., Marchini, D., and Sanderson, D., 1983, Possible Quaternary movement on the Meers Fault: Oklahoma Geology Notes, v. 43, no. 5, p. 124–133.

Donovan, R. N., Bridges, S., and Bridges, K., 1985, The formation of the Meers Valley, southwestern Oklahoma: Geological Society of America Abstracts with Programs, v. 7, no. 3, p. 156.

Evans, O. F., 1929, Old beach markings in the western Wichita Mountains: Journal of Geology, v. 37, p. 76–82.

Gibson, R. G., III, 1981, Petrography and geochemistry of mafic enclaves in the Reformatory Granite, western Wichita Mountains, Oklahoma [B.S. thesis]: Meadville, Pennsylvania, Allegheny College, 41 p.

Gilbert, M. C., 1982a, Stop 6; French Lake Dam, *in* Gilbert, M. C., and Donovan, R. N., eds., Geology of the eastern Wichita Mountains southwestern Oklahoma: Oklahoma Geological Survey Guidebook 21, p. 130–134.

—— , 1982b, Stop 4; Mount Scott Campground, *in* Gilbert, M. C., and Donovan, R. N., eds., Geology of the eastern Wichita Mountains southwestern Oklahoma: Oklahoma Geological Survey Guidebook 21, p. 118–123.

—— , 1983a, The Meers Fault; Unusual aspects and possible tectonic consequences: Geological Society of America Abstracts with Programs, v. 15, no. 1, p. 1.

—— , 1983b, Timing and chemistry of igneous events associated with the Southern Oklahoma Aulacogen: Tectonophysics, v. 94, p. 439–455.

—— , 1984, Comments on structure within the igneous core, Wichita Mountains crustal block: Oklahoma City Geological Society, Technical proceedings of the 1981 AAPG Mid-Continent Regional meeting, p. 177–190.

Gilbert, M. C., and Donovan, R. N., 1982, Geology of the eastern Wichita Mountains, southwestern Oklahoma: Oklahoma Geological Survey Guidebook 21, 160 p.

—— , 1984, Recent developments in the Wichita Mountains, Guidebook for Field Trip #1, South-Central Section Annual Meeting, Geological Society of America: College Station, Texas, A&M University, Department of Geology, 101 p., 1 plate.

Ham, W. E., Denison, R. E., and Merritt, C. A., 1964, Basement rocks and structural evolution of southern Oklahoma: Oklahoma Geological Survey Bulletin 95, 302 p., 5 separate plates.

Harlton, B. H., 1963, Frontal Wichita Fault system of southwestern Oklahoma: American Association of Petroleum Geologists Bulletin, v. 47, p. 1552–1580.

Havens, J. S., 1977, Reconnaissance of the water resources of the Lawton quadrangle southwestern Oklahoma: Oklahoma Geological Survey Hydrologic Atlas 6, Sheet 1.

Hoffman, M. G., 1930, Geology and petrology of the Wichita Mountains: Oklahoma Geological Survey Bulletin 52, 83 p.

Johnson, K. S., and Denison, R. E., 1973, Igneous geology of the Wichita Mountains and economic geology of Permian rocks in southwestern Oklahoma (Guidebook for Field Trip No. 6, Geological Society of America Annual Meeting): Oklahoma Geological Survey, SP73-2, 33 p.

Lambert, D. D., and Unruh, D. M., 1984, Rb-Sr and Sm-Nd isotopic study of the Glen Mountains Layered Complex, Wichita Mountains, Oklahoma: Geological Society of America Abstracts with Programs, v. 16, no. 2, p. 105.

Merritt, C. A., 1948, Meers quartzite: Oklahoma Academy of Science Proceedings, v. 28, p. 76–77.

—— , 1958, Igneous geology of the Lake Altus area, Oklahoma: Oklahoma Geological Survey Bulletin 76, 70 p.

—— , 1965, Mt. Scott Granite, Wichita Mountains, Oklahoma: Oklahoma Geology Notes, v. 25, p. 263–272.

—— , 1967, Names and relative ages of granites and rhyolites in the Wichita Mountains, Oklahoma: Oklahoma Geology Notes, v. 27, p. 45–53.

Miser, H. D., 1954, Geologic map of Oklahoma: U.S. Geological Survey and Oklahoma Geological Survey, scale 1:500,000.

Myers, J. D., Gilbert, M. C., and Loiselle, M. C., 1981, Geochemistry of the Cambrian Wichita Granite Group and revisions of its lithostratigraphy: Oklahoma Geology Notes, v. 41, p. 172–195.

Powell, B. N., and Fischer, J. F., Plutonic igneous geology of the Wichita magnetic province, Oklahoma (Guidebook for Field Trip No. 2, South-Central Section Geological Society of America Annual Meeting): Oklahoma Geological Survey, SP76-1, 35 p.

Powell, B. N., Gilbert, M. C., and Fischer, J. F., 1980, Lithostratigraphic classification of basement rocks of the Wichita province, Oklahoma: Geological Society of America Bulletin, Part I Summary, v. 91, p. 509–514; Part II, v. 91, no. 9, p. 1875–1994.

Powell, B. N., and Phelps, D. W., 1977, Igneous cumulates of the Wichita province and their tectonic implications: Geology, v. 5, p. 52–56.

Taylor, C. H., 1915, Granites of Oklahoma: Oklahoma Geological Survey Bulletin 20, 108 p.

Pennsylvanian deformation and Cambro-Ordovician sedimentation in the Blue Creek Canyon, Slick Hills, southwestern Oklahoma

R. N. Donovan, *Geology Department, Texas Christian University, Fort Worth, Texas 76129*
D. Ragland, M. Rafalowski, D. McConnell, and W. Beauchamp, *Geology Department, Oklahoma State University, Stillwater, Oklahoma 74078*
W. R. Marcini and D. J. Sanderson, *Geology Department, Queens University of Belfast, Belfast, Northern Ireland*

LOCATION

Blue Creek Canyon is located approximately 15 mi (24 km) north-northwest of Lawton in southwestern Oklahoma (T.4N.,R.13W.). Oklahoma 58 (Fig. 1) runs through the canyon. The area can be most easily reached from Lawton by taking the Medicine Park exit from the H. E. Bailey Turnpike a few miles north of Lawton.

The entire area described here is part of the Kimbell Ranch, owned by Mr. and Mrs. David Kimbell of Wichita Falls. The ranch house is situated at the south end of the canyon (Fig. 2) and is managed by Charlie Bob and Dixie Oliver. Both the Olivers and the Kimbells have greatly encouraged the visits of geologists, and several local universities use the area as a student training ground. Permission to visit should be sought from Mr. Oliver, and the usual common sense and courtesy extended. The ranch address is Star Route A, Box 124, Lawton, OK 73501. The Olivers carry their interest in geology to a remarkable level; during monitoring of the recently active Meers Fault, they had a seismograph installed in their bedroom as part of a regional network.

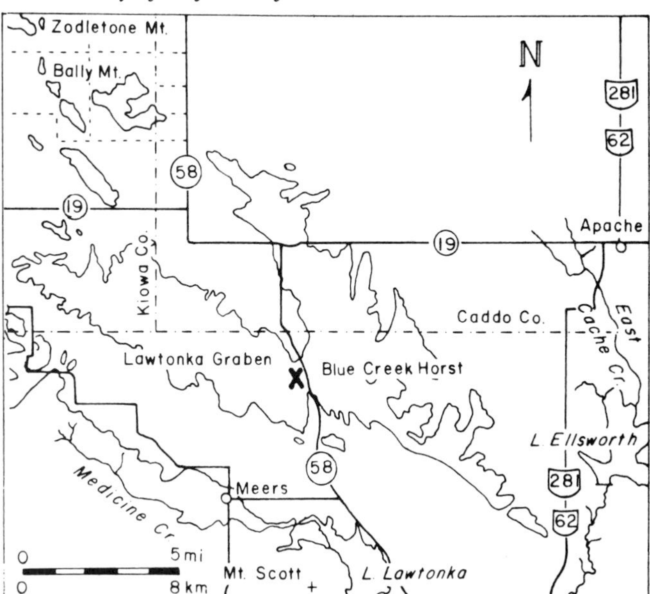

Figure 1. Locations of Blue Creek Canyon.

The descriptions in this field guide are for a series of walks that start approximately 0.5 mi (0.8 km) north of the ranch house at Red Hill (a small hill of Carlton Rhyolite capped by Post Oak Conglomerate which is immediately west of the road) (Fig. 2). The walks are all rough and fairly hilly; exposure is more or less continuous. The area is well drained even after heavy rain, although some difficulty may occur in crossing Blue Creek at such times. Take water on the longer trips (especially in summer when the temperature on the limestone may be 120+°F (49+°C), and beware of rattlesnakes.

SIGNIFICANCE

The relatively small area around the Blue Creek Canyon contains a remarkable number and variety of interesting sites. For this part of the world, exposure quality is very good to excellent. In general terms the exposures illustrate several aspects of the development of the Southern Oklahoma aulacogen: (1) the initial rift-related igneous activity, (2) early sedimentation in a basin with a decaying thermal imprint; (3) late Paleozoic deformation, (4) burial of the resulting Permian relief.

The oldest rocks in the area are part of the Cambrian Carlton Rhyolite Group. The lavas were intruded by a number of diabase dikes, gently tilted, and cut by a number of north-south faults before they were reduced to a range of low hills by late Cambrian (Franconian) time. The subsequent marine transgression (Fig. 3) was initially siliciclastic in character (the Reagan Formation), but eventually a highly productive carbonate "factory" developed (the Honey Creek Formation). By this time the rhyolite land surface had become an archipelago of small rocky islands. These islands were buried by the time deposition of the Arbuckle Group commenced. The Arbuckle Group is more than 5,500 ft (1,680 m) thick in the southern Oklahoma aulacogen and is one of the world's great platform carbonate sequences. (An unbroken section of these rocks at Bally Mountain is described elsewhere in this guide.) Succeeding Lower Paleozoic rocks (the Simpson, Viola Group, Sylvan, and Hunton) were deposited before deformation began in late Mississippian time.

This deformation, which concluded in early Permian time, resulted in dismemberment of the aulacogen into a series of NW-SE trending basins (e.g., Anadarko, Hollis, and Ardmore) and uplifts (e.g., Wichita, Arbuckle, and Criner Hills). As the deformation concluded, the surface of the uplift areas was sculpted by erosion into irregular hills that were subsequently buried by Permian sediments. At present these hills are in various stages of exhumation from beneath the Permian cover. In general terms the Permian rocks are conglomerates close to the old hills (the Post Oak Conglomerate facies) but fine outward into the ancient valleys and basins.

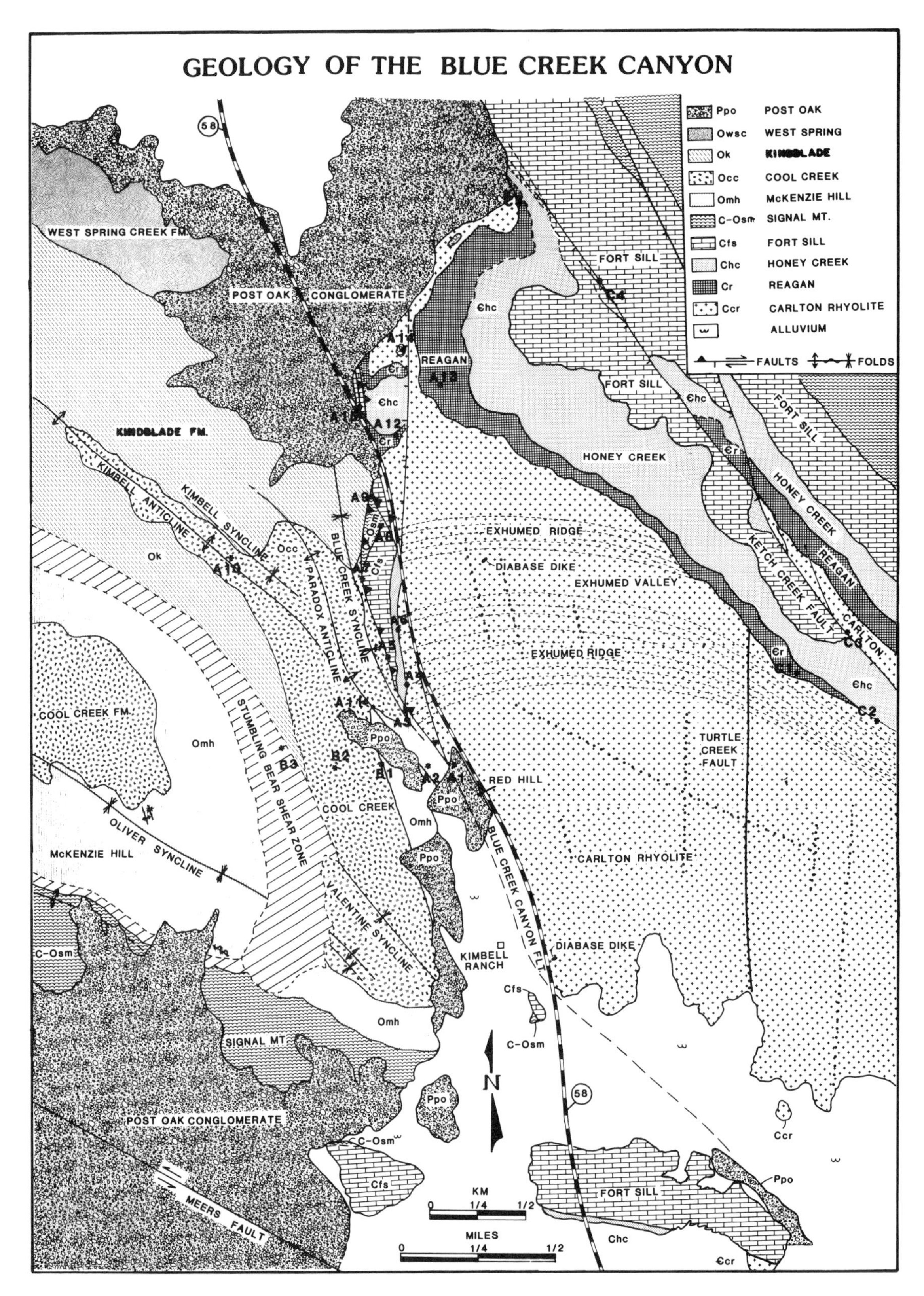

Figure 2. Geological map of Blue Creek Canyon area. A1–A15, B1–B3, and C1–C5 are suggested stops on three excursions.

STRUCTURAL OUTLINE

In a regional context the area forms part of the WNW-ESE trending Wichita Frontal fault zone which constitutes the hinge between the Anadarko basin (to the north) and the Wichita uplift. The zone was the locus for Pennsylvanian tectonism during the late deformation stage of the development of the Southern Oklahoma aulacogen (Donovan, 1982; Gilbert and Donovan, 1984). The principal character of this deformation was transpressive with left-lateral motion along the length of the zone. The zone is broken up into a number of blocks bounded by major faults (e.g., the Meers, Blue Creek Canyon, and Mountain View Faults).

The Blue Creek Canyon fault is the only one of these major structures whose Pennsylvanian character can be examined at the surface. It is an oblique high-angle reverse/left-lateral structure, trending just west of north (170–350) in the canyon area. It has a stratigraphic downthrow to the west that decreases northward from 2,400 ft (730 m) to 1,800 ft (550 m) along the length of the canyon (this variation in downthrow is not regionally significant but simply reflects fold geometry in the western "footwall" block).

The north-trending segment of the fault exposed in the canyon is anomalous in trend. To north and south, the fault bends until it is almost parallel to the regional WNW-ESE trend. Anomalous fault trends of similar direction are common in both the Wichita and Arbuckle areas. Such segments may have utilized an early north-south fracture trend that is pre-Reagan/post–Carlton Rhyolite in age.

The major faults in the Wichita Frontal zone define blocks of terrain that have undergone differing amounts of deformation. In the present instance the Blue Creek Canyon fault separates the "Blue Creek horst," to the east, from the "Lawtonka graben" (terminology of Harlton, 1972). The "horst" is a relatively undisturbed terrain dominated by homoclinal dips of moderate angles to the NE. The "graben," on the other hand, has been subjected to a "squeeze play" between the "horst" to the NE and the Wichita Mountains (to the SW across the Meers fault). The result is a structural mayhem of refolded reverse faults, rotated en-echelon folds, and brittle shears that together reflect intense left-lateral transpression and constitute the most intensely deformed terrain exposed in either the Wichitas or the Arbuckles (Donovan, 1982, 1984; Beauchamp, 1983; McConnell, 1983; Marcini, 1986).

In the small segment of the Lawtonka "graben" exposed in the Blue Creek Canyon, the dominating structures are a group of four major folds that trend WNW-ESE and plunge consistently northward. These folds are the Blue Creek Syncline, the Paradox anticline, and the Kimbell fold pair (Fig. 2). The latter folds have a lopsided "rabbit ear" relationship to the structurally lower Paradox fold. A fifth large fold, the Blue Creek anticline, is located in the "hanging wall" block east of the Blue Creek Canyon fault. A feature of interest is that the Carlton Rhyolite is quite clearly involved in the folding. Farther to the west is the enigmatic Stumbling Bear shear zone, a region of intensively deformed rock across which there is a stratigraphic discontinuity of up to 2,200 ft (670 m). This structure has recently been interpreted as an early thrust that was subsequently subject to considerable left-lateral shear (Marchini, 1986).

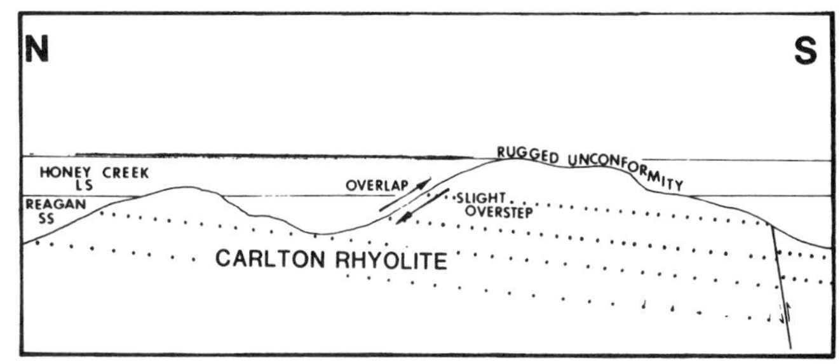

Figure 3. Generalized stratigraphic relationships between the Carlton Rhyolite and the Timbered Hills Group (from Rafalowski, 1984).

WALKS FROM RED HILL

Site 80. First Excursion: Stops A1–A15, Fig. 2

A1. The Permian Post Oak Conglomerate facies unconformably overlies the Carlton Rhyolite at the eastern end of Red Hill, whereas at the western end of the hill it overlies the McKenzie Hill Formation of the Arbuckle Group. Fracturing in the rhyolite is probably due to the Blue Creek Canyon fault, which can be projected to pass under Red Hill close to the road (and thus accounts for the differing unconformable relationships at either end of the hill [Donovan and others, 1982]). The Permian deposit is a breccio-conglomerate consisting of pebble-sized clasts, crudely bedded in layers derived either from the west (in which case the clasts are Ordovician limestone) or from the east of the canyon (in which case the clasts are rhyolite or Cambrian limestone). The top of the hill is a good vantage point for viewing the Ordovician terrain; the conspicuous anticline in the foreground is the Paradox anticline.

A2. After crossing Blue Creek (with care), the Paradox anticline can be examined in detail. The lithology here is the upper, cherty part of the McKenzie Hill Formation. In common with other major folds in the area, the Paradox anticline is a complex northward-plunging fold of variable tightness (30–130°) and with fairly sharply defined axes ('E' classification, Hudleston, 1973). All the large folds are basically parallel (Class I, Ramsay, 1967), and as a result their axial traces are displaced by layer-parallel slippage in a number of places. One of these slippage décollements occurs on the eastern limb of the anticline between A2 and A3, where a group of small-scale drag folds showing asymmetry and locally developed cleavage is confined to a layer about 20 ft (6 m) thick between undisturbed beds in the McKenzie Hill Formation.

A3. Here a gully marks the position of the Thatcher Creek member (Ragland and Donovan, 1985), a quartz-rich cross-

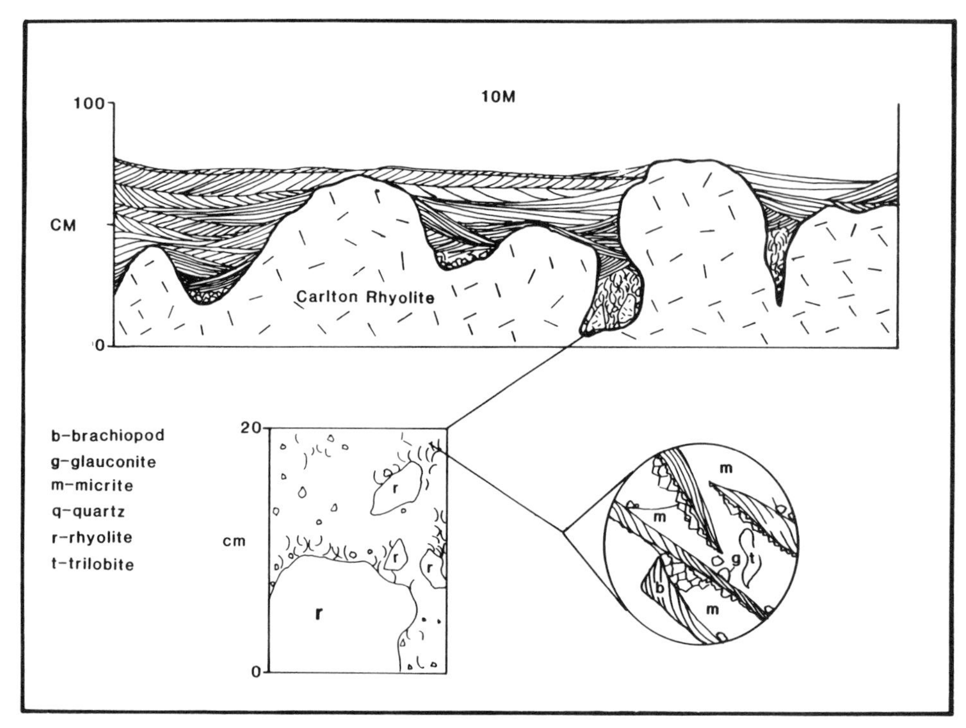

Figure 4. Interpretation of the Carlton Rhyolite–Honey Creek unconformity illustrating rock pedestals and fissures on wave-cut platform. Insets show petrographic details (from Rafalowski, 1984).

bedded limestone that marks the base of the Cool Creek Formation. This member is a critical stratigraphic marker in unraveling the complexities of the Lawtonka "graben," maintaining its character over many miles of outcrop; here it can be followed around the axis of the Blue Creek Syncline. The Blue Creek Canyon fault runs through the trees to the east, and as the fault is approached, the beds become overturned by as much as 30°. The route from A3 to A4 involves entering the trees to the NW, crossing the hidden fault trace, and scrambling over some scrappy orange-brown outcrops of Carlton Rhyolite to the unconformity with the Honey Creek Formation.

A4. The Reagan Formation is missing at this unconformity; outcrops along the length of the canyon (e.g., at A6) indicate that this area was an island until late in the history of the Honey Creek Formation. The exposure at A4 is interesting in that a stabilized shell bank consisting of orthid brachiopod valves formed above the thin basal breccia (no hammers here, please—there is plenty of loose material). Small calcite/hematite stromatolites are also present; the section has been interpreted as a sheltered coastline deposit (Donovan and Rafalowski, 1984).

A5. A short distance above the unconformity, the Honey Creek passes up into the Fort Sill, lowest formation of the Arbuckle Group (a general log of these rocks is given in the accompanying guide to Bally Mountain). To the west the Fort Sill outcrop is truncated by the Blue Creek Canyon fault trace. This trace is a conspicuous linear depression that can easily be traced, particularly where the upper massive member of the Fort Sill Formation is faulted against the Cool Creek Formation, as the former is a ridge-forming element. The fault plane dips to the east at a high angle, indicating a reverse element of movement. In addition, prominent phacoidal shearing suggests a left-lateral component of motion.

A6. The unconformity between the rhyolite is well-exposed at a small prospect pit to the northeast of A5 (Fig. 4). The unconformity is most irregular and appears to be a wave-cut platform with numerous fissures and pedestals that was covered by a detrital coarse-grained carbonate sand composed mostly of broken primitive echinoids. This exposure has been interpreted as an exposed shoreline setting on the windward side of an island jutting from the Honey Creek sea (Donovan and Rafalowski, 1984).

A7. From the unconformity, the hanging wall block can be recrossed in a NW direction to rejoin the fault trace at A7. Here the uppermost massive member of the Fort Sill Formation is overlain by the thinly bedded and more poorly exposed Signal Mountain Formation. This contact has been displaced by a number of E-W faults that downthrow to the south. The Cambro-Ordovician boundary is approximately 330 ft (100 m) above the base of the Signal Mountain Formation.

A8. From A7 it is instructive to follow the strike of the Signal Mountain Formation in a NE direction. It will be clear that a major space problem is created by a change in strike (from N-S to NE-SW); as a result the Signal Mountain Formation appears to be on a collision course with the Carlton Rhyolite. However, at A8, a major N-S trending fault cuts the section and emplaces a thin sliver of upper Fort Sill Formation between the

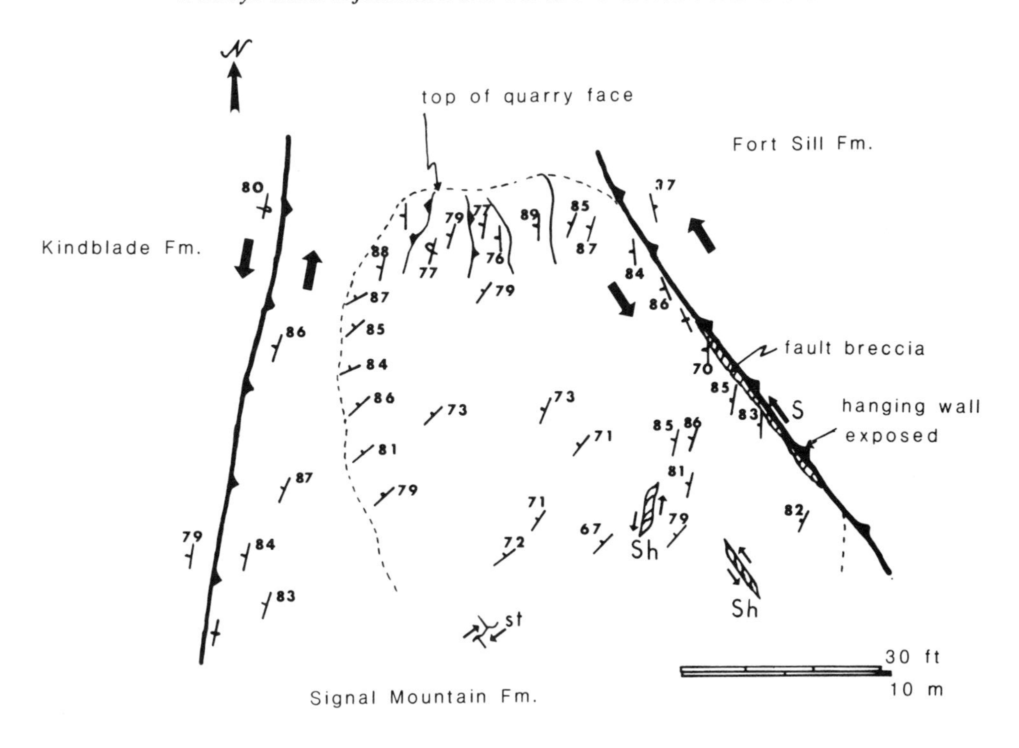

Figure 5. Sketch illustrating structural deformation in quarry (location A9). Note "squeeze" effect between converging oblique left-lateral, high-angle reverse faults.

Signal Mountain and the rhyolite. A second fault between the Fort Sill and the rhyolite can be inferred to run more or less beneath the existing road (Fig. 2). The pond next to the road (which is home to beavers and snapping turtles) is fed by springs emerging from this fault plane.

A9. The fault at A8 can easily be traced northward into an old quarry excavated when the road through the canyon was made. The quarry was excavated into Signal Mountain Formation in a zone that was crushed between the Blue Creek Canyon fault (to the west) and the "pond" fault (Fig. 5). The hanging wall of the latter fault is exposed on the eastern wall of the quarry and shows oblique slickensides and gouging suggestive of left-lateral transpression. Left-lateral shear phacoids are exposed on the floor of the quarry.

A10. Immediately north of the quarry the two major faults almost come together. However, the inferred contact is hidden under the highway. From this point a walk of 0.5 mi (0.8 km) west to the top of the hill on the horizon crosses all four of the major folds in the area (Fig. 2). The most photogenic of these folds is the Kimbell anticline at the top of the hill (where else?). The view from the hilltop (Kimbell Mountain) is an impressive one. To the west lies the contorted terrain of the Lawtonka "graben"; to the north is the valley eroded along the Blue Creek Canyon fault; to the east lies the Blue Creek "horst" where the Cambrian Timbered Hills Group unconformably overlies Carlton Rhyolite (the conspicuous peak is Ring Top Mountain); to the south the igneous Wichita Mountains rise across the Meers Valley.

A11. The Red Hill can easily be seen from the top of the hill at A10. The recommended return route is along the axis of the Kimbell anticline until this fold merges into the western limb of the Paradox anticline. The "space problems" of parallel fold geometry are clearly apparent on this walk—nowhere more so than at A11 where the Paradox anticline is spectacularly well exposed. The paradox here is provided by an interplay of plunge and hill slope that results in the fold having the convincing appearance of a syncline with a very tight axis. Sitting astride this axis and looking generally northward is a good place to reflect on the geometries of the four major folds in the area. Because of the consistent plunge to the north, their form can be studied through a stratigraphic thickness of over 3,000 ft (914 m). From this point it is an easy walk back to the Red Hill.

A12. From the Red Hill drive northward along Oklahoma 58 for 3,000 ft (910 m) to a point 300 ft (91 m) north of the pond described in A8. Go through the gate to the east of the road, walk north 600 ft (182 m), and cross Blue Creek. In this area, flow-banded Carlton Rhyolite is well exposed. Above the rhyolite lies the Reagan sandstone, which here comprises a 5-ft (1.5-m) thick basal conglomerate consisting of pebbles of rhyolite cemented by hematite. Above this is a 2-ft (60 cm) lithic sandstone followed by about 8 ft (2.3 mF) of highly glauconitic cross-bedded green-sand, which is interbedded with a thin rhyolite pebble conglomerate. Throughout the Slick Hills the greensand is everywhere the top unit of the Reagan, consisting of up to 60 percent fine sand-sized pellets of glauconite, plus quartz, rhyolite, and broken phosphatic shells of inarticulate brachiopods, all either set in a matrix of iron-rich illite or (toward the top) cemented by calcite.

The passage from the Reagan into the Honey Creek Formation (marked by the incoming of lenses of cross-bedded coarse grainstones) is located in a prospect pit a few yards to the northeast. Such prospect pits are a common feature in the area. Some of the later ones were uneconomic attempts to mine iron ore (hematite is an abundant cement in parts of the Reagan), while others seem to have resulted from a confusion of glauconite with malachite.

A13. By crossing an obvious fault-controlled gully to the northeast, a second exposure of Reagan can be examined in A13. The basal facies is a very coarse grained lithic sandstone (which at one time was quarried as a building stone), overlying which is a coarse-grained, cross-bedded (sets up to 2 ft; 60 cm) quartz arenite, some shale (mostly covered), and the upper greensand unit. This sequence is about 60 ft (18 m) thick and passes up into a full sequence of Honey Creek limestone on the slopes of Ring Top Mountain to the east.

A14. Six hundred feet (180 m) to the NW of A13 lies an enigmatic exposure of Post Oak Conglomerate consisting of angular boulders of the Fort Sill Formation up to 20 ft (6 m) in diameter. The deposit, which rests unconformably on the Carlton Rhyolite, is one of a number of similar deposits in the Slick Hills, all of which seem to be related to nearby fault scarps (Donovan and others, 1982).

A15. From A14, cross the small stream that runs north of the conglomerate and follow the track that runs down the stream valley to the main road. The north slope of the valley is a more typical exposure of the Post Oak (pebble conglomerate); in the valley bottom, close to the road, are near vertical exposures of the Fort Sill Formation caught up in the most northerly exposures of the Blue Creek Canyon fault. On rejoining Oklahoma 58, turn south for the pond (A8). After about 600 ft (180 m) examine the ditch to the east of the road (A15), where Post Oak Conglomerate, composed of diverse types of locally derived clasts up to 2 ft (60 cm) in diameter, rests on an irregular surface of Fort Sill Limestone. The relationships here suggest that Blue Creek Canyon was first sculpted in Permian times. Pebble imbrication hints that drainage at this spot was to the north (not south, as now). Of particular interest are the coarsely crystalline fibrous calcite cements located close to or on the unconformity. These cements are coatings up to 2 in (5 cm) thick that record initial vadose conditions (i.e., flowstone, precipitation) followed by phreatic crystal growth, suggesting that the water table gradually rose through a highly porous gravel.

Second Excursion: Stops B1–B3, Fig. 2

B1. After crossing Blue Creek as for stop A2, bear northwest up the valley of the perennial Thatcher Creek. The first exposures encountered are impressive cliffs carved in the boundstones and associated sediments that form the top part of the McKenzie Hill Formation. A band rich in the brachiopod *Finkelnbergia* is present but different to find. More obvious are silicified sponges and a variety of cherts, including burrow-fills and nodules. Above the McKenzie Hill, where the valley widens and runs parallel to strike, is the type locality of the Thatcher Creek Member (B1, Fig. 2; Fig. 6; Ragland and Donovan, 1985; see also A3).

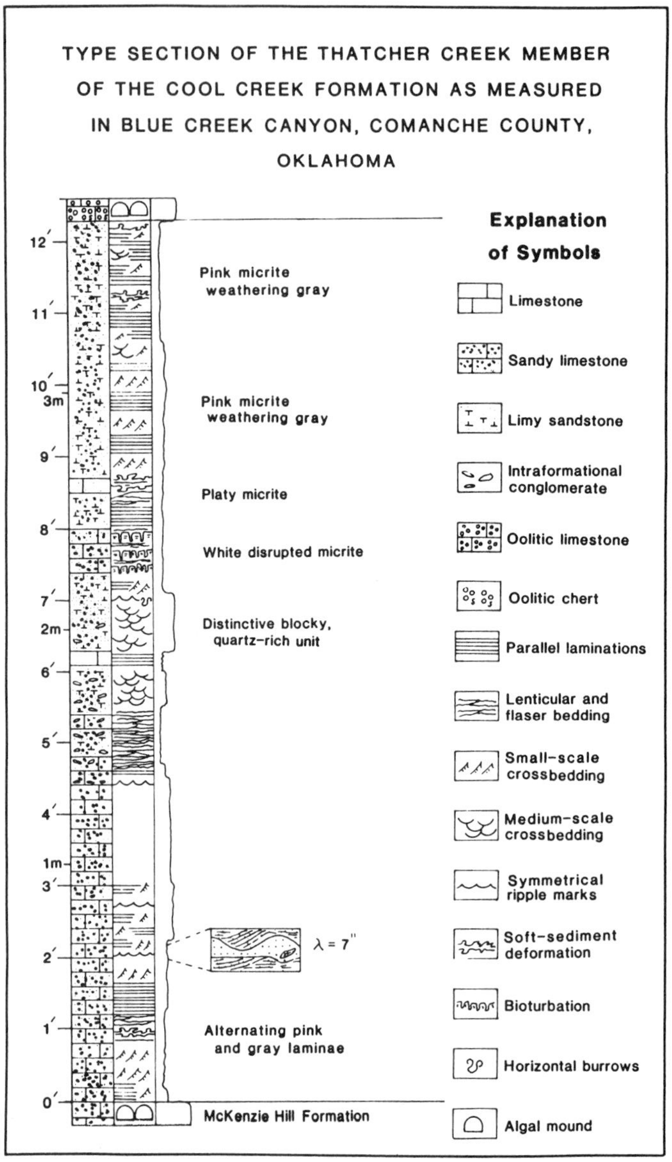

Figure 6. Type log of Thatcher Creek member of Cool Creek Formation in Thatcher Creek (location B1) (from Ragland and Donovan, 1985).

B2. Above B1 the Cool Creek Formation is well exposed and quite interesting, particularly on the lefthand (southern) side of the valley (Ragland, 1983). At a point due south of the first major junction in the valley, just above the valley rim, is the termination of an algal boundstone reef (B2; Fig. 7).

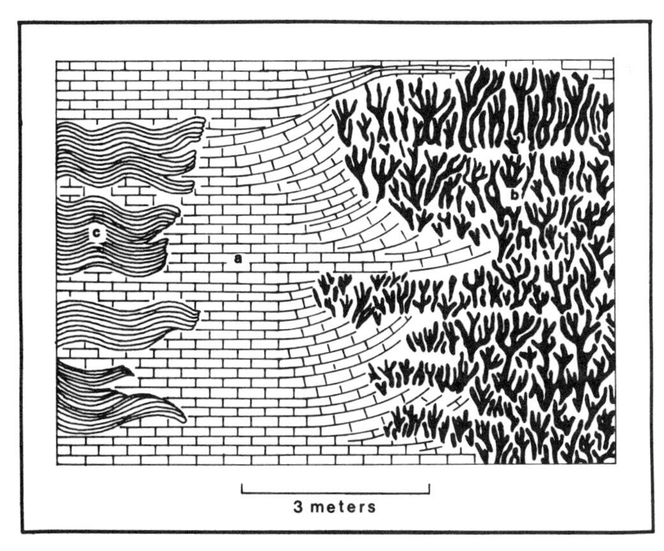

Figure 7. Schematic representation of the termination of an algal boundstone reef in the Cool Creek Formation at location B2. A compactional draping of laminated lime mudstone and intraformational conglomerates (a) against a bioherm (b); (c) is an algal mat interbedded with micrite and intraformational conglomerate (from Ragland, 1983).

B3. From B2, follow the minor lefthand (southern) valley due west across a homoclinal sequence of Cool Creek and Kindblade rocks disturbed only slightly by minor folding and fracturing to B3, where the Stumbling Bear shear zone is encountered. Within the shear zone the rocks are of Signal Mountain and McKenzie Hill age, indicating a stratigraphic discontinuity of over 2,000 ft (610 m). The zone, which extends throughout the Slick Hills, is up to 1,800n ft (550 m) wide and is probably the most structurally complex terrain exposed in Oklahoma. Dismembered folds, small faults, left-lateral shear belts, and pervasive pressure solution cleavage are well developed throughout. The intensity of deformation gradually decreases to the west into relatively undeformed McKenzie Hill rocks.

Site 81. Third Excursion: Stops C1–C5, Fig. 2

C1. Enter the gate on the east side of Oklahoma 58 at the Red Hill and walk up the rough track across the Carlton Rhyolite to a farm pond. From the pond climb the rhyolite hill to the northwest. On the other side of the hill lies a strike parallel valley which closes to north and south and is eroded in the Reagan Formation. There are several good exposures of the valley, particularly at C1. Here a tan-colored lenticular quartz arenite displays medium scale cross-bedding, reactivation surfaces, and a fine display of *Skolithos* trace fossils (Donovan, 1984). Some herringbone cross-bedding is present in subjacent sandstones. The whole sequence appears to record a storm-tide interplay in a shallow marine environment.

C2. The closure of the valley to the southeast of C1 is an expression of buried Cambrian relief that is best appreciated by walking the rhyolite-sediment unconformity to C2. From a maximum thickness of 120 ft (36 m), the Reagan gradually thins until it is completely overlapped by the Honey Creek. Among many interesting features are an ankerite band at the contact of the Honey Creek and Reagan and wedges of rhyolite conglomerate and breccia shed from the gradually disappearing rhyolite island (Donovan, 1984).

C3. The view east from C3 is across the valley of Ketch Creek, which is eroded along a fault with a stratigraphic downthrow of 500 ft (150 m) to the west at this point, decreasing to zero to the NNW. The fault bifurcates and gradually passes into a number of small fractures and west-facing monoclines in this direction. The dip of the fault plane wavers on both sides of vertical but is always steep; some springs are associated with the fault about 600 ft (180 m) north of C3.

C4. A walk along the Ketch Creek fault from C3 to C4 shows that while most of the related deformation (including some tight folding) is located in the downthrown (western) block, the stratigraphic disparity is reduced by the gradual appearance of younger beds in the eastern block. On this walk the view to the northeast encompasses almost 5,000 ft (1,500 m) of unbroken Arbuckle Group carbonates dipping northeast at angles of 20–30°.

C5. The final stop on this walk is a spectacular double unconformity involving the Reagan–Carlton Rhyolite relationship and the Post Oak Conglomerate and Lower Paleozoic sequence. The latter unconformity is especially impressive; one side of a steep-sided Permian valley is clearly exposed, and the Post Oak clasts marked fine upward and outward from the unconformity. From here the best route back to Oklahoma 58 is to walk along the rhyolite outcrop to A14; good examples of flow folding can be found in the rhyolite.

REFERENCES

Beauchamp, W. H., 1983, The structural geology of the southern Slick Hills, Oklahoma [M.S. thesis]: Oklahoma State University, 119 p.

Donovan, R. N., 1982, Geology of Blue Creek Canyon, Wichita Mountains area, *in* Gilbert, M. C., and Donovan, R. N., eds., Geology of the Eastern Wichita Mountains, Southwestern Oklahoma: Oklahoma Geological Survey, Guidebook 21, p. 65–77.

——, 1984, The geology of the Blue Creek Canyon area, *in* Gilbert, M. C., and Donovan, R. N., eds., Recent developments in the Wichita Mountains, Guidebook for Field Trip No. 1: 18th Annual Meeting, Geological Society of America South-Central Section, Richardson, Texas, p. 40–101.

Donovan, R. N., and Rafalowski, M. B., 1984, The nature of the unconformity between the Honey Creek Limestone and the Carlton Rhyolite in the Slick Hills of southwestern Oklahoma Geological Society of America, Abstracts with Programs, v. 16, no. 2, p. 82.

Donovan, R. N., Babaei, A., and Sanderson, D. J., 1982, Blue Creek Canyon, *in* Gilbert, M. C., and Donovan, R. N., eds., Geology of the Eastern Wichita Mountains, Southwestern Oklahoma: Oklahoma Geological Survey, Guidebook 21, p. 148–153.

Gilbert, M. C., and Donovan, R. N., 1984, Recent developments in the Wichita Mountains, Guidebook for Field Trip No. 1: 18th Annual Meeting, Geological Society of America South-Central Section, Richardson, Texas, 101 p.

Harlton, B. H., 1972, Faulted fold belts of southern Anadarko basin adjacent to frontal Wichitas: American Association of Petroleum Geologists Bulletin, v. 56, p. 1544–1551.

Hudleston, P. J., 1973, Fold morphology and some geometrical implications of theories of fold development: Tectonophysics, v. 16, p. 1–46.

McConnell, D., 1983, The mapping and interpretation of the structure of the northern Slick Hills, southwest Oklahoma [M.S. thesis]: Oklahoma State University, 131 p.

Marcini, W.R.D., 1986, Transpression: an application to the Slick Hills, SW Oklahoma [Ph.D. thesis]: Belfast, Northern Ireland, Queens University of Belfast, 286 p.

Rafalowski, M. B., 1984, Sedimentary geology of the late Cambrian Honey Creek and Fort Sill Formations as exposed in the Slick Hills of southwestern Oklahoma [M.S. thesis]: Oklahoma State University, 147 p.

Ragland, D. A., 1983, Sedimentary geology of the Ordovician Cool Creek Formation as it is exposed in the Wichita Mountains of southwestern Oklahoma [M.S. thesis]: Oklahoma State University, 170 p.

Ragland, D. A., and Donovan, R. N., 1985, The Thatcher Creek member, basal unit of the Cool Creek Formation, in southern Oklahoma: Oklahoma Geology Notes, v. 45, no. 3, p. 83–91.

Ramsay, J. G., 1967, Folding and fracturing of rocks: New York, McGraw-Hill Book Company, 560 p.

Great Salt Plains and hourglass selenite crystals, Salt Fork of the Arkansas River, northwest Oklahoma

Kenneth S. Johnson, *Oklahoma Geological Survey, University of Oklahoma, Norman, Oklahoma 73019*

LOCATION

The Great Salt Plains comprise about 25 mi^2 (64 km^2) of flat, barren land, naturally encrusted with salt (halite), along the Salt Fork of the Arkansas River (Fig. 1). The salt plains, which also embrace the area where unique hourglass selenite crystals are currently growing, are almost entirely enclosed within the Great Salt Plains National Wildlife Refuge. Located in central Alfalfa County, Oklahoma, the land is about 4 mi (6.4 km) east of Cherokee and just northwest of Jet. Free access by car or bus is permitted at all times during the year to the salt flats in the vicinity of the observation tower on the west (site 1, Fig. 1) and to the dam and state park area in the east (including picnic and camping facilities and cabins). Also, free access by car or bus to the selenite crystal area (site 2, Fig. 1) is permitted through only one gate, which is unlocked from 8 A.M. until 5 P.M. only on Saturdays, Sundays, and holidays from April 1 to October 15 each year. The crystal area is about 1 mi (1.6 km) northeast of this gate and can be reached safely by driving along the specified road across the salt flats. Access to other parts of the Great Salt Plains is restricted by fences; quicksand, at scattered locations on the salt flats, makes trespass in nonauthorized areas hazardous. The area is covered by four 7½-minute quadrangle maps: Cherokee S, Cherokee N, Jet, and Manchester SW. For information concerning the National Wildlife Refuge and crystal-collecting area, contact the Refuge Manager, Great Salt Plains National Wildlife Refuge, Route 1, Box 76, Jet, Oklahoma 73749 (phone: 405/626-4794). For information on state park facilities, contact the Park Superintendent, Great Salt Plains State Park, Route 1, Box 28, Jet, Oklahoma 73761 (phone: 405/626-4731).

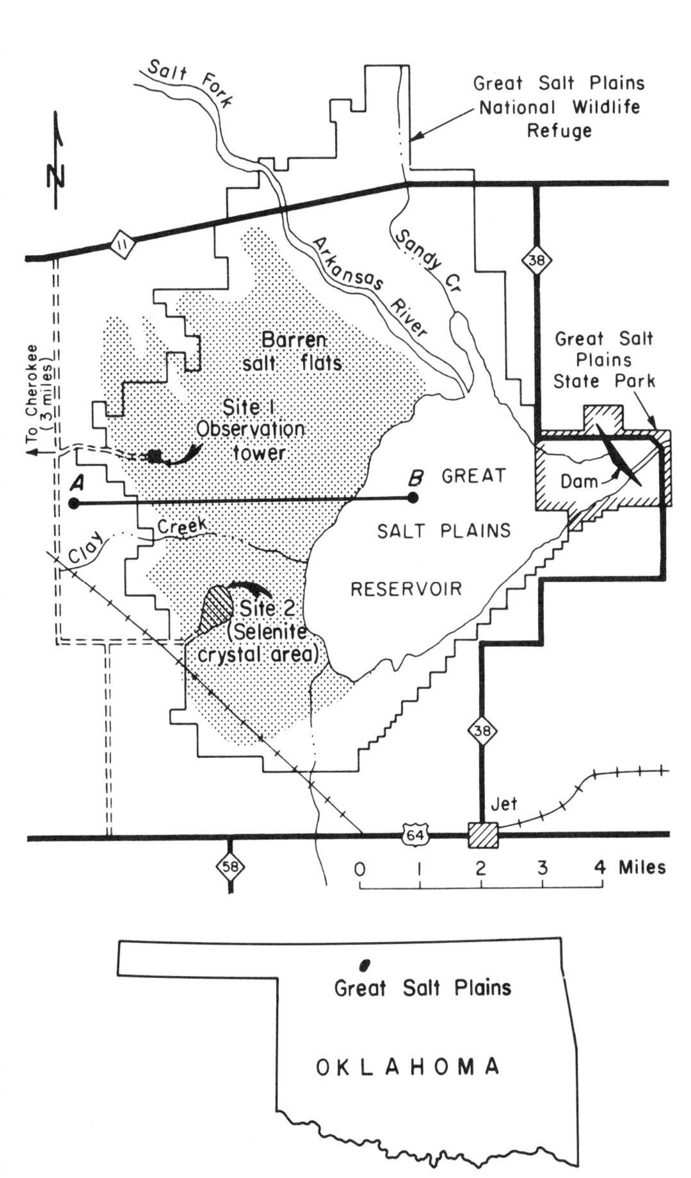

Figure 1. Location map for sites 1 and 2 at Great Salt Plains. Cross section A-B shown in Figure 2.

SIGNIFICANCE

There are a number of sites on the east flank of the Permian Basin where Permian salts are being dissolved and high-salinity natural brines are being emitted at the surface. Great Salt Plains is the largest of the resulting salt flats in north Texas, Oklahoma, or Kansas; it is the only salt plain located on public lands. In addition, it is the only location at which hourglass selenite crystals are growing today and can be collected by geologists and the general public. The current description is modified from Johnson (1972).

SITE INFORMATION

Quaternary deposits forming the salt plains are generally 10 to 25 ft (3 to 7.5 m) thick. They consist of alluvial and lacustrine sediments laid down on an irregular Permian bedrock eroded by

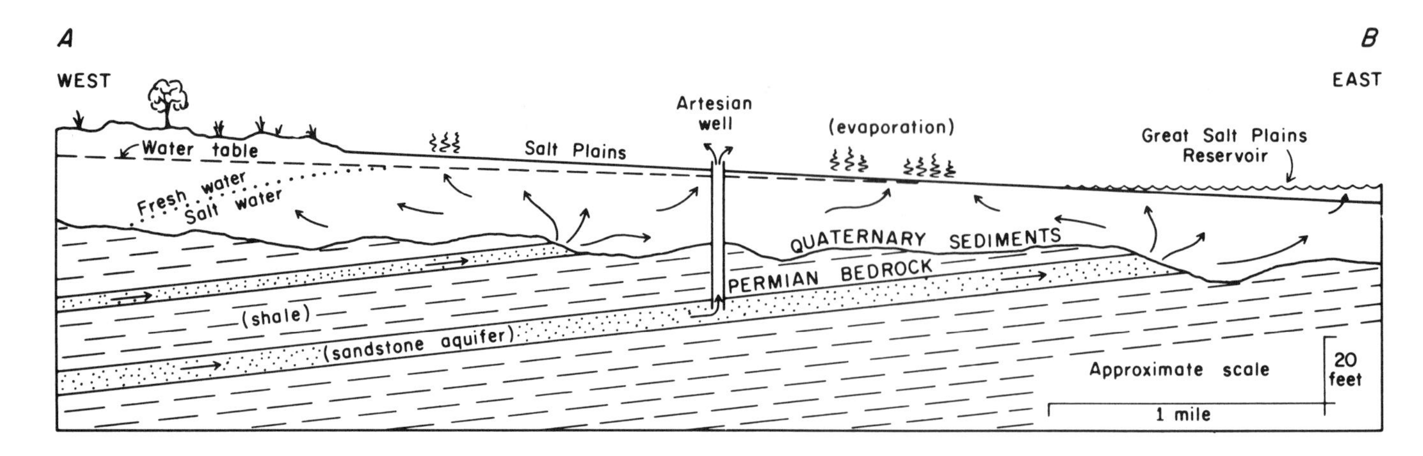

Figure 2. Schematic cross section A-B showing movement of salt water (arrows) from bedrock aquifers into overlying Quaternary sediments at Great Salt Plains. See Figure 1 for location of cross section.

streams and rivers, probably in late Pleistocene time. These unconsolidated sediments occur in thin layers deposited over large areas of the salt plains by three means: (1) by streams, which in time of flood, spread their water and sediment in a fanlike manner upon reaching the salt flats; (2) by intermittent lakes formed when water was backed up temporarily behind the narrow water gap through which the Salt Fork of the Arkansas River leaves the salt plains on the east; and (3) by wind, which tends to redistribute sediment and bevel surface irregularities. The present surface appears horizontal, but in fact it slopes toward the reservoir at a rate of 4 to 8 ft/mi (0.75 to 1.5 m/km).

Salt water moves laterally and upward under artesian conditions through several sandstone aquifers in the bedrock Hennessey Formation and is being discharged into the bottom of Quaternary deposits (Fig. 2). Natural brine in the bedrock aquifers and in Quaternary sediments has a sodium chloride content generally ranging from about 150 to 250 grams per kilogram (g/kg) of brine, or about 15 to 25 percent by weight. The only other ions in significant concentration are calcium (0.15 g/kg), magnesium (0.1 g/kg), and sulfate (0.7 g/kg). A thin crust of salt forms on the plains, usually after several days of dry weather. The salt is precipitated as water is evaporated from brine drawn to the surface of the salt flats by capillary action.

Hydrogeological studies show that natural dissolution of bedded Permian salts occurs at many places in western Oklahoma and adjacent areas (Johnson, 1970, 1981; Johnson and Denison, 1973). Meteoric water apparently seeps into the ground at unknown locations northwest, west, or southwest of Great Salt Plains; migrates downward and laterally to subsurface salt beds that are present on the north flank of the Anadarko Basin; and dissolves the salt to form brine. The resulting brine is then forced laterally and upward by hydrostatic pressure through aquifers, or through fractures in aquitards, until it is discharged at the surface beneath the salt plain. The salt being dissolved most likely is bedded salt in the Permian Cimarron Evaporites at a depth of 300 to 400 ft (90 to 120 m) beneath the plain (these strata are deeper and thicker farther west and southwest), or small salt crystals and finely disseminated salt masses present in small quantities in porous sandstones through which the ground water moves.

Of special interest at Great Salt Plains are the unique hourglass selenite crystals that can be collected during certain times of the year. The ground water here is highly mineralized; concentrations of sodium chloride, calcium, and sulfate are already near saturation level. Owing to the high dissolved-solids content of the water and the high rate of evaporation in the area, salt and gypsum crystals are continually precipitated on and just below the surface of the salt plains.

One variety of gypsum, selenite, is hydrous calcium sulfate ($CaSO_4{\cdot}2H_2O$). The interior of selenite crystals at Great Salt Plains contains a ghostlike hourglass form consisting of sand, silt, and clay particles incorporated within the crystal as it grows (Fig. 3). As a crystal grows, loose particles are enveloped only at the ends, whereas particles adjacent to the sides of the crystal are merely pushed aside. Apparently the bond between molecules forming the smooth faces along the sides of the crystal is so strong that each new layer of molecules forces foreign particles away from the crystal. On the other hand, the bond between molecules along crystal faces developing at the ends is not strong enough to push foreign matter aside, and therefore new layers envelop the particles. The brown color is acquired with the brown sediment enclosed in the crystal.

Crystals are formed just below the surface, seldom more than 2 ft (0.6 m) deep. They occur only in those few parts of the salt plains where the gypsum and salt concentrations in the brine are favorable for selenite precipitation. The selenite occurs as individual crystals and in clusters, with some individual crystals reaching 7 in (18 cm) long, and complete clusters weighing as

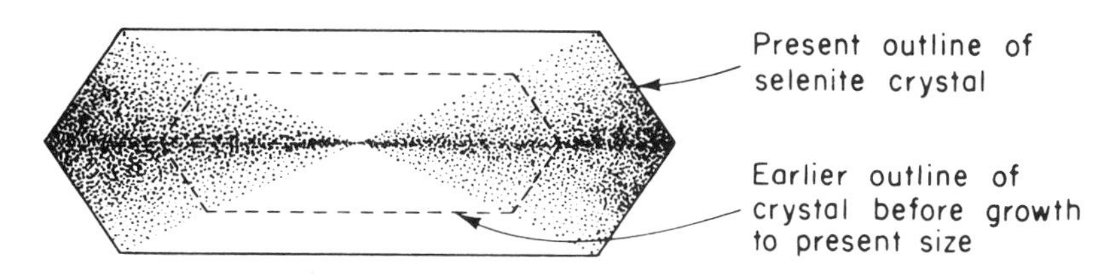

Figure 3. Drawing of hourglass selenite crystal illustrating two stages of growth and incorporation of sand, silt, and clay at ends of crystal.

much as 38 lb (17 kg). Crystals can enclose sticks, rocks, bones, and cocklebures that are part of the host sediment.

From May 1966 to May 1969, members of the Enid Gem and Mineral Society conducted a study of the shapes and growth rates of selenite crystals they had "planted" in small tracts of the Great Salt Plains from which all other crystals had been removed. Results of their study indicate that under favorable climatic conditions (cool, wet springs and hot, dry summers) crystals can grow as much as 26 percent larger within a year. The quality of crystals depends in part on the grain size of the host sediment, with the clearest crystals forming in clay and more cloudy ones forming in silt and sand.

One prolific area of the Great Salt Plains has been set aside as a place where the public may dig for crystals during certain times of the year (site 2, Fig. 1). A brochure from the Great Salt Plains National Wildlife Refuge recommends the following procedures: dig a hole about 2 ft (60 cm) across and 2 ft (60 cm) deep; allow it to fill with water that seeps in from below (water may need to be supplied); splash water against the sides of the hole and gently wash the soil away from the crystals until they are free. At first the crystals are very fragile and must be gently placed where the sun and wind will dry them; when dry, they are quite hard and can be handled with normal care.

Owing to hazardous conditions on some parts of the plains, cars should be driven to the digging area on the salt flats only along prescribed lanes. Also, the white salt surface increases the possibility of sunburn, so sunglasses and protective clothing should be worn. Bring plenty of drinking water and extra water to wash the salt water off all equipment and tools to prevent corrosion.

The dam that created Great Salt Plains Reservoir was completed in 1941. The lake covers approximately 14 mi^2 (36 km^2) of land that was once part of the salt plain. The water is less than 15 ft (4.5 m) deep in most places but is nearly 20 ft (6 m) deep at the eastern end near the dam. At this end, particularly just below the dam, there are excellent exposures of flat-lying Hennessey strata (red-brown shales and light gray sandstones) similar to the rocks covered by the salt plains.

The salinity of water in the lake is normally 2 to 15 g/kg of water (considerably less than the 30 g/kg of normal sea water). The lake reaches higher salinities when brief but heavy rains on the adjacent salt flats dissolve the salt and wash it into the lake, or when the lake volume is greatly reduced through evaporation while the amount of salt in the lake remains the same or increases. Lake water salinity is diluted when large quantities of fresh water flow into the reservoir. Although the salinity varies, the amount of salt (NaCl) that flows out of the reservoir through the Salt Fork of the Arkansas River averages about 3,000 tons per day.

Much attention has been focused on this area recently because of the adverse effect of these brines on water quality downstream from the salt plains. One remedial measure being considered by the U.S. Army Corps of Engineers involves rechanneling the Salt Fork of the Arkansas River around the salt plains and the reservoir, allowing the dam to impound and isolate permanently brine and any water which flows across the salt flats.

An alternate salt plain worth visiting, Big Salt Plain, is located near Freedom on the Cimarron River, about 50 mi (80 km) west of Great Salt Plain (Johnson, 1970, 1972, 1981). Here, brine formed by dissolution of the Permian Flowerpot salt is being evaporated in large earthen solar pans to produce commercial salt. Cargill Salt Company acquired the property in 1984 and anticipates producing several hundred thousand tons of salt annually.

REFERENCES CITED

Johnson, K. S., 1970, Salt produced by solar evaporation on Big Salt Plain, Woods County, Oklahoma: Oklahoma Geology Notes, v. 30, p. 47–54.

—— , 1972, Guidebook for geologic field trips in Oklahoma; Book II, northwest Oklahoma: Oklahoma Geological Survey Educational Publication 3, 42 p.

—— , 1981, Dissolution of salt on the east flank of the Permian Basin in the southwestern U.S.A.: Journal of Hydrology, v. 54, p. 75–93.

Johnson, K. S., and Denison, R. E., 1973, Igneous geology of the Wichita Mountains and economic geology of Permian rocks in southwest Oklahoma: Oklahoma Geological Survey Special Publication 73-2, Guidebook for Field Trip No. 6 at Annual Meeting (Dallas, 1973) of Geological Society of America, 33 p.

Shelf to slope facies of the Wapanucka Formation (lower-Middle Pennsylvanian), frontal Ouachita Mountains, Oklahoma

Robert C. Grayson, Jr., Department of Geology, Baylor University, Waco, Texas 76798
Patrick K. Sutherland, School of Geology and Geophysics, University of Oklahoma, Norman, Oklahoma 73019

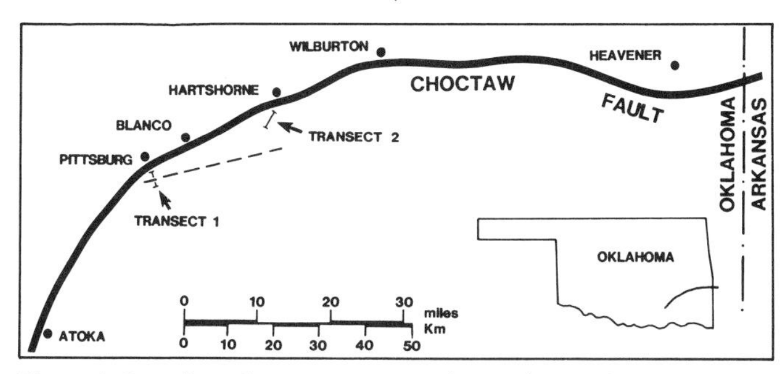

Figure 1. Location of transects 1 and 2 in relation to Choctaw Fault. Dashed line represents estimated location of outer margin of shelf during deposition of lower limestone member.

LOCATION

Two transects examine shelf-to-slope facies relations of the Wapanucka Formation in southeastern Oklahoma near the towns of Pittsburg and Hartshorne (Fig. 1). Detailed directions to several localities along each transect are given in the text.

SIGNIFICANCE

The frontal Ouachita Mountains in Oklahoma provide an exceptional opportunity to examine a shelf-to-slope facies change in the Wapanucka Formation. This is possible because the Wapanucka Formation is repeated by faulting three to five times within a distance of 1.5 to 2 mi (2.5 to 3 km). These fault-slice exposures parallel the Choctaw Fault for a distance of about 55 mi (90 km) along the frontal zone of the Ouachita Mountains (Figs. 1, 2). Stratigraphic relationships are complicated, however, by the fact that the frontal faults cut obliquely across strata representing depositional strike on the outer margin of the lower-middle Pennsylvanian shelf (Fig. 1). Two transects spaced along the narrow outcrop belt are to be examined (Fig. 2), and each transect consists of three or four localities (Figs. 3, 4).

TRANSECT 1

The Wapanucka Formation consists of four members (Fig. 5), which record basinward progradation of the lower-middle Pennsylvanian shelf and shelf margin. The dashed line in Figure 1 represents an estimate of the margin separating shelf facies of the lower limestone member from slope facies of the Chickachoc Chert member. This transition observed in transect 1 is marked by significant basinward changes in facies (Fig. 3). Accumulation of the middle shale member, which drastically thickens basinward from 15 to 288 ft (4.5 to 88 m), substantially increased the basinward extent of the shelf. As a consequence, exposures of the upper sandstone limestone member consist predominantly of shelf facies that thicken irregularly but only moderately basinward and exhibit less striking basinward facies changes than those seen in the lower limestone member.

The Wapanucka Formation has traditionally been included entirely within the Morrowan Series, but detailed studies of the conodont faunas (Grayson, 1979, 1984) show that the highest units are Atokan in age (Fig. 5) and represent the lateral equivalent of the Lower Atokan Spiro Sandstone in the subsurface Arkoma Basin to the north (Fig. 6).

Locality 1, Measured Section 303B (Kiowa East 2)

Location. The section is measured along an oil company service road located immediately east of a mostly abandoned section-line gravel road. The road intersects Oklahoma 63 about 1 mi (1.6 km) east of the intersection of Oklahoma 63 with U.S. 69 in Kiowa. Drive south on the section-line road approximately 1 mi (1.6 km), then bear left at the fork in the road for about 0.1 mi (0.2 km). The section begins on the east side of the service road immediately above the steep incline up Limestone Ridge, W½,NW¼,NW¼,Sec.31,T.3N.,R.13E., Kiowa Quadrangle, Pittsburg County, Oklahoma (Fig. 2).

Description. The visitor should examine the light-gray micritic limestone of the lower limestone member (units 5 through 11) in order to be able to compare these with better exposures of this member in transect 2 (Fig. 4) and to contrast them with the darker weathering and deeper water spiculiferous limestones of the Chickachoc Chert Member that are well exposed at locality 2 (Fig. 3).

The upper sandstone limestone member is particularly well exposed at locality 1. Progradation cycles consist of spiculitic limestone, micritic limestone, and calcarenite. This member accumulated on the outer shelf in generally shallow water, where coarse clastic influx from the east or northeast controlled facies distribution and character. At this locality the lithologic succession (Fig. 3) is interpreted in terms of a generalized offshore linear sand bar model (Walker, 1984). High-energy bar facies consist of coarsening upward tabular cross-bedded bioclastic calcarenite or sandstone, whereas bar-margin facies generally lack cross-bedding. Interbar accumulations are represented by low-energy deposits including algal bank micritic limestone, spiculitic limestone, and shale. The marked lateral changes in lithology exhibited by the upper sandstone-limestone (Fig. 3) probably reflect strike-oriented facies belts on the shelf. Approximately 560 ft (170 m) to the west, on the north wall of a small quarry (section 17 of Grayson, 1979), fossiliferous bar calcarenite is well exposed. The most common fossils include carbonized plant re-

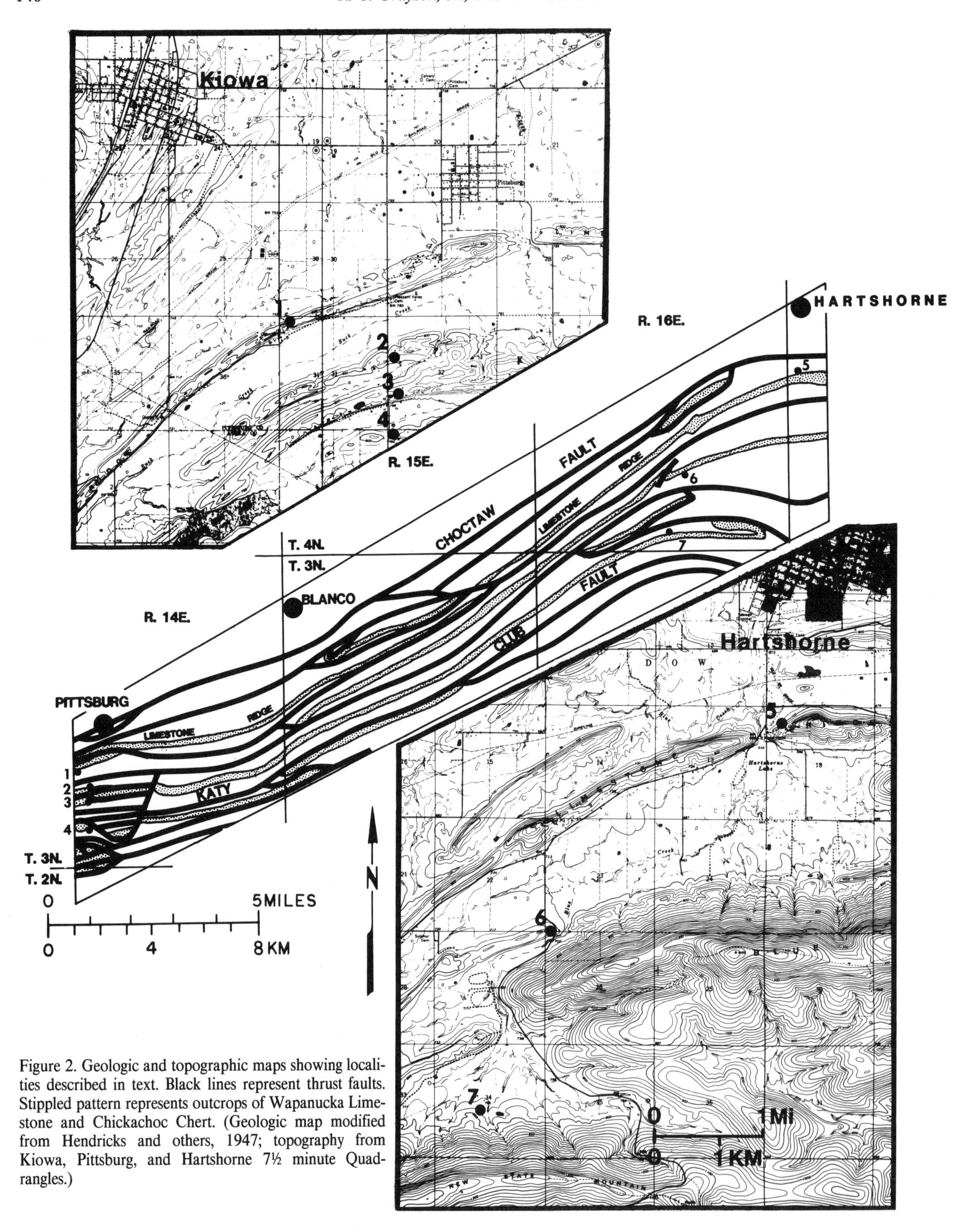

Figure 2. Geologic and topographic maps showing localities described in text. Black lines represent thrust faults. Stippled pattern represents outcrops of Wapanucka Limestone and Chickachoc Chert. (Geologic map modified from Hendricks and others, 1947; topography from Kiowa, Pittsburg, and Hartshorne 7½ minute Quadrangles.)

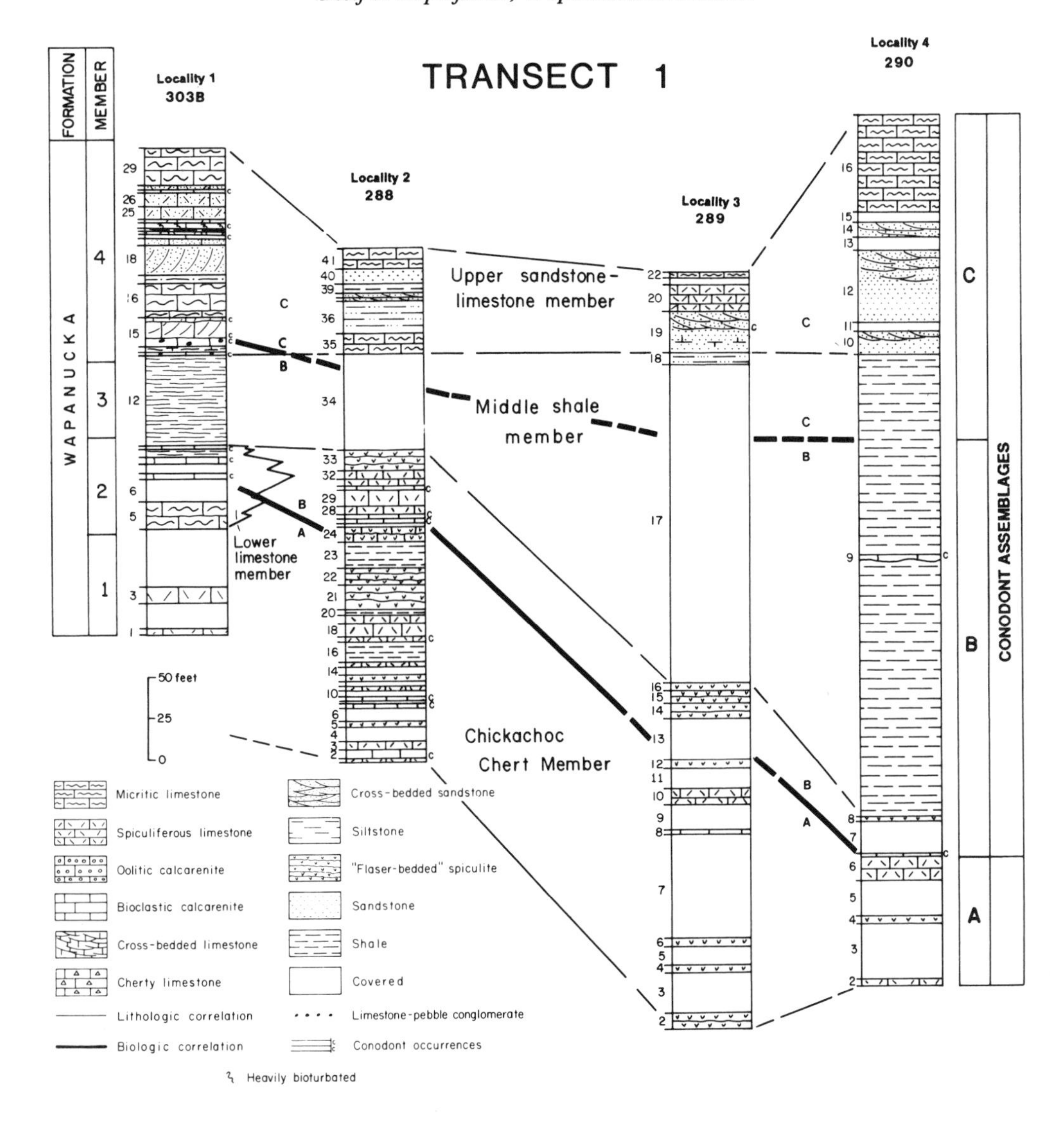

Figure 3. Stratigraphic cross section for transect 1, showing basinward facies changes. Numbers below locality numbers represent University of Oklahoma measured sections. Compare with Figures 1 and 2 for location and scale.

mains and ammonoid cephalopods. Limestone clasts that are spherical to oblate disk-shaped are also present.

Stop 2, Measured Section 288 (Pleasant Valley 1)

Location. Return to Oklahoma 63, turn east (right), drive 1.0 mi (1.6 km) to the section-line road. Turn south (right), proceed about 1.8 mi (2.9 km) to the first roadcut in the ridge south of Limestone Ridge: SW¼,NW¼,Sec.32,T.3N.,R.14E., Pittsburg Quadrangle, Pittsburg County, Oklahoma (Fig. 2).

Description. Compared to the previous locality, the lower part of the Wapanucka Formation at this stop is represented by an assemblage of facies with distinctly deeper water characteristics. This assemblage is typical of the Chickachoc Chert Member and consists of spiculite, spiculitic limestone, shale, and allodapic calcarenite units (units 2, 7, and 9) that exhibit partial Bouma sequences (Tc, Td). The Chickachoc Chert consists of two regressive cycles of deposition (units 1 to 20, 21 to 33), but the cycles are not as clearly distinguishable as those seen in the laterally equivalent lower limestone (Fig. 4, Locality 5). Exposures of the Chickachoc Chert located basinward of this locality (Fig. 3) record a comparatively constant water depth.

Facies changes are also obvious in the upper sandstone-limestone compared to the previous locality. Terrigenous clastic units are relatively more common and occur in two coarsening upward sequences that are typical of offshore linear sand bar successions. In contrast, carbonate facies are limited to algal bank micritic limestone that accumulated in interbar settings.

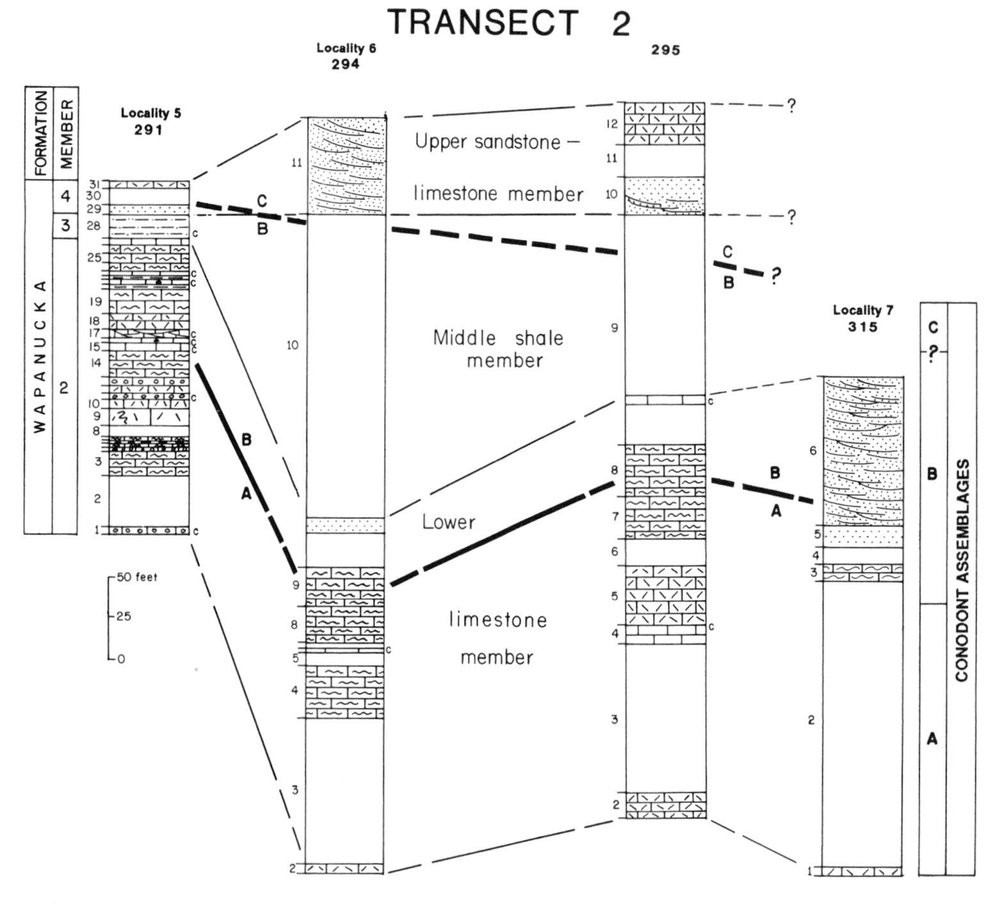

Figure 4. Stratigraphic cross section for transect 2. Compare with Figures 1 and 2 for location and scale.

Locality 3, Measured Section 289 (Pleasant Valley 2)

Location. This section is located in the drainage ditch mostly on the east side of section-line road 0.2 mi (0.3 km) south of Locality 2, along the west margin of SW¼,Sec.32,T.-3N.,R.14E., Pittsburg Quadrangle, Pittsburg County, Oklahoma. It is located on the second fault ridge south of Limestone Ridge.

Description. No obvious changes in sea level are recorded by the lithologic successions in the Chickachoc Chert between Localities 2 and 3. At Locality 3 the middle shale is not exposed, but the covered interval presumed to represent that member is substantially thicker compared to that at the previous stops (Fig. 3).

The lower algal build-up in the upper sandstone-limestone at Locality 2 has pinched out, and its stratigraphic position is occupied by a coarsening upward terrigenous clastic sequence. In addition to the "high" algal bank micritic limestone (unit 22), interbar carbonate facies at this locality include a spiculitic limestone unit.

Locality 4, Measured Section 290 (Pleasant Valley 3)

Location. This section was measured on the east side of the section-line road, 0.3 mi (0.5 km) south of Locality 3, in the NW¼,NW¼,Sec.5,T.2N.,R.14E., Pittsburg Quadrangle, Pittsburg County, Oklahoma.

Description. The stratigraphic and lithologic succession at this locality is comparable to that at Locality 3. However, the middle shale is markedly thicker, and comparatively well exposed. The upper sandstone-limestone consists of multiple sand bodies, indicating a local sand depocenter. The "high" algal bank limestone contains zones of the solitary rugose coral *Konickophyllum* in presumed life position.

TRANSECT 2

The three localities to be examined in transect 2 demonstrate the following stratigraphic relationships: (1) the main part of the Wapanucka consists of the lower limestone member (no Chickachoc Chert), since all localities represent a shelf environment; (2) the lower limestone member includes cross-bedded sandstones with an eastward source (Locality 7); (3) the upper sandstone-limestone member, in this more eastern area, includes a higher percentage of sandstone than was observed in transect 1. East of transect 2 there is a further increase in sand content in both the lower and upper shelf sequences.

Locality 5, Measured Section 291 (Hartshorne Quarry)

Location. This section, located in the NW¼,NW¼,Sec.18, T.4N.,R.17E., Hartshorne Quadrangle, Pittsburg County, Oklahoma, was measured on the western wall of an abandoned rock quarry (Fig. 2). The section is reached by driving south on 7th Street, in Hartshorne, approximately 1.5 mi (2.4 km) to the park area just north of the spillway on Hartshorne Lake. Walk east-northeast from the park, across a small pasture, and then ascend Limestone Ridge along an overgrown field road and walk about 0.25 mi (0.4 km) to the "floor" of the quarry.

Description. Several progradational platform carbonate depositional cycles can be recognized in the lower limestone member of the Wapanucka Formation at the Hartshorne Quarry section (Locality 5, Fig. 4). The lower cycle (units 1 through 8) is not well exposed and is represented by only the upper, mostly shallow-water part of the cycle. The middle cycle includes unit 9, a dark-colored, millimeter-laminated spiculitic limestone, and extends up-section to the top of unit 17. The upper cycle begins with unit 18 and extends upward to the base of the middle shale (unit 28). Each regressive cycle begins with spiculite or spiculitic limestone, and consists of similar lithologies and vertical succession of facies (Fig. 4). Lateral facies relationships and lithologic character suggest that spiculitic limestone represents a relatively low-oxygen, deeper water facies that mostly accumulated on the outer margin of the shelf (Newell, 1957), although spiculitic lithologies are known to have occurred in shallower water shelf settings, influenced by nearby deltas.

Algal bank micritic limestone is an important constituent of the lower limestone at this locality (Fig. 4). Two types of algal material provided the framework for micrite accumulation: (1) loose blade-like thalli of the red algae *Archaeolithophyllum,* and (2) the more common and rigid tubular red algae *Donezella.* By analogy to recent carbonate algal mudbank accumulations, *Archaeolithophyllum* and *Donezella* skeletal mud banks probably developed in shallow-water, low-energy, moderately circulated environments on the shelf. Algal banks periodically built up to sea level but generally were entirely subtidal, and may have initiated growth in water depths as great as 50 to 65 ft (15 to 20 m).

A relatively diverse suite of calcarenite lithologies occurs in the Hartshorne Quarry succession (Fig. 4). The majority of these can be interpreted in terms of a generalized calcarenite bar model. In an idealized calcarenite bar succession, the vertical sequence is indicative of shoaling conditions. A typical sequence might include a basal bioclastic calcarenite (packstone) grading through bioclastic grainstone to oolitic grainstone.

Locality 6, Measured Section 294 (Blue Creek)

Location. This section, on Mr. Blanchard's property, was measured adjacent to the winding country road and west of the Blanchard's fence line. The section begins to the north of the house near the road, passes well in front of the house, and ends in the valley of Blue Creek south of the house. Permission for access to the locality should be obtained at the Blanchard house, which can be reached by driving south on 7th Street from Hartshorne approximately 4.9 mi (7.9 km). The section extends from SW¼,SW¼,SW¼,Sec.23 south to NW¼,NW¼,NW¼,Sec.26,T.-4N.,R.16E., Hartshorne Quadrangle, Pittsburg County, Oklahoma.

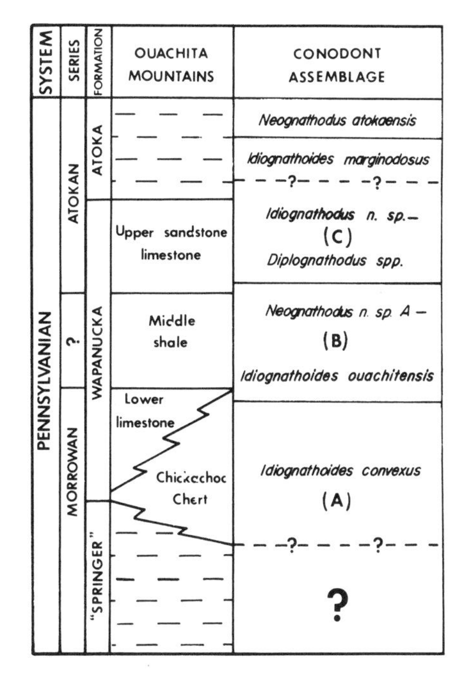

Figure 5. Lithostratigraphic and biostratigraphic subdivisions of Wapanucka Formation, Frontal Ouachita Mountains, Oklahoma.

Description. The correlation of measured units in the lower limestone member between Hartshorne Quarry (Locality 5) and Blue Creek (this Locality) is uncertain, owing to the lack of diagnostic conodont faunas and a marked change in facies. Unit 6 (Fig. 4) is a thin, interlaminated spiculitic limestone and bioclastic calcarenite that may correlate with the lowest spiculitic unit (unit 9, Locality 5, Fig. 4) in the Hartshorne Quarry section.

Low-energy, algal limestone is the predominant lithology of the lower limestone at this locality. Algal structures are generally difficult to discern but *Donzella* (tubular algae) boundstone can be observed at the top of unit 9. Spiculitic units (units 2 and 6) are relatively rare, and calcarenite development is limited to a thin interval at the base of unit 4. Approximately 5 ft (1.5 m) of siltstone is poorly exposed in Blanchard's driveway near the access road.

The middle shale and upper sandstone-limestone at this locality represent a considerably thickened stratigraphic interval when compared to the Hartshorne Quarry sequence. Most of the increased thickness is due to an expanded middle shale interval (unit 10) that is concealed by vegetation. The upper sandstone-limestone member consists of about 60 ft (20 m) of well-sorted, fine-grained sandstone. The sequence exhibits a slight coarsening upward trend that is also marked by a corresponding increase in thickness of sandstone layers. The upper, thicker bedded units

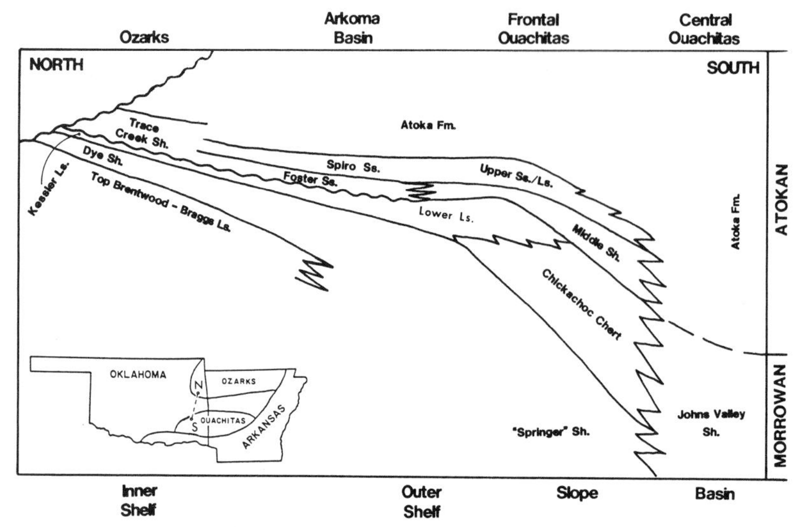

Figure 6. North-south depositional reconstruction during Late Morrowan and Early Atokan time; pre-Ouachita orogeny, Oklahoma. Not to scale; total present-day north-south distance 130 km (80 mi).

commonly contain large-scale tabular cross-bedding. This vertical sequence is indicative of prograding marine offshore shelf bars (Grayson, 1980) and is comparable to the marine-bar sequence of Exum and Harms (1968).

Locality 7, Measured Section 295 (New State Mountain)

Location. This section is located in NE¼,SW¼,Sec.34,T.-4N.,R.16E., Hartshorne Quadrangle, Pittsburg County, Oklahoma. It is located on the Blue Valley Ranch and can be reached by driving south from Locality 6 on winding country road for a distance of about 1.5 mi (2.4 km). At that point turn west into the ranch entrance and obtain permission either at foreman's house, located in valley about one-half mile to the west, or at ranch owner's house, on the ridge top, up a winding road to southwest. The section begins in a pasture at the base of the ridge immediately below (north of) the ranch owner's house and continues up the ridge slope to the south to unit 3; it is then offset about 500 ft (150 m) to the east. The best exposure of unit 6 is on the ridge top beside the road that leads to the ranch owner's house.

Description. This exposure was originally considered to represent the lower limestone, middle shale, and upper sandstone-limestone members (Grayson, 1979, p. 71–72). Subsequent study has cast some doubt on this interpretation. No identifiable fossils have been recovered from this sequence, but it now appears possible that unit 6 represents a facies of the lower limestone member rather than the upper sandstone-limestone, as was originally interpreted by Grayson (1979, stop 37). This supposition is based on the common occurrence of cross-bedded sandstones in the equivalent of the lower limestone farther to the east. In either event, the feature of greatest interest at this locality is the thick interval (50 ft; 15 m) of cross-bedded sandstone exposed in a chain of pinnacle-like blocks beside the road along the top of the ridge. Two sandstone facies are distinguished, based on the presence or absence of tabular cross-bedding. The tabular cross-bedded sandstone facies is well exposed (unit 6) and consists of multiple sets of cross-bedded layers ranging in thickness from about 1.5 to 3 ft (0.5 to 1.5 m). The contact between sets of cross-bedded sandstone is sharp and subparallel. Foreset laminations indicate a dominant basinward transport direction. This subfacies is interpreted as quartz-sand accumulations at the crest of an offshore shelf bar. Sandstone facies lacking tabular cross-bedding (unit 5) probably were deposited at the margin of the bar, in slightly deeper, less turbulent water.

REFERENCES CITED

Exum, F. A., and Harms, J. C., 1968, Comparison of marine-bar with valley-fill stratigraphic traps, western Nebraska: *in* Rocky Mountains—Breaking barrier boundaries: American Association of Petroleum Geologists Bulletin, v. 52, no. 10, p. 1851–1868.

Grayson, R. C., Jr., 1979, Stop descriptions–fifth day, *in* Sutherland, P. K., and Manger, W. L., eds., Mississippian–Pennsylvanian Shelf-to-Basin Transition, Ozark and Ouachita Regions, Oklahoma Arkansas: Oklahoma Geological Survey Guidebook 19, p. 67–76.

——, 1980, The Stratigraphy of the Wapanucka Formation (Lower Pennsylvanian) Along the Frontal Margin of the Ouachita Mountains, Oklahoma [Ph.D. thesis]: Norman, University of Oklahoma, 320 p.

——, 1984, Morrowan and Atokan (Pennsylvanian) conodonts from the northeastern margin of the Arbuckle Mountains, Southern Oklahoma, *in* Sutherland, P. K., and Manger, W. L., eds., The Atokan Series (Pennsylvanian) and Its Boundaries; A Symposium: Oklahoma Geological Survey Bulletin 136, p. 41–64.

Hendricks, T. A., Gardner, L. S., Knechtel, M. M., and Averitt, P., 1947, Geology of the Western Part of the Ouachita Mountains of Oklahoma: U.S. Geological Survey Oil and Gas Investigations Preliminary Map 66, scale 1:42,240, 3 sheets.

Newell, N. D., 1957, Permian reefs in the Guadalupe Mountains area; Treatise on marine ecology and paleoecology: Geological Society of America Memoir 67, v. 2, p. 407–436.

Walker, R. G., 1984, Shelf and shallow marine sands, *in* Walker, R. G., ed., Facies Models, 2nd ed.: Geosciences Canada, Reprint Series 1, Geological Association of Canada, p. 141–170.

Depositional and deformational characteristics of the Atoka Formation, Arkoma Basin, and Ouachita frontal thrust belt, Oklahoma

David W. Houseknecht and Michael B. Underwood, Department of Geology, University of Missouri, Columbia, Missouri 65211

LOCATION

Sedimentary facies and structure of the Atoka Formation of the Arkoma basin and Ouachita frontal thrust belt are illustrated by a north-south transect in LeFlore County, Oklahoma.

Locality 1 (94°29′32″W.; 35°12′30″N.; Hackett Quadrangle) is a roadcut on Oklahoma 112, 1.5 mi (2.4 km) south of the village of Pocola (Fig. 1). The Atoka Formation crops out along the Backbone anticline, which is the expression of a thrust fault that ramps to the surface just north of this exposure. The exposed strata are correlative with the "Alma" sandstone of subsurface terminology; in this part of the basin, the Alma is located about 2,650 ft (800 m) below the top and 4,650 ft (1,400 m) above the base of the Atoka Formation. The strata dip 45° to the south.

Locality 2 (94°38′06″W.; 34°49′36″N.; Hodgen Quadrangle) is a roadcut on U.S. 270/59, 1.1 mi (1.8 km) south of the village of Hodgen (Fig. 1). This north-dipping Atoka section is exposed just south of the Choctaw thrust fault (south of the Choctaw fault, the Atoka has been renamed the Lynn Mountain Formation in certain recent publications).

Locality 3 (94°37′37″W.; 34°44′46″N.; Big Cedar Quadrangle) is a roadcut on U.S. 270/59 just south of the village of Stapp. These strata have traditionally been placed within the Johns Valley Shale (Seely, 1963; Hart, 1963), but recent mapping indicates that the turbidites and debris-flow deposits are simply one of several facies associations within the Lynn Mountain Formation (Poole, 1985; McDonald, 1986). The section is slightly overturned to the north.

Locality 4 (94°41′05″W.; 34°45′52″N.; Hodgen Quadrangle) is an exposure along the south side of Holson Valley Road 2.2 mi (3.5 km) west of the intersection with U.S. 270/59. Here, the Lynn Mountain Formation crops out just south of the Ti Valley thrust fault (Fig. 1).

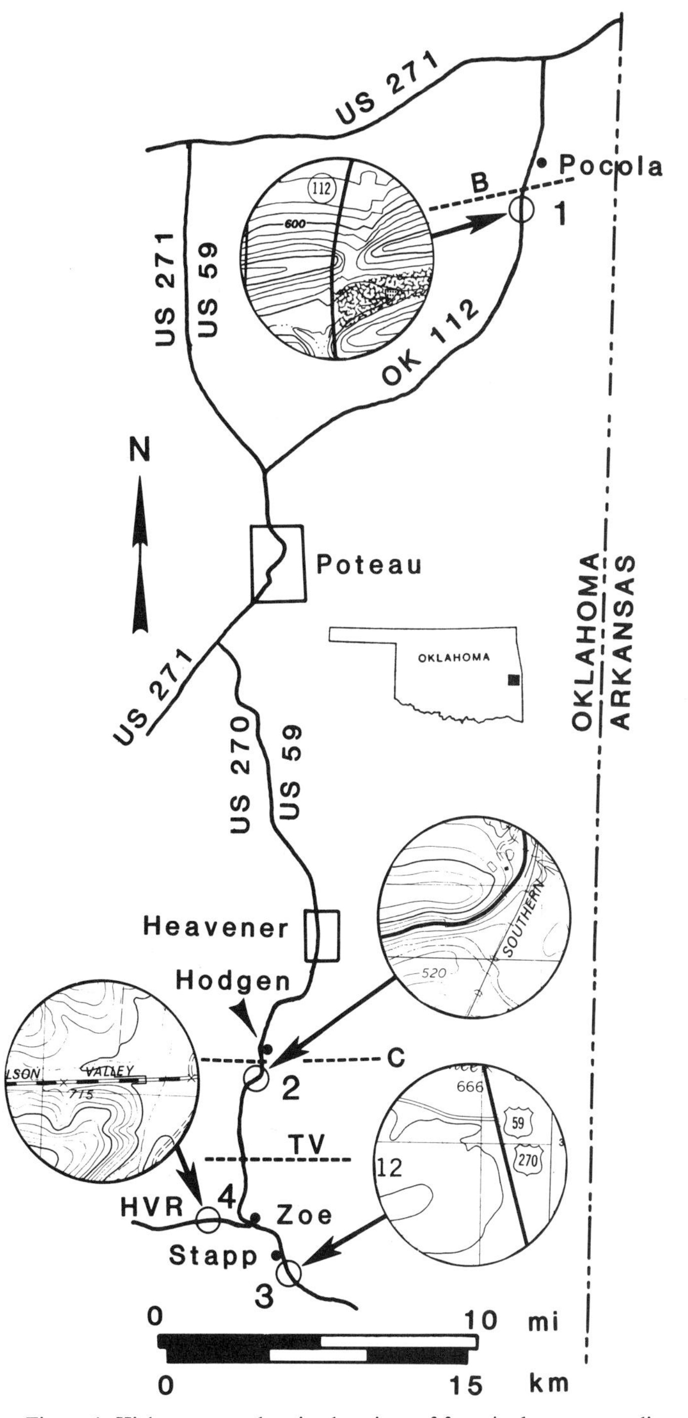

Figure 1. Highway map showing locations of four Atoka outcrops discussed in text. Dashed lines denote surface traces of three thrust faults: B, Backbone; C, Choctaw; TV, Ti Valley. HVR, Holson Valley Road.

SIGNIFICANCE

Atokan strata of the Arkoma basin and Ouachita frontal thrust belt represent sediment that was both deposited and deformed during convergent tectonic activity along the Ouachita orogenic belt. Older Carboniferous strata of the region accumulated in a basin characterized by bathymetrically distinct shelf and deep-marine environments (Fig. 2); in contrast, Atokan strata were deposited in a rapidly subsiding basin characterized by a southward gradation from shoreline to deep-marine environments. This important change in basin character resulted from the breakdown of the shelf by normal faults, apparently induced by

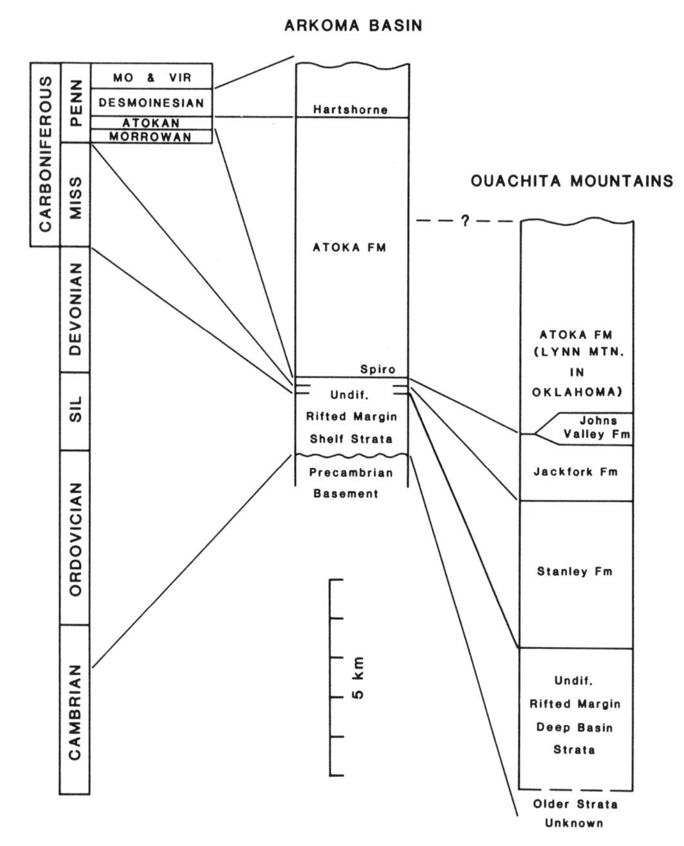

Figure 2. Generalized stratigraphy of Arkoma basin and Ouachita Mountains.

obduction of the Ouachita accretionary prism onto the southern margin of North American crust (Houseknecht, 1986). These normal faults are fundamental tectonic features that offset crystalline basement and the entire pre-Atokan stratigraphic section; the faults are manifested as syndepositional "growth faults" within the Atoka Formation, with thickness increases of more than 0.6 mi (1 km) across faults being common. This transect allows examination of both shoreline and deep-marine facies, as well as an assemblage of structural fabrics suggestive of tectonic deformation during subduction-accretion.

LOCATIONS

Locality 1

This roadcut exposes a 660-ft (200-m) section that represents a complete progradational shoreline sequence that displays well-preserved physical and biogenic sedimentary structures (the latter have been described by Chamberlain, 1978).

The lower 400 ft (127 m) of section includes four major sequences that coarsen upward from silty shale to ripple bedded sandstone intercalated with shale (Fig. 3). Trace fossils are variably abundant throughout; they range from sparse to so abundant that primary sedimentary structures are completely obliterated. The uppermost sequence culminates in a sandstone bed that represents the base of several fining upward sequences preserved in this exposure. The lowermost coarsening upward sequence represents deposition in shelf through lower shoreface environments during progradation of the shoreline. The overlying sequences closely mimic the upper portion of the basal sequence; they were likely the result of successive progradational events, episodic subsidence events, or variations in sea level. A modern analogue of these coarsening upward sequences has been described in the Gulf of Gaeta, Italy, by Reineck and Singh (1975), where the transition from shelf mud to lower shoreface sand represents a decrease in water depth from 600 ft to 15 ft (200 to 5 m).

The upper 230 ft (70 m) of section is characterized by repeated fining upward sequences. The lower part of each sequence is a sandstone that displays ripple, flaser, and swash bedding. Each basal sandstone grades upward into ripple-bedded sandstone intercalated with shale. A few of the sequences continue to fine upward into sandy and silty shale. Trace fossils are locally abundant throughout these fining upward sequences, with U-shaped burrows especially conspicuous within the sandstone beds. All of the physical and biogenic sedimentary structures noted above indicate that these fining upward sequences were deposited on tidal flats, with lower sand flat, middle mixed flat, and upper mud flat subfacies well defined. A modern analogue of these facies has been described by Evans (1975).

Locality 2

This roadcut displays several important elements of the Atokan turbidite basin. Lithofacies near the top of the exposed section (Fig. 4) include thin-bedded turbidites (facies C/D) and hemipelagic shale (facies G of Mutti and Ricci Lucchi, 1972). Partial Bouma sequences and trace fossils are displayed prominently, and flute casts indicate west-directed flow. Deposition of these beds probably occurred within the distal portion of a large, elongate submarine fan; more proximal facies are evident in coeval strata of west-central Arkansas. Working downsection at Locality 2, the thin-bedded turbidites are underlain successively by a zone of intense stratal disruption; a thick interval of shale; and a unit of fractured, massive facies B sandstone (Fig. 5A). This sand probably entered the basin from the north via fault-controlled submarine channels that flowed across the continental slope (Houseknecht, 1986); axial and transverse flows were thus intercalated on the deep basin floor.

The style of stratal disruption at Locality 2 (including web structure) is characteristic of this corridor of the frontal thrust belt. The deformation is discussed in the description of Locality 4.

Locality 3

Most of the turbidite lithofacies of Mutti and Ricci Lucchi (1972) is featured in the roadcut section (Fig. 4). Deposits include a prominent lens of clast-supported conglomerate (facies A), contorted, matrix-supported debris-flow deposits (facies F), hemipelagic shale (facies G), interbeds of shale and sandstone turbidites (facies C and D), and massive siltstone. The overall depositional environment probably contained relatively small distributary channels, associated levees, interchannel muds, and overbank or crevasse-splay deposits.

The origin of limestone clasts within the conglomerate and

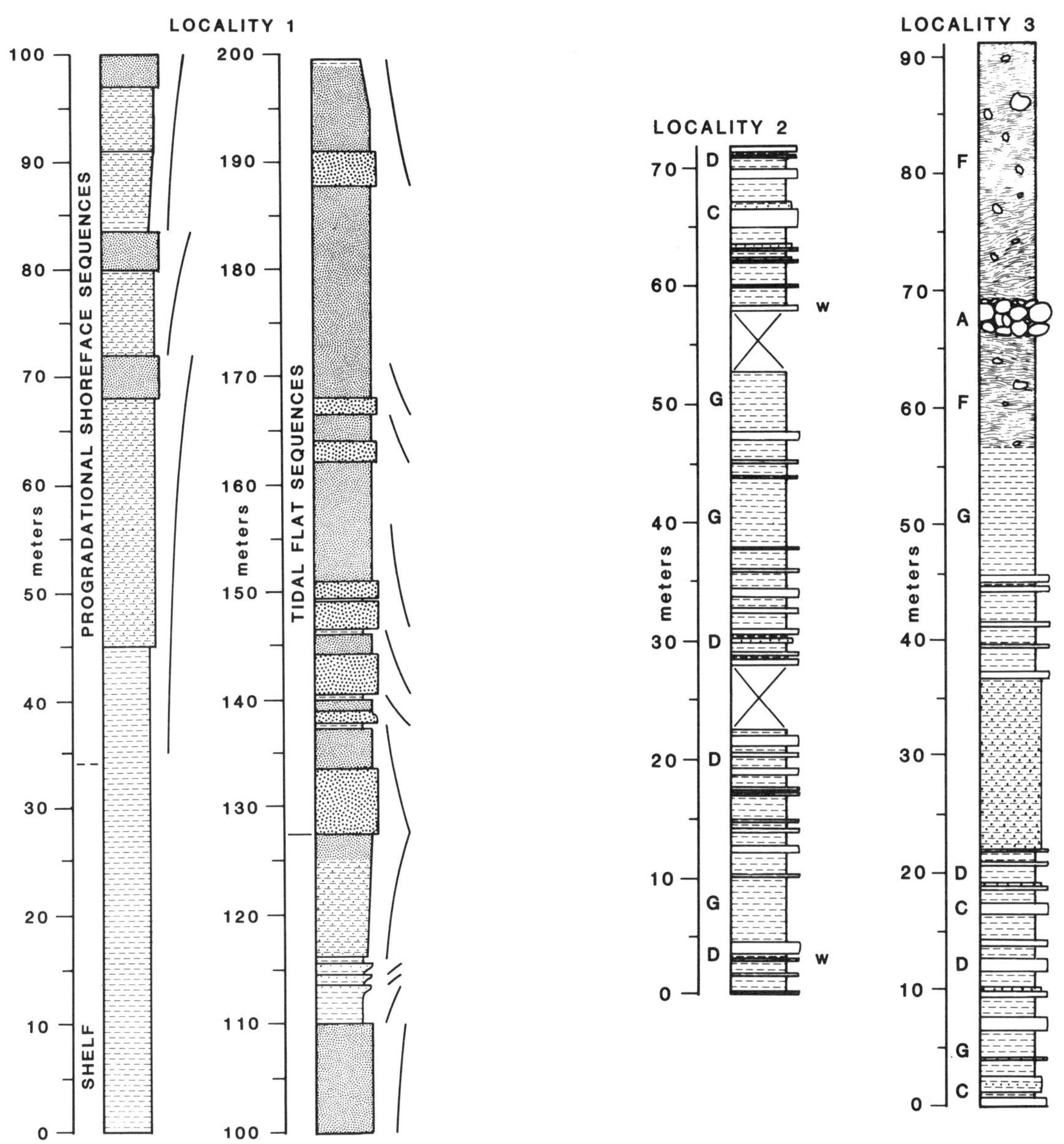

Figure 3. Measured section for locality 1. Key to symbols: shale, silty shale; silt, ripple-bedded siltstone intercalated with silty shale; fine sand, ripple-bedded sandstone intercalated with silty shale; coarse sand, ripple- and swash-bedded sandstone. Curved lines: upward deflection to right denotes coarsening upward sequence; upward deflection to left denotes fining upward sequence.

Figure 4. Measured sections for localities 2 and 3. Key to symbols: shale, shale; broken wavy lines, scaly shale; silt, siltstone; unpatterned, sandstone. Capital letters to left of columns denote facies designations of Mutti and Ricci Lucchi (1972); lower-case w's to right of columns denote presence of web structure (sections from McDonald, 1986).

debris-flow deposits represents one of the interesting controversies of Ouachita geology. Shideler (1970) recognized faunas ranging from upper Cambrian to lower Pennsylvanian, and he correlated lithologies with specific stratigraphic units of the Arbuckle, Ozark, and Ouachita facies. Exposure of rocks along fault scarps within a northern source terrane has been inferred (Shideler, 1970), but the presence of Ouachita-facies lithologies is problematic because rocks of that affinity do not crop out north of the Ti Valley fault. The possibility remains that source rocks were actually located to the south of the Ouachita trough, and mass wasting may have occurred along the walls of submarine canyons rather than fault scarps. Regardless of the source, the debris flows became intermixed within the axial turbidite system of the deep-marine basin.

Locality 4

A variety of structural fabrics defines the first phase of stratal disruption within the frontal zone of the Ouachitas; many of these are exposed along the roadcut. Phase-one deformational features include disharmonic folds, commonly with steeply plunging hinge lines and/or tight-to-isoclinal interlimb angles; many F_1

A

B

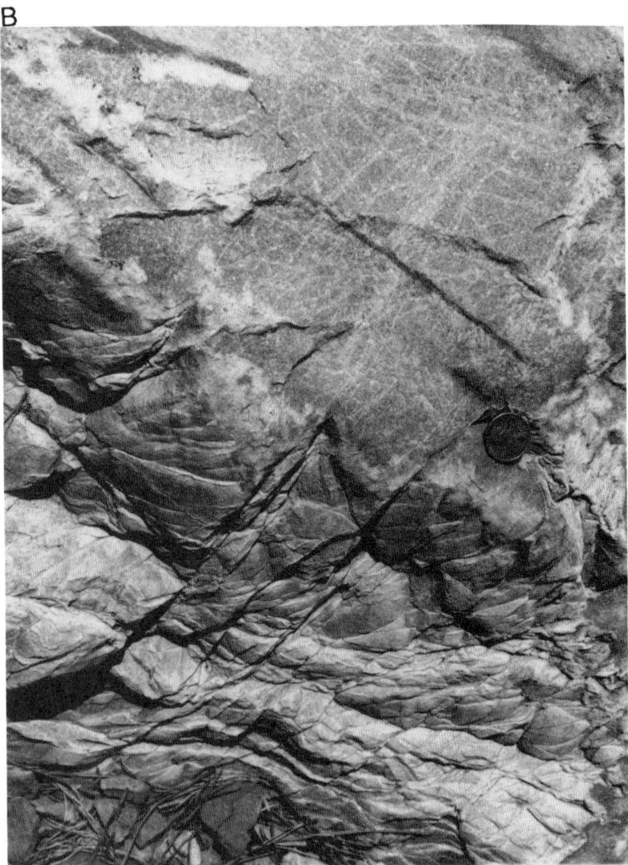

Figure 5. A: Thick-bedded sandstone located downsection from sequence illustrated in Figure 4, locality 2. These beds are assigned to facies B of Mutti and Ricci Lucchi (1972) and probably represent influx of sand from north side of turbidite basin. B: Example of web structure, a common structural feature within this portion of frontal thrust belt. Base of this bed is offset, but surfaces also penetrate into bed interior.

axial planes diverge significantly from the dominant east-west trend of F_2 folds. Some F_1 folds are completely overturned (e.g., the small antiformal syncline exposed on the east side of U.S. 270/59 at the intersection with Holson Valley Road). Other prominent D_1 structural features include scaly shale, bedding-parallel extension (pinch-and-swell, boudinage, necking), isolated sandstone phacoids, and web structure. Web structure is defined by an anastomosing network of slip surfaces in sandstone beds (Fig. 5B), and sedimentary structures such as flute casts are typically offset. Thin sections show that the slip surfaces are actually zones of grain-size reduction, grain breakage, and clay concentration. Evidently, web structure is produced by simultaneous fluid escape and cataclasis, under conditions of high confining pressure and elevated pore-fluid pressure.

Interpretations of the initial phase of deformation vary between gravitational failure of soft sediment (submarine slides) and tectonic deformation. The presence of cataclastic web structure, however, shows that the "soft sediment" deformation was caused by tectonic stresses; these stresses were probably generated during subduction-accretion, although perhaps at a relatively shallow position beneath the sea floor. Superimposed on the D_1 structural fabrics are regional-scale F_2 folds and associated north-verging thrusts such as the Ti Valley and Choctaw faults. The style of D_2 deformation is consistent with a collisional foreland setting; the thrusting occurred during obduction of the Ouachita accretionary prism onto the southern edge of North America.

REFERENCES CITED

Chamberlain, C. K., ed., 1978, A Guidebook to the Trace Fossils and Paleoecology of the Ouachita Geosyncline: Tulsa, Society of Economic Paleontologists and Mineralogists, 68 p.

Evans, G., 1975, Intertidal flat deposits of the Wash, western margin of the North Sea, *in* Ginsburg, R. N., ed., Tidal Deposits: New York, Springer-Verlag, p. 13–20.

Hart, O. D., 1963, Geology of the eastern part of Winding Stair Range, LeFlore County, Oklahoma: Oklahoma Geological Survey, Bulletin 103, 86 p.

Houseknecht, D. W., 1986, Evolution from passive margin to foreland basin; The Atoka Formation of the Arkoma basin south-central U.S.A., *in* Allen, P.A. and Homewood, P., eds., Foreland Basins: International Association of Sedimentologists, Special Publication 8, p. 183–201.

McDonald, K. W., 1986, Structural Style, Thermal Maturity, and Sedimentary Facies of the Frontal Ouachita thrust belt, LeFlore County, Eastern Oklahoma [M.S. thesis]: Columbia, University of Missouri, 211 p.

Mutti, E., and Ricci Lucchi, F., 1972, Turbidites of the northern Appenines; Introduction to facies analysis: International Geology Review, v. 20 (1978), p. 125–166.

Poole, L. A., 1985, Sedimentology, Structural Style, and Thermal Maturity of the Lynn Mountain Formation, Frontal Ouachitas, Latimer and LeFlore Counties, Oklahoma [M.S. thesis]: Columbia, University of Missouri, 189 p.

Reineck, H.-E., and Singh, I. B., 1975, Depositional Sedimentary Environments—with Reference to Terrigenous Clastics: New York, Springer-Verlag, 439 p.

Seely, D. R., 1963, Structure and Stratigraphy of the Rich Mountain area, Oklahoma and Arkansas: Oklahoma Geological Survey, Bulletin 101, 167 p.

Shideler, G. L., 1970, Provenance of Johns Valley boulders in Late Paleozoic Ouachita facies, southeastern Oklahoma and southwestern Arkansas: American Association of Petroleum Geologists Bulletin, v. 54, p. 789–806.

Carboniferous flysch, Ouachita Mountains, southeastern Oklahoma; Big Cedar-Kiamichi Mountain section

R. J. Moiola, *Mobil Research and Development Corporation, Dallas, Texas 75381*
G. Briggs, *Department of Marine, Earth, and Atmospheric Sciences, North Carolina State University, Raleigh, North Carolina 27650*
G. Shanmugam, *Mobil Research and Development Corporation, Dallas, Texas 75381*

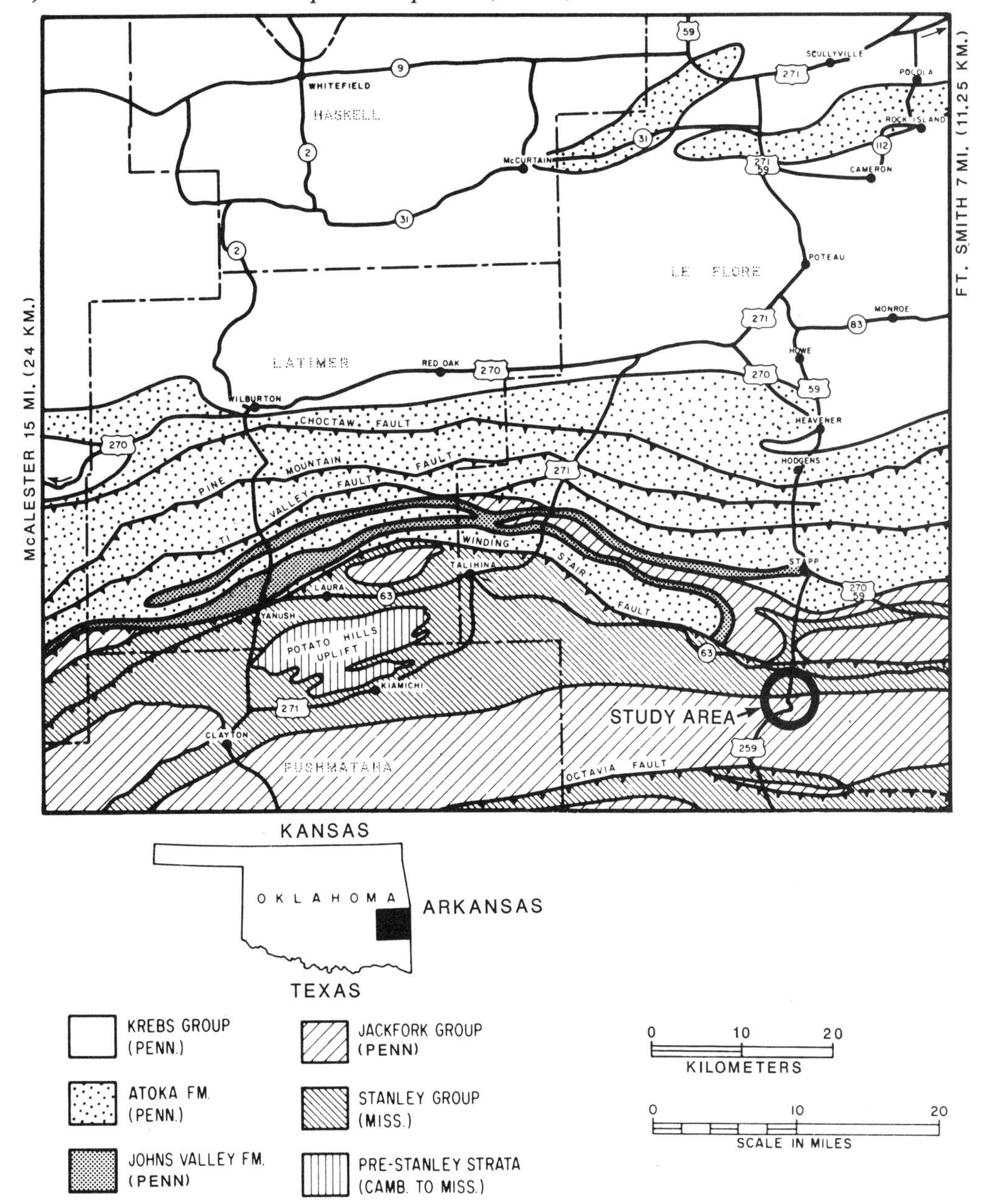

Figure 1. Generalized geologic map showing location of Big Cedar–Kiamichi Mountain section. (Modified from Briggs and others, 1975).

LOCATION

Located in LeFlore County (Big Cedar and Octavia 7½-minute Quadrangles), the Big Cedar–Kiamichi Mountain section is continuously exposed in roadcuts that parallel U.S. 259 (Figs. 1, 2, 3). Commencing 2.6 mi (4.2 km) south of the Big Cedar intersection of Oklahoma 63 and U.S. 259, at the contact between the Stanley and Jackfork Groups (Fig. 3), an uninterrupted sequence of Jackfork is superbly exposed as the road climbs up to and beyond the summit of Kiamichi Mountain. Reinterpretation

(Briggs, 1973, and this text) of the section Cline and Moretti (1956) measured at the time U.S. 259 was under construction indicates that 6,604 ft (2,013 m) of Jackfork are present (Fig. 2). The roadcuts are accessible all year, but caution must be exercised by parking cars well off the road, being careful of traffic, and avoiding dangerous areas with high rock faces or areas that may have loose or falling rocks.

SIGNIFICANCE

The Big Cedar–Kiamichi Mountain stratigraphic section exposes more than 6,500 ft (2,000 m) of rhythmically bedded sandstones and shales that were deposited in a deep-water setting during Late Mississippian–Early Pennsylvanian time. These strata, which are typical of the Ouachita flysch succession, elegantly display the physical/biogenic sedimentary structures and bedding styles characteristic of sediments deposited by turbidity-current and hemipelagic processes. Additionally, the section contains the most complete and undisturbed sequence of the Jackfork Group in the Ouachita foldbelt.

INTRODUCTION

The Ouachita Mountains of Oklahoma and Arkansas are part of a predominantly buried foldbelt that extends for approximately 1,240 mi (2,000 km) from the southern Appalachians to the border between southwest Texas and northern Mexico. Topographically, they comprise a series of complexly folded and faulted, east-west–trending ridges and valleys that extend from Little Rock, Arkansas westward about 186 mi (300 km) to McAlester, Oklahoma.

In southeastern Oklahoma, Carboniferous (Upper Mississippian–Lower Pennsylvanian) deep-water sandstones and shales form a thick flysch succession that is exposed in a series of northward-directed thrust slices (Fig. 1). The northernmost thrust, the Choctaw Fault, separates the foldbelt from the Arkoma Basin, and the Ti Valley Fault forms the boundary between the frontal and central Ouachitas. Strata present in the frontal Ouachitas are transitional between the flysch succession to the south and shallow-water/shelf facies to the north.

More than 22,960 ft (7,000 m) of interbedded deep-water sandstones (turbidites) and shales are exposed in the central Ouachitas. In ascending order, this succession comprises the Late Mississippian Stanley Group (11,000 ft; 3,350 m) and the Early Pennsylvanian Jackfork Group (6,600 ft; 2,000 m), Johns Valley Formation (650 ft; 200 m), and Atoka Formation (6,800 ft; 2,070 m) (Fig. 2). Lewis Cline (1970) and his colleagues (e.g., Briggs and Cline, 1967; Chamberlain, 1971; Morris, 1974) demonstrated that these strata, which exhibit characteristics analagous to the classic flysch of the Alpine and Carpathian areas of Europe, were deposited primarily by turbidity current and hemipelagic processes in bathyal to abyssal water depths. Paleocurrent measurements indicate that the dominant transport direction for the succession was westward, parallel to the axis of the Ouachita basin. Recently, Moiola and Shanmugam (1984) concluded that the Jackfork Group, as well as the flysch succession in general, was deposited in a series of longitudinal submarine fan complexes that prograded westward, essentially parallel to the axis of what was apparently a remnant ocean basin.

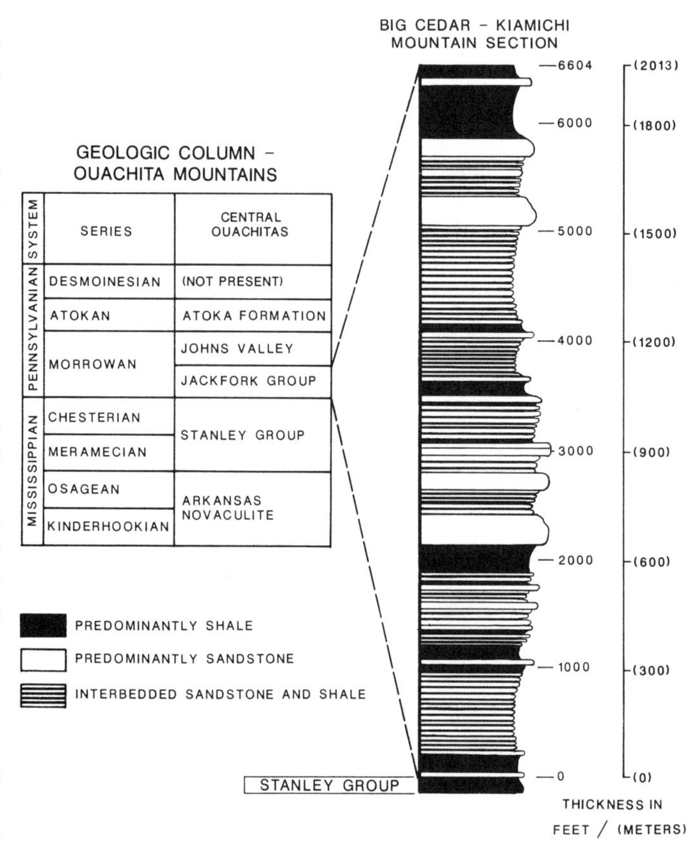

Figure 2. Geologic column for the central Ouachita Mountains, southeastern Oklahoma, with graphic column of Jackfork Group in Big Cedar–Kiamichi Mountain area. (Jackfork column based on section measured by Cline and Moretti, 1956.)

DESCRIPTION

The Big Cedar–Kiamichi Mountain section forms the northern limb of the Lynn Mountain Syncline (Fig. 3). Like other synclines that characterize the central Ouachitas, it is a long, gently plunging, asymmetrical feature whose southern and northern limbs are bounded by the Octavia and Windingstairs faults, respectively (Briggs, 1973). The southward-dipping strata of the northern limb include the valley-forming shales of the Stanley Group and the more resistant ridge-forming sandstones of the Jackfork. In general, the central Ouachitas consist of breached anticlinal valleys separated by long, linear synclinal ridges.

In Oklahoma, the Stanley Group has been divided into the Tenmile Creek, Moyers and Chickasaw Creek Formations, and the Jackfork Group has been divided into the Wildhorse Mountain, Prairie Mountain, Markham Mill, Wesley, and Game Re-

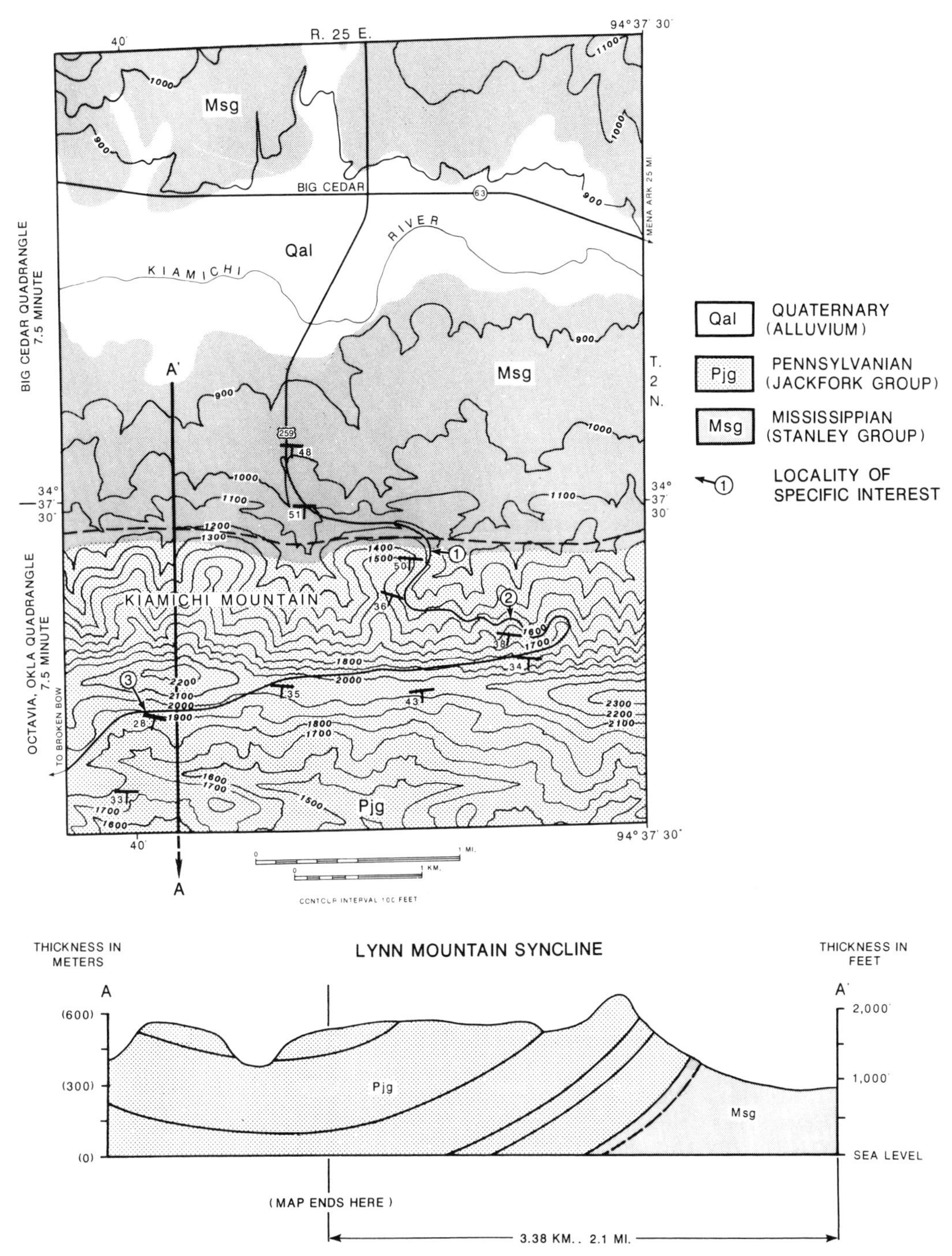

Figure 3. Geologic index map for Big Cedar–Kiamichi Mountain area and cross section through Lynn Mountain syncline. (Modified from Briggs, 1973.)

fuge Formations. Because these formations are quite similar overall, their boundaries are difficult to recognize (siliceous shale/chert intervals have been used). Consequently, for the purposes of this guide, we have not subdivided the Stanley and Jackfork. It is also difficult to determine the exact location of the Mississippian-Pennsylvanian boundary in the Ouachita succession; presumably, it falls somewhere in the lowermost part of the Jackfork Group. For simplicity, however, we have assigned the Jackfork to the Pennsylvanian and the Stanley to the Mississippian (Fig. 2).

The Stanley comprises only a small portion of the Big Cedar–Kiamichi Mountain section. Interbedded turbidite sandstones and hemipelagic shales are well exposed 1.6 mi (2.6 km) south of the Big Cedar intersection of Oklahoma 63 and U.S. 259. The uppermost part of the Stanley consists of dark-gray shales with interbeds of black siliceous shale and chert. This interval (dashed on map) occurs in conformable contact with the Jackfork at Locality 1 (Fig. 3), 2.6 mi (4.2 km) south of the Big Cedar intersection. Chert layers in the interval, especially at the contact with the Jackfork, commonly contain abundant white specks composed of siliceous sponge spicules and radiolarians. This "marker" interval is overlain by 6,604 ft (2,013 m) of rhythmically bedded Jackfork sandstones and shales.

Sandstones of the Jackfork Group consist of 85 to 95 percent quartz, 3 to 5 percent feldspar, and 2 to 3 percent rock fragments. Overall, they are less feldspathic, better sorted, and contain considerably less matrix than sandstones of the underlying Stanley. Associated shales, which are the products of both turbidity current and hemipelagic sedimentation, vary from carbonaceous to siliceous/cherty types.

The Jackfork sandstone beds exhibit all of the features typical of turbidites. Specifically, they have sharp bases, are graded, contain partial to complete Bouma sequences and display a wide variety of sole marks (e.g., flute casts, load casts, grooves, prod marks, etc.). Beds containing complete Bouma sequences (divisions A to E), however, are rare. Most beds begin with division B (parallel lamination) or C (ripple or convolute lamination). In addition, most beds consist of very fine to medium sand; consequently, when A divisions are present they are either thin and subtly graded or thick and massive. Many of the beds contain dish and pillar structures. According to Lowe and LoPiccolo (1974), who studied numerous samples from this section, these structures are postdepositional features that formed as a result of compaction and dewatering.

Unamalgamated sandstone beds are typically 2 in to 3 ft (5 cm to 1 m) thick and rarely pinch out with the limits of exposure. In terms of trends in bedding thickness, the section is characterized by packages of beds that crudely "thicken upward" (lobes). The thickness of these cycles is highly variable. At locality 2, which is 3.3 mi (5.3 km) south of the Big Cedar intersection (Fig. 3), several thickening upward cycles are displayed. They average 36 ft (11.0 m) in thickness and contain sandstone beds with Bouma BC, B, or C divisions. Irregular thinning upward cycles (channels) of variable thickness are also present in the section, but are much less common.

Using the facies association scheme of Mutti and Ricci Lucchi (1972), Moiola and Shanmugam (1984) concluded that the Jackfork in the Kiamichi Mountain area predominantly consists of outer fan depositional lobe deposits. These authors visualized a longitudinal submarine fan setting as the depositional framework for the Jackfork and for the flysch succession in general. Their regional facies analysis of the Jackfork indicates that sediments, derived primarily from the southern Appalachians and Illinois Basin, formed elongate submarine fan systems that prograded westward, essentially parallel to the axis of the Ouachita basin. For the Jackfork, a generalized east-west facies tract comprises inner fan deposits in the vicinity of Little Rock, Arkansas; middle fan deposits near Arkadelphia, Arkansas; and outer fan deposits in the Kiamichi Mountain area.

Olistostromes (slump and debris flow deposits) are important components of the flysch succession. Often containing "exotic" extrabasinal clasts (e.g., limestone boulders) derived predominantly from older facies (Arbuckle and Ozark) north of the basin, they are characteristic of the Johns Valley Formation, although they occur throughout the succession and in all parts of the Ouachita fan systems (Moiola and Shanmugam, 1984). In the Big Cedar–Kiamichi Mountain section, a slump interval containing rolled to contorted intrabasinal sandstone and shale masses can be seen 0.4 mi (0.6 km) south of Locality 2.

Trace fossils of the *Nereites* facies occur throughout the flysch succession and are the primary evidence for assigning a deep-water (bathyal/abyssal) setting to these strata (Chamberlain, 1971). At Locality 3, 5.8 mi (9.4 km) south of Big Cedar (Fig. 3), the Jackfork contains a diverse *Nereites* assemblage. Among the more distinctive traces present are *Lophoctenium, Phycosiphon, Scalarituba, Neonereites,* and *Helminthopsis.* The Big Cedar–Kiamichi Mountain section measured by Cline and Moretti (1956) ends 3.5 mi (5.6 km) south of this locality.

REFERENCES CITED

Briggs, G., 1973, Geology of the Eastern Part of the Lynn Mountain Syncline, LeFlore County, Oklahoma: Oklahoma Geological Survey Circular 75, 34 p.

Briggs, G., and Cline, L. M., 1967, Paleocurrents and source areas of Late Paleozoic sediments of the Ouachita Mountains, southeastern Oklahoma: Journal of Sedimentary Petrology, v. 37, p. 985–100.

Briggs, G., McBride, E. F., and Moiola, R. J., eds., 1975, Sedimentology of Paleozoic Flysch and Associated Deposits, Ouachita Mountains–Arkoma Basin, Oklahoma: Dallas Geological Society Guidebook, 128 p.

Chamberlain, C. K., 1971, Bathymetry and paleoecology of Ouachita geosyncline of southeastern Oklahoma as determined from trace fossils: American Association Petroleum Geologists, v. 55, p. 34–50.

Cline, L. M., 1970, Sedimentary features of Late Paleozoic flysch, Ouachita Mountains, Oklahoma, *in* Lajoie, J., ed., Flysch Sedimentology in North America: Geological Association Canada Special Paper 7, p. 85–101.

Cline, L. M., and Moretti, F. J., 1956, Description and correlation of two complete stratigraphic sections of the Jackfork Sandstone in Kiamichi Mountains, central Ouachita Mountains, Oklahoma: Oklahoma Geological Survey Circular 41, 20 p.

Lowe, D. R., and LoPiccolo, R. D., 1974, The characteristics and origins of dish and pillar structures: Journal of Sedimentary Petrology, v. 44, p. 484–501.

Moiola, R. J., and Shanmugam, G., 1984, Submarine fan sedimentation, Ouachita Mountains, Arkansas and Oklahoma: Gulf Coast Association Geological Societies Transactions, v. 34, p. 175–182.

Morris, R. C., 1974, Sedimentary and tectonic history of the Ouachita Mountains, *in* Dickinson, W. R., ed., Tectonics and Sedimentation: Society of Economic Paleontologists and Mineralogists Special Publication 22, p. 120–142.

Mutti, E., and Ricci Lucchi, R., 1972, Turbidites of the northern Appenines; Introduction to facies analysis: International Geology Review, v. 20 (1978), p. 125–166.

Turner Falls Park; Pleistocene tufa and travertine and Ordovician platform carbonates, Arbuckle Mountains, southern Oklahoma

R. N. Donovan, Department of Geology, Texas Christian University, Fort Worth, Texas 76129
D. A. Ragland and D. Schaefer, Department of Geology, Oklahoma State University, Stillwater, Oklahoma 74078

LOCATION

Turner Falls Park is located in the Arbuckle Mountains approximately 3 mi (4.8 km) south of Davis on U.S. 77. The park is shown on all highway maps of Oklahoma, and the easiest access is from appropriate exits on I-35 (Fig. 1). There is a small charge for entering the park.

The best-known view of Turner Falls is from the overview gift shop on U.S. 77 (free parking available). From here all the sites described below can easily be visited, although the descent to the falls is via rather steep steps. The less agile who wish to view the falls closely can enter the park via the automobile entrance (which is located approximately 0.5 mi [0.8 km] north of the gift shop). From here it is an easy walk to the foot of the falls. Alternatively it is possible to park above the falls; this is an easy way to examine the top of the old travertine deposit.

No hammers please; the exposures (which are highly photogenic) could easily be ruined.

SIGNIFICANCE

The composite travertine and tufa edifice at Turner Falls is a superb example of a limestone deposited by a stream that has been periodically supersaturated with respect to calcium carbonate. Many details of travertine and tufa texture are well exposed. The history of the deposit records climatic fluctuations during the Pleistocene.

The exposures of the Arbuckle Group (specifically the Ordovician Cool Creek Formation) here are the finest in Oklahoma and illustrate many aspects of carbonate platform sediments. Particularly noteworthy are the varieties of stromatolites and intraformational conglomerates. A distinctive collapse breccia, perhaps indicative of vanished evaporites, is well exposed, and several forms of chert are present.

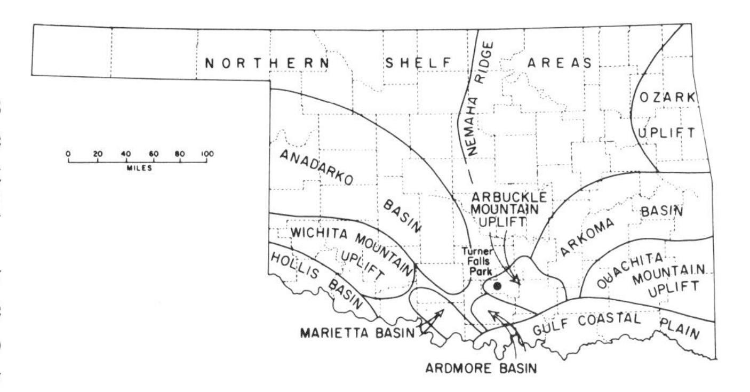

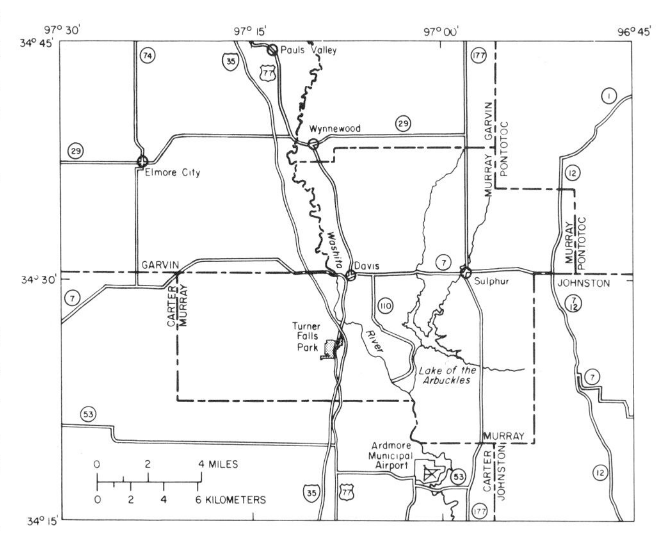

Figure 1. General location of Turner Falls in the Arbuckle Mountains, Murray County, Oklahoma.

37—PLEISTOCENE TUFA AND TRAVERTINE DEPOSITS: SITE INFORMATION

Turner Falls is the most impressive waterfall in Oklahoma. It is of unusual geological interest because the tumbling waters of Honey Creek have deposited a complex edifice of calcium carbonate (Ham, 1973); as a result the waterfall scarp is not receding upstream (as is the case with most waterfalls) but has advanced downstream as an impressive cliff of travertine and tufa.

Three large springs flowing from orifices in the limestones of the Arbuckle Group are the principal sources of Honey Creek. The water in the creek is hard and supersaturated with respect to calcite (Fairchild, 1984). As a result the creek bed is covered with precipitated travertine and tufa deposits (Fig. 2). In addition to Turner Falls, there are many lesser falls in the valley, each with a travertine coat. Similar falls also occur on nearby Falls Creek (e.g., Upper and Lower Prices falls). These travertine deposits form as a result of both physicochemical factors (i.e., evaporation) and biological induction (i.e., photosynthesis by algae [*Vaucheria* sp.] and moss) (Emig, 1917; Johnson and McCasland, 1971).

Emig, in 1917, recognized that the Turner Falls travertine was a complex deposit that effectively and eloquently recorded fluctuations in the Pleistocene climate. Five distinct stages can be recognized in the evolution of the deposit. The first (Stage 1) was the initial dissection of the area by vigorous streams with steep gradients. During this phase the Honey Creek Valley, including

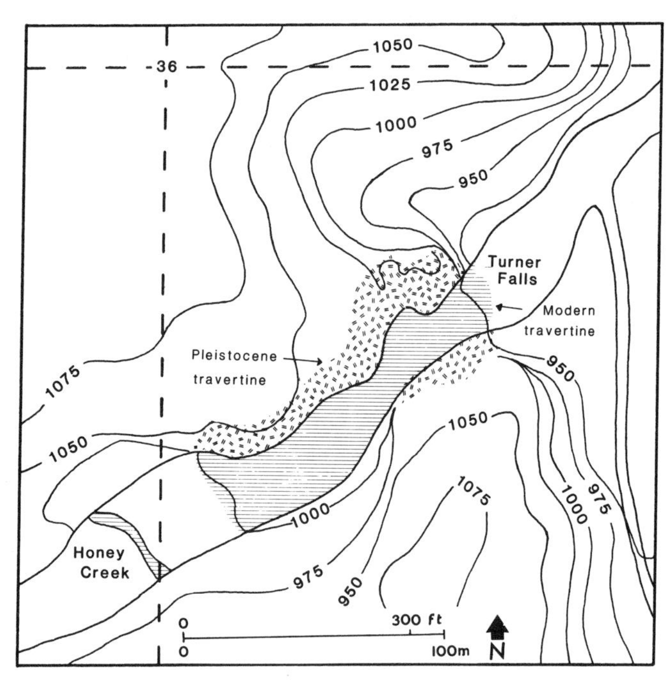

Figure 2. Distribution of travertine at Turner Falls.

Figure 3. Suggested restoration of Turner Falls during greatest phase of travertine production (based on Emig, 1917). The hill in the left background (to the southwest) is an outcrop of Colbert Rhyolite. Otherwise the topography is incised in the Arbuckle Group.

the imposing gorge downstream from the falls, was first excavated. The long profile of the (Pleistocene) Honey Creek was undoubtedly most irregular, reflecting the numerous faults and variations in dip found in the bedrock.

Subsequently the climate presumably became warmer and drier, calcite saturation increased, and as a result a major accumulation of travertine took place (Stage 2). This travertine deposit gradually grew upward and downstream until Honey Creek tumbled over a travertine cliff approximately 100 ft (30 m) high and 300 ft (90 m) wide (Fig. 3). Because of irregularities in the stream bed, the base of this deposit has a complicated unconformable relationship with the underlying McKenzie Hill Limestone of the Arbuckle Group (Fig. 4). The unconformity is marked by a discontinuous pebble conglomerate containing clasts of Colbert rhyolite, Arbuckle Group limestone, and reworked travertine. The overlying travertine is a textural chaos of banded travertine and moss tufa. Good examples of thick travertine tubes, originally deposited around now vanished tree logs, are common. The dominant fabric, which emphasizes the downstream progradation of the deposit, is formed by beds of petrified moss (*Didymodon tophaceus* and *Philonotis calcares*) that dip downstream at angles of up to 85°. These beds, which vary in thickness from 1 to 4 in (2 to 10 cm), can persist for as much as 33 ft (10 m). Within each bed the individual moss plant casts are horizontal (Fig. 4). On the modern falls, similar moss does not grow in areas of maximum water flow; hence it appears that the bed boundaries between the petrified moss colonies are an unusual type of disconformity, recording changes in stream volume or slight lateral migrations of the waterfall that controlled moss growth.

The older travertine is extremely porous, containing an elaborate cave system that can be examined by the agile (care should be taken; the largest cave is best entered from the top of the deposit). These caves, which are not solution phenomena but original cavities, probably formed behind large tree trunks that were cemented into the travertine, forming obstructions in the waterfall (similar features are presently forming in the modern deposit).

There is some evidence of speleothem phenomena within the older travertine. For example, flowstone and crude stalactites have precipitated on the walls of cavities, and some cave popcorn is present near the base of the deposit. It is clear from the evidence of horizontal layers of travertine and boxwork structures (calcified mudcracks) that at one time small pools existed near or on the basal unconformity. These speleothemic deposits may have formed during a semi-arid stage when the flow of Honey Creek was greatly diminished and probably intermittent (Stage 3). Supporting evidence for this stage can be found 2 mi (3.2 km) north of Turner Falls (on the north side of U.S. 77, adjacent to I-35) where a road cut through a hill formed of the Mississippian Sycamore Limestone clearly shows a well-developed calcrete (caliche) deposit on the hillside beneath the modern regolith.

Subsequently (Stage 4), presumably during a wetter and cooler stage, carbonate precipitation ceased, and Honey Creek eroded its own deposit, eventually cutting a few feet into the underlying bedrock in places and forming the steep-sided gorge above the modern waterfall (Fig. 5). Supporting evidence for this destructive stage in the valley's evolution can be found about a mile downstream (outside, the park boundary, in the grounds of the Cedar Vale cafe), where Honey Creek has cut through its own carbonate-cemented Pleistocene gravels.

Eventually (Stage 5) the climate became warmer and drier, and Honey Creek re-established its constructive mode. The modern travertine-tufa screen at the falls is the result.

The whole travertine-tufa edifice at Turner Falls is an eloquent expression of how climatic variations, as opposed to tec-

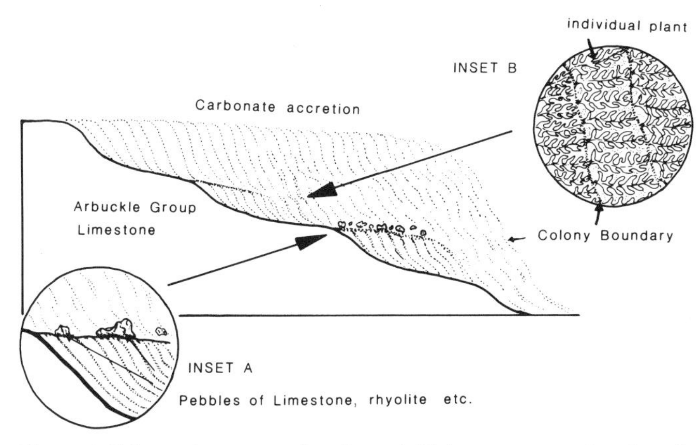

Figure 4. Schematic cross section through Pleistocene travertine deposit showing complex progradational relationships. Inset A: details of intra-travertine unconformity. Inset B: Detail of moss tufa showing horizontal growth of individual plants and steeply dipping disconformable surface of colonies (the latter are essentially casts of the waterfall surface).

Figure 5. Sketch of the modern Turner Falls showing the old Pleistocene travertine screen in the background, the later incised gorge, and the modern travertine construction in the foreground.

tonic or other adjustments to base level, can control fluvial erosion. The old travertine screen affords an unusual opportunity for the geologist to view such a deposit in great intimacy: literally inside and out (and without getting wet!). The modern waterfall, in addition to offering a compelling uniformitarian analogy, is a place of great beauty. The interplay of water and tan-colored tufa is enhanced by the unusual plants that thrive in the misty environment. Particularly noteworthy are the rare Venus maidenhair fern (*Adiantum Capillus-Veneris*), the purple-stemmed Cliffbrake (*Pellaea atropurpurea*), and the edible water cress (*Sisymbrium nasturtium-aquaticum*).

38—ORDOVICIAN PLATFORM CARBONATES: SITE INFORMATION

The Ordovician Cool Creek Formation is part of the Cambro-Ordovician Arbuckle Group, a thick accumulation of platform carbonates whose equivalents are present throughout most of the South Central region north and west of the Ouachita-Marathon front. In the Arbuckle Mountains, approximately 6,000 ft (1,800 m) of carbonates were preserved in the NW-SE trending Southern Oklahoma aulacogen.

The lower boundary of the formation with the McKenzie Hill Formation is marked by the incoming of abundant quartz sand into the sequence (the Thatcher Creek Member; Ragland and Donovan, 1985a). The upper boundary with the Kindblade Formation is marked by the incoming of the enigmatic gastropod genus *Ceratopea.*

There are two fine and characteristic sections near the gift shop. The stratigraphically lower one (which starts about 300 ft [90 m] above the base of the Cool Creek Formation) can be closely examined from the steps which lead down to the falls from the gift shop (Fig. 6). The best exposures are those worn smooth by the passing of tourists! The second section is a road cut precisely opposite the gift shop (Fig. 7). Both sections display similar lithologies: algal boundstones, intraformational conglomerates, quartz-rich oolitic grainstones, peloidal limestones, lime mudrocks, and heterolithic units. Chertification and dolomitization have affected these lithologies in varying degrees. An intraformational breccia is present in the lower section (Ragland and Donovan, 1985b).

The most common lithology is varieties of algal boundstone. The external forms present include encrustations, mats, mounds, and reefs. Internal organization into several growth forms of stromatolite and thrombolite (e.g., dentritic, cylindrical columnar, turbinate) is apparent (Fig. 8). A good three-dimensional appreciation of algal mounds can be obtained by examining exposures in the lower section at the first bend going down the steps (beware of the cliff edge).

About 10 percent of the formation consists of intraformational conglomerates, forming erosive-based beds up to 12 in (30 cm) thick. Most of the clasts are varieties of lime mud, but they also include fragments of encrusting stromatolites (Fig. 8). The majority of the clasts are disc- or rod-shaped. As a result, they have been sorted by storms, currents, and waves into distinctive packing arrangements: random, flat, imbricate, and vertical (Fig. 9). Textures observed in the conglomerates vary from mud supported to open framework. Sand-sized quartz grains and ooids are commonly observed in most of the conglomerates.

The intraformational conglomerates are the coarse end of a

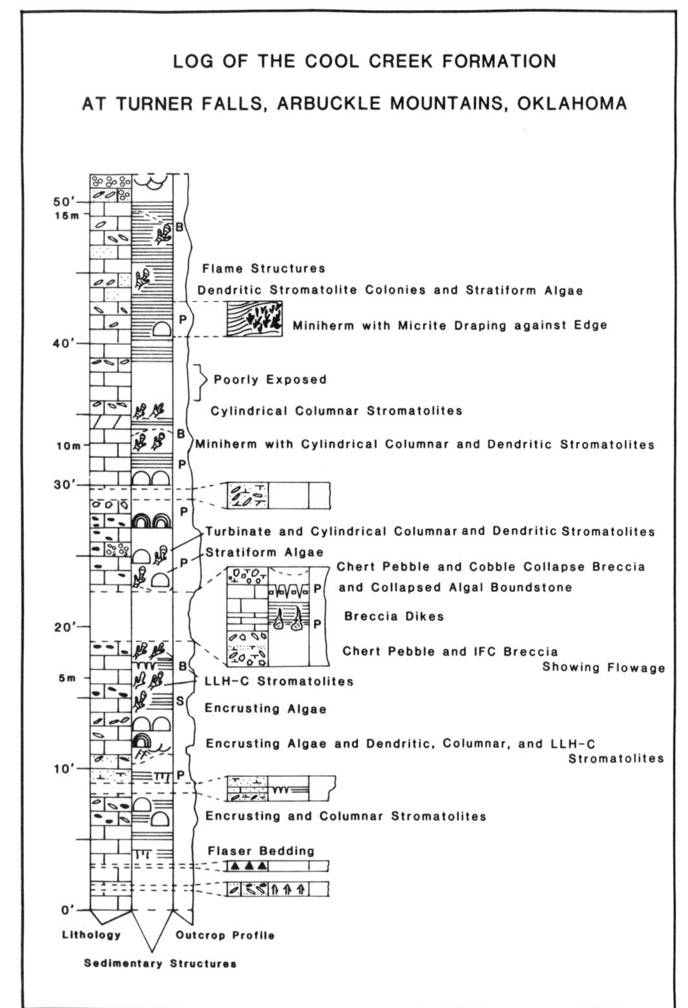

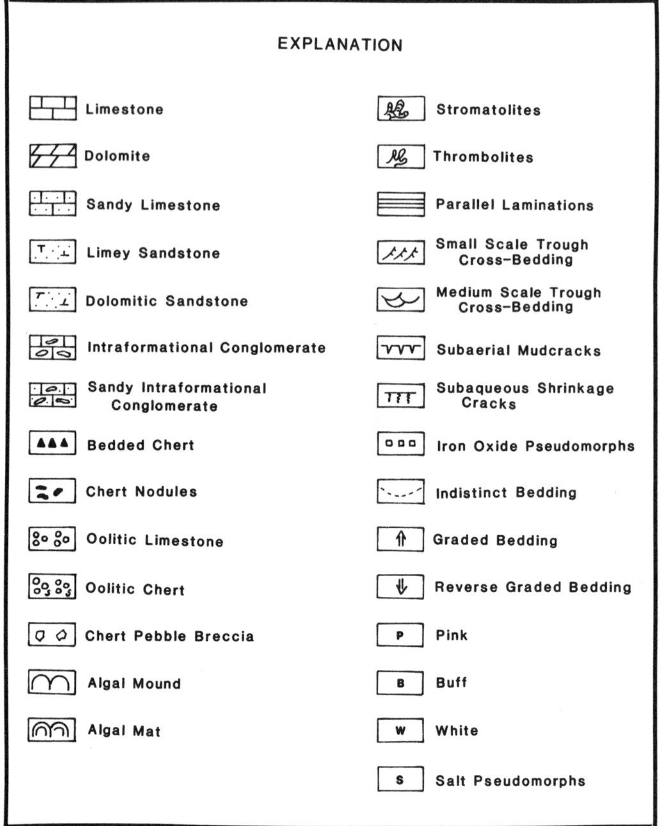

Figure 6. Stratigraphic section of Cool Creek Formation measured on cliff below gift shop at Turner Falls overlook (from Ragland and Donovan, 1985b).

spectrum of detrital limestones that includes grainstones, packstones, wackestones, and mudstones. The principal allochems in the grainstones are ooids and lime-mud intraclasts; less common skeletal fragments include trilobites, brachiopods, and (particularly) gastropods. Oval peloids (probably fecal pellets) are the most common allochems in the mud-supported textures. Variable amounts (from none up to 50 percent) of quartz (plus a little microcline) are present throughout both sections. All of the detrital lithologies are thin bedded, particularly the mud-supported textures (many bed boundaries have been stylolitized). Some of the grainstones evidence current action in the form of small trough crossbeds (maximum set thickness 3 in; 8 cm). Other grainstones are packed around and on top of algal boundstones, suggesting that growth of the latter was terminated when they were overwhelmed by migrating sand shoals. Some of the mud-supported textures are graded; otherwise lamination and bioturbation (including both vertical and horizontal burrows) are the principal sedimentary structures.

The heterolithic units are characterized by variations of flaser bedding. Alternating layers consist of lime mud and quartz-rich grainstones. The latter are characterized by small-scale cross-bedding, some of which can be related to both asymmetrical and symmetrical ripple marks. Units of this type are cut by both subaerial and subaqueous mudcracks (Ragland and Donovan, 1985b); they are also the most likely lithology to have been dolomitized.

Chert is present at several horizons and may have replaced any of the lithologies described above. In addition, the grainstones show some evidence of silica cementation. A variety of sizes of nodules is found, up to 3 ft (1 m) in length. Some of the cherts contain molds and pseudomorphs of a salt that appears to have been gypsum. Significantly, SEM studies have confirmed the presence of very small crystals of anhydrite and celestite in the pseudomorph-bearing cherts (Ragland and Donovan, 1984). The best place to see salt molds is approximately half way down the steps just before reaching a flat terrace (where the weary may rest on the ascent).

The collapse breccia is best exposed by the side of the steps about 30 ft (9 m) lower than the terrace noted above, at a point where the cliff slightly overhangs the trail. The breccia consists of angular blocks, up to 3 ft (1 m) across, that appear to be the result of gravity-controlled collapse of well-lithified rock (even chert

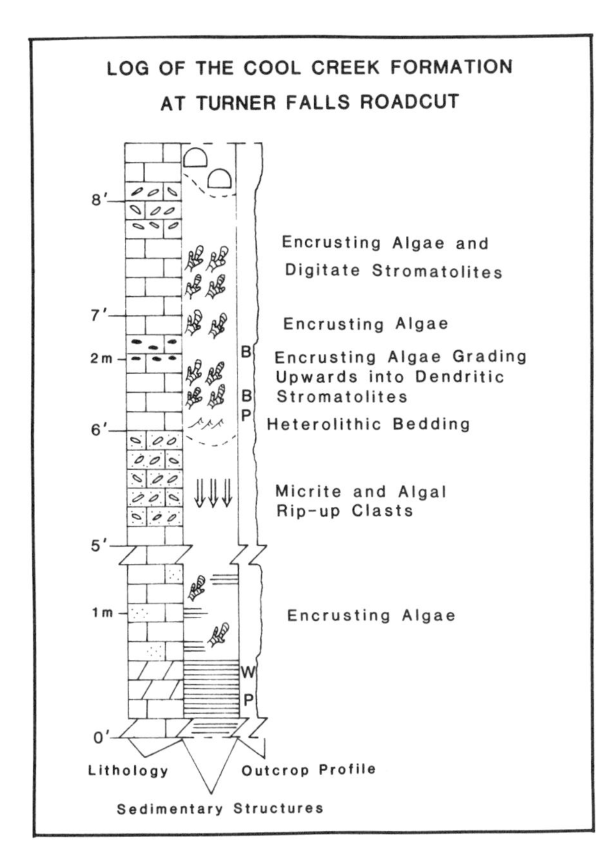

Figure 7. Stratigraphic section of Cool Creek Formation measured along U.S. 77 across from gift shop at Turner Falls overlook (from Ragland and Donovan, 1985b).

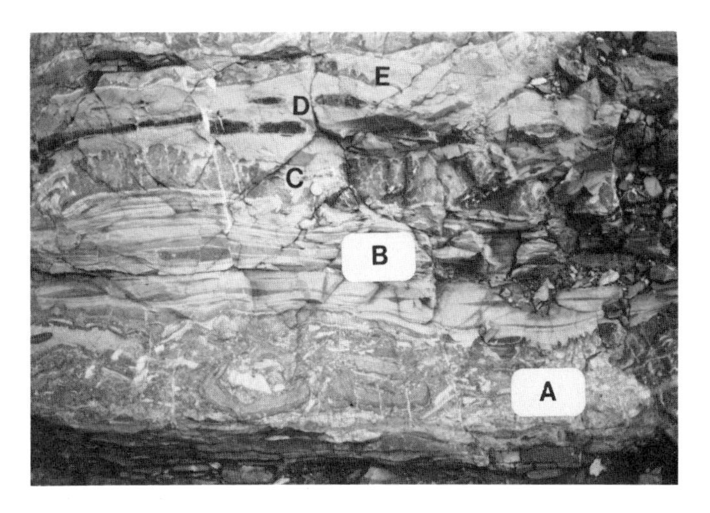

Figure 8. Part of section measured along U.S. 77, across from Turner Falls overlook. Base of unit is marked by a "storm deposit" (A) of intraformational clasts of lime mud, quartz sand, and fragments of algal boundstone. This deposit is overlain by a heterolithic unit (B) of interbedded micrite, dolomicrite, and quartz silt. Encrusting algae began growth after stabilization of the heterolithic unit and branched upward in a dendritic manner (C). A relatively homogenous micrite overlying the boundstone contains secondary chert nodules (D). Another thin and discontinuous boundstone (E) appears near the top of the unit (from Ragland and Donovan, 1985b).

nodules have been fractured). Three zones are present (Fig. 10): The lowest shows signs of liquifaction (including the upward injection of clastic dikes), the middle displays listric normal faulting in a heterolithic unit, and the upper comprises an intensively fractured algal boundstone. The fractures have been cemented by sparite (Fig. 11). Most fractures and faults have a north-northwest trend, and the overall sense of collapse is to the west-southwest. The breccia may be the result of subsurface dissolution of an evaporite-rich layer at a time when (to judge from the limit of brecciation) at least 12 ft (4 m) of lithified sediment was present above the zone (Ragland and Donovan, 1985b).

Figure 9. Vertically packed intraformational conglomerate (IFC) of lime and clasts. Block has been set upside down into north wall of gift shop. The IFC was first cemented as a hardground and then eroded to form a planar surface on which laminated micrite was deposited (from Ragland and Donovan, 1985b).

Summary

Within the confines of the Southern Oklahoma aulacogen, the Arbuckle Group is one of the world's thickest accumulations of carbonates. The exceptional exposures of the Cool Creek Formation at Turner Falls elegantly display those features of the formation that have led several authors to reconstruct the environment as a highly productive shallow marine carbonate platform in an arid or semi-arid setting (e.g., Ham, 1973; St. John and Eby, 1978; Donovan, 1982; Ragland and Donovan, 1985b).

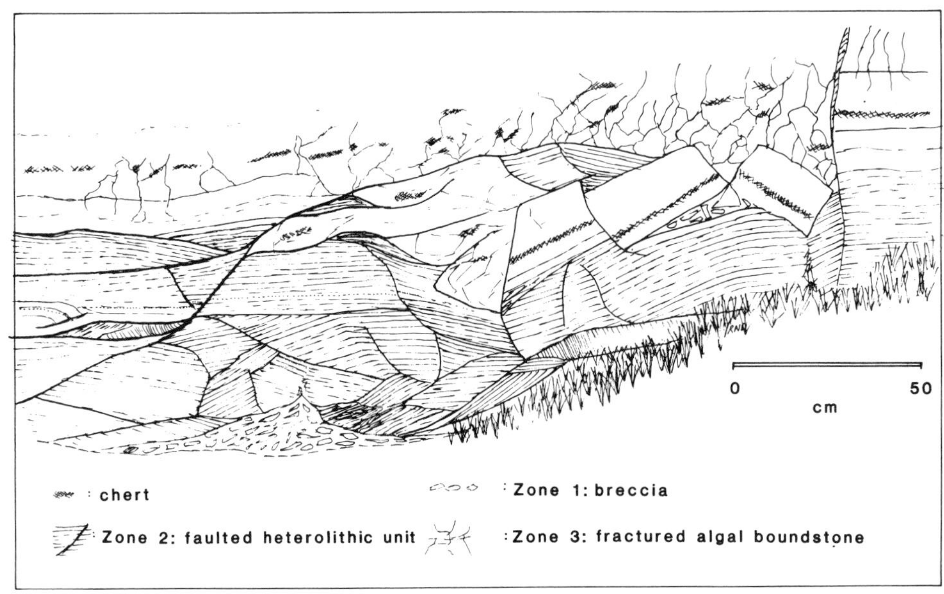

Figure 10. Sketch of part of intraformational breccia showing style of deformation.

In detail, the exposures evidence the dominant role of blue green algae in constructing boundstones. The differing forms of boundstone are probably a response to varying water depth. The intraformational conglomerate pebbles probably owed their origin to storms; some of these deposits were subsequently sorted by waves and tidal currents. Similarly, the broad spectrum of detrital carbonate sandstones seems to indicate an extensive interplay between storms, waves, and weak tidal currents. The subtle evidence for evaporites plus the regular occurrence of thin, stratigraphically persistent dolomite suggest that emergent sabkha-type conditions existed during periods of relatively low sea level. During early diagenesis, ground waters of complex and variable chemistry migrated through the sediments, as is evidenced by silica cementation and replacement, sulphate pseudomorphs, and the intraformational breccia.

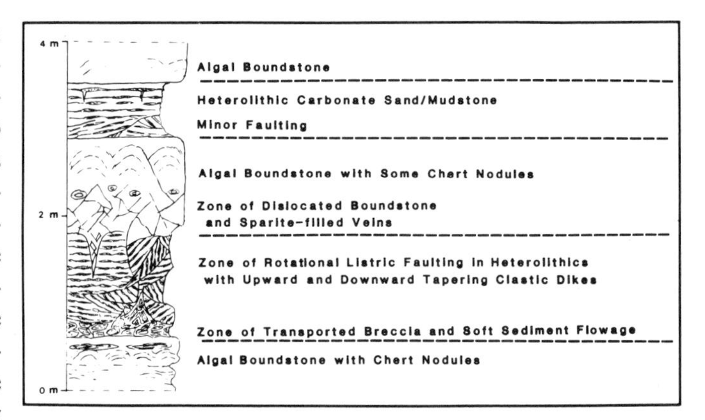

Figure 11. Schematic representation of deformation zones seen in intraformational breccia.

REFERENCES

Donovan, R. N., 1982, Geology of Blue Creek Canyon, Wichita Mountains area, *in* Gilbert, M. C., and Donovan, R. N., eds., Geology of the eastern Wichita Mountains, southwestern Oklahoma: Oklahoma Geological Survey Guidebook 21, p. 65–77.

Emig, W. H., 1917, Travertine deposits of Oklahoma: Oklahoma Geological Survey Bulletin 29.

Fairchild, R. W., 1984, Springs in the Arbuckle Mountain Area, South-Central Oklahoma: Oklahoma Geology Notes, v. 44, no. 1, p. 4–11.

——, 1973, Regional geology of the Arbuckle Mountains, Oklahoma: Geological Society of America 1973 Annual Meeting, Dallas, Texas, Guidebook for Field Trip No. 5.

Johnson, K. S., and McCasland, W., 1971, Highway geology in the Arbuckle Mountains and Ardmore area, southern Oklahoma: Oklahoma Department of Highways 22nd Annual Highway Geology Symposium Field-Trip Guidebook.

Ragland, D. A., and Donovan, R. N., 1984, Evidence supporting an evaporitic shallow marine paleoenvironment for the Cool Creek Formation of Oklahoma: Geological Society of America Abstracts with Programs, v. 16, p. 629.

——, 1985a, The Thatcher Creek Member; Basal unit of the Cool Creek Formation in Southern Oklahoma: Oklahoma Geology Notes, v. 45, no. 3, p. 84–91.

——, 1985b, The Cool Creek Formation (Ordovician) at Turner Falls in the Arbuckle Mountains of southern Oklahoma: Oklahoma Geology Notes, v. 45, no. 4, p. 132–148.

St. John, J. W., Jr., and Eby, D. E., 1978, Peritidal carbonates and evidence for vanished evaporites in the Lower Ordovician Cool Creek Formation, Arbuckle Mountains, Oklahoma: Gulf Coast Association of Geological Societies Transactions, v. 28, pt. 2, p. 589–599.

Pennsylvanian conglomerates in the Arbuckle Mountains, southern Oklahoma

R. N. Donovan, *Geology Department, Texas Christian University, Fort Worth, Texas 76129*
W. D. Heinlen, *2508 E. 88th St. South, #9, Tulsa, Oklahoma 74137*

LOCATION

Two formations illustrate the characteristics of Pennsylvanian conglomerates in the Arbuckle Mountains of southern Oklahoma: the Vanoss Formation and the Collings Ranch Conglomerate.

The Vanoss Formation is exposed on Bromide Hill in Platt National Park which is located on the southern edge of Sulphur and is easily reached by good roads from all directions (Fig. 1). To gain access to the area, follow U.S. 177 south from Sulphur into the park. Approximately 900 ft (300 m) inside the park boundary, immediately after crossing Rock Creek, turn west and follow the perimeter drive for about 1 mi (1.4 km) to Bromide Pavilion (Fig. 2).

The Collings Ranch Conglomerate is exposed on both sides of I-35 and in the median, approximately halfway between the Davis and Turner Falls turnoffs, and forms part of the magnificent Arbuckle Mountains road cuts (Fig. 1) (see also Fay, this volume). A rest area, which provides a lay person's explanation of the geology, is conveniently located at the southern margin of the Conglomerate on the south lane of I-35. The principal outcrops are between Interstate Stations 2600 and 2630 as designated by the Oklahoma Department of Highways.

SIGNIFICANCE

Both conglomerates record some of the latest events in the evolution of the Southern Oklahoma aulacogen. During Pennsylvanian time the ancient faults that had developed during the early evolution of the aulacogen were reactivated (Wickham *in* Wickham and Denison, 1978). In Wickham's view, the stresses affecting the region appear to have been resolved into a complex of left-lateral wrench movements that were associated both with localized compression (transpression) and extension (transtension) (see Tapp, this volume). The Collings Ranch Conglomerate, which is the earlier of the two deposits, accumulated in what appears to have been a small, closed transtensional pull-apart basin bounded by major faults. Brown (1982) has challenged the wrench interpretation, suggesting an enhanced role for pure compression and low-angle reverse faulting. Subsequently the conglomerate was affected by further tectonic pulses and as a result bears a significant deformational imprint.

By contrast, the relatively undisturbed conglomerates of the younger Vanoss Formation appear to be among the final detrital products of Pennsylvanian deformation. The conglomerates were deposited as fringing alluvial fans on the northern flank of the Arbuckle Mountains. These fans debouched onto an alluvial plain that stretched many miles to the north; the climate appears to have been semi-arid at this time (Al-Shaieb and others, 1977).

VANOSS FORMATION

Platt National Park History

The town of Sulphur grew up around cold-water mineral springs as a spa for those who drank and bathed in the sulphur-, bromine- and iron-rich waters in a whimsical attempt to improve their health. Previously the banks of Rock Creek had been a favorite spring and summer camping ground of the Chickasaw and Choctaw Indians. In 1902 the two tribes ceded the area of the park to the government in order to prevent commercial exploitation and to ensure that the springs could be used "by all men for perpetuity" (Barker and Jameson, 1975). The national park came into existence in 1906; by 1930 "taking the waters" was a major industry (at one time visitors were limited to a single gallon from Bromide Spring, so great was the demand for this stimulating beverage). Under the auspices of the New Deal, the park was landscaped between 1933 and 1940. Subsequent changes have been minor; nowadays the area is incorporated into the Chickasaw National Recreation Area.

Bromide Hill

Bromide Hill rises steeply 140 ft (40 m) above Rock Creek; it is the finest viewpoint in the park. The hill is built of the Vanoss Formation, which is well exposed by the side of the park trail (Fig. 3). However, in order to view the calcretes near the base of the section, it is necessary to scramble a few yards eastward along a rough trail from the first bend on the "official" footpath. Similarly the upper parts of the section (including the Carla Member [Fig. 4]) can only be examined by climbing the small broken cliff that abuts the path about two-thirds of the way up the hill (although care should be taken here, the exercise is recommended, as the exposure shows excellent facies relationships). Once the top of the hill is reached, one can continue along the path to the perimeter road, which eventually returns to the Bromide Pavilion (Fig. 2). En route, a second exposure of the Vanoss Formation can be examined in a road cut immediately south of Rock Creek (Fig. 5).

Geology: Details of the Exposure

The Vanoss Formation is of late Pennsylvanian (Virgilian) age and consists in its entirety of a lower conglomerate member that is 650 ft (200 m) thick and an upper shale member, up to 900 ft (275 m) thick (Ham, 1973). In northern Oklahoma the Vanoss Formation disconformably overlies the Pennsylvanian

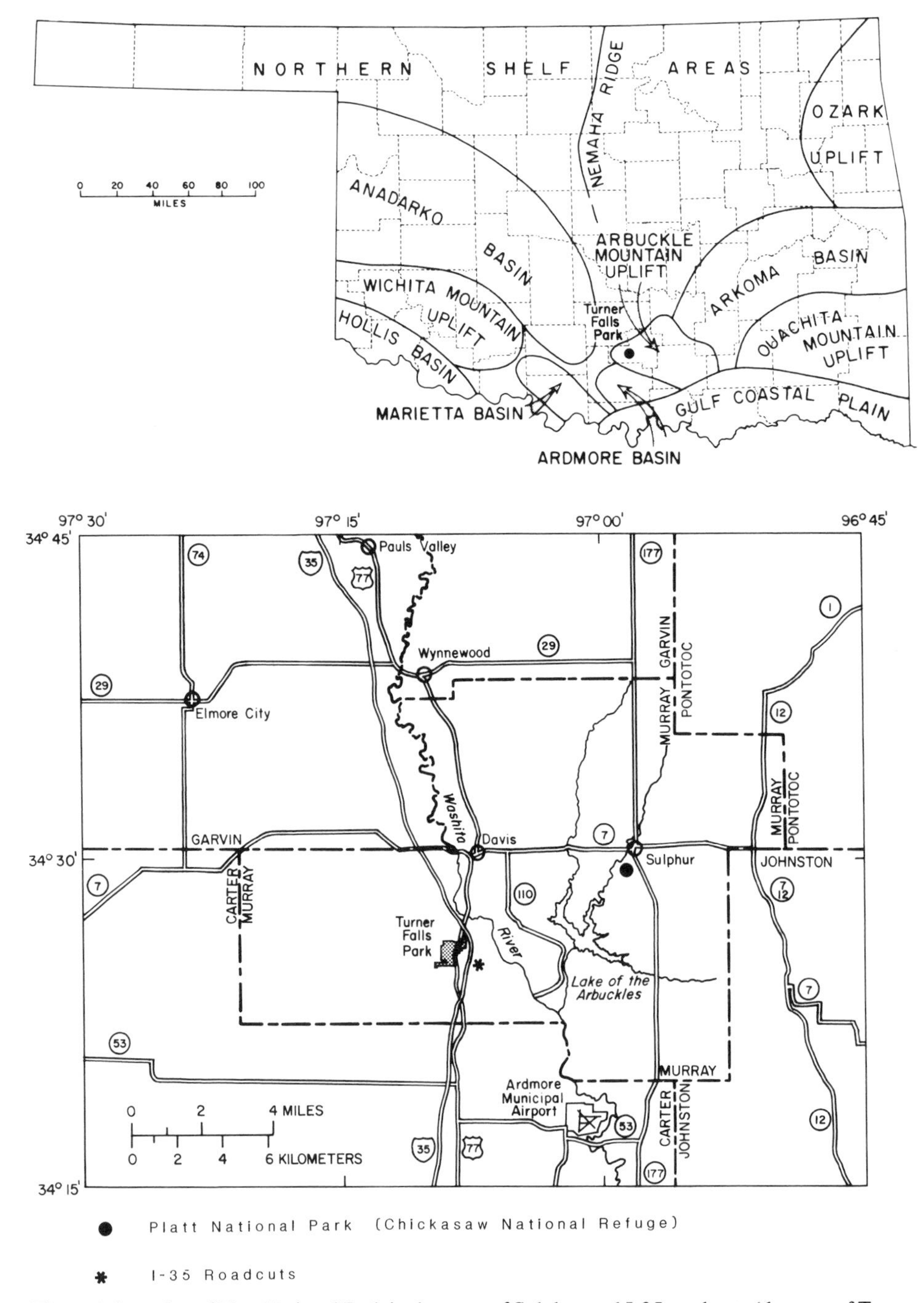

Figure 1. Location of Platt National Park in the town of Sulphur and I-35 road cuts (due east of Turner Falls Park).

Ada Formation, but in the Arbuckle region the conglomerate member of the formation oversteps folded older Paleozoic rocks with great unconformity. All the pebbles found in the various outcrops of the Vanoss could have been derived from rock types presently exposed in the Arbuckle Mountains. This relationship, coupled with the northward disappearance of the conglomerate facies, suggests a clear "cause and effect" relationship between Arbuckle tectonism, uplift, and the conglomerates.

The most obvious lithology seen on Bromide Hill consists of conglomerates composed of carbonate pebbles and cobbles. These are interbedded with brown and red-brown siliclastic sandstones and with red, buff, and green siltstones and mudstones. The conglomerate beds have lenticular geometries; many show markedly erosive bases; multistoried stacking of beds is commonly observed. Although individual beds usually appear massive at first sight, in places it is possible to see slight variations in pebble size and indistinct layering that together indicate that the conglomerates were deposited as gravel bars by rivers. (An example of this occurs in the large exposure at the first bend of the trail, 22 ft [7 m] above the base of the section given in Fig. 3.) Textures observed in the conglomerates are grain supported; most pebbles are roughly spherical and subangular to well rounded. In some beds sand-sized quartz grains are present between the pebbles (these may record an infiltration into gravel bars). Most conglomerates are cemented by calcite. However, near the base of one or two beds, a mud matrix is present (e.g., at 75 ft [23 m] on Fig. 3); this mud was probably derived from erosion of the underlying beds.

The finer beds seen in the hillside are laterally more persistent than the conglomerate beds (except where truncated by the latter). The sandstones show parallel lamination and small scale cross-bedding; they appear to be low-energy fluvial deposits.

At a number of horizons, calcrete (caliche) deposits are present. Calcretes are carbonates of pedogenic origin that today form in arid or semi-arid regions of the world. They form by the regular addition of calcium carbonate to the soil profile, and a sequence of textures from irregular nodules to more massive limestones develops with age (Al-Shaieb and others, 1977). Their occurrence in the Vanoss suggests that the climate was semi-arid on occasion and that long periods of geomorphologic stability occurred.

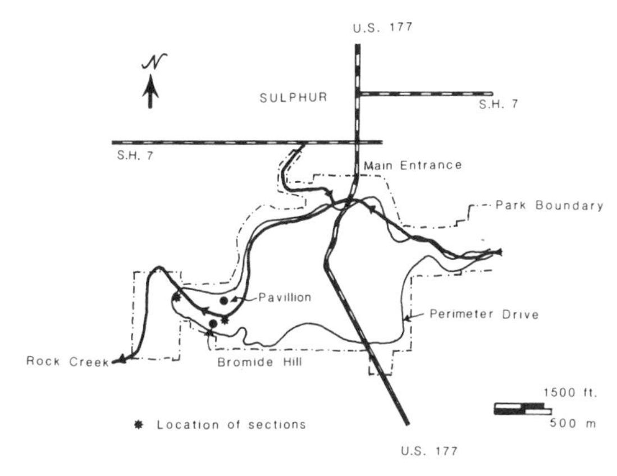

Figure 2. Location of Vanoss exposures in Platt National Park.

The interbedding of conglomerates with finer sediments, coupled with the presence of calcretes, reflects an interesting dichotomy in the development of the Vanoss. On the one hand, the finer sediments and the calcretes suggest a quiet low-energy setting: probably a broad alluvial plain crisscrossed by shallow braided river courses. On the other hand, the conglomerates bespeak an active alluvial fan system with considerable erosive potential.

The most eloquent example of this dichotomy is provided

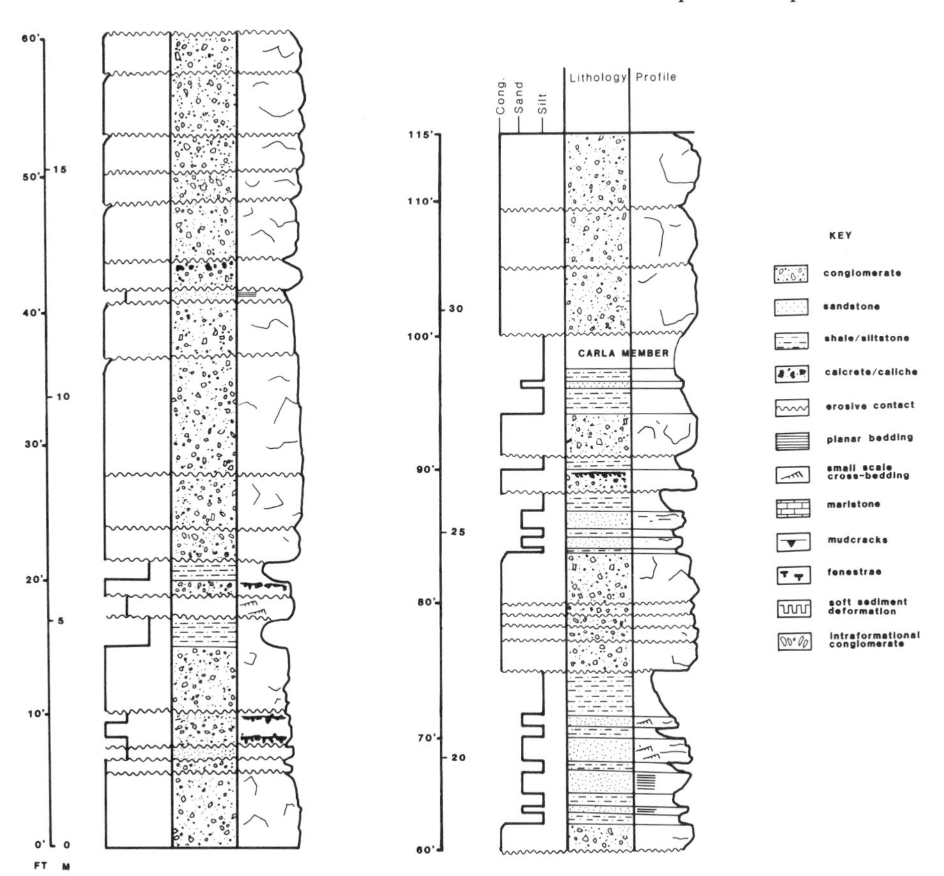

Figure 3. Measured section along park trail on Bromide Hill in Platt National Park.

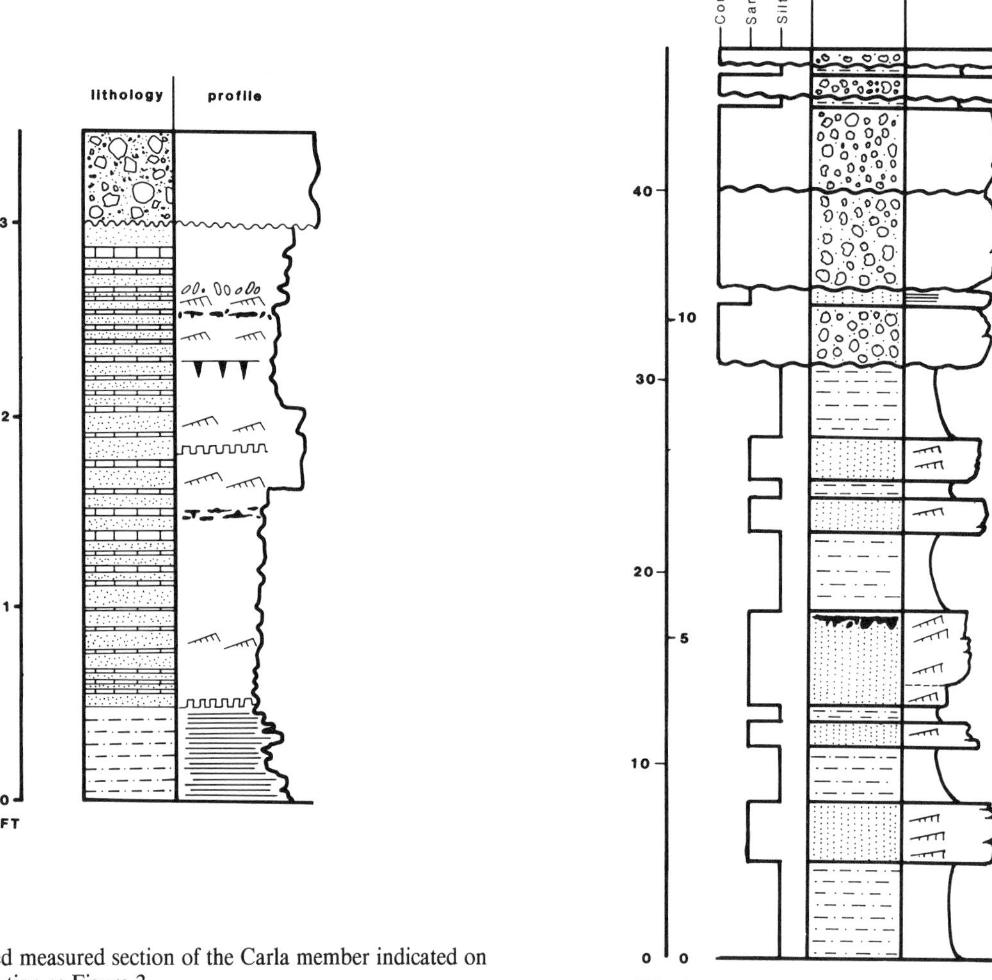

Figure 4. Detailed measured section of the Carla member indicated on Figure 3. Explanation as Figure 3.

Figure 5. Measured section taken from a road cut on the perimeter road, Platt National Park at the western end of Bromide Hill. Explanation as Figure 3.

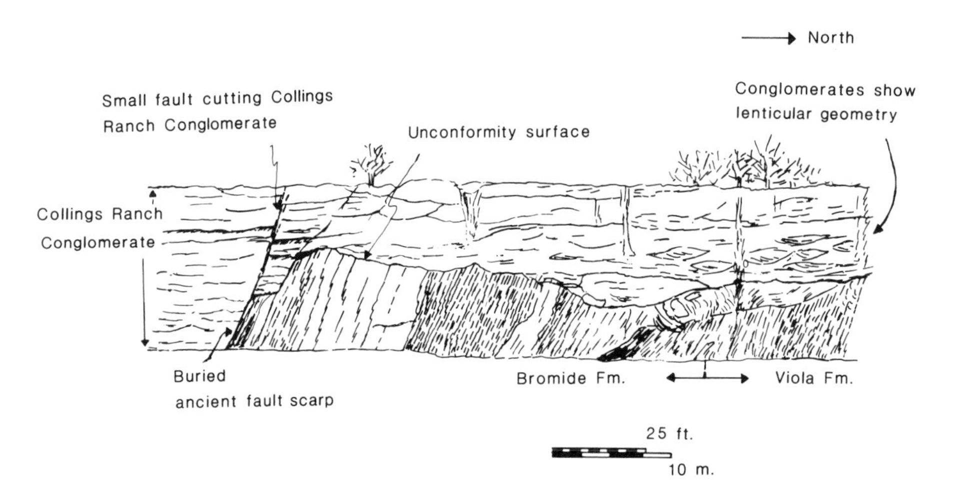

Figure 6. Sketch of unconformity between Collings Ranch Conglomerate and Lower Paleozoic rocks as seen in a road cut on the western side of the southbound lane of I-35 in the Arbuckle Mountains.

by the newly designated "Carla member." This member consists of thin interbeds of very fine light brown quartzose sandstone and pale buff marlstone. The latter consists of micrite with minor clays and other siliclastics. Textures present in the carbonate layers include laminae (? varves), fenestrae, and mudcracks. The sandstones exhibit small-scale cross-bedding and have abrupt or slightly erosive bases; one of these has cut down into an underlying micrite and disrupted it into intraformational clasts. This short sequence is interpreted as the record of a small, shallow, and relatively calm lake. The outcrop of the member is limited to a small area on the south of the exposure (this is the right-hand side when facing the cliff). To the north the member has been completely eroded out by the overlying conglomerate layer.

Two explanations, or an interplay between two explanations, seem appropriate to account for the abrupt changes in character of the Vanoss of Bromide Hill. Either there was frequent avulsion of small imbricate alluvial fans, or periodic rejuvenation of a nearby fault scarp reactivated these fans.

COLLINGS RANCH CONGLOMERATE

Description

Traveling southward from the Davis turnoff on I-35, the first exposure of the Conglomerate is seen on the western road cut of the southbound lane, on the southern slope of the second large hill south of the turnoff. The field relationships present show the Conglomerate overlying the Ordovician Bromide and Viola Formations with great angular unconformity. The underlying beds (which "young" northward) are vertical to slightly overturned (Fay, 1973), and dip of the conglomerate is slight (Fig. 6). The unconformity surface is irregular; near the southern edge of the exposure it falls abruptly beneath road level. This abrupt fall has been interpreted as a fault scarp buried by conglomerate deposition (Fay, 1973). The overlying pebble and cobble conglomerate (to which access is difficult) clearly displays lenticular beds that appear to be gravel bars.

About 1,500 ft (450 m) south of the unconformity, good exposures of the conglomerate can be closely examined in median road cuts. The entire exposure here forms part of a highly asymmetric syncline with a faulted steep (northern) limb. The best view of this fold is in the eastern road cut of the northbound lane of the interstate.

In detail, the conglomerate comprises stacked pebble and cobble conglomerates forming lenticular beds whose contacts are either erosional scour and fill structures or are defined by impersistent surfaces of red mudstone. Much of the deposit appears to have formed as gravel bars deposited by flashy streams. Where present, erosional channels show imbricate structure (Pybas and Cemen, 1986). Glahn and Laury (1985) recognized the importance of sieve deposits in which fine sediment (mostly red mud and silt) has infiltrated between pebble clasts. They also record minor debris flows.

Clasts found in the conglomerate all appear to have been derived from Lower Paleozoic sedimentary sources, particularly the Arbuckle Group (Ham, 1973). This suggests either that the igneous core of the Arbuckles was not exposed at this time or that it did not outcrop within the drainage area of the conglomerate.

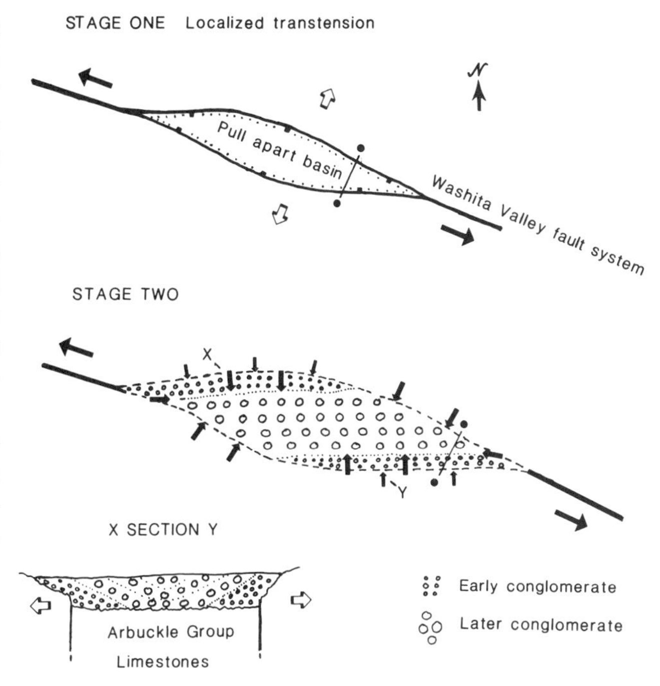

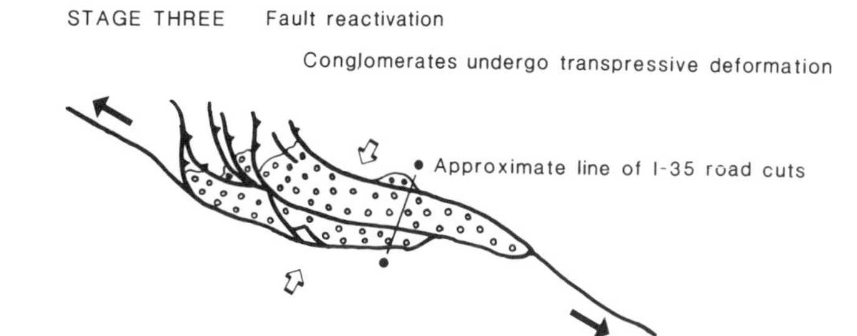

Figure 7. Stages in the development of the Collings Ranch Conglomerate according to the left lateral wrench movement model. Stage one: opening of a small pull-apart basin at a bend in the Washita Valley system. Stage two: syntectonic infill of the basin by the Conglomerate. Note that the geometry of the opening basin leads to illusory sediment thicknesses when logged along a horizontal plane. Stage three: reactivation of the fault system causes compression of the Conglomerate (which has effectively "plugged the hole" created by the bend in the fault system).

Close examination of the conglomerates reveals that much pressure solution has taken place between carbonate clasts; spectacular cases of up to 0.75 in (2 cm) of pebble cannibalism can be found. The orientation of these solution surfaces varies from horizontal to vertical. Glahn and Laury (1985) suggest that this pressure solution was a response to tectonically induced stress with a strong horizontal component. Pressure solution predated partial cementation of the deposit by coarse calcite cement (there may be a cause and effect relationship between the two events).

As noted above, the conglomerate has been folded into a faulted asymmetric syncline that trends approximately NW-SE. This flexural slip fold may have formed prior to cementation of the deposit (Pybas and Cemen, 1986). The vertical fault (trend WNW-ESE) which forms the edge of the steep northern limb of the fold is associated with phacoidal fracture patterns suggestive of left lateral motion. A similar sense of displacement is given by slickensides associated with numerous small faults that cut the deposit (Glahn and Laury, 1985).

The Collings Ranch Conglomerate was deposited in a grabenlike basin, trending WNW-ESE, which was several miles in length and up to 0.5 mi (0.8 km) in width. Wickham (in Wickham and Denison, 1978, p. 100–102) has noted that the basin is intimately associated with the major Wichita Valley fault system and has suggested that it formed as a left-lateral wrench pull-apart depression located along a bend or leftward step in the major fault (Fig. 7). If this is the case, then the apparent maximum measured thickness of 3,000 ft (900 m) assigned to the Conglomerate (Ham, 1973) may be illusory, particularly as the maximum vertical thickness that can be measured at one place is only 300 ft (90 m). The deposit may in fact consist of a series of small overlapping (imbricate) fans that record a continuous syntectonic response as the basin gradually opened.

An alternative explanation is that the Collings Ranch was deposited in a true graben formed during crestal extension of the Arbuckle anticline (Brown and Reneer, 1985). Subsequent to its initiation, the graben was rotated to an offcrest position as the anticline tightened. This explanation is compatible with Ham's (1954) ideas and his original estimate of the thickness of the conglomerate.

Subsequently the fault system reactivated and transpressed the basin fill, causing prelithification folding and faulting plus quantitatively significant pressure solution between rotating limestone clasts. It is these reactivated faults that presently form most of the boundary of the Conglomerate exposure except in a few areas, including that so conveniently exposed during the I-35 excavation where the fault scarp of one of the earlier faults was buried (Fig. 6).

REFERENCES

Al-Shaieb, Z., Shelton, J. W., Donovan, R. N., Hanson, R. E., May, R. T., Hansen, C. A., Morrison, C. M., White, S. J., and Adams, S. R., 1977, Evaluation of uranium potential in selected Pennsylvanian and Permian units and igneous rocks in southwestern and southern Oklahoma: Grand Junction, Colorado, Final report for Bendix Field Engineering Corporation.

Brown, T. A. and Reneer, B. V., 1985, The Collings Ranch Conglomerate, *in* Brown, W. G., Grayson, R. C., Jamison, W. H. and Altum, J. T., eds., Tectonism and sedimentation in the Arbuckle region, Southern Oklahoma Aulacogen: Baylor Geological Society, American Association of Petroleum Geologists Student Chapter publication, 43 p.

Brown, W. G., 1982, Shortcourse in structural geology (course notes): Southwestern Section, American Association of Petroleum Geologists, 79 p.

Barker, B. M., and Jameson, W. C., 1975, Platt National Park: Norman, University of Oklahoma Press.

Fay, R. O., 1973, Arbuckle anticline along Interstate Highway 35, *in* Ham, W. E., Regional geology of the Arbuckle Mountains, Oklahoma: Geological Society of America 1973 Annual Meeting, Dallas, Texas, Guidebook for Field Trip No. 5.

Glahn, T. E., and Laury, R. L., 1985, Sedimentation, diagenesis, and deformation of a Pennsylvanian conglomerate, Arbuckle Mountains, Southern Oklahoma: Geological Society of America Abstracts with Programs, v. 17, no. 3, p. 159.

Ham, W. E., 1954, Collings Ranch Conglomerate (Late Pennsylvanian) in the Arbuckle Mountains, Oklahoma: American Association of Petroleum Geologists Bulletin, v. 38, p. 2035–2045.

——, 1973, Regional geology of the Arbuckle Mountains, Oklahoma: Geological Society of America 1973 Annual Meeting, Dallas, Texas, Guidebook for Field Trip No. 5.

Pybas, K., and Cemen, I., 1986, Collings Ranch Conglomerate; An example of syn-depositional deformation in an extensional strike slip basin at the Arbuckle mountains, Oklahoma: Geological Society of America Abstracts with Programs, v. 18, no. 3, p. 260.

Wickham, J., and Denison, R., 1978, Structural style of the Arbuckle region: Geological Society of America, South Central Section, Guidebook for Field Trip No. 3.

Basal sandstone of the Oil Creek Formation in the quarry of the Pennsylvania Glass Sand Corporation, Johnston County, Oklahoma

J. G. McPherson, R. E. Denison, D. W. Kirkland, and D. M. Summers, Mobil Research and Development Corporation, Dallas Research Laboratory, P.O. Box 819047, Dallas, Texas 75381

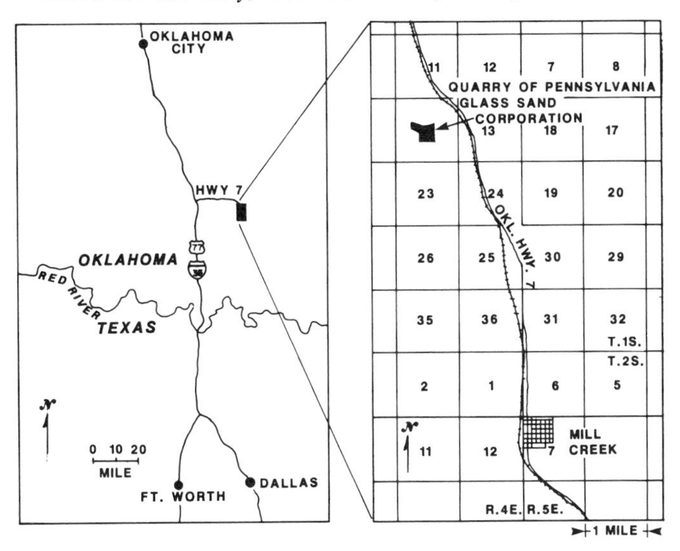

Figure 1. Map showing the location of the Oil Creek Formation sand quarry studied.

Figure 2. An overview (looking south) of the Pennsylvania Glass Sand quarry showing the basal sandstone (white) of the Oil Creek Formation occupying the lower half of the quarry face. The sequence is overlain by skeletal calcarenites (dark gray) of the upper Oil Creek Formation. Photograph taken September 1985.

LOCATION

The basal sandstone of the Oil Creek Formation is well exposed in a quarry located 5 mi (8 km) north northwest of the small town of Mill Creek in Johnston County, Oklahoma (Sec.14,T.1S.,R4E.) (Figs. 1 and 2). The quarry is owned by the Pennsylvania Glass Sand Corporation and is about 0.5 mi (0.8 km) west of Oklahoma 7, which passes through Mill Creek and connects with I-35 about 20 mi (32 km) to the west.

SIGNIFICANCE

The depositional environment of the basal sandstone of the Oil Creek Formation has been a matter of conjecture. In part, this is because on the weathered and dry quarry walls, especially on sunny days, few sedimentary features can be discerned. In addition, even under favorable conditions, internal stratification and sedimentary structures within the sandstone are difficult to recognize because the grain composition is essentially monomineralic and because the grains have a uniform size and shape. Another puzzling aspect of this Ordovician sandstone concerns the question of why it should be almost totally uncemented. Was it at one time well cemented and subsequently subjected to dissolution or has it remained uncemented throughout much of the Phanerozoic?

Precautions. Great caution should be used when visiting the workings. Permission to visit must be obtained from the superintendent of the quarry. The walls are up to 100 ft (30.5 m) high, and because the sandstone is poorly cemented and purposely undercut during hydraulic mining, they are very unstable. Avalanches are common. Geologic features should be investigated in low faces on the quarry floor. **Under no circumstances should the steep quarry walls be approached.** Failure to abide by this rule could result in injury or death and will result in suspension of permission to visit the quarry for future groups.

STRATIGRAPHIC SETTING

The Oil Creek Formation belongs to the Simpson Group of Middle Ordovician age. In the vicinity of the quarry the basal sandstone of the Oil Creek Formation is about 350 ft (107 m) thick (Lewis, 1982). North of the quarry are outcrops of the underlying West Spring Creek Formation of the Arbuckle Group, which consists of laminated fine-grained dolomites. Overlying the sandstone, and exposed near the top of the quarry, are skeletal calcarenites typical of the upper Oil Creek Formation. The Oil Creek at the quarry lies directly on dolomites of the Arbuckle Group, but some 5 mi (8 km) to the south, the Joins Formation is the basal unit of the Simpson Group (Fig. 3).

The Arbuckle Group is as much as 6,500 ft (1,980 m) thick

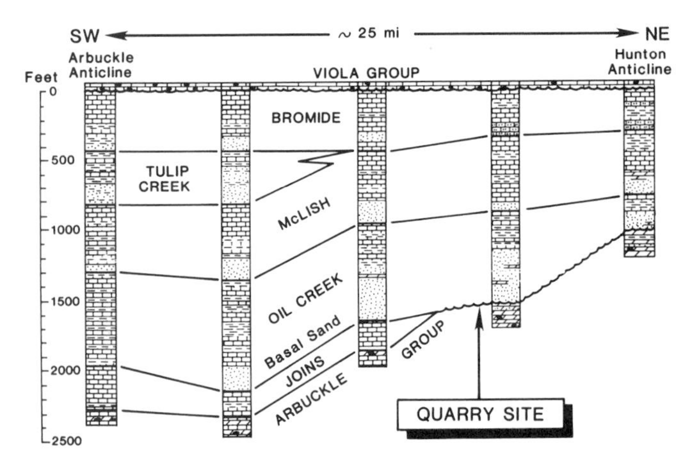

Figure 3. Stratigraphic relationship of the basal sandstone of the Oil Creek Formation to other formations of the Simpson Group (after Ham, 1955).

Figure 4. Vertical face (~50 ft; 16 m), wet from hydraulic mining, showing bedding within the basal sandstone. Note the high lateral continuity and considerable variation in thickness of beds. Photograph taken September 1985.

in the Arbuckle anticline (8 mi; 13 km to the west), but is probably no more than 4,500 ft (1,370 m) thick in the vicinity of the quarry. It is dominated by shallow-water carbonates with a low percentage of terrigenous material, and deposition was essentially continuous from Late Cambrian through earliest Middle Ordovician. The Late Cambrian–Early Ordovician was a time of extraordinarily stable carbonate-platform deposition unparalleled in the Phanerozoic of North America, for carbonate sediments of equivalent age were deposited over a vast area from New York to Arizona and Alabama to the Great Lakes.

Deposition of the sandstone at the base of the Oil Creek began a new and different geologic chapter in southern Oklahoma sedimentation. On this phenomenal carbonate platform a widespread sequence of Middle Ordovician limestones and clastics was deposited. Conspicuous in the sequence are sandstones of unusual purity. The wide distribution and the source of the sand have consumed much geologic speculation (the so-called "St. Peter problem"). There is a sandstone at the base of each formation above the Joins Formation within the Simpson Group. (In ascending order these are: the Oil Creek, the McLish, the Tulip Creek, and the Bromide [Fig. 3]). All the Simpson sandstones are economically important as sources or potential sources of glass sand, but they are best known as extraordinary reservoirs for oil and gas.

DESCRIPTION

The basal sandstone of the Oil Creek Formation, as observed on fresh, wet, vertical cuts of the quarry, consists of stacked beds, 1 to 10 in (3 to 25 cm) thick, each with a sharp or erosional base and an upward-increasing clay content (Fig. 4). The sand fraction displays no megascopic size grading. A thin (up to several millimeters) clay layer caps each bed, and some of this clay has been postdepositionally redistributed into the sand below. Commonly, individual beds of sandstone can be distinguished only by the clayey intercalations.

Bedding, where visible, is laterally continuous (over many tens of meters) for even the thinnest of beds (Fig. 4). Most of the sands appear to be parallel laminated (Fig. 5), although details of the bedding are difficult to discern. Ripple cross-lamination occurs in the upper part of some individual beds, but is not ubiquitous. Wavy and flaser bedding is prominent in one interval. Convex-upward bedding surfaces, suggestive of oscillatory- or combined-flow ripple cross-stratification, are present.

Neither large-scale cross-stratification nor bioturbation is observed. Lewis (1982) reports *Skolithos-Planolites* from one locality, but emphasizes that distinct biogenic structures are rare. Compactional dewatering structures (Fig. 5) (dish and pipe structures) are not common, but where present they attest to rapid sedimentation of water-charged sands. Small-scale loading structures occur at some clay-sand contacts. The sandstone beds have erosional lower-bounding surfaces, evidenced by common inclusion of rip-up clasts of the underlying clay beds. Amalgamation of individual sandstone-claystone couplets by complete erosion of the clay intervals produced sandstone units up to 30 in (80 cm) thick.

The sands are very fine- to fine-grained (graphic mean = 2.84 phi), well-sorted (inclusive graphic standard deviation = 0.36 phi), and are fine-skewed (inclusive graphic skewness = +0.13) (see also Lewis, 1982). Histograms and cumulative weight plots of the grain size distribution show a minor secondary sand mode. The detrital grains have very high sphericity and roundness (Fig. 6) and show grain-surface frosting (Denison and Ham, 1973; Lewis, 1982) as well as crescentic impact marks.

The sandstone is a quartzarenite with only 0.04 to 0.71 percent detrital feldspars (orthoclase, microcline, and albite) and 0.006 to 0.307 percent detrital and authigenic heavy minerals (most commonly pyrite, tourmaline, zircon, and garnet) in addition to small amounts (at most a few weight percent) of clay

Figure 5. A relatively thick, clay-free and parallel-laminated sandstone bed displaying dewatering structures and fluid-escape pipes at the base. Scale (middle right) interval in centimeters.

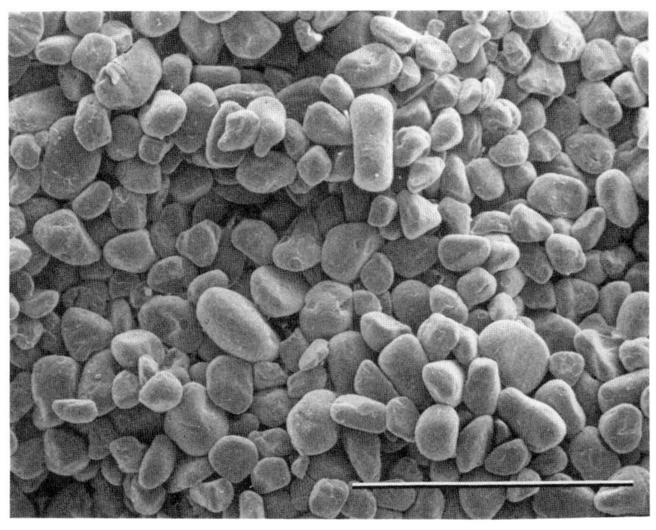

Figure 6. SEM photomicrograph of Oil Creek sandstone showing that grains are well rounded and have high sphericity. Scale bar equals 1 mm.

(Ham, 1945). The quartz grains are monocrystalline and unstrained; polycrystalline quartz grains are rare.

The clay is largely illite, but minor kaolinite and chlorite(?) are also present. The sandstone is commonly greenish in color, particularly toward the tops of individual beds. Where clay laminae and clasts are preserved, they are also green. The coloration of the sand is due to illite as partial to complete grain coatings 2 to 3 microns thick (Fig. 7; also Denison and Ham, 1973). In the whitest sands, illite is rare and occurs only as thin ridges around grain contacts and along grain surfaces. The kaolinite and chlorite(?) occur as rare pore-filling cements. The illitic clay in the upper portion of individual beds is interpreted as depositional, whereas the illitic coats on the detrital quartz grains are probably the result of postdepositional redistribution of clay from interbedded clay-rich horizons.

Even by inspection of hand specimens, the porosity and permeability of Oil Creek sandstones are clearly high. Plugs were dry cut from large samples collected in the quarry. The resulting porosities ranged from 28.5 to 35.3%. Point counting (500 counts) of pore space in seven thin sections gave results between 22.5 and 30.5%. Measurements on the least illitic (white) samples gave permeabilities (to air) between 2,000 and 2,500 md. The most illitic (green) samples had between about 800 and 1,150 md. The fragile nature of the sandstone suggests that these values may not necessarily represent the bulk of the rock. Nonetheless, the results demonstrate that by any standard the Oil Creek sandstones have exceptional reservoir quality.

The sandstone is friable, for the most part lacking appreciable cementation. A sample placed in water soon disaggregates into loose sand grains and a milky clay fraction held in suspension. However, where there is faulting in outcrop, extensive granulation has resulted in thin, harder, anastomosing seams of low porosity and permeability (Pittman, 1981). The observation that most contacts between detrital grains are point or tangential and only a minor percentage are shallow concavo-convex suggests that the sands have undergone mechanical compaction but little pressure solution (Fig. 8). Clean, linear fractures have developed in some quartz grains. Their origin is unknown and may be due to cleavage plane weaknesses (Wellendorf and Krinsley, 1980) or to crystallization stresses developed within the quartz crystals (Smalley, 1966). Cementation is rare and consists of incipient bonding by illite at points of contact between grains, some minor quartz probably derived from pressure solution (Fig. 9), and minor pods of carbonate. In a few areas on the floor of the quarry, sandstone well cemented by carbonate is found in blocks 3 ft (1 m) or more in diameter. In one block, the dolomite-to-calcite ratio is approximately 3 to 1, and in two samples from these blocks the carbonate cement has δC^{13} PDB values of $-6.08‰$ and $-5.68‰$, and δO^{18} PDB values of $-4.20‰$ and $-4.80‰$, indicating an origin other than marine.

SEDIMENTOLOGY

The Oil Creek sandstones in the Pennsylvania Glass Sand quarry are characterized by stacked sandstone–claystone beds. Such bedding sequences are a characteristic feature of storm-influenced sedimentation in both modern and ancient shelf settings (Hayes, 1967; Reineck and Singh, 1972; Brenchley and others, 1979; Kreisa, 1981; Aigner and Reineck, 1982; Brenchley and Newall, 1982; Nelson, 1982; Aigner, 1985). Fair-weather sedimentation produces an accumulation of clay. Storms not only erode previously deposited fines, but may advect large quantities of sand from the shore. The resultant storm sequence consists of relatively thick sands derived from wave action and erosion inshore, capped by thinner clays deposited during the post-storm stage. The upward fining, which is characteristic of the uppermost portion of each cycle, is a consequence of storm decay.

Whether the bulk of the sand in the storm layers was depos-

Figure 7. Photomicrograph of thin section in plane polarized light showing highly birefringent illite grain coats. Scale bar equals approximately 0.25 mm.

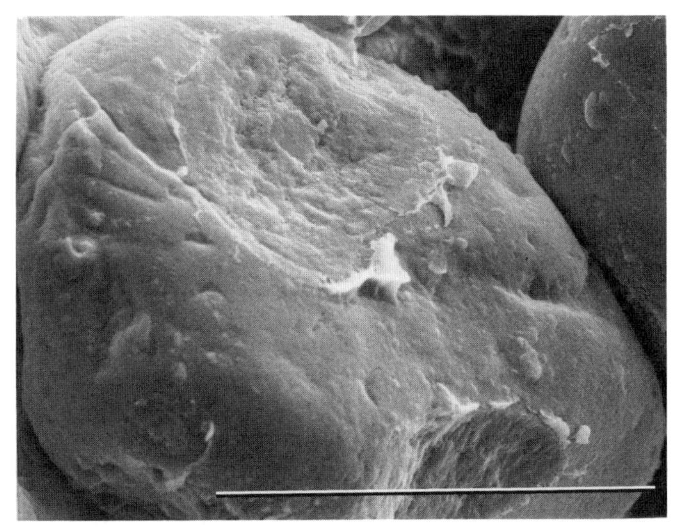

Figure 8. SEM photomicrograph of quartz grain showing pressure solution pits with remnant illite rim coatings. Some crescentic impact marks are present and suggestive of eolian transport. Scale bar equals 0.1 mm.

ited during the height of the storm (Allen, 1982), or as a product of the decay of the storm (Brenchley and others, 1979) is a matter of current debate. Storm erosion is evidenced by the erosional nature of the basal contact of the storm sand layers and locally derived claystone rip-ups. The high sandstone-to-mudstone ratio in the storm deposits and the predominance of parallel lamination in the sands are functions of high and sustained orbital velocities produced during storms of high intensity and long duration. These are the characteristics of storm deposits that have been investigated in near-shore, inner-shelf settings (Aigner and Reineck, 1982; Allen, 1982; Aigner, 1985) (Fig. 10). The ripple cross-lamination observed in the quarry is probably of current, combined-flow, and oscillatory-flow origin, although a distinction could not be made because of outcrop limitations. The absence of large-scale cross-stratification does not indicate an absence of currents of appreciable flow strength, but is merely a reflection of the lack of a suitable grain size from which to produce the bedforms (Middleton and Southard, 1985).

The anomalously low percentage of silt and clay seen in the basal sandstone is, at least in part, probably due to a prolonged period of eolian winnowing of the sediment that was supplied to the shelf. A similar explanation has recently been proposed for widespread Cambrian and Lower Ordovician orthoquartzites of North America (Dalrymple and others, 1985). The Ordovician coastal plain would have been a barren terrain devoid of vegetation and very susceptible to eolian winnowing. Further and direct evidence for an eolian influence comes from distinctive and characteristic eolian frosting and pitting of the quartz grain surfaces. In addition, the high sphericity and roundness of such fine-grained quartz is difficult, if not impossible, to produce by processes other than eolian abrasion. Furthermore, the secondary grain-size mode is compatible with an eolian source, in that bimodality is a characteristic feature of eolian sediments (Folk, 1968).

The great thickness of the basal sandstone (350 ft; 107 m) in the vicinity of the quarry suggests that the inner-shelf conditions were maintained for a prolonged period of time at this locality. The delicate balance between basinal subsidence rates, sediment supply rates and sea-level change was sustained in such a way as to maintain the inner-shelf setting.

The basal sandstone is overlain (without bedding plane discordance) by about a 300-ft-thick (100-m) sequence of alternating mudstone, limestone, and minor quartzarenite, representing the remainder of the Oil Creek Formation. This sequence is interpreted to represent shelf to tidal-flat sedimentation (Lewis, 1982). Although the contact between the upper Oil Creek and the basal sandstone has been classed as an intraformational unconformity (Lewis, 1982), we find no paleontologic or sedimentologic evidence for a significant time gap at the boundary. The transition is probably conformable and reflects a facies change to a mixed carbonate-clastic shelf.

DIAGENESIS

Within the storm layers, and particularly at the tops of the layers, are crystals of pyrite or their limonitic alteration products. The pyrite is apparently of early diagenetic origin. Within the upper part of each storm layer, some reactive organic matter of algal and microbial derivation, probably much of it within fecal pellets, accumulated. Molecular oxygen from the overlying sea water could not readily enter the storm sediments by diffusion because of the thin clay layer capping each cycle. Also, bioturbation was slower and shallower in pre-Devonian time than later (Thayer, 1983) and was accomplished by metazoans that lived on the surface of the substrate. Because of this limitation on bioturbation, molecular oxygen in solution was further restricted from entering the sediments of each storm cycle (see Demaison and

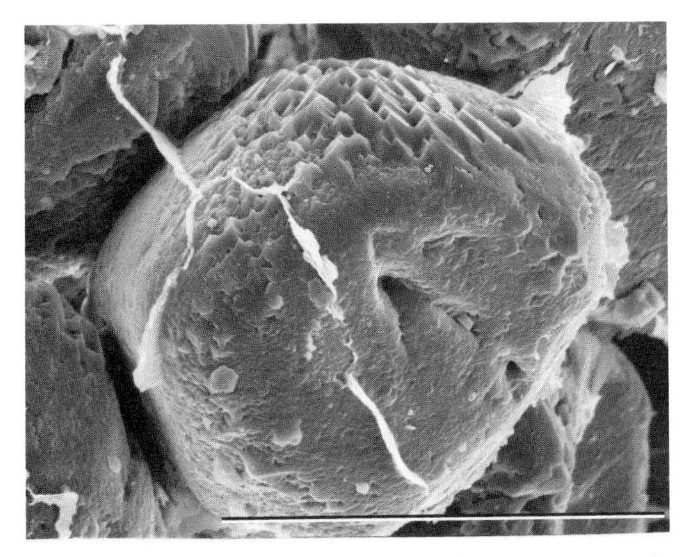

Figure 9. SEM photomicrograph showing incipient quartz overgrowths. The white bands are probably illite. Scale bar equals 0.1 mm.

Figure 10. Schematic diagram showing proximal and distal facies of typical storm-sand sequences in relation to water depth and distance from shore as the two most important environmental gradients controlling the characteristics of storm deposits. The basal sandstone of the Oil Creek Formation in the quarry is characterized as a storm deposit of an inner-shelf, near-shore setting, i.e., on the shoreward side of the transition zone in the above scheme (after Aigner and Reineck, 1982).

Moore, 1980). As oxygen in the sediments became depleted by reaction with organic matter, the remaining labile organic matter was free to react with trapped sulfate anions, via sulfate-reducing bacteria, to form hydrogen sulfide. Near the tops of the storm layers, this hydrogen sulfide, in turn, reacted with iron derived from the thin illitic capping layers to form the authigenic pyrite.

The Oil Creek sandstone, although of Ordovician age, is almost totally unlithified—sand peels can be made from outcrops. Pressure solution (chemical compaction) has been minimal because of the shallow depth of burial and the very high framework grain sphericity. The grains are virtually uncemented, and this absence of cement has been attributed to grain coating by illite that inhibited the formation of quartz overgrowths. This is supported by the observation that silica cement is found where the sand grains are broken and the illitic coats disrupted (Denison and Ham, 1973). Because no systematic difference occurs in degree of cementation between white (less illitic) and greenish (illitic) sands, the presence of illite is probably not the only important factor. Another contributor to the lack of cementation (and pressure solution) is the shallow maximum depth of burial (about 5,000 ft; 1,500 m).

A question remains concerning the possibility that the porosity in the Oil Creek sandstone is not primary but is secondary (i.e., the sandstone was cemented and has been decemented). If the sandstone was cemented at one time, then that cement has been nearly totally removed; we have observed no partially dissolved minerals that could be interpreted as the remnants of pore-filling cements. We did, however, observe large blocks of sandstone on the quarry floor that were cemented with dolomite and, to a lesser extent, with calcite. The isotopic values for this carbonate are not consistent with a marine origin. This dolomitic cement would be expected to be less soluble than a predominantly calcite cement and might represent a local relict of a more pervasive calcite cement. Grains in the uncemented Oil Creek are commonly frosted but not embayed, as one might expect if a preexisting carbonate cement had been removed. However, the carbonate that is now present does not obviously embay grains. A final consideration is that it is difficult to imagine a mechanism for massive removal of carbonate cement from a sandstone that occurs in a dominantly carbonate section. In summary, we have no compelling evidence that supports a secondary origin for the porosity of the Oil Creek sand.

HYPOTHESIZED REGIONAL HISTORY

Sea level reached a lowstand in early Middle Ordovician time, resulting in the draining of the vast shelf that had dominated early Paleozoic sedimentation (Ross, 1976). An ultra-regional disconformity formed. The vast exposed area must have been one of the more remarkable scenes in the Phanerozoic landscape—a broad and featureless carbonate plain with little or no organized drainage and no life other than algae and bacteria. The Oil Creek sand must have been transported across this vast platform from far (many hundreds of kilometers) to the north.

The Oil Creek sandstone is thickest at the northeastern margin of the southern Oklahoma aulacogen (near the quarry). Although the deep-rift phase of the aulacogen had ended in the Middle Cambrian, the area continued to be represented by a gentle linear zone of increased subsidence. While the shelf was for the most part exposed, marine deposition continued in the aulacogen, and is represented by shallow-water carbonates of the Joins Formation. The thickest Oil Creek sandstone was deposited near the northwest-trending Joins shoreline. The sandstone is thinner to the north and virtually disappears to the west. The sand may have been transported from the north across the emergent carbonate plain by eolian processes; ultimately the sand became concentrated near the Joins shoreline in a dune-beach complex.

As sea level rose again, the sand was reworked into an extensive diachronous transgressive blanket.

The Oil Creek is the oldest of the several pure sandstones that were deposited in North America during the Middle Ordovician. Time equivalents include the Everdon Dolomite of the southeast Missouri area and the sandy upper part of the Swan Peak Formation of Utah. The renowned St. Peter Sandstone that is found from northern Arkansas through southeastern Minnesota, the Harding in Colorado, the Winnipeg of the northern Great Plains, and the Eureka Quartzite of the western states are the equivalents of the upper part of the Simpson Group (Ross and others, 1982).

CONCLUSIONS

1. The basal sandstone of the Oil Creek Formation is characterized by stacked beds of pure quartzose sandstone. The bedding is a product of storm-influenced sedimentation in an inner-shelf setting. Each storm cycle consists of a sandstone bed (1 to 10 in; 3 to 25 cm thick) derived from wave action and erosion inshore, capped by a claystone layer (up to several millimeters in thickness) deposited during the post-storm stage.

2. Textural and compositional evidence suggest that the sand grains of the basal sandstone have passed through an eolian cycle. The sand was transported from the north to the coastal setting across a broad (hundreds of kilometers) emergent carbonate plain. Eolian processes may have provided an important mode of sand transport from the source area.

3. The basal sandstone is unconsolidated and has high porosity (30%) and high permeability (1–3 darcies). The sandstone has undergone minimal pressure solution because the quartzose framework grains are of such high sphericity and because of the shallow depth of burial; it is also almost totally uncemented. Where present, illitic grain-coatings, which resulted from depositional infiltration, probably protected framework grain surfaces from later quartz cementation.

REFERENCES CITED

Aigner, T., 1985, Storm depositional systems: New York, Springer-Verlag, 174 p.

Aigner, T., and Reineck, H. E., 1982, Proximialitiy [proximality] trends in modern storm sands from the Helgoland Bight (North Sea) and their implications for basin analysis: Senckenbergiana marit, v. 14, p. 183–215.

Allen, J.R.L., 1982, Sedimentary structures, their character, and physical basis, v. 2, *in* Developments in sedimentology, v. 30: Amsterdam, Elsevier, 663 p.

Brenchley, P. J. and Newall, G., 1982, Storm-influenced inner-shelf sand lobes in the Caradoc (Ordovician) of Shropshire, England: Journal of Sedimentary Petrology, v. 52, p. 1257–1269.

Brenchley, P. J., Newall, G., and Stanistreet, I. G., 1979, A storm surge origin for sandstone beds in an epicontinental platform sequence, Ordovician, Norway: Sedimentary Geology, v. 22, p. 185–217.

Dalrymple, R. W., Narbonne, G. M., and Smith, L., 1985, Eolian action and the distribution of Cambrian shales in North America: Geology, v. 13, p. 607–610.

Demaison, G. J., and Moore, G. T., 1980, Anoxic environments and oil source bed genesis: American Association of Petroleum Geologists Bulletin, v. 64, p. 1179–1209.

Denison, R. E., and Ham, W. E., 1973, Stop 7, Oil Creek Sandstone (Middle Ordovician) in quarry of Pennsylvania Glass Sand Corp. SW¼NE¼,Sec.6, T.2S,R.5E, Johnston County, *in* Ham, W. E., ed., Regional geology of the Arbuckle Mountains, Oklahoma, Guidebook for Field Trip No. 5, Norman, Oklahoma Geological Survey Special Publication 73-3, p. 49–51.

Folk, R. L., 1968, Bimodal supermature sandstones; Product of the desert floor: Proceedings of the 23rd International Geological Congress, Section 8, p. 9–32.

Ham, W. E., 1945, Geology and glass sand resources, central Arbuckle Mountains, Oklahoma: Oklahoma Geological Survey, Bulletin No. 65, 103 p.

—— , 1955, Field conference on geology of the Arbuckle Mountain region: Oklahoma Geological Survey Guidebook 3, 61 p., 1 map.

Hayes, M. O., 1967, Hurricanes as geologic agents; Case studies of Hurricane Carla, 1961, and Cindy, 1963: Texas University Bureau of Economic Geology Report of Investigation No. 61, 54 p.

Kreisa, R. D., 1981, Storm-generated sedimentary structures in subtidal marine facies with examples from the Middle and Upper Ordovician of southwestern Virginia: Journal of Sedimentary Petrology, v. 51, p. 823–848.

Lewis, R. D., 1982, Depositional environments and paleoecology of the Oil Creek Formation (Middle Ordovician), Arbuckle Mountains and Criner Hills, Oklahoma [Ph.D. thesis]: University of Texas at Austin, 351 p.

Middleton, G. V., and Southard, J. B., 1985, Mechanics of sediment movement: Society of Economic Paleontologists and Mineralogists Short Course No. 3, 401 p.

Nelson, C. H., 1982, Modern shallow-water graded sand layers from storm surges, Bering Shelf; A mimic of Bouma sequences and turbidite systems: Journal of Sedimentary Petrology, v. 52, p. 537–545.

Pittman, E. D., 1981, Effect of fault-related granulation on porosity and permeability of quartz sandstones, Simpson Group (Ordovician), Oklahoma: American Association of Petroleum Geologists Bulletin, v. 65, p. 2381–2387.

Reineck, H. E., and Singh, I. B., 1972, Depositional sedimentary environments: New York, Springer-Verlag, 439 p.

Ross, R. J., 1976, Ordovician sedimentation in the western United States: Rocky Mountain Association of Geologists Symposium Guidebook, p. 109–133.

Ross, R. J., and others, 1982, The Ordovician system in the United States: International Union of Geological Sciences Publication No. 12, 73 p.

Smalley, I. J., 1966, Formation of quartz sand: Nature, v. 211, p. 476–479.

Thayer, C. W., 1983, Sediment-mediated biological disturbance and the evolution of marine benthos, *in* Tevesz, M.J.S., and McCall, P. L., eds., Biotic interactions in recent and fossil benthic communities: New York, Plenum Press, p. 479–625.

Wellendorf, W., and Krinsley, D., 1980, The relation between the crystallography of quartz and upturned aeolian cleavage plates: Sedimentology, v. 27, p. 447–454.

ACKNOWLEDGMENTS

The Pennsylvania Glass Sand Corporation allowed us to investigate the quarry in Johnston County. We gratefully acknowledge this permission and especially appreciate the efforts and cooperation of Don Clark, Superintendent of the operation there. The field investigation of the Oil Creek in the quarry was a group effort by geologists of the Dallas Research Laboratory of Mobil Research and Development Corporation and we acknowledge the following colleagues: H. M. Chung, R. T. Clark, S. A. Dixon, J. T. Edwards, R. Evans, R. D. Kreisa, J. R. Markello, J. K. Sales, W. C. Seiler, M. L. Stockton, T. F. Tsui, J. M. Vizgirda, and C. M. Wall. We thank Mobil Research and Development Corporation for permission to publish this paper.

Middle Ordovician strata of the Arbuckle and Ouachita Mountains, Oklahoma; Contrasting lithofacies and biofacies deposited in southern Oklahoma Aulacogen and Ouachita Geosyncline

Stanley C. Finney, Department of Geological Sciences, California State University, Long Beach, California 90840

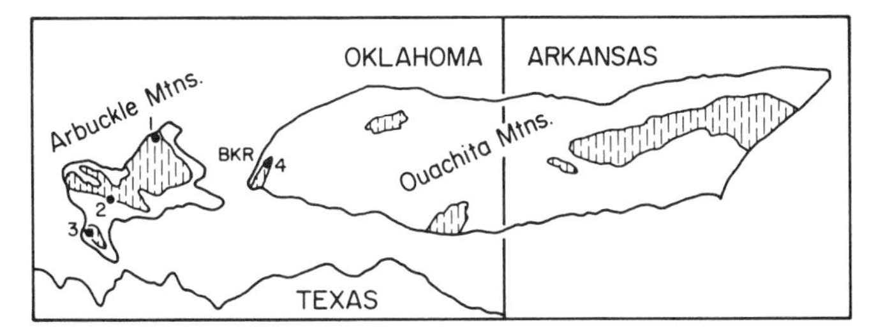

Figure 1. Map showing outcrops of Ordovician strata (dashed vertical lines) and localities in Arbuckle and Ouachita Mountains. BKR, Black Knob Ridge.

LOCATION

Middle Ordovician strata of the Arbuckle and Ouachita Mountains record markedly different depositional settings (Figs. 1, 2). Three localities in the Arbuckles and one in the Ouachitas demonstrate both the vertical sequences of lithofacies within each area and the contrasts in lithofacies and biofacies between the two areas.

Localities 1 through 3 (Fig. 1) are in the Arbuckle Mountain region. Locality 1 (Fig. 3) is in the northeastern part of the Arbuckles. It consists of a roadcut exposure along the west side of Oklahoma 99, about 3.5 mi (5.6 km) south of Fittstown, SW¼Sec.12,T.1N.,R.6E., Pontotoc County. It is readily accessible and requires no permission to visit. Locality 2 (Fig. 3) is a long roadcut along the west side of the southbound lanes of I-35 on the south flank of the Arbuckle Mountains. It is in the SW¼Sec.25,T.2S.,R1E., Carter County, is readily accessible, and requires no permission to visit. Locality 3 (Fig. 3) is at the north end of Criner Hills. It is a small quarry situated in the SW¼SE¼-Sec.9,T.5S.,R.1E., Carter County, readily accessible from I-35 and U.S. 70 by well-maintained gravel roads. The quarry is on private land. Permission for access can be obtained from the landowner, Ronald Burns of Ardmore, Oklahoma (405-657-8262).

Locality 4 (Figs. 1, 3) is at the western edge of the Ouachita Mountains. It is the Stringtown Quarry at the north end of Black Knob Ridge, directly north of the town of Stringtown, and situated in the center of the N½Sec.16,T.1S.,R12E., Atoka County. It is an active quarry, owned by the Amis Construction Company of Oklahoma City, and is readily accessible by paved roads. Access is generally limited to weekdays, and permission to visit must be obtained in advance (405-235-3555). Extra safety precautions, including the use of hard hats, should be taken because of the danger posed by the steep quarry walls.

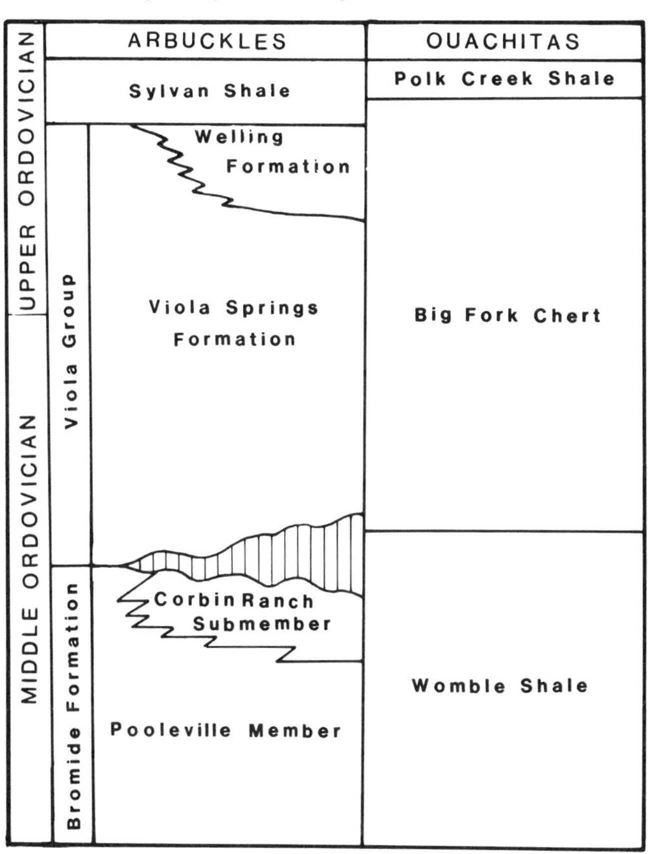

Figure 2. Middle to Upper Ordovician stratigraphy in Arbuckle and Ouachita Mountains.

SIGNIFICANCE

Middle Ordovician strata of the Arbuckle and Ouachita Mountains (Fig. 2) record markedly different depositional settings. The Arbuckle strata were deposited within and on the platforms bordering the Southern Oklahoma Aulacogen in environments ranging from the intertidal zone to water depths on the order of a few hundred meters. The Ouachita strata were deposited in the deep basin of the Ouachita Geosyncline off the southern margin of the North American craton and at water depths of 3,300 ft (1,000 m) or more. Slight differences in lithofacies and biofacies exist in the Arbuckle strata between sections representing deposition in the basin of the aulacogen (localities 2 and 3) and those representing deposition on the platforms bordering the aulacogen (locality 1).

Much greater contrasts exist, however, between the Ar-

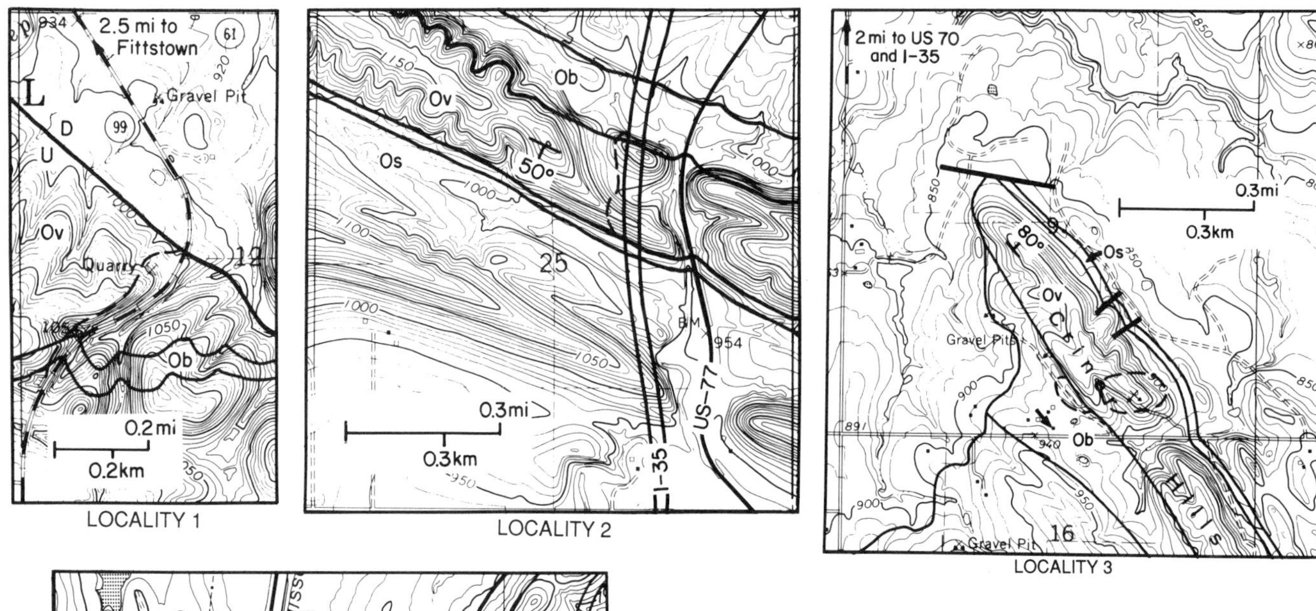

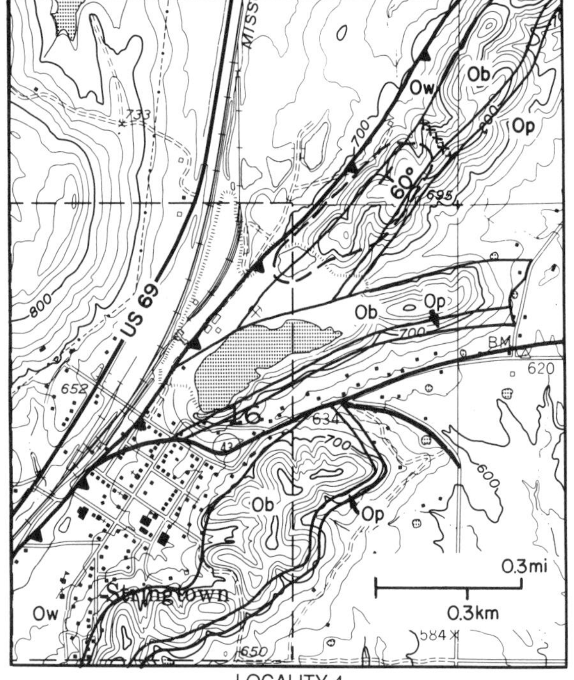

Figure 3. Maps of localities 1-4. Locality 1: U and D refer to movement along fault; dashed line marks top of roadcut. Ov, Viola Group; Ob, Bromide Formation. Locality 2: Labels as in locality 1. Os refers to Sylvan Shale; dashed line marks top of roadcut. Locality 3: Labels as in locality 1. Straight arrow points to landowner's house. Dashed lines denote margins of quarries. Locality 4: Ow, Womble Shale; Ob, Big Fork Chert; Op, Polk Creek Shale. Dashed line marks walls of quarry; row of X's indicates location of measured section.

buckle and Ouachita strata. Late Paleozoic thrusting brought the Ouachita facies onto the craton into close juxtaposition with the Arbuckle strata in eastern Oklahoma, thus accentuating the differences between the two regions. The transitional zone between the two geologic settings is probably buried beneath the Ouachitas. In spite of the contrasts in lithofacies and biofacies, graptolite biostratigraphy allows the lithostratigraphic units of the Arbuckles and Ouachitas to be accurately correlated not only to each other but also to those of other areas of North America and other continents (Finney, 1985, 1986). As a result, a major subsidence event in the Southern Oklahoma Aulacogen, marked by the contact between the Bromide Formation and Viola Group, has been shown to coincide with a eustatic drop in sea level that is recognizable in the Ouachita geosyncline by the contact between the Womble Formation and Big Fork Chert. These events were also contemporaneous with the emplacement of major allochthons (e.g., Taconic Allochthon and Hamburg Klippe) in the northern Appalachians (Finney and Bergström, 1985).

Locality 1 includes the oldest record of a Middle Ordovician subsidence event in the Southern Oklahoma Aulacogen. Here the upper part of the Bromide Formation and lower half of the Viola Group are well exposed in a long roadcut (Figs. 3, 4). The uppermost Bromide Formation is represented by the Corbin Ranch Submember of the Pooleville Member (Fig. 2). It consists of interbedded birdseye limestone and marl, and represents an intertidal to supratidal facies (Amsden and Sweet, 1983). The Corbin Ranch is sharply and directly overlain by the Viola Group, the basal part of which represents marine deposition of considerable depth (ca. 600 to 1,600 ft; 200 to 500 m). The lithofacies change at the Corbin Ranch–Viola contact might be interpreted as recording a transgressive event due to a eustatic rise in sea level. However, because a contemporaneous eustatic drop in sea level is recorded at many localities in the Ouachitas (see below) and other areas of North America (Finney and Bergström, 1985),

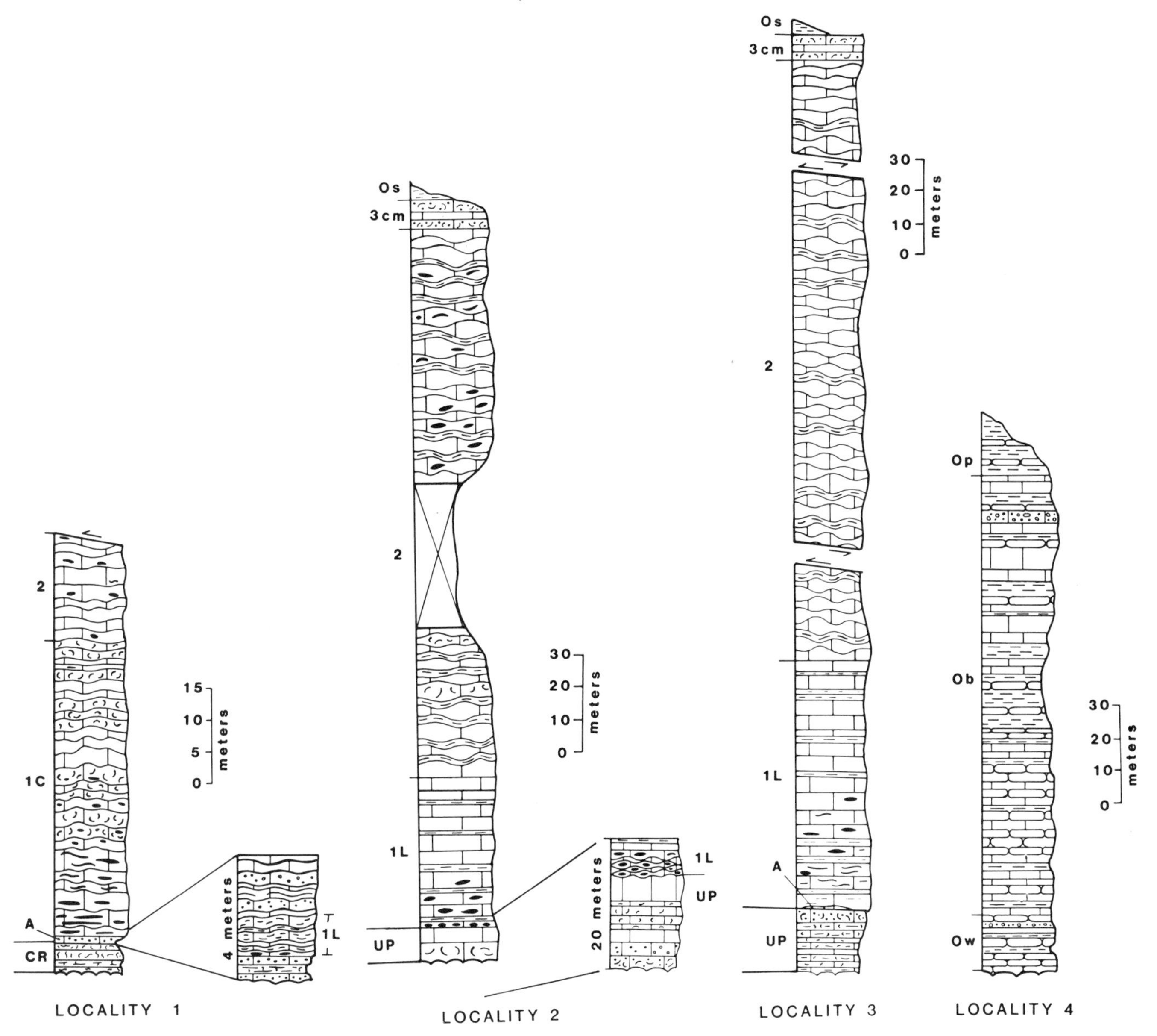

Figure 4. Stratigraphic columns for localities 1-4.
Locality 1: CR refers to Corbin Ranch submember of Pooleville Member of Bromide Formation. A is the basal three beds of Viola; 1C, 1L, and 2 are the lithofacies of Glaser (1965) and Alberstadt (1973).
Locality 2: UP refers to upper part of Pooleville Member; OS refers to Sylvan Shale. 1L and 2 (= Viola Springs Formation) and 3CM (= Welling Formation) are lithofacies of Glaser (1965) and Alberstadt (1973). (After Gentile, 1984.)
Locality 3. Labels as in localities 1 and 2.
Locality 4. Ow is Womble Shale; Ob is the Big Fork Chert; Op is the Polk Creek Shale.

subsidence is favored as a cause of the Corbin Ranch–Viola transgressive, lithofacies sequence.

Longman (1982a,b) demonstrated that the uppermost Bromide is composed of a number of lithofacies; these laterally intergrade with one another and record a variety of depositional environments at progressively greater depths from the northeast platform to the central basin of the Southern Oklahoma aulacogen. The presence of the Corbin Ranch submember indicates that Locality 1 was situated on the northeast platform when the uppermost Bromide Formation was deposited.

The Viola Group was long considered to consist of two formations: the lower Viola Limestone and the upper "Fernvale" Limestone. Glaser (1965) and Alberstadt (1973) recognized three lithofacies within the Viola Group: a lower unit 1, consisting of subunits 1L and 1C in the basin of the aulacogen and on the platform, respectively; a middle unit 2; and an upper unit 3, consisting of subunits 3CM and 3C in the basin and on the platform, respectively. Lithologic and paleontologic differences between units 1L and 1C and between 3CM and 3C are gradational across the Arbuckle Mountains, and reflect differences in the depth and water energy of the environments in which they were deposited. In the old terminology, units 1 and 2 composed the Viola Limestone, and unit 3 was equivalent to the Fernvale Limestone. Recently, the Viola Limestone was renamed the Viola Springs Formation, and the Fernvale was renamed the Welling Formation (Amsden, 1979; Amsden and Sweet, 1983). Sedimentologic analyses indicate that, in ascending order, the Viola Group represents a sequence of environments from anaerobic to dysaerobic to aerobic (Galvin, 1983; Grammer, 1983). This shallowing upward sequence was produced as sedimentation filled in the basin of deposition.

The upper part of the Viola Group has been faulted out at Locality 1 (Fig. 4). Only unit 1 and part of unit 2 are present. Unit 1 is composed in large part of subunit 1C, which is characterized by skeletal calcisiltites and calcarenites. However, its base includes 7 ft (2 m) of siliceous, laminated calcareous mudstones referrable to subunit 1L. These laminated mudstones, representing a deep, possibly anaerobic environment of deposition, do not rest directly upon the Corbin Ranch Submember. They are separated from it by two 2- to 4-m (5- to 10-cm)-thick beds of packstones and grainstones composed of coarse crinoidal skeletal fragments and separated by a 2-in (5-cm)-thick shale bed. These three beds are included in the Viola Group; they undoubtedly represent transitional environments between the supratidal to intertidal environment of the Corbin Ranch and the deep basinal environment of subunit 1L. The relatively rapid change in lithofacies across the Corbin Ranch–Viola contact attests to the rapidity of the subsidence event.

Graptolites are abundant and very well preserved in the lowest 6 ft (2 m) of the Viola Group. They are referrable to the *bicornis* subzone (Finney, 1986) and indicate that the subsidence event initiating deposition of the Viola Group began on the northeast platform at an earlier time than in the basin of the aulacogen situated farther to the southeast (see Localities 2 and 3). When the northeast platform subsided to great depth, the area that had been the basin of the aulacogen was at shallow to moderate depth and still receiving Bromide sediment. Considering that a eustatic drop in sea level was occurring at this time (see Locality 4), the subsidence had to be much greater than the sea-level drop to produce the transgressive lithofacies sequences.

Locality 2 (Figs. 3, 4) displays an excellent, complete section of the Viola Group that records a depositional history typical for the central basin of the aulacogen. Locality 2 also exposes an upper Bromide lithofacies that was deposited on a slope leading into the basin of the aulacogen: further demonstration of the pattern of subsidence in the aulacogen.

The upper 50 ft (15 m) of the Pooleville Member of the Bromide Formation consists of medium to massive beds of fossiliferous and nonfossiliferous, micritic limestones. Longman (1982b) referred them to his diverse fauna biomicrite and limited fauna biomicrite lithofacies, which he interpreted as being deposited on the slope leading from the northeast platform down into the basin of the aulacogen.

The Pooleville-Viola contact is sharp and distinct. The top of the highest bed in the Pooleville is an extensively burrowed biomicrite with abundant, coarse, skeletal fragments of receptaculitids, bryozoans, brachiopods, and cepholopods. The immediately overlying bed is the first in a distinctive 7-ft (2.1-m)-thick interval of wavy-bedded, organic-rich, laminated, cherty, phosphatic, medium-grained, calcarenitic packstones and interbedded graptolitic shales referrable to subunit 1L.

The overlying 140 ft (44 m) of subunit 1L consists of planar bedded, siliceous, laminated, calcareous mudstone with interbeds of shale and bedded or nodular chert. Graptolites are common. Except for a few beds rich in cryptolithid and isotelinid trilobites and an occasional conularid, other macrofossils are extremely rare. The lack of bioturbation and the organic-rich, phosphatic nature of the lower Viola, especially of the lowest 7 ft (2.1 m) attest to the anaerobic environment of deposition.

Unit 2 is 550 ft (167 m) thick, and consists of wavy-bedded, medium- to coarse-grained skeletal wackestones and packstones with thin shale interbeds. Graptolites are less common than in subunit 1L, but other macrofossils are common, including cryptolithid and isotelinid trilobites, orthoconic cephalopods, bryozoans, brachiopods, echinoderms, ostracods, and conularids. Bioturbation in the form of horizontal burrows appears and increases upwards through unit 2, attesting to a dysaerobic to aerobic environment.

Unit 3 (Welling Formation) is only 25 ft (7^{+} m) thick and transitional in lithology between subunits 3C and 3CM. It consists of coarse calcarenites and calcarenitic mudstones rich in skeletal fragments of a variety of shelly fossils. Bioturbation consists of vertical burrows. Unit 3 was deposited in an aerobic environment that was at a shallower depth and of a higher energy than those in which lower parts of the Viola were deposited.

Locality 2 demonstrates both the subsidence event and the subsequent filling of the basin to a moderate water depth. The subsidence event affected a locality that had previously been on a slope at moderate depth between the basin of the aulacogen to the south (Locality 3) and the northeast platform (Locality 1). The abrupt change in lithofacies at the Pooleville-Viola contact attests to the extent and rapidity of the subsidence event. Graptolites from the base of subunit 1L are referrable to the base of the *amplexicaulis* zone (Finney, 1986), which indicates that subsidence at Locality 2 was later than at Locality 1. The subsequent basin filling produced 720 ft (220 m) of compacted sediment and lasted until well into the Upper Ordovician.

Locality 3 (Figs. 3, 4) records changes in lithofacies accompanying the subsidence event in an area that was in the central basin of the aulacogen during Bromide deposition. Much of the lower Viola Group is poorly exposed on a hillside, but the upper Pooleville and the Pooleville-Viola contact are excellently exposed in a recently excavated shallow quarry that extends along the contact.

The upper 35 ft (11 m) of the Pooleville is a biomicrite rich in whole and broken fossils of trilobites, brachiopods, bryozoans, gastropods, cephalopods, and receptaculitids. Beneath this, the Pooleville is composed of evenly bedded couplets of calcareous shale and sparsely fossiliferous micrites. These represent Longman's (1982b) diverse fauna biomicrite and basinal lithofacies and were deposited in the central basin of the aulacogen as it existed at that time.

The Pooleville-Viola contact is exposed at a number of places along the top of the east wall of the quarry. The uppermost bed of the Pooleville is very fossiliferous. It is overlain by a bed of medium- to coarse-grained, crinoidal, skeletal packstone that varies in thickness from 0 to 8 in (20 cm). It, in turn, is directly overlain by siliceous, laminated, medium-bedded, graptolite-rich biomicrites referrable to subunit 1L of the Viola. The abrupt contact again attests to the rapidity of the subsidence event. At this locality, the crinoidal packstone, assigned to the basal Viola, represents facies and environments of deposition transitional between the upper Pooleville, deposited in a basin of moderate depth (see Longman, 1982b), and the lower Viola, deposited in a basin of much greater depth. Rare graptolites from the upper Pooleville are equivalent in age (*bicornis* subzone) to those from the basal Viola at Locality 1 (Finney, 1986), indicating that Pooleville deposition continued at Locality 3 while subsidence and Viola deposition was occurring at Locality 1. Graptolites from the basal Viola are referrable to the *amplexicaulis* zone, demonstrating that the subsidence event was contemporaneous at Localities 2 and 3.

The Viola Group is more than 500 ft (150 m) thick at Locality 3. The upper two-thirds are well exposed in another quarry to the east of the one exposing the Pooleville-Viola contact. Unit 2 and subunit 3CM are well exposed, but the section is broken by many faults.

The similarities and differences in the Viola Group between Localities 1, 2, and 3 indicate that, during the Middle Ordovician, a basin subsided initially in the region of Locality 1; then, a short time later, a possibly separate and adjacent basin subsided to an even greater depth in the region of Localities 2 and 3. Before subsidence, Locality 1 was situated on a supratidal to intertidal platform adjacent to the aulacogen. Locality 2 was on a south-facing slope leading into the basin of the aulacogen, and Locality 3 was in the central basin of the aulacogen but at a much shallower depth than after the subsidence event.

Locality 4 (Figs. 3 and 4) provides a complete exposure of the upper Womble Shale and Big Fork Chert, which represent Middle Ordovician deposition in the Ouachita Geosyncline. The best area for studying the section is high on the north wall of the quarry. The Womble–Big Fork contact extends along the west wall of the quarry, and the Big Fork–Polk Creek contact extends along the east wall. Only the upper few meters of the Womble and lower few meters of the Polk Creek Shale are exposed in the north quarry wall. The Big Fork is completely exposed for its full thickness of 470 ft (143 m).

Graptolite biostratigraphy indicates that the Womble–Big Fork contact is near the boundary between the *bicornis* subzone and the *amplexicaulis* zone (i.e., it correlates with the Bromide-Viola contact in the Arbuckles), and the Big Fork–Polk Creek contact correlates closely to the contact between the Viola Group and Sylvan Shale in the Arbuckles (Fig. 2).

The uppermost Womble Shale, which correlates with the upper Bromide and the lower Viola at Locality 1 and with the upper Bromide at Localities 2 and 3, is very different in both lithofacies and biofacies from correlative Arbuckle strata. It is composed of siliceous shales and bedded cherts. It includes several thin beds of limestone conglomerate 3.3 to 6.6 ft (1 to 2 m) below its top. Graptolites are extremely abundant throughout the formation; other fossils are rare.

The Big Fork Chert, which is largely correlative with and sharply differs from the Viola Group, is composed of relatively uniformly interbedded limestones, siliceous shales, and bedded cherts. Siliceous shales and bedded chert characterize the entire section from the upper Womble to the lower Polk Creek. The exact upper and lower boundaries of the Big Fork were not designated when the formation was first defined and described in the core of the Ouachitas (Purdue, 1909; Miser and Purdue, 1929), nor when it was studied at Black Knob Ridge (Hendricks and others, and 1937). Because the presence of limestones characterizes the interval referred to as the Big Fork Chert, the lower and upper boundaries are placed at the lowest and highest limestone beds in the section, respectively. The limestones are graded and medium bedded, fine- to coarse-grained, skeletal calcarenites that comprise at least 50 percent of the section. Skeletal fragments include pelmatazoans and brachiopods.

Three medium-to-thick beds of limestone conglomerate occur in the upper Big Fork. Clasts are angular, graded, and self-supporting. A variety of clasts indicative of an extra-basinal, shallower water environment include: micrites and biomicrites similar to lithologies found in the Bromide Formation and Viola Group, biosparites, skeletal fragments of brachiopods, trilobites, bryozoans, fine-grained clastics, and quartz grains. The graded calcarenites and limestones conglomerates occur in single beds that sharply overlie thin beds of siliceous shale and are overlain gradationally by bedded chert. The bases of the limestone conglomerate beds often show features of load deformation. Radiolarians and sponge spicules occur in all lithologies, especially in the cherts, and may have been the source of the siliceous sediment. Complete graptolite specimens and chitinozoans are common and well preserved in the limestones and are common in the shales. Except as skeletal debris in the calcarenites and conglomerates, other fossils are rare. Evidence of bioturbation is absent.

The lithofacies and biofacies of the Middle Ordovician strata in the Ouachitas are markedly different from those in the Arbuckles, attesting to the marked differences in paleogeographic and depositional setting. However, they are very similar to those of the Middle Ordovician sequence in the Marathon region of West Texas, which also was deposited in the Ouachita Geosyncline (Berry, 1960; Flawn and others, 1961). In Texas, the Woods Hollow Shale is overlain by the Maravillas Chert. Limestone conglomerates, graded calcarenites, and bedded chert characterize the Maravillas. McBride (1969, 1970) has interpreted the conglomerates and grade calcarenites as representing turbidity currents and debris flows that carried shallow-water clasts into the deep Ouachita basin. A similar interpretation can be made for the limestones in the Big Fork at Black Knob Ridge. The conglomerates are much more abundant, thicker, and coarser grained in the Marathon region than at Black Knob Ridge suggesting that the Marathon region represents a site closer to the edge of the craton than the Black Knob Ridge locality.

An unconformity may exist between the Woods Hollow and Maravillas in the Marathon region (Bergström, 1978). Graptolite biostratigraphy indicates that no such hiatus is present between the Womble and Big Fork and that the Womble–Big Fork contact correlates with a level represented by the possible break in the Marathon sequence (Finney, 1986). For this reason, the similar changes in lithofacies at the base of the Big Fork and Maravillas are considered to represent a similar geologic event contemporaneous in the two areas. Finney (1986; Finney and Bertström, 1985) and Bergström (1978) favor a eustatic drop in sea level. The correlation of the Womble–Big Fork contact with the Bromide-Viola contact indicates that the sea-level drop was contemporaneous with paleogeographic changes in the Arbuckle strata that are best interpreted as being a rapid subsidence event. It is interesting to note that the eustatic and subsidence events were closely contemporaneous with the emplacement of the Taconic Allochthon and Hamburg Klippe in the northern Appalachians.

REFERENCES CITED

Alberstadt, L. P., 1973, Articulate brachiopods of the Viola Formation (Ordovician) in the Arbuckle Mountains, Oklahoma: Oklahoma Geological Survey Bulletin 117, 90 p.

Amsden, T. W., 1979, Welling Formation, new name for Upper Ordovician unit in eastern Oklahoma (formerly called "Fernvale"): American Association of Petroleum Geologists Bulletin 63, p. 1135–1138.

Amsden, T. W., and Sweet, W. C., 1983, Upper Bromide Formation and Viola Group (Middle and Upper Ordovician) in eastern Oklahoma: Oklahoma Geological Survey Bulletin 132, 76 p.

Bergström, S. M., 1978, Middle and Upper Ordovician conodont and graptolite biostratigraphy of the Marathon, Texas graptolite zone reference standard: Palaeontology, v. 21, pt. 4, p. 723–758.

Berry, W.B.N., 1960, Graptolite Faunas of the Marathon Region, West Texas: University of Texas Publication 6005, 179 p.

Finney, S. C., 1985, A re-evaluation of the upper Middle Ordovician graptolite zonation of North America: Geological Society of America Abstracts with Programs, v. 17, no. 1, p. 19.

——, 1986, Correlation of graptolite biofacies, subsidence, and eustatic events in Ordovician of Arbuckle and Ouachita Mountains, Oklahoma and Arkansas: Palaios 1 (in press).

Finney, S. C., and Bergström, S. M., 1985, Regional contemporaneity of eustatic, subsidence, and tectonic events in the Middle-Upper Ordovician of the Appalachian and Ouachita orogens and the Southern Oklahoma aulacogen: Geological Society of America Abstracts with Program, v. 17, no. 7, p. 583.

Flawn, P. T., Goldstein, A., Jr., King, P. B., and Weaver, C. E., 1961, The Ouachita System: University of Texas Publication 6120, 401 p.

Galvin, P., 1983, Deep-to-shallow carbonate ramp transition in Viola Limestone (Ordovician) southwest Arbuckle Mountains, Oklahoma: Bulletin of American Association of Petroleum Geologists, v. 67, no. 3, p. 466–467.

Gentile, L. F., 1984, Sedimentology and graptolite biostratigraphy of the Viola Group (Ordovician), Arbuckle Mountains and Criner Hills, Oklahoma [M.S. thesis]: Stillwater, Oklahoma State University, 104 p.

Glaser, G. C., 1965, Lithostratigraphy and Carbonate Petrology of the Viola Group (Ordovician), Arbuckle Mountains, South-central Oklahoma [Ph.D. thesis]: Stillwater, University of Oklahoma, 197 p.

Grammer, G., 1983, Depositional History and Diagenesis of the Viola Limestone (Ordovician), Southeastern Arbuckle Mountains, Oklahoma [M.S. thesis]: Dallas, Texas, Southern Methodist University, 156 p.

Hendricks, T. A., Knechtel, M. M., and Bridge, J., 1937, Geology of Black Knob Ridge, Oklahoma: Bulletin of American Association of Petroleum Geologists, v. 21, p. 1–29.

Longman, M. W., 1982a, Depositional setting and regional characteristics, *in* Sprinkle, J., ed., Echinoderm Faunas from the Bromide Formation (Middle Ordovician) of Oklahoma: Lawrence, Kansas, University of Kansas Paleontological Contributions, Monograph 1, p. 6–10.

——, 1982b, Depositional environments, *in* Sprinkle, J., ed., Echinoderm Faunas from the Bromide Formation (Middle Ordovician) of Oklahoma: Lawrence, Kansas, University of Kansas Paleontological Contributions, Monograph 1, p. 17–30.

McBride, E. F., 1969, Stratigraphy and Sedimentology of the Woods Hollow Formation (Middle Ordovician), Trans-Pecos, Texas: Geological Society of America Bulletin, v. 80, p. 2287–2302.

——, 1970, Stratigraphy and origin of Maravillas Formation (Upper Ordovician), West Texas: Bulletin of American Association of Petroleum Geologists, v. 54, no. 9, p. 1719–1745.

Miser, H. D., and Purdue, A. H., 1929, Geology of the DeQueen and Caddo Gap Quadrangles, Arkansas: U.S. Geological Survey Professional Bulletin 808, 195 p.

Purdue, A. H., 1909, The Slates of Arkansas: Arkansas Geological Survey, p. 1–95.

Structural styles in the Arbuckle Mountains, southern Oklahoma

J. Bryan Tapp, Department of Geosciences, The University of Tulsa, Tulsa, Oklahoma 74104

LOCATION

This chapter describes structural features of outcrops in the Arbuckle Mountains, southern Oklahoma, along I-35 and near Arbuckle Lake. The stops are arranged as though the reader is driving south from Oklahoma City toward Dallas. For a description of stratigraphic stops along this same route, and for a generalized location map and cross sections, the reader is referred to the preceding chapter by Fay. The body of the discussion of outcrops evaluates evidence that the structures and sediments seen in the Arbuckle Mountains formed in an aulacogen that has undergone left-lateral-wrench deformation. All of the outcrops along I-35 are public domain. Parking along the interstate is tolerated as long as vehicles are pulled well off of the roadway and groups or individuals do not cross the highway. If a large group is to examine the exposures, it would be best to park in the scenic turnouts and walk along the highway shoulders. The outcrop described near Arbuckle Lake is in a county-owned quarry that is usually inactive on weekends.

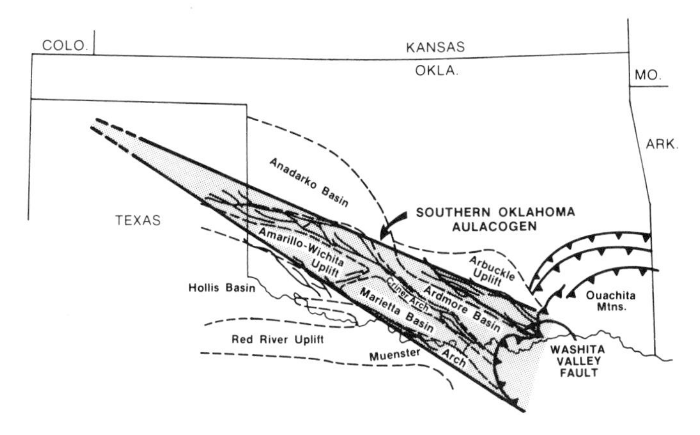

Figure 1. Generalized map of region containing the Southern Oklahoma Aulacogen and part of the Ouachita system. The map shows the relationship of the Arbuckle-Wichita trend to the Ouachita foldbelt. Rocks in the Ouachita system are deep marine clastics and chert, and represent an ancient continental margin that formed in Cambrian, and was deformed by closure (continent-continent collision) in Pennsylvanian time. The Arbuckle-Wichita trend represents an aulacogen that rifted in Cambrian and was deformed in Pennsylvanian time. The rocks in the aulacogen are predominantly shallow marine carbonates with minor terrestrial clastics and conglomerates.

SIGNIFICANCE

The Arbuckle Mountains preserve one of the thickest sections of Paleozoic sediments in the central United States. The structural style and stratigraphy of the Arbuckles are the expression of one of the best preserved and best exposed Paleozoic aulacogens in the world. The Arbuckle-Wichita trend formed during the Wilson cycle that formed the Ouachita Mountains. The style of deformation in the trend is associated with a left-lateral wrench. This text discusses the features associated with aulacogens and strike-slip faults as seen in the Arbuckle Mountains.

GEOLOGY OF THE ARBUCKLE MOUNTAINS

Paleozoic Stratigraphy

The Arbuckle-Wichita trend consists of paired structural basins and uplifts along a NW–SE trend that intersects the Ouachita Mountain disturbed belt at nearly a right angle (Fig. 1). The sediments in the Arbuckle-Wichita trend include shallow-water carbonates with interspersed marine carbonate clastics, terrestrial clastics, and conglomerates. Thicknesses of formations within the Arbuckle-Wichita Trend are greater than thicknesses of equivalent formations north or south of the trend. Rocks exposed in the Arbuckle Mountain region include Precambrian granitic basement to Cretaceous carbonates and clastics (refer to Fay, this volume, for a listing of Paleozoic strata exposed on the Arbuckle uplift).

Kay (1951) classified the Arbuckle-Wichita trend as a zeugeosyncline based on the type and extent of Pennsylvanian conglomerates. Ham and others (1964) stated that the Arbuckle-Wichita trend evolved through three separate geosynclinal stages to reach its final configuration. These stages included an early taphrogeosynclinal stage followed by a miogeosynclinal stage, and culminated by a zeugeosynclinal stage. Shatskiy (1946) described the Dnieper-Donets basin of the Russian platform and the Arbuckle-Wichita system as type aulacogens; long lived, fault bounded troughs in an otherwise stable platform that intersect a mountain belt or a continental margin at a high angle, and contain an abnormally thick section of terrestrial and marine sediments. The tectonic significance of the Southern Oklahoma Aulacogen was first recognized by Hoffman and others (1974).

Rocks in the Arbuckle region were deposited in a trough that formed in early Cambrian and persisted through late Pennsylvanian time (Ham and others, 1964). Early Cambrian rocks include 7,000 ft (2,133 m) of volcanic and volcaniclastic rocks (Colbert rhyolite porphyry). These are overlain by the Reagan sandstone that grades from terrestrial to marine to the east (Ham, 1969). The Colbert porphyry and the Reagan sandstone represent the rifting phase of the aulacogen. The Reagan sandstone is overlain by 8,800 ft (2,440 m) of late Cambrian through Ordovician

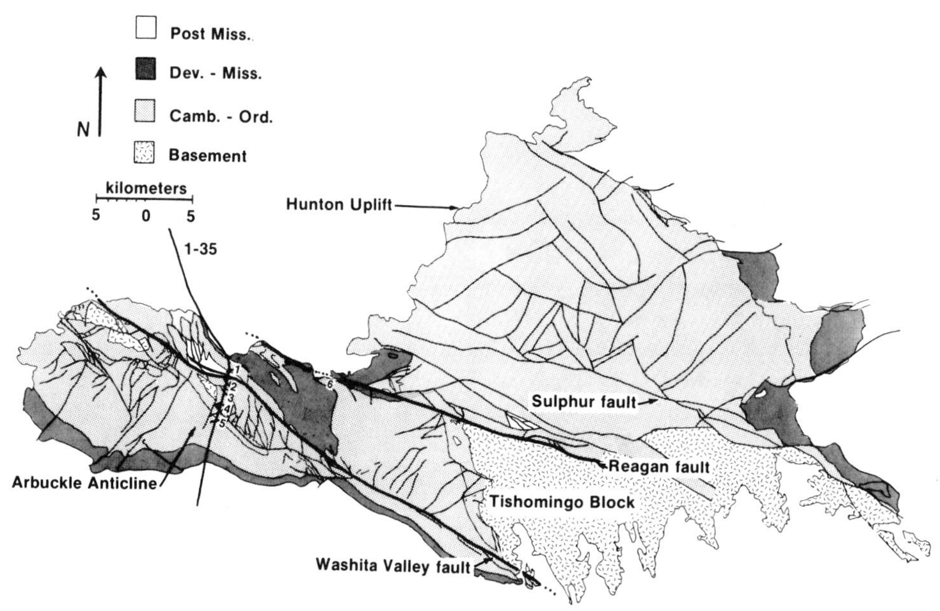

Figure 2. Generalized geological map of the Arbuckle region. The stops described in this paper are shown as well. The map shows the major structural elements in the Arbuckle region.

carbonates of the Arbuckle Group. These carbonates are shallow water limestones and dolostones. The Arbuckle Group was deposited in a deepening trough where sedimentation rates paced subsidence rates. The subsidence phase of the aulacogen is preserved in the Arbuckle Group carbonates and the Devonian through Mississippian sediments. The sediments above the Arbuckle Group consist of Ordovician through Mississippian clastics and limestones that accumulated to a thickness of 5,500 ft (1,680 m). Clastics are dominant in Devonian and Mississippian rocks.

Subsidence of the aulacogen was initially rapid, and may have been accommodated by motion along faults (Feinstein, 1981). Subsidence slowed after Cool Creek time, possibly due to locking of faults (Feinstein, 1981). Thicknesses of Cambrian and Ordovician strata in the aulacogen are significantly greater than thickness of equivalent strata north and south of the aulacogen. The Washita Valley fault seems to mark the northern boundary of the aulacogen; sediments are thinner north of the fault (Ham, 1969).

The deformation phase of the aulacogen is preserved by Pennsylvanian conglomerates and clastics in the Arbuckle region and Ardmore basin. Approximately 13,000 ft (4,000 m) of Pennsylvanian clastics and conglomerates accumulated in the Ardmore basin. These sediments were derived from local uplifts, and many of the Pennsylvanian conglomerates are restricted to single basins (Ham, 1969).

Surface Structures in the Arbuckle Mountains: The Wrench Model

Wrench faults are large-scale, strike-slip faults that are generally nearly vertical. Offset on these faults involves all rock units, including basement. Interpretation of the surface structures in the Arbuckle Mountains has been problematic. The structures present include small thrust complexes in upper Ordovician strata, wrench related close to tight, upright-plunging symmetric folds, high angle normal and reverse faults, and strike-slip faults in middle and lower Ordovician strata (Fig. 2). Deformation in the region has been related to wrench tectonics (Wickham and Denison, 1978). Recent seismic and drilling data suggest that the Arbuckle Uplift may be allochthonous. The presence of thrusting in wrench systems has been documented in the vicinity of restraining bends, and in flower structures (Harding, 1985). As a result, industrial geologists and geophysicists have interpreted the Arbuckle feature as a flower structure.

The Arbuckle uplift displays a range of deformation intensity in a region of roughly 1,000 mi^2 (2,600 km^2). There are three prominent regions of differing deformation, each separated by high angle faults that strike approximately N60°E (Fig. 2). These areas are the Hunton arch, Tishomingo block, both on the platform, and Arbuckle anticline in the aulacogen.

The Hunton arch consists of gently deformed lower Paleozoic carbonates, separated from the Tishomingo block by the Sulphur fault, Mill Creek fault, Mill Creek syncline, and Reagan fault. Studies in the Mill Creek syncline, and along the Reagan and Sulphur faults have shown that the style of deformation in this area is consistent with a wrench fault model (Luke, 1975; Haas, 1981; Islam and Nielsen, 1984). The Hunton arch has been interpreted as the oldest structural feature in the Arbuckles, resulting from epeirogenic uplift during Desmoinian and Missourian (Ham, 1969); however, it seems more likely that the lack of intense deformation may be related to its position relative to the aulacogen.

The Tishomingo block, bounded on the north by the Reagan fault and on the south by the Washita Valley fault, has been interpreted as a broad, open fold with a N-S axis. Detailed structural analysis of the Reagan and Washita Valley faults has shown that both are wrench faults (Booth, 1981; Haas, 1981). Luke (1975) analyzed fractures and faults associated with the Reagan fault, and found that the fracture sets fell into the expected synthetic and antithetic groupings of a wrench fault. Later work by Haas (1981) on the Reagan fault verified this, and also showed that the dominant slickenside orientation was horizontal.

Deformation in the Tishomingo block increases up-section from the Arbuckle Group into the Hunton anticline quarry, White Mound structures, Dougherty anticline, Vines dome, and Southwest Davis structures. These features are most likely related to wrench style deformation (Wickham and Dennison, 1978; Wiltse, 1978; Booth, 1981; Haas, 1981). The change in deformation intensity may be related to competency contrasts of the deforming medium or to a restraining bend in the Washita Valley fault (Wiltse, 1978).

The Arbuckle anticline is bounded on the north by the Washita Valley fault, and on the south by the Ardmore basin. The Arbuckle anticline and its associated minor folds have been interpreted as forming in two events that first formed the main anticline and its secondary folds in middle Virgilian, and the Dougherty anticline in late Virgilian (Ham, 1969). The Arbuckle anticline is within the range of axial orientations proposed for a primary fold in a left-lateral wrench system.

Deformation in the Arbuckle Mountain region occurred in late Pennsylvanian (Ham, 1969); although minor deformation may have occurred in Devonian as is shown by minor unconformities in the Hunton Group, and the presence of a local conglomerate at the base of the Woodford on US 77. The structural style seen in the Arbuckle Mountains is consistent with a wrench fault. The structural evidence gathered so far supports the interpretation that the dominant deformation style in the Arbuckle region is horizontal left-lateral wrench. Offset on the Washita Valley fault has been suggested to be from 40 mi (64 km), based on offsets in the basal Oil Creek and McLish sands (Tanner, 1967), to 20 mi (32 km), based on offset of a Hunton thin (Carter, 1979).

Deformation in the vicinity of the Washita Valley fault in the Arbuckle Group is limited to a narrow zone of shear, folding, and fractures at the edge of the open Arbuckle and Sycamore Creek anticlines, while deformation in the overlying Simpson and younger rocks is in a broader zone of folding and thrusting north of the fault. This change may be due to a widening of the fault zone near the syntectonic erosion surface, a change in material competency up section, a restraining bend, or a combination of these factors. The older strata of the Arbuckle Group are more uniform and more competent as a unit than the overlying interlayered sands, shales, and carbonates of the Simpson and younger groups. Deformation in the more competent units may be taken up by more gentle flexure overall, along with deformation in narrow shear zones near the fault, while the Simpson and younger rocks can fold and thrust more easily due to the interlayered shales and sands.

DESCRIPTION OF SELECTED OUTCROPS

Stop 1. Ordovician Viola Formation and Bromide Formation, Pennsylvanian Collings Ranch conglomerate. S½Sec.30 to N½Sec.31,T.1S.,R.2E., along I-35, Turner Falls 7½-minute Quadrangle. Milepost 50. West side of interstate, southbound lane.

This roadcut displays a variety of features. The outcrop is on the northern limb of the Arbuckle anticline. Bedding in the Bromide Formation is overturned, dipping approximately 85 degrees to the south. The Viola Formation shows a wide range of structural dips. Traversing the Viola Formation shows several folds and faults (Fig. 3a). Some of the faults are curved, being nearly vertical at the roadbed level, and becoming nearly horizontal toward the top of the outcrop. These curved faults may be similar to the antithetic faults that turn into thrusts suggested by Wiltse (1978) in the Southwest Davis area. Beds under the nose of the syncline (Fig. 3a) are nearly vertical, showing a folded fault separating the layers comprising the fold and the vertical layers beneath the fold. The presence of compressional deformation is consistent with a wrench system in the vicinity of restraining bends or en echelon faults (Crowell, 1974). Folding and faulting are seen in the Viola on both large and small scale. Fault gouge can be seen in the northern part of the outcrop, as well as extensile rotation of small scale blocks to the north of the fault.

Deformation in this outcrop is accommodated by a mixture of brittle failure, bedding plane slip, and minor pressure solution. Folds in the outcrop are gentle to tight, with vertical to inclined fold hinge surfaces. Folding was accomplished primarily by slip along bedding plane surfaces. Bedding plane slip alone, however, was not the only mechanism responsible for the formation of the folds seen. Some fold hinges exposed in the outcrop are thickened by small overlapping thrusts, and are more fractured than the limbs (Fig. 3b). Tectonic pressure solution surfaces that are subparallel with the hinge surface of the host fold can be found in some hinges. All of the folds in these outcrops were formed in the brittle field of rock deformation. There is little indication of plastic deformation (deformed fossils, flattened grains, etc.) other than twinning of some calcite spar.

To the south, there is a conformable contact between the Viola and the Poolville member of the Bromide Formation. The angular unconformity between the Pennsylvanian Collings Ranch conglomerate and the Ordovician Bromide and Viola Formations is exposed at this locality. The normal fault near the south end of the outcrop is a splay of the Washita Valley fault. The Collings Ranch is limited to a half graben interpreted as a pull-apart basin (Booth, 1981), located in a releasing bend of the Washita Valley fault (Fig. 2). The Collings Ranch conglomerate is composed of cobbles and boulders sourced primarily from the Arbuckle Group. The Collings Ranch conglomerate has been described by Ham (1954, 1969) and Dunham (1955) as a fanglomerate, and may be as old as Desmoinesian (Wickham and Dennison, 1978). The depositional sequence of the Collings Ranch has not been worked out in detail.

The Collings Ranch crops out from here to the Scenic Turn-

Figure 3. a) West wall of roadcut at Stop 1. A large syncline is seen in the exposure, which is in the Voila Formation (note graduate student for scale). Several faults, some nearly vertical at road level, which flatten near the top of the roadcut, are also seen. The large syncline rests directly on top of vertical beds of the Viola, suggesting that the syncline has been thrusted into its present position, and that the thrust fault has been folded. b) Closeup of a small fold hinge collected from the upper Viola at the northern end of the roadcut. The fold hinge is fractured and contains overlapping thrusts that may have moved sequentially to accommodate folding. These folds are similar to folds found in the Hunton Formation at Stop 6. The orientation marker was horizontal in the field, and is 1.18 in (3 cm) long. d) Fault related deformation in the Kindblade Formation. The photograph shows a small scale flower structure. The features are the result of motion of a minor fault. Several folds can be seen which are intimately related to faults. The faults vary in orientation from nearly vertical to nearly horizontal. The rocks are fractured, and contain abundant pressure solution surfaces. The hammer in the center of the photograph is 10 in (25 cm) long. e) Flexural slip folds in the McKenzie Hill Formation (note graduate student for scale). The photograph is of the west wall of the roadcut. The folding seen may be related to motion on the Chapman Ranch fault. The style and extent of folding are dependent on bedding thickness and the presence of shales in this portion of the McKenzie Hill. A small flower structure (?) can be seen across the Interstate in the Royer Formation. f) The photograph shows a channel sand in the Cool Creek Formation directly overlying a breccia horizon. The outcrop is east of the Interstate, between the roadway and the first clump of trees as you look toward the single (southernmost) radio tower. The lens cap is 2.16 in (54 mm) in diameter.

out, approximately 0.25 mi (402 m) to the south. A walk from this outcrop to the next affords a look at a large open syncline that was effected by later faulting. The deformation of the Collings Ranch is minor, producing an open syncline; the trend of the syncline in the Collings Ranch is close to that of the Washita Valley fault. At milepost 50, undeformed Collings Ranch can be seen unconformably overlying deformed Viola, indicating that the Viola was deformed prior to the deposition of the Collings Ranch. Folding and faulting of the Collings Ranch, therefore, must have been produced by a later event.

Stop 2. Kindblade megabreccia. NE¼,NE¼,SW¼,Sec.31, T.1S.,R.2E., first Scenic Turnout on I-35, southbound. Turner Falls 7½-minute Quadrangle. Milepost 49.7. Approximately 0.25 mi (400 m) south of Stop 1.

The chaotic blocks of Kindblade, seen in the exposures at the southern end of the scenic turnout, are within what has been termed the I-35 megabreccia. The I-35 megabreccia is limited to the immediate area of the roadcuts, extending no more than 300 ft (90 m) to the west and 1,000 ft (300 m) to the east. The unit is approximately 300 ft (90 m) wide at the widest point. The megabreccia can be seen on both the eastern and western wall of the cut. The unit can also be seen on the eastern side of the I-35 southbound lane roadcut and on the western and eastern walls of the northbound lanes of I-35.

This unit has been interpreted as a tectonic breccia resulting from overthrusting (Fay, 1969), as a solution collapse feature along closely spaced tension joints (Tapp, 1978), and as a collapse feature at the unsupported trailing edge of a gravity-glide thrust (Phillips, 1983). Clasts in the unit have a random orientation, based on bedding orientation measurements within the megabreccia (Tapp, 1978). Clays (consisting of kaolinite, illite, and minor montmorillonite) in the unit surrounding the limestone blocks are identical to clays within the limestone blocks and were probably derived directly from the Kindblade. Travertine has not been identified in any samples from the unit, and the clay material surrounding the blocks is devoid of palynomorphs. This feature is distinctly different from sink holes present in the Royer Formation to the south, and within Ordovician units on the Hunton Uplift. Rocks surrounding the megabreccia are fractured, with the orientation of the primary fractures consistent with extension in the vicinity of a releasing bend. There are no documented thrust faults present within the region of the megabreccia.

Stop 3. Small scale fault structures in Kindblade Formation. SW¼SE¼SW¼Sec.31,T.2S.,R.2E. Along I-35 in the southbound lane between the Scenic Turnout and Turner Falls exit. Turner Falls 7½-minute Quadrangle. The exposure is in the west wall. Approximately 650 ft (200 m) south of Stop 2.

This exposure (Fig. 3c) shows fault related folding in the Kindblade Formation. In this outcrop, several sub-horizontal to horizontal faults branching from a nearly vertical fault can be seen. The rocks in the immediate vicinity of the faults have been tightly folded and fractured. Rocks north and south of the feature are not deformed. The feature is a small-scale example of a flower structure. The narrowness of the deformation zone is undoubtedly related to the relatively small displacement on the fault. It may, however, also be a function of the material as well as the depth of burial during deformation. At Stop 1, the Viola Formation was severely fractured and contained several faults and folds. The primary deformation mechanism was brittle failure of the rock material. At this stop, the Kindblade Formation is fractured, but there is a much greater degree of deformation by pressure solution. The outcrop contains several tectonic stylolites that are sub-parallel to parallel with local fold hinges. The structural differences between the outcrops at Stop 1 and this outcrop may be related to depth of burial (deformation at shallower levels will be more brittle and to the relative resistance to deformation of the two units.

The Kindblade Formation is in the middle of the Arbuckle Group, a massive carbonate that should deform as a single, thick unit. As such, it was difficult to deform the entire group into tight folds. In addition to the thickness and material properties of the surrounding units, the Kindblade was at significantly greater depths during deformation. Consequently, the material response was controlled more by pressure solution and possibly by plastic mechanisms.

The Viola Formation is a thinly bedded carbonate surrounded by sandstones and shales. As a result, the Viola deformed as a multilayer unit embedded in a more ductile medium, folding into tighter folds involving the entire unit. The perceived intensity of deformation in the Arbuckle Mountains increases upsection in the thinner limestone units that are surrounded by sandstones and shales. The deeper units are folded into open folds, with only narrow zones of deformation by a more plastic material response adjacent to the fault. The upper units were folded and faulted by more brittle processes into a wider zone of deformation.

Stop 4. Small scale folds in the McKenzie Hill Formation. c SW½Sec.6,T.2S,.R.2E. along I-35 in the southbound lane 330 ft (100 m) before the Turner Falls exit. Turner Falls 7½-minute Quadrangle. The exposure is in the west wall.

There are several small scale folds in the McKenzie Hill Formation at this locality (Fig. 3d). These folds are flexural-slip folds. The dominant mechanisms of folding is bedding-plane slip and brittle failure. Pressure solution surfaces have been found in samples from this locality. The folds show signs of thrust stacking in the hinge of the folds. The folding here is probably related to the proximity to the Chapman Ranch fault to the south. Fay (1969) interpreted the Chapman Ranch fault as a bedding plane thrust that became steeper with depth. Recent work by John Bohon (personal communication) has shown that the Chapman Ranch fault is vertical at the surface, and may be part of a flower structure.

Stop 5. Bedded breccias in the Cool Creek Formation. E½E½E½SE¼Sec.13,T.2S.,R.1E. along I-35. Turner Falls 7½-minute Quadrangle. The outcrops are along the west wall of the southbound lane roadcut between the exit and entrance for the Scenic Turnout.

The outcrop along the roadcut shows a series of bedded breccia deposits in the Cool Creek Formation. The breccias are composed dominantly of chert fragments with limestone fragments as the secondary constituent. The breccias contain some quartz and rare feldspar, ooids and pellets, and are typically overlain by sand stringers (Fig. 3e). The upper and lower contacts are sedimentary. Bedded breccias in the Cool Creek are continuous around the Arbuckle anticline. Locally the clast size in breccia units reaches boulder size. In some areas, the breccias pinch out into a sandy, cherty limestone. The breccias exposed in this roadcut have been interpreted as shoaling upward carbonate cycles reflecting cyclic changes in water depth controlled by motion on the Washita Valley fault (Smith, 1981). A separate breccia to the south of the roadcut was described by Tapp (1978) who interpreted the breccia as a clastic carbonate deposit derived from uplift of source on an active fault during Cool Creek deposition.

The presence of the breccias in the Cool Creek Formation coincides with the most rapid subsidence in the aulacogen Feinstein (1981). Sedimentation of the Cool Creek breccias may be related to cyclic subsidence of the aulacogen along bounding faults. Cycles are also present in the Kindblade Formation (Tenney, 1984); however, the cycles are not as well defined. The change in cycles from the Cool Creek to the Kindblade is coincident with a decrease in subsidence rate in the aulacogen during Kindblade time. As proposed, a typical cycle in the Cook Creek involves uplift of a source by elastic rebound of the upthrown side of the fault, followed by erosion and transportation of debris into the aulacogen, where shallow water carbonate deposition was ongoing. Downwarping of the aulacogen may have been cyclic, with regional subsidence interrupted by localized faulting and rebound when stress on the bounding fault exceeded the strength of the material.

Stop 6. Hunton anticline quarry. cSec.31,T.1S.,R.3E., Dougherty 7½-minute Quadrangle. From Stop 5, drive south to the Springer exit. At the Springer exit, turn back north on I-35, and drive to the Davis (Oklahoma 77) exit. Turn toward the east, and drive toward Davis (northeast). At the intersection of Oklahoma 7, turn east and drive through Davis. Approximately 0.75 mi (1.2 km) east of Davis, turn south on Oklahoma 110. Continue south on Oklahoma 110 for approximately 3 mi (4.8 km). The road will split off to the south and east. Turn to the east and drive for approximately 3 mi (4.8 km) until the road splits to the east or south. Turn to the south and drive approximately 0.25 mi (400 m). There, a side road crosses Rock Creek and continues on the south side of Arbuckle Lake. The quarry is approximately 0.5 mi (800 m) east along this road.

The limestones in the quarry wall are in the Siluro-Devonian Hunton Group. The upper quarry wall is in the Bois d'Arc Formation. The lower bench is in the Henryhouse-Haragan formations. The quarry, which was opened for road chat and limestone rip rap, exposes an example of the effect of bedding thickness and material competence on the folding process. The folds in the Woodford (less competent, thinner bedded) are tight, with narrow, closely spaced hinges. (Note: The outcrop of Woodford has been removed for road material, refer to Fig. 3f.) The upper fold in the Bois d' Arc is open with a broad hinge, and has been cut by a minor thrust. Folds in the Henryhouse-Haragan are flexural-slip, where the fold hinge zone migrates over the outcrop from the base to the top. The hinge zone of folds in the Henryhouse-Haragan is formed by intersecting small scale thrusts.

REFERENCES

Booth, S. L., 1981, Structural analysis of portions of the Washita Valley fault zone, Arbuckle Mountains, Oklahoma: Shale Shaker, v. 31, p. 107–120.

Carter, D. W., 1979, A study of strike-slip movement along the Washita Valley Fault, Arbuckle Mountains, Oklahoma: Shale Shaker, v. 30, p. 80–108.

Crowell, J. C., 1974, Sedimentation along the San Andreas Fault, California, *in* Dott, R. H., and Shaver, R. H., eds., Modern and Ancient Geosynclinal Sedimentation: Society of Economic Paleontologists and Mineralogists Special Publication 19, p. 292–303.

Dunham, R. J., 1955, Pennsylvanian conglomerates, structure, and orogenic history of Lake Classen Area, Arbuckle Mountains, Oklahoma: American Association of Petroleum Geologists Bulletin, v. 39, no. 1, p. 1–30.

Fay, R. O., 1969, Geology of the Arbuckle Mountains along I-35: Ardmore Geological Society.

Feinstein, S., 1981, Subsidence and thermal history of the Southern Oklahoma Aulacogen; Implications for petroleum exploration: American Association of Petroleum Geologists Bulletin, v. 65, p. 2521–2533.

Haas, E. A., 1981, Structural analysis of a portion of the Reagan fault zone Murray County, Oklahoma: Shale Shaker, v. 31, p. 93–104.

Ham, W. E., 1954, Collings Ranch Conglomerate, Late Pennsylvanian in the Arbuckle Mountains, Oklahoma: American Association of Petroleum Geologists Bulletin, v. 38, p. 2035–2045.

—— , 1969, Regional geology of the Arbuckle Mountain region: Oklahoma Geological Survey Guide Book 17, 52 p.

Ham, W. E., Dennison, R. E., and Merritt, C. A., 1964, Basement rocks and structural evolution of southern Oklahoma: Oklahoma Geological Survey Bulletin 95, 302 p.

Harding, T. P., 1985, Seismic characteristics and identification of negative flower structures, positive flower structures, and positive structural inversion: American Association of Petroleum Geologists Bulletin, v. 69, p. 582–600.

Hoffman, P., Dewey, J. F., and Burke, K., 1974, Aulacogens and their genetic relation to geosynclines, with a Proterozoic example from Great Slave Lake, Canada, *in* Dott, R. H., and Shaver, R. H., eds., Modern and Ancient Geosynclinal Sedimentation: Society of Economic Paleontologists and Mineralogists Special Publication 19, p. 38–55.

Islam, Q., and Nielsen, K. C., 1984, Detailed fault history in the east and central portions of the Arbuckle Mountains, Oklahoma: Geological Society of America Programs with Abstracts, v. 16, no. 3, p. 87.

Kay, M., 1951, North American geosynclines: Geological Society of America Memoir 48, 143 p.

Luke, R. F., 1975, Structure of the eastern part of the Mill Creek syncline [M.S. thesis]: Norman, University of Oklahoma, 59 p.

Phillips, E. H., 1983, Gravity slide thrusting and folded faults in western Arbuckle Mountains and vicinity, Southern Oklahoma: American Association of Petroleum Geologists Bulletin, v. 67, p. 1363-1390.

Shatskiy, N. S., 1946, Great Donbass and Wichita System; Comparative tectonics of old platforms: SSSR Izvestia, Seria Geologie no. 6, p. 57–90.

Smith, M., 1981, Recurring shoaling-upward sequences of a carbonate tidal flat, Cool Creek Formation (lower Ordovician), Arbuckle Mountains, Oklahoma [M.S. thesis]: Madison, University of Wisconsin, 71 p.

Tanner, J. H., 1967, Wrench fault movements along Washita Valley fault, Arbuckle Mountain area, Oklahoma: American Association of Petroleum Geologists Bulletin, v. 51, p. 126–141.

Tapp, J. B., 1978, Breccias and megabreccias of the Arbuckle Mountains, Southern Oklahoma Aulacogen, Oklahoma [M.S. thesis]: Norman, University of Oklahoma, 126 p.

Tenney, C., 1984, Facies analysis of the Kindblade Formation, Upper Arbuckle Group, Southern Oklahoma [M.S. thesis]: Norman, University of Oklahoma.

Wickham, J. S., and Dennison, R., 1978, Structural style of the Arbuckle region: Geological Society of America South Central Section Field Trip no. 3 Guidebook, 111 p.

Wiltse, E. W., 1978, Surface and subsurface study of the Southwest Davis Oil Field, Sec. 11 and 14, T1S,R1E, Murray County, Oklahoma [M.S. thesis]: Norman, University of Oklahoma, 72 p.

ACKNOWLEDGMENTS

Support from the American Chemical Society Petroleum Research Fund on Grant #16872-G2, and by The University of Tulsa Faculty Research Program is gratefully acknowledged. John Cook served as a field assistant during the 1985 field season, and assisted in the preparation of samples. The overall presentation of this field guide benefited from discussion with colleagues. I alone, however, am responsible for any omission of fact or logic in this presentation.

I-35 Roadcuts; Geology of Paleozoic strata in the Arbuckle Mountains of southern Oklahoma

Robert O. Fay, Oklahoma Geological Survey, Norman, Oklahoma 73019

LOCATION

The field-trip site consists of excellent exposures of Paleozoic strata in roadcuts along I-35 through the Arbuckle Mountains of south-central Oklahoma (Figs. 1 and 2). The exposures extend from 11 to 18 mi (18 to 29 km) north of the city of Ardmore, and are in parts of Murray and Carter Counties.

Access to the roadcuts is open at all times, but extreme caution must be exercised because the exposures are along the right-of-way of a major 4-lane divided highway. Parking should be on the far right side of the shoulder, and one should not walk upon or cross the highway. If a large group is to examine the exposures, it would be best to park vehicles in the scenic turnouts and hike along the highway shoulders to examine the roadcuts.

SIGNIFICANCE

The Arbuckle Mountains area is a classic site for exposure of the Paleozoic sequence that underlies most of the Midcontinent region. About 10,000 ft (3,050 m) of Cambrian through Mississippian strata, mostly carbonates, are exposed in a homoclinal sequence dipping off the south flank of the Arbuckle Anticline; these same strata are also well exposed in a series of fault blocks on the north flank of the anticline. These outcrops constitute the best exposure of this fossiliferous marine sequence in all of the Midcontinent, and the roadcuts are visited every year by hundreds of geologists.

STRATIGRAPHY

I-35 cuts north-south through 7 mi (11 km) of steeply dipping rocks of the Arbuckle Anticline in the Western Arbuckle Mountains, exposing about 10,000 ft (3,050 m) of Cambrian through Mississippian rocks, with successively older rocks toward the middle or top of the anticline (Figs. 1 and 2). General references on the stratigraphy and geology of the area include those by Ham and McKinley (1954), Ham (1955), Ham, Denison, and Merritt (1964), Ham and others (1969), Hart (1974), Denison (1982), Johnson and others (1984), and Fay (1986).

Basement Rocks

The oldest rock in the vicinity of I-35 is the Colbert Rhyolite Porphyry (525 Ma, Middle Cambrian). It is not exposed along the highway but crops out in the East Timbered Hills, near the highest part of the mountains 1,377 ft (420 m) high, where antenna towers top the hills. Older Precambrian rocks, about 1.2 Ga, occur in the eastern Arbuckles; they are termed Tishomingo and Troy Granites and Blue River Gneiss. Please refer to Ham, Denison, and Merritt (1964) for an understanding of the basement rocks.

Cambrian-Ordovician Rocks

Unconformably above the rhyolite and granite is the Reagan Sandstone (Late Cambrian), about 240 ft (73 m) thick, dipping away from outcrops of the Colbert. The Reagan and the overlying Honey Creek Limestone, together about 105 ft (32 m) thick, are not exposed along the highway. The Reagan and Honey Creek belong to the Timbered Hills Group. Recently Stitt (1971) studied these rocks and described the trilobites.

The Arbuckle Group, next above, consists of 6,800 ft (2,073 m) of carbonates of Late Cambrian-Early Ordovician age, exposed for 5 mi (8 km) along the highway. These are best seen on the south flank, and are named from bottom to top: Fort Sill Limestone (155 ft; 47 m, thick, mostly covered), Royer Dolomite (717 ft; 219 m), Signal Mountain Limestone (415 ft; 126 m), Butterly Dolomite (300 ft; 91 m), McKenzie Hill Limestone (900 ft; 274 m), Cool Creek Formation (1,300 ft; 396 m), Kindblade Formation (1,440 ft; 439 m), and West Spring Creek Formation (1,528 ft; 466 m). For published data see Ham (1955), McHugh (1964), Ham and others (1969), Stitt (1971), Toomey and Nitecki (1979), and Fay (1986).

The Simpson Group, next above, is a fossiliferous sequence of Middle Ordovician formations about 2,400 ft (732 m) thick, named, from bottom to top, the Joins (294 ft; 90 m), Oil Creek (747 ft; 228 m), McLish (475 ft; 145 m), Tulip Creek (395 ft; 120 m), and Bromide (420 ft; 128 m) Formations. The latter four are cyclic, with a sandstone at the base, a shale in the middle, and a limestone at the top. The Oil Creek and Bromide contain many shelly fossils. The sandstones are major oil reservoirs in Oklahoma, such as the Oklahoma City Field. Some basic references are Decker, Merritt, and Harris (1931), Ham (1955), Herndon (1965), Ham and others (1969), and Fay (1986).

The Viola Group and overlying Sylvan Shale, of Middle to Late Ordovician age, are next above the Simpson. The Viola is limestone, about 700 ft (213 m) thick, with graptolites and chert in the lower part. The Sylvan is about 300 ft (91 m) thick, and erodes into a valley next to the high Viola hills. For more details refer to Alberstadt (1973), Amsden (1975, 1980), Ham (1955), Ham and others (1969), and Fay (1986).

Silurian-Mississippian Rocks

The Hunton Group (130 to 230 ft; 40 to 70 m) and overlying Woodford Shale (280 ft; 85 m) are next above the Sylvan. The Hunton carbonates are Late Ordovician through Early Devonian in age, with a shelly fauna. The Woodford is black shale and chert, with palynomorphs and fossil wood, of Late Devonian-Early Mississippian age. For published studies of these rocks, see Amsden (1960, 1975, 1980), Ham and others (1969), and Fay (1986).

Next above is the Sycamore Limestone (220 to 370 ft; 67 to 113 m), forming the first line of hills cut by the highway (Fig. 3a).

The Sycamore is Mississippian in age and has been described by Ham (1950), Ham and others (1969), and Fay (1986).

The overlying Delaware Creek Shale (425 ft; 130 m) and next overlying Goddard Formation are exposed along Tulip Creek, west of the west lane on the south flank. They are Late Mississippian in age. The Delaware Creek was formerly called the Caney Shale or Mississippian Caney, and the Goddard was called the Pennsylvanian Caney or Lower Springer. The Goddard now includes all shales and sandstones of Mississippian age above the Delaware Creek, up to the Lake Ardmore Sandstone. For more data see Elias (1956), Ham and others (1969), and Straka (1972).

Pennsylvanian Rocks

The Collings Ranch Conglomerate (3,000 ft; 914 m) is a red-bed limestone conglomerate of Late Pennsylvanian age, unconformably overlying the Viola-Bromide Formations on the north flank of the mountains, north of the turnouts (Fig. 3b). This unconformity dates the Late Pennsylvanian time of major uplift of the Arbuckle Mountains. The conglomerate is slightly folded and faulted, showing that there was later movement. Refer to Ham (1954, 1955), Tomlinson and McBee (1959), and Ham and others (1969) for information on the conglomerates.

ROAD LOG FROM SOUTH TO NORTH

The best approach to the mountains is from the south, where the stratigraphic section is undisturbed and dips about 50 degrees southwestward. A series of brass markers 4 in (10 cm) in diameter have been placed about 6 ft (2 m) above ground level in the rocks throughout the highway cuts; these brass markers contain stratigraphic information and are further described in a guidebook by Fay (1986). The first markers are on the west side of the west lane, on the south flank of the mountains. It is best to park vehicles off the shoulder and walk north on the west side.

Milepost 44, West Lane, Sycamore-Bromide Section. The first hills seen at Milepost 44 are formed by the Sycamore Limestone, of Mississippian age. The Sycamore is 370 ft (113 m) thick, comprised of several limestones and shales, with a green glauconitic shale at the base (Fig. 3a). A brass marker is 13 ft (4 m) below the top of the upper limestone. The Sycamore is a good oil reservoir where the rock is fractured.

The Woodford Shale (290 ft; 88 m) is composed of black shale and chert, with some phosphatic concretions eroding into a valley. Brass markers are 9 ft (2.7 m) below the top and 1 ft (0.3 m) below the base. The Woodford is a good source rock for oil.

The underlying Hunton Group (229 ft; 70 m) is composed of Bois d'Arc Limestone (8 ft; 2.4 m, at top), Haragan Marlstone (26 ft; 8 m, Lower Devonian), Henryhouse Marlstone (170 ft; 52 m, Silurian), Clarita Limestone (12 ft; 3.7 m), and Cochrane Limestone (13 ft; 4 m). On the old highway (U.S. 77) to the east, the underlying Keel Oolite is a few inches thick, and is Late Ordovician in age. A brass marker is in the Cochrane, 3 ft (1 m) above the base. The Hunton is a good reservoir rock where fractured or dolomitized.

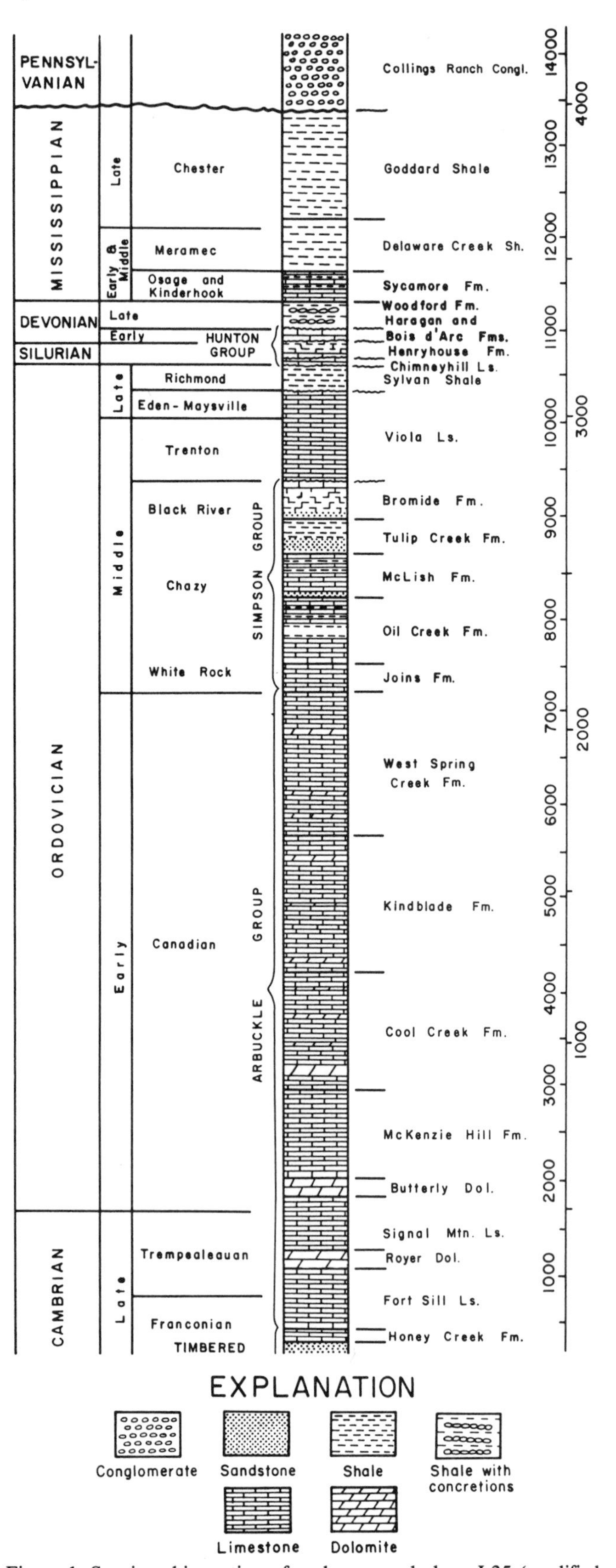

Figure 1. Stratigraphic section of rocks exposed along I-35 (modified from Ham and others, 1969).

The Sylvan Shale (305 ft; 93 m) is poorly exposed in a valley. It is Late Ordovician in age and contains graptolites and microfossils where exposed.

The Viola Group is 684 ft (208 m) thick, composed of limestone eroding into two prominent ridges, with the upper part containing trilobites and brachiopods of Late Ordovician age and the lower part containing chert and graptolites of Middle Ordovician (Trenton) age. Brass markers have been placed 15 ft (4.6 m) below the top and 1 ft (0.3 m) below the base on the west side of the west lane. The Viola is a good reservoir rock where fractured.

The Simpson Group, of Middle Ordovician age, is 2,400 ft (732 m) thick and is mostly covered. The upper Bromide and most of the Oil Creek Formations are exposed. The Pooleville Member (120 ft; 37 m) of the Bromide Formation is exposed for about 15 ft (4.6 m) below the limonitic contact with the Viola. It is a dense limestone, comprising supratidal micritic layers underlain by a massive burrowed calcilutite with *Ischadites.* The underlying Mountain Lake Member (300 ft; 91 m) is mostly shale and sandstone, with many fossils, but is only exposed on the adjoining private property where permission to collect must be granted.

Milepost 45, East Lane, Simpson-Upper Arbuckle Groups. The Tulip Creek (395 ft; 120 m) and McLish (475 ft; 145 m) Formations are mostly covered.

The Oil Creek Formation (747 ft; 228 m) is exposed at Milepost 45, on the east side of the east lane, where the rocks are brownish shaly fossiliferous limestones of Chazyan age. There is a brass marker 49 ft (15 m) above the base, a short distance north of Milepost 45. The Simpson sandstones form low ridges covered with scrub oak and are excellent reservoir rocks for oil in Oklahoma. The basal Oil Creek sandstone is gradational into sandy cross-bedded limestone here, but is 120 ft (37 m) thick one mile to the east.

The Joins Formation (294 ft; 90 m) is mostly limestone and is covered.

The next outcrops to the north are in the Arbuckle Group (Lower Ordovician), West Spring Creek Formation, beginning in an algal bed about 45 ft (14 m) below the top, about 0.15 mi (0.25 km) north of Milepost 45.

The West Spring Creek Formation (1,528 ft; 466 m) contains many different carbonates of diverse origin (supratidal, intertidal, and subtidal). There is a brass marker 67 ft (20 m) below the top on the east side of the east lane, and another is in a brachiopod-gastropod zone (C-D-P Zone), 284 ft (87 m) below the top, just above a 6-ft (2-m) thick sandstone bed. In this area are exposures of red beds, oolites, and algal beds or stromatolites of shallow-water origin. About half-way down the section is a bluish shaly limestone with *Didymograptus.* The basal brownish silty dolomite is a good reservoir rock, termed the Brown Zone.

The Kindblade Formation (1,440 ft; 439 m) is next below, beginning about 0.6 mi (1 km) north of Milepost 45. It is mostly a bluish-gray, even-bedded limestone and dolomite, forming tombstone topography, and containing fossil sponges of deeper-water origin. There is a brass marker 2 ft (0.6 m) below the top on the east side of the east lane. Near the north end of the east turnout, about 0.1 mi (0.16 km) south of Milepost 46 on the east side of the east lane, is a small anticline with a brass marker about 100 ft (30 m) above the base of the Kindblade. The underlying Cool Creek Formation is faulted out here, and the contact is covered.

Milepost 46, West Lane, Cool Creek-Royer Section. The Cool Creek Formation (1,300 ft, 396 m) is similar to the West Spring Creek, with many diverse carbonates. Only the middle part is exposed, on the west side of the west lane. A brass marker is near the center of the outcrop, about 700 ft (213 m) below the top, about 0.2 mi (0.3 km) north of Milepost 46.

The McKenzie Hill Formation (900 ft; 274 m), Butterly Dolomite (297 ft; 90 m), and Signal Mountain Limestone (415 ft, 126 m) are mostly covered. The Butterly forms a jagged outcrop of coarsely granular dolomite, with seemingly many small faults, and with limonite-zinc mineralization. The Cambrian-Ordovician boundary is at the top of the Signal Mountain Formation. The Signal Mountain-Royer contact is about 0.8 mi (1.3 km) north of Milepost 46, on the west side of the west lane, with a brass marker at the contact. The Royer is a coarse-grained pinkish tectonic dolomite, with many joints and caves.

Milepost 47, West Lane, Royer-Fort Sill Section. Near Milepost 47, on the west side of the west lane, there is a brass marker about 150 ft (46 m) above the Royer-Fort Sill contact. About 0.1 mi (0.16 km) north of Milepost 47, in the gully on the west, the Fort Sill Limestone is exposed, with the Chapman Ranch Thrust Fault below. This fault is along a bedding plane in the Fort Sill and has at least 1.5 mi (2.4 km) of displacement, depending upon the angle of the fault. The Cattle Pens Overpass is about 0.4 mi (0.64 km) north of Milepost 47, in the Royer Dolomite. This is near the crest of the Arbuckle Anticline.

Milepost 48, Both Lanes, Fort Sill-Cool Creek Section. Near Milepost 48, on either side of the highway, are 2 brass markers near the Fort Sill-Royer contact, with the thin-bedded Fort Sill Limestone being folded and faulted. About 0.5 mi (0.8 km) north of Milepost 48, the dip is reversed to the northeast, in the Royer Dolomite. A short distance north there is a brass marker in the Signal Mountain Limestone on the west side of the west lane, and a brass marker in the Butterly Dolomite on the east side of the east lane. Farther north, much of the McKenzie Hill and Cool Creek are faulted out of the section.

Milepost 49, West Lane, Kindblade-West Spring Creek-Collings Ranch. At Milepost 49, the Kindblade is dipping northward. At both turnouts the Kindblade is well exposed, with a tectonic breccia also well exposed on the west turnout just below the West Spring Creek Formation. There is a brass marker 52 ft (16 m) below the Kindblade top on the west turnout and another 25 ft (7.6 m) below the Kindblade top on the east turnout.

At the overlook on the west turnout, there is a metal sign and cross section explaining the geology of the Arbuckle Mountains. There is also a granite marker explaining the first use of seismic reflection for oil exploration. A short distance north of the entrance ramp is a brass marker in the Collings Ranch Conglom-

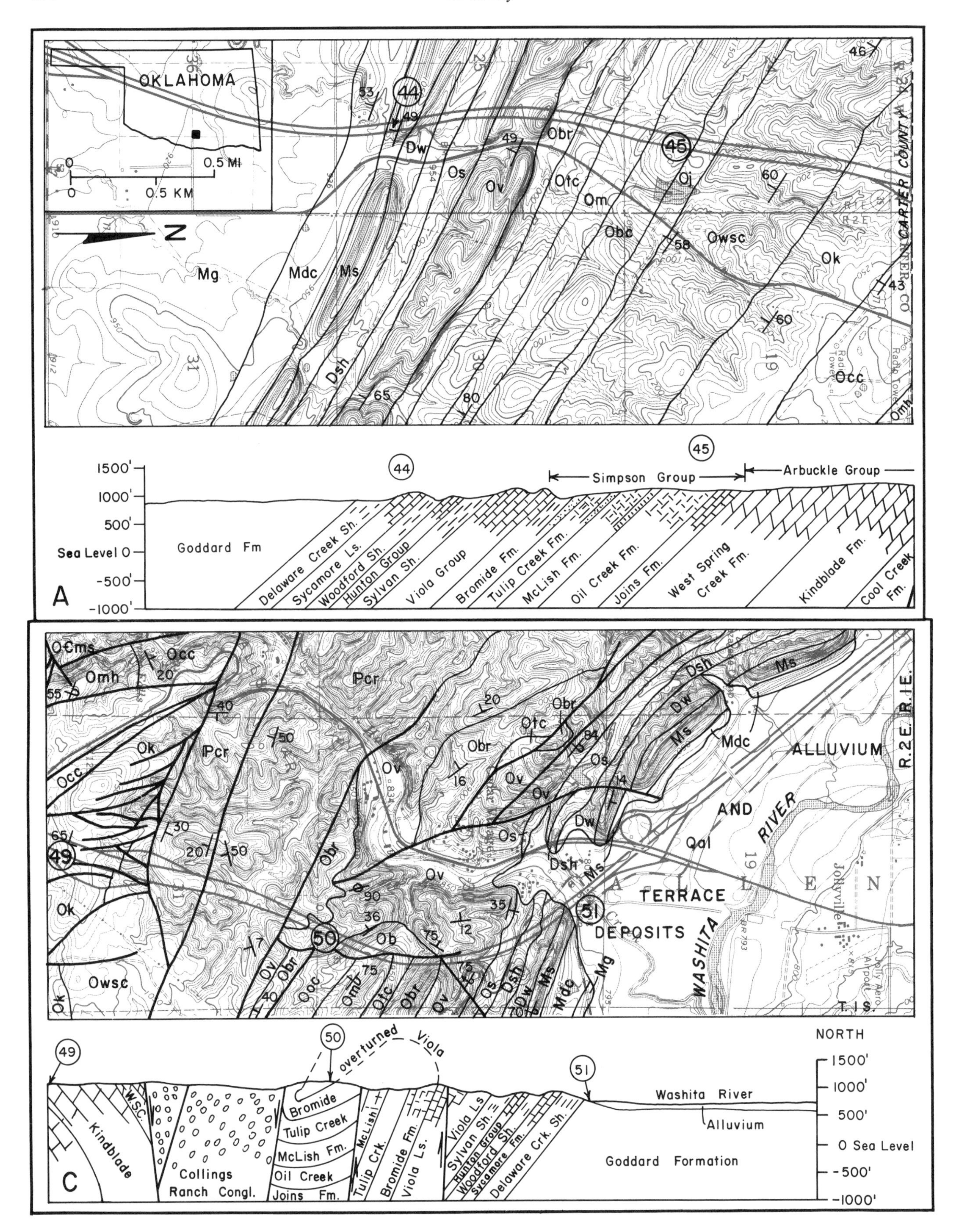
OKLAHOMA
0 0.5 MI
0 0.5 KM
N
CARTER COUNTY
R.2E. R.1E.
44
45
Mg
Mdc
Ms
Dsh
Dw
Os
Ov
Obr
Otc
Om
Obc
Oj
Owsc
Ok
Occ
Omh
1500'
1000'
500'
Sea Level 0
-500'
-1000'
A
Simpson Group
Arbuckle Group
Goddard Fm
Delaware Creek Sh.
Sycamore Ls.
Woodford Sh.
Hunton Group
Sylvan Sh.
Viola Group
Bromide Fm.
Tulip Creek Fm.
McLish Fm.
Oil Creek Fm.
Joins Fm.
West Spring Creek Fm.
Kindblade Fm.
Cool Creek Fm.
O€ms
Pcr
Ob
Ooc
Omb
Qal
49
50
51
ALLUVIUM
AND
RIVER
TERRACE
DEPOSITS
WASHITA
ALLEN
Jollyville
T.1S.
R.2E. R.1E.
NORTH
overturned Viola
Kindblade
W.S.C.
Collings Ranch Congl.
Bromide
Tulip Creek
McLish Fm.
Oil Creek
Joins Fm.
McLish
Tulip Crk.
Bromide Fm.
Viola Ls.
Viola Ls
Sylvan Sh.
Hunton Group
Woodford Sh.
Sycamore Fm.
Delaware Crk. Sh.
Washita River
Alluvium
Goddard Formation
C
1500'
1000'
500'
0 Sea Level
-500'
-1000'

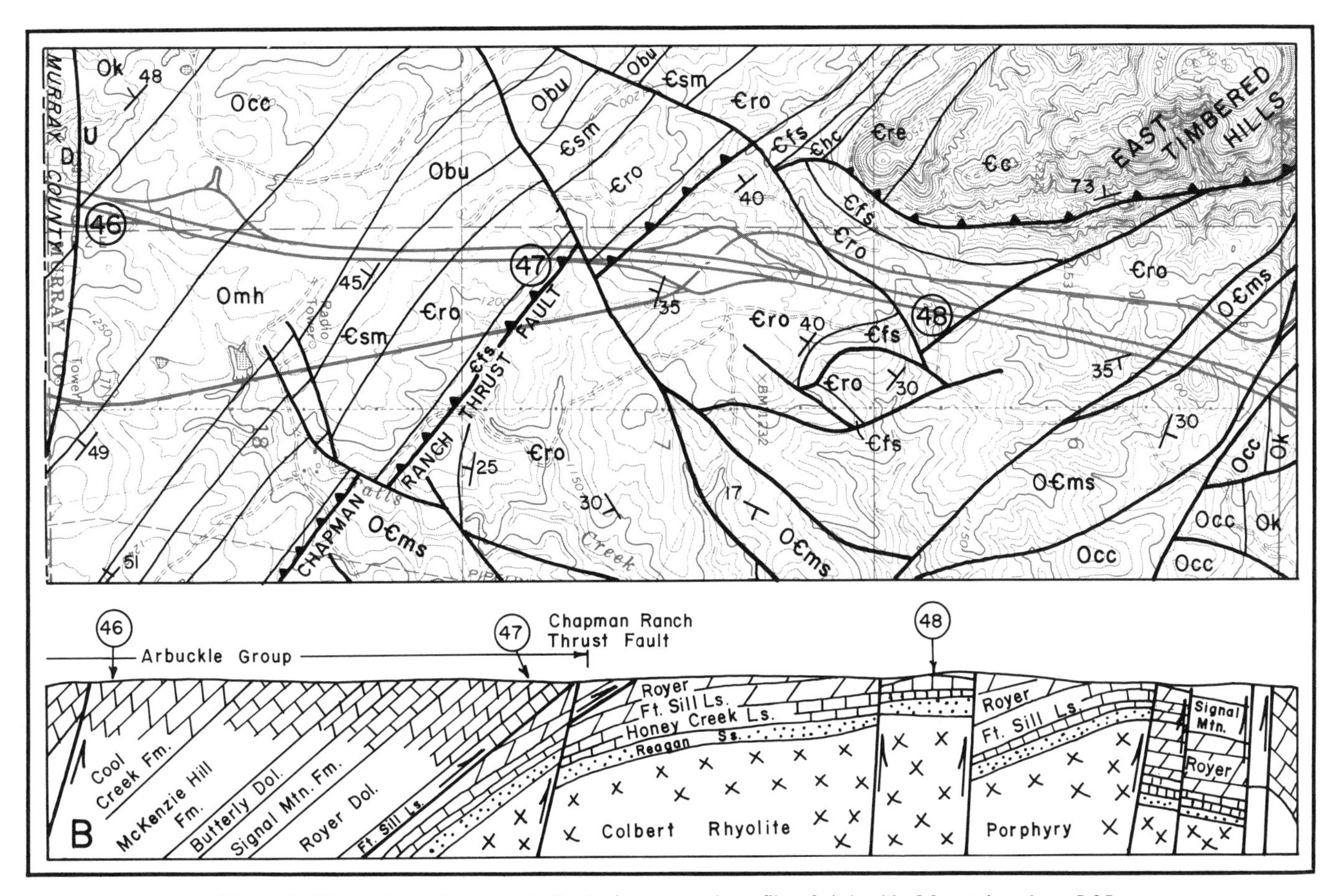

Figure 2 (this and previous page). Geologic map and profile of Arbuckle Mountains along I-35, south-central Oklahoma. A is to the south; C is to the north.

erate, about 70 ft (21 m) north of the fault with the West Spring Creek Formation.

About 600 ft (183 m) farther north, on the east side of the east lane, is another brass marker in the synclinal axis of the Collings Ranch Conglomerate.

About 0.9 mi (1.4 km) north of Milepost 49 is the north fault of the Collings Ranch Graben, with the Collings Ranch Conglomerate next to the Bromide Formation and with some of the conglomerate unconformity above the upturned edges of the Bromide-Viola sequence (Fig. 3b). Several brass markers have been placed here in the Bromide, on the west side of the west lane. A deep cut exposes a syncline in the Viola Group, with the north flank dipping southward above the Bromide Formation. A brass marker is in the Viola, about 175 ft (53 m) above the base on the west side of the west lane.

Milepost 50, East Lane, Viola-Sycamore Section. A short distance south of Milepost 50, there is a brass marker 15 ft (4.6 m) below the top of the Pooleville Limestone Member of the Bromide Formation. Near Milepost 50 is a brass marker 6 ft (1.8 m) below the top of the Mountain Lake Member of the Bromide Formation. Many fossils occur in this member. A short distance north is a brass marker in the basal Bromide sandstone, about 53 ft (16 m) above the base. Along the highway, a north-south fault separates the Simpson beds on the east from the Viola-Bromide beds on the west. The beds on the east are overturned, dipping steeply southward instead of normally northward. The beds on the west are part of a faulted anticline, with the Viola-Bromide dipping 30 degrees south on the south flank, in a normal sequence, but the north flank is overturned in places and has southward dip.

On the east side of the east line, about 0.2 mi (0.32 km) north of Milepost 50, there is a brass marker in the McLish Formation, about 2 ft (0.6 m) below the basal Tulip Creek sandstone.

The next outcrop to the north is a 156-ft (48-m) cut in the Viola-Bromide, with a brass marker in the Viola 2 ft (0.6 m) above the base and about 0.4 mi (0.64 km) north of Milepost 50. Many fossils occur in the Bromide shale here, where the beds are slumped (Fig. 3c).

The overpass above State Highway 77D is in the Hunton Group (132 ft, 40 m), with the Woodford (274 ft, 84 m) exposed on the north side of 77D, steeply overturned to the south. Farther north is the Sycamore Limestone ridge (221 ft; 67 m), on the north flank of the mountains, with a brass marker 71 ft (22 m) above the base at about 0.9 mi (1.4 km) north of Milepost 50.

The overpass of U.S. Highway 77 to Turner Falls is a short

Figure 3 (a). View looking northwest at Sycamore Limestone (Mississippian) on the south flank of the Arbuckle Mountains. Shale to left is shale break below upper Sycamore limestone, near Milepost 44, not seen here. (b) View looking west at Collings Ranch Conglomerate (Pennsylvanian) resting unconformably upon upturned edges of Bromide-Viola rocks, north flank of Arbuckle Mountains. (c) View looking southeast at Viola limestone (overturned), dipping 75° SW, on east side of north flank of Arbuckle Anticline. On right are shaly beds of underlying middle Bromide Formation (Ordovician).

distance to the northwest, about 0.2 mi (0.3 km) north of Milepost 51.

U.S. 77, Viola-Sycamore Section. About 0.2 to 0.4 mi (0.3 to 0.6 km) south of the I-35 underpass below U.S. 77, along the old highway, are new outcrops of Sycamore-Hunton-Sylvan rocks. Brass markers have been placed in the Viola 24 ft (7.3 m) below the top, in the Keel Limestone Member of the Hunton 2 ft (0.6 m) above the base, and in the Sycamore 110 ft (34 m) above the base. This is the best place to see the Keel, Cochrane, Clarita, Henryhouse, and Haragan Formations of the Hunton Group. Farther east, on Oklahoma 77D, the Woodford-Haragan contact is exposed. Many crinoid roots (*Camaraocrinus*) have been collected from the Haragan in this area.

About 1 mi (1.6 km) south along U.S. 77 is Turner Falls Park, where Honey Creek flows over a travertine waterfall built up by algae. This is the most popular tourist attraction in the State of Oklahoma.

REFERENCES

Alberstadt, L. P., 1973, Articulate brachiopods of the Viola Formation (Ordovician) in the Arbuckle Mountains, Oklahoma: Oklahoma Geological Survey, Bulletin 117, 90 p.

Amsden, T. W., 1960, Hunton stratigraphy, Part 6, *in* Stratigraphy and paleontology of the Hunton Group in the Arbuckle Mountain region: Oklahoma Geological Survey, Bulletin 84, 311 p.

——, 1975, Hunton Group (Late Ordovician, Silurian, and Early Devonian) in the Anadarko Basin of Oklahoma: Oklahoma Geological Survey, Bulletin 121, 214 p.

——, 1980, Hunton Group (Late Ordovician, Silurian, and Early Devonian) in the Arkoma Basin of Oklahoma: Oklahoma Geological Survey, Bulletin 129, 136 p.

Decker, C. E., Merritt, C. A., and Harris, R. W., 1931, The stratigraphy and physical characteristics of the Simpson Group: Oklahoma Geological Survey, Bulletin 55, 99 p.

Denison, R. E., 1982, Geologic cross section from the Arbuckle Mountains to the Muenster Arch, southern Oklahoma and Texas: Geological Society of America, Map and Chart Series MC-28, 8 p.

Elias, M. K., 1956, Upper Mississippian and Lower Pennsylvanian formations of south-central Oklahoma, *in* Petroleum Geology of southern Oklahoma, v. 1: American Association of Petroleum Geologists, p. 56–134.

Fay, R. O., 1986, Geology of the Arbuckle Mountains along Interstate Highway 35, southern Oklahoma: Oklahoma Geological Survey, Guidebook 23, (in press).

Ham, W. E., 1954, Collings Ranch Conglomerate, Late Pennsylvanian, in Arbuckle Mountains, Oklahoma: American Association of Petroleum Geologists Bulletin, v. 38, p. 2035–2045.

——, 1955, Regional stratigraphy and structure of the Arbuckle Mountain region: Oklahoma Geological Survey, Guidebook 3, 31 p.

Ham, W. E., and McKinley, M. E., 1954, Geologic map and sections of the Arbuckle Mountains: Oklahoma Geological Survey, Map A-2.

Ham, W. E., Denison, R. E., and Merritt, C. A., 1964, Basement rocks and structural evolution of southern Oklahoma: Oklahoma Geological Survey, Bulletin 95, 302 p.

Ham, W. E., and others, 1969, Regional geology of the Arbuckle Mountains, Oklahoma: Oklahoma Geological Survey, Guidebook 17, 52 p.

Hart, D. L., Jr., 1974, Reconnaissance of the water resources of the Ardmore and Sherman Quadrangles, southern Oklahoma: Oklahoma Geological Survey and U.S. Geological Survey, Hydrologic Atlas 3, HA-3, 4 maps, including geologic map, scale 1:250,000.

Herndon, T., ed., 1965, Symposium on the Simpson: Tulsa Geological Society Digest, v. 33, 298 p.

Johnson, K. S., Burchfield, M. R., and Harrison, W. E., 1984, Guidebook for Arbuckle Mountain Field Trip, southern Oklahoma, during UNITAR conference on development of shallow oil and gas resources: Oklahoma Geological Survey, Special Publication 84-1, 21 p.

McHugh, J. W., ed., 1964, Symposium on the Arbuckle: Tulsa Geological Society Digest, v. 32, p. 35–158.

Stitt, J. H., 1971, Late Cambrian and earliest Ordovician trilobites, Timbered Hills and lower Arbuckle Groups, western Arbuckle Mountains, Murray County, Oklahoma: Oklahoma Geological Survey, Bulletin 110, 83 p.

Straka, J. J., II, 1972, Conodont evidence of age of Goddard and Springer Formations, Ardmore Basin, Oklahoma: American Association of Petroleum Geologists Bulletin, v. 56, no. 6, p. 1087–1099.

Tomlinson, C. W., and McBee, W., Jr., 1959, Pennsylvanian sediments and orogenies of Ardmore District, Oklahoma, *in* Petroleum Geology of southern Oklahoma, v. 2: American Association of Petroleum Geologists, p. 3–52.

Toomey, D. F., and Nitecki, M. H., 1979, Organic buildups in the Lower Ordovician (Canadian) of Texas and Oklahoma: Fieldiana, New Series, v. 2, 181 p.

Lithostratigraphy and depositional environments of the Springer and lower Golf Course Formations, Ardmore Basin, Oklahoma

Frederick B. Meek, R. Douglas Elmore, and Patrick K. Sutherland, School of Geology and Geophysics, University of Oklahoma, Norman, Oklahoma 73019

LOCATION

The two lower members of the Springer Formation are well exposed in the vicinity of City Lake, located northwest of Ardmore, Oklahoma. The upper member of the Springer Formation and the lower member of the Golf Course Formation are well exposed along Phillips Creek, northwest of Springer, Oklahoma (Fig. 1). Detailed directions to these localities are given in the text.

SIGNIFICANCE

The Springer and lower Golf Course formations are a Mississippian-Pennsylvanian interval of sandstone and shale in the Ardmore and Anadarko structural basins in southern Oklahoma. The sequence is of particular interest because the sandstones are important hydrocarbon reservoirs. Despite this, little has been published on the depositional character of this sequence. The only area in which these units outcrop is in the Ardmore Basin (Fig. 1). The facies are here described and interpreted, with a particular emphasis on the coarsening-upward sequences of the major sandstone intervals. The sequence is also of interest because it was deposited in the Southern Oklahoma Aulacogen during the deformational stage of this trough, as first described by Ham and others (1964). As a result, the distribution and characteristics of the sediments were controlled, at least in part, by structural deformation.

GENERAL SETTING

The Ardmore structural basin is an elongate northwest-southeast–trending basin in south-central Oklahoma. It is a deep trough 15 to 27 mi (24 to 43 km) wide and lies between the Arbuckle uplift to the north, and the Wichita–Criner Hills uplift to the southwest. To the northwest, the Ardmore Basin merges with the southeast extension of the Anadarko Basin. To the southeast, the basin meets the thrusts of the Ouachita front but is buried below the Cretaceous coastal plain.

The Ardmore Basin was once a part of a larger northwest-southeast–trending trough extending through southwest Oklahoma, the Southern Oklahoma Aulacogen (Ham and others, 1964; Wickham, 1978). The aulacogen today consists of a number of northwest-southeast–trending uplifts and basins. The Anadarko, Ardmore, and Marietta basins, and the Arbuckle, Criner, and Wichita uplifts are the result of a series of Pennsylvanian orogenies that occurred during the deformational, or final, stage of development of the aulacogen. The Springer and lower Golf Course formations were deposited during this period.

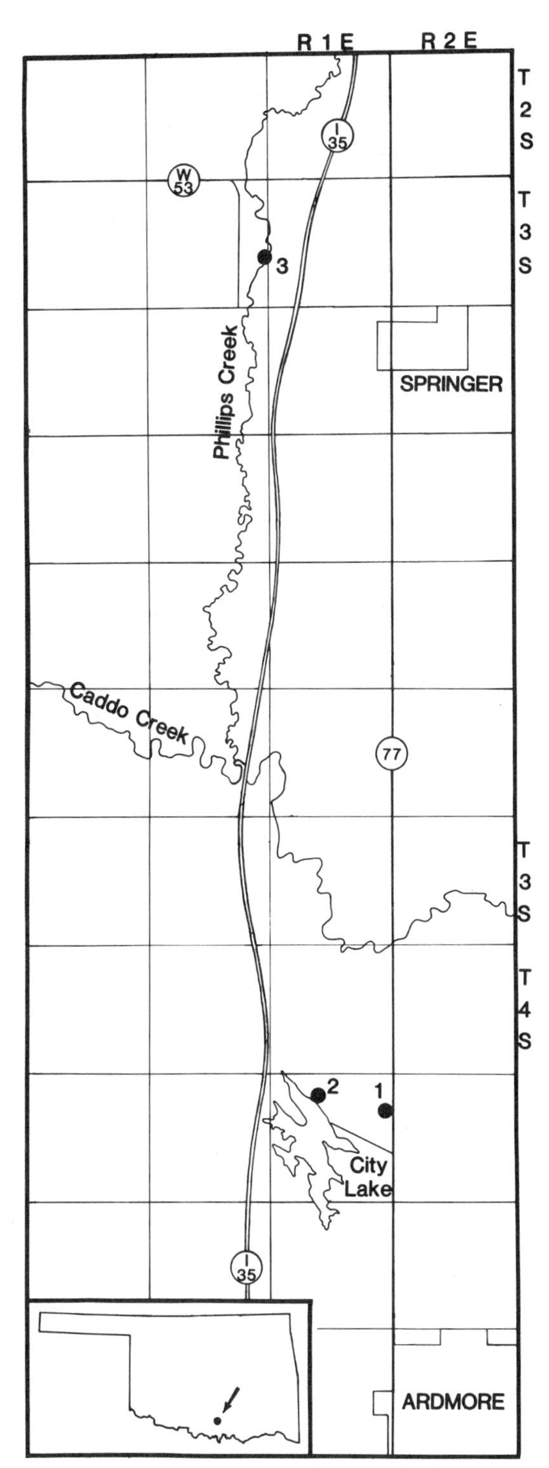

Figure 1. Map showing localities described in text.

The Springer and Golf Course formations comprise up to 3,600 ft (1,100 m) of thick marine shales with thin sandstones and extremely rare limestones in the outcrop area in Carter County. Individual sandstone beds are generally discontinuous, but four sandy zones can be recognized: the Rod Club, the Overbrook, and the Lake Ardmore members of the Springer Formation, and the Primrose, the basal member of the overlying Golf Course Formation. These strata range in age from Late Mississippian (Chesterian) through Early Pennsylvanian (Morrowan) (Fig. 2). The units as described here are based on exposures north of Ardmore and northwest of Springer, Oklahoma. Possibly two of the three Springer sandstones crop out in the southern part of the basin, but these units have not as yet been correlated with certainty with those north of Ardmore.

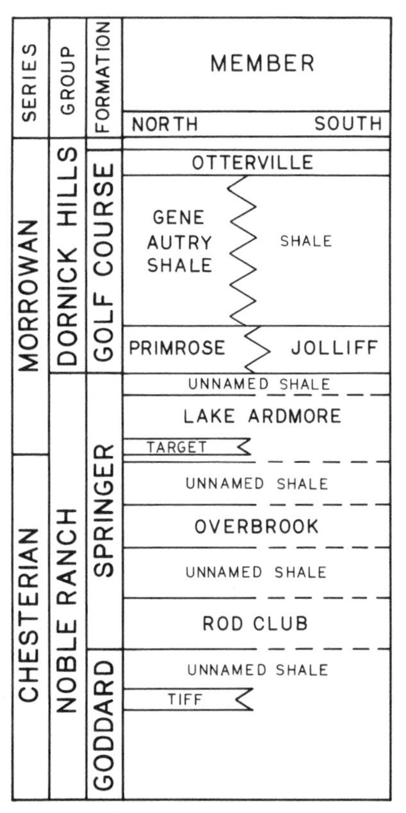

Figure 2. Table of formations and members for late Mississippian–early Pennsylvanian units in the Ardmore Basin.

FACIES

Six principal facies are found in the Springer and lower Golf Course Formations in the outcrop area in Carter County (Table 1): shale (B), sandstone (A), rippled glauconitic sandstone (G), thin-bedded siltstone-shale (C), thin-bedded sandstone (D), and cross-bedded sandstone (E) (Fig. 3). The latter three facies occur in coarsening and thickening upward sequences separated by thick intervals of the shale facies. Minor facies found in the Springer Formation include skeletal and oolitic packstones, massive sandstone, conglomeratic calclithites, and carbonate mudstone. Because these minor facies are rare, they will not be described here, but details are provided in Meek (1983).

Progradational sequences composed of thin-bedded siltstone-shale (C), thin-bedded sandstone (D), and cross-bedded sandstone (E) are the most common units in the study area (Fig. 3). The sequences are 13 to 150 ft (4 to 45 m) thick and generally become thinner higher up in the section. Correlations in the field and on logs suggest the progradational sequences have flat tops and bases, and are encased in shale. Paleocurrent data indicate a dominant transport direction southeastward, away from the continental interior. In the subsurface, the Springer sandstones become thinner and shaley toward the southeast, and the shales thicken (Tomlinson and McBee, 1959). This implies a southeastward-dipping trough with a major source of sediment to the northwest at the time of Springer deposition.

The progradational sequences in the Springer Formation in the outcrop area are similar to delta-front or lower strandplain sequences. Marine fossils and glauconite, although not invariably present, are found throughout the Springer sequences and strongly suggest a marine environment. The absence of abundant channel or delta-plain deposits at the top of the progradational sequences suggest deposition in a delta-front setting, with the more fully developed delta complexes being located in the subsurface to the northwest.

STRATIGRAPHIC AND DEPOSITIONAL HISTORY

The facies of the Springer and lower Golf Course formations are not randomly distributed; there is a distinct change in the types and abundance of facies from the base to the top of the studied interval. Deeper water facies are generally found lower in the section while shallow-water facies occur higher in the section; hence, the interval reflects an overall shallowing upward trend. In this section each of the formations and their sandstone members are briefly described in terms of which facies are present, followed by a discussion of the depositional history. Thick dark-gray shales (B) separate the different sandstone members.

Goddard Formation

Immediately below the Springer Formation is the Chesterian Goddard Formation (Fig. 2). The Goddard consists of approximately 2,200 ft (670 m) of soft dark-gray shales with rare light-buff to rusty-buff sandstones. The shales and sandstones are similar to those of the overlying Springer, except that the sandstones are fewer and thinner. The unit probably represents deposition in a basin or slope setting. The Goddard is thicker and less calcareous than contemporaneous deposits on the shelf to the north, indicating downwarping.

Springer Formation

Rod Club Member. The Rod Club Member consists of a

TABLE 1. SUMMARY OF FACIES

	Description and Interpretation					
Facies	A	B	C	D	E	G
Lithology	Well-sorted, fine-grained quartz arenites	Fissile, noncalcareous shale	Interbedded shale and siltstone	Well-sorted, fine-grained quartz-arenites with some shale	Well-sorted, medium-grained quartz arenites	Fine-grained, well-sorted glauconitic sandstones
Color	Green-gray	Dark-gray	Shale is dark-gray; siltstone is buff	Light-gray	Light-gray to buff	Light-gray
Bedding/ Structure	Beds 1-3 cm thick with erosional bases and rare flutes; Bouma sequences (A-B-C-E-) common; soft sediment deformational features (contorted beds, microfaults) also common	Thick beds (up to 250 m); comprises 90 % of strata in Springer	Thin-bedded (average 1 cm thick), rhythmically interbedded	Thin-bedded sandstone intervals (with some shale partings) 3-20 cm thick, laterally persistent. Parallel to wavy bedding. Ripple and low-angle planar cross-stratification present	Massive to thick-bedded with ripple and trough cross-stratification; interference ripple patterns and contorted bedding also found	From thin interbedded sandstones and shales to massive sandstones; wavy bedding with ripple cross-stratification
Other features	Sandstone contains minor amounts of ooids, glauconite, and fossil debris	Siderite concretions	Hematite ironstone concretions and a few burrows	Abundant horizontal and vertical burrows; mold faunas of brachiopods and crinoids	Wood detritus (_Calamites_ sp.), rare burrows, and some fossils (Crinoids, brachiopods, bryozoans); glauconite also present	Sponge spicules common; also crinoids and brachiopods
Interpretation	Deposition from turbidity currents on an unstable slope	Deposition from suspension in low-energy, moderate to deep marine setting	Moderate to deep water, probably below normal wave base; prodelta	Shallow-water marine shelf: distal delta (between distributary mouth bars)	Distributary mouth bars	Shallow marine shelf setting

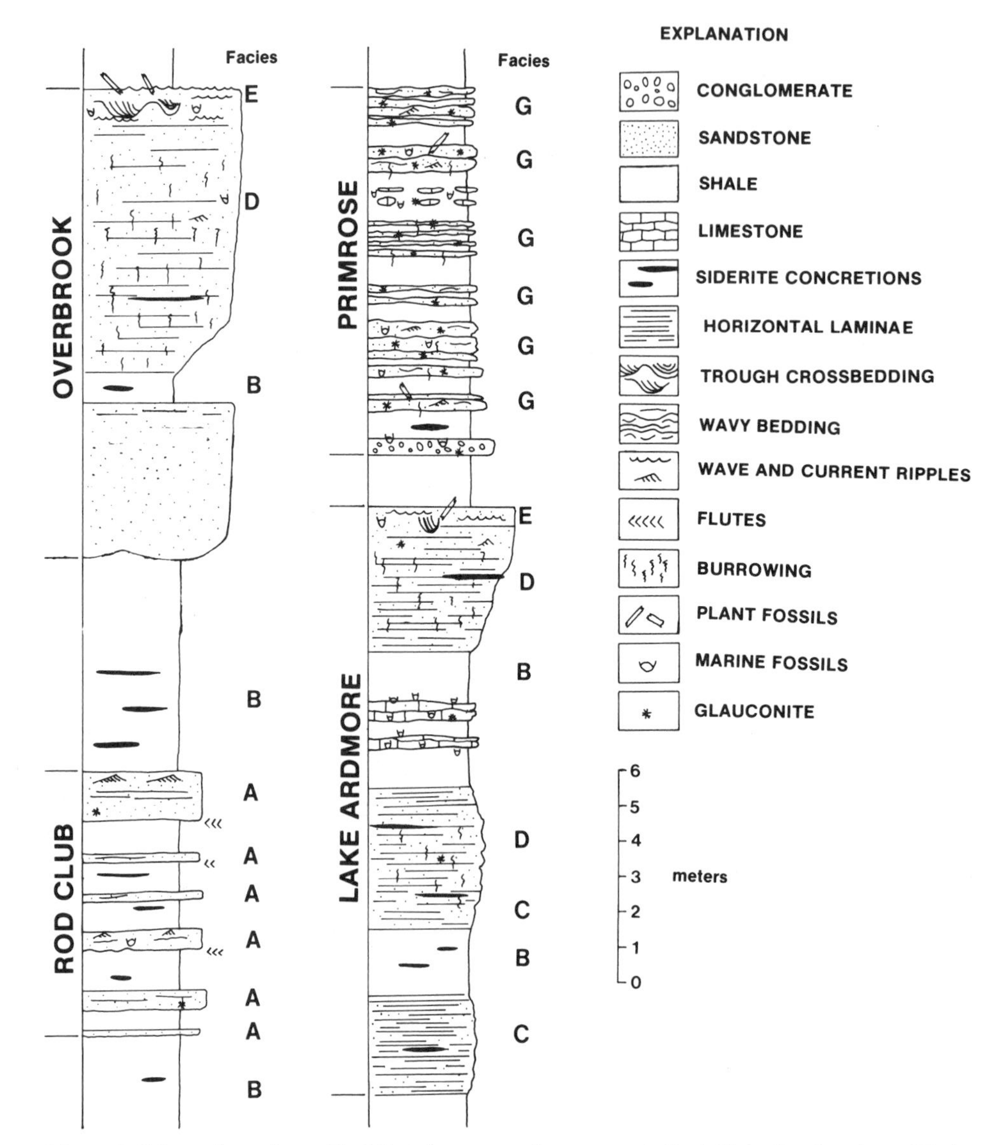

Figure 3. Schematic sections of bedding character and structures in Rod Club, Overbrook, and Lake Ardmore Members of the Springer Formation and Primrose Member of the Golf Course Formation. Facies designations are shown to the right. Sequences are diagrammatic and do not show the thick shales that separate each member.

discontinuous zone up to 490 ft (149 m) thick of interbedded green-gray sandstones and dark-gray shales. The stratigraphic position and conodont fauna (Straka, 1972) place the Rod Club in the Chesterian Series.

In the lower Rod Club, two facies are present: the sandstone (A) and shale (B) facies. The former are turbidites, whereas the latter probably record deposition from suspension. The soft sediment microfaults and various other soft sediment deformation features indicate the turbidites were deposited on an unstable slope. Paleocurrent data indicate the turbidites were transported toward the southeast, or down the axis of the aulacogen. These turbidites are not present in younger members of the Springer. The upper part of the Rod Club contains progradational sequences that probably record deltaic deposition.

Overbrook Member. This member lies 800 to 1,000 ft (240 to 300 m) above the Rod Club. It consists of discontinuous sandstone lenses and the interval is up to 150 ft (45 m) thick. Conodont recoveries are poor from this unit, but they possibly indicate a late Chesterian age (Lane and Straka, 1974).

The outcropping Overbrook sandstones may be divided into an upper and lower sandstone, separated by about 250 ft (76 m) of shale. The lower sandstone has coarsening upward sequences but is massive and apparently channeled. The upper Overbrook Sandstone is a typical coarsening upward deltaic sequence. Paleocurrent data and the isopach indicate a sediment source to the northwest.

Lake Ardmore Member. This member lies 500 to 700 ft (150 to 210 m) above the Overbrook. It is an interval of discontinuous sandstones interbedded with shales similar to the Overbrook Member. The total Lake Ardmore interval includes typically three sandstones with the total interval up to 500 ft (150 m) thick. Progradational sequences are the dominant feature in the Lake Ardmore Member. The sandstones are commonly noncalcareous, silica-cemented, porous, and thin-bedded to massive. Also occurring in the Lake Ardmore Member is the Target limestone lentil, a thin, mixed skeletal packstone that occurs in a limited area in the northeastern part of the basin, in the shale between the lowermost and second sandstones of the member. Conodonts from the Target limestone indicate a lowermost Morrowan age for the Lake Ardmore Member (Straka, 1972).

The paleocurrent data for the Lake Ardmore Member suggest sediment transport from the north and northwest. The Lake Ardmore sandstones thicken and become most numerous toward the north, and are thinnest along the Caddo Anticline, in the center of the study area. Overlying the Lake Ardmore Member is a dark-gray shale unit that varies from 70 to 500 ft (20 to 150 m) thick.

In the subsurface to the northwest, a fourth Springer sandstone zone called the Markham occurs above the Lake Ardmore equivalent (Aldredge) and below the Primrose sandstone. The absence of the Markham sandstone in the outcrop area is probably due to the sandstone facies having been replaced to the southeast by shale.

Golf Course Formation

Primrose Member. Tomlinson and McBee (1959) accounted for the absence of the Primrose Member of the Golf Course Formation south of Ardmore by truncation below a presumed regional unconformity at the base of the Jolliff Limestone Member (Fig. 2). In turn, the Jolliff, which is absent in the north, was assumed to correlate with some horizon in the shale above the Primrose in the north. It has recently been established that the Jolliff Member in the south is, in fact, a lateral facies of the Primrose in the north (Robert C. Grayson, Jr., personal communication, 1985). Grayson reported that the Jolliff contains the same three middle Morrowan conodont zones described by Straka (1972) from the Primrose member in the northern part of the basin. In ascending order, they are: *Neognathodus symmetricus, N. bassleri,* and *Idiognathodus sinuosis* zones.

The Primrose Member lies 70 to 500 ft (20 to 150 m) above the Lake Ardmore sandstones. It consists of 100 to 360 ft (30 to 109 m) of thin, light-gray, calcareous sandstones and shales with a few thin limestones. The dominant facies in the Primrose is the Rippled Glauconitic Sandstone Facies (G), which is thought to represent deposition in a shallow shelf setting. The Primrose differs from the underlying Springer sandstones in being distinctly more calcareous and in having fairly common small chert and limestone rock fragments. The change in mineralogy is not abrupt; rather, a sandstone bed with 1 percent chert fragments may lie between two sandstone beds without fragments. This suggests that two or more source areas were providing sediment at the same time. The source for the sands is presumed to have continued from the north and northwest, but the chert and limestone fragments were almost certainly derived from the Criner-Wichita uplift to the southwest, as were sediments for the Jolliff Member south of Ardmore.

Locality 1, Rod Club Member, U.S. Highway 77

Location. This outcrop is in the roadcut on the west side of U.S. 77, 1.8 mi (2.9 km) north of the junction with U.S. 70 in Ardmore, Oklahoma: NE¼,NE¼,SE¼,NE¼,Sec.12,T.4S.,R.1E., Ardmore West Quadrangle, Carter County, Oklahoma. Equivalent to units 4 to 6 in measured section 395 of Meek (1983).

Description. The shales in the Rod Club Member are not well exposed at this locality, but the sandstone layers show the unique features that are characteristic of the member. The member is 259 ft (79 m) thick in this section. The beds strike northwest-southeast and dip steeply southwest. Rod Club sandstones (facies A) and shales (facies B) are exposed in the partly grass-covered slope. The shales are dark-gray, fissile, and noncalcareous. The sandstones are green-gray, graded, fine-grained quartzarenites. The sandstones are 4 to 16 in (10 to 40 cm) thick, and some contain partial Bouma sequences. Soft-sediment microfaults (<1 cm displacement) are common. Flutes are rare but present in some beds. A coarsening upward sequence (units 6a, 6b, painted on the rocks) is poorly exposed at the south end of the hill.

The presence of graded beds, erosive lower contacts, flutes, and Bouma sequences suggest the facies A sandstones were deposited by turbidity currents. The microfaults are interpreted as soft-sediment deformational structures that formed as a result of deposition on an unstable slope.

Locality 2, Overbrook Member, Spillway at Dam on City Lake

Location. Drive 1.4 mi (2.25 km) north on U.S. 77 from the junction with U.S. 70 in Ardmore, then turn left (west) and drive 0.5 mi (0.8 km) to Ardmore City Filtration Plant. Park and ask at the plant for permission to walk northwest along the dam to the spillway, SE¼,NE¼,NW¼,Sec. 12,T.4S.,R.1E. The strata strike northwest-southeast and dip 80 to 90° to the southwest. It is equivalent to the middle part of Measured Section 395 of Meek (1983). The unit numbers are painted on the rocks.

Description. The Overbrook Member, 131 ft (40 m) thick, is superbly exposed at this locality. A coarsening upward sequence is seen, which consists of facies B (unit 8) overlain by facies D (unit 9). The sandstones are light-buff and contain numerous burrows and ripples. The upper 56 ft (17 m) (unit 9c) contains several smaller coarsening upward cycles. The upper part of a progradational sequence (facies E) is not well exposed at this locality. The sands are interpreted as representing the distal part of a deltaic sequence.

Locality 3, Lake Ardmore and Primrose Members, Phillips Creek

Location. Drive 11 mi (17.5 km) north on I-35 from U.S. 70 at Ardmore to the Oklahoma 53 exit and turn west; drive 0.7 mi (1.1 km) and turn left on first southbound road (this road is not on a section line); drive 0.2 mi (0.3 km) and stop at the first farm house (the west side of the road) and ask permission to enter the property; then drive southward an additional 0.3 mi (0.5 km). Walk 0.25 mi (0.4 km) due east to Phillips Creek. The base of Lake Ardmore is exposed in the creek bed at this point, whereas the base of the Primrose is found in the creek bed 0.2 mi (0.3 km) downstream (south). Exposures of the two members extend for 0.5 mi (0.8 km) along the creek bed to an east-west road at the south margin of Section 2. The base of Lake Ardmore in SW¼,SW¼NW¼,Sec. 1 and the top of the Primrose is in SW¼,SE¼,SE¼,Sec2,T.3S.,R.1E., Springer Quadrangle, Carter County, Oklahoma. The strata strike northwest and dip at 8 to 35° to the south-southwest. These beds are equivalent to units 15 through 44 in Measured Section 397 of Meek (1983).

Description. The Lake Ardmore and Primrose Members are well exposed at this locality. The Lake Ardmore consists predominantly of gray calcareous sandstones and siltstones with some shale lenses. The sandstones consist of facies D (units 15 and 21) and facies C (unit 22). Burrowing is common in some beds. The sands are calcite-cemented quartzarenites. The deposits are interpreted as representing a distal deltaic setting, perhaps delta-front sheet sands.

The Primrose consists of light-gray, calcareous glauconitic sandstones, siltstones, and shales. The unit consists predominantly of facies G (units 29, 36, 39) and some facies C (unit 42). Wavy bedding and ripple cross-stratification is common. Larger scale cross-stratification similar to hummocky cross-stratification may be present in some beds (unit 36), and may indicate storm processes. Herringbone cross-stratification, suggesting a tidal influence, may also be present (e.g., unit 39). The Primrose sandstones probably record deposition in a distal, marine-dominated part of a delta, or perhaps on shallow marine bars in a shelf setting.

REFERENCES CITED

Ham, W. E., Denison, R. E., and Merritt, C. A., 1964, Basement Rocks and Structural Evolution of Southern Oklahoma: Oklahoma Geological Survey Bulletin 95, 302 p.

Lane, R. H., and Straka, J. J., II, 1974, Late Mississippian and Early Pennsylvanian Conodonts, Arkansas and Oklahoma: Geological Society of America Special Paper 152, 139 p.

Meek, F. B., 1983, The Lithostratigraphy and Depositional Environments of the Springer and Lower Golf Course Formations (Mississippian-Pennsylvanian) in the Ardmore Basin, Oklahoma [M.S. thesis]: Norman, University of Oklahoma, 212 p.

Straka, J. J., 1972, Conodont evidence of Goddard and Springer formations, Ardmore Basin, Oklahoma: American Association of Petroleum Geologists Bulletin, v. 56, p. 1087–1099.

Tomlinson, C. W., and McBee, W., Jr., 1959, Pennsylvanian sediments and orogenies of Ardmore District, Oklahoma, *in* Mayes, J. W., Westheimer, J. M., Tomlinson, C. W., and Putman, D. M., eds., Petroleum Geology of Southern Oklahoma: American Association of Petroleum Geologists, v. 2, p. 3–52.

Wickham, J. S., 1978, The Southern Oklahoma Aulacogen, *in* Structural Style of the Arbuckle Region: Geological Society of America Guidebook, South-Central Secton Field Trip 3, p. 9–41.

Beavers Bend State Park, Broken Bow Uplift, Oklahoma

K. C. Nielsen, Program in Geosciences, University of Texas at Dallas, Richardson, Texas 75083-0688

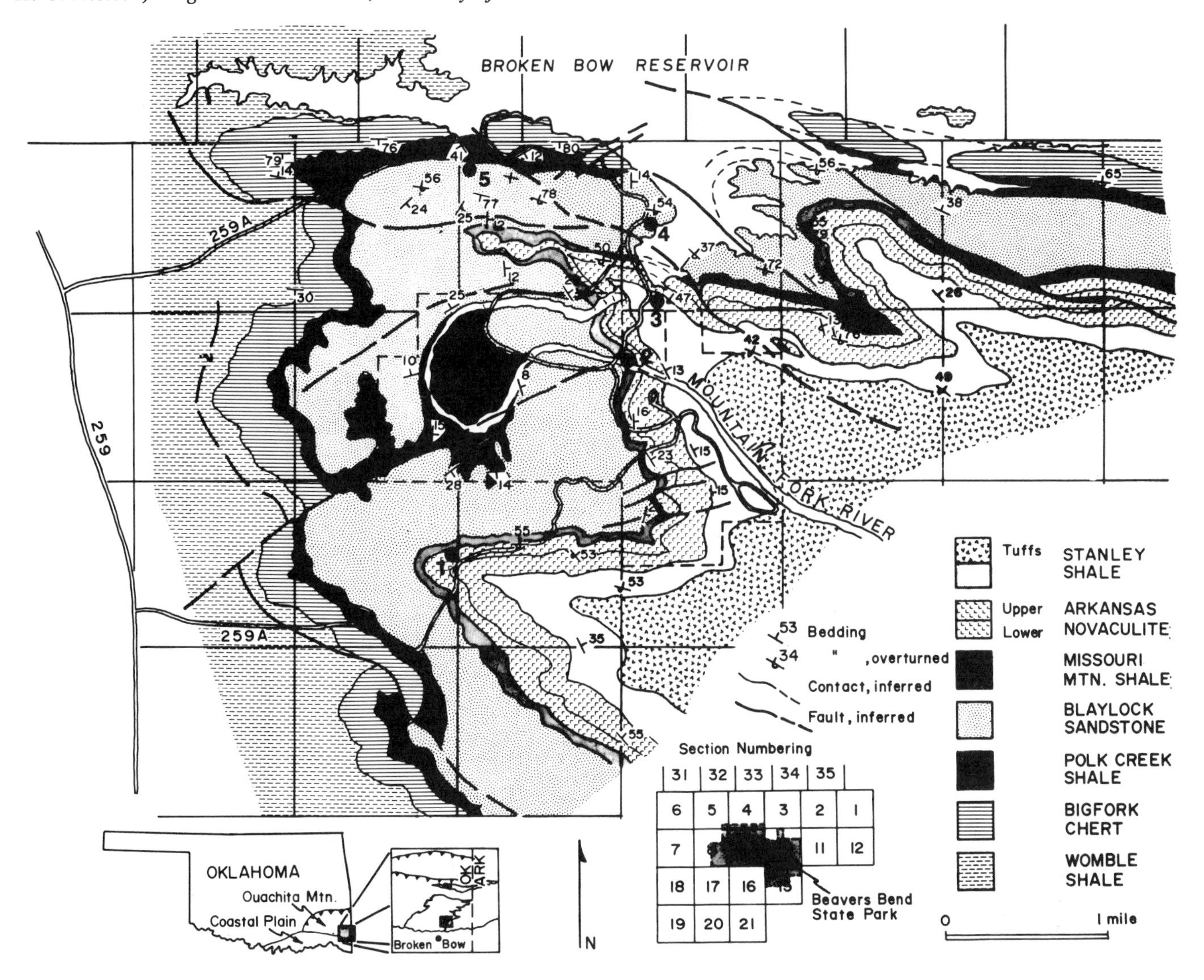

Figure 1. Location and geological map of the Beavers Bend State Park within southeastern Oklahoma. Stops are indicated by dots.

LOCATION

Beavers Bend State Park is located approximately 10 mi (16 km) north of Broken Bow, on U.S. 259A. In this park the Mountain Fork River has cut an entrenched meander loop into the Middle Paleozoic flank sequence of the Broken Bow Uplift (Fig. 1). As such, this is a convenient locality to examine the stratigraphic sequence and the structural style of the central uplift of the Ouachita Mountains of Oklahoma.

Within the park, U.S. 259A forms a loop east of U.S. 259 (Fig. 1). There are several good hiking trails; therefore, many of these stops will involve light hiking in order to examine the sequence. Also, detailed maps are provided in Figures 1 and 3 in the hope that interested people will hike independently to study the anatomy of these structures.

The stratigraphic sequence (Site 45) is examined in two stages. The Womble Shale through the Blaylock Sandstone are discussed at the Stephens Gap Area (Locality 5). The Blaylock Sandstone up to the Lower Stanley Shale is reviewed near the Power Station (Locality 2).

The structural sequence (Site 46) is discussed at various locations: the earliest folding at the Dam Overlook (Locality 4), the Dam (Locality 3), and the Southern Ridge (Locality 1); the second folding events at Stephens Gap Area (Locality 5), the

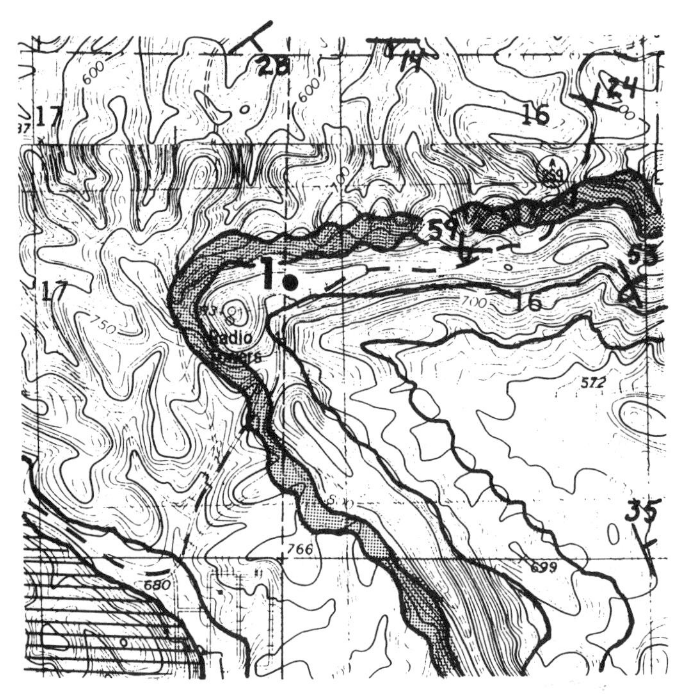

Figure 2. Geology at Locality 1. Missouri Mountain Shale and Bigfork Chert are shown. See Figure 1 for legend.

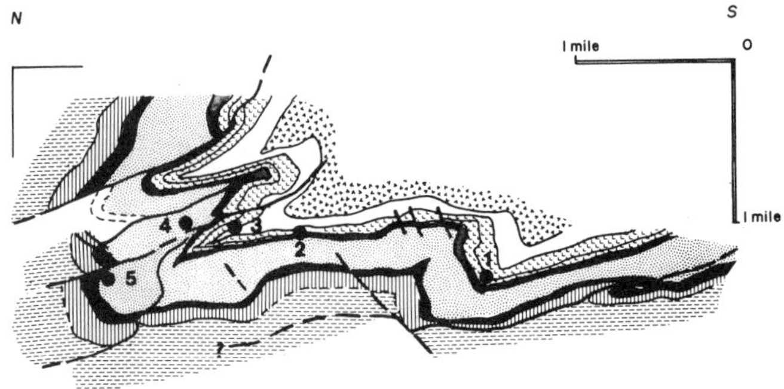

Figure 3. "Down-plunge" view of the Beavers Bend State Park area. Units are the same as in Figure 1. Relative locations of stops are shown by dots.

Dam Overlook (Locality 4), and the Power Station (Locality 2); the third event at Stephens Gap (Locality 5) and the Dam (Locality 3).

SIGNIFICANCE

The Broken Bow uplift is part of the core zone of the Ouachita Mountains. This zone of highly deformed low-grade metasediments extends northeastward into Arkansas. To the north, Carboniferous age rocks are involved in large-scale northerly directed thrusts. To the south, Mesozoic sediments rest unconformably on a thick sequence of deformed Carboniferous sediments (Lillie and others, 1983). The exposures in Beavers Bend State Park include units from Middle Ordovician through middle Mississippian and therefore straddle the stratigraphic boundary between preflysch and flysch deposition within the Ouachita Mountains (Viele, 1973). Structurally, evidence for three ductile deformation events is preserved, recording a significant change in shortening direction during the Carboniferous orogeny.

Topographically, the stratigraphy is characterized by ridge lines of chert or sandstone (Arkansas Novaculite, Bigfork Chert, or Blaylock Sandstone) with the interlayered shales forming valleys. Maximum relief in this area is approximately 600 ft (180 m). Thick vegetation covers the area; low areas consist of virtually inpenetrable tangles of briars. Welcome to the Ouachita Mountains. Bring bug repellent.

LOCALITY DESCRIPTIONS

This traverse will follow U.S. 259A, beginning at the southern intersection of U.S. 259A and 259 (SW¼,Sec.18, T.5S.,R.25E.), approximately 6.5 mi (10.4 km) north of Broken Bow, Oklahoma (Fig. 1). This drive proceeds due east 1.7 mi (2.7 km) and then northeastward 0.6 mi (1 km) up onto a major ridge line in this area. Along this route we travel up section from the Womble Shale to the lower portion of the Arkansas Novaculite (Fig. 1). The subdued relief over the first mile (1.6 km) and thick forest cover are typical of the shale exposures, particularly in areas of shallow dip. Where the road curves and climbs up over a small hill in the SW¼ of Sec. 17 (Fig. 1), you cross the Bigfork Chert. Watch the abrupt turn to the northeast. From here to the ridge crest, you climb up through the Blaylock Sandstone (there are exposures along the roadside), and the Missouri Mountain Shale into the Arkansas Novaculite (Fig. 2). There is a parking area at the top of the ridge, but watch the traffic as it is a hard left turn.

Locality 1 (Figs. 1, 2, 3; center, section line 17/16,T.5S.,R.25E.) is in a parking area at the west end of a sharp ridge trending east-west through Sec. 16 and 15 (Fig. 2). During winter months, the view to the north reveals an irregular topography associated with sandstone exposures and the larger more linear ridges of the cherts. In the summer, the view is more restricted, but the Broken Bow Reservoir and the dam 2 mi (3 km) away are visible slightly east of north. This view is across the large fold sequence we are examining (Fig. 1). To help visualize this large structure, a "down-plunge" view is constructed from the data in Figure 1 (Fig. 3). The relative position of each stop is shown in this cross section.

Examination of the large blocks of float at Locality 1 will reveal massively bedded white to gray-black chert (Fig. 2). These blocks are characteristic of the dominant ridges and come from the lower Arkansas Novaculite. Structurally, you are just north of the hinge of a large syncline (Fig. 3). This section of chert is vertical or slightly overturned. To the southeast of this ridge, younger Stanley Shale is exposed, and to the north is a large area of Blaylock Sandstone and older units. Generally in this area, east-west–trending ridges are overturned and the more irregular and subdued northwest-trending ridges are upright. As shown in the cross section, these folds are asymmetric with relatively planar limbs and angular hinges. These folds are gently plunging to the east (091°/14°). The sense of rotation is clockwise (southerly verging) when looking down plunge (Fig. 3). This southerly directed transport direction has provided one of the interesting structural problems within the otherwise northerly directed thrust sequence of the Ouachita Mountains.

Driving eastward along the ridge (~0.5 mi; 0.8 km) there are several good views to the south and of the hinge area of the large syncline. Just north of the center of Sec. 16, we will drop off the ridge and cross the Womble Shale and into the Blaylock Sandstone (Fig. 2). Within 0.25 mi (0.4 km) of the ridge line, we cross the axial trace of the anticline. Careful examination of the cross section (Fig. 3) reveals the inclined orientation of this large anticline and of all the folds in this area (apparent dip of axial surface is 20°N). At the state park gate, we are in upright, gently dipping Blaylock Sandstone; continuing northeast, we travel up section, crossing the lower novaculite contact in the SW¼ of Sec. 10. The location of this contact is a little difficult to find, but the massive white chert can be seen on the east side of the road just as the road slopes downhill toward the main park area. Proceed to the stop sign west of the park museum. At this point you are at the contact of the Arkansas Novaculite and the Stanley Shale. The campground to the southeast is in the Stanley Shale, and good exposures of the lower tuff sequences can be seen in the hillsides at the southeast end of the camp area. Turn left at the stop sign and drive west up a very gentle slope of the Arkansas Novaculite. At the top of the hill a thin outcrop of white novaculite can be seen overlying a section of Missouri Mountain Shale. You will follow the road down a small valley, which has Blaylock Sandstone on the west and Missouri Mountain Shale in the stream bank. Turn right at the intersection and proceed across the river; park either by the swimming area to the west (left) or in the park area to the east (right).

Locality 2 (Figs. 1, 4) is a traverse through the upper stratigraphic sequence of the Beavers Bend Park. The units include the upper contact of the Blaylock Sandstone just west of the swimming area, the Missouri Mountain Shale, the Arkansas Novaculite (at the road and behind the Power Station), the Stanley Shale at the east end of the low ridge, and finally the Beavers Bend and Hatton tuffs exposed on the slopes of Rattlesnake Bluff. The total distance is almost 1 mi (1.6 km). The reader is encouraged to read the unit descriptions in Honess (1923), Miser and Purdue (1929), Spradlin (1959), and Flawn and others (1961).

To begin the traverse in the oldest units, walk about 800 ft (230 m) west of U.S. 259A. North of you there will be a small valley cut into Devils Backbone (Fig. 4). This valley has developed at the contact between the Blaylock Sandstone and Missouri Mountain Shale. The contact between these units is gradational with the gray-green fine- to medium-grained sandstone layers giving way to shale. The shale is normally greenish to olive drab, but in this area, it is more commonly dark gray to black and slaty in character. Total thickness is 83 ft (~25 m) in this area (Spradlin, 1959). Another locality for continuous exposure of the Missouri Mountain Shale is north of the dam (south-

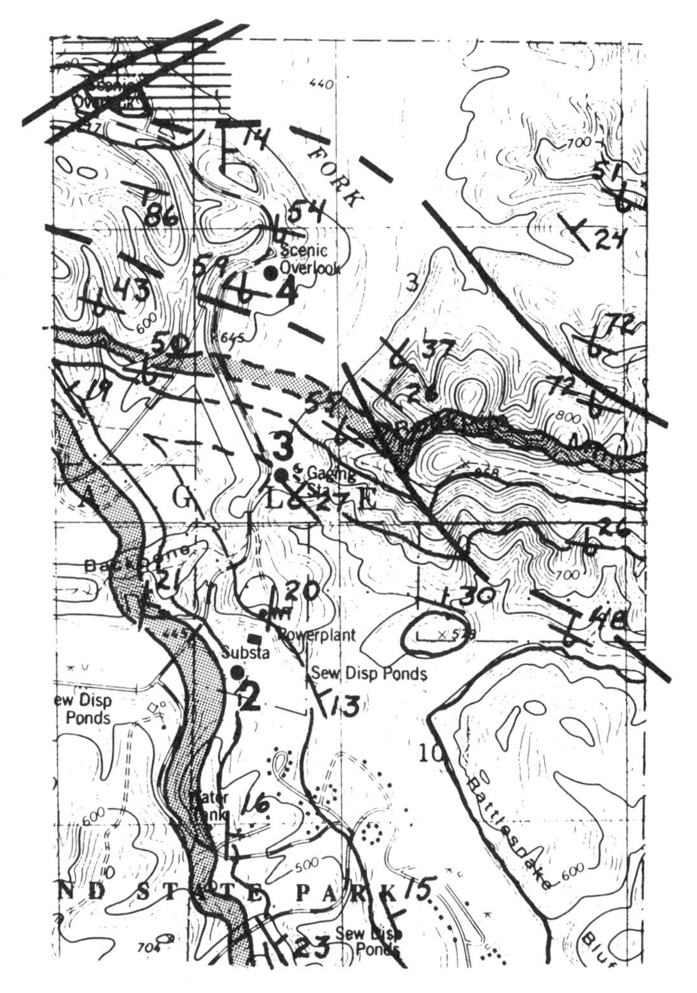

Figure 4. Geology at Localities 2, 3, 4. Missouri Mountain Shale and Bigfork Chert are shown. See Figure 1 for legend.

central Sec.3,T5S,R25E) along the lakeshore (Fig. 1). The stratigraphic position of the Missouri Mountain Shale is presumed to be Silurian (Flawn and others, 1961). A basal conglomerate has been described in Arkansas (Miser and Purdue, 1929), but no basal sequence has been documented here, and Honess (1923) considered these two units conformable. Because of the very gentle dip, this thin unit is exposed in the cliff face nearly all the way back to the road. Cleavage in this sequence of rocks is essentially parallel to bedding. Keep this idea in mind when you move to Locality 3.

The contact between the Arkansas Novaculite and Missouri Mountain Shale is fairly abrupt, with black siliceous shale overlain by the massively bedded white to gray chert. This contact can be seen along the slope north of the swimming area at this locality. Proceeding to the northeast corner of the parking area and just at the foot of the road bed, the first massive chert can be observed. Interestingly, this section of white chert is relatively thin when compared to the section north of the dam. While this author does not have a detailed measured section in this area, the estimated thickness for entire novaculite is 180 to 200 ft (55 to 60 m). When compared to the measured section north of the dam (312 ft; 95 m; Spradlin, 1959) there does appear to be a significant thinning. The cross section illustrates this thinning and relates it to structural position (Fig. 3). The upright limbs of the folds appear to be thinner. The interpretation of this relationship will be discussed at the last stop, Locality 5. Because of the shallow dip, the novaculite sequence is exposed along the full length of the cliff behind the power station.

The Arkansas Novaculite is the most distinctive formation in the Ouachita Mountains. Traditionally, this formation is considered to have three divisions. The lower division is characterized by the massive white novaculite. At this locality, this division is thin and covered by the road (U.S. 259A). Spradlin (1959) measured ~116 ft (35 m) for this division north of the dam. The rock name "novaculite" refers to this dense, even textured, light-colored, cryptocrystalline siliceous rock. The formation name includes a wide range of lithologies.

The middle division consists of interbedded shales and cherts, which range in thickness from 2 to 5 in (≤13 cm). The shales are dark green to black and generally very siliceous. A few layers within this sequence are graphitic, and some (tuffaceous?) sandy sections are observed. The cherts range from olive to gray. Whereas this entire exposure appears relatively flat and undeformed, a faulted fold can be seen in the middle division approximately 300 ft (90 m) east of the road. The fault is a bedding-plane fault, and the small associated anticline indicates slip toward the south. This fold actually deforms the regional cleavage. Exposures of the middle division extend to the east side of the power station.

The upper division of the Arkansas Novaculite at this locality is distinctive in that two thin carbonate units can be observed (Murgatroyd, 1980). East of the power station, the cliff exposure is mainly the upper division. The two carbonate units are apparent because of the cavities developed in the weathered surface. Each unit is approximately 10 to 20 ft (3 to 6 m) thick. Walking to the east, take time to look closely at these units, which reveal very irregular replacement textures. The carbonate occurs as thin laminations, small elongate irregular pods or large nodules. Weathering accentuates the lithologic differences between the chert and the carbonate. The upper carbonate section appears to have a higher percentage (50 percent) of carbonate; whereas the lower section (20 percent) is more silicified. Associated with these carbonates are massive sandstone layers, interbedded chert and shale, very thin conglomeratic layers, and an upper section of calcareous chert (Murgatroyd, 1980). In other portions of the Ouachita Mountains, the upper division is described as similar to the lower division, a massive chert but with distinctive tripolitic chert horizons.

At the eastern end of the low ridge, there is a wide area (~1,400 ft; 425 m) of nearly horizontal black fissile shale. This shale is the basal Stanley Formation. The contact with the Arkansas Novaculite appears gradational. Examination of the black shales behind the water tank above the power station reveals significant sulfide accumulation and evidence for soft-sediment deformation. Higher in the section, volcaniclastic sands are interlayered with these black shales. Continuing the traverse toward the east, across the low area, the next sequence of rocks is the lowest of five major tuff sequences (Niem, 1977) within the Stanley Formation. Two of the units are exposed here. Walking up the western flank of Rattlesnake Bluff, you first cross through the Beavers Bend Tuff and then, at the top of the hill, the Hatton Tuff. These units are thickest here in the southeast portion of the uplift. Niem (1977) used that information and the grain-size distribution to argue for a southern source for these tuffs. The tuffs are rhyodacitic in composition, and each tuff has a thick lower unstratified pumiceous vitric crystal base overlain by a thin-bedded pumiceous tuff and an upper massive fine-grained siliceous vitric tuff. There is an interval between the tuffs with quartzose and feldspathic turbidites. At the top of the bluff, large flattened pumice fragments are altered to chlorite, forming a very distinctive texture. Looking back to the west, you can see back down section and across the stratigraphic boundary between the flysch sequence of Carboniferous age and the preflysch lower Paloezoic sequence. In walking back to the cars you will pass this boundary in the upper division of the Arkansas Novaculite.

From Locality 2, return to U.S. 259A and head north up the hill. The vuggy, reddish-weathering, upper division of the novaculite is exposed where the road curves toward the east. Beyond this curve, the road turns north again and the black shales of the lower Stanley Formation crop out. At the eastern end of the dam, turn right into the large parking area.

Locality 3 (Figs. 1, 3, 4) is located just north of the axial trace of a large syncline (Figs. 1, 3). To the south, units of the lower Stanley Formation are tightly folded and the shales have eroded to form the narrow valley north of the large water tank. To the north, the Arkansas Novaculite is rotated such that it is overturned, dipping moderately to the north (Fig. 4). The distinctive upper division of the novaculite sequence can be seen in the reddish-weathering bank below the gaging station, just north of the parking area. From Figure 3 the degree of overturning and the very sharp hinge of this syncline are apparent.

Examine the sequence of shale and sandstones of the lowest Stanley Shale exposed at the east end of the parking lot. Here, thin fine-grained sandstone layers are interlayered with a black shale sequence. Weathering turns these black shales to a light brown or buff color. Slaty cleavage is well developed here, dipping gently toward the northeast. The dip of the bedding is variable, but steeper than cleavage, indicating an overturned limb of a major fold (Fig. 3).

From here we will walk up the road approximately 240 ft (75 m) and down to the left into the first cove along the east shoreline of the reservoir (Fig. 4). Careful examination of the siliceous shales along the dirt road reveals a series of open, upright folds spaced approximately 6 to 15 ft (2 to 5 m) apart. These folds are asymmetric with the steep limb toward the northwest. The axial surfaces for these folds are steeply dipping toward the southeast (60°/65°SE). There is a cleavage associated with these folds; it varies from a crenulation to a rough ("fracture") character. A set of these data yield an average orientation of 60°/68°SE for this cleavage. These folds refold the slaty cleavage and reveal a younger ductile event.

On a larger scale these later folds can be seen in subtle flexures of the novaculite ridges (fig. 1). In the NW¼ of Sec. 11 and east-central sec. 16, two inconsistent fold patterns are seen in the overturned limbs. On the upright limbs of early folds, more gentle warps are developed (i.e., west section line, Sec. 10). In Sec. 4, the axial trace of the principal syncline is deflected from a due east to a southeast trend (Fig. 1). Also the Polk Creek Shale forms a broad structural dome in the NW¼ of Sec. 9. This uplifted section is indicated in the cross section. A simple model for this superposition is illustrated in Figure 5. The weak superposition of northeast-trending folds on the preexisting south verging folds creates slightly asymmetric, steeper plunging folds on the overturned limbs and more symmetric, shallower plunging folds on the upright limbs. On an even larger scale the northeast trend of the Broken Bow uplift is believed to be related to this late stage superposition; in other words, the uplift is a structural dome (Fig. 1). Using the plunge of the early folds and the northeast trend of the flanks, an estimated wavelength of 25 mi (40 km) and an amplitude of 2.5 to 5 mi (4 to 8 km) was proposed for the Broken Bow uplift (Nielsen, 1982). COCORP data suggest similar magnitudes for the Benton uplift to the east in Arkansas (Lillie and others, 1983).

From Locality 3, drive northwest across the dam and stop and the overlook on the west side of the lake.

Locality 4 (Figs. 1, 3, 4) is located on the shoreline of Broken Bow Reservoir at the west end of the dam. Looking east, you can see an extremely well-developed sequence of tight, overturned folds in the Blaylock Sandstone. A representative set of bedding and cleavage data (Fig. 6) reveals the geometry of these early folds. These lower-hemisphere projections show that the shallow dipping upright limb dips gently to the northeast, while the stepper overturned limb is oriented nearly east-west. This agrees with the earlier statements about the ridge lines (see Locality 1). The hinge zones on the folds are sharp as you see across the lake. The plunge is very gentle toward the east. The cleavage data for these folds are effectively parallel to the upright limb (see Locality 2), indicating that these structures are either strongly asymmetric or that the slaty cleavage has been modified or produced by a later event. Feenstra and Wickham (1975) have argued that some of the slaty cleavage is younger than the folds you see across the lake and that this younger slaty cleavage transects these folds. The next stop will investigate both the remaining stratigraphic sequence and the remaining elements of the deformation sequence.

Continuing northwest on U.S. 259A, over the next 1 mi (1.6 km) we cross through a complex zone (Fig. 4). The first outcrop on the northeast side of the road is the Blaylock Sandstone, which is steeply dipping (NW¼, Sec. 3). If you follow these units to the north side of the knoll you will find the layers and the early fold set rotated to a very shallow dip. A low-angle fault brings Bigfork

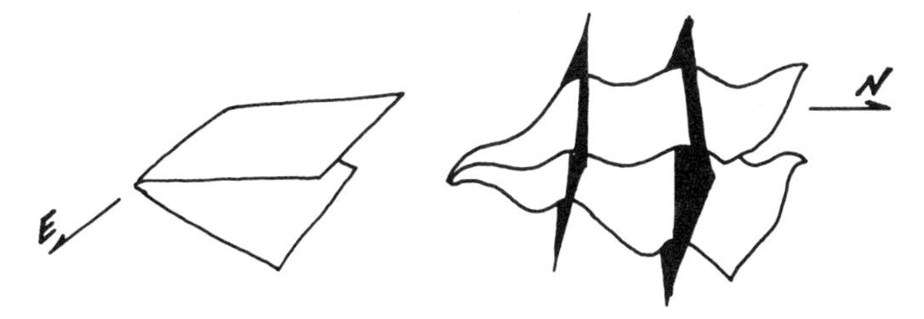

Figure 5. Fold model for the youngest folding event (F_3).

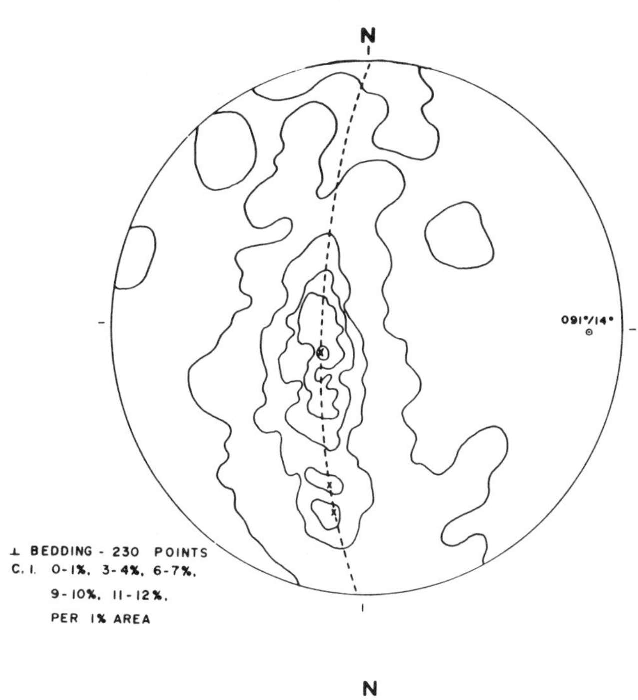

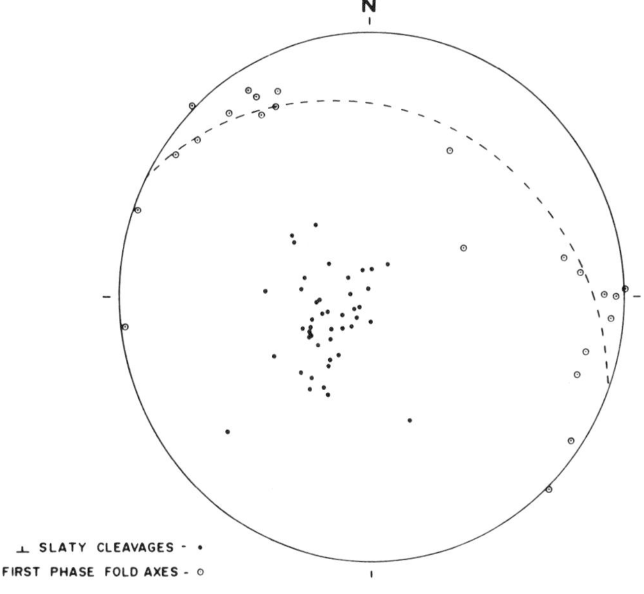

Figure 6. Lower hemisphere stereonets reveal geometry of earliest folding events. See text for explanation.

Chert nearly into contact with the Blaylock Sandstone, cutting out the Polk Creek Shale. At the turn out to the scenic overlook (NE¼, Sec. 4), exposures of the Polk Creek can be seen dipping toward the south. This area has yielded, surprisingly, some excellent graptolite specimens. From such specimens the Polk Creek Shale is assigned to the latest Ordovician (Spradlin, 1959). A series of northeast-trending high-angle faults cut across the Bigfork Chert ridge line north of the road. Proceeding west, cross over the spillway at Stephens Gap and park in the area below the spillway (NW corner, Sec. 4).

Locality 5 (Figs. 1, 3, 7) is in a strike valley of the Polk Creek Shale. Looking west, the Bigfork Chert is to the north, and the hills to the south are the Blaylock Sandstone. This valley is just north of the axial trace of a large syncline. Here, you examine the stratigraphy from the Blaylock Sandstone down to the Womble Shale. In addition, you complete the description of the folding events. Cross the small stream south of the parking area and proceed south along the dirt road (~1,300 ft; 400 m). The road will take you into an abandoned quarry within the Blaylock Sandstone. This thick accumulation of fine- to medium-grained sandstone is interlayered with black shale. The quarry runs east-west, parallel to the strike of the nearly vertical units. Bottom markings are apparent on the south wall of the quarry, indicating a southerly younging direction. These current markings, plus a few from the lakeshore in Sec. 3, suggest an east-to-west current (Fig. 8A). Recent work by Worrell (1984) suggests a more northwesterly current direction and a southern source for these sands. Examination of the internal stratification of the sand layers reveals locally well-developed Bouma sequences. The thickness of this sandstone is 885 ft (270 m) in the Beavers Bend State Park (Spradlin, 1959). Similar to the tuffs in the Stanley Formation, the Blaylock Sandstone thins toward the north and northwest. Lower Paleozoic sequences exposed in the Potato Hills 50 mi (80 km) to the northwest have no stratigraphic equivalent flysch sequence. Nor is there any equivalent at Black Knob Ridge near Atoka (~ 82 mi; 130 km to the west). This large submarine fan was apparently building from the south or southeast toward the west. It is interlayered between the two major chert sequences and represents some tectonic activity in the east during the Late Ordovician or Early Silurian.

Of particular structural interest are the mesoscopic fold elements in the quarry. Large buckles are seen in the quarry wall. These are open folds with horizontal axial surfaces. In cross-sectional view, a rough cleavage or parting can be seen associated with these "anomalous" folds (Feenstra, 1974). A slaty cleavage is also apparent in the black shale intervals. This cleavage generally dips to the *south,* an unusual orientation in light of the major folds (Fig. 3). As you walk out of the quarry, continue for a short distance southwest along the new quarry road. You will notice that the bedding and cleavage quickly turn to a very shallow dip toward the southeast (Figs. 7, 8B). In other words, you have walked across the axis of a very sharp synclinal hinge, which folds both bedding and cleavage. From Figure 8B, it is apparent that this later folding is approximatly coaxial with the earlier folds (see

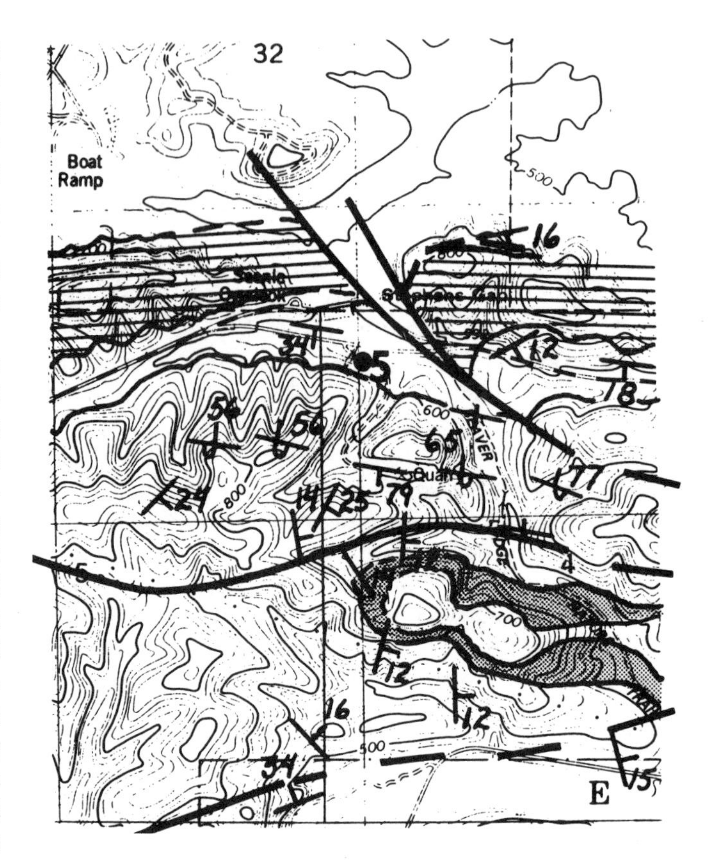

Figure 7. Geology at Locality 5. Missouri Mountain Shale and Bigfork Chert are shown. See Figure 1 for legend.

Fig. 6). This synclinal trace has been followed west-northwest to the NW¼ of Sec. 6 (Fig. 1) and east-southeast halfway across Sec. 4. This second-generation syncline is north of the earlier (principal) syncline cored by the novaculite (Fig. 1). A fault separates these two folds (Fig. 3).

Returning along the road to the north, you cross the gradational contact into the Polk Creek Shale. The lower Blaylock Sandstone is Late Ordovician based on graptolite fauna, while the upper portion is believed to be Silurian (Spradlin, 1959). Along this road and in the quarry, a few trace fossils have been observed. These feeding tracks are very common in certain locations (lakeshore, central Sec. 3). On the basis of these trace fossils, Chamberlain (1971) suggested a deep-water environment for the turbidites.

Some of the best outcrops of the Polk Creek Shale are exposed in the larger stream east of the parking area. These black graptolitic shales contain significant amounts of pyrite, which weathers reddish brown. Interbedded are layers of chert and some calcareous layers. At this locality many of these interlayers are boudinaged or broken by faults that dip toward the south or southwest. The slaty cleavage is nearly parallel to the few bedding measurements. In the stream bank northeast of the parking area, well-developed crenulation cleavage and kinklike folds are

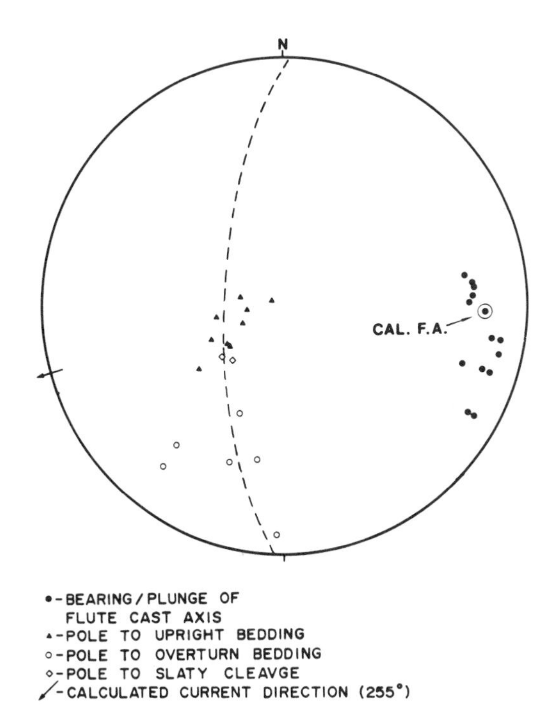

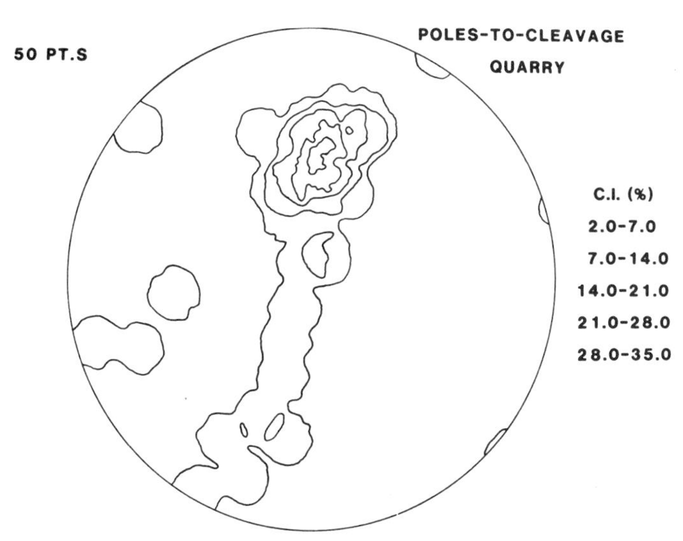

Figure 8. A. Lower hemisphere stereonets illustrating current directions in the Blaylock Sandstone, Beavers Bend State Park. B. Lower hemisphere stereonets illustrating the folding of cleavage in the Stephens Gap area (Locality 5).

strongly developed. These trend northeasterly and correspond to the youngest folding observed at Locality 3.

The Bigfork Chert and Womble Shale can be observed by climbing up the hill on the northeast side of the stream (into the SE¼,Sec.32,T4S,R25E). The Bigfork Chert is not well exposed; however, portions of the lower Bigfork Chert can be observed along the secondary road down to the lake. At the main highway, the section consists of highly deformed shales and sandstones. While the bedding and cleavage generally dips toward the south, the section is significantly disrupted by low-angle faults. Around the curve and down the hill, a section of black chert crops out. The angular weathered blocks of chert are commonly the only type of Bigfork exposure found. These chert layers are usually 4 in to a few feet (< 1 m) thick. Interlayered with the cherts are siliceous limestones and graptolitic shales, which grade upward into the Polk Creek Shales. Because of the faulting, this section is incomplete.

As this road turns back to the north and the lake access area, the exposures contain sandstones and shales, interpreted to be part of the Womble Shale sequence. Typically, the Womble sequence consists of soft, schistose, gray-green fine- to medium-grained sandstones and argillaceous shales (Spradlin, 1959). Farther away from the contact with the Bigfork Chert, the Womble Shale is more characteristically black, thin-bedded hard shale, which is frequently cut by quartz veins in this area (look north across the lake at the shoreline near the marina). For the adventurous, a scramble up and over the point to the lakeshore northeast of this road will reveal another distinctive lithology in the Womble Shale. Here you will find a conglomerate. The clasts are dominantly sandstone, subangular to subrounded, and generally matrix supported. This author has interpreted these conglomerates to be some type of debris flow, while Dix and Casey (1985) have suggested a tectonic origin, a type of broken formation. Some of these clasts are quite large (~ 2 to 3 ft; 1 m long). These clasts are also generally elongate in a southeasterly direction. A walk to the east along the lakeshore will reveal a variety of folds, and just west of the section line (32/33) a low-angle fault disrupts the section with indications of motion toward the southeast. Work in the older sequences by Hubert (1984) and Dix (1985) suggests that the flank sequence may be structurally detached from those older rocks. This is in agreement with Miser's (1929) earlier interpretation in which he suggested that the entire flank sequence is part of a younger-over-older fault.

CONCLUSIONS

As you struggle back to the parking area, think about these summary ideas. First, the stratigraphic sequence you have observed includes a major tectonostratigraphic contact. At the Arkansas Novaculite/Stanley Shale contact there is a major change. The rate of basin subsidence and the rate of sediment deposition increased dramatically. The Carboniferous section includes conservatively 40,000 ft (12,000 m) of flysch. The preflysch sequence is significantly less (~ 6,300 ft; 2,000 m) and represents approximately twice as much time (Flawn and others, 1961). While this general pattern is well established, you have seen—particularly in the Blaylock Sandstone and secondarily in the Womble Shale—evidence for some tectonic activity in the Ouachita basin during the Early Silurian and Middle Ordovician respectively.

Second, the structural data outlined here support three ductile events (Fig. 9). The first folds are the dominant structures, the southerly verging overturned folds (Fig. 9A). The regionally developed slaty cleavage is associated with this event. Postdating these initial folds and probably representing a continuation of the same event, the early folds were refolded and the slaty cleavage rotated. Associated with this refolding, this author suggests that most of the low-angle faults seen in the cross section were developed (Fig. 9B). This refolding appears coaxial in the flank rocks on the southeast side of the uplift. This is not the case on the northwest flank. The low-angle faults tend to detach the flank sequence from the older rocks and to thicken the sequence. Associated with the tectonic thickening is the later flattening, which produces a reoriented slaty cleavage (particularly in the upright limbs of the early folds), possible thinning of the upright stratigraphic sequences, development of open buckles in the steeply dipping overturned limbs, a poorly developed horizontal cleavage on the steeply dipping limbs, and possible complex motion on preexisting faults. For example, vertically oriented shortening would tend to produce normal slip on the low-angle faults in Figure 9C. Similarly, the large syncline in which Locality 5 is found is likely to have been flattened and pushed back to the north. The final event is the late crenulation and northeast-trending folds associated with the major Broken Bow uplift (Nielsen, 1979). The moderately southeast-dipping faults (Fig. 3) are related to this final event.

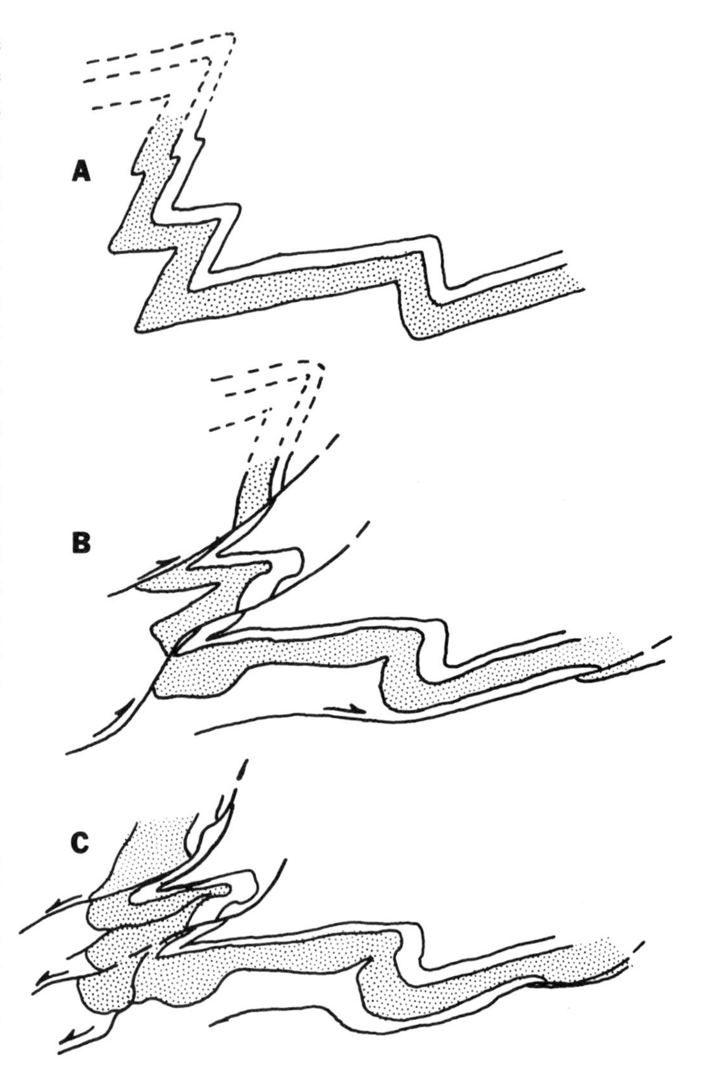

Figure 9. Summary of folding events. See text for explanation.

REFERENCES CITED

Chamberlain, C. K., 1971, Bathymetry and paleoecology of Ouachita geosyncline of southeastern Oklahoma as determined from trace fossils: American Association of Petroleum Geologists Bulletin, v. 55, p. 34–50.

Dix, M. C., 1985, Sequential deformation of Ordovician slates in the Stephens Gap area, eastern Broken Bow Uplift, Oklahoma Ouachita Mountains: Geological Society of America Abstracts with Programs, v. 17, p. 156.

Dix, M. C., and Casey, J. F., 1985, Tectonically broken formation in the Womble "Shale" of the Broken Bow Uplift and its association with the Glover fault: Geological Society of America Abstracts with Programs, v. 17, p. 156.

Feenstra, R., 1974, Evolution of folds in the Blaylock Formation (Silurian), Ouachita Mountains, southeastern Oklahoma [M.S. thesis]: Norman, University of Oklahoma, 77 p.

Feenstra, R., and Wickham, J., 1975, Evolution of folds around the Broken Bow Uplift, Ouachita Mountains, southeastern Oklahoma: American Association of Petroleum Geologists Bulletin, v. 59, p. 974–985.

Flawn, P. T., Goldstein, A., King, P. B., and Weaver, C. E., 1961, The Ouachita System: Bureau of Economic Geology, University of Texas Publication 6120, 401 p.

Honess, C. W., 1923, Geology of the southern Ouachita Mountains of Oklahoma: Oklahoma Geological Survey Bulletin 32, Pt. I, 278 p.

Hubert, L. M., 1984, Structure and lithologic variation in the central core of the Broken Bow Uplift, Ouachita Mountains, Oklahoma [M.S. thesis]: Richardson, University of Texas at Dallas, 112 p.

Lillie, R. J., and 7 others, 1983, Crustal structure of Ouachita Mountains, Arkansas; A model based on integration of COCORP reflection profiles and regional geophysical data: American Assocation of Petroleum Geologists Bulletin, v. 67, p. 907–931.

Miser, H. D., 1929, Structure of the Ouachita Mountains of Oklahoma and Arkansas: Oklahoma Geological Survey Bulletin 50, 30 p.

Miser, H. D., and Purdue, A. H., 1929, Geology of the DeQueen and Caddo Gap quadrangles, Arkansas: U.S. Geological Survey Bulletin 808, 195 p.

Murgatroyd, C., 1980, Significance of silicified carbonate rocks near the Devonian-Mississippian boundary, Ouachita Mountains, Oklahoma: Shale Shaker, v. 31, no. 1, p. 1–16.

Nielsen, K. C., 1979, Late stage folding along the southeastern margin of the Broken Bow uplift, Oklahoma: Oklahoma Geology Notes, v. 39, p. 237–238.

——, 1982, Regional implications of late stage folding within the Broken Bow uplift, southeastern Oklahoma: Geological Society of America Abstracts with Programs, v. 14, p. 134.

Niem, A. R., 1977, Mississippian pyroclastic flow and ash-fall deposits in the deep-marine Ouachita flysch basin, Oklahoma and Arkansas: Geological Society of America Bulletin, v. 88, p. 49–61.

Spradlin, C. B., 1959, Geology of the Beavers Bend State Park area, McCurtain County, Oklahoma [M.S. thesis]: Norman, University of Oklahoma, 105 p.

Viele, G. W., 1973, Structure and tectonic history of the Ouachita Mountains, Arkansas, *in* Gravity and tectonics: New York, John Wiley and Sons, p. 361–377.

Worrell, D. G., 1984, Sedimentology and petrology of the Blaylock Sandstone (Silurian) of the Ouachita Mountains, Arkansas and Oklahoma [M.S. thesis]: Baton Rouge, Louisiana State University, 128 p.

U.T.D. Contribution No. 496

Beaver Dam, northwestern Arkansas

Walter L. Manger, Department of Geology, University of Arkansas, Fayetteville, Arkansas 72701

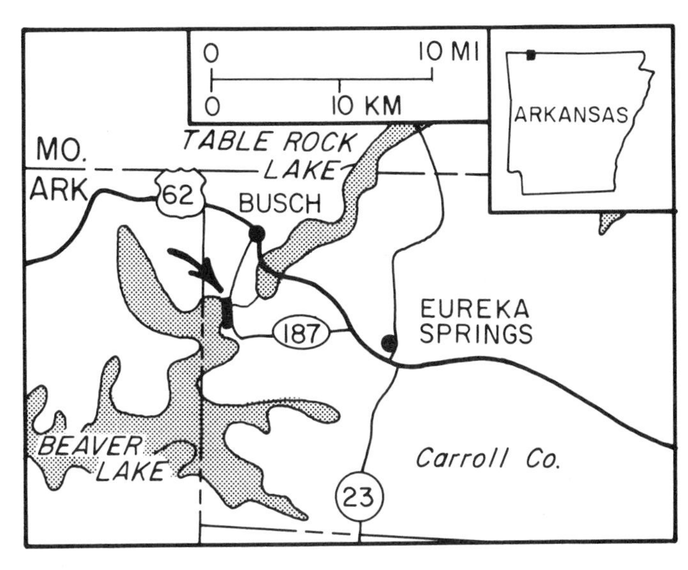

Figure 1. Index map of the northwestern part of Carroll County, Arkansas, showing location of the Beaver Dam section.

LOCATION

The Beaver Dam exposure is reached by taking Arkansas 187 south from its junction with U.S. 62 immediately south of Busch, Arkansas (Fig. 1). Arkansas 187 crosses Beaver Dam and a turnoff to the dam site overlook is located approximately 0.25 mi (0.4 km) south of the dam. Parking may be found at the overlook. This exposure occupies NW¼Sec.10,T.20N.,R.27W., Carroll County, Arkansas.

SIGNIFICANCE

Throughout the Paleozoic, the North American craton experienced periodic transgression and regression by shallow seas, leaving the stratigraphic record punctuated by unconformities. The Beaver Dam site, situated on the south flank of the Ozark Dome, displays an excellent example of Ordovician to Mississippian cratonic stratigraphy typical of the southern midcontinent region.

STRATIGRAPHIC SUCCESSION

The Beaver Dam site exposes portions of the Cotter and Powell Formations (Lower Ordovician), Clifty Sandstone (Middle Devonian), Chattanooga Shale (Upper Devonian), the St. Joe Limestone, and an incomplete section of the cherty limestones of the Boone Formation, both of Lower Mississippian age (Fig. 2).

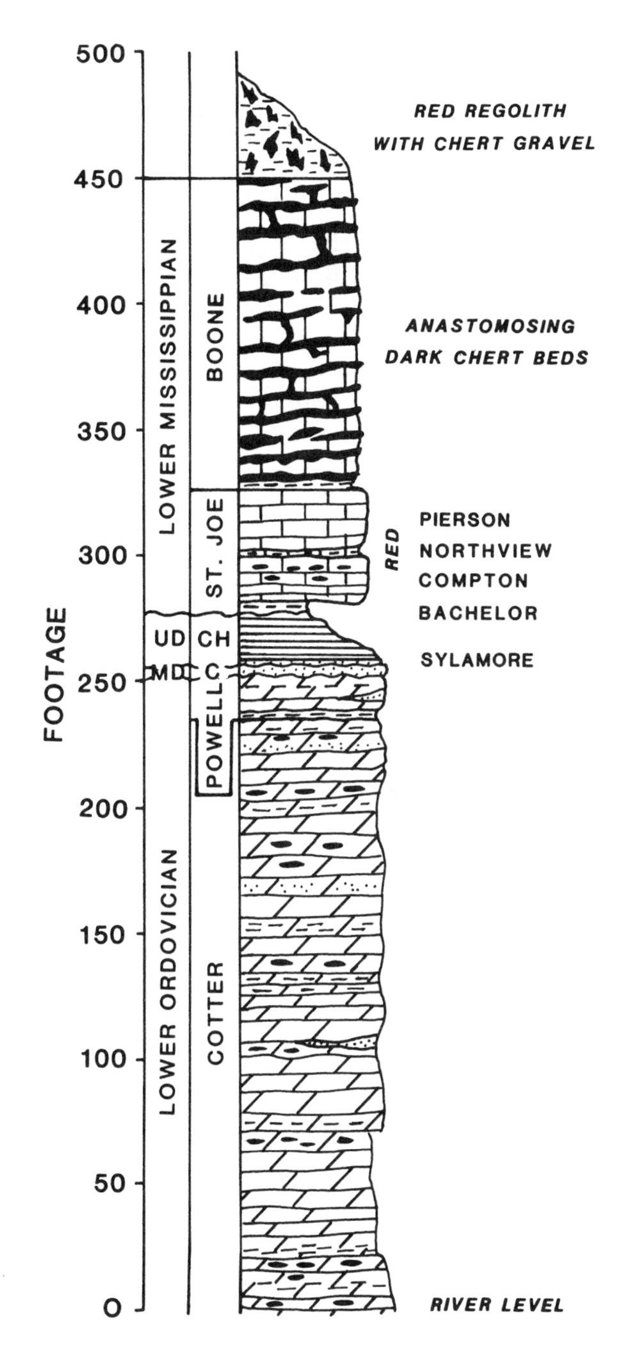

Figure 2. Generalized stratigraphic succession of formations exposed at the Beaver Dam Site. Standard lithologic symbols followed; UD, Upper Devonian; MD, Middle Devonian; CH, Chattanooga; C, Clifty. Section measured by U.S. Army Corps of Engineers, with modifications by J. D. McFarland III, Jeffrey Hall, Jeffrey Liner, and Krista Holland.

Figure 3. Section from river level to dam crest is entirely Lower Ordovician Cotter Formation (C). Thin remnant of Lower Ordovician Powell Formation may be seen beneath bench formed by Clifty Sandstone (arrow) of Middle Devonian age.

DESCRIPTION

The bulk of the exposure at Beaver Dam belongs to the Lower Ordovician Cotter Formation (Fig. 3). Here the Cotter is represented by nearly 250 ft (76 m) of monotonous, dark gray fresh to light brown weathered dolomite, with dark or banded chert as nodules and thin, discontinuous beds. The carbonates are dense and microcrystalline, occurring in thin to thick beds with laminated argillaceous, oolitic, stromatolitic, and quartz sand-bearing intervals developed sporadically through the section. Fossils are virtually nonexistent in the Cotter. Lenses or beds of intraformational breccia composed of chert and carbonate clasts, and orthoquartzitic sandstone are scattered through the succession as well. Conformably succeeding the Cotter is 8 to 10 ft (2.4 to 3.0 m) of Powell Formation. The Cotter-Powell contact is taken at the base of a persistent 1- to 2-ft-thick (0.3 to 0.6 m) dark gray shale (Fig. 4). The Powell is expressed as a light gray to very light brown, laminated, dolomudstone. At Beaver Dam, the Powell also lacks fossils, but in contrast to the Cotter, it lacks chert. A small orthoquartzite lense in the Powell may be seen at road level at the north end of the road cut. The Cotter-Powell interval reflects a series of shallowing upward cycles, indicating a change from shallow subtidal to supratidal environments.

The gently dipping Powell is succeeded unconformably by a massive sandstone ledge of the Middle Devonian Clifty Formation (Fig. 4). Here, the Clifty is a white to yellow-brown orthoquartzite comprising essentially a single bed. At its base, the Clifty contains a concentration of angular cavities interpreted as external molds of limestone clasts subsequently dissolved by ground water. The top of the Clifty is heavily burrowed and discolored. Conodonts from this sandstone ledge are Middle Devonian in age (Hall and Manger, 1978).

Unconformably overlying the Clifty Formation are thin beds of the Upper Devonian Sylamore Member Chattanooga Shale (Fig. 4). Beneath the dam site overlook, the Sylamore beds have been stripped away, but they are easily seen in the road-cut exposure (Fig. 4). There, the Sylamore consists of phosphatic, pebble-bearing orthoquartzite of medium yellow-brown color. Conodonts from this sandstone are Upper Devonian in age (Hall and Manger, 1978). The entire sand sequence between the Powell and Chattanooga Black Shale has been referred to the Sylamore Member by many previous investigators. However, the two units may be readily separated. Evidence for an unconformity between the Clifty and Sylamore is shown by an irregular, burrowed surface of contact, lithologic change, and difference in age of the two units. Thus, two sand units, not one, are present. Lumping these units together obscures a significant part of the geologic history of the Ozarks. The Sylamore is succeeded by typical black shale of the Chattanooga, although these beds are not well exposed here (Fig. 4).

The Mississippian Bachelor Member, St. Joe Limestone, overlies the Chattanooga Shale at the Beaver Dam site (Fig. 2). The basal contact is not exposed, but its approximate position is easily determined. The member also contains middle Kinderhookian conodonts at the dam site, but the precise age of the Chattanooga has not been determined here.

The remaining members of the St. Joe Limestone succeed the Bachelor conformably. Note that the reentrant of the Northview Member is well developed providing differentiation of the Compton and Pierson Members. Two striking differences are evident in the St. Joe at the dam site when compared to other exposures in northwest Arkansas. First, the Compton Member exhibits well-developed nodular chert in its upper part at Beaver Dam. Second, the Northview and Pierson Members exhibit carbonate beds that are medium red to pink. Red color predominates the St. Joe in its type region, north-central Arkansas.

The Boone Formation typically forms a rubble slope and the contact is not well exposed at the overlook. The St. Joe–Boone contact is drawn at the change to fine-grained carbonate packstones from gravel-sand size, packstone-wackestone lithologies of the St. Joe. It also coincides with the first "Boone-type" chert development and a shaley interval. An exposure of this contact is seen by returning to Arkansas 187 for approximately 1,500 ft (457 m) south of the overlook turnoff (Fig. 5). There the characteristic changes of the St. Joe–Boone contact occur through an interval of about 3.5 ft (1.1 m). Calcisiltites and fine-sand packstones appear abruptly above the last Pierson grainstones. These fine-grained carbonates contain small, irregular chert masses and become interbedded with medium brown shales, some as much as 6 in (15 cm) thick. Anastomosing chert beds of the "Boone"

Figure 4. Section exposed in road cut at south end of Beaver Dam. Solid arrows mark lower and upper boundaries of Powell Formation (Lower Ordovician). Open arrow denotes Sylamore Member, Chattanooga Shale (Upper Devonian), which unconformably succeeds Middle Devonian Clifty Sandstone. Chattanooga Shale forms grassed slope (C) at top of road cut.

Figure 5. Hammer marks contact of St. Joe (S) and Boone (B) limestones (Lower Mississippian) in vicinity of Beaver Dam Site. Here, typical lithologic changes used to mark boundary occur through a span of about 3.5 ft (1.1 m). Anastomosing, "Boone-type" chert appears at top of exposure.

type do not appear until the top of the exposure, several feet above the actual lithostratigraphic contact.

The Beaver Dam Site also affords an opportunity to examine the penecontemporaneous chert that is typical of the lower Boone Formation. As can be seen in the road cuts, this chert is dark gray to black, unfossiliferous, and occurs in nodules and beds that disrupt rather than follow the bedding. Drape and compaction features are common, indicating the early timing of this chert development. A view into the quarry near the top of the hill demonstrates the thin-bedded character and persistence of these Lower Boone chert-bearing carbonates. The Boone is capped by a distinctive red regolith composed of chert blocks in a clay matrix. The regolith is a Boone weathering product reflecting limestone solutioning by increased rainfall during portions of the Pleistocene Epoch.

REFERENCE CITED

Hall, J. D., and Manger, W. L., 1978, Devonian sandstone lithostratigraphy, northern Arkansas: Arkansas Academy of Science Proceedings, v 31, p. 47–49.

The Paleozoic rocks of the Ponca region, Buffalo National River, Arkansas

John David McFarland III, Arkansas Geological Commission, 3815 W. Roosevelt Road, Little Rock, Arkansas 72204

LOCATION

Three state highways provide the primary access to the Ponca region (Fig. 1). They are Arkansas 21, 43 (both north-south), and 74 (east-west). Three campgrounds are located in the area: Lost Valley, Steel Creek, and Kyle's Landing. These are all Buffalo National River campgrounds and designed for the tent camper. The roads leading into each camp are gravel and, in the case of Steel Creek and Kyle's Landing, involve steep grades. Each camp has excellent outcrops within easy walking distance.

The roadcuts and mines are easily accessible by passenger car and short walks. Although some of the river exposures may be reached by road or on foot, the ideal mode of travel is by canoe. Unfortunately, water levels typically allow canoe travel for only a few months in the winter and spring. The river in this area is rated class I-II white water, with fairly frequent, generally unobstructed rapids. During the floating season, canoes may be rented from any of several outfitters in Ponca.

The exposures along the River and around the campgrounds are within the boundaries of the Buffalo National River and are under the protection of the National Park Service. Please respect this reservation. Roadcuts in the immediate area but outside the park provide ample outcrops for sample collection.

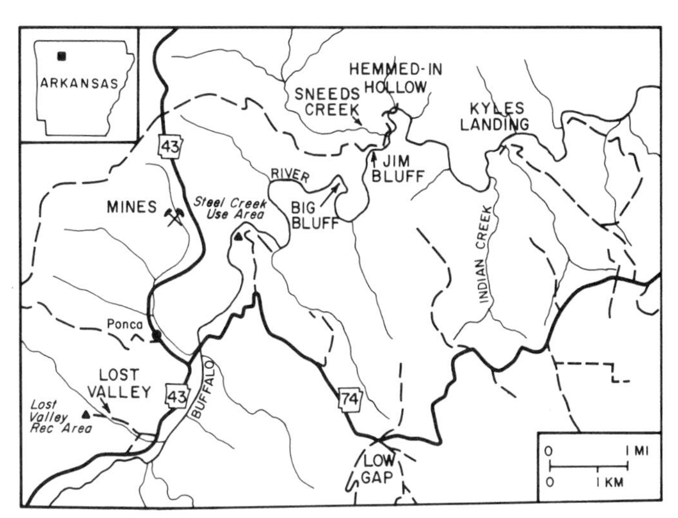

Figure 1. Map of part of northern Newton County, Arkansas. The sites mentioned in this guide are located in the central part of this map.

SIGNIFICANCE

In northern Newton County, Arkansas, the Buffalo National River has entrenched more than 1,200 ft (400 m) of flat-lying Paleozoic rocks along the north edge of the Boston Mountain plateau. Majestic bluffs, some more than 500 ft (150 m) high, cut from beds of sandstone, dolomite, and limestone of Ordovician and Mississippian ages, dominate the river bank exposures (Fig. 2a). Roads descending into the river valley reveal limestones, cherts, sandstones, siltstones, and shales of Mississippian and Pennsylvanian ages. Lead and zinc mines dating from Civil War days can be found around the town of Ponca, Arkansas. Ancient and modern karst features located in the area include sink holes, caves, sag structures, natural bridges, springs, and sinking streams.

The strata found in the area are, in general, representative of the north Arkansas stratigraphic section. Rock units found to the east and west of this area (but generally missing from the local section) are most of the upper Ordovician, almost all of the Silurian, and all of the Devonian. It appears that upper Ordovician thru Devonian sediments did collect across this area but were mostly removed by intervening periods of erosion prior to Mississippian time. Even so, isolated outcrops of these other rocks are found in the Ponca region.

This chapter provides general descriptions of various sites in the area that exhibit the greatest stratigraphic continuity, best outcrops, or most interesting features. For a more complete description or further information, contact the Arkansas Geological Commission.

STRATIGRAPHIC SUMMARY

Powell Dolomite. The early Ordovician-age Powell Dolomite is the oldest unit found along this part of the Buffalo River. The thickness of the Powell varies from 40 to almost 200 ft (12 to 60 cm) in the region but only the top 50 to 60 ft (15 to 20 m) are exposed in the Ponca area. The Powell is a light-gray, fine-grained, argillaceous dolomite with rare layers of concentrically banded nodular chert. The top of the Powell is marked where the typical Powell lithologies pass disconformably into the medium- to dark-gray, conglomeratic, dolomitic sandstone of the superjacent Everton Formation.

Everton Formation. The Middle Ordovician Everton Formation is about 250 to 350 ft (80 to 110 m) thick in this area and has been divided into several members by various workers. The most prominent of these members along the upper Buffalo is the 100-ft (30-m)-thick Newton Sandstone Member of the middle Everton. This sandstone, like other sandstones of the Everton, is a fine- to coarse-grained, well-rounded, frosted, often friable, occasionally well-sorted quartz arenite. Cementation of Everton sandstones is accomplished by dolomitic or calcitic carbonates or interlocking overgrowths of silica. In some beds the calcitic cement forms a poikilotopic texture. The Everton above and below the Newton Member consists of alternating beds of dolomite, limestone, and sandstone, none of which persist for any great distance laterally. Everton dolomites are very fine to coarse crystalline and often include variable amounts of limestone and sand. The limestones consist of calcilutites and calcarenites with variable amounts of dolomite and sand. Beds of quartz sandstone of the type found in the Newton occur throughout the intervals. A few discontinuous conglomerates, breccias, and cherts occur at various horizons in the Everton.

Plattin Limestone. There are only a few local sites where the Middle Ordovician Plattin Limestone is present. Lithologically, it varies from a dismicrite to a very fine grained dolomite. Most outcrops of Plattin in this area exhibit only a few inches to a few feet of limestone.

Fernvale Limestone. The Fernvale Limestone is an Upper Ordovician crinoidal calcirudite. It was only sporadically preserved by pre-Mississippian karst and erosion processes. Where preserved it rests unconformably on the Plattin or Everton Formations.

St. Joe Limestone. The early Mississippian–age (Kinderhookian-Osagean) St. Joe Limestone in this area is a 30- to 55-ft (9 to 17 m)-thick interval of very fine to coarse-grained, fossiliferous, tabular-bedded, argillaceous calcarenite with thin interbeds of calcareous shale. Throughout this region the St. Joe is distinguished by beds of brick-red limestones and shales. Nodules and/or thin beds of chert have been observed in some outcrops. The basal unit of the St. Joe is generally a conglomeratic quartz arenite beneath a thin, shaly limestone containing phosphate nodules. The pre-Mississippian erosion left a stratigraphically irregular surface. Almost every place where the Ordovician-basal Mississippian sequence is exposed in this area, its stratigraphic succession is different. In Newton County the St. Joe sits variously on Fernvale, Plattin, or Everton (Middle to Late Ordovician limestones, dolomites, and sandstones) but may overlie the Cason (an Ordovician-Silurian shaly unit) in a few places. In other parts of northern Arkansas, the St. Joe generally overlies a Devonian unit. The upper contact with the Boone Formation is conformable and generally marked by a thin shale and an increase in the volume of chert. This contact can be somewhat obscure in local areas where the chert content of the Boone is low.

Boone Formation. The Boone Formation (Osagean to Meramecian in age) consists of 250 to 350 ft (80 to 110 m) of fine- to coarse-textured, fossiliferous limestone, cherty limestone, and chert. The beds in the lower part of the Boone are generally fine-grained limestones with dark cherts (thought to be penecontemporaneous). Upper parts of the Boone exhibit coarser limestones and light-colored cherts (apparently diagenetic). Chert often predominates a typical Boone sequence, but it varies in volume both stratigraphically and spatially. In the vicinity of Indian Creek, near Kyles Landing, the Boone is primarily limestone with just a little nodular chert. Typically, the cherts of the St. Joe are nodular to smooth-bedded, whereas the cherts of the Boone tend to be irregular and anastomosing.

Batesville Formation. The Chesterian–age Batesville Formation consists of brown, calcareous, quartz arenites interfingered with a limestone member, the Hinesville. Although not very well exposed, the usual thickness of the Batesville in this area is 6 to 8 ft (2 to 3 m) but can range from being locally absent to more than 20 ft (6 m) in thickness. The Batesville rests unconformably on the Boone and grades into the overlying Fayetteville Shale.

Fayetteville Shale. The Fayetteville Shale (Mississippian, Chesterian age) is chiefly a black fissile claystone with locally abundant, sometimes septarian, clay-ironstone concretions. The upper Fayetteville frequently contains beds of dark-gray, fine-grained limestone. The unit is 150 to 250 ft (50 to 80 m) thick in this area but it is poorly exposed.

Pennsylvanian Strata. The Pennsylvanian portion of the local section is represented mostly by Morrowan age rocks of the Hale and Bloyd Formations. The Hale Formation is divided into two members—the older Cane Hill Member and the younger Prairie Grove Member. The Cane Hill is mostly silty shale with a few siltstone and sandstone beds. The Prairie Grove is a massively bedded, bluff-forming, calcareous sandstone. Its basal contact with the Cane Hill is sharp and unconformable. The Prairie Grove grades upward into a coarse bioclastic limestone that is lithologically similar to the overlying Brentwood Limestone Member of the Bloyd Formation. The contact between the Hale and the Bloyd is unconformable but obscure. The remainder of the Bloyd above the Brentwood is dominated by silty shales and thin sandstones interrupted by a bluff-forming, generally massive, fluviatile, cross-bedded quartz arenite informally known as the Gaither Mountain sandstone. Basal Atoka Series strata may cap some of the hills but its expression is minor and somewhat subjective.

SITE DESCRIPTIONS

Big Bluff. Big Bluff (Fig. 2b) is the highest bluff on the Buffalo National River, standing more than 500 ft (150 m) high. The basal 50 ft (15 m) or so is Powell Dolomite, which is mostly covered with vegetation and talus. Exposed above the Powell is about 60 ft (20 m) of lower Everton Formation dolomite, sandstone, and limestone. The massive 110-ft (35 m) thick Newton Sandstone Member of the Everton forms the lower bluff just above the riverside trees. The base of the vegetation break across the middle of the bluff marks the approximate top of the Newton. Overlying the Newton is about 100 ft (30 m) of upper Everton sandstone, dolomite, and limestone. The reentrant in the upper half of Big Bluff is known as the Goat Trail and is formed in the lower portion of the St. Joe Limestone and on a 4- to 6-ft (1.5 to 2 m) basal Mississippian sandstone. The Goat Trail may be reached from canoes by beaching at the downstream end of the Big Bluff pool, in the slough just above the exit shoal, then climbing across the talus slope to the base of the bluff and working around and up the south end of the bluff. The trail is indistinct until the top of the Newton Member. Travertine deposits are

a b c

Figure 2 (a). Typical outcrop along the upper Buffalo National River. This Everton Formation bluff is located adjacent to the Steel Creek campground and is composed of dolomites, sandstones, and limestones. (b) Big Bluff, thought to be the highest bluff between the Appalachian and Rocky Mountains. Rocks of Ordovician age make up the lower two-thirds of the bluff and underlie Mississippian deposits. Near the center of the figure, just below the basal Mississippian strata, a lens of post-Everton limestone is partially covered by travertine. Intraformational unconformity at base of Newton Member is visible to right of center just above riverside trees. (c) Hemmed-In Hollow Falls. At the terminus of a half-mile hike from the Buffalo National River, there is a vertical section of the Powell Dolomite with the Everton Formation above. The Powell-Everton contact is indicated on the figure. A spectacular 200-ft (60 m) waterfall develops here during the rainy season.

developed at various places along Goat Trail and are associated with fractures (joints) in the carbonates. Under Goat Trail, about midway across the bluff face, there is a lens of Plattin and Fernvale limestones (Ordovician). The St. Joe is about 53 ft (16 m) thick on Big Bluff and is succeeded by limestones and cherts of the Boone, which extend to the top of the bluff.

Jim Bluff. Jim Bluff (Fig. 1) is within the Jim Bluff graben and is stratigraphically equivalent to the Goat Trail sequence at Big Bluff. St. Joe Limestone makes up most of Jim Bluff. Massive sandstones at the base of the bluff are Everton, anomalously thick basal Mississippian sands, or some of each. The normal fault (downthrown to the north) that marks the south boundary of this graben strikes generally east-southeast. You can find the trace of this fault by working your way to the southwest end of the bluff where the fault truncates the bluff. The displacement of the fault is over 300 ft (90 m). Dip observed in this outcrop is thought to be drag-induced.

Sneeds Creek. A half-mile (0.8 km) up Sneeds Creek (Fig. 1) from its confluence with the Buffalo River is a long, wide, and reasonably flat slab of basal Mississippian or upper Everton sandstone known locally as Rocky Bottoms. The hike takes you past and along an old road that crosses Sneeds Creek on the edge of a bed of sandstone inclined to the west-southwest (upstream). The sandstone bed is cut off just downstream of the road-crossing by the normal fault (downthrown to the south) that defines the north edge of the Jim Bluff graben.

About 600 ft (200 m) farther upstream, this same fault trace can be seen again on the north side of the valley floor. A small drainage comes in there, over a rounded Everton sandstone bank. A shallow trench between the bank and the slab forming the valley floor has developed because of faster erosion of the rock in the fault zone. In addition, the slab has been bent by drag induced by the faulting. Immediately west of this point it is also apparent that the slab dips downstream. This incline is not due to drag but rather the overall geometry of Jim Bluff graben. The 3.5-mi (5.6 km)-long graben structure is hinged on each end and depressed more than 300 ft (90 m) in the middle. Beds of St. Joe limestone crop out on the south side of Rocky Bottoms.

Hemmed-In Hollow. Hemmed-In Hollow (Fig. 1) is a 0.5-mi (0.8-km)-deep box canyon reached by either of two trails; one starts at Sneeds Creek, the other starts at the mouth of the Hollow. Outcrops of Powell Dolomite occur along the stream that the trail crosses and follows within the Hollow. Local changes in the dip of the strata seem to be related to karst processes. Near the upper portions of the Hollow, chert beds assigned to the Powell are exposed in the main stream. The waterfall (Fig. 2c) at the head of the Hollow is about 200 ft (61 m) high and is reported to be the highest waterfall between the Appalachian and Rocky Mountains. The basal 50 to 60 ft (15–18 m) of this cliff face is Powell. The Everton Formation overlies the Powell and outcrops to the top of the cliff. The lip of the waterfall is in the Newton Sandstone Member. A slight trail leads up and off to the east from near the base of the waterfall and curls back across the cliff and behind the waterfall. This trail across the cliff face is in the Powell, a few feet below the Powell-Everton contact. The contact is marked by a basal Everton conglomeratic sandstone overlying argillaceous dolomite of the Powell.

A few yards upstream (on the Buffalo) from the mouth of Hemmed-In Hollow, a small collapse structure or sink is developed in the riverbed outcrops of the Everton.

About 0.5 mi (0.8 km) downstream from Hemmed-In Hollow, a sandstone-filled fossil sinkhole (Fig. 3) is present in a low bluff on the west side of the river. The sink is developed in the Everton. The top of the section is missing, making it difficult to say when the sink developed. The sandstone filling is "typical Ordovician" in character, suggesting an Ordovician (Everton?) age. However, it is also possible that this sink is post-Ordovician in age and filled with sand derived from Ordovician sandstones eroded as late as the Mississippian transgression.

Figure 3. Fossil sinkhole. This is a portion of a pre-Mississippian sinkhole complex developed in the Everton 0.5 mi (0.8 km) downstream from Hemmed-In Hollow.

Lost Valley. Lost Valley (Fig. 1) is fairly typical of the small karst valleys and hollows around this portion of the Buffalo National River. The trail to the upper end of Lost Valley passes by many karst features—springs, solution-enlarged fractures, a natural bridge in the St. Joe, an undercut bluff in the Boone, several waterfalls—and terminates at a small cave. Bring your flashlights.

About 0.5 mi (0.8 km) west of the entrance to Lost Valley on Arkansas 43/74 is a roadside quarry (private property) with a sequence of Everton, Fernvale, and St. Joe. The base of the St. Joe is represented by a shaly sandstone that is entrenched up to 3 ft (1 m) into the underlying Fernvale. Small deposits and films of bitumen have been found here in the top few feet of the Everton and in the lower part of the Fernvale.

Low Gap. About 485 ft (148 m) of almost continuous outcrop of Morrowan age strata are exposed along a dirt road ascending to the southwest from Arkansas 74 at Low Gap, Arkansas (Fig. 1). At the intersection is the Cane Hill Member of the Hale Formation. About 0.25 mi (0.4 km) from the intersection is the Prairie Grove Member of the Hale. The top of the Prairie Grove is placed above a thin shale overlying locally developed fine-grained limestone. The basal bed of the overlying Bloyd Formation (Brentwood Member) is a conglomeratic, calcareous, bioclastic, sandstone in sharp but irregular contact with the Prairie Grove. About 0.65 mi (1 km) from the intersection is a rather poor outcrop of Gaither Mountain sandstone, the middle Bloyd sandstone that generally forms a prominent bluff. The roadside exposures continue another 0.5 mi (0.8 km) to the top of the hill. The thick sandstone beds at the top of the roadcut exposures are thought to be basal Atokan deposits.

Mines and Mineralization. Lead and zinc have been mined sporadically in the Ponca area from Civil War days into the 1950s. More than 4,000 tons of galena, zinc carbonate, and zinc silicate concentrates have been produced from deposits in the Boone and Batesville Formations. The mineralized deposits in the Ponca district occur along the Ponca lineament, which is thought to be the surface expression of a Pennsylvanian– or Permian–age shear fracture related to the Ouachita orogeny. The Ponca lineament strikes north 30° east and can be traced on high-altitude photography or satellite imagery for more than 40 mi (60 km). Small faults associated with this lineament have been noted in and near many of the mines of this area. The most accessible mines are on private property just west of Arkansas 43 about 1.9 mi (3 km) north of the Ponca Post Office. There are several small drifts and spoil piles in this area. The mines are considered safe for the most part but do have areas of collapse—care should be taken.

SELECTED REFERENCES

Haley, B. R., and others, 1976, Geologic map of Arkansas: U.S. Geological Survey map.

Manger, W. L., 1985, Devonian-Lower Mississippian lithostratigraphy, northwest Arkansas: University of Arkansas Geology Department, 25 p.

McFarland, J. D., III, 1979, Geologic float guide on the upper Buffalo National River: Arkansas Geological Commission Guidebook 79-2.

McKnight, E. T., 1935, Zinc and lead deposits of northern Arkansas: U.S. Geological Survey Bulletin 853, 311 p.

Purdue, A. H., and Miser, H. D., 1916, Description of the Eureka Springs and Harrison quadrangles: U.S. Geological Survey Atlas, Folio 202.

Geology of the Buffalo River Valley in the vicinity of U.S. 65, Arkansas Ozarks

William W. Craig, Department of Earth Sciences, University of New Orleans, New Orleans, Louisiana 70148

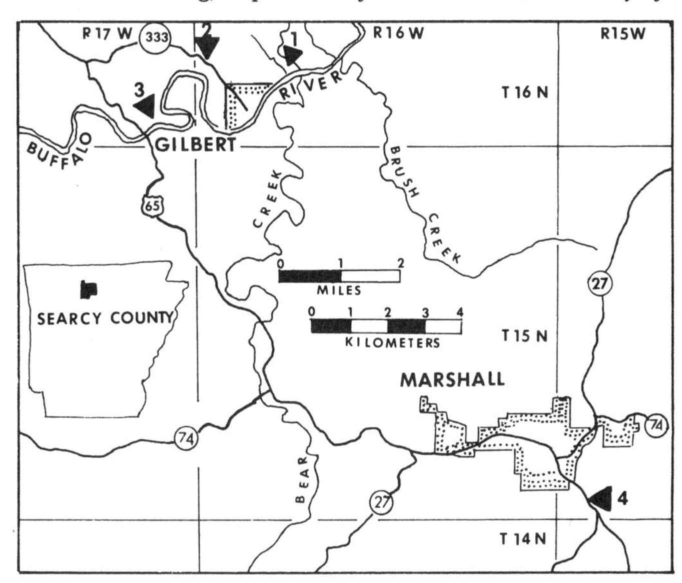

Figure 1. Location map. Black triangles show localities described here.

LOCATION

U.S. 65 is a major north-south artery across Arkansas. In north-central Arkansas it traverses the Ozark Plateau Province, which represents a stable Paleozoic platform located on the southwest flank of the Ozark Dome and north of the Ouachita trough. In Searcy County the highway crosses the deeply incised valley of the Buffalo River (Fig. 1) about two-thirds of the distance from Harrison to Marshall.

SIGNIFICANCE

Because the trend of U.S. 65 is approximately perpendicular to structural strike, its passage through the Buffalo River drainage system provides an excellent cross sectional view of the geomorphology, structure, and stratigraphy of the Ozark Plateaus (Fig. 2). The highway provides easy access to views of mature topography dissected by ingrown meanders of the Buffalo and its tributaries.

The Lower Paleozoic rocks exposed in the valleys of the Buffalo River and its tributaries provide an opportunity to examine the nature of supratidal to subtidal carbonate and quartz arenite associations characteristic of midcontinent rocks of this age. These rocks contrast with Upper Paleozoic strata in the Buffalo drainage, primarily in that the latter contain considerably more detrital material. This addition of detritus apparently reflects increased orogenic activity, which eventually culminated with the Ouachita Orogeny in the Late Paleozoic. McKnight (1935) gave an excellent comprehensive report on the region and Maher and Lantz (1952) provided detailed descriptions of the stratigraphic units.

GENERAL DESCRIPTION

The Buffalo River follows a meandering course of more than 120 mi (200 km) across the Springfield Plateau. The surface of the plateau is capped by the resistant chert-nodular Boone Formation of Mississippian (Osage-Meremac) age. The ingrown meanders of the Buffalo and its tributaries have incised deeply into the plateau to create a mature topography of rolling hills that forms the heart of the Arkansas Ozarks. Ordovician and Silurian sandstones and carbonates are well exposed in the Buffalo's valley walls, which are nearly 500 ft (150 m) high in places. Along much of the river's course this scenic landscape has been preserved as the Buffalo National River.

South of the Springfield Plateau is a higher plateau held up by rock of Pennsylvanian age. The north-facing escarpment of this plateau rises 600 to 1,000 ft (180 to 300 m) above the Springfield surface and contains in its face strata of Late Mississippian (Chester) age. Thorough dissection of the northern portion of this higher level by north-flowing tributaries to the Buffalo has produced the most rugged topography of the region. This portion of the Ozarks is named the Boston Mountains. Paleozoic rocks of the Springfield Plateau and the Boston Mountains dip gently southward toward the Arkansas Valley.

POINTS OF GEOLOGIC INTEREST

The easiest and most enjoyable way to see the Buffalo River country is by canoe, particularly if interest lies in the Lower Paleozoic stratigraphy of the region. Between the U.S. 65 bridge and Gilbert, gentle folds eventually bring all units between the Ordovician St. Peter Sandstone and Mississippian Boone Formation to river level. The best exposures in Arkansas of the Lower Silurian Brassfield Limestone occur along this stretch of the river.

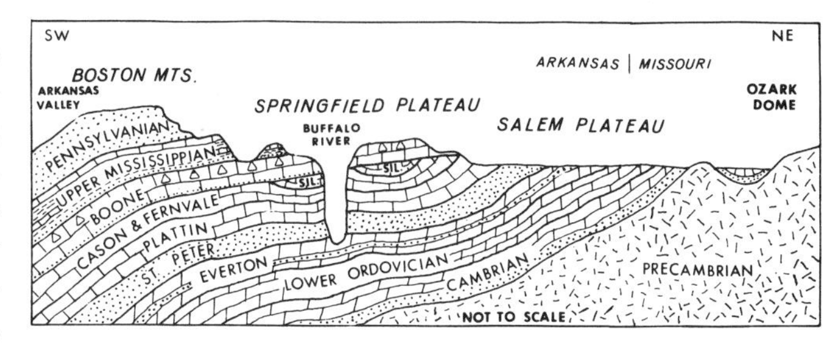

Figure 2. Geologic cross section through the Ozark Plateaus Province.

A map by Maher and Lantz (1953) serves as a guide. Unfortunately, units older than the St. Peter are not well exposed along the Buffalo River within the vicinity of U.S. 65. If interest is in pre–St. Peter strata, an excellent guide (McFarland, 1979) to a small stretch along the upper Buffalo River locates exposures of the Powell Dolomite and Everton Formation. Float outfitters at many places along the Buffalo River rent canoes and offer put-in, take-out, and car-shuttle services.

Four localities (Fig. 1) have been selected to illustrate the geology of the area. All are easily accessible from U.S. 65 by car, or a combination of car and short walk. The first three stops are within the boundaries of the Buffalo National River, administered by the National Park Service. Looking is fine, but if you plan extensive collecting, you should obtain permission from the ranger office in Harrison.

Locality 1 is along the old railroad grade on the north bank of the Buffalo River 1 mi (1.6 km) east of Gilbert, SW¼,SE¼-Sec.29,T.16N.,R.16W., Marshall 7½-minute Quadrangle. To reach the locality by land, take Arkansas 333 east from U.S. 65 to the Gilbert general store. Walk downstream (east) along the grade, which is easily located across from the store and about 100 ft (30 m) toward the river. The exposure is in the grade cut just across the first Buffalo tributary encountered. If on the river, pull in at the first south-flowing tributary beyond Gilbert and climb the tributary's east bank to the exposure.

The St. Peter sandstone is brought to the surface by a small flexure that imparts a slight westward dip at this locality. The St. Peter–Plattin contact is well exposed at the east end of the cut. About 70 ft (21 m) of Plattin occur between quartz arenite of the St. Peter below and approximately 20 ft (6 m) of Fernvale crinozoan grainstone above. Maher and Lantz (1953) mapped the Cason Shale here, but it is not well exposed. The hill is capped by the cherty limestone of the Boone Formation. The Plattin ranges from lime mudstone to intraclastic grainstone. Except for ostracodes, it is devoid of invertebrate fossils. Mud cracks, calcite pseudomorphs after halite, and stromatolites suggest that the Plattin here is dominantly intertidal to supratidal in origin (Deliz, 1984). The fossiliferous Fernvale represents shallow subtidal deposition. The absence of Joachim Dolomite between the St. Peter and Plattin, and Kimmswick Limestone between the Plattin and Fernvale, indicates an unconformable relationship between these units. However, geologists have argued over the meaning of the contacts between the Ordovician units above the St. Peter. For more details on these units, refer to the article on the post–St. Peter Ordovician succession at Allison in this guidebook.

Locality 2 is approximately 2.5 mi (4.2 km) east of U.S. 65 on Arkansas 333, just before the bridge over Dry Creek, NW¼SW¼Sec.30,T.16N.,R.16W., Maumee 7½-minute Quadrangle.

This is an excellent roadcut that exposes, in ascending order, 8 ft (2.4 m) of Cason Shale (Ordovician), 5 to 6 ft (1.5 to 1.8 m) of phosphatic sandstone of questionable formational assignment, 12 ft (3.7 m) of St. Joe Limestone Member of the Boone Formation, and several ft (m) of Boone Formation.

The stratigraphy of the Cason varies considerably across the state. In its type area in Independence County to the east, the name has been applied to an interval of heterogeneous rock types that range in age from Late Ordovician to Early Silurian. The problem has been addressed in detail by Craig (1969, 1975, 1984). The Cason in the Gilbert area, which is equivalent to only the lowest strata (Upper Ordovician) of the Cason in Independence County, has a sandy, slightly phosphate-conglomeratic base that grades up into silty, gray-green dolomitic shale. Only the upper dolomitic shale is exposed here. The lower portion of the Cason can be observed lying unconformably between the Fernvale Limestone and Boone Formation approximately 0.7 mi (1.2 km) east of this locality in a low roadcut and road ditch on the north side of 333 where the road climbs the hill into Gilbert. The upper part of the Cason at this latter locality contains unusual, large phosphate pebbles apparently reworked into the shale. The weathered nature of the outcrop requires shallow digging.

The phosphatic sandstone appears to grade up into the St. Joe Member at Locality 2. It was considered by Maher and Lantz (1953) to be a basal detrital phase of the Mississippian Boone Formation even though they collected from it a Late Devonian conodont fauna near here. Because of its Devonian age some geologists have assigned this unit to the Sylamore Sandstone, a name long-applied in northern Arkansas to a Late Devonian rock of similar lithic aspect. In other places in this general vicinity, phosphatic sandstone in this same stratigraphic position carries an Early Mississippian (Kinderhook) conodont fauna.

Whether this sandstone should be considered a Mississippian portion of the Sylamore or whether it represents a second sandstone distinct from the Sylamore is uncertain. The problems of the Late Devonian–Early Mississippian stratigraphy in northern Arkansas are discussed by Swanson and Landis (1962), Freeman and Schumacher (1969), Manger and Shanks (1976), Hall and Manger (1977), and Horner and Craig (1984).

The St. Joe Member is a clayey crinozoan-bryozoan packstone. It is overlain at Locality 2 by chert-bearing wackestone and lime mudstone of the upper Boone. These units are interpreted to have accumulated on an open shelf (Thompson and Fellows, 1970; McFarland, 1975; Horner, 1985).

An interesting aspect of this exposure is that a thin sliver of Silurian St. Clair Limestone 2 ft (0.6 m) present between the Cason and the Devonian sandstone in the western part of the roadcut disappears to the east. This exposure demonstrates the evenness of unconformable contacts in the Paleozoic of the Arkansas Ozarks and the disappearance by erosional truncation of major stratigraphic units over short distances.

Locality 3 is the first bluff on the north side of the Buffalo River approximately 0.25 mi (0.4 km) downstream (east) from the U.S. 65 bridge, NE¼SW¼Sec.36,T.16N.,R.17W., Marshall 7½-minute Quadrangle. The bluff is steep but climbable; it is most easily reached by boat. To reach the top of the bluff, turn east off U.S. 65 on the dirt road at the north end of the bridge. Follow the road, keeping right, for approximately 0.4 mi (0.7 km) through a deep tributary and up onto a high flat. Cross the

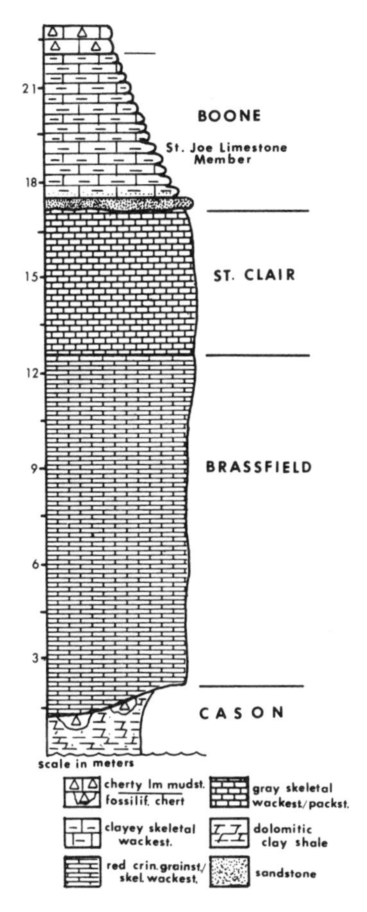

Figure 3. Section at U.S. 65 bridge over the Buffalo River.

flat toward the river on foot. The base of the bluff can be reached by turning east off U.S. 65 on the river access road at the south end of bridge and driving down to the gravel bar opposite the bluff. At the base of the bluff is one of the best swimming holes along the Buffalo River. Camping is allowed on the river bar, making this an excellent spot to combine geology with recreation.

The section is illustrated in Figure 3. About 5 ft (1.5 m) of Cason is exposed in the shallow cave at the base of the bluff. The large chert nodules in the upper foot are composed of fossil hash with oolite stringers.

Thirty-eight ft (11.5 m) of Lower Silurian Brassfield Limestone rest with irregularity on the Cason. The Brassfield in northern Arkansas is known principally from this area. It occurs at scattered localities, apparently preserved as erosional remnants beneath the St. Clair. The Brassfield is a pinkish-gray to red-mottled fossiliferous wackestone-packstone with minor lime mudstone and crinozoan grainstone. It is interpreted as a product of a protected subtidal environment.

The contact between the Brassfield and the Middle Silurian St. Clair Limestone is approximately three-fourths of the distance up the bluff (Fig. 4). The St. Clair is lithically similar to the Brassfield and is judged to have accumulated under similar environmental conditions. The two differ in that the St. Clair is gray

Figure 4. Exposure at locality 3. Cason (not visible) is exposed in cave at base of bluff. Brassfield–St. Clair contact marked by arrow. Slabby beds on top are St. Joe Limestone Member of the Boone Formation.

to pinkish-gray and lacks the mottled appearance of the Brassfield. The St. Clair also is coarser overall than the Brassfield, has a higher percentage of wackestone, and possesses more unfragmented fossils. There is, however, enough overlap in color and lithic characteristics to make the distinction between the two a subtle one. Their boundary is generally marked by the reddish, poorly washed grainstone of the Brassfield in contact with the gray to pinkish-gray wackestone/packstone of the St. Clair. The contact of the two limestones is sharp and even, exhibiting no marked irregularity. In many places it can be collected ("welded" contact), and in thin section shows truncation of Brassfield fossils by the base of the St. Clair. Similar contacts were reported by Freeman (1966) in the Ordovician of northern Arkansas. The Brassfield is absent over much of the area, so that the St. Clair rests directly on the Fernvale (Ordovician below the Cason). All these features suggest an unconformable relationship for the Brassfield and St. Clair. At this locality, the contact is stylolitic, a condition that provides no evidence on the depositional relationship of the two units.

In several places in northern Arkansas the St. Clair is followed conformably by the Lafferty Limestone. The Lafferty is a wackestone/mudstone similar to the wackestone interbedded with the coarser St. Clair. The St. Clair and Lafferty have been interpreted as a transgressive-regressive cycle, of which the Lafferty is the regressive part (Craig, 1969, 1984). The Lafferty is not present at this locality, apparently because it was removed by pre-Boone erosion. It is present above the St. Clair approximately 0.5 mi (0.8 km) downstream from stop 3, about midway around the large meander in the western half of section 36. The relationship between the Lafferty and overlying Boone at this latter locality is angular (Fig. 5). Subtle angular unconformities are characteristic of the contacts between many Ozark Paleozoic formations.

The top of the bluff at Locality 3 provides an excellent view of the incised Buffalo River Valley. The bluff is a cut bank on an ingrown meander of small radius. The meanders downstream show excellent asymmetrical cross sections characteristic of ingrown meanders.

Locality 4 is a magnificent roadcut along U.S. 65, 1 mi (1.7 km) south of its junction with Arkansas 27 in Marshall, SE¼,SE¼Sec.31,T.14N.,R.15W., Harriet 7½-minute Quadrangle. From the Buffalo River bridge to locality 4, U.S. 65 crosses the rolling uplands of the Springfield Plateau. Exposures along the road are mostly Boone Formation. Just south of Marshall the highway climbs the escarpment of the Boston Mountains. From Locality 4 the view to the south and southwest affords an excellent vista of the stripped-structural surface of the Springfield Plateau, capped by Boone. The isolated hills resting on the Springfield surface are outliers of the Upper Mississippian strata found in the face of the Boston Mountain escarpment. To the west, the Boston Mountains can be seen rising abruptly above the lower Springfield Plateau.

The roadcut exposes interbedded dark lime mudstone and black shale of the upper Fayetteville Shale overlain by light-colored bioclastic and oolitic carbonate of the lower Pitkin Limestone, both Late Mississippian in age. To the south, U.S. 65 passes through the Mississippian into the dominantly detrital rocks of the Pennsylvanian. More details on the rocks of Locality 4 and the Ozark Carboniferous in general are in McFarland (1979) and Sutherland and Manger (1977, 1979).

Figure 5. Angular contact between Lafferty and slabby beds of St. Joe Limestone Member of Boone, downstream from Locality 3.

REFERENCES CITED

Craig, W. W., 1969, Lithic and conodont succession of Silurian strata, Batesville district, Arkansas: Geological Society of America Bulletin, v. 80, p. 1621–1628.

—— , 1975, Stratigraphy and conodont faunas of the Cason Shale and the Kimmswick and Fernvale Limestones of northern Arkansas, *in* Wise, O. A., and Hendricks, K., eds., Contributions to the Geology of the Arkansas Ozarks: Arkansas Geological Commission, p. 61–95.

—— , 1984, Silurian stratigraphy of the Arkansas Ozarks, *in* McFarland, J. D., III, ed., Contributions to the Geology of Arkansas, v. 2: Arkansas Geological Commission Miscellaneous Publication 18-B, p. 5–32.

Deliz, M. J., 1984, Stratigraphy and petrology of the Plattin Limestone (Middle Ordovician) in Newton and Searcy Counties, Arkansas [M.S. thesis]: New Orleans, University of New Orleans, 259 p.

Freeman, T., 1966, "Petrographic" Unconformities in the Ordovician of northern Arkansas: Oklahoma Geology Notes, v. 26, no. 1, p. 21–28.

Freeman, T., and Schumacher, D., 1969, Qualitative pre-Sylamore (Devonian–Mississippian) physiography delineated by onlapping conodont zones, northern Arkansas: American Association of Petroleum Geologists Bulletin, v. 80, p. 2327–2334.

Hall, J. D., and Manger, W. L., 1977, Devonian sandstone lithostratigraphy, northern Arkansas: Arkansas Academy of Sciences Proceedings, v. 31, p. 47–49.

Horner, G. J., and Craig, W. W., 1984, The Sylamore Sandstone of north-central Arkansas, with emphasis on the origin of its phosphate, *in* McFarland, J. D., III, ed., Contributions to the Geology of Arkansas, v. 2: Arkansas Geological Commission, p. 51–85.

Horner, R. O., 1985, Petrography, diagenesis, and depositional environments, St. Joe Formation (Lower Mississippian), northern Arkansas [M.S. thesis]: New Orleans, University of New Orleans, 144 p.

Maher, J. C., and Lantz, R. J., 1952, Correlation and description of the Lower Paleozoic of Gilbert, Carver, and Marshall, Arkansas: U.S. Geological Survey Circular 160, 21 p.

—— , 1953, Geology of the Gilbert area, Searcy County, Arkansas: U.S. Geological Survey Oil and Gas Investigations Map OM 132.

Manger, W. L., and Shanks, J. L., 1976, Lower Mississippian lithostratigraphy, northern Arkansas: Arkansas Academy of Sciences, Proceedings, v. 30, p. 78–80.

McFarland, J. D., III, 1975, Lithostratigraphy and conodont biostratigraphy of the St. Joe Formation (Lower Mississippian), northwest Arkansas [M.S. thesis]: Fayetteville, University of Arkansas, 138 p.

—— , 1979, Geologic float guide on the Upper Buffalo River: Arkansas Geological Commission, 6 p.

McKnight, E. T., 1935, Zinc and lead deposits of northern Arkansas: U.S. Geological Survey Bulletin 853, 311 p.

Swanson, V. E., and Landis, E. R., 1962, Geology of a uranium-bearing black shale of Late Devonian age in north-central Arkansas: Arkansas Geological Commission Information Circular, no. 22, 16 p.

Sutherland, P. K., and Manger, W. L., eds., 1977, Mississippian–Pennsylvanian boundary in northeastern Oklahoma and northwestern Arkansas: Oklahoma Geological Survey, Guidebook 18, 185 p.

—— , eds., 1979, Ozark and Ouachita shelf-to-basin transition, Oklahoma–Arkansas: Oklahoma Geolgoical Survey, Guidebook 19, 81 p.

Thompson, T. L., and Fellows, L. D., 1970, Stratigraphy and conodont biostratigraphy of Kinderhookian and Osagean (Lower Mississippian) rocks of southwestern Missouri and adjacent areas: Missouri Geological Survey Report of Investigations 45, 263 p.

Post–St. Peter Ordovician strata in the vicinity of Allison, Stone County, Arkansas

William W. Craig, Department of Earth Sciences, University of New Orleans, New Orleans, Louisiana 70148
Michael J. Deliz, CLK Corporation, Suite 510, 1615 Poydras St., New Orleans, Louisiana 70112

LOCATION

Exposed at Allison are the Ordovician St. Peter Sandstone, Joachim Dolomite, Plattin, Kimmswick, and Fernvale Limestones, and Cason Shale, as well as the Mississippian Boone Formation. A geologic map of the vicinity of Allison is in McFarland and others (1979). The Allison section, along with related exposures, is included in Craig and others (1984). The outcrop consists of two obvious, closely spaced roadcuts, referred to by local geologists as Allison South and Allison West (Fig. 1). Allison South is on Arkansas 9, approximately 4 mi (6 km) north of Mountain View. Allison West is on Arkansas 14, approximately 0.5 mi (0.8 km) west of its junction with Arkansas 9, between the small community of Allison and the entrance to Blanchard Spring Caverns. The roadcuts are in Sec. 14 and 10,T.15N.,R.11W. of the Sylamore and Fifty Six 7½-minute Quadrangles, respectively.

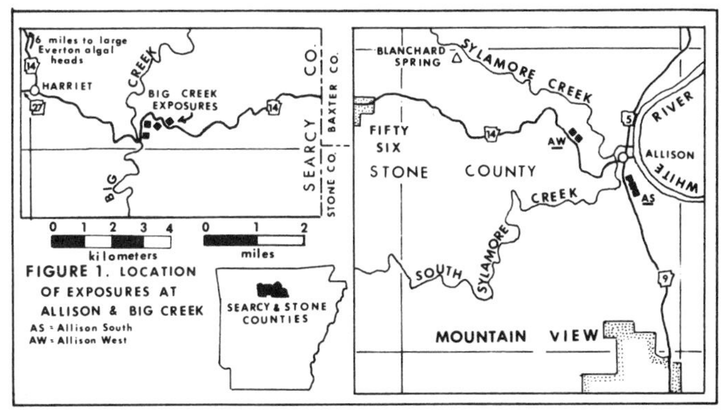

Figure 1. Location of exposures at Allison and Big Creek.

Both Allison localities are easily accessible. Neither Arkansas 9 nor 14 is heavily traveled. Ample shoulder space makes these locations ideal for relatively large groups. Parking at Allison South is better on the shoulder west of the highway. A turnout opposite Allison West allows convenient parking there. Allison South provides a nearly complete exposure from uppermost St. Peter Sandstone through Plattin Limestone. Higher rocks are better viewed at Allison West, which exhibits a complete section from uppermost Plattin into the lower part of the Boone Formation.

Geologically, the region is on the southwestern flank of the Ozark Dome. Regional dip of less than 1° to the south is interrupted by small folds that occur throughout the region, and down-to-the-south normal faults with displacements of a few meters to a few tens of meters. Physiographically, Allison lies at the southern margin of the Springfield Plateau subdivision of the Ozark Plateaus Province. The siliceous Boone Formation caps the Springfield Plateau, which is dissected by the White River and its tributaries, providing valleys 400 to 500 ft (120 to 150 m) below the plateau surface. It is within these valleys that Lower Paleozoic rocks are exposed. To the south, Ordovician rocks dip beneath the higher Boston Mountains, a dissected surface capped by Pennsylvanian sandstones. The north-facing escarpment of the Boston Mountains is traversed by Arkansas 9 south of Mountain View.

Blanchard Spring campground, which is maintained by the U.S. Forest Service, provides group camping and the most delightful scenery between the Appalachian and Rocky Mountains. Blanchard Spring Caverns, which are developed within the Plattin Limestone-Boone Formation interval, are well worth a tour.

SIGNIFICANCE

The two Allison sections provide the best exposure in the Arkansas Ozarks of the post–St. Peter Ordovician carbonate succession, which accumulated on a shelf to the north of the Ouachita trough. This succession of units records supratidal, intertidal, and subtidal environments in what appears to be an overall deepening upward succession. Abundant indicators of peritidal deposits make this sequence of strata ideal for study of shallow-water carbonates. In addition, the sections display well the controversial sharp, planar contacts that commonly separate Lower Paleozoic carbonate formations in the Ozarks.

POST–ST. PETER ORDOVICIAN SUCCESSION

Although there is relatively little disagreement on the interpretation of the environments of deposition of the sediments that comprise these strata, geologists have been divided on the historical meaning of the formational contacts. The controversy, which involves whether the contacts are conformable or unconformable, has appeared in the literature in a series of articles (Freeman, 1966a,b, 1972; Young and others, 1972a,b). Craig (1975a) reviewed the controversy in a summary of the history of investigations on the post–St. Peter rocks of the Arkansas Ozarks. Conformability of formational contacts is suggested by an apparent deepening upward succession, the logical consequence of landward migration of adjacent lithotopes. Lithofacies tracts outlined by Irwin (1965) serve as a model for these adjacent lithotopes. Evidence of truncation (mechanical or chemical) at formational contacts, however, indicates lithification and erosion

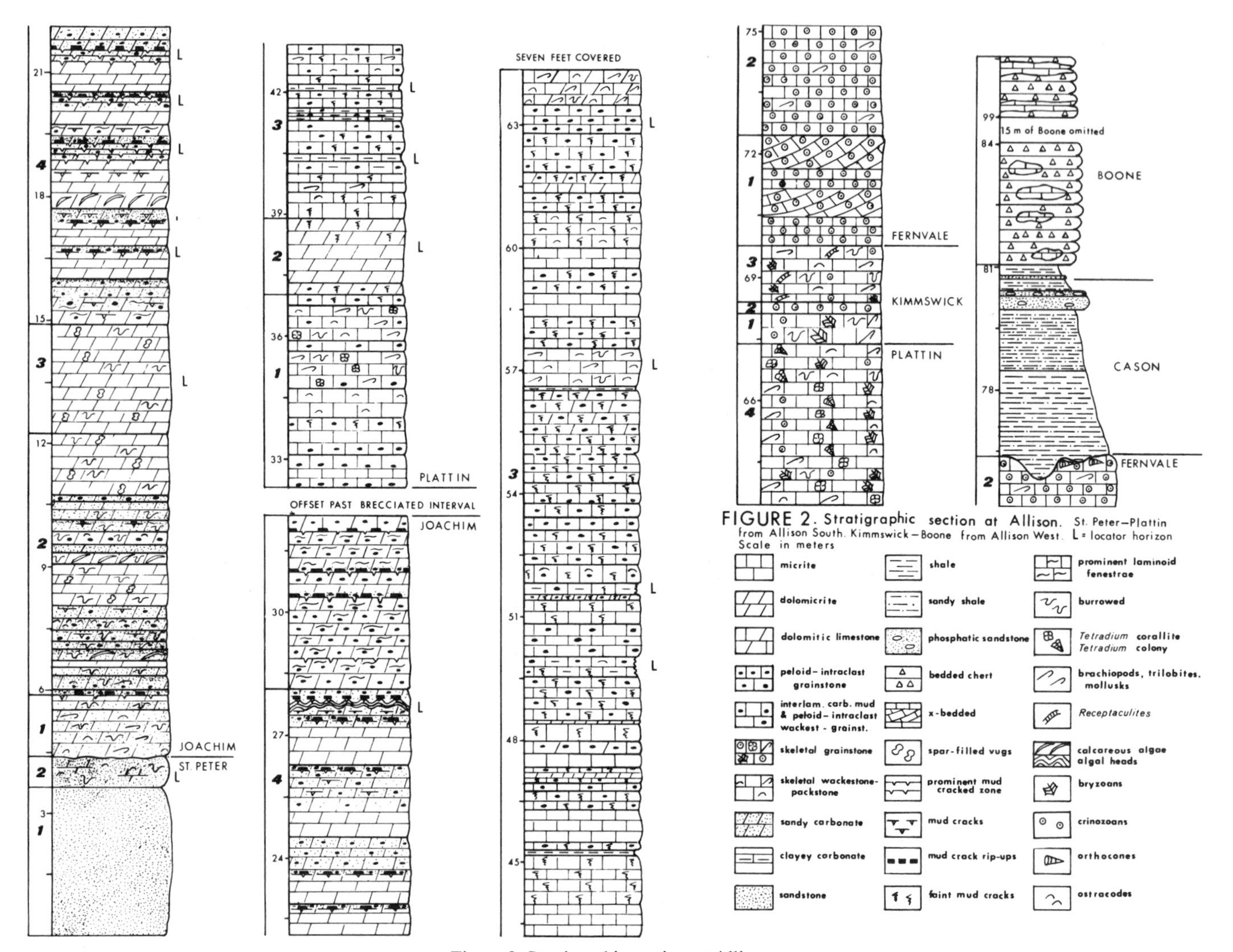

Figure 2. Stratigraphic section at Allison.

of each unit before deposition of the overlying unit. This erosion is evident on both large and small scale. Small-scale evidence includes truncation of fossils and burrows at the top of the units. Such features are best examined in thin sections and acetate peels across so-called "welded" contacts. Freeman (1966a) refers to these as "petrographic" unconformities. These surfaces bear some resemblance to the scalloped/planar erosional surfaces of Read and Grover (1977). However, the Arkansas surfaces do not seem to fit well into the Read-Grover model of development on a prograded, early cemented tidal flat or intertidal rock platform. Nor are they particularly suggestive of subtidal, penecontemporaneously cemented hardgrounds (Bathurst, 1975).

Large-scale truncation is manifest in slight angular discordance of contacts at a few locations. Surfaces of erosion (i.e., unconformities) separating units are further suggested by the distributions of stratigraphic units: each exhibits a regional variation in thickness that reflects little or no pattern relative to those of other units in the succession. Finally, certain units are locally absent: the Joachim and Kimmswick are missing from sections in Searcy County west of Allison, so there the Plattin rests directly on the St. Peter, and the Fernvale rests directly on the Plattin (see alternate stop 1).

We have not attempted a bed-by-bed interpretation of these Ordovician carbonates. Detailed descriptions of exposures at Allison have been provided by Rives (1977) for the Joachim, and by Jee (1981) for the Plattin. Inasmuch as such detail is beyond our purpose, we have not relied on these works in constructing our section. What we have attempted to do is to divide each formation into units with unifying sedimentologic features. These subunits are labeled with numbers along the left margin of our columnar section (Fig. 2). Tentative interpretations of these subunits are intended to serve only as stimuli for further discussion.

These rocks contain features far too abundant to describe in the space provided. To facilitate the integration of visitors' observations with those of our own, we have identified locator beds, marked by the letter L to the right of the stratigraphic column

(Fig. 2) in our measured sections. These are beds easily identified in the succession.

St. Peter Sandstone (Middle Ordovician)

The St. Peter is a quartz arenite that is widespread in the midcontinent region. It reaches a thickness of 200 ft (60 m) in eastern and central Arkansas, but thins toward the west. The unit is absent west of eastern Newton County. Where its base is exposed, it is seen to rest unconformably on a variety of underlying stratigraphic units. A description, from bottom to top, follows:

Subunit 1. Cross-bedded, fine- to medium-grained quartz arenite that is representative of the St. Peter throughout its occurrence; grains are well rounded; some sandy dolostone intraclasts. Interpretation: shallow-water, near-shore sand complex.

Subunit 2. Mottled light- to dark-gray sandy dolomicrite and dolomitic sandstone; sand in patches apparently concentrated by burrowing. Interpretation: transition between near-shore sand deposition (St. Peter type) to offshore carbonate accumulation (Joachim type), resulting from marine transgression.

Joachim Dolomite (Middle Ordovician)

The transitional nature of the Joachim–St. Peter contact indicates that the two units are conformable at this locality, even though the base of the Joachim is slightly undulatory. The Joachim is believed to have been a lime sediment that has been replaced by dolomite. Preservation of original limestone textures is excellent. The Joachim section here is complicated by numerous offsets across small faults. Careful tracing of beds is necessary to correlate across these displacements.

Subunit 1. Mottled light- to dark-gray dolomicrite, quartz sand in patches (burrowed), fossiliferous with ostracodes (locally abundant), snails, trilobites, and other shells; some thin laminae of sand, intraclasts, and smaller shell debris. Interpretation: shallow subtidal.

Subunit 2. Interbedded dolomicrite, intraclastic/bioclastic (ostracodes and molluscs) grainstone, and dolomicrite burrow-mottled with patches of grainstone and packstone. Dolomicrite faintly to more distinctly mud-cracked; large spar patches apparently resulted from filling of voids produced by burrowing; excellent laminoid fenestrae, in places truncated by grainstone; quartz sand distributed in intraclastic layers and as patches. Interpretation: low intertidal deposition of interlaminated grainstone and dolomicrite controlled by algal mat development, burrowed and scoured in places.

Subunit 3. Fine- to medium-grained mottled dolostone characterized by large spar patches with orange rims; texture enigmatic. The distinctive orange spar patches make this rock an excellent locator bed. Interpretation: probably low intertidal. We interpret the spar patches and mottling to be the result of burrowing, as in subunit 2.

Subunit 4. Sequences of faintly laminated dolomicrite, well-defined mud-cracked layers capped by prominent desiccation polygons, and sandy intraclastic "trash" zones with rip-up clasts from below (Figs. 3, 4). These zones are easily identified and make excellent locator horizons. Dolomicrite, interlaminated with fine intraclastic-peloidal grainstone/packstone, contains laminoid and irregular fenestrae and calcite pseudomorphs after halite. Subunit 4 is capped by an 8-in (20-cm) interval of stromatolite heads (LLH-C type), the largest of which measure 2 ft (0.6 m) across. Many of the stromatolites are asymmetric. Their mud-cracked tops are overlain by an intraclastic interval. Interpretation: we interpret the greater part of this unit as shallowing-upward sequences beginning with low intertidal cryptalgal-laminated dolomicrite and terminating with intensely desiccated dolomicrite of higher intertidal and/or supratidal zones. Sandy intraclastic rocks above the mud-cracked intervals probably represent a return to higher energy conditions near low mean tide and the initiation of another sequence. A cap of LLH-C–type stromatolites developed in the high-energy zone near mean low tide, and, as indicated by their mud-cracked tops, built up into the upper reaches of the intertidal zone.

Subunit 5. Intraclastic-peloidal packstone/grainstone with excellent laminoid fenestrae and thin intervals of mud-cracked dolomicrite. The upper few centimeters are only partially dolomitized. Interpretation: algal-mat–controlled sedimentation of the low- and middle-intertidal zones.

Plattin Limestone (Middle Ordovician)

The base of the Plattin is generally picked at the lowest occurrence of limestone. In the roadcut at Allison South, the contact between the two units is complicated by a 5-ft (1.5-m) interval of mixed lithic types (Fig. 5). Pieces of limestone isolated by solution joints are surrounded by dolomite, and layers of sandy limestone drape over truncated edges of beds within the interval. Some blocks are at angles to normal bedding, imparting a brecciated appearance to the interval. In the exposure on the platform above the roadcut, the brecciated zone departs from the contact and passes into the basal Plattin at a low angle to bedding. The best explanation for the interval seems to be a low-angle fault zone along which some solution and deposition has taken place. Where the brecciated interval does not coincide with the contact, the Joachim and Plattim are sharply separated by a stylolite.

At the place where the interval reaches road level, it consists of brecciated dolomite. Above the dolomite is a normal section of Plattin. Our section of Plattin begins at this point.

Subunit 1. Burrowed lime mudstone and skeletal wackestone with minor amounts of intraclastic-peloidal packstone/grainstone; skeletal fragments include ostracodes, *Tetradium* (coral), gastropods, and other shells; irregular fenestrae common. Interpretation: shallow subtidal.

Subunit 2. Laminated dolomicrite with faint mudcracks in upper portion; some thin laminae of peloidal packstone; ostracodes and very fine laminoid fenestrae; lower 1 ft (30 cm) is partially dolomitized intraclastic grainstone. Interpretation: intertidal zone.

Figure 3. Shallowing upward sequences in unit 4 of Joachim. Dark layers are laminated dolomicrite. Light layers are mud-cracked intervals capped by rip-up clasts. Between 55 and 75 ft (16.5 and 22.5 m) above base, Allison South.

Figure 4. Close-up of mud-cracked layers and capping rip-up clasts, unit 4 of Joachim Dolomite, 71 ft (21 m) above base of section, Allison South.

Figure 5. "Mixed" interval at Joachim-Plattin boundary, Allison South. Arrows on left mark base and top of interval; arrows on right mark pieces of limestone surrounded by dolomite.

Subunit 3. Random succession of: (1) finely interlaminated lime mudstone and peloidal packstone/grainstone, (2) intraclastic-peloidal packstone/grainstone, and (3) faintly laminated lime mudstone. The lime mudstone contains fine sparfilled mudcracks (polygons visible on some bed soles) and sheet cracks, calcite pseudomorphs after halite, and irregular and tubular fenestrae. Peloidal layers are densely packed and characterized by laminoid and irregular fenestrae. A very few fossiliferous wackestones with ostracodes and trilobites occur in the sequence. Also present are local dolomitic intervals. Clayey limestones and thin shales, which form less-resistant units in the section, serve as excellent locator beds. Interpretation: This major portion of the Plattin is

Figure 6. *Tetradium* colony in unit 4, Plattin Limestone, Allison West.

Figure 7. Closeup of Kimmswick-Fernvale contact, Allison West.

not characterized by specific successions as is true for the major portion of the Joachim. Nonetheless, we conclude that Plattin and Joachim rocks accumulated in essentially the same sedimentologic setting; that is, a tidal flat environment. We believe that the Plattin tidal flat contained a more varied hydrography than that of the Joachim, which included a complex of subtidal, intertidal, and supratidal subenvironments with tidal channels, levees, and ponds. Lateral migration of these subenvironments produced the random sequence of textures of subunit 3. The Plattin tidal flat was not subject to such intense desiccation as that of the Joachim and possibly existed under more humid conditions or accumulated lower on the tidal flat. Note: Higher rocks are better viewed at Allison West.

Subunit 4. Skeletal wackestone with an abundant, diversified biota. Leperditiid ostracodes and *Tetradium* colonies in growth position are obvious (Fig. 6). Please do not collect. Burrowed in places. Interpretation: shallow, protected subtidal environment.

Kimmswick Limestone (Middle Ordovician)

The Kimmswick Limestone is a fine- to coarse-grained bioclastic rock, most of which contains finely crystalline and micrite matrix. Its sharp contact with subunit 4 of the Plattin, from which it is distinguished by being more coarsely grained, is well displayed at Allison West.

Subunit 1. Skeletal packstone, dominated by crinozoans but containing a diversified fauna. Interpretation: protected subtidal.

Subunit 2. Crinozoan grainstone. Interpretation: open subtidal; probably represents a tongue of the protective barrier behind which most of the Kimmswick accumulated. Texturally, this rock is much like the overlying Fernvale and is difficult to distinguish from that formation in hand specimen.

Subunit 3. Skeletal wackestone/packstone with diversified fauna. Fine cross sections of *Receptaculites* and trepostomous bryozoans seen in hand specimen. Interpretation: protected subtidal.

Fernvale Limestone (Upper Ordovician)

The Fernvale is a crinozoan grainstone almost totally devoid of fine matrix. Its sharp contact with the Kimmswick, which is characteristic of the two units in the Arkansas Ozarks, is well displayed at Allison West (Fig. 7).

Subunit 1. Crinozoan grainstone with large-scale cross-bed sets (trough?). Interpretation: open, shallow subtidal.

Subunit 2. Crinozoan grainstone/minor packstone; finer grained than subunit 1, a little more diverse faunally, and planar bedded. Cross sections of orthocones can be seen on the upper surface of the Fernvale on top of the road cut. Interpretation: open subtidal, somewhat deeper water than subunit 1.

Note: For more detail on the Kimmswick and Fernvale, see Craig (1975b) and Craig and others (1984). The Kimmswick and Fernvale contain well-preserved Middle and Late Ordovician conodont faunas, respectively.

Cason Shale (Upper Ordovician)

Based on both lithology and conodonts, most of the detrital interval between the Fernvale and Boone at Allison West appears to belong the Cason Shale (Fig. 2). The Cason in this general vicinity is a silty clay shale with thin beds of phosphatic sandstone. Devonian-Mississippian phosphatic sandstone with some shale also occurs at the base of the Boone. Recent work suggests that this basal Boone sandstone is equivalent to the Sylamore Sandstone (Swanson and Landis, 1962; Freeman and Schumacher, 1969). The basal Boone detrital rock is characterized by an abundance of medium-grained, well-rounded quartz, whereas Cason quartz grains are dominantly silt to fine sand-size and angular. This change in grain size and shape occurs at the base of a 3-in (8-cm) sandstone at the 266-ft (81-m) level in our measured section. Furthermore, the shale below this thin sand includes a Late Ordovician conodont fauna, whereas conodonts from the sand are Early Mississippian. We have thus drawn the base of the Boone at the base of this thin sandstone, even though it rests without noticeable discordance on the shale below. The Cason is variable in lithology across the state. For details, see Craig (1975b, 1984) and Craig and others (1984).

Boone Formation (Lower Mississippian)

The Boone is a notable example of chert replacement of limestone. Residual patches of original fine-grained limestone are more common upward in the section. In the top 20 ft (6 m), laterally continuous beds of limestone occur. At nearby localities, the basal part of the Boone is a crinozoan limestone called the St. Joe Limestone Member. The Boone is interpreted as an open shelf deposit that accumulated below the wave base. Origin of the silica is unknown.

Alternate Stops

1. Arkansas 14 at Big Creek (BC, Fig. 1), starting just east of the bridge over Big Creek and proceding eastward, the following section can be seen in road cuts: Everton Dolomite (Middle Ordovician beneath the St. Peter), St. Peter Sandstone, Plattin Limestone, Fernvale Limestone, and Boone Formation with the St. Joe Limestone Member at its base. This exposure affords an excellent view of the pre–St. Peter unconformity. It also demonstrates the laterally discontinuous nature of units in the post–St. Peter carbonate succession, with both the Joachim and Kimmswick absent. A fault drops the St. Peter–Plattin contact to road level. According to Deliz (1984), a 115-ft (35-m) section of Plattin capped by subunit 4 occurs along the road. A "welded" Fernvale-Plattin contact is well exposed, with beds in the underlying Plattin at a slight angle to the base of the Fernvale. Careful examination demonstrates the removal of at least 12.5 ft (3.8 m) of Plattin prior to deposition of the Fernvale. Patches of Fernvale lithology in the upper Plattin are interpreted as fillings in solution cavities associated with karst at the top of the Plattin.

2. About 6 mi (10 km) north of Harriet on Arkansas 14, a roadcut exposure in the Everton Formation shows large stromatolite heads. Well worth the short drive.

REFERENCES CITED

Bathurst, R.G.C., 1975, Carbonate Sediments and Their Diagenesis, 2nd ed.; Developments in Sedimentology 12: New York, Elsevier, 658 p.

Craig, W. W., 1975a, History of investigations on the post–St. Peter Ordovician of northern Arkansas: The art of layer-cake geology, *in* Wise, O. A., and Hendricks, K., eds., Contributions to the Geology of the Arkansas Ozarks: Arkansas Geological Commission, p. 1–17.

—— , 1975b, Stratigraphy and conodont faunas of the Cason Shale and the Kimmswick and Fernvale Limestones of northern Arkansas, *in* Wise, O. A., and Hendricks, K., eds., Contributions to the Geology of the Arkansas Ozarks: Arkansas Geological Commission, p. 61–95.

—— , 1984, Silurian stratigraphy of the Arkansas Ozarks, *in* McFarland, J. D., III, ed., Contributions to the Geology of Arkansas, v. II: Arkansas Geological Commission, Miscellaneous Publication 18-B, p. 5–32.

Craig, W. W., Wise, O. A., and McFarland, J. D., III, 1984, A Guidebook to the Post–St. Peter Ordovician and the Silurian and Devonian Rocks of North-Central Arkansas: Guidebook, 2nd Annual Field Trip, Midcontinent Section of the Society of Economic Paleontologists and Mineralogists, Arkansas Geological Commission GB-84-1, 49 p.

Deliz, M. J., 1984, Stratigraphy and Petrology of the Plattin Limestone (Middle Ordovician) in Newton and Searcy Counties, Arkansas [M.S. thesis]: New Orleans, University of New Orleans, 259 p.

Freeman, T., 1966a, "Petrographic" unconformities in the Ordovician of northern Arkansas: Oklahoma Geology Notes, v. 26, no. 1, p. 21–28.

—— , 1966b, Petrology of the post–St. Peter Ordovician, northern Arkansas: Tulsa Geological Society Digest, v. 34, p. 82–98.

—— , 1972, Carbonate facies in Ordovician of northern Arkansas; Discussion: American Association of Petroleum Geologists Bulletin, v. 55, p. 2284–2287.

Freeman, T., and Schumacher, D., 1969, Qualitative pre-Sylamore (Devonian-Mississippian) physiography delineated by onlapping conodont zones, northern Arkansas: Geological Society of America Bulletin, v. 80, p. 2327–2334.

Irwin, M. L., 1965, General theory of eperic clear-water sedimentation: American Association of Petroleum Geologists Bulletin, v. 49, p. 445–459.

Jee, J. L., 1981, Stratigraphy and paleoenvironmental analysis of the Plattin Limestone (Middle Ordovician), White River region, Independence, Izard, and Stone Counties, Arkansas [M.S. thesis]: New Orleans, University of New Orleans, 168 p.

McFarland, J. D., III, Bush, W. V., Wise, O., and Holbrook, D., 1979, A Guidebook to the Ordovician-Mississippian Rocks of North-Central Arkansas: Guidebook, South-Central Section of the Geological Society of America, Little Rock, Arkansas Geological Commission, 25 p.

Read, J. F., and Grover, G. A., Jr., 1977, Scalloped and planar erosional surfaces, Middle Ordovician Limestones, Virginia: Journal of Sedimentary Petrology, v. 47, p. 956–972.

Rives, J. S., II, 1977, Paleoenvironmental Analysis of the Joachim (Middle Ordovician) in Northern Arkansas [M.S. thesis]: Baton Rouge, Louisiana State University, 166 p.

Swanson, V. E., and Landis, E. R., 1962, Geology of a uranium-bearing black shale of Late Devonian age in north-central Arkansas: Arkansas Geological Commission Information Circular, no. 22, 16 p.

Young, L. M., Fiddler, L. C., and Jones, R. W., 1972a, Carbonate facies in Ordovician of northern Arkansas: American Association of Petroleum Geologists, v. 56, p. 68–80.

—— , 1972b, Reply to Tom Freeman: American Association of Petroleum Geologists Bulletin, v. 56, p. 2287–2290.

Silurian rocks of northern Arkansas

O. A. Wise and W. M. Caplan, Arkansas Geological Commission, 3815 W. Roosevelt Road, Little Rock, Arkansas 72204

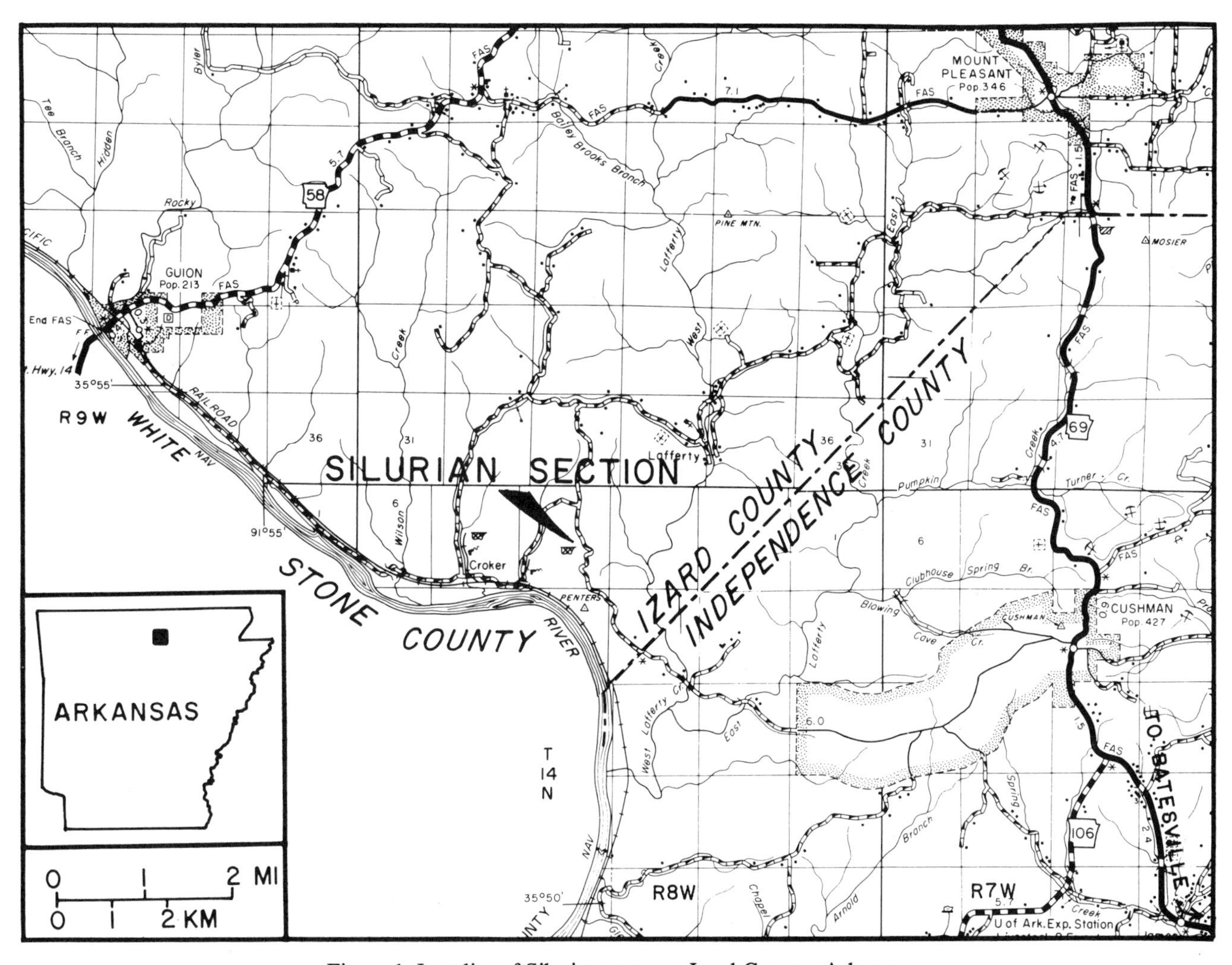

Figure 1. Locality of Silurian outcrop, Izard County, Arkansas.

LOCATION

From Batesville, Arkansas, in Independence County, follow Arkansas 69 north to Cushman (Fig. 1). Turn left (west) on the county road (paved) just beyond the Cushman General Store. If you consider this intersection the zero point of a road log, the pavement ends at about mile 3.5 (5.6 km). At mi 5.1 (8.2 km) you will drive past a good outcrop of Ordovician age limestones of the Plattin, Kimmswick, and Fernvale formations. Of particular interest here is the Plattin-Kimmswick contact, which is sharp and "welded" as well as angular. The Plattin occurs in a small synclinal fold that is truncated by the basal sandy Kimmswick. An orange-colored zone several yards up the road marks a fault that lowers the Fernvale to a position adjacent to the Kimmswick at road level. (See W. W. Craig's discussion of the lithologies of these units elsewhere in this volume.) At mi 7.0 (11.3 km) is the entrance to Love Hollow Quarry (Fig. 2). The outcrop described below can be found approximately 0.2 mi (0.3 km) north of Love Hollow.

The exposure consists of approximately 60 ft (18 m) of Brassfield, St. Clair, and Lafferty limestones. The shale on which the Brassfield rests at this locality is probably Ordovician age. The lithologic sequence consists of 4 ft (1.2 m) of spar-cemented, coarse fossil debris, overlain by 4 ft (1.2 m) of St. Clair, which in this locality is represented by a biomicrite. The contact at this point is slightly angular. The St. Clair is overlain by approximately 50 ft (15 m) of Lafferty alternating in color from cream to red. The rock is a grainy-looking micrite and may be slightly dolomitic. The weathered rock appears to be channeled and mud cracked. The upper contact with the overlying formation is obscured by chert float.

This exposure lies about a 0.25 mi (0.4 km) west of the type

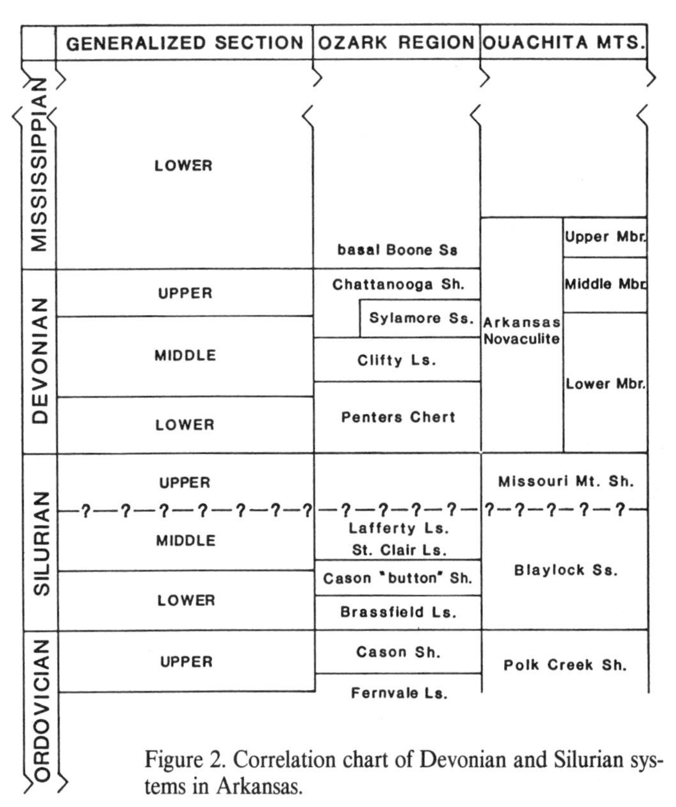

Figure 2. Correlation chart of Devonian and Silurian systems in Arkansas.

locality for the Lafferty, which was described by Miser (1922). One of the problems frequently encountered in the mapping of individual units in this area is the similarity of the spar-cemented calcirudites developed in the Fernvale, Brassfield and St. Clair formations from place to place. An incomplete stratigraphic section for this area is shown in Figure 2.

SIGNIFICANCE

Rocks of Silurian age are exposed in a few scattered outcrops in the Ozark region of north-central Arkansas. The outcrops are generally restricted to the Springfield Plateau physiographic subprovince of the Ozark Plateaus. The Silurian rocks are predominantly carbonates and have been assigned to the Brassfield Limestone (Ulrich, 1911), the St. Clair Limestone (Penrose, 1891), and the Lafferty Limestone (Miser, 1920; Fig. 2).

Recent field observations and paleontological data suggest that a part of the Cason Shale, previously considered as Ordovician by a number of workers, is also Silurian and is possibly a facies of the Brassfield Limestone (or vice versa).

The Silurian section of north Arkansas has for the most part been treated as an undifferentiated unit in the subsurface. It ranges in thickness from a feather edge (due to truncation) to more than 250 ft (75 m) in the Arkansas valley. The thickening appears to continue toward the Ouachitas where the Silurian is represented by over 1,200 ft (366 m) of clastic rocks. There is no information available at present indicating whether the nature of the change from Ozark to Ouachita facies is one of transition, overthrust, or separate basin deposition.

Brassfield Limestone. The Brassfield Limestone was named for exposures in Madison County, Kentucky, by Foerste (1905), and was identified in Arkansas by Ulrich in 1911.

Miser (1922) found Brassfield fossils (replaced by manganese oxide) at the Montgomery Mine 6 mi (9.6 km) northeast of Cushman, Independence County, Arkansas, and stated that they came from residual clays lying on the Cason Shale (Ordovician). He also mentioned exposures of Brassfield in the Yellville Quadrangle, Arkansas.

Maher and Lantz (1953) mapped exposures of Brassfield that attain a thickness of 26 ft (7.9 m) in the northern part of the Marshall Quadrangle in the Gilbert area of Searcy County, Arkansas. Here they described the formation as consisting of beds of coarsely crystalline fossiliferous limestone about 1 ft (0.3 m) thick, containing crinoids, brachiopods, trilobites, and an abundance of calcite-filled vugs. They stated that while the Brassfield resembled the St. Clair, there were recognizable lithologic differences.

Stripping operations at the Love Hollow Quarry, Izard County, Arkansas, have exposed a fossiliferous limestone on the north and south sides of the main quarry pit, within the Cason Shale. The exposure on the north side of the pit was removed before it could be sampled or described. The exposure on the south side occurs in the middle of a 10-ft-thick (3 m) bed of Cason Shale. At this latter exposure the limestone is 100 ft (30 m) long, lensing out to the east and west. It is 3 ft (1 m) thick at the thickest point, being equally divided between two rock types; the lower 1.5 ft consists of white oolitic limestone, and the upper 1.5 ft is a coarsely crystalline fossiliferous sparite quite similar to the Ordovician Fernvale Limestone that underlies the Cason Shale.

Amsden made a collection of fossils while examining this outcrop and states (written communication, 1967) that "the limestone bed in the Cason Shale at Love Hollow Quarry carries the brachiopod *Triplesia alata.*" This fossil had previously been described from the Brassfield of Arkansas by Ulrich and Cooper (Amsden and Roland, 1965).

Craig (written communication, 1967) states that conodonts collected at Love Hollow Quarry from the lower part of the Cason Shale indicate a Late Ordovician age; and that the conodonts from the "pelmatozoan limestone are definitely Early Silurian in age."

This would, in effect, make the shale mapped as Cason at Love Hollow in part Ordovician and in part Silurian, and presents the possibility that in some localities the limestone underlying that shale may be Silurian and not Fernvale (Upper Ordovician). It is also quite possible that the fossils found by Miser in the "residual clays" at the Montgomery Mine were in place in the weathered shale facies of this interval.

The question of the age of the Cason Shale is not a new subject. In the Arkansas Geological Survey Report for 1892 (p. 284), H. S. Williams states that the St. Clair Limestone, and the Cason Shale contain a Silurian fauna and that the "Polk

Figure 3. Slabby, clayey, fossiliferous lime mudstones of the Lafferty Limestone near the type section of the Lafferty.

Bayou Limestone" (now called Fernvale) contains an Ordovician fauna. Ulrich considered the Cason Shale age to be Ordovician in 1904, and Silurian in 1911.

Miser (1922) placed the Cason in the Ordovician. However, in 1964 (personal communication) he indicated that he was not satisfied with this placement, and thought a reexamination was in order.

The Brassfield Limestone is correlated with the Brassfield of Kentucky, Missouri, and Tennessee, and is equivalent or partially equivalent, to the Chimneyhill Formation of Olkahoma (Caplan, 1954).

The St. Clair Limestone. The St. Clair, as originally described by Penrose (1891), included the interval now mapped as Kimmswick Limestone (Ordovician) through Lafferty (Silurian). Williams (1892) redefined the St. Clair to include only that part now mapped as the St. Clair and Lafferty. The Lafferty, until separated by Miser (1920) was variously referred to as upper, or gray dense St. Clair.

Surface exposures are scattered in the north-central part of Arkansas principally between Batesville, Independence County, and Harrison, Boone County, and they attain a maximum outcrop thickness of 100 ft (30 m). The St. Clair is thought to be widely distributed in the subsurface, but thickness data are not readily available.

Usually, the St. Clair is characterized as being a gray to pink, fossiliferous limestone with a micritic matrix. Locally, the St. Clair contains an appreciable amount of sparry cement with spar-filled cavities and fossil molds. The St. Clair is quarried, 2 mi (3 km) north of Batesville, as a commercial marble. The pink and gray color of the host rock contrasts with the white calcite-filled vugs, fractures, and fossil molds, making a very attractive finishing stone.

A white oolitic limestone unit is sometimes present at the base of the St. Clair. This unit was identified by Maher and Lantz (1953) on the surface at St. Clair Springs, Independence County, and in the subsurface from the town of Marshall (deep water well), Searcy County. It is also present at Blanchard Springs, Stone County.

The St. Clair is thin- to medium-bedded but generally occurs as a single, massive, sharp-faced ledge on the outcrop. The St. Clair is a hard, brittle, resistant rock and, when struck, frequently causes the hammer to ring as if striking steel.

The St. Clair Limestone is correlated with the Laurel Limestone of Tennessee and is a partial equivalent of the Chimneyhill Formation of Oklahoma (Caplan, 1954).

Lafferty Limestone. The Lafferty Limestone (Fig. 3) was named for exposures at Tate Springs 1.25 mi (2 km) north of Penters Bluff Station on the Missouri Pacific Railroad, and 0.5 mi (0.8 km) west of West Lafferty Creek (Miser, 1920).

The Lafferty at the type locality is about 85 ft (26 m) thick (its maximum thickness), and is an earthy, compact, gray to red, thin to medium, even-bedded limestone. The reddish color, so conspicuous at the type locality, varies both laterally and vertically, and is thought to be related to the weathering of the rocks.

The Lafferty is considered to be unconformable with the overlying formations and appears to be disconformable with the underlying St. Clair Limestone. The contact with the St. Clair is characterized by a bedding break and an abrupt lithologic change from the micrite of the Lafferty.

The Lafferty is thought to be equivalent to, or partially equivalent to, the Henryhouse Shale of Oklahoma and to the Bainbridge Limestone of Missouri (Caplan, 1954).

REFERENCES CITED

Amsden, T. W., and Roland, T. L., 1965, Silurian stratigraphy of northeastern Oklahoma: Oklahoma Geological Survey Bulletin 10-5, 174 p.

Caplan, W. M., 1954, Subsurface geology and related oil and gas possibilities of northeastern Arkansas: Arkansas Research and Development Commission Bulletin 20, 124 p.

Foerste, A. F., 1905, Silurian clays: Kentucky Geological Survey Bulletin 6, pt. 2, p. 143–178.

Maher, J. C., and Lantz, R. J., 1953, Geology of the Gilbert area, Searcy County, Arkansas: U.S. Geological Survey Oil and Gas Investigations Map OM-132, scale 1:12,000.

Miser, H. D., 1920, Preliminary report on the deposits of manganese ore in the Batesville district: U.S. Geological Survey Bulletin 715-G, p. 93–124.

—— , 1922, Deposits of manganese ore in the Batesville district, Arkansas: U.S. Geological Survey Bulletin 734, 273 p.

Penrose, R.A.F., Jr., 1891, Manganese, its uses, ores, and deposits: Arkansas Geological Survey Annual Report, v. 1, 641 p.

Ulrich, E. O., 1911, Revision of the Paleozoic Systems: Geological Society of America Bulletin, v. 22, p. 281–680.

Williams, H. S., 1892, The Paleozoic faunas of northern Arkansas: Arkansas Geological Survey Annual Report, v. 5, p. 268–362.

Crowley's Ridge, Arkansas*

M. J. Guccione, Department of Geology, University of Arkansas, Fayetteville, Arkansas 72701
W. L. Prior, Arkansas Geological Commission, Geology Center, 3815 W. Roosevelt Road, Little Rock, Arkansas 72204
E. M. Rutledge, Department of Agronomy, University of Arkansas, Fayetteville, Arkansas 72701

GEOMORPHOLOGY

Crowley's Ridge is an erosional remnant of unconsolidated Tertiary clay, silt, sand, and lignite capped by Pliocene-Pleistocene gravel, sand, and clay, and middle-to-late Pleistocene loess (Haley, 1976). Located in eastern Arkansas and southeast Missouri, Crowley's Ridge is no more than 11 mi (18 km) wide, but it extends for 186 mi (300 km) from Campbell, Missouri, to Helena, Arkansas (Fig. 1). The ridge stands 100 to 200 ft (30 to 60 m) higher than the adjoining Eastern and Western Lowlands.

Crowley's Ridge was a divide formed during the Pleistocene, as ancestors of the Mississippi River to the west and the Ohio River to the east of the Ridge eroded Coastal Plain sediments (Call, 1891; Fisk, 1944) (Fig. 2). At that time the two rivers joined near Simmesport, Louisiana, 404 mi (650 km) south of where they join today. Abundant glacial outwash and a rising sea level contributed to the aggradation of the ancestral Mississippi and Ohio Rivers, causing a decrease in their gradients. This decrease in gradient and tectonic activity along the New Madrid seismic zone may have caused the eastward shift of the Mississippi River, culminating in the cutting of the present channel through the divide at Thebes Gap. Today the Mississippi River joins the Ohio River near Cairo, Illinois, and flows on the east side of the ridge.

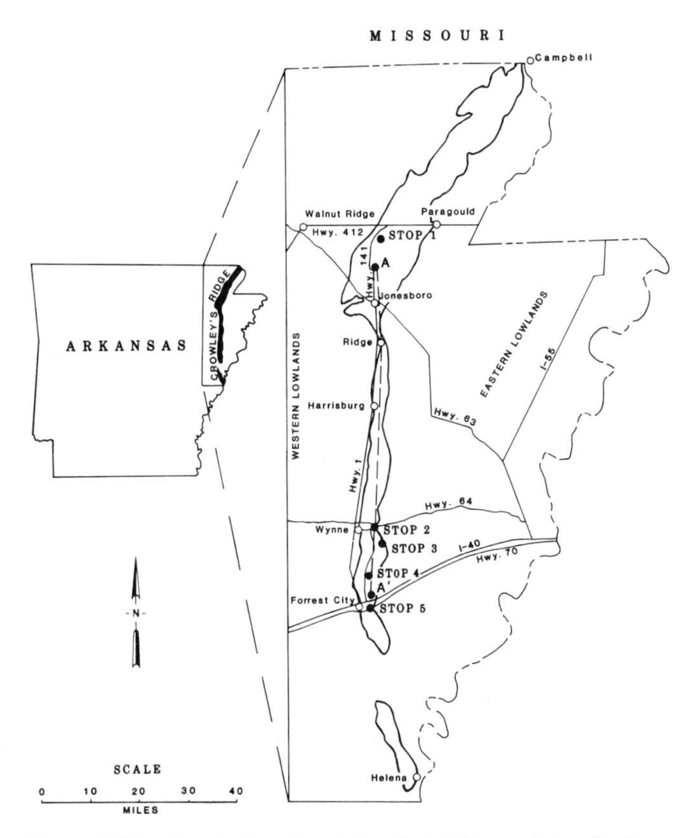

Figure 1. Map showing location of Crowley's Ridge and described sections in this field guide. Location of cross section (Fig. 3) is also shown.

GEOLOGIC SETTING AND STRATIGRAPHY

Crowley's Ridge is located in the northwestern section of the Mississippi River Embayment (Meissner, 1984). The embayment axis trends approximately parallel to the modern Mississippi River and plunges south toward the Gulf of Mexico. Cretaceous and Tertiary sediments in the Arkansas part of this basin dip 35 to 75 ft/mi (7 to 14 m/km) to the east-southeast toward the axis of the embayment (Meissner, 1984). The sediments also thicken toward the southeast due to greater rates of deposition and less subsequent erosion in that direction.

Three Tertiary basin-fill units, all Eocene, crop out on Crowley's Ridge (Fig. 3). As marine waters transgressed over the area, the sediments changed from dominantly fluvial to dominantly marine facies. The oldest unit, the Wilcox Group, is exposed along the flanks of the northern portion of the ridge from Campbell, Missouri, to Jonesboro, Arkansas. Wells indicate that the Wilcox is ~560 ft (170 m) thick in this area and that it thickens southward to ~775 ft (236 m) at the southern edge of the Ridge (Meissner, 1984). The sediments consist of interbedded sand, silt, clay, and some lignite deposited in a fluvial-deltaic environment.

The Claiborne Group overlies the Wilcox Group and is exposed along the flanks of the Ridge in the central part of Crowley's Ridge from Jonesboro, Arkansas, to Wynne, Arkansas (Fig. 1). The maximum thickness of the Claiborne Group is 690 ft (210 m) near the southern end of the Ridge (Meissner, 1984). The sediments consist of fine sand, silt, sandy clay, and minor lignite deposited in deltaic and nearshore marine environments.

The Jackson Group overlies the Claiborne Group and is exposed along the flanks of the southern portion of Crowley's Ridge from Wynne, Arkansas, to Forrest City (Fig. 1). Here the sediments are ~490 ft (150 m) thick and consist of sandy clay, silt, and glauconitic, fossiliferous, sandy clay deposited in a nearshore marine environment. The Jackson Group was deposited during the last marine transgression into Arkansas. The transgression was centered over the Desha Basin, which is a structural low trending east-west, south of Crowley's Ridge (Wilbert, 1953). An

*Published with permission of the Director of the Arkansas Agricultural Experimental Station.

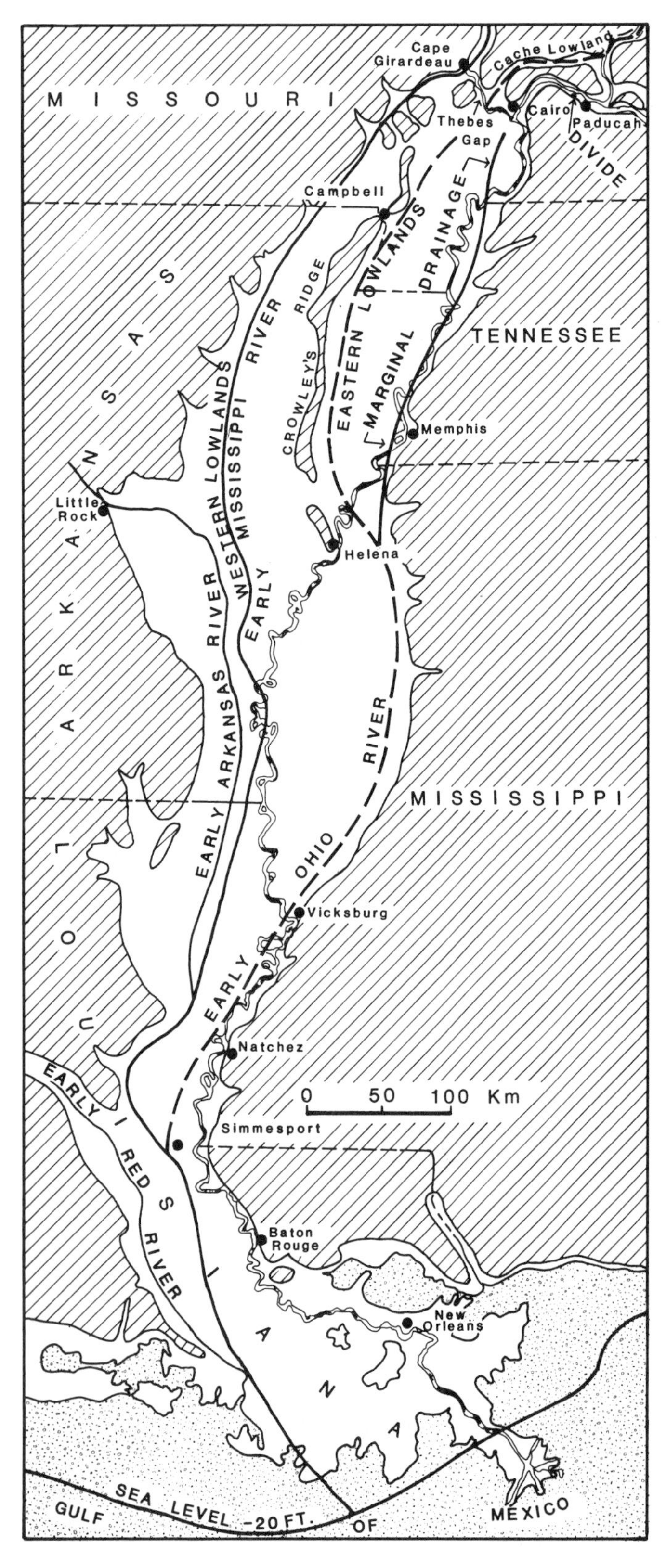

Figure 2. Map showing earlier stage relationships between Mississippi and Ohio Rivers. (modified after Fisk, 1944, Fig. 42).

undetermined thickness of Tertiary strata has been removed by erosion since the Eocene.

Overlying the Tertiary units on the Crowley's Ridge are Pliocene-Pleistocene gravels, sands, and silts and Pleistocene loesses (Fig. 3). The gravels and sands were deposited by braided streams that flowed down the Mississippi Valley. These strata, 0 to 125 ft (0 to 38 m) thick (Holbrook, 1980), dip to the south at a steeper gradient than those of the adjacent Holocene streams. The dominantly chert gravels are subrounded, imbricated, and weakly bedded, and may grade up to tabular cross-stratified sands and massive or laminated silts. These units may be blanket- or pod-shaped deposits. Two or three units are present at some localities, separated by moderately to well-developed, red, clayey soils. A similar, well-developed soil is commonly present in the upper portion of the gravel. On the basis of their lithology and stratigraphic position (Potter, 1955; Saucier, 1974), these deposits have been correlated with the Citronelle Formation and Lafayette Gravel, both of which occur adjacent to the Mississippi Embayment.

A silt, with a well-developed soil, is present above the gravels at some localities. At some locations this deposit may represent a fine-grained facies of the underlying sand and gravel, but elsewhere it can be a loess.

Middle and late Pleistocene loesses, wind-blown silts, overlie the Pliocene-Pleistocene gravel, sand, and silt. The loesses range from 0 to 140 ft (0 to 43 m) thick, generally thinning to the north (Call, 1891, p. 35, 45). Call (1891, p. 233) recognized two loesses, but more recent investigators (Wascher and others, 1947; Leighton and Willman, 1950; West and others, 1980) have recognized three loesses on most of the ridge. These units have been correlated with the Illinoian Loveland Silt and Wisconsinan Roxana Silt and Peoria Loess (Leighton and Willman, 1950; West and others, 1980) of the upper Mississippi Valley (Willman and Frye, 1970). This correlation is on the basis of stratigraphic position, relative thickness, degree of soil development, and colors of the soils and C horizons. The relatively thick Loveland Silt is overlain by a relatively thin Roxana Silt, which in turn is overlain by a relatively thick Peoria Loess.

Interbedded gravel, sand, and silt overlie the Tertiary units on the northern portion of Crowley's Ridge. The gravels, sands, and silts have been interpreted to be of various ages. Potter (1955) has identified them as Lafayette Gravel of Pliocene-Pleistocene age, and Haley (personal communication) considers them to be younger than the Peoria Loess on the central portion of Crowley's Ridge. Two of us (Guccione and Rutledge) find more evidence in support of Potter's correlation, because other workers (Wascher and others, 1974; Leighton and Willman, 1950; Thorp and Smith, 1952) have reported that the overlying loesses are continuous over the entire ridge, although thinning to the north. Our field investigations, although limited, support the presence of loesses on the northern portion of the ridge. For example, the highly weathered, red clay-rich soils developed in the gravels suggest to us that the gravels are unlikely to have been deposited since the late Pleistocene.

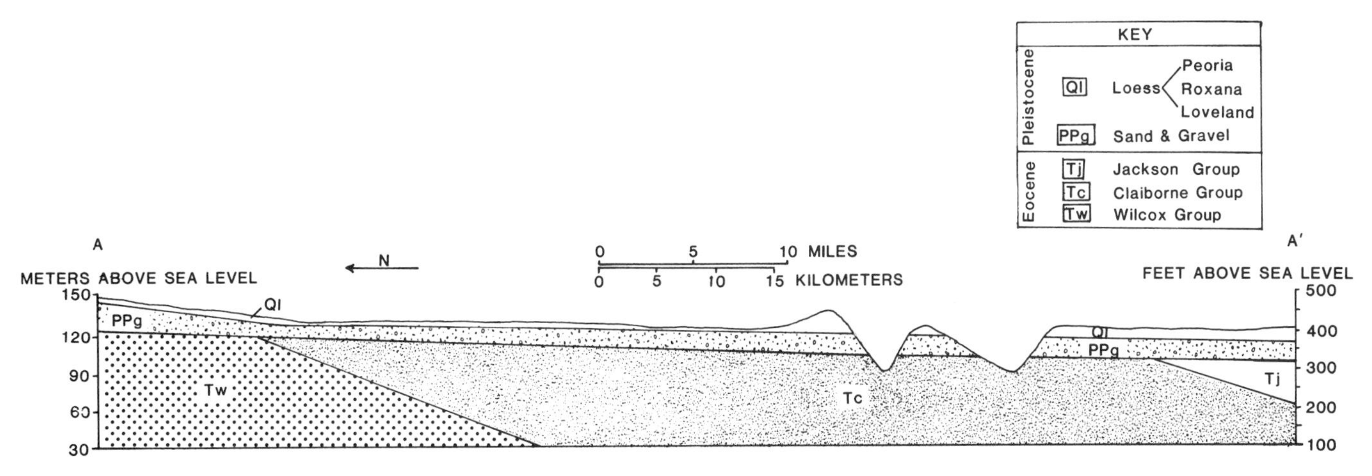

Figure 3. A north-south cross section of Crowley's Ridge. Location is shown in Figure 1.

The ages of the geomorphic features in eastern Arkansas are suggested by the nature and distribution of the Quaternary deposits. First, the Loveland Silt and Peoria Loess are present on high terraces in the Western Lowlands (Rutledge and others, 1985) as well as on Crowley's Ridge. Second, the source of the loess changed: the Loveland Silt was derived from both east and west of the ridge, but the source for the Roxana Silt and Peoria Loess is only from east of the ridge (West and others, 1980). The formation of the Western and Eastern Lowlands and Crowley's Ridge as a divide between these valleys of the ancestoral Mississippi and Ohio rivers predates deposition of Illinoian Loveland Silt. The diversion of the Mississippi River from the Western Lowland to the Eastern Lowland predates deposition of the Wisconsinan Roxana Silt. An alternate chronology is suggested by Haley (personal communication): the formation of the Western and Eastern Lowlands, the Ridge, and the diversion of the Mississippi River to the Eastern Lowlands postdate the gravels on the northern (Campbell, Missouri, to Jonesboro, Arkansas) portion of the ridge, which Haley interprets as younger than the Wisconsinan Peoria Loess.

DESCRIBED SECTIONS

The following sections are examples of the stratigraphy previously described:

Stop 1: Poplar Creek section. This stop is in SE¼,SW¼,SW¼,Sec.9,T.16N.,R.4E., Green County (Gainesville 15-minute Quadrangle). It can be reached by driving 1 mi (1.6 km) south of Walcott on Arkansas 141. At Mt. Zion Baptist Church turn east on a gravel road and continue 0.7 mi (1.1 km) to Mt. Zion Baptist Camp. Drive south on the gravel road 0.4 mi (0.6 km) to a low-water bridge. Walk 0.2 mi (0.3 km) upstream (east) to an outcrop on the north valley wall (Fig. 4). The landowner, William Oliver (501-539-6321), requests that visitors stay within the creek bed.

Forty-one ft (12 m) of the Wilcox Group are exposed at the base of the stream cut. The sediments are composed of alternating beds of fluvial sand and clay. The sand beds, 0.5 to 1.0 ft (0.2 to 0.3 m) thick, are medium-grained and planar or cross-stratified point bar deposits. The cross-stratified layers contain carbonaceous material and rip-up clasts of the underlying clay. The clay beds, 0.5 ft (0.2 m) thick, are white to maroon, sandy, and kaolinitic slack-water deposits. The sand units thin, and the clay units become thicker and more abundant in the upper part of the section.

Overlying the Wilcox Group is ~4 ft (1.2 m) of fluvial Pliocene-Pleistocene gravel and sand. At the base, a weakly bedded and imbricated gravel contains chert cobbles and boulders with a reddish-brown to orange, coarse-grained sand matrix. It grades upward to a weakly bedded, medium-grained sand, about 2 ft (0.6 m) thick. The sand laterally fines and thickens.

The upper unit is ~13 ft (4 m) of undifferentiated Pleistocene loess, which forms a vertical bluff. At the base, the brown noncalcareous silt contains disseminated pebbles and grains of sand, which are absent in the middle and upper part of the unit.

Figure 4. Photograph of Poplar Creek section looking west. The units exposed are Wilcox Group (Tw), Pliocene-Pleistocene sand and gravel (PPg), and Quaternary loess (Ql).

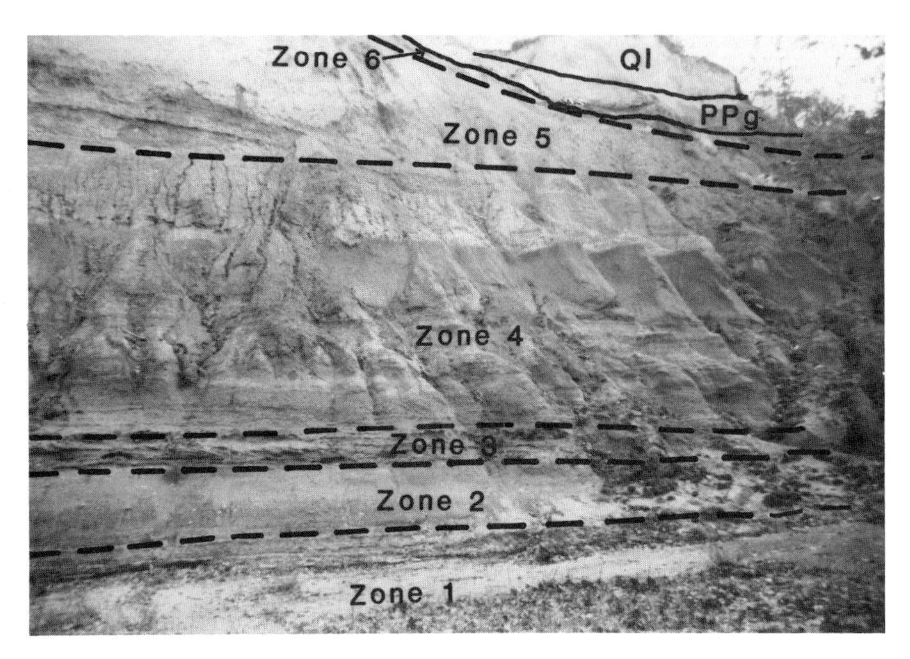

Figure 5. Photograph of Copperas Creek Section looking south. Units exposed are Claiborne Group (zones 1 through 6), Pliocene-Pleistocene sand and gravel (PPg), and Quaternary loess (Q1).

We (Guccione and Rutledge) correlate this whole sequence of gravel, sand, and silt with similar Pliocene-Pleistocene gravels and sands and Pleistocene loess found on the southern portion of the ridge. In contrast, B. R. Haley (personnal communication) considers this sequence to be younger than all the Pleistocene strata south of the town of Ridge, Arkansas.

Holocene deposits are exposed upstream (east) along the south bank, where gravel, sand, and silt with a weak (cambic) soil form a low terrace that is presently being eroded.

Stop 2: Copperas Creek section. This stop is in NE¼, SE¼,SE¼, Survey No. 498, Cross County (Whitmore 7½-minute Quadrangle). It can be reached by driving 0.4 mi (0.6 km) south of Arkansas 64 at Levesque, on a gravel road to Copperas Creek bridge. The outcrop is the south valley wall of Copperas Creek, just east of the bridge.

Approximately 81 ft (25 m) of the Claiborne Group are exposed at the base of a stream cut (Fig. 5). The sequence has been divided into six zones in ascending order.

Zone 1 is interlaminated clay and sand. The clay laminae contain fossil leaves and twigs. The sand laminae contain fossil root casts and burrows, suggesting that the environment of deposition was a delta or a coastal marsh. The clayey horizons supported vegetation and were buried by the sandy strata during recurrent flooding.

Zone 2 is ~4 ft (1.2 m) of mottled sandy clay with abundant burrows(?) and lignite fragments. The material may have been deposited during a larger flood event in which abundant clastics and transported pieces of wood (lignite) filled in the marsh.

Zone 3 is ~2 ft (0.6 m) of iron-cemented, cross-bedded friable sandstone with rip-up clay clasts and clay layers. This zone may have been deposited as a distributary channel.

Zone 4 is ~20 ft (6 m) of sandy clay. The lower portion of the zone is gradational into the underlying sand and contains selenite crystals. The sandy clay may have been deposited as a brackish water lagoon. The upper portion of zone 4 is fine sand. It may have been deposited where a prograding distributary channel or a bar entered the lagoon or bay.

Zone 5, the thickest layer, is ~25 ft (8 m) of mottled sandy clay and sand with some iron-cemented layers. These clays are the youngest unweathered Claiborne sediments and may have been deposited in a lagoon or bay environment that formed during a marine transgression.

Zone 6 is a weathering zone about 6 ft (2 m) thick, developed in the lagoonal sands and clays of zone 5. It is heavily iron-cemented. Overlying the Claiborne Group is ~7 ft (2 m) of fluvial Pliocene-Pleistocene gravels. The dominantly chert cobbles and boulders are embedded in a red, silty clay matrix. About 13 ft (4 m) of Pleistocene loess cap the exposure. The unit is a massive brown silt.

Stop 3: Wittsburg Section. This stop is in NE¼, NW¼,NW¼, Survey No. 494, Cross County (Whitmore 15-minute Quadrangle), about 3 mi (4.8 km) SSE of site 73CS01 of West and others (1980). It can be reached by driving 3.8 mi (6.1 km) south of stop 2 on a gravel road. At the gravel pits turn west on a gravel road and continue 0.5 mi (0.8 km) to a cut on the north side of the road (Fig. 6).

The oldest unit exposed in this roadcut is about 5 ft (1.5 m) of imbricated Pliocene-Pleistocene gravel. The dominantly chert

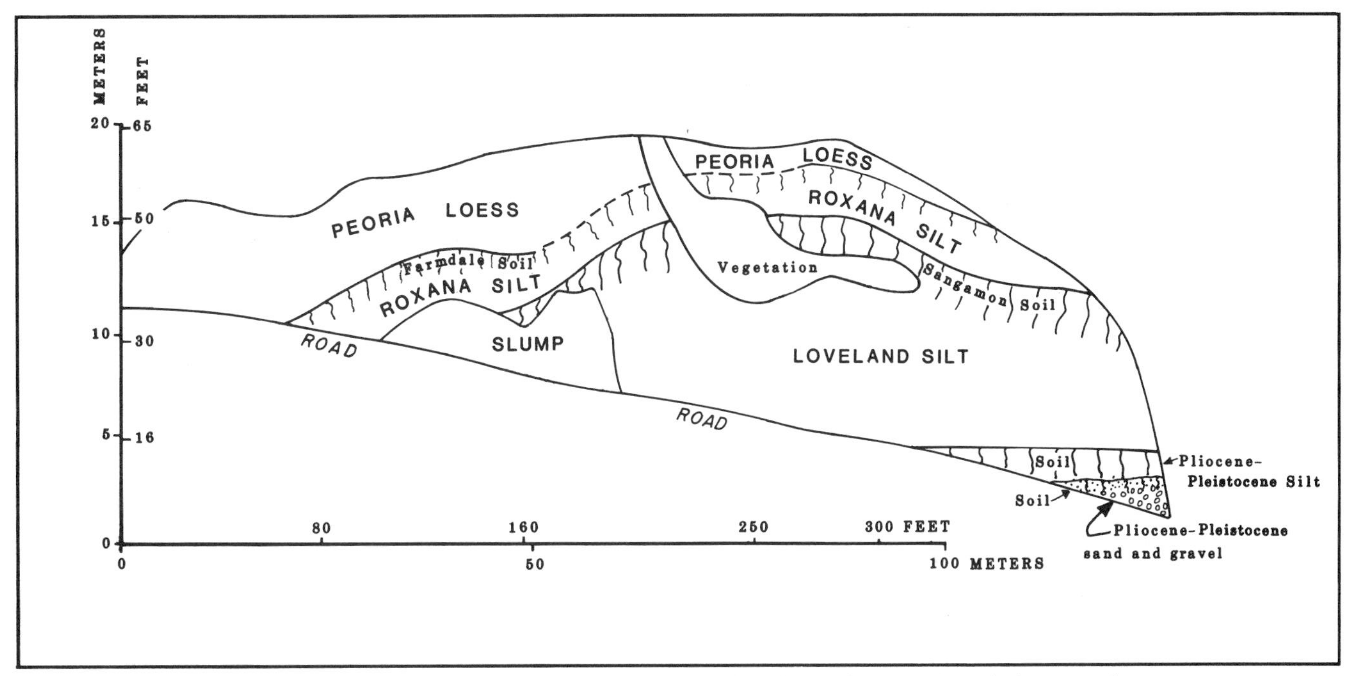

Figure 6. Diagram of the Wittsburg Section showing relationship of units exposed. Vertical exaggeration is 2.5×.

gravel has a brown patina and a yellowish-red matrix. The soil developed in the unit is partly truncated but is complete in the gravel pits south of the road. The gravel is underlain by, and probably laterally grades into, tabular cross-stratified, well-sorted sand. The sand is exposed in gullies and shallow pits east of the roadcut. The contact of this Pliocene-Pleistocene sand with the underlying sandy clay and lignite of the Claiborne Group is exposed near the east entrance of the gravel pits.

Overlying the gravel is ~5 ft (1.5 m) of a yellowish-red and gray, mottled, silty clay loam containing a well-developed soil. This is most likely a fine-grained facies of a fluvial deposit but may contain some loess component.

All three Pleistocene loesses are present above. The oldest loess, the Illinoian Loveland Silt, is unusually thick (34 ft or 10 m) at this exposure. The Loveland Silt is a massive brown silt loam weakly calcareous in some of its lower part, containing abundant gastropod shells. The Sangamon Soil developed in the Loveland Silt has a strong brown, silty clay loam B horizon that is ~7 ft (2 m) thick. The pedon is more developed than the ground soil in the Peoria Loess. The maximum duration of weathering of the ground soils on the southern portion of Crowley's Ridge is approximately 14,000 years.

The overlying middle Wisconsinan Roxana Silt is separated from the Loveland Silt by a gradational contact. It is also thicker (8 ft or 2.5 m) than that found on the adjoining uplands. The brown to dark-brown noncalcareous silt loam includes a weakly developed Farmdale Soil in its upper part.

The youngest loess, the late Wisconsinan Peoria Loess, is only ~15 ft (5 m) at this exposure because quarry operators have stripped approximately 6 ft (2 m) from the surface. It is a massive, noncalcareous, yellowish-brown silt loam, and is the parent material for most of the soils on the ridge.

Stop 4: Village Creek State Park section. This stop is in SW¼,Sec.7,T.16N.,R.4E., Cross County (Whitmore 15-minute Quadrangle). It can be reached by driving 2.5 mi (4.0 km) east of stop 3 on a gravel road. At Harris Chapel Baptist Church turn south on Arkansas 284 and continue 2.2 mi (3.5 km) to the entrance of Village Creek State Park (north of Forrest City and south of Wynne, Arkansas). Drive 1.8 mi (2.9 km) through the park to Lake Austell Dam. Walk south across the dam to the Military Road Trail Loop. The lower trail crosses the base of a landslide. The vertical scarp face is to the south (right).

The stop is of dual interest because the stratigraphy is exposed by a relatively recent and large landslide, similar to numerous slides that characterize the region (Call, 1891, p. 40). Initial movement of the slide occurred after the lake was impounded in 1978 and continued for several months. Most of the movement occurred during the first 45 days following heavy rains and coinciding with a local seismic event (McFarland and Stone, 1982).

The original slide area was 600 ft (183 m) long, but since has expanded 230 ft (70 m) farther south. Pressure ridges with tilted and uprooted trees, many of which have reoriented themselves to a vertical growth position, developed along the advancing toe of the slide.

Approximately 5 ft (1.5 m) of horizontally bedded Pliocene-Pleistocene gravel with a coarse sand matrix are exposed at the base of the scarp face. The gravel grades up into a 6-ft (3-m) fluvial unit of bedded sands and laminated silts containing a well-developed soil at the top.

The upper ~22 ft (7 m) of exposure comprise the three Quaternary loesses. The Loveland Silt below includes a well-developed Sangamon Soil, is overlain by the Roxana Silt with the weak Farmdale Soil, and the Peoria Loess—at the top of the exposure—contains the ground soil. Each loess is relatively thin, suggesting that the site was subject to erosion in the past, as it is today.

Stop 5: Crow Creek section. This stop is in SE¼, NE¼,SW¼,Sec.25,T.5N.,R.3E., St Francis County (Whitmore 15-minute Quadrangle). It can be reached by driving 2.9 mi (4.7 km) east of Arkansas 1 on U.S. 70 to Crow Creek bridge. The outcrop is the west valley wall of Crow Creek, just north of U.S. 70.

About 14 ft (4 m) of the marine Jackson Group are exposed at the base of the stream cut and along the stream bed. Two distinct units are recognizable. The lower unit is 5 to 8 ft (2 to 3 m) of gray clay with fine sand, mica, glauconite, and numerous fossils. Invertebrate fossils include pelecypod and gastropod shells and shell fragments; vertebrate fossils consist of shark's teeth. The clay was deposited in a lagoon or bay formed during the last marine transgression within Arkansas (Wilbert, 1953).

The upper unit of the Jackson Group is 4 to 7 ft (1 to 2 m) thick and ranges from brown, very fine, sandy silt to gray clay. It may be a deltaic crevasse splay or a sheet deposit that filled the bay or lagoon.

The gravels within the stream bed are reworked Pliocene-Pleistocene gravels that overlie the Tertiary units on Crowley's Ridge.

REFERENCES

Call, R. E., 1891, Annual Report of the Geological Survey of Arkansas for 1889, v. II. The Geology of Crowley's Ridge: Little Rock, Arkansas, Woodruff Printing Co., 283 p.

Clardy, B. F., 1979, Arkansas Lignite Investigations: Preliminary Report: Arkansas Geological Commission Miscellaneous Publication MP-15, 133 p.

Fisk, H. N., 1944, Geologic Investigation of the Alluvial Valley of the Lower Mississippi River: Vicksburg, Mississippi River Commission, Print 52, 78 p.

Haley, B. R., 1976, Geologic map of Arkansas: Arkansas Geological Commission and U.S. Geological Survey, scale 1:500,000.

Holbrook, D. F., 1980, Arkansas lignite investigations; Preliminary report: Arkansas Geological Commission, 157 p.

Leighton, M. L., and Willman, H. B., 1950, Loess formations of the Mississippi River Valley: Journal of Geology, v. 45, p. 577–601.

McFarland, J. D., III, and Stone, C. G., 1982, A case history of major landslide on Crowley's Ridge, Village Creek State Park, Arkansas, *in* McFarland, J. D., ed., Contributions to the Geology of Arkansas: Arkansas Geological Commission Miscellaneous Publication 18, p. 45–52.

Meissner, C. R., Jr., 1984, Stratigraphic Framework and Distribution of Lignite on Crowley's Ridge, Arkansas: Arkansas Geological Commission Information Circular 28-B, 14 p.

Potter, P. E., 1955, The petrology and origin of the Lafayette gravel: Journal of Geology, v. 63, no. 1, p. 1-38; no. 2, p. 115–132.

Rutledge, E. M., West, L. T., and Omakupt, M., 1985, Loess deposits on a Pleistocene age terrace in eastern Arkansas: Soil Science Society of America Journal, v. 49, p. 1231–1238.

Saucier, R. T., 1974, Quaternary geology of the lower Mississippi Valley: Arkansas Archeological Survey Research Series no. 6, 26 p.

Thorp, J., and Smith, H.T.U., 1952, Map of eolian deposits of the United States, Alaska, and parts of Canada: National Research Council and Geological Society of America, scale 1:2,500,000.

Wascher, H. L., Humbert, R. P., and Cady, J. G., 1947, Loess in the southern Mississippi Valley; Identification and distribution of the loess sheets: Soil Science Society of America Proceedings, v. 12, p. 389–399.

West, L. T., Rutledge, E. M., and Barber, D. M., 1980, Sources and properties of loess deposits on Crowley's Ridge in Arkansas: Soil Science Society of America Journal, v. 44, p. 353–358.

Wilbert, L. T., 1953, The Jacksonian Stage in southeastern Arkansas: Arkansas Geological Commission Bulletin 19, 125 p.

Willman, H. B., and Frye, J. C., 1970, Pleistocene stratigraphy of Illinois: Illinois Geological Survey Bulletin 94, 204 p.

Atokan stratigraphy of the Cherry Bend area, northwestern Arkansas

Gary D. Harris, Dean A. Ramsey, and Doy L. Zachry, Department of Geology, University of Arkansas, Fayetteville, Arkansas 72701

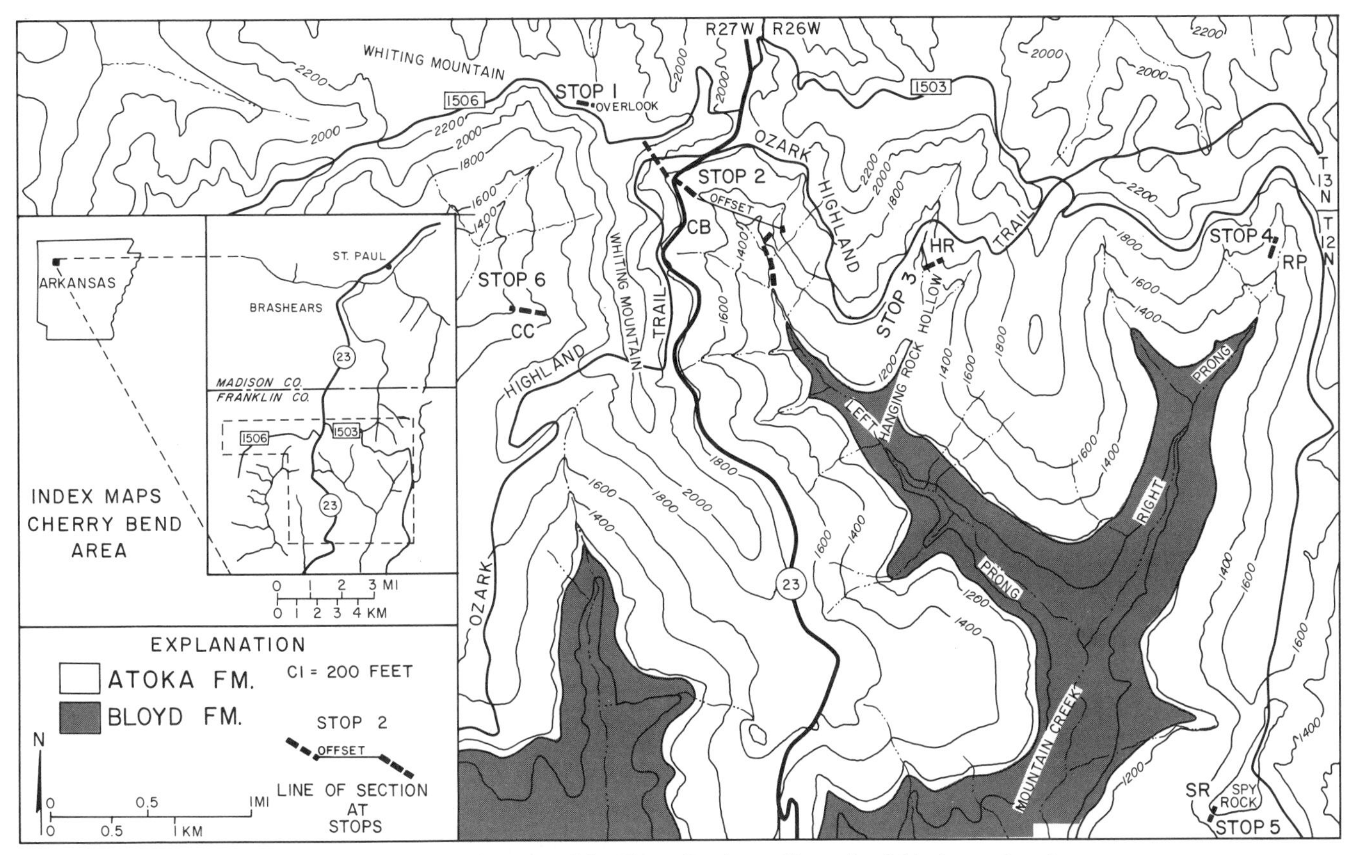

Figure 1. General location map of the Cherry Bend area illustrating field trip stop locations.

LOCATION

The Cherry Bend area is astride Arkansas 23 approximately 5 mi (8 km) south of Brashears, Arkansas (Fig. 1, inset), and is immediately north of the east-trending Cass Fault-Monocline System of the northern margin of the Arkoma Basin. Access to the various stops in the Cherry Bend area is from well-maintained forest service roads, and from Arkansas 23, as well as from the Ozark Highland Trail that transects the area from west to east (Fig. 1). The area is contained entirely within the Cass 7½-minute Quadrangle.

SIGNIFICANCE

Sandstone and shale units assigned to the lower and middle parts of the Atoka Formation are widely exposed in the Boston Mountains of northwestern Arkansas. South of the Cass Fault-Monocline System the units are continuous throughout the subsurface of the northern Arkoma Basin where they are overlain by upper Atoka strata. Sandstone units of the lower and middle Atoka succession contain important natural gas reservoirs in the Arkoma Basin, and have been assigned informal names by petroleum geologists active in the area. The names are applicable to units in outcrop north of the Cass Fault-Monocline.

The Cherry Bend area is deeply dissected by south-draining tributaries of the Mulberry River. Morrowan (early Pennsylvanian) strata assigned to the Bloyd and underlying Hale formations are exposed in the lower reaches of south-draining streams (Fig. 1). The Bloyd is overlain by the Atoka Formation (Early Pennsylvanian), which extends to the highest elevations in the area. Resistant units of the middle Atoka cap the hills at Cherry Bend and are the youngest Paleozoic units north of the Cass Fault-Monocline System. Upper Atoka units are confined to the Arkoma Basin south of the fault.

Field stops selected for the Cherry Bend area include a single overlook illustrating geomorphic features of the Boston Moun-

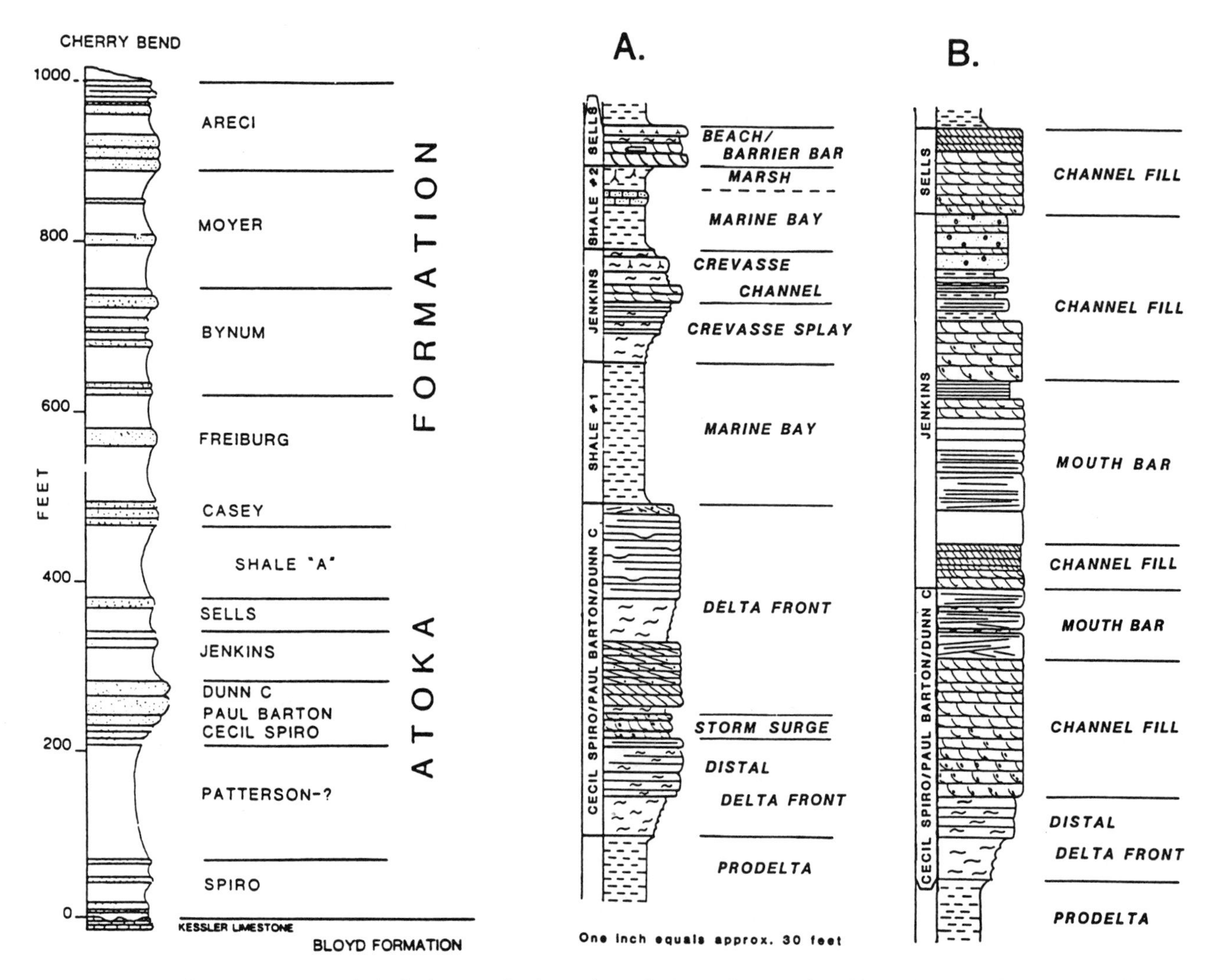

Figure 2. Stratigraphy section of lower and middle Atoka strata in the Cherry Bend section.

Figure 3. Generalized stratigraphic sections showing the succession of facies in "Cecil Series" strata. Section A is more characteristic of the northern part of the Cherry Bend area, and Section B of the southern.

tains, and five stratigraphic sections. One of the sections, Stop 2, illustrates the entire lower and middle Atoka succession. The remaining stops involving stratigraphic sections include only the upper part of the lower Atoka interval. This interval has been selected to illustrate various delta and delta-related facies associations established by correlation among the various sections. Stratigraphic and sedimentologic interpretations are the result of studies by Harris (1983) and Ramsey (1983) during thesis work at the University of Arkansas.

Stop 1. Stop 1 is a south-facing overlook along Forest Service Road 1506. It is reached by traveling west along the road from Arkansas 23. The stop is at an elevation of 2,100 ft (640 m) and overlooks the drainage basins of Cove Creek and Lindsey Branch. The Arkansas River, approximately 27 mi (43 km) to the south, is visible on clear days. The view illustrates the geomorphic character of the Boston Mountains in an area where relief is extreme and elevations at a near maximum.

Elevations in the Cherry Bend area range from approximately 900 ft (270 m) along Mountain Fork Creek where it leaves the area near its confluence with the Mulberry River to approximately 2,300 ft (700 m) on Whiting, Fly Gap, and Hare mountains (Fig. 1). Streams rise near the mountain crests, and stream gradients are steep across beds of the Atoka Formation. Most of the 1,400 ft (426 m) of relief in the area is achieved here. Stream gradients diminish southward across the shale successions of the Bloyd Formation in the lower valleys.

The area is intensely dissected but the mountain tops are flat and at approximately the same elevations. They form part of a formerly continuous erosion surface, the Boston Mountain Plateau, that was dissected during late Tertiary and Quaternary time. The surface has been variously interpreted as a peneplain and as a pediplain, and is part of a series of successively lower lying and less dissected surfaces to the north.

Stop 1 is on the Boston Mountains drainage divide. South-

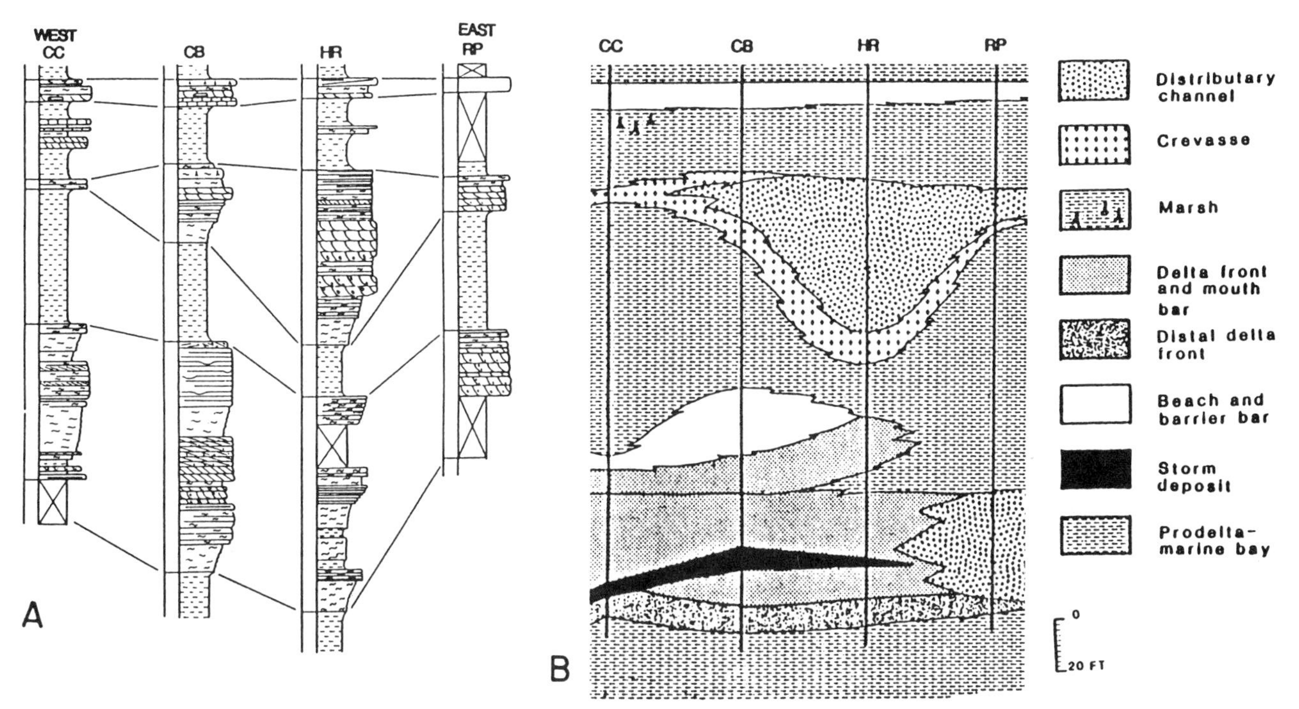

Figure 4. A. East-west lithostratigraphic cross section across the northern part of the Cherry Bend area. Datum is the top of the Sells Sandstone. The Cove Creek (CC), Cherry Bend (CB), Hanging Rock Hollow (HR), and Right Prong (RP) sections are illustrated. B. East-west cross section depicting an environmental interpretation of the facies illustrated in A.

flowing streams enter the Mulberry River, a tributary of the Arkansas River. North-flowing streams enter the East Fork of the White River.

Stop 2. The lower and middle parts of the Atoka Formation are composed of an alternating sequence of sandstone and shale units. Informal names for sandstone-shale packages as applied by petroleum geologists in the Arkoma Basin are used for the surface successions in the Cherry Bend area (Fig. 2). The Cherry Band section (CB) at Stop 2 affords a look at the entire Atoka sequence.

The base of the section is reached by descending into Left Prong Creek from the Ozark Highland Trail. An outcrop of the Kessler Limestone Member of the Bloyd Formation overlain by the basal sandstone unit of the Atoka Formation, the Spiro, occurs at an elevation of approximately 1,200 ft in the stream (Fig. 1). The boundary is an unconformity. The line of section extends upstream along Left Prong Creek and up a small tributary through the Cecil Spiro, Paul Barton, Dunn C, and Jenkins sandstones. The Cecil Spiro forms a striking bluff with its base clearly exposed. The interval between the base of the Spiro Sandstone and the base of the Cecil Spiro Sandstone is formally assigned to the Trace Creek Member of the Atoka, the only widely used and formally named member of the formation. The Patterson Sandstone within the Trace Creek Member is widely exposed in the Cherry Bend area but is absent in the Cherry Bend section. The Cecil Spiro, Paul Barton, and Dunn C sandstones are not differentiated in the Cherry Bend area.

Above the Jenkins Sandstone the section is offset to the northwest to a position just below Arkansas 23 at approximately 1,600 feet. The section continues upward through the Sells Sandstone to the highway. The line of section crosses the highway above the Sells at a sharp bend, Cherry Bend, and continues up the drainage to the top of Whiting Mountain intersecting Forest Service Road 1506 at the crest and near the top of the Areci Sandstone. An interval of shale above the Sells and below the Casey, Shale "A," is prominent and easily correlated throughout the Cherry Bend area and throughout the northern part of the Arkoma Basin. The base of this unit is the base of the middle Atoka succession. The interval from the base of the Cecil Spiro to the top of the Sells is the "Cecil Series" of Arkoma Basin terminology.

The middle Atoka Casey Sandstone is encountered immediately above Arkansas 23. The section continues upward through the Casey, Freiburg, Bynum, Moyer, and Areci sandstones ending on Forest Service Road 1506.

"Cecil Series." The "Cecil Series," Cecil Spiro through Sells, has been examined from a sedimentological standpoint in the Cherry Bend area. It is composed of an association of deltaic and delta-related facies. Specific facies identified and their stratigraphic associations are indicated in Figure 3.

Stops 2 through 6 are stratigraphic sections through the "Cecil Series." They are illustrated as parts of stratigraphic cross sections. Sections not indicated as stops are also included to supplement the stratigraphic information. Environmental interpreta-

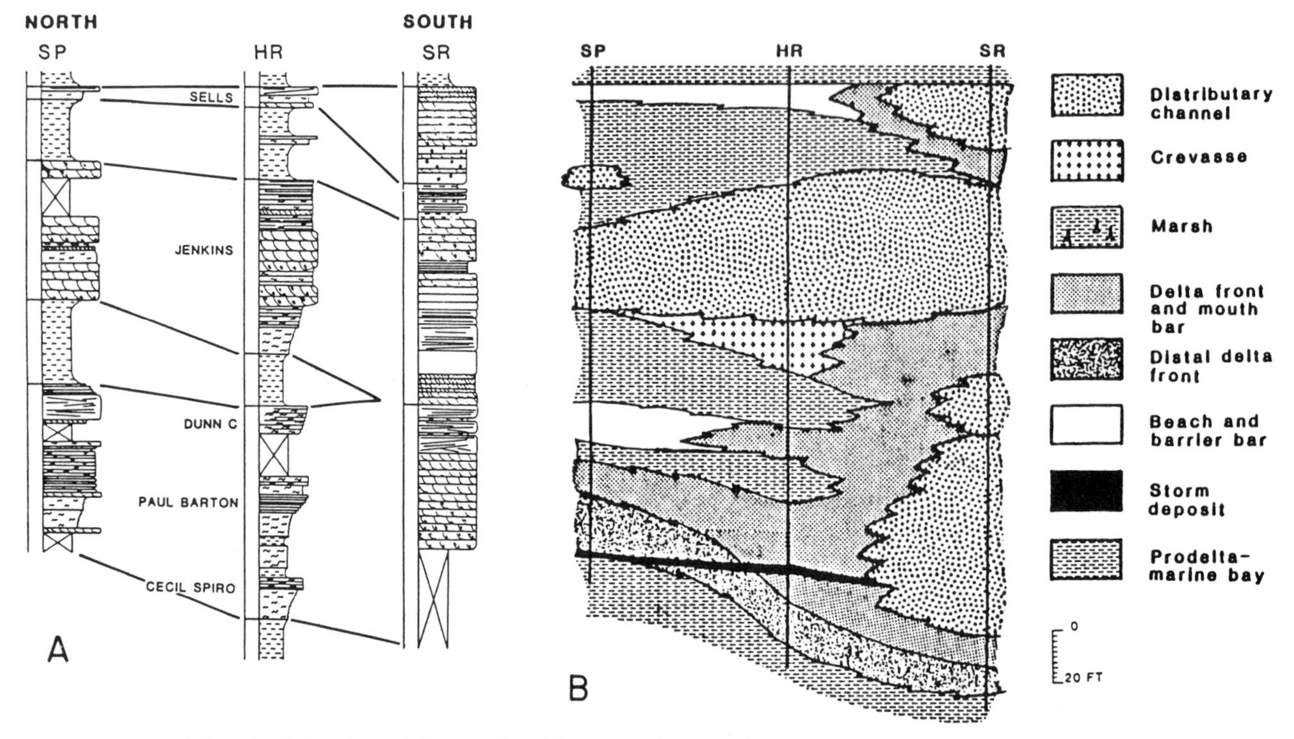

Figure 5. A. North-south lithostratigraphic cross section across the eastern part of the Cherry Bend area. Datum is the Sells Sandstone. The St. Paul (SP), Hanging Rock Hollow (HR), and Spy Rock (SP) sections are illustrated. The St. Paul section is immediately north of the Cherry Bend area at St. Paul, Arkansas. B. North-south cross section depicting an environmental interpretation of the facies illustrated in A.

tions for each cross section are presented as separate figures. Access to the Cherry Bend section through the "Cecil Series" is described under Stop 2.

Stop 3. Hanging Rock Hollow (HR) is reached by descending from the Ozark Highland Trail (Fig. 1). The base of the Cecil Spiro is at approximately 1,350 feet and is marked by a conspicuous bluff with a clearly exposed base. The section is illustrated as HR in Figures 4 and 5.

Stop 4. The Right Prong section (RP) is reached by walking along the Ozark Highland Trail from Forest Service Road 1503 and descending into the Right Prong of Mountain Creek (Fig. 1). The base of the Cecil Spiro is at approximately 1,500 feet. The section is illustrated in Figure 4.

Stop 5. Spy Rock Hollow (SR) is reached from Forest Service Road 1503. A spur of the Ozark Highland Trail extends from the road to the top of the section. The section is illustrated in Figure 5.

Stop 6. Cove Creek section (CC) is reached by a road that extends south along the crest of Whiting Mountain parallel to a natural gas pipeline (Fig. 1). Descend the second drainage to the west along the pipeline road. The section is illustrated in Figure 4.

Depositional Synthesis—"Cecil Series". A stratigraphic cross section (Fig. 4) from west to east and including the Cove Creek section (CC) of Stop 6, the Cherry Bend section (CB) of Stop 2, the Hanging Rock Hollow section (HH) of Stop 3, and the Right Prong section (RP) of Stop 4 indicates that the "Cecil Series" contains three laterally extensive but variable sandstone units separated by intervals of shale. The Cecil Spiro–Paul Barton and Dunn C intervals compose the lower sandstone unit, and the Jenkins and the Sells the upper two sandstone units. The north-south cross section (Fig. 5) also illustrates this subdivision with the intermediate shale units thinning toward the Spy Rock Hollow section. The sandstone intervals of the "Cecil Series" represent three distinct deltaic pulses separated by periods of abandonment, subsidence, and transgression by marine-bay or open-shelf environments. Environmental conclusions for facies encountered along the east-west cross section are also presented in Figure 4, and those for the north-south cross section in Figure 5.

REFERENCES CITED

Harris, G. D., 1983, Sedimentology and depositional history of a deltaic lower Atoka (Pennsylvanian) sandstone, northwestern Arkansas [Masters thesis]: University of Arkansas at Fayetteville, 154 p.

Ramsey, D. A., 1983, Lithostratigraphy and petrography of the Atoka Formation in the St. Paul area, northwest Arkansas [Masters thesis]: University of Arkansas at Fayetteville, 90 p.

Peyton Creek Road Cut, northern Arkansas

Walter L. Manger, Department of Geology, University of Arkansas, Fayetteville, Arkansas 72701

LOCATION

Road cut on east side of U.S. 65 beginning directly south of the Peyton Creek Bridge, 0.2 mi (0.3 km) south of the Searcy-Van Buren county line and 3.4 mi (5.5 km) south of Leslie. NE¼,Sec.11 and NW¼,Sec.12,T.13N.,R.15W., Van Buren County, Arkansas (Fig. 1).

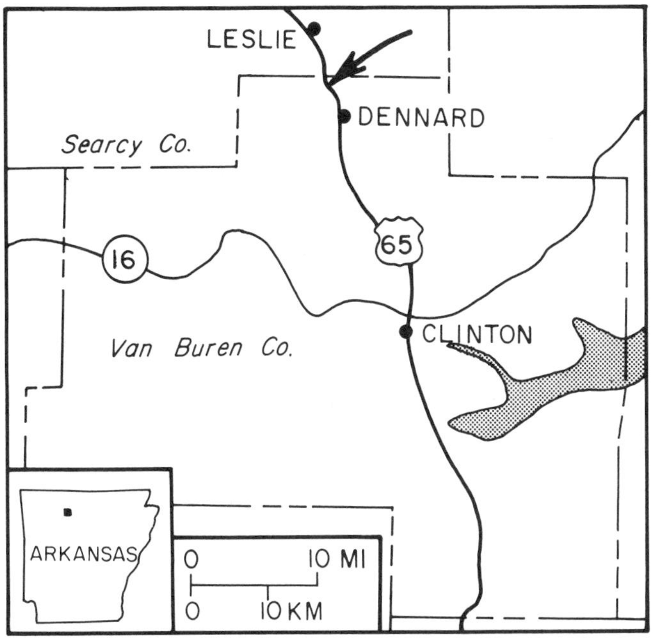

SIGNIFICANCE

The Imo Formation has figured prominently in analysis of the Mississippian-Pennsylvanian boundary in the southern mid-continent of North America and its intercontinental equivalents. This unit is the youngest biostratigraphically dated Mississippian unit in the southern mid-continent.

STRATIGRAPHIC SUCCESSION

The section begins in the upper Pitkin Limestone and exposes all of the Imo Formation, both of Chesterian (Mississippian) age. It ends in strata of presumed lower Pennsylvanian (Morrowan) age, probably referable to the Witts Springs Formation (Fig. 2).

Figure 1. Index map of Van Buren County, Arkansas, showing location of Peyton Creek road cut.

DESCRIPTION

The type locality for the Imo Formation is in Sulphur Springs Hollow, a tributary to Bear Creek, in adjacent Searcy County (Gordon, 1965). However, the Peyton Creek exposure has produced the bulk of the Imo fossils, and understanding of the age and correlation of the unit is based primarily on this exposure. This exposure has been described by Sutherland and Manger (1977, 1979) and the following discussion is updated from those publications.

The upper Pitkin Limestone comprises a series of oolitic crinozoan grainstones with minor dark shale probably deposited on an open shelf (Fig. 3). These beds are significant in that they have yielded biostratigraphically sensitive assemblages of conodonts and calcareous foraminifers. The conodonts, described by Lane (1967) and Lane and Straka (1974), are referable to the *Adetognathus unicornis* Zone of the standard Mississippian succession. This occurrence indicates equivalence of the Pitkin at Peyton Creek to the Grove Church Shale, and thus to the top of the type Chesterian Series (Lane, 1967). Smaller calcareous foraminifers from the Pitkin were reported by Brenckle (1977) to be dominated by eosigmoilinids characteristic of Mamet Zone 19, establishing the presence of that zone in the midcontinent for the first time. Recollection of the type Chesterian has produced similar foraminifer assemblages supporting the correlations based on conodonts (Brenckle and others, 1977). No ammonoids have been recovered from the Pitkin at this locality, but at Leslie, 5 mi (8 km) north, the Pitkin yields an ammonoid assemblage with *Eumorphoceras bisulcatum* and *Cravenoceras richardsonianum* (Saunders and others, 1977). This assemblage may be correlated intercontinentally with Arnsbergian zone E_2a of the standard Namurian succession.

The Imo succeeds the Pitkin with apparent conformity at the Peyton Creek locality. Use of the name Imo has been controversial and is unsettled. It was proposed and rejected on the same page of the same manuscript by Gordon (1965, p. 34). The name is not used officially by either the Arkansas Geological Commission (e.g., Haley, 1976) or the U.S. Geological Survey, but it has continued to appear in some recent literature (e.g., Manger and Sutherland, 1984; Mapes and Rexroad, 1986). The bulk of the formation is fossiliferous gray to black shale with scattered beds of sandstone and conglomeratic limestone representing a transgressive-regressive history (Fig. 4). Well-known ammonoid assemblages from the Imo are dominated by *Anthracoceras discus,* but include more sensitive taxa such as *Fayettevillea, Eumorphoceras, Delepinoceras,* and *Cravenoceras,* suggesting equivalency to the upper Arnsbergian Stage (E_2b-c; Saunders and others, 1977). The basal black shale of the Imo contains an assemblage of crushed ammonoids characterized by *Eumorphoceras bisulcatum erinense* (Fig. 3). This occurrence suggests that the

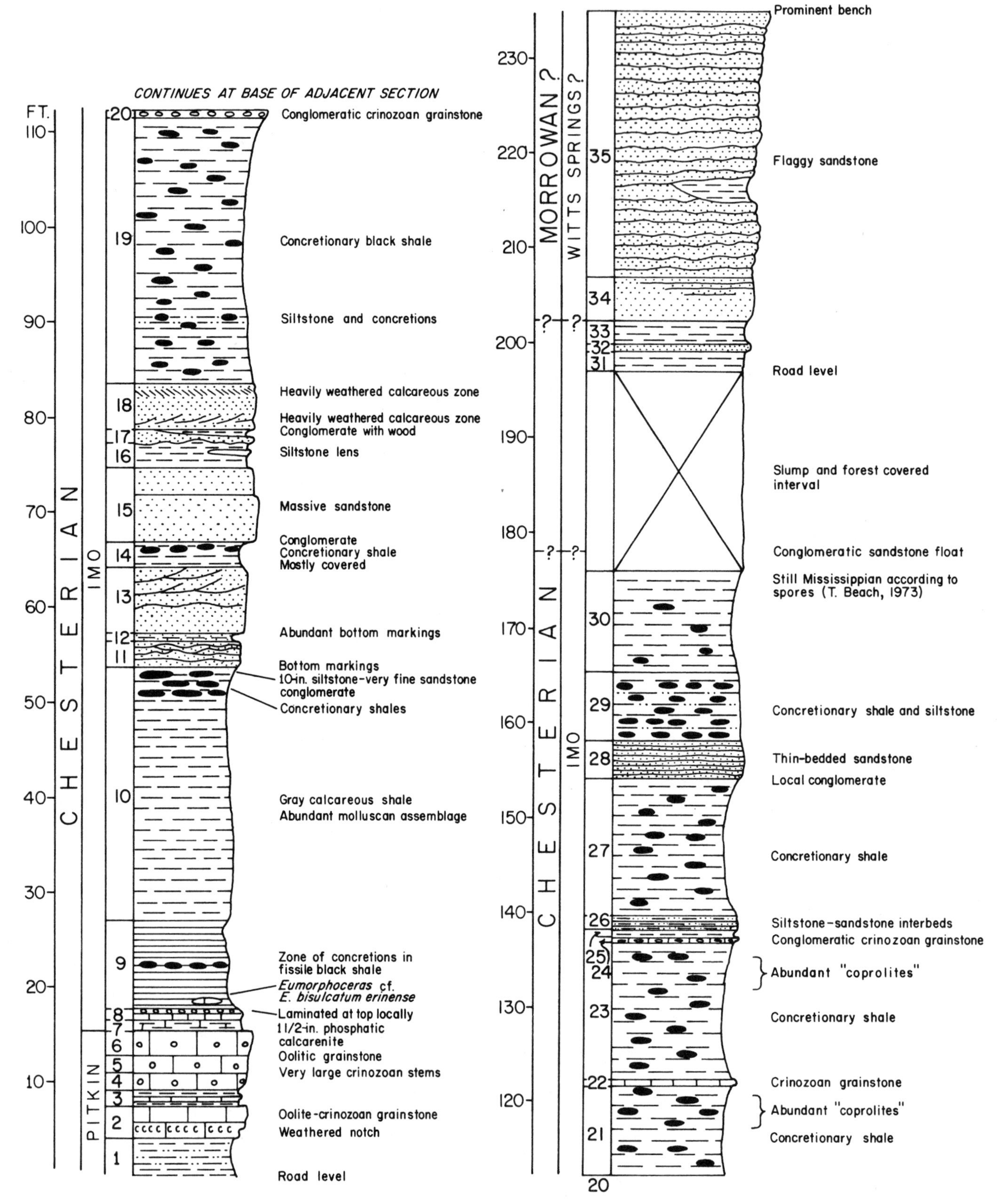

Figure 2. Generalized stratigraphic succession of the Mississippian Imo Formation exposed at Peyton Creek, Van Buren County, Arkansas (from Sutherland and Manger, 1979).

Figure 3. Oolitic grainstones of the Pitkin limestone (arrow) succeeded by black, fissile shale, gray, calcareous shale, and quartz sandstones of the lower Imo Formation. Numbered units correspond to those illustrated in Figure 2.

Figure 4. Dark shales and interbedded conglomeratic limestones and quartz sandstones in the upper Imo Formation. Unit 27 represents the highest shale with abundant fossils. The upper Imo is succeeded by quartz sandstones of presumed Pennsylvanian age referred questionably to the Witts Springs Formation (arrow).

lower portion of the Imo Formation is equivalent to E_2a of the standard Namurian succession (Saunders and others, 1977). Conodonts of the *Adetognathus unicornis* assemblage zone continue from the Pitkin through the primary ammonoid-bearing zone of the Imo (Lane, in Brenckle, 1977; Mapes and Rexroad, 1986). Limestones of the Imo yield Mamet Zone 19 foraminifers (Brenckle, 1977). Data from the ammonoids, conodonts, and calcareous foraminifers suggest that the Imo Formation is in part younger than the youngest type Chesterian Formation (Grove Church Shale; Manger and Sutherland, 1984).

At the Peyton Creek exposure, the highest dark shales in the road cut still yield Mississippian palynomorphs (Owens and others, 1984). Above this zone is an extensive covered interval, now enhanced by a run-away truck ramp! Conglomeratic sandstone float is common in the lower part of this covered interval, but no outcrops are known. The next exposures exhibit an apparently gradational contact of dark shale and quartz sandstone overlain by a thick sequence of flaggy sandstone. The top of the Imo Formation has not been defined and cannot be placed with assurance in the Peyton Creek road cut. It may fall within the covered interval at the horizon indicated by the conglomeratic sandstone, at the base of the first massive sandstone, or somewhere in between. The top of the Imo is also presumed to coincide with the Mississippian-Pennsylvanian boundary in this area.

Assignment of the sandstones above the Imo to a formation is unclear. As noted above, the name Imo was initially withdrawn by Gordon (1965), to be replaced by the term Cane Hill Formation of Chesterian and Morrowan age (Glick and others, 1964). Usage of Cane Hill as a formation in north-central Arkansas was questioned by Quinn (1966) and Saunders (1973) and has not been widely accepted. The Witts Springs Formation of Morrowan age was proposed for strata overlying the "Cane Hill Formation" in the north-central Arkansas (Glick and others, 1964). Since the relationship of the top of the Imo Formation to Morrowan strata has not been clearly established, the sandstone sequence above the Imo is assigned questionably to the Witts Springs to avoid confusion with the name Cane Hill (Fig. 2). Although it is common practice to show the Imo Formation overlain by the strata referred to the Cane Hill Member of the Hale Formation (type Morrowan), that relationship has never been demonstrated. In addition, the Mississippian-Pennsylvanian boundary in the region of Imo exposures in north-central Arkansas has never had a faunal assessment.

The Imo ammonoid assemblage has received the bulk of taxonomic and biostratigraphic attention (Furnish and others, 1964; McCaleb and others, 1964; Gordon, 1965, 1970; Saunders, 1966, 1973, 1975; Saunders and others, 1977; Manger and Quinn, 1972). However, the fauna is actually dominated by gastropods and bivalves; yet neither of these groups has received much taxonomic treatment. Published studies of other Imo fossils include crinoids (Burdick and Strimple, 1973), phyllocarids (Copeland, 1967), plant petrifactions (Taylor and Eggert, 1967) and palynomorphs (Sullivan and Mischell, 1971; Owens and others, 1984), and conodonts (Mapes and Rexroad, 1986).

REFERENCES CITED

Brenckle, P., 1977, Foraminifers and other calcareous microfossils from late Chesterian (Mississippian) strata of northern Arkansas, *in* Sutherland, P.K., and Manger, W.L., eds., Upper Chesterian–Morrowan stratigraphy and the Mississippian–Pennsylvanian boundary in northeastern Oklahoma and northwestern Arkansas: Oklahoma Geological Survey Guidebook 18, p. 73–88.

Brenckle, P., Lane, H.R., Manger, W.L., and Saunders, W.B., 1977, The Mississippian–Pennsylvanian boundary as an intercontinental biostratigraphic datum: Newsletters on Stratigraphy, v. 6, p. 106–116.

Burdick, D.W., and Strimple, H.L., 1973, New late Mississippian crinoids from northern Arkansas: Journal of Paleontology, v. 47, p. 231–243.

Copeland, M.J., 1967, A new species of *Dithyocaris* (Phyllocarida) from the Imo Formation, Upper Mississippian of Arkansas: Journal of Paleontology, v. 41, p. 1195–1196.

Furnish, W.M., Quinn, J.H., and McCaleb, J.A., 1964, The Upper Mississippian ammonoid *Delepinoceras* in North America: Palaeontology, v. 7, p. 173–180.

Glick, E.E., Frezon, S.E., and Gordon, M., Jr., 1964, Witts Springs Formation of Morrow age in the Snowball Quadrangle, north-central Arkansas: U.S. Geological Survey Bulletin 1194–D, 16 p.

Gordon, M., Jr., 1965, Carboniferous cephalopods of Arkansas: U.S. Geological Survey Professional Paper 460, 322 p.

—— , 1970, Carboniferous ammonoid zones of the south-central and western United States: Compte Rendu, Sixteme Congres International de Stratigraphie et de Geologie du Carbonifere, Sheffield, 1967, v. 2, p. 817–826.

Haley, B.R., 1976, Geologic map of Arkansas: Arkansas Geological Commission and U.S. Geological Survey, scale 1:500,000.

Lane, H.R., 1967, Uppermost Mississippian and Lower Pennsylvanian conodonts from the type Morrowan region, Arkansas: Journal of Paleontology, v. 41, p. 921–942.

Lane, H.R., and Straka, J.J., 1974, Late Mississippian and early Pennsylvanian conodonts, Arkansas and Oklahoma: Geological Society of America Special Paper 152, 144 p.

McCaleb, J.A., Quinn, J.H., and Furnish, W.M., 1964, The ammonoid family Girtyoceratidae in the southern midcontinent: Oklahoma Geological Survey Circular 67, 41 p.

Manger, W.L., and Quinn, J.H., 1972, Carboniferous dimorphoceratid ammonoids from northern Arkansas: Journal of Paleontology, v. 46, p. 303–314.

Manger, W.L., and Sutherland, P.K., 1984, The Mississippian–Pennsylvanian boundary in the southern midcontinent, United States: Compte Rendu, Neuvieme Congres International de Stratigraphie et de Geologie du Carbonifere, Urbana, 1979, v. 2, p. 369–376.

Mapes, R.H., and Rexroad, C.B., 1986, Conodonts from the Imo Formation (Upper Chesterian), north-central Arkansas: Geologica et Palaeontologica, v. 20, p. 113–123.

Owens, B., Loboziak, S., and Coquel, R., 1984, Late Mississippian–early Pennsylvanian miospore assemblages from northern Arkansas: Compte Rendu, Neuvieme Congres International de Stratigraphie et de Geologie du Carbonifere, v. 2, p. 385–390.

Quinn, J.H., 1966, The Pitkin and superjacent formations in northern Arkansas: Shale Shaker, v. 17, p. 2–12.

Saunders, W.B., 1966, New goniatite ammonid from the late Mississippian of Arkansas: Oklahoma Geology Notes, v. 26, p. 43–48.

—— , 1973, Upper Mississippian ammonoids from Arkansas and Oklahoma: Geological Society of America Special Paper 145, 110 p.

—— , 1975, The Upper Mississippian *Eumorphoceras richardsoni-Cravenoceras friscoense* ammonoid assemblage, North American midcontinent: Compte Rendu, Septieme Congres International de Stratigraphie et de Geologie du Carbonifere, Krefeld, 1971, v. 4, p. 201–207.

Saunders, W.B., Manger, W.L., and Gordon, M., Jr., 1977, Upper Mississippian and lower and middle Pennsylvanian ammonid biostratigraphy of northern Arkansas, *in* Sutherland, P.K., and Manger, W.L., eds., Upper Chesterian–Morrowan stratigraphy and the Mississippian–Pennsylvanian boundary in northeastern Oklahoma and northwestern Arkansas: Oklahoma Geological Survey Guidebook 18, p. 117–137.

Sullivan, H.J., and Mischell, D.R., 1971, The Mississippian–Pennsylvanian boundary and its correlation with Europe: Compte Rendu, Sixteme Congres International de Stratigraphie et de Geologie du Carbonifere, Sheffield, 1967, v. 4, p. 1533–1540.

Sutherland, P.K., and Manger, W.L., eds., 1977, Mississippian–Pennsylvanian boundary in northeastern Oklahoma and northwestern Arkansas: Oklahoma Geology Survey Guidebook 18, 185 p.

—— , 1979, Mississippian–Pennsylvanian shelf-to-basin transition Ozark and Ouachita regions, Oklahoma and Arkansas: Oklahoma Geological Survey Guidebook 19, 81 p.

Taylor, T.N., and Eggert, D.A., 1967, Petrified plants from the upper Mississippian (Chester Series) of Arkansas: American Microscopical Society Transactions, v. 84, p. 412–416.

Atoka Formation in the Lee Creek area, Arkansas

J. D. McFarland III, Arkansas Geological Commission, 3815 W. Roosevelt Road, Little Rock, Arkansas 72204

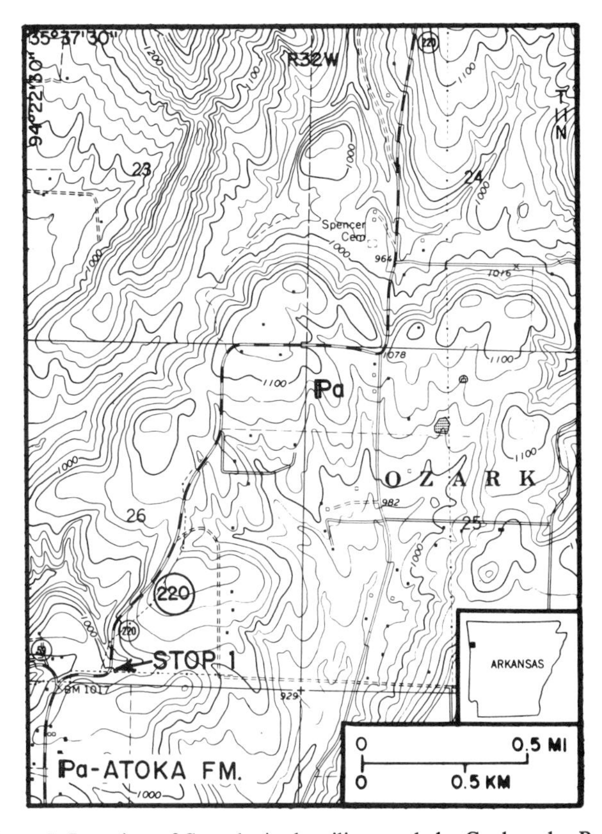

Figure 1. Location of Stop 1; Atoka siliceous shale. Geology by Boyd Haley.

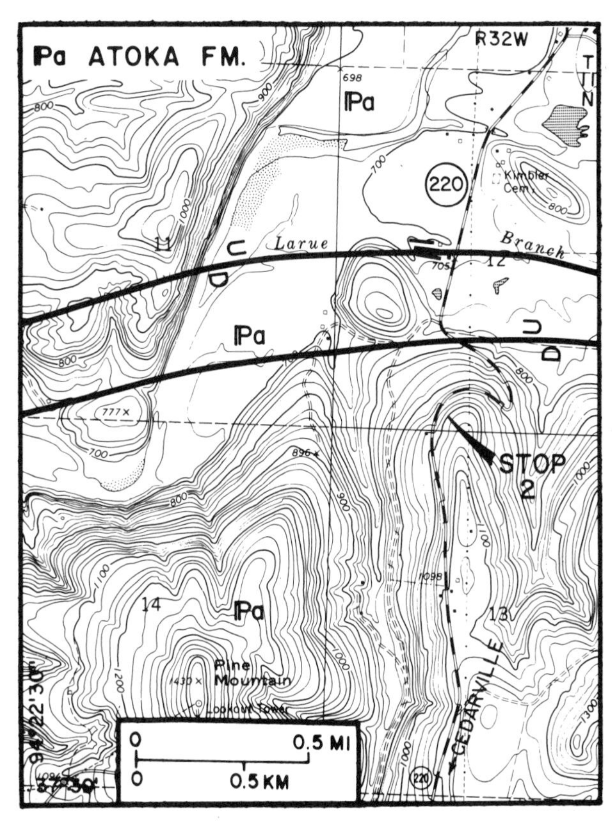

Figure 2. Location map for Stop 2; Atoka shale, siltstone, and sandstone along east side of road. Geology by Boyd Haley.

LOCATION

The outcrops described here are found along Arkansas 220 in northern Crawford County. Start at the intersection of Arkansas 59 and 220. Drive north on Arkansas 220 for 0.1 mi (0.16 km) to Stop 1.

SIGNIFICANCE

The Atoka Formation is composed of about 3,500 ft (1,060 m) of shallow water clastics in the Boston Mountains of Arkansas. In the Arkansas Valley and frontal Ouachitas to the south it thickens to over 25,000 ft (7,575 m). In the Arkansas Valley the Atoka produces natural gas from several horizons; it is, therefore, useful to examine these shallow-water rocks as a preliminary to the Atoka outcrops of the valley.

SITE DESCRIPTIONS

Stop 1. (Fig. 1) Siliceous shale of the Atoka. Outcrops on both sides of the road. (This stop is on a dangerous curve.) This zone of siliceous shale has been traced eastward about 40 mi (65 km) from the Oklahoma line to north of Ozark, Arkansas. It is about 1,200 ft (365 m) above the base of the Atoka Formation. The more this rock is weathered the more siliceous or cherty it appears, which may be one reason it has never been noted in logging well samples. Continue north on Arkansas 220 for 3.2 mi (5.1 km).

Stop 2. (Fig. 2) Atoka shale, siltstone, and sandstone to the east. You may examine these outcrops as you descend this hill. Excellent exposures of distributary channel deposits and the underlying fringe deposits are found here. The crossbedded limy sandstone at the top of the massive, distributary channel units can be variously interpreted as part of the channel deposits, a beach sand, or an offshore bar deposit. Continue north on Arkansas 220 for 6.1 mi (9.8 km).

Stop 3. (Fig. 3) Lee Creek. Atoka sandstone with plant and marine fossils. Walk to the outcrop on south side of the stream to the east of the road (Fig. 4). These rocks are interpreted as delta fringe deposits (Bush and others, 1978) with the more sandy parts

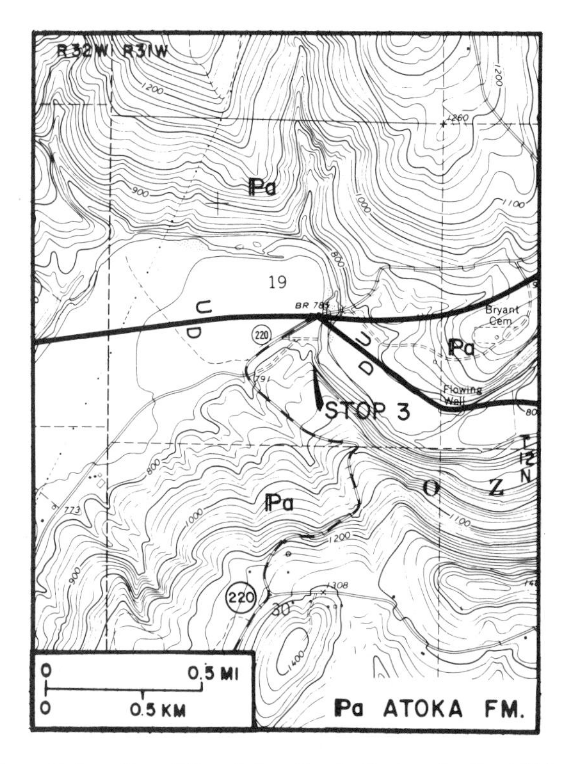

Figure 3. Location of Stop 3; Lee Creek. Geology by Boyd Haley.

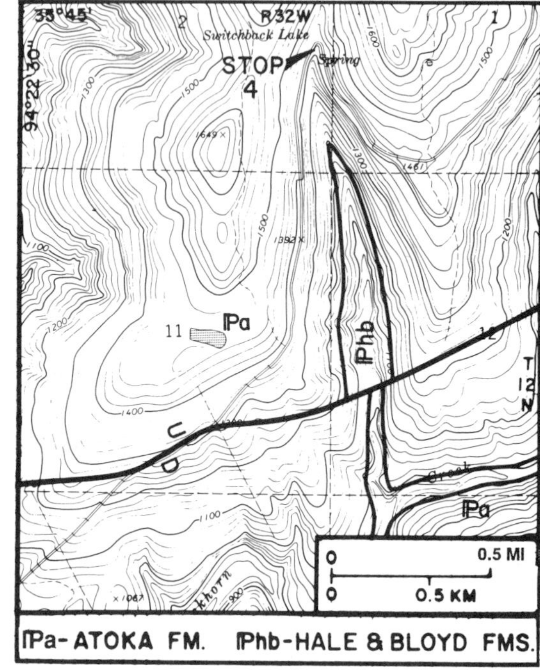

Figure 5. Location of Stop 4; Switchback Lake. Pa = Atoka Formation; Phb = Hale and Boyd formations. Geology by Boyd Haley.

Figure 4. Lee Creek, Atoka Formation. Delta fringe deposits. The more shaly rocks below were deposited farther from the front of the prograding delta than the overlying sandy sequence.

Figure 6. Switchback Lake, Atoka Formation. Delta fringe deposits overlain by crossbedded distributary channel deposits.

being deposited closer to the delta front (shoreline). If all the sediments exposed at this stop are fringe deposits, then the delta front never reached this area; however, the plant fragments would seem to indicate its nearby presence.

Continue north across the Lee Creek bridge and up the hill on the gravel road. At 0.7 mi (1 km) north of the bridge a road enters from the right, continue north (stay left). At 5 mi (8.1 km) past the bridge another road enters from the right, stay left. You will reach Switchback Lake 6.4 mi (10.3 km) past the bridge.

Stop 4. (Fig. 5) Switchback Lake. Two sequences of Atoka sandstone are exposed at this location. Each can be classified as a part of a prograding delta (Bush and others, 1978). The lower sequence (Fig. 6) is capped distributary channel deposits, and the upper sequence is capped by either offshore bar or river mouth deposits. Note the mound of travertine under the cliff, also the abundant brachiopods near the base of the cliff-forming sandstone. Retrace your steps to return to the paved highway.

REFERENCE CITED

Bush, W. V., Haley, B. R., Stone, C. G., and McFarland, J. D., III, 1978, A guidebook to the Atoka Formation in Arkansas: Arkansas Geological Commission Guidebook 78-1, 62 p.

Horsehead Lake spillway, Arkansas

J. D. McFarland III, Arkansas Geological Commission, 3815 W. Roosevelt Road, Little Rock, Arkansas 72204

LOCATION

Horsehead Lake is a small U.S. Forest Service lake located in western Johnson County, Arkansas, at the juncture of the Boston Mountains and the Arkansas River Valley (Fig. 1). To reach the lake, take Arkansas 164 north from I-40 (the Hunt-Coal Hill exit). From downtown Hunt, continue 2. 8 mi (4.5 km) north on Arkansas 164 to the entrance to Horsehead Lake. Walk across the dam to the spillway cut.

SIGNIFICANCE

The strata here represent various facies of a deltaic depositional system. The high hills on the skyline north of Horsehead Lake are underlain by the lower part of the Middle Pennsylvanian Atoka Formation. Near the northern end of the lake is a set of east–west-trending normal faults (the Mulberry fault system), downthrown to the south with about 2,500 ft (760 m) of displacement. South of the fault system are small exposures of deltaic and fluvial sequences of the Middle Pennsylvanian upper Atoka Formation, the Hartshorne Sandstone, and the McAlester Formation. The most striking exposures are at the west end of the dam where a southward-tilted sandstone hogback has been dissected by the spillway cut.

SITE DESCRIPTION

The rocks exposed at the north end of the spillway near the dam are in the upper part of the Hartshorne Sandstone. In the area of Horsehead Lake, the Hartshorne Sandstone consists of lenticular units of thinning-upward, ripple-marked, and cross-bedded sandstone intercalated with a few very thin beds of siltstone and shale. These units have a sharp channel-type lower contact as does the Hartshorne Sandstone where it unconformably overlies the Atoka Formation. The above criteria lead to the conclusion that in the vicinity of Horsehead Lake most of the rocks in the Hartshorne Sandstone were deposited in a meandering stream depositional environment.

The shale and the overlying Lower Hartshorne Coal bed (Fig. 2) were deposited in the marsh and swamp environment of a delta plain. The unit of sandstone above the Lower Hartshorne Coal thins to the east and south of the area of the spillway and the shale and coal bed above the unit (south of Fig. 2) becomes correspondingly closer to the Lower Hartshorne Coal. The beds of sandstone in this unit thin upward in thickness and the sandstone is extremely cross-laminated and ripple-marked. Most of the ripple marks are oriented to indicate a southeastward current flow during the deposition of the sediments in the unit. Therefore, the sediments in the unit could have been deposited in the channel of a crevasse through the natural levee of a stream to the north or northwest.

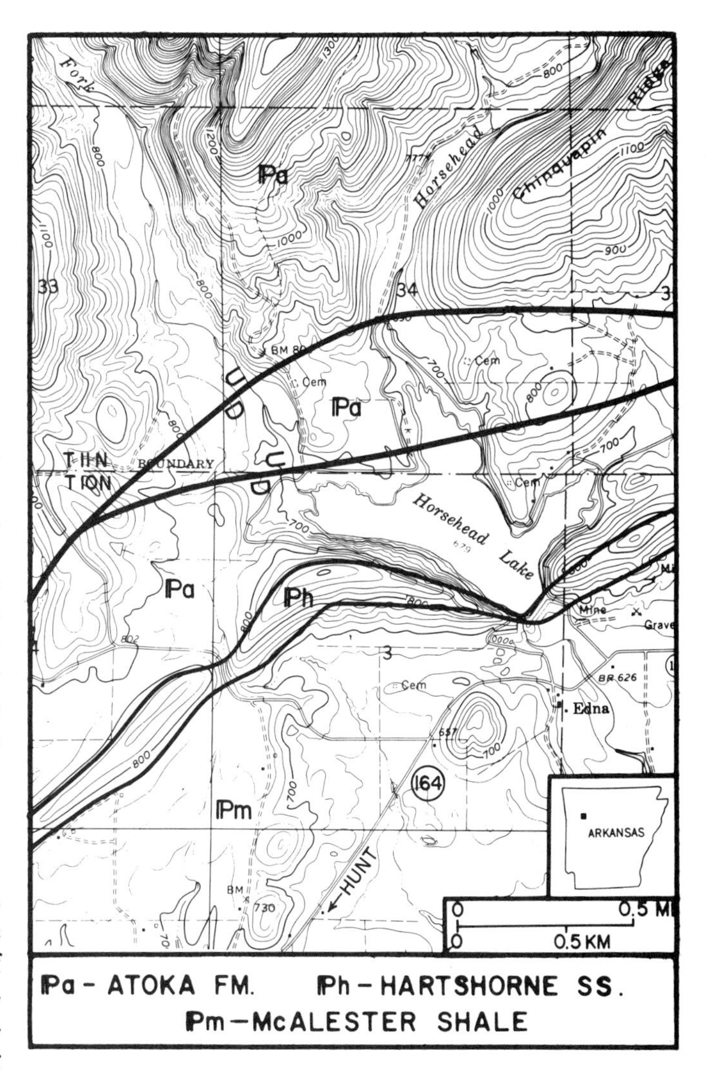

Figure 1. Geologic map of the Horsehead Lake area. Pm = McAlester Shale; Ph = Hartshorne Sandstone; Pa = Atoka Formation. Base from Hunt 7½-minute Quadrangle. Geology by Boyd Haley.

These crevasse channel deposits are overlain by shale and an overlying coal bed that are indicative of a delta plain depositional environment. If the crevasse channel deposits continue to thin to the south and east, and the two coal beds continue to get closer together, then the thin bed of shale in the Lower Hartshorne Coal bed near Clarksville, about 8 mi (13 km) to the south, could represent the outer fringe of the crevasse channel deposits.

The cross-laminated dark gray argillaceous siltstone and the very silty shale overlying the upper coal bed contain beds and lenses of siderite and probably represent sediments deposited in a shallow-water brackish to marine depositional environment.

Figure 2. Spillway at Horsehead Lake. The sandstone beds in the lower part of this sequence are at the top of the Hartshorne Sandstone and represent stream channel deposits. Above this lie delta plain sediments, the Lower Hartshorne Coal bed, and crevasse channel deposits of the McAlester Formation.

The rank of coal in the Arkansas Valley Coal Field ranges from low-volatile bituminous coal to semianthracite. The dry, mineral-matter-free fixed carbon content of the coal at this outcrop is about 81.3 percent (Bush and others, 1980). This percentage of fixed carbon places the coal in the rank classification of low-volatile bituminous coal, which has a range of dry, mineral-matter-free fixed carbon of 78 percent or more and less than 86 percent.

REFERENCE CITED

Bush, W. V., Haley, B. R., Stone, C. G., Holbrook, D. F., and McFarland, J. D., III, 1980, A guidebook to the geology of the Arkansas Paleozoic area: Arkansas Geological Commission Guidebook 77-1 (revised, 1980), 79 p.

Blue Mountain Dam and Magazine Mountain, Arkansas

Richard R. Cohoon and Victor K. Vere, Department of Physical Sciences-Geology, Arkansas Tech University, Russellville, Arkansas 72801

LOCATION

Blue Mountain Dam can be reached by driving east 17 mi (27 km) from Booneville, Arkansas, on Arkansas 10 to Waveland, and then turning south on Arkansas 309 for 0.8 mi (1.3 km) to the entrance to the Waveland Recreation Area. Turn west on the entrance road and follow the road over the spillway bridge and dam to the outcrops at the south abutment of the dam (Fig. 1).

Magazine Mountain can be reached by continuing east 6.4 mi (10 km) from Waveland on Arkansas 10 to Havana. At Havana, take Arkansas 309 north 4.9 mi (8 km) to Stop 2 on the flanks of Magazine Mountain (Fig. 2). To reach Stop 3, continue along Arkansas 309 for 3.6 mi (5.8 km). Stop 4 is 0.5 mi (0.9 km) farther along the road, at the top of the mountain.

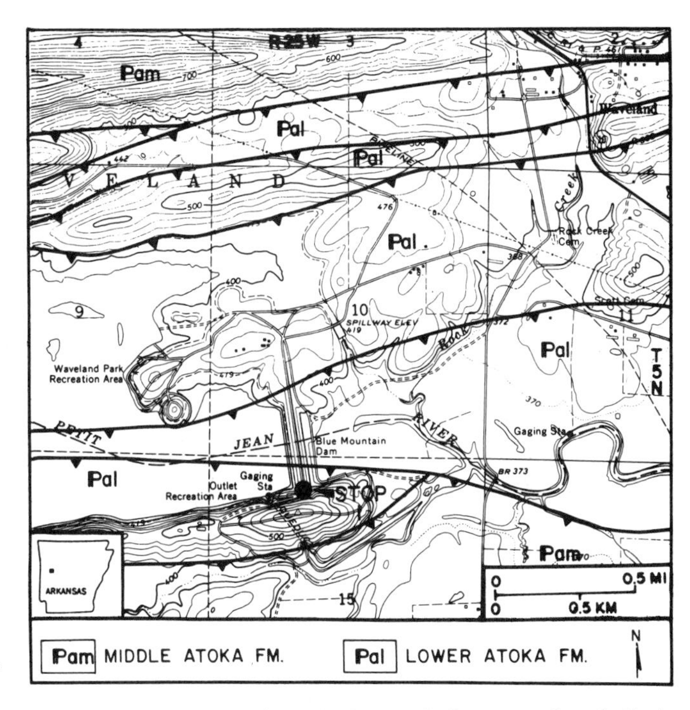

Figure 1. Geologic map of the Blue Mountain Dam area, Stop 1. Geology by Boyd Haley.

SIGNIFICANCE

The sites described here show strata of the Atoka Formation of lower Middle Pennsylvanian–Atokan age; the Hartshorne Sandstone, the McAlester Formation, and a lower sandstone of the Savanna Formation—all of which are Middle Pennsylvanian–Des Moinesian in age–can also be seen here. In a distance of approximately 22 mi (35 km), you will be able to study sedimentary rocks deposited on a continental slope by turbidity currents. These strata were formed in most of the depositional environments of a typical deltaic sequence, including: prodelta, outer fringe, inner fringe, distributary channel, and delta plain–lake. Additionally, the structure of this portion of the northern Ouachita Mountains and Arkansas Valley (transition zone) is described, as well as some spectacular vistas.

SITE DESCRIPTIONS

Stop 1, Blue Mountain Dam. Lower Atoka Formation submarine inner fan, slope, and/or canyon channel deposition (Bush and others, 1977, p. 44; and Stone and McFarland, 1981, p. 55–61). Park at the south end of the dam and walk from the east end of the cut toward the flood gate control building to the west, staying on the upper level road.

The Atoka Formation, a series of alternating sandstones, siltstones, and shales, thickens southward from approximately 3,500 ft (1,100 m) in the northern Arkansas Valley to more than 25,000 ft (7,600 m) in the northern Ouachita Mountains. Depositional environments of the Atoka in this area range from deltaic and shallow-water marine in the northern Arkansas Valley to deep-water marine in the southern Arkansas Valley and northern Ouachita Mountains (Stone, 1968, p. 9).

The Atoka strata are approximately 19,000 ft (5,800 m) thick in this area, and the exposure at Stop 1 is about 9,000 ft (2,700 m) stratigraphically lower than the Hartshorne Sandstone (Stone and McFarland, 1981, p. 55). The strata of the lower Atoka Formation exposed at this locality are silty, clayey sandstones, clayey siltstones, and silty shales. These sediments constitute a turbidite sequence deposited in the transitional zone between shallow-water shelf deposits to the north and deeper flysch facies to the south. Many sedimentary textures and structures typical of turbidites are present; for example, graded bedding, bottom marks, convolute bedding, load casts, channels, contorted structures, lenticular beds, slide masses, and trace fossils are quite conspicuous (Stone and McFarland, 1981, p. 55).

Considering the textures and structures represented, it is logical to propose that such relationships could be developed in a submarine fan on the low-dip upper edge of the continental slope. High-energy turbidity currents could have eroded the channels, which were later filled with sand by lower energy currents. The sand-filled channel may then have been covered by sheets of sediment resulting in the typical overlying turbidite sequences.

Forces active during the Ouachita orogeny resulted in folding, thrust faulting, tilting of blocks between faults, fracturing, jointing, and rock cleavage; all of which are displayed in the vicinity of Stop 1.

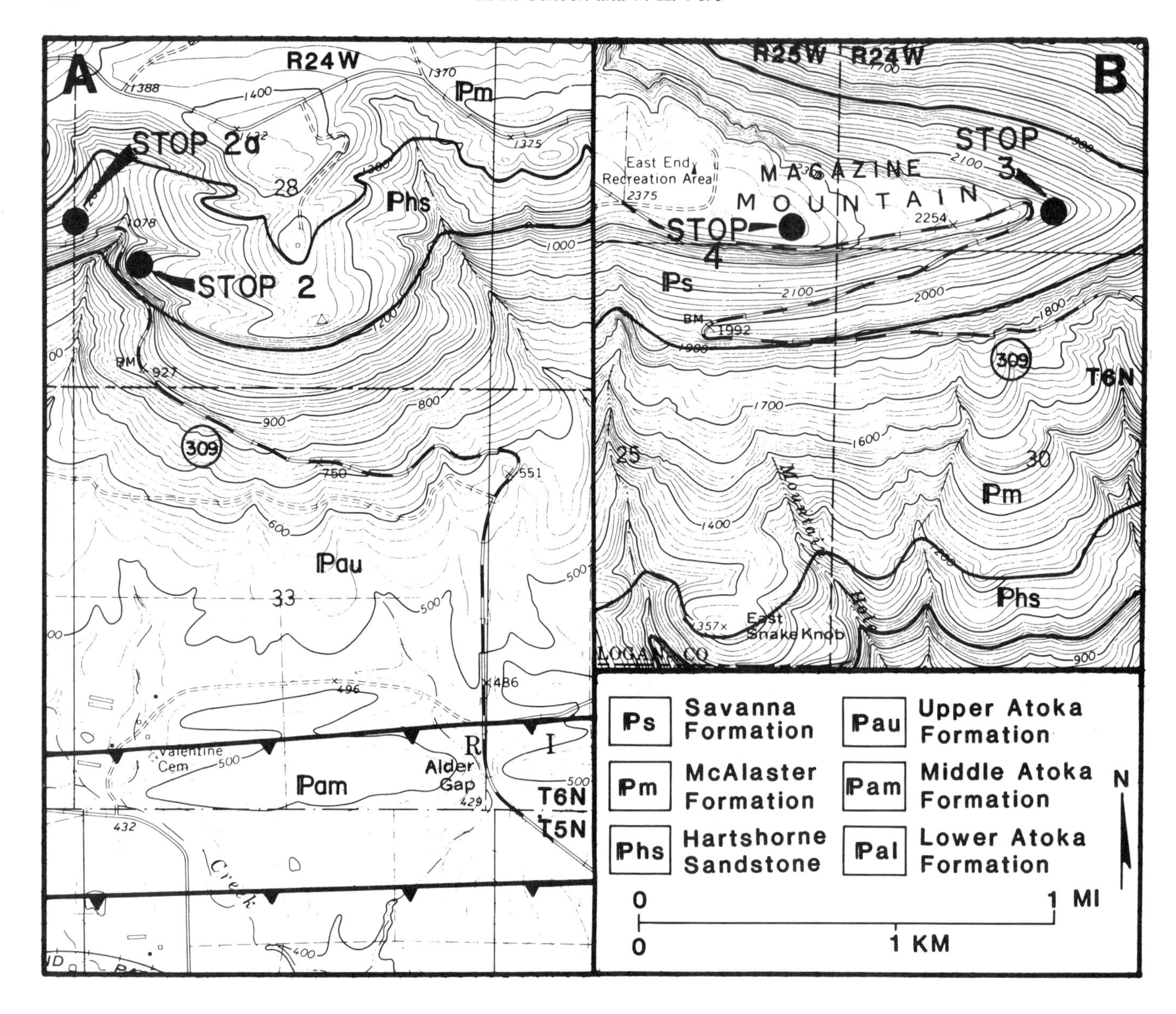

Figure 2. Geologic maps of the Magazine Mountain area, Stops 2 to 4. Geology by Boyd Haley.

As you walk to the west, notice the convoluted to nearly vertical bedding associated with some strata adjacent to the channel fill sandstone (Fig. 3). The absence of changes in thickness of the strata indicates that the beds acted as a unit during a submarine slump-block movement. The channel fill material consists of fine- to medium-grained, silty, clayey, quartz sandstone that is massively bedded and displays various bottom markings, such as load casts. A study of the channel geometry suggests that perhaps two channels are present or a change in channel direction occurred.

The channel sandstone thickens and thins toward the west and continues to show many sedimentary structures characteristic of soft-sediment deformation. Near the flood-gate control building a channel fill sandstone is seen near road level. In this same area, at the top of the cut, a sandstone lens, possibly caused by soft-sediment sliding, is surrounded by shales. At the west end of the cut near the control building, note the deformed flute casts on the bottoms of the slumped sandstone beds and the deformed bedding in the thinly bedded silty shale below the channel. Approximately 15 ft (4.5 m) above the chain-link fence a sandstone dike (Fig. 4) transects the shales, offering additional evidence of the soft-sediment deformation. Numerous features in this exposure, such as graded bedding, thinning, and reduction of grain sizes progressively upward, indicate that the strata were deposited by turbidity currents and the deformed flute casts appear to indicate a southerly current flow direction.

Along the lower road, which leads toward the lake, submarine-slump masses of shale and siltstone crop out (Fig. 5).

Figure 3. Stop 1. Sandstone-filled submarine-fan channel in shale and siltstone of the slope facies. Upper portion of the lower Atoka Formation at Blue Mountain Dam (photo by M. Satterfield).

Figure 5. Stop 1. Submarine-slump masses of shale and siltstone in the upper portion of the lower Atoka Formation at Blue Mountain Dam (photo by M. Satterfield).

Figure 4. Stop 1. Sandstone dike in turbidite sequence of the lower Atoka Formation at Blue Mountain Dam.

Figure 6. Stop 2. Contact of the Hartshorne Sandstone and the upper Atoka Formation on Arkansas 309.

Notice that the dip and strike of the slump units is not the same—a discrepancy caused by semiindependent movement of each block. Another interesting feature in this exposure is produced by a combination of fracturing and spheroidal weathering conditions that produce "dike-like" zones in the strata.

En route to Stop 2, 2.2 mi (3.5 km) north of Havana, note the nearly vertical standing sandstone beds of the middle part of the Atoka Formation at Alder Gap. These sandstones, known as the "traceable three" (Stone, 1968, p. 5), are bounded on the north and south by east-west–trending thrust faults (Fig. 2) that separate the lower, middle, and upper parts of the Atoka Formation at this locality. The middle part of the formation is approximately 6,000 ft (1,800 m) thick; the lower half is composed of a silty shale sequence. The shale is overlain by three mappable, thin-bedded, fine-grained, micaceous, silty sandstone units. Sedimentary structures and textures seen in the lower portion of the unit suggest turbidity current deposition on the upper edge of a continental slope. The mappable sandstones were probably deposited as sand sheets or fans in the inner fringe zone of a deltaic depositional system (Stone, 1968, p. 5–8).

Stop 2. Contact of the Atoka Formation and the Hartshorne Sandstone. Park in the wide shoulder area to the right side of the road just before the sharp curve to the left. Walk downhill along the road to the Atoka-Hartshorne contact.

At this locality the upper part of the Atoka Formation consists of thin-bedded, silty, sandy gray shale and is less resistant to weathering and erosion than the overlying ledge-forming Hartshorne Sandstone (Fig. 6). Numerous coalified fossil plant frag-

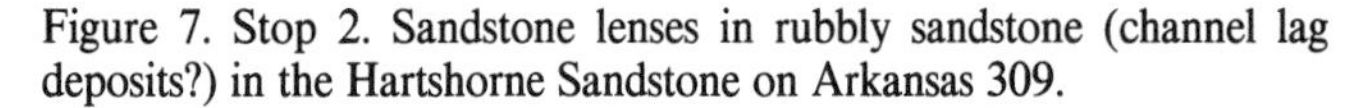

Figure 7. Stop 2. Sandstone lenses in rubbly sandstone (channel lag deposits?) in the Hartshorne Sandstone on Arkansas 309.

Figure 8. Stop 2a. Flame structures indicating soft-sediment deformation in the Hartshorne Sandstone on Arkansas 309.

ments occur at the contact, especially on the underside of the Hartshorne Sandstone.

The upper part of the Atoka Formation thins toward the north and contains several thin coal beds and sandstone-filled channels. The upper part of the formation generally contains a higher proportion of gray-black shale (90 percent of the total thickness) than other parts of the Atoka Formation, with only about 10 percent of the upper part being fine-grained silty sandstone. All evidence seems to indicate that the upper Atoka units were deposited in a shallow-water marine/delta plain environment.

The massively bedded Hartstone is a medium-grained, clean, light gray quartz sandstone, which develops prominent limonitic staining upon weathering. As you proceed uphill along the road (up the stratigraphic section), note the planar cross-bedding and the rubbly sandstone lenses (channels?), some of which are several tens of feet wide (Fig. 7). Several massive sandstone lenses are found in some of the rubble zones. Evidence of fluvial deposition, such as foreset planar cross-bedding, rubble zones (channel lag deposits?), and plant fossils, indicates that the Hartshorne Sandstone was deposited in stream (distributary) channels.

According to recent studies by Houseknecht (1981) and Houseknecht and Iannacchione (1982), the Hartshorne Sandstone and associated facies were deposited in a meandering, east-west–trending, westward-prograding deltaic system. The thick (up to 250 ft; 75 m) distributary channel "shoestring" sandstones interfinger with the thinner interdistributary-bay facies.

The strata strike in an east-west direction and dip approximately 6° to the north toward the axis of the Mount Magazine syncline. Prominent, nearly vertical, joint sets strike approximately east-west and north-south.

Stop 2a. "Flame" structures in the Hartshorne Sandstone. Walk uphill along the road from the parking area approximately 600 ft (180 m). Prominent "flame" structures are seen at the end of the uphill walk behind some small pine trees to the right of the road.

Observe the Hartshorne outcrops during the walk and notice such features as channeling, cross-bedding, and soft-sediment deformation "flame" structures (Fig. 8)—which offer additional evidence of a shallow water, distributary-channel depositional environment.

Between Stops 2 and 3, the highway crosses the McAlester Formation, which overlies the Hartshorne Sandstone conformably. McAlester strata, approximately 1,300 ft (400 m) thick in an area south of Mount Magazine, thin to about 600 ft (180 m) in areas north of the mountain. The McAlester Formation is mostly shale, but contains minor amounts of siltstone, sandstone, and coal. The shale is usually dark gray, fissile, micaceous, silty, and may contain beds of siltstone and sandstone. Siltstone of the McAlester is almost always micaceous and may contain fine sand at some localities. Bedding in the formation is thin, irregular, and often lenticular. Eroded tops of beds are usually ripple marked and often contain plant fossil fragments. Sandstones are usually lenticular, filling channels, and may vary in thickness from a few feet to a hundred feet in less than a mile. The sandstone beds range from massive to irregular to convolute and often show foreset cross-bedding and soft-sediment deformation structures.

Stop 3. Savanna Formation (Fig. 2). Stop on sharp "hairpin" curve to the left, park cars on the wide shoulder to the right of the road, and walk downhill to the exposure of sandstone.

Here, ledge-forming sandstone of the Savanna Formation that caps Mount Magazine is exposed. The sandstone is massively bedded and shows cross-bedding and bioturbation, but does not display the channel scour and fill structure seen elsewhere. At other localities, especially on the north-facing escarpment of the mountain, soft-sediment deformation features, ripple marks, and current lineations are abundantly exposed.

The Savanna Formation is approximately 2,200 ft (670 m) thick in the Arkansas Valley Section of the Ouachita Mountains Province. The formation consists of shale, silty shale, siltstone, and very fine- to fine-grained sandstone. At some localities the sandstone is silty and may contain medium- to coarse-grained, rounded quartz sand. The siltstones and sandstones contain ripple marks and fossil plant fragments. Bedding varies from thin to massive, regular to irregular and may include cross-bedding. At some localities, channeling is obvious, as are soft-sediment deformation structures, cross-bedding, current ripples, plant fossil fragments, and clay inclusions—all evidence of distributary-channel sedimentation. In the vicinity of Paris, Arkansas, Short Mountain and Horseshoe Mountain are capped by erosional remnants of a sandstone-filled channel in siltstone and shale (Haley, 1961, p. 8). According to Haley (personal communication, 1984), the sandstone of the Savanna Formation that forms the caprock of Mount Magazine is a distributary channel-fill of extraordinarily large size.

Figure 9. Stop 4. View to the southwest from the south rim of Mount Magazine. The Savanna Formation sandstone caprock is visible in the lower right corner of the photo. The Petit Jean River valley and Blue Mountain Lake are seen in the middle portion of the picture. Structurally, this region is part of the Ranger anticline and the lower, middle, and upper members of the Atoka Formation are exposed. On the horizon the Savanna Formation forms the caprock of the mesalike mountains devloped along the axis of the Poteau syncline.

Stop 4. Scenic overlook at the top of Mount Magazine (south rim). Park in the overlook parking area and walk to the south edge of the overlook.

Structurally, Mount Magazine is an asymmetrical, east-west–trending syncline. The south limb of the syncline dips more steeply than the north limb to a common axis. This mesalike mountain is the highest elevation in Arkansas (elevation 2,753 ft; 839 m).

From this vantage point there is a magnificent view of the Poteau syncline (its axis lies about 10 mi; 16 km distant) and of the Ouachita Mountains farther south (Fig. 9). The west-southwest–trending axis of the Poteau syncline is marked by a series of mountains capped by the Savanna Formation, an excellent expression of topographic inversion. Mountains from east to west, respectively, include White Oak, Round, Pilot Knob, East Poteau, and Poteau. The Hartshorne Sandstone crops out as a series of ridges outlining limbs of the Poteau syncline. The low ridges at the base of Mount Magazine (mapped as part of the Ranger anticline) and the anticlinal Dutch Creek Mountain seen to the southeast of the Poteau syncline are composed of units of the middle and lower Atoka Formation. Numerous east-west–trending thrust faults either form or parallel the contact between the middle and lower members of the Atoka in the area between Mount Magazine and Havana, Arkansas (Haley, 1976).

Glick (in Haley and others, 1979, p. 3) suggests that large-magnitude growth faults developed along a northeast-trending zone in west-central Arkansas during Middle Pennsylvanian time. The activity of these faults influenced depositional environments, resulting in shallow-water shelf facies being deposited north of the fault zone and deep-water turbidite facies developing to the south of the zone. A similar fault system, in northeastern Arkansas, may be related to the Reelfoot rift system. The faults in west-central Arkansas could have had a similar relationship to a rift located south of the Ozark dome. Early pulses of the Ouachita orogeny affected rocks of Des Moinesian age but the main phase of the orogeny did not begin until Late Pennsylvanian time.

Proceed north on Arkansas 309 and bear to the right at 1.9 mi (3 km) from the last stop. Travel 1.5 mi (2.4 km) and turn right onto Overlook Drive. After traveling another 0.7 mi (1.1 km) turn left into Ross Hollow overlook parking area.

Stop 5. Ross Hollow overlook. Boulders at the base of the cliff form "rock streams" on the northern slopes of Mount Magazine. These rock streams appear to be moving slowly downslope and are best developed at valley reentrants in Savanna Sandstone ledge exposures where the northerly trending joints are most numerous. Drainage patterns are joint controlled and the combination of jointing with the concentration of water along the joints produces areas of more intense weathering, erosion, and mass wasting at the heads of the rock streams.

To the north, in the foreground, can be seen the broad Hartshorne Sandstone bench of Mount Magazine. Farther to the north (near Paris, Arkansas, about 9 mi [14 km] distant) Savanna Sandstone can be seen again as a ledge-former capping Short and Horseshoe mountains where the sandstone is preserved along the axis of the Paris syncline. The surrounding low-lying areas are underlain by shales and siltstones of the Savanna Formation.

On the far-distant northern horizon, the Boston Mountains can be seen rising sharply above the Arkansas valley where steeply-dipping, east-west–trending normal faults have repositioned the Atoka Formation to a higher elevation. The sandstones of the Atoka are the mountain-forming strata of the southern Boston Mountains.

As you return to the parking area, note the "parting lineations" (current lineations) on the Savanna fieldstone steps and walkways.

REFERENCES CITED

Bush, W. V., Haley, B. R., Stone, C. G., Holbrook, D. F., and McFarland, J. D., 1977, A guidebook to the geology of the Arkansas Paleozoic area: Arkansas Geological Commission Guidebook 77-1, p. 43–48 (revised 1980).

Haley, B. R., 1961, Geology of the Paris Quadrangle, Logan County, Arkansas: Arkansas Geological Commission Information Circular 20-B, p. 6–8.

—— , 1976, Geologic map of Arkansas: Arkansas Geological Commission, scale 1:500,000.

Haley, B. R., Glick, E. E., Caplan, W. M., Holbrook, D. F., and Stone, C. G., 1979, The Mississippian and Pennsylvanian (Carboniferous) systems in the United States; Arkansas: U.S. Geological Survey Professional Paper 1110-0, 14 p.

Houseknecht, D. W., 1981, High-constructive, tidally influenced deltaic sedimentation in Arkoma Basin; Desmoinesian Hartshorne Sandstone [abs.]: American Association of Petroleum Geologists Bulletin, v. 65, p. 1499.

Houseknecht, D. W., and Iannacchione, A. T., 1982, Anticipating facies-related coal mining problems in Hartshorne Formation, Arkoma Basin: American Association of Petroleum Geologists Bulletin, v. 66, p. 923–930.

Stone, C. G., 1968, The Atoka Formation in north-central Arkansas: Arkansas Geological Commission, 11 p.

Stone, C. G., and McFarland, J. D., 1981, Field guide to the Paleozoic rocks of the Ouachita Mountain and Arkansas Valley provinces, Arkansas: Arkansas Geological Commission Guidebook 81-1, p. 37–61.

Late Paleozoic rocks, eastern Arkoma basin, central Arkansas Valley Province

P. L. Kehler, *University of Arkansas at Little Rock, Little Rock, Arkansas 72204*

LOCATION

A late Paleozoic section, as well as structures and geomorphology typical of the Arkansas Valley Province and the Frontal Ouachita Mountain Province are demonstrated by a transect extending from near Morrilton, Conway County, Arkansas to near Perryville, Perry County, Arkansas (Fig. 1).

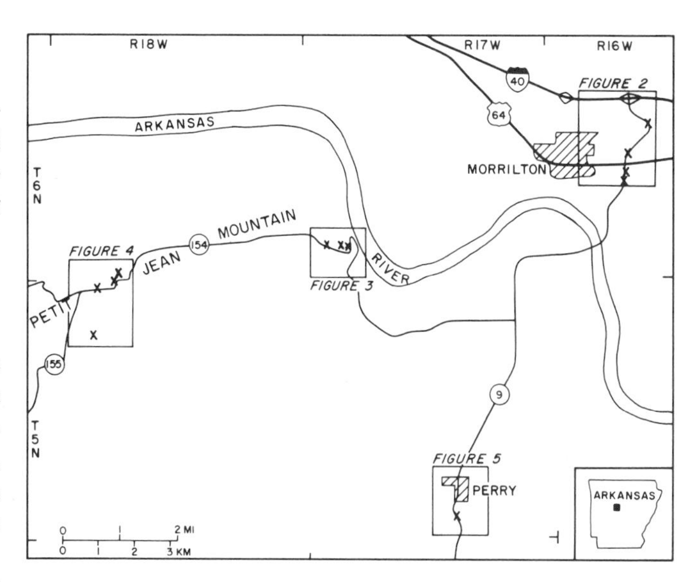

Figure 1. Index map showing locations of more detailed locality maps.

SIGNIFICANCE

Stratigraphically, this transect begins in the lower part of the upper Atoka Formation of Atokan age at the Morrillton anticline and continues upward through the upper Atoka into the Hartshorne Sandstone of Desmoinesian age at Petit Jean Mountain. Crossing the Ross Creek fault, the northernmost thrust fault of the Ouachita Mountain Province, it culminates in the lower Atoka on Perry Mountain.

The Pennsylvanian section in this part of the midcontinent includes shales, siltstones, and sandstones of a deltaic facies and of submarine fan facies. The total section is incredibly thick, displaying a maximum exceeding 20,000 ft (6,000 m) in the subsurface south of Morrilton. These rocks represent sediments deposited on a shelf edge and a marine slope on the north side of a closing trough between the North American plate and a southern plate. Major thrusting has disrupted and imbricated the rocks at the southern terminus of this transect.

Structurally, the transect begins on one of the typical anticlines of the Arkansas Valley, long and narrow compared to the broad intervening synclines. These folds plunge gently toward the west or southwest and are cut by normal faults, many of which dip southward. Subsurface data indicate that some of these appear to have been active during Atokan time. The transect terminates on a thrust-faulted structure, the northernmost of several imbricated slices on the northern edge of the Ouachita Mountain Province.

Geomorphically, the transect begins in the central part of the easternmost Arkoma Basin, which in west-central Arkansas is commonly known as the Arkansas Valley Province. The province includes broad valleys interspersed with long parallel ridges. The trends of these ridges and valleys reflect the underlying folds. The structural plunge also dictates the changes in trend of these features. The major drainage in the region is provided by the Arkansas River, a stream with large meanders, broad floodplains, and several obvious terraces. This river has been modified by a series of locks, dams, and groins. The transect ends in the northern portion of the Frontal Ouachita Mountains where prominent long ridges and narrow valleys are controlled by underlying imbricate thrust slices. Trellis drainage is evident in the tributaries of such streams as the Fourche La Fave. The elevations of the ridge crests are curiously accordant.

Land use is clearly a function of the stratigraphy and structure. Farmland occupies the floodplains, grazing land is found on the synclinal valleys and lower slopes, and timberland is found on the anticlinal ridges and the resistant ridges in the various thrust slices. The long, parallel ridges in the Frontal Ouachitas stand in sharp contrast to anticlinal and synclinal ridges of the Arkansas Valley.

STRATIGRAPHY

A variety of classic deltaic sequences is displayed in roadcuts near Morrilton. Excellent fluvial deposits are exposed on top of Petit Jean Mountain, and slope-fan deposits are exposed on top of Perry Mountain.

In order to interpret ancient delta sequences it is imperative to understand the manner of formation and distribution of the various deltaic environments and the dynamics of each. The details of the dynamic processes and detailed discussion of each of the delta environments are beyond the scope of this chapter, but are reviewed thoroughly by Wright and others (1974), Coleman and Wright (1975), Coleman (1976), Scruton (1960), and by Morgan (1970a and b), among many others. Significant deltaic accumulation of sediment requires a major river system carrying substantial quantities of clastic sediments to a coast and inner

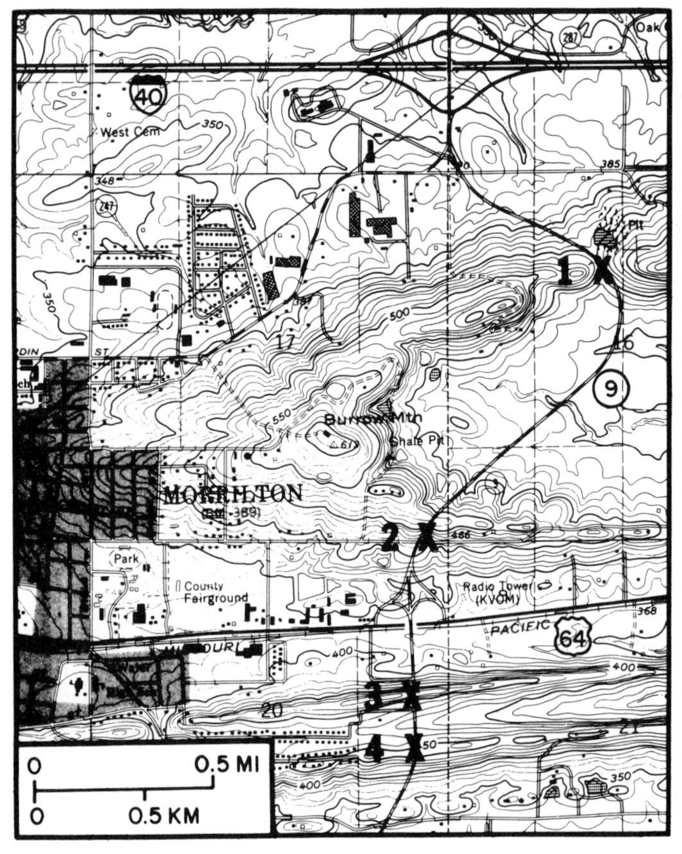

Figure 2. Part of Morrilton East 7½-minute Quadrangle showing locations of the four roadcuts of Locality 1.

shelf more rapidly than it can be redistributed by marine processes (Coleman, 1976). Deltas are most simply defined as subaqueous and subaerial coastal deposits molded by waves, currents, and tides found along deltaic coasts and on delta plains. The late Paleozoic deposits represent distributary channels, river-mouth bars, interdistributary marshes and bays, tidal flats and ridges, beaches, dunes, and evaporite flats. Modern deltas are found in any climate and in a variety of settings where waves, tides, currents, and river discharges interact.

The following localities display stratigraphic sequences that provide an excellent opportunity for the application of various delta models.

LOCALITY 1. UPPER ATOKA FORMATION, MORRILTON, ARKANSAS

Four excellent roadcuts are included in this locality, which are located in Figure 2. These roadcuts are found in Sec. 16, 17, and 20, T.6N.,R.16W., Morrilton East Quadrangle, Conway County, Arkansas.

Accessible by car or bus, these cuts are located east of Morrilton, Arkansas, on Arkansas 9 Bypass, the first of which is 0.7 mi (1.1 km) south of the Arkansas 9 and I-40 interchange.

The roadcuts are located in resistant units in the upper Atoka Formation on the flanks of the Morrilton anticline. The northernmost cut (first going southward on Arkansas 9) is on the northern limb, and the other three are in the southern limb of that fold. This fold is one of several narrow anticlines and broad synclines in the Arkansas Valley Province and contains several coarsening-upward and thickening-upward stratigraphic sequences typical of deltaic cycles. Sedimentary features and structures found in these rocks include load features, scour channels, crossbedding, shale and sandstone clasts, plant fragments that are often coalified, shallow-water trace fossils, water or gas expulsion structures, bioturbation features, flaser bedding, and ripple marks. Occasional thin carbonate units containing invertebrate fossils are found in these sequences.

The roadcuts are identified herein as 1, 2, 3, and 4 from north to south along Arkansas 9, and are briefly described as follows:

Roadcut 1. These rocks are dipping northerly and, viewing upsection, one may observe marine shales and siltstones, prodelta shales and siltstones, sandstone of an inner fringe aspect (LeBlanc, 1977), cross-bedded sandstone representing a river-mouth bar, even-bedded sandstone of an inner fringe aspect, and sandstone representing a possible distributary channel. Small faults are also exposed (Stone and McFarland, 1981).

Roadcut 2. The rocks here are dipping southerly and, observing upsection, one may inspect a delta sequence of gray-black silty shale and thin gray flaser-bedded siltstone, followed by silty sandstone comprising a prodelta and outer fringe stratigraphic sequence. Above that is a sequence of thin-bedded sandstone of an inner fringe aspect, and crossbedded sandstones representing a distributary channel, which in turn is overlain disconformably by a shale most likely representing a clay plug or crevasse filling, and finally topped by fringe sandstones. Bioturbation evidence is found in the lower deltaic packet. Pyrite and its oxidation product melanterite are notable near the base of the inner fringe sandstone (Stone and McFarland, 1981). These rocks are herein considered to be equivalent to those exposed in Roadcut 1. Several delta models might be tested here. One model by LeBlanc (1977) is used as an example for stratigraphic interpretation.

Roadcut 3. Dipping southward, an almost complete coarsening-upward delta cycle is exposed (Fig. 3). Viewing upsection one may first observe a black marine shale containing a thin fossiliferous limestone, a gray to black silty shale of a prodelta aspect, silty sandstone and shale representing the outer fringe, even-bedded sandstones representing the inner fringe, and a thin fossiliferous sandstone followed by a black silty marine shale. Evidence of bioturbation may be observed, as well as trace fossils (*Conostichus*); a water or gas expulsion feature is present on the east side of the highway in sandstones of the inner fringe packet (Stone and McFarland, 1981).

Roadcut 4. This last exposure traveling southward displays some now-familiar sequences and some new sequences relating to

Figure 3. An almost complete delta cycle (view to west) exposed in the roadcut just south of U.S. 64 at Locality 1. A black marine shale is overlain by a prodelta silty shale, silty sandstones and shales of an outer fringe, even-bedded sandstones of an inner fringe, then a fossiliferous sandstone topped by a black, silty marine shale. A distributary-channel sandstone is not present.

deltaic cycles. Again viewing upsection, the sandstone and shale intervals may represent a prodelta with overlying outer fringe rocks, a distributary channel sandstone, a thin transgressive marine unit at the top, inner fringe strata followed by a river-mouth bar sequence, inner fringe rocks, and lastly a distributary channel (Stone and McFarland, 1981).

The outer fringe interval contains small ripples and flaser bedding. Inner fringe sandstones and siltstones are evenly bedded. The distributary channel sandstones show load features, scour channels, shale and sandstone clasts, and crossbedding. The distributary channels are distinguished from the river-mouth bar sandstones by the crossbedding style. Single direction crossbeds may indicate channels, and crossbeds with multiple directions may indicate river-mouth bars (LeBlanc, 1977).

Sandstones included in these Atokan sequences are especially important in the subsurface of the Arkansas Valley Province as traps for natural gas. The entire Atoka interval is more than 20,000 ft (6,000 m) thick south of this locality. This thickness and lateral variation of the Atoka Formation in this area certainly indicates the need for more detailed studies and more regional mapping. Additional mappable units may be required to formulate a better understanding of the framework of southern North America during Atokan time.

LOCALITY 2. PETIT JEAN MOUNTAIN

Petit Jean Mountain is located in Perry and Conway counties on the Adona, Atkins, Casa, and Morrilton West quadrangles. Several localities are available on the mountain (Figs. 4 and 5) and may be reached from the east by Arkansas 9 and Arkansas 154 west from Oppelo; from the west by Arkansas 7 and Arkansas 154 east from Centerville; or from the south on Arkansas 10 and Arkansas 155 north from Casa.

General Geology of the Area. The mountain is located in the southeastern portion of the Arkansas Valley Province (eastern Arkoma Basin) between the gently domed and normal-faulted Ozark Province to the north and the complexly folded and thrust-faulted Ouachita Mountains to the south. It is a large, flat-topped erosional remnant preserving Pennsylvnian strata of the upper Atoka Formation and the Hartshorne Sandstone, and a portion of

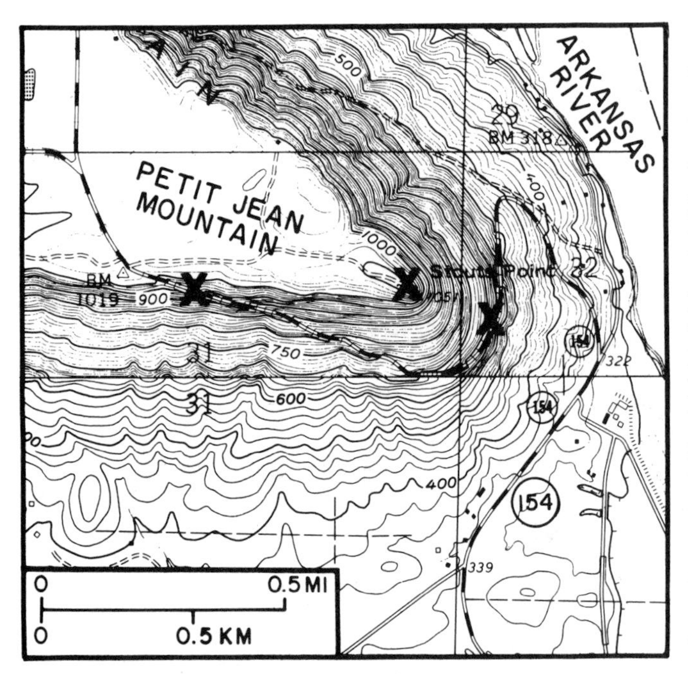

Figure 4. Part of Morrilton West 7½-minute Quadrangle showing locations of the roadcuts and Stouts Point Overlook of Locality 2a on the east side of Petit Jean Mountain.

the McAlester Formation to the south in Round Mountain. Important deformation of the Ouachitas occurred in Late Pennsylvanian time, affecting the Arkansas Valley Province by producing open and upright folds with gentle plunges. The mountain is part of a set of broad folds forming an irregular bowl-shaped structure: a syncline underlying Petit Jean Mountain, the Ada Valley anticline, and a syncline underlying Rose Creek Mountain to the south, all of which plunge gently toward the southwest.

Several sedimentologic, stratigraphic, and geomorphic features are beautifully exposed on Petit Jean Mountain. A view eastward from Stouts Point (Fig. 4) includes many features associated with a modern stream system such as a floodplain, meanders, meander scars, oxbows, terraces, and the effects of a managed stream system. To the north the fault-bounded Boston plateau of the Ozark Province is visible on the horizon. To the south, one can view the Frontal Ouachita ridges where the lower Atoka Formation is thrust northward over the middle and upper Atoka Formation. East and west the broad structures comprising the Arkansas Valley Province are visible.

The stratigraphy exposed on Petit Jean Mountain is representative of much of the Arkansas Valley, which experienced considerable sedimentary filling preceding the closing of the Ouachita trough in the late Paleozoic Ouachita orogeny.

The Atoka, Hartshorne, and McAlester Formations are Atokan and Desmoinesian in age and consist of sequences of alternating sandstones and shales with a few coal beds. Fluvial, deltaic, shallow marine, and paludal environments are represented in these sequences.

The geomorphology of the mountain records a late Tertiary and Quaternary history that is yet to be fully documented and understood. Cedar Falls (Adona Quadrangle, SE¼,Sec.32, T.6N.,R.18W.) is a spectacular example of a nickpoint. Headward erosion is enhanced by removal of the underlying Atoka shales and by collapse and rockfall in the Hartshorne Sandstone at the lip of the falls. The numerous caves and overhangs on the valley walls of Cedar Creek are most likely the result of the undercutting and subsequent erosion by the ancestral Cedar Creek operating at higher, now-abandoned levels. Rock slides and rock falls have produced the tremendous amount of boulder debris along the present creek below the falls. The effects of Pleistocene climates is unknown in this region, but they may have played an important role in the weathering and mass wasting observed on the mountain. The effects of possible arid-humid alternations are also unknown.

Along the Arkansas River—west, north, and east of the mountain—are several terrace levels. Shells from a terrace about 30 ft (9 m) above the present river floodplain have given C^{14} data of 38,000 years. Some of the gravels contain clasts of rock types not found in Arkansas. It is easy to speculate that the Arkansas River has been a major drainage outlet during the Pleistocene and late Tertiary.

East Side Petit Jean Mountain. (Locality 2a). Ascending Petit Jean Mountain from the east (Fig. 4) on Arkansas 154, there are a few hundred feet of relatively flat-lying shales, siltstones, and sandstones (with a thin coal bed) of the upper Atoka Formation (Morrilton West, NW¼,Sec.32,T.6N.,R.17W.). Channeled surfaces and coarsening-upward and thickening-upward sequences indicative of deltaic environments are well exposed in the roadcuts on the west side of the highway. This portion of the upper Atoka contains several deltaic sequences, incompletely preserved and variably superposed, representing delta platform, delta slope, and prodelta facies. Near the top of the hill (NE¼, Sec.31,T6N,R17W) an obvious disconformity is displayed between the Atoka and the Hartshorne Formation. This unconformity very likely has regional importance. The Hartshorne is thick-bedded sandstone exceeding 200 ft (61 m) in thickness. Crossbedding of several styles is well exposed in these roadcuts near the top of the mountain. Paleocurrent analysis indicates a northeasterly source for the fluvial units. These sandstones represent a meandering fluvial system, or perhaps a fluvial system alternating between meandering and braided conditions, whereas in western Arkansas the Hartshorne displays more obvious deltaic facies.

Petit Jean State Park. (Locality 2b) Most of the park is underlain by the Hartshorne Sandstone, which is typically composed of massive to thick-bedded sandstones and thin siltstones and shales. Spectacular bluffs all around the mountain and in the canyon of Cedar Creek are formed by this unit. Along Cedar Creek, beneath Cedar Falls, the disconformable contact between the Hartshorne and Atoka is exposed (Fig. 5) (Adona Quadrangle, SE¼,Sec.32,T.6N.,R.18W.). West along Cedar Creek the Hartshorne is approximately 350 ft (107 m) thick. The valley of

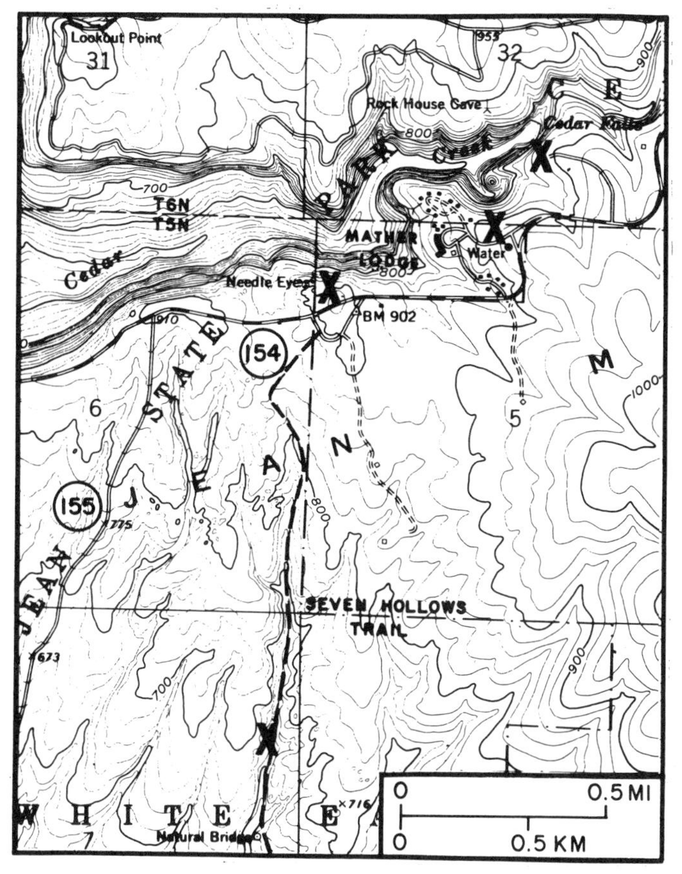

Figure 5. Part of Adona 7½-minute Quadrangle showing locations of outcrops for Locality 2b in Petit Jean State Park.

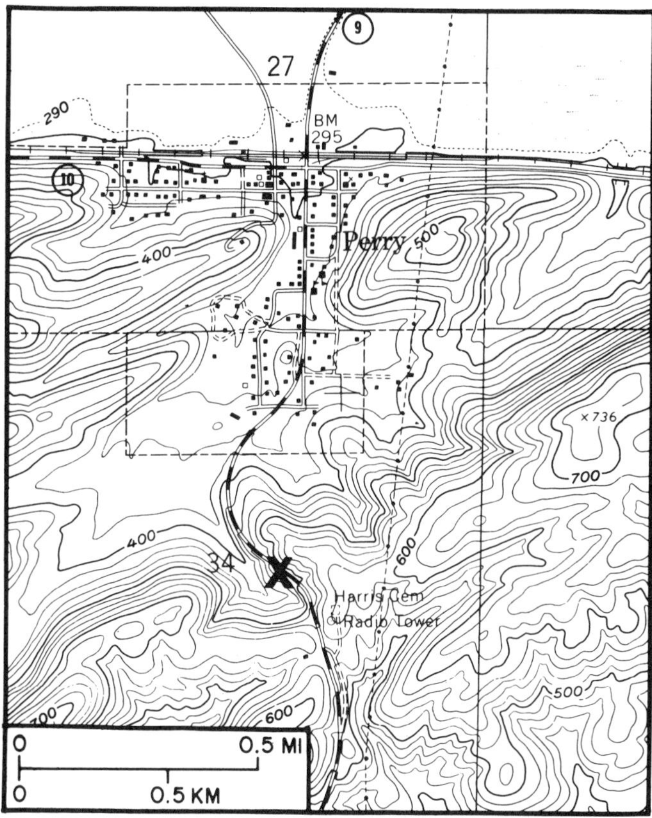

Figure 6. Part of Perryville 7½-minute Quadrangle showing location of outcrop for Locality 3.

Cedar Creek below the falls is cut into the nonresistant siltstones and shales of the upper Atoka Formation. In both the Hartshorne and the Atoka, abundant plant fragments and thin coal streaks can be found. In an abandoned quarry west of the State Park the upper Atoka contains abundant *Sigillaria.* The overlying McAlester Formation is exposed along the southern boundary of the park.

Seven Hollows Trail (Adona Quadrangle, Secs. 5, 6, 7, and 8,T.5N.,R.18W.) leads to many outstanding outcrops of the Hartshorne Sandstone (Fig. 5), which display fluvial crossbedding and features related to differential weathering and erosion performed by obsequent streams, observable north of Natural Bridge. Periglacial conditions may have contributed to the etching of almost unbelievable shapes and forms. The so-called "turtle rocks," observable east of Mather Lodge, and near Needle Eye (also called Bear Cave area), are the weathered tops of sets and cosets of crossbedded sandstones etched by weathering along joints. Spectacular spheroidal forms etched in the outcrop surfaces are produced by resistant iron oxide rings and oval layers, perhaps produced by some process analogous to those that produce liesegang rings. Boxwork is also visible on the surfaces of many outcrops along the Seven Hollows Trail and at several overlooks within the park.

LOCALITY 3: PERRY MOUNTAIN

Perry Mountain is located in Perry County, Arkansas on the Perryville Quadrangle. The outcrop is a roadcut on Arkansas 9 and 10 in the NE¼NW¼,Sec.34, T.5N.,R.17W., approximately 2.4 mi (3.8 km) north of Perryville and 1.1 mi (1.7 km) south of Perry (Fig. 6). The location is 12.8 mi (20.6 km) south of I-40 on Arkansas 9. Traffic is quite heavy here, and the best parking is at a roadside rest at the crest of the mountain south of the roadcuts, which are on the east side of the highway.

The lower Atoka Formation (Stone, 1968) exposes repetitions of southward-dipping subgraywackes, micaceous sandstones and siltstones, and black to dark gray shales. The subgraywackes are thin to thick bedded, and exhibit graded bedding, convolute bedding, and bottom markings. The siltstones are ash-colored and rich in coalified plant fragments. The shales contain siderite concretions. Trace fossils indicative of deep-water deposition are fairly common. Fining-upward and coarsening-upward sequences are represented in these exposures, which are characteristic of flysch sequences. Turbidite models (Bouma, 1962) and submarine fan models (Walker, 1978) are applicable to these outcrops. Few of the Bouma sequences are complete, and submarine fan channels and lobes are observable here, as well as evidence for soft-sediment deformation. A mid-fan position is most likely indicated by these rocks, but more study would facilitate our understanding of the exact nature of this record of a late Paleozoic continental rise and its relationship to the Ouachita trough. The source of the sediments appears to be eastward.

The lower Atoka Formation in this area may be 13,000 ft thick (4,000 m) (Stone, 1968). It is thrust northward over the upper Atoka north of here, at Perry along the Ross Creek fault, which strikes east–west approximately parallel to the Rock Island railroad tracks. The displacement required is approximately 15,000 (4,600 m). South of this locality the Atoka is imbricated along northward-directed thrusts, and the underlying Johns Valley Shale and the Jackfork Formation are in turn thrust over the Atoka along the "Y" City fault and related thrusts.

South of this locality, approximately 0.75 mi (1.2 km) on Arkansas 9 and 10 (NE¼,Sec.3,T.4N.,R.17W.) on the east side of the road, a lamprophyric breccia is poorly exposed. There are many tabular lamprophyric intrusions found in the Ouachita region that are probably of Late Cretaceous age. This particular breccia contains sedimentary rock fragments, altered chlorite, and lamprophyric fragments, which are poorly exposed in this roadcut.

SUMMARY

Pennsylvanian rocks in the eastern Arkansas Valley and northern Ouachita Mountains represent a thick pile of clastic sediments deposited in deltaic environments and submarine fan complexes on the south-facing slope of a closing ocean. The resulting peripheral foreland basin controlled the geographic extent of these thick deposits, which stand in sharp contrast to the deeper-water fine clastics and cherts to the south and the carbonate-dominated rocks to the north. The provenance of this thick section is in general recognized, but the specific points of sediment input and their direct relationship to Appalachian–Ouachita tectonics remain unresolved.

The geomorphology of this region clearly dictates the land usage. The topography and drainage is controlled by the underlying structures in this Arkansas Valley–Frontal Ouachita transition zone.

REFERENCES CITED

Bouma, A. H., 1962, Sedimentology of some flysch deposits; A graphic approach to facies interpretation: Amsterdam, Elsevier, 168 p.

Coleman, J. M., 1976, Deltas; Processes of deposition and models for exploration: Champaign, Illinois, Continuing Education Publishing Co., 92 p.

Coleman, J. M., and Wright, L. D., 1975, Modern river deltas; Variability of processes and sand bodies, *in* Broussard, M. L., ed., Deltas, models for exploration, 2nd ed.: Houston, Texas, Houston Geological Society, p. 99–150.

LeBlanc, R. J., 1977, Distribution and continuity of sandstone reservoirs, Parts I and II: Journal of Petroleum Technology, v. 29, p. 776–804.

Morgan, J. P., 1970a, Deltas; A resume: Journal of Geological Education, v. 18, no. 3, p. 107–117.

—— , 1970b, Depositional processes and products in the deltaic environment: Society of Economic Paleontologists and Mineralogists Special Publication 15, p. 31–47.

Scruton, P. C., 1960, Delta building and the deltaic sequence, *in* Recent sediment, northwest Gulf of Mexico: American Association of Petroleum Geologists, p. 82–102.

Stone, C. G., 1968, The Atoka Formation in north-central Arkansas: Arkansas Geological Commission Miscellaneous Publication, 24 p.

Stone, C. G., and McFarland, J. D., III, 1981, Field guide to the Paleozoic rocks of the Ouachita Mountain and Arkansas Valley provinces, Arkansas: Arkansas Geological Commission Guidebook 81-1, 140 p.

Walker, R. G., 1978, Deep-water sandstone facies and ancient submarine fans; Models for exploration for stratigraphic traps: American Association of Petroleum Geologists Bulletin, v. 62, p. 932–966.

Wright, L. D., Coleman, J. M., and Erickson, M. W., 1974, Analysis of major river systems and their deltas; Morphologic and process comparisons: Louisiana State University, Coastal Studies Institute Technical Report 156, 114 p.

I-430 bypass, Little Rock, Arkansas

J. D. McFarland III, Arkansas Geological Commission, 3815 W. Roosevelt Road, Little Rock, Arkansas 72204

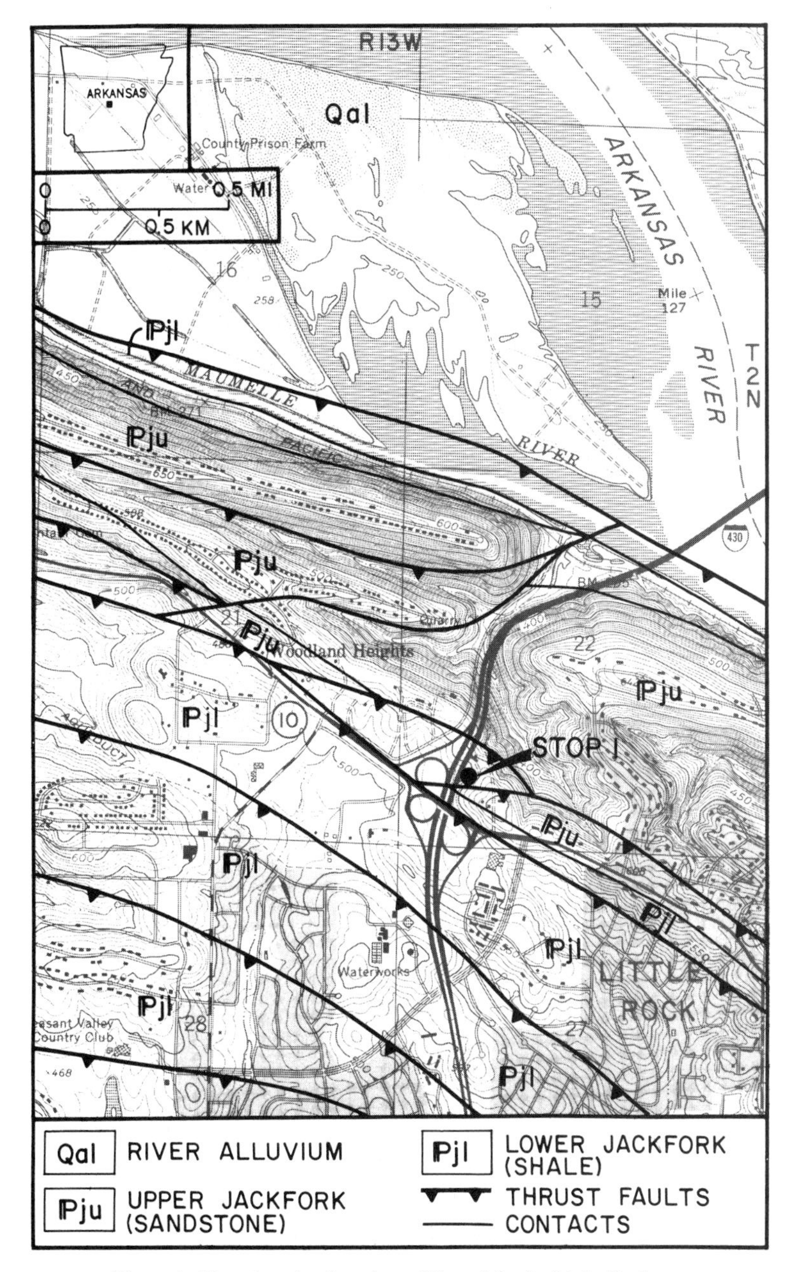

Figure 1. Map showing location of Stop 1 in the Little Rock area.

LOCATION AND SIGNIFICANCE

The I-430 bypass around the west end of Little Rock, Arkansas, offers a unique section of Paleozoic age rocks demonstrating the various facies and structural styles in the eastern Ouachita Mountains (Stone and McFarland, 1981). The sections are easy to reach; just drive to Little Rock, Arkansas, on the interstate system and follow the signs to I-430. The site descriptions are arranged from north to south, starting about 0.8 mi (1.3 km) south of the I-430 bridge over the Arkansas River.

SITE DESCRIPTIONS

Stop 1 (Fig. 1): Jackfork Sandstone at junction of Arkansas 10 with I-430. This large roadcut is in the black shale of the middle Jackfork Sandstone (Fig. 2) and massively bedded,

Figure 2. Interbedded sandstone and shale of the middle Jackfork Sandstone in the vicinity of Stop 1. Stratigraphic up is to the right. The cleavage and milky quartz veins dip gently to the north.

gray, quartzitic sandstone of the basal part of the upper Jackfork. These rather intensely sheared and, in part, thrust-faulted sequences are on the south flank of the Big Rock syncline. Submarine fan channels are present in the sandstone sequences, but these rocks were probably deposited a little farther down the slope of the fan compared to the same sequence at the abandoned Big Rock Quarry in North Little Rock as evidenced by the lack of deep-cutting channels and less abundant intraformational conglomerates. Structural features caused by tectonic deformation combined with soft sediment deformation are present at a number of places in this exposure.

The folds are upright and overturned to the southwest and have nearly horizontal hingelines. Throughout most of the area south of this stop, the Jackfork and older strata are complexly folded and faulted. Nearly all of the rocks, fault planes, and cleavages dip to the north. The structure of these rocks is thought to be a product of a combination of geologic events—namely: (1) southward slumping of soft sediments in a submarine continental slope depositional environment; (2) northward stacking of several thrust-fault slices; (3) several periods of folding; and (4) backfolding and faulting caused in part by 'piling up' and crowding at the toe of the larger northward-moving thrust plates.

Hydrothermal quartz veins of late Paleozoic age are common in the area, and some of them contain rectorite, cookeite, pyrite, and other minerals. Dickite is present in some of the fault zones.

Stop 2 (Fig. 3): Polk Creek Shale, Bigfork Chert, and Womble Shale at the junction of I-430 and Colonel Glenn Road. From north to south, rocks of the Ordovician Polk Creek Shale, Bigfork Chert (Fig. 4), and Womble Shale are exposed in this roadcut. All of the rocks are sheared and tightly folded. Northward-dipping thrust faults are present and are marked by small quartz veins and slickensides. A large block of the Lower Division of the Arkansas Novaculite (lower Devonian) has been thrust over the Womble Shale on the small hill southwest of the roadcut. Two alkalic igneous dikes (now altered to clay) of Cretaceous age (about 90 Ma) are present on the east side of the roadcut. Graptolites of Middle Ordovician age have been collected from shale in the upper part of the Womble. Graptolites of Late Ordovician age have been found in the Polk Creek in nearby localities.

Two opinions have been suggested to explain the complex structure of the rocks at Stops 2 and 3. One proposes gravitational sliding and cascading of rock units northward from a series of overturned nappes. The other proposes a series of northward-moving thrust plates with an overall southward increase in the structural complexity of the rocks in each thrust plate. The southward overturning of the rocks and of the fault planes south

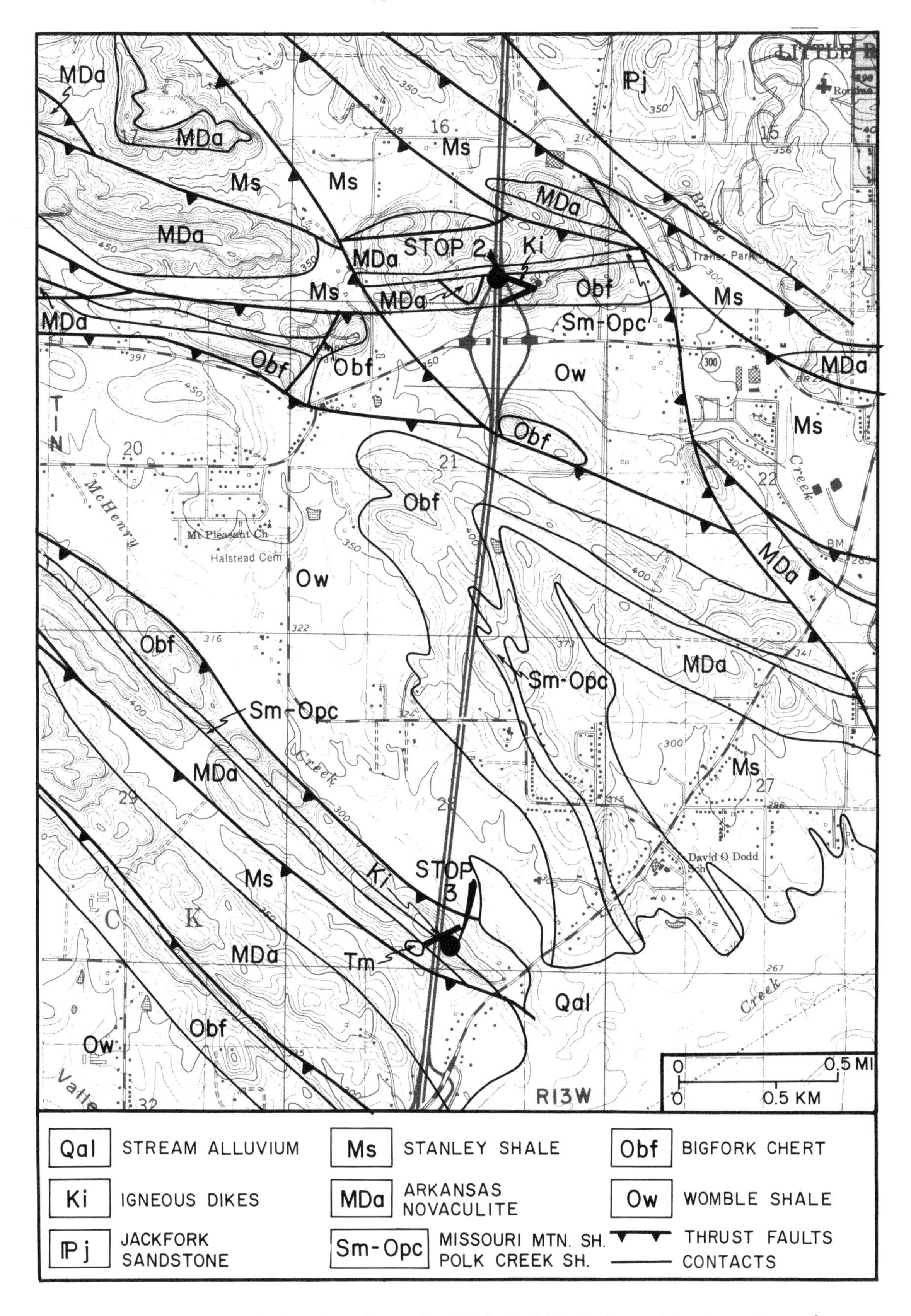

Figure 3. Map showing locations of Stops 2 and 3 in the Little Rock area. Tm = tiny outcrop of Paleocene Midway Group.

Figure 4. Chert and siliceous shale of the Bigfork Chert with two dissecting altered alkalic dikes exposed on the east side of I-430 just north of Colonel Glenn Road.

of Stop 1 can be attributed to backfolding at the toe of each thrust plate and also of subsequent thrust faulting. Still later faulting is suggested by the outlier of Novaculite overlying Womble at this stop 2.

Stop 3 (Fig. 3): Bigfork Chert, Polk Creek Shale, Missouri Mountain Shale, and Middle and Lower Divisions of Arkansas Novaculite 0.4 mi (0.65 km) north of the junction of I-430 and Arkansas 5. From north to south along this exposure the rocks consist of: chert and siliceous shale of the Bigfork Chert; black shale and dark gray chert of the Polk Creek Shale; gray and tan shale with chert and novaculite of the Missouri Mountain Shale; massive bedded, in part tripolitic, very light gray novaculite and chert of the Lower Division and black siliceous shale and chert of the Middle Division of the Arkansas Novaculite. The rocks are overturned to the south, and the style of folding varies with lithology. Pervasive northward-dipping cleavage is especially evident in the Bigfork Chert. Some northward-dipping thrust faults with quartz veins and minor gouge are present. An alkalic igneous dike (altered to clay) of early Late Cretaceous age (about 90 Ma) dissects the Middle Division of the Arkansas Novaculite; both are overlain by a remnant gravel, sand, and clay that may be of Paleocene age (Midway Group).

Keller and others (1985) have determined that the diameter of the polygonal triple point grains developed during recrystallization of the chert or novaculite increases as a function of the degree of metamorphism. The triple-point grains in the chert and novaculite at this stop average about 40 microns in diameter and are among the coarsest observed in chert or novaculite from the Ouachita Mountains.

REFERENCES CITED

Keller, W. D., Stone, C. G., and Hoersch, A. L., 1985, Textures of Paleozoic chert and novaculite in the Ouachita Mountains of Arkansas and Oklahoma and their geological significance: Geological Society of America Bulletin, v. 96, p. 1353–1363.

Stone, C. G., and McFarland, J. D., III, 1981, Field guide to the Paleozoic rocks of the Ouachita Mountain and Arkansas Valley provinces, Arkansas: Arkansas Geological Commission Guidebook 81-1, 140 p.

Igneous rocks at Granite Mountain and Magnet Cove, Arkansas

J. Michael Howard, *Arkansas Geological Commission, 3815 W. Roosevelt Road, Little Rock, Arkansas 72204*
Kenneth F. Steele, *Department of Geology, University of Arkansas, Fayetteville, Arkansas 72201*

LOCATION

The Granite Mountain area (Fig. 1) is located to the south of Little Rock, Arkansas, just south of the interchange between U.S. 65/167 and I-30/630. The exposures at Magnet Cove are located along or near Arkansas 51 west of I-30, about 40 miles southwest of Little Rock.

SIGNIFICANCE

Although the igneous rocks of Arkansas are insignificant in terms of their area of surface exposure (less than 18 mi^2 [46 km^3], they are highly significant because of their variety of unusual rocks and economic importance. Most of the intrusions are small, thin lamprophyre or syenite dikes and sills. Even the largest, Granite Mountain, has exposures totaling less than 7 mi^2 (18 km^2). They can be divided into two groups, those composed of nepheline syenite (Granite Mountain in Pulaski County and the Bauxite area in Saline County) and those that are syenite complexes (Magnet Cove in Hot Spring County and Potash Sulfur Springs in Garland County). Granite Mountain and Magnet Cove have been dated by K/Ar and Rb/Sr methods at 86 ± 3 to 91 ± 5 Ma and 95 ± 5 to 99 ± 10 Ma, respectively (Naeser and Paul, 1969; Zartman and others, 1967). All of the igneous activity is thought to be related chemically and in time (Steele and Wagner, 1979). Granite Mountain and Magnet Cove have been chosen for this field guide because they are excellent representatives of the unusual igneous rocks of Arkansas and have outcrops which are accessible by passenger car.

GRANITE MOUNTAIN

The outcrops to be examined are the largest exposures of igneous rock (syenite) in Arkansas, covering some 6 mi^2 (15.6 km^2) south of Little Rock. This area is shown on the Little Rock 7½-minute Quadrangle as Granite Mountain (Fig. 1) and in older literature as Fourche Mountain. Granite Mountain is the dome or boss of a large buried batholith that was intruded during Late Cretaceous time into intensely folded Paleozoic sedimentary rocks. Paleocene (Midway Group) and Eocene (Wilcox Group) sediments cover low-lying portions of the batholith. Bauxite deposits mined in Pulaski and Saline Counties were developed by lateritic weathering of syenite (Gordon and others, 1958).

Notable outcrop features include jointing and spheroidal residual boulders formed by weathering along joints, small segregation veins, rare xenoliths, fracture-filling fluorite, trachytic textures, and subtle zones of flow banding. Collectable rock types are "blue" syenite (pulaskite), "gray" syenite (nepheline syenite) and trachyte.

The following rock outcrops are accessible by passenger car and bus and are located on public domain. The best time of the year to visit is late fall or winter. To examine these exposures, begin by turning onto Pine Bluff Highway (U.S. 65–167) from I-30.

Mileage

0.0: Stop 1. Mile marker 1 on U.S. 65–167 (south-bound lane). Massive pulaskite is exposed for approximately 500 ft (150 m) along this roadcut. Williams (1891) described pulaskite as a rock made up of orthoclase, pyroxene (var. diopside and aegirine), amphibole (var. arfvedsonite), and a little eleolite (nepheline) or its decomposition product analcime. Johannsen (1938) defined pulaskite as "a granular to trachytoid rock of Family 2113, and as consisting essentially of a potash-feldspar, usually cryptoperthite, and a small amount of nepheline and pyribole." These rocks are now classified as foid-bearing alkali-feldspar syenite and foid syenite in the nomenclature of Streckeisen and others (1973). Megascopically, the rock is gray, varying from dark-bluish-gray to light-gray. The structure is semi-porphyritic with large feldspars forming the principal part of the rock. Between the larger feldspars is a finer grained second generation of feldspar and an occasional biotite flake or crystal of hornblende or augite. Microscopically, the texture is between hypautomorphic-granular and granito-porphyritic. This rock makes up the greater part of Granite Mountain, particularly the ridges and higher slopes. Spheroidally weathered boulders are seen in the upper part of the roadcut and on the slope above the roadcut. Rare miarolitic cavities containing zeolites are present.

1.2 mi (1.9 km): Stop 2. Approximately 330 ft (100 m) of pulaskite and nepheline syenite are exposed on this hill roadcut. Near the center of the outcrop is a fluorite-filled joint, an expression of late degassing mineralization. Subtle banding near the south end of the exposure is indicative of near-margin flowage. Radial fractures seen here were caused by blasting for highway construction.

1.4 mi (2.2 km): Photo Stop A. Granite Mountain Quarry Plant 2, operated by McGeorge Construction Company is visible to the west (right) of U.S. 65–167. The plant and quarry is one of several active syenite quarry operations on this dome. Total production of syenite has been more than 5 million tons per year since 1973. The two largest quarry operations in Arkansas, this one and Minnesota Mining and Manufacturing Company's (3M) Big Rock Quarry, are located on the Granite Mountain syenite. Products are used for riprap, concrete aggregate, road material (SB2), roofing granules, and confined construction fill (fines).

1.6 mi (2.6 km): Stop 3. Dixon Road (Arkansas 338) exit

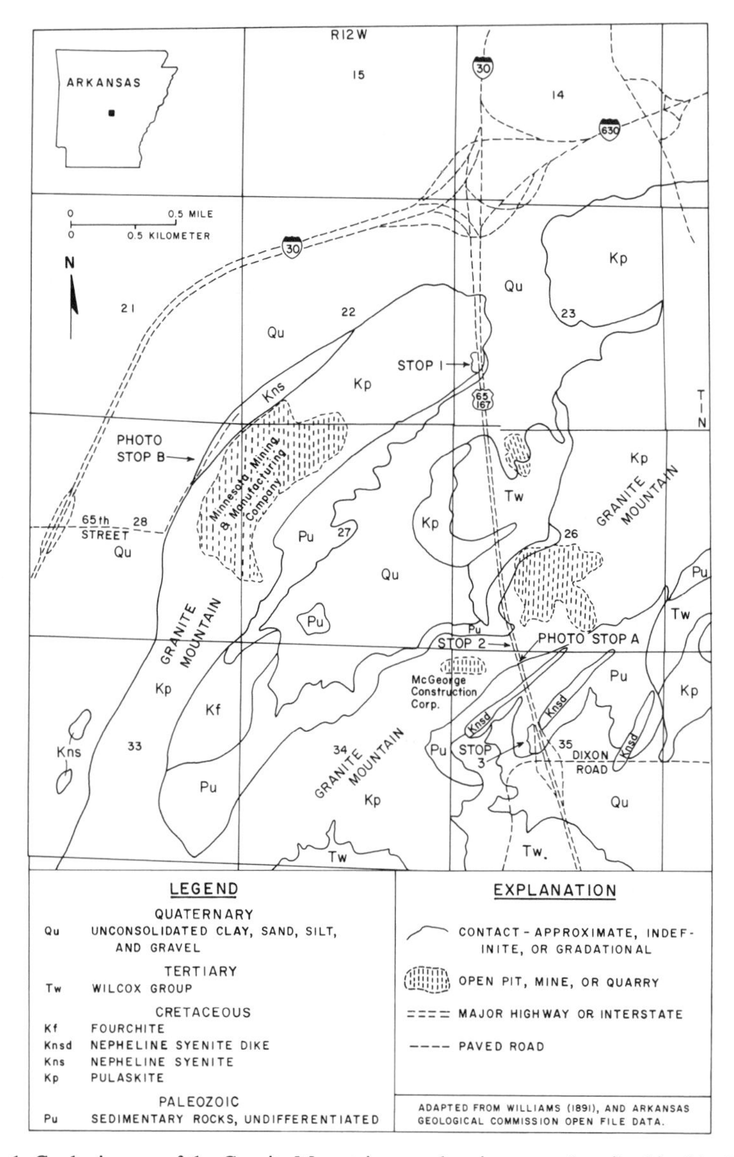

Figure 1. Geologic map of the Granite Mountain area showing stops described in this chapter.

cut. This outcrop extends approximately 600 ft (180 m) on both sides of the exit ramp. Brownish-gray syenite, with partially resorbed, zoned "phenocrystic" feldspars set in a matrix of smaller feldspar crystals, is exposed near the middle of the outcrop on the west side of the road. Near the north end of the outcrop, syenite dike rock grades into a trachytic textured chill margin consisting of feldspar set in a fine-grained matrix. Metasediments are exposed at the north end of the outcrop. Early workers interpreted this rock as a dike, but recent examination suggests that this "syenite" may represent a fenitized roof pendant.

1.8 mi (2.9 km). Turn left on Arkansas 338.

1.9 mi (3 km). Turn left back onto U.S. 65 North.

2.4 mi to 3.0 mi (3.8 to 4.8 km). Syenite outcrop on east side of U.S. 65 North.

4.2 mi (6.7 km). Exit left onto I-30 West toward Texarkana.

6.9 mi (11 km). Exit freeway at exit 135–65th Street.

7.1 mi (11.3 km). Left (east) on 65th Street. The rise ahead is syenite.

7.9 mi (12.6 km). Turn left (north) onto Arch Street Pike.

8.2 mi (13.1 km): Photo Stop B. The main offices of 3M's Big Rock Operations and view of conveyor system.

The primary crusher (grizzly) on the hillside takes quarry feed, crushes it, then transports material by way of conveyor belt to stockpiles in Fourche Creek bottom to the west or to hoppers where the material is loaded into rail cars for shipping to 3M's Little Rock roofing granule plant.

Return to I-30, turn south (left) toward Texarkana.

MAGNET COVE

Magnet Cove, a 4.6-mi^2 (12-km^2) syenite complex, has been

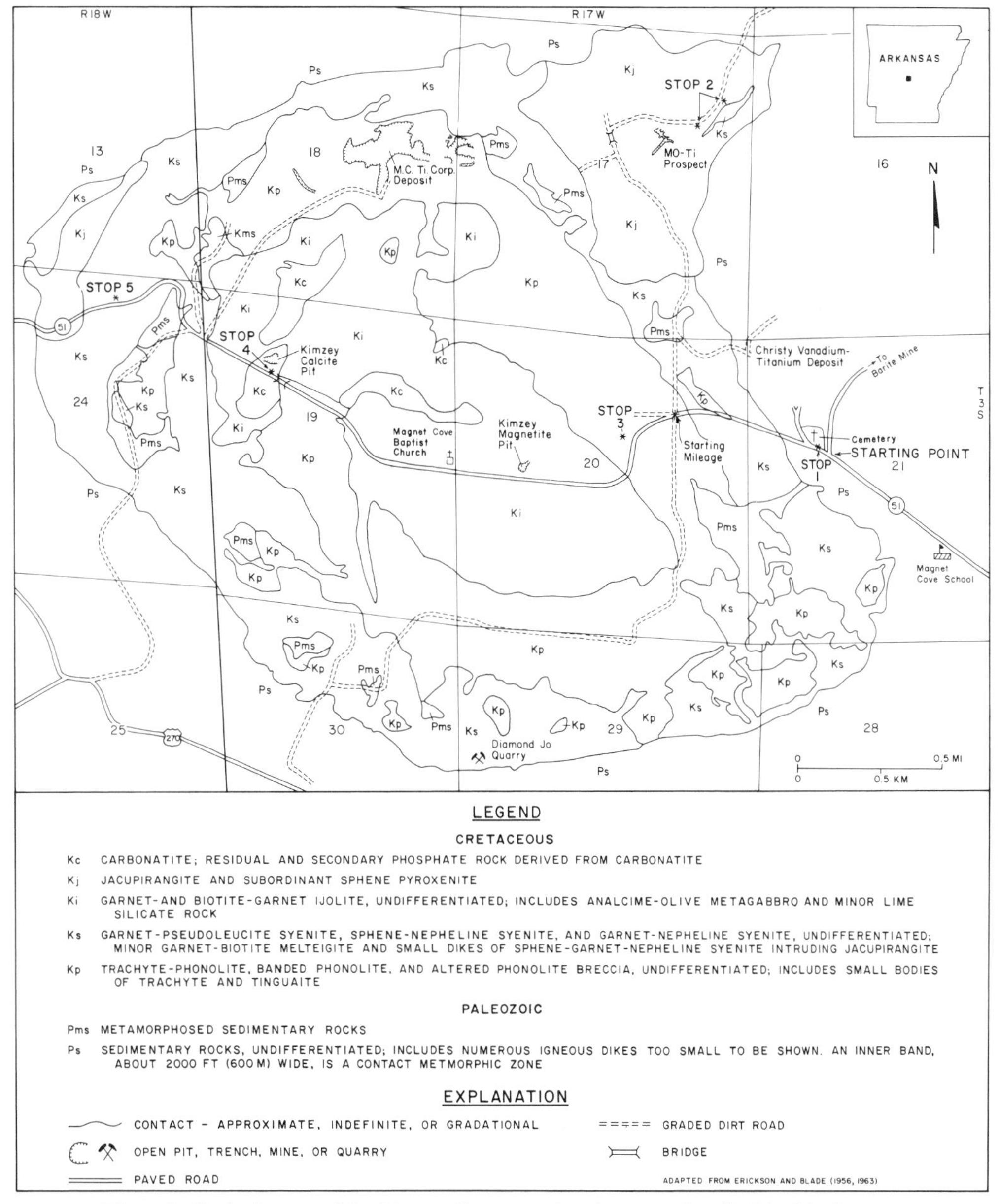

Figure 2. Geologic map of the Magnet Cove area showing stops described in this chapter.

interpreted as a ring dike (Fig. 2) consisting of (1) an inner core of carbonatite and ijolite, (2) an intermediate ring of trachyte-phonolite, and (3) an outer ring of nepheline syenite and jacupirangite (Erickson and Blade, 1963). Magnetite for iron ore, rutile for welding rod coatings, and calcite for agricultural lime have been mined and the Union Carbide Corporation has recently discovered vanadium reserves ajdacent to this complex.

Directions: Take Exit 98 B from I-30 onto U.S. 270 West toward Hot Springs; at the junction of U.S. 270 West and Arkansas 51 (0.7 mi; 1.1 km) take Arkansas 51 (right). The Magnet Cove school sign is on the left at 4.3 mi (6.9 km). At 5.4 mi (8.6 km) is the starting mileage point (Fig. 2) at the intersection of Arkansas 51 with a county dirt road , 0.5 mi (0.8 km) west of Magnet Cemetery.

Stop 1. 0.5 mi (0.8 km) east of starting point. Outcrop of metamorphosed Arkansas Novaculite. The Arkansas Novaculite (Devonian-Mississippian) is exposed in roadcuts due south of the Magnet cemetery on the north side of Arkansas 51. At this stop the novaculite exhibits textural effects from contact metamorphism related to the adjacent Magnet Cove intrusion. Hand specimens have a fine sugary texture and are crumbly owing to recrystallization of the rock and a lack of cementing material. Keller and others (1977, 1983) investigated the Arkansas Novaculite from several areas in Arkansas and Oklahoma and, using scanning electron microscope techniques, discovered that the average background grain size for the novaculite approximately 2.5 mi (4.2 km) from the Magnet Cove intrusion is about 5 μm, increasing to about 45 μm near the contact. Jackson and Nichols (personal communication, 1973) showed a fluid inclusion temperature gradient from about 200°C in quartz approximately 2.5 mi (4.2 km) from the pluton to about 440°C at this location. This coarser texture is developed on the initial finer texture, the finer being the result of older regional metamorphic effects and the coarser being an igneous contact effect. Occasional quartz veins crosscut the outcrop and are thought to represent silica mobilized by alkaline fluids that penetrated the fractured novaculite. In this vicinity, smoky quartz veins with brookite (TiO_2) and rutile (TiO_2) are present. The titanium mineralization is thought to represent contributions to the host rock from the intrusive fluids.

Stop 2. North on dirt road from starting point. At 0.2 mi (0.3 km), bear left at intersection; cross the iron bridge at 1 mi (1.6 km). Outcrops are in Cove Creek, 0.3 to 0.4 mi (0.5 to 0.6 km) east (upstream) from bridge. Jacupirangite is exposed on the south side of Cove Creek (across the creek) and is inaccessible without waders during cold weather. However, jacupirangite (plus other rocks) is also exposed approximately 500 ft (150 m) downstream and is accessible except during extreme high water conditions. The jacupirangite has been intruded by a variety of other rocks including ijolite, fourchite, tinguaite, melteigite and phonolite. Another rock type, meladiorite, is present as the result of assimilation of country rock by the jacupirangite. Locating jacupirangite at the downstream site is difficult because of the relatively large volume of other rocks; however, much of the jacupirangite can be located quickly by using a magnet. The jacupirangite is a fine to medium-grained holocrystalline rock containing at least 50 percent pyroxene and 2 to 25 percent magnetite-ilmenite. Apatite, biotite, sphene, garnet and perovskite are also present, sometimes comprising as much as 10 percent of the rock.

Stop 3. West on Arkansas 51, 0.2 mi (0.3 km) from starting point. This stop is located on the north side of a natural gas line cut on the west side of the highway. The rocks are largely holocrystalline and aphanitic, with some having a microporphyritic texture. Erickson and Blade (1963) referred to these rocks as trachytes and phonolites. They are primarily composed of sodic orthoclase, sodic plagioclase, nepheline, analcime, sodalite group minerals, diopside, biotite, garnet, hornblende, and aegirine. Small sedimentary and igneous xenoliths are present. Some of the trachyte boulders exhibit a definite banding that is apparently due to flow. About 200 ft (60 m) west of the highway, ptygmatic injection features are visible.

Magnet Cove Baptist Church. (0.9 m (1.4 km) from starting point. Ijolite is rarely exposed; however, a boulder may be found in the small stream between Arkansas 51 and Magnet Cove Baptist Church. The ijolite ranges from a medium- to very coarse-grained, feldspar-free rock composed mostly of nepheline, pyroxene, garnet, biotite, and magnetite. Carbonatite veins, some phases containing large amounts of bluish biotite, are sometimes exposed in the stream bed.

Stop 4. West on Arkansas 51, 1.6 mi (2.6 km) from starting point and 0.7 mi (1.1 km) west of Magnet Cove Baptist Church. Carbonatite is exposed on the north side of the road, and a residual boulder of nepheline syenite pegmatite is present on the south side of the road near the west end of the bridge. The carbonatite consists of medium- to coarse-grained calcite and contains the following accessory minerals: light-yellowish-green, radiating, acicular carbonate-apatite; light-brownish, granular monticellite; greenish-black crystals of biotite; dull, black, granular (occasionally octahedral) crystals of magnetite; reddish-bronze cubes of pyrite; and shiny black octahedrons of perovskite. The nepheline syenite pegmatite is medium to coarse grained, and contains aegirine crystals up to 6 in (15 cm) long. Ruby-colored eudialyte crystals as much as 1 in (2.5 cm) across have been reported for this rock type.

Stop 5. West on Arkansas 51, 2.4 mi (3.8 km) from starting point and 1.5 mi (2.4 km) west of Magnet Cove Baptist Church. All of the rocks described at this stop are exposed on the north side of the highway. At the crest of the west rim of Magnet Cove is a large boulder of coarse-grained sphene nepheline syenite containing megascopically visible pyroxene, hornblende, feldspar, nepheline, and sphene. Both cognate and foreign xenoliths are present and are apparently aligned parallel to flow lineation. About 300 ft (90 m) to the east of the crest of the hill, fine-grained sphene nepheline syenite containing phenocrysts of pseudoleucite is exposed. The xenoliths at this location are generally larger than those at the crest of the hill, and one outcrop consists of a large block of fine-grained metamorphosed, sulfide-bearing shale surrounded by fine-grained sphene nepheline syenite. Slightly farther to the east, spheroidal boulders of igneous rocks in clay (saprolite) are present.

REFERENCES CITED

Erickson, R. L., and Blade, L. V., 1963, Geochemistry and Petrology of the Alkalic Igneous Complex at Magnet Cove, Arkansas: U.S. Geological Survey Professional Paper 425, 95 p.

Gordon, M., Tracey, J. L., and Ellis, M. W., 1958, Geology of the Arkansas Bauxite Region: U.S. Geological Survey Professional Paper 299, 268 p.

Johannsen, A., 1938, A Descriptive Petrography of the Igneous Rocks, v. 4: University of Chicago Press, 523 p.

Keller, W. D., Stone, C. G., and Hoersch, A. L., 1983, Textures of chert and novaculite; An exploration guide: Transactions of the Gulf Coast Association of Geological Societies, v. 33, p. 129.

Keller, W. D., Viele, G. W., and Johnson, C. H., 1977, Texture of Arkansas Novaculite indicates thermally induced metamorphism: Journal of Sedimentary Petrology, v. 47, no. 2, p. 834–843.

Naeser, C. W., and Paul, H., 1969, Fission-track annealing in apatite and sphene: Journal of Geophysical Research, v. 74, p. 705–710.

Steele, K. F., and Wagner, G. H., 1969, Relationship of the Murfreesboro Kimberlite and other igneous rocks of Arkansas, U.S.A., *in* Boyd, F. R., and Meyer, H.O.A., eds., Kiberlites, Diatremes, and Diamonds; Their Geology, Petrology, and Geochemistry: American Geophysical Union, p. 393–399.

Streckeisen, A. L., and 19 others, 1973, Plutonic rocks, classification and nomenclature recommended by the IUGS Subcommission on the Systematics of Igneous Rocks: Geotimes, v. 18, no. 10, p. 26–30.

Williams, J. F., 1981, The Igneous Rocks of Arkansas: Annual Report of the Geological Survey of Arkansas for 1890, v. 11, 457 p.

Zartman, R. E., Brock, M. R., Heyl, A. V., and Thomas, H. H., 1967, K/Ar and Rb/Sr ages of some alkalic intrusive rocks from central and eastern Arkansas: American Journal of Science, v. 265, p. 848–870.

Geological features at Hot Springs, Arkansas

J. D. McFarland III, Arkansas Geological Commission, 3815 W. Roosevelt Road, Little Rock, Arkansas 72204

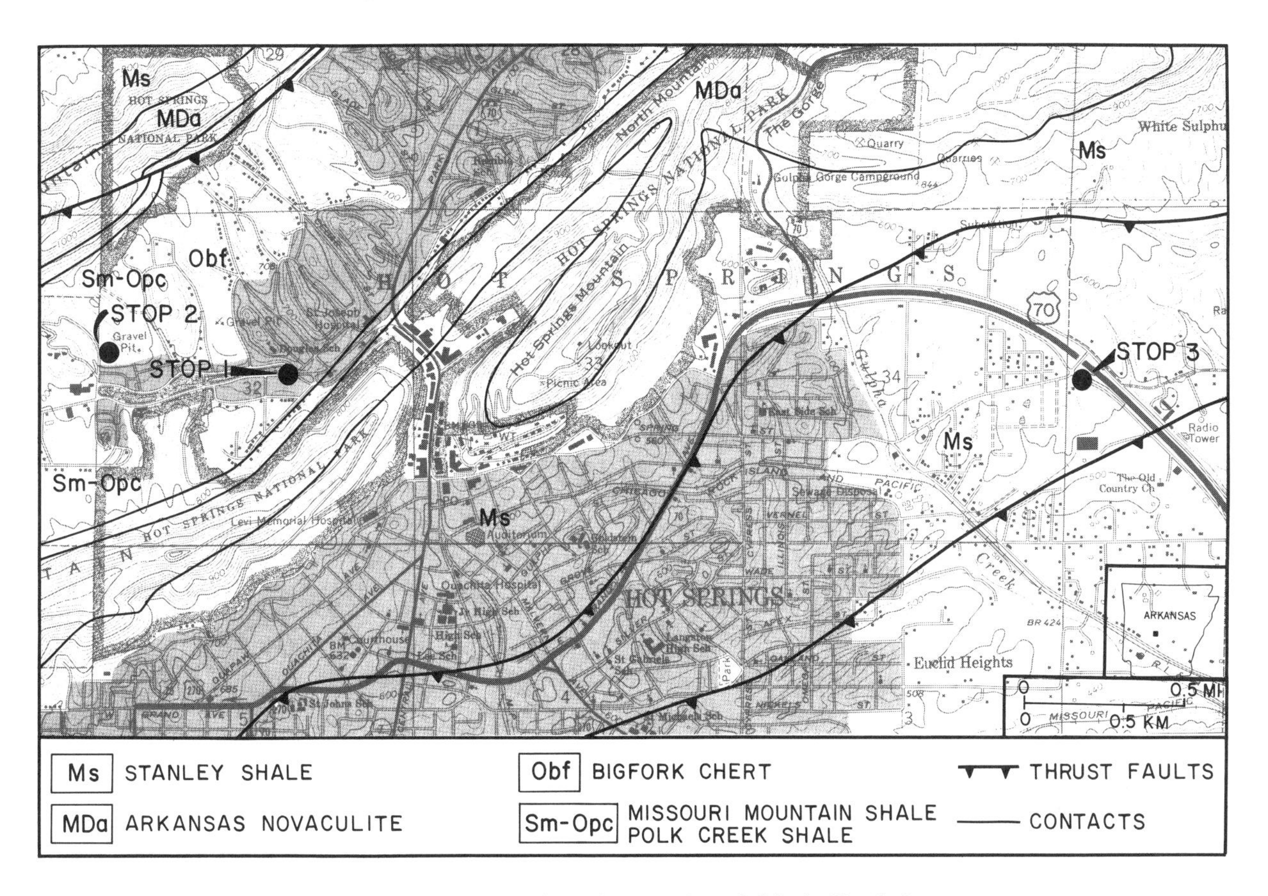

Figure 1. Map showing locations of Stops 1 through 3 in the Hot Springs area.

LOCATION AND SIGNIFICANCE

The area in and around Hot Springs National Park offers many fine exposures of Ordovician through Mississippian rocks along with some recent formations particular to the thermal springs (Stone and Bush, 1982). These strata are often steeply inclined to overturned, with many tight folds and numerous thrust faults formed by the late Paleozoic orogeny that uplifted the Ouachita Mountains. These rocks often contain quartz veins, small Cretaceous-age igneous dikes, and both hot and cold springs. The units to be examined in this area include the Middle Ordovician age Bigfork Chert, the Upper Ordovician age Polk Creek Shale, the Silurian-age Missouri Mountain Shale, the Devonian-Mississippian Arkansas Novaculite, and the Mississippian-age Stanley Shale (including the Hot Springs Sandstone Member). Roads that lead up to the crest of Hot Springs Mountain and West Mountain have several lookouts that provide views of the surrounding countryside. A high tower on Hot Springs Mountain is an excellent place to view the wide Mazarn Basin (a synclinorium underlain by Mississippian age Stanley Shale), the distant Trap Mountains (formed mostly by the Devonian-Mississippian Arkansas Novaculite), the Zigzag Mountains (Arkansas Novaculite and other formations), and the "core' area of the Ouachita Mountains (an anticlinorium with extensive exposures of Ordovician sediments).

SITE DESCRIPTIONS

Stop 1. (Fig. 1) Hot Springs National Park. This is a brief walking tour of the hot springs located off Central Avenue in downtown Hot Springs, Arkansas. The Grand Promenade Trail provides convenient access to hot springs, and exposures of tufa rock and Hot Springs Sandstone. Nearby there is a fine museum at the Hot Springs National Park Headquarters. This was the first designated federal reservation (1832) and became the eighteenth national park (1921) in the United States.

The U.S. Park Service regulates the use of the water from the hot springs that issue from fractures and joints in the Hot Spring Sandstone Member of the Stanley Shale along the base and slope of Hot Springs Mountain.

Hot Springs Mountain is a westward plunging, faulted, southward overturned anticlinal ridge of the Zigzag Mountain subprovince. American Indian tribes, Spanish conquistadors, early settlers, and modern man have all exploited the therapeutic properties of the springs. "Tah-ne-co," the place of the hot waters, was the Caddo Indian name for the site. About 50 of the original 71 springs produce hot water. The temperature range is from 95.4° to more than 147°F (35° to 65°C). The hot spring water is slightly radioactive, apparently caused by radon gas. A soil-and-vegetation–covered gray calcareous tufa formed by the hot springs covers an area of 20 acres (8 hectares) and in places is 6 to 8 ft (2 to 2.5 m) thick. Measurements of the hot springs' flow to the central collection reservoir have been made periodically since 1970 by the U.S. Geological Survey. These measurements indicate a range in spring flow of 750,000 to 950,000 gpd with an average flow of about 825,000 gpd.

The tritium and carbon-14 analyses of the spring water indicate a mixture of a very small amount of water less than 20 years old and a preponderance of water about 4,400 years old.

A study of the geochemical data, rate of flow, and geological structure of the region by the U.S. and Arkansas geological surveys (Bedinger and others, 1979) supported the concept that virtually all of the hot springs water is of local meteoric origin. Recharge to the hot springs artesian flow system is by infiltration of rainfall in the outcrop areas (north and east) of the Bigfork Chert and the Arkansas Novaculite. The water moves slowly to depth where it is heated by contact with rocks of high temperature. Highly permeable zones related to jointing or faulting collect the heated water in the aquifer and provide avenues for water to travel rapidly to the surface.

Stop 2. (Fig. 1) City Quarry in Bigfork Chert. Folded and faulted rocks in the middle part of the Bigfork Chert are exposed in a quarry north of the Weyerhaeuser Company office at 810 Whittington Avenue, Hot Springs, Arkansas. The quarry is near the axis of the southwestward-plunging Hot Springs anticline. Massive novaculite of the Lower Division (Devonian) of the Arkansas Novaculite forms the ridges to the north and south.

In the quarry, the Bigfork consists of very thin-bedded and often graded, calcareous (often decalcified) siltstone, gray chert, and minor beds or laminations of siliceous to carbonaceous shale. The basal silty part of each interbedded sequence likely represents minor influxes of fine clastics and bioclastics transported into the Ouachita trough by turbidity and bottom currents; and the overlying cherts and siliceous shales represent slowly deposited deep-water pelagic accumulations. Stylolites are often present in the calcareous siltstones and indicate a significant removal and thinning of the section.

The Bigfork is complexly folded in the quarry with a large box fold and associated kink, chevron, and buckle folds inclined both to the south and north. Most of the strata dip to the north, indicating a dominant southward vergence. North-dipping cleavage often refracts across the cherts and calcareous siltstones. Flowage of rock can be seen in some of the tight fold hinges. Several apparently small reverse faults dissect the rocks and they often contain gouge (dickite?)-impregnated chert breccia.

The Bigfork Chert is the most reliable aquifer in the Ouachita Mountains. Clear or small chalybeate (iron-rich) springs are often present in the basal outcrops. The water moves along joints, fractures, and bedding planes in the thin-bedded basal sequence.

Stop 3. (Fig. 1) Lower Stanley Shale east of Hot Springs. The lower part of the Stanley Shale is well exposed at several localities east of Hot Springs along U.S. 70. At an outcrop on the south side of U.S. 70 near an overpass about 1 mi (1.6 km) east of the turnoff to Gulpha Gorge Campground, the rocks consist of black shale or slate, thin-bedded siltstone, thin- to thick-bedded subgraywacke and graywacke, chert, and a few cone-in-cone concretions. The maximum thickness of the Stanley is approximately 9,500 ft (2,900 m); these rocks are about 1,200 ft (370 m) above the base. Most of the rocks are overturned to the south into the Mazarn Basin and dip steeply to the north. The lower Stanley is complexly folded, with both flexural-flow and slip types, and exhibits at least two generations of cleavage. Thrust faults cutting the sequences are indicated by numerous slickenside zones with quartz veins and dickite coating on the rock surface.

Sedimentary features include bottom marks, ripple marks, graded bedding, cross-laminations, debris flows, loading and slumping, and bioturbations. Many of the beds thicken and coarsen upward and are believed to represent lobe sequences of an outer submarine fan depositional environment. Some sandstone packets thin upward, indicative of a submarine fan channel depositional environment. These sands were likely derived from sources to the south and southeast. The sands were initially carried to the north and northwest down deep-water submarine fans and subsequently carried by turbidity currents to the west down the Ouachita trough. The clastic units show both structural boudinage and sedimentary pull-aparts. The weathered Stanley Shale at the top of the roadcut is characteristically greenish brown.

The middle and lower parts of the Stanley Shale are equivalent to the Tenmile Creek Formation of the lower part of the Stanley Group in the Ouachita Mountains of Oklahoma. In Arkansas and Oklahoma there are chert intervals in the lower, middle, and upper parts of the Stanley that represent reliable markers. Sandstones in the Stanley are often tuffaceous in Arkansas, but the acidic volcaniclastic beds of the Hatton and Mud Creek Tuff lentils, which are prominent in the lower 1,500 ft (460 m) in southwestern Arkansas and Oklahoma, are rarely present in this area. Miser (1934) and Niem (1977) have shown that the tuffs were derived from a source south or southwest of Broken Bow, Oklahoma.

REFERENCES CITED

Bedinger, M. S., Pearson, F. J., Jr., Reed, J. E., Sniegocki, R. T., and Stone, C. G., 1979, The waters of the Hot Springs National Park, Arkansas; Their nature and origin: U.S. Geological Survey Professional Paper 1044-C, 33 p.

Miser, H. D., 1934, Carboniferous rocks of the Ouachita Mountains: American Association of Petroleum Geologists Bulletin, v. 18, p. 834–843.

Niem, A. R., 1977, Mississippian pyroclastic flow and ashfall deposits in the deep-marine Ouachita flysch basin, Oklahoma and Arkansas: Geological Society of America Bulletin, v. 88, p. 49–61.

Stone, C. G., and Bush, W. V., 1982, Guidebook to the geology of the eastern Ouachita Mountains, Arkansas: Arkansas Geological Commission Guidebook 82-2, 24 p.

The Eocene Jackson Group at White Bluff, Arkansas

George W. Colton and William V. Bush, Arkansas Geological Commission, 3815 W. Roosevelt Road, Little Rock, Arkansas 72204

LOCATION

White Bluff is immediately north of Lock and Dam No. 5, about 15 mi (24 km) upriver from the city of Pine Bluff, Jefferson County, Arkansas. The best method to reach the bluff is by boat along the Arkansas River. Launching ramps are available at the Tar Camp Public Use Area, approximately 1.5 mi (2.4 km) upriver from White Bluff. Tar Camp Public Use Area is accessible via Redfield River Road from Redfield, which is on Arkansas 365. U.S. 65 parallels Arkansas 365 and the two are connected by Arkansas 46 (Fig. 1). Access to the top of the cliffs from the west is not recommended because of their steepness and the uncertain footing on the poorly consolidated materials. Other excellent exposures of the Jackson are present in less-steep cliffs on the west bank of the river at Red Bluff and South Red Bluff, approximately 2 mi (3.2 km) to the north (Fig. 1).

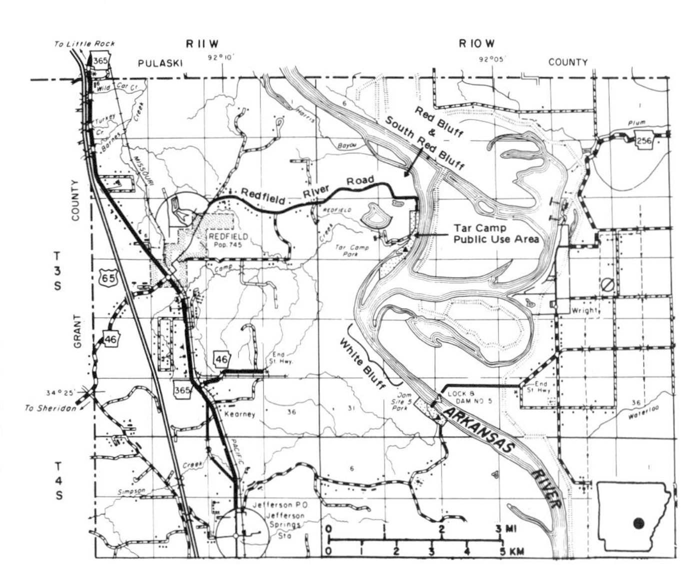

Figure 1. Location map of White Bluff, Jefferson County, Arkansas.

SIGNIFICANCE

The most extensive exposures of the Jackson Group in Arkansas occur along the west bank of the Arkansas River at White Bluff. The White Bluff exposures are described by Wilbert (1953) as a classic locality for study of the Jacksonian Stage and as the type locality for the White Bluff Formation, one of the two formations composing the Jacksonian. The bluff stretches for more than 2 mi (3 km) and ranges from 40 to 60 ft (12 to 18 m) in height.

GENERAL SETTING

The Jackson Group, of late Eocene age, is the youngest Tertiary unit reported in Arkansas, and underlies much of the southeastern and east-central parts of the state. Its presence at the surface is limited to two areas separated by nearly 90 mi (145 km). The first, and by far the smaller, outcrop area is a narrow belt along the east side of Crowley's Ridge, entirely within St. Francis County. The second area is a highly irregular north–south belt covering sizeable parts of Grant, Jefferson, Cleveland, Lincoln, Bradley, and Drew counties, and a small part of northern Ashley County (Haley and others, 1976).

Relatively little work has been done on the Jackson Group in the subsurface beneath its cover of younger Quaternary fluvial sediments. Apparently it thickens to the east and south from the two outcrop areas as it dips gently toward the axis of the Mississippi embayment. A map by Wilbert (1953, Fig. 2) shows the known subsurface extent of the group in Arkansas and in nearby parts of Tennessee, Mississippi, and Louisiana.

The Jackson Group overlies the Claiborne Group, also of Eocene age. The contact is gradational in a regional sense, but local unconformities have been noted. A pronounced unconformity separates the outcropping Jackson from the overlying Quaternary fluvial deposits.

Facies Relationships. An excellent study of the Jackson Group in southeastern Arkansas by Wilbert (1953) showed that the group comprises a complex assemblage of marine and nonmarine facies. They were deposited during the earlier transgression and later regression of a body of marine water occupying the Desha Basin in southeastern Arkansas and part of nearby Mississippi.

The dominantly marine deposits of the Jackson Group were designated as the White Bluff Formation by Wilbert (1953). It constitutes the bulk of the group in the southern part of the outcrop belt and is subdivided into three members based on lithology and faunal assemblages. Dominantly nonmarine beds were placed in the Redfield Formation (Wilbert, 1953), which Wilbert considered sufficiently homogeneous that further division into members was unnecessary. The Redfield Formation was named by Wilbert after the town of that name 3 to 4 mi (5 to 6 km) northwest of White Bluff (Fig. 1). The Redfield is thickest in the northern part of the outcrop belt. Owing to the transgressive–regressive nature of the deposits constituting the Jackson, the two formations are in large part time equivalent, and one thickens at the expense of the other.

THE WHITE BLUFF LOCALITY

The Jackson Group is not completely exposed at the White Bluff locality (Fig. 2). The exposures start at water level shortly above the base of the White Bluff Formation, extend upward across the contact between the White Bluff and Redfield and into the Redfield Formation. Owing to the gentle, regional southward dip in this part of the Mississippi embayment, the stratigraphically lowest beds of the White Bluff Formation are present at the north end of the exposures. An unknown but probably considerable part of the Redfield has been removed by erosion. Here, as elsewhere in southeastern Arkansas, the Redfield is truncated by Quaternary sediments.

The White Bluff Formation is represented only by the Pastoria Sand Member. The other two members of the formation, the Caney Point Marl and Rison Clay, which were mapped by Wilbert in the southern part of the outcrop belt, are absent in the cliffs at White Bluff. They represent the more offshore facies of the White Bluff Formation. The Caney Point disappears by intertonguing to the north with the near-shore Pastoria Member and with the overlying, nonmarine Redfield Formation.

The following descriptions are restricted largely to the lithology of the units exposed at White Bluff. Those interested in the fauna of these deposits are referred to Wilbert (1953), who described the faunal content in considerable detail.

Pastoria Sand Member of the White Bluff Formation. The Pastoria consists of irregularly interbedded sand, silt, and clay. The proportions of each change laterally along the cliff; lensing and intertonguing are common and clearly visible. Typically the sand is dark greenish gray, argillaceous, and silty and is composed largely of very fine angular quartz grains. The greenish cast is caused by glauconite, which is widely disseminated in many of the sands. Well-preserved mollusc shells are fairly abundant, and a few forams and ostracods are present in some beds. Clay in the Pastoria ranges from medium to dark gray and is typically sandy and silty. Stringers of sand, some of which contain mollusc remains and isolated brachiopod valves, are common in some clay units. Several of the clay units contain small spherical concretions and carbonaceous plant debris. Silt, which is less abundant than sand or clay, is typically greenish gray, may contain scattered calcareous concretions, and may be fossiliferous. Bedding characteristics, fairly abundant plant debris, and a relatively impoverished fauna (as compared to the laterally equivalent Rison Member of the White Bluff Formation) suggest that the Pastoria accumulated in a near-shore environment of reduced salinity.

Redfield Formation. The Redfield occupies the upper part of the cliffs at White Bluff. It consists largely of silt and sand and contains little clay. Consequently, it stands out in sharp contrast to the clay-rich Pastoria Member below. The Redfield is also distinguished by scattered stringers of lignite and much scattered plant debris. The silt is yellow-gray, evenly bedded, and may contain some interbedded clay. Leaf imprints and shards of carbonaceous plant debris are abundant. In the middle part of the exposure along White Bluff, a 5-in (13-cm) bed of lignite is present at the base of the upper silt unit of the Redfield. It extends for nearly a mile along the outcrop, and is a prominent marker bed about 17 ft (5 m) below the highest visible outcropping beds

Formation	Member	Thickness	Description
REDFIELD FORMATION		17	Silt, yellowish-gray, interbedded with silty clay; sandier near base; leaf imprints abundant.
		0.4	Lignite
		20	Sand, light gray, crossbedded, with irregular layers of lignitic silt and clay; lower 8 feet grades northwestward into silt, yellowish-gray, and dark silty clay; leaf imprints abundant in silt and clay.
		10	Sand, light-gray, interbedded with dark-gray, lignitic clay; grades northwestward into silt, yellowish-gray, and dark silty clay; leaf imprints abundant.
WHITE BLUFF FORMATION	Pastoria Sand Member	2	Clay, medium-gray; sand stringers containing marine fossils present.
		21-30	Sand, dark greenish-gray, argillaceous and slightly glauconitic; abundant well preserved mollusc shells.
		6-15	Clay, dark-gray, blocky, with stringers of silt and sand containing mollusc shells; some irregularly distributed small spherical concretions.
		7	Clay, dark-gray, lignitic, silty and sandy; stringers of glauconite and fragmented fossil debris common.
		7	Silt, greenish-gray, interbedded with sand; calcareous concretions present in some sand beds.

Figure 2. Jacksonian strata exposed along White Bluff, Jefferson County, Arkansas. Figures show thickness in feet. Adapted from Wilbert (1953).

of the formation. Sand in the Redfield is light gray, thinly and irregularly bedded, and locally crossbedded. Most beds interfinger laterally with silt. Leaf impressions are present but are not abundant. In addition to the general absence of marine fossils, the abundance of sand (in part crossbedded) and the presence of appreciable lignitic material and leaf imprints, all indicate a nonmarine but aqueous environment of deposition—probably in very shallow water on or near a delta.

REFERENCES CITED

Haley, B. R. and others, 1976, Geologic map of Arkansas: U.S. Geological Survey map, scale 1:500,000.

Wilbert, L. J., Jr., 1953, The Jacksonian Stage in southeastern Arkansas: Arkansas Division of Geology Bulletin 19, 125 p.

Lower Stanley Shale and Arkansas Novaculite, western Mazarn Basin and Caddo Gap, Ouachita Mountains, Arkansas

Jay Zimmerman and James Timothy Ford*, *Department of Geology, Southern Illinois University at Carbondale, Carbondale, Illinois 62901*

INTRODUCTION

Elements of the sedimentology and structural geology of two of the most prominent lithologic units in the Arkansas Ouachitas, the Arkansas Novaculite (Devonian-Mississippian), and the Stanley Shale (Mississippian), are illustrated along a 6.1 mi (9.8 km) highway traverse between the village of Glenwood and Caddo Gap, Arkansas. The outcrops described below serve as an introduction to the two broad facies groups present in the Ouachita Mountains and to the structural style that dominates major parts of the range.

LOCALITY AND ACCESS

The traverse is located in the southern part of the Benton Uplift of west central Arkansas, about 50 mi (80 km) southwest of Hot Springs and 110 mi (180 km) southwest of Little Rock. It begins in northeastern Pike County (Glenwood 7½-minute Quadrangle) and terminates in southeastern Montgomery County (Caddo Gap 7½-minute Quadrangle) (Fig. 1). All stops except one are in roadcuts along Arkansas 8-27, providing direct access by motor vehicle. Stop 3 (Fig. 1) requires an easy walk of about 0.25 mi (0.4 km) along footpaths and railroad tracks but presents parking problems for larger groups (see below).

GENERAL GEOLOGY

From Glenwood to Caddo Gap, the traverse proceeds down-section through shales, siltstones, and graywackes of lower Stanley Shale flysch to the base of the Arkansas Novaculite. The gradational contact between the two units is interpreted as a transition from an earlier (Lower Paleozoic through Osagean) starved basin sequence to a succeeding (Meramecian through middle Atokan) flysch, marked by an increase in rate of deposition and the emerging predominance of clastic over biochemical sedimentation (Gordon and Stone, 1977).

Within the Benton Uplift, relatively non-resistent rocks of the Stanley Shale crop out across the Mazarn Basin (Miser and Purdue, 1929), an east-west trending, thrust faulted, and folded structural basin that narrows to the west and is surrounded on all sides by ridges of resistant Arkansas Novaculite. Detailed stratigraphy of the Stanley Shale in the western part of the Mazarn Basin has not been established. Most workers assign the rocks exposed in this area to the lower part of the unit, probably equivalent to the Tenmile Creek Formation defined elsewhere in the Arkansas and Oklahoma Ouachitas.

Throughout the basin, the lower Stanley Shale sequence has been repeated across east-west trending, south-dipping thrust faults produced during northward tectonic transport in middle to late Pennsylvanian time. Synchronous, north-verging, monoclinic, mesoscopic folds can be seen at several locations along the road traverse. Macroscopic folding occurs at the leading edges (tip lines) of thrust faults. Elements of the general character of large-scale, thrust related folding are illustrated in Figure 2a. All of the bedding attitudes were measured at outcrops described in the following section. Notice that most beds dip moderately or steeply to the south and that aggregate folding has a shallow (β = 20°) westward plunge. More widely distributed data from the surrounding area (Zimmerman and others, 1984, Fig. 3) indicate that macrofold plunges vary in amount and direction in this part of the western Mazarn Basin but probably do not exceed 40°. The somewhat conical pattern of poles to bedding surfaces in Figure 2a is caused by plotting elements from several differently oriented macrofolds on the same net. It does not necessarily indicate that fold geometry is non-cylindrical.

From Caddo Gap northward into the central Benton Uplift, the attitudes of structural elements suggest that large scale, south-directed (clockwise looking east) bulk rotation of a major part of the section occurred relatively late in or subsequent to the period of thrust transport (Zimmerman and others, 1982).

B. R. Haley and C. G. Stone (see Stone and McFarland, 1981, Plate 35) mapped the traces of ten thrust faults between Glenwood and Caddo Gap, primarily from air photo interpretation of local and regional outcrop patterns (Fig. 1). Among their criteria for recognizing thrust traces in outcrop are dip reversals, disrupted bedding and networks of coarse, quartz-filled veins. Only those faults for which there is direct evidence in roadside exposures have been included in the descriptions below.

ROAD TRAVERSE FROM GLENWOOD TO CADDO GAP

The traverse begins at the junction of Arkansas 8-27 and U.S. 70B in Glenwood (corner of 1st Street and Broadway near the railroad crossing). All subsequent mileages are measured from this junction.

Follow Arkansas 8-27 northwest to **STOP 1** (1.35 mi; 2.2 km), the roadcut opposite the Glenwood Christian Church (Fig. 1). There is room to park on the shoulder, but beware of traffic along the highway!

*Present address: Sohio Petroleum Company, Two Lincoln Centre, Suite 1000/LB03, Dallas, Texas 75240.

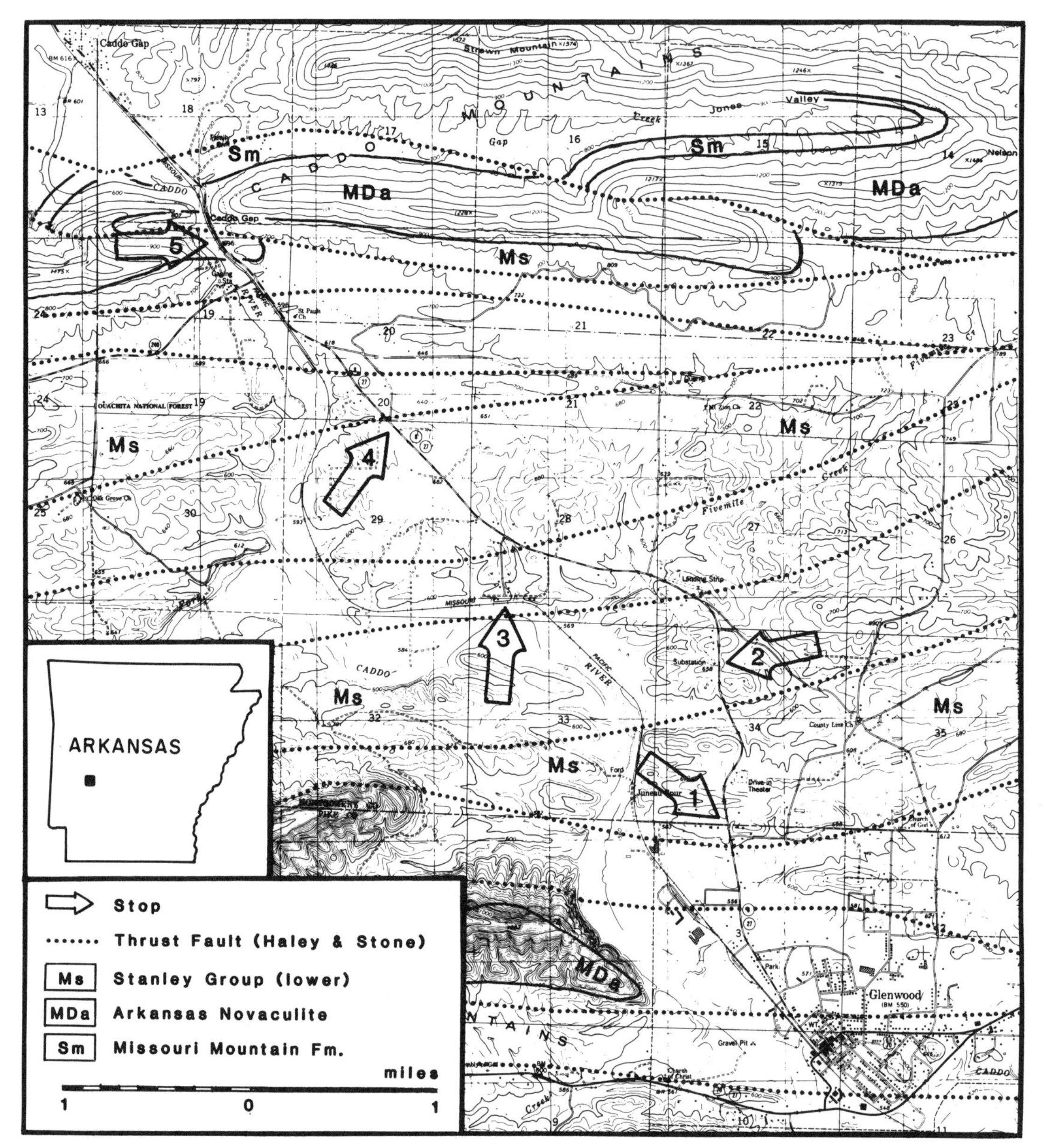

Figure 1. Map of the highway traverse between Glenwood (lower right) and Caddo Gap (upper left). Base: Glenwood and Caddo Gap 7½-minute Quadrangles. Geology after Haley and Stone (see Stone and McFarland, 1981, Plate 35).

Sandstones and siltstones of the lower Stanley Shale are best exposed on the east side of the roadcut. Beds strike east-west, dip steeply to vertically, and top to the south. The tan to orange colors of weathered sandstone and olive-brown of weathered shale are typical of Stanley rocks. Fresh sandstones are usually gray, and fresh shales are typically dark gray to black.

Plant fragments can be found in numerous lower Stanley siltstones. J. R. Jennings (personal communication, 1985) has identified *Calamites* and *Stigmaria* at this outcrop.

Several high-angle faults with minor displacement cut the section at this location. The most prominent is a normal fault that dips about 50° southward and displaces a sequence comprising (from north to south) thick sandstone, a thin shale and a thin sandstone bed. The estimated dip separation component along this fault is about 12 ft (4 m).

Proceeding northward from the Glenwood Christian Church roadcut, note the steeply south-dipping, medium to thickly bedded lower Stanley sandstones and subordinate shales exposed on the east side of the road at the entrance to the drive-in theater (mi 1.50; km 2.4). Cross-stratification and tool marks indicate that beds top to the south.

At mile 1.90 (km 3.1), cross the trace of a thrust fault mapped

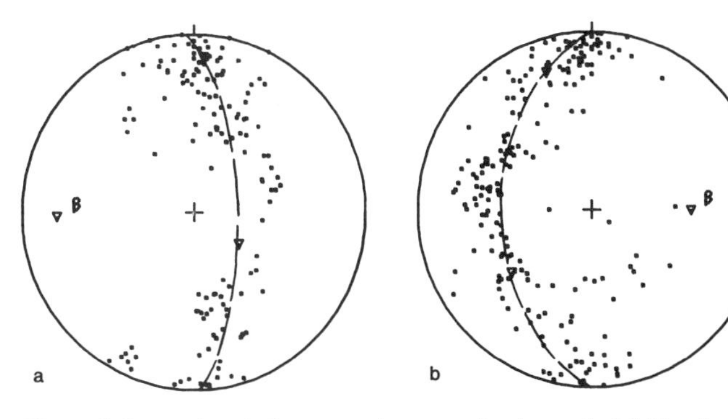

Figure 2. Lower hemisphere, equal area projections of: a) 143 poles to bedding surfaces in the lower Stanley Shale along Arkansas Highway 8-27 between Glenwood and Caddo Gap; and b) 200 poles to bedding surfaces in the Arkansas Novaculite at Caddo Gap. The representative, statistical macroscopic fold axis (β) for Stanley rocks plunges 20°/269°, and the equivalent mesoscopic fold axis for the Novaculite plunges 42°/090°. Inverted triangles are average (eigen-) vectors calculated from bedding plane pole orientation matrices.

by Haley and Stone. Contorted, north-south striking shales with quartz veins can be seen in drainage ditches and adjacent embankments along the road.

STOP 2 is located at mi 2.20 (km 3.5). Park on the east side of the highway opposite the electric power substation (Fig. 1).

At this stop, rocks of the lower Stanley Shale can be examined for a distance from about 420 ft (128 m) south to about 360 ft (110 m) north of the substation.

The southern outcrops (east side of the road) include relatively fresh, dark colored, fine-grained graywackes and shales. Pencil-fractures in the shales suggest incipient slaty cleavage development.

A north-verging, gently westward plunging syncline can be seen in the sandstone outcrop directly adjacent to the substation (west side of road). This fold is partly covered and poorly exposed but typical of many mesoscopic structural elements in the Stanley Shale.

Very thickly bedded sandstones crop out on the east side of the highway north of the substation. Bedding strikes east-west and dips to the north, indicating reversal of the steep southward dips more typical of this part of the Mazarn Basin. Horizontal thrust faults with very small displacements occur at the northern end of the outcrop.

Continuing the traverse, cross the trace of a thrust fault at mi 2.45 (km 3.9), just south of the old, unused landing strip (Fig. 1) as the highway curves toward the west-northwest. Quartz veins are exposed and dip variations in the sandstones can be seen in the drainage ditch.

Those planning to visit Stop 3 should note that the turn-off is the second dirt road to the left past the west end of the landing strip.

At mi 2.70 (km 4.3) Stanley sandstones and shales crop out over a distance of more than 200 ft (61 m) along the southwest side of the road. Beds are upright, dip moderately to the southwest, and are warped about a westward-plunging hinge. Quartz veins and slickensided surfaces are present at the northwest end of this series of outcrops.

Several small, north-verging, northwest plunging folds are exposed in shales, siliceous shales, and cross-laminated siltstones on the southwest side of the highway at mi 2.80 (km 4.5). This is a down-section continuation of the previous outcrop.

The trace of yet another thrust fault mapped by Haley and Stone is crossed at mi 3.50 (km. 5.6). A few quartz veins and bedding attitude variations can be found in the drainage ditches.

STOP 3. This stop presents parking problems. We suggest that groups traveling by bus or utilizing more than three automobiles or carryall-type vehicles delete STOP 3 from their itinerary and proceed to STOP 4. At mi 3.60 (km 5.8), turn left and proceed southward on the unpaved road for 0.3 mi (0.5 km) to the dead end in a small residential area. Park at the dead end, being careful that vehicles do not obstruct local residential traffic. Proceed on foot westward along unpaved lane for about 600 ft (180 m). Turn left and follow path to south to the Missouri Pacific railroad tracks, a distance of about 260 ft (80 m).

Despite minor access problems, **STOP 3** provides a much better opportunity to observe specific features of Stanley Shale rocks and to gain insight into the sedimentological character of the unit than the less extensive and more highly weathered outcrops along Arkansas 8-27.

Excellent exposures on the south-facing railway cut above the Caddo River permit examination of approximately 50 ft (15 m) of moderately (about 45°) northwestward-dipping, fresh sandstone, shale, and siltstone of the upper (?) lower Stanley Shale. The outcrop illustrates numerous features associated with flysch-like sedimentary rocks including graded bedding, flute casts, tool marks of various types, load structures, and horizontal filled burrows. Although these outcrops have not been studied in detail, preliminary data from flute cast orientations suggest a predominantly south to southwesterly paleocurrent direction.

Several normal faults with probable maximum dip separation of about 12 ft (4 m) and local drag folding cut the sequence at this location.

Return to the intersection with Arkansas 8-27 at mi 3.60 (km 5.8) and turn left (west) to continue the traverse toward Caddo Gap. *Note that the side excursion to Stop 3 has not been counted in the mileage from Glenwood.*

The route continues to cross east-west trending ridges consisting of sandstone-rich packets in the Stanley and intervening low, shale-dominant valleys incised by small tributary streams of the Caddo River. From the higher elevations along the road, there are good views of the Caddo Mountains to the north and the Cossatot Range to the south. Both of these small ranges are defined by exposures of folded and thrust-faulted Arkansas Novaculite. As the Mazarn Basin narrows westward, the Caddo and Cossatot Mountains merge into a single, complex set of west-northwest trending ridges (Haley and others, 1976), and the geographical, if not the geological, distinction between them is lost.

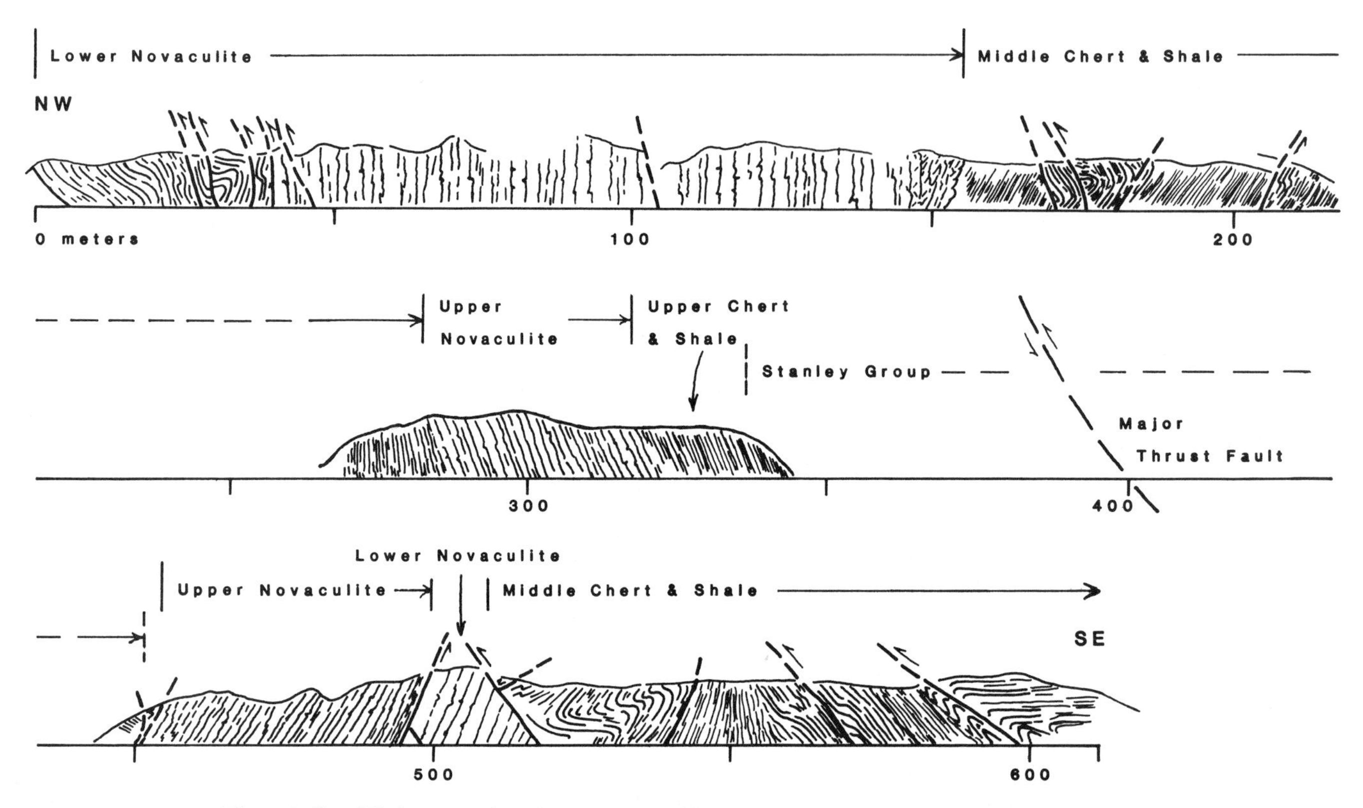

Figure 3. Simplified cross section of the sequence of Arkansas Novaculite exposures at Caddo Gap. The section was compiled previous to the start of a highway construction project in late 1985 (see text) and should be used solely as a general guide to structural style and major structural elements.

STOP 4 is on the east side of the road at mi 4.50 (km 7.2), opposite a small pond and between two farm roads that enter the highway from the east. Limited parking space is available on the shoulder of the road.

Thinly to thickly bedded sandstones and dark gray Stanley shales strike north-northeast and dip vertically over most of the outcrop. Limited geopetal indicators suggest that beds top to the north. A south-dipping normal fault offsets a thick shale unit near the southern edge of the outcrop.

Part of the hinge area of a north-verging, inclined, mesoscopic fold with well developed axial-plane cleavage is exposed at the north end of the roadcut. Cleavage in folded rocks of the Stanley Shale in the western Mazarn Basin is comparatively rare. It has been better, if only locally, developed in Ordovician through Silurian units in the Cossatot and Caddo Mountains. Well developed slaty cleavage becomes a common structural element only in the north-central and northern parts of the Benton Uplift.

The trace of a regional-scale thrust fault is crossed at mi 4.65 (km 7.5). This fault defines the northern boundary of the Hopper Belt (see Stone and Haley, 1984, Plate 2), one of the major tectonic subdivisions of the Ouachita Mountains according to the model proposed by Haley and Stone (1982). Bedding attitudes change radically in its vicinity.

At mi 4.80 (km 7.7), thinly to thickly bedded sandstones and interstratified shales crop out on the east side of the road. Beds strike approximately N60°W and dip 80°N with tops to the north. Compare these with attitudes of strata on the west side and about 137 ft (50 m) farther northward. These beds strike about N80°E, dip 60°S, and top to the south. This abrupt change in bedding attitude suggests that at this location the highway either follows a tear fault or trends across the hinge area of a large, relatively tight, north-verging, steeply inclined, southeast-plunging, mesoscopic fold.

At mi 5.20 (km 8.4), pass the entrance to the Bird and Son Granule Plant. Arkansas 240 enters from the west at mi 5.60 (km 9.0), and the southernmost outcrops of Arkansas Novaculite (middle chert and shale member) are reached at mi 5.70 (km 9.2).

STOP 5. In November 1985, a major highway construction project was begun at Caddo Gap in order to straighten several dangerous curves. Parts of the original sequence of outcrops have been altered during the course of the construction, and some details of the following descriptions may no longer apply. The fact remains that the Caddo Gap roadcuts comprise, arguably, the best and most easily accessible sections through the Arkansas Novaculite extant in the Arkansas Ouachitas. Although construction may have obliterated or defaced specific mesoscopic structural elements referred to below, it cannot remove major structures or alter definitive aspects of structural style. Because of the recent construction, we are unable to suggest specific parking areas. Formerly, a picnic ground on the east bank of the river west of the road at mi 5.85 (km 9.4) was used for this purpose. If parking in the immediate vicinity of the exposures is impossible, vehicles can be left just south of the Arkansas 240 intersection at mi 5.60 (km 9.2).

Caddo Gap (a village of the same name is located 1 mi (1.6 km) farther north) was formed by downcutting of the Caddo River through the Caddo Mountains, a macroscopically folded and thrust faulted, generally east-west trending ridge of Arkansas Novaculite. The Novaculite, together with minor amounts of the conformably underlying Missouri Mountain Formation (Silurian) and conformably overlying Stanley Shale, is exposed for about 2,000 ft (609 m) along the east side of the highway. Additional outcrops of poorer quality occur along the Missouri Pacific railroad tracks between highway and river.

The Arkansas Novaculite in this area is about 931 ft (284 m) thick, and a complete or nearly complete section can be examined along the northern part of the outcrop sequence. Miser and Purdue (1926) divided the formation into lower, middle, and upper divisions. More recently Sholes (1977) proposed a five-part division comprising (1) lower chert and shale, (2) lower novaculite, (3) middle chert and shale, (4) upper novaculite, and (5) upper chert and shale members. All except the first can be recognized easily at Caddo Gap. Owing to disagreement about the thickness of the lower chert and shale member at this location (compare Lowe, 1976, p. 2105, and Sholes, 1977, p. 140 and 142), we will refer to the oldest parts of the formation as the "lower novaculite division."

Figure 3 is a simplified, measured cross section of the outcrop sequence. Because it was drawn from a photo mosaic taken prior to the new highway construction, it should be considered no more than a general guide to the relative locations of major structures and subdivisions of the formation.

At the scale of the exposures, the structural style is dominated by open, upright to reclined mesoscopic kink folds with an average plunge of 42° to the east (Fig. 2b). Class 1C and 1B folds are most common in novaculite and chert, while class 3 folds are typical of interbedded shale (Zimmerman, 1984). These structures are best developed in the lower part of the lower novaculite division and in the middle chert and shale member. Overall parallel geometry and slickensided bed surfaces in the hinge areas of several folds suggest that the novaculite folded by flexural slip, although thickening of some hinges and limbs indicates local viscosity variation during deformation. Flow of shale into hinge areas dismembered and/or disharmonically folded thin chert layers, producing small-scale folds that were flattened by continued flexure of the larger kinks. Most folds originated as north-verging, monoclinic structures synchronous with the northward thrust transport that characterized the Ouachita orogenic belt. The entire section, however, has undergone subsequent clockwise rotation (looking east), and as a result, the longer back limbs of most large kink folds are vertical, short middle limbs are horizontal, and hinge surfaces dip to the north. Although comparatively rare, south-verging kinks are present in the sequence. While some may have formed prior to the more typical north-verging structures, others are clearly conjugate to and synchronous with the latter.

Small, bedding-parallel thrust faults, typically with no more than a few meters displacement, are common in the less thickly bedded parts of the sequence. These thrusts often parallel the vertical back limbs of kink folds, but in some cases they cut across hinge areas, obscuring fold geometry. Some reclined kinks, particularly those near the southern end of the outcrops, are spatially and perhaps genetically related to bedding plane thrust faults.

At least one thrust fault with displacement greater than 330 ft (100 m) cuts across the sequence, dividing it into a (northern) footwall block—containing an essentially complete section of Arkansas Novaculite—and a (southern) hanging wall block in which members of the formation are juxtaposed across subsidiary faults (Fig. 3).

The best way to view the formation at Caddo Gap is to begin at the north end of the outcrop and proceed southward along the highway on foot. Because of high-speed truck and automobile traffic, pedestrians should stay as far off the roadbed as possible. This traverse permits examination of the nearly unbroken section of Arkansas Novaculite, from base to top, in the footwall block of the major thrust (Fig. 3). Except where modified by faults and mesoscopic folds, beds dip steeply and top to the south. The contact between the novaculite and light-colored shales of the underlying Missouri Mountain Formation is gradational. At this location, the lower division is about 410 ft (125 m) thick and consists primarily of medium to thickly bedded, slightly silty, spiculitic, pelletal novaculite with thin beds and laminae of shale and quartz arenite (Sholes, 1977).

Large kink folds with height/width ratios of about 2.6 and maximum short-limb height of about 30 ft (9 m) are the most prominent structural elements. Antivergent folds, vertically-plunging folds, and bedding-parallel, back limb thrusts are also exposed in the lowermost 80 ft (25 m) of the division.

Bed thickness increases markedly in the upper part of the lower novaculite. Where well developed, as at Caddo Gap, such very thickly bedded, massive and heavily jointed layers are major ridge forming units in the Arkansas Ouachitas. Examples of Lowe's (1976) "nonglacial varves" can be found in this part of the section. The brecciated zone immediately beneath the contact with the middle chert and shale member occurs on a regional scale (C. G. Stone, personal communication, 1982).

An abrupt change from brecciated, massive novaculite to very thinly to thinly bedded dark chert and black to greenish shale marks the contact between the lower division and the 363 ft (111 m) thick middle chert and shale member. Broad, low, domal features on the uppermost bedding surface of the lower division rocks are of uncertain origin but may be related to differential loading soon after deposition.

Kink folds in the middle member are typically smaller than those in the lower division but have equivalent geometry. The appearances of some folds have been modified by faults that cut across hinge areas.

A dramatic change in lithology and surface appearance, together with an increase in bedding thickness mark the contact between the middle chert and shale and rocks of the upper novaculite member. The latter comprise about 115 ft (35 m) of white to gray, spiculitic and pelletal novaculite (Sholes, 1977). Its rough, "tripolitic" surface texture and the presence of numerous, small, flattened, solution (?) cavities are probably due to leaching of carbonate crystals or grains.

The absence of exposed folds in the upper novaculite may be the result of the influence of increased bedding thickness on dominant wavelength. Spacing between hinge areas increases with bed thickness and may, in this case, exceed the dimensions of the outcrop.

The upper novaculite is abruptly succeeded by approximately 35 ft (11 m) of greenish-gray to tan, thinly interbedded, upper chert and shale that, in turn, grade up-section (southward) into relatively soft, typically olive-green rocks of the lower Stanley Shale.

A gap in the outcrop sequence opposite the former low-water bridge across the Caddo River marks the trace of the major thrust (Figs. 1 and 3) that carried an internally faulted, southern (hanging wall) block northward over the footwall block just traversed. The present attitude of the fault plane is probably steep but has not been verified by direct observation. Scattered exposures of lower Stanley rocks are present in this interval.

The hanging wall block of the major thrust has been cut by a number of faults that bring slices of various members of the Arkansas Novaculite into juxtaposition out of their normal depositional sequence. From the base of the block southward, the rocks encountered consist of elements of the lower Stanley Shale, middle chert and shale, upper novaculite, upper lower novaculite, and middle chert and shale, in that order (Fig. 3). There is either direct or indirect evidence of faulting (slickensides, stratigraphic juxtaposition, and/or abnormal dimensions of units) along all of these contacts.

The last well-exposed rocks of the hanging wall block (part of the middle chert and shale member) contain a number of large and small, moderately eastward-plunging, open kinks that are similar to folds in the footwall block in both geometry and spatial orientation. This geometry is maintained but the degree of plunge increases to nearly vertical in a series of reclined kink folds near the southern end of the outcrop. At least two reclined folds directly overlie bedding-parallel, back-limb thrusts, and it is possible that their plunges were substantially increased by rotation along these faults at a relatively late stage in the sequence of deformative events.

A south-dipping thrust fault separates the series of reclined kinks from gently warped, essentially horizontal rocks that mark the final exposures at Caddo Gap. From the last outcrops of the middle chert and shale member to the intersection of Arkansas 8-27 and 240, float from the upper novaculite member and scattered outcrops of Stanley shales can be found along the roadside.

REFERENCES CITED

Gordon, M., Jr., and Stone, C. G., 1977, Correlation of the Carboniferous rocks of the Ouachita trough with those of the adjacent foreland, *in* Stone, C. G., ed., Symposium on the geology of the Ouachita Mountains, v. 1: Little Rock, Arkansas Geological Commission, p. 70–91.

Haley, B. R., and Stone, C. G., 1982, Structural framework of the Ouachita Mountains, Arkansas: Geological Society of America Abstracts with Programs, v. 14, no. 3, p. 113.

Haley, B. R., Glick, E. E., Bush, W. V., Clardy, B. F., Stone, C. G., and others, 1976, Geologic Map of Arkansas: Little Rock, Arkansas Geological Commission and U.S. Geological Survey, scale, 1:500,000.

Lowe, D. R., 1976, Nonglacial varves in Lower Member of Arkansas Novaculite (Devonian), Arkansas and Oklahoma: American Association of Petroleum Geologists Bulletin, v. 60, no. 12, p. 2103–2116.

Miser, H. D., and Purdue, A. H., 1929, Geology of the De Queen and Caddo Gap Quadrangles, Arkansas: U.S. Geological Survey Bulletin 808, 195 p.

Sholes, M. A., 1977, Arkansas Novaculite stratigraphy, *in* Stone, C. G., ed., Symposium on the geology of the Ouachita Mountains, v. 1: Little Rock, Arkansas Geological Commission, p. 139–145.

Stone, C. G., and Haley, B. R., 1984, A guidebook to the geology of the central and southern Ouachita Mountains, Arkansas: Little Rock, Arkansas Geological Commission Guidebook 84-2, 131 p.

Stone, C. G., and McFarland, J. D., III, 1981, Field guide to the Paleozoic rocks of the Ouachita Mountain and Arkansas Valley Provinces, Arkansas: Little Rock, Arkansas Geological Commission Guidebook 81-1, 140 p.

Zimmerman, J., 1984, Geometry and origin of folds and faults in the Arkansas Novaculite at Caddo Gap, *in* Stone, C. G., and Haley, B. R., A guidebook to the geology of the central and southern Ouachita Mountains, Arkansas: Little Rock, Arkansas Geological Commission Guidebook 84-2, p. 111–115.

Zimmerman, J., Roeder, D. H., Morris, R. C., and Evansin, D. P., 1982, Geologic section across the Ouachita Mountains, western Arkansas: Geological Society of America Map and Chart Series, MC 28-Q, scale, 1:250,000.

Zimmerman, J., Sverdrup, K. A., Evansin, D. P., and Ragan, V. S., 1984, Reconnaissance structural geology in the western Mazarn Basin, southern Benton uplift, Arkansas, *in* McFarland, J. D., III, and Bush, W. V., eds., Contributions to the geology of Arkansas, v. II: Little Rock, Arkansas Geological Commission Miscellaneous Publication 18-B, p. 161–177.

ACKNOWLEDGMENTS

The authors gratefully acknowledge the Arkansas Geological Commission, Southern Illinois University at Carbondale, and Sohio Petroleum Company for support during various phases of our work. Discussions both on and off the outcrop with D. P. Evansin, B. R. Haley, J. R. Jennings, C. G. Stone, J. E. Utgaard, and G. W. Viele have improved our understanding of the geology and the quality of this manuscript, but we accept full responsibility for any errors in fact or interpretation.

Turbidite exposures near DeGray Lake, southwestern Arkansas

J. D. McFarland III, Arkansas Geological Commission, 3815 W. Roosevelt Road, Little Rock, Arkansas 72204

LOCATION

DeGray Lake may be reached by taking Exit 78 from I-30 just north of Arkadelphia, Arkansas, and proceeding north on Arkansas 7. The access road to the dam area is approximately 3 mi (4.8 km) from the exit.

SIGNIFICANCE

Carboniferous rocks form most of the exposures along the Athens Plateau of the southern Ouachita Mountains. In the DeGray Lake area there are several sequences of Stanley Shale (mostly Chesterian), Jackfork Sandstone (mostly Morrowan), and Johns Valley Shale (Morrowan) that illustrate turbidite deposition, with source areas often indicated to the south and/or east. Two of the more accessible sites have been chosen here for a brief discussion of the environments of deposition (modified from Stone and McFarland, 1981).

SITE DESCRIPTIONS

Stop 1. (Fig. 1) Sequence of middle and upper Jackfork Sandstone at DeGray Lake spillway. Park at the DeGray Lake access area about 0.5 mi (0.8 km) east of the visitor's center near the dam; walk to the east and then to the south along the spillway cut. The sequence of Jackfork Sandstone exposed here (Fig. 2) offers the student a rare opportunity to examine a wide variety of flysch and submarine fan deposition in a limited area. The Jackfork is about 6,000 ft (1,800 m) thick in this area, but only 1,000 ft (300 m) of it are exposed in this cut.

The rock exposed here is a sequence of southward-dipping, fine-grained, quartzitic sandstone, subgraywacke, gray siltstone, and black shale in the middle and upper Jackfork Sandstone. The sandstones contain Bouma sequences, graded bedding, load structures, dish and pillar structures, bottom marks, ripple marks, broad scours, clay balls, and other features. Slurry and slump intervals of probable intraformational origin are present, and one unusual zone contains small sandstone cobbles, chert pebbles, iron carbonate concretions, clay balls, and other lithologies in a shale matrix. Sandstone olistoliths ("glumps and gloops" in the local terminology) occur in a few intervals, especially in the shale sequences below the upper more massive sandstone at the south end of the spillway. Coalified plant fragments are quite common in some of the debris-flow deposits represented by the siltstone "blue beds." A few invertebrate fossil remains can be found at various places along the exposure.

At the south end of the spillway (Fig. 3) there are at least 15 cycles of granule or "grit" deposition in a 200-ft (60-m) sequence. These "grit" beds are commonly graded and have some small to rather large scour features. The granules are composed of quartz and metaquartzite.

The north end of the spillway contains mostly upward-thinning and upward-fining sandstone sequences and are thought to represent middle submarine fan channel deposits. Near the middle of the spillway there are several thickening and upward-coarsening fan lobe sequences. At least one upward-thickening sequence proceeds into an upward-thinning dying lobe sequence. These intervals in the middle of the spillway represent a thick regressive or prograding section of outer fan lobe or possible midfan interchannel deposits. The exposures at the south end of the spillway represent upward-thinning, high-energy, probable middle fan channels. It is thought that these deposits were supplied from deltas and through submarine canyons down the slopes from two major areas: (1) to the northeast (Illinois basin); and, (2) to the east (southern Appalachians).

Some small tear faults cut the rocks and contain milky quartz, dickite, and traces of cinnabar. In this southern Ouachita Mountain area, pervasive shearing, cleavage, boudinage, and other structural deformation features are absent. Viele (1973) notes that along the northern margin of the Ouachita Mountains there is evidence for northern tectonic transport. Here on the south side there is evidence for the same direction of movement. However, in the lower Paleozoics, we have abundant evidence of southward movement. This is but one of the many Ouachita problems.

Studies by R. C. Morris (1971) indicated that the Athens Plateau Jackfork section is about 70 percent wackes and arenites and that the sandstones have a mean grain size of 2.8 phi (sd = 1.0 phi). He found an average of 86 percent framework grains (dominantly siliceous resistates but with minor feldspars and unstable lithic fragments). In addition, he noted carbonaceous plant remains, crinoid columnals, shale chips, and small pockets of ferric clay residue. Further studies by Morris (1974) showed that the turbidity currents that deposited the Jackfork in this area were flowing in a westward direction down the axis of an elongate marine trough.

Stop 2. Johns Valley Shale north of Caddo Valley. This exposure is on a minor side road approximately 1 mi (1.6 km) south of the intersection of the access road to DeGray Lake and Arkansas 7. The outcrop is about 300 ft (90 m) southeast of Arkansas 7 and is in southward-dipping lower Johns Valley Shale. The north end of the roadcut is a few hundred feet from the apparent contact with the underlying Jackfork Sandstone, but there is a thrust fault between the two units. At the north end of the exposure the rocks consist of several thinning and fining-upward sequences of brown, silty, micaceous subgraywacke with

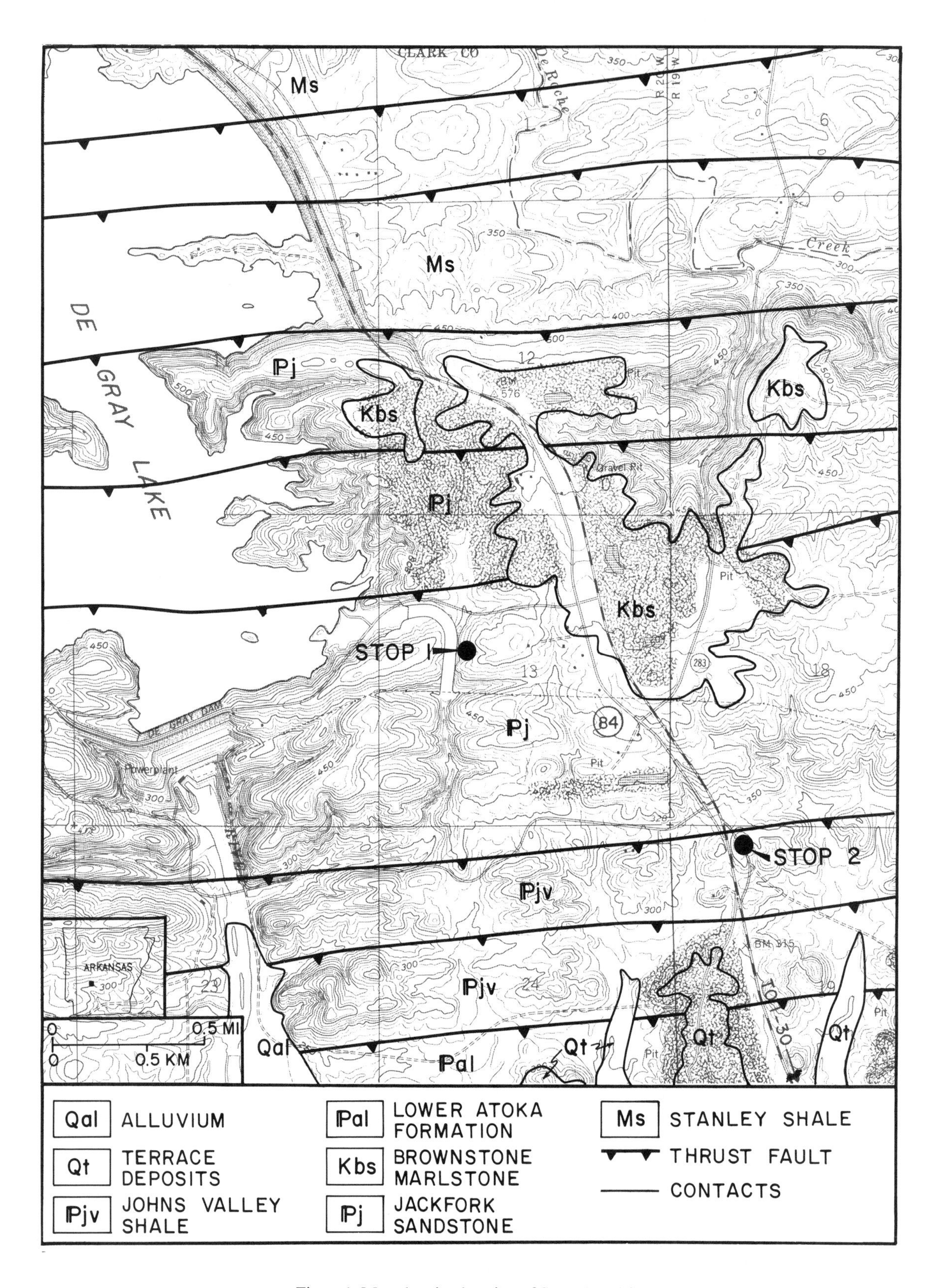

Figure 1. Map showing location of Stops 1 and 2.

Figure 2. Upper Jackfork Sandstone at the south end of DeGray Lake spillway. This section is a massive quartzitic sandstone sequence deposited in channels by proximal submarine fan processes.

some thin beds of siltstone and silty gray shale. Near the middle of the roadcut there are contorted and rubbly sandstone sequences, some with sandstone, chert, and other exotic clasts that, for the most part, represent sedimentary slumps. The slumping of these rocks is believed to be from the south to the north. There are also some later tectonic faults dissecting these slumped beds that have slickensides and dickite. At the south end of the exposure there are many thin-bedded intervals of sandstone and shale. Bottom marks, Bouma sequences, graded bedding, ball and pillow structures, and other features are common. Trace fossils are present and appear to be of deep-water type.

REFERENCES CITED

Morris, R. C., 1971, Stratigraphy and sedimentology of the Jackfork Group, Arkansas: American Association of Petroleum Geologists Bulletin, v. 55, p. 387–402.

——, 1974, Carboniferous rocks of the Ouachita Mountains of Arkansas; A study of the facies patterns along the slope and axis of a flysch trough, *in* Briggs, G., ed., Carboniferous of the southeastern United States: Geological Society of America Special Paper 148, p. 241–242.

Stone, C. G., and McFarland, J. D., III, 1981, Field guide to the Paleozoic rocks of the Ouachita Mountain and Arkansas valley provinces, Arkansas: Arkansas Geological Commission Guidebook 81-1, 140 p.

Viele, G. W., 1973, Structure and tectonic history of the Ouachita Mountains, Arkansas, *in* DeJong, K., and Scholten, R., eds., Gravity and tectonics: New York, Wiley and Sons, p. 361–377.

Figure 3. Upper Jackfork Sandstone at the south end of DeGray Lake spillway. A high energy, graded granule "grit" sandstone interval of an individual submarine fan channel.

The upper Jackfork Section, Mile Post 81, I-30, Arkadelphia, Arkansas

Denise M. Stone, Amoco Production Company, P.O. Box 3092, Houston, Texas 77253
David N. Lumsden, Department of Geological Sciences, Memphis State University, Memphis, Tennessee 38152
Charles G. Stone, Arkansas Geological Commission, 3815 West Roosevelt Road, Little Rock, Arkansas 72204

LOCATION

A readily accessible 440-ft (132 m) section of the upper Jackfork Sandstone crops out along I-30 near Arkadelphia, Arkansas. The outcrop is on the west side of I-30 between Exits 83 (Friendship) and 78 (Arkadelphia, Hot Springs) at the Mile Post 81 marker (Fig. 1).

SIGNIFICANCE

This section is an excellent example of a Ouachita facies turbidite sequence. The following text presents a measured stratigraphic section of this exposure and an interpretation of its depositional environment (Stone and others, 1982). The section has been subdivided on the megascopic appearance of rock type, grain size, sedimentary structures, vertical distribution of bed thicknesses, and cyclicity. Based on these sedimentary attributes, subdivisions are then assigned to an appropriate submarine fan subenvironment.

The Jackfork Sandstone is Pennsylvanian (Morrowan) in age. It overlies the Mississippian-age Stanley Shale (mostly Chesterian) and is overlain by the Pennsylvanian-age Johns Valley Shale (Morrowan). The Jackfork forms a portion of a thick sequence of Carboniferous-age clastic sediments that was deposited as a shelf-to-slope complex in the Ouachita Trough and environs. Morris (1971, 1974), Thomson and LeBlanc (1975), Stone and others (1982), and many others have established that the Ouachita facies represent deep marine continental slope rise sediments deposited largely by submarine fan processes.

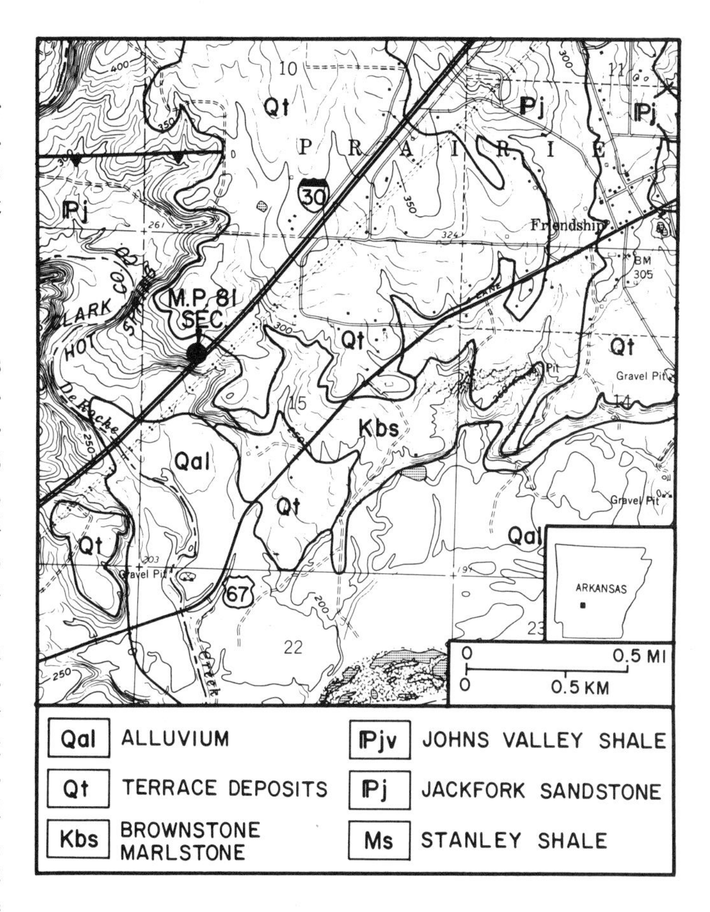

Figure 1. Map showing location of upper Jackfork sandstone outcrops at Mile Post 81, I-30, near Arkadelphia, Arkansas (Modified from Stone and McFarland, 1981, plate 45).

GEOLOGY OF THE SECTION

The Mile Post 81 section has a true vertical thickness of 440 ft (133 m). It dips 43° to the south and strikes N78°E. The entire section is in the upper Jackfork Sandstone, although neither top nor bottom formational contacts are present in this exposure. Boyd Haley (personal communication, 1981) indicated that this section lies stratigraphically 700 ft (212 m) above the DeGray Dam Spillway section located 10 mi (16 km) to the west and 500 to 600 ft (152 m to 182 m) below the Johns Valley Shale.

A measured columnar section of the Mile Post 81 outcrop is presented in Figure 2. Two lithologies dominate the section: dark gray to blue-gray shales and buff to light brown sandstones. Measurement started at the northeastern end of the outcrop at the base of a massive bed of conglomeratic sandstone that contains abundant shale clasts toward the top. This begins what has been interpreted to be nine submarine fan channels and one depositional lobe. Each fan channel is composed of at least two of the following stratigraphic subdivisions: active channel fill, partially abandoned channel fill, and abandoned channel fill. The top of the measured section, to the southwest, is marked by prominent reddish breccia along a fault plane. Approximately 50 ft (15 m) of somewhat distorted sandstones and shales are present above this fault. Two intermittent springs are also present, one at 130 ft (40 m) and a second at 315 ft (96 m) above the base. Six reverse faults are observed but appear to be of local extent. Two prominent joint sets perpendicular to bedding form a diamond pattern on exposed bedding surfaces.

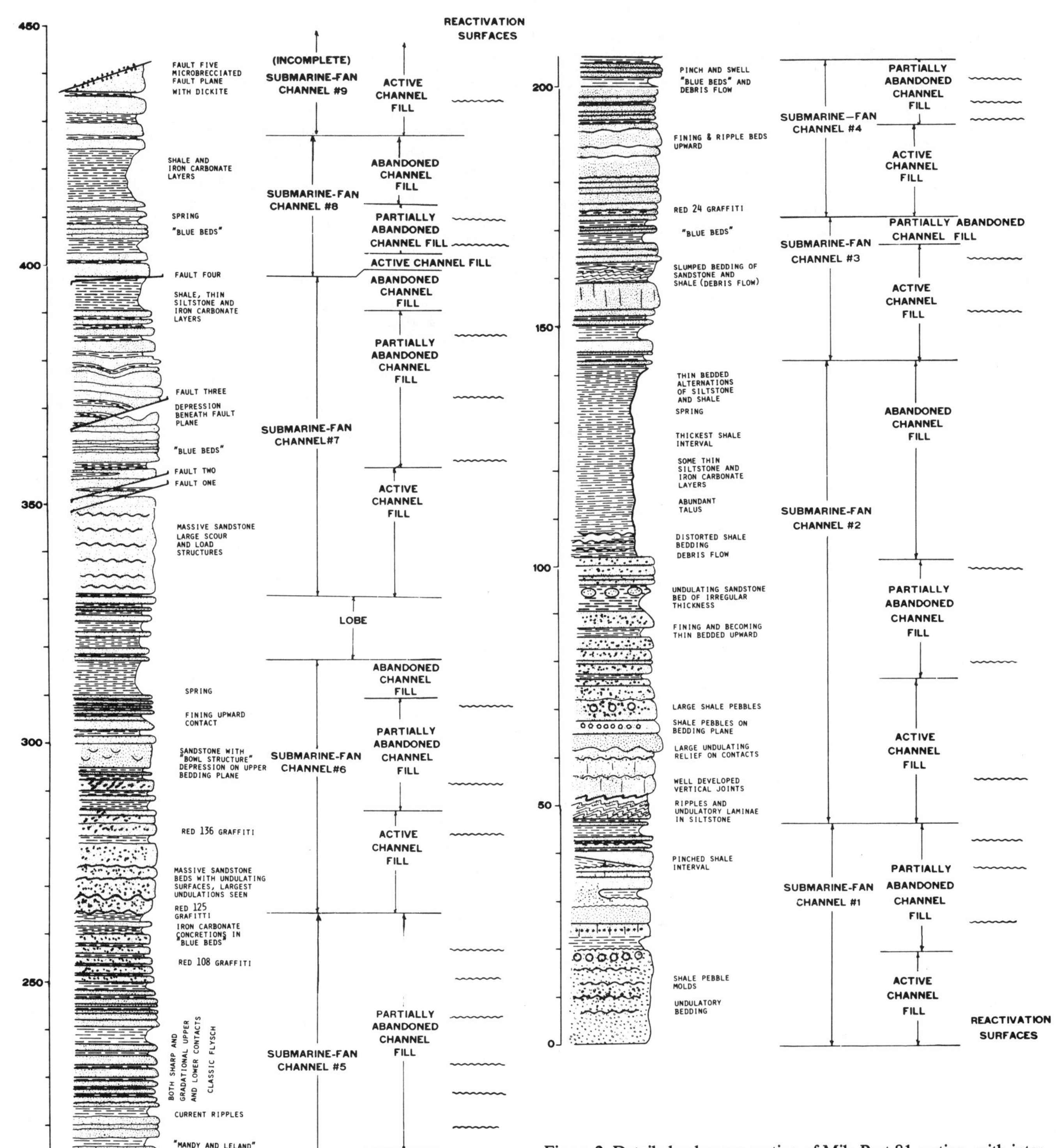

Figure 2. Detailed columnar section of Mile Post 81 section, with interpretation of submarine-fan depositional environments. Labeled subsequences represent subjective interpretations of pattern of vertical variation in sandstone/shale ratio, and are intended as guides to interpretation, not as formal unit designations.

RECOGNITION OF ENVIRONMENTS

The character and vertical variation of lithologies at Mile Post 81 is readily interpretable using Walker's (1978) model. This model combines the pioneer work of Bouma (1962) with recent stratigraphic studies of Mutti (1974) and Ricci Lucchi (1975) and the work on modern depositional systems such as that of Normark (1978). Steeply sloping submarine canyons are incised into the slope shelf adjacent to the area of submarine fan deposition, and are floored with slumped debris flow deposits. Sediment-charged currents that spasmodically surge out of these canyons initially erode the underlying soft sediments; then, as their energy wanes, they deposit a sequence of sediments whose grain size and sedimentary structures depend on where in the fan the sediments ultimately come to rest. The lateral pattern of sediments migrates with time, building the fan upward and forming a vertical succession of facies that can be used to interpret which portion of the fan occupied the outcrop location at a particular time.

Each submarine fan channel may be generally described as a fining-upward sequence of interbedded sandstone and shale. Active channel fill that occupies the base of each fan channel is characterized by thick, massive sandstone units containing sedimentary features indicative of high-energy turbidity flow. From the base toward the top of each fan channel, the massive sandstone units thin and become interbedded with siltstone and shale. These interbedded deposits, the accumulation of partially abandoned channel fill, reflect a decrease in the sand-carrying capacity of the currents and an increase in the amount of silt and clay material supplied to the environment. Finally, the uppermost portion of the fan channel is the abandoned channel fill. It is recognized by thick accumulations of fissile shale with thin siltstone and iron carbonate layers. Thick turbidite sandstones are virtually absent in this portion of the fan.

Sandstones of the active and partially abandoned channel fill are medium- to fine-grained, moderately sorted sublitharenites. They form approximately 58 percent of the Mile Post 81 section. Within the active channel fill, individual sandstone beds with sharp upper and lower contacts are commonly stacked in thicknesses of 20 to 30 ft (6 to 9 m) (Fig. 2). Bedding is typically planar but may have considerable relief. This relief is most likely due to scour, indicating that each sand bed was deposited by a single advancing current. Irregular ripples and large enigmatic "bowl structures" are also present.

High-energy erosional features at lithologic contacts and/or bedding planes are labeled reactivation surfaces (Fig. 2). These surfaces are evidence of a sudden change in the erosional and/or depositional power of the turbidity current. Deposition was in a sense "reactivated" from nonexistence or a slower rate by the introduction of one advancing current. This sudden surge of suspended sediment eroded a fine undisturbed substrate, leaving scour marks to be quickly filled by new material. The lateral extent of each reactivation surface, although unseen, directly corresponds to the lateral extent of each turbidity current.

Although shale beds are rather sparse in accumulations of active channel fill, some sandstone beds do contain large shale clasts. These are clasts of partially consolidated shale substrate that were ripped up, transported, and eventually redeposited within the body of the current transported sediment. During transport they acquired their roundness and present shape. At the Mile Post 81 exposure, many shale clasts themselves have been removed by weathering and erosion, and only molds remain. Typically they are located toward the top of sandstone beds and are up to 6 in (15 cm) in diameter.

Also occurring above some of these sandstones and grading upward into the shales are rather numerous intervals of chaotic sandstone and shale, often with carbonized plant fragments that are considered to be of debris flow origin. These beds are called "blue beds" because of their characteristic fresh color, although upon weathering they become light brown.

The number and thickness of shale beds noticeably increases in the upper portion of each submarine fan channel. These fine-grained quartz-rich shales form 42 percent of the measured section. They can be broadly divided into two groups: thin-bedded shales and thick intervals of fissile shales. Thin-bedded shales occur as the upper portions of graded beds; more specifically, dark gray beds 1 to 2 ft (30 to 61 cm) thick fine upward from sandstone to silty shale to thin-bedded shale. Deposits of partially abandoned channel fill are identified by these graded beds. Some exhibit partial Bouma sequences (Bouma, 1962). The base of the sandstones are typically in sharp contact with the subjacent shales and often exhibit sole marks. These marks are much smaller in size, however, than those of the active channel. Cross-laminations are also present but diminish as each bed fines upward. Current and interference ripple marks may be seen where bedding planes of silty shale are exposed.

The dark gray fissile shales are found in continuous thicknesses of 10 ft (3 m) or more. These shales are easily recognized because they weather to splinters that form large talus accumulations at the base of each exposure. Interbeds of slightly more resistant siltstone and siderite are common. A low-energy marine environment, relatively out of the path of turbidity current flow, is believed to be the site of deposition for these thick shales. Very slow hemipelagic sedimentation over a long period was responsible for much of their accumulation.

Between 320 and 330 ft (97 and 100 m), an interval of coarsening upward sediment is present. Thick, dark shales are interrupted by several thickening upward sand layers. We have assumed this thickening pattern is that of an anomalous depositional lobe.

SUMMARY

The 440 ft of upper Jackfork Sandstone exposed at Mile Post 81 has been interpreted as nine submarine fan channels and one depositional lobe. It represents an accumulation of deep marine continental slope/rise sediments deposited in Pennsylvanian time as part of the Ouachita facies. The channel deposits are characterized as fining upward sequences of interbedded sandstone and shale. Each has been subdivided vertically as active channel fill, partially abandoned channel fill, and abandoned channel fill, depending on depositional energy.

REFERENCES CITED

Bouma, A. H., 1962, Sedimentology of some flysch deposits: New York, Elsevier Publishing Company, 168 p.

Morris, R. C., 1971, Stratigraphy and sedimentology of the Jackfork Group, Arkansas: American Association of Petroleum Geologists Bulletin, v. 55, p. 387–402.

——, 1974, Carboniferous rocks of the Ouachita Mountains, Arkansas; A study of facies patterns along the unstable slope and axis of a flysch trough, *in* Briggs, G., ed., Symposium on the Carboniferous rocks of the southeastern United States: Geological Society of America Special Paper 148, p. 241–279.

Mutti, E., 1974, Examples of ancient deep-sea fan deposits from Circum-Mediterannean geosynclines, *in* Modern and ancient geosynclinal sedimentation: Society of Economic Paleontologists and Mineralogists Special Publication 19, p. 92–105.

Normark, W. R., 1978, Fan valley channels and depositional lobes in modern submarine fans; Characters for recognition of sandy turbidite environments: American Association of Petroleum Geologists Bulletin, v. 62, p. 912–931.

Ricci Lucchi, F., 1975, Depositional cycles in two turbidite formations of northern Appennines (Italy): Journal of Sedimentary Petrology, v. 45, p. 3–43.

Stone, C. G., and McFarland, J. D., III, 1981, Field guide to the Paleozoic rocks of the Ouachita Mountains and Arkansas Valley Provinces, Arkansas: Arkansas Geological Commission Guidebook 81-1, 140 p.

Stone, D. M., Lumsden, D. N., and Stone, C. G., 1982, The upper Jackfork section, mile post 81, I-30, Arkadelphia, Arkansas, I: Arkansas Geological Commission Miscellaneous Publication 18 and 18B, p. 147–160.

Thomson, A., and LeBlanc, R. J., 1975, Carboniferous deep-sea fan facies of Arkansas and Oklahoma: Geological Society of America Abstracts with Programs, v. 7, p. 1298–1299.

Walker, R. G., 1978, Deep-water sandstone facies and ancient submarine fans: Models for exploration for stratigraphic traps: American Association of Petroleum Geologists Bulletin, v. 62, p. 932–966.

The DeQueen Formation at the old Highland Quarry, Arkansas

J. D. McFarland III, Arkansas Geological Commission, 3815 W. Roosevelt Road, Little Rock, Arkansas 72204

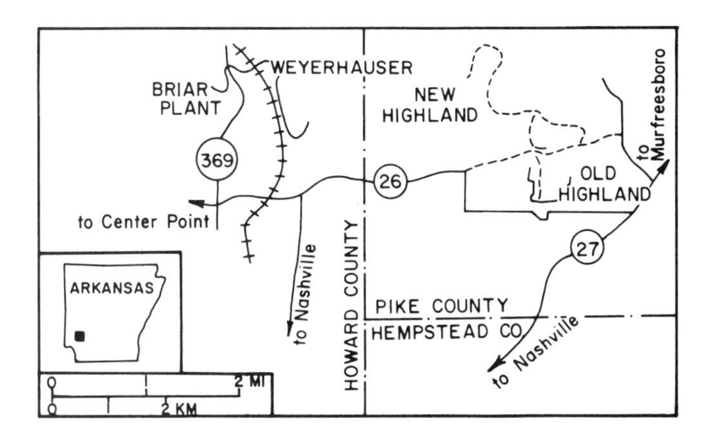

Figure 1. Map showing location of old Highland Quarry.

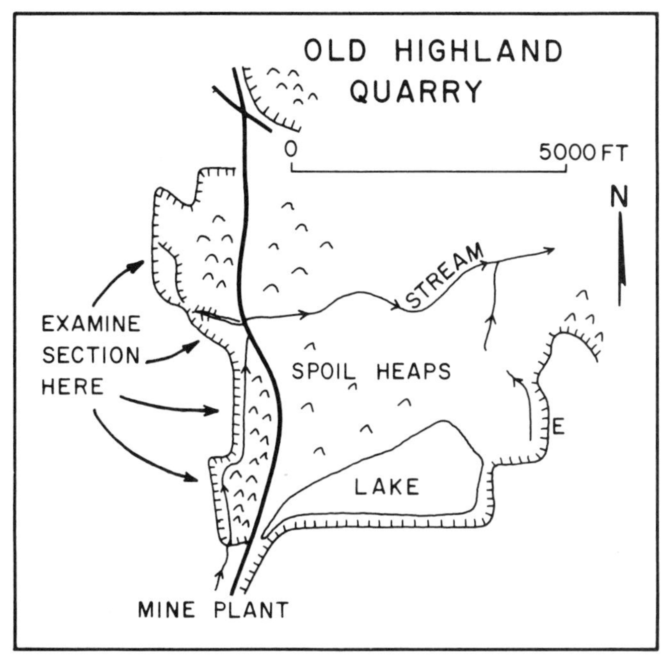

Figure 2. Old Highland Quarry. You will enter from the south. The stratigraphic section (Fig. 3) is generalized from the west wall of this quarry (after Lock and others, 1984).

LOCATION

To reach the Highland Quarry, start at the junction of Arkansas 24 and 27 at the north edge of Nashville, Arkansas. Drive north on Arkansas 27 (toward Murfreesboro) for 8.6 mi (13.9 km) and turn left (west) on Arkansas 26 (Fig. 1). The gravels at this junction are units of the Tokio Formation (Upper Cretaceous). Drive 1.6 mi (2.6 km) to the west on Arkansas 26 to the entrance of the quarry. Enter the quarry and ask permission at the mine office. The most convenient exposures are on the west wall of the quarry (Figs. 2, 3). This end of the quarry has been inactive for several years, but if you follow the road through the quarry for 2 mi (3.2 km) or so you will reach more recently developed pits.

SIGNIFICANCE

The DeQueen Formation is one of more economically important formations of the Lower Cretaceous Trinity Group in Arkansas. It outcrops in a narrow band across southwest Arkansas, generally through Pike, Howard, and Sevier counties. It is composed of limestones, gypsum, mudstones, and minor sands. The gypsum is mined by the Weyerhaeuser Company at Briar, Arkansas, less than 5 mi (8 km) to the west of the Highland pit. The Highland Quarry was mined commercially for gypsum in the past and currently offers the most accessible exposures of the unit.

DISCUSSION

The DeQueen Formation was named by Miser and Purdue (1918) for exposures near DeQueen, Arkansas. Recently, Lock and others (1983) have described in greater detail the sedimentology and paleontology of the DeQueen Formation in this area. The following is the abstract of that paper.

"The mixed evaporite/carbonate/terrigenous clastic sediments of the De Queen Formation, in southwestern Arkansas, were deposited at the landward margin of a broad shallow lagoon formed behind the Glen Rose reef. About sixty percent of the sedimentary volume consists of mudstone, silt and sand with brackish water to hypersaline ostracod faunas, believed to result from influx of flood waters from the Ouachita highlands located a few miles to the north. The lower part of the formation contains discontinuous beds of gypsum, ranging in thickness from a few centimeters to composite beds of over three meters, and displaying mosaic structure with vertically oriented, elongate nodules. These beds, which are lenticular, are interpreted to result from subaqueous precipitation of vertical selenite crystals (subsequently recrystallized) in discrete ponds and pools on microtidal-range mud flats. Intrastratal growth of gypsum nodules and displacive halite occurred at the margins of the pools.

"The upper part of the formation contains no gypsum beds, but halite pseudomorphs at the base of and within some of the thin limestones suggest the presence of supratidal brine pools. Several minor unconformities exist, of which one has a regional extent and is underlain by red-brown mudstones. Algal-mat lamination, lenticular gypsum pseudomorphs (an intrastratal growth form), and syneresis cracks occur in the limestones. A supratidal environment is envisaged from a significant proportion of the time of deposition.

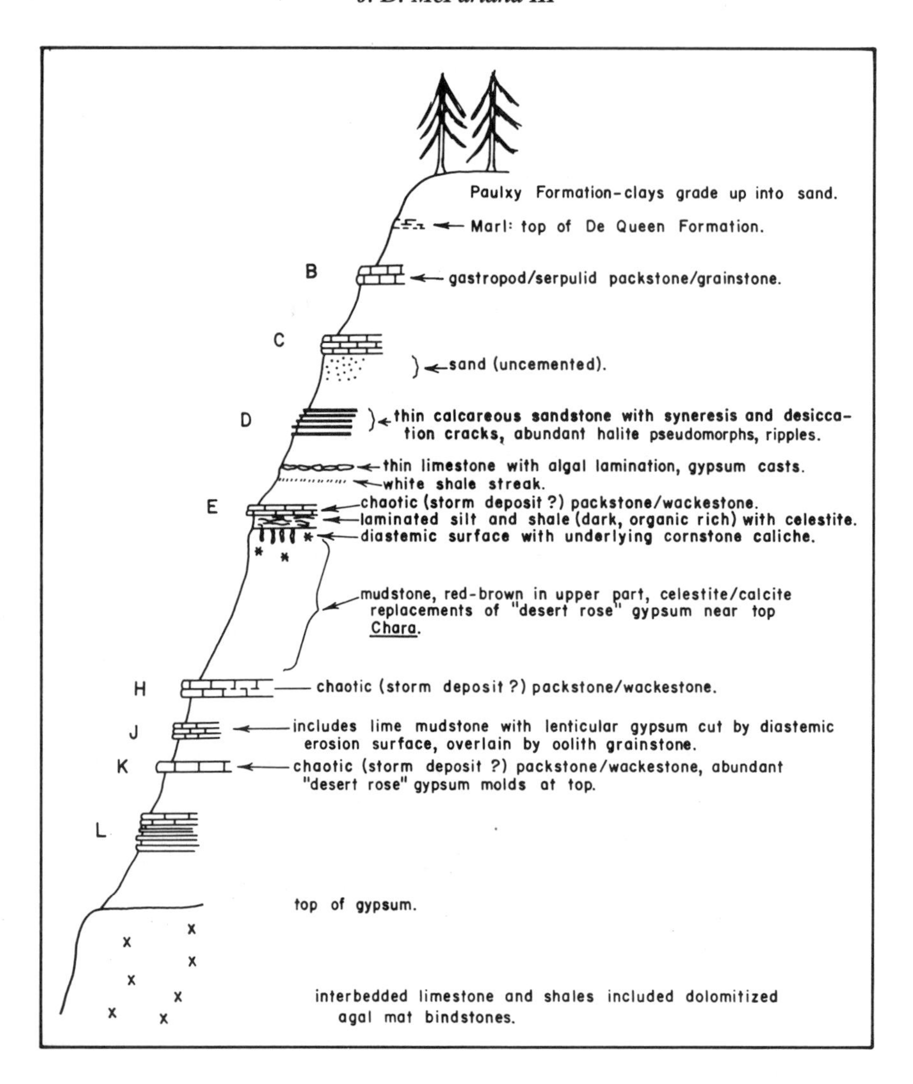

Figure 3. Generalized sketch of stratigraphic section exposed in west wall of old Highland Quarry, not to scale (after Lock and others, 1984).

"The limestones generally have a restricted fauna of ostracods, bivalves, cerithid gastropods, serpulid worms, and miliolid foraminifera, and range in texture from lime mudstones to grainstones. The most abundant grain types are pellets, superficial ooliths, and terrigenous quartz. A paucity of dolomite is a striking feature. Some of the thinner bedded units are rippled, and some ripples were truncated during periods of emergence. The limestones are believed to represent periods of shallow water, slightly hypersaline to slightly hyposaline conditions of variable energy.

"The regressive trend displayed by these two divisions continued with deposition of the overlying formation. The uppermost three meters of the De Queen consist of mudstone with a thin marl at the top. A conformable contact exists with the overlying Antlers Formation, which has basal mudstones becoming more silt- and sand-rich upwards and finally giving way to the typical Antlers (Paluxy) sands" (Lock and others, 1983, p. 145).

In Weyerhaeuser's gypsum mines at Briar, Arkansas, in the same sequence of rocks, Jeffrey Pittman discovered an extensive exposure of dinosaur footprints (Pittman and Gillett, 1984). The prints, numbering in the thousands, are from at least two kinds of large sauropods. Some of the prints are 2 ft (0.61 m) across, and several sets were made by the active stride of individual animals.

REFERENCES CITED

Lock, B. E., Darling, B. K., and Rex, I. D., 1983, Marginal marine evaporites, Lower Cretaceous of Arkansas: Gulf Coast Association of Geological Societies Transactions, v. 33, p. 145–152.

Lock, B. E., and others, 1984, Cretaceous rocks of southwest Arkansas: Gulf Coast Association of Geological Societies Transactions, Guidebook, 26 p.

Miser, H. D., and Purdue, A. H., 1918, Gravel deposits of the Caddo Gap and DeQueen quadrangles, Arkansas: U.S. Geological Survey Bulletin 690-B, p. 15–29.

Pittman, J. G., and Gillett, D. D., 1984, Tracking the Arkansas dinosaurs: The Arkansas Naturalist, v. 2, no. 3, p. 1–12.

The Claiborne Group in southwest Arkansas

William Lee Prior and Quin Baber, Arkansas Geological Commission, Little Rock, Arkansas 72204

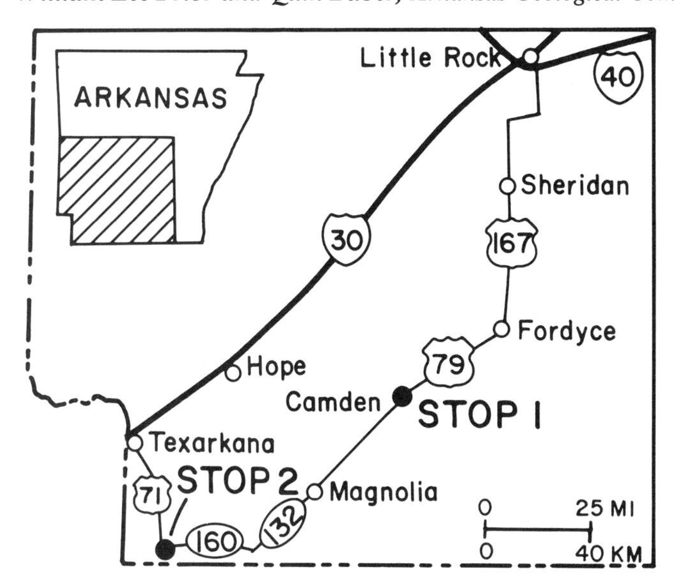

Figure 1. Index map of southwestern Arkansas showing location of Stops 1 and 2.

TERTIARY DEPOSITIONAL ENVIRONMENTS OF ARKANSAS LIGNITE

Although the middle Eocene age Claiborne Group of southwest Arkansas covers a large area, outcrops are sparse. The easily erodable Claiborne is composed of unconsolidated sand, silt, clay, and lignite. It has a maximum thickness of 1,500 ft (457 m) in Arkansas, generally strikes to the northeast, and dips an average 60 ft to the mi (11 m to the km) to the southeast. The exposures at Camden and Red River in southwestern Arkansas (Fig. 1) display features of a deltaic plain environment. The two exposures are 75 mi (120 km) apart and are probably of different parts of the Claiborne section. Currently, the Claiborne Group is interpreted as having accumulated in fluvial-deltaic system. Studies by Bernard and others (1970) have established a classification scheme of sedimentary environments within deltas. Texture, fabric, sedimentary features, geometry, and trends of sediments associated with deltas are controlled by geologic processes and the availability and influx of different sediment types. Based on these criteria, delta depositional environments have been divided into (1) delta plain with natural levees, distributary channels, interdistributary marshs, and flood basins; (2) inner fringe with river-mouth bars; (3) outer fringe; (4) prodelta; and (5) open marine (Figs. 2 and 3). These environments are discussed within the context of high sediment influx, low energy type of coastal delta (birdfoot, lobate) interpretation of the Claiborne Depositional system, as opposed to low sediment, high coastal energy deltas (arcuate, estuarine).

The delta plain is the principal subaerial part of a delta and includes distributary and tidal channels, natural levees, and interdistributary areas or flood basins (swamps, marshes, lakes, bays, and gathering streams) (LeBlanc, 1975). Distributary channels are irregular, divergent streams flowing away from the main stream and not returning to it. As these channels diverge and migrate, old channels become abandoned and are filled in with fine-grained sediment and organic material, generally as fining upward sequences (Fig. 3). The typical distributary channel sequence grades upward from a fine- to medium-grained, usually cross-bedded sand into locally lignitic and commonly laminated silt and clay. Rapidly abandoned channels such as oxbow lakes are mainly filled with clay and silt with very little sand; these fine-grained channel fillings are called "clay plugs." Distributary channel deposits in map view are relatively narrow, arcuate, sinuous, or, rarely, straight.

Bordering the distributary channels are natural levees, which are very low, subdued ridges. The slope toward the channel is very steep and the slope away from the river is very gentle. The deposits are of sand, silt, and clay, laminated and cross-bedded on a small scale, mottled and disturbed by weathering, roots, and animal burrows.

Interdistributary areas of flood basins are low, flat, relatively featureless, and poorly drained. They lie adjacent to and between the slightly higher distributary channels or natural levees and are subject to flooding during over-bank flood stages. Flood-basin sediments consist of fine material, largely clay, silt, and very fine-grained sand (Bernard and others, 1970). Marshes and swamps are indicated by a predominantly clay sequence with a high organic content and commonly containing calcareous and ferruginous nodules. Also present between the distributary channels are lakes and bays that contain fine-grained deposits or ripple-laminated sand, silt, and clay and may contain freshwater, brackish, or marine faunas and floras. Restricted lakes or bays are commonly algal-laminated and may contain calcareous nodules and gypsum crystals. Flood basins of abandoned deltas subside because of compaction and permit the accumulation of thick sections of peat and organic-rich silt and clay. Further subsidence may lead to deposition of lagoonal or marine sediment (Bernard and others, 1970).

The inner fringe environment is the shallowest part of the subaqueous deltaic area. It includes fringing beaches, river mouth bars, and lagoons. Fringing beaches are low, narrow, and elongated and parallel the shoreline, usually with a steep seaward side (foreshore) and a gentle landward slope (backshore). The sediments normally consist of fine well-sorted sand, with abundant shells and fragments, (Bernard and others, 1970). River mouth bar deposits consist of symmetrical to asymmetrical crescent-shaped masses of sand and silt with small-scale to medium-scale (giant ripple) cross-bedding and with some horizontal laminations. Lagoons are areas of open water (commonly marine but some are brackish), which form behind river mouth bars or in

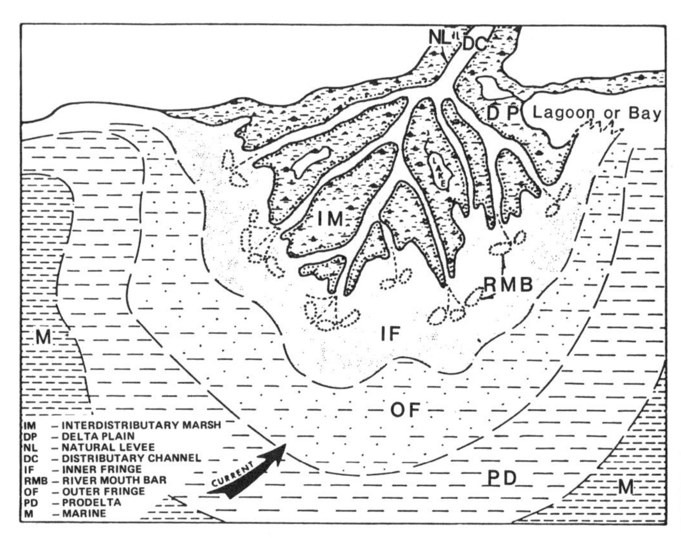

Figure 2. Depositional environments typical of modern deltas. Adapted from LeBlanc (1972).

Figure 3. Typical sequences of deltaic deposits. From LeBlanc (unpublished).

interdistributary areas where sedimentation rates are relatively low. Sediments are silt and clay and can contain a high organic content if vegetation takes root.

Lying seaward of the inner fringe environment is the outer fringe. The sediments of the outer fringe environment are finer grained (with less than 50 percent sand and silt) than those of the inner fringe. For the most part they are deposited below the wave base, are finely laminated, and are not burrowed by organisms.

The prodelta sedimentary environment is characterized by fine-grained riverborn clay deposited as a broad fan on the floor of the sea. The deposits contain a sparse fauna, are commonly laminated with flaser sedimentary features, frequently massive, but rarely burrowed or churned by organisms.

The normal marine environment shows the least influence of riverborn sediment. Open marine sedimentary sequences are clays, which have a normal marine fauna, and are generally heavily bioturbated. Glauconite, carbonates and other precipitated marine sediments are present.

FIELD STOPS

Stop 1. The outcrop at Camden, Ouachita County, Arkansas (Fig. 4) is located in the NE¼ of Sec.27,T.13S.,R.17W., northwest of the intersection of U.S. 79 and 79B and Arkansas 4. The hillside cut (Fig. 5) was made for a shopping center and is easily reached by car. The section exposed is 25 ft (7.6 m) high, 105 ft (32 m) long, and is composed mostly of interlaminated sands and clays in the lower Claiborne.

The outcrop has been divided into six zones. The lowest, Zone 1, is 3 (1 m) thick and consists of very fine sand with some silt and mica. Close examination reveals that it is cross-bedded and probably is a distributary channel fill.

Zone 2, which is present only in the southern part of the exposure, is composed of interlayered lignitic clay and fine, angular sand with flaser bedding. Sulfates are leaching from the clay layers. The unit grades upward and laterally into Zone 3.

Zone 3 thickens to the north (right) at the expense of Zone 2. Where it is in contact with Zone 1, the boundary appears to be an erosional contact. Like Zone 4 above, it is composed of alternating ripple laminations of clay and sand, but contains a smaller percentage of sand than does the overlying unit. In both zones the lignitic clays are 0.125 to 2 in (3 to 50 mm) thick, interlaminated with very fine sands, 0.25 to 3 in (6 to 76 mm) thick (Fig. 6).

The clay intervals are heavily stained with organic matter

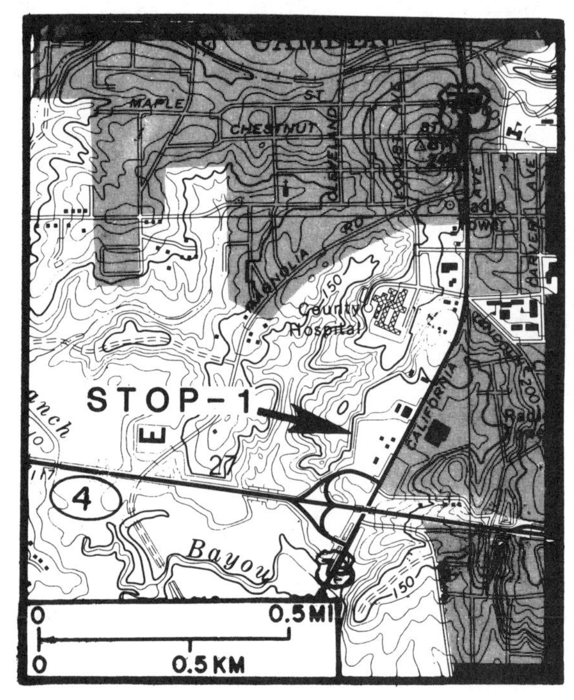

Figure 4. Detailed location map for Stop 1.

Figure 5. East facing exposure of Claiborne Group at Stop 1, Zones 1–6 are marked.

Figure 6. Interlaminated clays and sands in Zone 2 at Stop 1.

and burrowed. The organic material is made up of very fine particles and many represent periods of plant growth in an interdistributary lagoon or lake. Intervals of clastic deposition from nearby distributary channels are marked by the layers of sand between the organic clays.

Zone 5 consists of gray sandy clay and is 4 ft (1.2 m) thick. At other localities in the area, this zone contains shark and other fish fossils. The sandy clay may have accumulated in a lagoon, in the inner fringe, or in a lake that had some connection to the marine environment. No invertebrates or microfauna have been found at this outcrop, but some gypsum is present. Lignite logs and amber occur elsewhere in the vicinity indicating that forests or swamps of some type were close by to provide the organic matter. It is in these interdistributary swamps that the lignite in the Claiborne would be formed and these are the areas that hold the best potential for economic deposits within the Claiborne.

Zone 6 is composed of 4 ft (1.2 m) of highly weathered, interlayered fine sand and clay containing several ironstone layers. This zone is similar to zones 2, 3, and 4 and marks a return to a depositional environment similar to these lower zones.

Stop 2. The Spring Bank outcrop is located on the west bank of the Sulphur River in the NW¼ of Sec.24,T.19S.,R.26W., Miller County, Arkansas (Fig. 7). The exposure is accessible by car to within 150 ft (46 m). After parking, walk to the stream gauging station at the abandoned Arkansas 160 Spring Bank Ferry Landing. The outcrop is 200 ft (61 m) long; 54 ft (16 m) are exposed above normal river level (elevation 210 ft; 64 m). The exposure has been divided into three zones (A, B, and C) in ascending order.

The Spring Bank outcrop exhibits two depositional regimes of a deltaic depositional system-flood basin (zones A and C) and distributary channel (zone B). The sediments occupying the lower and upper portions of the section are indicative of flood basin deposition as evidenced by the fining upward sequences and bioturbation. The middle portion of the section contains festoon cross-bedding and is interpreted as representing a distributary channel.

Zone A, the lowest 22 ft (6.7 m) of the exposure, consists of laminated silt with minor fine micaceous sand and burrowed lignitic clay layers. The lower part contains pyrite nodules along with sulfates derived from weathering processes.

Zone B, above is 8 ft (2.4 m) thick. The lower 2 ft (0.6 m) exhibits festoon cross-bedding. The remaining 6 ft (1.8 m) are composed of parallel laminations of fine-grained subangular quartz sand and contain abundant plant fragments. This portion of the section is largely covered by an irregular iron concretionary sand drape resulting from the weathering of sulfides in the upper portion of the zone.

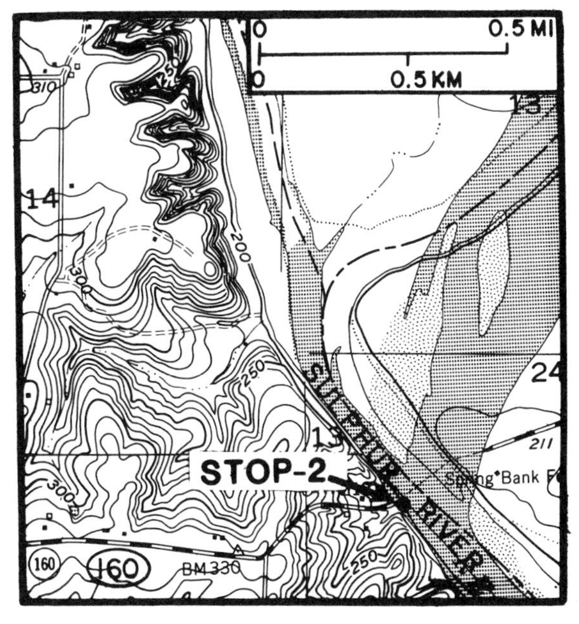

Figure 7. Location of Stop 2 on the western bank of Sulphur River at the old gauging station.

Zone C comprises the uppermost 24 ft (7.3 m) of the outcrop and consists of very fine quartz sand with intermittent silt and lignitic clay layers. Small pieces of lignitic wood and weathered pyrite nodules are randomly scattered within this zone. The very top of the section contains sand-filled burrows found in thin clayey lignitic layers.

REFERENCES CITED

Bernard, H. A., Major, C. F., Jr., Parrett, B. S., and LeBlanc, R. J., Sr., 1970, Recent sediments of southeast Texas; A field guide to the Brazos alluvial and deltaic plants and the Galveston barrier island complex: University of Texas Bureau of Economic Geology Guidebook 11, 132 p.

Clardy, B. F., 1979, Arkansas lignite investigations, a preliminary report: Arkansas Geological Commission Miscellaneous Publication MP-15, 133 p.

LeBlanc, R. J., 1972, Geometry of sandstone bodies, *in* Cook, T. D., ed., Underground waste management and environmental implications: American Association of Petroleum Geologists Memoir 18, p. 133–189.

—— , 1975, Significant studies of modern and ancient deltaic sediments, *in* Broussard, M. L., ed., Deltas, models for exploration: Houston Geological Society, p. 13–85.

—— , 1977, Distribution and continuity of sandstone reservoirs, Parts 1 and 2: Journal of Petroleum Technology, v. 29, p. 776–804.

Morton, R. A., and McGowan, J. H., 1980, Modern depositional environments of the Texas Coast: Bureau of Economic Geology, University of Texas at Austin Bureau of Economic Geology Guidebook 20, 156 p.

Taff, J. A., 1980, Preliminary report on the Camden coal field of southwestern Arkansas: 21st Annual Report, U.S. Geological Survey, p. 320–322.

The Lingos formation, western Rolling Plains of Texas

S. Christopher Caran and Robert W. Baumgardner, Jr., Bureau of Economic Geology, The University of Texas at Austin, Box X, University Station, Austin, Texas 78713-7508

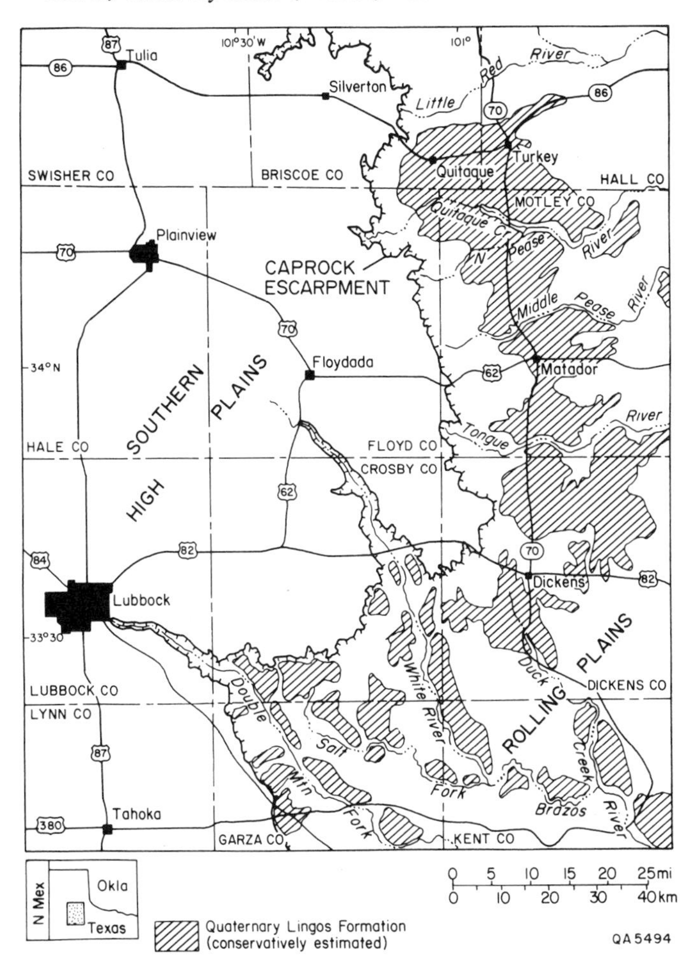

Figure 1. Areal extent of the Quaternary Lingos formation in the western Rolling Plains of Texas. Areal geologic maps published before 1984 do not depict these deposits.

LOCATION

Lingos strata (Figs. 1, 2) are well exposed at localities in Hall, Briscoe, and Floyd counties, Texas. Except where the formation is dissected or where remnants of bedrock project through it, the Lingos terrane is a smooth continuous geomorphic surface sloping southeastward approximately 80 ft/mi (15 m/km) near the escarpment and 40 ft/mi (7.5 m/km) farther east. The route between the Field Guide localities described here crosses this geomorphic surface, within the area covered by the Plainview sheet of the Geologic Atlas of Texas (Barnes, 1968). However, none of the deposits now recognized as the Lingos formation are shown on the atlas or other maps published before 1984. Topographic maps (7½-minute) covering the localities in this Field Guide are: Turkey, Folley, Quitaque and Lake Theo.

SIGNIFICANCE

The western Rolling Plains of Texas are mantled with thick, coarse- to fine-grained, late Quaternary clastic sediments and lacustrine marls (Caran and others, 1987). These deposits are here informally designated the Lingos formation (Figs. 1 and 2). The Lingos formation discontinuously covers more than 3,000 mi^2 (7,800 km^2), forming a clastic apron that extends as much as 50 mi (80 km) eastward from the Caprock Escarpment. Thickness of these deposits exceeds 250 ft (76 m) locally. The thickest sections fill broad karstic subsidence basins produced by dissolution of Permian salt at depths of 400 to 800 ft (120 to 240 m) (Gustavson and others, 1982; Caran and others, 1987). Lingos sediments constitute a significant freshwater aquifer that is the major source of water for communities and agriculture in the western Rolling Plains.

At most sites the Lingos formation comprises a sequence of three principal genetic facies: (1) basal tabular lithosomes of locally fossiliferous sand and gravel, deposited by alluvial fans; (2) medial lenses of calcareous, fossiliferous, lacustrine clay and marl; and (3) uppermost sheet and channel deposits of mostly nonfossiliferous eolian sand and silt, stacked paleosols, and fluvial silt, sand, and gravel, containing sparse archaeological material (Caran and others, 1987) (Fig. 3). Locally, facies 1 overlies a thin relict Quaternary saprolite developed on Permian siltstone. Chronologic control is afforded by more than 50 finite radiocarbon dates (Caran and Baumgardner, 1986a), as well as by fossil and archaeological remains. The best known vertebrate fossil locality is the Quitaque local fauna site of Dalquest (1964, 1986) and Caran and Baumgardner (1986b).

LOCALITY 1

The first site is the Fort Worth and Denver Railroad cut beneath Texas 70, approximately 0.5 mi (0.8 km) north of Turkey, southwestern Hall County (34°24′21″N,100°53′38″W, Turkey 7½-minute Quadrangle) (Fig. 2: Locality 1). Pedestrian access to the cut is along an unpaved road immediately northeast of the Texas 70 bridge. This locality is described by Baumgardner and Caran (1986a). Quaternary strata are exposed primarily south of the bridge. Permian bedrock (Quartermaster Formation) crops out approximately 0.5 mi (0.8 km) to the north, where it is locally disrupted by small folds, joints, and thrust faults. Structural disturbance probably resulted from karstic subsidence owing to dissolution of evaporites at depth.

In the eastern wall of the cut, 80 ft (25 m) south of the highway bridge, the exposed Quaternary section is approximately 19 ft (6 m) thick. Beds lowest in this section (site A of Baumgardner and Caran, 1986a) are slightly clayey sandy silts that are

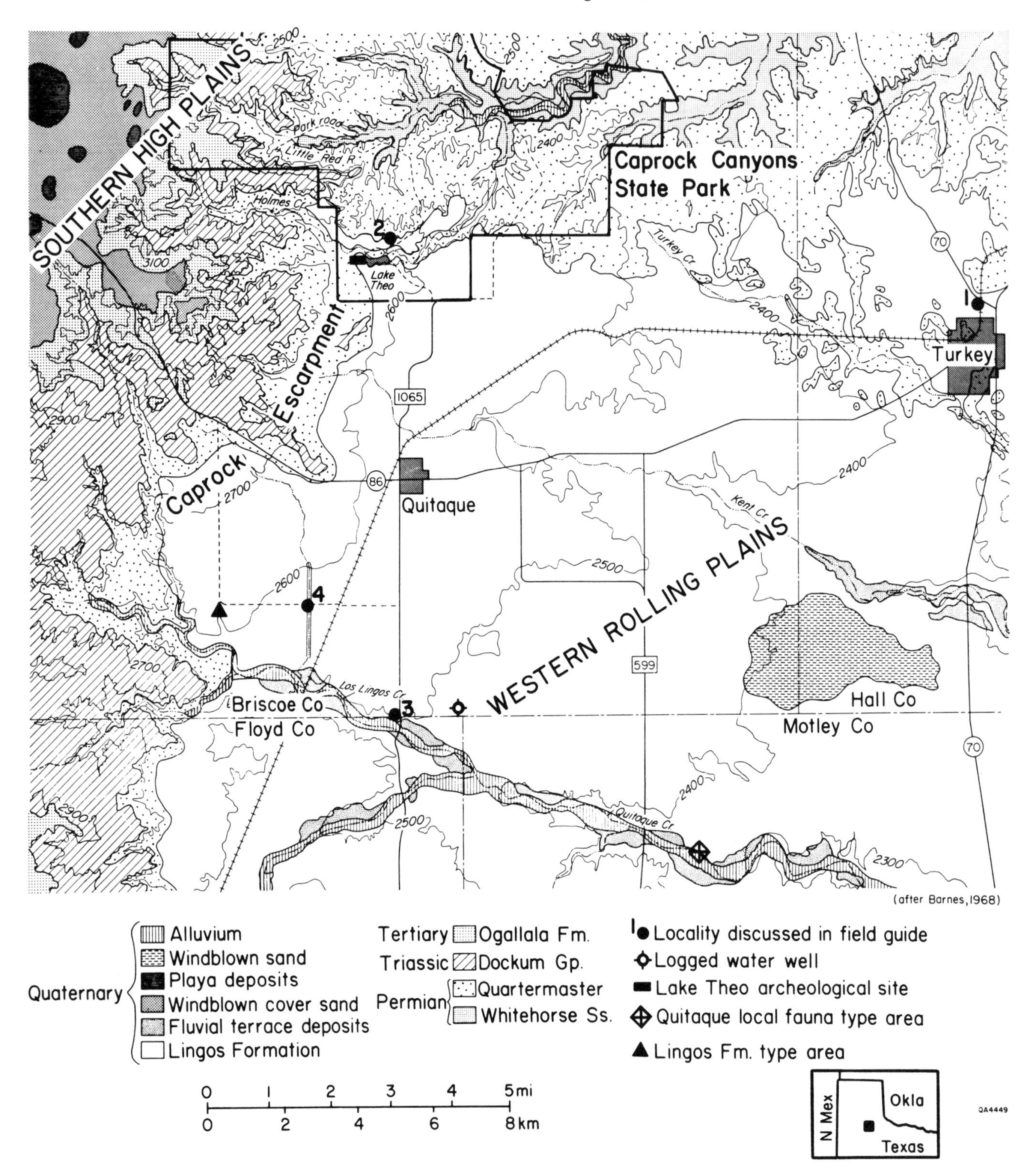

Figure 2. Areal geologic map of part of the western Rolling Plains. Numbered localities and other sites mentioned in the text are denoted symbolically. Map is based on Barnes (1968) but has been revised extensively, particularly with regard to Quaternary deposits. The area shown lies in the northeastern part of the region represented in Figure 1, in Briscoe, Hall, Floyd, and Motley counties. Contour lines are topographic, in feet above mean sea level.

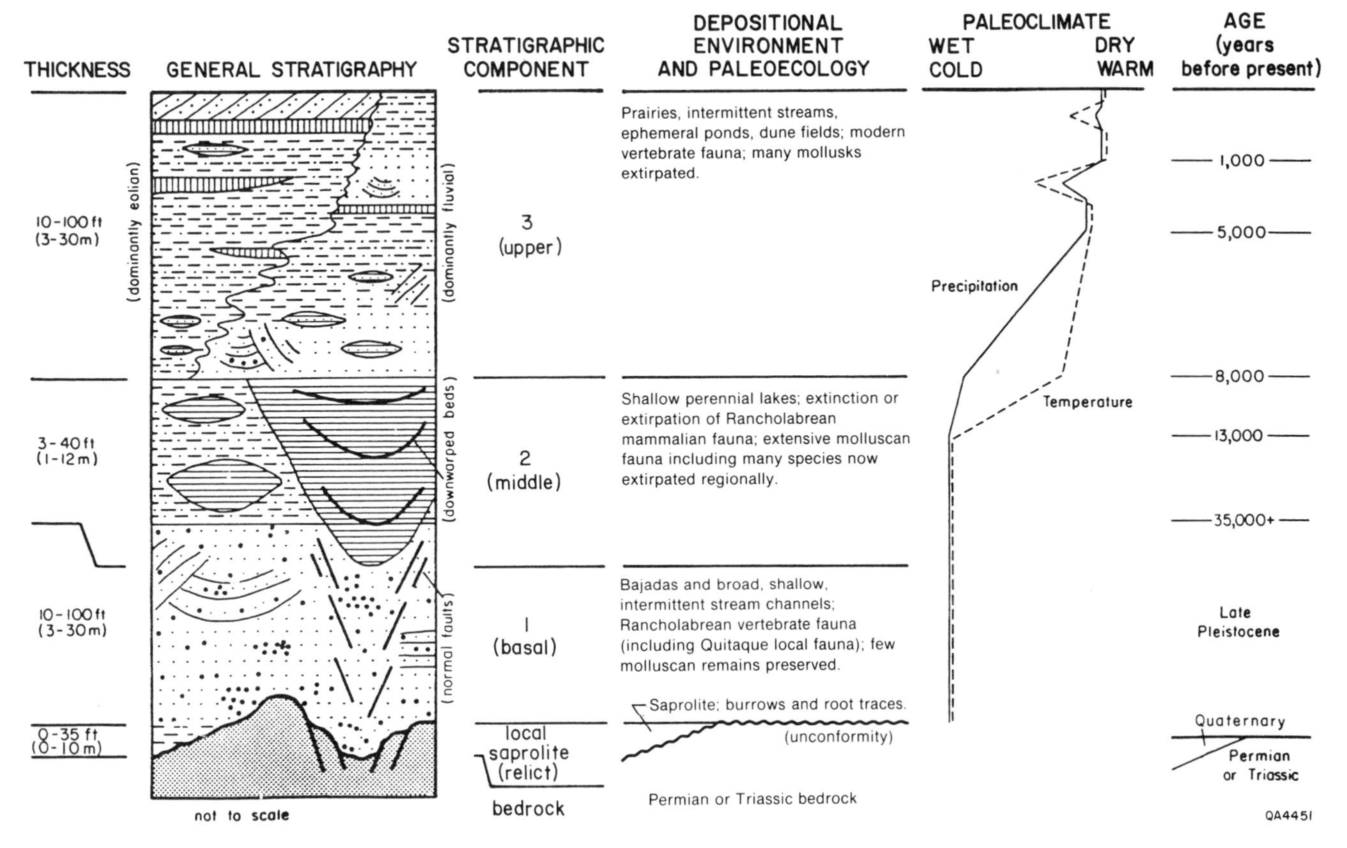

Figure 3. Generalized composite section, paleoenvironmental interpretation, and approximate chronostratigraphy of the Lingos formation.

downwarped and jointed locally. Subsidence created a small structural basin that gradually became isolated from sources of coarse sediment, although deposition of fine sediment continued. Ultimately, 6 ft (2 m) of highly calcareous light gray fossiliferous lacustrine clay and silty clayey marl were deposited. Fossils from these deposits include terrestrial and aquatic gastropods, pelecypods, ostracodes, charophytes, and diatoms. Evidently, these same strata were sampled by Frye and Leonard (1963, p. 34 and Fig. 2).

Part of the lacustrine sequence (facies 2) and the upturned edges of underlying beds are eroded locally. A variably indurated silty sandy fluvial gravel covers this erosion surface. The gravel generally is overlain by a prominent paleosol consisting of red to dark reddish brown clayey and fine sandy silt or silty clay. This paleosol and thin soils that overlie it are pedogenically modified fine-grained fluvial and eolian deposits. The oldest radiocarbon-dated deposit is an abandoned-channel fill of very dark gray sandy silty clay exposed in the western wall of the cut 100 ft (30 m) south of the bridge. Organic humates from these strata were dated 12,710 ± 140 yr B.P. (Caran and Baumgardner, 1986a). The channel fill clearly postdates most of the extant section in the eastern wall.

From Locality 1, drive 0.3 mi (0.5 km) southeast along Texas 70 to its junction with Texas 86. Turn south (right) for 0.8 mi (1.3 km) to the village of Turkey, where Texas 70 and 86 diverge (Fig. 2). Turn west (right) and continue on Texas 86 across the broad rolling geomorphic surface that characterizes the Lingos terrane. This surface has been aggrading through addition of eolian sediment during much of the Holocene.

Approximately 5.0 mi (8.0 km) west of the divergence of Texas 70 and 86 (in Turkey), near a drive-in theater on the southern side of the road, a water well penetrates 245 ft (75 m) of mostly water-saturated Quaternary sediment overlying Permian red beds (Popkin, 1973, Table 6: well number BL-12-41-101). Gravel that appears to be an eroded remnant of Lingos facies 1 is exposed in the road cut approximately 0.6 mi (1 km) west of the theater, across from Valley School. Irregularity of the Permian subcrop in this area probably resulted from differential subsidence.

From the Valley School road cut, continue west on Texas 86 for 4.1 mi (6.6 km) to its intersection with Farm-to-Market Road (FM) 1065 in Quitaque. Turn north (right) onto FM 1065 and travel approximately 2.9 mi (4.7 km) to the marked turnoff leading to Caprock Canyons State Park. Take the north (left) fork and travel 0.9 mi (1.4 km) north and northwest on the park road to the headquarters building. A small entrance fee is required.

Park regulations forbid sample collecting and alteration of outcrops.

From the headquarters building, continue 0.1 mi (0.2 km) northwest on the main park road to its intersection with a partly paved secondary road leading west. Approximately 0.7 mi (1.1 km) westward along this secondary road, a historical marker designates the Lake Theo archaeological site (41B170) (Harrison and Killen, 1978; Johnson and others, 1982) (Fig. 2), where Paleoindians butchered now extinct *Bison antiquus.* From the intersection of the main and secondary park roads, continue north on the main road and travel 0.3 mi (0.5 km) across the Lake Theo dam, to the intersection of another park road leading southeast and east. Turn southeast (right) onto that road and travel approximately 0.25 mi (0.4 km) to the outdoor interpretive pavilion.

LOCALITY 2

Exhibits at the interpretive shelter (34°25′10″N.,101° 03′34″W, Lake Theo 7½-minute Quadrangle) (Fig. 2: Locality 2) illustrate the geologic and human history of the park. A small amphitheater is located immediately north of the shelter on a point of land projecting into the valley of the Little Red River. Southwest of the amphitheater, the broad geomorphic surface on the Lingos sediment apron is seen along topographic and depositional strike. This surface slopes gently southeastward in the general direction of depositional dip. Less than 0.5 mi (0.8 km) southeast of the amphitheater, the present lower reach of Holmes Creek captured the upper reach, carving a deep, steep-walled canyon. Stream capture and incision during the middle to late Holocene isolated the narrow remnant of the original geomorphic surface on which the interpretive center stands. A thin cover of Quaternary sediment overlies the Permian Quartermaster Formation in the cliff faces immediately west and east of the shelter. The narrow reentrant canyon to the east has been cut to within a few tens of feet from a meander loop of Holmes Creek. In the future, stream piracy probably will divert drainage from upper Holmes Creek northward through this reentrant.

Quaternary deposits have been stripped away north of the amphitheater, exposing contorted Permian bedrock in the valley of Little Red River. The variable dip of these folded Permian strata resulted from karstic subsidence (Goldstein and Collins, 1984). To the west and northwest, the Lingos formation is relatively undissected and covers part of the Triassic Dockum Formation (Fig. 2). The thick light-colored strata on the horizon to the west and southwest are part of the Tertiary Ogallala Formation.

Return to Quitaque along the park roads and FM 1065. From the intersection of FM 1065 and Texas 86, continue south on 1065 for 3.5 mi (5.6 km) to Locality 3. Approximately 2.7 mi (4.3 km) south of Quitaque, observe the vegetation-stabilized dune fields west of 1065. This small tract of brush-invaded rangeland preserves a sample of the original rolling topography that existed prior to widespread cultivation.

LOCALITY 3

Locality 3 is a deep steep-walled gully on the H. E. Blair, Jr., farm, at the Briscoe-Floyd County line, 80 ft (25 m) west of FM 1065 (34°18′44″N, 101°03′33″W, Quitaque 7½-minute Quadrangle) (Fig. 2: Locality 3). The gully is private property, and prior authorization by the landowner is required for access. However, most of the section can be seen from vantage points within the public right-of-way along the roadside. The gully walls are unstable at some points: **Approach the gully cautiously!**

This site is described by Caran and Baumgardner (1986c). The gully formed when drainage was diverted from FM 1065, exposing part of facies 3, the uppermost stratigraphic interval in the Lingos formation. Aggregate thickness of strata within the gully is 40 ft (12 m). Beds lowest in the composite section are exposed at the southern end. These units consist of cross-bedded braided-stream deposits of sand and gravel forming cut-and-fill sequences. Vertebrate remains are rare in these strata. Overlying the beds of sand and gravel is a wide lens of horizontally bedded clayey fine-sandy silt deposited within an abandoned channel. Strata higher in the local section, at a depth of approximately 3 ft (1 m), were radiocarbon dated. The oldest date obtained is 1,230 ± 90 yr B.P. The dated units comprise floodplain and eolian deposits modified by pedogenesis.

A water well on the Blair farm 0.95 mi (1.5 km) east of this site penetrates 176 ft (54 m) of Quaternary strata overlying Permian bedrock (Caran and Baumgardner, 1986c) (Fig. 2). All three major facies of the Lingos formation are represented. The composite section measured in the gully west of FM 1065 appears to be correlative with the upper part of the well section.

From Locality 3, travel northward along FM 1065 for 1.6 mi (2.6 km) to the intersection of an unpaved, unnamed county road leading westward. Turn west (left) onto that road. **Caution:** Parts of this road may be impassable to two-wheel-drive vehicles during wet weather. Continue west for 1.5 mi (2.4 km) to Locality 4.

LOCALITY 4

Locality 4 is the western side of a long deep ditch on the Rusty Henson farm in southeastern Briscoe County, south of the county-road bridge (34°20′13″N, 101°05′07″W, Quitaque 7½-minute Quadrangle) (Fig. 2: Locality 4). The ditch is private property, and prior authorization by the landowner is required for access. However, most of the section can be seen from the road bridge and from vantage points within the public right-of-way. The walls are unstable at some points: **Approach the ditch cautiously!**

This site is described by Baumgardner and Caran (1986b). The ditch was excavated to divert drainage from a railway to the east (Fig. 2). Almost 20 ft (6 m) of facies 3 is exposed at this site. The lowest part of the local section predominantly consists of fluvial (floodplain) sand. These deposits are overlain by a wide irregular lens of silty clay filling a shallow abandoned channel.

The radiocarbon age of beds at the top of this lens is 1,950 ± 80 yr, indicating rapid accumulation of the overlying 11.5 ft (3.5 m) of sediment (eolian and overbank fluvial sandy silts). All strata in this section, particularly those overlying the lens of silty clay, have been modified pedogenically. Soil horizons have been superimposed on some deposits and formed during deposition of others.

From Locality 4, continue west on the unnamed county road for 1.5 mi (2.4 km). At that point, a deep wide gully extends southward from the shoulder of the road. The property west of this gully is the D. L. Smith farm, also known as the Lazy J Ranch, in southeastern Briscoe County. Little of the gully can be seen from the roadway, and access requires prior authorization by the landowner. The 0.7-mi (1.1-km) long gully is the most complete exposed section of the Lingos formation and is the proposed type area (Caran and Baumgardner, 1986d). Permian bedrock crops out near the confluence of this gully and Los Lingos Creek for which the formation is named. The local composite Quaternary section is 155 ft (47 m) thick and includes a thick basal saprolite, subsidence-deformed sandy gravels (facies 1), extensive fossiliferous lacustrine deposits (facies 2) filling a well-defined subsidence basin, and a moderately thick fluvial and eolian sequence (facies 3).

The unpaved county road turns northward at the point where it approaches the gully. Continue north on this road for 2.7 mi (4.3 km) to the second large bridge. Multicolored Triassic Dockum siltstone and sandstone are exposed beneath a residual veneer of Lingos sediment. Continue north on the county road for 0.1 mi (0.2 km) to its intersection with Texas 86. The Tertiary Ogallala Formation caps the escarpment west of this intersection, and Dockum sandstone is exposed in road cuts to the east. Turn southeastward (right) onto Texas 86 and travel 3.2 mi (5.1 km) southeastward and eastward to the intersection of Texas 86 and FM 1065 in Quitaque. En route, observe the updip limit of the Lingos sediment apron to the north, surrounding low hills that are local remnants of Triassic and Permian bedrock. The Lingos formation thickens rapidly to the south and east. At Quitaque, Lingos sediments are approximately 50 ft (15 m) thick.

REFERENCES CITED

Barnes, V. E., project director, 1968, Geologic atlas of Texas, Plainview sheet: Austin, The University of Texas, Bureau of Economic Geology, scale 1:250,000.

Baumgardner, R. W., Jr., and Caran, S. C., 1986a, Stop 11, Measured section, Fort Worth and Denver Railroad cut, *in* Gustavson, T. C., ed., Geomorphology and Quaternary stratigraphy of the Rolling Plains, Texas Panhandle: Austin, The University of Texas, Bureau of Economic Geology Guidebook 22, p. 47–55.

—— , 1986b, Stop 15, Measured section, Henson farm near Quitaque, Texas, *in* Gustavson, T. C., ed., Geomorphology and Quaternary stratigraphy of the Rolling Plains, Texas Panhandle: Austin, The University of Texas, Bureau of Economic Geology Guidebook 22, p. 67–72.

Caran, S. C., and Baumgardner, R. W., Jr., 1986a, Appendix; Summary of radiocarbon dates, western Rolling Plains of Texas, *in* Gustavson, T. C., ed., Geomorphology and Quaternary stratigraphy of the Rolling Plains, Texas Panhandle: Austin, The University of Texas, Bureau of Economic Geology Guidebook 22, p. 90–97.

—— , 1986b, Stop 12A, Quitaque Creek section, *in* Gustavson, T. C., ed., Geomorphology and Quaternary stratigraphy of the Rolling Plains, Texas Panhandle: Austin, The University of Texas, Bureau of Economic Geology Guidebook 22, p. 56–57.

—— , 1986c, Stop 13, Measured section, Blair farm, Quitaque, Texas, *in* Gustavson, T. C., ed., Geomorphology and Quaternary stratigraphy of the Rolling Plains, Texas Panhandle: Austin, The University of Texas, Bureau of Economic Geology Guidebook 22, p. 60–62.

—— , 1986d, Stop 14, Measured section, Smith farm, Quitaque, Texas, *in* Gustavson, T. C., ed., Geomorphology and Quaternary stratigraphy of the Rolling Plains, Texas Panhandle: Austin, The University of Texas, Bureau of Economic Geology Guidebook 22, p. 63–66.

Caran, S. C., Baumgardner, R. W., Jr., McGookey, D. A., Gustavson, T. C., and Neck, R. W., 1987, Quaternary stratigraphy of the western Rolling Plains of Texas; Preliminary findings: Lincoln, Nebraska Academy of Sciences, Institute for Tertiary-Quaternary Studies Ter-Qua Symposium Series, v. 2 (in press).

Dalquest, W. W., 1964, A new Pleistocene local fauna from Motley County, Texas: Kansas Academy of Science Transactions, v. 67, no. 3, p. 499–505.

—— , 1986, Stop 12B, Vertebrate fossils from a strath terrace of Quitaque Creek, Motley County, *in* Gustavson, T. C., ed., Geomorphology and Quaternary stratigraphy of the Rollings Plains, Texas Panhandle: Austin, The University of Texas, Bureau of Economic Geology Guidebook 22, p. 58–59.

Frye, J. C., and Leonard, A. B., 1963, Pleistocene geology of Red River Basin in Texas: Austin, The University of Texas, Bureau of Economic Geology Report of Investigations no. 49, 48 p.

Goldstein, A. G., and Collins, E. W., 1984, Deformation of Permian strata overlying a zone of salt dissolution and collapse in the Texas Panhandle: Geology, v. 12, no. 5, p. 314–317.

Gustavson, T. C., Simpkins, W. W., Alhades, A., and Hoadley, A., 1982, Evaporite dissolution and development of karst features on the Rolling Plains of the Texas Panhandle: Earth Surface Processes and Landforms, v. 7, p. 545–563.

Harrison, B. R., and Killen, K. L., 1978, Lake Theo; A stratified, Early Man bison butchering and camp site, Briscoe County, Texas: Canyon, Texas, Panhandle-Plains Historical Museum Special Archeological Report 1, 108 p. and Addendum.

Johnson, E., Holliday, V. T., and Neck, R. W., 1982, Lake Theo; Late Quaternary environmental data and new Plainview (Paleoindian) date: North American Archeologist, v. 3, no. 2, p. 113–137.

Popkin, B. P., 1973, Ground-water resources of Hall and eastern Briscoe Counties, Texas: Austin, Texas Water Development Board Report 167, 84 p.

ACKNOWLEDGMENTS

The cooperation of landowners in the western Rolling Plains is gratefully acknowledged. Particular thanks go to H. E. Blair, Jr., Houston; R. Henson, Quitaque; D. Smith, Quitaque; and the Texas Parks and Wildlife Department. The manuscript was typed by R. Wilson and G. Zeikus. Figures were drafted by R. L. Dillon and K. Prewitt. The following individuals reviewed the manuscript and offered helpful remarks: E. C. Bingler, Deputy Director, J. R. DuBar, Technical Editor, J. Raney, T. C. Gustavson, and T. F. Hentz of The University of Texas at Austin, Bureau of Economic Geology; K. S. Johnson of the Oklahoma Geological Survey; and J. H. Peck of Stone and Webster Engineering Corporation. Research and preparation of this paper were supported by the United States Department of Energy under contract number DE-AC97-83WM46651.

Late Pleistocene and Holocene stratigraphy, Southern High Plains of Texas

Vance T. Holliday, Department of Geography, Science Hall, University of Wisconsin, Madison, Wisconsin 53706

LOCATION

Begin in Crosbyton, Texas (Fig. 1). Take U.S. 82 west 35 mi (56 km) to Lubbock. On the northeast side of Lubbock at the intersection of U.S. 82 (which is also U.S. 62 along this stretch of the road) and Loop 289, turn right (north) off of U.S. 82 and follow Loop 289 to the north and west. Cross over I-27/U.S. 84, and take the University Avenue exit. Stay on the frontage road and continue west, parallel to the loop. Cross University Avenue and continue west with the large Texas Instruments (TI) plant to the north (right). Just past the TI plant the road drops into Yellowhouse Draw. The Lubbock Lake site, is in the trees to the northwest, about 300 ft (100 m) away. On the floor of the draw, take the only paved road north (right). Follow it for 0.3 mi (500 m) as it curves to the left along the edge of the draw. A chain-link fence enclosing most of the trees becomes visible. This is Lubbock Lake.

SIGNIFICANCE

Excellent examples of Pleistocene and Holocene stratigraphy are exposed in the east-central part of the Southern High Plains of Texas.

A trip from Crosbyton to Lubbock, which is well into the High Plains proper, allows excellent views of the generally flat, featureless topography of the region. The route also passes examples of landforms that provide the little topographic relief there is in the region, including numerous playas, a few dunes, and several draws. The highway goes near or through a number of shallow basins with ephemeral lakes or playas. There are thousands of these basins on the Southern High Plains, probably resulting mostly from wind deflation in late Pleistocene and Holocene time (Reeves, 1966; Holliday, 1985a). Typically, lacustrine sediment several meters thick is found on the floors of the playas (Holliday, 1985a). Along the highway, in the area of Lorenzo and Idalou (17 and 26 mi; 27 and 42 km west of Crosbyton, respectively), are some large dunes on the lee (east) side of the playas. These dunes are silty, calcareous deposits derived from adjacent playas that had calcareous lacustrine sediments (Holliday, 1985a). Late Pleistocene and Holocene valley fills, commonly found in ephemeral drainages or draws of the region, are exposed at the Lubbock Lake archaeological site (Fig. 1).

Lubbock Lake is the most intensively studied late Quaternary site on the High Plains (Johnson, 1987). The stratigraphy at the site is generally representative of that for all draws in the central part of the Southern High Plains.

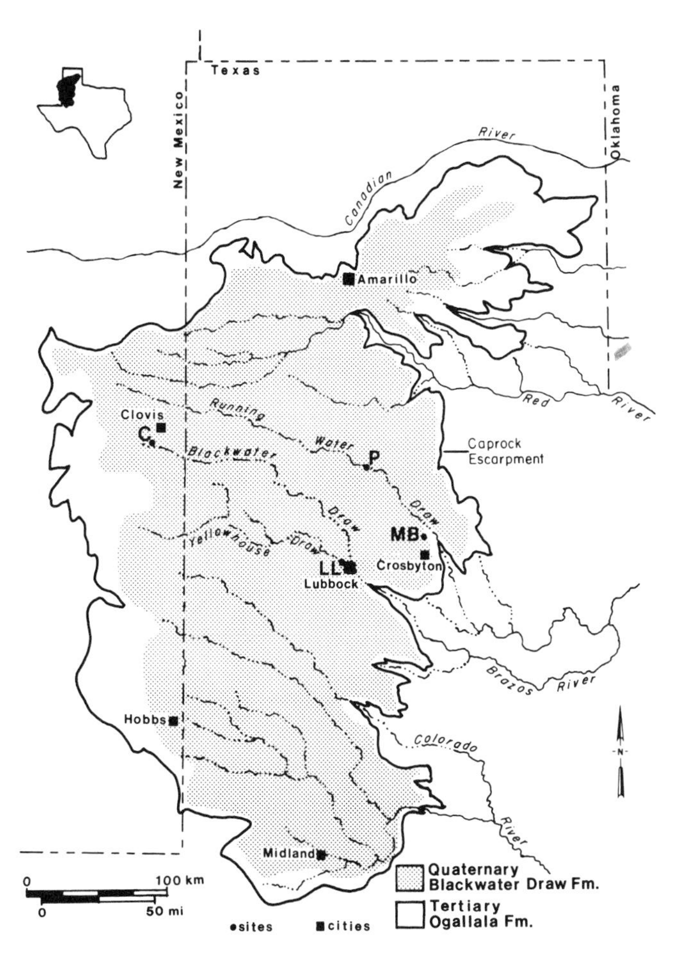

Figure 1. Map of Southern High Plains with principal regional surficial geologic units (Ogallala and Blackwater Draw Formations) and principal physiographic and cultural features, including those mentioned in text. (Key to sites: LL = Lubbock Lake; P = Plainview; C = Clovis)

LUBBOCK LAKE

Lubbock Lake (Figs. 1, 2) (33°31′13.5″N, 101°53′31.5″W; Lubbock West 7½-minute Quadrangle) is a well-stratified archaeological site composed of a thick sequence of sediments set in an entrenched meander of Yellowhouse Draw, a tributary of the Brazos River. The site, a State Archaeological Site and National Historic Landmark, covers 300 acres (120 ha) and contains a virtually complete geological, biological, and cultural record spanning the last 11,000 yr. It is owned by the city of Lubbock,

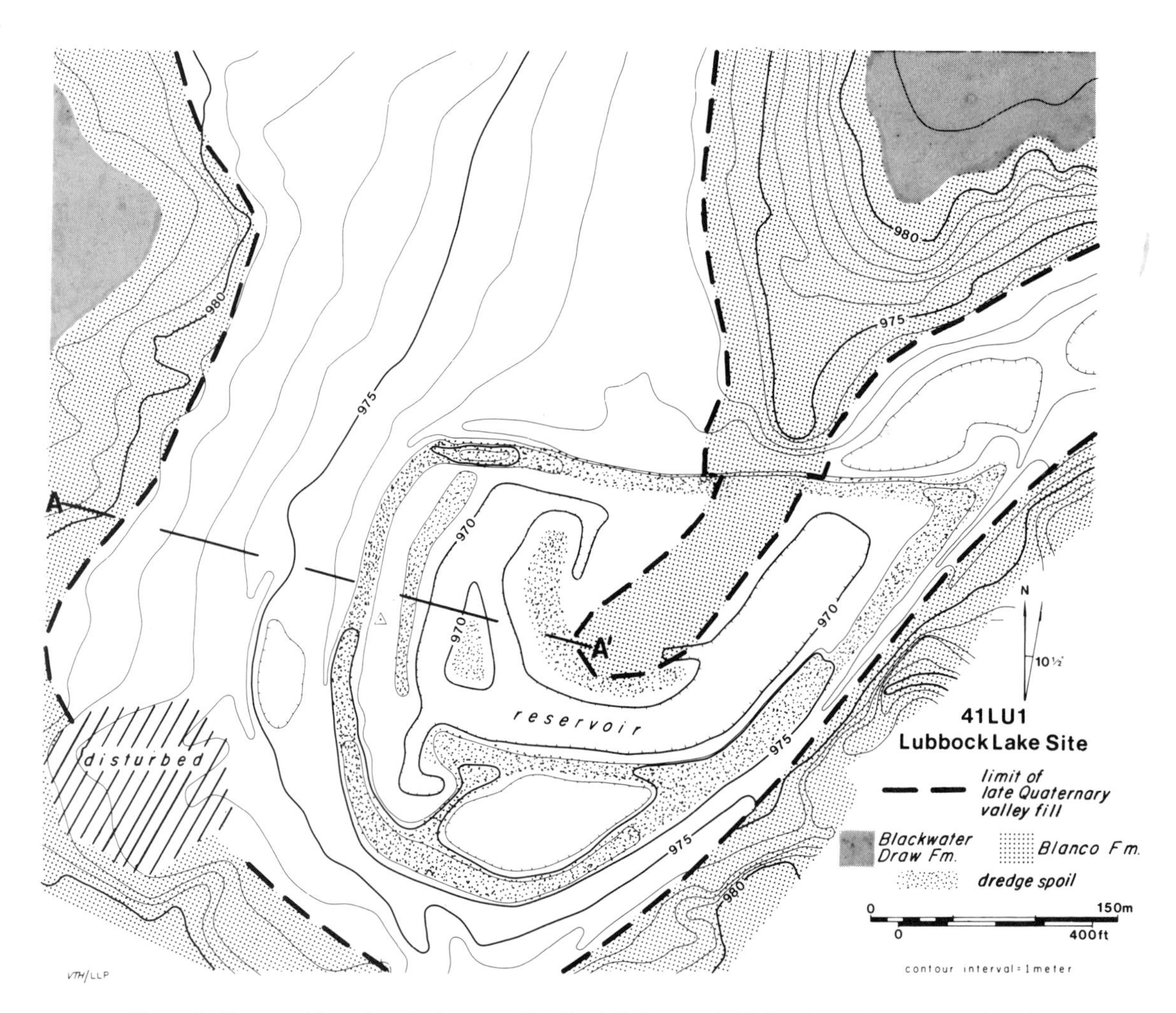

Figure 2. Topographic and geologic map of Lubbock Lake area in Yellowhouse Draw, with line-of-secton for Figure 4.

and is leased and managed by Texas Tech University. Tours of the site are given during the summer excavation season.

The site was discovered in 1936 during excavation of a U-shaped reservoir along the inside of the meander (Fig. 2). The excavations exposed the late Quaternary valley fill of Yellowhouse Draw that contains abundant debris of human occupation. Archaeological investigations were conducted intermittently at the site beginning shortly after its discovery. In 1972 the Lubbock Lake Project began, under the auspices of The Museum of Texas Tech University, which is continuing research at the site and other localities in the region. The results of much of the research are presented by Johnson (1987). The stratigraphy, soils, and geochronology of the site are described by Holliday (1985b-d) and Holliday and others (1983, 1985).

The local bedrock in the area of the site is weakly consolidated sandstone and dolimitic marl of the Blanco Formation. At the top of the unit is a Stage IV–V pedogenic calcrete, similar to that exposed at Mt. Blanco. These Blanco sediments were deposited in a basin that was separate from the basin in the Mt. Blanco area. The Blanco beds are well exposed immediately north of the entrance gate to the main excavation area of the site and along the paved road leading to the fence along the east side of the draw (Fig. 2). The Blanco outcrops on the east side of the draw were heavily quarried for road metal in the 1950s.

Yellowhouse Draw cut through the Blackwater Draw Formation and into the Blanco Formation in the late Pleistocene. One stage of this downcutting is indicated by a strath terrace cut on the Blanco Formation in the area of the entrance gate to Lubbock Lake (Fig. 2). Aggradation along the draw began at the end of the Pleistocene and continued intermittently throughout

TABLE 1. GENERALIZED DESCRIPTIONS OF STRATA 1 THROUGH 5 AT LUBBOCK LAKE*

Stratum	Valley Axis Facies	Valley Margin Facies
5	Substratum 5Bℓ: up to 1 m thick; gray to very dark gray (5YR5/1 to 3/1, dry); clay; weakly stratified. Substratum 5Aℓ: same as 5B.	Substratum 5B: 10-25 cm thick; brown (e.g., 7.5YR5/3, dry); sandy clay loam to sandy loam, interbedded with common sand and gravel lenses. Substratum 5A: 30-75 cm thick; brown (e.g., 7.5YR5/3, dry); sandy clay loam to sandy loam interbedded with few sand and gravel lenses.
4	Substratum 4Bℓ: same as 5B Substratum 4A: less than 1 m thick; olive gray (2.5YR hues); laminated to massive, often cross-bedded, well-sorted, loamy fine sand to sandy clay loam interbedded with blocky to granular, somewhat more organic clay to clay loam.	Substratum 4B: 1-3+ m thick; brown (e.g., 7.5YR5/4, dry); sandy clay loam to sandy loam. No valley margin equivalent of 4A.
3	Substratum 3ℓ: 30-100+ cm thick; white (10YR7/1, dry); massive to platy, friable, silty clay to silty clay loam.	Substratum 3e: 30-100+ cm thick; light brown (7.5YR7/3, dry); sandy loam.
2	Substratum 2F: up to 30 cm thick; light gray (e.g., 2.5YR7/2, dry); sandy loam. Substratum 2B: 30-80 cm thick; gray (e.g., 10YR5/1, dry); loam to silty clay loam to clay; locally abundant silicified roots; few lenses of diatomite. Substratum 2A: 3-100 cm thick; light gray (10YR7/1, dry) diatomite interbedded with gray (e.g., 10YR5/1, dry) silt to clay.	No facies variation noted in 2F. Substratum 2s (facies of 2A and 2B): up to 2 m thick; gray (e.g., 2.5Y7/2, dry); silty clay interbedded with light gray (E.g., 2.5Y7/2, dry) sandy clay. Substratum 2e (facies of 2A and 2B); up to 2 m thick; pale brown (e.g., 10YR6/3, dry); sandy clay loam.
1	Highly variable, stratigraphic subdivisions reflect local lithologic changes and can occur individually or in various combinations. Substratum 1C: up to 50 cm thick; light gray (e.g., 2.5YR7/2, dry), massive sandy clay to clay. Substratum 1B: up to 1 m thick; light gray (e.g., 2.5Y7/2, dry) loose, cross-bedded, sand to loamy sand, with lenses of carbonate gravel (with clasts up to 2 cm in diameter). Substratum 1A: up to 1.5 m thick; massive carbonate gravel with clasts up to 5 cm in diameter and lenses of cross-bedded sand to loamy sand.	

*Facies are not necessarily time equivalents. Modified from Holliday (1985b).

the Holocene. Five principal strata (numbered 1 through 5, oldest to youngest, Figs. 3, 4; Table 1) and five soils (named) are identified at the site.

The oldest valley fill at the site is stratum 1, a deposit of gravel, sand, and clay exposed low in the walls of the reservoir cut. These sediments were laid down by a meandering stream under somewhat cooler and more moist conditions than today. The beginning date of this alluviation is not known, but it terminated about 11 ka. The earliest cultural remains at the site are found in stratum 1. About 11.1 ka (Clovis cultural age), the local inhabitants butchered a variety of extinct mammals along point bars of the stream. The animals include extinct bison (*Bison antiquus*), mammoth (*Mammuthus columbi*), camel (*Camelops hesternus*), horse (*Equus francisi* and *Equus mexicanus*), short-faced bear (*Arctodus simus*), and giant armadillo (*Holmesina septentrionale*). The latter two finds are the first documented in association with humans and the youngest reported in the paleontological literature.

Conformably above stratum 1 are lake and marsh deposits of stratum 2. The bedded diatomite in the lower portion of the unit is a distinct marker bed in the walls of the reservoir. Stratum 2 accumulated from 11 to about 8.5 ka. A marsh soil (Firstview Soil) then developed in the top of the deposit from 8.5 to about 6.4 ka. Most of the Paleoindian occupation of the site took place during stratum 2 sedimentation, including the Folsom culture (10.5 to 10.2 ka), for which the site is best known. Most of the cultural features are composed of the butchered remains of *Bison antiquus* with associated stone tools and projectile points. The early Archaic cultural period occurred during formation of the Firstview Soil.

Stratum 3 conformably overlies stratum 2. This deposit is composed of highly calcareous lacustrine sediment along the valley axis, and sandy eolian material along the valley margin. Stratum 3 was deposited between 6.4 and about 5.5 ka and represents the first of two significant episodes of warm, dry climate for the region. Formation of the Yellowhouse Soil (5.5 to

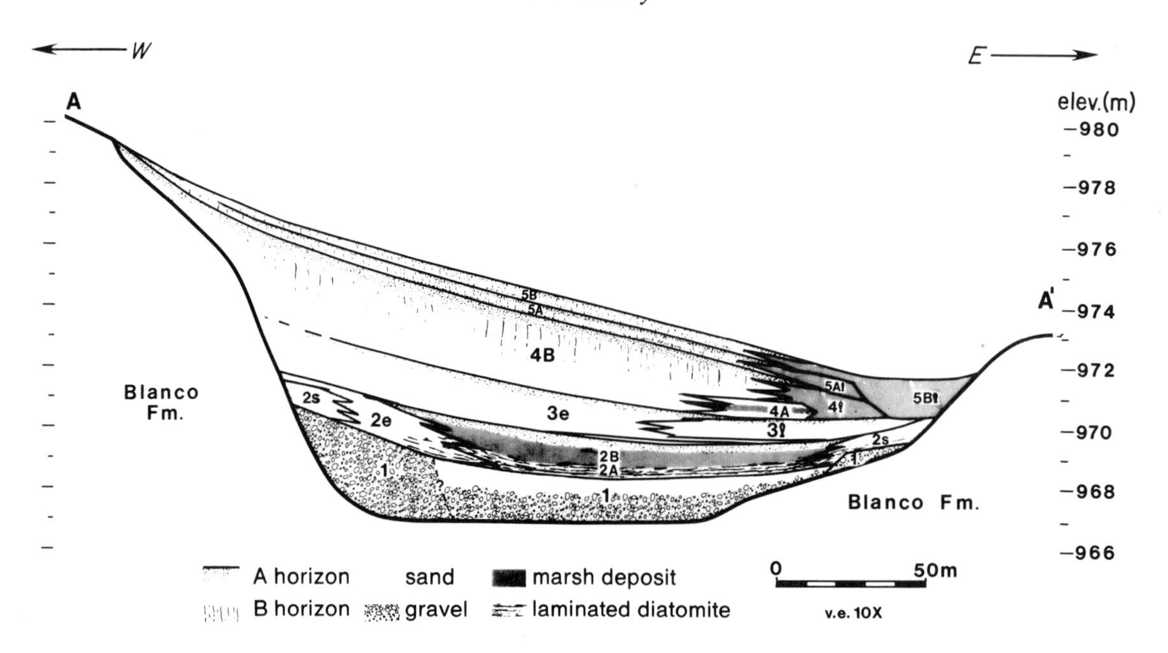

Figure 3. Late Quaternary stratigraphy of Yellowhouse Draw at Lubbock Lake (modified from Holliday, 1985b). Subdivisions of strata indicate either vertical sequence (such as A, B) or facies (s = shore, e = eolian, *l* = lacustrine).

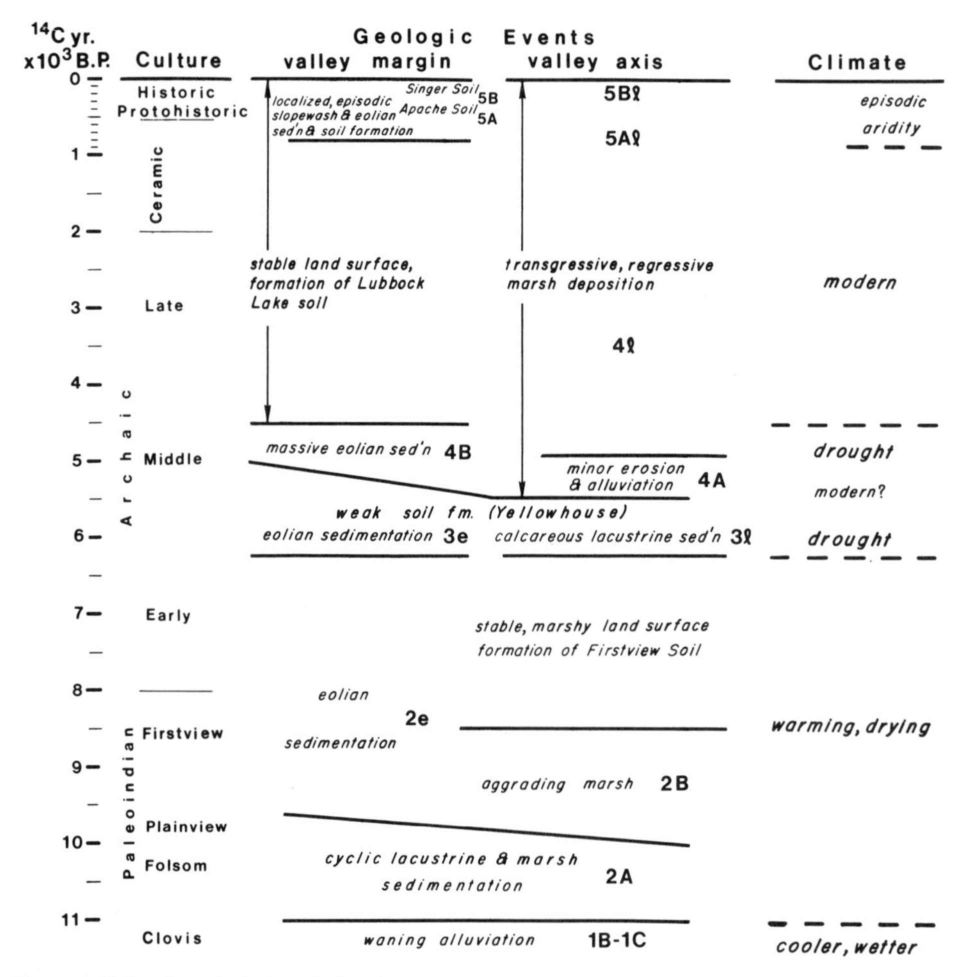

Figure 4. Cultural, geological, and climatic sequence at Lubbock Lake (modified from Holliday, 1985b).

5 ka in upper stratum 3 marks a brief, relatively cooler, wetter respite during the mid-Holocene dry interval.

Stratum 4 is an extensive deposit of sandy eolian material overlying stratum 3 and denoting the culmination of middle Holocene drough between 5.0 and 4.5 ka. Relatively more moist and cooler conditions prevailed beginning about 4.5 ka and represented by the Lubbock Lake soil, which is well developed in upper stratum 4 with a distinct A horizon and calcic horizon. Along much of the draw, stratum 4 is the most recent valley fill, and the Lubbock Lake Soil is the modern surface soil. Marsh clays were deposited along the valley axis during deposition of stratum 4 and formation of the Lubbock Lake Soil, indicating that the draw was never dry in this reach during middle and late Holocene time. Middle Archaic cultures occupied the site throughout the middle Holocene drought, and late Archaic cultures followed during the period of Lubbock Lake Soil pedogenesis. Most of the Archaic occupation features are accumulations of camping debris and butchered remains of modern bison (*Bison bison*).

Locally, stratum 4 is covered by stratum 5, which accumulated intermittently over the past 1,000 yr. Stratum 5 is composed of slopewash sand and gravel and eolian sand along the valley margin and marsh clay along the valley axis. The Apache and Singer Soils developed in stratum 5 and denote short-lived periods of nondeposition. The youngest of the stratum 5 marsh clays are probably related to a spring-fed lake that existed at the site throughout Historic times (it was noted by the early Spanish explorers), but disappeared in the early part of this century because of mining of the ground water. There are considerable accumulations of camping debris and butchered bison bone in stratum 5, the result of intense occupation during the late Ceramic, Protohistoric, and Historic cultural periods. The Apache were the Protohistoric and early Historic occupants of the region, replaced by the Comanche in the 18th century. Anglo-American settlements around the site in the 1880s mark the founding of Lubbock.

The sedimentology, invertebrate, and vertebrate paleontology, and paleobotany of Lubbock Lake provide an excellent record of late Quaternary climate change (Fig. 4). Furthermore, the stratigraphy and geochronology of the valley fill at Lubbock Lake is similar to that reported from other draw localities in the region (Fig. 5). Therefore, the record of late Quaternary climatic change at Lubbock appears to reflect regional climatic change.

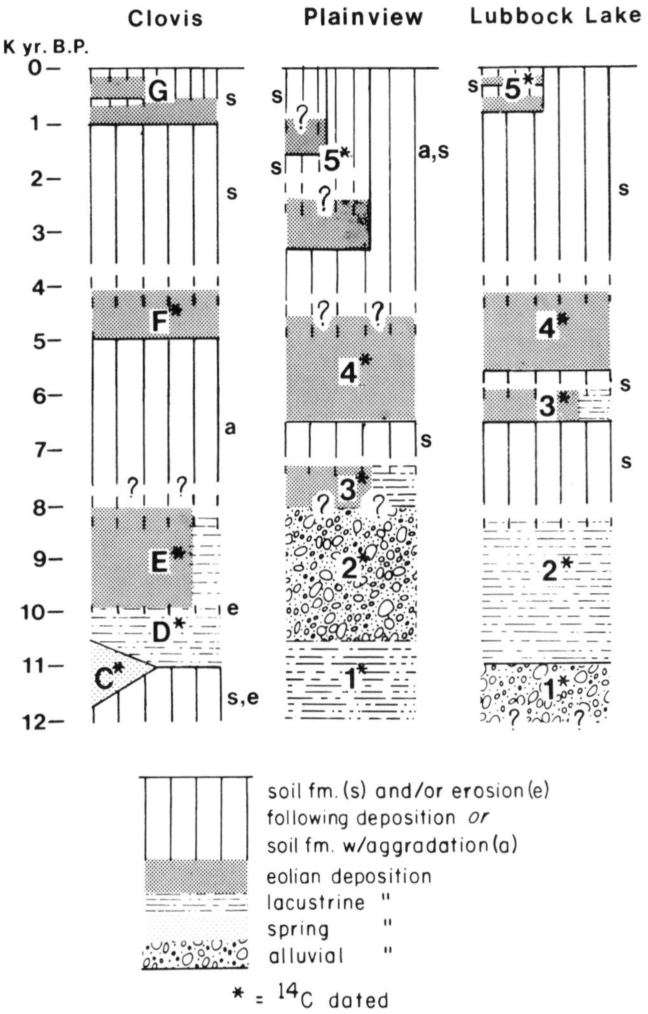

Figure 5. Stratigraphy at Lubbock Lake compared with that reported from Plainview Site in middle Running Water Draw (Fig. 1) (Holliday, 1985e, 1986) and Clovis site (Blackwater Draw Locality 1) in upper Blackwater Draw (Fig. 1) (Haynes and Agogino, 1966; Haynes, 1975; Holliday, 1985e).

REFERENCES CITED

Haynes, C. V., Jr., 1975, Pleistocene and Recent stratigraphy, *in* Wendorf, F., and Hester, J. J., eds., Late Pleistocene environments of the Southern High Plains: Publication of the Ft. Burgwin Research Center, v. 9, p. 57–96.

Haynes, C. V., Jr., and Agogino, G. A., 1966, Prehistoric springs and geochronology of the Clovis site, New Mexico: American Antiquity, v. 31, p. 812–821.

Holliday, V. T., 1985a, Holocene soil-geomorphological relations in a semiarid environment; The Southern High Plains of Texas, *in* Boardman, J., ed., Soils and Quaternary landscape evolution: Chichester, United Kingdom, John Wiley & Sons, p. 325–357.

——, 1985b, Archaeological geology of the Lubbock Lake site, Southern High Plains of Texas: Geological Society of America Bulletin, v. 96, p. 1483–1492.

——, 1985c, Early Holocene soils at the Lubbock Lake archaeological site, Texas: Catena, v. 12, p. 61–78.

——, 1985d, Morphology of late Holocene soils at the Lubbock Lake archaeological site, Texas: Soil Science Society of America Journal, v. 49,

p. 938–946.
——, 1985e, New data on the stratigraphy and pedology of the Clovis and Plainview sites, Southern High Plains: Quaternary Research, v. 23, p. 388–402.
——, 1986, Late Quaternary stratigraphy of the Plainview site and middle Running Water Draw, *in* Holliday, V. T., ed., Guidebook to the archaeological geology of classic Paleoindian sites on the Southern High Plains, Texas, and New Mexico: Geological Society of America Guidebook, Department of Geography, Texas A&M University, p. 60–70.
Holliday, V. T., Johnson, E., Haas, H., and Stuckenrath, R., 1983, Radiocarbon ages from the Lubbock Lake site, 1950–1980; Framework for cultural and ecological change on the Southern High Plains: Plains Anthropologist, v. 28, p. 165–182.
——, 1985, Radiocarbon ages from the Lubbock Lake site, 1981–1984: Plains Anthropologist, v. 30, p. 227–291.
Johnson, E., ed., 1987, Lubbock Lake; Late Quaternary Studies on the High Plains of Texas: College Station, Texas A&M University Press (in press).
Reeves, C. C., Jr., 1966, Pluvial lake basins of west Texas: Journal of Geology, v. 74, p. 269–291.

ACKNOWLEDGMENTS

This chapter benefited greatly by reviews from Thomas C. Gustavson (University of Texas) and John W. Hawley (New Mexico Bureau of Mines and Mineral Resources). The research at Lubbock Lake was supported by Eileen Johnson (director, Lubbock Lake Landmark, Texas Tech University) with funding from the National Science Foundation (SOC-14857, BNS76-12006, BNS76-12006-A01, BNS78-11155), the West Texas Museum Association, Texas Tech University, the Institute of Museum Research, and the Museum of Texas Tech University.

Ogallala and post-Ogallala sediments of the Southern High Plains, Blanco Canyon and Mt. Blanco, Texas

Paul N. Dolliver, *1100 Geomap Lane, Plano, Texas 75074*
Vance T. Holliday, *Department of Geography, Science Hall, University of Wisconsin, Madison, Wisconsin 53706*

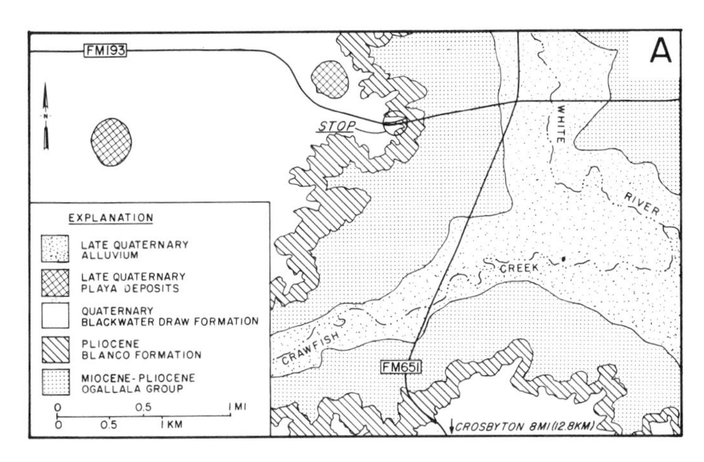

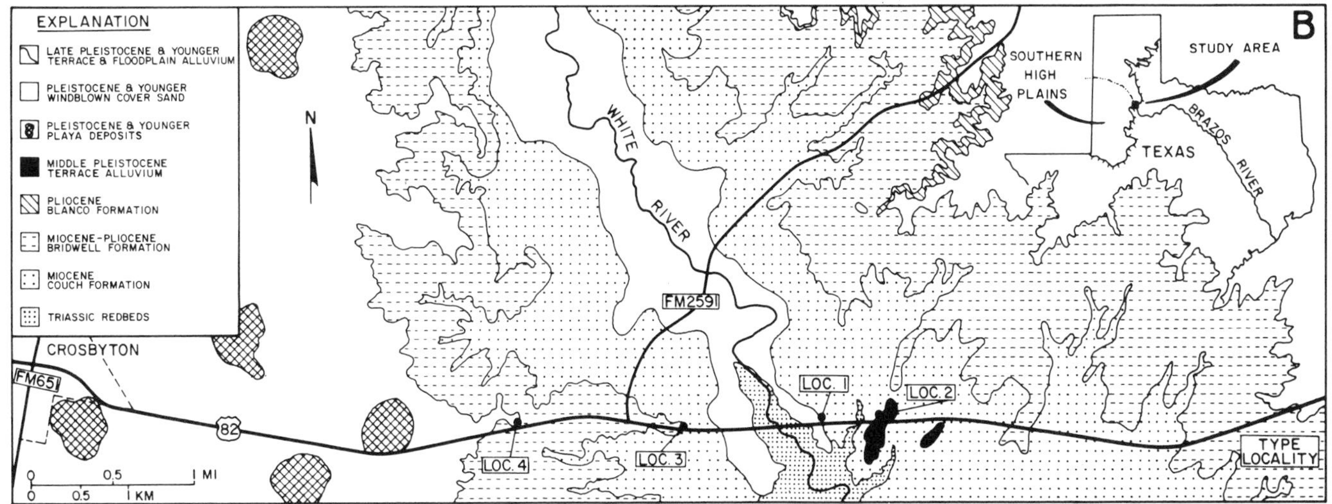

Figure 1. Geographic and geologic setting of described Mt. Blanco (A) and Blanco Canyon (B) localities, Floydada SE, Mt. Blanco, and Crosbyton 7½-minute Quadrangles. Outcrop limits are adapted from Barnes (1967) and Winkler (1985). The Couch-Bridwell type locality (on private property) is described by Evans (1949) and Winkler (1985).

LOCATION

Blanco Canyon and Mount Blanco are located along the eastern Caprock Escarpment of the Southern High Plains (Llano Estacado) in the Texas Panhandle near Crosbyton, Texas.

The Blanco Canyon exposures are along U.S. 82 where it crosses lower Blanco Canyon east of Crosbyton (Fig. 1). Within a distance of 5.5 mi (8.9 km), the highway descends from the High Plains through the Bridwell and Couch Formations (Ogallala Group) to Triassic red beds before climbing back onto the High Plains surface. The route passes near the Bridwell and Couch type sections and is flanked by several excellent roadside exposures. Outcrops at four Blanco Canyon localities are discussed in this chapter. Collectively, they document the buildup and subsequent dissection of the Southern High Plains since late Miocene time.

Mt. Blanco can be reached by travelling north from Crosbyton on Farm Road 651 across the High Plains surface (Blackwater Draw Formation). At 7.8 mi (12.5 km) the road descends into Blanco Canyon. Below the escarpment the highway crosses the mouth of Crawfish Creek, a major reentrant of Blanco Canyon. As the road descends the side of the canyon, the white beds of the Blanco Formation are apparent to the left and right just below the reddish sands of the Blackwater Draw Formation. After descending into the canyon, turn left (west) on Farm Road 193 and drive 0.7 mi (1.1 km) to Mt. Blanco, which is the conical erosional remnant on the south side of the road.

SIGNIFICANCE

Blanco Canyon is a deep reentrant into the southeast margin of the Southern High Plains where the White River, an upper tributary to the Salt Fork of the Brazos River, has excavated the canyon 6 mi wide (9.7 km) at its mouth and up to 450 ft (137 m)

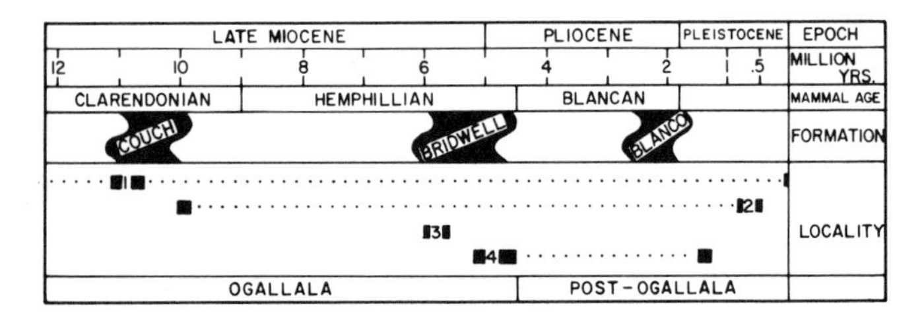

Figure 2. Chronostratigraphy of sites representing the Ogallala and post-Ogallala history of the Blanco Canyon area. The chart is based upon information from the following sources: Boellstorff, 1976; Hawley and others, 1976; Reeves, 1976; Lindsay and others, 1984; Tedford and Hunter, 1984; and Winkler, 1985.

below the High Plains surface, exposing Triassic red beds beneath Ogallala and post-Ogallala sediments (Figs. 2, 3). These sediments first attracted the attention of vertebrate paleontologists in the early 1890s, when remains of late Tertiary mammals were discovered (Schultz, 1972, 1977). Geologists in recent years have examined Blanco Canyon outcrops to assemble an exceptionally detailed, though inherently incomplete (Fig. 2), record of High Plains history.

The Mount Blanco site is the type locality for the Blanco Formation and the Blancan Local Fauna, which is the type fauna of the Blancan Land Mammal Age of North America (Evans and Meade, 1945; Meade, 1945; Evans, 1948; Schultz, 1977). The overlying Blackwater Draw Formation is the most extensive Quaternary deposit of the Southern High Plains. Within this formation are several buried soils and the 1.4-Ma Guaje Ash, which dates the lower Blackwater Draw Formation and provides an upper age for the Blanco Formation (Gustavson and Holliday, 1985).

BLANCO CANYON

Locality 1 consists of three exposures near the roadside park at Silver Falls, south of U.S. 82 and 4.75 mi (7.6 km) east of downtown Crosbyton, Texas (the intersection of U.S. 82 and Texas 652; Fig. 1). The falls occur where White River flows over resistant ledges of Triassic sandstone. Sand and gravel overlying the Triassic is either terraced alluvium from White River or, farther from the river, lower Couch Formation (basal Ogallala Group).

A roadcut north of U.S. 82 and opposite the eastbound exit from Silver Falls Park exposes 18 ft (5.5 m) of unconsolidated greenish gray to tan sand, silt, and gravel. Near the west end of the outcrop, massively bedded sand and silt contains abundant fossil shells of snails that lived on a well-watered flood plain during late Pleistocene (early Wisconsinan) time (Frye and Leonard, 1957). The gravel is largely concentrated in two laterally adjacent cross-stratified channel fill sequences. Most of the gravel is caliche, with some (less than 10 percent) quartzose material and a few mud balls. The late Pleistocene alluvium terminates upward in a smooth, essentially undissected terrace that is approximately 45 ft (13.7 m) above White River. Terraces of comparable age and landscape position floor virtually every valley of appreciable size that indents the eastern Llano Estacado (Barnes, 1967).

U.S. 82 climbs east of the early Wisconsinan terrace through older sand and gravel deposits of the lower Couch Formation. Two-tenths of a mile (0.3 km) east of the terrace, a roadcut west of a deep gully and south of the highway exposes 6 ft (1.8 m) of Couch gravel. Unlike the late Pleistocene gravel, Couch gravel contains few caliche pebbles or other material of local origin. Most of the clasts are quartzite, chert, granite, and other crystalline rocks derived from the Southern Rocky Mountains, more than 275 mi (443 km) to the northwest (Reeves, 1984; Dolliver, 1984). The east side of the gully rises to a 35-ft-high (10.7 m) roadcut. The lower 25 ft (7.6 m) of the cut is tan lower Couch sand unconformably overlying Triassic red beds. Lenses and stringers of gravel and mud balls are interbedded with the horizontally bedded to cross-stratified sand. The sequence is capped by up to 8 ft (2.4 m) of predominantly caliche gravel representing later depositional episode (see Locality 2).

Lower Couch gravel and mud-ball units commonly contain bones and teeth of late Miocene (early Clarendonian) mammals (Winkler, 1984, 1985). These remains date the beginnings of Ogallala alluviation on what is now the Great Plains, a physiographic province that extends from South Dakota to west Texas. Uplift of the Rocky Mountains and adjacent plains mobilized sediment that accumulated on a vast bedrock plain as alluvial fans, surficial veneers, and basal valley and basin fills (Seni, 1980; Hawley, 1984; Reeves, 1984). Ephemeral streams flowing across an irregular hummocky valley floor deposited older Ogallala sediments in the Blanco Canyon area (Knowles and others, 1984). The valley continued to fill with channel and overbank sediments, but also with progressively larger volumes of clayey eolian silt and sand (see Locality 3).

Locality 2 is a prominent roadcut of unconsolidated sand and gravel on the north side of U.S. 82, 0.1 mi (0.16 km) east of the most easterly Locality 1 exposure and 0.5 mi (0.8 km) east of White River (Fig. 1). The outcrop resembles terraced alluvium at Locality 1, though on a larger scale and for an earlier Pleistocene depositional cycle. Locality 2 culminates in a terrace that is about 60 ft (18 m) above its late Pleistocene counterpart and slightly more than 100 ft (30.5 m) above White River.

Alluvial fill at Locality 2 is stratigraphically continuous, with the caliche gravel capping lower Couch sand at Locality 1. The basal erosional contact noted at the first locality also crops out at the east end of Locality 2, where gravel lenses laterally abut a caliche horizon near the top of the lower Couch. Caliche gravel is concentrated in the lower 12 ft (3.7 m) of the outcrop. Average maximum clast size is significantly greater (by about 75 percent) than that of other Ogallala and post-Ogallala gravel. Approximately 27 ft (8.2 m) of sand and silt are interbedded with and overlie the gravel. Gravel laminae and reddish tan, pink, and greenish gray color banding highlight horizontal beds and cross-

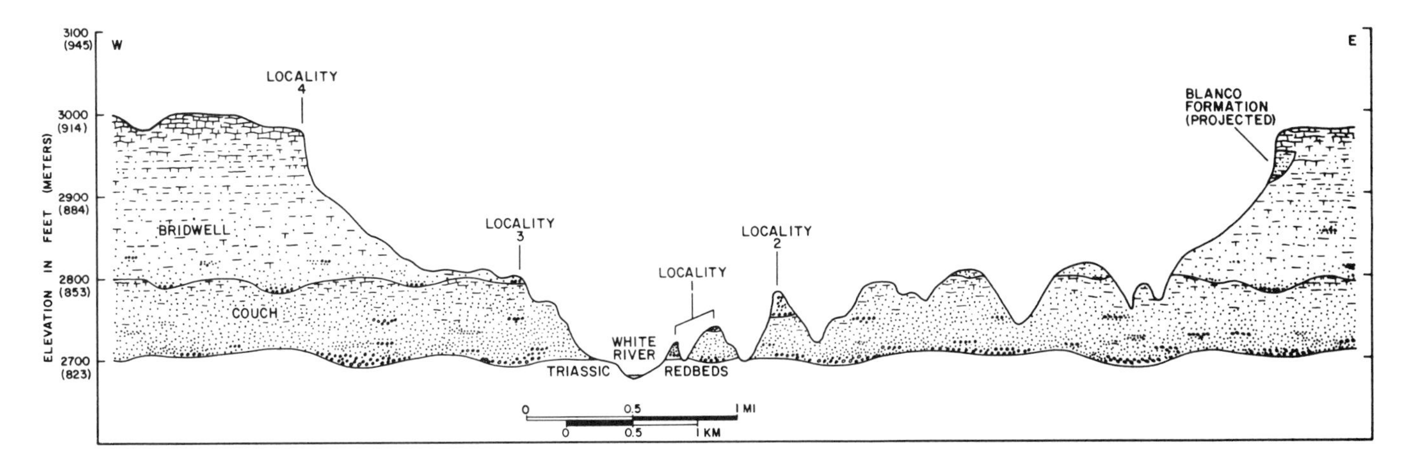

Figure 3. Generalized section across Blanco Canyon, emphasizing both the regional aspects of Ogallala stratigraphy and the magnitude of Pleistocene High Plains dissection. Line of section coincides with the route of U.S. 82 east of Crosbyton, Texas.

strata within the silty sand. Photographs published more than 20 years ago show the east end of Locality 2 culminating in up to 25 ft (7.6 m) of caliche gravel channel fill (Frye and Leonard, 1957, 1965). Only 6 ft (1.8 m) of this alluvium remains; the rest apparently has been excavated by the Texas Highway Department.

Fossils of aquatic and terrestrial snails, along with sparse clams and ostracods, are found within the sand at Locality 2. These creatures thrived during middle Pleistocene time on what was probably a well-watered wooded floodplain (Frye and Leonard, 1957; Pierce, 1975). The flood plain was built by a braided stream that was substantially larger than the modern White River. Today, White River originates on the New Mexico High Plains as Running Water Draw (Fig. 1), a small ephemeral stream occupying an anomalously large valley. The valley may have formerly contained the early-to-middle Pleistocene Portales River, a major drainage fed by montane headwaters and the increased runoff spawned by Pleistocene glacial climates (Reeves, 1972, 1976; Hawley and others, 1976; Gustavson and Finley, 1985). An ancestral Pecos River, migrating headward (northward) along the southern Rockies, pirated the Portales River during or soon after alluviation at Locality 2 (Pierce, 1975).

Locality 3 is a roadcut in the lower Bridwell Formation on the north side of U.S. 82, 1.2 mi (1.9 km) west of Locality 2 and 4 mi (6.4 km) east of downtown Crosbyton. After crossing White River, westbound U.S. 82 proceeds up section through nearly 100 ft (30 m) of the Couch Formation. The upper Couch (Crosbyton Member) is 45 ft (13.7 m) of light pinkish gray calcareous and clayey sand that typically weathers to form badlands, like those east of Locality 3. The locality itself is topographically less than 25 ft (7.6 m) above the middle Pleistocene terrace at Locality 2 (Fig. 3).

The lower Bridwell at Locality 3 is a 12 ft (3.7 m) section of unconsolidated, reddish clay-rich silty sand enveloping two gravel and sand-filled channels. Clasts within the channel fill are mostly well-rounded caliche, with a few quartzose pebbles and mud balls. The bases of both channels are just above road level, which here coincides with the approximate contact between the lower Bridwell and upper Couch formations. A dense caliche locally marks this contact and forms a prominent topographic bench (Fig. 3). Sparse, mostly weathered vertebrate fossils are scattered throughout the section, but tend to be concentrated within the channel fill. These remains are the chief basis for assigning a late Miocene (early Hemphillian) age to the lower Bridwell (Winkler, 1984).

Wind transported and deposited most sediment of the upper Couch Formation (Winkler, 1985). Caliche soils developed when the rate of sediment accumulation slowed. Couch deposition eventually ceased altogether, to be followed by a protracted period of erosion and deep stream dissection (Fig. 2). The resulting break in the sedimentary record is traceable throughout much of the southern Great Plains and coincides with renewed regional uplift of the southern Rockies (Winkler, 1985). Sedimentation resumed at Locality 3 with early Bridwell eolian deposition and channel filling. Mostly fine-grained Bridwell sediments continued to fill in local topographic lows throughout the Blanco Canyon area during alternating periods of eolian and ephemeral stream deposition (see Locality 4).

A pleasant diversion from the Blanco Canyon transect may be made between Localities 3 and 4. FM 2591 joins U.S. 82 0.4 mi (0.6 km) west of Locality 3. A 3-mi (4.8 km) drive northeast (to the right) from the junction leads obliquely across Blanco Canyon before climbing onto the High Plains (Fig. 1). White River is crossed where it occupies a 0.25 mi (0.4 km) wide-wooded flood plain. Locales such as this have provided shelter, food, wood, and water to humans on the High Plains for more than 10,000 years (see Holliday, this volume). Prior to irrigation, springs issuing from the lower Ogallala fed White River and several other High Plains border streams. Early pioneers referred to White River as the Freshwater Fork of the Brazos, to distinguish it from brackish streams common on the red-bed plains to the east (Rathjen, 1973). Geologists have noted artesian discharge from the Ogallala as being an important influence on retreat of the Caprock Escarpment, chiefly through the processes of sapping and seepage erosion (Fenneman, 1931; Osterkamp and Wood, 1984).

Flowing ground water also has modified the High Plains by dissolving subsurface Permian salt beds. Removal of up to several hundred feet of salt has induced extensive late Cenozoic surface subsidence, mostly along the periphery of the Southern High Plains (Gustavson and Finley, 1985). Blanco Canyon is west of this peripheral zone of dissolution and has undergone comparatively minor subsidence. FM 2591, near where it crests the High Plains, passes what may be indirect evidence of post-Ogallala surface collapse. Roadcuts at the base of the "caprock" caliche are in sand and carbonate of the late Pliocene Blanco Formation. Blanco sediments accumulated in playa lake basins that are situated over areas of thin Permian salt, suggesting dissolution-induced subsidence may have influenced basin location. The position of shallow Blancan lake basins in turn apparently guided early incision of Blanco Canyon (Brand, 1974; Gustavson and Finley, 1985).

Locality 4 is a prominent upper Bridwell roadcut on the north side of U.S. 82 east of where the highway emerges from Blanco Canyon onto the High Plains. The site is 0.7 mi (1.1 km) west of the intersection of U.S. 82 and FM 2591, 1 mi (1.6 km) west of Locality 3, and 3 mi (4.8 km) east of downtown Crosbyton. Aggregate thickness of the Bridwell Formation from its base at Locality 3 to its top marked by the "caprock" caliche at Locality 4 is approximately 165 ft (50 m); the upper 50 to 55 ft (15 to 17 m) are exposed in the Locality 4 roadcut. The total Bridwell thickness is about average for the Blanco Canyon area (Evans and Brand, 1956). Locally, the formation thickens where it fills valleys cut into the underlying Couch and thins over pre-Bridwell topographic highs (Evans, 1949; Winkler, 1985).

The lower 20 ft (6 m) of section at Locality 4 is reddish brown, horizontally-bedded sand and caliche gravel containing several lenses of cross-stratified channel fill. Mud balls are abundant in the fill. The uppermost 30 to 35 ft (9 to 11 m) of Bridwell is composed of red clay-rich silty sand containing two thick (10 to 13 ft; 3 to 4 m) caliche beds that are traceable for the length of the outcrop. The upper caliche bed grades into 20 ft (6 m) of overlying dense "caprock" caliche.

Upper Bridwell channel sands near road level at Locality 4 yield remains of late Miocene-early Pliocene (Hemphillian) age mammals (Winkler, 1985). The animals once inhabited a broad plain of coalescent alluviation built by aggrading ephemeral streams and influxes of windblown sandy loess (Schultz, 1977; Winkler, 1985). Eolian processes and caliche soil formation prevailed as stream valleys filled and divides were covered. Buildup of the Southern High Plains finally ended in development of a subsoil lime accumulation that through several Plio-Pleistocene cycles of exposure, dissolution, and redeposition became the resistant "caprock" caliche (Reeves, 1970).

Eolian sand of the Pleistocene Blackwater Draw Formation thinly mantles the crest of the Locality 4 roadcut. This areally extensive "cover sand" is both the source of agricultural productivity on the High Plains and a reason for the region's extraordinarily low relief. The top of Locality 4 is also an excellent vantage point from which to contemplate the late Miocene to Recent history of Blanco Canyon and the Southern High Plains.

MT. BLANCO

The Blanco Formation is about 60 ft (20 m) thick at Mt. Blanco, with the upper section exposed in the roadcuts. Interbedded sand and dolomitic marl of the lower Blanco Formation, including the bone-bearing zone and the Blanco Ash, are exposed in the badlands south of the road (note: this is private property—*do not enter*). The basal erosional contact between the white Blanco Formation and the red silty sand of the Ogallala Formation is also visible from the road.

The Blanco "beds," particularly those containing vertebrate remains, have been investigated by geologists and paleontologists since the end of the last century. Schultz (1977) has summarized these investigations, focusing on various interpretations of the age and origin of the Blanco Formation. Pierce (1973, 1974) conducted the most recent and comprehensive geological study of the unit.

Several lines of evidence support a late Pliocene age for the Blanco Formation at Mt. Blanco, if the age of 1.6 Ma is accepted for the Plio-Pleistocene boundary (e.g., Berggren and others, 1985; Aguirre and Pasini, 1985). A volcanic ash, probably derived from eruptions in the Pacific Northwest and termed the "Blanco Ash," overlies the sandy bone beds south of Farm Road 193 and has yielded a glass-shard fission track age of 2.32 ± 0.15 Ma (Izett and others, 1972; Izett, 1981). Paleomagnetic studies by Lindsay and others (1975) indicate that the time of Blanco deposition at the type locality was between 2.4 and 1.4 Ma. Sands of the Blackwater Draw Formation disconformably overlie the Blanco Formation at Mt. Blanco and contain the 1.4 Ma Guaje Ash (see below). The contact is marked by a Stage IV–V pedogenic calcrete developed in sediments of the Blanco Formation (after Machette, 1985), which probably took several hundred thousand years to form (Machette, 1985). Therefore, Blanco deposition probably ended prior to 1.6 Ma.

The Blanco Formation at Mt. Blanco occupies a broad basin inset into the Ogallala Formation. Clasts of the Ogallala "caprock caliche" in basal Blanco deposits indicate that the basin developed long after Ogallala deposition ceased. The Blanco Formation is as much as 90 ft (27 m) thick near the center of the basin, where it is composed primarily of very fine sand, dolomicrite, and magnesium-rich clay. Diatomite and volcanic ash occur locally. Near the basin margin, Blanco deposits thin to a feather edge and typically consist of limestone and coarse clastics, including caliche gravel. Fossil vertebrates, diatoms, ostracodes, and pollen from the lower Blanco indicate deposition during a semi-arid to arid (possibly preglacial) climate and in a seasonal to semi-permanent playa that existed between frequent periods of desiccation (Pierce, 1973, 1974).

The origins of the Blanco basin are unclear. Pierce (1974) has suggested that the basin was a stream channel enlarged by deflation. He also noted that "the caprock caliche . . . dips toward the [Blanco] basin . . . before pinching out" (Pierce, 1974, p. 16). This suggests solution-induced subsidence, which influenced development of large Pleistocene basins elsewhere in the region (Gustavson and Finley, 1985; Gustavson, 1986). Basins containing Blanco or other Plio-Pleistocene–age lake and playa deposits are common along reentrants of the eastern margin of the Southern High Plains.

The roadcuts along Farm Road 193 expose 23 to 26 ft (7 to 8 m) of sand and carbonate of the upper Blanco Formation. The lower 9 to 12 ft (3 to 4 m) of the cut are unconsolidated white sand. Most of the famous Blancan vertebrate fossils were collected from equivalents of this sand, where it directly underlies the Blanco Ash and diatomite. Laminated white, silty to sandy lacustrine carbonates (commonly contorted) and pink sands compose the uppermost 12 ft (4 m) of the Blanco Formation, with the previous mentioned pedogenic calcrete at the top of the unit.

Above the Blanco beds are reddish-brown, sandy sediments of the Blackwater Draw Formation, which cover most of the Southern High Plains. The Blackwater Draw Formation is the "Illinoisan cover sand" of Frye and Leonard (1957). Reeves (1976) renamed the deposits, and Gustavson and Holliday (1985) demonstrated that the sediments accumulated throughout most of the Quaternary, based in part on evidence from the Mt. Blanco section. At Mt. Blanco the 1.4-Ma Guaje Ash, derived from the Jemez volcanic field in New Mexico, provides a minimum date for the beginning of deposition of the Blackwater Draw Formation. At a locality near Tulia, Texas, the Yellowstone-derived 0.6 Ma Lava Creek ash is exposed within the Blackwater Draw Formation (Izett and Wilcox, 1982).

The Blackwater Draw Formation is an eolian deposit with numerous buried soils. It grades in texture from sandy in the southwestern part of the Southern High Plains to a silty clay in the northeast. No primary sedimentary structures are preserved. Most of the field characteristics of the Blackwater Draw Formation are the result of soil-forming processes. The buried and surface soils are quite similar, with reddish-brown (2.5 YR to 5 YR) hues, considerable translocated clay, strong prismatic structure, and distinct zones of calcium carbonate accumulation (calcic horizons).

At Mt. Blanco the Blackwater Draw Formation is about 15 ft (5 m) thick (Fig. 2). The Guaje Ash is a prominent ledge-forming layer in the lower portion of the formation. Below the ash is a buried soil about 1 m thick that rests on the Blanco calcrete. Above the ash there are at least two buried soils and the modern surface soil, each more than 1 m thick. Besides color, the most distinctive feature of these soils is the calcic horizons, which form relatively prominent ledges with sharp upper boundaries and diffuse lower boundaries. These zones are most noticeable in the cut south of the road (Fig. 4).

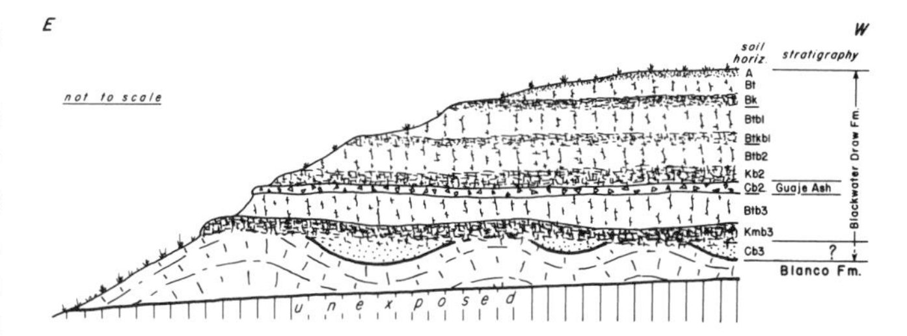

Figure 4. Generalized stratigraphic cross section of south side of roadcut at Mt. Blanco section (looking south).

SUMMARY

Blanco Canyon and the sediment it contains are evidence for the late Tertiary buildup and subsequent dissection of the Southern High Plains. Eleven million years ago, the area was a rolling subhumid parkland savanna that supported animals whose diversity rivaled that of today's African savannas. Horses, camels, deer, and rhinos were among the wildlife that probably frequented watering holes along intermittent stream courses (Winkler, 1985). Repeated flooding of these streams and periodic dust storms progressively built the Ogallala plain between extended periods of landscape stability and erosion. Animal variety diminished throughout late Miocene and Pliocene time as the climate gradually became drier and more seasonal (Webb, 1984). Late Pliocene grazing mammals and their predators watered at Blancan playa lakes that dotted a semiarid short-grass steppe resembling that described by early explorers and pioneers (Webb, 1931; Rathjen, 1973).

Conditions on the Southern High Plains changed dramatically less than 2 m.y. ago with the onset of cyclic Pleistocene glaciation far to the north and west. Cooler temperatures and probably greater rainfall sustained rivers that sharply incised the Ogallala plain. Two-thirds of the depth of Blanco Canyon was cut in roughly 1 m.y. (Fig. 2, 3). Like the processes that built the High Plains, those causing their dissection were probably most effective during infrequent flood events. Blanco Canyon, contrasted with the pre-Pleistocene geologic record that it exposes, exemplifies the profound imact of cyclic Pleistocene climatic change upon the landscape.

REFERENCES CITED

Aguirre, E., and Pasini, G., 1985, The Pliocene-Pleistocene boundary: Episodes, v. 8, p. 116–120.

Barnes, V. E., 1967, Lubbock sheet: The University of Texas at Austin, Bureau of Economic Geology Geologic Atlas of Texas, scale 1:250,000.

Berggren, W. A., Kent, V., Flynn, J., and Van Couvering, A., 1985, Cenozoic geochronology: Geological Society of America Bulletin, v. 96, p. 1407–1418.

Boellstorff, J. D., 1976, The succession of late Cenozoic volcanic ashes in the

Great Plains; A progress report: Kansas Geological Survey Guidebook Series, v. 1, p. 37–71.

Brand, J. P., 1974, Guidebook to the Mesozoic and Cenozoic geology of the southern Llano Estacado: Lubbock Geological Society, 72 p.

Dolliver, P. N., 1984, Cenozoic evolution of the Canadian River basin: Waco, Texas, Baylor Geological Studies Bulletin 42, 96 p.

Evans, G. L., 1948, Geology of the Blanco beds of West Texas, *in* Colbert, E. H., ed., Pleistocene of the Great Plains: Geological Society of America Bulletin, v. 59, p. 617–619.

—— , 1949, Upper Cenozoic of the High Plains, *in* Evans, G. L., and others, Cenozoic geology of the Llano Estacado and Rio Grande Valley: West Texas Geological Society and New Mexico Geological Society Field Trip Guidebook no. 2, p. 1–22.

Evans, G. L., and Brand, J. P., 1956, Eastern Llano Estacado and adjoining Osage Plains: West Texas Geological Society and Lubbock Geological Society Field Trip Guidebook, 99 p.

Evans, G. L., and Meade, G. E., 1945, Quaternary of the Texas High Plains: University of Texas Publication 4401, p. 485–507.

Fenneman, N. M., 1931, Physiography of western United States: New York, McGraw-Hill Book Company, Incorporated, 534 p.

Frye, J. C. and Leonard, A. B., 1957, Studies of Cenozoic geology along eastern margin of Texas High Plains, Armstrong to Howard counties: The University of Texas at Austin, Bureau of Economic Geology Report of Investigations no. 32, 62 p.

—— , 1965, Quaternary of the southern Great Plains, *in* Wright, H. E., Jr., The Quaternary of the United States: Princeton, New Jersey, Princeton University Press, p. 203–216.

Gustavson, T. C., 1986, A possible origin of the Tule basin, *in* Gustavson, T. C., ed., Geomorphology and Quaternary stratigraphy of the Rolling Plains, Texas Panhandle: University of Texas, Bureau of Economic Geology Guidebook 22, p. 73–78.

Gustavson, T. C., and Finley, R. J., 1985, Late Cenozoic geomorphic evolution of the Texas Panhandle and northeastern New Mexico; Case studies of structural controls of drainage development: The University of Texas at Austin, Bureau of Economic Geology Report of Investigations no. 148, 42 p.

Gustavson, T. C., and Holliday, V. T., 1985, Depositional architecture of the Quaternary Blackwater Draw and Tertiary Ogallala Formations, Texas Panhandle and eastern New Mexico: University of Texas, Bureau of Economic Geology Open File Report, OF-WTWI-1985-23, 92 p.

Hawley, J. W., 1984, The Ogallala Formation in eastern New Mexico, *in* Whetstone, G. A., ed., Proceedings of the Ogallala Aquifer symposium II: Lubbock, Texas Tech. University, p. 157–176.

Hawley, J. W., Bachman, G. O., and Manley, K., 1976, Quaternary stratigraphy in the Basin and Range and Great Plains provinces, New Mexico and western Texas, *in* Mahaney, W. C., ed., Quaternary stratigraphy of North America: Stroudsburg, Pennsylvania, Dowden, Hutchinson and Ross, Incorporated, p. 235–274.

Izett, G. A., 1981, Volcanic ash beds; Recorders of upper Cenozoic silicic pyroclastic volcanism in the western United States: Journal of Geophysical Research, v. 86, p. 10200–10222.

Izett, G. A., and Wilcox, R. E., 1982, Map showing localities and inferred distributions of the Huckleberry Ridge, Mesa Falls, and Lava Creek ash beds (Pearlette family ash beds) of Pliocene and Pleistocene age in the western United States: U.S. Geological Survey Miscellaneous investigations Series Map I-1325, scale 1:4,000,000.

Izett, G. A., Wilcox, R. E., and Borchardt, G. A., 1972, Correlation of a volcanic ash bed in Pleistocene deposits near Mount Blanco, Texas, with the Guaje pumice beds of the Jemez Mountains, New Mexico: Quaternary Research, v. 2, p. 554–578.

Knowles, T., Nordstrom, P., and Klemt, W. B., 1984, Evaluating the groundwater resources of the High Plains of Texas, v. I: Texas Department of Water Resources Report 288, 177 p.

Lindsay, E. H., Johnson, N. M., and Opdyke, N. D., 1975, Preliminary correlation of North American land mammal ages and geomagnetic chronology, *in* Smith, G. R., and Friedland, N. E., eds., Studies on Cenozoic paleontology and stratigraphy in honor of Claude W. Hibbard: University of Michigan, Museum of Paleontology, p. 111–119.

Lindsay, E. H., Opdyke, N. D., and Johnson, N. M., 1984, Blancan-Hemphilian land mammal ages and late Cenozoic mammal dispersal events: Annual Review of Earth and Planetary Sciences, v. 12, p. 445–488.

Machette, M. N., 1985, Calcic soils of the Southwestern United States, *in* Weide, D. L., ed., Soils and Quaternary geology of the Southwestern United States: Geological Society of America Special Paper 203, p. 1–21.

Meade, G. E., 1945, The Blanco fauna: University of Texas Publication 4401, p. 509–556.

Osterkamp, W. R., and Wood, W. W., 1984, Development and escarpment retreat of the Southern High Plains, *in* Whetstone, G. A., ed., Proceedings of the Ogallala Aquifer Symposium II: Lubbock, Texas Tech. University, p. 177–193.

Pierce, H. G., 1973, The Blanco beds; Mineralogy and paleoecology of an ancient playa [M.S. thesis]: Lubbock, Texas Tech University, 93 p.

—— , 1974, The Blanco beds, *in* Brand, J. P., ed., Mesozoic and Cenozoic geology of the Southern Llano Estacado: Texas Tech University, Department of Geosciences, p. 9–16.

—— , 1975, Diversity of late Cenozoic gastropods on the Southern High Plains [Ph.D. thesis]: Lubbock, Texas Tech. University, 267 p.

Rathjen, F. W., 1973, The Texas Panhandle frontier: Austin, University of Texas Press, 286 p.

Reeves, C. C., Jr., 1970, Origin, classification, and geologic history of caliche on the Southern High Plains, Texas and eastern New Mexico: Journal of Geology, v. 78, p. 352–362.

—— , 1972, Tertiary-Quaternary stratigraphy and geomorphology of west Texas and southeastern New Mexico, *in* Kelley, V. C., and Trauger, F. D., eds., Guidebook of east-central New Mexico: New Mexico Geological Society, p. 108–117.

—— , 1976, Quaternary stratigraphy and geologic history of Southern High Plains, Texas and New Mexico, *in* Mahaney, W. C., ed., Quaternary stratigraphy of North America: Stroudsburg, Pennsylvania, Dowden, Hutchinson and Ross, Incorporated, p. 213–234.

—— , 1984, The Ogallala depositional mystery, *in* Whetstone, G. A., ed., Proceedings of the Ogallala Aquifer Symposium II: Lubbock, Texas Tech. University, p. 129–156.

Schultz, G. E., 1972, Vertebrate paleontology of the Southern High Plains, *in* Kelley, V. C., and Trauger, F. D., eds., Guidebook of east-central New Mexico: New Mexico Geological Society, p. 129–133.

—— , 1977, Field conference on late Cenozoic biostratigraphy of the Texas Panhandle and adjacent Oklahoma: West Texas State University, Killgore Research Center Special Publication no. 1, 160 p.

Seni, S. J., 1980, Sand-body geometry and depositional systems, Ogallala Formation, Texas: The University of Texas at Austin, Bureau of Economic Geology Report of Investigations no. 105, 36 p.

Tedford, R. H., and Hunter, M. E., 1984, Miocene marine-nonmarine correlations, Atlantic and Gulf coastal plains, North America: Palaeogeography, Palaeoclimatology, Palaeoecology, v. 47, p. 129–151.

Webb, S. D., 1984, Ten million years of mammal extinctions in North America, *in* Martin, P. S., and Klein, R. G., eds., Quaternary extinctions; A prehistoric revolution: Tucson, University of Arizona Press, p. 189–210.

Webb, W. P., 1931, The Great Plains: New York, Grosset and Dunlap, 525 p.

Winkler, D. A., 1984, Vertebrate biostratigraphy of the Ogallala Group (Miocene), Southern High Plains, Texas: Geological Society of America Abstracts with Programs, v. 16, p. 698.

—— , 1985, Stratigraphy, vertebrate paleontology, and depositional history of the Ogallala Group in Blanco and Yellowhouse Canyons, northwestern Texas [Ph.D. thesis]: The University of Texas at Austin, 243 p.

ACKNOWLEGMENTS

Some of the research at the Mt. Blanco section was supported by the Bureau of Economic Geology, The University of Texas at Austin, with funds from the U.S. Department of Energy.

The Triassic section of the West Texas High Plains

Ted Gawloski, Mitchell Energy Corporation, 200 North Loraine, Midland, Texas 79701

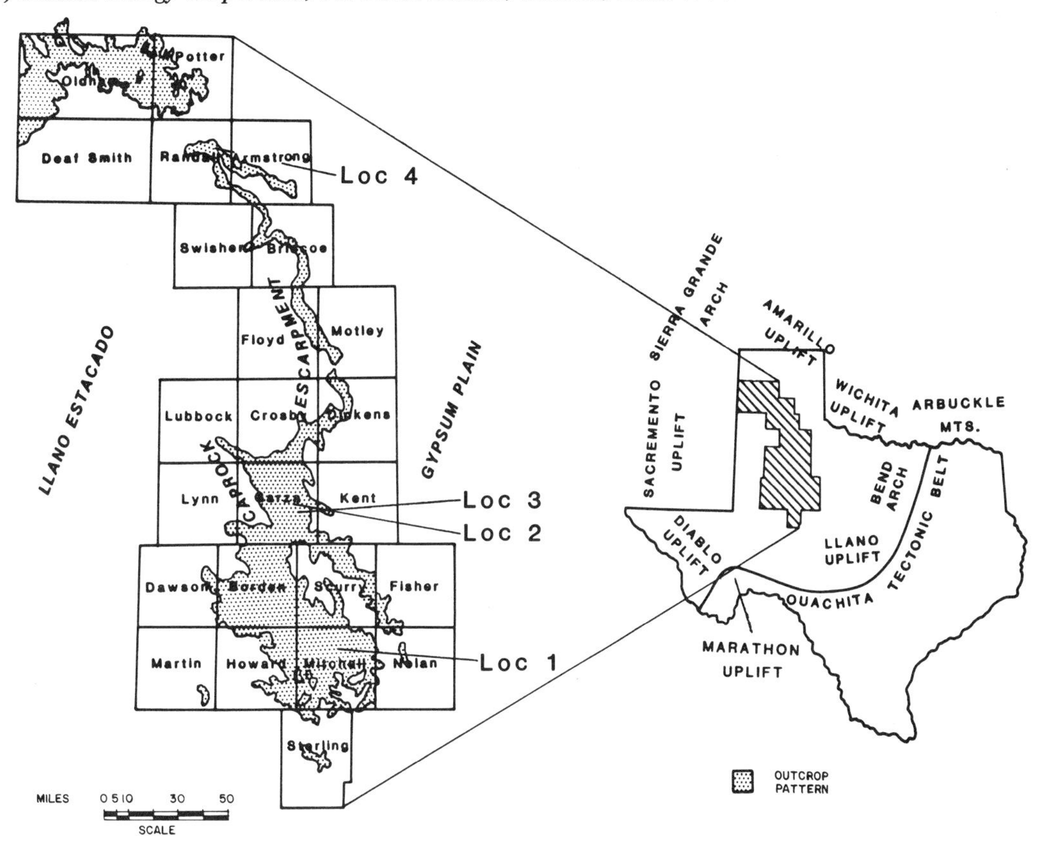

Figure 1. Outcrop and locality map of the Upper Triassic Dockum Group showing the major tectonic and physiographic provinces of the region.

LOCATION

The Triassic section of the West Texas High Plains is represented by the rocks of the Dockum Group. This section consists of up to 2,000 ft (610 m) of complex clastic sequence composed of conglomerates, sandstones, and shales. These sediments were deposited by a variety of terrigenous depositional systems into a broad, shallow, fluvial-lacustrine basin stretching across what is now the Panhandle region of Texas. The location and geometry of the Dockum Basin probably reflect Paleozoic structural elements lying in an area roughly defined by the Midland Basin. It is bounded to the north by the Sierra Grande Arch and Amarillo Uplift and to the south by the Llano and Marathon Uplifts (Fig. 1). Most of the Triassic deposits of West Texas occur beneath younger deposits of the Llano Estacado. These rocks crop out along the Caprock Escarpment from Oldham County southward to Sterling County (Fig. 1). Terminology applied to the Dockum Group of West Texas has undergone considerable evolution. Presently, the Dockum Group of the West Texas High Plains is divided into two formations. These are, from oldest to youngest, the Tecovas and Trujillo formations (Fig. 2). In the southern outcrop areas, however, the formational divisions are difficult to identify.

The traverse begins in rocks of the Dockum Group along the southern portion of the outcrop belt (locality 1) (Fig. 1) and proceeds northward along the Caprock Escarpment (localities 2 and 3) (Fig. 1), and finally to Palo Duro Canyon where thick sections of the Tecovas and Trujillo formations occur (locality 4) (Fig. 1).

SIGNIFICANCE

The Triassic section (Dockum Group) of the West Texas High Plains comprises a unique and complex red-bed sequence consisting of variegated sands, shales, and conglomerates. These upper Triassic rocks unconformably overlie the red shales of the Upper Permian Quatermaster Group. This contact represents a time gap of about 35 m.y. The Dockum Group is overlain by the Ogallala Group of Pliocene (Tertiary) age and to a lesser extent by lower Cretaceous limestones and shales. Figure 2 shows the lithologic and stratigraphic character of the various units that are associated with rocks of the Dockum Group.

The landscape of the outcrop belt clearly reflects the geo-

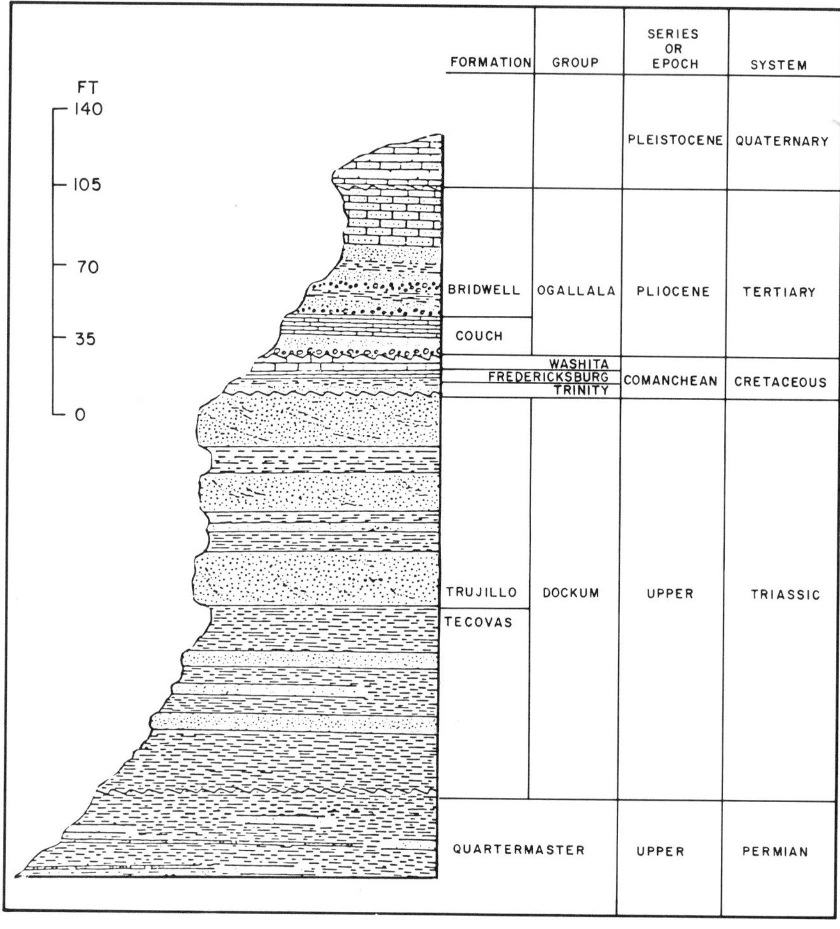

Figure 2. Composite columnar section of the Southern High Plains showing the lithologic and stratigraphic relationships of the Upper Permian, Upper Triassic, Cretaceous, and Tertiary sediments.

morphic association of the Dockum rocks to the overlying Cretaceous and/or Pliocene deposits. Any of the exposures of the Dockum rocks found along the eastern escarpment edge (Caprock) of the West Texas High Plains are representative of the general character of the Triassic stratigraphic section.

Dockum Outcrop—Mitchell County, Texas

This traverse begins at the intersection of I-20 and Texas 208 in Colorado City, Texas. Proceed south on Texas 208 to the junction with Texas 163, 1.7 mi (2.8 km). Turn left on Texas 163 and continue traveling south to locality 1, 8.4 mi (13.5 km).

Locality 1 is on the southern end of the Caprock Escarpment, just south of the present Colorado River drainage.

Locality 1 (Fig. 1) is exposed in a roadcut on both sides of the highway at the crest of a broad escarpment. In general, the section consists predominantly of discontinuous lenticular sandstones and conglomerates interbedded with red and green mudstones, siltstones and shales, divided into several lithologic units described (from bottom to top) in the following paragraph.

The lowermost unit is composed of red shale topped by a thin, light green, clay layer. This is overlain by a dark brown-gray, dense, micaceous, crossbedded, lenticular sand and conglomerate layer. The conglomerate has clasts of chert, sand and siltstone up to 3 in (8 cm) in diameter. Overlying this unit is a buff, medium-grained micaceous sandstone with large-scale crossbedding. This next unit consists of several alternating beds of dark red channel sandstones, siltstones, and shales. The sandstones are fine grained, silty, and exhibit excellent forset-crossbeds. The overlying unit contains numerous lenticular conglomerate and sandstone beds with well developed cut and fill structures, cut into the underlying siltstones and shales (Fig. 3). The next unit is a massive channel-fill sandstone approximately 120 ft (37 m) across and 5 to 7 ft (1.5 to 2 m) thick (Fig. 4), composed of fine- to medium-grained silty sandstone, exposed on both sides of the road. The overlying units consist of red sandstones interbedded with thin lenticular conglomerates. Above this zone is a 2 to 3 ft (0.6 to 0.9 m) dark brown, fine-grained, clean, resistant sandstone bed. The uppermost unit consists of fine-grained sandstone and siltstone lenses interbedded with red shales.

Figure 3. Locality 1. Well developed cut-and-fill structure characteristic of Dockum sedimentation is shown at the top of the roadcut. The deposits consist of several conglomerate and sandstone layers that have cut into the underlying shales and siltstones.

Figure 4. Locality 1. A massive sandstone channel-fill approximately 120 ft (37 m) across and 5 to 7 ft (1.6 to 2.1 m) thick is exposed along both sides of the roadcut. The channel is composed almost entirely of fine to medium-grained silty sandstone and is in sharp contact with the underlying shale and sandstone layers.

Dockum Outcrops—Garza County, Texas

This traverse begins at the intersection of U.S. 380 and U.S. 84 in Post, Texas. The traverse extends southeast along U.S. Highway 84.

Localities 2 and 3 also occur along the Caprock Escarpment where Dockum rocks form the smooth red slopes.

Locality 2 (Fig. 1) is 3.1 mi (5 km) south of Post on U.S. 84 in a roadcut on both sides of the highway. Here the Dockum consists of approximately 25 ft (7.5 m) of red and green shales overlain by 6 to 8 ft (1.8 to 2.4 m) of light green-gray-brown coarse-grained, crossbedded, silty channel sandstones and conglomerates. The sandstones and conglomerates form a resistant ledge with numerous slump blocks. Similar sections of Dockum rocks crop out 8 mi (13 km) and 9.3 mi (15 km) south of Post on U.S. 84.

Locality 3 (Fig. 1) is 16.7 mi (27 km) south of Post (2 mi; 3.2 km south of Justiceburg) on U.S. 84. Here the Dockum consists of an excellent channel-fill sandstone and conglomerate sequence cut into the adjacent red and green shales and clays. The channel-fill contains a basal conglomerate with clasts composed of weathered sandstone and siltstone up to 1.5 in (3.75 cm) in diameter, overlain by a light gray, medium-grained silty sandstone with abundant medium-scale crossbed sets and minor gravel lag deposits. Other well-exposed Dockum outcrops occur 1 mi (1.6 km) and 1.7 mi (2.7 km) south of locality 3 on U.S. 84.

Dockum Outcrops—Palo Duro Canyon State Park, Randall County, Texas

This traverse begins at the ranger headquarters to Palo Duro Canyon State Park 12.5 mi (20 km) east of Canyon, Texas, on Texas 217.

Locality 4 (Fig. 1) is at the scenic overlook 0.8 mi (1.3 km) from the park headquarters on Park Road 5. This locality gives an excellent overview of the canyon and offers a unique opportunity to view the Dockum rocks in association with the underlying Permian and overlying Pliocene strata (Fig. 2). Figure 5 is a view south from the overlook toward Timber Mesa and the Spanish Skirts illustrating the contacts between Permian Quartermaster Group (Pa), Triassic Tecovas Formation (TRtec), Triassic Trujillo Formation (TRtru), and Tertiary Ogallala Group (Tog), which are easily distinguished along the canyon slopes. The oldest rock group exposed in the canyon is the Upper Permian Quartermaster Group, composed of brick-red to vermillion shales interbedded with lenses of soft red sandstone and clay. The lower red shales contain a considerable amount of gypsum in white to pink bands of satin spar. The Quartermaster shales weather to form a typical badland topography. The overlying Tecovas shales form a relatively smooth slope easily distinguished from the gullied, steeper slopes of the Permian shales beneath them (Fig. 5). The Tecovas Formation is about 200 ft (60 m) thick and is composed of a lower lavender, gray, and yellow shale sequence; a middle bed of white crossbedded friable sandstone; and an upper unit of orange shale, which underlies the more massive Trujillo sandstones (Matthews, 1969). The Trujillo Formation at Palo Duro Canyon consists of prominant ledges of sandstone and conglomerate interbedded with red and gray shales and clays (Fig. 6). The sandstones are reddish brown-gray, fine to medium grained, crossbedded, and micaceous. Many of the sands are lenticular and are characteristic of cut-and-fill channel sedimentation. The overlying Ogallala rocks are composed primarily of unconsolidated fine-to-coarse-grained calcareous sand, gravel, and lenses of clay and silt (Underwood and others, 1977). A consolidated caliche bed forms the upper few feet of Ogallala. The Ogallala sediments are the principle water source for the Southern High Plains, a vital resource for an otherwise semi-arid region. A close-up view of the stratigraphy and sedimentation of the Pliocene, Triassic, and Permian rock groups can be seen as you proceed down the canyon slope on Park Road 5 (Fig. 6).

SUMMARY

The rocks of the Dockum Group were deposited in a broad, shallow fluvial-lacustrine basin during semi-arid to arid conditions, by complex depositional systems that derived sediments not from a single source but from multiple provenance areas (Gawloski, 1983). These source areas are probably related to slight rejuvenation of old Paleozoic structural elements (Fig. 1) resulting in increased erosion of Permian and older Triassic red beds. The Dockum lacustrine basin formed regional base levels that rose and fell in response to climatic changes, maintaining a delicate balance between erosion and deposition (McGowan and others, 1979). When climatic conditions were more humid, and lake levels and lake areas were at their maximum, prograding deltas and associated meandering streams constituted the principle depositional systems. With a shift toward more arid conditions, base level and lake size decreased and valleys were scoured. Lacustrine deposits, fan deltas, and local braided streams became the dominant depositional types (McGowen and others, 1979). The character of the Dockum and other Triassic rocks of Texas strongly suggests semi-arid deposition typical of the present-day low-latitude desert regions (Gawloski, 1983). During Triassic time, arid climates dominated a broad region of the global land mass, a part of which now occupies the High Plains of West Texas.

Figure 5. Locality 4. View of the Spanish Skirts and Timber Creek Canyon area of Palo Duro Canyon looking south from the scenic overlook. The stratigraphic contacts between the Permian Quartermaster Group (Pq), Tecovas Formation (Ꞧtec), Trujillo Formation (Ꞧtru), and Ogallala Group (Tog) are easily distinguished in the canyon.

Figure 6. Locality 4. The Tecovas-Trujillo contact is exposed along the canyon wall on Park Road 5, Palo Duro Canyon State Park. The more massive sandstones and conglomerate lenses of the Trujillo Formation are in sharp contrast to the underlying orange shales of the Tecovas Formation.

REFERENCES CITED

Adams, J. E., 1929, Triassic of West Texas: American Association of Petroleum Geologist Bulletin, v. 13, no. 8, p. 1045–1055.

Adkins, W. S., 1933, The Mesozoic systems in Texas, *in* The geology of Texas, v. 1, Stratigraphy: University of Texas Bureau of Economic Geology Bulletin 3232, p. 239–518.

Cazeau, C. J., 1962, Upper Triassic deposits of West Texas and northeastern New Mexico [Ph.D. thesis]: University of North Carolina, 94 p.

Cramer, S. L., 1973, Paleocurrent study of the Upper Triassic sandstones, Texas High Plains [M.S. thesis]: Canyon, West Texas State University, 31 p.

Gawloski, T. F., 1983, Stratigraphy and environmental significance of the continental Triassic rocks of Texas: Baylor Geological Studies Bulletin 41, 47 p.

Gould, C. N., 1906, The Geology and water resources of the eastern portion of the panhandle of Texas: U.S. Geological Survey Water Supply Paper 154, 64 p.

Kiatta, H. W., 1906, A provenance study of the Triassic deposits of northwestern Texas [M.S. thesis]: Lubbock, Texas Tech College, 63 p.

Matthews, W. H., 1969, The geologic story of Palo Duro Canyon, Texas: University of Texas Bureau of Economic Geology Guidebook 8, 51 p.

Maxwell, R. A., 1971, Geologic and historic guide to the State Parks of Texas: University of Texas Bureau of Economic Geology Guidebook 10.

McGowen, J. H., Granata, G. E., and Seni, S. J., 1979, Depositional framework of the lower Dockum Group (Triassic), Texas Panhandle: University of Texas Bureau of Economic Geology Report of Investigations 97, 60 p.

Patton, L. T., 1923, The Geology of Potter County: University of Texas Bureau of Economic Geology Bulletin 2330, 184 p.

Roth, R., 1961, Origin of siliceous Dockum conglomerates, west Texas: San Angelo Geological Societies, Upper Permian to Pliocene, San Angelo Area Field Trip Guidebook, p. 50–53.

Sidwell, R., 1943, Triassic sediments in west Texas and eastern New Mexico: Journal of Sedimentary Petrology, v. 15, no. 2, p. 50–54.

Underwood, J. R., and Hood, C. H., 1977, Geology of Palo Duro Canyon and vicinity, Randall County, Texas: Panhandle Geological Society Guidebook, Palo Duro Field Trip 70 p.

Permian strata of North-Central Texas

***James O. Jones**, Division of Earth and Physical Sciences, The University of Texas at San Antonio, San Antonio, Texas 78285-0663*
***Tucker F. Hentz**, Bureau of Economic Geology, The University of Texas at Austin, University Station, Box X, Austin, Texas 78713-7508*

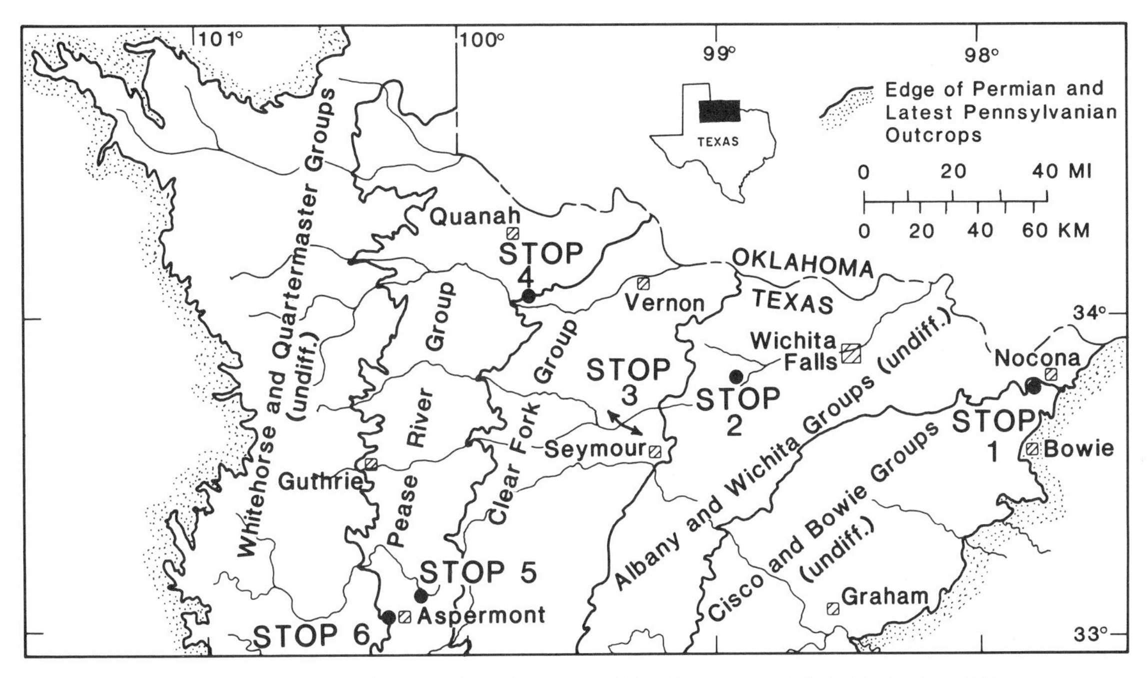

Figure 1. Generalized geologic map of northern part of Permian outcrop belt in North-Central Texas showing stops.

LOCATION

Permian sedimentary rocks of North-Central Texas are exposed in a northeast-striking outcrop belt that extends from overlying Cretaceous strata composing the Edwards Plateau in the south to the Red River and Panhandle region in the north. The gently inclined (0.5°), west-dipping Permian strata, over 5,500 ft (1,680 m) of section, are represented in this chapter as a transect of about 180 mi (290 km) in the northern part of the outcrop area (Fig. 1).

GEOLOGIC CONTEXT

Permian strata exposed in North-Central Texas were deposited on the Eastern Shelf of the Midland Basin. Most major tectonic elements that influenced sedimentation of the Eastern Shelf and adjoining regions during the Permian Period had developed by Virgilian and Wolfcampian time (Oriel and others, 1967). The Ouachita Foldbelt and uptilted Pennsylvanian strata of the eastern Fort Worth Basin to the south and east, the Arbuckle Mountains to the northeast, and the Wichita Mountains to the north acted as the principal sources of sediment to the southwest-sloping shelf. Permian strata of the Eastern Shelf are underlain by lithologically similar rocks of the Pennsylvanian System; precise placement of the chronostratigraphic boundary has historically been problematic (Gupta, 1977).

The stratigraphic sequence in this transect includes fluviogenic continental red beds in the eastern half (Stops 1 through 3) and mostly tidal-flat, deltaic, and sabkha units in the western half (Stops 4 through 6). This gross facies succession collectively records progressive infilling of the Midland Basin and adjacent shelf areas from relatively great bathymetric relief during Cisco/Bowie deposition (Wolfcampian) to a shallow evaporite basin during Blaine and Whitehorse deposition (Leonardian and Guadalupian). Stratigraphic units discussed in this chapter follow the usage of Hentz and Brown (1987) (Fig. 2).

Eastern transect region—coastal plain facies. The Bowie and Wichita Groups of the eastern part of the transect constitute continental beds composed predominantly of reddish-brown mudstone and regionally extensive cuesta-forming sandstone beds (Hentz, 1987). These deposits represent upslope facies of equivalent fluvial-deltaic and open-shelf marine Cisco and Albany Groups exposed south of the transect area (Fig. 2). Lateral gradation between the respective facies tracts coincides

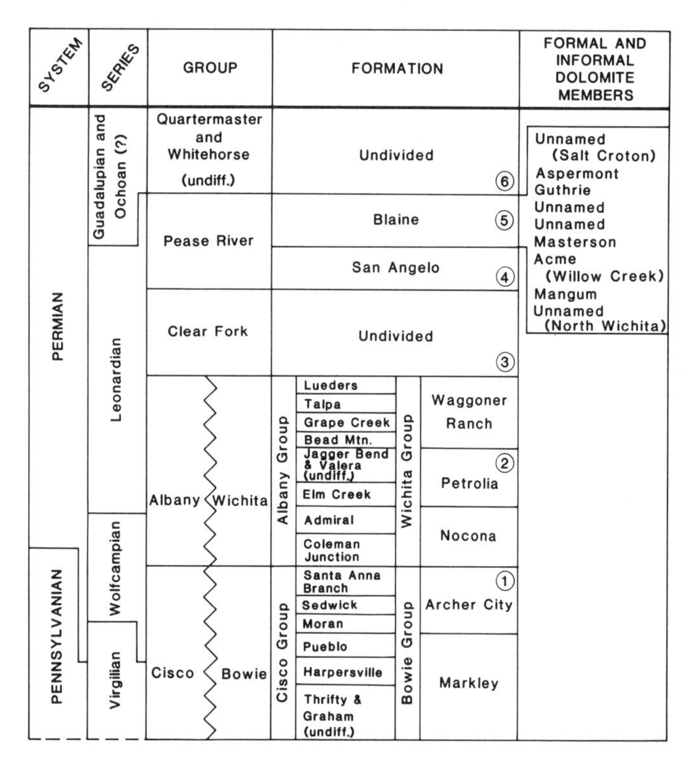

Figure 2. Permian stratigraphic units represented in transect area. Numbers show stratigraphic position of each stop.

approximately with the Brazos River (Fig. 1). Facies of the lower piedmont to updip margin of the upper coastal plain of the Bowie and lower Wichita Groups in easternmost (source-proximal) outcrops comprise regionally interstratified mixed-load-meandering and sandy-braided fluvial facies. Deposits of meandering rivers dominate upper coastal plain facies best developed in the Bowie and lower Wichita Groups in central and western areas of exposure and of the middle Wichita Group in northeastern outcrops. Meanderbelt systems (Stop 1) were the primary sediment transport routes to generally high-constructive delta systems that prograded across the Eastern Shelf, at times to the shelf edge of the Midland Basin (Brown and others, 1987). Floodplain facies associated with the upper coastal plain province include overbank, crevasse splay, and fossiliferous lacustrine (floodplain pond, backswamp) deposits.

Lower coastal plain facies typify southwestern (source-distal) exposures of the middle and upper Wichita Group (Hentz, 1987). Mud-rich fluvial channel-fill bodies record sedimentation of high-sinuosity, suspended-load meandering rivers (Stop 2). Floodplain facies are similar to those of the upper coastal plain; small ephemeral channel systems are common throughout the entire coastal plain. Thin variegated zones and abundant, locally bedded calcareous nodules in red mudstone (overbank facies) probably represent paleosols. Regionally interstratified tidal mud-flat facies contain locally preserved, pervasively desiccation-cracked red mudstone; lenticular storm-berm accumulations of intraformational calcareous nodules, marine invertebrates, and bones/teeth of terrestrial and aquatic vertebrates; and small sand- and mud-filled tidal or distributary channels of ill-defined muddy delta systems that existed during middle to late Wichita time. Thin beds of argillaceous limestone of the upper Wichita Group that interfinger with the tidal-flat and distal alluvial-plain sediments are nearshore equivalents of thick, open-shelf limestone units present in the subsurface and exposed to the south in the upper Albany Group.

Progressing westward along the transect, rocks of the lower to middle Clear Fork Group represent a marked progradational phase of strictly continental deposition (Olson, 1958). This red bed succession consists of mostly mudstone and siltstone with lesser amounts of sandstone and intraformational conglomerate. In the subsurface, these deposits grade basinward into coastal evaporites and peritidal dolomites (Handford, 1980). Exposed coastal plain paleoenvironments are similar to those of the underlying Wichita Group and include locally well-developed meanderbelt exposures of mixed- and suspended-load river systems (Stop 3). Lower Permian red beds of principally the Wichita and Clear Fork Groups contain the oldest assemblages of abundant primitive reptiles and amphibians in the world; badland exposures of Baylor and Archer Counties are classic collecting areas (Romer, 1935).

Western transect region—deltaic, tidal-flat, and sabkha facies. A significantly different depositional regime was present in the western province of exposed Permian strata in North-Central Texas. Marine and paralic mudstone, sandstone, dolomite, and gypsum are dominant lithologies in the upper Clear Fork, Pease River, and Whitehorse Groups. Component dolomite, sandstone, and gypsum members form moderate to prominent northeast-trending escarpments beginning at the base of the San Angelo Formation.

Interbedded red gypsiferous mudstone, dolomite, and thin, discontinuous gypsum of the uppermost Clear Fork Group recording probable tidal-flat and sabkha environments are unconformably overlain by deposits of regionally extensive high-constructive deltaic and sand-dominated tidal-flat systems (Duncan Member) (Stop 4) and mud-rich tidal-flat systems (Flowerpot Member) of the San Angelo Formation (Smith, 1974). Deltaic facies tracts are characterized by a suite of subfacies, including prodelta, delta-front, distributary channel and mouth bar, and delta-plain deposits. Tidal sandflat deposits and sandy tidal channel-fill bodies compose the dominant facies of the sand-rich tidal-flat systems; the mud-rich tidal-flat systems consist of laminated algal-bound shale, dolomitic shale with gypsum and barite nodules, thin ripple-bedded sandstone (swash zone facies), and lenticular sandy tidal channel fills. Locally within the San Angelo Formation, there are zones of stratabound copper mineralization (covellite, chalcocite, azurite, and malachite) within organic-rich tidal channel-fill sandstones and algal-mat mudstone facies (Smith, 1974).

The Blaine Formation is a sequence of dominantly red and gray mudstone interstratified with numerous regionally continuous beds of dolomite and nodular gypsum (Roth, 1945; Jones, 1971). The base of the formation coincides with the first occurrence of bedded gypsum. Mudstone and dolomite facies represent transgressive tidal-flat systems, whereas nodular gypsum and gypsiferous mudstone facies record sabkha deposition (Stop 5) (Smith, 1974; Jones, 1984). However, Fracasso and Hovorka (1986), in an investigation of the strata equivalent to the Blaine in the subsurface of the Palo Duro Basin, indicate that strict application of the sabkha model is not entirely appropriate. They interpreted evaporite deposition to have occurred mostly subaqueously on an expansive shallow-water shelf. Facies successions are cyclic, indicating recurrent periods of eustatic sea level change and/or systematic changes in rates of regional basin subsidence.

In Childress, Cottle, and King Counties, the base of the overlying Whitehorse Group is the Childress Dolomite (Roth, 1937), which thins southward and pinches out in southern King County. The Childress Gypsum, a gypsum bed at about the same horizon, marks the group contact in this area (Stop 6). This unit can be traced southward for several counties. Progressing westward, extensive red beds of the Whitehorse and Quartermaster Groups consist of sabkha and tidal-flat mudstone, sandstone, dolomite, and gypsum.

STOP DESCRIPTIONS

For a more precise stratigraphic perspective, the reader should refer to the 1:250,000-scale Geologic Atlas of Texas map sheets covering the transect area (Eifler, 1967, 1968; Hentz and Brown, 1987; McGowen and others, in preparation). County highway maps are available from the Texas State Department of Highways and Public Transportation (P.O. Box 5051, Austin, TX 78763).

Stop 1 (97°47'45" W, 33°47'30"N). This stop is located 4.0 mi (6.5 km) west of the intersection of U.S. 82 and Texas 175 in Nocona, Montague County (Fig. 3).

This roadcut exposure of a fluvial channel fill occurs in sandstone member 9 of the uppermost Archer City Formation (Bowie Group) (Hentz and Brown, 1987). This lower and middle point bar accumulation is approximately 15 ft (5 m) thick and extends about 550 ft (170 m) along the highway. Fine-grained upper point bar deposits are not represented at the top of this outlier. The exposure represents a single depositional unit; elsewhere, younger channel fills are locally incised into this unit, a common aspect of the proximal fluvial facies. Moderate sinuosity of channel courses is revealed by locally exhumed channel fills exposed on broad dip slopes. Maximum thickness of this facies rarely exceeds 25 ft (8 m).

Sedimentary features exhibited at this stop are indicative of the piedmont to upper coastal plain mixed-load meanderbelts. In contrast to the suspended-load channel-fill bodies of the lower coastal plain (Stop 2), point bars are predominantly fine- to very coarse-grained quartz sandstone with minor chert and feldspar. Mud-clast conglomerate is common near the channel base. Sedimentary structures are medium to large scale and include trough crossbeds, tabular crossbeds, planar lamination, low-angle foreset crossbeds, and cross-lamination exposed mostly in the upper part of the roadcut. The channel margin (cutbank) and red floodplain mudstone are exposed at the western end of the roadcut.

The most striking features of this outcrop are numerous inclined (to the west) parting surfaces within the point bar deposit. These surfaces commonly extend throughout the channel fill and represent lateral accretion surfaces that indicate apparent paleochannel migration toward the west. Within sandstone accretion units bounded by adjacent surfaces, grain size and scale of sedimentary structures generally fine upward. Pervasive soft-sediment deformation has imparted a wavy appearance to accretion surfaces. Because this exposure coincides approximately with the axis of meander-loop migration and lateral accretion surfaces are well developed, paleochannel dimensions can be calculated using methods presented in Ethridge and Schumm (1978). Conservative estimates (because the upper point bar is not preserved) of bankfull channel width and depth (215 ft [66 m] and 11.5 ft [3.5 m]), meander wave length (2,480 ft [756 m]), and meander amplitude (1,000 ft [305 m]) document the moderate size of this Permian river.

Stop 2 (98°56'03" W, 33°49'32"N). To locate this stop, proceed 5.2 mi (8.4 km) north on FM 2846 from its intersection with U.S. 82/277 at Dundee, northwest Archer County (Fig. 4). Turn left (west) and travel 2.7 mi (4.3 km) on FM 1180 and cross the bridge over the Wichita River below Diversion Reservoir dam. Stop 2 is a roadcut on the left (west) side of FM 1180.

This sandstone- and mudstone-bearing fluvial channel fill occurs in the upper Petrolia Formation (Wichita Group) approximately 60 ft (18 m) below the cuesta-supporting Beaverburk Limestone, a regionally continuous marker bed at the base of the Waggoner Ranch Formation. The meanderbelt, of which this exposure is part, crops out discontinuously for at least 6.2 mi (10 km) northward along the escarpment. The roadcut is about 17 ft (5.2 m) high and over 130 ft (40 m) long.

Sedimentary characteristics of this channel fill are representative of deposits of lower coastal plain suspended-load rivers found in the Wichita Group. These characteristics clearly distinguish this facies from the upslope-equivalent mixed-load meanderbelts from the group's northern area of outcrop. The dominant large-scale fabric of the outcrop is inclined (5–10°) thin interbeds of cross-laminated and planar-laminated very fine sandstone and silty red mudstone, which represent lateral accretion surfaces and record apparent meander-loop migration toward the northwest. The entire exposure is composed of three, superimposed, erosionally discordant point bar units (Fig. 4), suggesting episodic accretion and adjustment of channel course, probably during major floods. Elsewhere in the Wichita Group, this facies locally exhibits regularly spaced and depositionally concordant accretion surfaces, indicating relatively continuous lateral growth of the point bar. Point bar sequences of the Wichita Group range from 6.5 ft (2 m) to 11.5 ft (3.5 m) thick. Smaller, secondary channel-fill

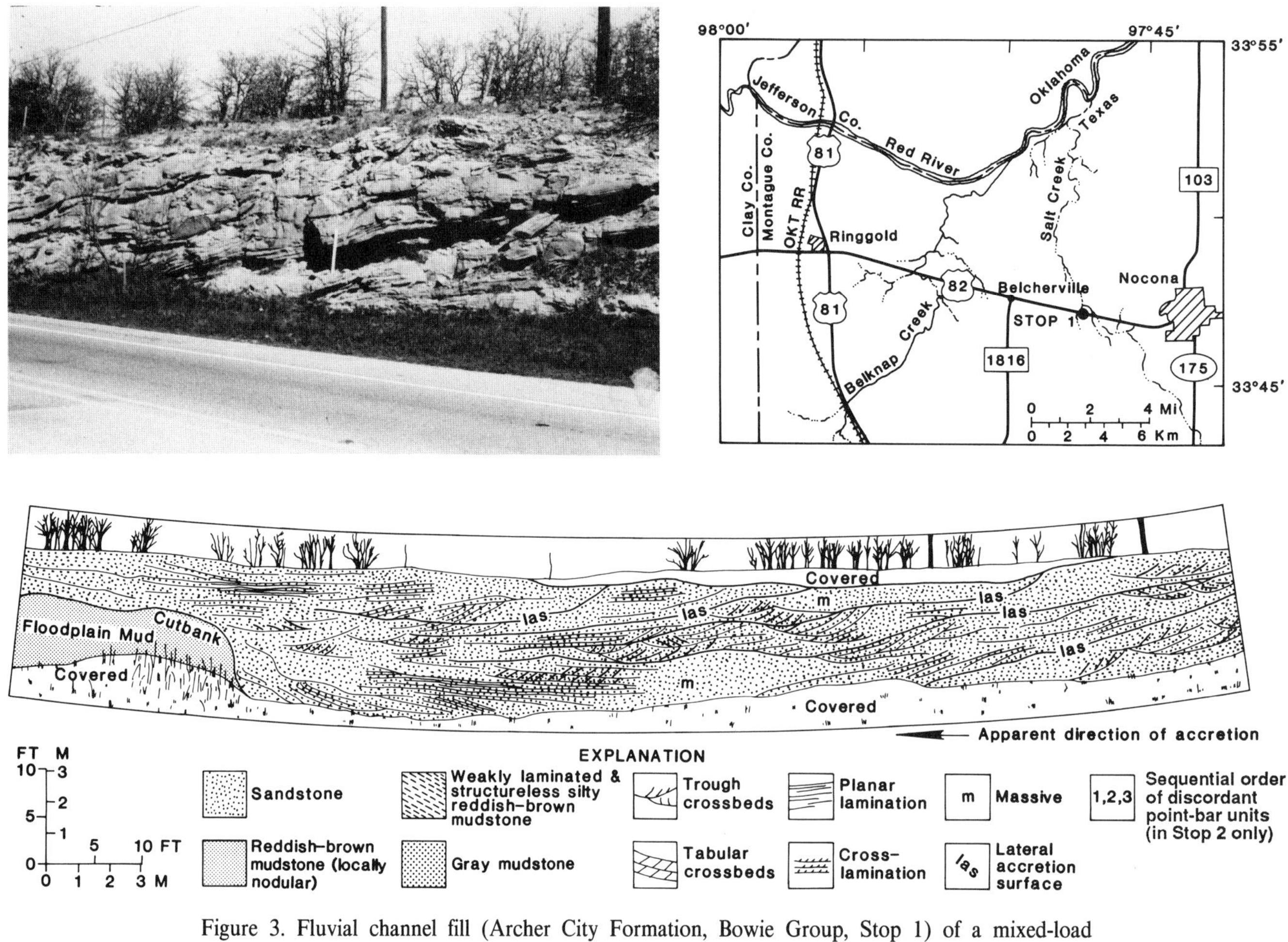

Figure 3. Fluvial channel fill (Archer City Formation, Bowie Group, Stop 1) of a mixed-load meandering river system characteristic of piedmont and upper coastal plain provinces of the early Permian (Wolfcampian) Eastern Shelf of the Midland Basin. Western part of roadcut is illustrated in diagram.

bodies are superimposed on the lower (scour fill) and upper (chute channel) parts of point bar units. Scour fills are interpreted to represent runnels that developed on the channel floor during periods of low discharge. The channel base is incompletely exposed at Stop 2, but other examples exhibit medium- and large-scale trough crossbeds. Mudstone intraclasts are locally abundant in the lower part of point bar sequences; intraformational calcareous nodules are ubiquitous channel-base lag constituents.

Paleochannel dimensions cannot be calculated because neither the channel base nor the full length of individual point bar units is exposed. However, these muddy river systems were undoubtedly smaller than the sandy systems represented at Stop 1. Partial exhumation of channel courses in badland exposures strongly suggests high sinuosity.

Lacustrine gray mudstone and laminated siltstone above the point bar deposits (Fig. 4) record the final stages of infilling of the subsequently abandoned channel. In contrast to the inclined bedding of the point bar facies, overlying red floodplain mudstones are horizontally stratified.

Stop 3 (99°17'38"W, 33°35'13"N). This stop comprises several outcrops of sandstone fluvial channels and floodplain deposits along FM 1919 northwest of Seymour, Baylor County (Fig. 5). The first notable exposure occurs 6.0 mi (9.7 km) north of the intersection of FM 1919 and U.S. 82/277, immediately north of Seymour. Traveling northwest for the succeeding 25 mi (40 km), there are almost continuous exposures along both sides of the road.

Stop 3 is in the middle Clear Fork Group and begins with a roadcut of sandstone member 5 (Hentz and Brown, 1987). This roadcut documents about 35 ft (11 m) of section and is 175 ft (35 m) long. Dark reddish-brown sandstone contains medium- to large-scale trough crossbeds and planar bedding in the lower part of the exposure. Medium-scale cut-and-fill scours and trough crossbeds occur in the upper and western half of the outcrop. This

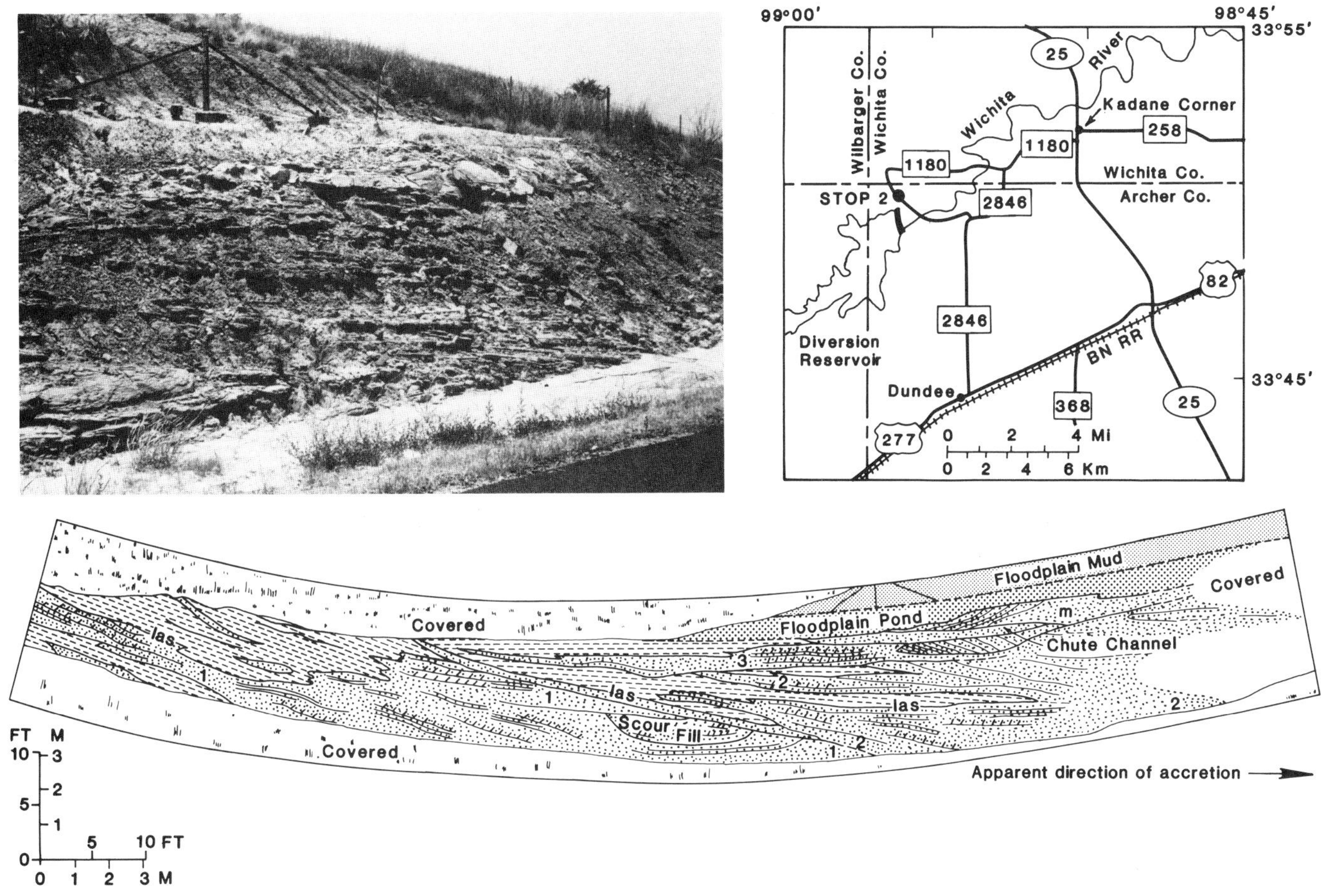

Figure 4. Fluvial channel fill (Petrolia Formation, Wichita Group, Stop 2) representative of suspended-load meandering rivers that drained the lower coastal plain of the Eastern Shelf during Leonardian time. See Figure 3 for explanation of symbols used in outcrop diagram.

sandstone sequence, exposed along the escarpment to the north, generally fines upward in grain size. It represents a point bar sequence of a mixed-load meandering river similar to that displayed at Stop 1. Sandstone 5 is one of several mappable, cuesta-forming sandstone beds in the lower and middle Clear Fork Group in the unit's northern area of exposure.

Proceed northwest on FM 1919 for about 10 mi (16.1 km) to a well-developed badland area. On both sides of the road are well-exposed floodplain and channel-fill facies of small, variable-discharge, high-sinuosity suspended-load meandering rivers (Edwards and others, 1983). Exhumed deposits of 23 point bars extend over a section of about 160 ft (50 m); two roadcuts display excellent vertical sections. These deposits are similar in both morphology and dimension to the meanderbelt of Stop 2 and probably represent the same fluvial environment; Stop 3 provides an areal perspective of these mud-rich systems.

These point bar deposits are exposed as arcuate accretionary ridges of interbedded ripple cross-laminated very fine sandstone and silty mudstone. Inclined sandstone beds (lateral accretion surfaces) display abundant lamination traces, interference and symmetrical oscillation ripples, desiccation cracks, erosional terraces, erosional runnels oriented parallel to the dip of accretionary surfaces, and trace fossils (arthropod tracks, *Scolicia* trails, and *Thallasinoides* burrows) (Edwards and others, 1983). Locally developed, broad, shallow scour fills containing planar laminations and small-scale trough crossbeds record erosion of the upper point bar during high discharge (chute channels?). At channel bases are lag conglomerates with clasts of intraformational mudstone and calcareous nodules. Medium-scale trough and planar crossbeds are locally developed. Channel-fills are 6.5 to 10 ft (2 to 3 m) thick.

An exceptional roadcut in the northwestern part of the locality area illustrates several of the point bar features. Approximate paleochannel dimensions calculated from this exposure (Ethridge and Schumm, 1978) are 79 ft (24 m) and 8.5 ft (2.6 m) for bankfull channel width and depth, respectively.

Stop 4 (99°43'09" W, 34°06'31" N). This stop, in and near Copper Breaks State Park, Hardeman County, provides outstand-

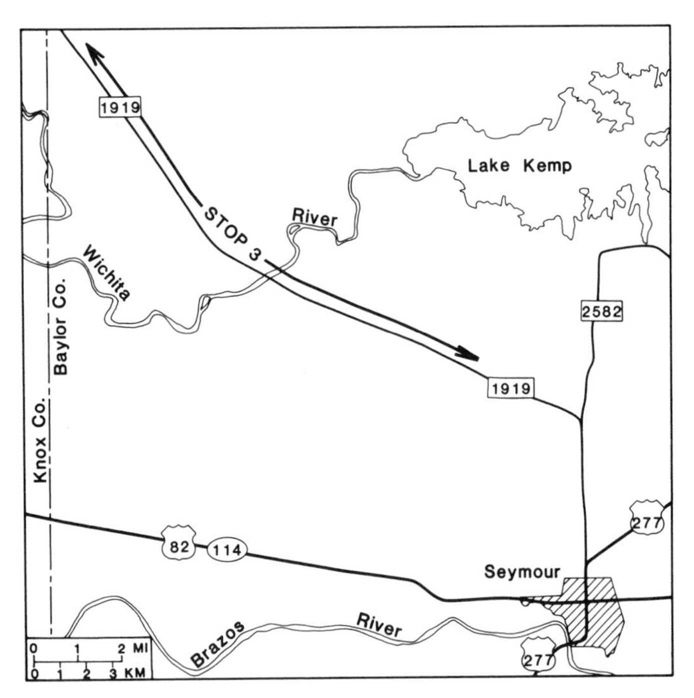

Figure 5. Location map of Stop 3, Clear Fork Group, Baylor County, Texas.

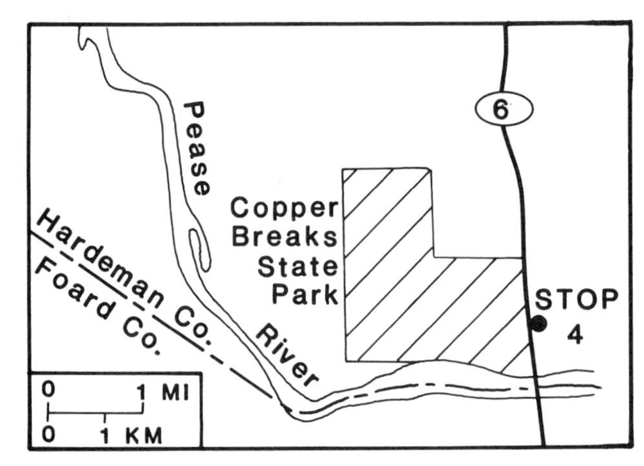

Figure 6. Location map of Stop 4, San Angelo Formation (Pease River Group), Hardeman County, Texas.

ing exposure of the lower part of the San Angelo Formation (Duncan Member) of the Pease River Group. The base of the formation is the first bedded sandstone 0.75 mi (1.2 km) north of the northern bridge abutment at the Pease River along Texas 6 (Fig. 6). Note copper mineralization near this contact. Entrance to the park is 0.3 mi (0.5 km) farther north.

A distinct escarpment is visible northwest of the Pease River bridge. The base of the slope-supporting sandstone at the top of the escarpment is the approximate Pease River–Clear Fork Group contact. In addition to roadcuts along Texas 6, a complete deltaic progradational sequence is well exposed near the top of the escarpment at various places within the southern part of the park (Smith, 1974).

Smith (1974) provides a paleogeographic overview and locations and descriptions of deltaic and tidal-flat facies within the park. Starting below the escarpment-supporting San Angelo sandstone, the thickly bedded, red, reddish-brown, and grayish-green mudstone of the upper Clear Fork records mud-rich tidal-flat deposition. Overlying the mudstone are sand-rich tidal-flat deposits composed of 4 to 7 ft (1.2 to 2.1 m) of sandy mudstone and lenticular (high width-to-thickness ratios), locally gypsum-cemented tidal channel-fill sandstone bodies with flaser beds, horizontal beds, and low-angle trough crossbeds. Above these rocks, high-constructive deltaic facies are superposed in a classic progradational sequence. Prodelta facies, conformably overlying the tidal-flat beds, comprise about 6 ft (1.8 m) of well-bedded reddish-brown mudstone lacking gypsum nodules and selenite seams characteristic of the underlying units. Sheet-like sandstone beds 10 to 12 ft (3 to 3.7 m) thick, exhibiting clay/sand interlamination, and trough crossbeds/ripple cross-lamination represent delta-front and distributary mouth bar facies, respectively. Distributary channel-fill sandstone bodies up to 30 ft (9 m) thick and 200 ft (61 m) wide locally were eroded into underlying deposits. Medium- to large-scale trough crossbeds are the principal sedimentary structures, although pervasive soft-sediment deformation (flow-roll structures) is common.

Stop 5 (100°08′39″W, 33°13′32″N). To get to this site, proceed 4.7 mi (7.6 km) north on FM 1263 from its intersection with U.S. 83 in Aspermont, Stonewall County, and then turn east on a paved road and travel 5.0 mi (8.1 km) to the bridge over the Salt Fork of the Brazos River (Fig. 7). At the south end of the bridge, turn west (upstream) for about 1,000 ft (320 m) to a bluff.

The middle Blaine Formation (Pease River Group) begins at the base of the cliff. The river bed may be legally traversed; however, the bluff is private property and trespassing is forbidden. Return to the paved road and proceed to the top of the hill to view the remainder of the section.

This stop illustrates the small-scale cyclicity of sabkha and tidal mud-flat systems characteristic of much of the Blaine (Fig. 8). Over 60 ft (18 m) of alternating tidal mud-flat mudstones and dolomites, algal-mat (cyanobacterial) deposits, and sabkha nodular gypsum and gypsiferous mudstone are exposed in the lower part of the section. Within mudstone units are sparse lenticular sandstone beds, interpreted as small tidal channel-fill bodies or tidal-flat bars. Thin dolomite beds within this lower interval are also largely lenticular. Gypsum beds are composed of a "chicken-wire" framework of nodular alabaster with locally red and gray clay matrix. Some gypsum beds exhibit enterolithic layers alternating with mudstone beds. Small symmetrical ripple marks at the base of some gypsum beds formed as casts from structures on the underlying mudstone. Secondary

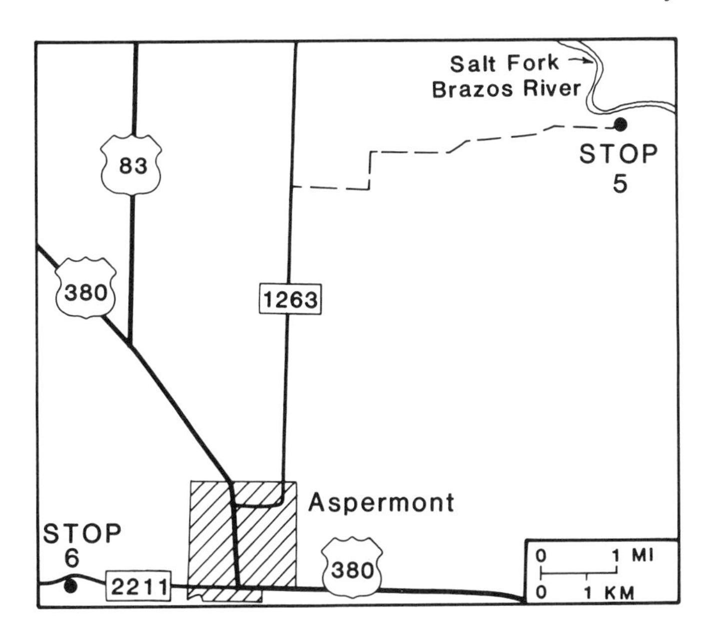

Figure 7. Location map for Stops 5 and 6 in the Pease River and Whitehorse Groups, Stonewall County, Texas.

Figure 8. Well-bedded mudstone, gypsum, and dolomite strata of the Blaine Formation represented at Stop 5, Stonewall County, Texas.

deposits of satin spar, either interbedded or in discordant fracture fills, are common throughout the Blaine Formation (Jones, 1971).

Increasing thickness of dolomite beds in the upper 115 ft (35 m) of the section is typical of the higher parts of the middle Blaine Formation. Two informally named dolomite members (North Wichita and Willow Creek) and the Acme Dolomite Member are present (Jones, 1971). The two informal units are regionally discontinuous but traceable for about 80 mi (130 km) along strike. The Acme Dolomite at the top of the section contains two thin beds of coarse fossil fragments and a coquina of pelecypods (*Schizodus sp.*). Dolomite members throughout the Blaine have locally occurring assemblages of abundant large ammonoids and nautiloids, gastropods, and pelecypods (Clifton, 1944). These accumulations in otherwise fossil-poor dolomite beds were probably concentrated as a result of storm surge in a broad, shallow nearshore zone. Halite hopper casts and molds averaging about 0.5 inch (1.3 cm) in diameter are present in some Blaine dolomites (Smith, 1974). Thickest dolomite units in the formation occur along the western part of the outcrop belt (Stonewall to western Hardeman County) and probably record the transitional zone between the tidal flat/sabkha and open marine regions.

Stop 6 (100°15'39" W, 33°08'08" N). To locate this stop, proceed west from Aspermont, Stonewall County, on FM 2211 for 2.0 mi (3.2 km) from the intersection with U.S. 83/380 (Fig. 7). The locality begins on the south side of the road and continues west to the top of the escarpment. Cliffs beyond the fences are private property.

The broad dip slope extending from east of Aspermont to this stop is developed on the top of the Aspermont Dolomite Member of the Blaine Formation (Fig. 2). Stop 6 in part comprises about 65 ft (20 m) of gypsiferous mudstone of the upper Blaine (Dog Creek Shale of Oklahoma) above this dolomite marker bed.

The Childress Gypsum of the Whitehorse Group is the interval of interest at this locality and is the unit upon which the distinct escarpment is developed. The Childress Gypsum and equivalent Childress Dolomite to the north form the basal unit of the Whitehorse Group. A regional unconformity has been interpreted to exist between the Whitehorse and Pease River Groups (Roth, 1937). About 15 ft (4.6 m) of gypsum with 3.0 ft (0.9 m) of red mudstone interlaced with satin spar are exposed near the middle of the Childress Gypsum. Ripple marks on the tops of weathered mudstones are preserved as casts on the bases of gypsum beds. Thick basal gypsum beds contain little clay and display distinct crossbeds on weathered surfaces; foreset laminae are enhanced by ooids and rounded fossil fragments. Crossbeds demonstrate a detrital origin for some gypsum beds. The Childress Gypsum probably represents a primarily subaqueous evaporite sequence deposited basinward of sabkha and tidal mud-flat facies similar to those exposed in underlying strata.

REFERENCES

Brown, L. F., Jr., Solis Iriarte, R. F., and Johns, D. A., 1987, Regional depositional systems, Upper Pennsylvanian and Lower Permian Systems, North-Central Texas: The University of Texas at Austin, Bureau of Economic Geology Report of Investigations (in press).

Clifton, R. L., 1944, Paleoecology and environments inferred for some marginal middle Permian marine strata: American Association of Petroleum Geologists Bulletin, v. 28, p. 1021–1031.

Edwards, M. B., Erickson, K. A., and Kier, R. S., 1983, Paleochannel geometry and flow patterns determined from exhumed Permian point bars in North-Central Texas: Journal of Sedimentary Petrology, v. 53, no. 4, p. 1261–1270.

Eifler, G. K., 1967, Lubbock sheet, *in* Bureau of Economic Geology Geologic Atlas of Texas: The University of Texas at Austin, scale 1:250,000.

——, 1968, Plainview sheet, *in* Bureau of Economic Geology Geologic Atlas of Texas: The University of Texas at Austin, scale 1:250,000.

Ethridge, F. G., and Schumm, S. A., 1978, Reconstructing paleochannel morphologic and flow characteristics; Methodology, limitations, and assessment, *in* Miall, A. D., ed., Fluvial sedimentology: Canadian Society of Petroleum Geologists Memoir 5, p. 703–721.

Fracasso, M. A., and Hovorka, S. D., 1986, Cyclicity in the middle Permian San Andres Formation, Palo Duro Basin, Texas panhandle: The University of Texas at Austin, Bureau of Economic Geology Report of Investigations No. 156, 48 p.

Gupta, S., 1977, Miofloral succession and interpretation of the base of the Permian System in the Eastern Shelf of North Central Texas, U.S.A.: Review of Palaeobotany and Palynology, v. 24, p. 49–66.

Handford, C. R., 1980, Lower Permian facies of the Palo Duro Basin, Texas; Depositional systems, shelf-margin evolution, paleogeography, and petroleum potential: The University of Texas at Austin, Bureau of Economic Geology Report of Investigations No. 102, 31 p.

Hentz, T. F., 1987, Lithostratigraphy and paleoenvironments of upper Paleozoic continental red beds, North-Central Texas; Bowie (new) and Wichita (revised) Groups: The University of Texas at Austin, Bureau of Economic Geology Report of Investigations No. 170 (in press).

Hentz, T. F., and Brown, L. F., Jr., 1987, Wichita Falls–Lawton sheet, *in* Bureau of Economic Geology Geologic Atlas of Texas: The University of Texas at Austin, scale 1:250,000.

Jones, J. O., 1971, The Blaine Formation of north Texas [Ph.D. dissertation]: Iowa City, The University of Iowa, 173 p.

——, 1984, Marine dominated coastal sabkha and tidal-flat deposition of the Blaine Formation, Permian Basin, Texas [abs.]: 27th International Geological Congress, Moscow, U.S.S.R., v. 2, p. 84.

McGowen, J. H., Hentz, T. F., Owen, D. E., and McGowen, M. K., Sherman sheet, *in* Bureau of Economic Geology Geologic Atlas of Texas: The University of Texas at Austin, scale 1:250,000 (in preparation).

Oriel, S., Myers, D. A., and Crosby, E. J., 1967, Paleotectonic investigations of the Permian System in the United States: U.S. Geological Survey Professional Paper 515, Chapter C, 80 p.

Olson, E. C., 1958, Fauna of the Vale and Choza; 14, summary, review and integration of the geology and the faunas: Fieldiana (Geology), v. 10, p. 397–448.

Romer, A. S., 1935, Early history of Texas redbeds vertebrates: Geological Society of America Bulletin, v. 46, p. 1597–1658.

Roth, R., 1937, Custer Formation of Texas: American Association of Petroleum Geologists Bulletin, v. 21, p. 421–474.

——, 1945, Permian Pease River Group of Texas: Geological Society of America Bulletin, v. 56, p. 893–908.

Smith, G. E., 1974, Depositional systems, San Angelo Formation (Permian), north Texas, facies control of red bed copper mineralization: The University of Texas at Austin, Bureau of Economic Geology Report of Investigations No. 80, 74 p.

Middle and late Pennsylvanian rocks, North-Central Texas

E. L. "Jack" Trice, *Marshall Exploration, P.O. Box 1689, Marshall, Texas 75670*
Robert C. Grayson, Jr., *Department of Geology, Baylor University, Waco, Texas 76798*

LOCATION

The middle and late Pennsylvanian (Desmoinesian-Missourian) section of north-central Texas is illustrated by a transect extending from eastern Parker County northwestward to northeastern Palo Pinto County (Fig. 1).

The transect begins in the Dobbs Valley Member of the Mingus Formation (middle Strawn Group, Desmoinesian; Localities 1 and 2) and crosses through the Brazos River Formation and Mineral Wells Formation (upper Strawn Group, Desmoinesian-Missourian; Localities 3 and 4), Palo Pinto Formation, Wolf Mountain Shale, and into the Winchell Limestone (Canyon Group, Missourian; Localities 5 and 6; Figs. 1, 2).

Strawn and Canyon rocks generally reflect the cyclic depositional nature of the Pennsylvanian. The vertical succession of these rocks illustrates the final filling of the Fort Worth Basin and cyclic fluvial/deltaic progradation and marine transgression of the epeiric sea over the Eastern Shelf (Fig. 3). A shelf edge into the Midland Basin developed in Missourian time. Slope and basin depositional systems are recognized in the subsurface in Missourian and Virgilian time. Sources for the clastics were the Quachita Mountains and Southern Oklahoma Mountains. This transect illustrates some of the major clastic and carbonate depositional systems in the Strawn and Canyon Groups.

Geomorphically, this transect lies within the Palo Pinto country of the Central Great Plains province of the central United States (Hill, 1900). Throughout this transect, the major depositional systems of the middle and late Pennsylvanian are illustrated, as well as the relationship of the geology and the resulting landscape.

SIGNIFICANCE

Coal, clay, limestone, and hydrocarbons have made the middle and late Pennsylvanian of North-Central Texas economically important since the late 1800s. Coal is mined from the fluvial/deltaic facies of the Strawn and Canyon Groups. Clay is mined from delta plain facies of the Strawn for bricks, ceramic, and other clay products. Limestones of the Strawn and Canyon Groups are used for aggregate. Many of the early oil plays were for Strawn fluvial/deltaic sandstone facies and later for Canyon carbonate bank (reef) facies. Because of its economic significance, the geology of this region has been studied extensively. The interested reader is referred to Kier and others (1980), for a general discussion of the Pennsylvanian in Texas. Brown and Wermund (1969) provide a more detailed description of late Pennsylvanian shelf sediments in north-central Texas, and the work of Brown and others (1973) gives detailed descriptions of depositional systems in the middle and late Pennsylvanian of north-central Texas.

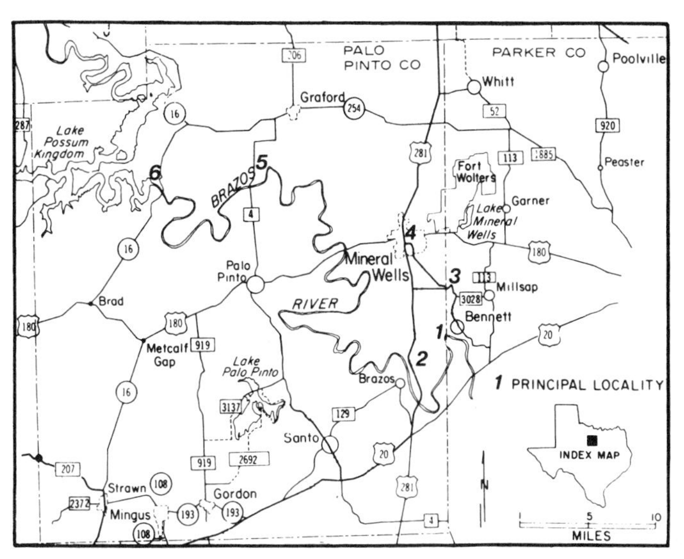

Figure 1. Map showing distribution of localities 1 through 6.

The Strawn and Canyon are characterized by cyclic sequences containing fluvial/deltaic sandstone, interbedded sandstone and shale, shale, interbedded limestone and shale, and algal bank limestone. At least 10 major cycles of fluvial/deltaic progradation terminated by marine transgression can be recognized from the Grindstone Creek Formation to the Winchell Limestone Formation (Fig. 4).

Landscape clearly reflects stratigraphy and structure. The middle and late Pennsylvanian rocks gently dip to the west-northwest. This structure allows the outcropping edge of resistant strata to form steep escarpments and gentle dip slopes, resulting in the development of northeast-southwest–trending cuestas. Stratigraphy within the Strawn and Canyon has a critical role in the development and prominence of these cuestas. Where a thick resistant limestone or sandtone crops out, a prominent cuesta has formed.

MIDDLE AND LATE PENNSYLVANIAN SECTION

The traverse begins at the intersection of I-20 and FM Road 113, approximately 5 mi (8 km) southeast of Bennett, eastern Parker County (Fig. 1). This is approximately 60 mi (96 km) west of Fort Worth on I-20. From here, the transect extends northwest to Bennett. Follow FM road 113 north to the Bennett Road (no highway number) and on to Bennett. The transect begins on the Buck Creek Sandstone Member of the Grindstone Creek Formation (Fig. 2) and continues stratigraphically upward into the Mingus Formation. Locality 1 consists of two railroad cuts on the main line of the Texas and Pacific Railroad just west

System	Series	Group	Formation	Member or Informal Units	Field Locality
Pennsylvanian System	Missouri Series	Canyon Group	Home Creek Limestone	Kisinger Sandstone	
			Colony Creek Shale	Unnamed sandstone	
			Ranger Limestone		
			Placid Shale	Unnamed sandstone	
			Winchell = Chico Ridge Limestone	Devil's Den Limestone	6
			Wolf Mountain Shale	Unnamed sandstone Rock Hill Limestone Lake Bridgeport Shale Unnamed sandstone	
			Palo Pinto Formation	Wiles Limestone = Willow Point Limestone Bridgeport Coal Oran Sandstone Unnamed limestone Wynn Limestone	5
	Missouri Series / Des Moines Series	Strawn Group	Mineral Wells Formation	Keechi Creek Shale Turkey Creek Sandstone Salesville Shale Unnamed sandstone Dog Bend Limestone Lake Pinto Sandstone East Mountain Shale Village Bend Limestone Unnamed sandstone Hog Mountain Sandstone	4
	Des Moines Series		Brazos River Formation		3
			Mingus Formation	Thurber Coal Unnamed sandstone Goen Limestone Dobbs Valley Sandstone Santo Limestone	1, 2
			Grindstone Creek Formation	Buck Creek Sandstone Brannon Bridge Limestone	
			Lazy Bend Formation	Steussy Shale Meek Bend Limestone Hill's Creek Shale	

Figure 2. Stratigraphic column of Strawn and Canyon Groups showing stratigraphic position of field localities. Modified from Brown and others, 1973.

Figure 3. Tectonic elements in Texas and southern Oklahoma.

of Bennett, Texas. The most distant of the cuts is 2.1 mi (3.4 km) west of the town (Fig. 1).

LOCATIONS

Locality 1 (Figs. 1, 5).

This is Stop 5 of Cleaves (1973). Access to this locality is obtained by a lengthy walk along the tracks. No prior permission is necessary, but remember to keep a constant watch for trains and stay off the fences between the tracks and the cliffs. (These are devices for detecting rubble that has rolled onto the tracks, and climbing on the fences could trigger a false danger signal at the railroad dispatcher's office in Fort Worth.)

Lobate deltas prograded westward and southwestward across eastern palo Pinto County during deposition of the Dobbs Valley Member (Cleaves, 1973). The Dobbs Valley Member crops out in a belt from western Parker County to west-central Erath County.

This locality is particularly significant because all the major facies of the prodelta and delta front are present in a single vertical sequence. The distal prodelta facies of the lowest Mingus Shale is overlain by the complete vertical sequence of sandstone facies of the Dobbs Valley Member characteristic of a small delta lobe (Cleaves, 1973). The most prominent units include a prodelta facies with slumped fragments of the delta front, a thin bedded, growth-faulted delta-front sandstone facies, and a massive channel mouth bar facies. The relationship between faulting and direction of deltaic progradation is readily seen.

The eastern cut exposes delta-front facies, which were marginal to the main axis of progradation. The prodelta facies is composed of siltstone and very fine sandstone flags. The best developed flow rolls in this area occur at the east end of this south-facing exposure. The delta-front sandstones are very fine grained and have a massive appearance that is largely due to intensive burrowing. The exposure lacks channel mouth bar facies.

The western exposure provides a complete vertical sequence of the progradation of a small lobe of a lobate delta. The basal 30 ft (9 m) of the cliff comprises proximal prodelta facies similar to those in the eastern exposure. There is an increase of total sand upward through the section. The lower part of the delta front is composed of thin beds of unburrowed, wave-rippled, very fine sandstone. Higher in the delta-front sequence, individual sandstones are fine- to medium-grained and thick-bedded. Horizontal laminations are the most prominent sedimentary structure. Small- and medium-scale trough cross-beds are also present, as well as

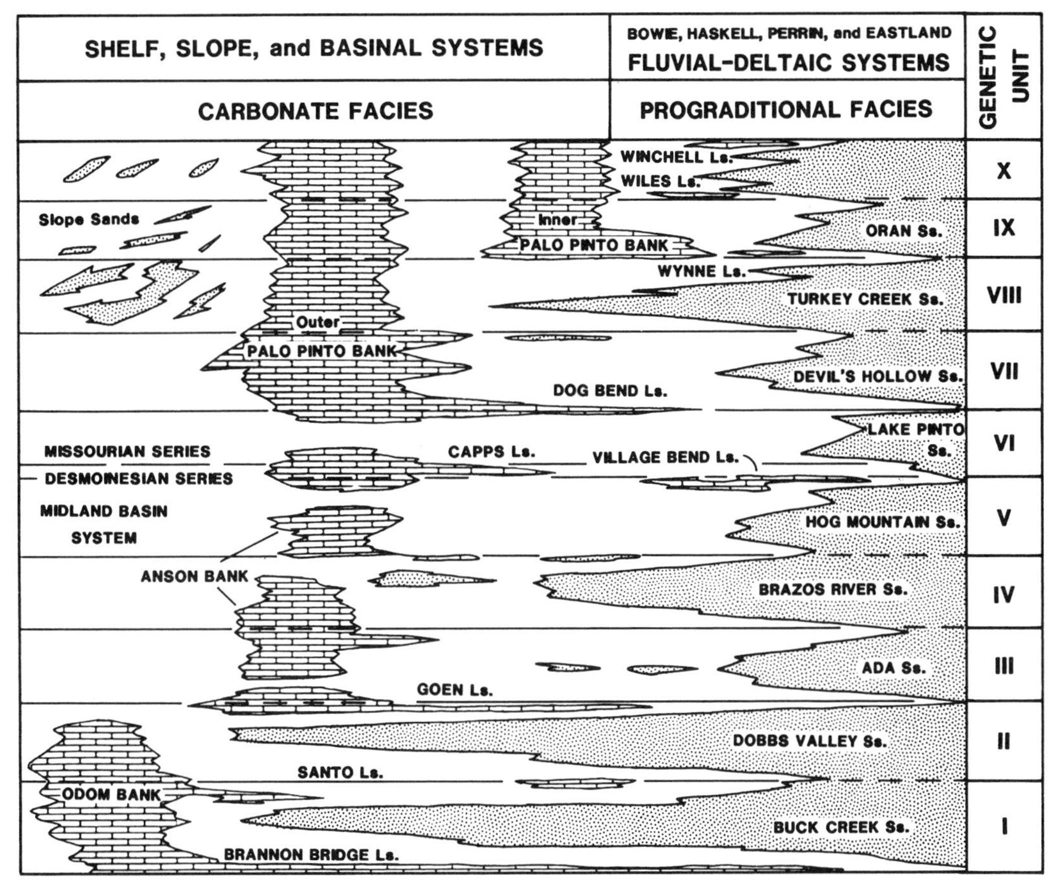

Figure 4. Deltaic cycles of Strawn and Canyon Groups, north-central Texas. Adapted from Cleaves, 1973.

oscillation ripples. Deposition of the horizontally laminated units took place in shallow water adjacent to the channel mouth bar. The upper part of the delta front shows less marine reworking than the prodelta–lower delta front. The channel mouth bar facies comprises massive, very well sorted blocky sandstone beds that lack significant shale breaks. Horizontal lamination is the dominant sedimentary structure, large-scale trough cross-beds cut into the bar at many points along the outcrop. Growth faulting is evident in the delta front and channel mouth bar facies. The fault plane is concave toward the direction of progradation, and individual sandstones are thicker on the downthrown side.

Locality 2 (Figs. 1, 2)

This is Stop 6 of Cleaves (1973). The locality exposes a thin distributary channel-fill sandstone cut into interdistributary-bay mudstones near the top of the Dobbs Valley member of the Mingus Formation (Fig. 2). It is a roadcut on the east side of U.S. 281, approximately 8.8 mi (14 km) south of Mineral Wells. Overlying this distributary sandstone is a thin veneer of fossiliferous, marine-reworked sandstones and sandy limestone, which is termed the Goen Limestone. The exposure of marine shale and Goen limestone at Goen Cemetery is a prolific Strawn fossil locality. The Thurber coal occurs within this interval to the north in central Parker County and to the southwest near Strawn.

This stop illustrates the internal sedimentary structures and geometry of channel fill. The base of the channel fill is undeformed and the vertical sequence is well preserved. The facies present at this stop include a thin, unfossiliferous mudstone at the base of the cut; the distributary fill sandstone and delta plain mudstones. The base of the distributary has a sharp, scoured contact with the underlying mudstone. Large-scale trough cross-beds contain abundant pebble-sized clay clasts in the bottom 5 ft (1.6 m) channel fill. Higher in the section, the fine-grained sandstones lack coarse clasts and contain small-scale trough cross-beds and ripples. Climbing ripples are more numerous away from the center of the channel. The abrupt change from sandstone to mudstone at the top of the section may indicate rapid distributary abandonment (Cleaves, 1973).

The transect continues northward on U.S. 281 to FM Road 1195. From Stop 2, continue over the valley floor in the upper Mingus Shale to the Brazos River Formation. Sandstones and chert pebble conglomerates are the resistive unit forming this northwest-southwest trending cuesta. Travel on the back side of the cuesta to FM Road 1195, turn right (east), and proceed to the cuesta face.

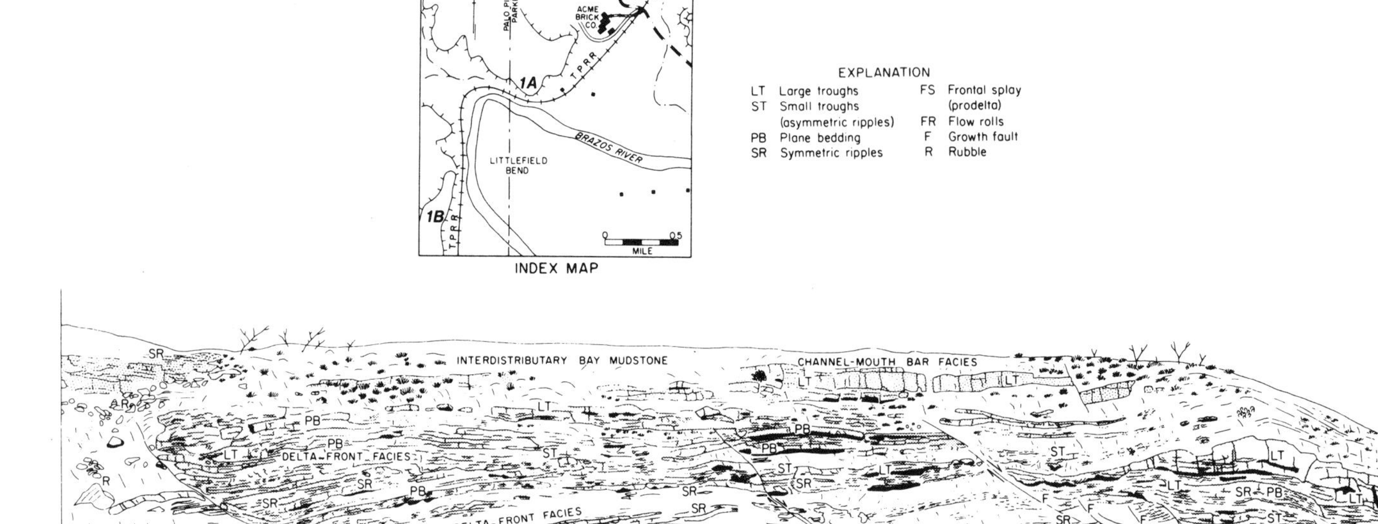

Figure 5. Field locality 1. Deltaic facies, Dobbs Valley Sandstone. Modified from Brown and others, 1973.

Locality 3 (Figs. 1, 2)

This is optional Stop D of Cleaves (1973). This locality exposes facies of the next major cycle of deltaic progradation, the Brazos River Formation. It is located 5.6 mi (9 km) east of Mineral Wells on FM road 3028, just south of the Mineral Wells airport, and just east of the junction of FM road 3028 and FM Road 1195 in a roadcut through the cuesta face. The base of the cuesta is composed of prodelta facies. This is overlain by thick-bedded delta-front sandstones, which are overlain by delta plain chert conglomerates. The delta-front sandstones are fine-grained, moderately sorted sandstones that exhibit trough cross-beds, horizontal laminations, and ripple marks. The uppermost units are chert pebble conglomerates that have large-scale trough cross-beds. The delta front exhibits an overall coarsening upward sequence, whereas the meander belt deposits exhibit a fining upward sequence. These large-scale troughs are overlain by smaller scale troughs and then ripples and/or parallel bedding.

The transect doubles back to the northwest to Mineral Wells, going back on FM road 3028; turn right (north) on FM Road 1195. Continue north on FM Road 1195 on the Brazos River cuesta and cross into the Mineral Wells Formation. The dissected hills to the west are the Hog Mountain Sandstone Member of the Mineral Wells Formation. At the intersection of FM Road 1195 and Texas 180, turn left (west) and drive into Mineral Wells on shales of the Mineral Wells Formation. Notice the cuestas formed by sandstone members of the Mineral Wells Formation to the north. Proceed west on Texas 180 to the Roadway Garage, turn in, and park.

Locality 4 (Figs. 1, 2)

This locality exposes the middle of the Mineral Wells Formation and is particularly significant because it contains the Desmoinesian-Missourian (Pennsylvanian) boundary. The cuesta behind the Roadway Garage is part of East Mountain, the type locality for East Mountain Shale. The cuesta exposes shales, sandstones, fossiliferous shales, the Village Bend Limestone Member, more shales, and the Lake Pinto Sandstone member of the Mineral Wells Formation. The Village Bend Limestone is best exposed on a small bench about 12 ft (4 m) above the lowest sandstone exposed and is about 1.5 to 2 ft (0.3 to 0.6 m) thick. The Lake Pinto Sandstone, exposed at the top of East Mountain, is a thick-bedded, moderately well sorted, fine-grained sandstone with a channeled base. This locality is locality 181-T-9 of Plummer and Hornberger (1935). Stauffer and Plummer (1932) and Trice (1984) sampled this exposure for conodonts. Numerous workers have sampled this interval for fusulinids. Fusulinid and conodont data indicate the Desmoinesian-Missourian boundary lies in the shale interval above the Village Bend Limestone and below the Lake Pinto Sandstone (Cheney, 1940; Plummer, 1950; Shelton, 1958; Laury 1962; Trice, 1984).

The transect continues west on Texas 180 toward Palo Pinto, Texas, traveling across small cuestas formed by sandstone members in the upper Minerals Wells Formation and then across a finger of the first cuesta in the Canyon Group, formed on the Palo Pinto Limestone. Turn right (north) at Palo Pinto on FM Road 4, travel north along the surface of Palo Pinto Limestones to the Brazos River.

Locality 5 (Figs. 1, 2)

This is Stop 3 of Wermund (1969). It is located on the north bank of the Brazos River below the bridge on FM Road 4. This locality exposes the lowermost member of the Palo Pinto Limestone, the Wynne Limestone, below a cuesta capped by the uppermost member of the Palo Pinto Formation, the Wiles Limestone. The limestones exposed are bank limestones distant from the bank core and are representative of the easternmost extent of Palo Pinto bank limestone where fingers of a thick subsurface bank intersect the outcrop (Wermund, 1969). The core of the Palo Pinto Limestone bank is about 30 mi (48 km) northeast and downdip from here. About 22 ft (6.6 m) of limestone is exposed on the cuesta face. The lowest strata exposed are about 6 ft (1.8 m) of light-gray, thin-bedded algal wackestone and mudstone. Phylloid algae and fusulinids are the most common allochems. The massive 3.5 ft (1 m) bed overlying the lowest limestones is an osagid-algal grainstone and contains *Syringopora* heads. In some cases, fusulinids drape over the *Syringopora* heads. The next sequence of beds is 4 ft (1.2 m) of phylloid algal packstone and wackestone. Rare fusulinids and *Syringopora* are present. These beds form an inclined, less-resistant bench below a nearly vertical wall of about 7 ft (2.1 m) of limestone. The lowest of these beds are mudstones, which have fractures filled with sparry calcite. Overlying beds are composed of algal-bryozoan packstone. Fragments of phylloid algae and fenestrate bryozoans are the most common allochems and the number of bryozoans increase in the upper beds.

Continue north on FM Road 4 to the intersection of Texas 254, turn left (west) on Texas 254, continue to the intersection of Texas 254 and 16. Take a left (south) on Texas 16 and proceed to the Morris Shephard Dam of Possum Kingdom Lake, Palo Pinto County. After leaving Stop 5, the road travels obliquely across the remnants of the dip slope developed on the Wiles Limestone, then over the valley underlain by the Wolf Mountain Shale. The prominent cuesta formed by the Winchell Limestone can be seen in the distance after turning left (west), on Texas 254. Locality 6 is on this cuesta face.

Locality 6 (Figs. 1, 2)

This is Stop 1 of Wermund (1969). It is located in a south-flowing intermittent stream, 0.25 mi (0.4 km) east of the Morris Shephard Dam of Possum Kingdom Lake, and on the north side of the Brazos River, Palo Pinto County, Texas. Walk east on the Power House Road, then north up the stream bed to FM Road 2353.

The basal limestones are thin-bedded, with common mudstone lenses. These beds are transitional from the underlying terrigenous rocks upward into the higher limestones. Phylloid algae increase in abundance as the carbonate environment becomes dominant. As the terrigenous sediments decrease, the osagiid packstones-grainstones become more common. In measured sections of bank facies in this area, this facies commonly makes up one-third of the limestones (Wermund, 1969). This facies is thickly bedded, compared to the algal-bioclastic wackestones-packstones. The bedding of the osagiid packstones-grainstones is also more lenticular. They appear as superposed lenses, interpreted as having been deposited by currents or waves in a high-energy environment. The lenses may have been deposited as large ripples (Wermund, 1969). The middle part of the section is an example of extensive lime mud flat deposition. This is shown by some 20-25 ft (6-8 m) of mudstones and crinoidal-algal wackestones.

At times, the bank limestones grew above the sea floor as mounds. Evidence of this at this locality includes draping of strata over a crinoidal grainstone-packstone that contains small *Chaetetes*-like pods, and nearly 15 ft (5 m) of dipping beds in the upper part of the section.

REFERENCES CITED

This bibliography contains references cited in this paper, plus others useful for background reading.

Brown, L. F., Jr., and Goodson, J. L., 1972, Abilene Sheet, *in* Barnes, V. E., Geologic Atlas of Texas: Austin, University of Texas, Bureau of Economic Geology.

Brown, L. F., Jr., and Wermund, E. G., eds., 1969, A Guidebook to the Late Pennsylvanian Shelf Sediments, North Central Texas: Dallas Geological Society, 69 p.

Brown, L. F., Jr., Cleaves, A. W., II, and Erxleben, A. W., 1973, Pennsylvanian Depositional Systems in North-Central Texas; A Guide for Interpreting Terrigenous Clastic Facies in a Cratonic Basin: Austin, University Texas, Bureau of Economic Geology Guidebook 14, 122 p.

Cheney, M. G., 1929, Stratigraphic and structural studies in north-central Texas: Bureau of Economic Geology No. 2913, 27 p.

—— , 1940, Geology of north-central Texas: American Association Petroleum Geologists Bulletin, v. 24, p. 65–118.

—— , 1947, Pennsylvanian classification and correlation problems in north-central Texas: Journal of Geology, v. 55, p. 202–219.

Cheney, M. G., and Goss, L. F., 1952, Tectonics of central Texas: American Association Petroleum Geologists Bulletin, v. 36, p. 2237–2265.

Cheney, M. G., and others, 1945, Classification of Mississippian and Pennsylvanian rocks of North America: American Association Petroleum Geologists Bulletin, v. 29, p. 125–169.

Cleaves, A. W., II, 1973, Depositional systems of the upper Strawn Group of north-central Texas, *in* Brown, L. F., Jr., Cleaves, A. W., II, and Erxleben, A. W., Pennsylvanian Depositional Systems in North-Central Texas; A guide for Interpreting Terrigenous Clastic Facies in a Cratonic Basin: Austin, University Texas, Bureau of Economic Geology Guidebook 14, p. 31–42.

Hill, R. T., 1900, Physical geology of the Texas Region: U.S. Geological Survey, Topographic Atlas, folio 3, 12 p.

Kier, R. S., Brown, L. F., and McBride, E. F., 1980, The Mississippian and Pennsylvanian (Carboniferous) Systems in the United States–Texas: Austin, University Texas, Bureau of Economic Geology Circular 80-14, 45 p.

Laury, R. L., 1962, Geology of the type area, Canyon Group, north-central Texas: Graduate Research Center Journal, v. 30, no. 3, p. 107–180.

Plummer, F. B., 1950, The Carboniferous rocks of the Llano region of central Texas: Texas University Publication no. 4329, 120 p.

Plummer, F. B., and Hornberger, J., 1935, Geology of Palo Pinto County, Texas: University Texas Bulletin 3594, 240 p.

Plummer, F. B., and Moore, R. C., 1921, Stratigraphy of the Pennsylvanian formations of north-central Texas: University of Texas Bulletin, 2132, 237 p.

Shelton, J. W., 1958, Strawn-Canyon boundary in north central Texas: Geological Society American Bulletin, v. 69, no. 12, pt. 1, p. 1515–1524.

Stauffer, C. R., and Plummer, H. J., 1932, Texas Pennsylvanian conodonts and their stratigraphic relations: University Texas Bulletin 3201, p. 13–20.

Trice, E. L., 1984, Conodont Biostratigraphy and Stratigraphic Relationships of the Strawn Group (Pennsylvanian) Colorado and Brazos River Valleys, Central Texas [M.S. thesis]: Waco, Texas, Baylor University, 140 p.

Turner, G. L., 1957, Paleozoic stratigraphy of the Fort Worth Basin: Abilene and Fort Worth Geological Society Guidebook, 1957 Joint Field Trip, p. 57–78.

Wermund, E. G., 1969, Field trip first day, *in* Brown, L. F., Jr., and Wermund, E. G., eds., A guidebook to the late Pennsylvanian shelf sediments, north-central Texas: Dallas Geological Society, p. 40–51.

Wermund, E. G., 1975, Upper Pennsylvanian Limestone Banks, North-Central Texas: Austin, University Texas, Bureau of Economic Geology Circular 75-3, 34 p.

The Comanchean Section of the Trinity shelf, central Texas

O. T. Hayward, Geology Department, Baylor University, Waco, Texas 76798

LOCATION

This traverse is a composite of six localities (Fig. 1)—four in the Trinity Group and one each in the Fredericksburg and Washita Groups (Fig. 2)—exposed along a route extending from near Eastland (Eastland County) on the west to a point near Blum (northwestern Hill County) on the east, a straight-line distance of about 75 mi (120 km). The section transects the eastern margin of the Trinity Shelf and ends near the western margin of the East Texas Basin (Fig. 3). The section extends from the Western Cross Timbers across the Glen Rose Prairie, the Lampasas Cut Plain, and the Washita Prairies, which together make up the Grand Prairies of Texas (Hill, 1900) (Fig. 4).

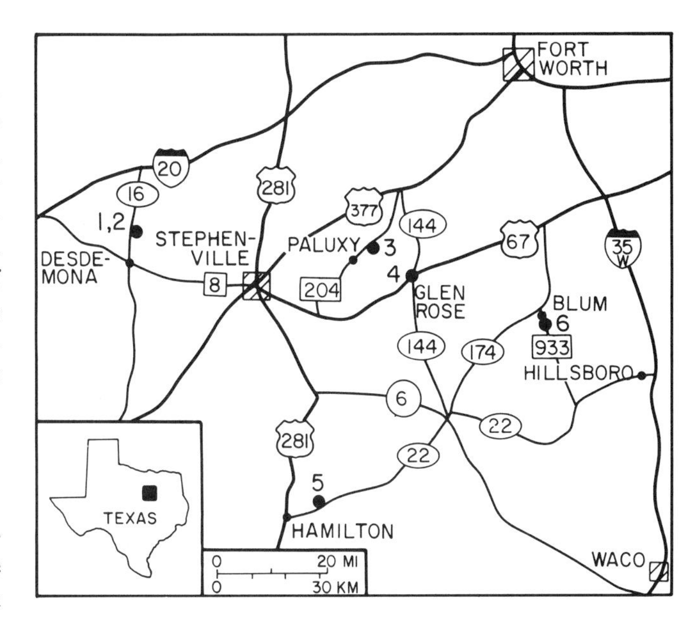

Figure 1. Index map showing distribution of localities along the Comanchean traverse.

SIGNIFICANCE

In the region described in this chapter, Hill (1901) established the type Comanchean Series (Figs. 1, 2) as well as the Trinity "Division," composed of "basement sands" and Glen Rose Limestone. Fredericksburg and Washita, the names for the other two "Divisions", were brought from distant areas. He emphasized that the Trinity Division comprises rocks of great lateral and vertical variability, the products of deposition in a shallow embayment, the Trinity Shelf (Lemons, 1987) (Fig. 3), whereas the Fredericksburg and Washita Divisions of the platform are widely uniform products of major transgression and regression across an almost featureless Cretaceous littoral (Figs. 2, 4).

The region of this transect has been subject to only minor structural disturbances since at least the beginning of Cretaceous time, so the rocks are in almost their original depositional attitudes. Thus the section exists in a degree of preservation that allows us to see facies in perhaps the world's best laboratory for the study of Early Cretaceous epeiric marine deposition.

LOCALITY 1

This locality (Figs. 2, 4) (98°32′49″W; 32°21′03″N; Desdemona 7½-minute Quadrangle) is on Texas 16, 5.5 mi (8.9 km) north of its intersection with Texas 8 in Desdemona, Eastland County, at a bridge over a small tributary of South Palo Pinto Creek. The locality extends southward 1.5 mi (2.4 km) to the road cut at the crest of the hill.

About 190 ft (58 m) of the lower Antlers Formation (Fig. 2), here about 240 ft (73 m) thick, is exposed in a series of small outcrops in the ditches on both sides of the highway. The outcrop begins with crinoidal limestone of the Pennsylvanian (Strawn) Mineral Wells Formation, overlain by 60 ft (20 m) of quartzose pea-gravel conglomerate and sand of the basal Antlers Formation, in turn overlain by massively bedded sandstone, siltstone, and red sandy mudstone. At the crest of the hill is a well-exposed section of interbedded sand, silty shale (some layers of which are highly carbonaceous) and at least one thin silty algal limestone capped by friable "packsand" about 20 ft (6 m) thick.

Antlers rocks rest with marked unconformity on the Wichita Paleoplain (Hill, 1901), initially a surface of subaerial erosion carved largely in Late Paleozoic rocks. These paleoplain landscapes were characterized by ridge-and-valley topography with relief as great as 200 ft (60 m) (Atchley, 1986). From the crest of the hill this exhumed topography is clearly visible to the north.

The earliest Cretaceous deposits were fluvial valley fills, chiefly mature sands and gravels of point bar and channel bar origin, bordered by silty red muds of floodbasin deposits (Boone, 1968; Hall, 1976; Ford, 1987). All such sediments are indicative of mixed-load streams of small size that originated far to the west—perhaps as far west as central Utah (Atchley, 1986). Basal gravels at most sites are interstratified with clean, well-sorted quartzose sands of fluvial to marine origin, suggesting that initial Cretaceous transgression was pulsed. Each pulse of transgression moved the shoreline westward along paleovalleys, inundating earlier fluvial deposits (Atchley, 1986; Ford, 1987; Lemons, 1987). Coarse fluvial deposits sporadically prograded over the marginal fluvial and marine sands to create an interstratification of coarse fluvial, fine fluvial, and marine clastics characteristic of the lowest Antlers/Twin Mountains Formations (Fig. 2).

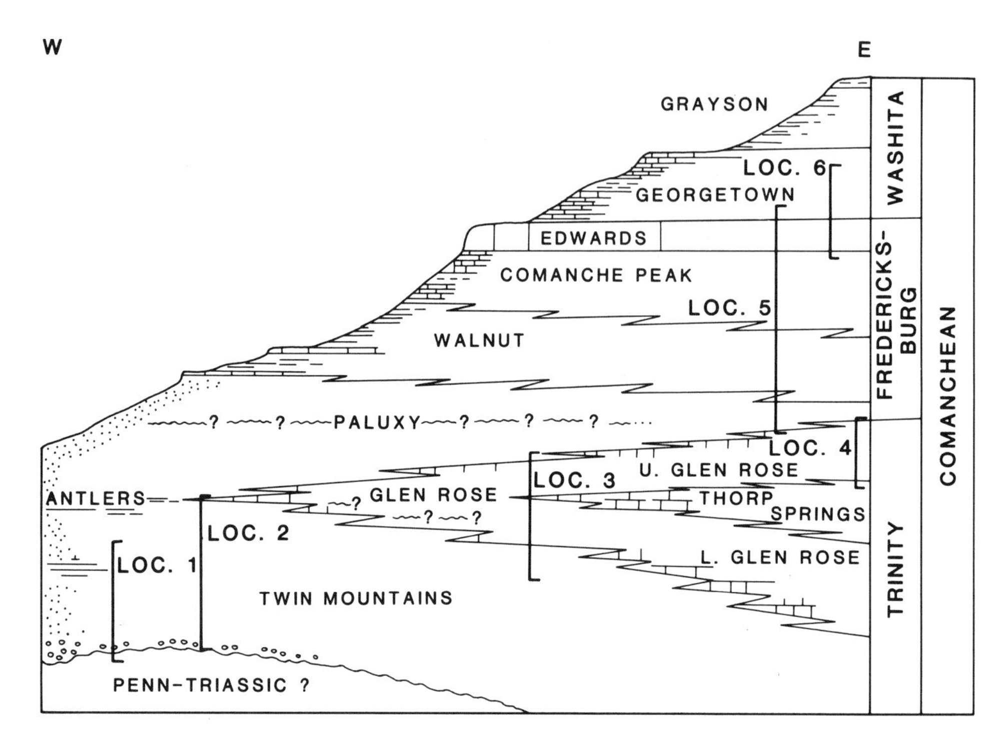

Figure 2. The Comanchean section of the Trinity shelf (after Lemons, 1987).

LOCALITY 2

This locality (Figs. 2, 4) (98°30′35″W; 32°20′57″N; Desdemona 7½-minute Quadrangle) is on an unnumbered county road, about 2.2 mi (3.5 km) due east of Locality 1, in a small road cut at the crest of a prominent north-facing escarpment, about 1 mi (1.6 km) south of Rattlesnake Mountain.

Two 6-in (13-cm) beds of rippled, sandy, extremely hard algal limestone of the upper Glen Rose Formation, separated by about 3 ft (1 m) of friable silty sand, are exposed in the road cut. Similar limestones, forming a continuous scarp, can be traced northeastward to Decatur, in Wise County, 78 mi (125 km) from Locality 2. This line marks the approximate western limit of Glen Rose marine carbonate deposition (Fig. 3). North of the road cut at Locality 2, and extending northward along the road about 2.2 mi (3.5 km), is a section about 220 ft (70 m) thick of Twin Mountains sand (Figs. 2, 4). The lower part of this section is equivalent to the section seen at Locality 1.

As transgression continued to drive the shoreline westward across a platform largely leveled by infilling of original topography by Twin Mountains sediments, Glen Rose Limestone overlapped marginal marine clastics, which in turn overlapped fluvial conglomerates and floodbasin muds. Thus, from west to east in any correlative stratigraphic sequence, the section grades from fluvial or marginal marine to marine, and at any one location, the same gradation exists vertically (Boone, 1968; Ford, 1987).

LOCALITY 3

This locality (Figs. 2, 4) (97°54′13″W; 32°16′28″N; Paluxy 7½-minute Quadrangle) is about 40 mi (62 km) due east from Locality 1, on FM 204, Hood County, at the bridge over the Paluxy River, and extends northward along the highway 1.3 mi (2.1 km) to the flat at the crest of the hill, exposing a total thickness of about 200 ft (68 m).

The section includes massive "packsands" grading upward to silty muds, about 100 ft (33 m) thick, of the Twin Mountains Formation. These are overlain by about 12 ft (4 m) of massive nodular carbonate bank limestone of the Thorp Springs Member of the Glen Rose Formation, in turn overlain by siltstones, nodular silty limestones, and thin dense hard limestones of the Upper Glen Rose Formation, about 100 ft (33 m) thick. The obvious increase in thickness in the Glen Rose section has occurred over a distance of 40 mi (65 km), a rate of thickening of about 4 ft/mi (0.75 m/km), which is characteristic of the Trinity rocks on the Trinity Shelf.

Maximum Trinity transgression apparently occurred during the deposition of Upper Glen Rose rocks (Davis, 1974; Ford, 1987), although along the Paluxy Valley, Thorp Springs rocks

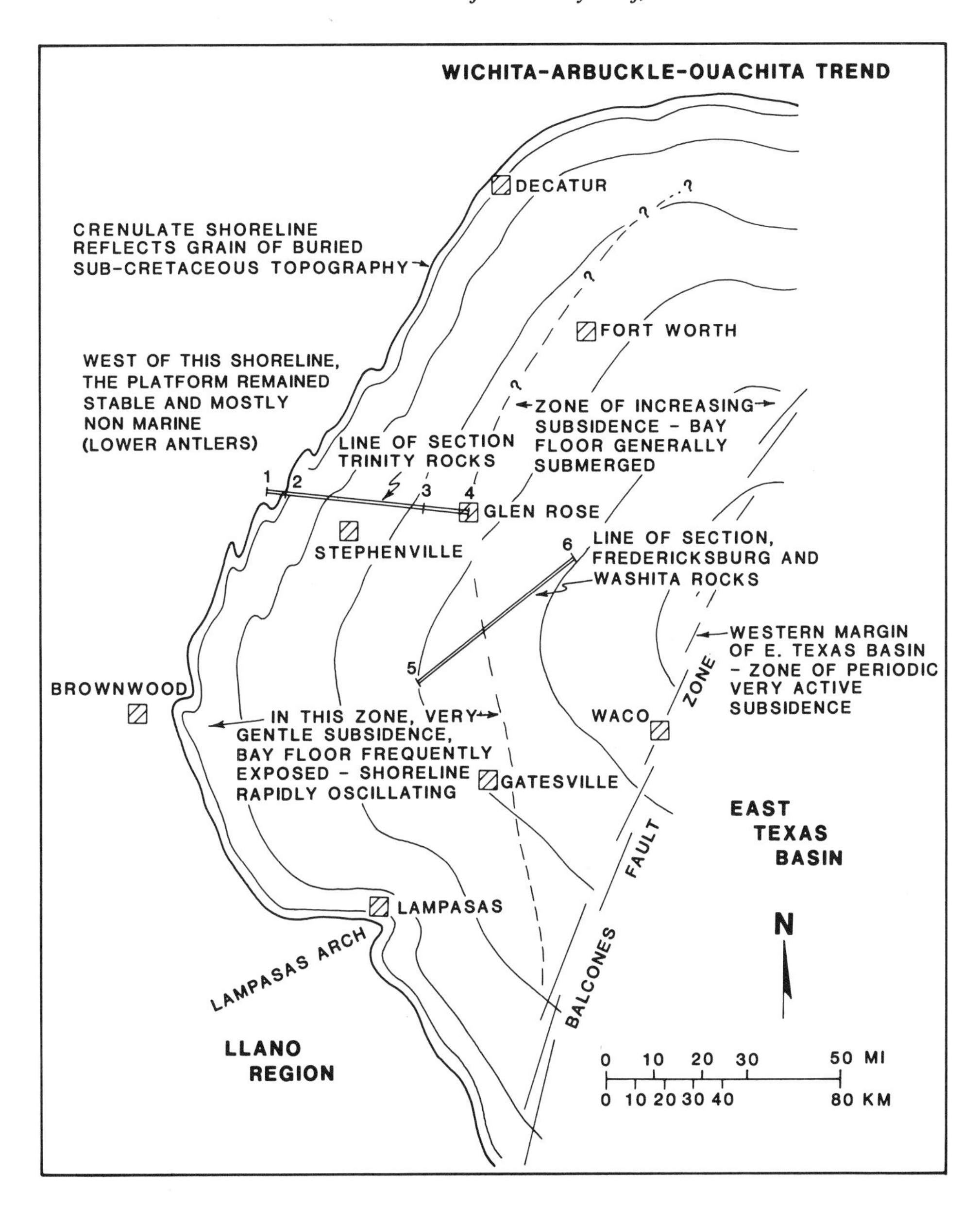

Figure 3. Trinity shelf at the time of maximum Glen Rose transgression. The shelf, the most stable part of the larger Central Texas Platform (Rose, 1972), acted as a shallow bay during most of Trinity deposition, subsiding gently with each major subsidence of the East Texas Basin (Lemons, 1987). It became a depositional platform during late Fredericksburg deposition (Ford, 1987). Coupling between the East Texas Basin and the Trinity shelf again became important during Washita deposition (Lemons, 1987).

represent a period of local transgression possibly related to local downwarping of the eastern margin of the Trinity shelf (Ford, 1987; Lemons, 1987). There is a great lithologic contrast between Localities 3 and 4. The upper Twin Mountains siliciclastic section of Locality 3 becomes the much-thickened Glen Rose carbonate section of Locality 4 in a distance of about 12 mi (19 km).

LOCALITY 4

This locality, in Glen Rose (Somervell County) (Figs. 2, 4) (97°44′50′; 32°14′17″N; Glen Rose 7½-minute East Quadrangle), is in the city park on the Paluxy River, 0.7 mi (1.1 km) east along Texas 144 from the flashing light in the center of town.

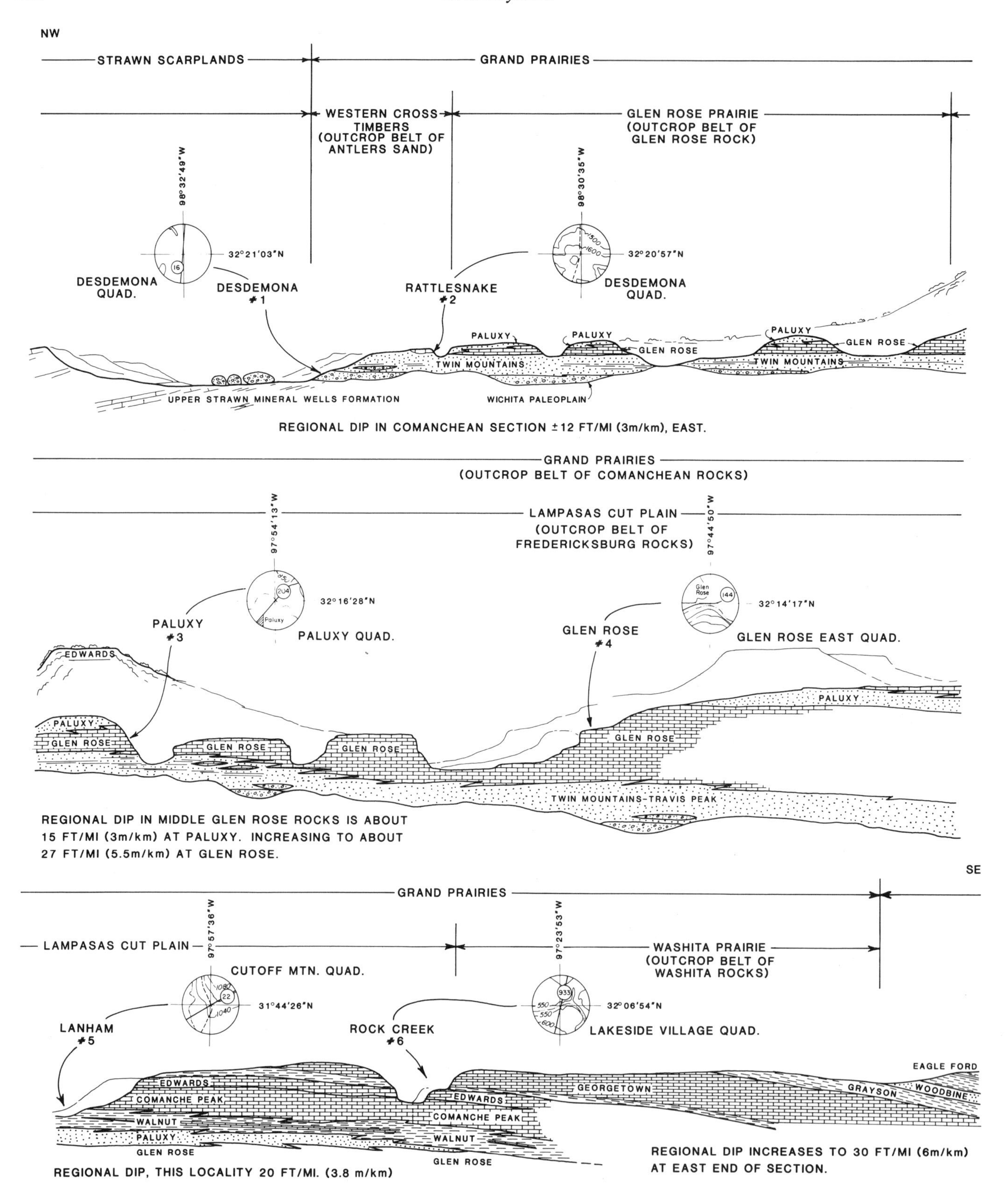

Figure 4. Diagrammatic west-to-east section of the Comanchean rocks on the Trinity Shelf, central Texas Platform. Localities 1–4 show downdip facies variations in Trinity rocks, from a totally land-derived-clastic section at Locality 1 to a marine carbonate section at Locality 4. Fredericksburg and Washita sections (Localities 5 and 6) show these groups in developments typical of most of the Trinity Shelf.

The large mushroomlike limestone joint blocks just above river level are of the Thorp Springs Member of the Glen Rose Formation, correlative with the same unit at Locality 3 (Fig. 2). The bluff on the opposite side of the river, about 150 ft (50 m) high, consists almost entirely of limestones of the Upper Glen Rose Member. The increased thickness and dominance of bank limestones is characteristic of the rocks of the transition zone between the Trinity Shelf and the more marine East Texas Basin during Trinity deposition (Fig. 3). Latest Glen Rose deposition was marked by a regression, represented here in uppermost Glen Rose rocks by interstratified sands and sandy limestones. The Glen Rose is overlain by Paluxy Sand, lowermost formation of the Fredericksburg Group (Fig. 2).

LOCALITY 5

This locality (Figs. 2, 4) (97°58′36″W; 31°44′26″N; Cutoff Mountain and Fairy 7½-minute Quadrangles) is centered near Lanham in a roadside park on the north side of Texas 22, at the intersection of an unnumbered county road and the highway, about 4.4 mi (7.1 km) east of the center of Hamilton, Hamilton County, and about 42 mi (68 km) south-southwest of Locality 4. The entire Fredericksburg section is present in this vicinity, though it must be pieced together from generally small partial sections in a 3.3 mi (5.3 km) traverse from west to east along Texas 22.

The section begins at the first small road cut on the south side of Texas 22, 0.5 mi (0.8 km) west of the roadside park. Exposed here is the uppermost Glen Rose Limestone, which is veneered with a layer of limonite and limonite-cemented sand. This is the weathered remnant of pyrite beds, which in the region of this transect commonly mark the top of the Glen Rose section (Corwin, 1982). The next road cut, 0.6 mi (1 km) west of the roadside park, on the north side of the highway, is of upper Paluxy Sand, here a massive upper shoreface sand 10 ft (3.3 m) thick, capped by a bed of extremely hard, ripple-marked coquinoid (*Texigryphaea*) limestone, the lowermost bed of the Walnut Clay. Eastward from the roadside park 2.5 mi (4 km) to the base of the prominent divide, the road ascends through Walnut Clay, exposed principally as oyster beds, and occasional hard limestones. From the base of the divide, the highway ascends through an excellent road cut in Comanche Peak Limestone, Edwards Limestone, and lowermost Georgetown Formation (Figs. 2, 4). The total Fredericksburg Section is 260 ft (85 m) thick.

The Paluxy sand (lowermost member of the Fredericksburg Group) grades upward from sandy limestones of the Upper Glen Rose through laminated muddy silts and sands into massive bar and shoreface sands, in places marked by dinosaur trackways and carbonized leaf and twig impressions. Nodular caliches here indicate temporary exposure (Amsbury, 1967). Paluxy Sand, shown as the lowermost formation of the Fredericksburg Group, thus represents the final regressive deposit of the Trinity section and the initial transgressive deposit of the Fredericksburg section (Amsbury, 1967; Owen, 1979; Corwin, 1982).

In the line of this section, although the Walnut Clay consists mostly of clay, the exposed parts are typically dense, fossiliferous, ripple-marked limestones, extremely fossiliferous marls and chalky limestones, some thin shales, and massive oyster banks, all of which are areally extensive. These rocks accumulated in rapidly changing shallow-water environments ranging from estuarine and brackish to normal epeiric marine (Flatt, 1976).

The Walnut Clay grades upward through a transition zone 10 ft (3 m) or more thick, into marly, nodular limestone of the Comanche Peak Formation. No new lithologies are represented, but the mix of lithologies changes to favor carbonates and more normal epeiric marine communities and perhaps to indicate more distant shorelines and greater transgression (Corwin, 1982).

The contact between the Comanche Peak Formation and the overlying Edwards Limestone is one of the most conspicuous in the Comanchean section. Typically it is an abrupt, apparently conformable transition from marly nodular Comanche Peak Limestones to the massive rudist-bearing biohermal and biostromal reeflike limestones of the Edwards Formation (Hertel, 1986; Corwin, 1982). On the Trinity Shelf the Edwards Limestone is characterized by four particularly significant features: (1) uniformity in thickness over the vast area of the shelf, (2) a composition essentially devoid of land-derived clastics (in contrast to formations below and above), (3) facies uniformity for a distance of at least 300 mi (480 km) along depositional strike across the platform to the northwest, and (4) marked facies contrast from reeflike limestones on the northeast to cherty dolomites and evaporites on the southwest (Corwin, 1982).

These combine to suggest that Edwards deposition may have been the product of sudden transgression and was completed in a very short time. This conclusion is supported by the presence of individual bioherms that penetrate almost the full thickness of Edwards Limestone (Corwin, 1982). Thus the Fredericksburg Group appears to be the result of a continuing series of pulsed transgressions, culminating in the final abrupt transgressive pulse of the Edwards Limestone. Although on a regional basis Washita transgression to the northwest was greater (D. Amsbury, oral communication, 1987), the most extensive transgression of the Trinity shelf during Comanchean time appears to have been that of Edwards Limestone (Corwin, 1982). Although there were later extensive transgressions and regressions, the overall aspect of post-Edwards rocks on the Trinity Shelf is that of regression.

The contact between the Fredericksburg and Washita groups is abrupt. Dense, reef-like rudist limestones, lime grainstones, and lime mudstones are overlain by the marly marine clay of the lowermost Georgetown Formation. This (Kiamichi) clay thins southward to pinch out near Waco and was clearly derived from the north. Generally the contact is considered unconformable, but there is no clear evidence of truncation. Apparently the period of exposure was a brief one.

LOCALITY 6

This locality (Figs. 2, 4) (97°23′55″W; 32°06′54″N; Lakeside Village 7½-minute Quadrangle) is at the bridge over Rock

Creek, on FM 933, 1.9 mi (3.1 km) south of Blum (Hill County) and 42 mi (68 km) northeast of Locality 5. The exposure is upstream along Rock Creek on a steep slope 100 ft (30 m) high in the Corps of Engineers-designated public area of Lake Whitney.

Here, the Edwards Limestone (16 ft, 5+ m) and the lower Georgetown Formation (60 ft, 21 m) are exposed along the creek, on the slope, and in the ditches beside the road on the south side of the bridge. The Grayson Marl, uppermost formation of the Washita Group in this region, is present (but poorly exposed) in the highlands to the east, and the top of the Grayson (the top of the Comanchean Section) is poorly exposed at Locality 1 of the Gulfian Section (Hayward, site 75, this volume).

The Georgetown Formation is typically a series of marly, nodular, chalky limestones, separated by thinner calcareous shales. Even in its most calcareous facies it shows strong admixtures of land-derived clays that thicken and become more coarsely clastic northward, indicating a northern source. The uppermost Georgetown limestone (Mainstreet) grades conformably upward through an interval of perhaps 4 ft (1.2 m) into the overlying Grayson Marl, a chalky, fossiliferous marl to slightly calcareous, ocherous, clay-shale.

The Grayson, consisting almost entirely of land-derived clastics, may mark a more prominent local regressive episode in the long regressive history of post-Edwards deposition on the Trinity shelf.

SUMMARY

In summary, the platform Comanchean Section illustrates the combined, but difficult to differentiate, effects of eustasy and tectonically controlled marine transgression over an erosional landscape of generally low relief and general structural stability. Deposition of Trinity rocks appears to have been largely regulated by the structural behavior of the East Texas Basin and its coupled western margin, the Trinity Shelf (Fig. 3), although superimposed upon this were the effects of eustasy. Deposition of Fredericksburg rocks was far more regional, for they indicate the effects of major transgression, culminating in the Edwards Limestone, across an essentially flat shelf. The Washita Group, characterized by the Georgetown Formation, with its increase in land-derived clastics, followed by the Grayson Marl, mostly land derived, marks the beginning of the "Great Regression" (Jackson, 1983) and dominance of clastics over carbonates, which yet continues. The volume of mud represented by these later deposits, and the marked change in depositional style, presents one of the most interesting problems of Cretaceous history on the Trinity shelf.

REFERENCES CITED

Amsbury, D. L., 1967, Caliche soil profiles in Lower Cretaceous rocks of central Texas: Geological Society of America Program with Abstracts, p. 4.

Atchley, S. C., 1986, The pre-Cretaceous surface in central, north, and west Texas; The study of an unconformity [M.S. thesis]: Waco, Texas, Baylor University, 233 p.

Boone, P. A., 1968, Stratigraphy of the basal Trinity (Lower Cretaceous) sands of central Texas: Baylor Geological Studies Bulletin 15, 64 p.

Corwin, L. W., 1982, Stratigraphy of the Fredericksburg Group north of the Colorado River: Baylor Geological Studies Bulletin 40, 64 p.

Davis, K. W., 1974, Stratigraphy and depositional environments of the Glen Rose Formation, north-central Texas: Baylor Geological Studies Bulletin 26, 43 p.

Flatt, C. D., 1976, Origin and significance of the oyster banks in the Walnut Clay Formation, central Texas: Baylor Geological Studies Bulletin 30, 47 p.

Ford, M. E., 1987, The stratigraphy of the Trinity rocks north of the Colorado River [M.S. thesis]: Waco, Texas, Baylor University, 351 p.

Hall, W. D., 1976, Hydrologic significance of depositional systems and facies in Lower Cretaceous sandstones, north-central Texas: University of Texas Bureau of Economic Geology Geological Circular 76-1, 29 p.

Hertel, C. O., 1981, The nature of the Comanche Peak–Edwards contact, Lampasas Cut Plain, central Texas [B.S. thesis]: Waco, Texas, Baylor University, 163 p.

Hill, R. T., 1900, Physical geography of the Texas region: U.S. Geological Survey Topographic Atlas of the United States, Folio 3, 12 p. plus maps.

—— , 1901, Geography and geology of the Grand and Black Prairies of Texas: U.S. Geological Survey, 21st Annual Report, part 7, 666 p.

Jackson, R. L., 1983, The stratigraphy of the Gulfian Series (Upper Cretaceous) east-central Texas [M.S. thesis]: Waco Texas, Baylor University, 103 p.

Lemons, D., 1987, The structural evolution of the Trinity shelf, central Texas [M.S. thesis]: Waco, Texas, Baylor University, 301 p.

Owen, M. T., 1979, The Paluxy sand in north-central Texas: Baylor Geological Studies Bulletin 36, 36 p.

Rose, P. R., 1972, Edwards Group, surface and sub-surface, central Texas: University of Texas Bureau of Economic Geology Report of Investigation 74, 198 p.

Gulfian rocks, western margin of the East Texas Basin

*O. T. **Hayward**, Department of Geology, Baylor University, Waco, Texas 76798*

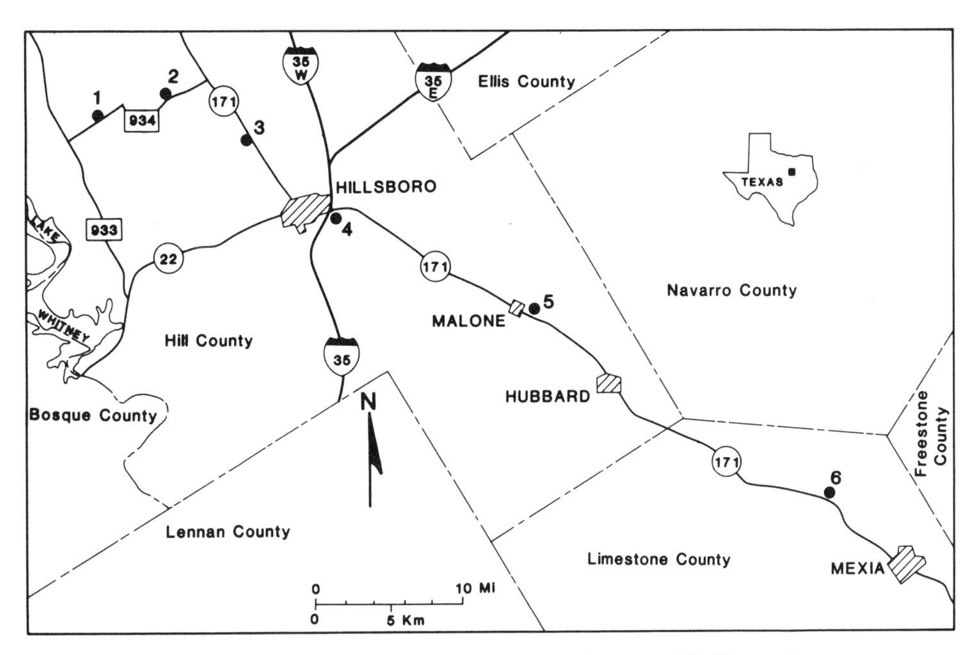

Figure 1. Index map showing localities along Traverse of Gulfian rocks.

LOCATION

The Gulfian section of Central Texas is illustrated by a transect extending from northwestern Hill County southeastward to northeastern Limestone County (Figs. 1 to 6).

This transect begins in latest Washita rocks of Comanchean age (locality 1; Figs. 1, 3), and crosses in succession the outcrop belts of Woodbine Sand (localities 1 and 2; Fig. 3); Eagle Ford Group (locality 3; Fig. 4); Austin Chalk (locality 4; Fig. 4); Taylor Group (locality 5; Fig. 5); and Navarro Group, all of Gulfian age, to end at the Midway Group (locality 6; Fig. 6) of Paleocene Age (Hill, 1901; Adkins, 1933; Bureau of Economic Geology, 1970, 1972). Structurally, it originates near the eastern margin of the Central Texas Platform (Fig. 3), and trends downdip across the northern end of the Balcones Fault Zone (Fig. 4) into the East Texas Basin (Fig. 3), to terminate in the Mexia Fault Zone (Fig. 6; Oliver, 1971). Geomorphically it begins at the eastern margin of the Grand Prairies (Fig. 3), and crosses the Eastern Cross Timbers and the Black Prairies (Figs. 4, 5, 6) to end at the western margin of the East Texas Timber Belt (Fig. 6; Hill, 1901). Throughout this route, the geology, landscape, history, and land use are clearly related.

SIGNIFICANCE

The Gulfian section (Fig. 2) consists of generally fossiliferous marine mudrocks with minor limestones and sandstones, all of which thicken rapidly eastward into the East Texas Basin.

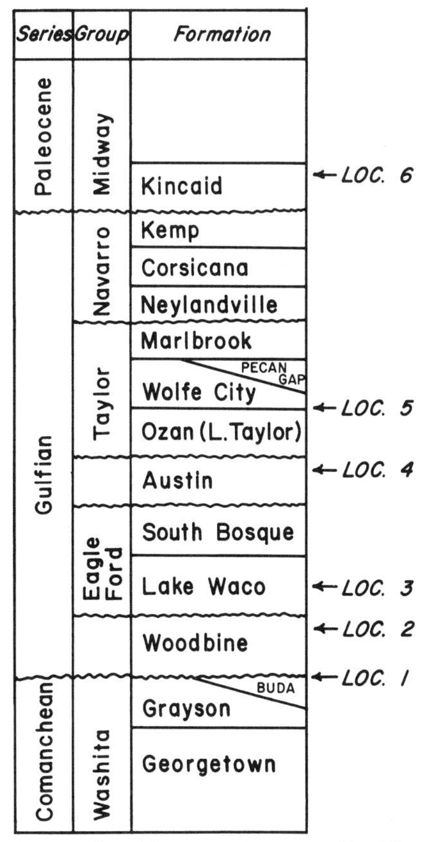

Figure 2. Gulfian stratigraphic nomenclature used in this study. It conforms to that of the Geologic Atlas of Texas (Bureau of Economic Geology, 1970, 1972).

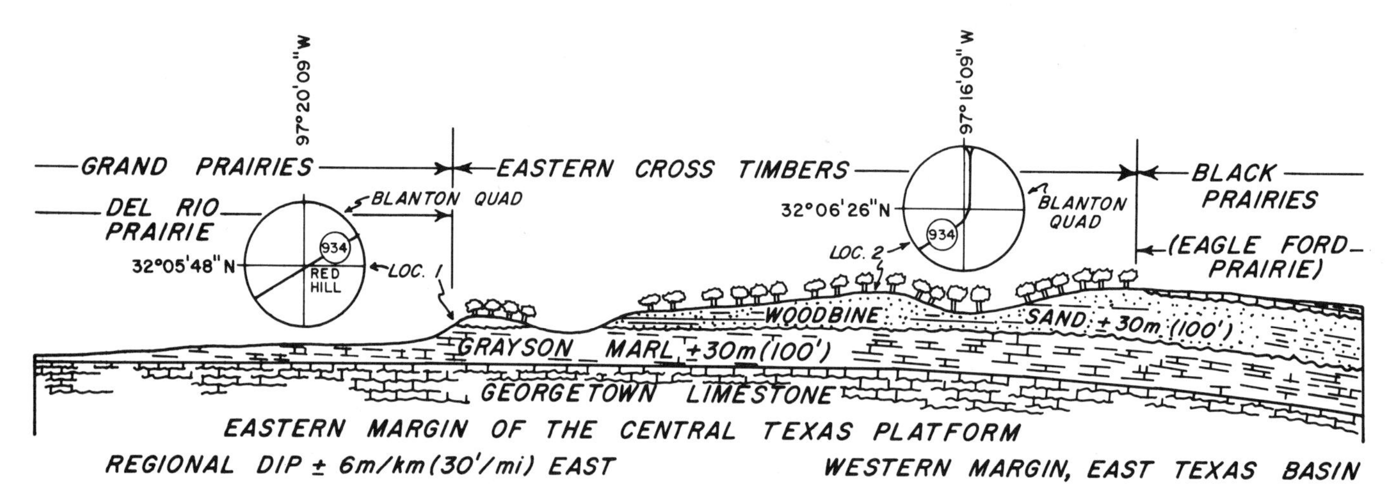

Figure 3. Diagrammatic section of Gulfian rocks, segment 1. Section not drawn to scale, but is designed to show visual relationships between landforms and geology. Figures 2, 3, 4, 5 join to form a continuous dip section.

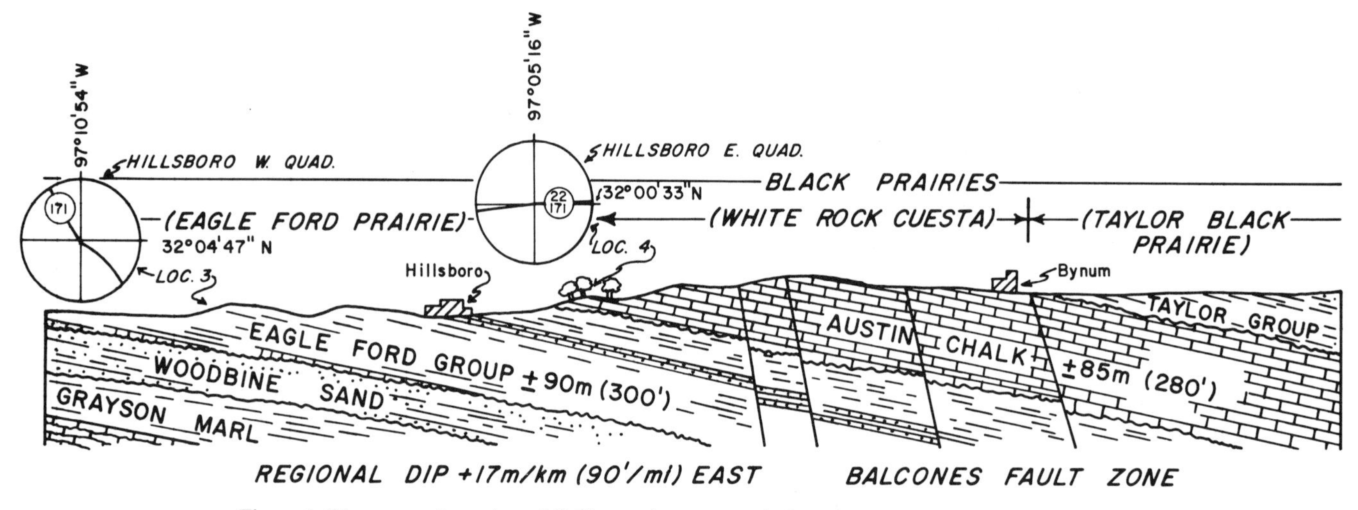

Figure 4. Diagrammatic section of Gulfian rocks, segment 2. See Figure 3 caption for description.

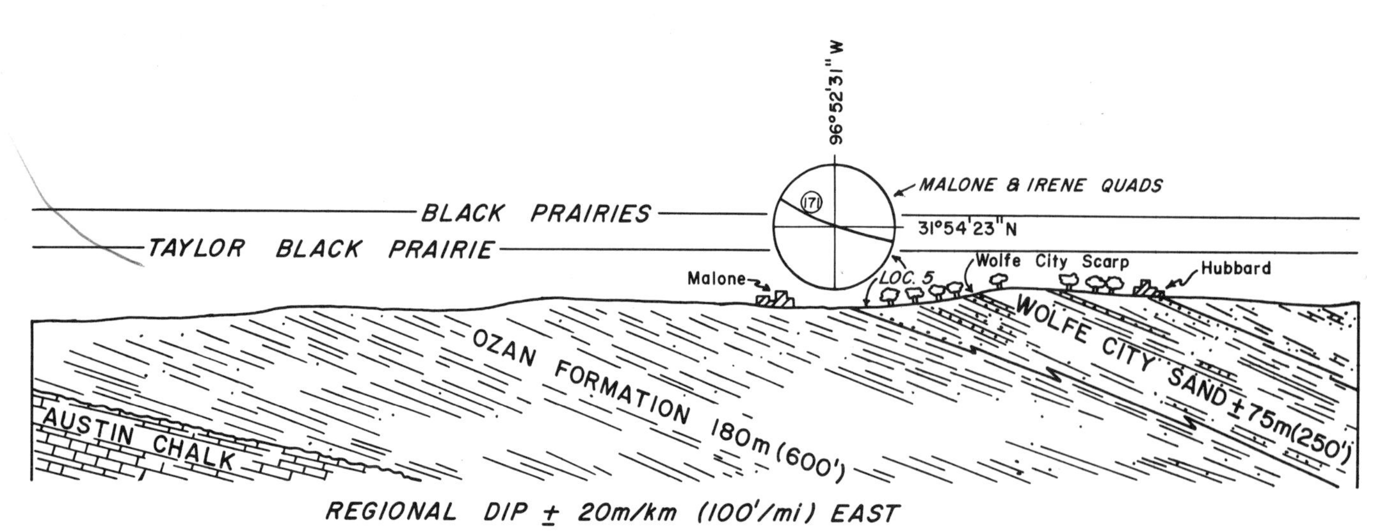

Figure 5. Diagrammatic section of Gulfian rocks, segment 3. See Figure 3 for description.

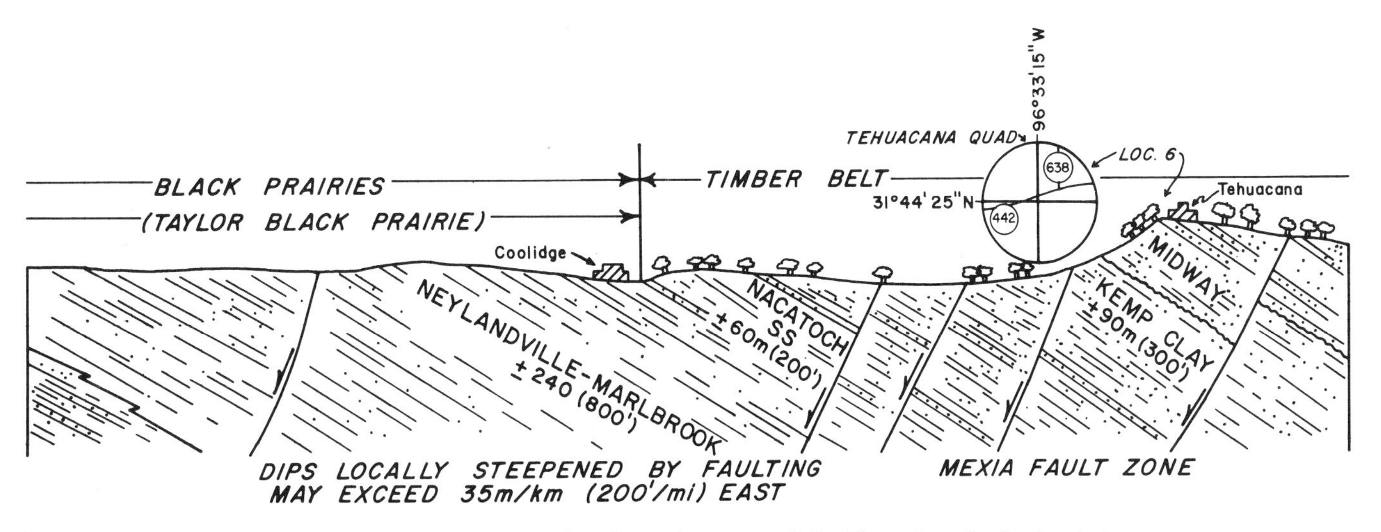

Figure 6. Diagrammatic section of Gulfian rocks, segment 4. See Figure 3 caption for description.

Within this region the outcrops are small, scattered, and unimpressive, yet because of the general homogeneity of major units, even these small exposures illustrate well the general character of the stratigraphic section.

Landscape clearly reflects stratigraphy and structure. From west to east the buttes and mesas of the Comanchean outcrop belt, with a vegetation cover of savannah, grassland, and pasture, give way to a Gulfian landscape of inconspicuous west-facing cuestas, now chiefly in farmland (Hill, 1901).

GULFIAN SECTION

This traverse begins at the intersection of FM 3049 and FM 934, 4.6 mi (7.4 km) southeast of Blum, northwestern Hill County (97°20′51″W.; 32°05′27″N. Blanton 7½-minute Quadrangle) (Fig. 1; Bureau of Economic Geology, 1972). The traverse extends southeast along FM 934.

Locality 1 of the traverse is on the eastern margin of the near-flat Del Rio Prairie, the easternmost margin of the Grand Prairies (Hill 1901, Plate 1).

Locality 1 (Figs. 1, 3) (97°20′09″W.; 32°05′48″N., Blanton 7½-minute Quadrangle) is on FM 934 0.8 mi (1.3 km) northeast of the starting point, in a low roadcut on the left (north) side of the road at the crest of gentle west-facing wooded escarpment. At this locality are 14 ft (4.6 m) of poorly exposed section, including the Woodbine/Grayson contact and lowermost Woodbine Sands.

Here, at the western margin of the Eastern Cross Timbers (Fig. 3), are a series of prominent, wooded, west-facing cuestas veneered with sandy soils that developed on the basal Woodbine Sand. This belt, which was designated on the earliest maps of Texas as a most conspicuous physiographic marker, is still heavily timbered, and is composed of a distinctive landscape of savannah and Post Oak forest. The contact between the Comanchean and Gulfian rocks in this line of section is unconformable. The unconformity is marked by buff to ocherous muddy sands of the Woodbine Group that rest on gray marly clays of the Grayson Formation. The thin Buda Limestone, apparently originally present, was stripped by erosion prior to initial deposition of Woodbine Sands. While this unconformity is everywhere present along the outcrop belt, it diminishes in magnitude downdip in the subsurface (Adkins and Lozo, 1951). This is another of the peculiarities of the Gulfian Section: numerous and significant basin-margin unconformities, which become less conspicuous or disappear basinward.

Locality 2 (Figs. 1, 3) (97°16′09″W.; 32°06′26″N., Blanton 7½-minute Quadrangle) is on FM 934, on the west valley wall of Aquilla Creek, 1.9 mi (3.0 km) northeast of the junction between FM 934 and FM 309. The outcrop is a prominent roadcut on the left (north) side of the road, in a sharp descending turn to the left. At this locality, 26 ft (8 m) of well-exposed section range from jarosite-stained shaly silt upward through thick, massively bedded, very friable sands typical of the upper Woodbine over much of the East Texas Basin.

In the line of section, the Woodbine Group is of coastal barrier and prodelta-shelf facies, derived from deltaic input from the northeast (Oliver, 1971). Lower Woodbine Sands are typically fine, well-sorted quartzose sandstones, often glauconitic, properties characteristic of the lower shoreface (locality 1). More massively bedded and cleaner upper shoreface sands with parallel bedding and silty interbeds are present in the Upper Woodbine (locality 2, Fig. 3).

The contact between the Woodbine and Eagle Ford Groups is not exposed along this transect, but it too is a basin-margin unconformity, typical of the Central Texas Gulfian Section. Though the actual contact is not visible, the geomorphic effect of the transition from Woodbine to Eagle Ford rocks is readily visible in the landscape. Woodlands of the Cross Timbers, developed on Woodbine rocks, end abruptly, and the treeless, heavily agricultural Black Prairies (Figs. 3, 4, 5, 6) stretch away to the eastern horizon, in one of the most conspicuous land-use contrasts in the Gulfian outcrop belt. The Eagle Ford Prairie, westernmost of the Black Prairies, was originally in tall grass, but the deep, dark, rich soils soon attracted farmers; today it is difficult to find even an acre of that sea of grass, which was the Texas tall grass prairie (Hill, 1901).

Locality 3 (Figs. 1, 4) (97°10′54″W.; 32°04′47″N., Hillsboro West 7½-minute Quadrangle) is on FM 171, 4.3 mi (7 km) southeast of Osceola (junction of FM 934 and FM 171), in a roadcut on the right (southwest) side of road, on the south valley wall of a small tributary of Hackberry Creek. At this locality are 30 ft (10 m) of well-exposed lower Eagle Ford rocks.

The section consists of buff-weathering, dark blue-gray, highly calcareous bentonitic shales, most often very thinly laminated; it is interbedded with very thin discontinuous limestone beds that are richly fossiliferous and contain both invertebrate and vertebrate remains. A particularly significant aspect of the Eagle Ford rocks, especially conspicuous in this exposure, consists of numerous, highly distinctive thin (1 to 5 cm) bentonite seams, each of which apparently represents a single ash fall. Associated with these bentonitic sections are finely laminated shale beds of ash-like appearance, very highly calcareous with fragments of carbonized wood, vertebrate remains (bones, teeth, scales), and abundant planktonic foraminifers, but with very few benthonic forms. In contrast, dense, thin limestone beds in the lower and middle Eagle Ford are formed largely of spicules, shells and shell debris of *Inoceramus* and other benthonic forms, with occasional small ammonites (Adkins and Lozo, 1951).

This highly calcareous lower Eagle Ford section grades upward into more massive, dark blue-gray, and much less calcareous shales of the upper Eagle Ford Group, which are not exposed in the line of section.

The various depositional environments of Eagle Ford rocks have been the subject of arguments for many years. However, the presence of the following features tends to suggest shallow shelf, platform, or lagoonal accumulation in which periodic stagnation resulted in toxic bottom conditions (Silver, 1963): marked unconformities at base and top, internal unconformities, abundant fragments of carbonized wood from land plants; and alternating sequences of undisturbed laminites (indicative of anoxic conditions), with clean fossiliferous calcarenites (indicative of oxygenated conditions). Isopachs of Eagle Ford rocks of the East Texas Basin indicate deltaic input from the north. The very limited sand development suggests that these were mud-dominated deltas, similar to other Gulfian mudrock deltas (M. A. Surles, Jr., 1985, oral communication).

In the landscape, the contact between the Eagle Ford Group and the overlying Austin Chalk (another of the basin margin unconformities so common in the Gulfian Section) is marked by the White Rock Cuesta (Fig. 4; Hill, 1901). This prominent west-facing escarpment marks the western margin of the White Rock Prairie (Fig. 4), a middle subprovince of the Black Prairies (Figs. 3, 4, 5, 6). The White Rock Cuesta is the westernmost of the coastal escarpments of Texas, and, locally at least, is considered to be the surface indication of the western margin of the East Texas Basin. This cuesta also marks the general trend of the Balcones Fault Zone and has the further distinction of being "home" to about one-fifth of the population of Texas. Denison, Dallas, Waco, Austin, San Antonio, and dozens of smaller towns in between, are centered on this narrow physiographic feature, which nowhere is more than 15 mi (24 km) wide. This unique association of geology and cities has no other parallel in Texas.

Locality 4 (Figs. 1, 4) (97°05′16″W.; 32°00′33″N., Hillsboro East 7½-minute Quadrangle) is on FM 171, 0.5 mi (0.75 km) east of the intersection of Texas 171 and I-35, at the entrance to Hill Junior College, and extending east in bar ditches on both sides of Texas 171 for 0.5 mi (0.75 km). In this stretch is a continuous exposure of Austin Chalk about 100 ft (30 m) thick, beginning very near the Austin/Eagle Ford contact. Additional exposures are conspicuous in drainages, bar ditches, quarries, and fields for the next 6 mi (10 km).

Austin Chalk is an anomaly in the Gulfian Section, a limestone in a major sequence dominated by mudrocks. In the line of section, it consists of the lowermost part of the Austin Group of north Texas. Upper units were removed by erosion prior to deposition of Taylor Marl (Stephenson, 1937). Despite its lithology and thickness over most of its outcrop area, the Chalk tends to be exposed only in very small outcrops in roadside ditches; it is in somewhat greater thicknesses in streamcuts and in quarries. It consists of interstratified, massive white chalk and gray chalky marl in beds up to 3+ ft (1 m) thick. Throughout its outcrop area it is most notable for uniformity in bedding, texture, and composition. Fossils include *Inoceramus,* often very large (up to 3 ft or 1 m in diameter), *Gryphaea,* some rudists and abundant microfossils (largely cocolithophorids) (Scholle, 1978, p. 17-18). Bed-for-bed correlation is possible for 100 mi (160 km) along strike and for 40 mi (60 km) down dip. Conspicuous among the minor depositional features of the lower Austin group are numerous channel cut-and-fill structures up to 100 ft (30 m) wide, possibly of tidal origin, that are particularly well-exposed in Dallas and Waco.

Areal uniformity of the formation and the abrupt reduction in land-derived clastics at the Austin/Eagle Ford contact suggest that Austin deposition was the product of major transgression out of the shallow East Texas Basin over a nearly flat Late Cretaceous landscape. Clastic sediments, chiefly clays in the marl beds, were derived from the north, though the source was distant from this line of section. There is considerable argument about Austin depositional environments, but in the beginning, at least, Austin seas were apparently shallow, perhaps no more than 33 ft (10 m) deep with clear, oxygenated waters. Nowhere along this line of section is there evidence of near-shore conditions during Austin deposition, though near-shore sands and muds occur about 60 mi (100 km) to the northeast.

The contact between Austin Chalk and overlying Taylor Marl is markedly unconformable. Both the middle and upper formations of the Austin Group, present in Dallas and Austin, are missing along the line of section. Correlations just downdip from the outcrop belt suggest that the upper Austin contact is a product of erosional truncation in which more than 200 ft (60 m) of section was removed.

Physiographic contrast between White Rock Prairie of the Austin Chalk outcrop belt and Taylor Black Prairie of the Taylor Outcrop belt is a subtle one. The most conspicuous indications of this transition are the existence of exposed chalk along stream courses and the presence of white-to-black mottled soils on divides, both of which characterize the Austin Chalk outcrop belt. These give way to inconspicuous exposures of gray-buff marl on streams, and generally dark soils in the Taylor outcrop area.

Taylor Black Prairie is typically gently rolling farmland, veneered with very deep black soil. This was once the largest of the prairie lands of Texas, covered by tall grass so high at flowering that it would hide a man on a horse. Its fertility was recognized early, and it became the base of the Texas "Cotton Empire." While it is still almost entirely agricultural, grains have replaced cotton as the major crop.

The Taylor Group is by far the thickest and most widely exposed of Gulfian units. It is also difficult to study, for Taylor outcrops are rare and almost always small. Yet in terms of sedimentary volumes this is the most impressive of all the Cretaceous formations (Beall, 1964).

Locality 5 (Figs. 1, 5) (96°52′31″W.; 31°54′23″N.; Malone 7½-minute Quadrangle) is on FM 171, 1.25 mi (2 km) southeast of the junction of FM 171 and FM 308 (Malone), southeastern Hill County, in a series of roadcuts on the right (southwest) side of the road, extending for 0.24 mi (0.4 km). In this series of roadcuts are about 60 ft (18 m) of total section, of which about half is exposed.

The section begins in the uppermost Ozan Formation (generally referred to as Lower Taylor Marl) and extends upward into the overlying Wolfe City (Fig. 2). The Ozan formation is dark olive-gray, fissile to blocky, calcareous, fossiliferous shale, which becomes slightly silty upward in the section. Limonite seams, originally pyrite, are prominent near the Wolfe City Sand contact. The contact with the Wolfe City Sand is gradational, but here is marked by the abrupt occurrence of thin beds of silty sand in silty shale near the crest of the Wolfe City Escarpment (Fig. 5). This section becomes more sandy upward, although in the line of section it is significantly more shale than sand.

Isopach character, clay mineralogy, and sand distribution and petrology suggest that Taylor deposition was largely a product of delta and pro-delta mud deposition that continued into latest Cretaceous time. For the Ozan formation, sediment input points were on the north and for the Wolfe City Sand from the west. Both of these units thicken into the East Texas Basin (Beall, 1964).

Pecan Gap Chalk, a chalky basinal marl that overlies the Wolfe City Sand, is not exposed along the line of section, nor is the overlying Neylandville Marl, though along this transect the combined thickness of these two formations exceeds 800 ft (250 m).

The Taylor-Navarro contact is unconformable (Fig. 2); it consists of sandy clay on silty clay, and therefore is difficult to distinguish in the field, either on the basis of physical stratigraphy or geomorphology.

Navarro Group consists of several named units, all marly, silty, and marine, which are in physical stratigraphy almost indistinguishable from each other. It is equally difficult to distinguish from the underlying Taylor marls and from the lowermost Tertiary Midway section (Figs. 2, 6) that immediately overlies it.

The general trend in lithology of Upper Cretaceous rocks, from Taylor upward, is toward more sandy and more nearshore deposition. Outcrops of uppermost Gulfian formations are very rare, and the best of these commonly expose only a few feet of rock. Along the line of this section, although latest Cretaceous rocks are present, they are not exposed. However, some fresh silty fossiliferous clays of latest Gulfian age can be obtained by digging in roadside ditches near the crests of gentle hills.

Locality 6 (Figs. 1, 6) (96°33′15″W.; 31°44′25″N.; Tehuacana 7½-minute Quadrangle) is on the west side of the town of Tehuacana, eastern Limestone County, on F.M. 442, and 0.2 mi (0.3 km) east of the junction with FM 171, in a roadcut along the right (south) side of road. The outcrop continues to the top of the Tehuacana Escarpment (Fig. 6). Somewhere near the base of the scarp face is the contact between uppermost Navarro silty clays and Lower Midway clayey silts. The top of the scarp is marked by 40 ft (12 m) of Tehuacana "limestone" (principally sand) of the upper Midway Group. To the east of Tehuacana this section is repeated several times by down-to-the-west faulting of the Mexia Fault Zone (Fig. 6).

The history of latest Gulfian deposition is that of increasingly sandy, mud-dominated sedimentation into a subsiding East Texas Basin from inputs along a prograding western shoreline. This sequence continues eastward to the outcrop belt of Eocene Mid-Wilcox rocks, the first truly alluvial sediments now exposed at the surface.

SUMMARY

Gulfian rocks generally reflect continuing regression or progradation interrupted by a few minor transgressive pulses, a history that has continued to the present time. Gulfian sediments accumulated to thicknesses exceeding 3,500 ft (1,000 m). Each subsidence of the East Texas Basin was apparently accompanied by renewed uplift in provenance areas to the west and north, yielding a continuing supply of clastic sediments. Thus, the section is characterized by great thicknesses of land-derived clays, silts, and sands, in which Austin Chalk constitutes an anomaly. Gulfian rocks accumulated in and adjacent to the subsiding East Texas Basin, in which the volume of muds was enormous, in marked contrast to the thin, carbonate-dominated platform deposits of the Comanchean section to the west. While sediment input points are generally recognized, the specific provenance of this great volume of mud, together with its correlatives in other areas, constitutes one of the more interesting geologic problems of the Gulfian and later sections.

The Gulfian section is further characterized by numerous minor basin-margin unconformities, and, although the section is a thick one, nowhere in this transect is there clear evidence of deep-water sedimentation.

Finally, there is a clear Gulfian-related geomorphology: even minor lithologic contrasts between units are geomorphically significant, resulting in unique topographies, vegetation, and soils. And each geomorphic province has acted as a control on land use, and hence on human history.

REFERENCES

This list contains the references cited in this paper, plus others useful for background reading on regional relationships.

Adkins, W. S., 1933, The Mesozoic Systems in Texas, *in* Adkins, Sellards, and Plummer, The Geology of Texas: University of Texas Bureau of Economic Geology Bulletin 3232, p. 239–518.

Adkins, W. S., and F. E. Lozo, 1951, Stratigraphy of the Woodbine and Eagle Ford, Waco Area, Texas, *in* Lozo, F. E., ed., The Woodbine and Adjacent Strata of the Waco Area, Central Texas: Southern Methodist University Press, Fondren Science Series, no. 4, p. 105–176.

Beall, A. O., 1964, Stratigraphy of the Taylor Formation, East–Central Texas: Baylor Geological Studies, Bulletin no. 6, 35 p.

Bureau of Economic Geology, 1970, Geologic Atlas of Texas, Waco Sheet: Bureau of Economic Geology, University of Texas, 10 p., and map, scale 1:250,000.

——, 1972, Geologic Atlas of Texas, Dallas Sheet: Bureau of Economic Geology, University of Texas, 11 p., and map, scale 1:250,000.

Hill, R. T., 1901, Geography and Geology of the Grand and Black Prairies, Texas: U.S. Geological Survey 21st Annual Report, part VII, 666 p.

Oliver, W. B., 1971, Depositional Systems in the Woodbine Formation (Upper Cretaceous), Northeast Texas: University of Texas, Bureau of Economic Geology Report of Investigations, no. 73, 28 p.

Pessagno, O. L., Jr., 1969, Upper Cretaceous Stratigraphy of the Western Gulf Coast Area of Mexico, Texas, and Arkansas: Geological Society of America Memoir 111, 119 p.

Scholle, P. A., 1978, Carbonate Rock Constituents, Textures, Cements, and Porosities: American Association of Petroleum Geologists Memoir 27, 241 p.

Silver, B. A., 1963, The Bluebonnet Member of the Lake Waco Formation (Upper Cretaceous), Central Texas; A Lagoonal Deposit: Baylor Geological Studies Bulletin, no. 4, 47 p.

Stehli, G. G., Creath, W. B., Upshaw, C. F., Forgotson, J. M., 1972, Depositional history of Gulfian Cretaceous of East Texas Embayment: American Association of Petroleum Geologists Bulletin, v. 56, p. 38–67.

Stephenson, L. W., 1937, Stratigraphic Relations of the Austin, Taylor, and Equivalent Formations in Texas: U.S. Geological Survey Professional Paper no. 186-G, p. 135–146.

The Claiborne Group of East Texas

Patricia S. Sharp, *Department of Geology, Stephen F. Austin State University, Nacogdoches, Texas 75962*
Austin A. Sartin, *Department of Geology, Centenary College, Shreveport, Louisiana 71134*

LOCATION

An ancestral coastal plain was the principal setting for the deposition of the Claiborne Group. Units within the Claiborne dip to the southeast at one to two degrees. Generally the Claiborne formations crop out in an arcuate pattern across the east Texas area, with the older formations cropping out farther to the north and west while the younger formations crop out nearer to the present coast.

The ancestral coastal plains area of Texas consists of several major structural features: Sabine Uplift in extreme East Texas, Tyler Basin in east Texas, San Marcos Arch in central Texas, and Rio Grande Embayment in southern Texas (Fig. 1). Formation of these structural features influenced the deposition of the Claiborne units. High areas, such as the Sabine Uplift and the San Marcos Arch, caused rivers to reroute and diverge their courses. This resulted in minimal amounts of material being deposited in those areas of uplift. Subsidence created the basinal features such as the Tyler Basin and Rio Grande Embayment, and great thicknesses of material were deposited in these susiding lows. Normal faulting has vertically displaced the Claiborne units throughout the area.

Claiborne exposures discussed here are located within approximately 6 mi (10 km) of Jacksonville, Texas, in Cherokee County (Fig. 2). Outcrops that primarily expose the Queen City and Weches formations are easily accessible along U.S. 69 north and south of Jacksonville and along U.S. 79 west of Jacksonville. In several localities, upper and lower contacts of the Queen City and Weches formations are visible. The formations are laterally continuous and relatively easy to trace.

SIGNIFICANCE

The Claiborne Group is a thick sequence of sedimentary strata deposited during the Eocene Epoch (Fig. 3). Fluviatile, deltaic, and marine accumulations were responsible for the formation of the deposits. Sediments occurred in distinctive marine and nonmarine cycles, and the Claiborne Group represents at least three cycles of deposition, with each cycle having a transgressive and regressive phase. Cycle 1: transgressive Reklaw Formation, regressive Queen City Formation. Cycle 2: transgressive Weches Formation, regressive Sparta Formation. Cycle 3: transgressive Cook Mountain Formation, regressive Yegua Formation.

In many respects the transgressive units within each cycle are similar, as are the regressive units. This guide will use the Weches Formation as a representative transgressive unit, and the Queen City Formation as a representative regressive unit. Even though every formation in the Claiborne has its own special identity, the attention to these two units will adequately reveal the cyclic sedimentation patterns of the Eocene.

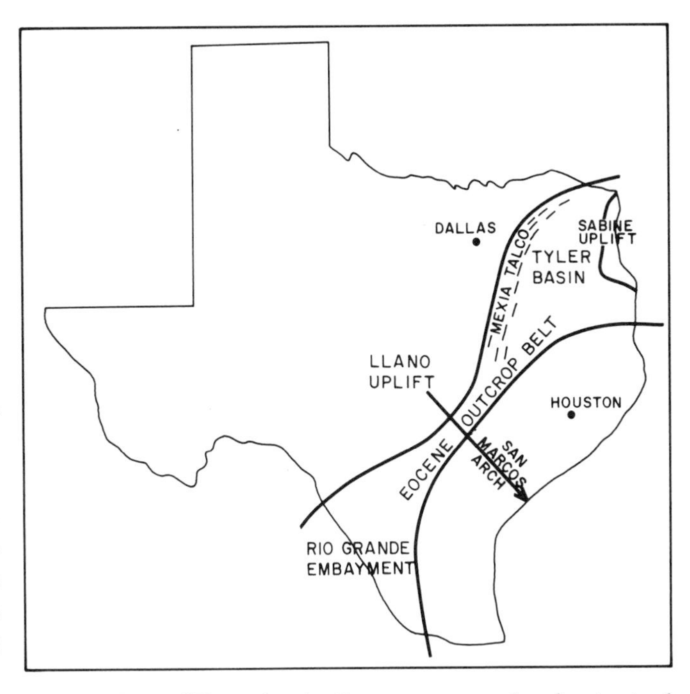

Figure 1. Map of Texas showing Eocene outcrop and major structural features (from Brown and others, 1969).

The Claiborne Group

Locality 1 (95°17'; 32°2.5', Bullard Quadrangle). Locality 1 is located on U.S. 69, 2 mi (3.3 km) north of the intersection of U.S. 69 and 79 in Jacksonville. The roadcut is exposed on the east side of the highway. At this locality are 36 ft (11 m) of well-exposed sandstone layers alternating with silt and clay beds (Fig. 4).

The lowermost unit (unit 1) is composed of very fine-grained sandstone and siltstone, and it is believed to be a shoreface facies. It is thought that this material was deposited as a result of offshore marine sediment being reworked by longshore currents. This produced trough crossbeds (unit 2), which paralleled the paleocoastline of the Gulf of Mexico during the Eocene (Hobday, 1980). Burrows have been observed in the lower portion of this facies unit and have been described as *Ophiomorpha* burrows by Hobday (1980). These burrows, it is generally agreed, are used to indicate littoral to shallow-water environments.

Above this unit are two sets of hummocky crossbeds (units 3 and 5) separated by a one-meter interval of silty clay (unit 4). This term is applied to undulating sets of low-angle crossbeds having low degrees of dip. The lowermost set probably formed below fair-weather base as a result of storm processes, and the

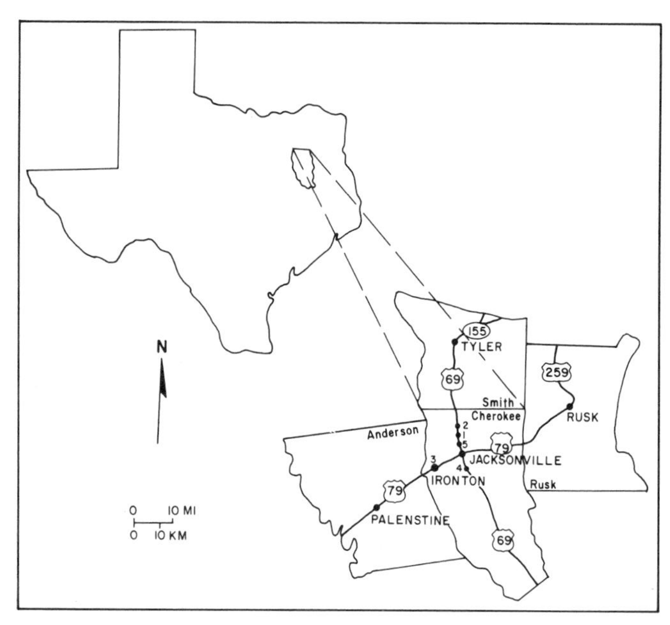

Figure 2. Index map showing location of study area and outcrop localities (modified from Hobday, 1980).

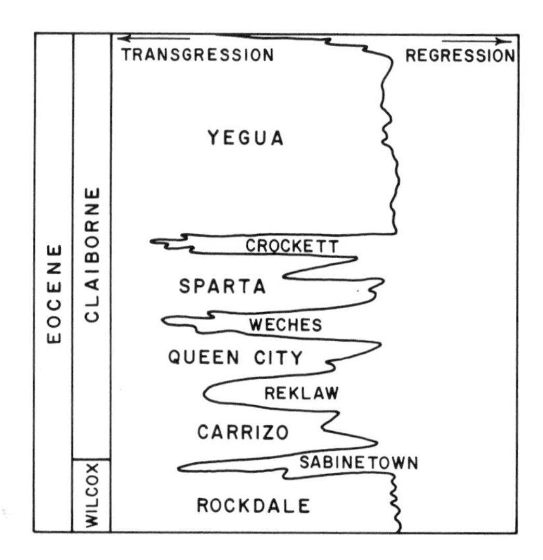

Figure 3. Claiborne Formation in east Texas showing cyclic sedimentation (modifieid from diagram by H. B. Stenzel in Sellards and others, 1932b).

sediment would be little disturbed by currents after deposition. Many of the grains in the hummocky crossbeds are sand-size clay rip-up clasts. The second set of hummocky crossbeds (unit 5) is believed to have formed in a similar manner. However, perhaps due to rapid deposition and a coastline with a low degree of slope, a barrier island formed. It was probably below mean high tide but served as a restriction to incoming marine water. As a result, a tidal flat formed behind the barrier and thus produced the overlying facies (unit 6). The unit is approximately 30 ft (9 m) thick and shows a decrease in grain size from fine sand to silt and clay-size material. Flaser, lenticular, and wavy bedding are common. The environment was one of low energy and allowed finer-grain sizes to be deposited. Alternating influxes of sand-rich sediment occurred due to higher-energy invasions from the sea. These influxes were probably due to channels that were cut as storm waves pounded the barrier.

Locality 2 (95°17'; 32°3', Bullard Quadrangle). Locality 2 is 1.1 mi (1.8 km) north of locality 1 and is a continuation of the vertical section at locality 1. Approximately 6.5 ft (2.0 m) of Queen City is exposed at the south end of this section. At locality 2 the sandstone is massive; it is thought to have been deposited as a result of the reworking of the barrier island system. Overlying the light-colored, massive Queen City sandstone is approximately 3 m of interbedded dark lignitic shales and thin sandstone stringers. The upper part of this interbedded unit is burrowed and represents the induration of the coastal marsh by a marine transgression. Overlying this transition zone at locality 2 are approximately 36 ft (11 m) of "glauconitic" clay (Weches Formation), which was deposited under marine conditions. Locality 2 contains an entire Weches succession and is capped by less than 2 m of Sparta sand. Large subhorizontal burrows can be observed in the lower part of the Weches. The botryoidal appearance of the uppermost ironstone layer is characteristic of the contact between the Weches and Sparta formations.

Locality 3 (95°22'30'; 31°55', Jacksonville Quadrangle). Locality 3 is located at Ironton on U.S. 79, 6.6 mi (11 km) southwest of Jacksonville. The outcrop consists of crossbedded Queen City deposits that were deposited as a result of inferred flood-tidal delta conditions (Hobday, 1980). The foresets show a wide range of inclination from subhorizontal to almost 40°, and there are numerous clay drapes and rip-up clasts. Several small channels have truncated the surrounding beds and can be observed near the top of the outcrop. The northeastern end of the section is overlain by the marine-dominated Weches Formation. The exposure is unique in that it shows, rather completely, the marine transgression (Weches Formation) across a barrier-dominated coastline (Queen City Formation).

Locality 4 (95°15'; 31°56', Rusk Quadrangle). Locality 4, located 2.6 mi (4.3 km) south of Jacksonville on U.S. 69, is dominated by several unique sedimentary structures in the Queen City Formation. The most diagnostic are diapir-like structures that are approximately 1.6 ft (0.5 m) in height and 2.5 ft (0.75 m) in diameter. Mean grain size of the structure is clay to silt-size, but they protrude upward into coarser sand-size material. Clay and silt were probably rapidly deposited with much water between the grains. A sudden shock could have resulted in rearrangement and liquefaction of particles. This would enable the sediments to react to the forces of gravity and to those of forced injection. As a result, random irregular shapes of the diapiric structures were produced. At locality 4, normal faults cut across the outcrop,

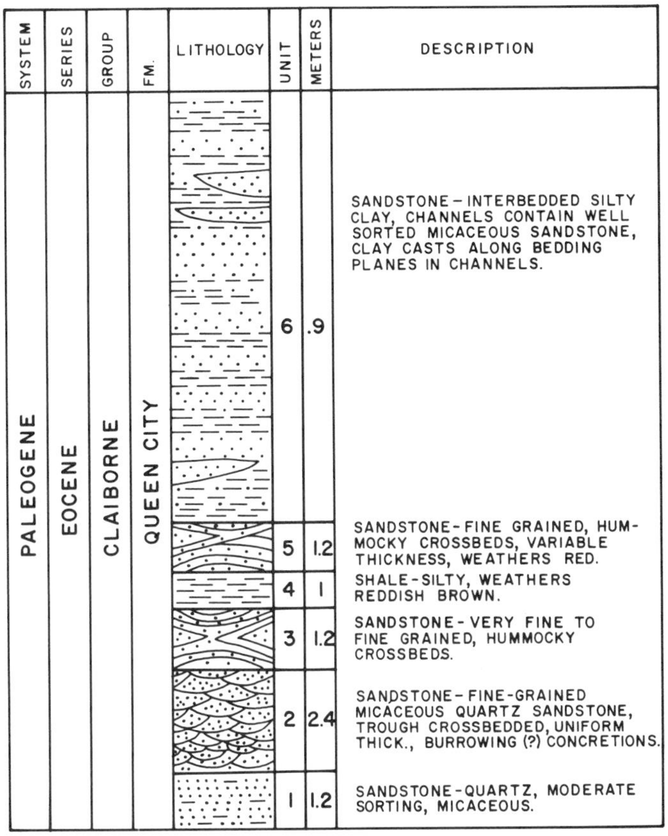

System	Series	Group	Fm.	Unit	Meters	Description
Paleogene	Eocene	Claiborne	Queen City	6	.9	SANDSTONE - INTERBEDDED SILTY CLAY, CHANNELS CONTAIN WELL SORTED MICACEOUS SANDSTONE, CLAY CASTS ALONG BEDDING PLANES IN CHANNELS.
				5	1.2	SANDSTONE - FINE GRAINED, HUMMOCKY CROSSBEDS, VARIABLE THICKNESS, WEATHERS RED.
				4	1	SHALE - SILTY, WEATHERS REDDISH BROWN.
				3	1.2	SANDSTONE - VERY FINE TO FINE GRAINED, HUMMOCKY CROSSBEDS.
				2	2.4	SANDSTONE - FINE-GRAINED MICACEOUS QUARTZ SANDSTONE, TROUGH CROSSBEDDED, UNIFORM THICK., BURROWING (?) CONCRETIONS.
				1	1.2	SANDSTONE - QUARTZ, MODERATE SORTING, MICACEOUS.

Figure 4. Stratigraphic section of the Queen City Formation at locality 1.

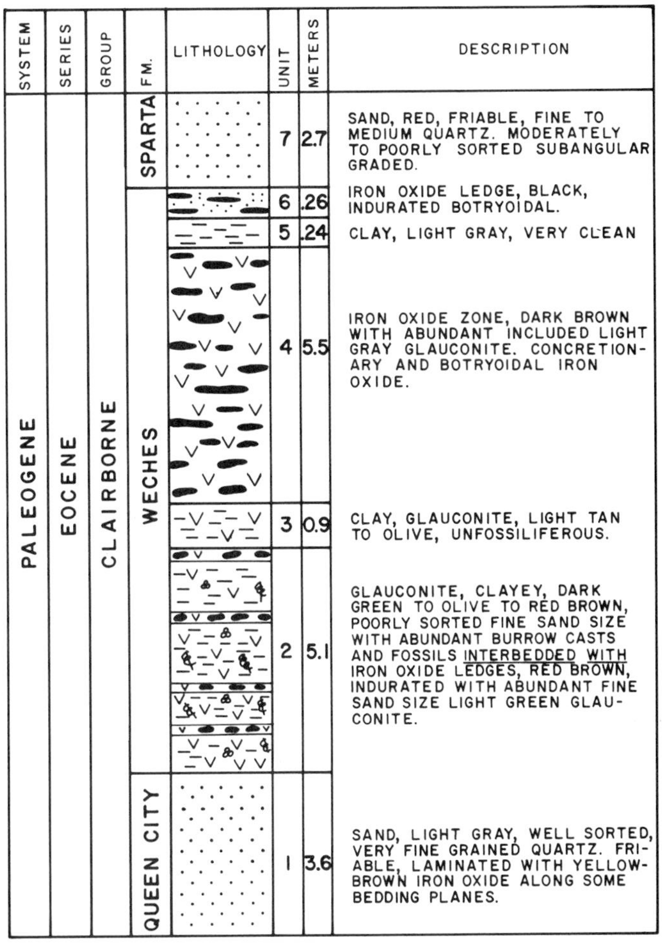

System	Series	Group	Fm.	Unit	Meters	Description
Paleogene	Eocene	Claiborne	Sparta	7	2.7	SAND, RED, FRIABLE, FINE TO MEDIUM QUARTZ. MODERATELY TO POORLY SORTED SUBANGULAR GRADED.
			Weches	6	.26	IRON OXIDE LEDGE, BLACK, INDURATED BOTRYOIDAL.
				5	.24	CLAY, LIGHT GRAY, VERY CLEAN
				4	5.5	IRON OXIDE ZONE, DARK BROWN WITH ABUNDANT INCLUDED LIGHT GRAY GLAUCONITE. CONCRETIONARY AND BOTRYOIDAL IRON OXIDE.
				3	0.9	CLAY, GLAUCONITE, LIGHT TAN TO OLIVE, UNFOSSILIFEROUS.
				2	5.1	GLAUCONITE, CLAYEY, DARK GREEN TO OLIVE TO RED BROWN, POORLY SORTED FINE SAND SIZE WITH ABUNDANT BURROW CASTS AND FOSSILS INTERBEDDED WITH IRON OXIDE LEDGES, RED BROWN, INDURATED WITH ABUNDANT FINE SAND SIZE LIGHT GREEN GLAUCONITE.
			Queen City	1	3.6	SAND, LIGHT GRAY, WELL SORTED, VERY FINE GRAINED QUARTZ. FRIABLE, LAMINATED WITH YELLOW-BROWN IRON OXIDE ALONG SOME BEDDING PLANES.

Figure 5. Stratigraphic section of the Weches Formation at locality 5 (modified from Brown and others, 1969).

causing the Queen City to be in lateral contact with the Weches Formation. They are part of a much larger system of faults (Mt. Enterprise Fault System), which extends to the southwest and to the east. This network of faults is clearly mapped on the Palestine Sheet (Flawn, 1965) showing the arcuate pattern through this area.

Locality 5 (95°17′; 32°2′, Bullard Quadrangle). Locality 5 is 1.7 mi (2.9 km) north of the intersecton of U.S. 69 and 79 in Jacksonville. The roadcut is exposed on the east side of the highway and is easily located because it is the underlying bedrock for a roadside park known as "Love's Lookout." The outcrop (Fig. 5) includes the upper part of the Queen City (unit 1), the entire Weches (units 2, 3, 4, 5, and 6), and the lower portion of the Sparta (unit 7). The lower half of the Weches Formation (unit 2) consists primarily of green "glauconitic" shale interbedded with abundant ironstone ledges. Burrow casts and fossils are common. The ironstone ledges are probably due to leaching of soluble ingredients in the "glauconite." The iron is in the form of hematite, siderite, or "limonite" and is redeposited, forming the ledges (Sellard and others, 1932a). Resistant ironstone caps hills and produces flat-topped, steep hills, which are typical in east Texas. Fossils in the lower half of the exposure are unaltered animal remains primarily from the Phylum Mollusca. Some of the common pelecypods found in the Weches are *Lutetia, Venericardia, Nucula,* and *Vokesula;* some of the common gastropods are *Dirocerithium* and *Turritella.* Overlying unit 2 is a 3-ft-layer (0.9 m) (unit 3) of nonfossiliferous, glauconitic clay. Unit 4 is a thick ferruginous zone, which contains layers of botryoidal and concretionary structures. The "glauconite" in this zone has undergone extensive weathering and transformation to the reddish colored iron-rich minerals (hematite, "limonite" or siderite). The upper unit is therefore easy to distinguish from the lower, "glauconite"-rich, green-colored unit. The Weches, at this locality, is capped by a 0.8-ft-thick (0.25 m) (unit 5) clay zone that is overlain by a botryoidal, ferruginous ledge (unit 6). Lone Star Steel (Lone Star, Texas) has actively mined these iron-rich deposits of the Weches Formation (Brown and others, 1969). The uppermost layer (unit 7) consists of fine- to medium-grained quartz sandstone and is interpreted to be the fluvial-dominated Sparta Formation.

The environment of deposition for the Weches Formation was probably shallow marine, which deepened at this locality through the first transgressive cycle (Sellard and others, 1932a). Generally, the Weches was deposited in an offshore environment with good circulation. This allowed for clear waters and little coarse sediment accumulation. The formation of the "glauconite" in this unit is questioned by many and is continually debated and discussed by geochemists.

This outcrop, locality 5, can be correlated effectively with locality 2, located 1.9 mi (3 km) north on U.S. 69. Locality 2 has a similar lithology, fossil content, and structure to locality 5, and the interpreted environments of deposition are compatible.

The northeastern part of the Ironton outcrop, locality 3, is environmentally similar to locality 5. The transition zone between the alternating sand and carbonaceous clays (Queen City) is a very dark lignitic sand and clay with flaser bedding. These basal carbonaceous deposits are interpreted to be back–barrier marsh deposits, while the overlying glauconite burrowed clays and sands (Weches) are interpreted to be nearshore (estuarine) marine deposits.

REFERENCES CITED

Brown, T., Newland, L., Campbell, D., and Ehlmann, A., 1969, Field excursion of East Texas; Clay, glauconite, ironstone deposits, 18th Clay Conference Guidebook: Austin, University of Texas Bureau of Economic Geology, no. 9.

Flawn, P. T., 1965, Geologic atlas of Texas; Palestine sheet: Austin, University of Texas Bureau of Economic Geology.

Hobday, D., 1980, Middle Eocene coastal plain and nearshore deposits of East Texas; A field guide to the Queen City Formation and related papers: Gulf Coast Section, Society of Economic Paleontologists and Mineralogists, 95 p.

Sellards, E. H., Adkins, W. S., and Plummer, F. B., 1932a, Geology of Texas: University of Texas Bulletin, v. 1, p. 635–651.

——, 1932b, Geology of Texas: University of Texas Bulletin 3232, v. 1, p. 528.

Cambrian algal reefs of the upper Wilberns Formation, central Texas, The Camp San Saba locality

Wayne M. Ahr, Geology Department, Texas A & M University, College Station, Texas 77843

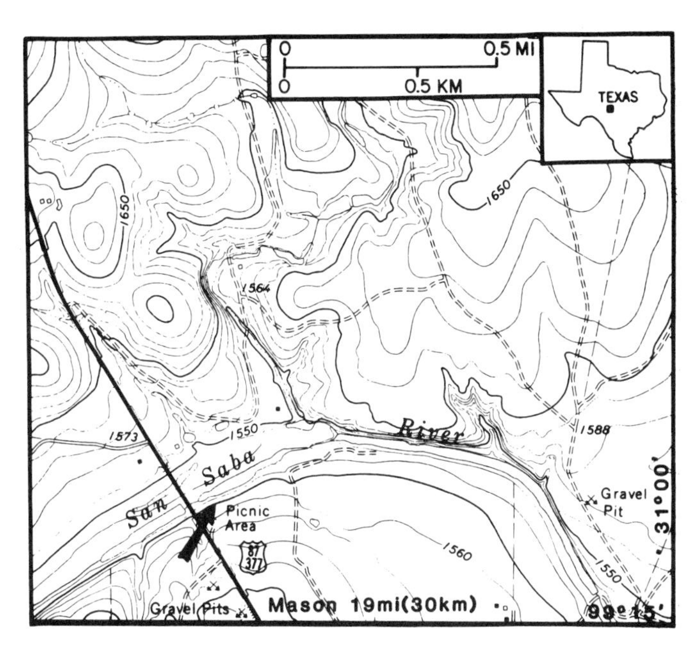

Figure 1. The location map for the Camp San Saba algal reef outcrop. (From the Brady South 7½-minute Quadrangle, 1970.) The parking area referred to in the text is shown on this figure as a "picnic area."

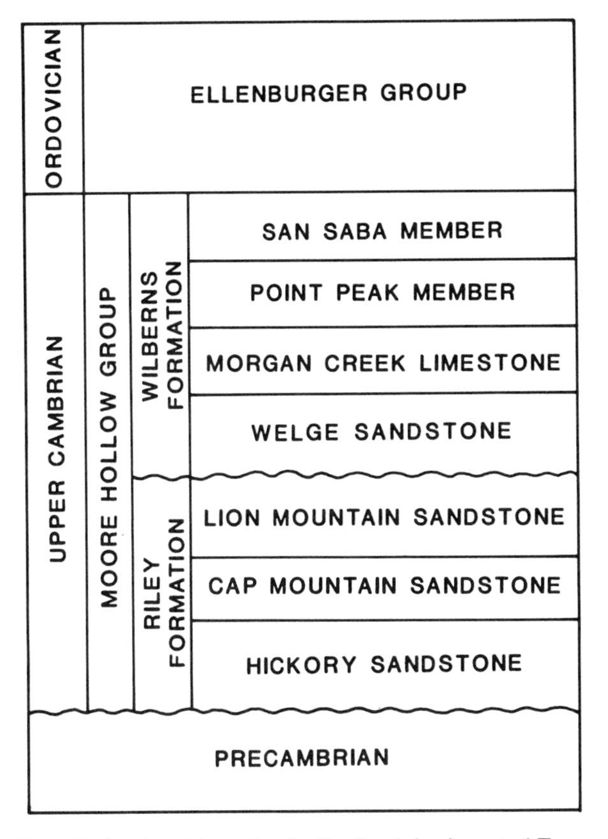

Figure 2. Stratigraphic section for the Cambrian in central Texas.

LOCATION

The famous algal reefs of the upper Wilberns Formation in central Texas are well illustrated by the section at the Camp San Saba crossing on the San Saba River, McCulloch County, Texas (Figs. 1, 2). These reefs are entirely within the San Saba Member of the Upper Cambrian Wilberns Formation (Barnes and Bell, 1977).

SIGNIFICANCE

Algal bioherms and clusters of bioherms, loosely termed reefs, are present in the Morgan Creek, Point Peak, and San Saba members of the Wilberns Formation. One of the more photogenic bioherms (not at the Camp San Saba locality) of the Point Peak Member has been illustrated in several publications, including the stratigraphy text by Dunbar and Rodgers (1957, p. 180). Virtually all of the Wilberns bioherms and reefs are limestone, and have not been severely altered by dolomitization or other kinds of fabric-destructive diagenesis. The preservation of depositional fabrics and even fossil algae is unusually good for such ancient rocks. Four different kinds of fossil algae (*Girvanella, Epiphyton, Renalcis,* and *Nuia*) and several growth forms of the alga *Girvanella* have been described from these algal limestones (Ahr, 1971). The microscopic algae and algal growth forms represent specific kinds of environmental settings; consequently, one can infer the growth history of an algal reef by identifying the sequence of algal growth forms from bottom to top of a bioherm or reef.

Most of the microscopic features have parallel macroscopic structures that are visible on the outcrop and may be identified as both non-laminated, laminated, or headlike (stromatolitic). In general, the nonlaminated and thickly laminated, non-headlike structures represent growth under quiescent conditions, whereas the stromatolitic features represent environments with more vigorous current or wave activity.

Most of the Wilberns bioherms and reefs are exposed on private land or in outcrops that are not easily accessible from major highways. The photogenic bioherms in the Point Peak Member of the Wilberns, for example, are on private land and

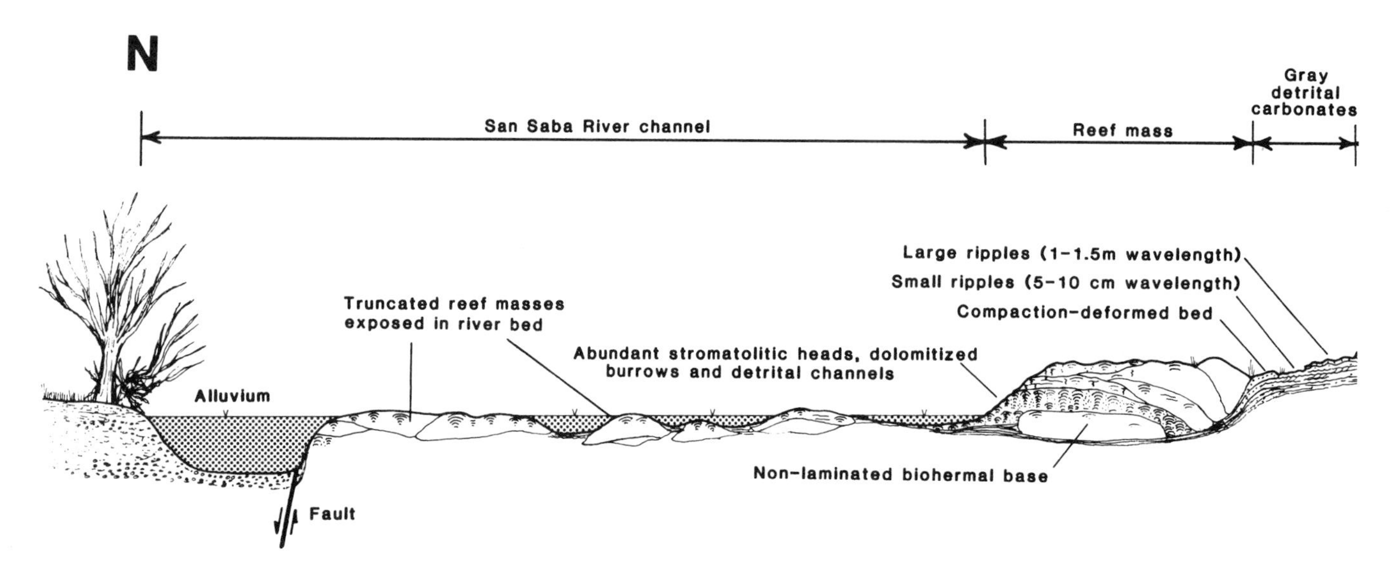

Figure 3 (this and facing page). A diagrammatic cross section from the roadside rest area at the U.S. 87 bridge looking north across the San Saba River. The rest area is near the upper right corner of the section. Two reef masses and an exposure of rippled, lime sands are shown, along with several additional algal buildups partially submerged in the river. A small fault, associated with a deep pool in the river, is shown at the lower left.

are exposed in a cliff face about 40 ft (12 m) above a fast-flowing river. The Camp San Saba exposure, on the other hand, has excellent bedding plane and cross-sectional exposures of the reefs and some associated detrital carbonate strata along several hundred meters of open river bank. All of the main features of the Texas Cambrian algal reefs may be seen there.

THE CAMP SAN SABA SECTION

The Camp San Saba section is exposed in the San Saba River bed and along the south bank of the river at the U.S. 87 bridge (99°16′10″W; 31°00′13″N, Brady South 7½-minute Quadrangle; Fig. 1) 18.7 mi (30 km) north of the courthouse in the city center of Mason, Texas. A roadside rest area with paved parking spaces and picnic tables is on the east side of U.S. 87 at the San Saba River bridge. The algal reefs are exposed along the river just below the rest area.

The San Saba exposure below the rest area extends for about 280 ft (85 m) from the rest area northward into the San Saba River (Fig. 3). The underlying Point Peak Member of the Wilberns Formation is not visible at this location; however, Barnes and Bell (1977) place the algal reefs in the lowermost San Saba very near the base of the member. The outcrop is continuous along the river to the west, and some excellent exposures are present under the U.S. 87 bridge.

Looking northward from the rest area, one can see several algal bioherms, detrital carbonate strata and a small fault along the 280-ft (85-m) exposure (Fig. 4). The bioherms are ellipsoidal in shape, about 40 by 25 ft (12 by 7.5 m), with long axes oriented about N35°E. In the river bed, the bioherms may appear as elliptical outlines, if they have been deeply weathered, or as small domes protruding out of the water. The detrital rocks are mainly lime sands consisting of broken trilobite and brachiopod skeletons.

The noticeable orientation of the algal bioherms is interpreted to be the result of currents in the Cambrian sea that streamlined the shape of the growing algal masses. Notice that there are several sets of asymmetrical, current ripples in the lime sands (Fig. 3) just beneath and between the bioherms nearest the parking area. These ripples have curved crests aligned approximately perpendicular to the long axes of the overlying algal "ellipsoids," suggesting that there may have been a relatively prolonged, northeast-to-southwest current regime at this locality.

ALGAL STRUCTURES IN THE REEFS

There are four basic "structures," or patterns of growth, visible in the algal buildups: (1) non-laminated, or "massive" structure; (2) laminated, sheet-like layers of variable thicknesses; (3) sheetlike masses with scattered domical features or poorly developed stromatolites; and (4) stromatolites. The four algal structures represent, from 1 through 4, respectively, development in progressively more current or wave-swept water. Stromatolites in these outcrops represent vigorous water movement and presumably, the shallowest water environment.

Red or yellow ochre mottling indicates weathered dolomite;

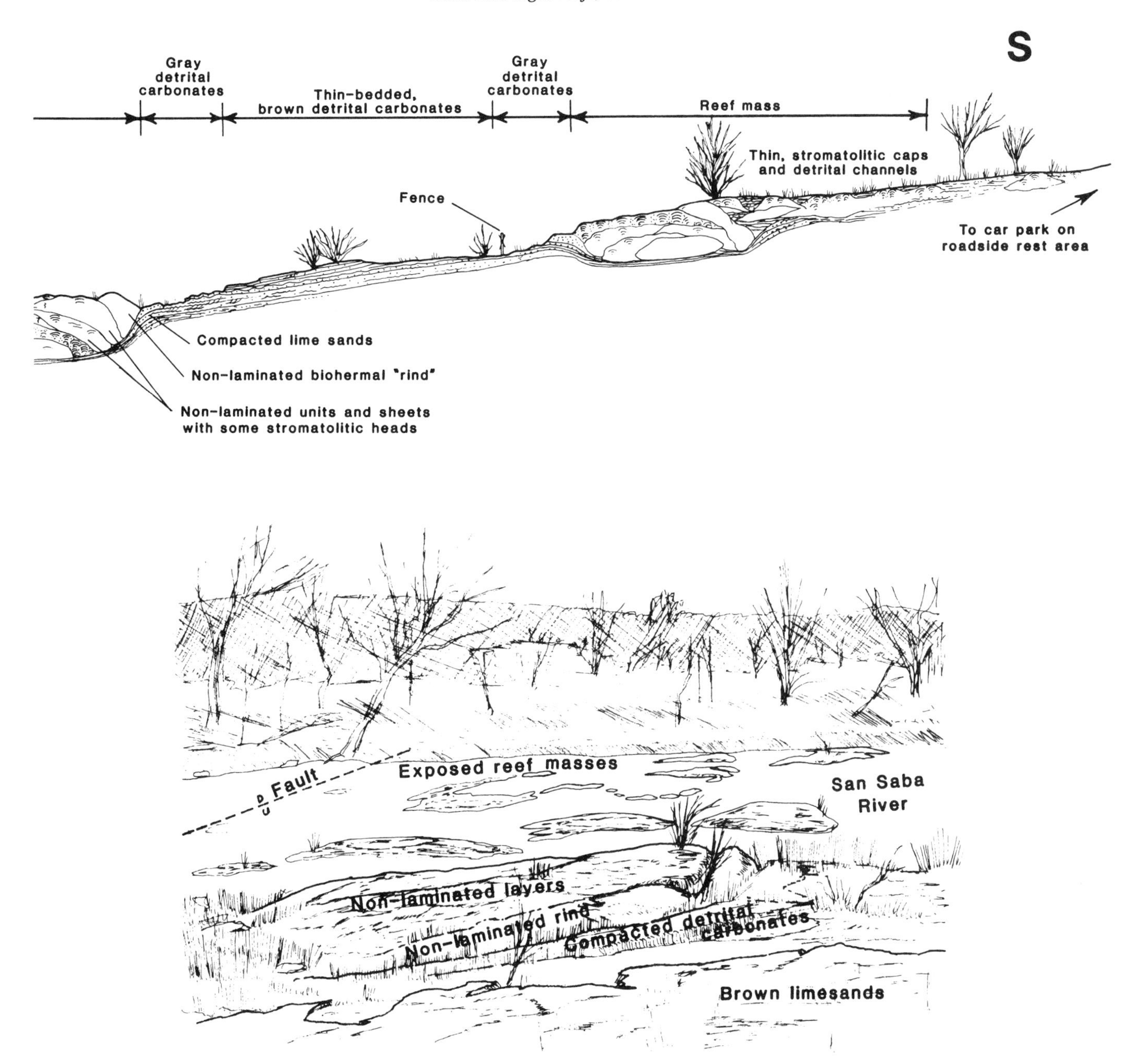

Figure 4. An ink sketch of the outcrop surface just north of the car park and picnic area showing the appearance of the algal reef masses, the rippled lime sands, and the fault across the San Saba River. The view is looking towards the north from the car park.

close inspection will show that the mottles are in raised relief because the dolomite is more resistant to dissolution than the limestone, and the rust color is from iron removed during partial dissolution of the dolomite. Many of the mottles represent trails and burrows of some unknown organisms. The burrow fillings were dolomitized because they were originally more permeable than the muddy algal masses on which the burrowers were prowling.

With this in mind, consider the diagrammatic sketch of the biohermal complex at the water's edge just below the rest area (Fig. 5). By climbing down near the water's edge and looking back at the car park, one can see a stromatolitic unit in the buildup overlain by about four units of sheet-like masses that are partly non-laminated and partly contain poorly developed stromatolites. The top unit of the buildup is a gray, non-laminated, "massive" feature termed a rind. Rinds in this buildup and others

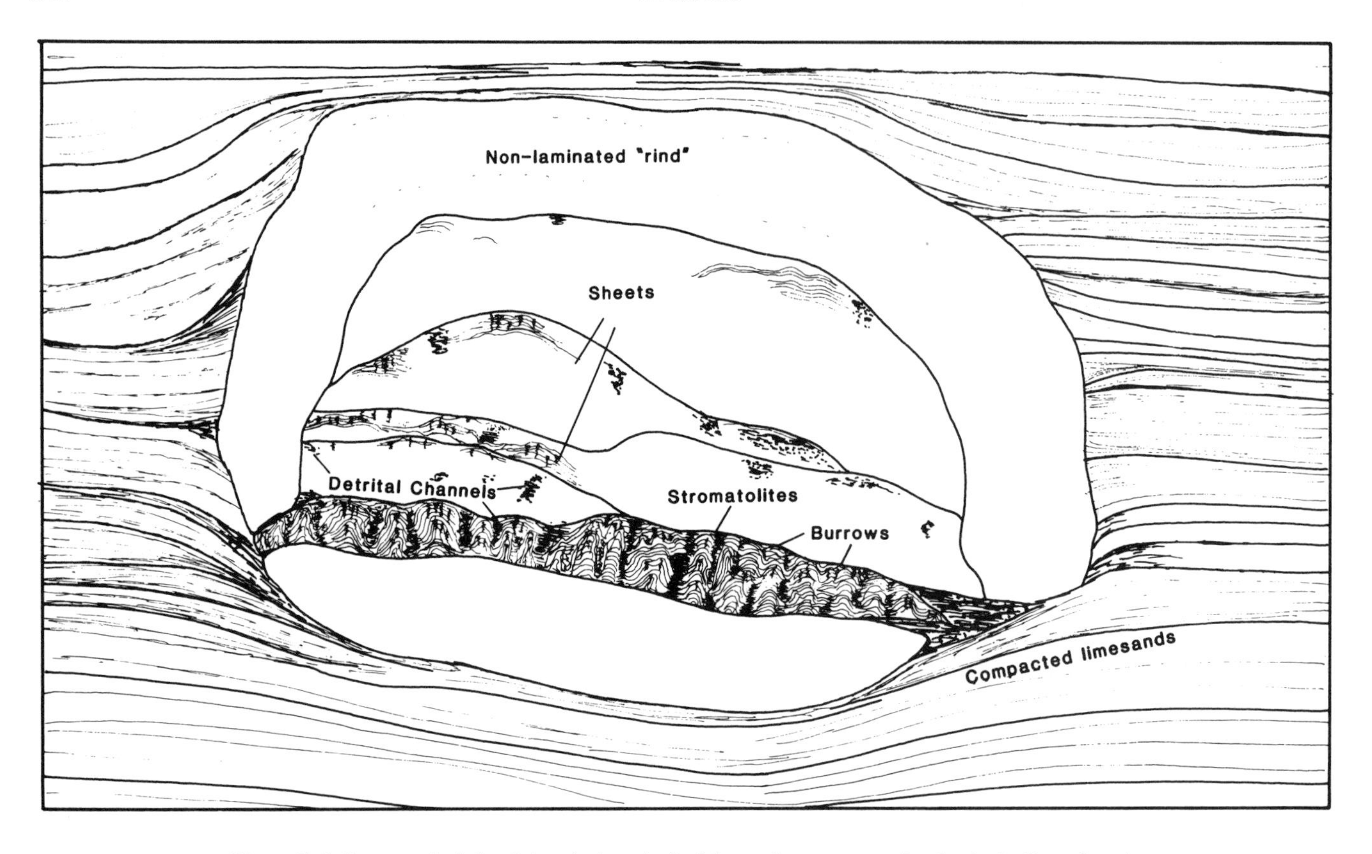

Figure 5. A diagramatical sketch (vertical section) of the reef mass at water's edge in the line of section shown in Figure 3. The buildup consists of a non-laminated basal unit, a stromatolitic unit, four units with sheet-like laminations, non-laminated portions and some poorly developed stromatolites, and a non-laminated capping unit termed the "rind." The entire mass is about 12.5 ft (3.75 m) thick.

in the area commonly began to develop on a bedding surface about halfway between the base and the top of the individual bioherms and then grew to cover the biohermal mass. This sequence of development suggests growth in a deepening upward environment or one in which water agitation became less pronounced as time went on, although many of the Wilberns reef complexes represent shoaling-upward sequences. Notice the "channels" of detritus (commonly dolomitized) between stromatolitic heads and the dolomitized carbonate between the individual laminae in each stromatolite. As the stromatolites compose only about one-fifth of the total buildup, it seems a misnomer to call these "stromatolitic bioherms." "Algal bioherms" is preferable.

Finally, as one stands on the bioherm at the river's edge and looks back at the car park and to the left (east), an expanse of rippled, lime sands is visible over about 75 ft (22.5 m) between bioherms. Close examination shows that the lime sands are older than the reef mass and are bent downward by compaction beneath each bioherm in the reef complex.

REFERENCES CITED

Ahr, W. M., 1971, Paleoenvironment, fossil algae, and algal structures in the Upper Cambrian of central Texas: Journal of Sedimentary Petrology, v. 41, p. 205–216.

Barnes, V. E., and Bell, W. C., 1977, The Moore Hollow Group of central Texas: University of Texas Bureau of Economic Geology Report of Investigations 88, 169 p.

Dunbar, C. O., and Rodgers, J., 1957, Principles of stratigraphy: New York, John Wiley and Sons, 356 p.

ACKNOWLEDGMENTS

The artwork in this chapter was done by Thomas Byrd, who, along with Stephen Lovell, helped check field localities.

Middle and Upper Pennsylvanian (Atokan-Missourian) strata in the Colorado River Valley of Central Texas

Robert C. Grayson, Jr., Department of Geology, Baylor University, Waco, Texas 76798
E. L. "Jack" Trice, Marshall Exploration, P.O. Box 1689, Marshall, Texas 75670

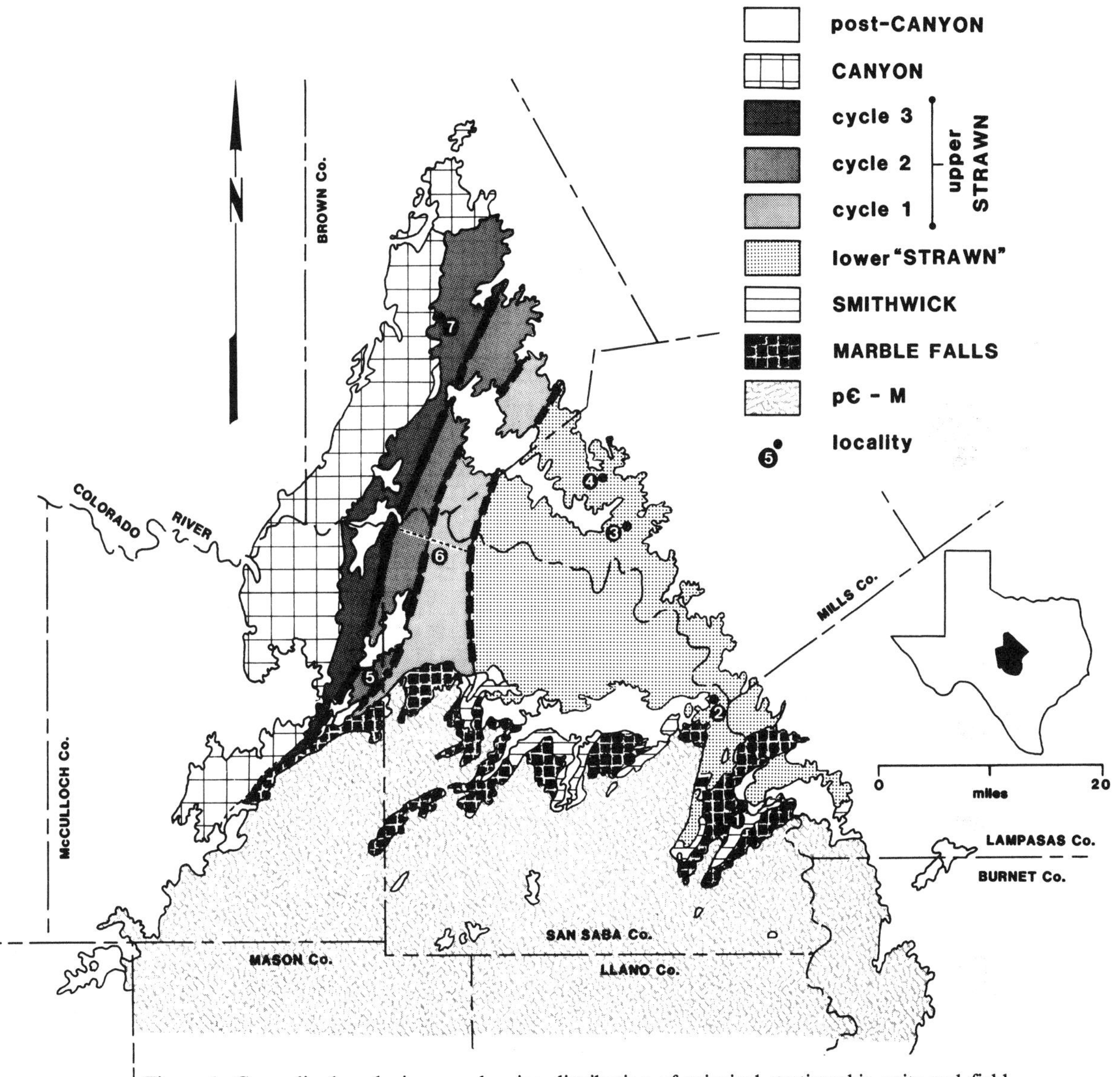

Figure 1. Generalized geologic map showing distribution of principal stratigraphic units and field localities. Geology adapted from Kier and others (1976, Brownwood Sheet).

LOCATION

The Middle and Upper Pennsylvanian stratigraphic sequence of Central Texas is exposed in a triangular shaped region that is bisected by the Colorado River (Fig. 1). This succession of strata (Fig. 2) is illustrated in this double-site guide. Site 78 begins in San Saba County near Bend, Texas, where exposures of the Marble Falls Limestone, Smithwick Shale, and lower "Strawn" Group are available for study. Three additional localities in San Saba and Mills counties permit special emphasis on the depositional character of the lower "Strawn" Group. The upper Strawn Group, Site 79, is seen along a transect across the outcrop belt in San Saba and McCulloch Counties. Single localities at opposite ends of the outcrop belt in San Saba and Brown Counties (Fig. 1) further document the various lithologies and stratigraphic relationships of the upper unit.

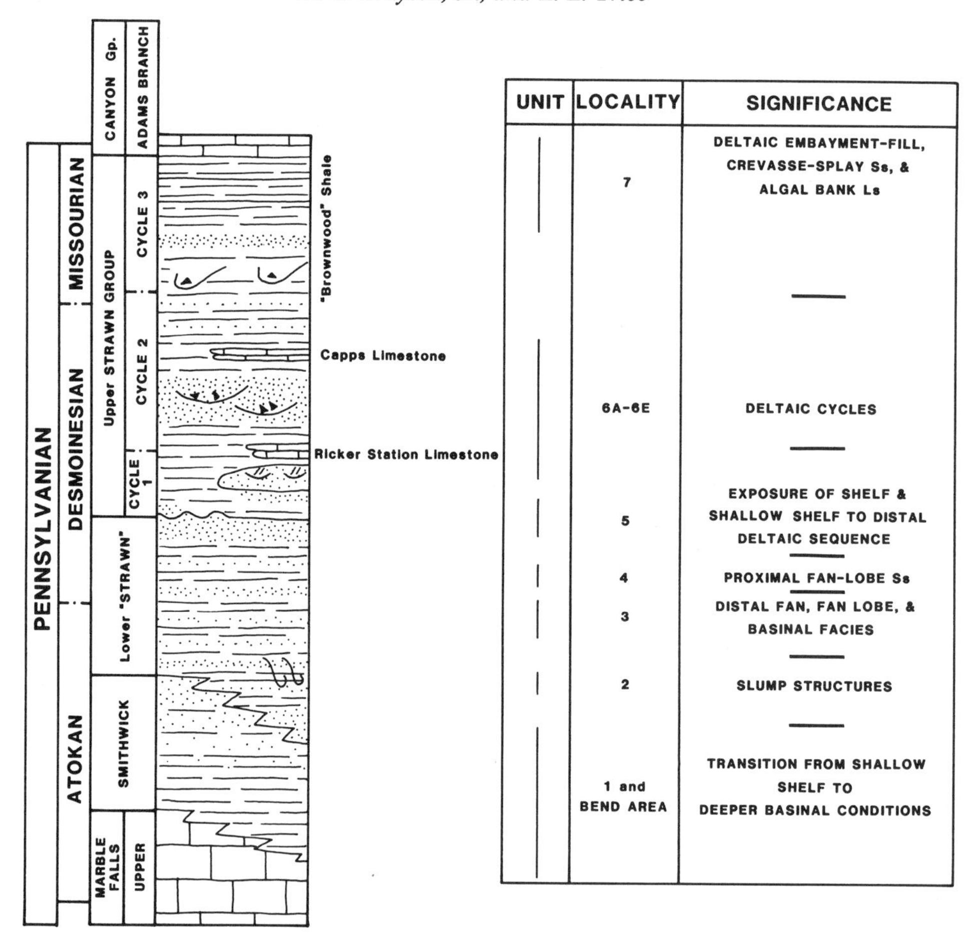

Figure 2. Stratigraphic interval and facies interpretation of units seen at field localities.

SIGNIFICANCE

Middle and Upper Pennsylvanian strata exposed in the Colorado River Valley and along the margin of the Llano Uplift were deposited in the Fort Worth Basin and on the Concho Platform (Fig. 3). The Llano Uplift, a dome cored by Precambrian igneous and metamorphic rocks (Cloud and Barnes, 1948), was not a significant tectonic element, although it served as a buttress that limited subsidence (Weaver, 1956).

The stratigraphic succession shown in this double-site guide records marked subsidence of the Fort Worth Basin and filling of the basin, Site 78, and westward progradation of fluvial-deltaic complexes across the almost filled basin, Site 79. The Marble Falls Limestone represents platform carbonate facies (Kier and others, 1979) that accumulated as a lateral equivalent to deeper water, platform margin and basinal Smithwick Shale (Namy, 1969; Grayson and others, 1985). As the eastern edge of the Concho Platform subsided during westward migration of the basin, Smithwick Shale was deposited on former sites of Marble Falls carbonate sedimentation. In the deeper part of the basin, clastic influx from the north established lower "Strawn" Group basin-fill shale and submarine fan (ramp) complexes. As the basin filled, upper Strawn Group fluvial-deltaic sediments and associated shelf deposits prograded across the Concho Platform and onlapped positive features such as the northwestern Llano area (Fig. 3). In Missourian time, the "filled" Fort Worth Basin and Concho Platform became, in effect, the Eastern Shelf to the now rapidly subsiding Midland Basin. Diminished paleogradients and reduced volumes of clastics resulted in widespread deposition of shelf carbonates and algal bank accumulations of the Canyon Group (Kier and others, 1979).

STOP DESCRIPTIONS

78–Basin and Basin Fill Successions

Locality 1—Bend, Texas. Several stratigraphic units are accessible for study in the vicinity of Bend, Texas (Fig. 4). Here the Middle Pennsylvanian strata record the change from carbonate platform to terrigenous clastic basinal depositional systems.

An excellent exposure of the upper Marble Falls Limestone occurs along Cherokee Creek to the east of Bend (Bell, 1957). Approximately 1.4 mi (2.4 km) southwestward from Bend on FM Road 580, an unimproved gravel road to the northwest (right) leads down to exposures of the lower Smithwick Shale located on the southwest bank of the Colorado River (Merrill,

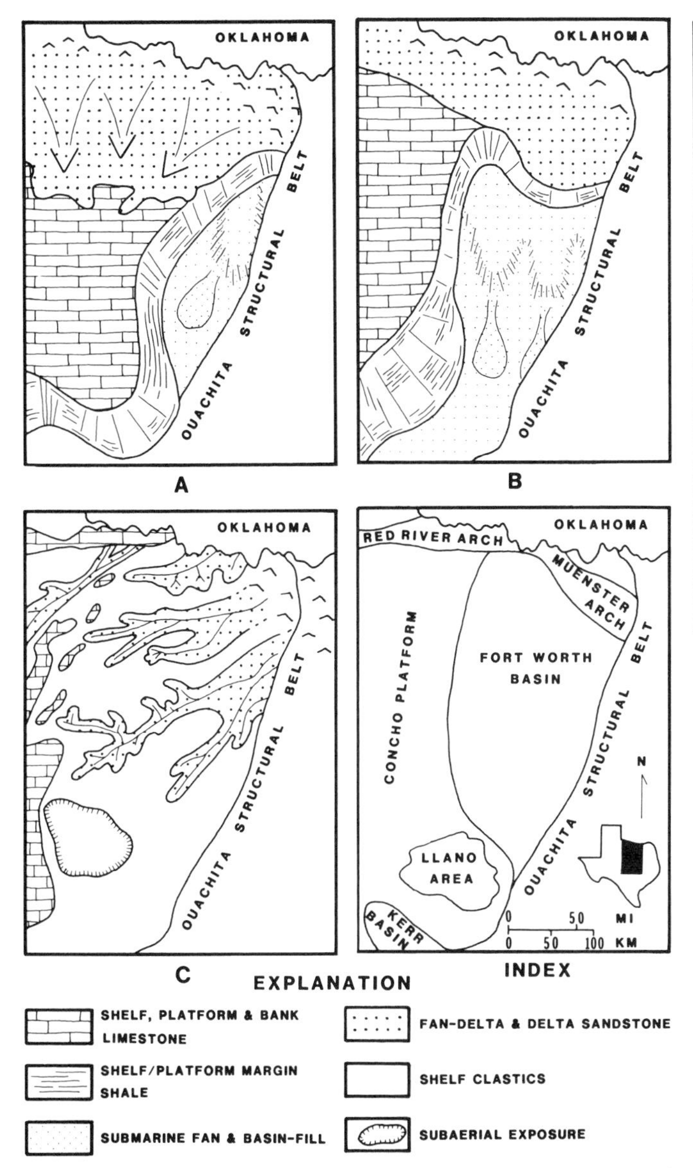

Figure 3. Generalized paleogeographic maps illustrating the evolution of Middle Pennsylvanian depositional systems: A) Early to Late Atokan time, B) Late Atokan to Early Desmoinesian time, C) Middle to Late Desmoinesian time. Sources: Turner (1957), Cleaves (1975), Crosby and Mapel (1975), and Thompson (1982).

1980, Stop 8). The upper Smithwick Shale can be examined along FM Road 501, which intersects FM Road 580 about 0.5 mi (0.8 km) to the southwest of the first stop. The uppermost Smithwick Shale and basal lower "Strawn" Group can be seen along the Jo-el Fishing Camp road about 2.9 mi (4.8 km) to the northwest from FM Road 501 off FM Road 580. Turn northeast (right) on to the gravel road that ascends the hill exposing upper Smithwick Shale. At the top of the hill, turn left and follow the principal road past a house where the road begins to descend toward the Colorado River; lower "Strawn" sandstones with partial Bouma sequences and good sole marks are exposed.

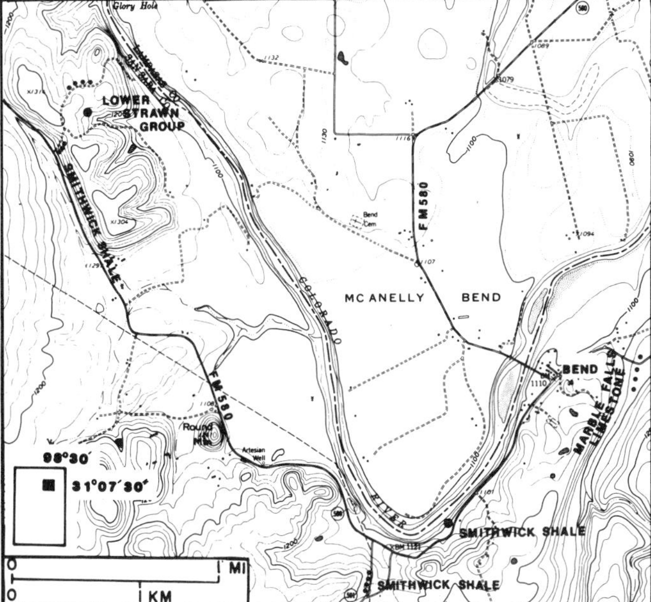

Figure 4. Location map showing some significant Middle Pennsylvanian exposures in the Bend, Texas region, San Saba County, within the Bend Quadrangle.

The Marble Falls Limestone (Morrowan-Atokan) consists of a variety of platform carbonate facies including calcarenite bars, algal bank limestone, and platform margin spiculite (Kier and others, 1979). The overlying Smithwick is about 250 ft (76 m) thick and consists of dark colored mudrock that is divisible into two general lithologic units (Grayson and others, 1985). The lower Smithwick is predominantly black fissile shale deposited under dysaerobic conditions along the subsiding margin of the Concho Platform. The upper Smithwick is lighter colored, silty shale that contains ironstone concretions, and records subsidence and the establishment of more basinal conditions. The succeeding thickening and coarsening upward sequence of lower "Strawn" Group shale, siltstone, and sandstone represents the initial progradation of basin-fill submarine fan facies.

Locality 2—Gulf Colorado and Santa Fe Railroad. Proceed northwestward on FM Road 580 to its intersection with U.S. 190, turn east (right) on U.S. 190 and drive about 3.5 mi (5.8 km) to the intersection from the north (left) with Harris Cemetery Road (Fig. 5). Turn north (right) on the Harris Cemetery Road and proceed to the railroad crossing. Walk to the east (right), about 0.5 mi (0.8 km) along the track to the top of hill where lower "Strawn" Group subaqueous gravity slump deposits are exposed in a vertical cut.

Slump structures have been recognized in the lower "Strawn" at this stop, as well as Blue Bluff, Elliot Creek (Pavlovic, 1958), and several undescribed localities. In most instances, the slump structures are located along the projected trends of faults or horst blocks mapped along the northern margin of the Llano Uplift (Kier and others, 1976). Presumably these structures

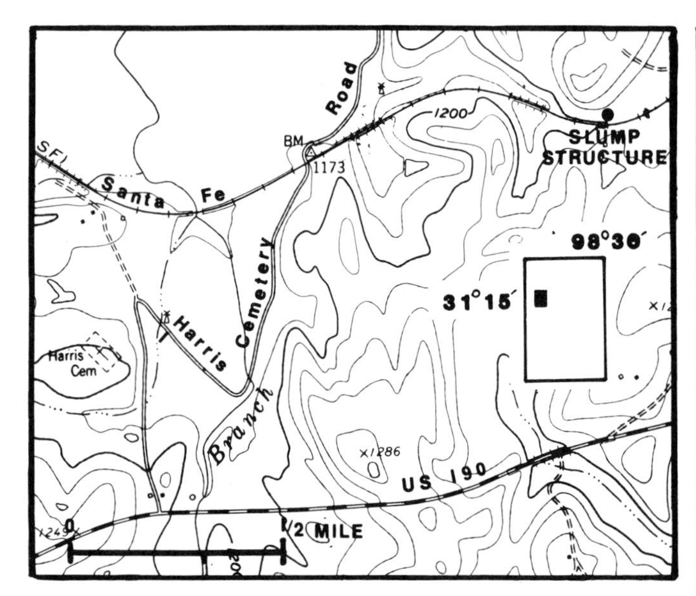

Figure 5. Location map for field locality 2, submarine sediment gravity slump structures in the lower "Strawn" Group, San Saba County, within the Wolf Ridge Quadrangle.

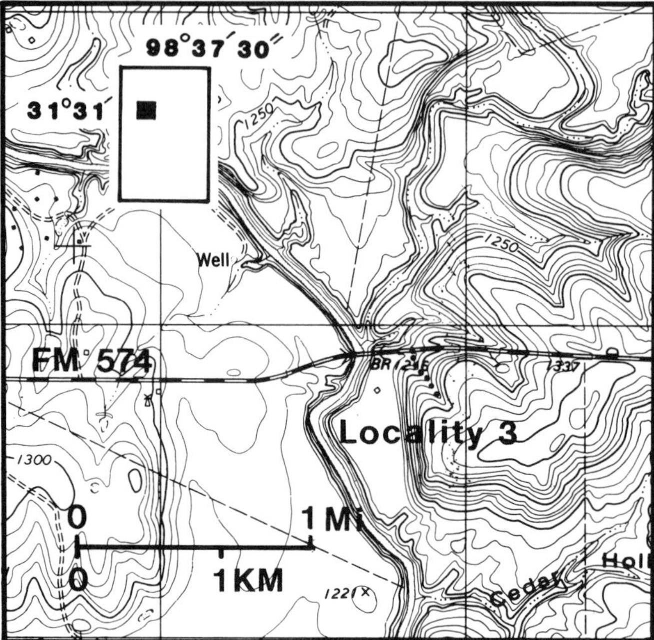

Figure 6. Location map for locality 3, Mills County, within the Big Valley North Quadrangle.

were active during accumulation of the lower "Strawn" Group, and slumping may have occurred as a response to movements along those faults.

Lower "Strawn" gravity slumps consist of complexly deformed sequences of shale interbedded with siltstone and sandstone. At this locality, zones of deformed strata are apparently separated by fault planes or detachment surfaces. Some deformation features that can be observed include isoclinal overturned folds, slump overfolds, jelly-roll structures, and micro-faulted sandstone masses.

Locality 3—Pecan Bayou 1. Return to U.S. 190, turn west (right), and proceed to the intersection with Texas 16 in San Saba. Turn north (right) on Texas 16 and continue ahead to the intersection with U.S. 183 in Goldthwaite, Texas. Turn north (left) onto U.S. 183. In Goldthwaite, take FM Road 574 to the southwest (left) for about 8.7 mi (14.5 km) to the bridge over Pecan Bayou. Locality 3 is on the east side of Pecan Bayou at the base of the hill on the south (left) side of FM Road 574 (Fig. 6).

Based on two current models of submarine fan depositional systems (Fig. 7), lower "Strawn" Group sandstone sequences exhibit sedimentary structures and bedding styles characteristic of middle fan environments. Trice and Grayson (1985) note the absence of slope deposits and feeder channel sequences in outcrop. They suggest, based on comparisons to Mutti and Ricci-Lucchi's (1978) model, these sequences might be expected to occur in the subsurface to the north. An equally possible relationship, at that time, is that lower "Strawn" middle fan facies associations were deposited on a submarine ramp proximal to deltaic complexes (Fig. 7).

The exposure at this locality consists of about 80 ft (24 m) of sandstone and shale, which represents lower-middle fan and basinal facies associations. The lowest 10 ft (3 m) interval of nearly continuous sandstone layers exhibits a vertical sequence consisting of two genetic units indicative of a distal, middle fan depositional lobe (Figs. 7 and 8). The lower genetic unit exhibits mostly amalgamated sandstone layers that thicken upward and contain partial Bouma sequences. This interval is interpreted to represent a more distal facies compared to the overlying sequence of apparently structureless sandstone layers, which exhibit a slight thinning-upward sequence. The thinning-upward sequence could represent a broad shallow channel-fill succession.

The upper and thickest portion of the section consists predominantly of shale with thin sandstone layers that contain abundant deeper water trace fossils and excellent sole marks including flute casts (Fig. 8). This interval probably represents outer fan and basinal depositional environments.

Locality 4—Pecan Bayou 2. Continue west on FM Road 574 for about 6.5 mi (10.9 km) to the intersection from the north (right) with FM Road 573. Take FM Road 573 for about 4.8 mi (8 km), to exposures located just past the bridge over Pecan Bayou (Fig. 9).

At this locality a progradational sequence of a proximal depositional lobe is exceptionally well exposed (see Fig. 12). The sequence exhibits two genetic intervals comparable to those seen at the previous locality, but probably represents accumulation in a more proximal middle fan environment. The lower, partially amalgamated, thick, even-bedded sandstone layers exhibit crude graded bedding and partial Bouma sequences. In contrast, the overlying unit consists of lenticular, channel-fill sandstone bodies. On the north side of the road, an abandoned thalweg filled with

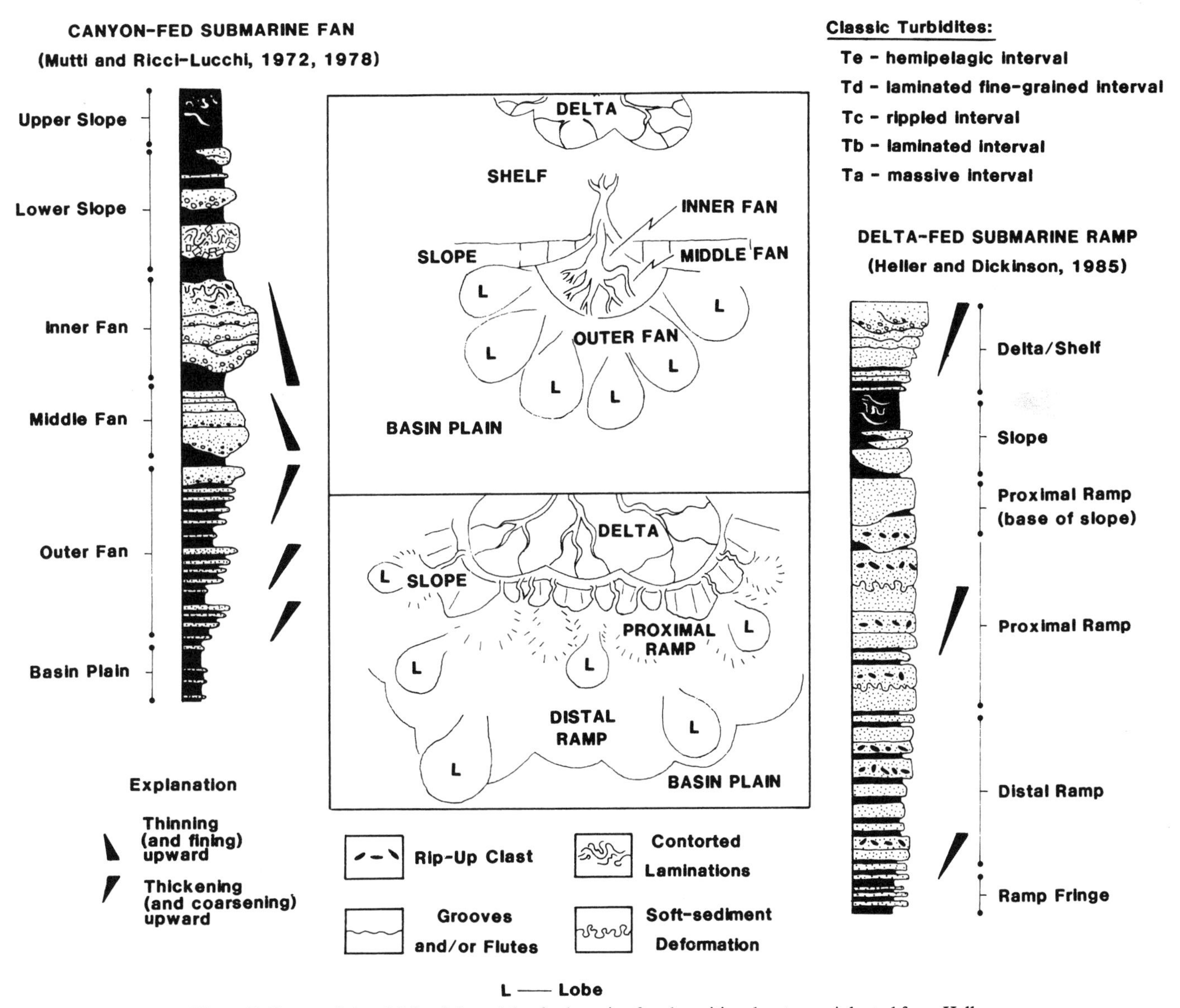

Figure 7. Canyon-fed and delta-fed models of submarine fan depositional systems. Adapted from Heller and Dickinson (1985).

pelagic shale is present in the upper-most portion of the channel-fill sequence.

79–Fluvial-Deltaic and Shelf Successions

Locality 5—Hall West. From locality 4, return to FM Road 574 and turn west (right). Continue on FM Road 574 to its intersection with Texas 45, turn south (left) on Texas 45 and continue to the intersection with FM Road 2997. Turn west (right) on FM Road 2997, and continue to its intersection with an unnumbered gravel road entering from the west (right). This road is located just north of Hall, Texas (Fig. 10). Turn west (right) and follow the road about 0.8 mi (1.4 km) to the point where the road turns sharply northward (right). Park and walk southward across the Gulf, Colorado, and Santa Fe railroad track into the bed of Richland Springs Creek. The main part of the exposure is located on a 20 ft (6 m) bluff immediately south of the railroad track in the creek bed. The upper part of the section is exposed across the track to the north, through a gate, past a dry stock tank, and then northward up a small gulley to the edge of a cultivated field. An exposure of the Marble Falls Limestone can also be seen in the bed of Richland Springs Creek about 0.5 mi (0.8 km) southeastward from the lowest exposures of the upper Strawn Group (Fig. 10).

At this locality, the stratigraphic sequence consists of Marble Falls Limestone succeeded by a covered interval, and then an upper Strawn Group succession that consists of shallow marine shelf to distal deltaic facies. This succession of stratigraphic units

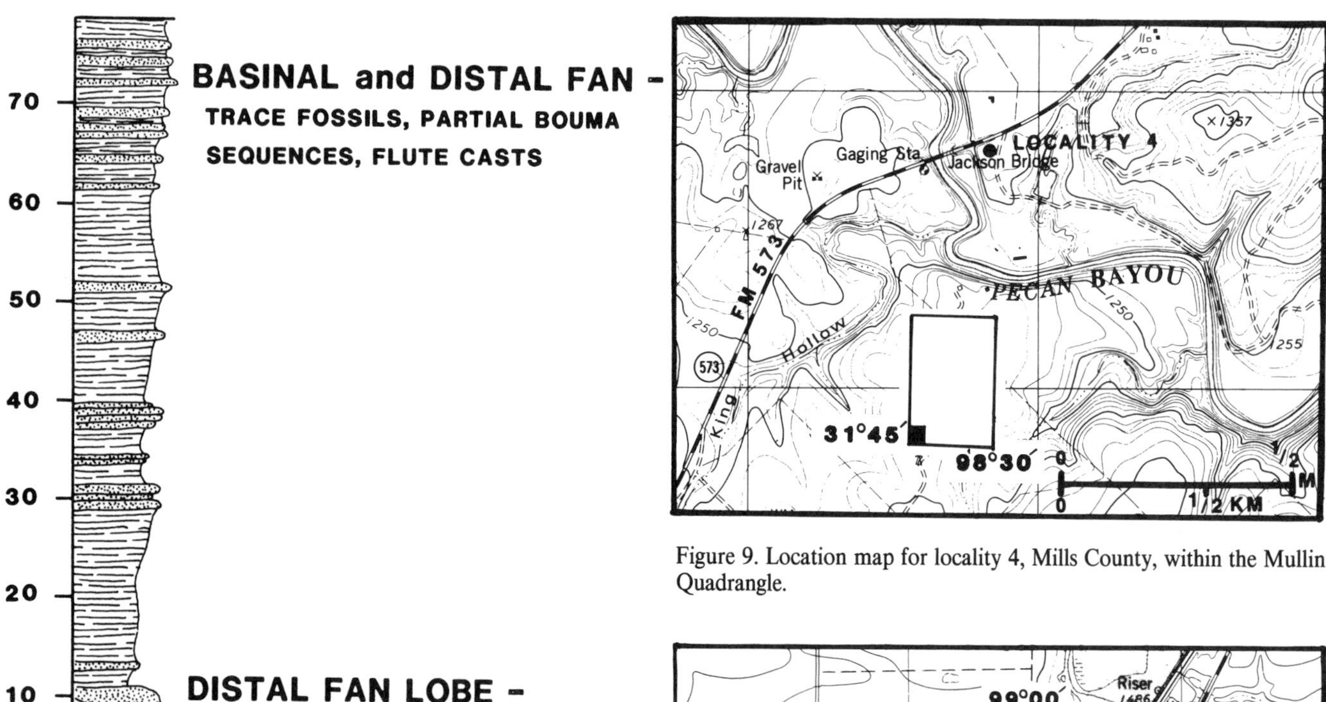

Figure 8. Columnar section for locality 3.

Figure 9. Location map for locality 4, Mills County, within the Mullin Quadrangle.

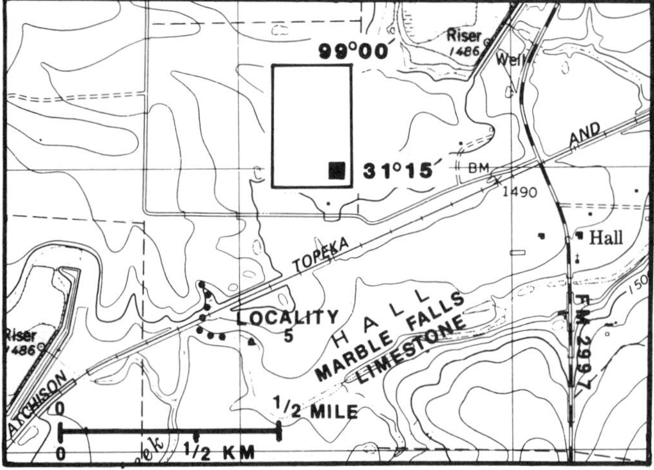

Figure 10. Location map for locality 5, San Saba County, within the Hall Quadrangle.

and facies is markedly different from that examined in the vicinity of Bend, Texas (Locality 1). In addition, the time-stratigraphic record at this locality also differs from that represented near Bend. In the Hall area, Grayson and others (1985) demonstrate by conodont biostratigraphic analysis that a substantial gap or hiatus exists at the contact of upper Strawn Group strata with older Smithwick Shale or Marble Falls Limestone. Apparently the horst block on the northwestern margin of the Llano Uplift (Fig. 1) was structurally active penecontemporaneous with deposition (Freeman and Wilde, 1964; Grayson and others, 1985; Trice and Grayson, 1985). By Late Atokan time sufficient uplift and lack of subsidence had occurred in the Hall area to preclude accumulation of lower "Strawn" deeper water facies. Prior to onlap of upper Strawn Group strata, the Hall area was subaerially exposed (Fig. 3C) and locally eroded down to the upper Marble Falls Limestone. Erosion of the Marble Falls provided the source for local accumulations of limestone-cobble conglomerates in the Strawn Group (Freeman and Wilde, 1964; Grayson and others, 1985).

Locality 6—Upper Strawn Group transect. Return to Texas 45, turn north (left) and proceed to the intersection with FM Road 765 and turn west (left). Figure 11 shows the location of several exposures that demonstrate the character of two upper Strawn deltaic cycles in an area where subsidence was more continuous and block faulting had little obvious effect on sedimentation. In the Colorado River Valley, upper Strawn Group deltaic depositional cycles are comparable to those described by Cleaves (1975) in the Brazos River Valley. A vertical sequence of an idealized cycle is shown as Figure 12.

From its intersection with Texas 45, FM Road 765 crosses the contact between the lower "Strawn" and upper Strawn Groups. In this area the contact is gradational and is marked by the change from deeper water sandstone bodies to prodelta and delta-front deposits of the first upper Strawn deltaic cycle (Localities 6A and 6B). The top of the first deltaic cycle is seemingly incised by the basal chert conglomeratic unit of the second deltaic cycle. Deltaic deposits of the second cycle exposed along FM Road 765 are represented by fluvial channel-fill facies (Localities 6C and 6D). The Capps Limestone, a transgressive shelf carbonate near the top of the second cycle can be seen just north of FM Road 765 (Locality 6E).

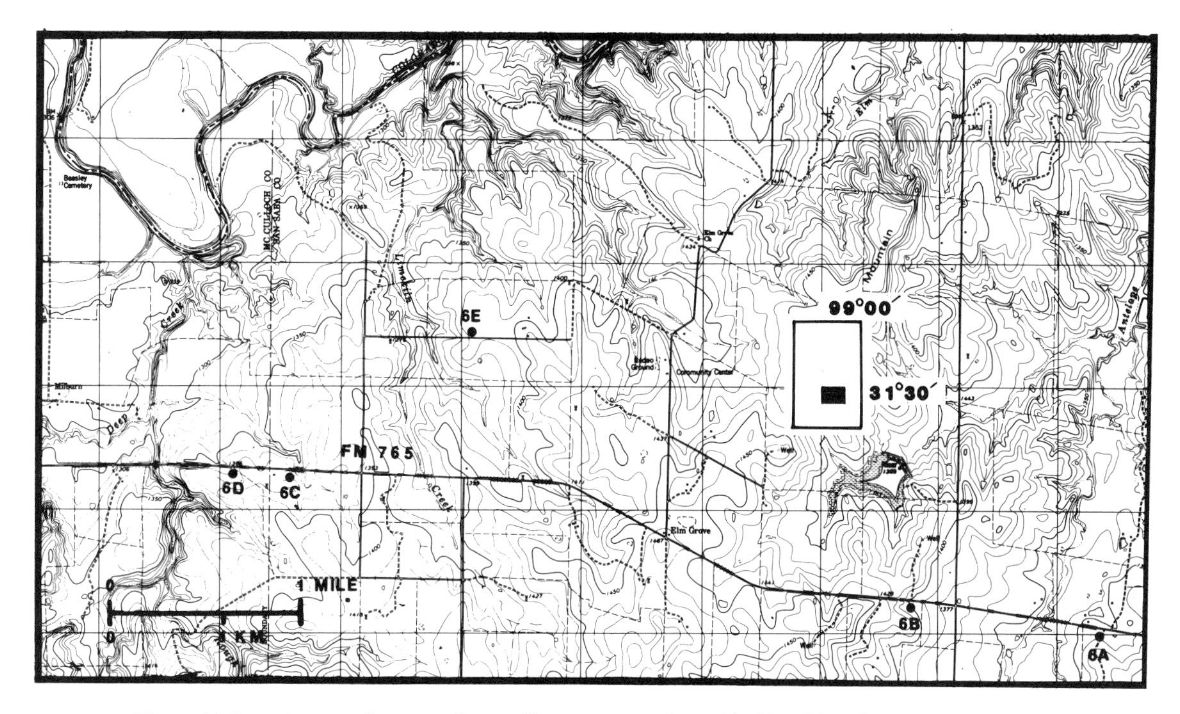

Figure 11. Location map for upper Strawn Group transect (locs. 6A-6E), all localities except 6E are low relief exposures on Fm 765, within the Elm Grove Quadrangle, San Saba and McCulloch Counties.

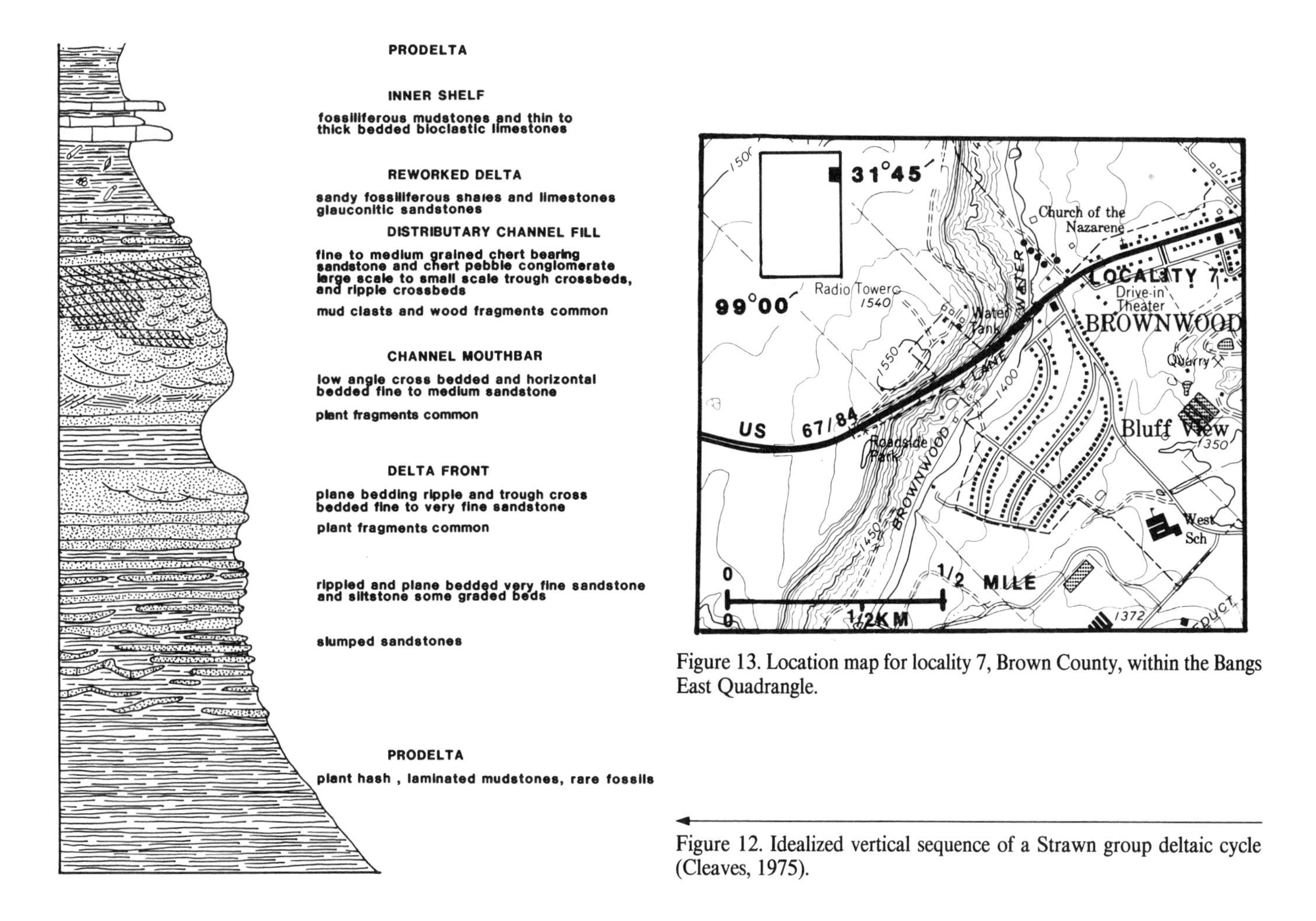

Figure 13. Location map for locality 7, Brown County, within the Bangs East Quadrangle.

Figure 12. Idealized vertical sequence of a Strawn group deltaic cycle (Cleaves, 1975).

Locality 7—Western Hills. Proceed westward on FM Road 765, past excellent exposures of the Cedarton Shale and Winchell Limestone of the Canyon Group, to Texas 377 and then turn north (right). Continue on Texas 377 to its junction with U.S. 67/84 in the northern part of Brownwood, Texas. Turn west (left) on U.S. 67/84, and drive westward about 1 mi (1.6 km) to the entrance of the Western Hills subdivision. Locality 7 is located up the escarpment to the north (right) and along the Western Hills private road (Fig. 13).

The stratigraphic succession at this locality includes part of the highest portion of the upper Strawn Group (Cycle 3) and the basal Adams Branch Limestone of the Canyon Group. Eargle (1960) referred the Missourian-age Strawn Group exposures at this locality to the Brownwood Shale, a name that has not been widely accepted.

The lower, well-exposed portion of this sequence consists of cross-bedded and bioturbated sandstone succeeded by gray and maroon-colored shale. This succession is thought to represent crevasse-splay sandstone and deltaic-embayment fill based on Brown's (1973, Fig. 22) deltaic cyclothem model. The overlying units, beginning with yellowish, fossiliferous shale, are mostly covered up to the Adams Branch Limestone. This interval probably represents a fine-clastic marine embayment and shallow shelf algal bank limestone.

REFERENCES

Bell, W. C., 1957, Road log for second day: Abilene and Fort Worth Geological Society, Guidebook, Oct. 1957, p. 43–46.

Brown, L. F., Jr., 1973, Pennsylvanian rocks of north and west-central Texas; An introduction, *in* Brown, L. F., Jr., Cleaves, A. W., II, and Erxleben, A. W., eds., Pennsylvanian depositional systems in north-central Texas; A guide for interpreting terrigenous clastic facies in a cratonic basin: Austin, University of Texas, Bureau of Economic Geology Guidebook 14, p. 1–9.

Cleaves, A. W., II, 1975, Upper Desmoinesian-Lower Missourian depositional systems (Pennsylvanian), north-central Texas [Ph.D. thesis]: Austin, University of Texas, 256 p.

Cloud, P. E., Jr., and Barnes, V. E., 1948, The Ellenburger Group of central Texas: Austin, University of Texas, Bureau of Economic Geology Publication 4621, 478 p.

Crosby, E. J., and Mapel, W. J., 1975, Paleotectonic investigation of the Pennsylvanian System in the United States, Part 1; Introduction and regional analyses of the Pennslyvanian System, central and west Texas: U.S. Geological Survey Professional Paper 853-K, p. 196–232.

Eargle, D. H., 1960, Stratigraphy of Pennsylvanian and Lower Permian rocks in Brown and Coleman Counties, Texas: U.S. Geological Survey Professional Paper 315-D, p. 55–77.

Freeman, T. J., and Wilde, G. L., 1964, Age and stratigraphic implications of a major fault in the Llano region, central Texas: American Association of Petroleum Geologists Bulletin, v. 48, no. 5, p. 714–718.

Grayson, R. C., Jr., Trice, E. L. (Jack), III, and Westergaard, E. H., 1985, Significance of some Middle Atokan to Early Missourian conodont faunas from the Llano Uplift and Colorado River Valley, Texas: Southwest Section of the American Association of Petroleum Geologists 1985 Transactions, p. 117–131.

Heller, P. L., and Dickinson, W. R., 1985, Submarine ramp facies model for delta-fed, sand-rich turbidite systems: American Association of Petroleum Geologists Bulletin, v. 69, no. 6, p. 960–976.

Kier, R. S., Brown, L. F., Jr., and McBride, E. F., 1979, The Mississippian and Pennsylvanian (Carboniferous) Systems in the United States—Texas: U.S. Geological Survey Professional Paper 1110-S, 45 p.

Kier, R. S., Harwood, P., Brown, L. F., Jr., and Goodman, J. L., 1976, Brownwood Sheet, *in* Barnes, V. E., project director, Geologic Atlas of Texas: Austin, University of Texas, Bureau of Economic Geology.

Merrill, G. K., 1980, Road log-Day two, *in* Geology of the Llano region, central Texas: Guidebook to the Annual Field Trip of the West Texas Geological Society, Oct. 19–21, 1980, p. 160–199.

Mutti, E., and Ricci-Lucchi, F., 1978, Turbidites of northern Appenines, introduction to facies analysis: American Geological Institute Report, Series 3, p. 125–167.

Namy, J. N., 1969, Stratigraphy of the Marble Falls Group, southeast Burnett County, Texas [Ph.D. thesis]: Austin, University of Texas, 385 p.

Pavlovic, R., 1958, Pennsylvanian outcrops of significance, Mills County, Texas: American Association of Petroleum Geologists Bulletin, v. 42, p. 888–892.

Thompson, D. M., 1982, Atoka Group (Lower to Middle Pennsylvanian), northern Fort Worth Basin, Texas; Terrigenous depositional systems, diagenesis, and reservoir distribution, and quality: Austin, University of Texas, Bureau of Economic Geology, Report of Investigation no. 125, 62 p.

Trice, E. L. (Jack), III, and Grayson, R. C., Jr., 1985, Depositional systems and stratigraphic relationships of the Strawn Group (Pennsylvanian), Colorado River Valley, central Texas: Southwest Section of the American Association of Petroleum Geologists 1985 Transactions, p. 192–205.

Weaver, O. D., 1956, Introduction to the Fort Worth Basin, *in* Steward, W. S., ed., Symposium of the Fort Worth Basin area and field study of the Hill Creek beds of the Lower Strawn, southwestern Parker County, Texas: Permian Basin Section, Society of Economic and Paleontological Mineralogists, Spring meeting and Field Symposium, p. 10–18.

Paleozoic strata of the Llano region, Central Texas

Robert S. Kier, *Robert Kier Consulting, Colina West, 8834 Capital of Texas Highway North, Suite 230, Austin, Texas 78759*

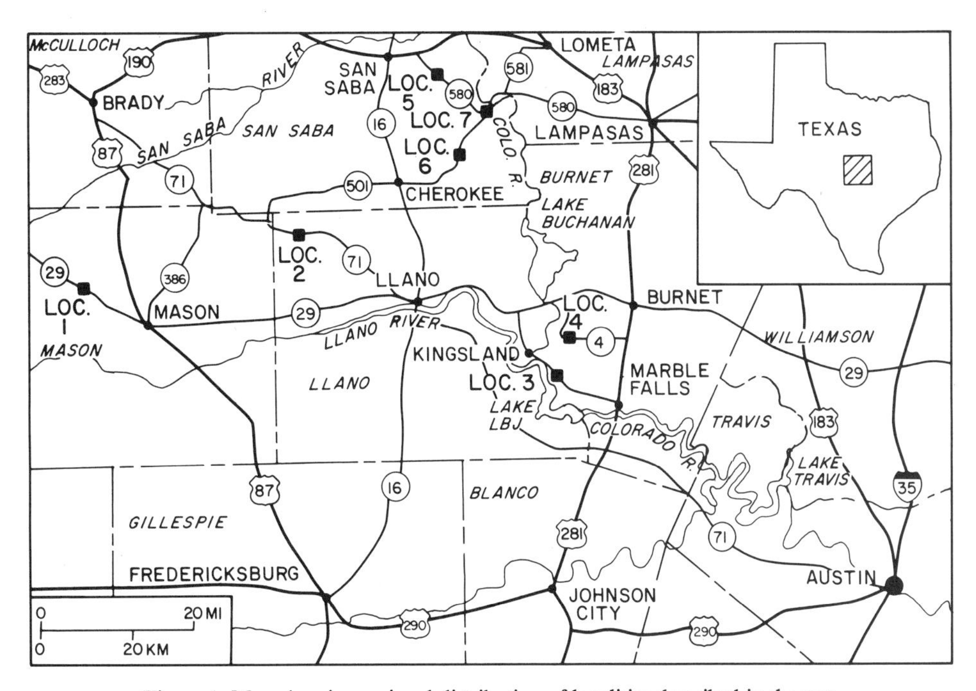

Figure 1. Map showing regional distribution of localities described in the text.

LOCATION

The Paleozoic rocks of central Texas occur in the numerous fault blocks that compose the Llano uplift, also known as the Central Mineral Region, a structural high in the Precambrian basement centered in Llano County, northwest of Austin (Fig. 1). The predominant trend of the fault blocks is southwest to northeast. For the most part, the Paleozoic rocks are preserved in grabens. Few fault blocks are large enough, however, to display the entire Paleozoic section; thus, no single transect can be made that will allow a view of the entire section. Rather, the Paleozoic section must be examined at a number of isolated localities from which a regional picture can then be formulated. Rapid facies changes in some of the formations compound the difficulty of piecing together a representative section. The Brownwood and Llano sheets of the Geologic Atlas of Texas (Barnes and others, 1976, 1981) cover the entire area of interest.

Most of the localities are along roadcuts where access presents no problem; possible restrictions on public access to locality 7 are noted. Because of access problems, no localities are presented for the thin, patchy Upper Ordovician, Silurian, and Devonian formations, few of which are in normal stratigraphic position. Most of these are preserved in collapse structures, sinks, and crack fillings.

INTRODUCTION

The Paleozoic strata of the Llano uplift comprise rocks ranging in age from Middle Cambrian through Early Pennsylvanian (Fig. 2). Cretaceous rocks of the Edwards Plateau and the Lampasas Cut Plain ring and overlie the Paleozoic and Precambrian rocks on the east, south, and west sides of the Llano uplift, and on much of the north side. Younger Paleozoic rocks are exposed north of the uplift where Cretaceous rocks are absent.

Erosion since Cretaceous time has resulted in a topographic inversion, so that the oldest and structurally highest rocks tend to occur at the lowest topographic elevations. Where the Cretaceous rocks rim the Llano uplift, a sharp topographic rise or bluff is common.

Most of the Llano region is in the Colorado River basin; a small part of the area is in the Guadalupe River basin to the south.

SITE 80—EARLY PALEOZOIC

Moore Hollow Group, Riley Formation

Locality 1, Basal Unit of Hickory Sandstone Member (Fig. 3). South side of Texas 29, approximately 7.4 mi (12. 3 km) west of Mason, 1.4 mi (2.3 km) northwest of Grit, and 0.4 mi (0.6 km) northwest of Honey Creek (Mason County; lat 30°48′01″N, long 99°19′58″W; Grit 7½-minute Quadrangle). Here Hickory Sandstone unconformably overlies Packsaddle

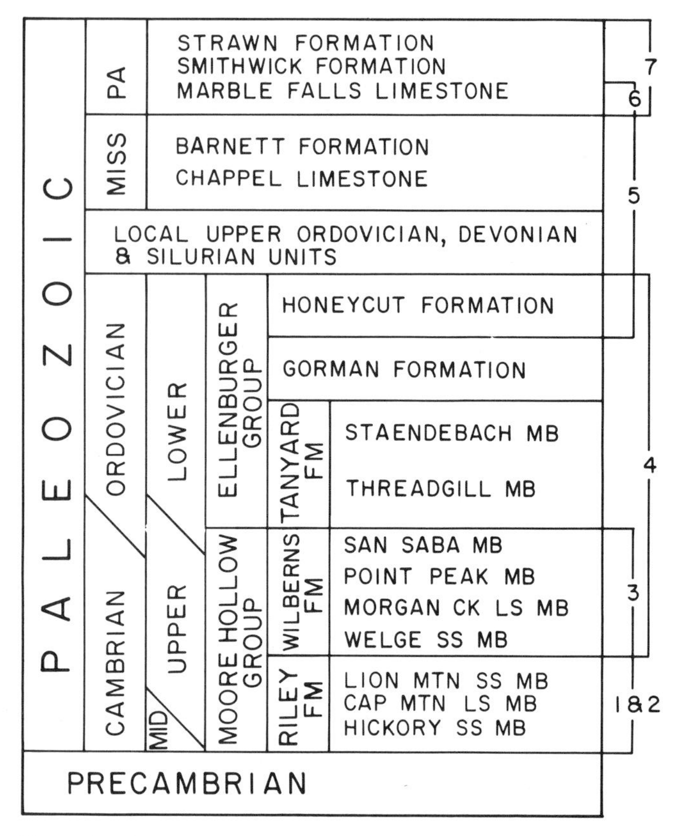

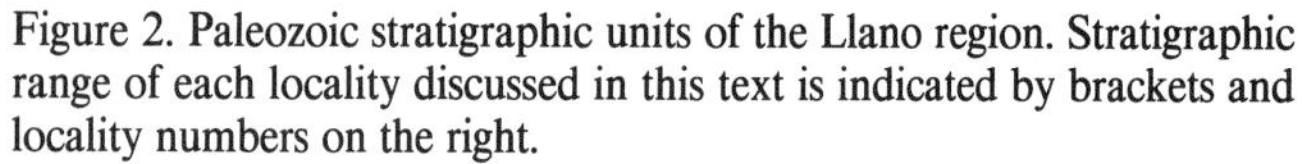
Figure 2. Paleozoic stratigraphic units of the Llano region. Stratigraphic range of each locality discussed in this text is indicated by brackets and locality numbers on the right.

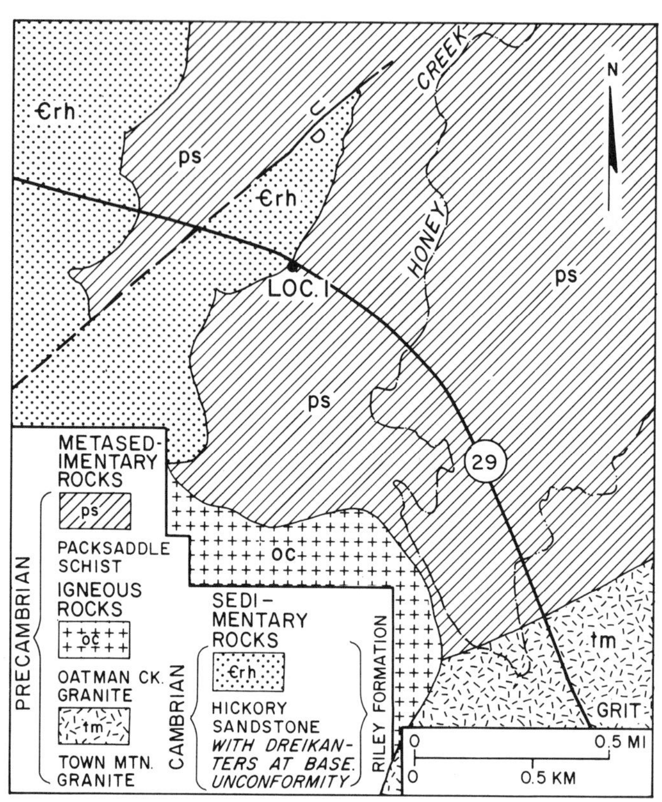

Figure 3. Geologic map of the Grit area (Locality 1), Mason County, Texas. Modified from Barnes and Bell, 1954.

Schist. Throughout much of the Llano area, the lower part of the Hickory Sandstone is typically a friable, poorly sorted, rounded to subrounded, fine-grained, quartz-cemented sandstone locally containing rounded, but unweathered feldspars derived from the Precambrian gneisses and granites. Feldspathic conglomerates occur elsewhere at or near the base of the formation. Pebbles and cobbles in the base of the flat-lying Hickory Sandstone are wind abraded and faceted. In the northwestern part of the Llano region, the middle part of the Hickory is argillaceous, silty, and micaceous. The upper part of the Hickory is typically a dark-red, friable, well-rounded, medium- to coarse-grained sandstone. Iron oxide is the principal cement. The purity and uniform grain size of the Hickory make it valuable as a "frac" sand, which is still mined to the north toward Brady.

The incorporation of ventifacts and unweathered feldspars into the Hickory indicates that the marine transgression during which the Hickory was deposited crossed a dry, windy, and possibly cold, erosion surface developed on the underlying Precambrian rocks. The time-transgressive nature of the Hickory Sandstone and overlying members of the Riley Formation indicates that the marine transgression occurred from southeast to northwest (Fig. 4). The thickness of the Hickory in the Llano region ranges from approximately 276 to 470 ft (84 to 143 m) and is controlled by the paleotopography on the underlying Precambrian surface and by lateral gradation with the overlying Cap Mountain Limestone. Variations in the position of the basal unconformity suggest that the topographic relief on the Precambrian erosion surface was comparable to the modern relief, approximately 800 ft (244 m). In the subsurface to the north and west of the Llano region, the Hickory Sandstone thins and is absent approximately 60 mi (100 km) northwest of Brady in McCulloch County.

The Hickory Sandstone in this area also serves as an aquifer. Ground water is pumped in the outcrop area for irrigation purposes. The city of Brady (McCulloch County) depends on artesian ground water from the Hickory aquifer for its municipal supply, and the city of San Angelo (Tom Green County) holds land in reserve for future ground water development (most ground water in Texas is a private-property right).

Locality 2, Upper Unit of Hickory Sandstone Member (Fig. 5). Texas 71, between Valley Spring and Pontotoc in Llano County (lat 30°52′32″N, long 98°53′44″W; Pontotoc 7½-minute Quadrangle). Here the Hickory Sandstone is cemented by hematite. The iron content ranges up to 14%, but production is

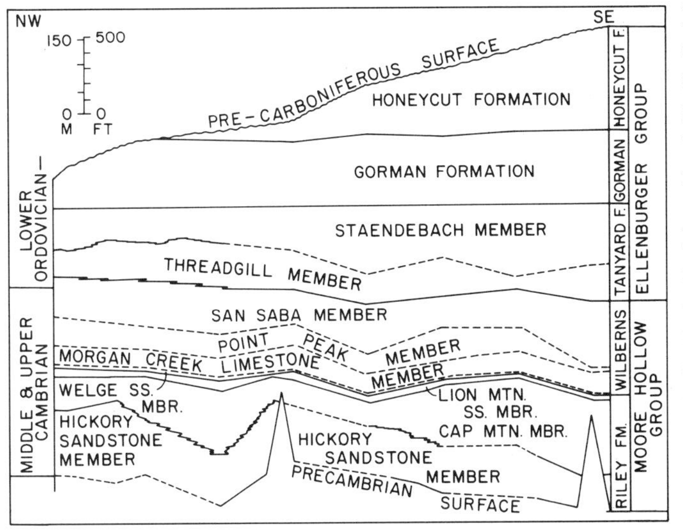

Figure 4. Diagrammatic representation of the Moore Hollow and Ellenburger Groups in central Texas. Modified from Barnes and Bell, 1977.

not currently economical. The state highway department has quarried Hickory Sandstone at this locality as base-coarse material for highway construction (Barnes and Schoenfield, 1964).

Moore Hollow Group, Riley and Wilberns Formations

Locality 3, Upper part of the Riley Formation and lower part of the Wilberns Formation (Fig. 6). Roadcut, northeast side of Farm to Market Road 1431 at Hoover Point overlooking Lake Lyndon B. Johnson (formerly called Granite Shoals Lake), approximately 3.7 mi (6 km) east of Kingsland (Llano County) and 1.3 mi (2.2 km) east of the Llano-Burnet County line at the Colorado River (lat 30°38′46″N, long 98°24′55″W, Kingsland 7½-minute Quadrangle). The Cap Mountain Limestone Member of the Riley Formation and the Lion Mountain Sandstone Member of the Wilberns Formation can easily be seen at the road level; the Welge Sandstone Member and the Morgan Creek Limestone Member of the Riley Formation can be seen but cannot be reached for close observation at the roadcut.

Hoover Point is the southwestern terminus of a graben structure that preserves Cambrian and Ordovician rocks of the Moore Hollow and Ellenburger Groups. The Paleozoic rocks are more resistant than the surrounding Precambrian rocks and form a topographic high called Backbone Ridge. The low, rounded mountain about 5 mi (8 km) west of this locality is Packsaddle Mountain. Packsaddle Schist crops out at the base of the mountain, but Packsaddle Mountain actually comprises Cambrian rocks. The higher mountains in the distance beyond Packsaddle Mountain are the Riley Mountains.

Cap Mountain Limestone Member. The lowest stratigraphic unit exposed at the road cut is the Cap Mountain Limestone Member of the Riley Formation. Only the upper limy and glauconitic facies of the Cap Mountain Limestone can be seen in the roadcut. The lower sandy and limy facies and the middle silty facies cannot be easily seen at this locality because they are below the parking area at the roadcut (the cliff below the road is very steep and dangerous).

The boundary between the Hickory Sandstone and the Cap Mountain Limestone is gradational and time transgressive to the northwest (Fig. 4), and it is reflected by displacement of the quartz or hematitic cement by calcite cement and by a change from quartzose sandstone to limestone. Although the boundary is sometimes difficult to discern by casual observation at the outcrop, it is marked by a change in the vegetation from deciduous oaks to cedar and is readily apparent on aerial photographs. The Cap Mountain Limestone varies in thickness from 90 ft (27.4 m) to 411 ft (125.3 m) in outcrop, and thins to absence in the subsurface to the north and west.

Lion Mountain Sandstone Member. The Lion Mountain Sandstone Member of the Riley Formation is the most accessible and spectacular stratigraphic unit at Hoover Point. The Lion Mountain typically consists of coarse-grained, dusky-green to grayish-olive-green, cross-bedded, glauconitic sandstone containing lenses of white, glauconitic trilobite coquinite and phosphatic brachiopods. Hematite nodules are common in the soil that develops on the Lion Mountain Sandstone in the northwestern part of the Llano region, and vegetative cover is sparse, consisting of widely spaced live oak mottes. The thickness of the Lion Mountain Sandstone varies from 29 ft (8.8 m) in Burnet and western Mason counties to 69 ft (21 m) in southeastern Mason County. The member is slightly more than 33 ft thick (10 m) at Hoover Point. The Lion Mountain thickens in the subsurface to the west toward the source area and thins to the east.

Welge Sandstone Member. The Welge Sandstone Member, the basal member of the Wilberns Formation, unconformably overlies the Lion Mountain Sandstone. Although well exposed at Hoover Point, it is out of reach at the Hoover Point roadcut. The Welge Sandstone is typically a medium- to coarse-grained, dark yellowish-brown, well-sorted quartz sandstone. Exposures of the member are generally sparsely fossiliferous and nonglauconitic, although the glauconite content increases to the south and east. Typically, the Welge forms a heavily vegetated scarp. In the subsurface to the southeast, the Welge grades into a green glauconitic sand and is indistinguishable from the underlying Lion Mountain Sandstone. The unconformity between the Lion Mountain Sandstone and the Welge Sandstone was apparently rather limited in extent, perhaps confined to the Llano uplift, and possibly of short duration. The Welge Sandstone ranges in thickness from 11 ft (3.4 m) to 30 ft (9.1 m); at Hoover Point, the Welge is approximately 13 ft thick (4 m).

Morgan Creek Limestone Member. The pink, sandy lower part of the overlying Morgan Creek Limestone Member of the Wilberns Formation is well exposed at Hoover Point but inaccessible at the top of the roadcut; however, a few blocks have been used instead of a guardrail to mark the southwestern side

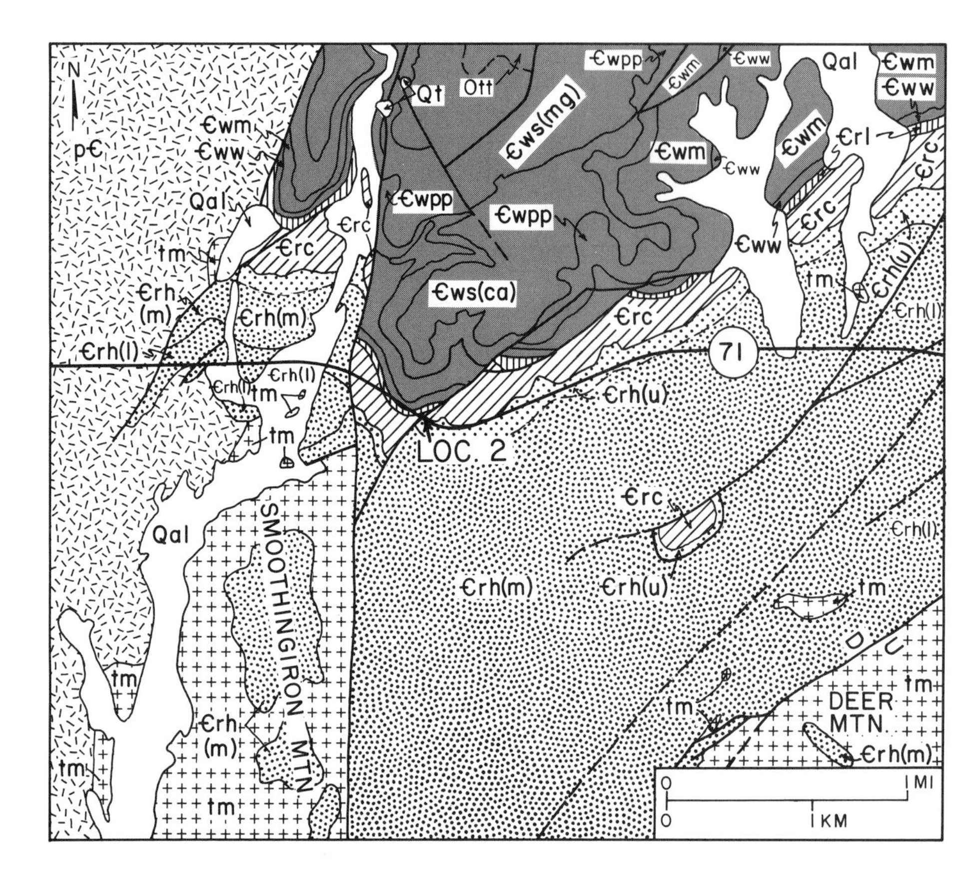

Figure 5. Geologic map of the Valley Springs area (Locality 2), Llano County, Texas. Modified from Barnes and Schoenfield, 1964. Precambrian: Valley Spring Gneiss = P€, random dashes; Town Mountain Granite = tm, crosses. Riley Formation: Hickory Sandstone Member = €rh(l,m,u), dots; Cap Mountain Limestone Member = €rc, diagonal lines; Lion Mountain Sandstone Member = €rl, vertical lines. Wilberns Formation: Welge Sandstone Member = €ww; Morgan Creek Limestone Member = €wm; Point Peak Member = €wpp; San Saba Member = €ws(ca), calcitic facies, €ws(mg), dolomitic facies; Tanyard Formation: Threadgill Member = Ott; all gray. Quaternary without pattern.

of the parking area adjacent to the highway. The boundary between the Welge and Morgan Creek is gradational.

Point Peak and San Saba Members. Although the Point Peak and San Saba Members of the Wilberns Formation crop out above the Morgan Creek Limestone at Hoover Point, they are not accessible but can be better viewed at Locality 4, farther to the northeast on Backbone Ridge. Except for the exposures at the roadcut, all other outcrops in the Hoover Point area are on private land that should not be entered without permission.

Moore Hollow Group, Wilberns Formation, Ellenburger Group, Tanyard and Honeycut Formations

Locality 4 (Fig. 7). Roadcuts along Park Road 4 begin at the overlook west of Longhorn Caverns, in Longhorn Cavern State Park, a little more than 2.5 mi (4 km) west of the park entrance (east side). The roadcuts continue downhill to the base of Backbone Ridge (Burnet County, lat 30°41′48″N, long 98°22′23″W, Longhorn Cavern 7½-minute Quadrangle). Rocks exposed here that are easily viewed from the park road include the upper part of the Morgan Creek Limestone, the Point Peak, and the San Saba Members of the Wilberns Formation and the lower part of the Tanyard and the Honeycut Formations, two of the three formations of the Ellenburger Group. The section here also is known as the Wedge section, because of the shape of the fault block between the Roaring Springs fault at the overlook (down to the east) and an unnamed fault farther to the west (also down to the east). The stair-stepped arrangement of these faults has preserved a thin wedge of the Wilberns Formation and the lower part of the Ellenburger Group; most of Backbone Ridge north of Hoover Point is underlain only by the Ellenburger Group. The fault block is tilted about 25° to the northeast. In contrast to the rocks of the Moore Hollow Group, rocks of the Ellenburger Group contain almost no glauconite and are sparsely fossiliferous.

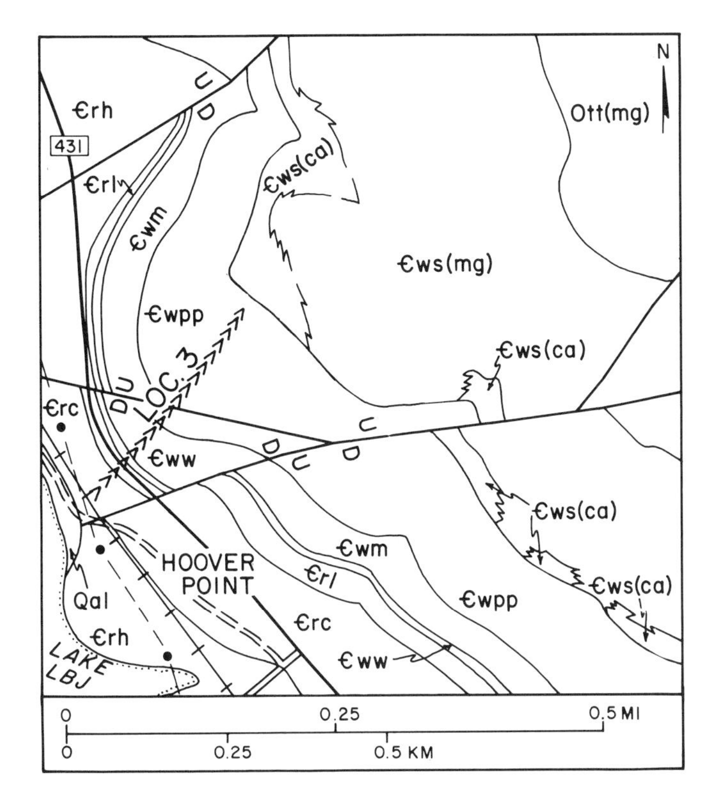

Figure 6. Geologic map of the Hoover Point area (Locality 3), Burnet County, Texas. Modified from Barnes and others, 1972. Symbols for Cambrian units same as for Figure 5. Symbol for Ordovician unit same as for Figure 7.

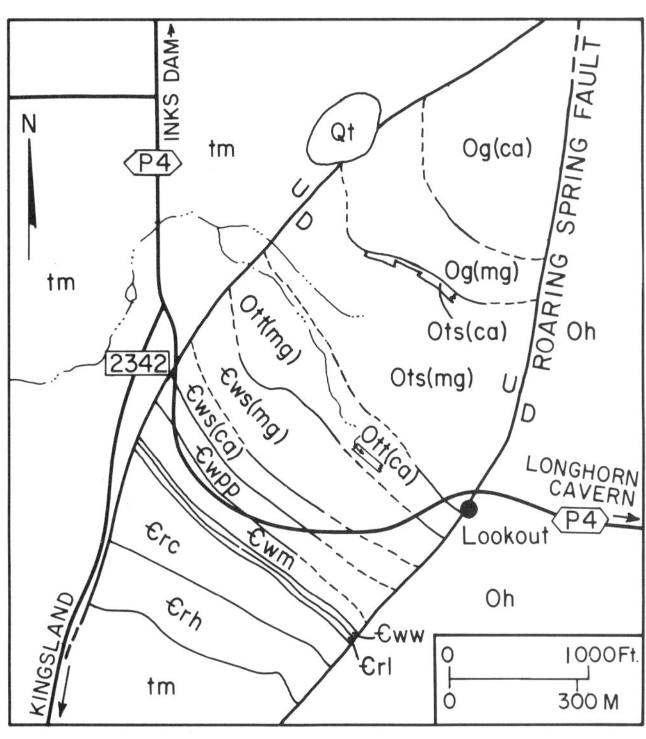

Figure 7. Geologic map of the fault wedge area (Locality 4), west of Longhorn Cavern, Burnet County, Texas. Modified from Merrill, 1980a; original mapping presented in Cloud and Barnes, 1948. Symbols for Precambrian and Cambrian units same as for Figure 5. Ordovician: Tanyard Formation, Threadgill Member = Ott, Staendebach Member = Ots; Gorman Formation = Og; Honeycut Formation = Oh. (ca) = calcitic facies; (mg) = dolomitic facies.

If the section is walked down the hill, beginning at the overlook, the view is down section, that is, beginning with the highest stratigraphic unit, or the opposite sense of the descriptions provided below. The Gorman Formation is also present in the Wedge section, but does not crop out along the park road. Longhorn Cavern has been developed entirely within the Gorman Formation, and it can best be viewed by touring the caverns. Alternatively, the Gorman Formation can be seen by visiting Gorman Falls, which is reached by taking the county road east then south from Bend, Texas, just east of the last locality described in this section of the field guide, and following the signs. The Tanyard Formation, most of the Gorman Formation, and the lower part of the Honeycut Formation also are exposed in a ditch along Texas 16, approximately 2.1 mi (3.5 km) north of Cherokee and 4 mi (6.7 km) south of San Saba.

Moore Hollow Group, Wilberns Formation

Morgan Creek Limestone Member. The Morgan Creek Limestone Member of the Wilberns Formation consists of coarse-grained, greenish-gray to light olive-gray, glauconitic limestone in the lower part of the member; bedding is thick but thins upward. The lower part of the member is commonly sandy and pinkish to reddish where it grades into the underlying Welge. Near the middle of the member, bedding is thin to medium; dark greenish-gray, silty and argillaceous, fine-grained limestone is interbedded with coarser grained limestone. The upper part of the member consists of coarse-grained, glauconitic limestone interbedded with thick-bedded, dark greenish-gray, silty, fine-grained limestone. Oolites are common in the Morgan Creek; selective dolomitization of the ooids and other small allochems occurs locally. Stromatolitic bioherms are common in the upper third of the Morgan Creek. The Morgan Creek Limestone ranges in thickness from 114 to 143 ft (34.7 to 43.6 m). Vegetation is aligned with bedding. The boundary with the overlying Point Peak Member is gradational; where the Morgan Creek Limestone becomes increasingly dolomitic in the subsurface to the south and the west, the Morgan Creek is difficult to distinguish from the Point Peak. Farther to the west and to the northwest, sandstone is the lateral equivalent of the Morgan Creek Limestone.

Point Peak Member. The Point Peak Member is principally thin-bedded, very light olive-gray, argillaceous, glauconitic, calcareous siltstone, and light olive-gray to yellowish-gray, fine-grained, argillaceous, silty limestone. Siltstone predominates in the lower part of the member; increasing amounts of limestone are present higher in the member. Varicolored intraformational conglomerate composed of thin, flat, subrounded, silty limestone clasts is common in the upper part of the member. Very light grayish-green, stromatolitic limestone also occurs in the upper part of the member, locally dominating the entire upper half of the member (e.g., east of White's Crossing of the Llano River: access is difficult; also in the Riley Mountains). The Point Peak Member is generally not resistant and weathers to vegetated flat benches and gentle slopes. The member averages 150 ft thick (45.7 m) over most of the Llano region. The boundary with the overlying San Saba Member is gradational and commonly arbitrary. The Point Peak thins to the northeast as the San Saba thickens, suggesting a time-transgressive facies change; bioherms extend across the boundary between the two members.

San Saba Member. The San Saba Member consists of both limestone and dolomite and varies considerably throughout the Llano region. The limestone is mostly thin- to thick-bedded, fine- to medium-grained, and glauconitic. The color of the limestone is various shades of gray, including yellowish-gray, olive-gray, and greenish-gray. Vegetation is aligned with the bedding. The dolomite is either medium-bedded and fine-grained, or thick-bedded and coarse-grained, and contains abundant chert. The dolomite also can be various shades of gray and yellowish-gray, but also can be pinkish-gray or mottled with red and purple. Limestone predominates in the western part of the Llano region. Where both limestone and dolomite occur, the dolomite is generally higher in the section. Stromatolites are common in the western part of the region, mostly near the base of the member; some of the stromatolites originated during deposition of the Point Peak and continued to grow throughout deposition of the San Saba. In the western part of the Llano region, about 55 to 70 ft (16.8 to 21.3 m) of medium-grained, well-rounded sandstone occurs in the member. The thickness of the San Saba Member generally varies from about 280 to about 325 ft (85.3 to 99.1 m). Anomalously thick sections occur locally. The boundary with the overlying Ellenburger Group is gradational. The upper boundary of the San Saba Member of the Wilberns Formation is time transgressive; the upper part of the unit is Early Ordovician in the western part of the Llano region, whereas the lower part of the Ellenburger is Late Cambrian age in the eastern part of the region.

Ellenburger Group, Tanyard Formation

Threadgill Member. In the eastern part of the Llano region, as seen in the Wedge section, the Threadgill Member of the Tanyard Formation consists of gray to light brownish-gray, medium- to coarse-grained dolomite. Locally, the dolomite is vuggy, and lenses of limestone occur. The transition from limestone to dolomite is abrupt. In the western part of the Llano region, the Threadgill Member becomes thin-bedded, pearl-gray, argillaceous and silty limestone. Generally, the Threadgill limestone contains little chert, but some allochems are selectively dolomitized. The thickness of the Threadgill Member ranges from 91 ft (27.7 m) in the eastern part of the Llano region to 294 ft (89.6 m) in the western part of the region. In the subsurface farther west, the Threadgill Member can no longer be separately identified. The San Saba Member of the Wilberns Formation becomes the lateral equivalent of the Threadgill and the upper member of the Tanyard Formation rests directly on the Wilberns Formation.

Staendebach Member. The Staendebach is the upper member of the Tanyard Formation. The member consists predominantly of light-gray to light brownish-gray, fine- to medium-grained, cherty dolomite. The chert is slightly dolomoldic and oolitic and is opaque to slightly translucent. In the western part of the Llano region, the member is entirely dolomitic. In the eastern part of the region, parts of the upper third of the member contain gray, cherty limestone. The chert contains small, elongated, aggregated algal bodies that resemble worm castings.

Gorman Formation. The Gorman Formation is a highly variable mixture of limestone and dolomite. In general, the lower part consists of beige to cinnamon-pink, microgranular to very fine-grained dolomite. In outcrop the basal dolomitic unit weathers to a gray to beige, blocky ledge that is a conspicuous marker in the Ellenburger. The upper limestone unit is composed of thin- to thick-bedded, aphanitic, gray limestone. In the eastern part of the Llano region, a microgranular dolomite facies occurs within the limestone beds, and well-rounded quartz sand is scattered at various levels throughout the formation. Where a complete section of the Gorman is present, it ranges from 430 to 500 ft thick (131.1 to 152.4 m); however, post-Ellenburger erosion has commonly removed much of the formation. In the subsurface, the Gorman Formation is the principal water-bearing formation in the Ellenburger Group, although all the Ellenburger formations can be productive.

Honeycut Formation. In the eastern part of the Llano region, the Honeycut Formation can be divided into three informal units. The lowest unit consists of alternating light-gray limestone; gray, fine- to medium-grained dolomite; and brown to pearl-gray microgranular dolomite. The middle unit consists predominantly of brown, microgranular dolomite. The upper unit of the Honeycut is predominantly light-gray to yellowish-gray limestone. The Honeycut is a maximum of 678 ft (206.6 m) thick. In the western part of the Llano region, post-Ellenburger erosion and truncation has removed all of the Honeycut Formation (Fig. 4) and locally, near Hext, all of the Gorman and Tanyard Formations as well.

SITE 81—LATE PALEOZOIC

Chappel Limestone and Barnett Formation

Locality 5 (Fig. 8). Roadcut on south side of Farm Road 1031, 2.7 mi (4.3 km) from the courthouse in San Saba (San

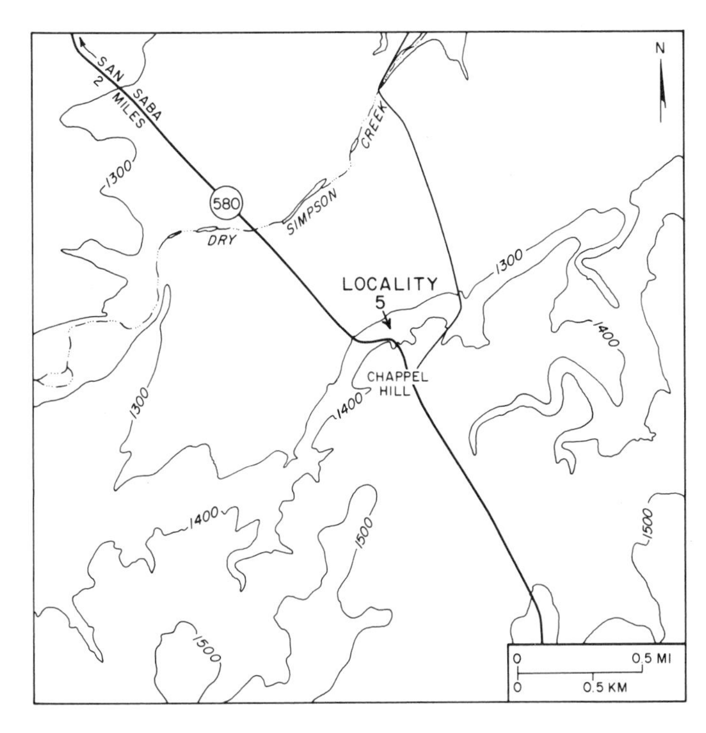

Figure 8. Type locality of Chappel Limestone (Locality 5), San Saba County, Texas. Modified from Windle, 1980.

Saba County; lat 31°09′57″N, long 98°41′30″W; San Saba 7½-minute Quadrangle). Three formations are displayed here; the Honeycut Formation of the Ellenburger Group, the Chappel Limestone, and the Barnett Shale. The formations are exposed in a cut where the road between San Saba and Chappel ascends the fault-line scarp between Smithwick Shale to the north and Ellenburger limestone and dolomite to the south. Because of differential weathering, the younger Smithwick is topographically lower than the older Ellenburger rocks and underlies the valley of Simpson Creek, a tributary of the San Saba River.

The Honeycut is Early Ordovician in age; the Chappel and the Barnett are considered to be Mississippian in age. Solution of the Honeycut during post-depositional erosion left a thin chert regolith that was incorporated into a basal transgressive breccia. Hass (1953, 1959) placed this in the Houy Formation (Ives Breccia); others appear to have lumped the basal breccia with the Chappel (Ellison, 1980). The amount of section removed during the Ordovician–Mississippian hiatus is not known.

Chappel Limestone. The Chappel Limestone generally consists of light olive-gray to pinkish-gray or light yellowish-brown, fine- to coarse-grained, packed, crinoidal biosparite and biomicrite. The crinoid fragments are mostly broken and disarticulated and commonly rounded. Other allochems include ostracodes, algae, foraminifers, brachiopods, trilobites, and conodonts. Locally, the Chappel contains glauconite. The conodonts include both Mississippian forms (Middle and Late Mississippian) and Devonian forms. The Devonian forms are probably reworked from pre-Chappel deposits. Locally, near the base of the formation, medium-grained angular to rounded, quartz-sand grains occur. For the most part, the Chappel is thin bedded, and the exposures of the formation seldom amount to more than 1 to 2 ft (0.3 to 0.6 m) in thickness. Although some exposures of the Chappel can be traced for miles, others seem to be isolated occurrences. In places, the Chappel seems to have accumulated in pre- and syn-depositional sinks in the underlying Ellenburger; in other places, the Chappel is now preserved in collapse structures formed in the Ellenburger after the Chappel was deposited. Accumulations of the Chappel in the sinks and collapse structures have amounted to as much as 50 ft (15 m). At this locality, the Chappel is approximately 5 ft thick (1.5 m); it is one of the thickest occurrences of the Chappel found outside a sinkhole or collapse structure. The calcareous Chappel appears to be analogous to quartz sand strandline deposits. Relief on the pre-Chappel Ellenburger erosion surface was very low.

Barnett Shale. The Barnett consists of shale and various kinds of limestone. At and in the vicinity of this locality, the Barnett Shale is black or dark gray to dark brown and very thinly laminated. Thin brachiopod and goniatite coquinites occur in the shale. Small to large (up to 8 ft; 2.75 m) ellipsoidal, black microsparite concretions are common in the upper part of the formation. The concretions tend to be laterally persistent and appear to have originated from thin beds of dark microgranular limestone. Small coiled to very large straight-shelled cephalopods are preserved in the Barnett concretions. Both the shale and the concretions are petroliferous; freshly broken concretions give off a strong odor. Uranium also occurs in the Barnett. Locally, the base of the Barnett is a finely laminated calcareous siltstone. The top of the Barnett is marked by fine- to coarse-grained, packed, goniatite- and pellet-bearing, phosphatic oomicrite. For the most part, the shale is poorly exposed and has been altered to caliche. Outcrops of Barnett Shale are commonly marked by mesquite-covered or tilled benches occurring between the underlying Ellenburger limestone and dolomite and the overlying Marble Falls Limestone. At this locality, the Barnett Shale is about 50 ft thick (15 m). The type locality of the Barnett Shale, which was probably mis-located when it was originally described, is about 3 mi (5 km) northeast of this locality. It cannot be visited because it is on private land.

To the west of San Saba, the lower part of the Barnett is light-colored, clayey shale that contains small concretions and, near the top of the lower part, phosphatic limestone. Farther to the west, near Bracy, the lower part of the Barnett contains increasing amounts of thin, phosphatic, and glauconitic limestone. The upper part of the Barnett is a grayish-black to yellowish-brown, fine- to coarse-grained, packed, glauconitic, and phosphatic biomicrite and micrite. The Barnett Shale is probably absent near Mason, at the far western side of the Llano uplift. To the south, near the town of Marble Falls, the Barnett is a light-colored, concretion-bearing shale capped by phosphatic limestone.

The Barnett was probably deposited in a sediment-starved,

restricted environment developed on the west side of the Fort Worth basin. Euxinic conditions are reflected by the dark petroliferous shale and general paucity of benthonic fauna. Lack of clastic influx is suggested by the thin beds, inferred length of time over which the Barnett apparently accumulated, and the abundance of limestone near the basin margins. The coquinas probably developed during temporary cessations of euxinic conditions. Near the margins of the basin, and as the basin filled, phosphatic limestone became common. The phosphatic limestone near the top of the Barnett indicates shoaling of the basin and return to normal-marine conditions.

Marble Falls Limestone

Locality 6 (Fig. 9). Roadcut and exposures marginal to the highway, extending from 1.8 to 2.8 mi (3 to 4.6 km) south of the community of Chappel on Texas FM-501 and 0.1 mi (0.16 km) south of Cherokee Creek (San Saba County; lat 31°01′52″N, long 98°34′38″W, Bend 7½-minute Quadrangle). To view the section from bottom to top, proceed to the south end of the exposures at the top of the second hill, where the road begins to bend to the south-southwest.

The Marble Falls consists of a variety of cherty and noncherty limestones interbedded with shale. Except near the town of Marble Falls, the formation can be divided into two informal members separated by an unconformity. The lower part of the Marble Falls is predominantly light to dark chert limestone and thin shale beds. Principal limestone types include algal biomicrite and biosparite, oosparite, spiculitic biomicrite, pelmicrite, micrite, and mixed skeletal-fragment biomicrite and biosparite. Coral and algal biolites occur locally. There has been little diagenetic alteration of the limestone, although at the surface the shale has commonly been altered to caliche.

It would seem that deposition of the Marble Falls Formation began with the establishment of open-marine conditions over the Llano uplift. Lithologically, the Barnett–Marble Falls boundary is commonly marked by a heavily vegetated, resistant limestone ledge. The nature and age of the Barnett–Marble Falls boundary, however, has been subject to considerable discussion, and the issue is still not resolved. Because the Barnett–Marble Falls boundary approximates the Mississippian/Pennsylvanian boundary, it commonly has been assumed that the boundary should be an unconformity. But most of those who have mapped the Marble Falls in detail believed that the boundary is conformable. On the basis of conodonts, the Mississippian/Pennsylvanian boundary, marked by a hiatus, appears to occur between the Barnett and the Marble Falls in the eastern part of the Llano region, but within the Barnett to the west. The "missing" conodont zones, thought to indicate an unconformity (Liner and others, 1979), are probably absent because of the restricted environment in which the Barnett was deposited (Merrill, 1980b).

The distribution of the different limestone types in the lower Marble Falls is complex and there is a great deal of local variation. In general, however, it appears that the facies were arranged in a semicircular pattern about the structurally high Llano uplift. The uplift formed the core of a carbonate platform where vertical accretion dominated over lateral spread of the individual facies, and facies deposited in higher energy environments expanded at the expense of facies accumulating in lower energy environments. Depositional relief was as much as 30 ft (9 m), and was a significant factor in controlling facies characteristics. With time, the platform built upward to sea level, and facies patterns near the end of the deposition of the lower part of the Marble Falls Formation indicate net regression. The lower part of the Marble Falls Formation is about 100 ft thick (30 m) but ranges from approximately 66 to 150 ft (21 to 45 m).

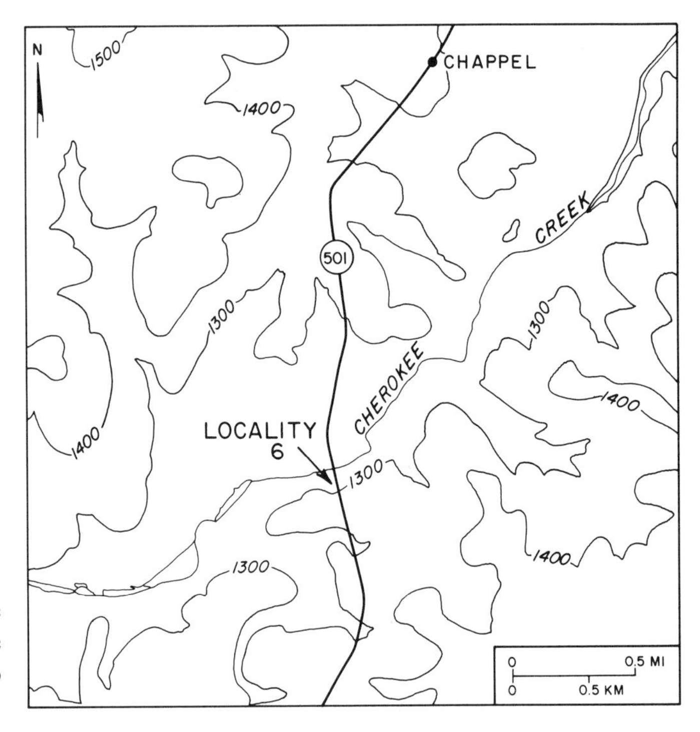

Figure 9. Location of the Cherokee Creek section of the Marble Falls Limestone (Locality 6), San Saba County, Texas. Modified from Merrill, 1980a.

The upper part of the Marble Falls consists predominantly of light to dark algal biomicrite, siliceous spiculitic biomicrite, and shale. In contrast to the lower part of the Marble Falls, facies patterns are oriented in a north-south direction, individual facies are thin and widespread, and depositional relief was very low. Although higher energy facies become more common upward and lateral facies shifts are common, the patterns record net migration to the west. Paleontological data support the conclusion that the upper part of the Marble Falls was time transgressive to the west and that the hiatus between the lower and upper parts of the Marble Falls was of increased duration to the west (Fig. 10). Locally, the unconformity is marked by channels cut

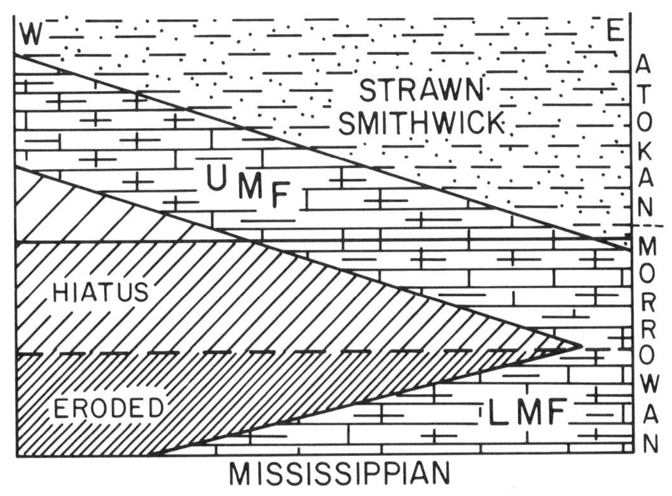

Figure 10. Age relations of the lower and upper parts of the Marble Falls Limestone (from Kier, 1980).

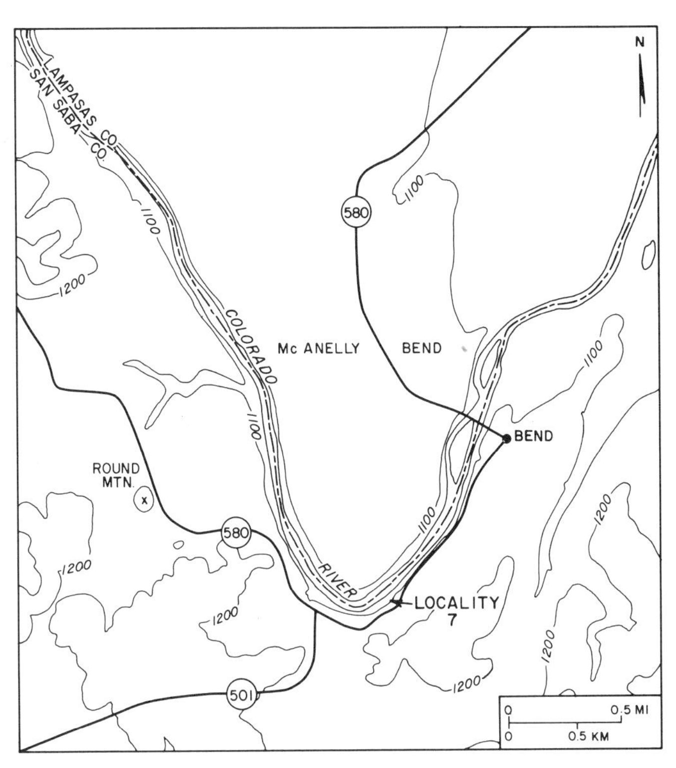

Figure 11. Location of the Bend Dump area (Locality 7), San Saba County, Texas. Modified from Merrill, 1980a.

into the lower Marble Falls. Near Mason, it appears that erosion has removed much of the lower part of the Marble Falls and, possibly, some of the Barnett Shale.

The upper part of the Marble Falls was deposited as algal buildups and calcarenite shoals; shale and spiculite were deposited in the topographically low intershoal depressions. Whereas much of the Llano uplift was being eroded after deposition of the lower Marble Falls, deposition of shale and spiculite continued on the submerged slope areas (near the town of Marble Falls). With renewed transgression, shale and spiculite moved over the erosion surface, and algal buildups and shoals were reestablished on the old platform.

Because of the wide variation in limestone types in the Marble Falls, only a few of them can be viewed at any one section. In the roadcut, however, several of the dominant facies types are exposed. At the south end of the roadcut (topographic high), oolitic calcarenite occurs. This facies makes up nearly 80% of the lower part of Marble Falls in this area. The oolite facies is analogous to the oolitic shoals of the Bahamian Platform. Walking north, toward Cherokee Creek, the dominant lithologies are the spiculitic limestone and shale of the lower part of the upper Marble Falls. The covered intervals are presumed to be shale; this part of the Marble Falls Formation is recessive and underlies the topographic low between the two hills. Foraminiferal biosparite occurs locally. The brownish, mottled coloration of the spiculitic limestone in the lower part of the upper Marble Falls Limestone distinguishes it from gray limestone common in the lower Marble Falls and higher in the upper Marble Falls. At the top of the first rise south of Cherokee Creek, algal biomicrite and a lesser amount of calcarenite are exposed. These are the high-energy facies of the upper Marble Falls. Higher in the section, but farther downhill, spiculitic biomicrite and shale are exposed.

Locality 7 (Fig. 11). The locality, also known as the Bend Dump, is between the road and the Colorado River 0.5 mi (0.8 km) east of the intersection of Farm Roads 501 and 580 (from San Saba) and approximately 0.75 mi (1.2 km) west of the community of Bend, where FM-580 crosses the Colorado River (San Saba County; lat 31°05′42″N, long 98°30′57″W, Bend 7½-minute quadrangle). The locality is probably on private property, but whose is unknown. For the most part, access presents no problem, but should there be a chain across the drive leading to the locality, proceed at your own risk. At this locality, the top of the Marble Falls Limestone and the Smithwick Shale can be examined in detail; sandstone of the Strawn Formation caps the hills to the south, west, and northwest. Conglomerate in this area is either Strawn or basal Cretaceous; outcrops must be examined closely to distinguish the two.

Top of Marble Falls Formation, Smithwick Shale and Strawn Group. For convenience, the top of the Marble Falls Limestone is placed at the highest, most prominent, and most resistant limestone. Here the Marble Falls Formation consists of dark, spiculitic limestone and interbedded black, fissile, calcareous shale. Upward transition from the Marble Falls to the Smithwick Shale is obvious. The uppermost limestone contains cephalopods (difficult to collect) and trace fossils (rooster tails and Caudi-gaili). A small, blue-green algal bioherm is at the south end of the outcrop.

The overlying Smithwick Shale is soft, black, poorly fossiliferous, and fairly homogenous. Nearby, there are thin siltstone and sandstone beds and iron-rich concretions; siltstone and sandstone beds increase in abundance upward. Fossils are rare, but do occur. Sedimentary bedforms, ripple marks, flute casts, groove casts, and slump features are common on the bedding surfaces of the sandstone beds, especially on the east side of the Llano uplift. Plant remains also can be found.

Smithwick is as thick as 100 ft (30 m) on the north side of the uplift and as much as 375 ft thick (121 m) on the east side of the uplift. In general, the Smithwick thins westward and locally is absent. For the most part, the Smithwick Shale weathers so easily that it is covered. Smithwick "flats" are common except where the overlying Strawn supports a ridge.

Broken, weathered float derived from the Strawn sandstone capping the hills can be picked up along the FM 580 roadside. The sandstone near this locality is a fine- to coarse-grained, moderately sorted, subangular to rounded quartz sand, commonly stained by hematite. The boundary between the Smithwick and the Strawn is gradational, but for convenience the base of the Strawn in this area is placed at the lowest resistant sandstone bed. No single bed is persistent; thus, the boundary appears to shift in stratigraphic position from place to place. Shale in the Strawn is similar to the Smithwick Shale and can be distinguished only by stratigraphic position.

The Smithwick and Strawn represent prodelta and delta-front facies and distributary mouth bar facies and flood-basin facies, respectively, of deltas the prograded across the Fort Worth basin from the Ouachita Mountains to the east, and impinged on the Llano uplift. With continued subsidence and deltaic progradation across the uplift, carbonate facies of the upper Marble Falls Formation were driven westward. Where subsidence was less, Strawn sandstones were deposited directly on the Marble Falls Formation.

REFERENCES CITED

Barnes, V. E., and Bell, W. C., 1954, Cambrian field trip–Llano area (in honor of West Texas Geological Society): San Angelo Geological Society Field Conference, 139 p.

——, 1977, The Moore Hollow Group of central Texas: University of Texas, Bureau of Economic Geology, Report of Investigations no. 88, 169 p.

Barnes, V. E., and Schoenfield, D. A., 1964, Potential low-grade iron ore and hydraulic-fracturing sand in Cambrian sandstones, northwestern Llano region, Texas: University of Texas at Austin, Bureau of Economic Geology, Report of Investigations no. 53, 58 p.

Barnes, V. E., and 7 others, 1959, Stratigraphy of the pre-Simpson Paleozoic subsurface rocks of Texas and southeast New Mexico: University of Texas at Austin Publication 5924, 836 p.

Barnes, V. E., and 6 others, 1972, Geology of the Llano region and Austin area, field excursion: The University of Texas at Austin, Bureau of Economic Geology, Guidebook no. 13, 77 p.

Barnes, V. E., and others, 1976, Brownwood sheet, geologic atlas of Texas: University of Texas at Austin, Bureau of Economic Geology, scale 1:250,000.

Barnes, V. E., and others, 1981, Llano sheet, geologic atlas of Texas: University of Texas at Austin, Bureau of Economic Geology, scale 1:250,000.

Cloud, P. E., Jr., and Barnes, V. E., 1948, The Ellenburger Group in central Texas: University of Texas at Austin Publication 4621, 473 p.

Ellison, S. P., 1980, Type locality of the Chappel Limestone (Mississippian) 2.4 miles southeast of San Saba, Texas: West Texas Geological Society Guidebook, Publication 80-73, p. 187–191.

Hass, W. H., 1953, Conodonts of the Barnett Formation of Texas: U.S. Geological Survey Professional Paper 243-F, p. 69–94.

——, 1959, Conodonts of the Chappel Limestone of Texas: U.S. Geological Survey Professional Paper 294-J, p. 365–399.

Kier, R. S., 1980, Depositional history of the Marble Falls Formation of the Llano region, central Texas, *in* Windle, D., ed., Geology of the Llano region, central Texas: West Texas Geological Society Guidebook, Publication 80-73, p. 59–75.

Liner, R. T., Manger, W. T., and Zachry, D. L., 1979, Conodont evidence for the Mississippian–Pennsylvanian boundary, northeastern Llano region, central Texas: Texas Journal of Science, v. 31, p. 309–317.

Merrill, G. K., 1980a, Roadlog, *in* Windle, D., ed., Geology of the Llano region, central Texas: West Texas Geological Society Guidebook, Publication 80-73, p. 116–203.

——, 1980b, Preliminary report on the restudy of conodonts from the Barnett Formation, *in* Windle, D., ed., Geology of the Llano region, central Texas: West Texas Geological Society Guidebook, Publication 80-73, p. 103–107.

ACKNOWLEDGMENT

In preparing the field guide to the Paleozoic strata of central Texas, I borrowed freely from the numerous other guidebooks that have been prepared for the area. These include an unpublished guidebook prepared by the late W. C. Bell for Humble Oil and Refining Company (Exxon); the Cambrian field trip—Llano area, prepared for the San Angelo Geological Society, for which Virgil E. Barnes and W. C. Bell (1954) were the leaders; Guidebook 13 of the University of Texas Bureau of Economic Geology, prepared by Barnes and others (1972); and especially the guidebook to the 1980 annual field trip of the West Texas Geological Society, assembled largely by Glenn K. Merrill (1980) and for which field trip I was one of several leaders. In addition, I used a report by Barnes and Schoenfield (1964) on potential low-grade iron ore and hydraulic fracturing ("frac") sand in Cambrian sandstones of the northwestern Llano region, and a report by Barnes and Bell (1977) on the Moore Hollow Group. Other excellent reports, upon which I did not directly rely, but which contain much information on the Llano region and the surrounding areas, are Cloud and Barnes (1948) and Barnes and others (1959). Because of space constraints, however, I have not attempted to provide a citation for every piece of information presented, nor to provide a complete bibliography for the area, which would be extensive. Rather, the pertinent references are largely contained in the citations mentioned above.

The Precambrian of Central Texas

Virgil E. Barnes, Bureau of Economic Geology, University of Texas at Austin, University Station, Box X, Austin, Texas 78713-7508

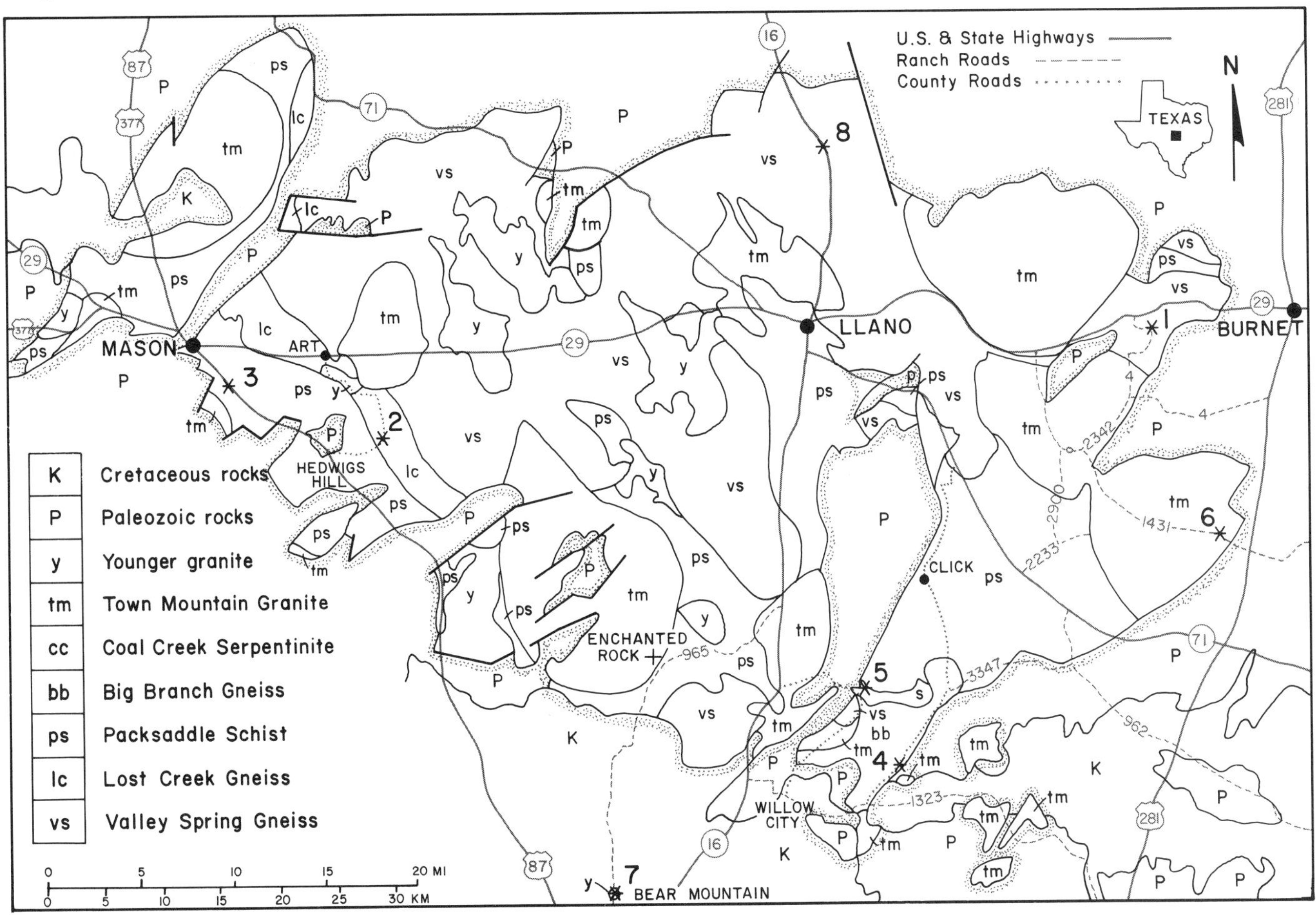

Figure 1. Map showing the Precambrian rocks of Llano region (Barnes, 1981) and the localities discussed in this chapter. The edge of the Phanerozoic cover is stippled.

LOCATION

"Precambrian rocks reach the surface in the Llano region of Central Texas in the highest part of a broad domal arch, the Llano uplift, which appears on a regional geologic map as an island of igneous and metamorphic rocks surrounded by Paleozoic and Cretaceous sedimentary rocks. The widest expanse of Precambrian rocks is about 65 mi (105 km), extending westward from the valley of the Colorado River through a subdued topographic basin drained by the Llano River. The broad, gentle basin carved into the Precambrian rocks is bordered by a discontinuous rim of flat-topped limestone hills which are the dissected edge of the Edwards Plateau. Within the basin and at its margins are erosional remnants and down-faulted blocks of Paleozoic rocks which form prominent hills, locally referred to as mountains." Clabaugh and McGehee (1972, p. 9).

The main Precambrian units present in the Llano region are listed in Table 1, and the distribution of some of these units is shown in Figure 1. This figure is derived from the Llano Sheet of the Geologic Atlas of Texas, scale 1:250,000 (Barnes, 1981). On that map (as well as Fig. 1), the Oatman Creek and Sixmile Granites (Table 1) are lumped as younger granitic intrusive rocks; locally, in the southeast part of the Llano Sheet, the Packsaddle Schist is subdivided into four formations (McGehee, 1979). Detailed 7½-minute quadrangle maps include most of the area in which the Packsaddle Schist has been subdivided (Barnes and McGehee, 1976; 1977a, b, c).

INTRODUCTION

The first significant study of the Precambrian rocks of the area was by Walcott (1884), who gave the name Llano Group to the metamorphic rocks. Mapping by U.S. Geological Survey geologists, headed by Sidney Paige (1911, 1912), led to the recognition of two major units in the Llano Group, for which he redefined names used by Comstock (1890). The lower unit is the

TABLE 1. PRECAMBRIAN STRATIGRAPHY OF THE LLANO REGION

Classification	Reference
Igneous rocks	
Llanite (quartz porphyry dikes)	Iddings, 1904
Sixmile Granite	Stenzel, 1932
Oatman Creek Granite	Stenzel, 1932
Town Mountain Granite	Stenzel, 1932
Metaigneous rocks	
Metagabbro and metadiorite (not named)	
Coal Creek Serpentinite	Barnes, 1940
Red Mountain Gneiss	Romberg and Barnes, 1949; Barnes and others, 1950
Big Branch Gneiss	Barnes, 1940
Metasedimentary rocks	
Llano Group	Walcot, 1884; Llano Supergroup, McGhee, 1979
Packsaddle Schist	Comstock, 1890; redefined by Paige, 1911
Click Formation	McGehee, 1979
Rough Ridge Formation	McGehee, 1979
Sandy Formation	McGehee, 1979
Honey Formation	McGehee, 1979
Lost Creek Gneiss	Ragland, 1960; redefined by Mutis-Duplat, 1982
Valley Spring Gneiss	Comstock, 1890; redefined by Paige, 1911

Valley Spring Gneiss, which is chiefly microcline-quartz gneiss with subordinate biotite and hornblende. The upper unit is the Packsaddle Schist, which includes graphite schist, leptite, marble, and amphibole schist.

West of the area mapped by Paige and others, a third unit has been identified between the Valley Spring Gneiss and the Packsaddle Schist. This unit is a distinctive augen gneiss, first noted by Barnes and others (1947, p. 121). Ragland (1960) mapped the northern occurrence of this gneiss, determined that it is metasedimentary, and named it Lost Creek Gneiss (see also Barnes and Schofield, 1964, p. 3–4). Garrison and others (1979) consider the Lost Creek Gneiss to be metarhyolitic.

Various igneous rocks intruded the Llano Group and became metamorphosed with it. Numerous bodies of serpentinite, soapstone, and other metamorphosed mafic igneous rocks, mostly south of the quadrangles mapped by Paige and others, have been mapped (Barnes, 1946, 1952a,b; Barnes and others, 1947; Clabaugh and Barnes, 1959). These rocks are discussed by Clabaugh and McGehee (1972).

Also widespread in the southern part of the region is a metamorphosed quartz diorite to which Barnes (1946) gave the name Big Branch Gneiss. Metamorphosed granite in the southeastern part of the region was recorded by Paige, and to this Romberg and Barnes (1949) assigned the name Red Mountain Gneiss. Clabaugh and Boyer (1961) found the granite gneiss to be slightly younger than the quartz diorite gneiss. The largest mafic body, Coal Creek Serpentinite; is 3.5 mi (5.6 km) long; contains amphibolite and talc schist inclusions from the Packsaddle; is cut by aplites, pegmatites, and quartz bodies derived from Town Mountain Granite; and appears to cross-cut Big Branch Gneiss (Barnes and others, 1950; Barnes, 1952a,b; Garrison and Mohr, 1984).

Large bodies of granite were intruded into the framework of folded metasedimentary rocks at the final stages of regional metamorphism. Paige distinguished two types, coarse-grained pink granite and fine-grained pink to gray granite, as well as dikes of granite porphyry and felsite. Stenzel (1932, 1935) proposed a more elaborate classification of the granites on the basis of color, grain size, and field relations. He named the oldest coarse-grained to porphyritic type Town Mountain Granite, the intermediate-age medium-grained gray to pink type Oatman Creek Granite, and the youngest fine-grained gray type Sixmile Granite.

The youngest granitic rock is a dike-like series of small irregular bodies of a distinctive granite porphyry to which Iddings (1904) gave the name llanite.

Age determinations on minerals from the Llano region were summarized by Flawn (1956, Table 4) and by Flawn and Muehlberger (1970, Table 3). Zartman (1962, 1964, 1965; Zartman and Wasserburg, 1962) made extensive determinations on whole rock samples and mineral separates from the granites and the Valley Spring Gneiss using rubidium-strontium (Rb-Sr) and potassium-argon (K-Ar) methods. Garrison and others (1979) also made Rb-Sr and K-Ar geochronologic and isotopic studies of Llano region rocks. These results indicate that the major granite bodies are 1,030 ± 30 million years old and the Valley Spring Gneiss is 1,120 ± 25 million years old. Garrison (1985) considered these Grenvillian-age rocks and their intrusions to be an island arc complex.

FIELD GUIDE SITES

Valley Spring Gneiss

Locality 1 (Fig. 2) (98°21′15″W.; 30°44′59″N., Longhorn Cavern 7½-minute Quadrangle) is along Park Road 4 about 330 ft (100 m) northwest of Spring Creek crossing in Inks Lake State Park.

The Valley Spring Gneiss is well exposed on hillsides adjacent to the eastern shore of Inks Lake. In the eastern part of the Llano region the Valley Spring Gneiss is predominantly microcline-quartz gneiss, with subordinate amounts of biotite, magnetite, epidote, and hornblende. Some layers are richer in biotite and therefore darker; a few lenses and layers are black amphibolite, but these were probably diabase dikes and other mafic igneous bodies intruded into the original rock before or during metamorphism. Farther west the Valley Spring Gneiss includes a greater variety of rocks, including one or more layers of marble (locally converted to wollastonite-garnet rock) and more numerous layers of dark schists and amphibolite. Its average composition closely approximates that of granite, whereas its distribution, layering, and structural features suggest that it was deposited as a sediment. Perhaps the simplest hypothesis of its origin is that the original rocks were ignimbrites and related rhyolitic volcanic rocks in which there were local mafic igneous rocks, calcareous tuffs, nonmarine limestone, and sediments de-

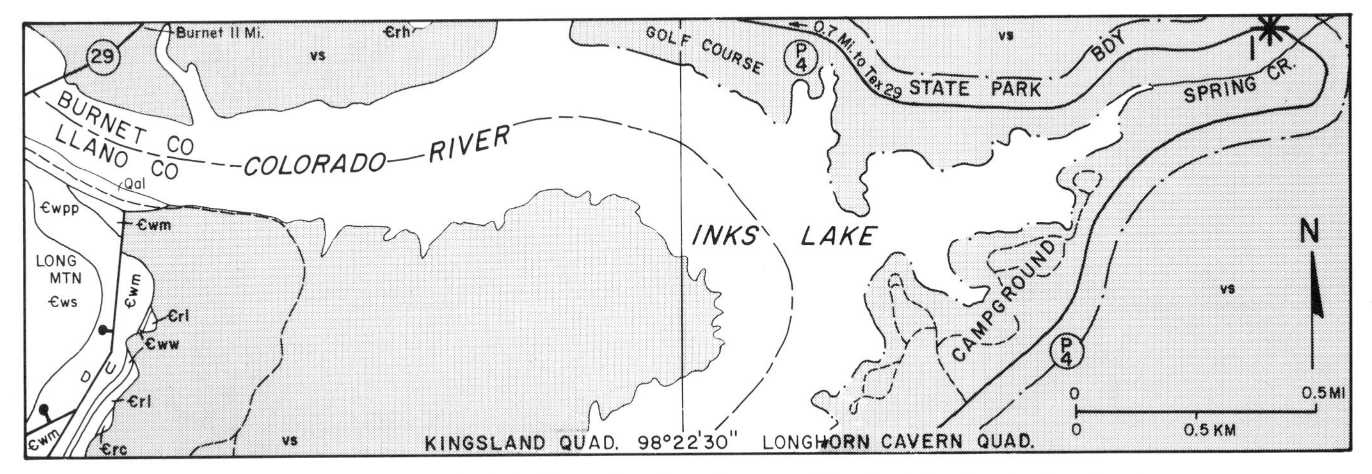

Figure 2. Geologic map of Inks Lake area, Longhorn Cavern and Kingsland Quadrangles, Burnet and Llano Counties, Texas (Barnes and McGehee, 1976). Key: Alluvium, Qal; Wilberns Formation—San Saba Member, €ws; Point Peak Member, €wpp; Morgan Creek Limestone Member, €wm; Welge Sandstone Member, €ww; Riley Formation—Lion Mountain Sandstone Member, €rl; Cap Mountain Limestone Member, €rc; Hickory Sandstone Member, €rh; Valley Spring Gneiss, vs.

rived from the volcanic rocks. Mutis-Duplat (1982) and Droddy (1978) estimated a thickness of 12,500 ft (3.8 km) for the Valley Spring Gneiss in the western part of the Llano region.

During metamorphism some of the gneiss appears to have undergone partial melting or to have been invaded intricately by granite magma to produce complex migmatites. Following the peak of metamorphism innumerable small pegmatites, aplites, and quartz veins, probably derived from the large granite plutons, were emplaced in the gneiss as well as in other Llano Supergroup rocks.

Of the Precambrian rocks, the Valley Spring Gneiss produces the roughest topography, especially in the eastern part of the Llano region. When the Cambrian sea encroached on the Llano region, the topography was equally rough with peaks standing more than 700 ft (200 m) above the surrounding lowlands. The Morgan Creek Limestone Member of the Wilberns Formation (Cambrian) rests in places directly on the Valley Spring Gneiss (Fig. 2). Within the Kingsland quadrangle the average thickness of the Cambrian rocks beneath the Morgan Creek is 715 ft (218 m).

Lost Creek Gneiss

Locality 2 (Figs. 3, 4) (99°3′38″W.; 30°40′47″N., Art 7½-minute Quadrangle) is 6.55 mi (10.7 km) by graded road southeast of Art at the Willow Creek crossing. This locality may also be reached from U.S. 87 at Hedwigs Hill, 0.4 mi (0.6 km) north of Llano River, by following a graded road eastward 3.9 mi (6.4 km).

The Lost Creek Gneiss in this exposure is overturned, dipping about 60° eastward. Another good roadside outcrop standing 10 ft (3 m) high is 0.4 mi (0.6 km) north of Willow Creek. Also, an almost continuous exposure is present between the road and Llano River at the windmill 0.4 mi (0.6 km) to the south. Hummocky Lost Creek Gneiss dots the broad expanse of Llano River downstream for about 0.6 mi (1 km). Outcrop continues upstream from the windmill 900 ft (300 m), at which point Lost Creek Gneiss and Packsaddle Schist are interlayered. Beyond this point outcrop is sporadic and the last one of Lost Creek Gneiss is about 600 ft (200 m) farther upstream.

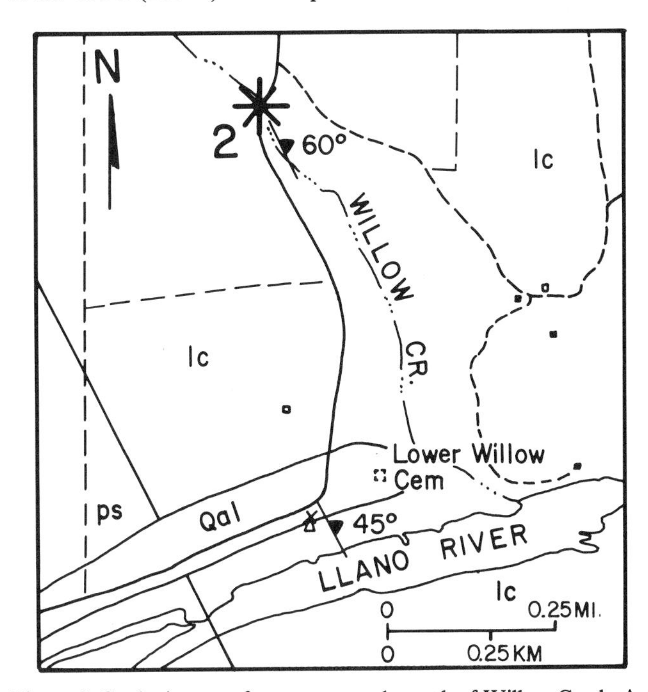

Figure 3. Geologic map of an area around mouth of Willow Creek, Art Quadrangle, Mason County, Texas (Barnes, 1981). Key: Alluvium, Qal; Packsaddle Schist, ps; Lost Creek Gneiss, lc.

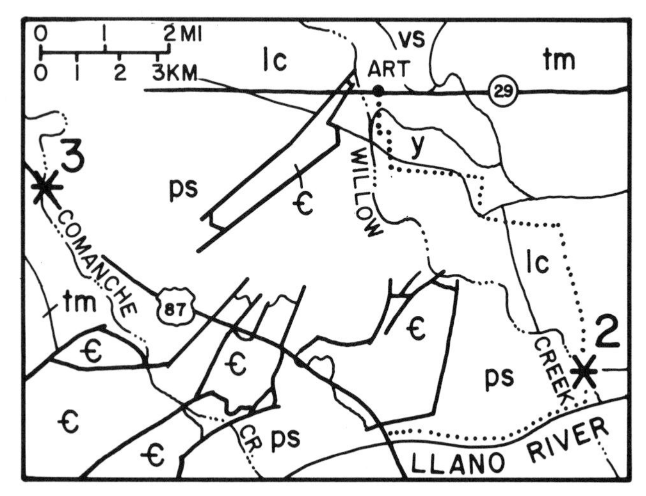

Figure 4. Geologic map of an area in central Mason County, Mason Quadrangle, Texas (Barnes, 1981). Key: Cambrian rocks, €; younger granite, y; Town Mountain Granite, tm; Packsaddle Schist, ps; Lost Creek Gneiss, lc; Valley Spring Gneiss, vs.

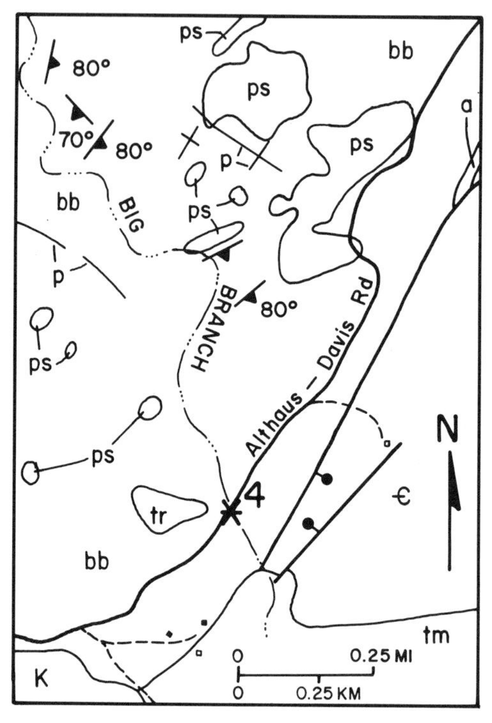

Figure 5. Geologic map of an area along Big Branch, Blowout Quadrangle, Gillespie County, Texas (Barnes, 1952b). Key: Cretaceous rocks, K; Cambrian rocks, €; aplite, a; pegmatite, p; Town Mountain Granite, tm; Big Branch Gneiss, bb; tremolite, tr; Packsaddle Schist, ps.

Mutis-Duplat (1982) described the Lost Creek Gneiss in the area he mapped as "medium- to coarse-grained, nonfoliated to very well foliated quartz-feldspar-hornblende-biotite gneiss grading into augen gneiss and migmatite." However, for most of its 2 mi (3 km) outcrop width along Llano River, the Lost Creek Gneiss is an augen gneiss with surprisingly few intrusions except near its contact with Packsaddle Schist. The original sample collected by Barnes and others (1947) is composed of quartz, oligoclase, biotite (in part altered to chlorite), a small amount of hornblende, and a little sphene deeply clouded by leucoxene; accessory minerals are magnetite, apatite, zircon, and pyrite.

Packsaddle Schist

Locality 3 (Fig. 4) (99°11′55″W.; 30°43′11″N., Mason 7½-minute Quadrangle) is 2.75 mi (4.5 km) southeast of Mason along U.S. 87 just short of Comanche Creek. Packsaddle Schist is exposed in a cut 10 ft (3 m) deep that extends along both sides of the highway for 0.2 mi (0.3 km). Hornblende schist and some biotite schist form most of the southeastern half of the exposure and leptite most of the rest with some interlayering of the two. Aplite and pegmatite intrusions are numerous and both the hornblende schist and leptite display intricate folding in places. The trace of foliation as seen on aerial photographs and a few foliation attitudes measured in the area indicate that locality 3 is on the northeastern limb of a southeastward-plunging overturned anticline.

McGehee (1979) subdivided the Packsaddle Schist in the southeastern part of the Llano region into four formations. The lower one, the Honey Formation, consists of graphite schist, marble, calc-silicate rock, leptite, quartz-feldspar gneiss, biotite schist, hornblende schist, and muscovite schist. Next is the Sandy Formation, an alternation of quartz-feldspar rock and hornblende schist. The third one, Rough Ridge Formation, is mostly gray quartz-feldspar rock; some muscovite schist occurs in the lower part, and some biotite gneiss with large cordierite porphyroblasts is in the upper part. The top formation, the Click Formation, has a basal member of actinolite schist, and the rest of the formation is light brown to pink leptite and quartz-feldspar schist.

Interlayered marble and hornblende schist is well exposed in Llano County along the Click road 0.8 mi (1.3 km) southwest of Texas 71 (Barnes and McGehee, 1977c).

Big Branch Gneiss

Locality 4 (Fig. 5) (98°35′49″W.; 30°25′4″N., Blowout 7½-minute Quadrangle) is in eastern Gillespie County near the Blanco County line where the paved Althaus-Davis county road crosses Big Branch, from which the Big Branch Gneiss received its name. This locality is 2.5 mi (4.1 km) northeast of RR 1323 at a point 5.5 mi (8.8 km) east of Willow City. As this county-maintained unfenced road passes through private property, visitors should confine their attention to the immediate sides of the road, to avoid trespassing.

The Big Branch Gneiss is quartz diorite. It is composed of plagioclase, quartz, hornblende, biotite, magnetite, sphene, apatite, and zircon. Some samples also contain microcline, as well as the secondary minerals chlorite, epidote, and albite.

The Big Branch Gneiss is well exposed adjacent to the west side of the road where inclusions of Packsaddle Schist can be seen. One reddish aplitic intrusion, in part linear with mostly internal cross-cutting quartz veins, contains dark inclusions and in places has filled in around breccia blocks of Big Branch Gneiss. A craggy exposure of Big Branch Gneiss along the road 0.12 to 0.18 mi (0.2 to 0.3 km) to the south is well foliated with some chevron-type deformation. The Big Branch Gneiss as a whole in this area is well foliated, and lineation pitches steeply in a direction about S10°E. Swarms of inclusions in some areas, mostly of Packsaddle Schist, are arranged parallel to the foliation. The Big Branch Gneiss has been feldspathized to augen gneiss locally near its contact with the Red Mountain Gneiss and is therefore older than the Red Mountain Gneiss (Clabaugh and Boyer, 1961).

Coal Creek Serpentinite

Locality 5 (Fig. 6) (98°37′50″W.; 30°28′59″N., Willow City and Blowout 7½-minute Quadrangles) is along a secondary road that crosses the Coal Creek Serpentinite mass. This locality is reached either by going north from Willow City 7.9 mi (13.0 km) on a paved county road or going east on the same road 4.9 mi (8.0 km) from Texas 16 at a point 18.8 mi (30.8 km) south of Llano, and then northeastward 0.4 mi (0.6 km) by a graded road. At this locality, likewise, the county-maintained road passes through private property and visitors should confine their attention to the immediate sides of the road. Permission must be obtained to visit the active quarry 0.3 mi (0.5 km) south of locality 5.

The serpentinite forms a very sparsely vegetated ridge about 3.5 mi (5.6 km) long in an east-west direction and about 0.65 mi (1 km) wide, with lobes extending northward at each end. The serpentinite mass pitches southward at an angle of approximately 45°. The original rock from which the serpentinite formed intruded Packsaddle Schist and Big Branch Gneiss, as evidenced by inclusions of both rock types in the serpentinite. The serpentinite consists dominantly of lizardite with minor magnetite, chromite, anthophyllite, and tremolite (Garrison and Mohr, 1984; Garrison, 1981).

The potential use for the serpentinite was discussed by Barnes and others (1950). Romberg and Barnes (1949), using gravity data, estimated 10^{10} metric tons of serpentinite to be present, of which about one billion tons is a readily available reserve.

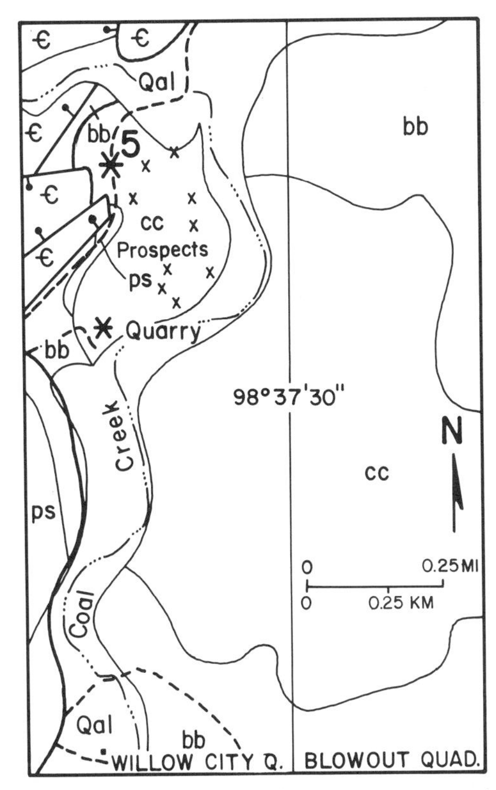

Figure 6. Geologic map of Coal Creek area near eastern end of Cutoff Gap, Willow City and Blowout Quadrangles, Gillespie County, Texas (Barnes, 1952a,b). Key: Alluvium, Qal; Cambrian rocks, €; Coal Creek Serpentinite, cc; Big Branch Gneiss, bb; Packsaddle Schist, ps.

POSTMETAMORPHIC IGNEOUS ROCKS

Town Mountain Granite

Locality 6 (Fig. 7) (98°17′55″W.; 30°35′32″N., Marble Falls 7½-minute Quadrangle) is along RR 1431 1.8 mi (2.9 km) west of U.S. 281 in Marble Falls. This locality is adjacent to a roadside park within sight of the quarry. A historical marker, situated in the park, gives the history of the Granite Mountain

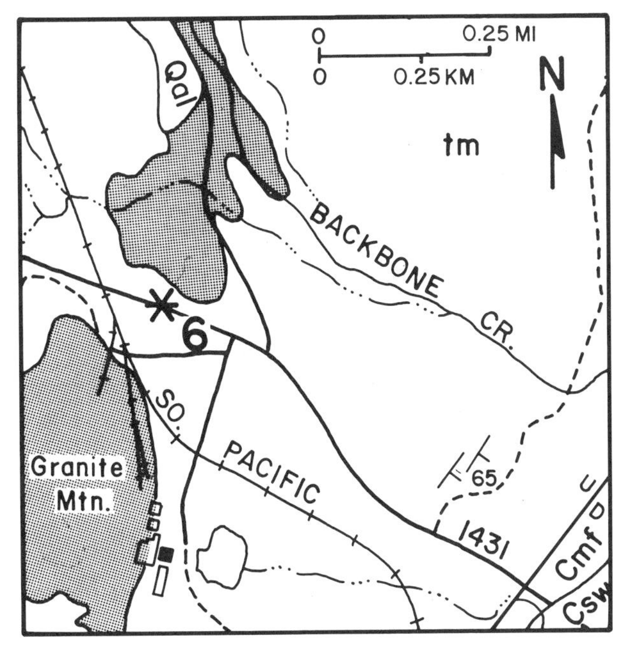

Figure 7. Geologic map of Granite Mountain area, Marble Falls Quadrangle, Burnet County, Texas (Barnes, 1982). Key: Alluvium, Qal; Smithwick Formation, Csw; Marble Falls Limestone, Cmf; Town Marble Granite (bare areas shown by stipple), tm.

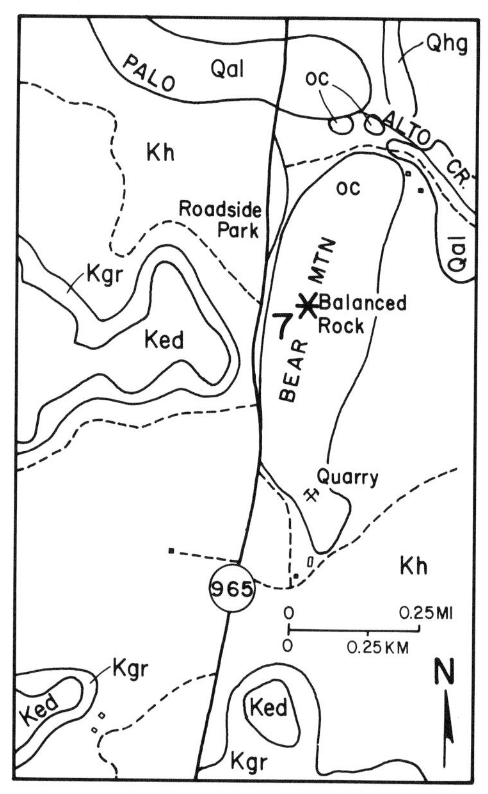

Figure 8. Geologic map of Bear Mountain area, Fredericksburg East Quadrangle, Gillespie County, Texas (Barnes, 1952c). Key: Alluvium, Qal; high gravel, Qhg; Edwards Limestone, Ked; Glen Rose Limestone, Kgr; Hensell Sand, Kh; Oatman Creek Granite, oc.

quarrying operations. A stile at the back of the park gives access to an area of bare granite from which a portion of the surface layer has been quarried, and where pegmatite veins, scarce xenoliths, and schlieren can be seen. The granite is coarsely crystalline and composed chiefly of microcline, plagioclase, and quartz, with minor amounts of biotite, hornblende, rutile, zircon, and allanite. A few of the large microcline crystals are complexly zoned and mantled by plagioclase.

This granite and the quarrying operation were described by Barnes and others (1947), and an update on quarrying methods and uses for the granite was published by Clabaugh and McGehee (1972).

Approximately 2 mi (3 km) north along the road east of locality 7, in Slaughter Gap (Barnes, 1982), Cambrian Hickory Sandstone east of the road has broken off in large blocks and is riding downslope on the decomposing Town Mountain Granite. About 0.75 mi (1.2 km) along RR 1431 to the east of locality 7 one of the major faults of the Llano region is excellently exposed, with Town Mountain Granite sharply in contact with upturned Pennsylvanian Marble Falls Limestone. The throw of the fault at this point is more than 3,300 ft (1,000 m).

For anyone interested in examining the surface of a large exfoliation dome of Town Mountain Granite, the Enchanted Rock State Natural Area in southeastern Llano County is recommended (McClay and Barnes, 1976). This area is accessible from RR 965 either by traveling 18 mi (30 km) north from Fredericksburg or west 8 mi (13 km) from Texas 16 at a point 14.7 mi (24.1 km) south of Llano County courthouse. A detailed map of the Enchanted Rock pluton (Hutchinson, 1956), of which Enchanted Rock is only a very small part, shows textural zones within the pluton and details of the metamorphic rocks into which the pluton was intruded.

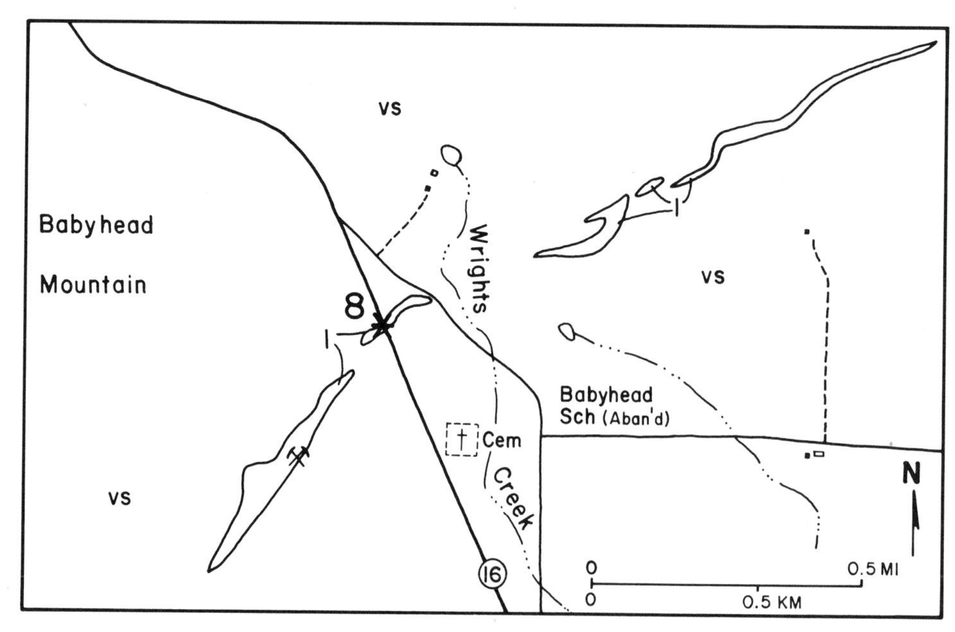

Figure 9. Geologic map of Babyhead area, Cherokee Quadrangle, Llano County, Texas (Barnes, unpublished data). Key: Llanite, l; Valley Spring Gneiss, vs.

Oatman Creek Granite

Locality 7 (Fig. 8) (98°51′24″W.; 30°19′40″N., Balanced Rock, Fredericksburg East 7½-minute Quadrangle) is 3.9 mi (6.4 km) north of U.S. 290 along the east side of RR 965. A picnic area is just north of the gap in the fence from which a marked trail leads to Balanced Rock on Bear Mountain.

The Oatman Creek granite is composed of plagioclase, microcline, quartz, and biotite, with a small amount of magnetite, fluorite, and apatite. The granite at Bear Mountain, used chiefly as a monumental and decorative stone, has been quarried for many years (Barnes and others, 1947; Barnes, 1952c). The surface of Bear Mountain is littered with large exfoliation boulders of granite. Balanced Rock, about 8 ft (2.4 m) in diameter, is now one of those boulders. It was blasted from its pedestal by vandals during the spring of 1986. Bear Mountain is a pre-Cretaceous monadnock that stands as high as the surface of the adjacent Edwards Plateau. About 180 ft (54 m) of the monadnock has been exhumed; the amount that remains buried is unknown.

An excellent overview of Enchanted Rock and the Llano basin can be obtained from the edge of the Edwards Plateau by continuing northward from Bear Mountain 10.2 mi (16.7 km).

Quartz-Feldspar Porphyry (Llanite) Dikes

Locality 8 (Fig. 9) (98°39′30″W.; 30°53′26″N., Cherokee 7½-minute Quadrangle) is in a cut at the crest of a sharp boulder-strewn ridge crossing Texas 16 east of Babyhead Mountain, Llano County (Barnes, unpublished data). The locality is 9.0 mi (14.4 km) north of the intersection of Texas 16 and 29 in Llano.

The porphyry (Iddings, 1904) is one of the youngest granitic rocks in the Llano region and forms a dike-like series of small irregular bodies of a distinctive quartz-feldspar porphyry that extends eastward from the area shown in Figure 9 to near Little Llano River, then southward to the Llano-Lone Grove road, a total of about 14 mi (22 km). The porphyry is characterized by phenocrysts of red feldspar and blue chatoyant quartz in a dark aphanitic groundmass. In the past the llanite from the quarry 0.3 mi (0.5 km) southwest of locality 8, marketed under the name opaline granite, was used as a monumental and ornamental stone. The llanite has a crushing strength of 37,800 pounds per square inch and is the strongest stone tested in the Llano region.

REFERENCES CITED

Barnes, V. E., 1940, Precambrian of Llano region with emphasis on tectonics and intrusives: Geological Society of America, Guidebook to Excursions in Connection with 53rd Annual Meeting, Austin, Texas, p. 44–55.

——, 1946, Soapstone and serpentine in the Central Mineral Region of Texas: University of Texas Bureau of Economic Geology Publication 4301, p. 55–92.

——, 1952a, Geology of the Willow City Quadrangle, Gillespie and Llano Counties, Texas: University of Texas Bureau of Economic Geology Geologic Quadrangle Map 4.

——, 1952b, Geology of the Blowout Quadrangle, Gillespie, Blanco, and Llano Counties, Texas: University of Texas Bureau of Economic Geology Geologic Quadrangle Map 5.

——, 1952c, Geology of Palo Alto Creek Quadrangle, Gillespie County, Texas: University of Texas Bureau of Economic Geology Geologic Quadrangle Map 8.

——, 1981, Geologic Atlas of Texas—Llano Sheet: University of Texas Bureau of Economic Geology.

——, 1982, Geology of the Marble Falls Quadrangle, Burnet and Llano Counties, Texas: University of Texas Bureau of Economic Geology Geologic Quadrangle Map 48.

——, unpublished data (on photograph CJC-6-138, USDA SCS): University of Texas Bureau of Economic Geology files.

Barnes, V. E., and McGehee, R. V., 1976, Geology of the Kingsland Quadrangle, Llano and Burnet Counties, Texas: University of Texas Bureau of Economic Geology Geologic Quadrangle Map 41.

——, 1977a, Geology of the Click Quadrangle, Llano and Blanco Counties, Texas: University of Texas Bureau of Economic Geology Geologic Quadrangle Map 43.

——, 1977b, Geology of the Dunman Mountain Quadrangle, Llano, Burnet, and Blanco Counties, Texas: University of Texas Bureau of Economic Geology Geologic Quadrangle Map 44.

——, 1977c, Geology of the Cap Mountain Quadrangle, Llano County, Texas: University of Texas Bureau of Economic Geology Geologic Quadrangle Map 45.

Barnes, V. E., and Schofield, D. A., 1964, Potential low-grade iron ore and hydraulic-fracturing sand in Cambrian sandstones, northwestern Llano region, Texas: University of Texas Bureau of Economic Geology Report of Investigations 53, 85 p.

Barnes, V. E., Dawson, R. F., and Parkinson, G. A., 1947, Building stones of central Texas: University of Texas Bureau of Economic Geology Publication 4246, 198 p.

Barnes, V. E., Shock, D. A., and Cunningham, W. A., 1950, Utilization of Texas serpentine: University of Texas Bureau of Economic Geology Publication 5020, 52 p.

Clabaugh, S. E., and Barnes, V. E., 1959, Vermiculite in central Texas: University of Texas Bureau of Economic Geology Report of Investigations 40, 32 p.

Clabaugh, S. E., and Boyer, R. E., 1961, Origin and structure of the Red Mountain Gneiss, Llano County, Texas: Texas Journal of Science, v. 13, p. 7–16.

Clabaugh, S. E., and McGehee, R. V., 1972, Precambrian rocks of Llano region, *in* Barnes, V. E., Bell, W. C., Clabaugh, S. E., Cloud, P. E., Jr., McGehee, R. V., Rodda, P. U., and Young, K., 1972, Geology of the Llano region and Austin area: University of Texas Bureau of Economic Geology Guidebook 13, p. 9–23.

Comstock, T. B., 1890, A preliminary report on the Central Mineral Region of Texas: Austin, Geological Survey of Texas, First Annual Report, p. 239–391.

Droddy, M. J., 1978, Metamorphic rocks of the Fly Gap Quadrangle, Mason County, Texas [Ph.D. thesis]: Austin, University of Texas.

Flawn, P. T., 1956, Basement rocks of Texas and southeast New Mexico: University of Texas Bureau of Economic Geology Publication 6505, 261 p.

Flawn, P. T., and Muehlberger, W. R., 1970, The Precambrian of the United States of America; South-Central United States, *in* The Geologic Systems, The Precambrian: Interscience, v. 4, p. 72–143.

Garrison, J. R., Jr., 1981, Coal Creek serpentinite, Llano uplift, Texas; A fragment of an incomplete Precambrian ophiolite: Geology, v. 9, p. 225–230.

——, 1985, Petrology, geochemistry, and origin of the Big Branch and Red Mountain Gneisses, southeastern Llano uplift, Texas: American Mineralogist, v. 70, p. 1151–1163.

Garrison, J. R., Jr., and Mohr, D., 1984, Geology of the Precambrian rocks of the Llano uplift, central Texas: Austin Geological Society Guidebook 5, 58 p.

Garrison, J. R., Jr., Long, L. E., and Richmann, D. L., 1979, Rb-Sr and K-Ar geochronologic and isotopic studies, Llano uplift, central Texas: Contributions to Mineralogy and Petrology, v. 69, p. 361–374.

Hutchinson, R. M., 1956, Structure and petrology of Enchanted Rock batholith, Llano and Gillespie Counties, Texas: Geological Society of America Bul-

letin, v. 67, p. 763–806.
Iddings, J. P., 1904, Quartz feldspar porphyry from Llano, Texas: Journal of Geology, v. 12, p. 225–231.
McClay, R., and Barnes, V., 1976, Field trip road log stops 1-5, October 30, *in* Economic Geology of South-Central Texas: San Antonio, South Texas Geological Society, p. 6–35.
McGehee, R. V., 1979, Precambrian rocks of the southeastern Llano region, Texas: University of Texas Bureau of Economic Geology Geological Circular 79-3, 36 p.
Mutis-Duplat, E., 1982, Geology of the Purdy Hill Quadrangle, Mason County, Texas: University of Texas Bureau of Economic Geology Geologic Quadrangle Map 52.
Paige, S., 1911, Mineral resources of the Llano-Burnet region, Texas: U.S. Geological Survey Bulletin 450, 103 p.
——, 1912, Description of the Llano and Burnet Quadrangles: U.S. Geological Survey Atlas, Llano-Burnet Folio, no. 183, 16 p.
Ragland, P. C., 1960, Geochemical and petrological studies of the Lost Creek Gneiss, Mason and McCulloch Counties, Texas [M.S. thesis]: Rice University, 99 p.
Romberg, F., and Barnes, V. E., 1949, Correlation of gravity observations with geology of the Coal Creek serpentine mass, Blanco and Gillespie Counties, Texas: Geophysics, v. 14, p. 151–161.
Stenzel, H. B., 1932, Precambrian of the Llano uplift, Texas: Geological Society of America Bulletin, v. 43, p. 143–144.
——, 1935, Precambrian structural conditions in the Llano region, *in* The Geology of Texas, v. II, Structural and Economic Geology: University of Texas Bureau of Economics Geology Bulletin 3401, p. 74–79.
Walcott, C. D., 1884, Notes on Paleozoic rocks of central Texas: American Journal of Science, 3rd ser., v. 28, p. 431–433.
Zartman, R. E., 1962, Rb^{87}-Sr^{87} and K^{40}-Ar^{40} ages of Precambrian rocks from the Llano uplift, Texas: Geological Society of America Program, 1962 Annual Meetings, p. 166A–167A.
——, 1964, A geochronologic study of the Lone Grove pluton from the Llano uplift, Texas: Journal of Petrology, v. 5, p. 359–408.
——, 1965, Rubidium-strontium age of some metamorphic rocks from the Llano uplift, Texas: Journal of Petrology, v. 6, p. 28–36.
Zartman, R. E., and Wasserburg, G. J., 1962, A geochronologic study of a granite pluton from the Llano uplift, Texas [abs.]: Journal of Geophysical Research, v. 67, p. 1664.

ACKNOWLEDGMENTS

I thank C. D. Henry and J. G. Price for their painstaking and helpful review of the manuscript for this chapter of the Guidebook.

Enchanted Rock dome, Llano and Gillespie counties, Texas

***Robert M. Hutchinson,** Department of Geology and Geological Engineering, Colorado School of Mines, Golden, Colorado 80401*

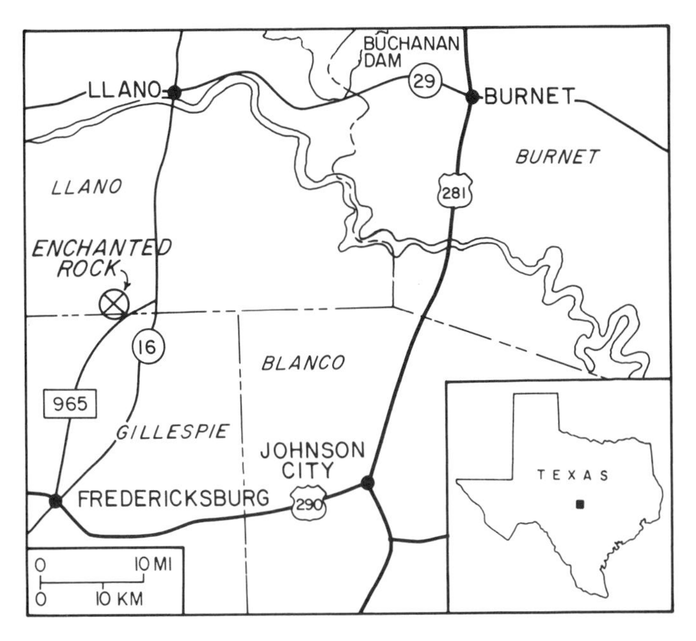

Figure 1. Map showing location of Enchanted Rock area.

Figure 2. Enchanted Rock Dome.

LOCATION

Enchanted Rock dome is centrally situated in Texas (Fig. 1) and is easily accessible via state asphalt roads. The site is maintained year long, and for a small fee the visitor can climb to the top of the dome. The site can be reached by driving 17 mi (27.4 km) north from Fredericksburg on Texas 965 or south from Llano 15 mi (24.1 km) to the intersection of Texas 965. From there it is 8 mi (12.9 km) to the site. The exfoliation dome is the largest of five domes extending in a line for 1.75 mi (2.82 km) along the west side of Sandy Creek. The dome covers an area 0.4 × 0.6 mi (0.64 × 0.96 km) and is 425 ft (129.5 m) high. Granite-tectonic features of the dome can easily be seen in half a day (Fig. 2).

SIGNIFICANCE

Enchanted Rock dome is an outstanding example of megascopic structural relations between (1) primary igneous flow structures, (2) mafic and felsic schlieren, (3) structurally significant pegmatite-aplite dikes that have occupied marginal fissures, (4) "tent-blister" type surficial geomorphological forms developed by unloading of the pluton by erosion, and (5) primary igneous fractures and joint sets related to emplacement and cooling of the Enchanted Rock batholith.

The dome is a small structural domain that is part of the 116 mi^2 (300 km^2) of Enchanted Rock batholith (Goldich, 1941; Hutchinson, 1956 and 1960) (Fig. 3). The regional study has provided information on (1) influence of tectonic setting on process of magma generation, (2) time-span relation of intrusion to regional tectonics, (3) type(s) and role of structural controls during rise and emplacement of granitic magma in a tectonically active structural domain, (4) crystallization and grain-to-grain autometamorphic changes in gross textural-mineralogical properties of the rocks, and (5) a solution to the so-called "room problem" in generation of magma at depth and creation of an opening that accommodated a rising magma column, as it adjusted itself structurally to the space and geometry available. The magma column came to rest in a structural position characteristic of the upper mesozonal–lower epizonal depth zone (Fig. 4).

SITE INFORMATION

Enchanted Rock dome is located on the extreme southeastern flank of Enchanted Rock batholith. The following structural features can be seen in a traverse from Sandy Creek up the southeast side to the top of the dome: (1) a near-vertical contact of porphyritic leucogranite with units of the Packsaddle Schist along Sandy Creek; (2) a 50-ft (15 m) wide mylonite zone along Sandy Creek; (3) vertical planar flow structure in the granite striking N40–50°E; (4) small- to medium-sized disc-shaped xenoliths of Packsaddle Schist subparallel to parallel with the planar

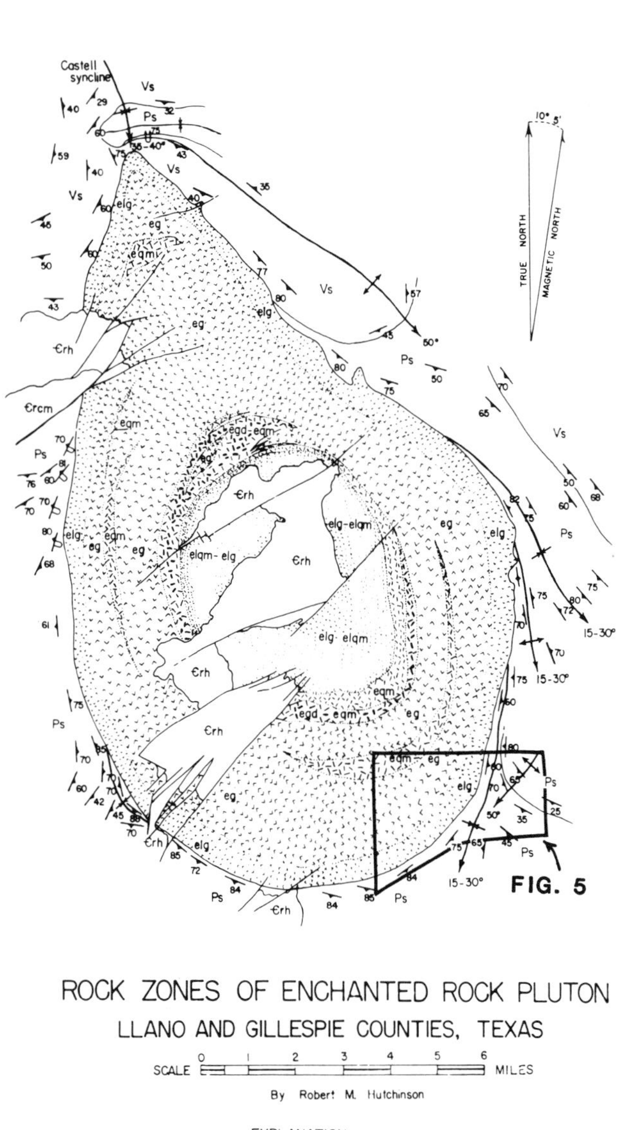

Figure 3. Rock zones of Enchanted Rock pluton and isometric block diagram of internal and wallrock structures.

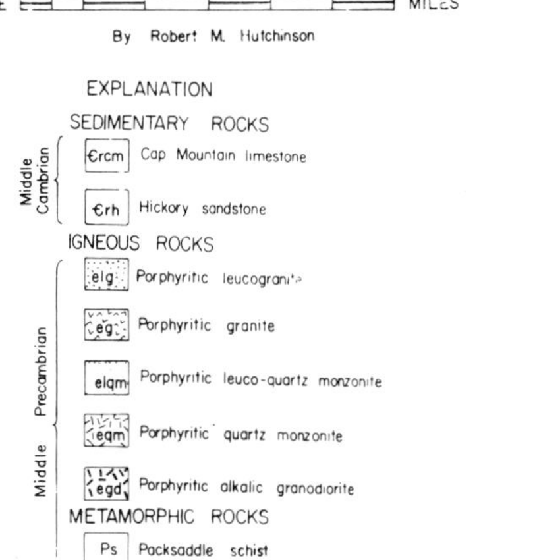

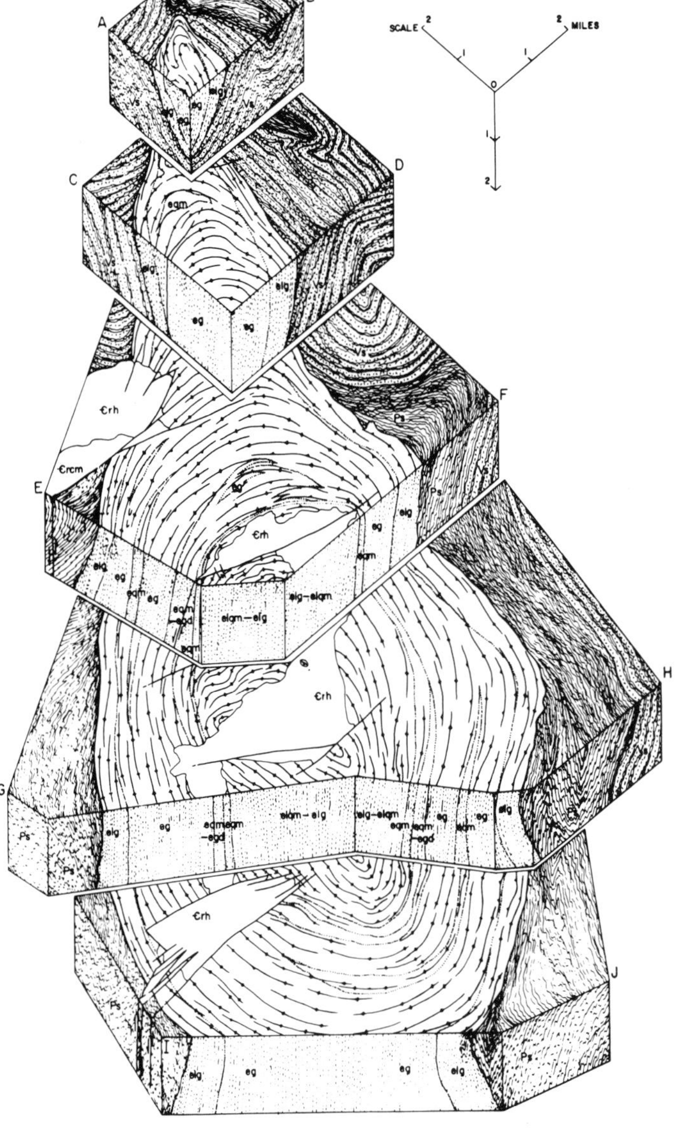

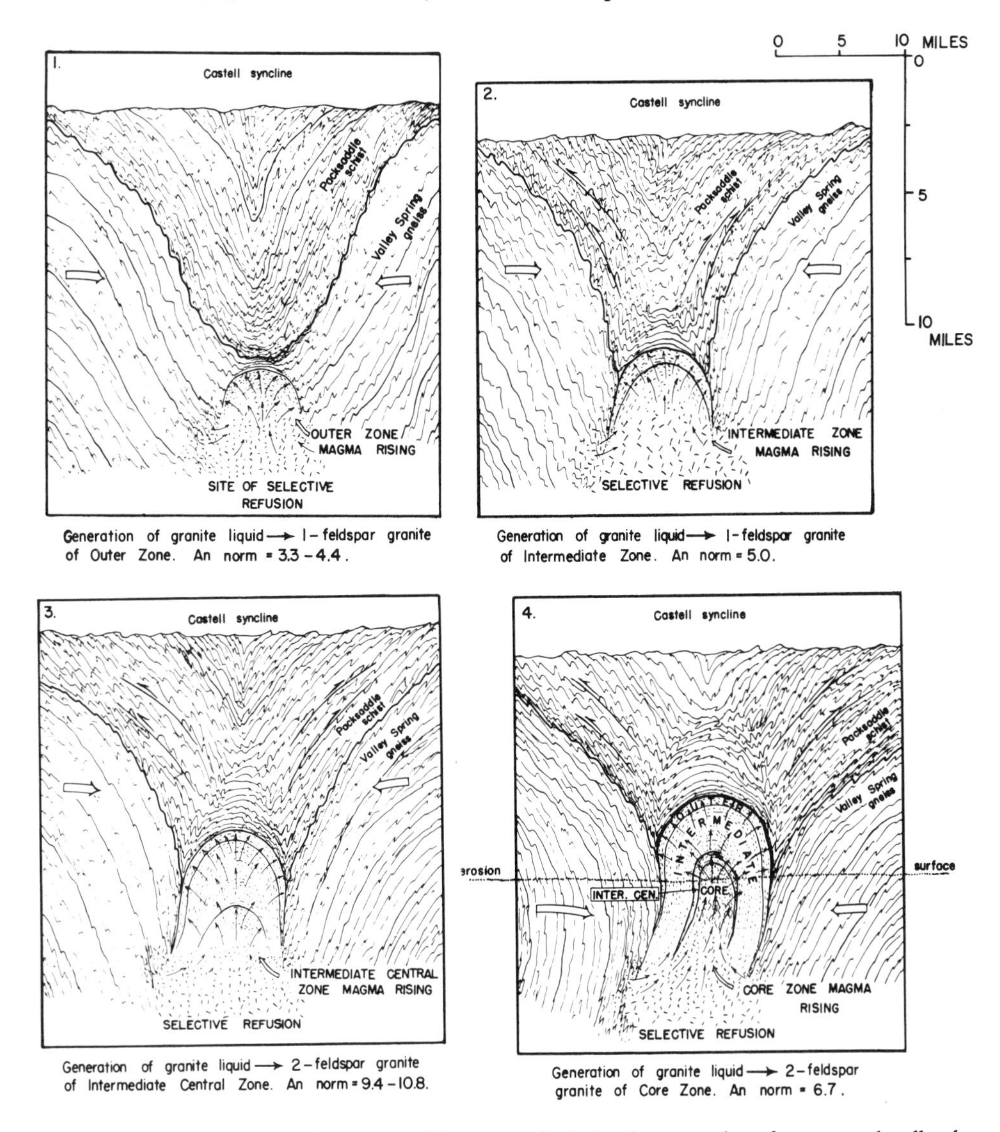

Figure 4. Schematic sketches showing possible structural relation for generation of magma and wallrock zones of Enchanted Rock batholith.

flow structure; (5) mafic biotite schlieren dipping 80° west that are parallel to planar flow structure, but which are cut by a gently westward dipping marginal fissure filled with pegmatite and aplite; (6) a granite "tent-blister" near the top on the north side of the dome; and (7) through-going primary joint sets of N45°E trending longitudinal (a-b) joints and N45–50°W trending a-c joints (Figs. 5 and 6). Narrow veins of quartz and irregular pods of quartz enrichment occur on the northeast side of the dome near Sandy Creek.

From the summit of Enchanted Rock one can view the north and west extension of the batholith for 10 to 15 mi (16 to 24 km). The dome contains many of the fabric features characteristic of the batholith. Sandy Creek is to the east entrenched along the contact of the granite and units of the Packsaddle Schist. The top of Enchanted Rock is a delightful place to each lunch and enjoy the view of the Llano country. During the rest of the day one can drive over most of the batholith via hard-bottomed gravel roads and observe the different lithologic zones of the batholith.

REFERENCES CITED

Goldich, S. S., 1941, Evolution of the central Texas granites: Journal of Geology, v. 49, p. 697–720.

Hutchinson, R. M., 1956, Structure and petrology of Enchanted Rock batholith, Llano and Gillespie counties, Texas: Geological Society of America Bulletin, v. 67, p. 763–806.

——, 1960, Petrotectonics and petrochemistry of late Precambrian batholiths of central Texas and the north end of Pikes Peak batholith, Colorado: Copenhagen, 21st International Geological Congress Proceedings, Part 14, p. 95–107.

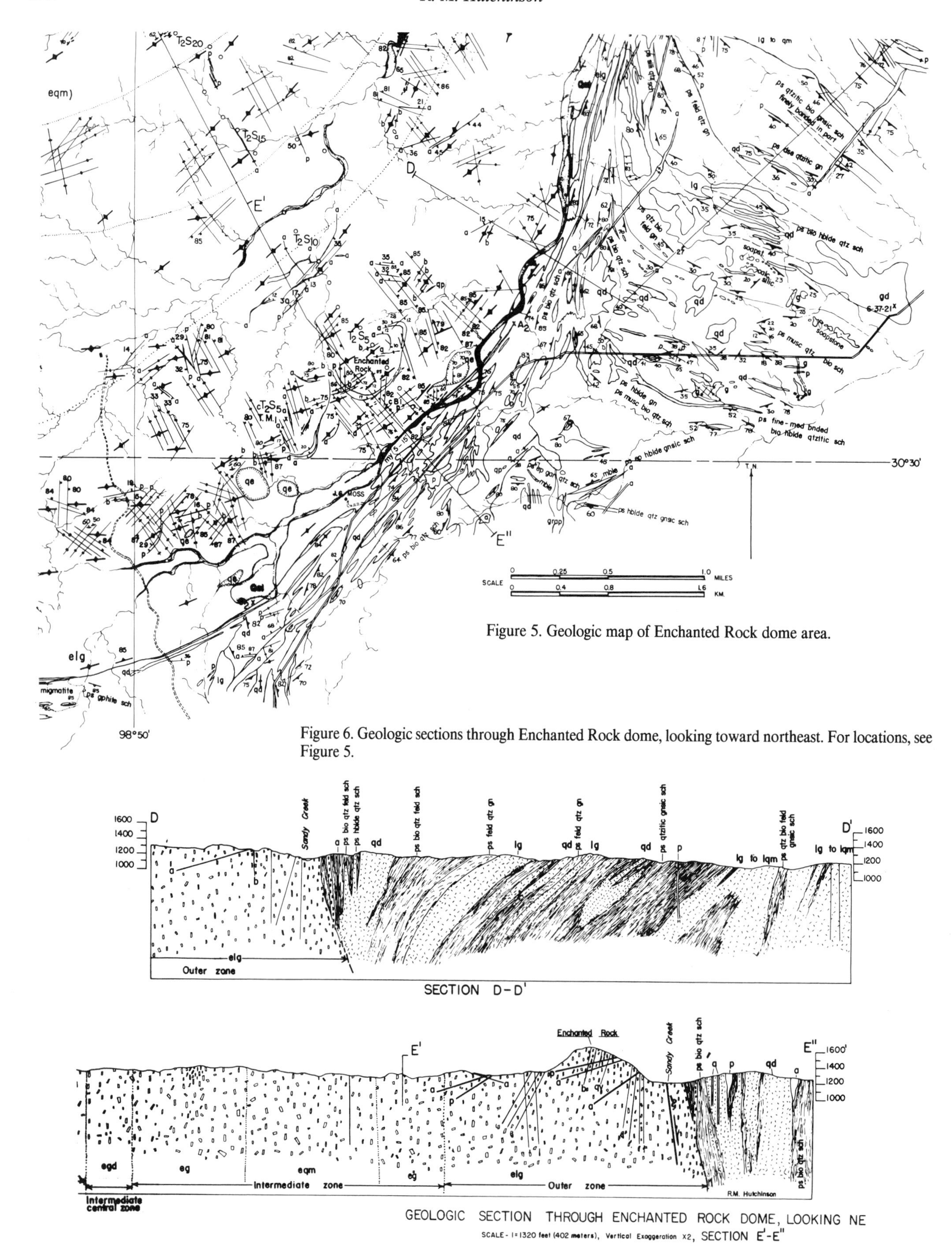

Figure 5. Geologic map of Enchanted Rock dome area.

Figure 6. Geologic sections through Enchanted Rock dome, looking toward northeast. For locations, see Figure 5.

The middle Comanchean section of Central Texas

David L. Amsbury, 1442 Brookwood Ct., Seabrook, Texas 77586

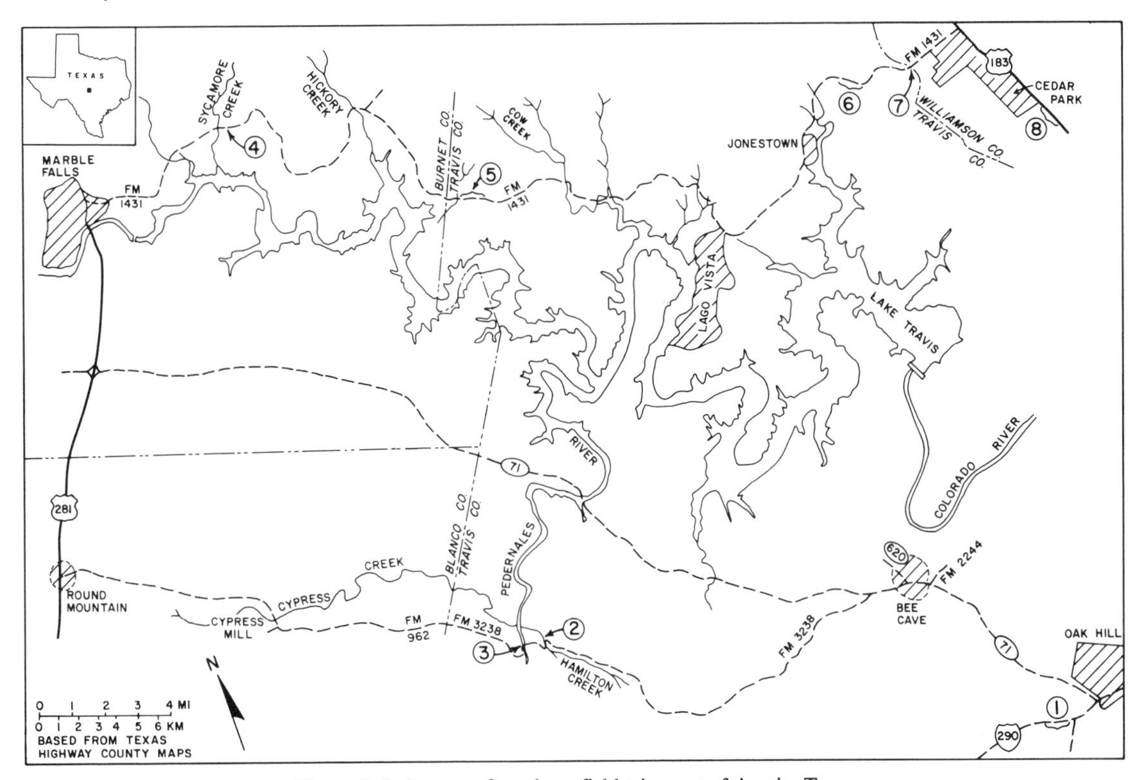

Figure 1. Index map for a loop field trip west of Austin, Texas.

LOCATION

The text describes outcrops of Comanchean strata around Lake Travis on the Colorado River between Austin and Marble Falls in central Texas (Fig. 1). The exposures selected are readily accessible and can provide a good understanding of the vertical sequence of Upper Aptian through Middle Albian rock types. Appreciation of the three-dimensional relationships will require study of the listed publications.

Many good exposures occur on private property within the region, but most are not indicated in the guide. The geologist in central Texas is well advised to *refrain from crossing any fences for any reason* without prior permission from the landowner (and land manager, if appropriate).

SIGNIFICANCE

Early strata (Trinity Division) of the Comanche Series lapped onto the craton in central Texas as Cretaceous seas expanded out of the ancestral Gulf of Mexico, and the craton in this area was covered with marine sediments during the Middle Albian (Fredericksburg Division). Cretaceous beds slowly filled pre-existing valleys as the craton was flooded; water depths were peritidal to perhaps 30 ft (10 m), as 650 to 1,000 ft (200 to 300 m) of relief was buried. The central Texas part of the craton not only was topographically higher than surrounding areas prior to inundation by Cretaceous seas, but subsequently subsided at a slower rate to form the San Marcos/Concho Arch. The sedimentologic and stratigraphic influence of the buried craton and arch is seen as facies changes in Fredericksburg rocks. Oölitic grainstone and rudist banks bordered the area on the northeast and southwest flanks and separated peritidal and evaporite rocks on the arch from marine mudstones and shale farther off the flanks.

The Lake Travis area contains the type or reference localities for the Sycamore, Hammett, Cow Creek, and Hensel formations, as well as several Trinity units not used in this guide. The Whitestone Lentil, Cedar Park Limestone, and Bee Cave Marl Member of the Walnut Formation were also named here. Outcrops in this area were seminal for modern concepts of central Texas Cretaceous sedimentation and stratigraphy.

SITE INFORMATION

Locality 1. Road cuts along the southeast side of U.S. 290 about 1.25 mi. (2 km) SW of the U.S. 290/Texas 71 intersection at Oak Hill, Travis County. Resistant upper Glen Rose carbonates (Fig. 2) occur in one of the westernmost fault blocks of the Balcones System (Barnes, 1981a).

Shallowing-upward sequences (James, 1984) are well exposed in the lower cuts. Such sequences, 3 to 10 ft (1 to 3 m) thick, are typical of the Glen Rose throughout its outcrop area and result in the stairstep topography displayed in hill slopes between localities 1 and 2. A typical sequence begins at an iron-stained hardground, overlain by intensely burrowed, clayey lime packstone containing agglutinated foraminifera, angular mollusk and echinoid fragmants, and an infauna of clams, snails, and echinoids. Grainstones formed of pellets, rounded shell fragments, miliolid foraminifera, and glauconite may occur as lenses within the packstone or in beds at the base or top. Succeeding beds typically are less-fossiliferous but burrowed dolomite that contains leached shell fragments in burrow fillings, and evenly laminated dolomite, rarely stromatolitic, capped by a hardground. About 500 ft (150 m) of these sequences accumulated to form the upper Glen Rose in the Austin area. Sequences in the lower Glen Rose below are thicker and contain rudist reef rocks.

Traverse to Locality 2. There are no post-Cretaceous faults west of the vicinity of Oak Hill within the area covered by the index map. Structural dip is less than 1° east or southeast. The Pedernales River and its tributaries flow through dissected alternating hard and soft beds of the Glen Rose Formation, then along the top of the Cow Creek Limestone (Fig. 2), and finally cut through to form cliffed, U-shaped valleys in the Hammett Shale and Sycamore Sand. Oak brush tends to grow along particularly marly beds on hillsides; prairie is being replaced by juniper brush elsewhere. Seeps and springs form during wet years at the bases of porous dolomite beds.

Locality 2. Hamilton Pool, on Hamilton Creek north of Ranch Road 3238, Travis County. This is a private resort in the process of becoming a state park; check with the management prior to collecting samples. Hamilton Creek flows through the Cow Creek Limestone between RR 3238 and the waterfall at the pool. The pool was excavated in Hammett Shale. Beach accretion beds that dip southeast are exposed in the creek; the rest of the Cow Creek and beds of upper Hammett Shale are exposed below the waterfall. Cretaceous caliche, red and green shale, and coarse subarkose of the Hensel Sand are exposed in road-metal quarries near the park entrance.

The vertical sequence of middle Trinity rocks is exposed well at Hamilton Pool; sedimentary structures are better exposed at other localities (Stricklin and others, 1971), most of which are inaccessible without permission from landowners. Gently dipping beach accretion beds in the upper Cow Creek overlie beds that contain large festoon cross-laminae formed by currents flowing parallel to the beaches. The rock is coarse lime grainstone composed mostly of well-rounded oyster and clam shell fragments. A middle silty member, locally dolomitized but deposited as fine-grained, well-rounded shell fragments and pellets, is exposed near the lip of the falls. This unit represents fine material winnowed from beaches and deposited offshore. Locally it contains abundant juveniles of the zonal ammonite *Dufrenoya justinae* (Upper Aptian). The falls are formed in the lower Cow Creek, a coarse shell-fragment grainstone containing abundant robust mollusk shells, including adult ammonites (Young, 1974). Inden (1974) and Inden and Moore (1984) interpreted the lower Cow Creek as

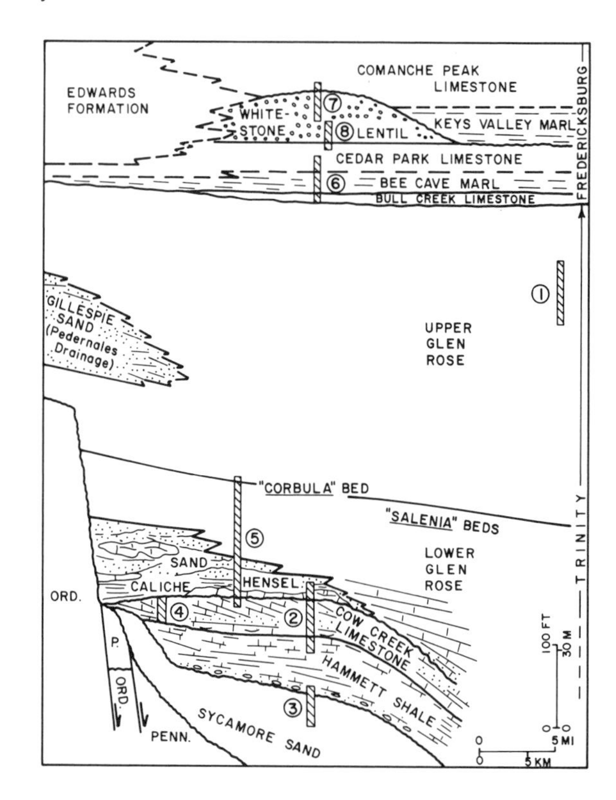

Figure 2. Diagrammatic west–east cross-section of the Trinity and Fredericksburg (Comanchean, Lower Cretaceous) strata west of Austin, Texas. Field trip stops described in the text are shown.

an offshore, open-ocean deposit contemporaneous with the prograding beach; Amsbury (1974) interpreted it as an open bay or sound deposit formed during transgression before the beach began to prograde.

The Hammett/Cow Creek boundary is placed for convenience at one of the shale beds near the top of the oyster-bearing, silty limestone beds under the overhang of the waterfall. Oyster biostromes, concretionary lime siltstones, and silty dolomite beds are well exposed there. Again, these beds may represent relatively deep, offshore equivalents of the overlying beds (Inden, 1974) or older lagoonal deposits equivalent to lime grainstone shoals farther offshore (Amsbury, 1974).

Locality 3. Hammett's Crossing. Road cut on the west side of the Pedernales River at the crossing of RR 3238, Travis County. Be careful to park in the widest area possible, and walk back down to the outcrop. The Sycamore "Sand" is mostly coarse, poorly sorted conglomerate and sand derived from Paleozoic and Precambrian rocks: clasts include dolomite, chert, limestone, feldspar, and quartz. Cement is dolomite spar and caliche (Amsbury, 1974). This unit probably is continuous with the subsurface Hosston Sandstone (Stricklin and others, 1971) rather than older, perhaps Triassic, rocks (Gawloski, 1983).

The basal bed of the Hammett Shale overlies an irregular surface at the top of the Sycamore. It is a resistant lime packstone 1 to 2 ft (0.3 to 0.6 m) thick containing oyster shells, molds of

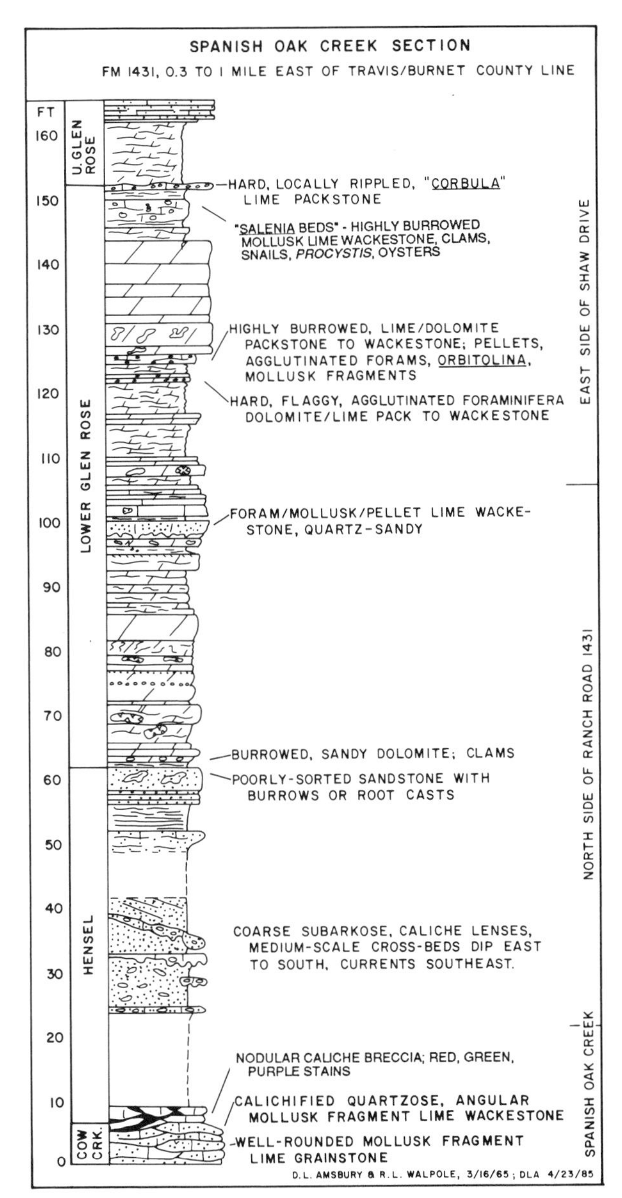

Figure 3. Measured section of the Hensel Sand and Lower Glen Rose Formation in the Spanish Oak Creek Section, Locality 5.

clams and snails, and scattered cobbles of carbonate rocks bored by clams. Fragments of echinoderms, blue-green and green calcareous algae, and corals were seen in thin section. This bed records the first open-marine water to flood the craton since the Pennsylvanian.

Traverse to Locality 4. Much of the route (Fig. 1) is across the top of an exhumed pre-Hensel landscape. At Cypress Mill, beach beds of the upper Cow Creek lap onto Pennsylvanian Marble Falls Limestone. West to Round Mountain and north to Marble Falls, much of the road runs on Ordovician Ellenburger Limestone (Barnes, 1981b). Glen Rose and Fredericksburg rocks on Shovel Mountain north of RR 962 are preserved on the divide between paleo-Pedernales and paleo-Colorado drainages; in places uppermost Glen Rose Limestone disconformably overlies Ellenburger Limestone.

About 1 mi (1.6 km) east of Marble Falls, RR 1431 turns abruptly north at the foot of a middle Pennsylvanian fault scarp between the Marble Falls Limestone and the younger Pennsylvanian Smithwick Shale; the road runs in Smithwick Shale to Locality 4.

Locality 4. Sycamore Creek Section (Stop 9 of Barnes and others, 1972). Road cut on the south side of RR 1431, 4.5 mi (7.2 km) east of the U.S. 281 intersection in Marble Falls. Park along the top of the cut—please do not block the ranch-house driveway.

The Cow Creek Limestone is only about 20 ft (6 m) thick, but a lower unit containing large oyster shells and other fossils, a middle terrigenous-rich unit, and an upper coarse lime grainstone unit are present. To the west the lower unit overlaps Hammett Shale onto the Sycamore Sand, the middle unit overlaps the lower one (Amsbury, 1974), and finally only the upper beach accretion beds rest on older rocks (Fig. 2).

Traverse to Locality 5. Most cuts along RR 1431 are in subarkose, shale, and caliche of the Hensel Sand; some are capped by basal Glen Rose beds. There are excellent exposures of the Cow Creek Limestone below some creek crossings; *do not trespass!*

Locality 5. Spanish Oak Creek Section. Road cuts along the north side of RR 1431, 16.5 mi (26.5 km) east of U.S. 281 in Marble Falls; and exposures along Shaw Drive south of RR 1431. The top of the Cow Creek was exposed in the bed of Spanish Oak Creek, but may not be accessible without trespassing.

Coarse, shell-fragment grainstone of the upper Cow Creek is commonly overlain by calichified angular-mollusk-fragment lime wackestone to packstone, and then by aphanitic nodular caliche with scattered birdseyes in the Colorado River drainage (Stricklin and others, 1971). The basal Hensel caliche is overlain by red, purple, and green shale containing nodules and beds of calcite and dolomite caliche; this unit is not exposed here but is exposed in nearby road cuts.

Coarse, gravelly subarkose in channel fills and more even beds overlie the clay section in most exposures of the Hensel Sand. Medium-scale cross-laminae indicate currents flowing southeast. The sand beds were weathered periodically; pebbles and cobbles of caliche are incorporated within successive gravels. The contact with the Glen Rose Formation is interbedded-gradational (Stricklin and others, 1971; Inden, 1974), but at most localities a reasonable contact can be picked. Here the boundary was picked at a thin clay shale above sandstone and below dolomite that contains molds of claims (Fig. 3).

Most of the lower Glen Rose at Spanish Oak Creek is dolomite. Large vugs lined by dog-tooth calcite may represent gypsum nodules. Distinctive fossiliferous beds, noted in Figure 3, were traced westward by Amsbury and R. L. Walpole in 1965 (unpublished) toward the pinchout. Judging from fair to poor exposures it is likely that the uppermost lower Glen Rose onlaps the Hensel onto Paleozoic rocks (Fig. 2). Similar relationships were mapped by V. E. Barnes in the Pedernales drainage. The *"Salenia"* beds and *"Corbula"* bed are very fossiliferous here and are nearly typical (Stricklin and others, 1971; Stricklin and Amsbury, 1974), but the reef units so well displayed in the Guadalupe and Blanco river drainages (Perkins, 1974) are represented only by peritidal dolomite. The *"Salenia"* beds elsewhere contain the Lower Albian zonal ammonite *Douvilleiceras mammillatum* (Young, 1974).

Traverse to Locality 6. Ranch Road 1431 ranges up and down from the Cow Creek through the Glen Rose. The Cow Creek Limestone on Cow Creek dips under the normal level of Lake Travis just upstream from Lago Vista, followed quickly by the *"Corbula"* bed (Barnes, 1981a); most of the cuts from Lago Vista to Locality 6 are in the upper Glen Rose.

Locality 6. (Stop 11 of Barnes and others, 1972.) Road cuts and ditches on the southwest side of RR 1431, 3.75 to 4 mi (6 to 6.5 km) west of the U.S. 183 intersection at Cedar Park (Fig. 1).

The basal Fredericksburg unit is white, nodular Bull Creek Limestone (Walnut Formation), which onlaps Paluxy Sand to the north (Moore, 1964). At Locality 6 the Bull Creek is about 28 ft (8.5 m) thick and is composed of intraclast lime packstone and molluskan lime wackestone to packstone; molluskan fossils are abundant. The resistant top is quartz-sandy and capped by a hardground (Moore, 1964).

A shallowing-upward sequence some 100 ft (30 m) thick overlies the Bull Creek Limestone. It begins with about 28 ft (9 m) of Bee Cave Marl (Walnut Formation), which grades into 25 to 30 ft (8 to 9 m) of Cedar Park Limestone (Walnut Formation). The sequence is capped in this area with as much as 40 ft (12 m) of oölith lime grainstone, the Whitestone Lentil (Walnut Formation, Moore, 1964; or Edwards Formation, Barnes and others, 1972).

Basal Bee Cave marls contain exogyrines, gryphaeids, ostreids, other mollusks, and echinoids. A resistant bed 12 to 14 ft (4 to 5 m) above the base contains abundant *Dictyoconus.* The Cedar Park Limestone is white, almost porcellaneous, mollusk-fragment lime wackestone; the upper part is interbedded with rudist lime packstone.

Locality 7. Quarries in the Whitestone Lentil east of RR 1431 and south of the Lime Creek Road, about 2 mi (3.2 km) southwest of the U.S. 183 intersection at Cedar Park (Stop 12 of Barnes and others, 1972). These quarries, active and inactive, are the property of Texas Quarries, Inc., Cedar Park, Texas. The company requires written permission and a release of liability in advance of entry.

About 20 ft (6 m) of the upper Whitestone is exposed. In tidal-channel depositional sequences, the lower part (marketed as Cordova Shell) contains abundant casts of the ribbed claim *Trigonia,* and the upper part (marketed as Cordova Cream) consists of festoon cross-laminated oölith lime grainstone. The Whitestone is capped by an iron-stained hardground bored by clams and plastered with exogyrines. Keys Valley Marl (Walnut Formation) of the next Fredericksburg depositional sequence overlies the hardground, and is overlain by nodular Comanche Peak Limestone. The marl contains mollusk shells, including oxytropidocerine ammonites (Middle Albian, Young, 1966).

Locality 8. If permission to enter the Whitestone quarries is not obtained, blocks of the material may be observed near the several public roads in the area. The basal Whitestone crops out in road cuts along the southwest side of U.S. 183, 1 to 2 mi (1.6 to 3.2 km) southeast of Cedar Park. Upper Cedar Park Limestone is overlain by rudist packstone and grainstone, and by oölith grainstone of the basal Whitestone Lentil.

REFERENCES CITED

Amsbury, D. L., 1974, Stratigraphic petrology of lower and middle Trinity rocks on the San Marcos Platform, south-central Texas, *in* Perkins, B. F., ed., Geoscience and man, v. VIII: Louisiana State University Press, p. 1–35.

Barnes, V. E., 1981a, Geologic atlas of Texas, Austin sheet: Bureau of Economic Geology, The University of Texas.

—— , 1981b, Geologic atlas of Texas, Llano sheet (Revised): Bureau of Economic Geology, The University of Texas.

Barnes, V. E., Bell, W. C., Clabaugh, S. E., Cloud, P. E., Jr., McGehee, R. V., Rodda, P. U., and Young, K. P., 1972, Geology of the Llano region and Austin area: Bureau of Economic Geology, The University of Texas, Guidebook 13, 154 p.

Gawloski, T., 1983, Stratigraphy and environmental significance of the continental Triassic rocks of Texas: Baylor Geological Studies Bulletin no. 41, 48 p.

James, N. P., 1984, Shallowing-upward sequences in carbonates, *in* Walker, R. G., ed., Facies models, Sec. Ed.: Geoscience Canada Reprint Series 1, p. 213–228.

Inden, R. F., 1974, Lithofacies and depositional model for a Trinity Cretaceous sequence, central Texas, *in* Perkins, B. F., ed., Geoscience and man, v. VIII: Louisiana State University Press, p. 37–52.

Inden, R. F., and Moore, C. H., Jr., 1983, Beach, *in* Scholle, P. A., and others, eds., Carbonate depositional environments: American Association of Petroleum Geologists Memoir 33, p. 211–265.

Moore, C. H., Jr., 1964, Stratigraphy of the Fredericksburg Division, south-central Texas: Bureau of Economic Geology, The University of Texas, Report of Investigations 52, 48 p.

Perkins, B. F., 1974, Paleoecology of a Rudist Reef Complex in the Comanche Cretaceous Glen Rose Limestone of central Texas, *in* Perkins, B. F., ed., Geoscience and man, v. VIII: Louisiana State University Press, p. 131–173.

Stricklin, F. L., Jr., Smith, C. I., and Lozo, F. E., 1971, Stratigraphy of Lower Cretaceous Trinity deposits of central Texas: Bureau of Economic Geology, The University of Texas, Report of Investigation 71, 63 p.

Stricklin, F. L., Jr., and Amsbury, D. L., 1974, Depositional environments on a low-relief carbonate shelf, middle Glen Rose Limestone, central Texas, *in* Perkins, B. F., ed., Geoscience and man, v. VIII: Louisiana State University Press, p. 53–66.

Young, K. P., 1966, Texas Mojsisovicziinae (Ammonoidea) and the zonation of the Fredericksburg: Geological Society of America Memoir 100, 225 p.

—— , 1974, Lower Albian and Aptian (Cretaceous) ammonites of Texas, *in* Perkins, B. F., ed., Geoscience and man, v. VIII: Louisiana State University Press, p. 175–228.

The lower Tertiary of the Texas Gulf Coast

Thomas E. Yancey and Elizabeth S. Yancey, Department of Geology, Texas A & M University, College Station, Texas 77843

LOCATION

The lower Tertiary section in the western Gulf Coast is most complete and best exposed in the bluffs of the Brazos River Valley (Figs. 1 and 2), and has received the most study of any area in the state. The low rolling hills of this region provide limited outcrops, so most good exposures are confined to high banks along the Brazos River. Strata dip toward the Gulf of Mexico at uniformly low dips, and exposures gradually shift upsection as one travels downriver. The regular progression upsection continues to the area of Navasota and Hempstead, where Quaternary sediments of the flat-lying coastal plain lap onto the youngest (Miocene age) of the more steeply dipping Tertiary formations. The Brazos Valley was one of the more heavily traveled routes by settlers and explorers, and the Stone City locality (site of the King's Highway [El Camino Real or Old Spanish Road—OSR] crossing of the Brazos River) was the source of material for the first study of Texas fossils.

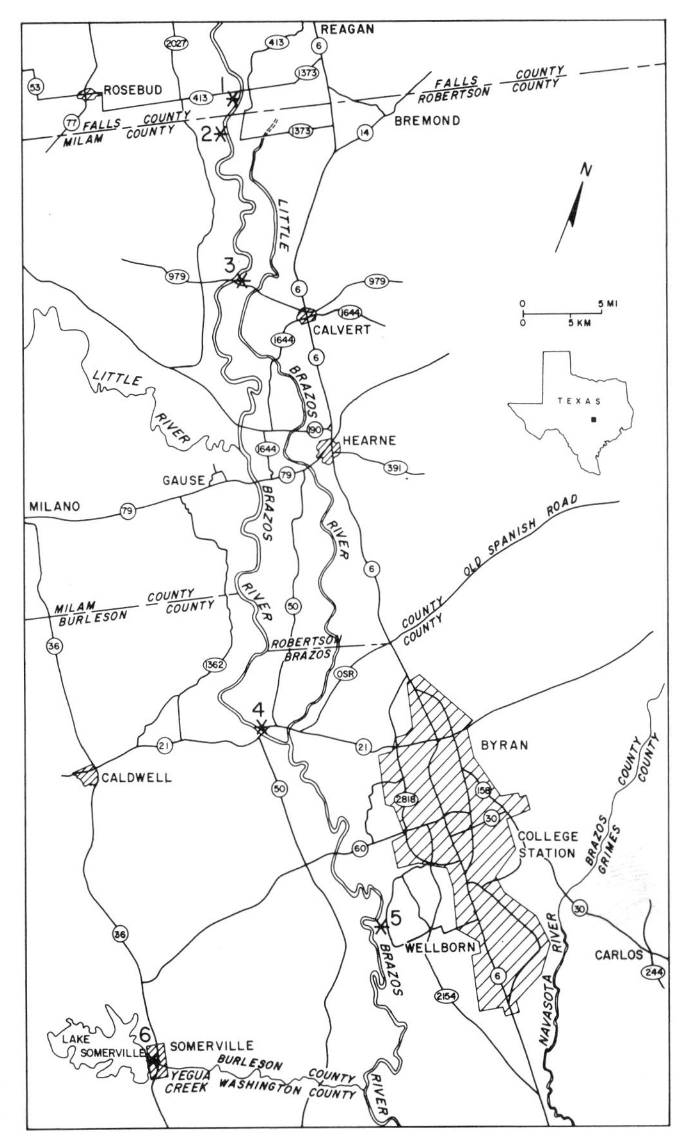

Figure 1. Map showing distribution of Localities 1 through 6.

DEPOSITIONAL SYSTEMS

Lower Tertiary strata are exposed in a wide band across Texas from the Louisiana border to the Rio Grande, and provide the best interval for examination of Cenozoic depositional systems. Cenozoic deposition is characterized by major transgressions and regressions along the Gulf margin, producing an alternation of nonmarine and shallow-marine sections in the exposure belt, and alternations of shallow-water marine and deep-water marine sections in more downdip locations (Israelsky, 1949).

Several major delta complexes occur in these units, commonly associated with river systems that have persisted in the same general location to the present day (such as the Brazos River system). The best known are the subsurface extensions of the Wilcox and Jackson groups, which contain important oil-producing sands. The exposure belt provides a view of the up-channel portions of these deltas. These units are economically important, holding major lignite reserves (Wilcox and Jackson groups), uranium reserves in south Texas, water aquifers (Wilcox Group), and much oil and gas in the subsurface.

SUPPLEMENTAL SOURCES OF INFORMATION

Outcrops of lower Tertiary strata in Texas have been visited many times on organized field trips run by geological societies, although there are few comprehensive studies of this part of the section. Detailed lithologic and stratigraphic studies of the Cretaceous–Tertiary contact section appear in Yancey (1984). The Midway Group is covered in detail by Gardner (1933), who discusses outcrop sections and stratigraphy of the Brazos River Valley area. The overlying Wilcox, Claiborne, and Jackson groups are most extensively described in articles contained in guidebooks printed by the geological societies. The focus of these guidebooks is mostly on individual sections within the outcrop belt. Persons wanting to visit the many additional good sections should consult the following guidebooks: Yancey (1984; an update of Kersey and Stanton, 1979, and Berg, 1970, and the best current overview); Atlee and others (1967, 1968); Rainwater and Zingula (1962); Smith (1958, 1959, 1962); Stuckey and others (1961); Perkins and Hobday (1980) for the Queen City Formation; and Russell (1960) for the Jackson Group.

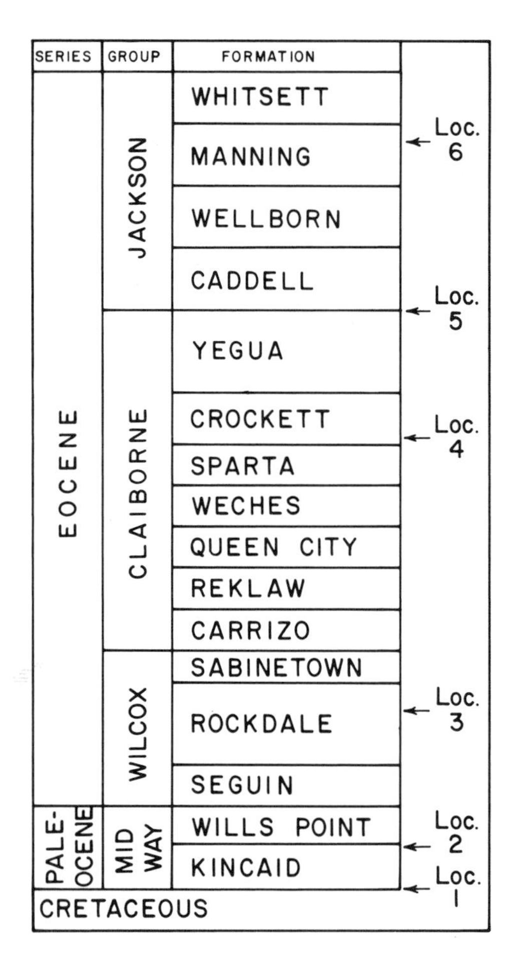

Figure 2. Lower Tertiary stratigraphic units of central Texas.

***Locality 1.* Cretaceous–Tertiary boundary on the Brazos River.** This section is exposed in the western banks of the Brazos River about 1,000 ft (300 m) downstream from the bridge on Texas FM 413, crossing over the Brazos River (Fig. 1). Texas FM 413 is the main road to Rosebud from Texas 6, and intersects with Texas 6 in Reagan, or via Texas FM 1373 from its intersection with Texas 6 at 2.3 mi (3.7 km) southeast of Reagan. From the bridge, access to the exposure is gained by walking along the river bank, or wading in the shallow river bottom. Watch out for snakes.

This is the best-studied marine section in North America of the Cretaceous–Tertiary contact, and contains a continuous depositional record of the latest Cretaceous. Recent studies (Yancey, 1984; Gartner and Jiang, 1985) have documented a complete record of coccolith zones, a good sequence of microfossils and macrofossils, and iridium anomalies at the Cretaceous–Tertiary contact.

There are 60 ft (18 m) or more of latest Cretaceous greenish gray clay shales exposed in the river bed (when water levels are low) below the bridge, where they form irregular shoals and ledges along the side of the river (Fig. 3A). The best exposures of these shales are in the river bed, since the bluffs are overgrown just below the bridge. The shales are massive, with indistinct bedding, and are fossiliferous throughout.

Immediately above the Cretaceous shales is a thin complex of resistant sandstones and siltstones that forms riffles in the river bed. This is a marker for the Cretaceous–Tertiary boundary, which earlier studies have placed at the base of the sandstone. Coccoliths recovered from this section show that the change from Cretaceous to Tertiary assemblages occurs at a level 15 cm above the top of the upper resistant unit, rather than at the base. The boundary interval is best exposed in the river bluff, which is partly overgrown and may have to be cleared somewhat to see the strata.

The resistant unit is mapped as the Littig Member of the Kincaid Formation, which elsewhere occurs at the base of the Paleocene. The unit is about 30 cm thick, and is composed of sandstone, siltstone, and mudstone, which occur in 4 subunits in this outcrop. The lowest subunit is a thin, nonresistant (5–20 cm) glauconitic sandstone containing common shells and phosphate nodules, similar to occurrences of the Littig elsewhere. This is overlain by a 10-cm-thick, fine-grained, rippled, calcareous sandstone, very resistant to erosion. The next subunit is a soft, sandy shale or mudstone 2–5 cm thick. The highest subunit is a 10- to 15-cm-thick hard siltstone or sandstone, also very resistant to erosion. The lower of the resistant subunits shows good examples of climbing ripples, and where the top has been stripped clear by water in the riverbed the surface shows extensive wavy bedding. These sands have been interpreted as the products of storm deposition.

The overlying sediments are shales and siltstones of the Pisgah Member of the Kincaid Formation. There are about 23 ft (7 m) exposed in the river bluff, consisting of greenish gray shale at the base, grading upward to a gray muddy siltstone that is moderately hard and very fossiliferous. (Discussion condensed from Yancey, 1984.)

***Locality 2.* Midway Group at Frost Bluff.** This interval is best exposed in Frost Bluff along the Brazos River, located on the west bank of the river a short distance downriver from the Cretaceous–Tertiary boundary locality (Fig. 1). The locality is reached by ranch roads from Texas FM 2027 (the Old River Road) at a turnoff 2.4 mi (3.8 km) south of the intersection of Texas FM 2027 and Texas FM 413. This is on the Ellison Ranch; an access fee of $2.00 per person can be paid at the gate. Drive on the main road track, bearing right at intersections until coming to a large pond, and drive across the pond dam to reach the river terrace level. This is 2.5 mi (4.0 km) from Texas 2027. Walk down the bluff to river level. This is a popular fishing area.

The Midway Group in Texas is composed of marine shales and thin sandstones deposited during a variable trend of shoaling, culminating in nonmarine deposits of the overlying Wilcox Group.

About 80 ft (25 m) (Fig. 3B) of marine Midway strata is exposed on the west side of the Brazos River in the river bluff 0.6 mi (1 km) downstream from the Milam-Falls county line, and referred to as Frost Bluff by Gardner (1933). Exposures extend

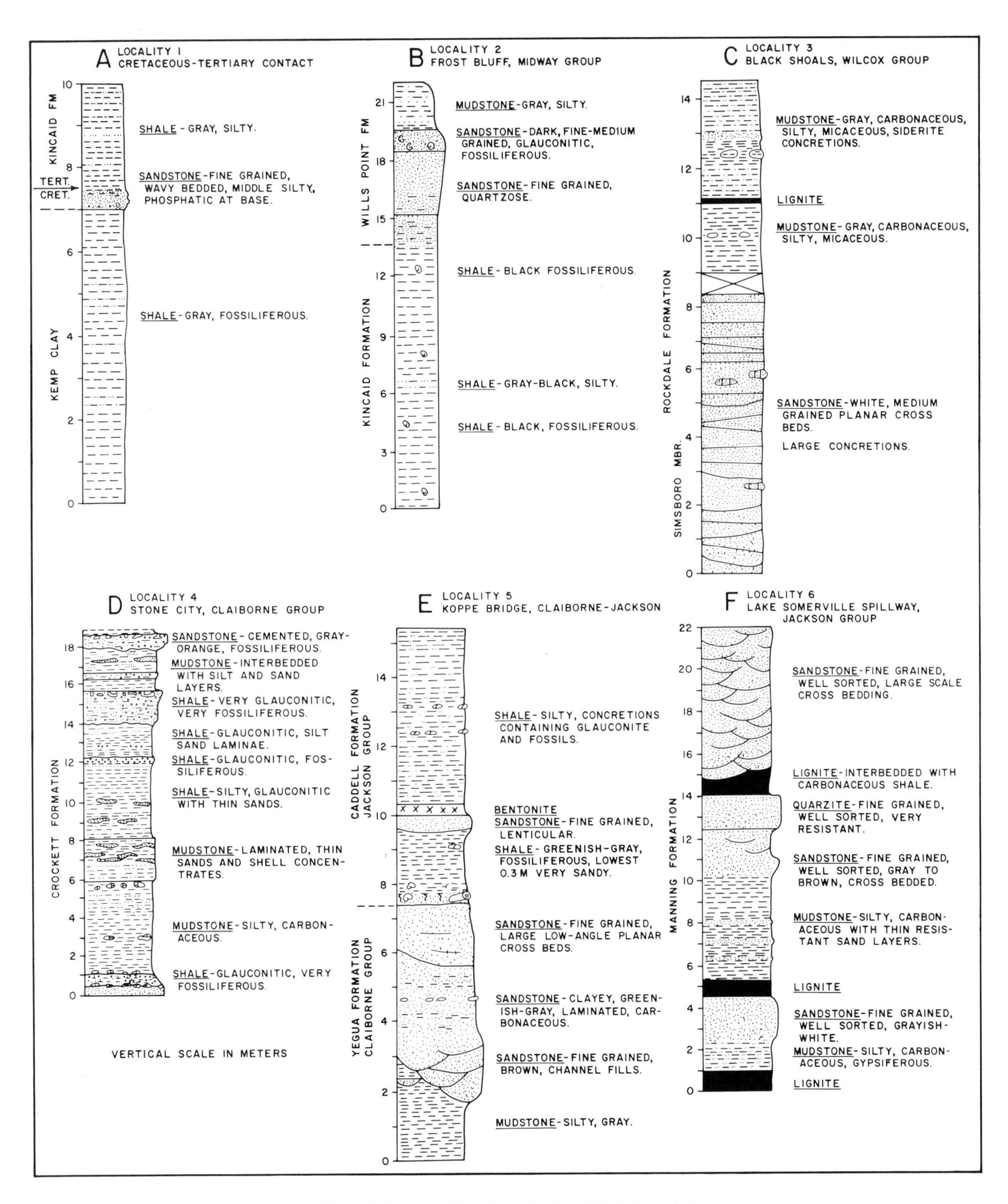

Figure 3. Stratigraphic columns for Localities 1 through 6.

along 0.3 mi (0.5 km) of the river, in two main areas of exposure separated by a gully choked with soil and vegetation. Strata dip gently in a downstream direction, with the lowest strata exposed in the upstream area, including exposures in the river bed accessible at low-water conditions, and higher strata exposed best in the downstream area. The sandstones of the upper part of the section form the top of the bluff, protecting the underlying strata from erosion, but as the sandstones dip down to the river level the exposures disappear beneath a soil cover. The highest strata are variably weathered, and the bluff is capped by a thin layer of Pleistocene gravel and sand.

The lower 56 ft (17 m) is Pisgah Member of the Kincaid Formation, while the upper 26 ft (8 m) of sandstone and mudstone is Mexia Member of the Wills Point Formation. At least half of the Pisgah is exposed in this bluff section, and consists of clay shale throughout. It is richly fossiliferous, with well-preserved shells. The formational boundary occurs in a gradational shale-to-sandstone sequence in which sand layers in the shale become more common until mud layers are lacking, and the boundary is placed at the level where layers of thick, well-sorted sand appear that are separated only by thin mud layers. Mexia sediments are varied, with quartzose sandstones at the base, glauconitic sandstones above, and blocky mudstones at the top. Although these sands are assigned to the Mexia by Gardner (1933), they may be better placed in the Tehuacana Member of the Kincaid Formation.

The strata are entirely marine, and were deposited in a shoaling-upward cycle ranging from deep-water shelf environment to a shallow-water nearshore environment. Deposition was continuous, with no evidence of significant gaps in the column. The glauconitic sand contains large heavy-shelled fossils indicative of shallow-water conditions. The mudstone overlying the glauconitic sands contains thin sand laminae and large and small fossil shells, showing a rapid change to quiet-water bay deposition. The glauconitic sandstones are variably lithified, and produce many large blocks that fall and accumulate at the foot of the cliff.

***Locality 3.* Wilcox Group at Black Shoals.** Good exposures of this interval are present at Black Shoals, in the bluffs on both sides of the Brazos River upstream and downstream from the Texas FM 979 bridge crossing over the river (Fig. 1). This is reached by turning onto Texas FM 979 at its intersection with Texas 6 on the north edge of Calvert, one of the oldest towns in the Brazos River Valley. Travel 5 mi (8 km) west to the bridge. A dirt road extends along part of the river bank.

The Wilcox Group is the oldest major unit of nonmarine deposits in the Tertiary section. Exposures of this unit throughout Texas are characteristically fluvial and swamp deposits, with a small amount of shore-zone deposits, and often contain large amounts of sandstone. Nonmarine deposits pinch out and change to marine deposits within 60 mi (100 km) downdip of the outcrop belt. The Wilcox is a major producer of lignite in Texas and, in the Brazos River Valley area, sands of the Wilcox (Simsboro Member) are important aquifers.

Carbonaceous mudstones and fluvial sands of the Simsboro Member of the Rockdale Formation are well exposed in the banks on both sides of the Brazos River (Fig. 3c). The sand unit forms a high ground on the northwest side of the river, and contains an abundance of large "cannonball" and flat slab concretions that weather out of the strata. Apart from the concretions, the sand is very loosely consolidated in outcrop, and is eroded by the river into steep bluffs.

The sands are nearly all medium graned, and of light color with a pepper-and-salt appearance. Quartz, feldspar, opaques, rock fragments, and mica are common components, and sand grains are conspicuously angular. The sands contain large-scale planar crossbeds throughout the bluff section. Crossbed sets range from several cm to 1 m in height, and uniformly dip in an upriver (northeast) direction. Crossbed laminae on foresets turn abruptly into bottomset laminae, which form a thin layer rarely more than a couple of centimeters thick. Medium-grained sand of about the same grain size as sand in foreset laminae is the dominant component of the bottomset sediments, along with many chunks of wood (now flattened).

These sands were deposited in the meanderbelt of a large river that fed the Brazos River Delta system of the Rockdale Formation. The large grain size and consistent northeast-dipping planar crossbedding suggest deposition from migrating megaripple dunes in a major river channel. Crossbed orientation is consistent through the section, and is parallel to the outcrop trend and perpendicular to the regional slope of the Eocene, implying deposition in a meander-loop of a larger river. This is also apparent in the consistency of crossbed type through the 8 m of section, and lack of interbeds or other sedimentary structures. Trough crossbedding is rare or lacking.

From the bridge site and downstream, carbonaceous mudstones and thin lignites and sandstones are exposed in the low bluffs of the river. This fine-grained interval also has zones of large concretions, formed with siderite cement, which weather out and concentrate in the river bed. Fine sands and mudstones exposed at the bridge abutments show much contorted bedding, and abundant large and small plant fragments. There is one thin (0.7 ft; 0.2 m) lignite bed in the section, and numerous very thin lignite streaks.

***Locality 4.* Stone City exposures of the Claiborne Group at Stone City Bluff.** This famous fossil locality is located on the south bank of the Brazos River at the Texas 21 bridge over the river (Fig. 1), which is the site of the Moseley Ferry crossing of the Old Spanish Road connecting San Antonio and Nacogdoches. Exposures of the formation occur around the abutments of the bridge, and in the bluff a short distance upriver. A foot path extends along the river bank. This locality is the best known of several richly fossiliferous localities of Claiborne strata in Texas, and was first described by Roemer (1848). The Claiborne in central and east Texas contains three main intervals of marine shales separated by nearshore and fluvial sands. The Stone City and overlying strata of the Crockett Formation were deposited during the last and most extensive of the Claiborne transgressions.

The Stone City section overlies fluvial to marginal marine sands of the Sparta Formation and is overlain by richly fossiliferous marine shales and sands of the Crockett Formation. A resistant cemented sandstone at the top of the bluff has been named the Moseley "Limestone," but it is not more calcareous than the underlying strata. The Stone City interval has been separated as a distinct formation from the lithologically similar overlying Crockett strata on the basis of an inferred unconformity between the two. The top of the Moseley contains sediment-filled burrows and a possible hardground, but there is no other evidence for an unconformity. Because of the supposed unconformity, the formation name, Crockett, was temporarily replaced by Cook Mountain, but the Stone City remains a member of the Crockett Formation. The underlying Sparta Formation crops out in the bed of the Brazos River at Stone City Bluff, but is exposed only at a very low water stage.

The Stone City section (Fig. 3D) consists of alternating beds of dark carbonaceous mudstones with discontinuous layers of sand and silt, and beds of fossiliferous, calcareous, glauconitic sands and muds. Greenish gray to brown mudstone occurs throughout, is the major rock type in the lower three-fourths of the section, and contains scattered woody plant fragments. Fossils are rare, occurring only in a few coquina lenses. Lenses and thin beds of laminated and cross-laminated sandstone are common within the mudstone, and show small-scale channeling and rippling.

Muddy glauconite pellet "sandstone," the other major rock type, is dominant in the upper one-fourth of the formation, interbedded with mudstone. Body fossils and trace fossils are abundant in the glauconitic sediment; intense bioturbation has destroyed any primary sedimentary structures.

The fauna of the Stone City Member consists largely of molluscs with fish otoliths and teeth, bryozoans, corals, crustaceans, foraminifers, and brachiopods as minor components. Diversity is high, with more than 100 taxa identified. The distribution of body and trace fossils is strongly correlated with that of the glauconite. In the mudstone, body fossils are restricted to small lenses and have probably been concentrated as lag deposits during occasional periods of higher-velocity water currents. Trace fossils are limited to small, straight horizontal to vertical tubes.

In contrast, body fossils in the glauconitic sediments are abundant, well preserved, and represent the original community living at the site. The fauna is dominated by corbulid and nuculid bivalves and turrid and naticid gastropods. Pervasive bioturbation in the glauconite beds was produced by crustaceans and polychaetes during relatively slow sediment deposition. Burrows of *Thallasinoides* and *Gyrolithes* occur in the upper part of the thickest glauconite bed.

Environments of deposition of Stone City sediments are varied, ranging from open shelf to brackish. The carbonaceous mudstones of the lower part are shallow, nearshore deposits, possibly deposited within a delta setting. The fossil-rich glauconitic beds contain a high-diversity fauna, containing turrid gastropods along with planktic foraminifera and planktic pteropod gastropods, suggesting open-marine, deeper-water environments, distant from shorelines or deltas. (Discussion modified from Stanton, in Berg [1970].)

***Locality 5.* Claiborne-Jackson contact at Koppe Bridge.** This locality is at the old Koppe Bridge site (Fig. 1) and immediately upstream, and is reached by taking Texas FM 2154 (either from its intersection with Texas 6 at 15 mi [24 km] southeast of College Station or from its intersection with Texas FM 2818 on the south side of College Station) to the village of Wellborn, turning west and crossing the railroad tracks onto Dowling Road, which joins with Koppe Bridge Road at this point. Follow Koppe Bridge Road southwest for 2.9 mi (4.6 km) to a sharp northwest turn in the road, continue on for another 0.9 mi (1.4 km) and turn onto a west-trending side road, which is the unmarked continuation of Koppe Bridge Road. At the end of this road (1.1 mi; 1.7 km) is the Koppe Bridge site.

The upper Claiborne and lower Jackson groups are exposed in bluffs along the east bank of the Brazos River, upstream from the old Koppe Bridge (Fig. 3E). The lower 23 ft (7 m) of the section is composed of sandstones and mudstones of the upper Yegua Formation, and the upper 26 ft (8 m) is part of the Caddell Formation of the basal Jackson Group. This section contains a complete record of marine transgression, with deposits ranging from fluvial to shore zone to offshore marine. This transgression starts at the base of the Jackson Group in this area of exposure, and is the lower of two major marine transgressions in the Jackson (the other one occurs at the top of the Jackson Group).

The lowermost 10 ft (3.3 m) of the Yegua is silty claystone that contains marine palynomorphs, some of which occur in Jackson shale above, but no foraminifera are found. The claystone is locally channeled by sandstone that is very fine grained (0.07 mm), crossbedded, and contains significant amounts of feldspar, rock fragments, and clay matrix. The channels range from small lenses 1.5 ft (0.5 m) thick to a large channel which replaces almost the entire 10 ft (3.3-m) interval.

Overlying beds are laminated, shaly sandstones that contain carbonaceous fragments, and the upper part of the Yegua is fine-grained sandstone (0.17 mm), with planar crossbeds in 15-to-30-cm sets. Maximum grain size of quartz is 0.25 mm, and rock fragments range from 3 to 18 percent. These sandstones are of fluvial origin and, together with the possible marine claystone below, indicate a prograding delta sequence at the top of the Yegua.

Immediately above the Yegua sandstones is a thin, greenish gray, clayey sandstone, 1 ft (0.3 m) thick, that has large vertical burrows, casts of bivalves and gastropods, some thin coaly lenses, and at least one large petrified log. This sandstone represents reworking of Yegua sand at the base of the Caddell marine transgression. The basal sand grades upward into greenish gray, glauconitic, and fossiliferous shales of the Caddell. These shales contain macrofauna and microfauna typical of Jacksonian faunas (Atlee and others, 1967). Thin sand lenses in the lower Caddell are very fine grained, and they are similar in composition to

Yegua sandstones. The lower Caddell shales were deposited in nearshore marine environments, or bays. Occurrences of calcareous foraminifera higher in the Caddell suggest deposition in deeper marine water (Atlee and others, 1967). (Discussion modified from Berg [1970].)

Locality 6. **Jackson Group at Lake Somerville Spillway.** This locality is located on the south side of the spillway for the dam forming Lake Somerville, on the edge of the town of Somerville on Texas 36 (Fig. 1). A road branches off Texas 36 on the south side of the town and leads across the dam. A parking lot is present at an overlook beside the spillway, and the section is exposed immediately below the parking area and on the north bank as well.

This section provides excellent exposures of lignites, swamp deposits, and fluvial sands that are characteristic of the Manning Formation of the Jackson Group. The Jackson is the younger of the two main Early Tertiary nonmarine sequences, and contains major lignite reserves at the Gibbons Creek lignite mine in Grimes County east of the Brazos River. This outcrop at Lake Somerville is representative of the formation where lignite is being mined.

The spillway section (Fig. 3F) contains interbedded sandy mudstones, lignites, and crossbedded sandstones, with sandstone predominating. The highest sandstones contain large tabular and trough crossbed sets, with a few thin beds of silty clay between sets. The sands are well-sorted fine-to-medium-grained sands, with scattered mud clasts in a few sets. The sandstone beneath the highest lignite is a well-cemented quartzite, which forms a ledge along the spillway. The source of silica is probably altered volcanic ash, which is common in this formation, and is also the probable source of silica for silicified wood found in strata at nearby Welsh Park. The quartzite in the spillway section posed unexpected excavation problems during construction of the dam.

The upper lignites are irregular and interbedded with carbonaceous clays. They contain many root traces in places, and some sand-filled burrows that are probably animal burrows. Sands underlying the lignites are cut with many root traces. The lignite at the base of section is the most laterally continuous lignite, but is not well exposed. Mudstones in the lower half of the section are quite sandy, with disseminated sand, and thin layers of sand scattered through them. Most of the section contains some gypsum.

The sandstone at the top of the section appears to be part of a large fluvial channel deposit. The lower sands are thinner and more tabular, and interbedded with lignites and sandy mudstones, suggesting deposition by periodic overbank flooding onto coastal plain swamps near large fluvial channels. These sands do not have clear indications of channel deposition.

The Manning Formation was deposited in a coastal plain setting. This encompassed large, sand-rich, fluvial-delta centers separated by low-energy plains and swamp environments, where peats accumulated along with muds and lesser amounts of sand. Occasional marine incursions deposited sheet sands (east of the Brazos River) across the plains. The laterally extensive lignites formed from forested swamps covering the plains between the deltas.

REFERENCES CITED

Atlee, W. A., Loep, K. J., Elsik, W. C., Zingula, R. P., Pointer, G. N., Ogden, J. C., 1967, Selected Cretaceous and Tertiary depositional environments: Tulsa, Olkahoma, Gulf Coast Section of Society of Economic Paleontologists and Mineralogists, Guidebook, 50 p.

Atlee, W. A., Elsik, W. C., Frazier, D. E., Zingula, R. P., 1968, Environments of deposition Wilcox Group: Houston, Texas, Houston Geological Society, Guidebook, 43 p.

Berg, R. R., ed., 1970, Outcrops of the Claiborne Group in the Brazos Valley, southeast Texas: 4th Annual Meeting, South-central Section Geological Society of America, Guidebook, College Station, Texas A&M University Department of Geology, 79 p.

Gardner, J., 1933, The Midway Group of Texas: The University of Texas Bulletin 3301, 403 p.

Gartner, S., and Jiang, M. J., 1985, The Cretaceous–Tertiary boundary in east-central Texas: Gulf Coast Association of Geological Societies Transactions, v. 35, p. 373–380.

Israelsky, M. C., 1949, Oscillation chart: American Association of Petroleum Geologists Bulletin, v. 33, p. 92–98.

Kersey, D. G., and Stanton, R. J., Jr., eds., 1979, Lower Tertiary of the Brazos River Valley: Houston, Texas, Houston Geological Society, Guidebook, 126 p.

Perkins, B. F., and Hobday, D. K., eds., 1980, Middle Eocene coastal plain and nearshore deposits of East Texas; A Field guide to the Queen City Formation and related papers: Tulsa, Oklahoma, Gulf Coast Section of Society of Economic Paleontologists and Mineralogists, Guidebook, 95 p.

Roemer, F., 1848, Contributions to the geology of Texas: American Journal of Science, ser. 2, v. 6, p. 21–28.

Rainwater, E. H., and Zingula, R. P., eds., 1962, Geology of the Gulf Coast and central Texas: Houston, Texas, Houston Geological Society, Guidebook, 319 p.

Russell, W. L., 1960, Jackson Group, Catahoula and Oakville formations and associated structures: Houston, Texas, Houston Geological Society and Gulf Coast Section of Society of Economic Paleontologists and Mineralogists, Guidebook, 50 p.

Smith, F. E., 1958, Upper and middle Tertiary of Brazos River Valley, Texas: Tulsa, Oklahoma, Gulf Coast Section of Society of Economic Paleontologists and Mineralogists and Houston Geological Society, Guidebook, 50 p.

Smith, F. E., 1959, Lower Tertiary and Upper Cretaceous of Brazos River Valley, Texas: Houston, Texas, Houston Geological Society and Gulf Coast Section of Society of Economic Paleontologists and Mineralogists, Guidebook, 54 p.

Smith, F. E., 1962, Upper Cretaceous and lower Tertiary rocks in east-central Texas: College Station and Waco, Texas A&M University and Baylor Geological society, Guidebook, 26 p.

Stuckey, C. W., Zingula, R. P., Rainwater, E. H., 1961, Middle Eocene of Houston Co., Texas: Tulsa, Oklahoma, Gulf Coast Section of Society of Economic Paleontologists and Mineralogists, Guidebook, 45 p.

Yancey, T. E., ed., 1984, The Cretaceous–Tertiary boundary and lower Tertiary of the Brazos River Valley: Tulsa, Oklahoma, American Association of Petroleum Geologists, Guidebook, 137 p.

The Catahoula Formation; A volcaniclastic unit in east Texas

Ernest B. Ledger, Department of Geology, Stephen F. Austin State University, Nacogdoches, Texas 75962

LOCATION

This site is near the dam at Sam Rayburn Reservoir, about 25 mi (40 km) southeast of Lufkin in northern Jasper County, east Texas (Fig. 1). Jasper County is in the Coastal Plain near the Louisiana border.

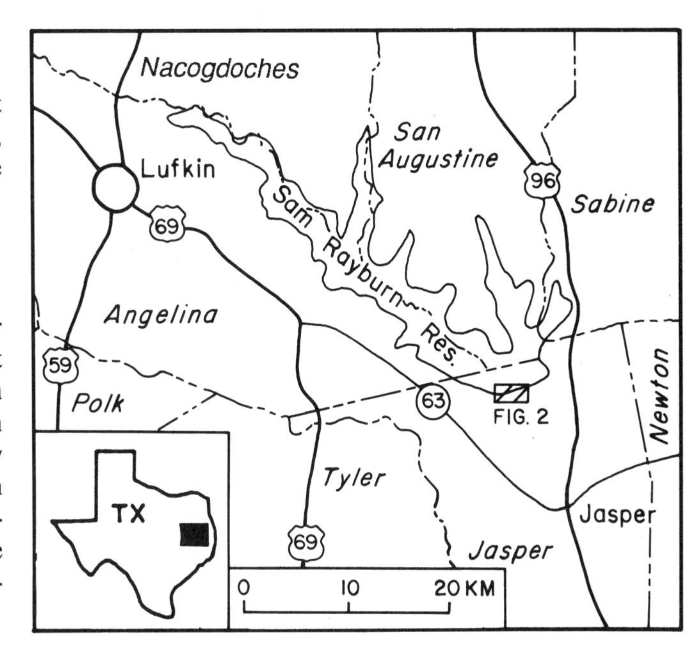

Figure 1. Map of part of eastern Texas showing location of Catahoula Formation outcrop at Sam Rayburn reservoir.

SIGNIFICANCE

The Oligocene/Miocene Catahoula Formation of the upper Texas coastal plain is a fluvial and lacustrine volcaniclastic unit composed of "normal" quartz-rich fluvial material mixed with distal rhyolitic air-fall ash from coeval volcanic source areas in Trans-Pecos Texas and northern Mexico. It consists of poorly sorted siltstones, sandstones, and mudstones. The silt-size fraction includes abundant, slightly altered, volcanic glass shards. Sand-size grains are mostly quartz, some of which are very coarse—the "rice sands". The clay-size fraction is montmorillonite, an alteration product of the glass.

The narrow outcrop belt of the Catahoula Formation, which is somewhat less than 300 ft (100 m) thick in Jasper County, extends from central Mississippi, through Louisiana and Texas parallel to the present coastline, and into eastern Mexico where it is not well studied.

Catahoula deposition is characterized by sporadic influxes of air-fall ash into low-gradient fluvial and coastal lake environments. Stratigraphic studies of the Catahoula Formation began as early as 1857, and the formation has since been studied by numerous stratigraphers and sedimentologists, including Hilgard, Penrose, Dumble, Veatch, and Deussen. Dumble (1918) gives a summary of this early work. Bailey (1926) named the lateral equivalent in the lower Texas coastal plain of south Texas the Gueydan, and along with Renick (1936) established the Catahoula as a mid-Tertiary fluvial unit with abundant volcaniclastic material. A more recent summary of stratigraphic nomenclature is given in Sheldt (1976). Other recent studies of the Catahoula are concerned with petrology, geochemistry, and uranium ore genesis.

Weeks and Eargle (1963) recognized that uranium in the south Texas uranium district was derived from local volcanic material that had been altered by pedogenesis. Differences in the details of sedimentation and weathering did not favor economic accumulations of uranium in east Texas (Ledger and others, 1984). In an extensive petrologic and field study of the Catahoula Formation in south Texas, McBride and others (1968) described the sedimentology and nature of diagenesis of sandstones and finer-grained beds. Also, they discuss in detail the evidence for the volcanic source area of the Catahoula Formation. McDowell (1978) determined K/Ar ages for many volcanic units of the Trans-Pecos area and found that the volcanic activity peaked in the Oligocene. Sheldt (1976) studied the petrology of outcropping Catahoula sandstones just west of the localities presented here. He estimated the time of deposition of the basal sandstones as Middle Oligocene based on eustatic sea level changes (Vail, 1975). Sheldt (1975) found the sandstones to be subarkoses with igneous rock fragments from both the Wichita Mountains of Oklahoma and the volcanic terrane of Trans-Pecos Texas. He found no volcanic glass shards in the sandstones because the high energy of the fluvial channel environment destroyed these fragile grains before deposition. Glass shards are more abundant (up to 90% of the sediment) in the siltstones, which are lake and floodplain deposits. The concept of depositional systems has been applied to the Catahoula Formation by Galloway (1977). On the basis of well logs, he recognizes two depositional systems in Texas, the Chita-Corrigan fluvial system in east Texas, and the Gueydan fluvial system of the Rio Grande Embayment. They are separated by the San Marcos Arch, which may have been breached during part of Catahoula deposition.

The volcanic material in the Catahoula Formation was certainly derived from a western source, most probably from Trans-Pecos Texas and northern Mexico, this being the closest source of appropriate age and chemical affinity. Caldera-type eruptive centers were most active during the Oligocene, and such explosive eruptions produced three units (Sparks and Walker, 1977). Two of the units are deposited locally in the source area, an ignimbrite

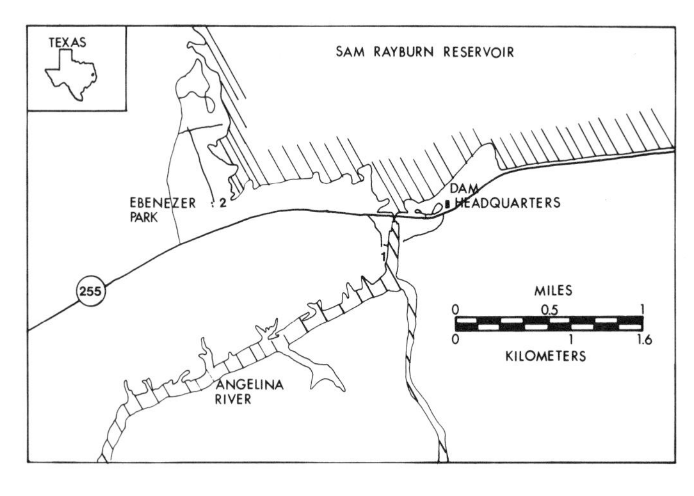

Figure 2. Catahoula Formation outcrop near the dam at Sam Rayburn Reservoir, northern Jasper County, southeastern Texas.

and a pumice-fall. The third unit, the co-ignimbrite ash-fall, is deposited over a wide area and may be measurably thick more than 620 mi (1,000 km) from the eruption. During an eruption, ascending magma consists of early-formed crystals suspended in a volatile-rich magma. The crystals are preferentially incorporated in the ignimbrite, while the magma freezes to volcanic glass and is incorporated in all three units. The co-ignimbrite ash-fall is glass-rich, and the constituent shards exhibit a "bubbly" morphology because of exsolution of volatiles during the eruption, which propels them high into the atmosphere. During Catahoula deposition, upper level winds carried ash clouds to the east. After deposition as a mantle over the paleosurface, the easily eroded ash washed into the fluvial systems. This resulted in about a five-fold increase in volume in the fluvial systems based on the areal ratio of floodplain area to total area in present day east Texas. Delivery of ash was sporadic, but geologically almost continual. Silty floodplain deposits were highly tuffaceous, so mudflows were common because of the buildup of ash by erosion of surrounding areas.

The only fossils found in the Catahoula are abundant plant fragments including petrified pine, oak, and palm logs, smaller reedy plants and palmetto, and diatoms in the lake deposits. Catahoula fossils need more study.

CATAHOULA FORMATION AT SAM RAYBURN RESERVOIR

The middle of the Catahoula section is exposed in the vicinity of the dam and spillway (Fig. 2, locality 1). The base is poorly exposed about 1.2 mi (2 km) to the northwest (Fig. 2, locality 2). Downstream to the south the upper part of the formation is homogeneously clayey and very poorly exposed. At the measured section (Fig. 3) it consists of overbank sandstones and tuffaceous siltstones and mudstones. The siltstones are particularly glass

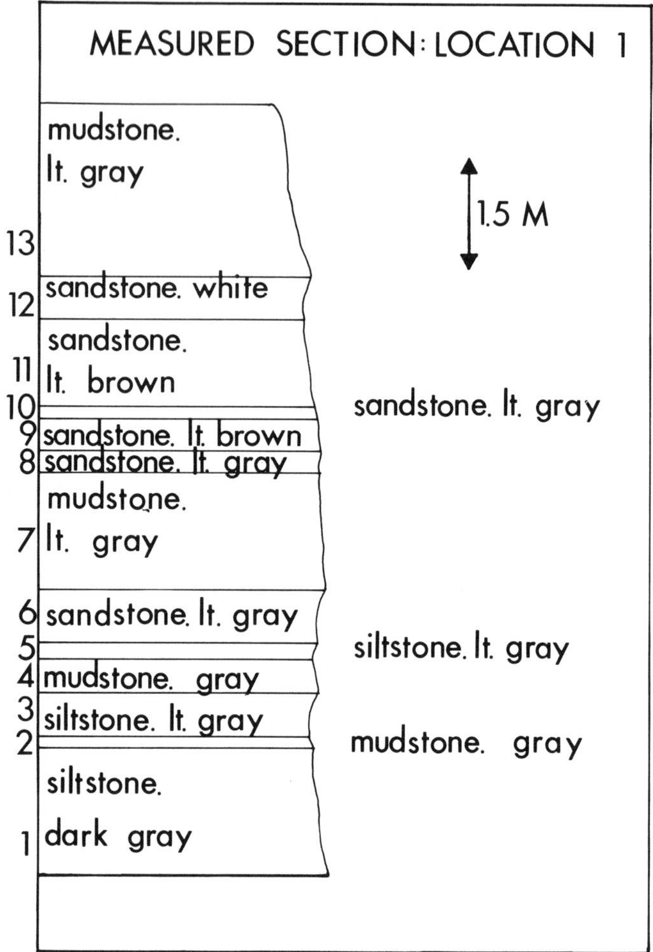

Figure 3. Measured section from the middle part of the Catahoula Formation, along the spillway south of the dam, Sam Rayburn Reservoir.

shard–rich and commonly contain diatoms, but most units are more than 50% silt-size volcanic glass shards exhibiting mostly bubble-wall and micropumice morphologies. A few shards are sand-size, but are difficult to see with the hand lens (Fig. 4). Details of glass share morphology are shown in Ledger (1981) and Ledger and others (1984).

The following is a brief description of the units in the measured section at the spillway just south the dam:

Unit 13. Mudstone; light gray, no color change with weathering, some sand and silt, lower contact conformable, 7.8 ft (2.4 m) thick.

Unit 12. Sandstone; quartzose, white, no color change with weathering, subangular to subrounded, some silica cementation, upper contact not exposed, 2 ft (0.6 m) thick.

Unit 11. Sandstone; light brown, poorly sorted, weathers lighter color, some silica cementation, lower contact conformable, 3.9 ft (1.2 m) thick.

Unit 10. Sandstone; light gray, no color change with weathering, subangular very fine sand, some plant fragments and iron-oxide cement (burrows?), lower contact conformable, 0.7 ft (0.2 m) thick.

Unit 9. Sandstone; light brown, poorly sorted, weathers tan, thinly bedded, subangular to subrounded, minor silica cementation, 1.6 ft (0.5 m) thick.

Unit 8. Sandstone; light gray, no color change with weathering, thinly bedded, very fine sand, some silica cementation, 1 ft (0.3 m) thick.

Unit 7. Mudstone; silty, light gray, weathers very light gray, thickly bedded, contact conformable, 4.9 ft (1.5 m) thick.

Unit 6. Sandstone, light gray, poorly sorted, no color change with weathering, very fine sand, minor feldspar and black chert, iron-oxide-cemented burrows, 2.6 ft (0.8 m) thick.

Unit 5. Siltstone; light gray, no color change with weathering, poorly sorted with very fine quartz sand, local iron-oxide cement, 0.7 ft (0.2 m) thick.

Unit 4. Mudstone; gray, weathers light gray, thinly to thickly bedded, thickness varies laterally, poorly sorted with some silt and sand, 1 to 3 ft (0.3 to 0.9 m) thick.

Unit 3. Siltstone; light gray, no color change with weathering, thinly bedded, poorly sorted, some quartz sand, local iron oxide, 2 ft (0.6 m) thick.

Unit 2. Mudstone; gray, weathers light gray, thinly bedded, some silt and sand, local iron oxide, 0.7 ft (0.2 m) thick.

Unit 1. Siltstone; dark gray, no color change with weathering, thickly bedded, local pyrite/marcasite nodules, lower contact not exposed, 5.9 ft (1.8 m) thick.

The lower part of the Catahoula Formation is poorly exposed along the beach to the north, and the contact with the underlying Whitsett Formation of the Eocene Jackson Group is near the point of land about 1.2 mi (2 km) north–northwest of the dam. About half way along this beach to the north is a small fluvial channel filled with almost pure volcanic ash. The delicate nature of the coarse silt to sand-size shards indicates that they were wind-transported from explosive eruptions in the volcanic source area and reworked only a very short distance in the small channel. The path of the channel appears to have been influenced by contemporaneous slumping of the not yet lithified Catahoula sediments, probably on the edge of a large flood plain that was choked with air-fall ash washed into the flood plain by smaller tributary channels.

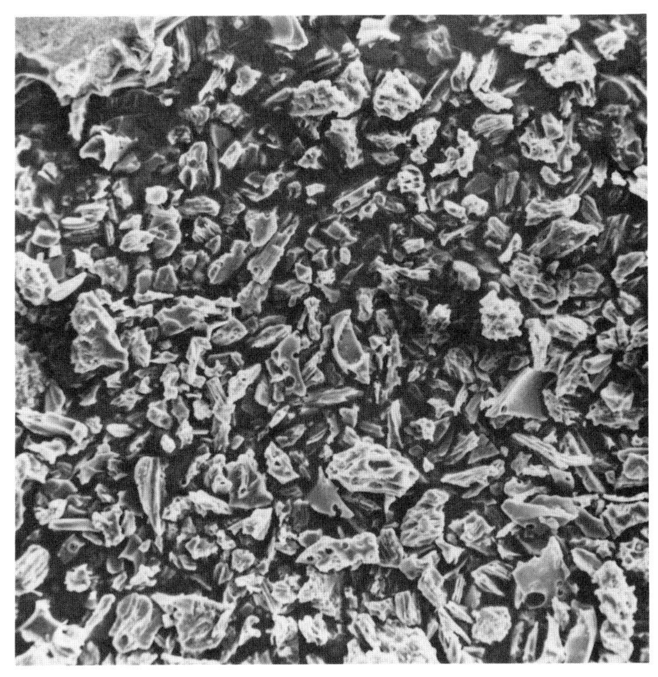

Figure 4. Scanning electron photomicrograph of Catahoula siltstone from Sam Rayburn Reservoir. Clay has been removed to show abundant glass shards. Field of view is 10 millimeters wide. Larger shards are about 100 micrometers.

REFERENCES CITED

Bailey, T. L., 1926, The Gueydan, a new middle Tertiary formation of the southwestern coastal plain of Texas: University of Texas Bulletin 2645, p. 1–187.

Dumble, E. T., 1918, The geology of east Texas: University of Texas Bulletin 1869, p. 1–388.

Galloway, W. E., 1977, Catahoula Formation of the Texas coastal plain: University of Texas at Austin, Bureau of Economic Geology Report of Investigations, no. 100, 81 p.

Ledger, E. B., 1981, Evaluation of the Catahoula Formation as a source rock for uranium mineralization, with emphasis on east Texas [Ph.D. thesis]: College Station, Texas A & M University, 263 p.

Ledger, E. B., Tieh, T. T., and Rowe, M. W., 1984, An evaluation of the Catahoula Formation as a uranium source rock in east Texas: Gulf Coast Association of Geological Societieis Transactions, v. 34, p. 99–108.

McBride, E. F., Linemann, W. L., and Freeman, P. S., 1968, Lithology and petrology of the Gueydan (Catahoula) Formation in south Texas: University of Texas at Austin, Bureau of Economic Geology Report of Investigations, v. 632, 122 p.

McDowell, F. W., 1978, Potassium-argon dating in the Trans-Pecos Texas volcanic field, *in* Walton, A. W., ed., Cenozoic geology of the Trans-Pecos volcanic field of Texas: Lawrence, University of Kansas, p. 9–18.

Renick, W. C., 1936, The Jackson Group and the Catahoula and Oakville Formation in a part of the Texas Gulf coastal plain: University of Texas Bulletin 3619, p. 1–104.

Sheldt, J. C., 1977, Petrology of the Catahoula sandstones of east Texas: Gulf Coast Association of Geological Societies Transactions, v. 27, p. 365–375.

Sparks, R.S.J., and Walker, C.P.L., 1977, The significance of vitric enriched air-fall ashes associated with crystal-enriched ignimbrites: Journal of Volcanology and Geothermal Research, v. 2, p. 329–341.

Weeks, A. D., and Eargle, D. H., 1963, Relation of diagenetic alteration and soil-forming processes to the uranium deposits of the southeast Texas coastal plain, *in* Clays and clay minerals; National Conference of Clays and Clay Minerals Proceedings: New York, MacMillan Co., v. 10, p. 23–41.

Vail, P. R., 1975, Eustatic cycles from seismic data for global stratigraphic analysis [abs.]: American Association of Petroleum Geologists Bulletin, v. 59, p. 2198–2199.

Precambrian and Paleozoic stratigraphy; Franklin Mountains, west Texas

David V. LeMone, Department of Geological Sciences, The University of Texas at El Paso, El Paso, Texas 79968

LOCATION AND ACCESS

The Franklin Mountains are located in far West Texas (El Paso County) and southernmost central New Mexico (Dona Ana County). The range, when including the North Franklin Mountains (north of Anthony Gap), is 23 mi (38 km) long and less than 5 mi (8 km) wide (Fig. 1). It consists of a linear series of north–south-trending, westward-tilted fault blocks, which are bounded on the east and west by major normal faults. Gravity slides are common throughout the range. The mountains are reasonable examples of the basin-and-range structural and physiographic type. The structure is distinctly different from the thrusted (autochthon and three allochthons) Cretaceous sequence of the northern Chihuahua trough of the Sierra de Juarez to the south in Mexico. The Texas lineament, the major structural feature separating these disparate northern and southern structural styles, if it exists, would pass through downtown El Paso.

Access to the area is by I-10. Three localities are suggested for examination (Fig. 1). Localities 1 and 2 are on paved, public thoroughfares. Locality 3 (Vinton Canyon) may be reached from I-10 at Westway, approximately 3 mi (4.8 km) south of the Texas–New Mexico line, by proceeding eastward on a dirt track toward the mountains. A high-clearance vehicle is required.

SIGNIFICANCE

The Franklin Mountains are a major key to the comprehension of the geology of this region; the contrast between the tectonic styles of this range and the Sierra de Juarez to the south is critical. The relatively undisturbed and continuous stratigraphic sequence is absolutely essential to the development of any regional analysis of this area of the southwestern United States and the adjacent portions of northern Mexico.

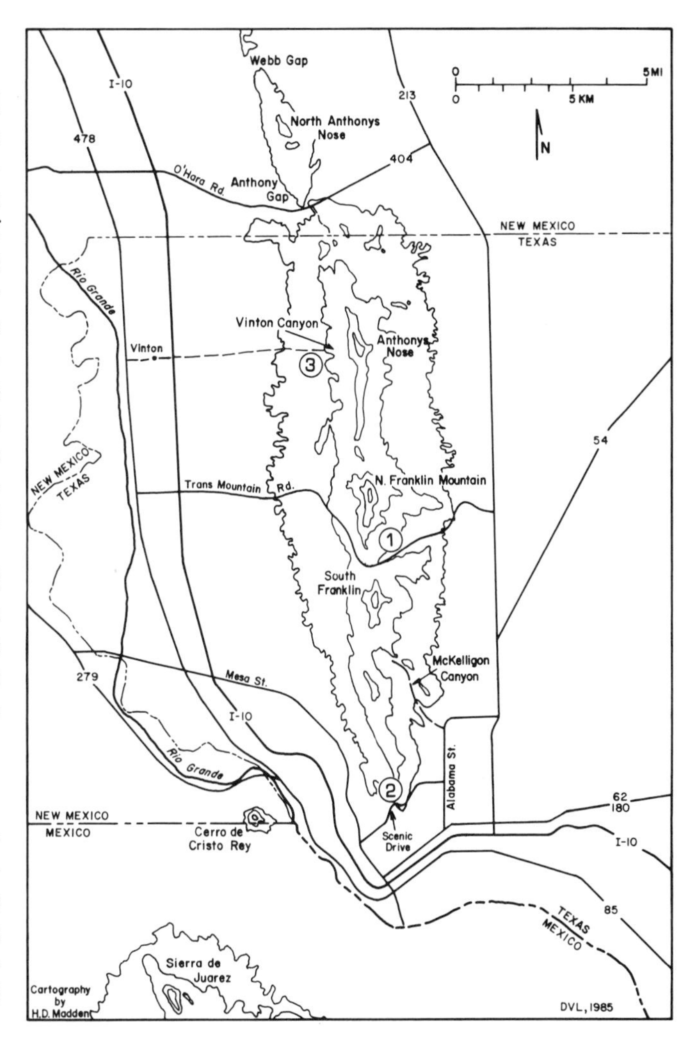

Figure 1. Index map showing places and localities (numbered) mentioned in the text.

SITE 87. PRECAMBRIAN STRATIGRAPHY

In the Franklin Mountains, Precambrian strata consist of six formations of metasediments and metaigneous rocks that are approximately 5,340 ft (1,628 m) thick and range in age from 995 to 950 Ma (Riphean). The intrusion of the Red Bluff Granite complex into Precambrian strata has been interpreted to be similar in development to the Nigerian (Niger Province) peralkaline and associated granitic ring complexes; a minimum of seven igneous phases of common magmatic origin are recognized (see Ray in LeMone, 1983).

Locality 1: Trans-Mountain Road. Precambrian sequence of the Eastern Slope

Exceptional and continuous exposures of the Precambrian are observed in Fusselman Canyon along the north side of Trans-Mountain Road (Fig. 1). A 4.75 mi (7.6 km) road guide is included from the intersection of Trans-Mountain Road and the North–South Freeway westward to the crest of Smuggler's Pass. Trans-Mountain Road is a public thoroughfare. It is part of the old U.S. Army Castner Artillery Range; do not disturb or pick up any shells or ordnance fragments as they may still be live. Exhibit more than normal care in climbing in this area, especially in descending the frequently unstable slopes.

The six recognized units of Precambrian (Riphean) sedimentary and volcanic strata of the Franklin Mountains are approximately 5,340 ft (1,628 m) thick. They have been intruded and

Geochronologic Units	Chronostratigraphic Units	Group	Lithostratigraphic Units	
PERMIAN	WOLFCAMPIAN	HUECO GROUP	ALACRAN MT. FM. CERRO ALTO LS. HUECO CANYON FM.	8
PENNSYLVANIAN	VIRGILIAN MISSOURIAN		PANTHER SEEP FM.	7
	DES MOINESIAN ATOKAN MORROWAN	MAGDALENA GROUP	BISHOP CAP FM. BERINO FM. LA TUNA FM.	6
MISSISSIPPIAN	CHESTERIAN		HELMS FM.	5
	MERAMECIAN		RANCHERIA FM.	
	OSAGE - MERAMEC		LAS CRUCES FM.	
DEVONIAN	UPPER		PERCHA SHALE	4
	MIDDLE		CANUTILLO FM.	
SILURIAN	NIAGARAN (MIDDLE) ALEXANDRIAN (LOWER)		FUSSELMAN DOL.	3
ORDOVICIAN	CINCINNATIAN (UPPER)	MONTOYA GROUP	CUTTER FM. ALEMAN FM. UPHAM DOL.	2
	CANADIAN (LOWER)	EL PASO GROUP	FLORIDA MTS. FM. SCENIC DRIVE FM. McKELLIGON FM. JOSE FM. VICTORIO HILLS FM. COOKS FM. SIERRITE FM.	1
CAMBRO-ORDOVICIAN	CROXIAN AND/OR CANADIAN		BLISS SANDSTONE	
YOUNGER PRECAMBRIAN	RIPHEAN	INTRUSIVES RIEBECKITE GRANITE RED BLUFF GRANITE COMPLEX MICROGRANITE SILLS	THUNDERBIRD GROUP – TOM MAYS PARK FM., SMUGGLERS PASS FM., CORONADO HILLS FM. LANORIA QUARTZITE MUNDY BRECCIA CASTNER MARBLE	

(LeMone 1982)

Figure 2. Paleozoic and Precambrian stratigraphy of the Franklin Mountains.

metamorphosed by the Red Bluff Granitic Complex (units 1 to 4; Fig. 3). Radiometric age dates for the Precambrian here range from 995 to 917 Ma; the most reasonable range would be from 995 to 953 Ma (Thomann, 1981; LeMone, 1983).

Seven references for review and additional pertinent references are recommended. Harbour (1972) and LeMone (1983) are recommended for general relationships for the entire Precambrian. Specific data and best discussions for the strata (listed in ascending order) are: Castner Marble (Hoffer, 1976); Castner stromatolites (Toomey, 1983); Mundy Breccia (Thomann and Hoffer, 1985); Lanoria Quartzite (Harbour, 1972); Thunderbird Group (Coronado Hills Conglomerate, Smuggler's Pass Formation, and Tom Mays Park Formation) (Thomann, 1981). The current and best interpretation of the Red Bluff Granite Complex in the area has been made by Ray (1982).

The oldest exposed Precambrian is the Castner Marble. Its stratotype was measured approximately 1,000 ft (305 m) north of Trans-Mountain Road in a roof pendant exposing the maximum thickness of the Castner Marble (1,390 ft; 424 m) and a complete section of the Mundy Breccia (Stops 5–7; Table 1), which is disconformably overlain by a small sequence of the Lanoria Quartzite. This roof pendant is underlain and overlain by microgranite sills and is bounded on the north side by a biotite granite intrusion, all of which are related to the Red Bluff Granite Complex. The Castner is a massive, medium- to coarse-grained marble alternating with layers of hornfels and intruded by hypabyssal diabase sills. Fine-grained clastics of the hornfels increase in quantity toward the top of the Castner. The well-preserved, typically ribbed, sequence of greenish gray hornfels and more easily eroded gray to light gray marbles also display excellent examples of load

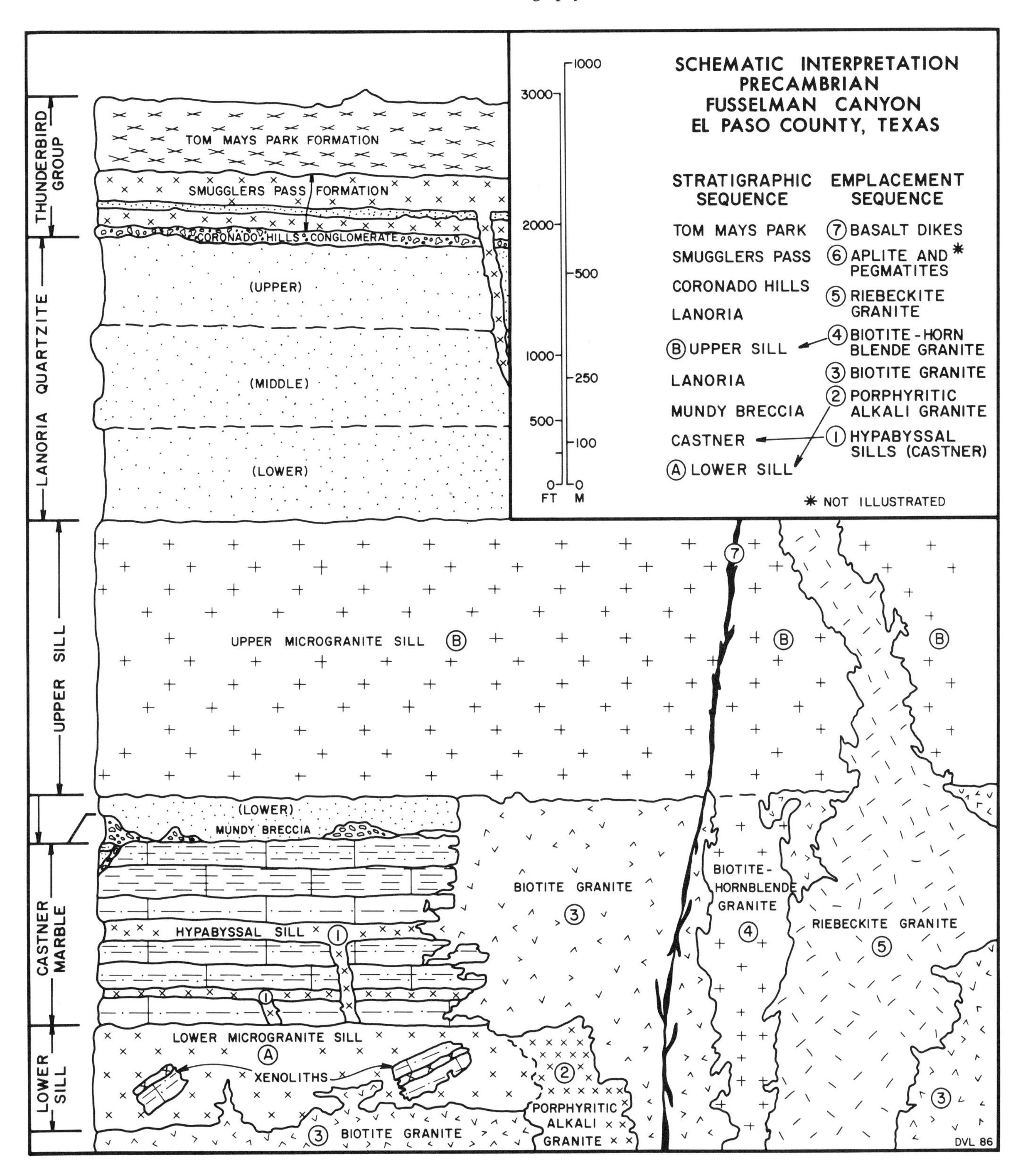

Figure 3. Schematic interpretation of the Precambrian geology along the Trans-Mountain Road.

TABLE 1. SIMPLIFIED ROAD LOG OF PRECAMBRIAN EXPOSURES ALONG THE TRANS-MOUNTAIN ROAD ON THE EAST SIDE OF THE FRANKLIN MOUNTAINS

Stop Number	Cumulative mi	(km)	Description
1	0.00	(0.0)	Intersection of Trans-Mountain Road and the North-South Freeway (old War Road).
2	0.65	(1.0)	Wilderness Park Museum. Road cuts in biotite granite primarily.
3	1.65	(2.6)	The Teardrop Roadcut. This spectacular roadcut (approximately 0.2 mi or 0.3 km long) displays a series of xenoliths composed of Castner Marble and its associated hypabyssal sills floating in granite. These rocks also display numerous light-colored aplite dikes as well as pegmatites. An excellent exposure of a riebeckite pegmatite is observed on the eastern end of this exposure. At the west end of the roadcut, a near-vertical dipping xenolith of the Castner Marble may be observed. Please exercise extreme caution at this roadcut outcrop, as well as the others on this transect of the Precambrian, as the back is very poor and frequently avalanches large blocks down on the roadway edge.
4	2.25	(3.6)	Lower microgranite sill. Estimated by Ray to be 815 ft (248 m) thick.
5	2.35	(3.8)	Base of the Castner stratotype. Note the sharp boundary between the lower microgranite sill and the injected base of the Castner. The basal whitish units, especially well exposed at the basal contact with the microgranite, seen north of road, are algal stromatolites. The near-black units, well exposed in the medial part of the section, are the Castner hypabyssal sills.
6	2.55	(4.1)	Parking area for the examination of the upper Castner. Note the excellent development of euhedral to subhedral garnets on bedding planes in the upper section. Note the increase of soft sediment deformation toward the top of the unit.
7	2.7	(4.3)	Mundy Breccia. The relationships of the Castner-Mundy, Mundy-Lanoria, Castner-Lanoria, and Lanoria-upper microgranite sill contacts require climbing to the north. The north side of this Castner-Lanoria block is intruded by biotite granite.
8	2.85	(4.6)	Upper microgranite sill. This sill is estimated to be at least 2,145 ft (654 m) thick (Ray, 1982).
9	3.4	(5.5)	Small roadcut displaying a typical outcrop of the Lanoria Quartzite.
10	4.45	(7.2)	Colluvial conglomerates of Pleistocene age. Across the road on the south side are the exposures of the roadcut into Smuggler's Pass. The basal conglomerates of the Coronado Hills Conglomerate of the Thunderbird Group may be observed on the east side of this long exposure. All three formations of the Thunderbird Group can be observed by walking through the roadcut. Please note the problems of major sliding caused by the jointing characteristics of the group.
11	4.75	(7.6)	Crest of Smuggler's Pass at 5,280 ft (1,600 m) (Milepost 7). Carefully veer over to left lane, cross median and oncoming traffic to Stop 12.
12	4.85	(7.8)	Overlook of the Mesilla Valley. Small limestone hills in the immediate foreground include the Alacran Mountain Formation of late Wolfcamp (early Permian) age, which is the uppermost formation of the Franklin Mountains Paleozoic section. The west-boundary fault zone is between the overlook and the light-colored hills of Permian carbonates. North Mount Franklin (7,192 ft; 2,193 m) lies immediately to the north; it is the highest point in the Franklin Mountains. Its peak consists of the Tom Mays Park Formation of the Thunderbird Group, and is the highest structural point in the state of Texas.

casts, soft sediment deformation, intraformational breccias, and edgewise conglomerates toward the top of the section. In addition, distinctive stromatolitic structures, mounding, bioturbation, oolites, ripple marks, and tepee structures are reported and best developed in the lower portion of this section. Castner stromatolite mounds range up to 5 ft (1.5 m) in height and 6 to 8 ft (1.8 to 2.4 m) in width. They are composed of relatively broad colonies of vertically stacked hemispheroidal stromatolites. Some of these structures clearly demonstrate a central column and V-shaped laminae typical of conophyton-type stromatolites. Laminated marbles near the base of the section are interpreted to be the remains of algal mats (Toomey, 1983).

The environment of deposition of the Castner is interpreted to have been high intertidal to shallow subtidal. It was deposited on a normal marine, mixed clastic-carbonate shelf. Correlation of this sequence to the Allamore Formation in the Van Horn area to the east is questionable at best, for a variety of reasons (e.g., age, stromatolite types, etc.). Hoffer (1976) has concluded that the sequence has been metamorphosed to the hornblende-hornfels and K-feldspar-cordierite-hornfels facies, which indicate a pressure to 1,000 bars and temperature that ranged from 515 to 670°C. An additional series of xenolithic blocks is observed at Stop 3 (Table 1), including a magnificent vertical exposure on the west end of that outcrop.

The overlying Mundy Breccia has been most recently interpreted as an epiclastic volcanic breccia as defined by Fisher in 1961 (Thomann and Hoffer, 1985). Reported thickness ranges from 0 to 250 ft (0 to 76 m). The breccia filled channels and produced differential loading depressions. Its original thickness is unknown as it was eroded prior to burial by sediments of the Lanoria Quartzite. The thick, viscous, mudflow breccia may have formed from either the erosion of a lava flow or from a shallowly emplaced sill. The unsorted, unstratified breccia cannot be positively interpreted as a single or a multiple flow unit nor can it be determined whether the breccia was a surface or subsurface flow (Thomann and Hoffer, 1985).

The overlying Lanoria Quartzite (2,600 ft; 793 m) (Stop 9; Table 1) is divided into three members (Harbor, 1972). The poorly exposed lower member consists of fine-grained quartzites, quartzitic siltstones, and "baked" shales (1,100 ft; 335 m). It is overlain by a distinctive, vertical cliff–forming, middle member quartzite (800 ft; 244 m), which is easily distinguished by a medial dark-weathering band that is in sharp contrast to the adjacent white-weathering quartzites. The poorly exposed, slope-forming upper member (700 ft; 213 m) consists of gray- to brown-weathering sandstones, siltstones, and shales. The Lanoria disconformably overlies an irregular topography of the Mundy Breccia and the Castner Marble. It is disconformably overlain by sediments and volcanics of the Thunderbird Group.

Thomann (1981), on the basis of detailed petrographic and stratigraphic analyses, established the Thunderbird Group. It consists of three formations (in ascending order): Coronado Hills, Smuggler's Pass, and Tom Mays Park. The stratotype section for the group is 3,000 ft (915 m) northeast of Smuggler's Pass on North Mount Franklin on the south side of an arroyo.

The basal Coronado Hills Conglomerate (35 to 90 ft; 11 to 27 m) consists of well-rounded cobbles and pebbles of quartzite, siltstone, shale, chert, jasper, ignimbrite, and trachyte. This section, which is well exposed along the south side of the easternmost part of the outcrop at Smuggler's Pass (Stop 10; Table 1), develops horizontal bedding, scour-fill structures, and cross-bedding.

The overlying Smugglers Pass Formation (460 ft; 140 m) consists of (in decreasing thicknesses): porphyritic trachytes (200 ft; 61 m), basal melanocratic porphyritic trachytes (150 ft; 46 m), tuffaceous sandstones and conglomerates (90 ft; 27 m), silicified ignimbrites (20 ft; 6 m), and a few thin (several cm), locally silicified, ash-fall deposits.

The youngest stratigraphic unit is the Tom Mays Park Formation (550 ft; 168 m), which was extensively eroded prior to Paleozoic deposition. It is an ignimbrite suite (953 ± 13 Ma) composed primarily of metamorphosed crystal-vitric tuffs. It also contains widely scattered porphyritic rhyolite dikes. Thomann (1981) considers the petrographically similar metavolcanics at the Pump Station Hills (70 mi; 113 km) east of El Paso to be correlative to the altered ignimbrites of the Tom Mays Park Formation. Both Ray (1982) and Thomann (1981) consider the strong possibility that there is a comagmatic relationship between the metamorphosed Thunderbird Group flows and ignimbrites and the intrusive granites of the Red Bluff Granite Complex.

The Red Bluff Granite was originally designated for exposures on the cliffs in the arroyo behind the outdoor amphitheater in McKelligon Canyon (Fig. 1) in the southern Franklin Mountains. Subsequent studies in the area of the Trans-Mountain Road by Ray (1982) refer to the intrusive phase of the Precambrian as a complex of associated rocks having a compatible mineralogy and geochemistry. He also suggests that the magma that intrudes the Precambrian strata (from the Castner Marble to the Tom Mays Park Formation) is related to incipient continental rifting. The concordantly injected, oldest, hypabyssal mafic sills that occur in the Castner Marble are not considered to be part of the Red Bluff Granite Complex.

Ray (1982) has proposed a relatively elegant model for the intrusive history of the Red Bluff Granite Complex. He recognizes three distinct, vertically differentiated phases from the parent magma of the Red Bluff Granite Complex. They are (in order of sequential emplacement): porphyritic alkali granite (most differentiated and derived from the upper levels of the magma chamber); biotite granite (intermediate in differentiation and initiated from intermediate levels in the magma chamber); and biotite-hornblende granite (least differentiated and developed from the lower levels of the magma chamber).

The upper and lower microgranite sills, in which the Castner Marble, Mundy Breccia, and lower member of the Lanoria Quartzite seem to be floating (Stops 6 and 7; Table 1), represent fractions derived from different levels of the magma chamber. The lower microgranite sill is interpreted to have developed during the early stages of porphyritic alkali granite intrusion from the upper levels of the magma chamber and was injected concordantly into the lower section of the Castner Marble. The upper microgranite sill, which was injected into the lower member of the Lanoria Quartzite, is interpreted by Ray (1982) to have been from a source in the lower levels of the magma chamber at the time of the intrusion of the biotite-hornblende granites.

He believes that the later riebeckite granite intrusion (unit 5; Fig. 2) has no direct genetic relationship to the other granite phases. He interprets it as being derived from a separate parent magma that intruded all other earlier phases of the complex. The numerous pegmatites and aplites of the area (Stop 3; Table 1) were late, residual fluids genetically related to the various phases of the magma. These late-stage intrusives were injected along an intricate network of fractures primarily developed by forces generated during intrusion. Basalt dikes of variable thickness (unit 7; Fig. 2) are interpreted to be the last phase of intrusive activity despite the fact that cross-cutting relationships with the riebeckite granite cannot be demonstrated. The interiors of the wider basalt dikes develop a diabasic texture.

One should be reminded, while examining the exposures

along the Trans-Mountain Road, that this thick suite of Precambrian metasediments and metavolcanics forms a distinct sequence of strata. This fact should be kept in mind constantly when interpreting geophysical reflection seismic data in the local region. Significant similarities between the emplacement of the Red Bluff Granite and other intrusions, such as the peralkaline and associated granite ring complexes of the Nigeria–Niger Province (Ray, 1982), offer interesting possible analogs for those involved in igneous petrology. Metamorphic petrologists will enjoy detailed examination of this Precambrian suite, especially the Castner Marble. Structural geologists will be especially interested in complex jointing patterns developed from the sequence of intrusions in the range. Investigation of the sedimentary structures and algal development of the Castner in comparison to temporal equivalents will be of interest to sedimentologists and paleophycologists. Finally, anyone contemplating regional studies of the Precambrian in this area should visit these well-exposed, easily accessible Franklin Mountain sequences.

SITE 88. PALEOZOIC STRATIGRAPHY

The Paleozoic sequence consists of 23 formations and is separated from Precambrian rocks by a major nonconformity (±450 Ma). The section is 8,910 ft (2,717 m) thick and ranges in age from Gasconadian (Early Ordovician) to Wolfcampian (Early Permian) (LeMone, 1983). The Paleozoic sediments of the Franklin Mountains may be divided into eight discrete sedimentation units (Fig. 2): (1) Early Ordovician Bliss Sandstone and El Paso Group; (2) Late Ordovician Montoya Group; (3) Early and Middle Silurian Fusselman Dolomite; (4) Middle Devonian Canutillo Formation and Late Devonian Percha Shale; (5) Late Mississippian Las Cruces Formation (Osage–Meramec), Rancheria Formation (Meramec), and Helms Formation (Chester); (6) Early and Middle Pennsylvanian Magdalena Group; (7) Late Pennsylvanian Panther Seep Formation; and (8) the Early Permian (Wolfcamp) Hueco Group.

Locality 2. Scenic Drive and Scenic Point. Paleozoic sedimentological units 1 and 2; Ordovician.

Park at Scenic Point and examine the disconformity between the Late Ordovician Montoya Group (Unit 2) (447 ft; 136 m) and the underlying Early Ordovician El Paso Group and Bliss Sandstone (Unit 1; Fig. 3). The cliff-forming Upham Formation (103 ft; 31 m) is a mottled (bioturbated), massive, normal marine, tropical, gray to brown-gray dolomite that contains a spectacular fauna of brachiopods, trilobites, gastropods, receptaculitid algae, corals (solitary and colonial), and a minimum of 14 genera of nautiloids. The formation may be examined by walking a short distance to the west.

The underlying Bliss–El Paso Group forms a time-transgressive unit that is older to the west and younger to the east. It has been progressively eroded to a feather edge in a regional angular unconformity that extends north into central New Mexico. Therefore, the section exposed here displays one of the most complete Canadian sections available in the area. It should be recalled that the next Paleozoic section south–southeast of El Paso is exposed at Placer de Guadalupe, Chihuahua, 200 mi (320 km) away, and is apparently unrelated to the Paleozoic sequence of the Franklin Mountains.

Seven of the ten recognized regional formations of the El Paso Group are exposed here. Taking care for traffic on the descent, one passes stratigraphically downward through the following units: the Florida Mountains Formation (39 ft; 12 m); the Scenic Drive Formation, including the Nameless Canyon Member (228 ft; 70 m) and Black Band Dolomite Member (60 ft; 18 m); the McKelligon Formation, including the Mounds Member (605 ft; 184 m) and Pistol Range Member (70 ft; 21 m); the Jose Formation (72 ft; 22 m); the Victorio Hills Formation, (290 ft; 88 m); the Cooks Formation (109 ft; 33 m); The Sierrite Formation (121 ft; 37 m); and the Bliss Sandstone (225 ft; 69 m). The base of the Paleozoic section is not seen on Scenic Drive. However, it may be easily observed in the McKelligon Canyon area (Fig. 1). The most notable features in this equatorial, shallow-water, subtidal to supratidal sequence that was deposited on a planar to irregular surface are: Scenic Drive Black Band Dolomite Member, supratidal sediments and clastics; the cyclic *Pulchrilamina spinosa* mounds of the Mounds Member of the McKelligon Formation; the distinct, orange-weathering, basal sand unit of the Pistol Range Member of McKelligon Formation; the cyclic digitate algae and sand units of the lower 54 ft (16.5 m) of the Jose Formation; and the upper glauconitic member of the Bliss Formation.

Locality 3: Vinton Canyon. Paleozoic Sedimentological Units 3–8: Silurian Through Permian

Vinton Canyon (Fig. 4) is distinguished by having stratotypes for the Fusselman, Canutillo, Las Cruces, Rancheria, La Tuna, Berino, and Bishop Cap formations (Fig. 2). The area is reached from I-10 by taking the Vinton–Westway Exit and turning east toward the Franklin Mountains at the service stations. The road is followed directly to the range and becomes a dirt track.

Approximately 3.5 mi (5.6 km) east of the stations, there is a low range of hills with quarries that contain excellent exposures of the Hueco Canyon Formation of the Hueco Group (Unit 8) (2,514 ft; 766 m). This locality contains outstanding specimens of phylloid algae and fusulinids, in combination with a great diversity of marine fossils.

The next 0.6 mi (1 km) is covered and is interpreted to be underlain by the Panther Seep Formation (Unit 7) (1,180 ft; 376 m). At the first outcrops to the south there are excellent exposures of the basal conglomerate of this formation (Fig. 4). Continue straight ahead to the road fork where the top formation of the Magdalena Group (Unit 6), the Desmoinesian Bishop Cap Formation (590 ft; 180 m), is exposed. Phylloid algae and an exceptional invertebrate fauna can be seen to the south. Continue ahead on the dirt track. Do not descend into the canyon. On the south wall of the canyon is an excellent exposure of a bed of small

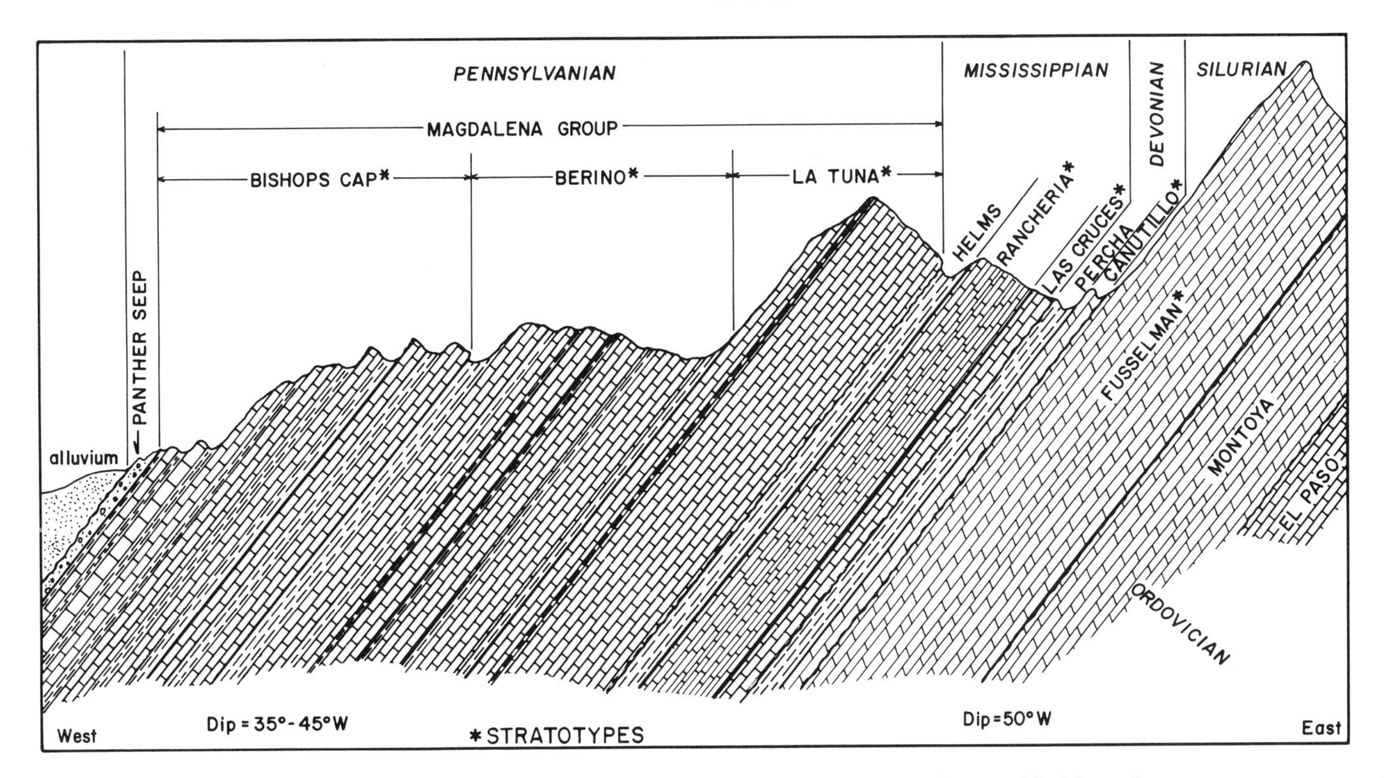

Figure 4. Schematic stratigraphic cross section at Vinton Canyon, west flank of the Franklin Mountains (modified from Nelson, 1940).

masses of the sclerosponge *Chaetetes milleporaceous.* The lowest beds exposed in the quarry below the parking area in the north wall of Vinton Arroyo are at the contact between the Atokan–Desmoinesian Berino (514 ft; 157 m) and the underlying Morrowan–Atokan La Tuna (505 ft; 154 m) formations. Walk up the canyon cut through the basal cliffs of the La Tuna, which are underlain by shales of the Mississippian sequence (Unit 5). The Helms–La Tuna disconformity may be examined by climbing to the dissected alluvial bench on the south wall of the canyon. Please use extreme care in examination of the cliff base.

The Mississippian sequence in the Franklin section does not include the Caballero Formation and the overlying Lake Valley Group (Kinderhook and Osage), which are so well exposed to the north in New Mexico. The sequence here is shallowing upward from the distal carbonate turbidites of the Las Cruces Formation to the shallow-water carbonates of the uppermost Helms Formation. The essentially shaly Helms Formation (Chester) (98 ft; 30 m) may be examined along the cut on the south wall of the canyon. Recross the arroyo to the north bank and continue on up the canyon to examine the underlying Rancheria Formation (Meramec). This 411-ft (124-m) formation is easily divisible into three members: upper and lower orange members and a light gray middle member. The basal Mississippian unit is the thin (79 ft; 24 m), distinct, Las Cruces Formation (Osage–Meramec). This very light gray cliff- and ridge-forming unit is exposed on the east-facing slope of the north-trending branch of the canyon arroyo. It is a light gray, fetid, dense, hard limestone, which is gray-black on fresh surfaces. It is interpreted to be composed of distal carbonate turbidites with minor siliciclastics deposited in a deep, restricted, marine basin. The Las Cruces is underlain by the Devonian Percha Shale (Late) and Canutillo Formation (Middle) (Unit 4).

The Percha Shale (140 ft; 42 m) is mostly covered, black, and weathers with limonite staining. By traversing north along the Las Cruces cliff front, one can see an exposure of the formation on the north side of a westward-trending arroyo. In the arroyo, below this sequence, there is an excellent exposure of the cherty, clayey, dolomitic Canutillo Formation of Middle Devonian age.

The underlying Early to Middle Silurian Fusselman Formation (1,000 ft; 305 m) (Unit 3) here forms the light-colored carbonate dip slope on the west flank of the main ridge of the Franklin Mountains. It is composed of a series of sequences of shoaling-upward carbonates that are typically capped by supratidal units. The upper contact of this karsted unit is typically stained to a reddish hue.

The sequence exposed in Vinton Canyon is primarily of carbonates with minor siliciclastics. It contains six of the eight sedimentological units of the Paleozoic. It is fossiliferous throughout the entire Silurian through Permian sequence, with the apparent exception of Unit 7, the Late Pennsylvanian Panther Seep Formation.

REFERENCES CITED

Harbour, R. L., 1972, Geology of the northern Franklin Mountains, Texas and New Mexico: U.S. Geological Survey Bulletin 1298, 129 p.

Hoffer, R. L., 1976, Contact metamorphism of the Precambrian Castner Marble, Franklin Mountains, El Paso County, Texas [M.S. thesis]: El Paso, University of Texas, 77 p.

LeMone, D. V., 1983, Stratigraphy of the Franklin Mountains, El Paso County, Texas, and Dona Ana County, New Mexico, *in* Delaware Basin Guidebook: West Texas Geological Society Publication No. 82-76, p. 42–72.

Nelson, L. A., 1940, Paleozoic stratigraphy of the Franklin Mountains, West Texas: Bulletin of the American Association of Petroleum Geologists, v. 24, p. 157–172.

Ray, D. R., 1982, Geology of the Precambrian Red Bluff Granite Complex, Fusselman Canyon Area, Franklin Mountains, Texas [M.S. thesis]: El Paso, University of Texas, 295 p.

Thomann, W. F., 1981, Ignimbrites, trachytes, and sedimentary rocks of the Precambrian Thunderbird Group, Franklin Mountains, Texas: Geological Society of America Bulletin, v. 92, p. 94–100.

Thomann, W. F., and Hoffer, J. M., 1985, Petrology and geochemistry of the Precambrian Mundy Breccia, Franklin Mountains, Texas: Texas Journal of Science, v. 36, p. 267–281.

Toomey, D. F., 1983, Stromatolites in the Precambrian Castner Marble, Fusselman Canyon, northern Franklin Mountains, West Texas, *in* Precambrian and Paleozoic algal carbonates, West Texas–Southern New Mexico: Colorado School of Mines Professional Contributions no. 11, p. 92–114.

The southern Guadalupe Mountains, Texas; Permian stratigraphy and Great Plains/Basin and Range structural transition

Cleavy L. McKnight, Department of Applied Earth Sciences, Stanford University, Stanford, California 94305

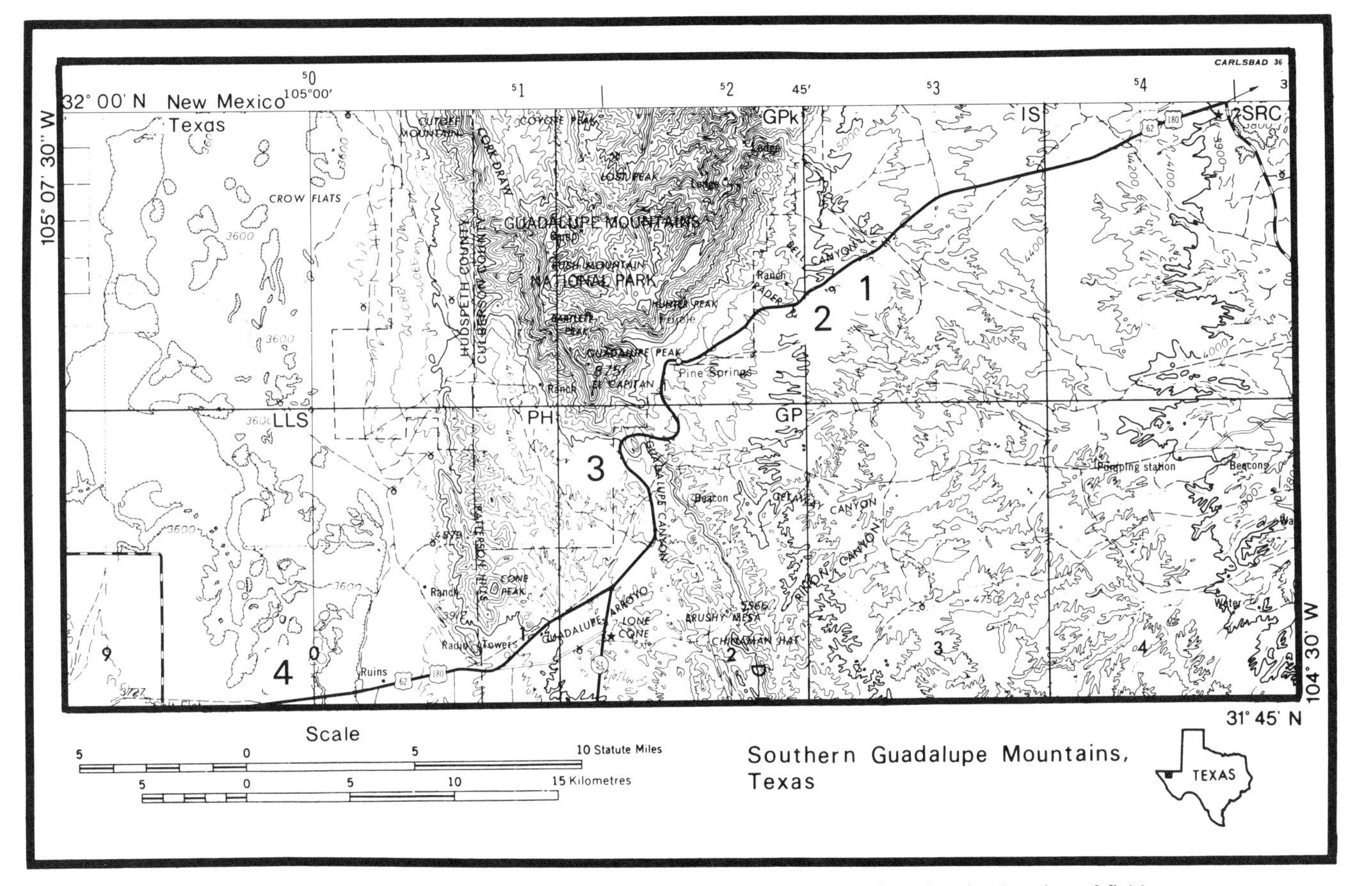

Figure 1. Topographic map of the southern Guadalupe Mountains region, showing location of field localities and 7½-minute topographic quadrangle maps traversed along the route. Numbers refer to stops described in the text. Abbreviations of 7½-minute quadrangle names are located in the northeast corner of each quadrangle. Key to 7½-minute quadrangles: SRC, South Rector Canyon; IS, Independence Spring; GPk, Guadalupe Peak; GP, Guadalupe Pass; PH, Patterson Hills; LLS, Linda Lake South.

LOCATION

The Guadalupe Mountains, easternmost of the block-faulted ranges of the Basin and Range at this latitude, expose a classic section of Permian shelf and basin carbonates and clastics (King, 1942, 1948). In a series of panoramic views and roadcuts the southern Guadalupe Mountains in Culberson and Hudspeth counties, Texas, display both the Permian stratigraphic transition of shelf clastics and evaporites to shelf-margin carbonate buildups to basinal clastics and the modern physiographic/structural transition from the relatively undisturbed, flat-lying sedimentary strata of the Great Plains to the tilted and block-faulted rocks of the Basin and Range. The exposures are all located along U.S. 62-180. Stop 1 is located 43 mi (69 km) southwest of Carlsbad, New Mexico; Stop 4, the final locality, is located 37 mi (59 km) farther west, or 96 mi (154 km) east of El Paso, Texas. The immediate field area is shown in Figure 1. The regional geography is shown on the Van Horn, Texas (1975), and Carlsbad, New Mexico (1972), 1° × 2° topographic sheets; geographic place names not shown in Figure 1 may be found on these two maps. The route traverses the South Rector Canyon (1973), Independence Spring (1973), Guadalupe Peak (1973), Guadalupe Pass (1973), Patterson Hills (1973), and Linda Lake South (provisional, 1984) 7½-minute Quadrangles, all in Texas.

INTRODUCTION

The Guadalupe Mountains represent the exhumed remnant of a shelf-edge carbonate buildup that fringed the Delaware Basin of west Texas and southeastern New Mexico in Permian time

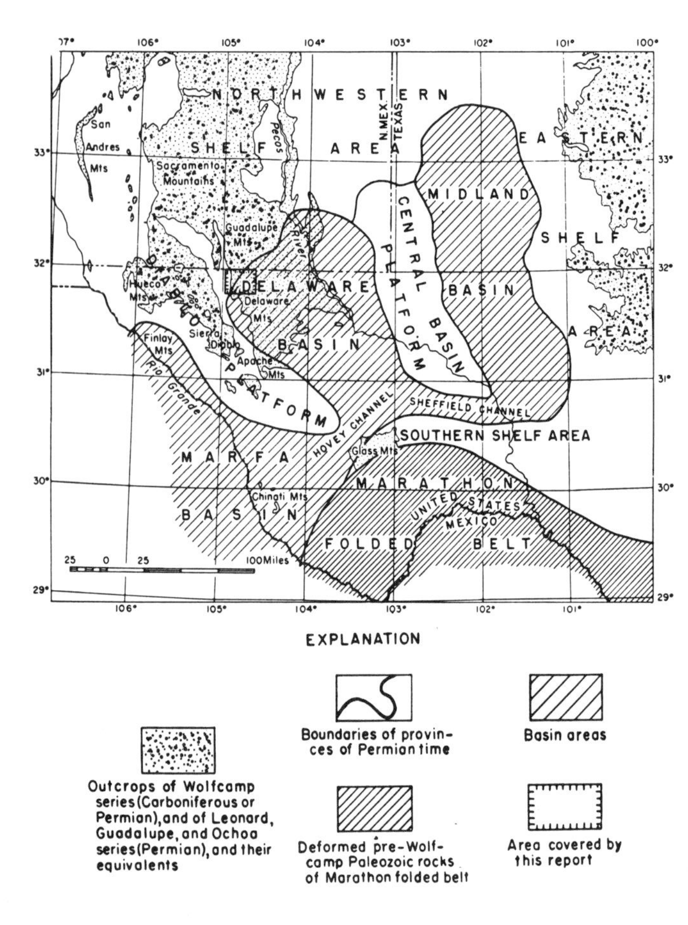

Figure 2. Index map showing location of field trip area (hachured box) in relation to paleogeographic provinces during Permian time (from King, 1948). Permian carbonates deposited around the margins of the Delaware Basin are exposed in the Guadalupe, Sierra Diablo, Apache, and Glass mountains of west Texas.

(Fig. 2). Their location adjacent to a major petroleum province, the Permian Basin, adds to their importance as a natural laboratory for the study of geomorphology, stratigraphy, structural geology, sedimentology, and paleogeography. A threefold division exists between the facies tracts of the shelf, shelf margin, and basin (Fig. 3); strata representative of each of these facies tracts can be seen either in roadcuts or in panoramic views of the eastern and western escarpments of the southern Guadalupe Mountains along this field-trip route.

The southern Guadalupe Mountains in Texas are located in Guadalupe Mountains National Park, and the field stops are along the main highway skirting the park. At each stop, **caution should be exercised when crossing the highway.** The stops are listed in the order in which they would be encountered during a drive southwest from Carlsbad, New Mexico. An additional field stop of related interest is the exposure of laminated Upper Permian (Ochoan) Castile Formation basin-filling evaporites described by Anderson and Kirkland (1987) in the companion Rocky Mountain Section Centennial Field Guide volume. The exposure is located 0.75 mi (1.2 km) north of the Texas–New Mexico state line on U.S. 62-180, and may be conveniently visited on the way from Carlsbad, New Mexico, to the first stop described in this paper.

The route begins in Quaternary alluvial fan and stream deposits shed from the eastern face of the Guadalupe Mountains toward the Pecos River to the east. The Guadalupe Mountains proper are an uplifted, gently northeast-dipping block uplift, bounded on the west by a zone of steep, down-to-the-west normal faults. Uplift of the range dates from Plio-Pleistocene time. The upper surface of the range is essentially a dip slope, modified extensively by fluvial erosion. Ground water, too, has played an important role in the geomorphic evolution of the range (McKnight, 1986), as evidenced by the numerous caves present in the mountains; Carlsbad Caverns in New Mexico is the largest.

The starting point for this trip is about 7.5 mi (12 km) from the eastern scarp face, or basinward of the Permian shelf margin. As the trip progresses, and the mountain front is approached at an oblique angle, progressively older Permian rocks of the Delaware Basin are exposed. Stop 1 is a roadcut in the black, anoxic basinal Lamar Limestone Member of the Permian Bell Canyon Formation (Fig. 3), with an opportunity to view the Permian Reef Escarpment in the western distance. Stop 2 exposes a submarine debris flow in the Rader Limestone Member of the Bell Canyon Formation in another roadcut. The route crosses the Permian shelf margin, where a Quaternary fault zone marks the modern structural transition from the Great Plains to the Basin and Range. Stop 3 presents a view of El Capitan, the southern promontory of the Guadalupe Mountains, and the Delaware Mountain Group basinal sands. The trip ends west of the mountains in the Salt Basin, a graben that forms the easternmost basin in the Basin and Range province. Stop 4 provides a panoramic view of the western fault escarpment of the Guadalupe Mountains, where both Permian sedimentary features and Basin and Range normal faulting are displayed. This stop also affords a chance to examine modern playa evaporites forming in the Salt Basin graben. In addition to the stops themselves, some geologic features along the road in the reaches between stops are noted.

ROUTE DESCRIPTION

For convenience in measuring mileage, the route begins at the junction of Texas Ranch Road 652 with U.S. 62-180, 0.1 mi (0.2 km) south of the New Mexico/Texas state line and 33 mi (53 km) southwest of downtown Carlsbad, New Mexico (Culberson County; 104°32′03″W; 31°59′58″N; elevation 3,940 ft or 1,201 m; South Rector Canyon Quadrangle).

Stop 1. Basinal Lamar Limestone. Drive west on U.S. 62-180 for 9.9 mi (15.9 km; Culberson County; 104°41′40″W; 31°58′07″N; elevation 4,730 ft or 1,442 m; Independence Spring Quadrangle). Park on the side of the road, just west of the roadcut.

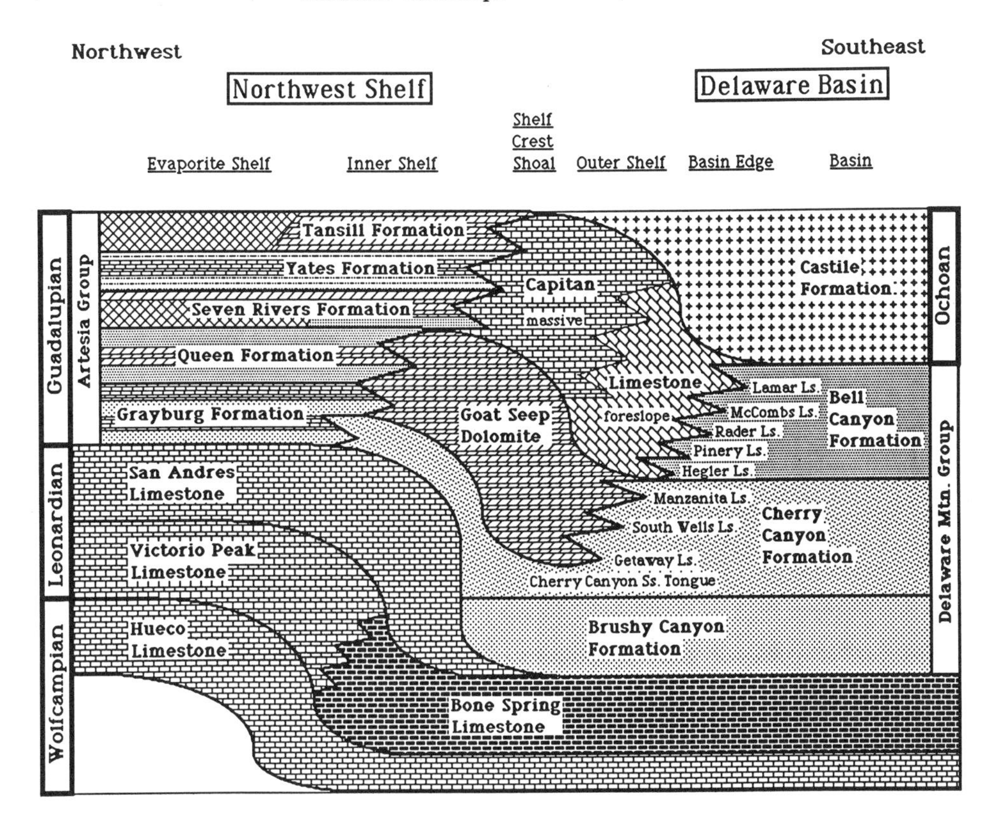

Figure 3. Diagrammatic northwest-southeast section across the northwestern margin of the Delaware Basin (after Pray and Crawford, 1984; and Sarg and Lehmann, 1986); no scale intended. Note the threefold division of facies from shelf to shelf margin to basin, represented during Guadalupian (late Permian) time by the Artesia Group, Capitan Limestone, and Delaware Mountains Group, respectively. Ochoan (uppermost Permian) sediments of the Castile Formation, shown only on the righthand portion of the diagram, are evaporites which filled the Delaware Basin as it became increasingly restricted and dried up. Representative depositional environments of various facies in the Guadalupian strata are shown along the top of the diagram.

The roadcut on both sides of the road exposes laminated, black lime mudstone and wackestone of the Lamar Limestone Member of the Bell Canyon Formation. The Bell Canyon is a basinal deposit, uppermost of three formations making up the Delaware Mountain Group. It is composed primarily of fine-grained siliciclastics and contains five intertonguing carbonate units (Fig. 3). The Lamar is the uppermost of these five members. The Lamar here contains few fossils. Soft-sediment deformation features are visible in several places, particularly near the western end of the roadcut. On a freshly broken surface, the rock smells of hydrogen sulfide. This scent, together with the black color and low fossil content of the rock, suggests that the bottom of the Delaware Basin was a stagnant, anoxic, reducing environment at the time of deposition of the Lamar Limestone Member. A burrowed zone near the base of the outcrop indicates a period of oxygenated bottom conditions (Babcock, 1977).

The top of the roadcut is a good vantage point from which to view the Permian Reef Escarpment as it is exposed along the eastern face of the Guadalupe Mountains 4 to 5 mi (6 to 8 km) to the northwest. Several features of this escarpment are important to the interpretation of the Permian stratigraphy and paleogeography of the region. The relatively flat surface at the top of the escarpment is developed on essentially flat-lying bedded deposits of backreef carbonates, clastics, and evaporites (Fig. 3). Bedding planes may be visible in the upper walls of McKittrick Canyon. The steep upper slopes of the escarpment are massive carbonate buildup ("reef") units of the Permian shelf edge. These are particularly apparent at the mouth of McKittrick Canyon. The massive, unbedded carbonate buildup composing the Capitan Limestone (Fig. 3) consists predominantly of algae (including the problematic *Archaeolithoporella*, which seems to share affinities with red algae), sponges, and the problematic fossil *Tubiphytes*; it was cemented in situ on the sea floor as it was deposited (Toomey and Babcock, 1983). Below the massive Capitan cliffs, the slope becomes gentler as it is developed on east- (basinward) dipping carbonate detritus shed off the growing reef mass. This carbonate debris grades into the almost flat-lying basinal clastics and carbonates of the Bell Canyon Formation. The Lamar Limestone seen in this roadcut can be traced up into the forereef slope at the base of the escarpment. The Lamar is correlated with the uppermost part of the massive Capitan Limestone at the shelf crest, and with the Tansill Formation of the shelf (Fig. 3). The Lamar Limestone tongue can be traced a total of about 17 mi (27 km) into the basin (Babcock, 1977). Most of the relief between this stop and the upper surface of the Guadalupe Mountains apparently reflects Permian depositional topography, perhaps somewhat accentuated by later tectonic uplift (Kelley, 1971, 1972; McKnight, 1986).

Stop 2. The Rader slide. Continue west on U.S. 62-180 for 2.4 mi (3.8 km) to another roadcut (Culberson County; 104°43′47″W; 31°55′54″N; elevation 4,900 ft or 1494 m; Independence Spring Quadrangle). Park on the side of the road, just east of the roadcut.

Both sides of this roadcut expose a 12-ft (4-m) debris-flow deposit of sand and carbonate boulders with overlying carbonate turbidites, all in the Rader Limestone, middle of the five carbonate members of the Bell Canyon Formation (Fig. 3). Unlike the Lamar Limestone at the previous stop, whose laminated character suggested quiet-water deposition, most of the Rader Limestone here was carried downslope rapidly from the shelf margin and deposited on the basin floor during essentially one mass-flow event (Beard, 1985). Limestone boulders (light gray) are conspicuous against the tan to yellow Bell Canyon sands and silts making up the debris-flow matrix. The debris-flow unit has a sharp basal contact with the underlying Bell Canyon sands, which are laminated and relatively undisturbed by the debris flow; locally, a large boulder at the base of the debris flow indents the underlying laminated sands (Beard, 1985). The flow unit is crudely graded, with limestone blocks from 3 to 10 ft (1 to 3 m) in diameter (one is 30 ft, or 10 m) at the base diminishing to clasts of only a few in at the top (Toomey and Babcock, 1983; Beard, 1985). Above the debris-flow unit is a 1- to 2-ft (0.3- to 0.6-m) sequence of thin, graded, dark carbonate turbidites. This sequence resembles the Lamar Limestone at Stop 1, but contains coarser and more abundant fossil grains. Most of the discrete carbonate units are separated by siliciclastic laminae, indicating that they were not deposited at the time of the thick debris flow (Beard, 1985).

The top of this roadcut again affords an excellent view of the Permian Reef Escarpment. This location is about 2 to 3 mi (3 to 5 km) from the escarpment, and the Rader Limestone debris flow has been traced more than 9 mi (15 km) into the basin (Koss, 1977). Carbonate boulders in the debris flow derive from high on the shelf edge, mostly in upper forereef and reef facies. Fossils indicative of normal marine conditions (brachiopods, crinoids, bryozoans, fusulinids, sponges, and calcareous algae) are present as allochems (Beard, 1985). The carbonate clasts were obviously lithified before being broken off and transported into the basin. The Rader Limestone Member has been correlated with the middle Capitan massive facies of the shelf margin and the backreef Yates Formation (Fig. 3). The size of the entrained carbonate boulders and the basinward extent of the Rader Limestone Member debris flow attest to the depositional relief present at the basin margin in Permian time, perhaps as much as 1,500 ft (460 m; Koss, 1977).

Approximately 6 mi (10 km) west of Stop 2, the route crosses the Border fault zone (King, 1948) separating the Great Plains physiographic province from the Basin and Range physiographic province. High-angle normal faults here trend northwest to southeast and are mostly downthrown to the west (Barnes, 1983). Faults are conspicuous in roadcuts 6.4 mi (10.2 km) west of Stop 2 (0.5 mi or 0.8 km west of the crest of Guadalupe Pass, elevation 5,695 feet or 1,736 m) and 8.9 mi (14.2 km) west of the stop.

Stop 3. El Capitan view. From Stop 2, continue southwest on U.S. 62-180 for 10.2 mi (16.3 km) to a pair of roadside rest areas (Culberson County; 104°50′39″W; 31°52′14″N; elevation 4,920 ft or 1,500 m; Guadalupe Pass Quadrangle). Park in the space provided at the roadside park.

A number of geologic features are visible from this scenic viewpoint. The most conspicuous is El Capitan (elevation 8,085 ft or 2,465 m), the southern promontory of the Guadalupe Mountains, located to the northwest. El Capitan has served as a landmark to travelers for centuries, from the southwestern Indians to the Spanish conquistadores in the 1500s to the Butterfield Overland Mail in 1858 to the present-day motorist and airline pilot heading for El Paso. The cliffs of El Capitan are developed in resistant, steeply dipping (10° to 20°) Capitan foreslope facies (Beard, 1985). The lower slopes are formed on basinal Cherry Canyon Sandstone, middle of the three formations of the Delaware Mountain Group (Fig. 3). The prominent topographic bench in the foreground is supported by a massive sandstone bed marking the top of the Brushy Canyon Formation, lowest formation of the Delaware Mountain Group (Fig. 3). Brushy Canyon Sandstone continues down the slopes beneath the bench, and underlies these roadside rest areas. Visual tracing of ledge-forming sandstone beds in the vicinity of Stop 3 reveals that many have overall lenticular shapes. A closer view of one of these beds, 0.5 mi (0.8 km) south of this stop (best seen in the western roadcut face), reveals a scoured channel margin, suggesting that much of the sand was deposited in submarine channels.

To the south and southeast of Stop 3, the west-facing fault escarpment of the Delaware Mountains, the southward continuation of the Border fault zone, is visible. Brushy Canyon Formation sandstone beds can be seen in the scarp face and in the hills to the west; Cherry Canyon Sandstone caps the east-dipping backslope of the cuesta.

To the west lies the Salt Basin, a north-south–trending graben. In the distance to the southwest are the Sierra Diablo Mountains, an east-facing upthrown fault block consisting of gently west-dipping pre-Capitan Permian strata deposited on the Delaware Basin margin (Beard, 1985).

Stop 4. The Salt Basin. Continue west from Stop 3 on U.S. 62-180 for 14.2 mi (22.8 km) to Texas Historic Marker near the center of the Salt Basin (Hudspeth County; 105°00′27″W;

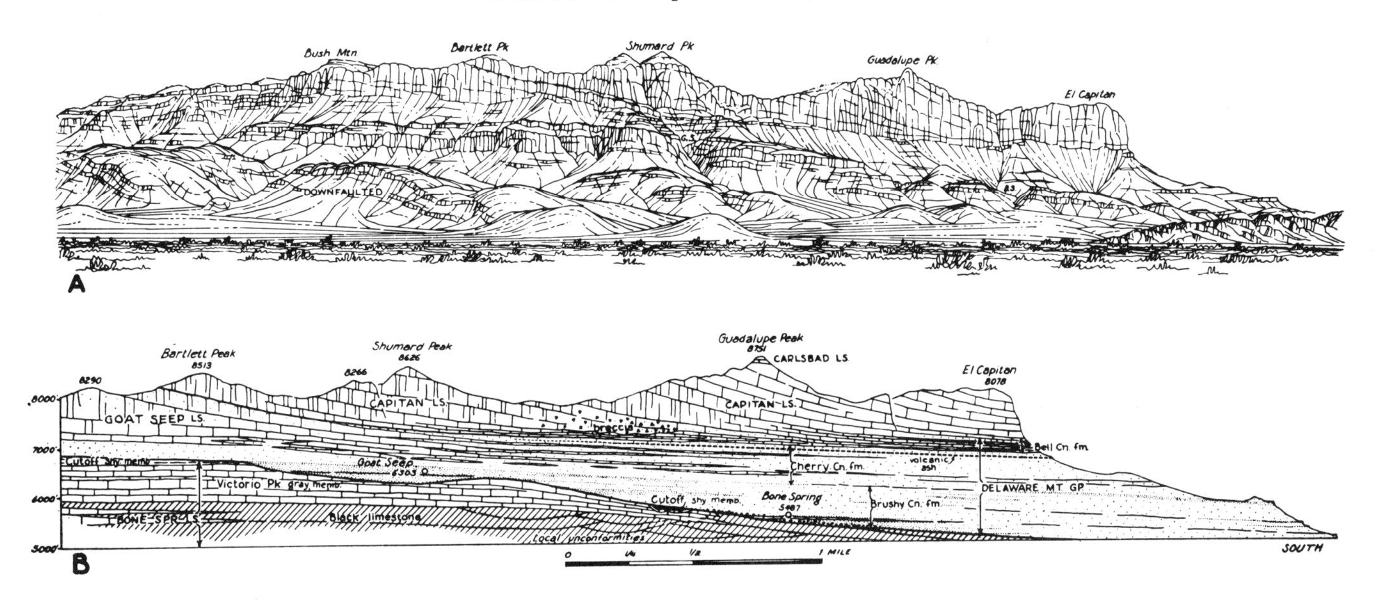

Figure 4. (A) Sketch of western escarpment of the southern Guadalupe Mountains formed by down-to-the-west Basin and Range normal faulting; and (B) diagrammatic section showing Permian strata visible in the scarp face (modified from King, 1942). View is from approximate location of Stop 4. Note bedded shelf strata of the Artesia Group capping Guadalupe Peak, foreslope bedding planes visible in Capitan Limestone, and older Goat Seep shelf-margin carbonates below Bush Mountain.

31°45′08″N; elevation 3,613 ft or 1,102 m; Linda Lake South Quadrangle). Park on the side of the road. (Note: at a distance of 4.7 mi [7.6 km] southwest of Stop 3, U.S. 62-180 is joined by Texas 54 from Van Horn, Texas, 54 mi [87 km] to the south. Stay on U.S. 62-180 in order to reach Stop 4.)

The Salt Basin is a north-south–trending graben, the youngest and most easterly portion of the Basin and Range province. The basin extends about 43 mi (69 km) north from here into New Mexico, and another 49 mi (78 km) south from here to the vicinity of Van Horn, Texas, where it bifurcates. Beginning at the Sierra Diablo Mountains, 18 mi (29 km) south of Stop 4, the southern end of the basin is offset to the east, "dogleg"-fashion, relative to its northern end. The eastern margin of the graben is formed by down-to-the-west normal faults of the Border fault zone. The western margin is formed by a parallel series of down-to-the-east normal faults along the eastern edge of the Diablo Plateau, a large, north-south–trending Basin and Range horst block, which reflects the southern extension of the buried Precambrian Pedernal Uplift (Kelley, 1971). The Salt Basin is actively subsiding, and faults of Quaternary age are found along its margins (Goetz, 1980). A map of Quaternary bolson fill in the graben shows a maximum of 2,000 ft (610 m) of sediment about 30 mi (48 km) south of this stop (Veldhuis and Keller, 1980). Because it lacks high heat flow and any association with young volcanism, Seager and Morgan (1979) considered the Salt Basin not a part of the "mature" Rio Grande Rift, but possibly a developing arm of the rift.

If there has been no recent rain, the surface of the salt flats at Stop 4 should be solid enough to walk out on. The white crystals covering the surface are primarily gypsum with some halite. The crystals do have the distinctive taste of halite. Gypsum dunes may be visible in some places. A hole dug in the sediment reveals a succession of thin, light gray to tan layers of mud separated by slightly darker laminae. About 2 to 3 ft (0.6 to 1.0 m) down, a thin, black horizon occurs, beneath which are darker gray laminated muds. The black horizon produces a hydrogen sulfide smell, indicating reducing conditions. The individual laminae in the mud presumably reflect individual rain storms in the mountains to the east. Rainwater dissolves Permian carbonate and evaporite minerals in the Guadalupe Mountains and transports its solution load, together with suspended clays and silts, into the Salt Basin. After a heavy rain, the Salt Flats may be a huge, shallow playa lake. Suspended particles settle out in the quiet lake waters, and, as the waters evaporate, minerals precipitate. Carbonate minerals identified by X-ray diffraction include calcite, aragonite, and dolomite (Friedman, 1966; Dunham, 1972). The Texas Historic Marker beside the highway commemorates the "Salt War" of 1877, when private U.S. interests attempted to claim the salt deposits that the Mexican residents of El Paso and Chihuahua, Mexico, had used freely for years (Boyd and Kreitler, 1986).

The western face of the Guadalupe Mountains is spectacularly displayed in the distance to the east of Stop 4. The fault escarpment provides an excellent profile view of the Permian carbonate shelf margin facies (Fig. 4). Guadalupe Peak (elevation 8,749 ft or 2,667 m), the highest point in Texas, is capped with flat-lying backreef strata above steeply dipping (up to 35°) Capitan foreslope beds (Beard, 1985). There is almost a mile of relief (5,136 ft or 1,566 m) between this stop and Guadalupe Peak. Facies of the older Permian Goat Seep reef (Fig. 3), over which the Capitan reef prograded, are visible below Bush Mountain (Fig. 4). Farther north, in New Mexico, are the Brokeoff Mountains, a faulted horst block of layered shelf carbonates and clastics.

To the northwest, near Dell City, Texas, are two small nepheline trachyte intrusions, Round Mountain and the Wilcox Hills. Round Mountain, the larger of the two, rises 177 ft (54 m) above the surrounding surface. These intrusions, part of the Oligocene northern Trans-Pecos magmatic province (Barker and others, 1977), are located near the western boundary of the Salt Basin graben, but predate its origin.

From this final stop it is 96 mi (154 km) west on U.S. 62-180 to El Paso, Texas.

SUMMARY

The southern Guadalupe Mountains of Texas exhibit in their eastern and western scarp faces important features of the shelf and shelf margin carbonates that formed the northwestern boundary of the deep Delaware Basin in late Permian time. Bedded backreef strata, massive shelf-edge carbonate buildups, and dipping foreslope detrital deposits display the vertical and lateral facies succession of a major progradational carbonate shelf margin. Roadcuts in the basinal facies provide examples of contemporaneous pelagic and current deposition of carbonates and clastics in the deep waters in front of the shelf. The northwest to southeast trending fault zone that produced the western escarpments of the Guadalupe and Delaware Mountains also marks the boundary between relatively flat-lying, structurally undisturbed sedimentary strata of the Great Plains physiographic province on the east and tilted, block-faulted rocks of the Basin and Range physiographic province on the west. The Salt Basin is a graben that is filling with bolson and playa sediments. Its continuing subsidence and Quaternary faulting along its margins are evidence of the active tectonics prevalent in the Basin and Range.

REFERENCES CITED

Anderson, R. Y., and Kirkland, D. W., 1987, Banded Castile evaporites, Delaware basin, New Mexico, *in* Beus, S. S., ed., Rocky Mountain Section of the Geological Society of America: Boulder, Colorado, Geological Society of America, Centennial Field Guide, v. 2, p. 455–458.

Babcock, L. C., 1977, Life in the Delaware Basin; The paleoecology of the Lamar Limestone, *in* Hileman, M. E., and Mazzullo, S. J., eds., Upper Guadalupian facies, Permian reef complex, Guadalupe Mountains, New Mexico and Texas: Permian Basin Section, Society of Economic Paleontologists and Mineralogists Publication 77–16, p. 357–389.

Barker, D. S., Long, L. E., Hoops, G. K., and Hodges, F. N., 1977, Petrology and Rb-Sr isotope geochemistry of intrusions in the Diablo Plateau, northern Trans-Pecos magmatic province, Texas and New Mexico: Geological Society of America Bulletin, v. 88, p. 1437–1446.

Barnes, V. E., 1983, Geologic atlas of Texas; Van Horn–El Paso sheet: The University of Texas at Austin, Bureau of Economic Geology, 12 p. and map, scale 1:250,000.

Beard, C., ed., 1985, Permian carbonate/clastic sedimentology, Guadalupe Mountains; Analogs for shelf and basin reservoirs: Permian Basin Section, Society of Economic Paleontologists and Mineralogists Publication 85–24, 151 p.

Boyd, F. M., and Kreitler, C. W., 1986, Hydrogeology of a gypsum playa, northern Salt Basin, Texas: The University of Texas at Austin, Bureau of Economic Geology Report of Investigations no. 158, 37 p.

Dunham, R. J., 1972, Capitan reef, New Mexico and Texas; Facts and questions to aid interpretation and group discussion: Permian Basin Section, Society of Economic Paleontologists and Mineralogists Publication 72–14, variously paged.

Friedman, G. M., 1966, Occurrence and origin of Quaternary dolomite of Salt Flat, west Texas: Journal of Sedimentary Petrology, v. 36, p. 263–267.

Goetz, L. K., 1980, Quaternary faulting in Salt Basin graben, west Texas, *in* Dickerson, P. W., and Hoffer, J. M., eds., Trans-Pecos region, southwestern New Mexico and west Texas: New Mexico Geological Society, 31st Field Conference, p. 83–92.

Kelley, V. C., 1971, Geology of the Pecos country, southeastern New Mexico: New Mexico Bureau of Mines and Mineral Resources Memoir 24, 75 p.

—— , 1972, Geometry and correlation along Permian Capitan escarpment, New Mexico and Texas: American Association of Petroleum Geologists Bulletin, v. 56, p. 2192–2212.

King, P. B., 1942, Permian of west Texas and southeastern New Mexico: American Association of Petroleum Geologists Bulletin, v. 26, p. 535–763.

—— , 1948, Geology of the southern Guadalupe Mountains, Texas: U.S. Geological Survey Professional Paper 215, 183 p.

Koss, G. M., 1977, Carbonate mass flow sequences of the Permian Delaware Basin, west Texas, *in* Hileman, M. E., and Mazzullo, S. J., eds., Upper Guadalupian facies, Permian reef complex, Guadalupe Mountains, New Mexico and Texas: Permian Basin Section, Society of Economic Paleontologists and Mineralogists Publication 77–16, p. 391–408.

McKnight, C. L., 1986, Descriptive geomorphology of the Guadalupe Mountains, south-central New Mexico and west Texas: Waco, Texas, Baylor University, Baylor Geological Studies Bulletin no. 43, 40 p.

Pray, L. C., and Crawford, G. A., 1984, A field guide to the geology of the Permian shelf margin-to-basin transition of the western escarpment, Guadalupe Mountains, Texas: Houston, Texas, Gulf Coast Section, Society of Economic Paleontologists and Mineralogists, variously paged.

Sarg, J. F., and Lehmann, P. J., 1986, Lower-middle Guadalupian facies and stratigraphy, San Andres/Grayburg formations, Permian Basin, Guadalupe Mountains, New Mexico, *in* Moore, G. E., and Wilde, G. L., eds., Lower and middle Guadalupian facies, stratigraphy, and reservoir geometries, San Andres/Grayburg formations, Guadalupe Mountains, New Mexico and Texas: Permian Basin Section, Society of Economic Paleontologists and Mineralogists Publication 86–25, p. 1–8.

Seager, W. R., and Morgan, P., 1979, Rio Grande rift in southern New Mexico, west Texas, and northern Chihuahua, *in* Riecker, R. E., ed., Rio Grande rift; Tectonics and magmatism: Washington, D.C., American Geophysical Union, p. 87–106.

Toomey, D. F., and Babcock, J. A., 1983, Precambrian and Paleozoic algal carbonates, west Texas–southern New Mexico: Golden, Colorado School of Mines Professional Contribution no. 11, 345 p.

Veldhuis, J. H., and Keller, G. R., 1980, An integrated geological and geophysical study of the Salt Basin graben, west Texas, *in* Dickerson, P. W., and Hoffer, J. M., eds., Trans-Pecos region, southeastern New Mexico and west Texas: New Mexico Geological Society, 31st Field Conference, p. 141–150.

Lower Cretaceous of western Trans-Pecos Texas

David L. Amsbury, *1422 Brookwood Court, Seabrook, Texas 77586*
Donald F. Reaser, *Department of Geology, The University of Texas at Arlington, Arlington, Texas 76019*

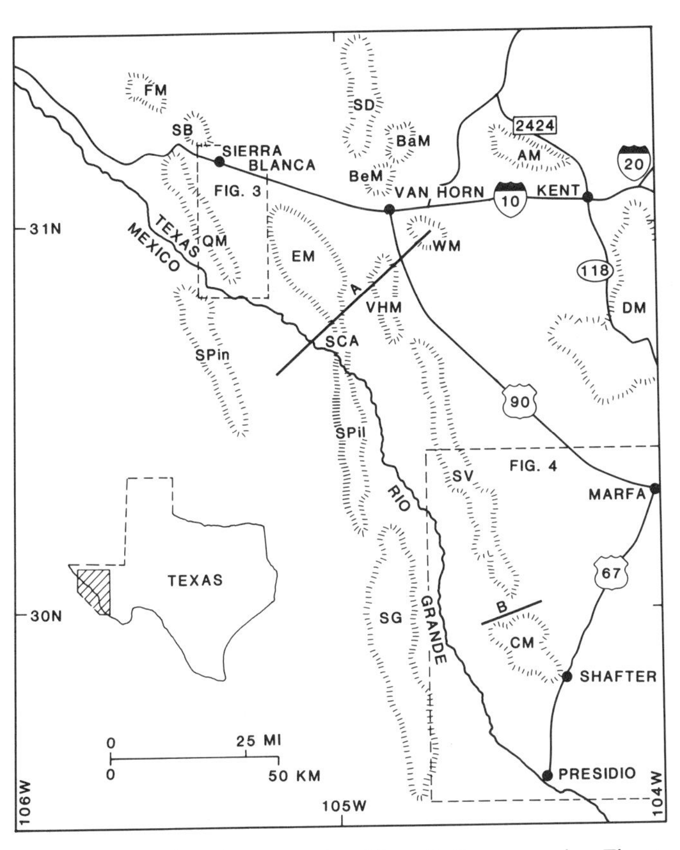

Figure 1. Index Map. A, cross section, Figure 2; B, cross section, Figure 5. Mountain Ranges: AM, Apache Mountains; BaM, Baylor Mountains; BeM, Beach Mountain; CM, Chinati Mountains; DM, Davis Mountains; EM, Eagle Mountains; FM, Finlay Mountains; QM, Quitman Mountains; SB, Sierra Blanca; SCA, Sierra Cola de Aguila ("Southern Indio Mountains"); SD, Sierra Diablo; SG, Sierra Grande; SPil, Sierra Pilares; SPin, Sierra del Piño; SV, Sierra Vieja; VHM, Van Horn Mountains; WM, Wylie Mountains.

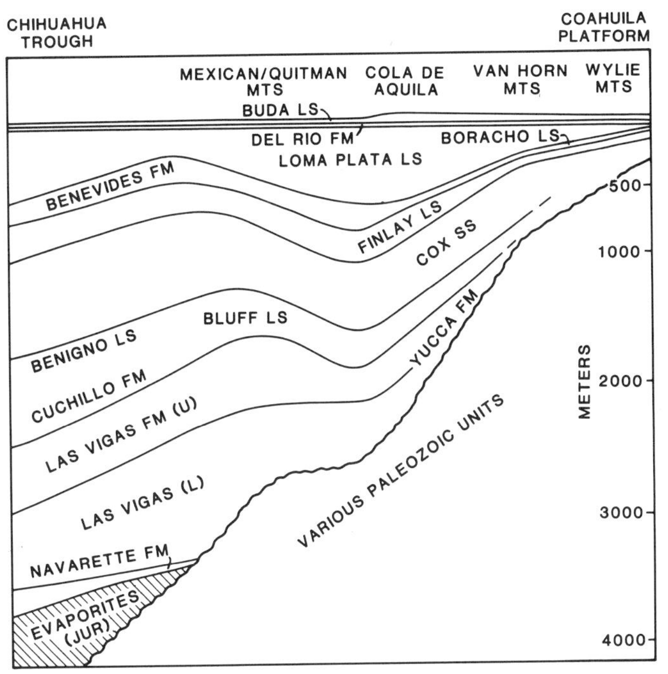

Figure 2. Cross section A, Figure 1. A diagrammatic trough-to-shelf transect showing major named units, and the tenfold thinning of Lower Cretaceous rocks from southwest to northeast. Modified from Haenggi ("1965" in 1966); reproduced as Figure F in SEPM (1970); compare with Adkins (1933, Fig. 15 and p. 296).

LOCATION

The area is accessible from I-10, U.S. 90, and (from Big Bend National Park) Texas 170. Outcrops of Lower Cretaceous rocks occur along public roads near Kent; south of Sierra Blanca in the Quitman Mountains; southwest of Marfa and north of the Chinati Mountains; and south of Shafter (Fig. 1).

SIGNIFICANCE

Lower Cretaceous strata in Trans-Pecos Texas and adjacent northern Mexico were deposited in the Chihuahua Trough along and westward from the Rio Grande (Fig. 1) and on the Diablo Platform from Marfa through Van Horn and westward through Sierra Blanca. By late Mesozoic time the southeastward end of the Diablo Platform had merged with the northwestern end of the Coahuila Platform (see Stevens, this volume). Within the Chihuahua Trough, Lower Cretaceous strata conformably overlie evaporites of probable Jurassic age; elsewhere Lower Cretaceous strata unconformably overlie Paleozoic rocks.

There is a tenfold difference in thickness of the Lower Cretaceous section between the Chihuahua Trough and the Diablo Platform (Fig. 2). Part of the difference represents filling of previously existing topographic lows, but much represents subsidence of the trough at a rate faster than the platform. Dominantly terrigenous clastic units at the base of the section lap onto the platform, so that basal clastic rocks on the platform are younger than those in the trough. In contrast, the overlying carbonate and shaly formations, sequences of beds, and even individual beds thin markedly from the trough to the platform.

Lower Cretaceous rocks in Trans-Pecos Texas are similar to those in other parts of the state. The top of the Coahuila Series (lower Aptian and older) is at about the Las Vigas/Cuchillo boundary. The rest of the units shown in Figure 2 form the Comanche Series (uppermost Aptian, Albian, and part of the Cenomanian). Gulfian shale, limestone, and sandstone are preserved within the Chihuahua Trough and locally along the northern front of the Davis Mountains.

The Las Vigas Formation (quartzose conglomerate and sandstone) is similar to the subsurface Hosston Formation in South Texas and in the East Texas Basin; the Yucca Formation (carbonate-cemented limestone conglomerate) resembles the outcropping Sycamore Sandstone west of Austin (see Amsbury, Chapter 84, this volume); much of the Cuchillo Formation and Bluff Limestone are very similar to the Glen Rose Limestone; the Cox Sandstone occupies the stratigraphic position of the Paluxy Sand in north-central Texas and of the Maxon Sandstone around the Marathon Uplift; the Finlay Limestone is continuous with the Edwards Group; the Benevides Formation resembles the Kiamichi Shale; the Loma Plata Limestone and the Boracho Formation are similar to the Georgetown Limestone; and the Del Rio Shale is continuous eastward, as is the Buda Limestone.

During the Laramide orogeny (poorly dated to Late Cretaceous and early Tertiary) rocks within the Chihuahua Trough were folded tightly and thrust northeastward onto the edge of the Diablo Platform. Evaporites underlying the trough section played a major role in the structural configuration (Gries, 1980). After deposition of a few kilometers of middle Tertiary volcanic rocks, Miocene block faulting and then extensive erosion formed the present topography, which consists of isolated ranges formed from Lower Cretaceous and middle Tertiary rocks, separated by wide valleys covered with younger material.

PLATFORM SECTION

Representative outcrops of the platform section, though with few accessible road cuts, may be seen 7.5 to 12 mi (12 to 19 km) north of Sierra Blanca, where Ranch Road 1111 passes through the east end of the Finlay Mountains.

Good roadside exposures of the thin (1,000 ft; 300 m) platform section occur along public roads near Kent (Brand and DeFord, 1958). Northward from I-10 at Kent along Ranch Road 2424: 0.5 mi (0.8 km)—Finlay Limestone; 0.9 mi (1.4 km)—Cox Sandstone/Finlay Limestone contact; 1.7 mi (2.7 km)—aphanitic, birdseye limestone of the Yearwood Formation (Yucca facies, perhaps equivalent to the Bluff Formation); and 2.0 mi (3.2 km)—dark brown, cross-bedded, chert-pebble–bearing sandstone of the Cox. See SEPM (1958, Fig. 3 and p. 7–13) for details of the structural geology of these outcrops.

Southward from I-10 at Kent along Texas 118: 0.6 mi (1 km—nodular but unfossiliferous limestone of the Levinson Member (= Kiamichi and lower Georgetown) of the Boracho Formation; 0.9 mi (1.1 km)—resistant wackestone of the Levinson containing toucasiids, gryphaeids, and agglutinated foraminifera; 2.0 mi (3.2 km)—resistant limestone of the San Martine Member (= upper Georgetown) of the Boracho; 2.5 mi (4 km)—San Martine/Buda Limestone contact, with less than a foot of calcareous sandstone referable to the Del Rio Formation; 2.7 mi (4.3 km)—Buda Limestone; and 2.9 mi (4.7 km)—flagstones of the Gulfian Boquillas Limestone. A superb exposure of the Levinson Member has been created on the south side of I-10, 3.0 mi (4.8 km) west of Kent and 0.7 mi (1.1 km) east of the Hurd Draw overpass. A few beds contain gryphaeids, echinoids, and casts of clams, snails, and ribbed upper Albian ammonites.

TABLE 1. SYNONYMS

Bluff Formation	This guidebook
Bluff Mesa Formation	Smith, 1940; Albritton and Smith, 1965
Bluff Formation	Underwood, 1963
Cuchillo Formation, plus Glen Rose Formation	Scott, 1940
Quitman bed	Adkins, 1933
Quitman Formation	Jones and Reaser, 1970
Las Vigas Formation	This guidebook, quartzose facies
Yucca Formation	This guidebook, carbonate conglomerate facies
Las Vigas Formation	Adkins, 1933 (downdip); Scott, 1940
Mountain Formation	Jones and Reaser, 1970
Yucca Formation	Adkins, 1933 (updip); Campbell, 1970, 1980 (all outcrops); Allbritton and Smith, 1965

SHELF-TO-TROUGH TRANSECT (HUDSPETH COUNTY)

The geology of this area is covered by the Marfa and the Van Horn/El Paso sheets of the Geological Atlas of Texas (Barnes, 1979b, 1983).

The town of Sierra Blanca lies approximately on the northeastern edge of thrust-faulted Lower Cretaceous rocks. Nomenclature of these rocks is complex; Table 1 supplies the terminology used herein with synonyms applied to the same strata by several authors. Figure 2 illustrates the stratigraphic relationships.

This section is based on SEPM (1970), unfortunately out of print as of 1986. Drive south on Ranch Road 1111 from its intersection with I-10 in Sierra Blanca (Fig. 3). Yucca Mesa is the prominence on Devil Ridge to the south, east of RR 1111. The Mesa is capped by about 1,300 ft (400 m) of Bluff Limestone overlying about 1,000 ft (330 m) of Yucca Formation.

0.8 to 1 mi (1.3 to 1.6 km). Road-cut exposures of the Cox Sandstone and Finlay Limestone, which also form Texan Mountain to the west. The Cox along Ranch Road 1111 is cross-bedded, medium- to fine-grained quartz sandstone containing interbeds of red and green shale. The Finlay Limestone is light gray, fossiliferous wackestone to packstone, but contains some beds of lime grainstone, marl, and quartz sandstone. The ⅛-inch (3 mm) foraminifer *Dictyoconus walnutensis* and 1-ft-long (30 cm) caprinids are prominent in beds near the top of the exposed section.

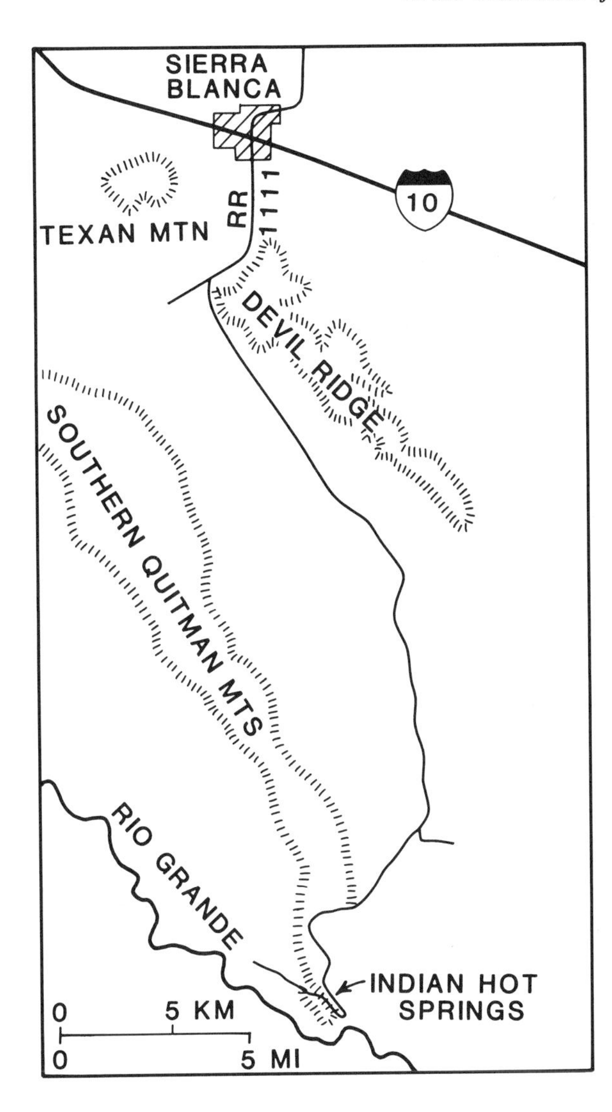

Figure 3. Route for the shelf-to-trough transect from the Diablo Platform into the Chihuahua Trough (Hudspeth County).

The Cox and Finlay formations represent the Fredericksburg Division in Trans-Pecos Texas. The lower part of the Cox is probably equivalent to the uppermost Glen Rose Formation (Trinity Division) of central Texas; despite diligent search by determined stratigraphers, no regionally useful dividing plane between the divisions has been found in Trans-Pecos. The Cox is about 660 ft (200 m) thick in the Finlay Mountains northwest of Sierra Blanca (Albritton and Smith, 1965; Brand and DeFord, 1958). In the Devil Ridge, Eagle, and Indio (Cola de Aguila) mountains to the southeast, the Cox ranges from 1,150 to 1,640 ft (350 to 500 m) thick (SEPM, 1958; Smith, 1940; Underwood, 1963). Quartzose sandstone dominates, but marl and limestone containing a marine molluscan fauna occur at most localities.

4.4 mi (7 km). Pavement ends; intersection of the Quitman Gap Road (right, southwest) and Red Light Draw Road (left, southeast). Continue to the left (southeast) on the Red Light Draw Road. Exposures of the Bluff Formation east of the intersection are typical. Resistant ledges of siliceous lime wackestone and rudist packstone protrude from a steep hillside nearly covered by lechuguilla, thorny shurbs, sotol, cactus, and some bunchgrass.

15.2 mi (24.5 km). Red Light windmills; low water crossing of Red Light Draw (Quitman Arroyo). You are traveling down the Red Light Bolson, which contains up to 820 ft (250 m of Tertiary(?) and Quaternary fill.

22.5 mi (36 km). Road forks after swinging southwest and passing outcrops of pinkish bolson fill deposits. Continue right (southwest) rather than taking the south fork, which goes to a ranch on the Rio Grande.

23.5 mi (38 km). Road forks in light-colored Tertiary ash-flow tuff. Keep right (southwest to west) toward Indian Hot Springs. You are now entering the main body of the southern Quitman Mountains. The Quitman range is the topographic expression of an eroded, northeast-verging, nearly recumbent anticline. Numerous minor folds are superimposed on the inverted limb of the major fold. At places, small Z- and S-shaped folds are common features adjacent to some large faults. The dominant deformational mechanism in this part of the range is flexural slip, that is, slippage along bedding surfaces.

26.9 mi (43 km). Quitman Summit. The road here runs along the Red Bull fault zone, a set of ENE–trending tear faults that cut Bluff and Las Vigas rocks. These two left-lateral faults offset Lower Cretaceous strata about 1.5 mi (2.4 km).

The middle member of the Bluff, exposed on both sides of the road, is about 650 ft (200 m) thick and is composed of shale, siltstone, limestone, and sandstone. It is highly fossiliferous and contains, among other mollusks, lower Albian ammonites (Scott, 1940; Young, 1974). Do not trespass away from the road; information on land access may be obtainable at Indian Hot Springs.

Southward, the member is well exposed for about 2.5 mi (4 km) along a prominent strike valley (Mayfield Canyon). Near the Rio Grande, Flint (1984) reported that the lower part of the member contains biostromes of exogyrids, and lenticular, cross-laminated quartz sandstone and siltstone characterized by sets of hummocky cross-stratification.

The Bluff Limestone in the Quitman Mountains is about 1,800 to 1,970 ft (550 to 600 m) thick (Jones and Reaser, 1970). Despite the fact that these beds have been thrust several tens of kilometers toward the Diablo Platform from the Chihuahua Trough (Reaser, 1982), the lithofacies is very similar to that of the equivalent platform sections in Trans-Pecos (and indeed to the Glen Rose of central Texas). Evidently, subsidence in the Chihuahua Trough, although rapid enough to allow deposition of a relatively thick set of deposits, was too slow to cause a major transgression and drastic facies change to occur during the late Aptian and early Albian.

27.0 mi (43 km). Road turns south and crosses the southernmost fault of the Red Bull zone. The high ridge on the left (east) exposes the lower member of the Bluff. The upper part of the member, which forms the west wall of Mayfield Canyon, is marked by a 6-ft-thick (2 m) coral biostrome (Martin, 1987). The upper member of the Bluff (not visible from this vantage point) forms the east wall of Mayfield Canyon. The lower part of the upper member is a rudist packstone to boundstone that forms a prominent 75-ft (23 m) escarpment.

28.6 mi (46 km). Steel pipe bridge. Beds of the upper member of the Las Vigas Formation are vertical on the east and west sides of the road. The Las Vigas is about 5,250 ft (1,600 m) thick in the southern Quitmans. The upper member here consists of calcsiltite, lime mudstone, marl, clay mudstone, and sandstone. Thin-bedded limestones contain charophytes, snails, turtles, and rare fish and dinosaur bones; petrified logs occur in some beds (Campbell, 1970).

29.9 mi (48.1 km). The ridge along which the road is built formed from resistant sandstone and interbedded shale of the middle member of the Las Vigas Formation. Campbell (1970, 1980) described multiple fining-upward depositional sequences, probably fluvial, from the west side of the ridge. Polymict conglomerate fills basal scours of the sequences, overlain by cross-bedded sandstone, and then by horizontally laminated siltstone and mudstone that commonly contains rib-and-furrow structure, burrows, trails, root molds, and carbonate nodules (Campbell, 1980, p. 161). Composition of the pebbles and sand grains is consistent with derivation from the Diablo Platform.

It is desirable to remain on or very close to the road. Aside from the hazards of trespassing, the slope is very steep, contains much loose rock, and is partly covered by lechuguilla and other sharp-tipped plants.

SHELF-EDGE TRANSECT (PRESIDIO COUNTY)

This part of the area is covered by the Marfa and Emory Peak–Presidio Sheets of the Geologic Atlas of Texas (Barnes, 1979a,b). There are no recent guidebooks for the Cretaceous rocks.

The suggested transect runs southwest from Marfa through Pinto Canyon, southeast to Presidio, and north through Shafter to Marfa (Fig. 4). Pinto Canyon is out of the way for travelers passing through Big Bend National Park, and the unpaved road can be difficult after extremely heavy rains; however, the trip is spectacular, the road is well maintained, and it is usually not difficult for *careful* drivers of passenger cars.

Drive south on Ranch Road 2810, which intersects U.S. 90 0.6 mi (1 km; 8 blocks) west of U.S. 67 in Marfa. Begin the mileage log at this intersection. The surface of the Marfa Plateau is formed on Quaternary and middle Tertiary gravel and silt, with outcrops of the youngest middle Tertiary basalt and tuff at some places. Drainage is to the Rio Grande.

32.1 mi (51.6 km). End of pavement.

34.6 mi (55.7 km). Horseshoe curve to the left; pull off on the right side of the road, before crossing the cattleguard into the Pinto Canyon Ranch (Fig. 5). The view westward illustrates major Lower Cretaceous outcrops. The blue mountains on the horizon are across the Rio Grande in Mexico. The broad hill mass in the middle distance, past the sharp peak of Tertiary intrusive rock, is formed from Bluff Limestone. The Lower

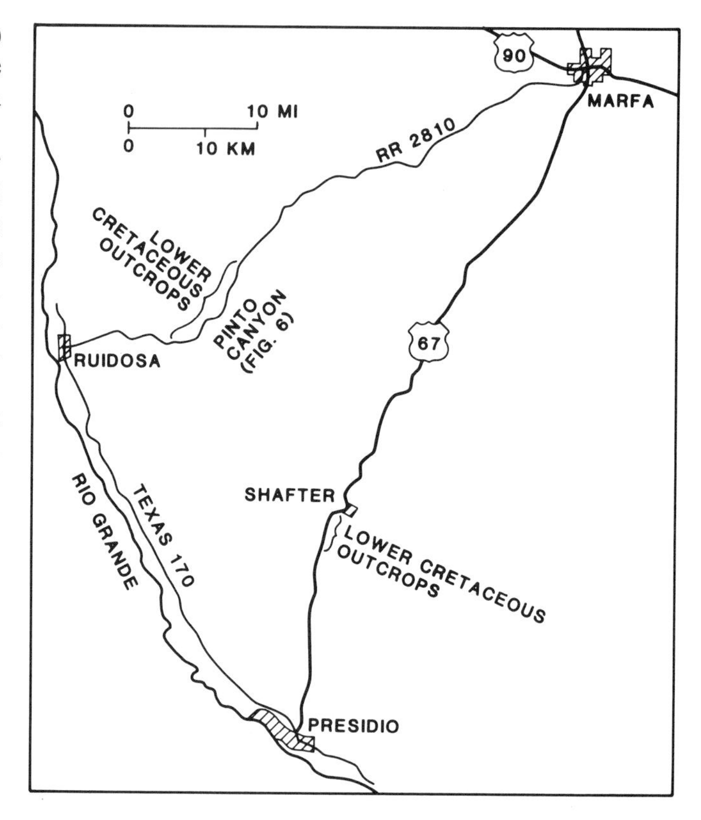

Figure 4. Route for the shelf-edge transect near the juncture of the Diablo Platform, Coahuila Platform, and Chihuahua Trough (Presidio County).

Member is about 260 ft (80 m) thick and is composed of unfossiliferous lime wackestone that forms light-colored ledges in the upper part of the hill. The Upper Member of the Bluff, which caps the hills and forms most of the slopes facing the viewer, is about 460 ft (140 m) thick and is composed of marl, nodular limestone, rudist (toucasiid) lime packstone, shale, and quartzose sandstone. It contains echinoids, the foraminifer *Orbitolina,* and mollusks simliar to those of the Glen Rose Formation in central Texas.

Lower slopes of the igneous peak and the canyon in the foreground are formed on reddish-brown Cox Sandstone. Light-colored cliffs of Finlay Limestone protrude from the canyon walls (Fig. 5). The Finlay Limestone, about 328 ft (100 m) thick (Fig. 6) is composed of an upper resistant member of lime mudstone to rudist-miliolid-pellet lime packstone; and a lower, less-resistant marly unit. The Lower Finlay slope merges with the Upper Member of the Cox, which is composed of shale, marl, nodular limestone, and quartose sandstone.

The Finlay is overlain by Tertiary lava and tuff along most of the north wall of the canyon in the foreground, but near the head of the canyon yellowish slopes on the Benevides Formation can be seen. The Benevides, about 164 ft (50 m) thick, is composed of shale, marl, and nodular limestone capped by a prominent bed of pellet-oölith-echinoid lime grainstone. It contains

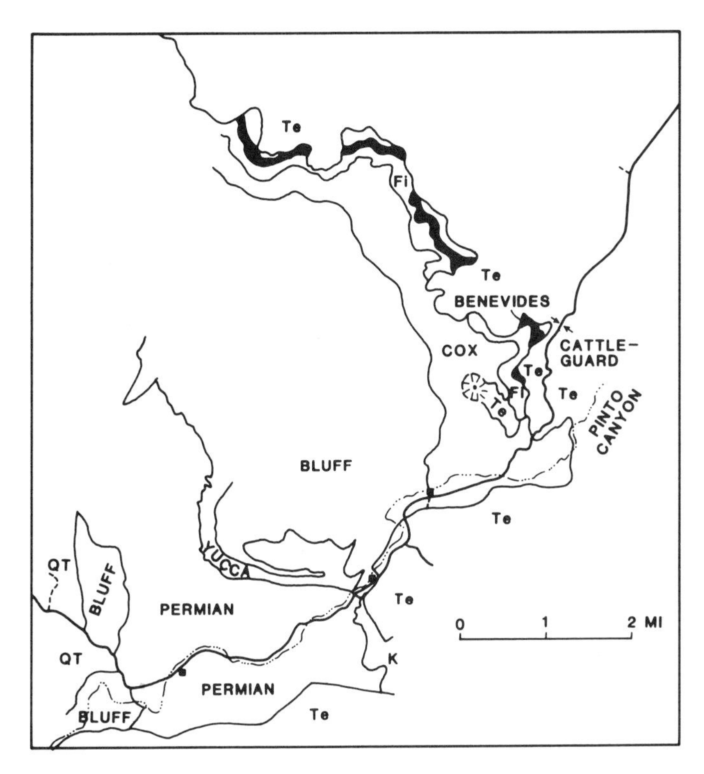

Figure 5. Route map and simplified geological map (from Amsbury, 1958) through Pinto Canyon (Presidio County). "Cattleguard" is the entrance to the Pinto Canyon Ranch; Fi = Finlay Limestone cliff; and the overlying unit shown in black is the Benevides Formation. Te = Tertiary undivided. K = Cretaceous undivided.

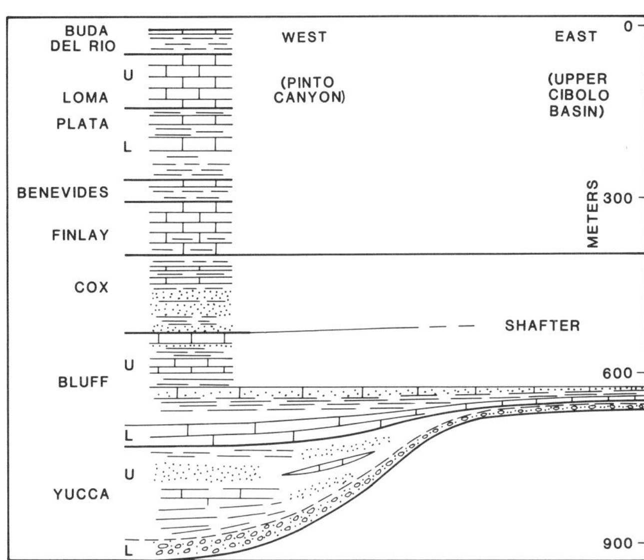

Figure 6. Cross section B, Figure 1. A diagrammatic shelf-edge transect showing major named units in the Pinto Canyon area. Data from Amsbury (1957, Figs. 7 and 8) and Amsbury (1958).

ammonites and other mollusks characteristic of the Kiamichi Shale in north-central Texas. There are no easily accessible outcrops of the Benevides and the overlying Loma Plata, Del Rio, and Buda formations anywhere in Presidio County.

36.0 mi (57.9 km). Pull off for another viewing spot in a curve toward the right. Pay strict attention to the No Trespassing signs; some landowners take them very seriously.

Light-colored ledges of the Lower Bluff form the mass of hills to the northwest, capped by brownish beds of the Upper Bluff. The Cox forms reddish-brown slopes to the northwest in the middle distance. To the north along the southern Vieja Rim, Finlay cliffs are overlain by yellowish brown Benevides slopes.

36.3 to 36.7 mi (58.4 to 59 km). Exposures of highly faulted Cox Sandstone in road cuts. The Cox is about 490 ft (150 m) thick and composed of cross-bedded brown sandstone, yellowish brown, softer sandstone, shale of various colors, and light gray, nodular limestone. A recent study by Keller (1982) indicated that Cox Sandstone beds in Pinto Canyon were deposited as a series of fan deltas.

38.0 mi (61.1 km). Primitive road to the north. In the hill north of the creek reddish brown Cox is faulted against gray limestone of the Lower Bluff. The basal Upper Bluff "caprock sequence" forms a butte to the northwest.

38.1 mi (61.3 km). Cattleguard. The basal Upper Bluff "caprock sequence" crops out on the hill straight ahead. Cox sandstone is exposed along the road.

38.6 mi (62.1 km). Primitive road to the left (road to the former Fred Shely Ranch house). Outcrops of faulted Upper Bluff ahead.

41.1 mi (66.1 km). Wide spot in the road. The relatively wide valley is cut in Permian rocks. Dark, cherty, contorted limestones are middle Guadalupian, above dark shale containing reddish weathering sandstone, of Wolfcampian age.

Northward from the road, low hills of Guadalupian cherty limestone protrude from the larger hill formed from Lower Cretaceous rocks. A light-colored cliff formed by the Lower Member of the Yucca marks the basal Cretaceous; a basal limestone-cobble conglomerate is overlain by reddish shale, sandstone, and nodular limestone of the Upper Member of the Yucca. Cliffs and ledges that cap the hill are outcrops of the Lower Member of the Bluff Limestone.

Figure 6 illustrates about 886 ft (270 m) of thinning of the basal Lower Cretaceous section from this area onto middle Guadalupian rocks in the upper Cibolo Basin, 6 mi (9.6 km) east-southeast. Intermediate sections demonstrated that onlap is supplemented by stratal thinning in the Yucca and Bluff formations (Amsbury, 1957); this area was the trough/platform transition zone during the late Aptian(?) and early Albian. Early to middle Albian sandstone beds thin and pinch out across the transition zone, so that the Upper Member of the Bluff and the Cox Sandstone merge into a single mappable unit, termed "Shafter Formation" by Rix (1953).

42.6 mi (68.5 km). Wide spot in the road on a high terrace. Directly to the south across the creek reddish beds of the Upper

Member of the Yucca are faulted against Wolfcampian and middle Guadalupian Permian rocks, and dark Upper Bluff rocks are faulted against the Yucca.

The basal Upper Bluff rocks exposed here represent a distinctive set of beds. They contain *Orbitolina,* oysters (*"Exogyra quitmanensis"*), echinoids, abundant casts of clams and snails, and the lower Albian ammonite *Douvilleiceras* (Young, 1974, p. 188). Amsbury (1957) termed this set of beds the "caprock sequence"; it forms the upper part of Rix's (1953) Presidio Formation. The sequence is capped by a distinctive brownish gray weathering, quartz–sandy lime grainstone. The sequence is about 246 ft (75 m) thick at the lower end of Pinto canyon, 164 ft (50 m) thick to the east in Pinto Canyon and at Shafter, and 82 ft (25 m) thick in the upper Cibolo Basin. It just might correlate with the *"Salenia"* beds of the middle Glen Rose in central Texas.

43.4 mi (69.8 km). Curve at telephone pole. The upper, cross-bedded, sandy limestone of the caprock sequence is exposed on the steep slope north of the road.

44.0 mi (70.8 km). Road to the Juan Benevides Ranch on the north side of the county road. The highest, pointed hill directly to the north across the ranch house has cliffs of the Upper Member of the Loma Plata Limestone, a slope formed in the Lower Member of the Loma Plata and the Benevides Formation, and a lower cliff of Finlay Limestone.

51.9 mi (83.6 km). Intersection of the Marfa–Pinto Canyon Road with Ranch Road 170 at Ruidosa. Continue to Presidio (40 mi [64.4 km] with no gasoline, water, or cold drinks).

0.0 mi (Begin new mileage series.) Intersection of Ranch Road 170 with U.S. 67 north of Presidio. Drive north on U.S. 67 toward Shafter.

14.2 mi (22.8 km). Beds of the Lower and Upper Members of the Finlay Limestone are exposed on the west side of the road. Park on the east side of the road on a curve and walk back. Mollusk lime wackestone of the Upper Finlay is faulted against reddish brown sandstone of the Shafter Formation (i.e., Cox).

14.5 mi (23.3 km). Low cut along the east side of the road. A massive lege of Finlay wackestone is interbedded with nodular limestone and marl.

14.7 to 15.2 mi (23.6 to 24.5 km). Outcrops of the Presidio Formation (Bluff). An oyster biostrome, probably part of the "caprock sequence," is faulted down at the north end of the road cut against cross-bedded, coarse-grained sandstone.

18.0 mi (29 km). Extensive outcrops of the Shafter Formation (Upper Bluff and Cox).

18.3 mi (29.4 km). Sandstone, shale, nodular limestone containing small oysters, and marl of the Shafter Formation.

18.6 mi (29.9 km). Road east into the village of Shafter. The extensive lead/silver mineralization here formed in Permian (probably upper Guadalupian) limestone.

REFERENCES CITED

Information about Cretaceous rocks in Trans-Pecos Texas is scattered; there is no detailed regional synthesis. Many field trip guidebooks are out of print, most M.A. theses and some Ph.D. theses remain unpublished, and most of the published Ph.D. theses were only in summary form to accompany geologic maps.

Adkins, W. S., 1933, The Mesozoic systems in Texas, *in* The geology of Texas; Vol. 1, Stratigraphy: University of Texas Bulletin 3232, p. 239–518.

Albritton, C. C., Jr., and Smith, J. F., Jr., 1965, Geology of the Sierra Blanca area, Hudspeth County, Texas: U.S. Geological Survey Professional Paper 479, 131 p.

Amsbury, D. L., 1957, Geology of the Pinto Canyon area, Presidio County, Texas [Ph.D. thesis]: University of Texas, 203 p.

—— , 1958, Geology of the Pinto Canyon area, Presidio County, Texas: University of Texas Bureau of Economic Geology Geological Map 22 (text printed on back of sheet), scale: 1:63,360.

Barnes, V. E., 1979a, Emory Peak–Presidio Sheet, Geological atlas of Texas: University of Texas Bureau of Economic Geology, scale 1:250,000.

—— , 1979b, Marfa Sheet, Geological atlas of Texas: University of Texas Bureau of Economic Geology, scale 1:250,000.

—— , 1983, Van Horn–El Paso Sheet (revised), Geological atlas of Texas: University of Texas Bureau of Economic Geology, scale 1:250,000.

Brand, J. P., and DeFord, R. K., 1958, Comanchean stratigraphy of Kent Quadrangle, Trans-Pecos Texas: American Association of Petroleum Geologists Bulletin, v. 42, p. 371–386.

Campbell, D. H., 1970, Depositional environments of the Mountain Facies in the Yucca Formation, southern Quitman Mountains, Hudspeth County, Texas: Society of Economic Paleontologists and Mineralogists 1970, p. 70–75 (also Ph.D. thesis, Texas A&M University, 1968, 323 p.).

—— , 1980, The Yucca Formation; Early Cretaceous continental and transitional environments, southern Quitman Mountains, Hudspeth County, Texas: New Mexico Geological Society Guidebook, 31st Field Conference, p. 159–168.

Flint, S. A., 1984, Paleoenvironments of the middle member of the Quitman Formation, Hudspeth County, Texas [M.S. thesis]: Alpine, Texas, Sul Ross State University, 216 p.

Gries, J. C., 1980, Laramide evaporite tectonics along the Texas–northern Chihuahua border, Trans-Pecos region: New Mexico Geological Society Guidebook, 31st Field Conference, p. 93–100.

Haenggi, W. T., 1966, Geology of El Cuervo area, northeastern Chihuahua, Mexico [Ph.D. thesis]: University of Texas, 403 p.

Jones, B. R., and Reaser, D. F., 1970, Geology of southern Quitman Mountains, Hudspeth County, Texas: University of Texas Bureau of Economic Geology Geological Quadrangle Map 39, 24 p. text, scale 1:48,000.

Keller, C. M., 1982, Depositional environment of the Lower Cretaceous Cox Formation, Pinto Canyon area, Presidio County, Texas [MA thesis]: Houston, Texas, University of Houston at Clear Lake City, 154 p.

Martin, B. S., 1987, Petrography and depositional environment of the lower member of the Cretaceous Quitman Formation, Hudspeth County, Texas [MS thesis]: Arlington, Texas, University of Texas at Arlington, 261 p.

Reaser, D. F., 1982, Geometry and deformational environment of the Cieneguilla-Quitman Range in northeastern Chihuahua, Mexico, and western Trans-Pecos Texas, USA: Rocky Mountain Association of Geologists, v. 1, p. 425–449.

Rix, C. C., 1953, Geology of the Chinati Mountains Quadrangle, Presidio County, Texas [Ph.D. thesis]: University of Texas, 188 p.

Scott, G., 1940, Cephalopods from the Cretaceous Trinity Group of the south-central United States: University of Texas Bulletin 3945, p. 969–1125.

SEPM, 1958, Cretaceous platform and geosynclines, Culberson and Hudspeth counties, Trans-Pecos Texas: Society of Economic Paleontologists and Mineralogists Guidebook 1958 Fieldtrip, 90 p.

—— , 1970, Geology of the southern Quitman Mountain area, Trans-Pecos Texas: Society of Economic Paleontologists and Mineralogists Guidebook 70-12, 127 p.

Smith, J. F., Jr., 1940, Stratigraphy and structure of the Devil Ridge area, Texas: Geological Society of America Bulletin, v. 51, p. 597–637.

Underwood, J. R., Jr., 1963, Geology of Eagle Mountains and vicinity, Hudspeth County, Texas: University of Texas Bureau of Economic Geology Geological Quadrangle Map 26, 32 p., text, scale 1:48,000.

Young, K. P., 1974, Lower Albian and Aptian (Cretaceous) ammonites of Texas, *in* Perkins, B. F., ed., Geoscience and man, Vol. 8: Baton Rouge, Louisana State University Press, p. 175–228.

The Davis Mountains volcanic field, west Texas

Don F. Parker, *Department of Geology, Baylor University, Waco, Texas 76798*

LOCATION

The Davis Mountains are located south of the junction of I-10 and I-20 and northwest of Alpine, predominantly in Jeff Davis County, southwestern Texas. The 14 field localities described here are keyed to the schematic cross sections of Figure 1. The field localities vary from roadcut exposures to hillside exposures to scenic panoramas; all were chosen to elucidate important features of Davis Mountains geology. A brief description of the location, the rock units involved, and chief references are given for each locality. K-Ar ages on Figure 1 were recalculated from Parker and McDowell (1979), using new standard decay constants (Steiger and Jager, 1977).

SIGNIFICANCE

The Davis Mountains are the largest volcanic area within the South-Central Section of the Geological Society of America. Cenozoic magmatism and tectonism associated with subduction along the southwestern margin of North America were responsible for a strong late Eocene–early Oligocene (~39–35 Ma) pulse of silicic volcanic activity in this area, which was followed during the Miocene by basin-and-range faulting and minor associated basaltic volcanism. The rocks of the Davis Mountains are unusual for this tectonic setting because their alkalic mineralogy and chemistry is more typical of intraplate or rift-related suites.

INFORMATION SOURCES

This abbreviated guide cannot be comprehensive in its treatment of all previous works or in field descriptions of the many exposed geologic units. Instead, the guide provides major sources of published geologic information; general descriptions of the rocks, their ages, and structures; and 14 publicly accessible field localities. The guide is intended to be used in conjunction with the references cited, particularly the two sheets (Marfa and Fort Stockton) of the Texas Atlas Project covering the area (Barnes, 1979, 1982). Useful topical references are summarized below.

Regional igneous geology: Barker (1977, 1979). Regional stratigraphy and isotopic ages: Henry and McDowell (1986), Henry and Price (1984), Maxwell and Dietrich (1970), Price and Henry (1984), Stevens and Stevens (1985), Wilson (1980). Davis Mountains stratigraphy, volcanology, and isotopic ages: Eifler (1951), Gibbon (1969), Parker (1983, 1986), Parker and McDowell (1979). Geologic maps: Anderson (1968), Barnes (1979, 1982). Roadlogs: Dickerson and Muehlberger (1985), Price and others (1986), Walton and Henry (1979).

GEOLOGIC SUMMARY

The Davis Mountains part of the Trans-Pecos magmatic province falls within the eastern, alkalic belt of Barker (1977). The rocks form a series from basaltic rocks (technically mostly hawaiite and mugearite) to trachyte, quartz trachyte and rhyolite, with rocks intermediate between basalt and trachyte poorly represented, and basaltic rocks subordinate in volume to the silicic rocks. Nepheline-bearing trachyte is present in some late intrusions. Some silicic rocks, particularly intrusions, are peralkalic and thereby possess sodic pyroxene and amphibole as mafic constituents.

The volcanic strata include silicic lava flows, ash-flow tuffs, volcaniclastic sedimentary rocks, and mafic lavas. Silicic lava units form broad shields; ash-flow tuffs are even more widely distributed. Some ash-flow tuffs that have secondary flow structures are difficult to distinguish from lava flows, particularly where pyroclastic textures have been destroyed by granophyric recrystallization (Parker and others, 1985). Volcaniclastic sedimentary units include reworked air-fall or ash-flow tuff, mudstone, sandstone, and conglomerate, and typically are interstratified with mafic lava flows. Mafic flows are much thinner (some flows only 3 to 6 ft; 1 to 2 m thick) than the silicic units, which can be as much as several hundreds of meters thick.

Volcanic centers identified for the Davis Mountains units include the dike swarms associated with the Paisano trachytic shield volcano (Parker, 1983), and the 12-mi-diameter (20 km) Buckhorn Caldera (Parker, 1986), which was the source of the widespread Gomez Ash-Flow Tuff. Many volcanic centers remain to be identified. Several large silicic flow units may represent shields similar to that of the Paisano Volcano. Several widespread ash-flow tuff sheets may have been erupted from caldera complexes in the central Davis Mountains west and northwest of Fort Davis.

LOCALITY DESCRIPTIONS

Locality 1. Gomez Peak, northern Davis Mountain. Gomez Peak is located on private land, but may be viewed for scores of miles on the eastern, northern, and northwestern flanks of the Davis Mountains. Gomez Peak is located on the northern rim of the Buckhorn Caldera (38 Ma), in which the Gomez Tuff accumulated to thicknesses up to 1,600 ft (500 m). The profile shown on Figure 1 may be viewed from the vicinity of Toyahvale, Texas from either U.S. 290 or Texas 17. References: Barnes (1979, 1982), Parker (1986), Price and others (1986).

Locality 2. Timber Mountain, northeastern Davis Mountains. View from Ranch Road 1832 (~9 mi or 14.5 km from Texas 17). A thick section of volcanic rocks on the southeastern tip of Timber Mountain exposes, in ascending order, the Huelster Formation (bedded tuff), the Star Mountain Formation (quartz trachyte lava), Gomez Tuff (ash-flow tuff), and Adobe Canyon

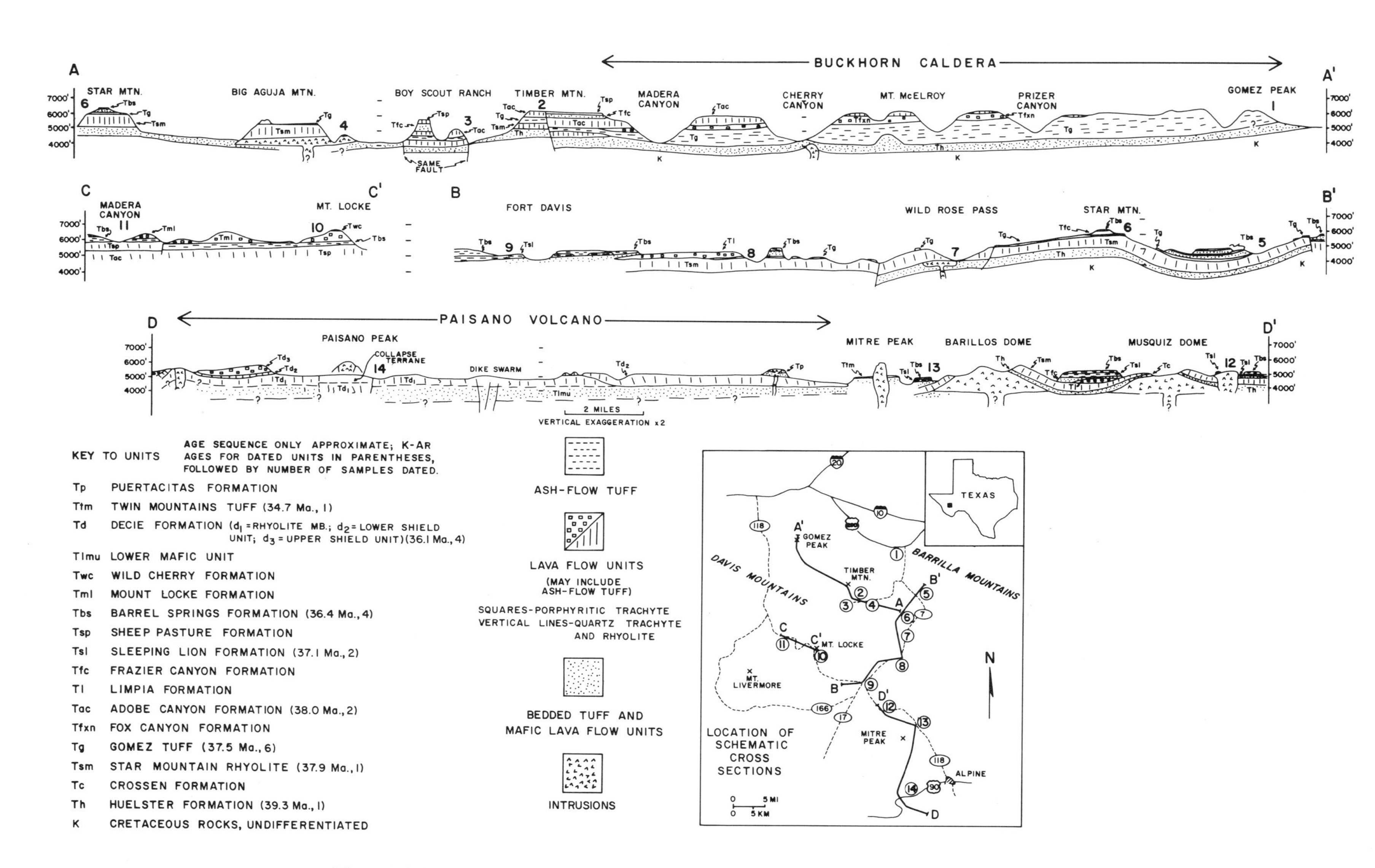

Figure 1. Schematic cross sections, Davis Mountains, Texas. In several areas, features slightly off from the line of section have been projected onto the section. Localities of geological interest, indicated by the bold numbers 1–14, are described in the text.

Formation (rhyolite lava). A basin-and-range fault that throws white, bedded tuff of the Frazier Canyon Formation against the Adobe Canyon rhyolite flows can easily be seen from several stops along Ranch Road 1832. Timber Mountain is located only 2 mi (3 km) from the southeastern edge of the Buckhorn Caldera. References: Barnes (1982), Price and others (1986).

Locality 3. Little Aguja Canyon. Ranch Road 1832 ends at the entrance to the Buffalo Trail Boy Scout Ranch (~11 mi or 18 km from Texas 17), which is private property. A northwest-trending basin-and-range fault, which the highway follows for several miles, has dropped volcanic rocks down to the southwest, and Little Aguja Creek has carved a deep canyon in these strata, which form precipitous cliffs surrounding the road. The section on the southwest side of the fault includes strata from the Star Mountain Formation at the base, up through the Gomez Tuff and Adobe Canyon formations. The section exposed on the upthrown side includes Cretaceous limestone and shale, and volcaniclastic sedimentary rock of the Huelster Formation. References: Barnes (1982), Price and others (1986).

Locality 4. Big Aguja Mountain. A peralkalic quartz trachyte laccolith that domed Star Mountain Formation flows crops out south of Ranch Road 1832 (~5 mi or 8 km from Texas 17). References: Barnes (1982), Gibbon (1969), Price and others (1986).

Locality 5. Roadcuts in Barrel Springs Formation along Texas 17. Texas 17 parallels the axis of the Rounsaville syncline for a number of kilometers. Cuestas formed by the outcrop of the Gomez Tuff and welded ash-flow tuff members of the Barrel Springs Formation crop out adjacent to the highway from about 7 to 13 mi (~11 to 21 km) south of the intersection with U.S. 290. Roadcuts allow inspection of the various welded tuff and bedded nonwelded tuff units within the Barrel Springs Formation, which most likely was derived from caldera sources in the central Davis Mountains. Folds, like this syncline, in the northeastern Davis Mountains and Barrilla Mountains were probably formed as drapes over basement faults, although other interpretations have been made. References: Barnes (1982), Eifler (1951), Pearson (1985), Price and others (1986).

Locality 6. Star Mountain. The type locality of the Star Mountain Formation may be viewed from several different angles from Texas 17. The Star Mountain Formation forms the steep cliffs on the flanks of the mountain; the Gomez Tuff and Barrel Springs Formation form thinner ledges on top. Three thick flows of Star Mountain Formation are visible on the flanks of Star Mountain and in exposures surrounding Wild Rose Pass. The Star Mountain Formation may be examined in roadcuts near the "dogleg" of Texas 17 east of Star Mountain, or in Limpia Canyon (see locality 8 below). References: Barnes (1982), Gibbon (1969), Eifler (1951), Price and others (1986).

Locality 7. Wild Rose Pass. An olivine diabase crops out in and around Wild Rose Pass on Texas 17 (~19 mi or 31 km south of the intersection with U.S. 290). Roadcuts contain fresh rock, which forms spheroidal boulders in more weathered material. The McCutcheon fault downdropped Gomez Tuff almost to road level south of the pass. References: Barnes (1982), Gibbon (1969), Price and others (1986).

Locality 8. Limpia Canyon. Texas 17 follows Limpia Canyon from Wild Rose Pass to Fort Davis. For many miles, the road stays in the upper part of the Star Mountain Formation (note the palisades formed by columnar jointing). Many good exposures of this unit occur in roadcuts from about 20 to 25 mi (32 to 40 km) south of U.S. 290. The Gomez Tuff forms a rare roadcut where the canyon broadens about 6.7 mi (10.8 km) west of Wild Rose Pass. References: Barnes (1982), Price and others (1986).

Locality 9. Fort Davis. Old Fort Davis is located at the junction of Texas 17 and 118; the modern town is located just west of this intersection. Fort Davis National Historic Site and the adjacent Davis Mountains State Park form the largest expanse of public land in the Davis Mountains. (However, samples may not be taken from either area without permission from authorities.) The Sleeping Lion Formation, an ash-flow tuff with secondary flow structures, forms the prominent cliffs behind the old fort. Roadcuts along Texas 118 east of the fort in Limpia Canyon expose stacked basaltic flows within the Frazier Canyon Formation and, nearer the entrance to Davis Mountains State Park, grey porphyritic rhyolite of the overlying Sleeping Lion Formation. References: Barnes (1982), Price and others (1986).

Locality 10. Mount Locke. A well-exposed section along Texas 118 west of Fort Davis on the flanks of Mount Locke exposes the top of the Sheep Pasture Formation (silicic lava?), the Barrel Springs Formation (ash-flow tuff), and the lower part of the Mount Locke Formation (trachyte lava flows). Nonwelded to slightly welded bedded tuff in between major units appears to be ash-flow tuff. McDonald Observatory, which maintains an 82-in and a 107-in telescope on Mount Locke, has a visitor center at the base of the mountain. References: Anderson (1968), Barnes (1979), Mattison (1979).

Locality 11. Madera Canyon road side park on Texas 118 west of Fort Davis. A good lunch stop with Ponderosa Pines. The stream-cut across from the park exposes massive breccia and vitrophyre within the top of the Sheep Pasture Formation. References: Anderson (1968), Barnes (1979).

Locality 12. Weston laccolith, roadcuts on Texas 118, approximately 4.5 mi (7.2 km) south of Fort Davis. Several roadcuts expose the green trachyte of the intrusion where the road drops into the Musquiz Canyon drainage. One cut exposes a tongue-like projection of the trachyte where it has flexed mafic lava and bedded tuff of the Frazier Canyon Formation; another contains the trachyte faulted against mafic lava. In exposures on private land off the road, the intrusion can be shown to have both flexed and faulted the Sleeping Lion Formation, the palisaded ash-flow tuff unit that crops out around Fort Davis and along Musquiz Canyon. The Gomez Tuff is locally present in the Musquiz Canyon area (Elkins, 1986, personal communication). References: Barnes (1982), Price and others (1986).

Locality 13. Barillos Dome overlook, intersection of Ranch Road 1837 with Texas 118. Barillos dome, a classic laccolith structure, has domed the Star Mountain Formation and associated units. Erosion has exposed a composite trachyte/rhyolite intrusive core. The Weston and Barillos laccoliths are two of about five such major structures in the Musquiz Canyon drainage area. References: Barnes (1982), Price and others (1986).

Locality 14. Roadcuts along U.S. 90 west of Alpine expose lava, ash-flow tuff, and dike rocks associated with the Paisano volcano. A complex area west of Paisano Peak may be a 3-mi-diameter (5 km) caldera collapse zone; it contains megabreccia and lahar deposits with large clasts of ash-flow tuff. References: Barnes (1982), Parker (1979, 1983), Price and others (1986).

REFERENCES CITED

Anderson, J. E., Jr., 1968, Igneous geology of the central Davis Mountains, Jeff Davis County, Texas: University of Texas at Austin Bureau of Economic Geology Quadrangle Map 36, scale 1:62,500.

Barker, D. S., 1977, Northern Trans-Pecos magmatic province; Introduction and comparison with the Kenya rift: Geological Society of America Bulletin, v. 88, p. 1421–1427.

——, 1979, Cenozoic magmatism in the Trans-Pecos province; Relation to the Rio Grande rift, *in* Riecker, R. E., ed., Rio Grande rift; Tectonics and magmatism: Washington, D.C., American Geophysical Union, p. 382–392.

Barnes, V. E., project director, 1979, Marfa Sheet, *in* Geologic Atlas of Texas: University of Texas at Austin Bureau of Economic Geology, map with explanatory text, scale 1:250,000.

——, 1982, Fort Stockton Sheet, *in* Geologic Atlas of Texas: University of Texas at Austin Bureau of Economic Geology, scale 1:250,000.

Dickerson, P. W., and Muehlberger, W. R., eds., 1985, Structure and tectonics of Trans-Pecos Texas: West Texas Geological Society Publication 85-81, 278 p.

Eifler, G. K., 1951, Geology of the Barrilla Mountains, Texas: Geological Society of America Bulletin, v. 62, p. 339–353.

Gibbon, D. L., 1969, Origin of the Star Mountain Rhyolite: Bulletin Volcanologique, v. 33, p. 438–474.

Henry, C. D., and McDowell, F. W., 1986, Geochronology of the Tertiary volcanic field of Trans-Pecos Texas, *in* Price, J. G., Henry, C. D., Parker, D. F., and Barker, D. S., eds., Igneous geology of Trans-Pecos Texas: The University of Texas at Austin, Bureau of Economic Geology Guidebook 23, p. 99–122.

Henry, C. D., and Price, J. G., 1984, Variations in caldera development in the Tertiary volcanic field of Trans-Pecos Texas: Journal of Geophysical Research, v. 89, p. 8765–8786.

Mattison, G. D., 1979, A reinterpretation of the Sheep Pasture tuffs, *in* Walton, A. W., and Henry, C. D., eds., Cenozoic Geology of the Trans-Pecos volcanic field of Texas: University of Texas at Austin, Bureau of Economic Geology Guidebook 19, p. 83–91.

Maxwell, R. A., and Dietrich, J. W., 1970, Correlation of Tertiary rock units, West Texas: University of Texas at Austin, Bureau of Economic Geology Report of Investigations 70, 34 p.

Parker, D. F., 1979, The Paisano volcano; Stratigraphy, age, and petrogenesis, *in* Walton, A. W., and Henry, C. D., eds., Cenozoic geology of the Trans-Pecos volcanic field of Texas: University of Texas at Austin, Bureau of Economic Geology Guidebook 19, p. 97–105.

——, 1983, Origin of the trachyte–quartz trachyte–peralkalic rhyolite suite of the Oligocene Paisano volcano, Trans-Pecos Texas: Geological Society of America Bulletin, v. 94, p. 614–629.

——, 1986, Stratigraphic, structural, and petrologic development of the Buckhorn caldera, northern Davis Mountains, Trans-Pecos Texas, *in* Price, J. G., Henry, C. D., Parker, D. F., and Barker, D. S., eds., Igneous geology of Trans-Pecos Texas: The University of Texas at Austin, Bureau of Economic Geology Guidebook 23, p. 286–302.

Parker, D. F., and McDowell, F. W., 1979, K-Ar geochronology of Oligocene volcanic rocks, Davis and Barrilla mountains, Texas: Geological Society of America Bulletin, Part I, v. 90, p. 1100–1110.

Parker, D. F., Price, J. G., Henry, C. D., Sanders, K. C., and Powell, K. H., 1985, Discrimination of widespread silicic lava flows from rheomorphic ash-flow tuffs, Trans-Pecos volcanic field, Texas: EOS, v. 66, p. 1125–1126.

Pearson, B. T., 1985, Tertiary structural trends along the northeast flank of the Davis Mountains, *in* Dickerson, P. W., and Muehlberger, W. R., eds., Structure and tectonics of Trans-Pecos Texas, West Texas Geological Society Publication 85-81, p. 153–156.

Price, J. G., and Henry, C. D., 1984, Stress orientations during Oligocene volcanism in Trans-Pecos Texas; Timing the transition from Laramide compression to basin and range extension: Geology, v. 12, p. 238–241.

Price, J. G., Henry, C. D., Parker, D. F., and Barker, D. S., eds., 1986, Igneous geology of Trans-Pecos Texas: University of Texas at Austin Bureau of Economic Geology Guidebook 23, 360 p.

Steiger, R. H., and Jager, E., 1977, Subcommission on geochronology; Convention on the use of decay constants in geo- and cosmochronology: Earth and Planetary Science Letters, v. 36, p. 359–362.

Stevens, J. B., and Stevens, M. S., 1985, Basin and range deformation and depositional timing, Trans-Pecos Texas, *in* Dickerson, P. W., and Muehlberger, W. R., eds., Structure and tectonics of Trans-Pecos Texas: West Texas Geological Society Publication 85-81, p. 157–163.

Walton, A. W., and Henry, C. D., eds., 1979, Cenozoic geology of the Trans-Pecos volcanic field of Texas: University of Texas at Austin Bureau of Economic Geology Guidebook 19, 193 p.

Wilson, J. A., 1980, Geochronology of the Trans-Pecos Texas volcanic field, *in* Dickerson, P. W., and Hoffer, J. M., eds., Trans-Pecos region: New Mexico Geological Society, 31st Field Conference Guidebook, p. 205–211.

Geology of the Marathon Uplift, west Texas

Earle F. McBride, *Department of Geological Sciences, University of Texas at Austin, Austin, Texas 78713*

LOCATION

Roadcuts provide the only exposures of public access in the Marathon Uplift. However, roadcuts on U.S. 385 south of the town of Marathon and on U.S. 90 east of Marathon display spectacular features. Mileages to the first five stops in this guide are given from the junction of U.S. 385 south and U.S. 90 at the west edge of Marathon; mileages to the six remaining stops are from the junction of U.S. 385 north and U.S. 90 (Fig. 1).

SIGNIFICANCE

The Marathon Uplift is a broad, domal uplift 78 mi (125 km) in diameter, of early Tertiary age. Erosion of Cretaceous and younger strata from the crest of the uplift produced the topographic Marathon Basin (Fig. 1) where, in an area 31 by 47 mi (50 by 75 km), deformed pre-Permian Paleozoic rocks are exposed that have a composite stratigraphic thickness of 16,400 ft (5,000 m). These Paleozoic rocks are part of a belt of deformed rocks that were deposited along the southeastern margin of North America during Paleozoic time, and which make up the Ouachita orogen. The orogen extends from Arkansas across Oklahoma and Texas, and has been traced almost to Mexico.

INTRODUCTION

Bedrock is exposed in the Marathon Uplift in northeast-trending ridges, rising up to 1,000 ft (300 m) above the valley floors, which are covered with Quaternary alluvium and colluvium. The latter support a variety of cacti, yucca, creosote bushes, cat-claw, grasses, and other desert plants. Rainfall averages 17 in (43 cm) per year. The uplift is bordered on the north and east by Cretaceous rocks of the Edwards Plateau; on the northwest by the Glass Mountains, whose Permian rocks rest unconformably on the Marathon Basin sequence; on the southwest by Cretaceous rocks and by the volcanic terrane of the Davis Mountains; and on the south by a dissected south-dipping homocline of Cretaceous rocks.

The stratigraphic sequence in the Marathon Uplift includes rocks ranging in age from Late Cambrian to Late Pennsylvanian (Fig. 2). The depositional history of the Paleozoic sequence in the Marathon region can be divided into two stages on the basis of rate of deposition. The initial stage, from Late Cambrian to Mississippian (Fig. 2), records the slow accumulation of only 3,100 ft (950 m) of rocks in a time span of nearly 170 m.y. Rocks of the initial stage are shale, limestone, chert, sandstone, conglomerate, and olistostromes of the Dagger Flat, Marathon, Alsate, Fort Peña, Woods Hollow, Maravillas, and Caballos Formations. Stratigraphic and sedimentological evidence indicates the presence of unconformities (debatably of submarine rather than subaerial origin) in places at the base of the Caballos Novaculite and the top of each of the novaculite members within the formation. Many previous workers have reported unconformities elsewhere in the section, but the evidence is debatable. No Silurian fossils have been identified. The second stage, from Mississippian through Pennsylvanian, records the rapid accumulation of 13,800 ft (4,200 m) of rock in only 60 m.y. This represents an order of magnitude increase in sedimentation rate. Rocks of the second stage are shale, sandstone, limestone, chert, conglomerate, and olistostromes of the Tesnus, Dimple, Haymond, and Gaptank Formations. The basic stratigraphy of the rocks was established by Baker and Bowman (1917) and King (1930, 1937).

Differences in interpretation of the processes and environments of deposition exist for nearly all the formations in the Marathon sequence, with the greatest debate concerning the

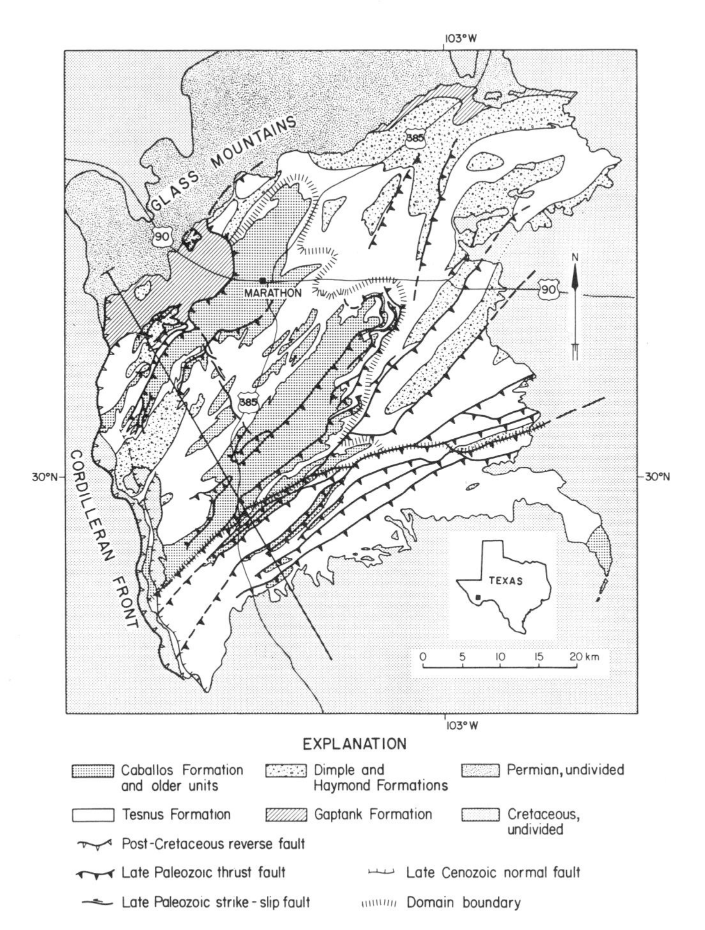

Figure 1. Generalized geologic map of the Marathon region showing structural domains described in text. Line of section is that of Figure 3. Adapted from Muehlberger and Tauvers (1988, Fig. 1).

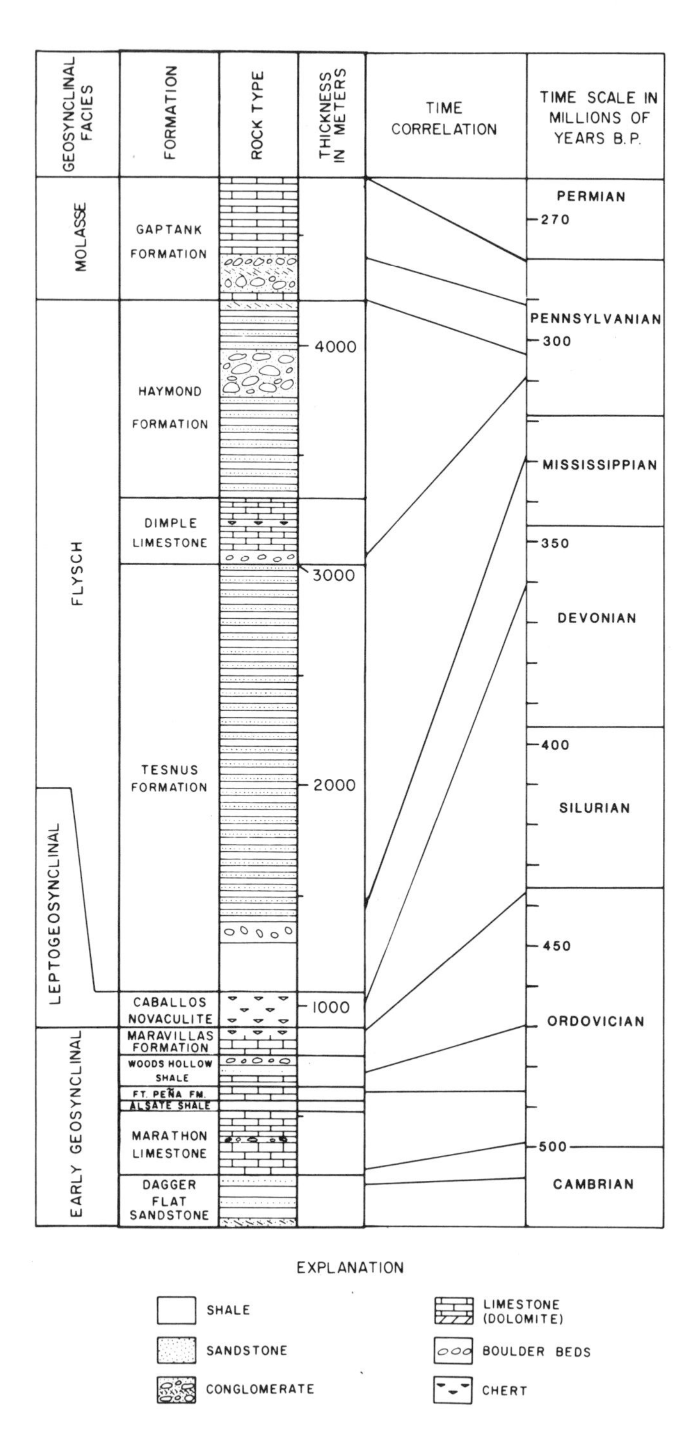

Figure 2. Stratigraphic section of pre-Permian strata in the Marathon Uplift. The early geosynclinal and leptogeosynclinal (starved basin) facies make up the deposits of the initial stage of deposition, whereas the flysch and molasse make up the late stage of deposition (see text). Tie lines from the time scale identify stratigraphic positions for which paleontological ages are known. Not all Ordovician ages are so identified.

depth of water during deposition. The majority of workers favor the interpretation that all pre-Gaptank rocks were deposited in submarine slope and basinal environments that were from hundreds to thousands of meters deep (e.g., Thomson, 1964; McBride, 1966; McBride and Folk, 1977). However, other workers have interpreted some formations to have been deposited in water only tens of meters deep or to be subaerial in origin (e.g., Fan and Shaw, 1956; Flores, 1972; Folk, 1973). There has been only minor debate about the source of detritus in the sequence. Stratigraphic and paleocurrent data demonstrate that most of the carbonate detritus in the entire sequence was derived from a source to the northwest of the basin, whereas terrigenous siliciclastic detritus was derived from a source to the east and southeast of the basin. The northwestern source probably was the edge of the North American craton, which, being overridden during thrusting, lay to the southeast of the present outcrop. The name "Llanoria" was popularized for the eastern land element by Dumble (1920), Miser (1921), and King (1937). Ideas on the composition and size of Llanoria have changed over the years. The continental margins of Africa and South America have been suggested (Keller and Cebull, 1973; Rowett and Walper, 1973), although a fragment of what is now Central America is also a possibility (A. Salvador, 1985, personal communication). Olistostromes, boulder-bearing mudstones deposited by submarine mass flows, are present in most of the stratigraphic units and attest to multiple episodes of uplift of basin-margin elements. Olistoliths include some rock types indigenous to the exposed Paleozoic sequence, but many were derived from unknown extrabasinal sources. Most olistoliths were derived from the North American craton during Cambrian to Pennsylvanian time; Llanoria shed olistoliths only during Early Pennsylvanian (Haymond) time.

The stratigraphy and sedimentology of each formation in the Marathon uplift are described in references cited in the following guidebooks: *Permian Basin Section, S.E.P.M.,* 1964 Field Trip Guidebook; also 1978 Guidebook; *Dallas Geological Society,* 1969 Guidebook, E. F. McBride, editor; and in McBride (1987).

The Ouachita sequence in this area underwent strong compressional deformation during Late Pennsylvanian and Early Permian time, during which the main structural features of the region formed (King, 1937; Ross, 1962; Ross and Ross, 1985). Wells drilled in the Marathon Uplift indicate that the entire exposed pre-Permian Paleozoic sequence is allochthonous (Ross, 1962; King, 1980). Sedimentological and structural evidence suggests that the allochthon has been transported at least 125 mi (200 km) northwestward from its root (Muehlberger and others, 1984). Deformation was probably continuous during this late Paleozoic orogeny, although there were probably pulses of stronger deformation (King, 1937; Ross and Ross, 1985; Tauvers, 1985). The area was subjected to compressional forces during the Laramide orogeny (80 to 55 Ma), when a high-angle reverse fault formed at the southwestern edge of the Uplift, and some Paleozoic fold trends within the Uplift were reoriented. The area was domed sometime during Tertiary time, when Paleozoic strike-slip

Figure 3. Generalized cross-section along line shown in Figure 1. Mts, Tesnus Formation; DCT, Dugout Creek overthrust; HHAT, Hell's Half Acre thrust fault.

faults underwent normal faulting, and numerous small igneous intrusions, chiefly of syenite, were intruded within and marginal to the Uplift. The main episode of intrusion in west Texas was from 38 to 15 Ma (Stevens and Stevens, 1985).

Paleozoic structural elements include several orders of folds, southeastward-dipping thrusts, northwest-trending strike-slip faults, and several joint sets. First-order folds are reflected by a synclinorium and adjacent flanking anticlinoria (Fig. 3). Other well-displayed structural features include duplex structures, folded thrusts, klippen, imbricate thrusts, and disharmonic folds. In spite of stratal shortening of 6 to 1 in the Warwick Hills (K. Coley, 1986, personal communication), the rocks of the Marathon Uplift are not metamorphosed.

Three major structural domains can be identified in the Marathon Uplift (Fig. 1). The western domain is highly folded and imbricated lower Paleozoic formations whose major structures are controlled chiefly by the actions of the relatively competent Maravillas/Caballos couplet and a major décollement within the underlying Woods Hollow shale. The eastern domain is dominated by large-scale folds of upper Paleozoic flysch formations (Fig. 2). Shale within the lower part of the Tesnus Formation served as the main décollement in this domain. The southern domain comprises all outcrops south of the Hells Half Acre thrust fault. Formations in this domain have unique facies and thicknesses (Muehlberger, 1978; Muehlberger and others, 1984; Muehlberger and Tauvers, 1988).

ROADCUTS SOUTH OF MARATHON

Roadcut. 3.5 mi (5.6 km) south on U.S. 385. At this locality the Caballos Novaculite is 436 ft (133 m) thick and includes the following members: Lower Chert, 1 ft (0.3 m); Lower Novaculite, 121 ft (37 m); Lower Chert and Shale, 144 ft (44 m); Upper Novaculite, 3 ft (1 m); Upper Chert and Shale, 164 ft (50 m). The Lower Chert and Lower Novaculite are exposed on the north side of the road; the Upper Chert and Shale are exposed on the south side of the road. The contact of the Caballos with the Maravillas is concealed, but the contact with the Tesnus is exposed. The Novaculite is milk-white to off-white chert composed of nonfibrous microcrystalline quartz. The white color is imparted by light reflections off microcrystal boundaries. The novaculite lacks detrital and chemical components that provide color to cherts in the other members of the formation (organic matter and manganese oxide = black, illite = green, hematite = red; limonite = brown). Novaculite in the Caballos has not been subjected to temperatures as high as novaculite in the Arkansas Novaculite, and it lacks the triple-junction fabric typical of the Arkansas Novaculite (Keller and others, 1977). Protolith sediment of the Caballos Novaculite was probalby carbonate mud (60 percent?) that contained abundant siliceous sponge spicules and moderately few radiolaria. Upon burial, the biogenic opaline tests dissolved and the silica completely replaced all carbonate grains. Sponge spicule ghosts are visible in thin section, and can also be seen with a hand lens in the gray-pigmented parts of some beds. No stratification is visible in the novaculite, and "beds" are separated by stylolite surfaces. (Did they form before or after silicification of the protolith?)

The Upper Chert and Shale Member on the south side of the road is composed of chert beds of various colors (mostly gray, green, and brown) from 0.4 in to 1 ft (1 to 30 cm) thick that are separated by shale beds less than 0.8 in (2 cm) thick. These cherts probably formed by the silicification of protolith sediment that was a mixture of carbonate mud, carbonate allochems, siliceous sponge spicules and radiolaria, and terrigenous clay. Detrital grain ghosts, now all silicified, are visible in thin sections of many samples. These detrital grains plus a few chert-granule conglomerate beds (0.4 to 4 in; 1 to 10 cm thick) attest to a significant contribution of coarse material to the chert and shale members. Coarse sediment of this type is absent from the novaculite beds. Radiolarian ghosts make up 30 percent of some samples and are more abundant than spicules. Shale beds are composed of illite clay with scattered radiolaria. Many chert and shale beds are deeply oxidized in this roadcut. Many chert beds have uneven thickness that is the result of differential compaction during silicification. The upper part of the Caballaos here has red shale beds several meters thick that are visible in the bar ditch. Olive-drab siliceous shale beds of the overlying Tesnus Formation are present above the red shale.

Manganese oxide coats fracture surfaces of many chert beds in the chert and shale members. Within the Upper Chert and Shale Member north of the road is an old manganese prospect pit in a particularly manganese-rich zone.

Road sign. 7.7 mi (12.3 km) south on U.S. 385. The Upper Novaculite Member forms the crest of Horse Mountain (elevation 5,012 ft; 7,528 m), from which the Caballos Novaculite gets its name. Straight ahead, at 12:00, is flat-topped Santiago Peak, a Tertiary syenite intrusion. At 2:00 is East Bourland Mountain, a breached, doubly plunging anticline whose flanks have conspicuous flatirons of Lower Novaculite. Topographic ridges in the uplift are formed chiefly by the Maravillas and Caballos Formations.

Roadside picnic area. 10.1 mi (16 km) south on U.S. 385. To the west is Simpson Spring Mountain, another doubly plunging anticline that has been breached. The Lower Novaculite Member of the Caballos forms flatirons at the base of the hill and forms the ridge crest also. A folded thrust runs along the valley in the Tesnus Formation just south of this site.

Roadcut. 11.5 mi (18.4 km) south on U.S. 385. This roadcut transects the limb of a fold that is overturned to the northwest at the site of a left-lateral strike-slip fault trending N20°W. A nearly complete section of the Maravillas Formation (Late Ordovician) is exposed on the north side of the road, whereas fault breccia of the strike-slip fault and chert and shale members of the Caballos Novaculite are exposed on the south side of the road. Quaternary travertine, locally cavernous, cements colluvium in the southern half of this wind gap.

The Maravillas here is 353 ft (108 m) thick and consists of 80% limestone, 18% chert, 2% shale, and a trace of conglomerate. All beds are black where fresh, due to the presence of organic matter, but the limestones weather light gray. Limestone beds range from 0.4 to 28 in (1 to 70 cm) thick and show laminations and graded bedding in places, especially at the base of sedimentation units. Limestones are nonporous beds composed chiefly of neomorphosed silt- and clay-sized grains with some sand-sized skeletal fragments, carbonate rock fragments, quartz and glauconite grains, and chert clasts. The limestone beds were deposited by turbidity currents derived from a shelf area to the northwest of the depositional site. Contorted chert beds in the upper third of the section formed by submarine slumping prior to chertification.

Chert nodules formed by the partial replacement of limestone beds by silica, and chert beds are probably fine-grained limestones that have been totally replaced by silica. Silica was derived chiefly from siliceous sponge spicules, and it selectively replaced fine-grained carbonate particles.

Roadside stop in Dagger Flat. 17 mi (27 km) south on U.S. 385. At 1:00 is Three-Mile Hill, which exposes the southeastern facies of the Caballos Novaculite. Here the Upper Novaculite is the thick novaculite at the hill crest, and the Lower Novaculite forms a small ledge (16 ft; 5 m thick) lower on the hillslope. The position of a southeast-dipping thrust fault is marked by a conspicuous brush line one-third of the way up the slope. The thrust plate, which includes the Woods Hollow through the lower part of the Tesnus Formation, rests on the Dagger Flat Formation.

ROADCUTS EAST OF MARATHON

Roadside stop. 9 mi (14 km) east of Marathon on U.S. 90. The Warwick Hills several kilometers to the south are intricately folded thrust sheets that have undergone crustal shortening as great as 6 to 1 (K. Coley, 1986, personal communication). Ridges are supported by the Maravillas and Caballos Formations.

Roadcut. 17.0 mi (27.2 km) east on U.S. 90. Nearly vertical beds of the Haymond Formation (Pennsylvanian) are exposed in this cut. The stratigraphic top is to the east. Sandstone beds (including siltstone) feature sharp basal surfaces, some with trace-fossils and flute and groove casts, and pass gradually into shale at their top surfaces. Limonite stains most fracture surfaces and obscures internal stratification, but Bouma A, B, and well-developed C divisions are visible on sawed surfaces. Sandstone beds are turbidites that pass upward to turbidite shale and then to hemipelagic shale, which cannot be distinguished in the field. At least one bed in the roadcut is a calcarenite (strongly limonitized) that has the same composition as turbidites in the underlying Dimple Formation. Shale beds contain minute amounts of plant detritus and bedding-parallel trace fossils. Paleocurrent data indicate turbidity currents travelled west and northwest at this site. The thin beds and scarcity of slump structures suggests that this facies of the Haymond, like much of the entire formation in outcrop, was deposited along the nearly flat basin floor far from the edge of the shelf. Thirteen of the thickest beds from the central 230 ft (70 m) of this section of the exposure can be correlated with beds exposed in a railroad cut 6.5 mi (10.5 km) south of this roadcut (but across a thrust fault; Dean and Anderson, 1967). There is a 46 percent decrease in the number of turbidite/shale couplets in this roadcut as compared to the more proximal railroad cut.

This outcrop provides a classic example of soil creep. Notice the change in dip of the sandstone beds in a strip 1.6 ft (0.5 m) thick across the top of the hill. The thin (< 0.4 in, 1 cm) white powdery "beds" scattered throughout the roadcut are Quaternary accumulations of caliche. A metal plaque on the south side of the road, entitled "Denuded Ouachita Rock Belt," describes part of the geologic history of the area.

Roadcut. 17.7 mi (28 km) east on U.S. 90. This roadcut exposes the Haymond thrust fault, which places the Tesnus on top of the Haymond Formation. Thin-bedded Haymond sandstones and shales, folded but coherent, are exposed at the west end of the cut, whereas thick-bedded Tesnus sandstones and shales are exposed at the east end of the cut. The central part of the cut is the thrust fault zone, in which sandstone beds have been folded, faulted, and deformed into boudins, and the shales intensely contorted. Microfaults cut the bottom surface of many beds.

This exposure provides a challenge for deciphering the degree of induration that the sediments had undergone at the time of deformation.

Spectacular boxwork structure is present in Haymond sandstone beds that have been case hardened along intersecting joint sets.

Roadcut. 18.1 mi (29 km) east on U.S. 90. The cut exposes a tight syncline and strongly sheared and boudinaged beds in the Tesnus Formation. The intensity of deformation, plus structural features similar to the previous roadcut, raise the possibility that there is an intraformational thrust at this site.

Roadcut. 18.5 mi (30 km) east on U.S. 90. This long exposure of steeply dipping beds in the Tesnus Formation reveals a variety of rock types, including sandstone, shale, mudstone, chert, and tuff, as well as a variety of features formed during soft-sediment deformation. Sandstone "beds" include turbidites (sharp base, gradational top, some faint laminations), clastic sills (sharp base and top, internally structureless to faintly laminated), and "beds" that could be either sills or turbidites. Some sills distinctly cut across bedding or form anastomosing bodies, and many have lateral or oblique dike offshoots into adjacent shale and mudstone beds. Other sills can only be distinguished from beds by the flute casts that occur on both sides of the body. Sandstone dikes were injected into shale beds at various angles to bedding. The dikes show a random or great circle girdle when poles are plotted on a stereo net (T. Diggs, 1986, personal communication). Dikes perpendicular to bedding have ptygmatic folds that indicate up to 50 percent compaction of the shale beds since dike injection.

Spheroidal weathering of mudstone beds is common in the exposure. These beds, most of which contain abundant land-plant fragments, formed as submarine mud flows. Hemipelagic shale beds are olive-colored, fissile clay beds with only a trace of silt.

Approximately 82 ft (25 m) from the east end of the exposure is a 60-ft-thick (18 m) sequence of tuff beds, which occurs in three cyclic units. Each cycle consists of structureless to graded tuff (mustard yellow), tuff with lenses of radiolarian-bearing siliceous shale, and fissile clay shale. Glass shards (visible with hand lens) and crystal grains have all been altered to clay minerals. The tuff beds are debris-flow deposits that originated upslope (east) of their site of deposition. siliceous shale beds were dismembered and incorporated into the debris flows (Imoto and McBride, 1986).

Present in several shale beds are disc-shaped concretions of siderite, now altered to limonite, with margins of cone-in-cone structure.

The Tesnus exposed in this cut was deposited on a fairly steep submarine slope in water deepr than storm base. Episodic shocks, probably generated by westward migration of Llanoria toward the North American craton, triggered the debris flows, slumps, and injection episodes.

Roadcut. 19.6 mi (32 km) east on U.S. 90. Steeply dipping, gray to black limestone, mudstone, and chert beds of the basinal facies of the Dimple Formation (Pennsylvanian; Thomson and Thomasson, 1969) are exposed in the blasted face south of the highway. Approximately the lower half of the Dimple is exposed in the cut. Shallower-water facies of the Dimple are exposed 12 mi (20 km) to the northwest. The gradational contact with the Tesnus Formation is visible in the stream cut at the west end of the exposure.

Limestone beds are turbidites that show Bouma A, B, and C divisions (graded, laminated, current rippled, and convolute laminated). Structureless calcitic and spiculitic mudstones that overlie limestone beds are the suspension deposits of the turbidity currents. Hemipelagic shale beds are thin, dark gray, and have few spicules. Meandering trails of bottom-feeding animals (*Fodinichnia*) are present in some shale beds. Thin black chert beds consist of microcrystalline quartz with abundant siliceous sponge spicules and a trace of organic matter that serves as the pigment. These beds are fine-grained, spicule-rich limestones that were replaced by silica derived from sponge spicules in the calcitic mudstone beds. Some radiolaria are present in the chert beds also. Silica has selectively replaced the fine-grained tops and fine-grained laminations in many of the graded limestone turbidite beds. Differential weathering of chertified convolute laminations in some of the thicker limestone beds forms spectacular patterns at the ground surface on top of the cut.

Limestone beds are nonporous and contain fossil fragments, carbonate rock fragments, chert clasts, quartz, glauconite, phosphate grains, and conodonts in their coarser fraction. The fine-grained tops of beds are composed of silt- and clay-sized carbonate grains. Carbonate and other detritus in the Dimple came from a source area to the west of this area. In the lowermost 200 ft (60 m) of the formation, limestone beds thicker than 20 in (50 cm) can be correlated bed-for-bed with those exposed in a roadcut 3.5 mi (5.6 km) west of this locality. The two localities were more than 6 mi (9.6 km) apart before folding and faulting of the strata. Limestone beds differ in thickness between the two exposures by less than 10 percent; two-thirds of the beds are thicker in the western (proximal) exposure.

REFERENCES CITED

Baker, C. L., and Bowman, W. F., 1917, Geologic exploration of the front range of Trans-Pecos Texas: University of Texas Bulletin 1753, p. 61–177.

Dean, W. E., Jr., and Anderson, R. Y., 1967, Correlations of turbidite strata in the Pennsylvanian Haymond Formation, Marathon region, Texas: Journal of Geology, v. 75, p. 59–75.

Dumble, E. T., 1920, The geology of East Texas: University of Texas Bulletin 1869 (August 10, 1918), 388 p.

Fan, P. H., and Shaw, D. E., 1956, The Tesnus Formation of Trans-Pecos Texas: Journal of Sedimentary Petrology, v. 26, p. 258–267.

Flores, R. M., 1972, Delta front-delta plain facies of the Pennsylvanian Haymond Formation, northeastern Marathon Basin, Texas: Geological Society of America Bulletin, v. 83, p. 3414–3424.

Folk, R. L., 1973, Evidence for peritidal deposition of Devonian Caballos Novaculite, Marathon Basin, Texas: American Association of Petroleum Geologists Bulletin, v. 57, p. 702–705.

Imoto, N., and McBride, E. F., 1986, Volcanism recorded in Carboniferous

Tesnus Formation, Marathon Basin, Texas: Geological Society of America Abstracts with Programs, v. 18, p. 643.

Keller, G. R., and Cebull, S. E., 1973, Plate tectonics and the Ouachita system in Texas, Oklahoma, and Arkansas: Geological Society of America Bulletin, v. 83, p. 1659–1666.

Keller, W. D., Viele, G. W., and Johnson, C. H., 1977, Texture of Arkansas novaculite indicates thermally induced metamorphism: Journal of Sedimentary petrology, v. 47, p. 834–843.

King, P. B., 1930, The geology of the Glass Mountains, Texas; Part I, Stratigraphy: University of Texas Bulletin 3038, 167 p.

—— , 1937, Geology of the Marathon region, Texas: U.S. Geological Survey Professional Paper 187 (published 1938), 148 p.

—— , 1980, Geology of the eastern part of the Marathon Basin, Texas: U.S. Geological Survey Professional Paper 1157, 40 p.

McBride, E. F., 1966, Sedimentary petrology and history of the Haymond Formation (Pennsylvanian), Marathon Basin, Texas: University of Texas Bureau of Economic Geology Report of Investigations 57, 101 p.

—— , 1969, Stratigraphy, sedimentary structures, and origin of flysch and preflysch rocks, Marathon Basin, Texas: Dallas Geological Society Guidebook, 104 p.

—— , 1987, Stratigraphy and sedimentary history of pre-Permian Paleozoic rocks of the Marathon Uplift, *in* Hatcher, R. D., Jr., Thomas, W. A., and Viele, G. W., The Appalachian and Ouachita regions; U.S.: Boulder, Colorado, Geological Society of America, The Geology of North America, v. F-2 (in press).

McBride, E. F., and Folk, R. L., 1977, The Caballos Novaculite revisited; Part II, Chert and shale members and synthesis: Journal of Sedimentary Petrology, v. 47, p. 1261–1286.

Miser, H. D., 1921, Llanoria, the Paleozoic land area in Louisiana and eastern Texas: American Journal of Science, v. 2, p. 61–89.

Muehlberger, W. R., 1978, Notes on the structural domains of the Marathon region: Permian Basin Section, Society of Economic Paleontologists and Mineralogists 1978 Field Conference Guidebook, Publication 78-17, p. 51–54.

Muehlberger, W. R., and Tauvers, P. R., 1988, Marathon fold-thrust belt update, *in* Hatcher, R. D., Jr., Thomas, W. A., and Viele, G. W., The Appalachian and Ouachita regions; U.S.: Boulder, Colorado Geological Society of America, The Geology of North America, v. F-2 (in press).

Muehlberger, W. R., DeMis, W. D., and Leason, J. O., 1984, Geologic crosssections, Marathon region, Trans-Pecos Texas: Geological Society of America Map and Chart Series MC-28T, 6 p. text, scale 1:250,000.

Permian Basin Section, Society of Economic Paleontologists and Mineralogists, 1964, The filling of the Marathon geosyncline: Field Trip Guidebook, Publication 64-9, 104 p.

—— , 1978, Tectonics and Paleozoic facies of the Marathon geosyncline, west Texas: Field Conference Guidebook, Publication 78-17, 271 p.

Ross, C. A., 1962, Permian tectonic history of Glass Mountains, Texas: American Association of Petroleum Geologists Bulletin, v. 46, p. 1728–1746.

Ross, C. A., and Ross, J. P., 1985, Paleozoic tectonics and sedimentation in west Texas, southern New Mexico, and southern Arizona, *in* Dickerson, P. W., and Muehlberger, W. R., eds., Structure and tectonics of Trans-Pecos Texas: West Texas Geological Society Publication 85-81, p. 221–230.

Rowett, C. L., and Walper, J. L., 1973, Plate tectonics and new proposed intercontinental reconstruction [abs.]: American Association of Petroleum Geologists Bulletin, v. 57, p. 422.

Stevens, J. B., and Stevens, M. S., 1985, Basin and range deformation and depositional timing, Trans-Pecos Texas, *in* Dickerson, P. W., and Muehlberger, W. R., eds., Structure and tectonics of Trans-Pecos Texas: West Texas Geological Society Publication 85-81, p. 157–164.

Tauvers, P. R., 1985, Décollement tectonics of the frontal zone, western domain, Marathon Basin, Texas, *in* Dickerson, P. W., and Muehlberger, W. R., eds., Structure and tectonics of Trans-Pecos, Texas: West Texas Geological Society Publication 85-81, p. 69–76.

Thomson, A., 1964, Genesis and bathymetric significance of the Caballos Novaculite, Marathon region, Texas: Permian Basin Section, Society of Economic Paleontologists and Mineralogists Field Trip Symposium and Guidebook, Publication 64-9, p. 12–16.

Thomson, A., and Thomasson, R. M., 1969, Shallow to deep-water facies development in the Dimple Limestone (Lower Pennsylvanian), Marathon region, Texas, *in* Friedman, G. M., ed., Depositional environments in carbonate rocks: Society of Economic Paleontologists and Mineralogists Special Publication 14, p. 57–77.

Persimmon Gap in Big Bend National Park, Texas; Ouachita facies and Cretaceous cover deformed in a Laramide overthrust

***Peter R. Tauvers and William R. Muehlberger,** Department of Geological Sciences, University of Texas at Austin, Austin Texas 78713-7909*

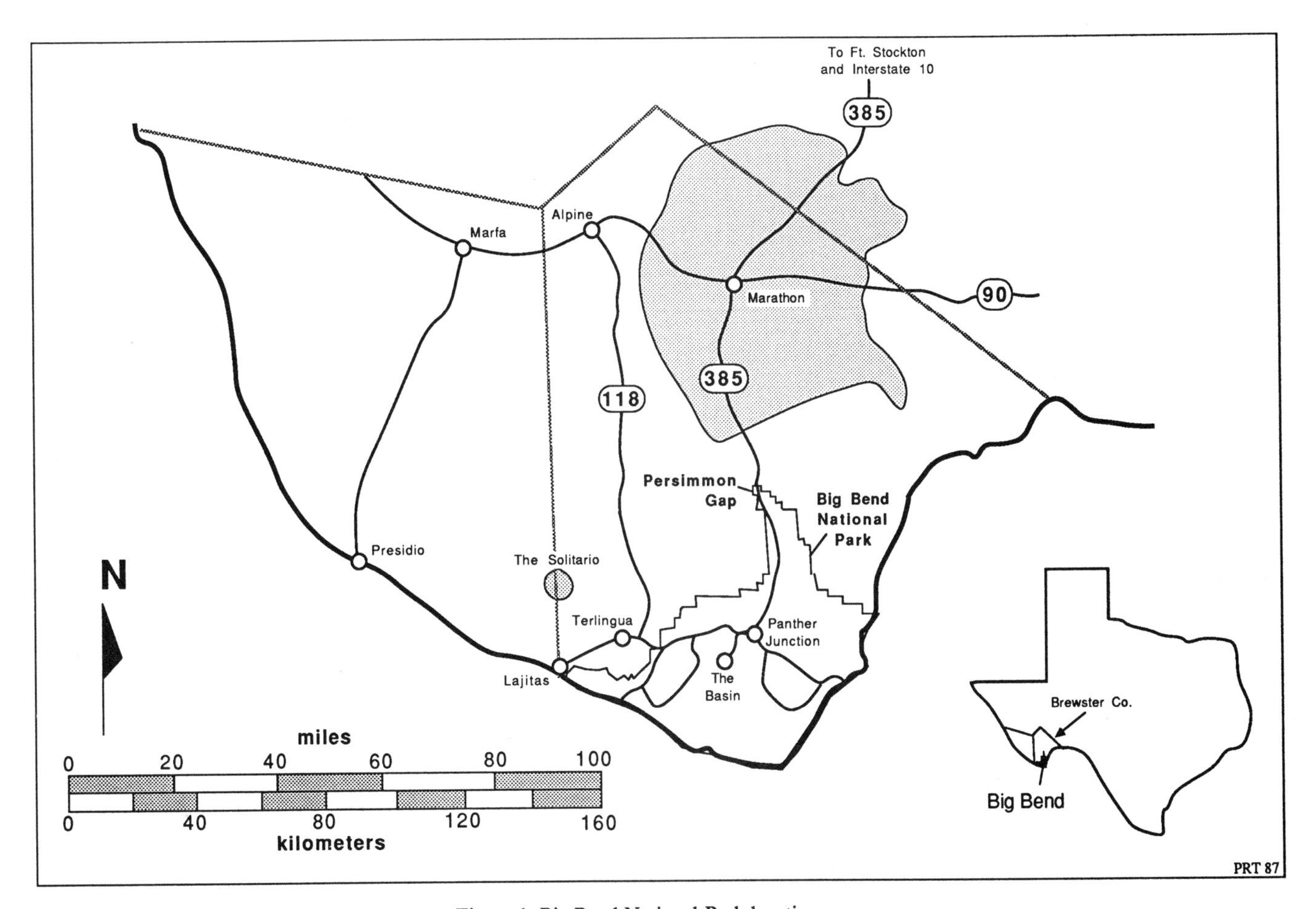

Figure 1. Big Bend National Park location map.

LOCATION

Persimmon Gap. A wind gap that was part of the historic Comanche Trail crosses the southern Santiago Mountains at the northern entrance to Big Bend National Park. The Persimmon Gap site is 42 mi (68 km) south of Marathon, Texas, along U.S. 385 (Fig. 1). The main outcrop is a prominent peak 0.5 mi (0.8 km) east of the highway that can be reached easily by walking up an arroyo into the hills (Fig. 2). Beware of cactus! Permission to walk off road should first be obtained at Persimmon Gap Ranger station 0.2 mi (0.3 km) south of the outcrop. The park has a nominal entrance fee.

SIGNIFICANCE

Persimmon Gap is unique in that the effects of three major orogenies are exposed. A Paleozoic inlier features the southernmost exposed thrust faults (Upper Ordovician over Lower Pennsylvanian) of the Ouachita system in Texas. These rocks and the Cretaceous cover (predominantly limestone) were involved in oblique overthrusting during Laramide left-slip transpression, locally creating a positive flower structure. Right-slip transtensional faulting of the present cycle of Basin and Range deformation is evident via downdropped fault blocks. The geologic history of the region is dominated by repeated movements of basement-rooted northwest-trending faults.

SITE INFORMATION

The north-trending gap and state highway obliquely cut across late Paleozoic, Laramide, and Basin-and-Range structural features, exposing allochthonous Ouachita facies rocks unconformably below a thick Cretaceous cover sequence. Hill (1900) and Baker and Bowman (1917) originally described the regional setting.

Paleozoic sequence

Paleozoic rocks in the Trans-Pecos region are exposed in the Marathon Basin uplift (King, 1937) and the intrusion-induced dome of the Solitario (Wilson, 1954; Calkins, 1986; Fig. 1). A narrow band of middle Paleozoic units outcrops from the southwest corner of the Marathon Basin along the trend of the Santiago Mountains into the Persimmon Gap region (Pearson and Greenlee, 1972; Maxwell and others, 1967; Fig. 3). The Paleozoic rocks are allochthonous with respect to the North American southern margin. They were telescoped onto the margin during the polyphase Early Mississippian to Early Permian Ouachita orogeny (Waterschoot van der Gracht, 1931; Flawn and others, 1961). Their exposure today is almost wholly due to erosion sometime after Laramide uplift in the Late Cretaceous–early Tertiary (Baker, 1934).

Three Paleozoic units are exposed beneath the Cretaceous cover in the Persimmon Gap area—the Upper Ordovician Maravillas Formation, the Devonian Caballos Formation, and the Lower Mississippian Tesnus Formation (Fig. 4).

Pre-flysch stratigraphy. The Paleozoic section begins with an unknown thickness of the typical black Maravillas thin-bedded chert. Individual chert beds range in thickness from 2 to 8 in (average 4 in; 5 to 20 cm, average 10 cm). The upper Maravillas is a 6.6-ft (2 m) interval of "leached and baked-appearing shale, varicolored, orange, yellow-brown, and white," with chert nodules at the base and a 1-ft thick (0.3 m) highly jointed chert bed at the top (Wilson, 1954, p. 2463). In contrast, thin, dark limestone layers are abundant in the Dog Canyon region about 2 mi (3 km) to the southeast. The overlying 28-ft thick (8.4 m) "Persimmon Gap shale" (Wilson, 1954) is an illite-bearing section of greenish, red-weathering shale with several 1- to 4 in-thick (3 to 10 cm) grey chert beds in the lower half. Wilson (1954, p. 2470) notes that the "Persimmon Gap shale" is "generally missing" in the Marathon Basin proper. However, Berry and Nielsen (1958, p. 2259) correlate the "Persimmon Gap Shale" with the ubiquitous middle chert member of the Caballos Formation of King (1937).

In the Marathon Basin and the Solitario, the next higher unit, the Caballos Formation, is generally dominated by one of two members of the ridge-forming Caballos Novaculite (Folk and McBride, 1976). The massive upper novaculite is up to 165 ft (50 m) in thickness within the Marathon Basin. In contrast, the total thickness of the Caballos Formation at Persimmon Gap is only 145 ft (44 m), and the single novaculite horizon found there

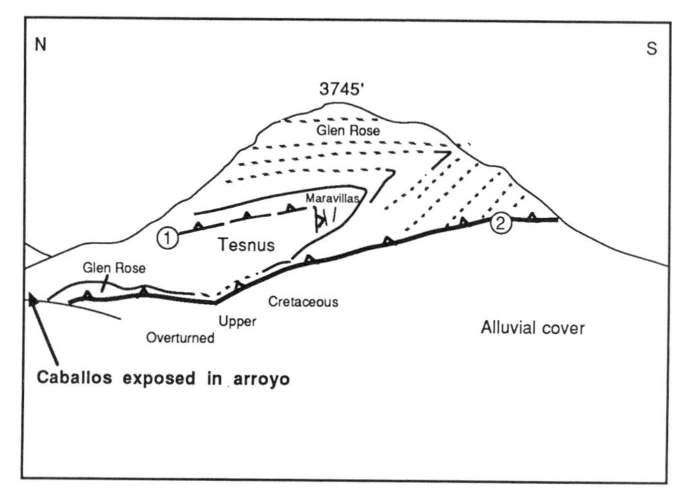

Figure 2. Photo and line drawing of 3,745 ft (1,141 m) peak at Persimmon Gap site (see arrow near range station in Fig. 5 for viewpoint and viewing direction). Outcrop can be reached by walking up arroyo north of hill. Two major episodes of thrust faulting are apparent: (1) thrust fault within Paleozoic rocks that placed Upper Ordovician Maravillas Formation over Lower Mississippian Tesnus Formation from southeast to northwest. This early fault is folded by (2) a major west-directed Laramide thrust fault. Lower Cretaceous Glen Rose beds immediately against Laramide thrust fault on hanging wall are overturned. See Figure 6b for cross section.

is only 33 ft (10 m; Wilson, 1954, p. 2462). The rest of the formation consists of thin-bedded 1- to 4 in-thick (3 to 10 cm) varicolored chert beds intercalated with brown siliceous shale.

Mississippian flysch. Onset of Alleghanian orogenic movements to the southeast of the Trans-Pecos region is recorded by the Lower Mississippian Tesnus Formation, which reaches a thickness of nearly 8,200 ft (2,500 m) in the Marathon Basin. At Persimmon Gap, these rocks consist of approximately 495 ft (150 m) of siliceous shale and highly jointed interbedded sandstone with irregular veinlets of white quartz. The section contains a

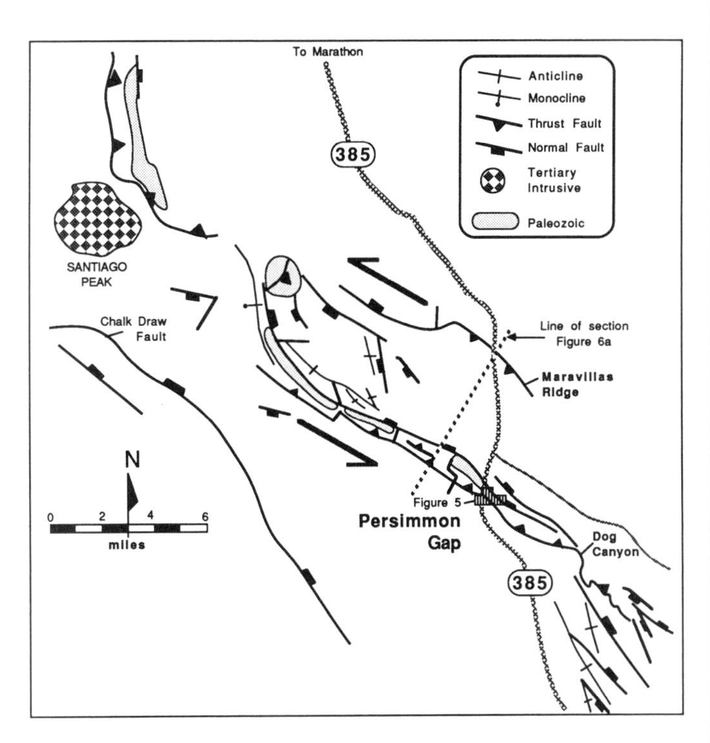

Figure 3. Major structural features and exposures of Paleozoic rocks within the Santiago Mountains. Large arrows illustrate Laramide left-lateral movement. Normal faults reflect the superposition of Basin and Range right-slip extension. After Cobb and Poth (1980).

Persimmon Gap Stratigraphy

Era	Series	Formations
CRETACEOUS	Gulfian	Aguja Fm. Pen Fm. Boquillas Fm. Buda Ls. Del Rio Clay
	Comanchean	Santa Elena Ls. (ridge former) = Georgetown of central TX Sue Peaks Fm. Del Carmen Ls. (ridge former) = Edwards of central TX Telephone Canyon Fm. Maxon Ss. Glen Rose Ls. (ridge former)
PALEOZOIC	Miss	Tesnus Formation (Flysch)
	Dev	Caballos Formation (One novaculite mbr plus chert)
	U.Ord	Maravillas Formation

Figure 4. Stratigraphic column. From Cobb and Poth (1980). See Maxwell and others (1967) for complete stratigraphic descriptions.

distinctive brown siliceous conglomerate at the base. The conglomerate is 8 in to 2 ft (0.2 to 0.6 m) thick with angular pebbles of gray and white novaculite (Maxwell and others, 1967, p. 27).

Cretaceous cover. Unconformably overlying the Paleozoic rocks here are the limestone, shale, and minor sandstone of the Comanchean and Gulfian series (Lower and Upper Cretaceous, respectively; Fig. 4). The Comanchean series is on the order of 1,640 ft (500 m) thick in the map area, whereas the Gulfian units are more than 990 ft (300 m) thick.

The lowest unit in the Comanchean section is a distinctive reddish brown pebble and cobble conglomerate. The clasts of the conglomerate are chert and novaculite derived from underlying Ouachita facies rocks. The conglomerate grades upward into minor sandstone and shale before giving way to the characteristic limestone and marl of the Glen Rose Formation. The fossiliferous Glen Rose is a major ridge former, locally as much as 660 ft (200 m) thick, and makes up the crest and southeast slope of the 3,745 ft (1,141 m) hill pictured in Figure 2. *REMINDER: specimen collecting is not permitted in any National Park.*

The foreslope of the 3,745 ft (1,141 m) hill exposes Upper Cretaceous (Gulfian) units in the footwall of the Laramide overthrust, namely the Pen and Aguja Formations (Figs. 2 and 5; Maxwell and others, 1967). The Pen Formation is principally a brown gypsiferous shale with large (up to 3 ft; 1 m diameter), disk-shaped calcareous concretions. It is perhaps 320 ft (100 m) thick in the Persimmon Gap area (Poth, 1979). The Aguja is the youngest Cretaceous unit in the area. It consists of medium- to coarse-grained cross-bedded sandstone of unknown thickness (Poth, 1979).

Structure

Paleozoic. The oldest structural features at Persimmon Gap are northeast-trending thrust faults within allochthonous Ouachita facies rocks. Figure 5 illustrates two Paleozoic thrust faults. Both have low southeast dips and fault Maravillas northwestward over Tesnus. The outcrop photo of Figure 2 shows the exposure of the structurally higher thrust fault within the main valley at Persimmon Gap. Note that this fault was later folded along with the Cretaceous cover.

Some of the shale at Persimmon Gap has been described as "schistose" (Lonsdale and others, 1955, p. 54) alluding to its

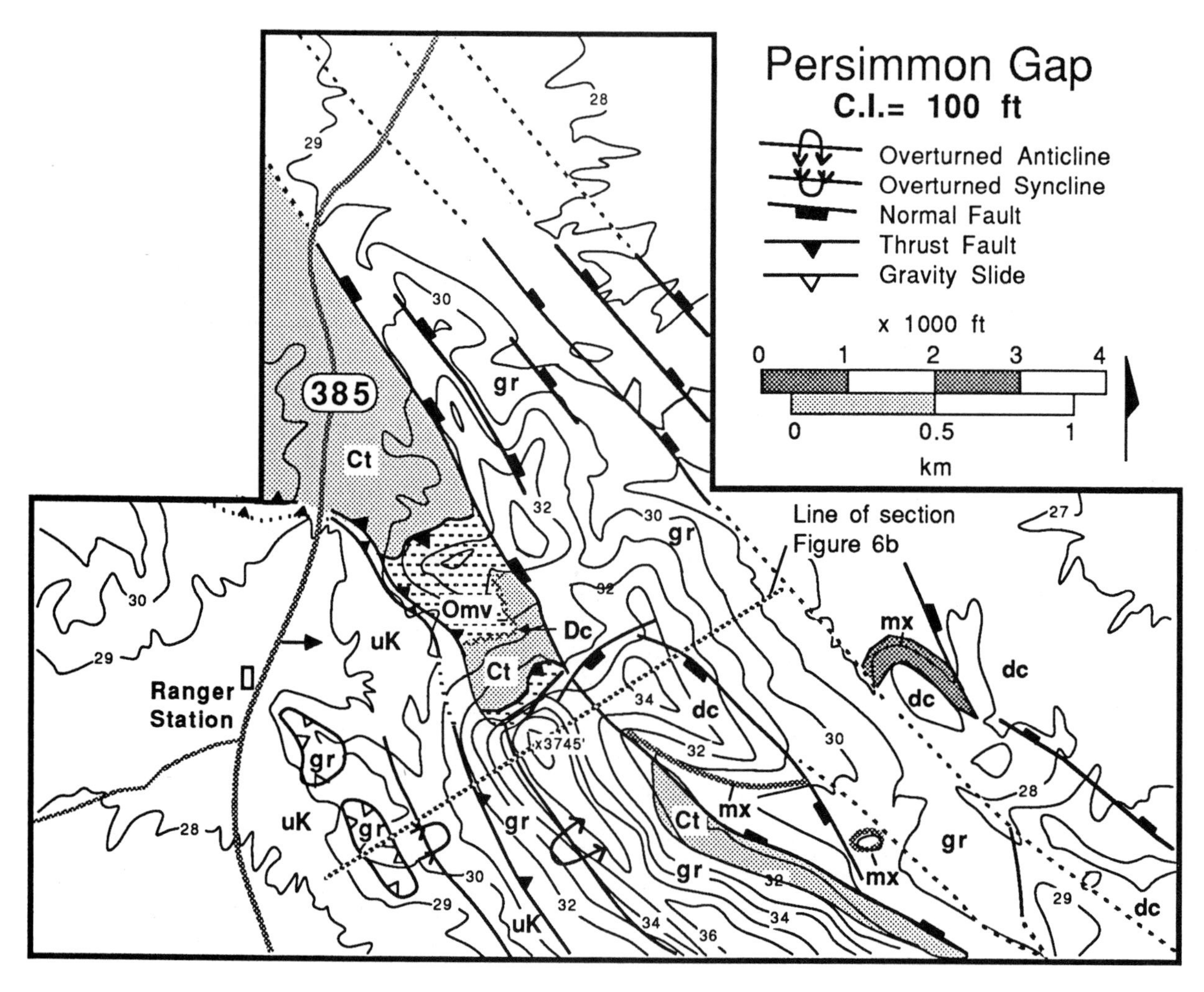

Figure 5. Persimmon Gap map. Arrow near Ranger station marks starting point for 0.3 mi (0.5 km) hike to the outcrop from U.S. 385. Paleozoic units: Omv, Maravillas Formation; Dc, Caballos Formation; Ct, Tesnus Formation. Cretaceous units: gr, Glen Rose; mx, Maxon Sandstone; dc, Del Carmen limestone; uK, Upper Cretaceous undifferentiated. Compiled from Maxwell and others (1967) and Poth (1979). Note N35°E trend in topographic breaks across region.

fracture cleavage and the recrystallization of its clay matrix to illite. These rocks, however, are still part of the frontal zone of the Ouachita system and are in the footwall relative to the greenschist facies rocks of the interior zone (known from subcrop relations) farther to the southeast (Flawn and others, 1961; Muehlberger and others, 1984).

Laramide. The timing of post-Paleozoic, pre–late Tertiary deformation in the Persimmon Gap region is not known precisely due to the lack of crosscutting relationships with syndeformational sediments or intrusives (Baker, 1934, p. 165). However, such deformation can be tied directly to two major Laramide events—the doming of the Marathon Basin uplift, and left-lateral movement along the northern boundary faults of the Texas lineament zone of Muehlberger (1980).

The west wall of the Marathon Basin is characterized by a major south-trending overthrust system in which upper Paleozoic rocks were thrust faulted over Cretaceous (King, 1937). South of Santiago Peak, this structure dramatically changes trend to the southeast within the southern Santiago Mountains before nearly resuming its former trend south of Dog Canyon (Fig. 3). This jog in the structural belt is along the northwest-trending Texas lineament.

The Marathon dome is asymmetric, being structurally higher in the west (King, 1937). Structural data from nearby regions to the south document east-northeast Laramide compression (Moustafa, 1983; Price and Henry, 1984), which resulted in the formation of major monoclines and overthrusting on high-angle reverse faults rooted in Precambrian basement. Monoclines north and south of Persimmon Gap have structural relief of 2,300 to 2,600 ft (700 to 800 m); reverse-faulted monoclines have

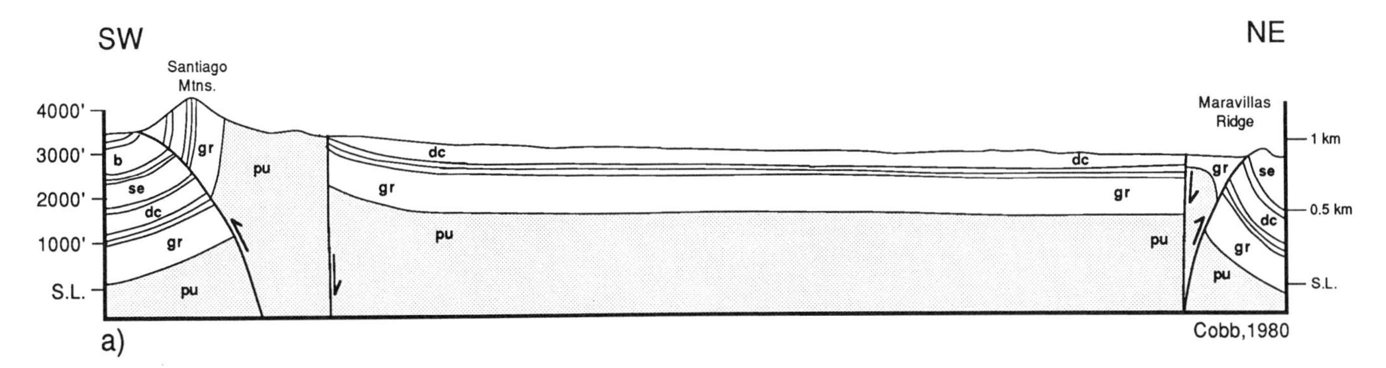

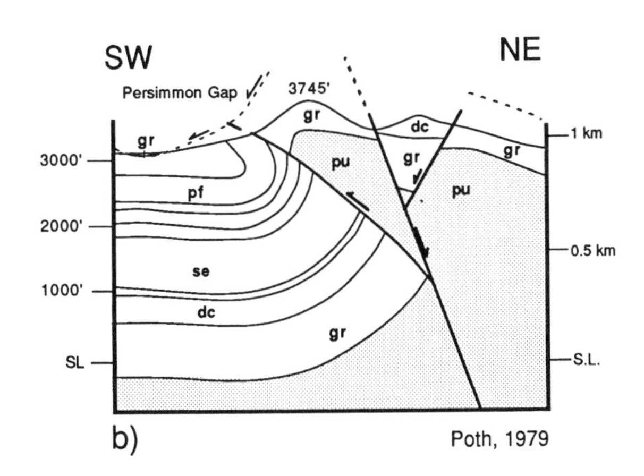

Figure 6. a) Structure section through positive flower structure (V/H=1). See Figure 3 for location. b) Structure section through hill in Figure 2 (V/H=1). See Figure 5 for location. Cross sections are incomplete because of lack of information concerning true depth to basement. Nor are they balanced as the plane-strain assumption typical of thrust belts does not apply in such a strike-slip setting. Arrows only show dip-slip component: reverse faults also have left-slip components; normal faults right-slip. The sections clearly show the superposition of Basin and Range normal faulting on the preexisting Laramide structure, which effectively cancels the Laramide heave in the hanging-wall block. Symbols: pu, upper Paleozoic rocks; Cretaceous units: gr, Glen Rose; dc, Del Carmen; se, Santa Elena; b, Buda, pf, Pen.

structural relief of about 3,000 ft (900 m; Fig. 6; Cobb and Poth, 1980).

From these kinds of relationships, Eifler (1943) concluded that the Santiago Mountains were underthrust below the Marathon region. Pearson and Greenlee (1972, p. 39) suggested, "within the mapped area the Santiago Mountain structures may be the result of gravity-sliding away from the central part of the Marathon Uplift."

In contrast, Cobb and Poth (1980) noted the acute angle between the Laramide compression direction and the trend of the Santiago Mountains and inferred a left-lateral transcurrent Laramide history. Major thrust faults are believed to have propagated vertically from basement before rolling over to horizontal in places along the present level of exposure. Tauvers (1986), noting the northeast-vergent Laramide overthrust along Maravillas ridge (Figs. 3 and 6a) parallel to the Persimmon Group structure, likened this transpressive system to a (positive) flower structure.

Cobb and Poth (1980) interpret several earlier-mapped "klippe" of Glen Rose beds southeast of the Persimmon Gap Ranger Station to be landslide blocks (Figs. 5 and 6b). These blocks now rest on Upper Cretaceous rocks.

Figure 3 shows the location of Dog Canyon, about 2 mi (3 km) east of U.S. 385 and southeast of Persimmon Gap, which exposes the Santa Elena Limestone in the tight synclinal part of the Laramide structure seen at Persimmon Gap. The visitor with time (and water) should walk into this canyon to observe the spectacular exposure of a tightly folded thick, brittle limestone sequence.

Basin and Range. Cobb and Poth (1980) documented the role of late-stage high-angle normal faults superimposed on Laramide structural features along the southern Santiago Mountains.

Northwest-trending, down-to-the-northeast normal faults are common in the map area (Fig. 5). The faults are known to cut mid-Tertiary basalt to the southeast (St. John, 1965) where they are part of a Basin and Range right-lateral transtensional system along this part of the Texas lineament. Cobb and Poth (1980, p. 72) described a decrease in dip-slip separation over a distance of 9.3 m (15 km) from 2,800 ft (850 m) in the northwest to about 200 ft (60 m) at Persimmon Gap. They conclude that the Laramide thrust fault and main Basin and Range normal fault are superimposed, born of the same basement structure at depth. The normal fault, however, remains nearly vertical to the present level of exposure and so is always found east of the east-dipping Laramide thrust fault traces (Figs. 5 and 6).

CONCLUSIONS

The Persimmon Gap site is an excellent example of superimposed deformation related to reactivation of deep-seated basement faults. Ouachita thrust faults are exposed today because of Laramide uplift along northwest-trending basement features followed by Basin and Range transtension. Examination of the topography shown in Figure 5 suggests a northeasterly basement control as well. Linear features trending N35°E transect the Santiago Range in this area. For example, the location of the northeast-oriented valley at Persimmon Gap, the southern wall of which is pictured in Figure 2, may be similarly controlled.

REFERENCES CITED

Baker, C. L., 1934, Major structural features of Trans-Pecos Texas, *in* The geology of Texas, Vol. II, Structural and economic geology: University of Texas Bulletin 3401, p. 137–214.

Baker, C. L., and Bowman, W. F., 1917, Geologic exploration of the southeastern Front Range of Trans-Pecos Texas: University of Texas Bulletin 1753, p. 60–177.

Berry, W.B.N., and Nielsen, H. M., 1958, Revision of the Caballos novaculite in Marathon region: American Association of Petroleum Geologists Bulletin, v. 42, p. 2254–2259.

Calkins, C., 1986, Stratigraphy and structure of the Solitario, *in* Geology of the Big Bend area and Solitario Dome, Texas: West Texas Geological Society 1986 Field Trip Guidebook, Publication 86-82, p. 91–96.

Cobb, R. C., 1980, Structural geology of the Santiago Mountains between Pine Mountain and Persimmon Gap, Trans-Pecos Texas [M.A. thesis]: University of Texas at Austin, 69 p.

Cobb, R. C., and Poth, S., 1980, Superposed deformation in the Santiago and northern Del Carmen mountains, Trans-Pecos, Texas: New Mexico Geological Society Guidebook, 31st Field Conference, Trans-Pecos Region, p. 71–75.

Eifler, G. K., 1943, Geology of the Santiago Peak Quadrangle, Texas: Geological Society of America Bulletin, v. 54, p. 1613–1644.

Flawn, P. T., Goldstein, A., King, P. B., and Weaver, C. E., 1961, The Ouachita System: Texas Bureau of Economic Geology Bulletin 6120, 401 p.

Folk, R. L., and McBride, E. F., 1976, The Caballos Novaculite revisited; Part I, Origin of novaculite members: Journal of Sedimentary Petrology, v. 46, p. 659–669.

Hill, R. T., 1900, Physical geography of the Texas region: U.S. Geological Survey Topographic Atlas, Folio 3, 12 p.

King, P. B., 1937, Geology of the Marathon region, Texas: U.S. Geological Survey Professional Paper 187, 148 p.

Lonsdale, J. T., Maxwell, R. A., and Wilson, J. A., 1955, Geology of Big Bend National Park: West Texas Geological Society Guidebook, 1955 Spring Field Trip, March 18–19, 1955, 142 p.

Maxwell, R. A., Lonsdale, J. T., Hazzard, R. T., and Wilson, J. A., 1967, Geology of Big Bend National Park, Brewster County, Texas: Texas Bureau of Economic Geology Publication 6711, 320 p.

Moustafa, A. R., 1983, Analysis of Laramide and younger deformation of a segment of the Big Bend region, Texas [Ph.D. thesis]: University of Texas at Austin, 278 p.

Muehlberger, W. R., 1980, The Texas lineament revisited: New Mexico Geological Society, 31st Field Conference Guidebook, Trans-Pecos region, p. 113–121.

Muehlberger, W. R., Demis, W. D., and Leason, J. O., 1984, Geologic cross-section, Marathon region, Trans-Pecos, Texas: Geological Society of America Map and Chart Series MC-28T, scale 1:250,000.

Pearson, B., and Greenlee, D. W., 1972, Paleozoic outcrops in the Santiago Mountains northwest of Persimmon Gap: West Texas Geological Society Guidebook 72-59, p. 39–41.

Poth, S., 1979, Structural transition between the Santiago and Del Carmen mountains in northern Big Bend National Park, Texas [M.A. thesis]: University of Texas at Austin, 56 p.

Price, J. G., and Henry, C. D., 1984, Stress orientations during Oligocene volcanism in Trans-Pecos, Texas; Timing the transition from Laramide compressions to Basin and Range extension: Geology, v. 12, p. 238–241.

St. John B., E., 1965, Structural geology of Black Gap area, Brewster County, Texas [Ph.D. thesis]: University of Texas at Austin, 132 p.

Tauvers, P. R., 1986, Tertiary deformation within the Marathon Basin, Texas: Geological Society of America Abstracts with Programs, v. 18, no. 2, p. 191.

Waterschoot van der Gracht, W. A. J. M. van, 1931, Permo-Carboniferous orogeny in south-central United States: American Association of Petroleum Geologists Bulletin, v. 15, p. 991–1057.

Wilson, J. L., 1954, Ordovician stratigraphy in Marathon folded belt, West Texas: American Association of Petroleum Geologists Bulletin, v. 38, p. 2455–2475.

ACKNOWLEDGMENTS

Patricia Wood Dickerson, Richard J. Erdlac Jr., and Michael O. Maler, reviewed the manuscript with a care that we hope is reflected in the final result.

Basal Gulfian and Comanchean Section, Anguila Fault Zone and Santa Elena Canyon, Big Bend National Park, Trans-Pecos Texas, and Chihuahua, Mexico

J. B. Stevens, Department of Geology, Lamar University, Box 10031, L.U. Station, Beaumont, Texas 77710

LOCATION

Santa Elena Canyon and the Anguila Fault Zone northwest of the canyon are occupied by the Rio Grande. These features separate Mesa de Anguila, the westernmost part of Big Bend National Park, Brewster County, Texas, from northeastern Chihuahua, Mexico (Fig. 1). The traverse begins on the river at Lajitas (reached by FM 170), southwestern Brewster County, about 2 mi (3.2 km) above the entry to an upper canyon, and ends at the lower end of Santa Elena Canyon, where Terlingua Creek joins the Rio Grande from the northwest. This location is accessible by a paved park road. The canyons are accessible only by river, a trip best made in a sturdy rubber raft. Travel through the canyons is strictly regulated by the Big Bend National Park administration, particularly with respect to safety equipment and camping. Permits may be obtained at a ranger station at Lajitas, or at park headquarters at Panther Junction. The simplest procedure is to arrange a guided tour through businesses in Lajitas, Terlingua Ghost Town 12.5 mi (20.1 km) east of Lajitas on FM 170, or Study Butte (junction of Texas 118 and FM 170) 16.9 mi (27.2 km) east. A permit is necessary to collect rock samples in any national park; even if the intent is to restrict sampling to the Mexican side of the river, it would be best to obtain such a permit.

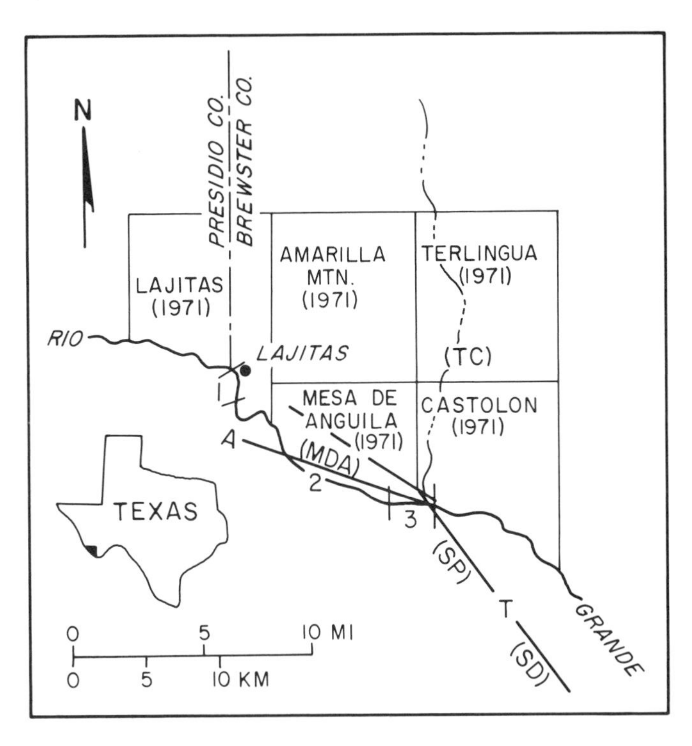

Figure 1. Locations of discussed reaches of Rio Grande (1 through 3) (Santa Elena Canyon); topographic quadrangles. MDA = Mesa de Anguila, SP = Sierra Ponce, SD = Sierra Diablo, TC = Terlingua Creek; trends of Anguila (A) and Terlingua (T) fault zones.

Santa Elena Canyon and the lesser (but still impressive) canyon immediately upstream provide access at and near river level to virtually continuous exposure, in a relatively simple structural setting, of basal Gulfian rocks and part or all of every Comanchean formation known from the park area (Fig. 2) southwest of the Santiago Mountains. Three reaches of the river are discussed. One can view, and examine successively, Rio Grande flood-plain and terrace sediments, basal Terlingua Group (Maxwell and others, 1967) resting disconformably on Washita Group, and, in the main canyon, rocks correlative with those of the Washita, Fredericksburg, and Trinity Groups of central Texas.

STRUCTURAL AND GEOMORPHIC SETTING: RIO GRANDE AND SANTA ELENA CANYON

The Rio Grande from El Paso southeast to the Big Bend is a young river where the Texas-Chihuahua border follows a series of fault zones. Fault zones striking northwest control the river from the Rio Grande Rift to midway between Presidio and Redford, along the southern boundary of Presidio County, and an east-west zone from there to Lajitas. From Lajitas the river wanders generally south and southeast for 7 mi (11.3 km) (10.8 mi or 17.4 km by river) within the Anguila Fault Zone, and then turns east into the more impressive Santa Elena Canyon. Subsurface and surface geological studies suggest episodic movement on these and related trends under a variety of regional stress orientations since 1,500 Ma. Peaks of activity occurred during late Paleozoic time, the Laramide Orogeny, and Basin and Range events (DeCamp, 1981, 1985; Goetz and Dickerson, 1985; Muehlberger, 1980; Pearson, 1981; Stevens and Stevens, 1985). Broadly, the Big Bend region is in a zone of wrench faulting between the Texas Lineament (in Texas, roughly along a line from El Paso to Uvalde) and a megashear to the south in Mexico (Longoria, 1985).

Laramide structures, present in the neighborhood of Mesa de Anguila most obviously as low-amplitude open folds, make only a subtle contribution to the topography (Fig. 3). Santa Elena

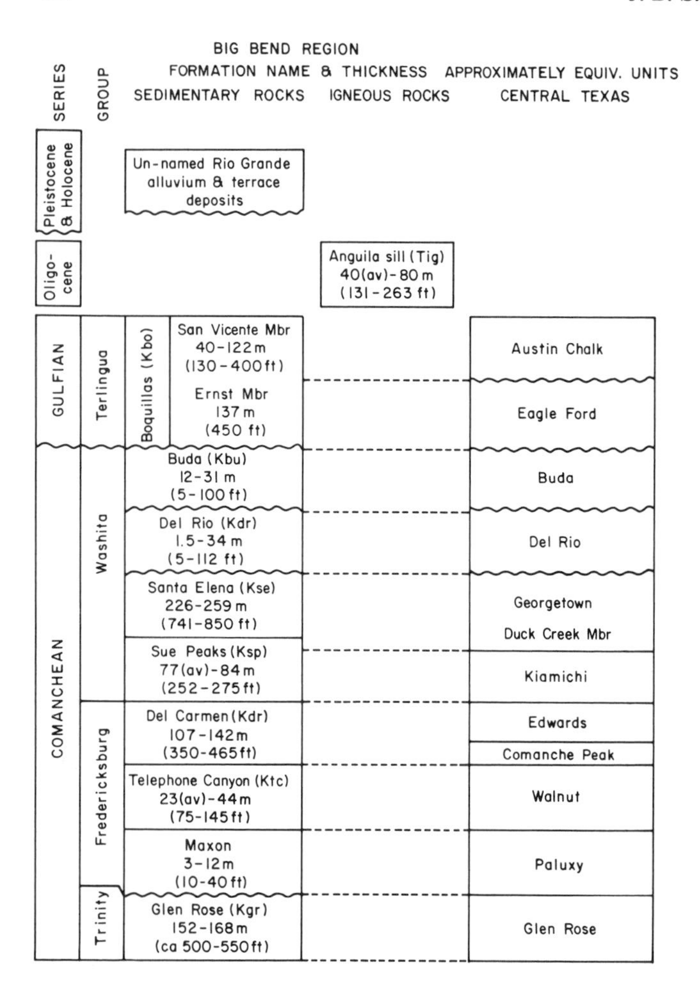

Figure 2. Comanchean and Gulfian stratigraphic nomenclature follows usage in Geological Atlas of Texas (Bureau of Economic Geology, 1979), based on that of Maxwell and others (1967). This figure provides key to formation symbols used in sections (Figs. 3-5).

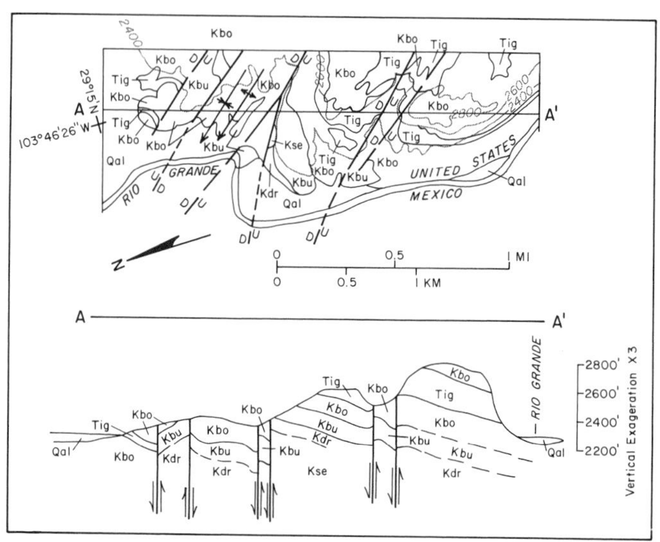

Figure 3. Reach 1 and Locality 1: strip map (simplified from map of DeCamp, 1981), north end, Mesa de Anguila, and diagrammatic section (oriented N16E) across the Anguila Fault Zone in lower part of reach. Section illustrates setting and relationships of Comanchean and Gulfian rocks, and Oligocene Anguila sill in Reach 1, and in area of Locality 1; also applies to upper part of Reach 2 where river crosses and recrosses parts of fault zone. Section is based on mapping done by DeCamp (1981). Symbols for rock units given in Figure 2.

Canyon is a zig-zag course cut along minor faults and joint zones that mark a structural sag across a northwest-striking block of resistant Comanchean limestones uplifted and tilted about 5° southwest by Basin and Range faulting. The block forms Mesa de Anguila, and, in Mexico, Sierra Ponce and Sierra Diablo. By earlier Quaternary time, lower parts of the region were largely filled with debris eroded from the uplifts and mid-Tertiary volcanic centers. Modern topographic relief in the area is the result of later Quaternary exhumation of Basin and Range and older volcanic topography (for example, the Bofecillos Mountains immediately northwest of Lajitas; the Chisos Mountains in Big Bend National Park) during development of the Rio Grande drainage. Occupation of Santa Elena Canyon by the Rio Grande may be the result of piracy.

PLEISTOCENE ALLUVIUM

Late Pleistocene and Holocene sediments along the river, composed of coarse, rounded gravel, very coarse to medium sand, and silt, have been of local economic importance. They provide the only arable land near a dependable supply of irrigation water in an area with an annual precipitation of only about 12 in (30.5 cm) and a climate that promotes evaporation of seven to eight times that amount. Early settlements were along the river or its larger tributaries, such as Terlingua Creek, at structurally controlled points, where development of a narrow alluvial plain was possible. The flood plain, when and where not actively cultivated, is densely covered by salt cedar, mesquite, cane, and scattered cottonwood trees.

Reach 1 (Figs. 1-3) (from 103°47′08″W, 29°15′39″N, to 103°46′26″W, 29°14′50″N, Lajitas Quadrangle) extends 1.3 mi (2.1 km) south-southeast to south along the river as it flows in a small complex graben, the northwest-dipping floor of which is covered with a thin mantle of Rio Grande alluvium. Most of this flood plain was cultivated before Lajitas became a resort town. Terrace deposits stand above the rapidly narrowing floodplain just south of Locality 1.

GULFIAN AND UPPER COMANCHEAN SECTION

The Boquillas Formation, basal Gulfian Series in the Big Bend region, is exposed above the flood plain in Reach 1, and along Reach 2 (Figs. 1, 4). The Boquillas Formation (Fig. 2),

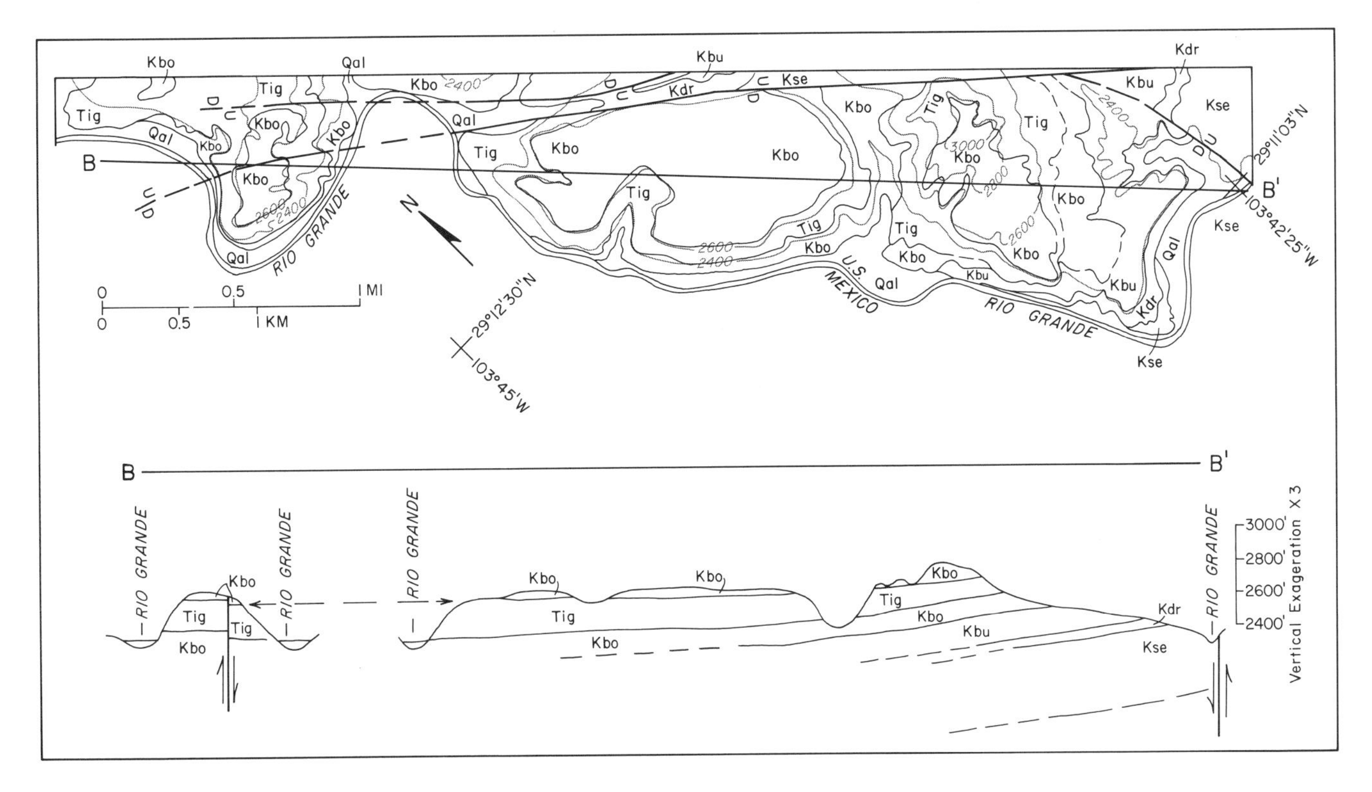

Figure 4. Reach 2: strip map and diagrammatic section (oriented N49W) along Anguila Fault Zone along southwest side of Mesa de Anguila. Map is simplified from, and section based on, work of DeCamp (1981). Symbols for rock units given in Figure 2.

named by Udden (1907), was extended and divided between the lower, Ernst, and the upper, San Vicente Member, by Maxwell and others (1967) on the basis of exposures in southeastern Big Bend National Park. Thin-bedded yellow limey shale and limestone found in many roadcuts for 2 mi (3.2 km) east of Lajitas along FM 170 is probably San Vicente member, although there are unresolved structural-stratigraphic problems in the area. All Gulfian rocks encountered along the river traverse belong to the Ernst Member of the Boquillas Formation roughly equivalent to Udden's (1907) original "Boquillas Flags" (Maxwell and others, 1967). The thinner bedded upper Ernst Member is separated from the lower part by the Oligocene (37.8 ± 1.0 Ma) olivine gabbro (hawaiitic?) Anguila sill, which averages about 131 ft (40 m) in thickness (DeCamp, 1981). The sill is found on Mesa de Anguila wherever there is sufficient exposure of the Ernst, and extends southwest into Chihuahua. The first Comanchean rocks encountered are the Del Rio and Buda Formations (Fig. 2), named from south Texas (Val Verde County: Hill and Vaughan, 1898) and central Texas, respectively (Hays County: Vaughan, 1900). These two units usually occur together, for Del Rio Clay is soon removed if derpived of a resistant cap of Buda Limestone.

Locality 1 (Figs. 1-3) [103°46′26″W, 29°14′28″N, Lajitas Quadrangle, 0.16 mi (0.25 km) southeast of the Rio Grande, on an unnamed ephemeral stream that enters the Texas side of the river 1.3 mi (2.1 km) below Lajitas at a point where the river turns abruptly west]. It is on the southwest side of a northwest-tilted graben at the base of the north face of a hill, part of a fault-line scarp, that rises about 220 ft (67 m) above the river (Fig. 3). Exposures of the Del Rio and Buda Formations, and the basal part of Ernst Member, Boquillas Formation, dip 16 to 19° to the south-southeast. Where a more westerly splay joins the main fault, there is a small outcrop of Santa Elena Limestone on which the Del Rio rests disconformably; sections of Del Rio and Buda, here about 102 ft (31 m) and 82 ft (25 m) thick, respectively, are complete.

The Del Rio consists mainly of a thinly laminated bluish to brownish-green shale that weathers brown or yellow-brown, and appears massive except on a very fresh exposure. There are sparse, thin (0.8 to 2.4 in or 2 to 6 cm) interbeds of tan silicified limestone that weather dull orange. The limestone beds contain a uniserial arenaceous foraminifer, *Cribratina texana* (Conrad) (commonly, 10 to more than 20 millimeters long) not abundant in any other formation. The limestone commonly has small ripples, both vertical and horizontal burrowing, and bedding planes with *Cribratina* tests oriented by flow. The top of the Del Rio Formation often contains thin, discontinuous beds of nodular limestone, some places apparently marking channels, that resemble marly parts of the overlying Buda Formation, but that may be caliches. There is a regional unconformity between the formations in central and south Texas (Smith, 1970). Here, the contact of the Del Rio with the Buda Formation is sharp, but shows little evidence of weathering or erosion, even where exposed for considerable distances (lower part of Reach 2).

The Buda has three distinct subequally thick subunits: lower and slightly thicker upper units of massive, thick-bedded (1.6 to 7.6 ft or 0.5 to 2.3 m) very pale gray porcellaneous wackestone and packstone, with conchoidal fracture, are separated by indistinctly bedded nodular marly wackestone. The nodular middle unit produces a distinct break in slope. Fossils (Maxwell and others, 1967, p. 52) are sparse at this and most localities. Both the Del Rio and Buda appear to have been deposited at or mainly slightly below wave base on a slowly subsiding shelf, with the difference in sediments supplied perhaps emphasized by an episode of nondeposition.

Mild folding of Boquillas and Buda Formations in the graben immediately north and northwest of the hill at Locality 1 is evidence of Laramide left lateral convergent wrench faulting, whereas the normal faults and the graben itself were produced by Basin and Range right transtension (DeCamp, 1981, 1985). Northwest-striking stylolites more or less normal to bedding in limestones of the area provide further evidence of Laramide transpression.

Reach 2 (Figs. 1, 3, 4) (from the lower end of Reach 1, Lajitas Quadrangle to 103°42′25″W, 29°11′03″N, Mesa de Anguila Quadrangle) extends 10.2 mi (16.4 km) in a southeasterly meandering course distorted by local structure, as the river follows the trend of the Anguila Fault Zone in comparatively easily eroded Ernst Member, Boquillas Formation. Normal faulting (Fig. 3) is prominent in the upper half of the reach, with throw on the faults increasing and then decreasing again (DeCamp, 1981); faulting is sparser in the lower half of the reach, and the walls of the canyon are steeper. The canyon of Reach 2 is not considered part of Santa Elena Canyon, but relief near the river along this reach, about 350 to 530 ft (107 to 162 m), provides excellent exposure of the Ernst, the Anguila sill, and 8.6 mi (13.8 km) below the beginning of the reach, the Buda, Del Rio, and the thick-ledgy top of the Santa Elena Formation. The distinct disconformity between the Del Rio and Santa Elena Formations can be seen in the last 1.1 mi (1.8 km) of Reach 2. According to Yates and Thompson (1959), the minor karst features in the top approximately 66 ft (20 m) of the Santa Elena developed before deposition of the Del Rio Formation.

There is marked lithologic contrast at the Buda-Boquillas (Comanchean-Gulfian) contact that can be seen in most of this area (Fig. 4). Yates and Thompson (1959) considered this contact conformable, but Maxwell and others (1967) attributed variations in thickness of the Buda in the Big Bend region to pre-Boquillas erosion. The basal 6.7 to 16 ft (2 to 5 m) of the Ernst Member, Boquillas Formation consists of thin-bedded, fine sandy and silty limestone, separated by silty, marly shales reddened at the surface. Mild folding and occasional crumpling of these layers may be soft sediment deformation (DeCamp, 1981). Upward, the unit—gray to very dark gray on a fresh surface—is characterized by a light tan–yellow weathering, darker near the top, which is again sandier. Limestone, poorly washed foraminiferal biomicrite in 0.3 to 3 ft (0.1 to 1 m) beds clearly dominates the lower third to half of the Ernst, but layers of marly shale become more important upward, accounting for half, or slightly more, of the section toward the top. The Boquillas Formation, and particularly the Ernst Member, closely resembles rhythmically bedded Upper Cretaceous pelagic carbonate deposits found in many parts of the world that are discussed as climatically induced products of Milankovitch cycles by the Research on Cretaceous Cycles Group (1986).

COMANCHEAN SECTION BELOW THE DEL RIO FORMATION

The Comanchean formations (Figs. 2, 5) encountered in the canyon below the Del Rio, are the Santa Elena, Sue Peaks, Del Carmen, Telephone Canyon, and Glen Rose Formations. Of these, all but the Glen Rose were named by Maxwell and others (1967) from localities in Big Bend National Park. The Maxon Formation (King, 1930), found in the Comanchean section around the Marathon Uplift and near the northeasternmost part of the park is not known from the western part of the park or adjacent areas (Maxwell and others, 1967). It may be closely related to (facies of ?) the Telephone Canyon Formation, and is included in Figure 2 for the sake of completeness. The Glen Rose Formation, Trinity Group, was named by Hill (1891) and described from a type locality in Somervell County, central Texas.

The thick, mainly massively bedded resistant siliceous limestones of the Santa Elena and Del Carmen Formations (Fig. 2) cap many topographic highs in the Big Bend region. In the western part of the park and adjacent areas, the Santa Elena is far better exposed than the Del Carmen Formation; the exposures of Del Carmen at river level in Santa Elena Canyon are the most easily accessible in the area. The marly, shaley Sue Peaks and Telephone Canyon Formations are usually poorly exposed in rubble-covered slopes below the more resistant units, and Santa Elena Canyon (Reach 3; Fig. 5) provides some of the best exposure of them in the region.

Reach 3 (Figs. 1, 5) (from the lower end of Reach 2, Mesa de Anguila Quadrangle, to 103°36′43″W, 29°09′55″N, Castolon Quadrangle) is Santa Elena Canyon, which extends 7.4 mi (12 km) between Mesa de Anguila and Sierra Ponce, with walls about 450 ft (140 m) high at the beginning of the reach, and 1,500 ft (460 m) high at the lower end. The fault at the beginning of Reach 3 is a branch of the fault crossed 1.4 mi (2.2 km) farther on. The second fault has displacement of only about 230 ft (70 m) but the rock is severely shattered. A portion of the canyon wall has collapsed to form the "Rock Slide," which in turn produces rapids that have claimed many canoes, kayaks, cameras, and at least one life. Unless one is with an experienced guide, it is best to portage around this point.

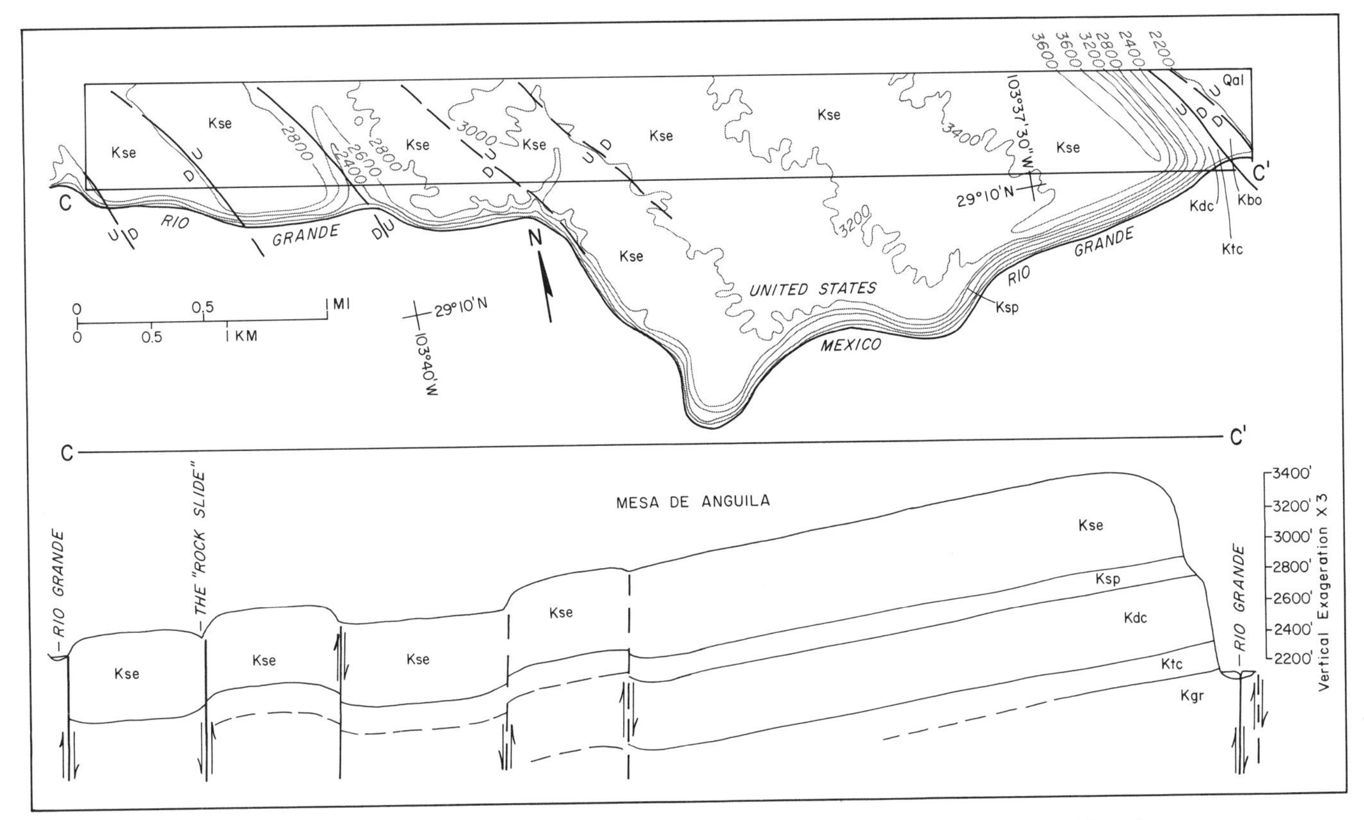

Figure 5. Reach 3: strip map and diagrammatic section (oriented N76W), across southern Mesa de Anguila (adjacent to northern Sierra Ponce), roughly parallel to Reach 3, show Comanchean rocks along Santa Elena Canyon. Map is simplified from, and the section based on, work of DeCamp (1981). Symbols for rock units given in Figure 2.

The Santa Elena Formation—hard, siliceous foraminiferal wackestone to packed oyster rudstone and rudist biohermal rock in massive 3 to 10 ft (1 to 3 m) beds—is at river level for the first 3.7 mi (6.0 km) of Reach 3. Pale gray on a fresh surface, the Santa Elena weathers darker gray (or brown where pyrite-bearing), and develops rillenstein on infrequently abraded surfaces. Light brown chert nodules, singly or in bedding plane groups, and silicified rudistids are common. The Del Carmen Formation, with river-level exposure beginning 5 mi (8.1 km) from the top of Reach 3, is quite similar to the Santa Elena, a source of frustration in structurally complex areas.

River-level exposure of the much thinner (Fig. 2) and less resistant Sue Peaks and Telephone Canyon Formations begins at 5.1 mi (8.2 km) and 6 mi (9.7 km), respectively, below the top of Reach 3. The fossiliferous (Maxwell and others, 1967, p. 44) gray and tan-gray shales, marls, and soft clayey limestones, alternating in 1 to 4 ft (0.2 to 1.2 m) beds of the Sue Peaks Formation belong to a distinctive lower member (Maxwell and others, 1967). An upper limestone member is either not present, or is not lithologically distinct from the Santa Elena (Maxwell and Dietrich, 1965, Fig. 41; Smith, 1970). The thinner Telephone Canyon Formation consists of tan-gray to gray marly wackestone and nodular wackestone.

Only the top 100 ft (30.5 m) of the resistant Glen Rose Formation, alternating thicker beds of pale gray miliolid packstone and wackestone and thinner beds of marly wackestone, is exposed in the last 0.6 mi (0.9 km) of the canyon (beginning 6.9 mi or 11 km beyond the top of Reach 3), which ends at the Terlingua Fault. This Basin and Range fault has a stratigraphical throw on the order of 3,900 ft (1,200 m), increasing to the southeast and decreasing to the northwest.

SUMMARY

The Comanchean section of the western Big Bend region represents continuing rudistid limestone deposition on a slowly subsiding shallow shelf well to the west of the more active shelf margin in northeastern Mexico and south-central Texas. Smith (1970) called the area the "western limb" of the Devils River depositional environment. Sporadic incursions of fine terrigenous sediments from the west or northwest may suggest occasional episodes of mild tectonism or lowering of sea level. This series of events culminated in subaerial exposure of the Santa Elena Formation. For a time, subsequent rapid transgression produced slightly deeper water deposition in basins with poorly oxygenated bottom waters. The change in the major supply of sediment was more striking than any previous one. It is not clear that regional subaerial exposure of the Big Bend region preceded or succeeded deposition of the Buda, but the contacts of the Buda do represent abrupt changes in depositional style, and, at the very least, unrecorded time.

Together with the disconformity at the base of the Del Rio, the shift in sedimentation probably represents a change in tectonic style. The Ernst Member, Boquillas Formation, marks the beginning of increased, and gradually more terrigenous, deposition that produced fairly uncomplicated regression (Boquillas and younger Cretaceous formations) on a still subsiding Coahuila platform. A few miles to the southwest, subsidence in the Chihuahua trough may have accelerated; there, upper Cretaceous stratigraphy is different, and the thickness of rocks much greater. Laramide and Basin and Range tectonism, and resequent rugged relief contrast strongly with quiet Cretaceous history.

REFERENCES

This list contains references cited in this paper and includes others useful for background reading on regional relationships.

Bureau of Economic Geology, 1979, Geologic atlas of Texas, Emory Peak–Presidio Sheet: Joshua William Beede Memorial Edition, University of Texas at Austin, Bureau of Economic Geology, 14 p., and map, scale 1:250,000.

DeCamp, D. W., 1981, Structural geology of Mesa de Anguila, Big Bend National Park, Trans–Pecos Texas [M.A. thesis]: University of Texas at Austin, 185 p.

——, 1985, Structural geology of Mesa de Anguila, Big Bend National Park, Texas, *in* Dickerson, P. W., and Muehlberger, W. R., eds., Structure and tectonics of Trans-Pecos Texas: Field Conference Guidebook, West Texas Geological Society Publication 85-81, p. 127–135.

Dickerson, P. W., 1985, Evidence for Late Cretaceous and Early Tertiary transpression in Trans-Pecos Texas and adjacent Mexico, *in* Dickerson, P. W., and Muehlberger, W. R., eds., Structure and tectonics of Trans-Pecos Texas: Field Conference Guidebook, West Texas Geological Society Publication 85-81, p. 185–194.

Goetz, L. K., and Dickerson, P. W., 1985, A Paleozoic transform margin in Arizona, New Mexico, West Texas, and Mexico, *in* Dickerson, P. W., and Muehlberger, W. R., eds., Structure and tectonics of Trans-Pecos Texas: Field Conference Guidebook, West Texas Geological Society Publication 85-81, p. 173–184.

Gries, J. C., and Haenggi, W. T., 1971, Structural evolution of the eastern Chihuahua Tectonic Belt, *in* Seewald, K., and Sundeen, D., eds., The geologic framework of the Chihuahua Tectonic Belt; Symposium honoring R. K. DeFord: West Texas Geological Society and The University of Texas at Austin, sponsors, p. 119–137.

Groat, C. G., 1972, Presidio Bolson, Trans-Pecos Texas and adjacent Mexico; Geology of a desert aquifer system: University of Texas at Austin, Bureau of Economic Geology Report of Investigations, no. 76, 46 p.

Henry, C. D., and Price, J. G., 1985, Summary of the tectonic development of Trans-Pecos Texas: University of Texas at Austin, Bureau of Economic Geology Miscellaneous Map no. 36, with text, 8 p.

Hill, R. T., 1891, The Comanche Series of the Texas-Arkansas region [with discussion by C. A. White and others]: Geological Society of America Bulletin, v. 2, p. 503–528.

Hill, R. T., and Vaughan, T. W., 1898, Geology of the Edwards Plateau and the Rio Grande Plain adjacent to Austin and San Antonio, Texas, with reference to the occurrence of underground waters: U.S. Geological Survey Eighteenth Annual Report, pt. 2, p. 193–321.

Hills, J. M., and Kottlowski, F. E., co-ordinators, 1983, Correlation of stratigraphic units in North America; Southwest/southwest Mid-Continent correlation chart: COSUNA, American Association of Petroleum Geologists.

King, P. B., 1935, Outline of structural development of Trans-Pecos Texas: American Association Petroleum Geologists Bulletin, v. 19, p. 221–261.

Longoria, J. F., 1985, Tectonic transpression in the Sierra Madre Oriental, northeastern Mexico; An alternative model: Geology, v. 13, p. 453–456.

Maxwell, R. A., 1968, The Big Bend of the Rio Grande: University of Texas at Austin, Bureau of Economic Geology Guidebook 7, 138 p.

Maxwell, R. A., and Dietrich, J. W., 1965, Geology of the Big Bend area, Texas: Field Trip Guidebook, West Texas Geological Society Publication 65-51, 216 p. (reprinted by the society, with additions, as Publication 72-59, 248 p.).

Maxwell, R. A., Lonsdale, J. T., Hazzard, R. T., and Wilson, J. A., 1967, Geology of Big Bend National Park, Brewster County, Texas: University of Texas at Austin, Bureau of Economic Geology Publication 6711, 320 p.

McKnight, J. F., 1970, Geologic map of Bofecillos Mountains area, Trans-Pecos Texas: University of Texas at Austin, Bureau of Economic Geology Quadrangle Map 37, with text, p. 1–36.

Muehlberger, W. R., 1980, Texas lineament revisited, *in* Dickerson, P. W., and Hoffer, J. M., eds., Trans-Pecos region: 31st Field Conference, Southeastern New Mexico and West Texas, New Mexico Geological Society, p. 113–121.

Pearson, B. T., 1981, Some structural problems of the Marfa Basin area, *in* Jons, R. D., and French, L. R., Jr., eds., Marathon-Marfa region of West Texas, symposium and guidebook: Permian Basin Section, Society of Economic Paleontologists and Mineralogists Publication 81-20, p. 59–73.

Research on Cretaceous Cycles Group, 1986, Rhythmic bedding in Upper Cretaceous pelagic carbonate sequences; Varying sedimentary response to climatic forcing: Geology, v. 14, p. 153–156.

Scholle, P. A., 1978, Carbonate rock constituents, textures, cements, and porosities: American Association of Petroleum Geologists Memoir 27, 241 p.

Smith, C. I., 1970, Lower Cretaceous stratigraphy, northern Coahuila, Mexico: University of Texas at Austin, Bureau of Economic Geology Report of Investigations no. 65, 101 p.

Stevens, J. B., and Stevens, M. S., 1985, Basin and range deformation and depositional timing, Trans-Pecos Texas, *in* Dickerson, P. W., and Muehlberger, W.R., eds., Structure and tectonics of Trans-Pecos Texas: Field Conference Guidebook, West Texas Geological Society Publication 85-81, p. 157–164.

Udden, J. A., 1907, A sketch of the geology of the Chisos country, Brewster County, Texas: University of Texas Bulletin 93, 101 p.

Vaughan, T. W., 1900, Description of the Uvalde Quadrangle: U.S. Geological Survey Geological Atlas, Folio 64 (Uvalde Folio), 7 p.

Wilson, J. L., 1975, Carbonate facies in geologic history: New York, Springer-Verlag, 471 p.

Yates, R. G., and Thompson, G. A., 1959, Geology and quicksilver deposits of the Terlingua District, Texas: U.S. Geological Survey Professional Paper 312, 114 p.

In addition, the following U.S. Geological Survey 7½-minute topographic quadrangles (1:24,000) were used for plotting the locality, and the ends of reaches along the Rio Grande: Castolon, 1971; Lajitas, 1971; Mesa de Anguila, 1971.

Mid-Tertiary and Pleistocene sections, Sotol Vista to Cerro Castellan, Big Bend National Park, southwestern Brewster County, Trans-Pecos Texas

James B. Stevens, Department of Geology, Lamar University, Box 10031, L.U. Station, Beaumont, Texas 77710

LOCATION

Texas 118 becomes the main park road at the western entrance to Big Bend National Park, in southwestern Brewster County, Trans-Pecos Texas. The traverse lies along Ross Maxwell Appreciation Scenic Drive, which turns off the main park road 9.4 mi (15.1 km) east of this entrance, and provides principal access to the southwestern part of the park. The traverse begins at Sotol Vista (Fig. 1), a 0.3 mi (0.5 km) loop off the drive 8.5 mi (13.7 km) from the turn, and ends at the eastern base of Cerro Castellan 9.4 mi (15.1 km) to the southwest. The traverse provides examples of the stratigraphy (Figs. 2, 3) and structure on which many conclusions about the eastern part of an intraplate boundary are based.

This traverse lies entirely within a national park; without the proper permits it is illegal to collect rocks, fossils, plants, or animals, and unwise to carry a rock hammer. Collecting permits should be arranged well in advance with the Resource Management Office, Big Bend National Park. It is also important to remember that this is desert country. There is no reliable supply of potable water along Ross Maxwell Appreciation Drive.

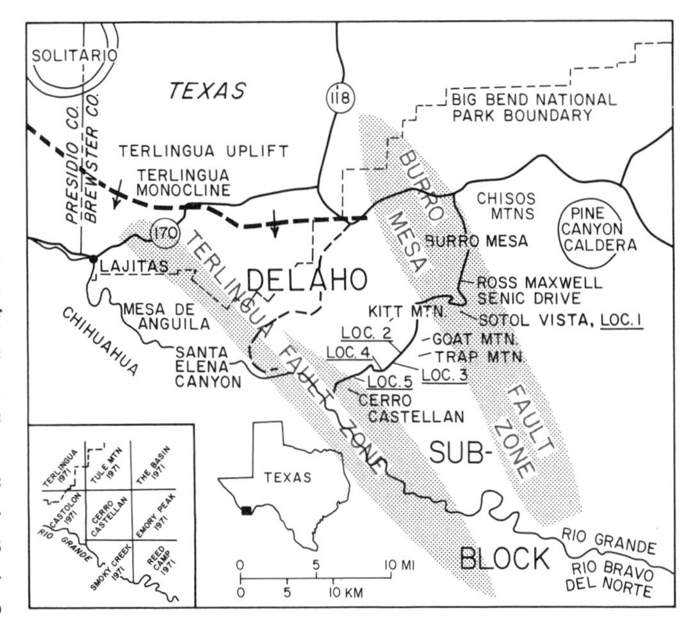

Figure 1. Map showing location of traverse and described localities, major geographic and structural features, and 7½-minute topographic quadrangles, southwestern part of Sunken Block, and Big Bend National Park, Texas.

Structural Setting: Southwestern Big Bend National Park

Udden (1907) discussed the Chisos Mountains and the lowlands around them, now the largest part of Big Bend National Park, as sitting in a "Sunken Block" bounded on the northeast by the Del Carmen and Santiago Mountains, and on the southwest by the face of the erosionally separated blocks called Mesa de Anguila (U.S.), Sierra Ponce, and Sierra Diablo (both in Mexico). King (1935) summarized a great deal of his own work, and published and unpublished work of Baker (1927, 1928), Adkins (1933), and R. E. King, among others, to produce a synthesis of the structure and tectonic setting of the Sunken Block. This synthesis was little questioned for 40 years.

Present views relate the Sunken Block to the Texas Lineament, either as lying within and near the zone of the lineament (Muehlberger, 1980; Muehlberger and Moustafa, 1984), or as a part of a larger second-order rhomb related to it (Moody and Hill, 1956; Stevens and Stevens, 1986). There is general agreement that structures in the area have been produced by a long series of episodes of regional compression and tension accompanied by (induced by?) wrench faulting, not always with the same sense of motion, between northern and Mexican parts of the North American plate (Muehlberger, 1980; Goetz and Dickerson, 1985; DeCamp, 1985; Dickerson, 1985; Ross, 1986). The result is an area in which the terms "Laramide" and "Basin and Range" are not applied to structures entirely similar to usual developments of these tectonic styles. From the later Eocene to about the end of the Oligocene (geochronological definitions of Berggren and others, 1985, are used throughout this report), the time of the major volcanic and intrusive activity in the area, there is little evidence of strong regional stresses. Right lateral divergent wrenching (Muehlberger, 1980) has occurred episodically for the past 20 m.y. (Stevens and Stevens, 1985). This continuing activity has reactivated parts of older faults, and produced high-angle oblique slip (apparently normal) faults, the dominant structures on this traverse.

The Chisos Mountains sit on a horst, defined on its west side by the complex north-northwest Burro Mesa fault zone (Fig. 1). The southwestern part of the park, and immediately adjacent Chihuahua, Mexico, lie within the most deeply subsided part of the Sunken Block, the Delaho Sub-block of Stevens and Stevens (1986), bounded on the southwest by the Terlingua fault zone. The traverse starts structurally high on the east side of the Delaho

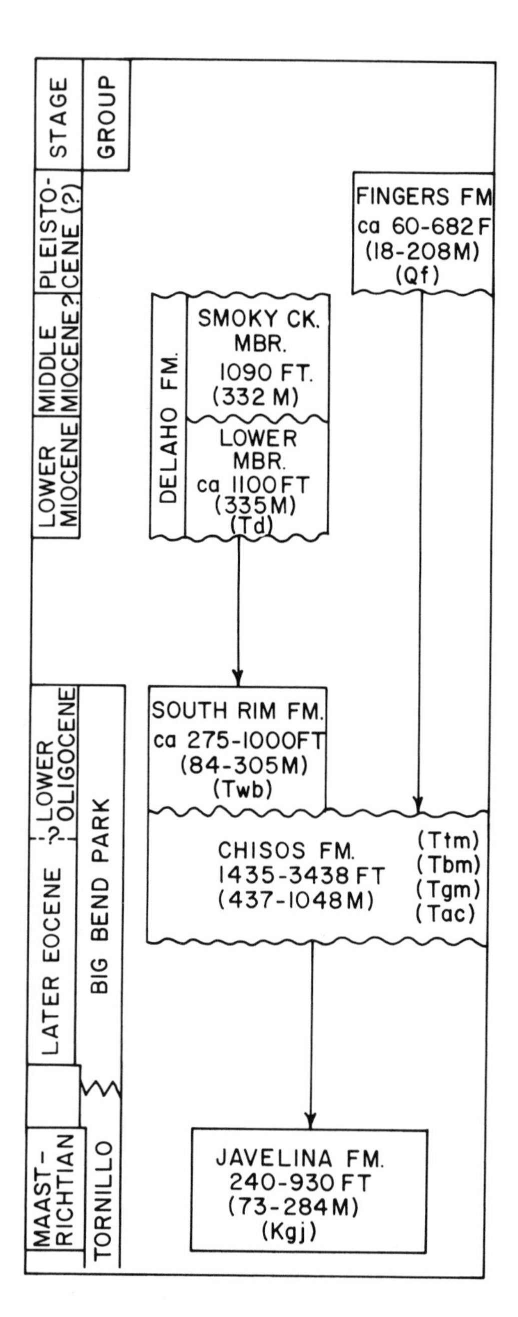

Figure 2. Series, formations, and groups exposed along traverse on Ross Maxwell Drive. Udden (1907) named the Chisos Formation ("Chisos beds") and used the name "Tornillo." Modern designations and descriptions of pre-Miocene units were given by Maxwell and others (1967). Miocene and younger units were named and described by Stevens (1969) and Stevens and others (1969). Symbols are given for units shown in the cross section (Fig. 4): Kgj, Javelina Formation; Tir, rhyolitic intrusions; Chisos Formation: Tac, Alamo Creek Member; Tgm, Goat Mountain Member; Tbm, Bee Mountain Member and Mule Ear Spring Member; Ttm, Tule Mountain Member; South Rim Formation: Twb, Wasp Spring Member and Burro Mesa Member; Td, lower member, Delaho Formation; Qf, Fingers Formation.

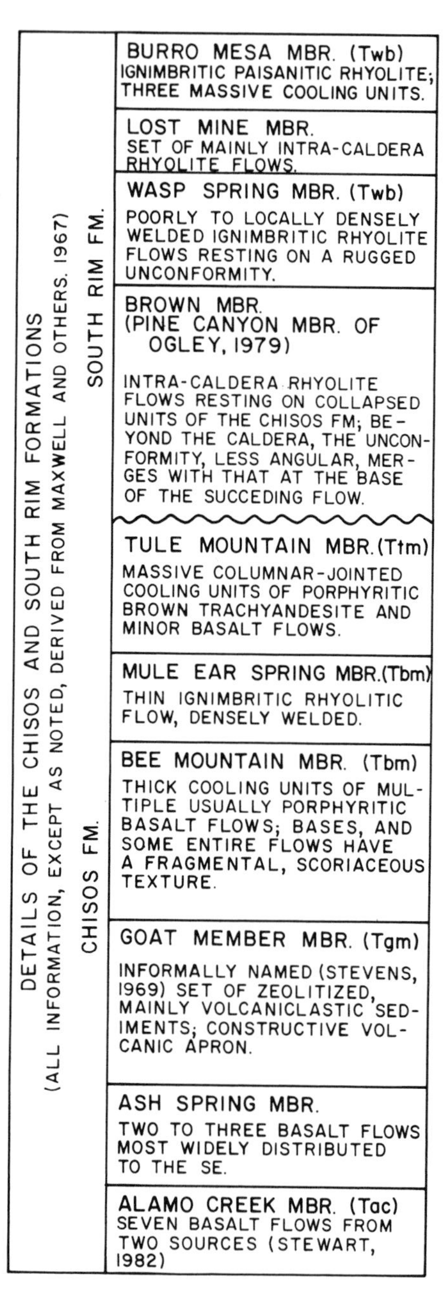

Figure 3. Members (named and described by Maxwell and others, 1967) of the Chisos and South Rim Formations. Symbols are given for units shown in the cross section (Fig. 4): Chisos Formation: Tac, Alamo Creek Member; Tgm, Goat Mountain member; Tbm, Bee Mountain Member and Mule Ear Spring Member; Ttm, Tule Mountain Member; South Rim Formation; Twb, Wasp Spring Member and Burro Mesa Member.

Sub-block, and crosses a series of faults, that strike northwesterly and are downdropped to the southwest, ending on the structurally lowest element (Figs. 1, 2). The difference in elevation between the two ends of the traverse, about 2,000 ft (600 m), is somewhat less than the structural relief.

Stratigraphy and Geologic History

Rocks of latest Cretaceous through probable Quaternary (Fig. 2) age are exposed along the traverse or within a fairly short hike from it. There is no accessible locality (or geographically orderly succession of localities) where the units can be examined in sequence. The following stratigraphic discussion offers the organization necessary to an understanding of the relationships of the localities.

The oldest rocks exposed in the Delaho Sub-block south of the extreme northern margin along the Terlingua Monocline (Fig. 1) are the Upper Cretaceous Aguja and overlying Javelina Formations; only the Javelina is exposed along the traverse. Regionally, deposition of the uppermost Cretaceous, and particularly the lowermost Tertiary, was restricted to a small area of eastern Big Bend National Park; in that area there is no pronounced unconformity. The pronounced unconformity lies between deposits representing a latest Cretaceous through early Eocene progression from shallow marine to continental deposition, and the onset of late Eocene through Oligocene volcanic episodes. This unconformity may represent the time of the greatest Laramide activity in the area. Laramide events are not otherwise dated in the Big Bend region.

Deposits of an east- to southeast-flowing late Eocene precursor of the Rio Conchos–Rio Grande system record the transition to volcanism in the Big Bend region. These deposits are found 31.1 mi (50 km) to the northwest of Sotol Vista (lower member of the Devil's Graveyard Formation, Stevens and others, 1984), and 19.9 mi (32 km) to the northeast (Big Yellow Member and the basal part of the upper member of the Canoe Formation, Maxwell and others, 1967; Rigsby, 1982, 1986).

Volcanic members of the Eocene-Oligocene Chisos Formation, with the exception of the Ash Spring Member, are widespread to the northwest beyond Lajitas, 28.6 mi (46 km) west-northwest of Sotol Vista (McKnight, 1970), and must once have covered the Mesa de Anguila–Sierra Ponce block to the west and southwest. They are not known from the eastern flanks of the Chisos Mountains, nor to the east or northeast of them. The Mexican part of the Sierra del Carmen, southeast of the Chisos Mountains, has not been investigated. The Alamo Creek Member, a 42-Ma basal unit of the Chisos Formation, rests on a surface of low relief, erosional along this traverse. Beneath this surface is the Javelina Formation in the area of the traverse, possible Canoe Formation to the north (Maxwell and others, 1967), and late Eocene river deposits near Lajitas. The Goat Mountain member (informal; Stevens, 1969) is a constructive volcanic apron, the major sedimentary part of the Chisos Formation, and contains the Eocene-Oligocene boundary, although not precisely located. A lull in volcanic activity resulted in some destruction of this apron and produced a second low-relief erosional surface, buried in rapid succession by younger members of the Chisos Formation (Figs. 2, 3), which served to further reduce relief in the area.

The main mass of the Chisos Mountains is closely connected with the South Rim Formation (Figs. 2, 3) and related rocks. Eruptions of Pine Canyon Caldera (Ogley, 1979; Henry and Price, 1984) in the northeastern part of the mountains produced the South Rim Formation about 33 to 32 Ma, or perhaps slightly less. The caldera is contemporaneous with doming of the Solitario and major volcanic activity in the Chinati Mountains. Locally, doming preceding eruption of the caldera produced a pronounced unconformity between the Chisos and Second Rim Formations. Regionally, deposition continued on a surface of remarkably low relief, as shown by the areal coverage (more than 3,000 mi^2 or 7,770 km^2) achieved by the superjacent 32 Ma Mitchell Mesa Rhyolite. Much of the rock in the northeastern Chisos Mountains is the product of intracaldera eruption and dome building. The Wasp Spring and Burro Mesa Members of the South Rim Formation escaped the caldera to the southwest into the area of this traverse. Probably in the range of 27 to 24 Ma, the Chisos Mountains pluton, chemically similar to the iron-rich peralkaline Pine Canyon Caldera rocks but also perhaps similar to younger peralkaline comenditic rocks in the area, domed the region west and south of the caldera in the Chisos Mountains and may have affected an area to the southwest. The thickness and distribution of the Eocene-Oligocene rocks suggests persistent sagging of the area west of the Chisos Mountains, but the sagging was accomplished without extensive faulting. First evidences of faulting occur in the Delaho Sub-block and others areas in this same range, 27 to 24 Ma.

About 23 Ma, deposition of the first alluvial fans (lower member, Delaho Formation; Stevens, 1969; Stevens and others, 1969) began on a terrane where the Chisos and South Rim Formations had already been brought into fault contact and bevelled. These first fans in the park and elsewhere (Stevens and Stevens, 1985) are dated both by vertebrate fossils and what are nearly the last volcanic flows (basaltic) in the region. Much thicker units, recording the principal episodes of high-angle faulting that produced the modern structural relief (20 to 18, 10 to 9 m.y. ago), are found in the park but are not easily accessible. Sediments recording earlier Pleistocene movement eventually partially buried the Chisos Mountains and completed filling of structural lows, such as the Delaho Sub-block. An example of these does occur in the traverse. Pedimentation during later Pleistocene organization of the Rio Grande drainage in the Big Bend area is the youngest series of geologic events yet proved. Faulting may continue on the east side of the Chisos Mountains.

Regionally, age determinations on intrusive rocks are late Eocene through late early Miocene. Those not successfully dated radiometrically usually have field relationships that make ages within that range most probable. Debate on the Mariscal sill, a folded igneous body with dates in the same range (Henry and

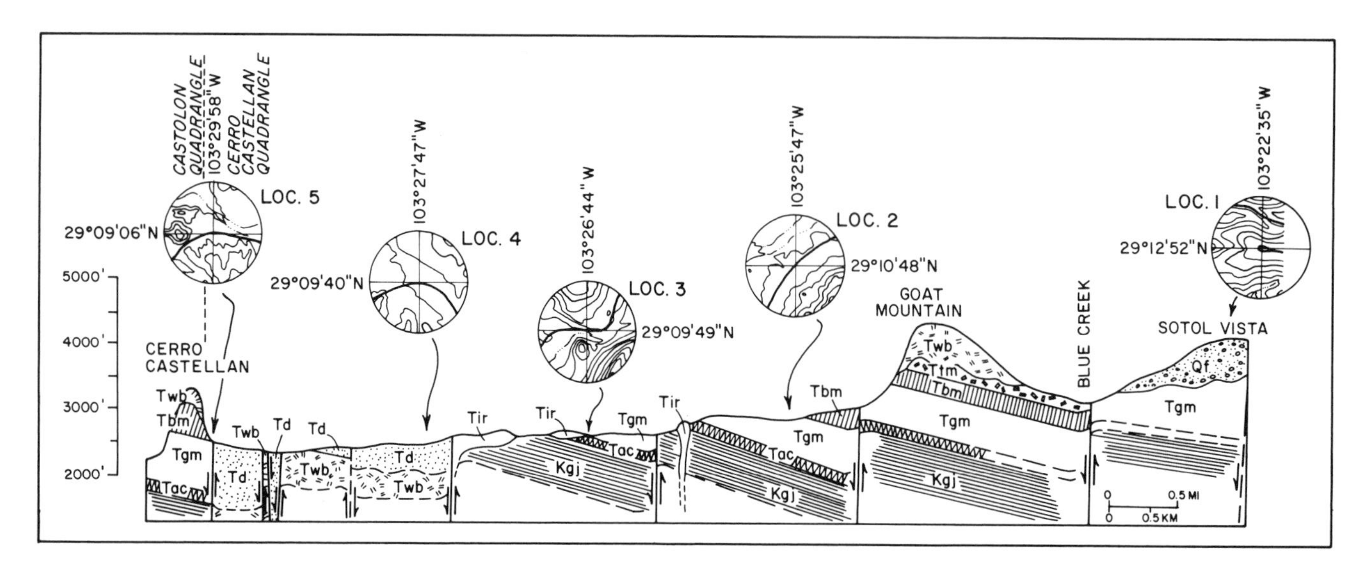

Figure 4. Diagrammatic cross section (modified from Stevens, 1969) along the traverse from Sotol Vista to Cerro Castellan, southwestern Big Bend National Park. Circular insets show topographic situations of the five localities described in the text. All insets are from the Cerro Castellan Quadrangle, except that for Locality 5, which was taken from both the Castolon and Cerro Castellan Quadrangles. Symbols for rock units are those given in the stratigraphic columns (Figs. 2, 3).

others, 1986), will continue. At present no intrusive rocks are known to be unequivocally related to Laramide events. In the area of the traverse there are only a few small rhyolitic intrusive bodies possibly related to emplacement of the Chisos Mountain pluton. Northwest of the traverse but within the Delaho Sub-block, at Rattlesnake and Peña Mountains, along the well maintained gravel road from Santa Elena Canyon to near the west entrance of the park, are accessible examples of older (Tertiary, probably later Eocene) syenodioritic intrusions (Carman and others, 1975; Carman and Cameron, 1986).

Geomorphology

Much-eroded fault-line scarps, expressed by resistant Oligocene volcanic flows, are prominent in the area. From the turn-off of the main road, the drive passes about 8.1 mi (13 km) along an obsequent fault-line scarp before climbing to Sotol Vista. Consequent fault line scarps are found along the traverse. Resistant Oligocene rhyolite flows (Burro Mesa Member, South Rim Formation) occupied a drainage system roughly approximated by the southwesterly trend of Ross Maxwell Drive; the result is a large-scale example of inversion of relief. The Oligocene valleys have been sliced transversely by high-angle faults, and softer rocks have been deeply eroded. Many of the peaks along the traverse locate parts of the Oligocene valleys. Badlands mark outcrop areas of finer mid-Tertiary sediments, and steep rounded hills are charactistic of the later Tertiary to Quaternary fanglomerates. The traverse also crosses a broad expanse of the best developed pediment in this area.

Locality 1 (Figs. 1-4) (103°22′35″W, 29°12′52″N, Cerro Castellan Quadrangle) is Sotol Vista, a viewing point and an exhibit on a short (0.3 mi; 0.5 km) loop that turns off Ross Maxwell Drive 8.5 mi (13.7 km) south of the junction with the main park road. Many aspects of the geology discussed above are visible from this point. In the distance to the west-southwest is the 1,500-ft (460 m) fault-line scarp of the Terlingua fault zone marking the southwest limit of the Sunken Block and the Delaho Sub-block, and cut by Santa Elena Canyon. On the skyline to the northwest is the Solitario rim. The Solitario sits on the north end of the Terlingua Uplift; the Terlingua Monocline across the southern part of the uplift marks the northern end of the Delaho Sub-block. Many of the peaks in a general southwesterly direction, and adjacent Burro Mesa (north-northwest), are capped by the Burro Mesa Member of the South Rim Formation, and mark the Oligocene valleys that ran down from the Pine Canyon Caldera (Fig. 1). These include Cerro Castellan, 8.5 mi; 13.6 km away (9.4 mi; 15.1 km by road), and nearby Goat Mountain (southwest); Trap Mountain (south-southwest); and Kitt Mountain (west). The peaks sit on successive downdropped fault blocks, so that although Cerro Castellan rises 1,000 ft (305 m) above its base; its highest point is still 250 ft (75 m) lower than the 4,250-ft (1,295 m) elevation of Sotol Vista.

To the northeast, the Chisos Mountains pluton forms the northern and northwestern parts of the Chisos Mountains; directly east, the Chisos Formation was contorted by intrusion of the pluton. The gravels that form the Fingers Hills and Sotol Vista, and extend onto the south end of Burro Mesa are pedimented and faulted remnants of the proximal facies of an earlier(?) Pleistocene alluvial fan, 680 ft (208 m) thick along the traverse.

Locality 2 (Figs. 1-4) (103°25′47″W, 29°10′48″N, Cerro Castellan Quadrangle) is the Goat Mountain exhibit and viewpoint on the south side of Ross Maxwell Drive at the base of Trap Mountain 7.4 mi (11.9 km) southwest of Locality 1. The lower part of the mountain is Chisos Formation. At the base are Goat Mountain member zeolitized volcaniclastic mudstone, fine to medium sandstone, and conglomerates, tilted 21° to 5° east (the dip decreasing away from the fault between Goat and Kitt Mountains). The Bee Mountain (basalt; more properly, mugearite), and Mule Ear Spring (pink, welded ash flow tuff) Members overlie the Goat Mountain sedimentary rocks, and in turn are overlain by a thin unit of sedimentary tuffs and the Tule Mountain Member, flows of coarsely to very coarsely porphyritic brown trachyandesite, with some approaching benmoreite composition (Carman and others, 1975).

The unconformity between the Chisos and South Rim Formations is spectacularly exposed in the west face of Goat Mountain. It cuts through the Tule Mountain Member to below the Mule Ear Spring Member. Above the unconformity is the Wasp Spring Member (poorly welded green-yellow and blue-gray ash flow breccia and minor sedimentary beds) of the South Rim Formation overlain by three thick, columnar-jointed cooling units of the Burro Mesa Member. The Burro Mesa Member is a paisanitic rhyolite (comendite) commonly showing strong flow foliation as a result of the interlamellar concentration of the navy blue to blue-green amphibole(?) that contrasts strongly with the pale gray aphyric groundmass. The unit was named "Burro Mesa Riebeckite Rhyolite" by Maxwell and others (1967), but there is now considerable doubt that the mineral that produces the characteristic appearance of this and several lithologically and chemically similar bodies of rock (Parker, 1983) in the Big Bend region is always, or even usually, riebeckite. A younger, but chemically similar, rhyolite intrusion cuts through the central part of the Goat Mountain exposure. Volcanic members of the Chisos Formation are generally basic and do not have known sources. The Mule Ear Spring Member (ash flow tuff) is exceptional in composition and mode of emplacement (ignimbritic) in having a satisfactory age determination (34 Ma), and, at least tentatively, a source at Sierra Quemada, a possible caldera about 3.8 mi (6.1 km) east-southeast of Sotol Vista (Henry and Price, 1984).

Locality 3 (Figs. 1-4) (103°26′44″W, 29°09′49″N, Cerro Castellan Quadrangle) is a roadcut exposing 30 ft (9 m) of unusually fresh basalt flows of the Alamo Creek Member, 2.6 mi (4.2 km) southwest of Locality 2. The Alamo Creek Member of the Chisos Formation consists of a suite of related(?) alkalic olivine basalt and mugearite flows (Carman and others, 1975). The query arises from a later study, in which Stewart (1982) found that at least seven flows, from two distinct but unlocated sources, are collectively referred to as the Alamo Creek Member. The Alamo Creek Member rests unconformably on red and tan gypsiferous clay of the Maastrichtian Javelina Formation.

About 0.2 mi (0.3 km) southwest of Locality 3, the Alamo Creek Member is overlain by a amphibole(?) rhyolite, very shallowly intrusive or extrusive, similar to the intrusions at Goat Mountain, but showing well-developed, sometimes exotic, columnar jointing. This unit can be traced to intrusions along a major fault 0.9 mi (1.5 km) west of Locality 3, where Ross Maxwell Drive turns northwest to parallel the fault. The fault brings the Burro Meso Member of the South Rim Formation in contact with the Javelina Formation, indicating vertical displacement of about 3,300 ft (1,000 m). The fault cuts the 23 Ma lower member of the Delaho Formation, but part of the movement may be roughly contemporaneous with the approximately 26 Ma intrusions.

Locality 4 (Figs. 1-4) (103°27′47″W, 29°09′40″N, Cerro Castellan Quadrangle) is near minor roadcuts in medial fan facies of the lower member of the Delaho Formation, 3.2 mi (5.2 km) west of Locality 3 on Ross Maxwell Drive. To the north across Blue Creek are pediment-gravel–capped cliffs of the same facies of the lower member. Sediments were derived from erosion of the South Rim Formation to the east and northeast. About 2.2 mi (3.6 km) south of here, near De la Ho Spring, there is a thin basaltic flow near the base of the Delaho Formation, dated at 23.3 ± 0.6 Ma, among the youngest dated evidences of volcanic activity in the park. The date is in excellent agreement with vertebrate biostratigraphic evidence (Stevens and others, 1969; Stevens, 1977) of earliest Miocene age for coeval sediments. The Delaho Formation rests here on the Wasp Spring and Burro Mesa Members of the South Rim Formation; on another fault block 5 mi (8 km) south, the unit rests on the Alamo Creek Member. The maximum preserved thickness of the lower member of the Delaho Formation, is about 1,000 ft (305 m). The younger Smoky Creek Member, not exposed near the traverse, is both thicker, and, in its proximal facies, much coarser than the lower member.

Locality 5 (Figs. 2-4) (103°29′58″W, 29°09′06″N, Cerro Castellan Quadrangle) (inset for Locality 5; includes a small part of Castolon Quadrangle) is a fault contact between the lower member of the Delaho Formation, and the Goat Mountain member of the Chisos Formation, 1.7 mi (2.7 km), southwest of Locality 4, at the base of Cerro Castellan. The Delaho Formation sits on the most deeply dropped fault block in the Delaho Subblock and rests on the Wasp Spring and Burro Meso Members of the South Rim Formation, the same stratigraphic units that occur near and at the top of Cerro Castellan. This locality has diverse points of interest. Pink-buff distal fan deposits of the lower member of the Delaho Formation, are exposed here and in roadcuts and the lower east-facing slopes of Cerro Castellan for 0.4 mi (0.6 km) to the south. It was with considerable surprise that John A. Wilson (now Professor Emeritus of Geology, the University of Texas at Austin) discovered, near Locality 5, earliest Miocene fossils in what was supposed, at the time, to be Quaternary alluvium. Here, white zeolitized sedimentary tuffs of the Goat Mountain member of the Chisos Formation, are nearly pure clinoptilolite. About 0.4 mi (0.6 km) to the northwest along the fault exposed there, small amphibole(?) rhyolite intrusives on the north side of the road closely mimic tree stumps; a larger intrusion of the same material (somewhat less glassy) has raised a small trap door of the Alamo Creek Member into a near vertical position.

REFERENCES CITED

Adkins, W. S., 1933, The Mesozoic System in Texas, *in* Sellards, E. H., Adkins, W. S., and Plummer, F. B., The geology of Texas; Volume 1, Stratigraphy: University of Texas at Austin Bulletin 3232, p. 239–518.

Baker, C. L., 1927, Exploratory geology of a part of southwestern Trans-Pecos Texas: University of Texas at Austin Bulletin 2745, 70 p.

——, 1928, Desert range tectonics of Trans-Pecos Texas: Pan-American Geologist v. 50, p. 341–373.

Berggren, W. A., Kent, D. V., Flynn, J. J., and Van Couvering, J. A., 1985, Cenozoic geochronology: Geologic Society of America Bulletin, v. 96, p. 1407–1418.

Carman, M. F., and Cameron, M. E., 1986, Petrology of shallow intrusive rocks near Study Butte, Texas, *in* Pausé, P. H., and Spears, R. G., eds., Geology of the Big Bend area, and Solitario dome, Texas: Spring 1986 Field Trip Guidebook, West Texas Geological Society Publication 86-82, p. 153–158.

Carman, M. F., Cameron, M. E., Gunn, B., Cameron, K. L., and Butler, J. F., 1975, Petrology of Rattlesnake Mountain sill, Big Bend National Park, Texas: Geological Society of America Bulletin, v. 86, p. 177–193.

DeCamp, D. W., 1985, Structural geology of Mesa de Anguila, Big Bend National Park, Texas, *in* Dickerson, P. W., and Muehlberger, W. R., eds., Structure and tectonics of Trans–Pecos Texas: Field Conference Guidebook, West Texas Geological Society Publication 85-81, p. 127–135.

Dickerson, P. W., 1985, Evidence for Late Cretaceous and Early Tertiary transpression in Trans–Pecos Texas and adjacent Mexico, *in* Dickerson, P. W., and Muehlberger, W. R., eds., Structure and tectonics of Trans-Pecos Texas: Field Conference Guidebook, West Texas Geological Society Publication 85-81, p. 185–194.

Goetz, L. K., and Dickerson, P. W., 1985, A Paleozoic transform margin in Arizona, New Mexico, West Texas, and Mexico, *in* Dickerson, P. W., and Muehlberger, W. R., eds., Structure and tectonics of Trans-Pecos Texas: Field Conference Guidebook, West Texas Geological Society Publication 85-81, p. 173–184.

Henry, C. D., and Price, J. G., 1984, Variations in caldera development in the Tertiary volcanic field of Trans-Pecos Texas: Journal of Geophysical Research, v. 89, no. B10, p. 8765–8786.

Henry, C. D., and others, 1986, Compilation of potassium-argon ages of Tertiary igneous rocks, Trans-Pecos Texas: University of Texas at Austin, Bureau of Economic Geology Geological Circular 86–2, 34 p.

King, P. B., 1935, Outline of structural development of Trans-Pecos Texas: American Association of Petroleum Geologists Bulletin, v. 19, p. 221–261.

Maxwell, R. A., Lonsdale, J. T., Hazzard, R. T., and Wilson, J. A., 1967, Geology of Big Bend National Park, Brewster County, Texas: University of Texas at Austin, Bureau of Economic Geology Publication 6711, 320 p.

McKnight, J. F., 1970, Geologic map of Bofecillos Mountains area, Trans-Pecos Texas: University of Texas at Austin, Bureau of Economic Geology Quadrangle Map 37, with text, p. 1–36.

Moody, J. D., and Hill, M. J., 1956, Wrench-fault tectonics: Geological Society of America Bulletin, v. 67, p. 1207–1246.

Muehlberger, W. R., 1980, Texas lineament revisited, *in* Dickerson, P. W., and Hoffer, J. M., eds., Trans-Pecos region: 31st Field Conference, southeastern New Mexico and West Texas, New Mexico Geological Society, p. 113–121.

Muehlberger, W. R., and Moustafa, A. R., 1984, Late Cenozoic pull-apart graben development, Big Bend region, Texas [abs.]: American Association of Petroleum Geologists Bulletin, v. 68, p. 510.

Ogley, D. S., 1979, Eruptive history of the Pine Canyon Caldera, Big Bend National Park, *in* Walton, A. W., and Henry, C. D., eds., Cenozoic geology of the Trans-Pecos Volcanic Field of Texas: University of Texas at Austin, Bureau of Economic Geology Guidebook 19, p. 67–71.

Parker, D. F., Jr., 1983, Origin of the trachyte-quartz trachyte-peralkalic rhyolite suite of the Oligocene Paisano volcano, Trans-Pecos Texas: Geological Society of America Bulletin, v. 94, p. 614–629.

Rigsby, C. A., 1982, Provenance and depositional environments of the Middle Eocene Canoe Formation, Big Bend National Park, Brewster County, Texax [M.S. thesis]: Baton Rouge, Louisiana State University and Agricultural and Mechanical College, 111 p.

——, 1986, The Big Yellow Sandstone; A sandy braided stream, *in* Pausé, P. H., and Spears, R. G., eds., Geology of the Big Bend area, and Solitario dome Texas: Spring 1986 Field Trip Guidebook, West Texas Geological Society Publication 86-82, p. 111–115.

Ross, C. A., 1986, Paleozoic evolution of southern margin of Permian Basin: Geological Society of America Bulletin, v. 97, p. 536–554.

Stevens, J. B., 1969, Geology of the Castolon area, Big Bend National Park, Brewster County, Texas [Ph.D. thesis]: Austin, The University of Texas, 129 p.

Stevens, J. B., and Stevens, M. S., 1985, Basin and range deformation and depositional timing, Trans–Pecos Texas, *in* Dickerson, P. W., and Muehlberger, W. R., eds., Structure and tectonics of Trans-Pecos Texas: Field Conference Guidebook, West Texas Geological Society Publication 85-81, p. 157–164.

——, 1986, An outline of the Cenozoic geologic history of the area around Big Bend National Park, *in* Stevens, J. B., Roberts, D. C., and Stevens, M. S., Field seminar on the Big Bend, Trans–Pecos region, Texas: Spring Field Trip Guidebook, Houston Geological Society, p. 45–73.

Stevens, J. B., Stevens, M. S., and Wilson, J. A., 1984, Devil's Graveyard Formation (new), Eocene and Oligocene age, Trans–Pecos Texas: University of Texas at Austin, Texas Memorial Museum Bulletin 32, 21 p.

Stevens, M. S., 1977, Further study of Castolon local fauna (early Miocene), Big Bend National Park, Texas: University of Texas at Austin, Texas Memorial Museum, The Pearce-Sellards Series, no. 28, p. 1–69,

Stevens, M. S., Stevens, J. B., and Dawson, M. R., 1969, New early Miocene formation and vertebrate local fauna, Big Bend National Park, Brewster County, Texas: University of Texas at Austin, Texas Memorial Museum, The Pearce-Sellards Series, no. 15, 53 p.

Stewart, R. M., 1982, A stratigraphic and petrologic characterization of the Alamo Creek Basalt, Big Bend National Park, Texas: Geological Society of America Abstracts with Programs, v. 14, p. 137.

Udden, J. A., 1907, A sketch of the geology of the Chisos country, Brewster County, Texas: University of Texas Bulletin 93, 101 p.

In addition, the following U.S. Geological Survey 7½-minute topographic quadrangles (1:24,000) were used for plotting the localities: Castolon, 1971; Cerro Castellan, 1971.

Dikes in Big Bend National Park; Petrologic and tectonic significance

Jonathan G. Price and Christopher D. Henry, Bureau of Economic Geology, The University of Texas at Austin, University Station, Box X, Austin, Texas 78713-7508

INTRODUCTION

Dikes in Big Bend National Park illustrate the petrologic variety and tectonic setting of Tertiary magmatism in Trans-Pecos Texas. Most igneous activity occurred between two major episodes in the tectonic history of western North America: (1) Late Cretaceous to early Eocene Laramide compression and (2) late Oligocene to recent Basin and Range extension. Volumetrically minor magmatism occurred during the initial stage of Basin and Range extension and perhaps during the final stage of Laramide deformation.

The general geology of the park, including aspects of stratigraphy, structure, petrology, and geomorphology, is best described in the treatise by Maxwell and others (1967). The 1:62,500-scale geologic map in Maxwell and others' (1967) volume is also available at many stores in the Big Bend region as part of a nontechnical geologic guide to Big Bend by Maxwell (1968). The 1:250,000-scale geologic map by Barnes (1979) covers the southern part of the Trans-Pecos region. A field trip guidebook by Maxwell and Dietrich (1972) complements these publications. More recent studies have added considerably to our understanding of the tectonics, volcanic stratigraphy, and petrology of the region. Research results are summarized in guidebooks edited by Dickerson and Hoffer (1980), Dickerson and Muehlberger (1985), and Price and others (1986). In addition, details of the tectonic and magmatic history of the Trans-Pecos region are provided by Henry and Price (1984, 1985, 1986), Price and Henry (1984), and Price and others (1987).

This chapter describes localities for examining dikes in Big Bend National Park (Fig. 1). These dikes represent two ends of the chemical spectrum of igneous rocks found in the Trans-Pecos region: rhyolites and nepheline-normative basalts. In addition, the dike orientations record both regional tectonic controls related to Laramide compression and local structural controls related to caldera development. A summary of Tertiary geologic history provides the framework for determining the significance of dike compositions and orientations.

TERTIARY TECTONIC AND MAGMATIC HISTORY OF TRANS-PECOS TEXAS

Laramide Deformation. Laramide compression, which affected most of western North America, caused folding, thrust faulting, and wrench faulting in the Trans-Pecos region. In the eastern part of the park, north-northwest-trending monoclines and thrust faults and left-lateral slip on west-northwest-striking fractures (the Texas lineament of Muehlberger, 1980) indicate east-northeast compression during Laramide deformation (Moustafa, 1983). Using occurrences of unconformities and conglomerates during the Cretaceous and early Tertiary, Wilson (1971) determined that Laramide deformation in the park began in the Cretaceous, peaked in the late Paleocene, and ended in the early Eocene, about 50 Ma.

Convergence of the Farallon and North American plates is the probable cause of Laramide crustal shortening throughout North America (Dickinson, 1981). Igneous activity during Laramide deformation was dominantly west of the Trans-Pecos region, closer to the paleotrench that lay off the coast of western North America (Coney and Reynolds, 1977; Damon and others, 1981). This magmatism was dominated by calc-alkaline activity typical of a continental arc tectonic setting.

Middle Eocene to Early Oligocene Magmatism. Igneous activity began in Trans-Pecos Texas about 47 Ma and perhaps as early as 57 Ma (Henry and McDowell, 1986; Henry and others, 1986). The igneous rocks in the park are representative of the Trans-Pecos region. Both intrusive and extrusive rocks are abundant (Table 1). Although some of the Upper Cretaceous sedimentary rocks in the park contain bentonitic clays ultimately derived from volcanic sources probably to the west in Mexico, the oldest rocks of local volcanic origin are basalt lavas of the middle Eocene Canoe Formation (Maxwell and others, 1967). Overlying the Canoe Formation are the Chisos and South Rim formations, both of which contain tuffaceous sedimentary rocks interlayered with volcanic rocks (Table 1). Large volumes of volcanic rocks were erupted from calderas throughout the Trans-Pecos region. Two calderas have been recognized in the park (Ogley, 1979; Henry and Price, 1984; Barker and others, 1986): (1) the Pine Canyon caldera, located east of the basin high in the Chisos Mountains (and source of the South Rim Formation), and (2) the Sierra Quemada caldera, located southwest of the basin and probable source of the Mule Ear Springs Tuff.

Two distinct petrologic suites characterize the igneous rocks of the Trans-Pecos region. Examples of each occur in the park: (1) a silica-undersaturated suite ranging along probable crystal fractionation paths from alkali basalt through hawaiite, mugearite, benmoreite, and trachyte to phonolite and (2) a silica-saturated to silica-oversaturated suite ranging from trachybasalt through trachyandesite and quartz trachyte to rhyolite. The nepheline-normative silica-undersaturated rocks are dominantly intrusive. Both suites of rocks are more alkaline (sodium and potassium richer and calcium poorer) than calc-alkaline rocks typical of continental and island arcs. Chemical variation within

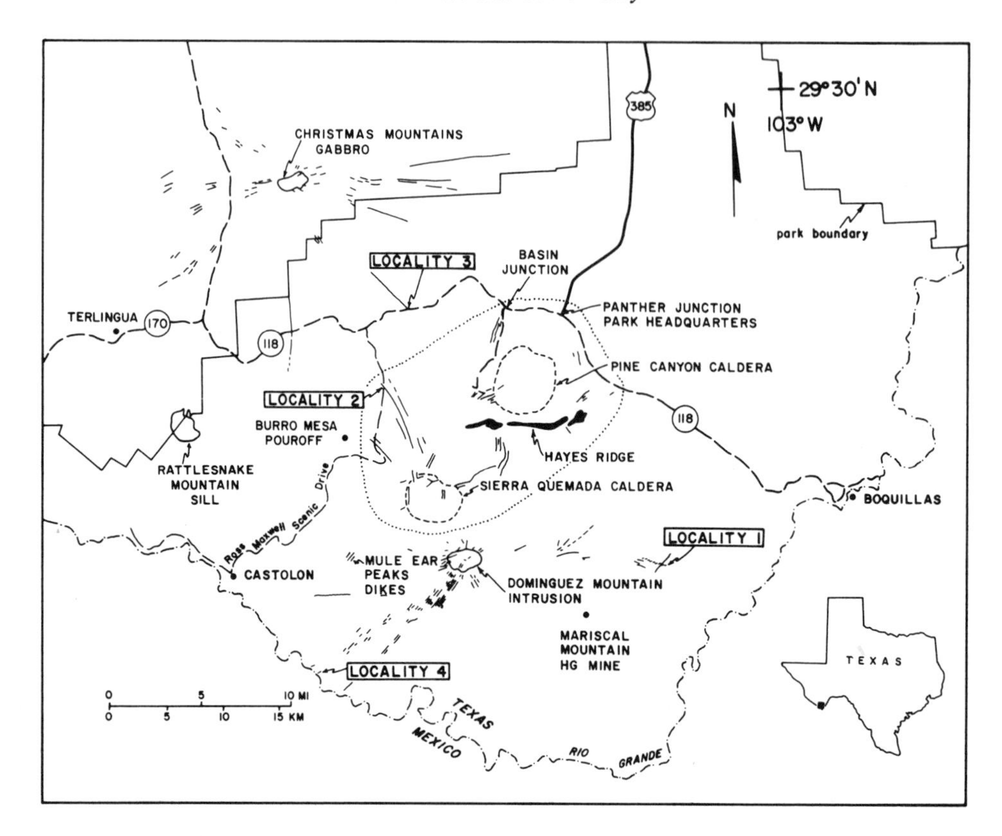

Figure 1. Major dikes in Big Bend National Park, modified from Maxwell and others (1967). Rhyolite porphyry dikes associated with the Pine Canyon and Sierra Quemada calderas are outlined with dots. Swarms of dominantly east-northeast-striking dikes emanate from the Christmas Mountains gabbro north of the park and from the Dominguez Mountain intrusion in the southern part of the park. Check at park headquarters at Panther Junction concerning road conditions to Localities 1 and 4.

the Trans–Pecos region, however, suggests a subduction origin for the igneous rocks. The middle Eocene to early Oligocene igneous rocks can be divided into an eastern alkalic belt and a western alkali-calcic (less alkaline) belt (Barker, 1977; Henry and Price, 1984). The north–northwest-trending boundary between the two belts parallels the paleotrench that lay off the west coast of Mexico. Trans–Pecos igneous rocks are at the eastern edge of a group of subduction-related rocks that range from Laramide-age calc-alkaline varieties to the west in northern Mexico, New Mexico, and Arizona to Eocene and Oligocene alkaline rocks in Big Bend National Park.

Dikes and veins that formed during middle Eocene to early Oligocene igneous activity in Texas are preferentially oriented east–northeast. Price and Henry (1984) interpreted this observation as evidence that the maximum principal compressive stress, σ_1, was east–northeast, essentially the same as during Laramide deformation. Although the middle Tertiary igneous rocks are generally not folded, the region was still under a mild state of compression. The orientation of en echelon dikes in Big Bend National Park, such as those at the first locality described in this section, further strengthens this link between magmatism and stresses residual from Laramide deformation. This structural evidence supports the contention that the main phase of Trans–Pecos igneous activity occurred in a continental arc tectonic setting.

Potassium-argon dates of igneous rocks in the vicinity of the park (Table 1) suggest pulses in magmatic activity. An early period from at least 47 to 39 Ma is dominated by basalts (the Alamo Creek Basalt and the Christmas Mountains gabbro and associated dikes). A later period of basaltic activity from about 37 to 33 Ma (Dominguez Mountain dike swarm, Ash Spring and Bee Mountain basalts) overlaps with a period of caldera eruptions from 34 to 32 Ma.

Basin and Range Extension. Orientations of dikes in the Sierra Rica caldera complex west of the park in Mexico indicate that about 31 Ma, stress orientations shifted from the directions characteristic of Laramide compression to those characteristic of early Basin and Range extension. Basin and Range normal faulting, which is locally well developed in the park, did not begin in Texas, however, until about 24 Ma (Henry and Price, 1986). The Sierra Rica caldera complex was the source of the San Carlos and Santana tuffs (Table 1), which crop out in Texas west of the park. During the period of early tension without significant extension from 31 to 24 Ma, minor igneous activity in the park yielded the

TABLE 1. TERTIARY VOLCANIC STRATIGRAPHY AND POTASSIUM-ARGON AGES OF IGNEOUS ROCKS, BIG BEND NATIONAL PARK AND VICINITY (from Henry and others, 1986)

Extrusive	Age (Ma, ± 1σ)	Intrusive
	20 to 24	Basin and Range dikes and flows, Black Gap (NE) and Terlingua areas (NW of park)
Santana Tuff (erupted from caldera in Chihuahua)	27.8 ± 0.6	
	28.2 ± 0.6	Mule Ear Peaks peralkaline rhyolite dike.
	28.6 ± 0.4	Rattlesnake Mountain syenogabbro sill.
	30.2 ± 1.4	Basalt dike in SE part of park
San Carlos Tuff (erupted from caldera in Chihuahua)	30.4 ± 0.7	
South Rim Formation (erupted from Pine Canyon caldera)		
Burro Mesa Rhyolite	31.7 ± 0.7	
Wasp Spring Member		
Lost Mine member	33.0 ± 0.7	
Boot Rock Member	32.9 ± 0.7	
Pine Canyon Rhyolite		
Chisos Rormation		
Tule Mountain Trachyandesite		
Mule Ear Springs Tuff (erupted from Sierra Quemada caldera)	33.9 ± 0.7	
Bee Mountain Basalt	34.5 ± 1.8	
Ash Spring Basalt	34.5 ± 1.7	
	33.4 ± 1.4	Basalt dikes of Dominguez Mountain swarm
	36.8 ± 1.7	
Tuff in Tuff Canyon	42.4 ± 0.7	
	42.4 ± 0.7	Christmas Mountains gabbro
	39.6 ± 1.2	Basalt and trachyte dikes of Christmas Mountains swarm
	39.9 ± 0.7	
	41.7 ± 1.2	
	42.6 ± 1.3	
	43.5 ± 1.8	
Alamo Creek Basalt (NW of park)	46.1 ± 3.1	
(SW part of park)	46.9 ± 1.1	
	47.2 ± 2.0	Basalt dike in central west part of park
	57.1 ± 5.6*	Basalt dike at Glenn Draw
Canoe Formation		

*Accuracy of this date is doubtful; the large error is due to a low radiogenic argon percentage.

Rattlesnake Mountain sill, dikes near Mule Ear Peaks (Table 1), and perhaps other undated intrusions.

Volumetrically minor basaltic magmatism was widespread in the Trans–Pecos region from about 24 to 17 Ma (Henry and Price, 1986). Examples near the park include dikes and plugs north of Terlingua and flows and dikes in the Black Gap area northeast of the park. A lava flow at the base of the Delaho Formation in the southwestern part of the park probably belongs to this same episode (J. B. Stevens, personal communication, 1985).

Early Basin and Range basaltic dikes have preferred north–northwest orientations throughout the region (Henry and Price, 1986). Stress orientations during early extension in Texas, therefore, were similar to those throughout the Basin and Range Province (Zoback and others, 1981); the least principal compressive stress, σ_3, or direction of extension, was east–northeast. Thus about 32 or 31 Ma, σ_3 shifted approximately 90° from north–northwest to east–northeast. Some dike injection occurred along northwest-striking fractures of the Texas lineament. Right-lateral oblique slip occurred along these fractures during extension (Muehlberger, 1980; Moustafa, 1983).

Another change in σ_3, from east–northeast to west–northwest, may have occurred in Texas about 10 Ma, as it did in other parts of the Basin and Range Province (Zoback and others, 1981). Timing for this change in Texas is poorly constrained, because igneous activity ceased about 17 Ma. Price and others (1985) argued that sandstone-hosted silver-copper veins near Van Horn formed when the region was undergoing northwest–southeast extension. These veins are unrelated to Tertiary igneous rocks and probably formed during Basin and Range extension. Abundant Quaternary fault scarps (Muehlberger and others, 1978; Seager, 1980; Henry and others, 1983) and a 1931 earthquake with a magnitude of approximately 6.0 (near Valentine, northwest of the park; Dumas and others, 1980) indicate that the Basin and Range Province in Texas is actively extending today.

LOCALITY 1. EN ECHELON DIKES NEAR GLENN DRAW

A set of en echelon dikes near Glenn Draw illustrates the stress orientations during the main phase of magmatism in Trans–Pecos Texas, between 47 and 32 Ma. The dikes are most easily

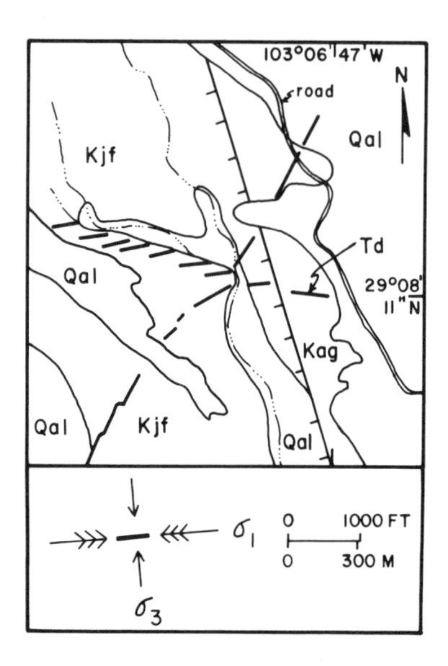

Figure 2. Geologic map of Glenn Draw area, Big Bend National Park, modified from Maxwell and others (1967). Symbols, in chronologic order: Kag = Cretaceous Aguja Formation; Kjf = Cretaceous Javelina Formation; Td = heavy lines = Tertiary dikes; line with tick marks = Basin and Range normal fault; Qal = Quaternary alluvium. Base map is the southwest corner of the San Vicente 7½-minute Quadrangle.

Figure 3. Low hills of en echelon dikes surrounded by clays and sandstones of the Upper Cretaceous Javelina Formation.

reached by the following route: drive from Panther Junction (park headquarters) 15.6 mi (25.1 km) southeast on the paved road toward Boquillas; turn right (south) onto the dirt road toward Castolon via the Mariscal Mountain mercury mine, 0.5 mi (0.8 km) west of and before crossing Tornillo Creek; travel 8.7 mi (14.0 km) on this road, which loops south, southwest, west, then northwest; 0.8 mi (1.3 km) past Rooneys Place ruins, turn right (north) onto the dirt road toward Glenn Spring, the back road to Panther Junction; travel 3.0 mi (4.8 km) northwest along this road, driving up a Quaternary gravel surface; park and walk 400 ft (120 m) to the southeastern end of the set of en echelon dikes (Fig. 2).

Two sets of basalt dikes occur at this locality, one with a general northeast strike and one consisting of east-northeast-striking en echelon dikes (Fig. 2). The dikes intrude sandstones and clays of the Upper Cretaceous Aguja and Javelina formations. Except where metamorphosed by dikes, the poorly consolidated sedimentary rocks weather more readily than the basalts, leaving dike segments standing as small ridges (Fig. 3). Relative ages of the dikes have not been determined. Judging from their petrographic similarity, we surmise that they may be contemporaneous. Henry and others (1986) reported a potassium-argon date of 57.1 ±5.6 Ma on plagioclase from a segment of the en echelon dikes, but the accuracy of this date is questionable (Table 1). Plagioclase from a similar porphyritic basalt (Maxwell and others, 1967, western end of Locality 358) yielded a date of 30.2 ±1.4 Ma. Maxwell and others (1967, Locality 174) described the dikes in this area as heavily weathered analcite basalts. Analcite is probably the hydrothermal alteration product of nepheline or interstitial glass. A relatively fresh sample of the en echelon dike (Sample H84-33 in Table 2) is chemically typical of the silica-undersaturated suite of igneous rocks in Trans-Pecos Texas. Such nepheline-normative alkali basalt could differentiate to phonolite.

The en echelon dikes at Glenn Draw are tectonically significant because they indicate orientations of σ_1 and σ_3. Individual dike segments strike generally east-northeast, parallel to σ_1 and perpendicular to σ_3. The dikes fill tension cracks along a left-lateral shear that trends west-northwest. En echelon dikes filling tension cracks along complementary northeast-trending right-lateral shears occur at Maxwell and others' (1967) Localities 170 (5.0 mi or 8.0 km southwest of here at the north end of Mariscal Mountain) and 358 (3.6 mi or 5.8 km southwest of here). The stress orientations indicated at Glenn Draw coincide with the regional pattern of stresses during the main phase of igneous activity in Trans-Pecos Texas. East-northeast compression at this time was probably residual from Laramide deformation.

LOCALITY 2. RHYOLITE DIKES ASSOCIATED WITH CALDERAS

Rhyolite dikes occur around and radial to the two calderas within the park (Fig. 1). The most accessible dikes are at Basin Junction, 3.2 mi (5.1 km) west of Panther Junction and at Locality 2 (29°16′36″N,103°22′08″W) along Ross Maxwell Scenic Drive. Take the following route to Locality 2: drive from Panther Junction 12.9 mi (20.8 km) west on Texas 118 to Ross Maxwell Scenic Drive, the paved road to Castolon; travel 3.4 mi (5.5 km) south on this road to the dike locality, a roadside rest in the southwestern part of the Basin 7½-minute Quadrangle.

On the ridge to the southeast, rhyolite dikes intrude tuffaceous sedimentary rocks and the Ash Spring Basalt Member of the Chisos Formation. The north-northwest-striking dikes appear to radiate from the Sierra Quemada caldera (Fig. 1). Petrographically similar quartz-alkali feldspar porphyry dikes are concentric to the Pine Canyon caldera. The Hayes Ridge dike, which is up to 2,200 ft (670 m) wide and 5.6 mi (9.0 km) long, and several other dikes are roughly parallel to the regional trend of east–northeast dikes, but overall these rhyolite dikes exhibit a radial pattern indicative of little tectonic control. The dikes have not yielded feldspar phenocrysts sufficiently fresh for potassium-argon dating. They were presumably intruded during the activity of the Sierra Quemada caldera (34 Ma) and the Pine Canyon caldera (33 to 32 Ma). Thus the dikes were emplaced near the end of the main phase of regional igneous activity between 47 and 32 Ma.

In contrast to the basalt dikes at Glenn Draw, these rhyolite dikes chemically belong to the silica-oversaturated suite. Highly differentiated rocks in the eastern alkalic belt of Trans-Pecos Texas are generally peralkaline. That is, relative to aluminum they contain more sodium and potassium than can be incorporated in feldspars. The peralkalinity of these and other rhyolites in the park is mineralogically expressed by the occurrence of sodic amphibole (arfvedsonite rather than the commonly used name riebeckite). The components acmite (ac) and sodium metasilicate (ns) appear as normative minerals calculated from the chemical analysis of these peralkaline rocks (Table 2). Few of the rhyolite dikes, such as Sample H84-32 (Table 2), are unaltered and unweathered enough for arfvedsonite to be present. Arfvedsonite generally occurs as fine grains that impart an overall bluish green color to the peralkaline rocks. The Burro Mesa Rhyolite, an extrusive rock that forms spectacular exposures at the Burro Mesa pouroff (8.1 mi or 13.0 km further south on this road) is a relatively fresh example of peralkaline rhyolite.

TABLE 2. CHEMICAL ANALYSES OF REPRESENTATIVE DIKES IN BIG BEND NATIONAL PARK

wt.%*	Porphyritic analcite basalt Sample No. H84-33 Glenn Draw	Rhyolite porphyry Sample No. H84-32 Hayes Ridge
SiO_2	45.61	75.23
TiO_2	3.81	0.22
Al_2O_3	17.23	10.54
Fe_2O_3	3.52	2.33
FeO	8.98	1.47
MnO	0.18	0.05
MgO	4.53	<0.03
CaO	10.41	0.46
Na_2O	3.58	4.82
K_2O	0.97	4.66
P_2O_5	0.98	<0.05
CO_2	<0.01	<0.01
H_2O^+	0.40	0.20
Total	100.20	99.98
	NORM	
Q	---	32.30
or	5.73	27.53
ab	26.52	28.28
an	28.08	---
ne	2.04	---
ac	---	6.74
ns	---	1.13
di	14.02	2.04
hy	---	1.35
ol	8.79	---
mt	5.10	---
il	7.24	0.42
ap	2.14	---
Total	99.66	99.79

*Analyses by Steven W. Tweedy, Mineral Studies Laboratory, Bureau of Economic Geology (quality assurance data provided upon request).

LOCALITIES 3 AND 4. DIKE SWARMS

The two main dike swarms in the park area exhibit preferential east–northeast strikes indicative of residual Laramide compression during the main phase of igneous activity. The Christmas Mountains swarm of basalt and trachyte dikes emanates from the Christmas Mountains gabbro (Fig. 1). With the exception of one 47-Ma north-striking dike (Laughlin and others, 1982), dates of these rocks are between 44 and 39 Ma (Table 1). An easily accessible dike of the Christmas Mountains swarm (Locality 139 of Maxwell and others, 1967, and Sample VCK-82-TX-028 of Laughlin and others, 1982) crops out at Locality 3 29°20′03″N,103°20′48″W) near Texas 118, 10.1 mi (16.3 km) west of Panther Junction or 2.8 mi (4.5 km) east of the intersection with Ross Maxwell Scenic Drive (Fig. 1). The dike is located in the northwestern part of the Basin 7½-minute Quadrangle. This nepheline-normative basaltic dike intrudes sandstones and clays of the Upper Cretaceous Aguja Formation.

A second swarm of basalt dikes emanates from the composite gabbro-syenite intrusion at Dominguez Mountain (Fig. 1). Laughlin and others (1982) dated two of these dikes at 36.8 ±1.7 and 33.4 ±1.4 Ma. The older dike at Locality 4 (29°03′25″N, 103°25′56″W, Locality 160 of Maxwell and others, 1967, and Sample VCK-82-TX-030 of Laughlin and others, 1982) can be reached by driving southeast from Castolon 9.7 mi (15.6 km) along the dirt road that follows the Rio Grande. The younger dike (Sample VCK-82-TX-031 of Laughlin and others, 1982) crops out 2.2 mi (3.5 km) farther down the same road. The dikes are located in the central part of the Smoky Creek 7½-minute Quadrangle. Both dikes intrude tuffaceous sedimentary rocks of the Chisos Formation. Laughlin and others (1982) indicated that the older dike is nepheline normative (silica undersaturated) whereas the younger dike is hypersthene normative (silica saturated). The close spatial association of undersaturated and saturated to oversaturated rocks is common in the Trans-Pecos magmatic province.

SUMMARY

Dikes of Big Bend National Park characteristically strike east–northeast, as do most middle Eocene to early Oligocene dikes in Trans-Pecos Texas. This regional trend suggests that stresses during the main phase of magmatism were residual from

Laramide deformation, which folded and faulted prevolcanic strata. Magmatism during later Basin and Range extension is characterized by north-northwest-striking dikes. The two chemically distinct suites of igneous rocks, silica-undersaturated and silica-oversaturated varieties, are well represented in the park. They are broadly contemporaneous and are commonly spatially associated.

REFERENCES CITED

Barker, D. S., 1977, Northern Trans-Pecos magmatic province; Introduction and comparison with the Kenya rift: Geological Society of America Bulletin, v. 88, p. 1421–1427.

Barker, D. S., Henry, C. D., and McDowell, F. W., 1986, Pine Canyon caldera, Big Bend National Park; A mildly peralkaline magmatic system with multiple shallow reservoirs, *in* Price, J. G., Henry, C. D., Parker, D. F., and Barker, D. S., eds., Igneous geology of Trans-Pecos Texas: The University of Texas at Austin, Bureau of Economic Geology Guidebook 23, p. 266–285.

Barnes, V. E., 1979, Emory Peak–Presidio sheet: The University of Texas at Austin, Bureau of Economic Geology, Geologic Atlas of Texas, scale 1:250,000.

Coney, P. J., and Reynolds, S. J., 1977, Cordilleran Benioff zones: Nature, v. 270, p. 403–406.

Damon, P. E., Shafiqullah, M., and Clark, K. F., 1981, Age trends in igneous activity in relation to metallogenesis in the southern Cordillera, *in* Dickinson, W. R., and Payne, W. D., eds., Relations of tectonics to ore deposits in the southern Cordillera: Arizona Geological Society Digest, v. 14, p. 137–154.

Dickerson, P. W., and Hoffer, J. M., eds., 1980, Trans-Pecos region, southeastern New Mexico and West Texas: New Mexico Geological Society 31st Field Conference Guidebook, 314 p.

Dickerson, P. W., and Muehlberger, W. R., eds., 1985, Structure and tectonics of Trans-Pecos Texas: West Texas Geological Society Publication 85-81, 278 p.

Dickinson, W. R., 1981, Plate tectonic evolution of the southern Cordillera, *in* Dickinson, W. R., and Payne, W. D., eds., Relations of tectonics to ore deposits in the southern Cordillera: Arizona Geological Society Digest, v. 14, p. 113–135.

Dumas, D. B., Dorman, H. J., and Latham, G. V., 1980, A reevaluation of the August 16, 1931, Texas earthquake: Bulletin of the Seismological Society of America, v. 70, p. 1171–1180.

Henry, C. D., and McDowell, F. W., 1986, Geochronology of magmatism in the Tertiary volcanic field, Trans-Pecos Texas, *in* Price, J. G., Henry, C. D., Parker, D. F., and Barker, D. S., eds., Igneous geology of Trans-Pecos Texas: The University of Texas at Austin, Bureau of Economic Geology Guidebook 23, p. 99–122.

Henry, C. D., and Price, J. G., 1984, Variations in caldera development in the Tertiary volcanic field of Trans-Pecos Texas: Journal of Geophysical Research, v. 89, no. B10, p. 8765–8786.

—— , 1985, Summary of the tectonic development of Trans-Pecos Texas: The University of Texas at Austin, Bureau of Economic Geology, Notes to accompany Miscellaneous Map 36, 8 p.

—— , 1986, Early Basin and Range development in Trans-Pecos Texas and adjacent Chihuahua; Magmatism and orientation, timing, and style of extension: Journal of Geophysical Research, v. 91, no. B6, p. 6213–6224.

Henry, C. D., Price, J. G., and McDowell, F. W., 1983, Presence of the Rio Grande rift in West Texas and Chihuahua, *in* Clark, K. F., and Goodell, P. C., eds., Geology and mineral resources of north-central Chihuahua: El Paso Geological Society, Field Conference Guidebook, p. 108–118.

Henry, C. D., McDowell, F. W., Price, J. G., and Smyth, R. C., 1986, Compilation of potassium-argon ages of Tertiary igneous rocks, Trans-Pecos Texas: The University of Texas at Austin, Bureau of Economic Geology Geological Circular 86-2, 34 p.

Laughlin, A. W., Kress, V. C., and Aldrich, M. J., 1982, K-Ar ages of dike rocks, Big Bend National Park, Texas: Isochron West, no. 35, p. 17–18.

Maxwell, R. A., 1968, The Big Bend of the Rio Grande, a guide to the rocks, geologic history, and settlers of the area of Big Bend National Park: The University of Texas at Austin, Bureau of Economic Geology Guidebook 7, 138 p.

Maxwell, R. A., and Dietrich, J. W., 1972, Geology of the Big Bend area, Texas; Field trip guidebook with road log and papers on natural history of the area: West Texas Geological Society Publication 72-59, 196 p.

Maxwell, R. A., Lonsdale, J. T., Hazzard, R. T., and Wilson, J. A., 1967, Geology of Big Bend National Park, Brewster County, Texas: University of Texas Publication 6711, 320 p.

Moustafa, A. R., 1983, Analysis of Laramide and younger deformation of a segment of the Big Bend region, Texas [Ph.D. thesis]: The University of Texas at Austin, 157 p.

Muehlberger, W. R., 1980, Texas Lineament; Revisited, *in* Dickerson, P. W., and Hoffer, J. M., eds., Trans-Pecos region, southeastern New Mexico and West Texas: New Mexico Geological Society 31st Field Conference Guidebook, p. 113–121.

Muehlberger, W. R., Belcher, R. C., and Goetz, L. K., 1978, Quaternary faulting in Trans-Pecos Texas: Geology, v. 6, p. 337–340.

Ogley, D. S., 1979, Eruptive history of the Pine Canyon caldera, Big Bend National Park, *in* Walton, A. W., and Henry, C. D., eds., Cenozoic geology of the Trans-Pecos volcanic field of Texas: The University of Texas at Austin, Bureau of Economic Geology Guidebook 19, p. 67–71.

Price, J. G., and Henry, C. D., 1984, Stress orientations during Oligocene volcanism in Trans-Pecos Texas; Timing the transition from Laramide compression to Basin and Range extension: Geology, v. 12, p. 238–241.

Price, J. G., Henry, C. D., Standen, A. R., and Posey, J. S., 1985, Origin of silver-copper-lead deposits in red-bed sequences of Trans-Pecos Texas; Tertiary mineralization in Precambrian, Permian, and Cretaceous sandstones: The University of Texas at Austin, Bureau of Economic Geology Report of Investigations No. 145, 65 p.

Price, J. G., Henry, C. D., Parker, D. F., and Barker, D. S., eds., 1986, Igneous geology of Trans-Pecos Texas: The University of Texas at Austin, Bureau of Economic Geology Guidebook 23, 360 p.

Price, J. G., Henry, C. D., Barker, D. S., and Parker, D. F., 1987, Alkalic rocks of contrasting tectonic settings in Trans-Pecos Texas: Geological Society of America Special Paper 215 (in press).

Seager, W. R., 1980, Quaternary fault system in the Tularosa and Hueco Basins, southern New Mexico and West Texas, *in* Dickerson, P. W., and Hoffer, J. M., eds., Trans-Pecos region, southeastern New Mexico and West Texas: New Mexico Geological Society 31st Field Conference Guidebook, p. 131–135.

Wilson, J. A., 1971, Vertebrate biostratigraphy of Trans-Pecos Texas and northern Mexico, *in* Seewald, K., and Sundeen, D., eds., The geologic framework of the Chihuahua tectonic belt: West Texas Geological Society, p. 157–166.

Zoback, M. L., Anderson, R. E., and Thompson, G. A., 1981, Cainozoic evolution of the state of stress and style of tectonism of the Basin and Range province of the western United States: Philosophical Transactions of the Royal Society of London, v. 300, p. 407–434.

ACKNOWLEDGMENTS

Research support was provided by grant number G1154148 from the U.S. Department of Interior, Bureau of Mines, to the Texas Mining and Mineral Resources Research Institute. Publication was authorized by the Director of the Bureau of Economic Geology.

The Anacacho Limestone of southwest Texas

R. W. Rodgers, Physical Science Department, Pan American University, 1201 W. University Drive, Edinburg, Texas 78539

LOCATION

The Anacacho Limestone represents an anomaly in the Gulfian-age rocks of Texas. Like the Austin Chalk that it unconformably overlies, the Anacacho is a limestone in a section dominated by terrigenous clastics and mudrocks; unlike the Austin Chalk (which is extensive throughout the margin of the Gulf Basin), the Anacacho is a small carbonate bank that extends less than 96 mi (160 km) along an east-west trend on the flanks of the Uvalde Salient. The formation thins perceptibly to the east toward San Antonio, where it is replaced by rocks of the Taylor Group. The formation reaches its maximum thickness on the flanks of the Uvalde Salient, and terminates abruptly at the western edge of the Anacacho Mountains. West of this termination the Anacacho is replaced by the Upson and San Miguel Formations.

From the Nueces River, west of the Uvalde Salient, westward to the Anacacho Mountains, the limestone outcrop is continuous for approximately 28 mi (46 km). The width of the outcrop varies from 2.4 mi (4 km) to 9 mi (15 km). East of the Uvalde Salient, outcrops of the Anacacho Limestone are discontinuous and scattered. Much of the section is either covered by alluvium or is lost along faults of the Balcones Fault Zone (Bureau of Economic Geology, 1974). This section, from Blanco Creek eastward to San Antonio, is approximately 60 mi (100 km) in length. However, outcrops in this section are limited in length from a few miles to several hundred feet. Exposures are mostly along the banks and beds of small ephemeral streams.

West of the Uvalde Salient the Anachacho Limestone forms a prominent north-facing cuesta locally called the Anacacho Mountains. To the west in Kinney County, the cuesta and dip slope are dissected by small ephemeral streams that flow southward from the crest. From the Kinney-Uvalde county line eastward, the outcrop pattern of this limestone in the Anacacho Mountains is broken by numerous exposures of igneous rocks that form rounded hills; the most prominent are Asphalt Mountain, Saddleback Mountain, Sulfur Mountain, and Nueces Hill along the Nueces River.

The outcrop area west of the salient contains the most complete sections for field studies. Unfortunately, most of the outcrops are unavailable or have limited access due to ranchers restricting access to their property. To the east of the salient, outcrops are more accessible, but are much less complete.

In addition to outcrops that are readily accessible, the traverse includes two localities which have limited or restricted access; they are important in illustrating the stratigraphy and depositional pattern of the Anacacho.

Stratigraphy

The Anacacho Limestone was first described by Hill and Vaughan (1898, p. 240), who wrote, "In Uvalde and Kinney Counties in the stratigraphic position occupied to the eastward by the Taylor Marls, is a series of hard yellow and white limestones with interbedded marls and occasional sandstone ledges for which the local name Anacacho Formation is proposed, after the locality of their characteristic occurrence, the Anacacho Mountains of Kinney County which are capped by this formation."

Subsequent investigations by Deussen (1924), Stephenson (1927), and Getzendanner (1930) supported the correlation with the Taylor Marls to the east and the Upson–San Miguel beds to the west. The contacts with the underlying Austin Chalk and overlying Escondido Formation were considered unconformable.

In 1956, Hazzard proposed a revision to the stratigraphy of the Anacacho Limestone. He subdivided the formation into three units: a Lower Anacacho Limestone, a Middle Milam Chalk, and an Upper Anacacho Limestone. Additionally, he proposed a new correlation for these units. He placed the Lower Anacacho Limestone with the Austin Group and proposed a conformable contact between the two units. He also correlated this lower Anacacho unit with the Gober Chalk of northeast Texas, based on the ammonites *Menabites* sp. and *Proplacenticeras* sp. (Hazzard, 1956). The contact between the Austin and Taylor Groups was then placed at a disconformity between the Lower Anacacho and the Milam Chalk. The Milam Chalk and Upper Anacacho were then correlated with the Pecan Gap Chalk of northeast Texas, based on the ammonites *Pachydiscus* sp., *Parapachydiscus* sp., and *Bostrychoceras* sp. (Hazzard, 1956). Durham (1957) disagreed with this interpretation; he placed the Austin-Taylor boundary at the contact between the Big House Chalk and the Upson Clay, noting similar disconformable surfaces in sections on the Rio Grande at Tequesquite Creek and in the western edge of the Anacacho Mountains. Harvill (1959), Brown (1965), and Wilson (1984) have all subsequently placed the Austin-Taylor, Santonian-Campanian boundary at the base of the Anacacho Limestone.

It is unfortunate that Hazzard placed the Austin-Taylor boundary at the base of the Milam Chalk, as this unit cannot be traced laterally for any great distance. It appears to have been replaced to the west by limestones and is not found at the western margin of the Anacacho Mountains.

Line of Section

The traverse begins on Texas 1572 approximately 9 mi (15 km) south of the intersection with U.S. 90. The traverse then extends along Texas 1572 and 90 to the Nueces River on the west flank of the Uvalde Salient.

Locality 1—Restricted access—(Figs. 1, 2) (100°18′23″W; 29°10′23″N Spofford Quadrangle) is on Texas 1572 opposite the entrance to Anacacho Ranch. The westernmost exposure of

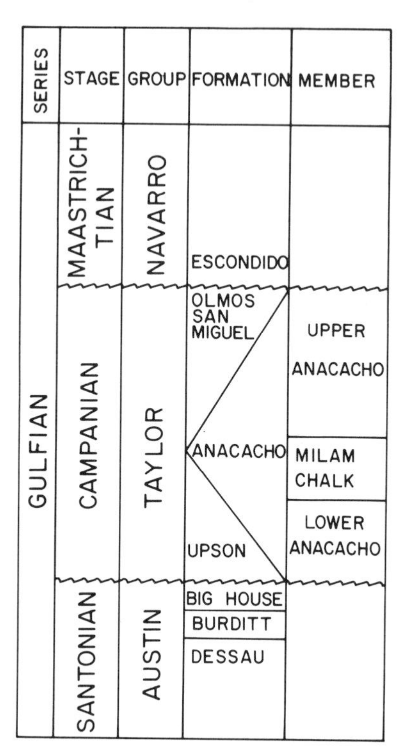

Figure 1. Gulfian stratigraphic nomenclature used in this paper conforms in part to that of the Geologic Atlas of Texas (Bureau of Economic Geology, 1974, 1977).

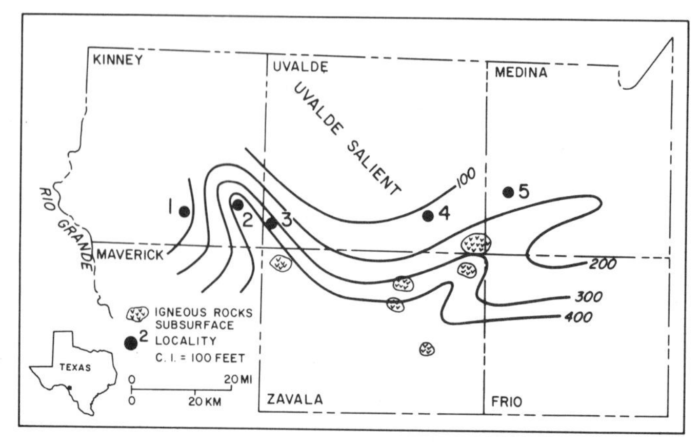

Figure 2. Isopach map of Anacacho Limestone illustrating reentrants on flanks of the Uvalde Salient.

the Anacacho Limestone is exposed along Elm Creek on the western margin of the Anacacho Mountains. Here, " . . . the Anacacho Limestone ends as abruptly on the west as if the Anacacho Sea there had washed against a stationary shoreline throughout the entire period" (Getzendanner, 1930, p. 1431).

The sequence at this locality consists of a broad, flat exposure of the Big House Chalk (Burditt), the uppermost unit of the Austin Group. Unconformably overlying this is approximately 30 ft (10 m) of cream- to brown-colored, nodular and sandy clays of the Upson Formation. Overlying the Upson are 138 ft (45 m) of massive bedded lime mudstones and packstones of the Anacacho Formation. Cross-bedded fossiliferous grainstones containing abundant *Pseudorbitoides israelskyi* form prominent bluffs along the creek. These are accretion cross-beds dipping to the northeast in channels that are cut into the underlying Upson Clay. The channels are on the margin of a reentrant on the southwest flank of the Uvalde Salient. The grainstones rapidly thicken toward the east to more than 300 ft (98 m) at the Kinney-Uvalde county line (Fig. 2).

Locality 2—Restricted access—(Figs. 1, 2) (100°08′30″W; 29°11′20″N Cline Quadrangle) is 3.2 mi (5.3 km) south of U.S. 90 and 1.9 mi (3.16 km) west of the Uvalde-Kinney county line in an east-facing scarp on the Milam Ranch. This is the type locality of the Milam Chalk proposed by Hazzard in 1956. Just to the west of this locality is the area first measured by Hill and Vaughan in 1898.

The broad exposure flat south of U.S. 90 is on the Big House Chalk (Burditt) of the Austin Group. The contact with the lower Anacacho unit is covered, but the lower Anacacho Milam Chalk contact is well exposed. Here, the upper surface of the lower Anacacho contains abundant *Exogyra ponderosa* in a prominent ledge that may be traced for some distance. Overlying this ledge are some 30 ft (10 m) of finely laminated, ashy?, locally cross-bedded lime mudstones of the Milam Chalk. Internal molds of numerous pelecypods are locally abundant in the chalk. The chalk may be traced along the outcrop by the distinctive band of vegetation which forms a dark band separating the lower and upper Anacacho limestone units. The outcrop pattern begins to thin to the west and cannot be traced as a lithologic unit into the western portion of the Anacacho Mountains. The Milam Chalk is overlain by some 240 ft (80 m) of fossiliferous lime packstones and grainstones containing abundant mollusks, echinoids, bryozoans, and foraminifera. Locally there are patches of what appear to be in situ accumulations of these fossils in the top of the Upper Anacacho unit.

Locality 3 (Figs. 1-3) (100°05′40″W; 29°09′40″N Cline Quadrangle) is 8.4 mi (14 km) south of U.S. 90 on Texas 1022. The outcrop is the mine quarry of White's Uvalde Mines at Dabney, Texas. Other abandoned quarries of the Uvalde Rock Asphalt Company may also be visited in this area.

The White's Mine exposes more than 150 ft (48 m) of Anacacho Limestone. Three distinct facies or units may be seen in the mine walls: an upper unit of approximately 36 ft (12 m) of cream-colored recrystallized lime mudstones and packstones that do not contain asphalt, a middle unit approximately 30 ft (10 m) thick containing fossiliferous lime packstones and grainstones that contain asphalt, and a lower unit approximately 90 ft (30 m) thick of coarse-grained fossiliferous lime grainstones containing abundant asphalt. The shell fragments in the lower unit are markedly coarser than those in unit 2 and also appear to increase in coarseness toward the southeast. This would support the hypothesis for sediment distribution by longshore currents moving from the southeast to the northwest (Fig. 2).

The middle unit is finer grained than the underlying thicker unit and contains planar cross-beds. The contact between this and the underlying unit is marked by a uniform layer of gray-green–colored volcanic ash which contains abundant pyrite. The ash layer acts as a barrier to the vertical movement of ground water, which seeps out of the mine walls above this contact. Oxidation of the pyrite results in extensive limonite staining of the lower unit. The base of the middle unit above the ash contains extensive burrows and local accumulations of fossils. One such accumulation consists of numerous molds of a turritellid gastropod. The molds are filled with asphalt and rimmed with pyrite, and the chambers are filled with a micritic mud matrix. Below the gastropod zone is a soft, poorly cemented coquina on which a thin and extensively bored layer of oyster shells is cemented.

The two asphalt-bearing units contain an abundant and varied fauna. In some cases the shells are too fragmented for positive identification. Some shells are whole or only partly broken; in some cases, there are local in situ accumulations. The fauna consists of mollusks, echinoderms, abundant benthonic foraminifera, calcareous red algae, and a few planktonic foraminifera. The molluscan fauna includes abundant pelecypods and gastropods, a few cephalopods, and some rudists including *Monopleura* sp.? and *Toucasia* sp. Unit 2 can be traced laterally to the west in Kinney County. Rocks with the same lithology and identical fauna containing asphalt indicate that these units are not just local accumulations.

Origin of the asphalt has long been an unresolved question, with proponents of in situ formation and petroleum residuum formation presenting arguments. Vaughan (1897) argued that heat from the adjacent basalt intrusions had driven the oil from organic-rich sections to accumulate in the porous limestones. Presenting several lines of reasoning, Utterback (1953) proposed a contemporaneous in situ origin of the asphalt. Harvill (1959) and Wilson (1984) used similar arguments concerning the occurrence of asphalt in secondary pore spaces to support the petroleum residuum hypothesis.

Evidence can be seen to support both hypotheses. "Plumes" of asphalt cutting across bedding planes with barren zones of limestone in between may be seen in the lower unit. Zones of uniform lithology saturated with asphalt adjacent to barren zones of slightly differing lithology and asphalt occupying secondary moldic and intergranular porosity zones would seem to indicate the migration of a liquid that had devolatilized. Yet there are zones, such as the previously mentioned gastropod zone of unit 2, that contain asphalt and that have no apparent interconnecting pore spaces. Studies of total carbon, soluble organic matter, and vitrinite reflectance indicate a migrated crude rather than in situ accumulation.

Locality 4 (Figs. 1-3) (99°32′45″W; 29°19′40″N, Blanco Lake Quadrangle) is 5.1 mi (8.5 km) east of Knippa on U.S. 90, and 0.5 mi (0.9 km) north of the Texas and New Orleans Railroad bridge on the east bank of Blanco Creek. From the bed of Blanco Creek to the top of the hill on the east bank are approximately 120 ft (40 m) of fossiliferous lime grainstones with a thick, interbedded volcanic sequence.

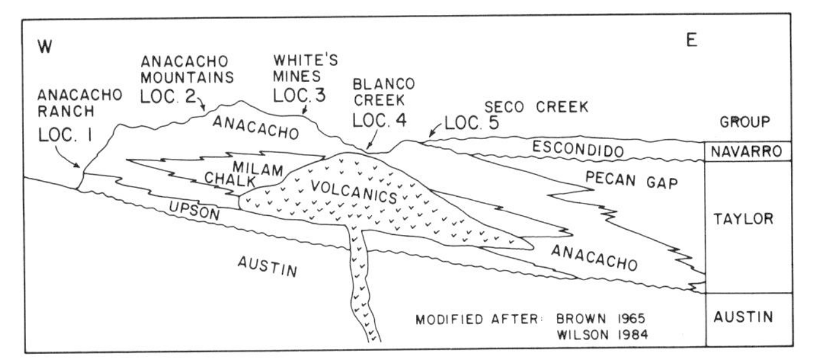

Figure 3. Diagrammatic section of the Anacacho Limestone and associated rocks illustrating approximate relationships at the section localities. Section not drawn to scale.

The lower limestone consists of approximately 22 ft (7 m) of fossiliferous lime grainstones containing abundant echinoids, bryozoan fragments, and benthonic foraminifera. Overlying the limestone is a sequence of evenly bedded gray-green–colored clays which appear to be volcanic in origin. The section thickens from less than 10 ft (3 m) in a section 0.6 mi (1 km) downstream from the bridge to more than 60 ft (20 m) 1 mi (1.5 km) upstream from the bridge. The evenly bedded clays contain abundant fragments and boulders of olivine basalt. Many of the fragments are vesicular, and the boulders are highly fractured. Interbedded with the olivine basalts are well-preserved specimens of *Exogyra ponderosa.* The section appears to be a volcanic mudflow associated with one of the nearby igneous masses.

Overlying the volcanic section are approximately 75 ft (25 m) of massive-bedded fossiliferous lime packstones and grainstones containing abundant foraminifera and bryozoan fragments. The limestone contains some asphalt filling intergranular pore spaces.

Zones of igneous rocks such as those seen on Blanco Creek have been encountered in numerous test holes in Uvalde and Medina Counties. A test hole near White's Mines encountered five zones ranging in thickness from 10 to 78 ft (3 to 26 m). Tests drilled by Shell Development Company penetrated zones of bedded volcanics and pyroclastic material, weathered basalts, and volcanic ash layers. To the south in Dimmit and Zavala counties, volcanic activity formed cones that grew to sea level. Shallow water on the flanks of the cones favored rudist reef development, which, with other marine fauna, supplied abundant shell material for reworking into shoals (Luttrell, 1977). The Project Mohole test well in southern Uvalde County encountered asphalt-bearing coarse grainstones and fractured basalts containing highly viscous oil.

All localities where asphalt has been reported in the Anacacho Limestone are within the arcuate belt of igneous rocks that rims the Uvalde Salient. Asphalt does not occur in the Anacacho Ranch area to the west in Kinney County, and is limited or does not occur to the east in Bexar County.

Locality 5 (Figs. 1-3) (99°17′00″W; 29°21′30″N D'Hanis Quadrangle) is 1.7 mi (2.7 km) north of D'Hanis along Texas

1726 on the right (west) bank of Seco Creek. This is also the section for U.S. Geological Survey Mesozoic Localities 7693 and 7694. Approximately 1 mi (1.6 km) upstream from this locality on the west bank of Seco Creek, the Anacacho Limestone is disconformably overlain by the Escondido Formation. This is the section for USGS Localities 7695 and 12909.

The first section consists of approximately 40 ft (13 m) of massive- to thin-bedded fossiliferous lime grainstones grading into massive- to thin-bedded packstones and lime mudstones. Biostromes containing abundant *Pycnodonte (Gryphaea) mutabilis* Morton and other mollusks are interbedded within the grainstones and marls. *Exogyra ponderosa* is abundant in the upper massive-bedded lime mudstone. The grainstones contain abundant asphalt occupying both moldic and intergranular pore spaces. Downstream to the south in Seco Creek is a section of approximately 30 ft (10 m) of hard, massive-bedded grainstones that may be Anacacho, but which Brown (1965) suggested may contain a fauna older than the lower Anacacho. This section is also slightly asphaltic.

The second section upstream on Seco Creek has restricted access, but contains one of the best exposed contacts between Upper Cretaceous beds in the Gulf Coastal Plain (Brown, 1965). Approximately 26 ft (8.5 m) of thinly bedded fossiliferous grainstones and mudstones grade upward into thinly bedded, slightly arenaceous limestone at the top of the Anacacho Limestone. These beds are the type locality for *Bostrychoceras secoense* Young and the large echinoid *Echinocorys texana* (Cragin). Disconformably overlying the Anacacho are approximately 30 ft (10 m) of yellow to brown, interbedded, silty shales and marls of the Escondido (Corsicana) Formation.

If Brown (1965) is correct in his interpretation of the Austin below the disconformity, the total Anachacho thickness at this locality is 66 ft (22 m). If the lower section is considered part of the Anacacho, the total section is less than 96 ft (32.5 m).

The traverse ends at this point as the Anacacho continues to thin rapidly toward the east (Fig. 2). Outcrops of the Anacacho Limestone east of Locality 5 are small and difficult to locate. Outcrops of a thinned Anacacho section are present in the quarries of the Longhorn and Alamo cement companies on the northeast side of San Antonio. Access to these outcrops, however, is very restricted. From the area of San Antonio on to the northeast the stratigraphic position of the Anacacho is mapped as the Pecan Gap Chalk of the Taylor Marls.

SUMMARY

The Anachacho Limestone represents a minor transgressive pulse in an otherwise regressive or progradational sequence during Gulfian time. The massive build-up of carbonate material flanking the Uvalde Salient is an anomaly in a sequence dominated by terrigenous clastic sediments.

The igneous rocks forming an arcuate band around the Uvalde Salient provided the shallow-water environment necessary for the development of marine organisms that supplied abundant shell material for reworking into shoals. Longshore currents transported this shell material to form shoals in reentrants flanking the salient (Fig. 2). Shell material was probably not transported much farther than the area of the present outcrop. Numerous basin margin disconformities, cross-bedding, and thinning of the section to the west and east indicate the local limited source of sediments for the Anacacho.

Igneous activity reached a peak during Anacacho deposition, forming a sea-floor topography that allowed rudist reef development with abundant associated marine organisms. This build-up was sufficient to create a lagoon shoreward of the reef facies in which the Milam Chalk was deposited. With the waning of volcanic activity, the shell material around the build-up was disbursed by longshore currents, bringing Anacacho Limestone deposition to a close.

REFERENCES CITED

Brown, N. K., 1965, Stratigraphy of upper Cretaceous beds in the vicinity of D'Hanis, Medina County, Texas, *in* Upper Cretaceous asphalt deposits of the Rio Grande embayment: Corpus Christi Geological Society Guidebook, p. 23–30.

Bureau of Economic Geology, 1974, Geologic atlas of Texas, San Antonio Sheet: University of Texas at Austin, Bureau of Economic Geology, 7 p. and map, scale 1:250,000.

——, 1977, Geologic atlas of Texas, Del Rio Sheet: University of Texas at Austin, Bureau of Economic Geology, map, scale 1:250,000.

Deussen, A., 1924, Geology of the coastal plain of Texas west of the Brazos River: U.S. Geological Survey Professional Paper 126.

Durham, C. O., Jr., 1957, The Austin Group in central Texas [Ph.D. thesis]: New York, Columbia University, Department of Geology, 130 p.

Getzendanner, F. M., 1930, Geological section of the Rio Grande Embayment and implied history: American Association of Petroleum Geologists Bulletin, v. 14, p. 1425–1437.

Harvill, L. L., 1959, Petrology of the Anacacho Limestone of southwest Texas: Gulf Coast Association of Geological Societies Transactions, v. 9, p. 161–165.

Hazzard, R. T., 1956, Cretaceous rocks south of Tarpley to Del Rio, *in* Four provinces field trip: San Angelo Geological Society Field Trip Guidebook, pt. III, p. 43–72.

Hill, R. T., and Vaughan, T. W., 1898, Geology of the Edwards Plateau and Rio Grande Plain adjacent to Austin and San Antonio, Texas: U.S. Geological Survey, 18th Annual Report, p. 199–321.

Luttrell, P. E., 1977, Carbonate facies distribution and diagenesis associated with volcanic cones in the Anacacho Limestone (Upper Cretaceous), Elaine Field, Dimmit (and Zavala) County, Texas: Bureau of Economic Geology Report of Investigations no. 89, p. 260.

Stephenson, L. W., 1927, Notes on the stratigraphy of the Upper Cretaceous formations of Texas and Arkansas: American Association of Petroleum Geologists Bulletin, v. 11, p. 1–17.

Utterback, D. D., 1953, Information on the origin of oil as obtained from the study of some asphaltic limestones: Gulf Coast Association of Geological Societies Transactions, v. 3, p. 115–126.

Vaughan, T. W., 1897, The asphalt deposits of western Texas: U.S. Geological Survey, 18th Annual Report, Pt. V, p. 929–935.

Wilson, D. H., 1984, Geology of the Anacacho Formation in southwest Texas, *in* Wilson, W. F., and Wilson, D. H., Meteor impact site, Anacacho asphalt deposits: South Texas Geological Society Field Trip Guidebook, p. 60–98.

The following maps have been used for plotting field localities: Blanco Lake, 1971, 1:24,000; Cline, 1949, 1:62,500; D'Hanis, 1970, 1:24,000; Spofford, 1942, 1:62,500.

Geological Society of America Centennial Field Guide—South-Central Section, 1988

Late Quaternary geology of the Texas coastal plain

Robert A. Morton, *Bureau of Economic Geology, The University of Texas at Austin, University Station, Box X, Austin, Texas 78713-7508*

INTRODUCTION

Modern depositional environments prominently displayed along the northwestern Gulf of Mexico served as the first widely accepted sedimentary facies models that have successfully guided exploration and production of a variety of energy and nonfuel resources (petroleum, lignite, uranium, geothermal energy, construction aggregate, ground water). Because these industrial activities both required and provided a tremendous data base, the depositional environments of the gulf coast are among the best known anywhere. The following sections of the guidebook describe selected geological features that typify late Pleistocene and Holocene sediments of the Texas coastal plain and illustrate the diversity of associated depositional environments. The described sites represent a spectrum of transgressive and regressive sequences coexisting along a microtidal storm-dominated coast undergoing a significant reduction in sediment supply and a slight sea-level rise.

Late Quaternary history. The Texas coastal plain is a broad, flat, depositional surface created by several rivers that eroded large volumes of sediment from remote areas of the state and adjacent mountainous regions and deposited the sediment as coalescing deltas in the Gulf of Mexico (Barton, 1930; Doering, 1935; Winker, 1979). It extends eastward and merges with the coastal plain of Louisiana, which has a similar morphology but a slightly different origin. Rather than being formed by several rivers and deltas of moderate size as was the Texas coast, the Louisiana coastal plain was mostly deposited by two large alluvial systems (Mississippi and Red rivers) that drained much of the continental interior.

Like present-day rivers, the load and discharge of these late Pleistocene streams were controlled by climate, drainage basin size, and composition of sediments they traversed. Nearly all Texas rivers continue to carry multicycle fine-grained sediments derived from Tertiary and Mesozoic rocks that crop out in their drainage basins. As a result of sediment recycling, coastal plain deposits are dominantly composed of mud with some concentrations of fine sand. The sand deposits represent depositional environments exposed to relatively high energy such as fluvial channels, barrier islands, and strandplains.

Fluctuations in sea level related to glacial and interglacial stages during the past 120,000 years have been chiefly responsible for shaping the Texas coastal plain. Although absolute ages of coastal plain deposits are unknown, the youngest Pleistocene depositional surface is assumed to be Sangamonian in age because that was the interglacial highstand of sea level that preceded the current highstand. These late Pleistocene deposits, known in Texas as the Beaumont Formation (Figs. 1–3), extend coastwise in a band that parallels the shoreline from Louisiana to Mexico.

About 90,000 years ago, as continental glaciers expanded, sea level began falling, and major rivers and coastal-plain streams began eroding valleys in response to lowered base level. The former continental shelf was exposed, subaerially eroded, and a soil profile formed on sediments that were subsequently deposited on the broadened coastal plain during the early and late Wisconsinan sea-level lowstands.

The Wisconsinan glacial stage was characterized by cooler temperatures and greater precipitation than today. Enlarged drainage basins and integrated drainage networks of the major coastal-plain streams combined with higher rainfall to form several large rivers that dominated the landscape. Increased runoff from the exposed uplands and increased erosion from the entrenched rivers provided abundant sediment for prograding the shelf margin by several major deltas (Suter and Berryhill, 1985). These deltas were subsequently abandoned and transgressed about 18,000 years ago as meltwaters from retreating glaciers caused a rapid rise in sea level. The post-glacial rise in sea level inundated the entrenched valleys with marine water causing a progressive landward retreat of the rivers, deltas, and adjacent shoreline.

Most coastal-plain rivers were unable to keep pace with the rapidly rising sea, and their valleys eventually became estuaries that aggraded mainly with bay center and bay margin mud and shell deposits. When the rate of sea-level rise diminished about 5,000 years ago, these rivers built small bayhead deltas that were confined by the entrenched valley walls. The largest rivers (Brazos–Colorado and Rio Grande; Figs. 1 and 3) filled their estuaries and constructed large oceanic deltas that prograded onto the inner continental shelf about 25 mi (40 km) seaward of the present shoreline (Price, 1958). Most of the remaining Texas Gulf shoreline was characterized by eroding mainland beaches or migrating transgressive barriers that were transformed into regressive barriers when sea level reached its present position.

Average sea level has remained nearly constant for the past few thousand years, although it may have oscillated within a range of a few meters. During that same period, however, the Gulf shoreline has undergone a remarkable change in morphology and position. A warmer and drier Holocene climate with resulting decreased sediment supply caused the large oceanic deltas to retreat. Attendant erosion of the delta shorelines and compaction of the delta plains contributed to the formation of young lagoons and transgressive barrier islands on the delta flanks (i.e., Laguna Madre and South Padre Island; Fig. 3).

Between the large Holocene deltas, the older transgressive barriers began to prograde (Fig. 2) under the influence of stable sea level and abundant sediment supply. Barrier progradation was

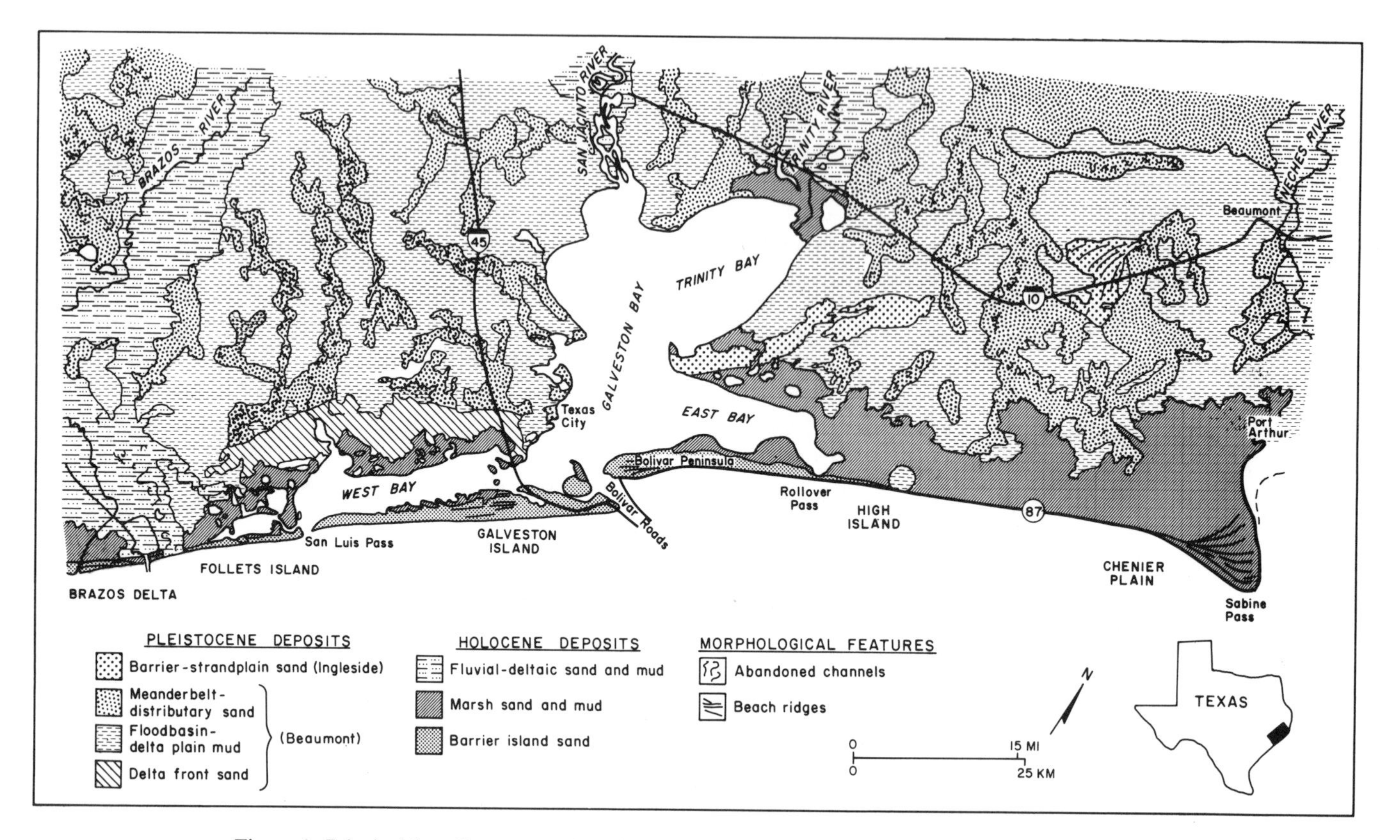

Figure 1. Principal late Quaternary depositional environments of the upper Texas coast. Geology is generalized from the map of Fisher and others (1972).

supplied by longshore transport of wave reworked fluvial sediment, updrift shoreline erosion, and erosion of underlying Pleistocene barrier-strandplain sands as the shoreface and inner shelf adjusted to an equilibrium profile. East of Galveston Bay (Fig. 1), a promontory composed of Pleistocene sediments continued to erode because of moderate wave energy and a deficit in sediment supply, while the adjacent beach near Sabine Pass prograded as a series of beach ridges and intervening swales. Natural reductions in sediment supply during the late Holocene have been further aggravated by recent human activities such as damming rivers and constructing jetties. The flood-control structures cause rivers to lose their sediment transporting capacity, whereas the jetties disrupt longshore drift and compartmentalize the coast, preventing sediment movement from one segment to another. As a result, some coastal segments that were formerly accreting, or were at least stable, have begun to erode.

Structural elements. The principal structural elements controlling positions of large-scale depositional systems in the western Gulf Basin throughout the Tertiary period have continued to influence sedimentation patterns during the Quaternary period. The largest river and delta systems occupied the more rapidly subsiding Houston and Rio Grande embayments of the upper and lower Texas coast (Figs. 1 and 3), while smaller deltas and wave-dominated shorelines were located over the more stable San Marcos Arch of the central coast (Fig. 2).

Tectonic deformation of the coastal plain is a slow process acting nearly uniformly to uplift inland areas and to cause seaward tilting of the lower coastal plain and continental shelf (Doering, 1935). Minor (<5 m) structural warping of late Pleistocene sediments along the coast has occurred at much lower rates than either sea-level fluctuations or dewatering and compaction of the sediments themselves.

Climate. The Texas coastal climate, which ranges from subhumid to semiarid, is greatly influenced by the systematic westward decrease in precipitation and increase in evapotranspiration. This strong climatic gradient has a dramatic affect on biological assemblages, landforms, and physical processes. For example, the average annual rainfall for the upper coast is 50 in (125 cm); as a result, nutrient-rich freshwater inflow to the bays supports lush saltwater marshes (Fig. 1) and large oyster reefs, but the higher rainfall also minimizes the influence of eolian processes. In far-south Texas, where evapotranspiration generally exceeds precipitation, coastal waters commonly have abnormally high salinites, and saltwater marshes represent only minor bay-margin environments; moreover, large active dune fields and wind-tidal flats dominate the nearshore landscape (Fig. 3).

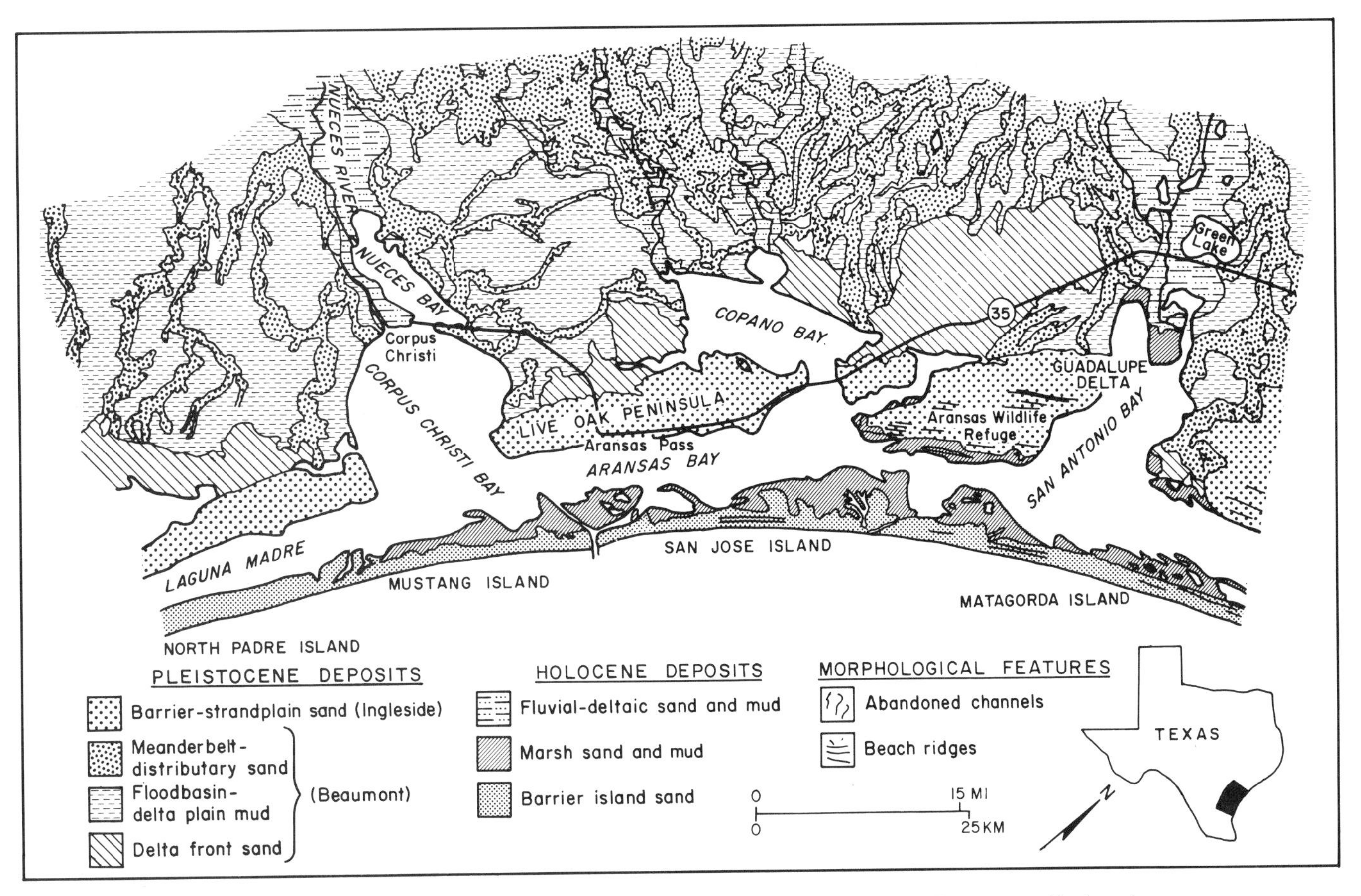

Figure 2. Principal late Quaternary depositional environments of the middle Texas coast. Geology is generalized from the map of McGowen and others (1976).

The dominant wind systems also affect the surficial distribution of coastal plain sediments. Most of the year (spring, summer, and fall) the wind blows from the southeast, but during the winter, strong northerly winds accompany the passage of cold polar air masses that can suddenly lower the temperature, rapidly change wind direction, and greatly increase sea state. Powerful wind-driven currents associated with hurricanes can also redistribute substantial quantities of surface sediment. These tropical cyclones originate at low latitudes in the Atlantic, travel westward, and recurve northward following paths that frequently cross the Texas coast. During the late Quaternary, extreme storms undoubtedly played an important role in reworking and redepositing sediment originally delivered to the Gulf by rivers.

Physiography. The surface of the late Pleistocene Beaumont Formation slopes gently seaward at 0.2 to 0.4 m/km. Vertical exposures of these sediments are rare because relief is extremely low, and dissection is not advanced except near major streams. Despite the lack of outcrops, sedimentary facies can be interpreted by the morphogenic features preserved on delta plain and alluvial plain surfaces. Natural levees, abandoned channels, and point-bar scrolls are evidence of the large mixed-load and suspended-load fluvial systems that constructed the coastal plain as they flowed to the Gulf. In addition, ridge and swale topography and sandy soils are ample proof of wave-dominated beaches and barriers that once formed the Gulf shoreline seaward of, or adjacent to, the mud-rich deltas.

Holocene alluvial-deltaic plains have depositional surfaces similar to the Beaumont surface except for a slightly lower regional dip of about 0.2 m/km. Also, they are primarily composed of fluvial-deltaic sand and mud that retain much of their original depositional relief and surficial texture. Soils observed in roadside ditches and cultivated fields all along the coast directly reflect the composition of underlying Pleistocene and Holocene sediments. Some types of vegetation also can be used to interpret sediment composition. For example, mesquite trees grow on muddy substrates, whereas live oak trees prefer sandy, well-drained soils. Furthermore, the sandiest soils are commonly uncultivated because they do not retain enough moisture to support crops.

The coastal plain is used primarily for ranching and agriculture. Important crops include feed stocks (milo, sorghum, corn), rice, sugar cane, and cotton. In addition to these products, citrus fruit and vegetables are grown using irrigation methods in the Rio

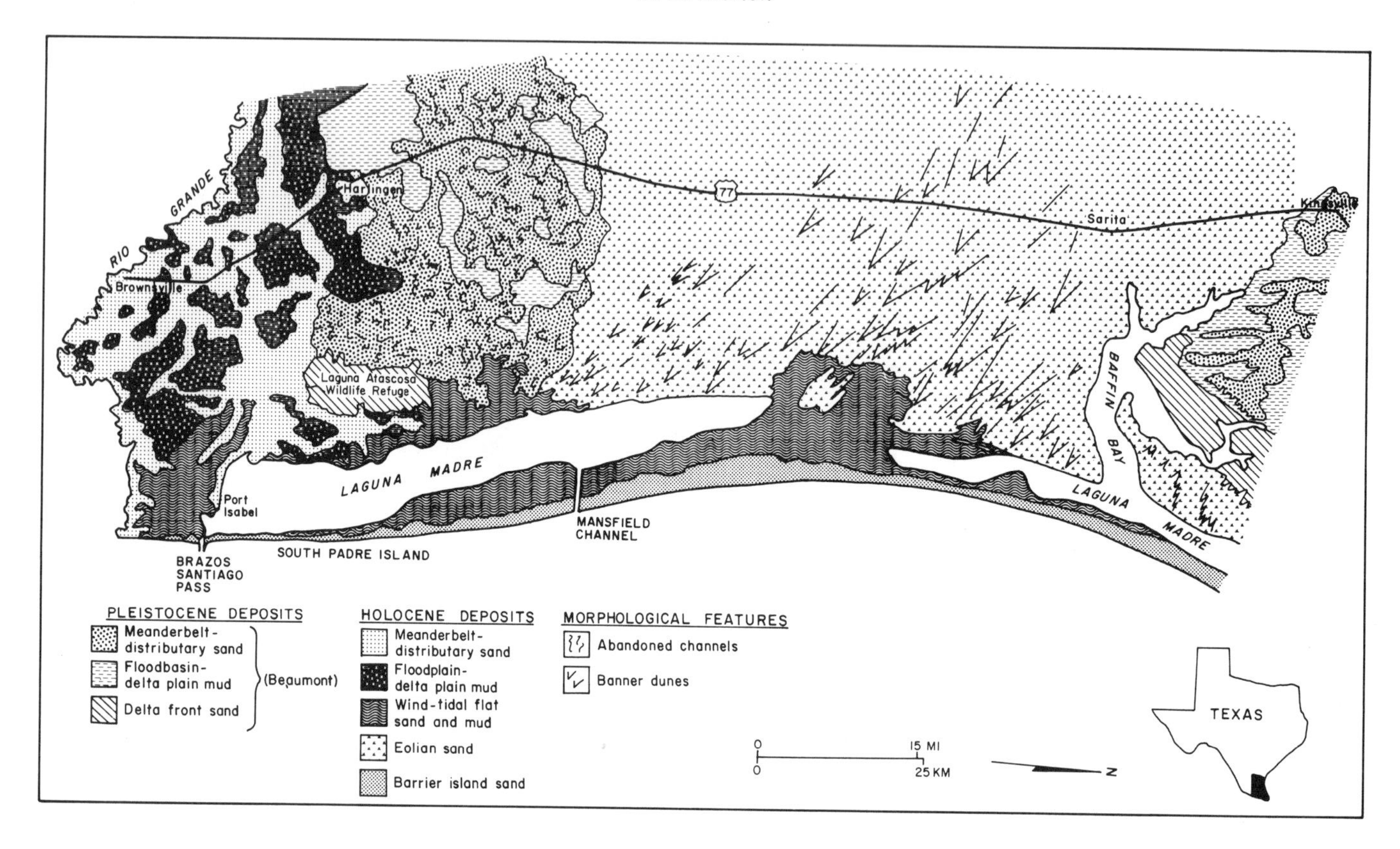

Figure 3. Principal late Quaternary depositional environments of the lower Texas coast. Geology is generalized from the maps of Brown and others (1977, 1980).

Grande Valley. Other land uses important to the area's commerce are oil and gas facilities, petrochemical plants, and numerous industrial sites relying on abundant supplies of energy and access to international shipping lanes.

SITE 98. GEOLOGY OF THE UPPER TEXAS COAST

In several ways the coastal plain of southeastern Texas (Fig. 1) is more similar to southwestern Louisiana than other coastal segments in Texas. It primarily consists of a broad expanse of marsh that merges with a chenier plain near the Gulf. Furthermore, it is one of the few areas where 1) shallow piercement salt domes punctuate the landscape, and 2) the Gulf shoreline is an eroding mainland beach.

Both I-10 and Texas 87 cross this region in an east–west direction (Fig. 1). The inland route (I-10) traverses Pleistocene sand and mud of fluvial–deltaic and strandplain origin that is densely forested in some places. Road traffic along this route is heavy and travels at high speed, preventing leisurely viewing of the monotonous landscape. Texas 87, on the other hand, offers a good view of the coastal plain and a free ferry ride to or from Galveston. The road parallels the Gulf, and along some stretches it coincides with the Gulf shoreline. When tides are abnormally high, this road floods and is temporarily closed to traffic.

The Chenier Plain. About 2 mi (3.2 km) southwest of the town of Sabine Pass (93°55′00″W, 29°42′35″N, Sabine Pass, Tex.–La. Quadrangle), Texas 87 follows a topographic high that is the narrow crest of an old beach ridge (Fig. 1). Marsh vegetation grows on the low, poorly drained, muddy substrates that parallel the beach ridge. Subtle changes in elevation and vegetation on both sides of the road delineate other beach ridges and intervening swales that comprise the chenier plain. The ridges rise about 6 to 10 ft (2 to 3 m) above sea level and are composed mainly of sand and shell. They follow arcuate paths converging to the southwest toward a common inflection point that marks the boundary between accreting shorelines and eroding shorelines over the past few thousand years.

Gould and McFarlan (1959) used shallow borings and radiocarbon dates to interpret the sedimentary facies and stratigraphic history of the chenier plain. They show a transgressive and regressive wedge of sediments overlying a soil zone that is also the Pleistocene-Holocene unconformity. The wedge thickens from 10 to 30 ft (3 to 10 m) and is progressively younger in a seaward direction. Vertical sequences within the wedge vary slightly, but normally consist of basal and upper layers of marsh or bay mud separated by intermediate layers of shoreface sand and mud. Shoreface deposits either grade upward into chenier sand and shell or are overlain by bay and tidal flat sand and mud. A thin but extensive layer of modern organic-rich mud (marsh) caps the sequence.

Shoreline composition and rate of seaward growth of the chenier plain depended on energy of Gulf waves and fluctuations in sediment supply. Shallow-water mudflats rapidly prograded when distributaries of the Mississippi Delta were located in southwestern Louisiana, but when those distributaries were abandoned, Gulf waves reworked the mudflats, concentrated the coarsest material (sand and shell), and deposited the beach ridges. Periodic repetition of these processes accounts for the alternating beach ridge and mudflat morphology. The chenier plain complex was prograding 10 to 15 ft/yr (3 to 4.5 m/yr) from 2,800 to about 250 years ago (Gould and McFarlan, 1959). During the past few hundred years, natural changes in sediment supply and human activities (jetty construction at Sabine Pass, channel maintenance of the Mississippi River) have caused both accretion and subsequent erosion of the modern marsh.

Erosional Gulf shoreline. Like many shorelines of the world, the upper Texas coast is eroding near Texas 87 (Fig. 1, 94°17′30″W, 29°35′00″N. Mud Lake Quadrangle) because sea level is rising relative to the land surface, and sediment supplied to the beach by longshore currents is less than sediment removed by wave energy (Morton and others, 1983). Either of these factors is capable of causing shoreline retreat, but together they produce average long-term erosion rates of more than 10 ft/yr (3 m/yr). Relative sea-level rise in the western Gulf Basin is primarily associated with compactional subsidence of young clastic sediments (Swanson and Thurlow, 1973), but there also is convincing evidence that global sea level is increasing. The volume of sand in the littoral drift system is small because nearby rivers empty into estuaries that trap the coarse fraction and prevent it from reaching the beach. Furthermore, the eroding Pleistocene delta-plain sediments contain little sand for beach nourishment.

Along this coastal segment the beach is composed of a thin veneer of highly mobile sand and shell overlying the eroded Holocene marsh mud. Driving on such a beach can be risky after the backbeach has been flooded and the underlying muds are saturated. For this reason, the road was built on the thicker and higher washover terrace, which provides a stable base for light traffic. Future road construction, necessitated by continued erosion, will be expensive because the washover terrace has been prevented from migrating landward with the beach. Consequently, the adjacent water-saturated marsh will not support road construction without hauling in additional material and building up the roadbed.

Texas 87 was relocated in 1930 because the road was destroyed by beach erosion. In addition to old road pavement cropping out on the beach, other signs of erosion include: a narrow, steep beach, absence of high sand dunes, and high concentrations of oyster shells, rock fragments, and caliche nodules. This gravel-size fraction is a lag deposit eroded from Pleistocene deltaic sediments exposed on the inner shelf. Other field evidence of erosion includes exposed foundations or septic tanks of beach houses and presence of oil field production equipment on the beach.

High Island salt dome. High Island (Fig. 1) is a small town located on a salt dome (94°23′00″W, 29°33′45″W; High Island Quadrangle) that rises about 25 ft (7.5 m) above the surrounding coastal plain. It can be reached by driving on Texas 87 to FM 124 and then north about 1 mi (1.6 km). The unusually high elevation protects the local community from hurricane flooding that periodically inundates the coast. Shallow piercement domes of thick deformed salt are common structural features in the Houston Embayment, but few domes beneath the coastal plain have undergone such recent downbuilding that they deform Pleistocene sediments and have prominent surficial expression. The hill created by the dome has steep slopes on its flanks and several sinkholes near the crest containing fresh water. At High Island, the caprock lies about 200 ft (60 m) below the surface.

It was the anomalous relief that got the attention of wildcatters who first drilled the dome for oil in 1901 (Halbouty, 1936). More than 132 millions bbls of oil and 350 bcf of gas have been recovered since production began in 1922. Similar salt domes near High Island (West Hackberry, Big Hill, Bryan Mound) are currently used to store extremely large volumes of oil as part of our nation's Strategic Petroleum Reserve. Huge underground caverns, created for this purpose by salt dissolution, provide impervious containers for the oil, which is pumped in and monitored using on-site facilities.

Bolivar Peninsula, a Holocene progradational barrier. Beginning at Rollover Pass (94°30′00″W, 29°30′30″N, Frozen Point Quadrangle) and extending along Texas 87 to its termination at Bolivar Roads (94°45′30″W, 29°21′00″N, Galveston Quadrangle), Bolivar Peninsula (Fig. 1) exhibits characteristics of many Holocene barriers that prograded seaward several thousand years ago after sea level reached its current position (Morton and McGowen, 1980). Barrier subenvironments are defined on the basis of morphology, biological assemblages, and active processes depending on the most important variable at present. A reason for using this classification scheme is that some subenvironments are relict features of a transgressive barrier that was associated with the rising-sea-level phase preceding the present stillstand. For example, large, fan-shaped expanses of backbarrier marsh with narrow, radiating channels were once active tidal delta/washover complexes. The abandoned fans are locally subsiding and eroding because sediment is no longer being deposited on the fan surface. Most of the subsidence is natural, but some of it near the community of Caplen (just west of Rollover Pass; Fig. 1) has been induced by deep hydrocarbon production. Prolonged fluid withdrawal and attendant pore-pressure decline have reactivated a growth fault, causing displacement of about 3 ft (1 m) at the surface.

Subenvironments encountered going across Bolivar Penin-

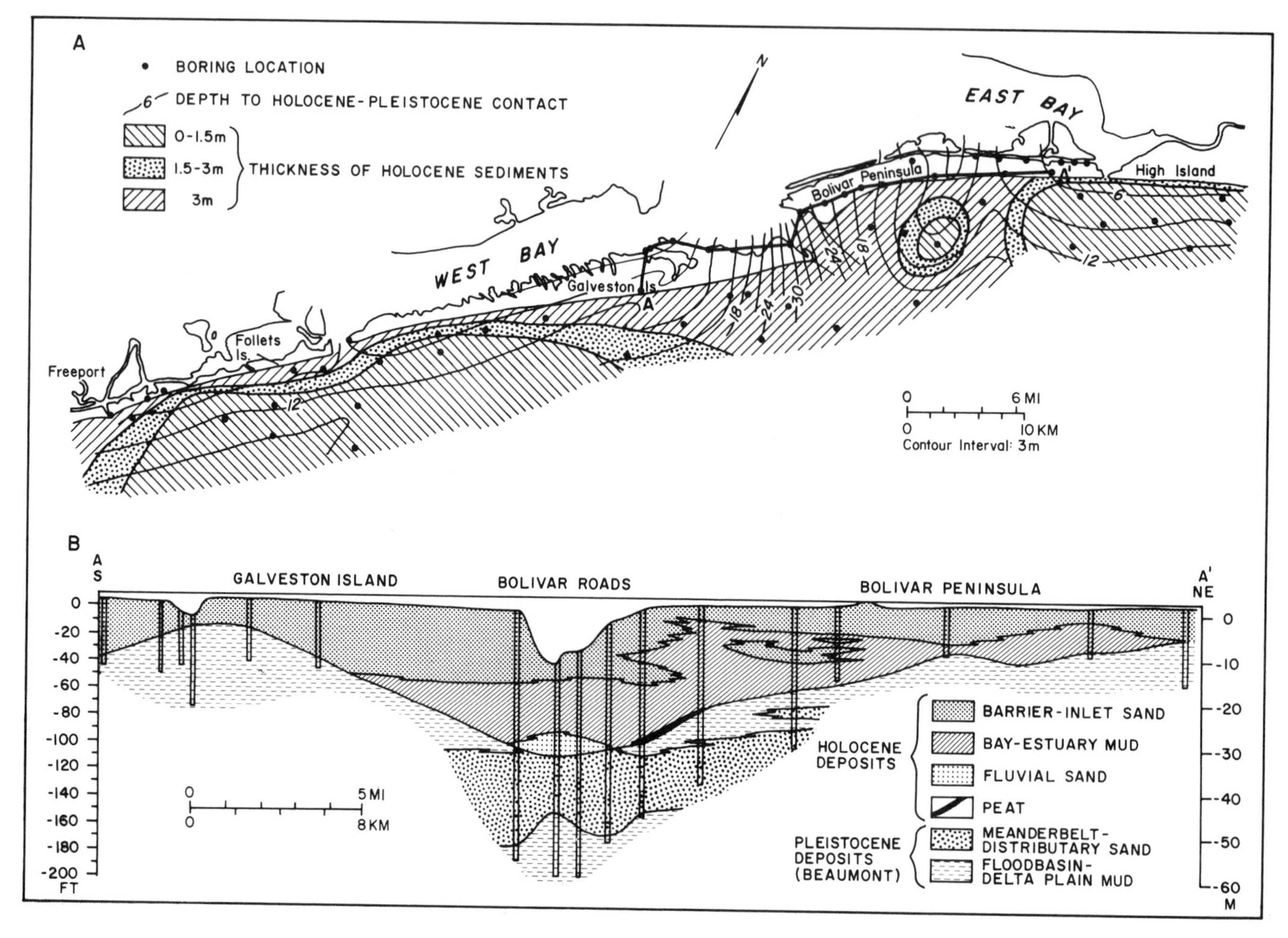

Figure 4. A. Isopach map and B. stratigraphic cross section showing the late Quaternary depositional sequences of the upper Texas coast composed of fluvial–deltaic, aggradational valley fill and regressive barrier-island sediments. General location shown in Figure 1. From Morton and Nummedal (1982).

sula from the bay to the Gulf are: bay margin shoals, backbarrier marsh, vegetated barrier flat, foreisland dunes, beach, and shoreface. Most of these units maintain a fairly uniform width along the barrier except for the vegetated barrier flat, which widens to the southwest. The grass-covered flat also has the most relief, due to the ridge and swale topography that reflects former positions of shorelines and storm berms.

The barrier is a relatively high (10 ft; 3 m) and broad (2 mi; 3.2 km) sandy lithosome bracketed between muddy marine sediments of the inner continental shelf and bay. The adjoining bay was partially restricted to tidal exchange before Rollover Pass was artificially opened in 1955. Before then, Bolivar Peninsula was attached to the mainland and grew seaward and to the southwest in the direction of net littoral drift. Consequently, the barrier is wider and thicker toward the natural tidal inlet at Bolivar Roads. Thickness of the barrier sand (Fig. 4) ranges from 15 to 30 ft (5 to 10 m) except at Bolivar Roads where sand reaches a depth of about 60 ft (18 m). Bolivar Peninsula and eastern Galveston Island both overlie Holocene muds deposited in estuarine and open-bay environments.

Galveston Bay, a flooded Pleistocene valley. The ferry traveling between Bolivar Peninsula and Galveston Island crosses the lower reaches of Galveston Bay and Bolivar Roads (94°45′30″W, 29°21′00″N, Galveston Quadrangle), which is the natural tidal inlet connecting the bay and Gulf of Mexico (Fig. 1). Since the late 1800s, this inlet has been dredged deeper and stabilized with jetties to allow entrance of deep-draft ships traveling to the ports of Galveston and Houston. Beneath the inlet and bay waters are a variety of late Pleistocene and Holocene sediments (Fig. 4) deposited during the late Wisconsinan sea-level lowstand and subsequent post-glacial rising sea level and highstand.

The oldest sediments penetrated beneath the barriers are fluvial sand and deltaic mud of the Beaumont Formation (Fig. 4).

An erosional unconformity separates these over-consolidated regressive sediments from the overlying regressive-transgressive sequence. Fluvial sands of the ancestral Trinity River occupy the base of the erosional valley, which is about 100 ft (30 m) below sea level. These regressive deposits grade upward into shelly and sandy muds that originated in estuarine and bay environments as the rising sea flooded the former entrenched valley. The modern barriers mainly represent the regressive depositional phase that began when sea level became stable.

Follets Island, a narrowing barrier. Compared to the higher and wider barriers to the east, Follets Island (Fig. 1, 95°08′00″W, 29°02′30″N, Christmas Point Quadrangle) is a narrow, low-relief barrier that is narrowing as both the gulf and bay shorelines erode. Evidence of the erosion can be seen as one travels between San Luis Pass and Freeport on Texas 3004. Island elevations are generally less than 5 ft (1.5 m) above sea level, and island width generally decreases from 0.5 mi (800 m) to 0.2 mi (300 m) in a southwesterly direction. Although the barrier is frequently inundated by storm surges, it is not migrating in a landward direction like a transgressive barrier because washover sediments do not reach the bay shoreline.

Storm washover of Follets Island generally occurs as sheet flooding rather than as confined flow through individual washover channels that are characteristic of some transgressive barriers. Consequently, sand transported onto the barrier flat is deposited slightly inland from the backbeach and forms a narrow washover terrace that parallels the island. This terrace terminates bayward in an avalanche face that controls the faunal assemblage and separates the backbarrier marsh from the grass-covered barrier flat. Gulf shoreline erosion prevents accumulation of large dunes on the eastern end of the island, but dune height and continuity both increase to the southwest where sand is locally supplied by the Brazos River. In 1983, these higher elevations were breached at several locations as bay water was driven against the dunes by Hurricane Alicia. After the storm, water draining from the bay to the Gulf enlarged the breaches and caused headward erosion of short channels that undermined the road in several places.

Unlike the adjacent barriers, Follets Island was created by subsidence of the Holocene Brazos Delta. As the delta sank, the lagoon extended laterally, flooding the delta plain and separating the beach ridge from the mainland. The barrier core is composed of sand that rests on stiff, delta-plain mud and grades landward into lagoon mud. Maximum thickness of the island is about 15 ft (4.5 m) on its eastern end; it thins westward and landward to about 3 ft (1 m) where it merges with the delta plain (Bernard and others, 1970).

The narrowing of Follets Island illustrates how barrier islands respond to a slight increase in sea level. It also indicates what will happen to currently stable barriers if sea level continues to rise as many scientists have predicted. If extant processes remain unchanged, the barrier will eventually become so narrow that washover will reach the lagoon, and the island will begin migrating landward. Such transgressive barriers are already shifting landward on some other segments of the Texas coast (Morton and others, 1983).

SITE 99. GEOLOGY OF THE MIDDLE TEXAS COAST

Guadalupe River and Delta. Modern coastal-plain rivers are relatively small underfit streams occupying much larger valleys that were broadened as base level rose and scalloped as the channels meandered from one valley wall to the other. The Holocene alluvial valley of the Guadalupe River is typical of rivers that (1) retreated while sea level was rising and then, (2) prograded a series of bayhead deltas within their confined valleys during the subsequent stillstand. Texas 35 crosses the Guadalupe alluvial valley at a location (Fig. 2, 96°50′00″W, 28°30′00″N. Austwell Quadrangle) that exposes wetlands and abandoned distributaries associated with former river channels (Hogs Bayou, Schwings Bayou), and delta lobes that were actively building into the upper reaches of San Antonio Bay about two thousand years ago (Shepard and others, 1960).

Alluvial channel deposits of the Guadalupe River near Victoria consist of mixed gravel and coarse sand at the base that grade vertically into floodbasin mud. Coarsest clastics are reworked late Tertiary and Quaternary gravels that locally blanket the older coastal plain sediments. The valley fill is about 4 mi (6.5 km) wide and about 25 ft (7.5 m) thick. Despite a coarse substratum, the surface is fine grained, commonly mud veneered, and lacks accretionary depositional features. Transported sediment rapidly decreases in size downstream so that lower alluvial plain and delta plain sediments are composed only of fine sand and mud. The Guadalupe River is classified as a suspended-load fluvial system because it primarily transports mud rather than sand.

Continuous deltaic sedimentation filled most of the lower alluvial valley a few feet above sea level. However, rapid progradation and incomplete valley filling left deltaic sediments surrounding some open bodies of water. These sediment bypass areas, such as Green Lake (Fig. 2), are common features of most delta systems. They diminish in size either by gradual deposition of backwater debris and floodbasin mud or rapidly by deposition of crevasse splay sand and mud.

The modern Guadalupe delta (96°49′00″W, 28°25′30″N, Austwell Quadrangle) can be reached from Texas 35 by a narrow shell road, located 500 ft (150 m) southwest of the Guadalupe River; the shell road follows the main river channel to its bifurcation on the upper delta plain. This river-dominated, shallow-water delta is typical of stable platform deltas constructed within intracratonic basins having low wave energy. Initial progradation of the modern lobe paralleled the valley axis, producing a highly elongate delta. Subsequent lobes were constructed in a counterclockwise direction as branching distributaries shifted the locus of deltaic deposition and changed the morphology to a lobate pattern because of the progressively shallower water. Sluggish discharge and frequent flooding caused by extensive log jams within distributary channels may have contributed to repeated channel avulsion and creation of new delta lobes. Most of the Guadalupe Delta is abandoned because sediment supply has been reduced, and much of the discharge has been artificially diverted into Mission Lake. Accompanying delta abandonment are (1) bay margin erosion, (2) delta plain subsidence, and (3) conversion of the delta plain to marsh and brackish-water lakes.

Total delta thickness, about 12 ft (3.5 m), corresponds to water depth in the bay during initial progradation. Despite this shallow depth, a variety of subenvironments can be recognized both at the surface and in the regressive deltaic sequence. Principal sedimentary facies from bottom to top include: bay, prodelta, distributary-mouth bar, and delta plain; the latter subenvironment is further subdivided into natural levee, active and abandoned distributary channels, marsh, interdistributary bay, and beach ridge (Donaldson and others, 1970). Most delta-plain environments can be seen from the wooden bridge about 4 mi (6.4 km) from Texas 35. The river (active distributary channel) is lined with trees growing at slightly higher elevations (natural levee) than the adjacent delta plain. Toward the south are low, brackish-water marshes and San Antonio Bay, as well as the Pleistocene escarpment. Looking toward the north, one can see extensive freshwater marshes, ponds, and another tree-lined levee flanking the North Guadalupe distributary channel.

Stable platform deltas exhibit certain unique characteristics including thin progradational facies (prodelta and delta-front deposits) as compared to relatively thick aggradational facies (channel-fill and delta-plain deposits). Furthermore, they commonly prograde into water that is shallower than scour depths of the distributary and alluvial channels. As a result of this relationship, some or all of the deltaic sediments are removed by erosion as the system advances seaward (Donaldson and others, 1970). Channel-fill and distributary-mouth bar deposits typically constitute the sand framework of these shallow-water fluvial–deltaic systems. About 0.5 mi (750 m) southwest of the Guadalupe River, Texas 35 crosses the Pleistocene–Holocene contact that marks the western wall of the entrenched valley. The difference in elevation between the Beaumont surface and the alluvial valley at the contact, which is about 30 ft (9 m), represents the unfilled portion of the eroded valley. It also serves as a rough, albeit slightly high, estimate of sea level during the preceding late Pleistocene highstand.

Ingleside Barrier–Strandplain. The Aransas Wildlife Refuge (Fig. 2) offers a unique opportunity to view modern marshes and bays as well as a late Pleistocene barrier-strandplain system known morphologically as the Blackjack or Live Oak Peninsula and stratigraphically as the Ingleside sand (Fig. 5). It can be reached from Texas 35 by following signs on Texas 239 and 2040 south of Tivoli. Maps for self-guided tours are available at the visitors center. At the refuge, juxtaposed Pleistocene and Holocene barrier-lagoon systems (Fig. 5) encompass a diverse range of aquatic, wetland, and upland environments that sustain a variety of plants and animals, including some former (alligator, pelican) and current (whooping crane, prairie chicken) endangered species. Other common animals include coyotes, deer, wild turkeys, javalinas, armadillos, and numerous shore birds.

Long linear swales between relict beach ridges, which can be seen within the refuge, are occupied by ponds surrounded by freshwater marshes with rushes and cattails. The ponds, which accumulate organic detritus and some mud, serve as important ecological niches in the wetlands system by providing sources of fresh water for the abundant wildlife. Many old, live oaks growing on the Ingleside sand are sculptured by the prevailing onshore wind. A fine salt spray carried by sea breezes inhibits windward growth of new vegetation causing the trees to grow asymmetrically in a downwind direction.

The mainland part of the refuge is situated on a strandline sand of Sangamonian age, which was first described as the Ingleside Barrier-Strandplain by Price (1933). This late Pleistocene strike-fed system formed between the major deltas where minor influx of fluvial sediments allowed the accumulation of thick barrier and strandplain sands. It exhibits accretionary beach ridges and contains a nearshore marine fauna like the adjacent modern barriers.

Remnants of the Ingleside shoreline can be traced from northeastern Mexico to southwestern Louisiana (Price, 1958). Along its extent it displays the same morphogenic features as the modern Texas Coast, including barrier islands, strandplains attached to the mainland, and erosional beaches seaward of alluvial plains (Winker, 1979). The Ingleside strandplain of the central Texas Coast (Figs. 2 and 5) consists of a broad (10 mi; 16 km), strike-aligned sand body that was deposited under wave-dominated marine conditions perhaps as much as 120,000 years ago. The sand was supplied by high-sinuosity rivers that meandered across the coastal plain and emptied into the Gulf 5 to 15 mi (8 to 25 km) inland from the present-day shoreline.

Sands of the Ingleside barrier-strandplain attain a maximum thickness of about 80 ft (25 m) (Fig. 5). They interfinger with deltaic muds (Beaumont Formation) in a landward direction and thin in a seaward direction. Their seaward limit has not been established, although they have been penetrated in borings beneath modern barriers of the central Texas coast (Fig. 5). The nearshore marine deposits are typically coarse- to fine-grained, yellow to brown sand, or dark gray, shelly sand that is intercalated with brown mud. The Ingleside sands overlie hard, blue to gray, sandy clay containing calcareous concretions.

Numerous authors have shown that the Ingleside is a multistory beach-shoreface complex deposited contemporaneously with deltaic muds of the Beaumont Formation (Fig. 5). The Ingleside sediments primarily aggraded in response to coastal plain subsidence and a highstand of sea level. However, during the last depositional phase, beach-ridges prograded as sea level fell. The lowered base level caused some Beaumont deltas to bury the Ingleside shoreline as they advanced seaward beyond the former strandline (Figs. 1 and 2).

The road from the refuge entrance to the visitors' center passes a brackish-water marsh of cattails. The marsh has accretionary morphology and partly fills a meander loop carved in the

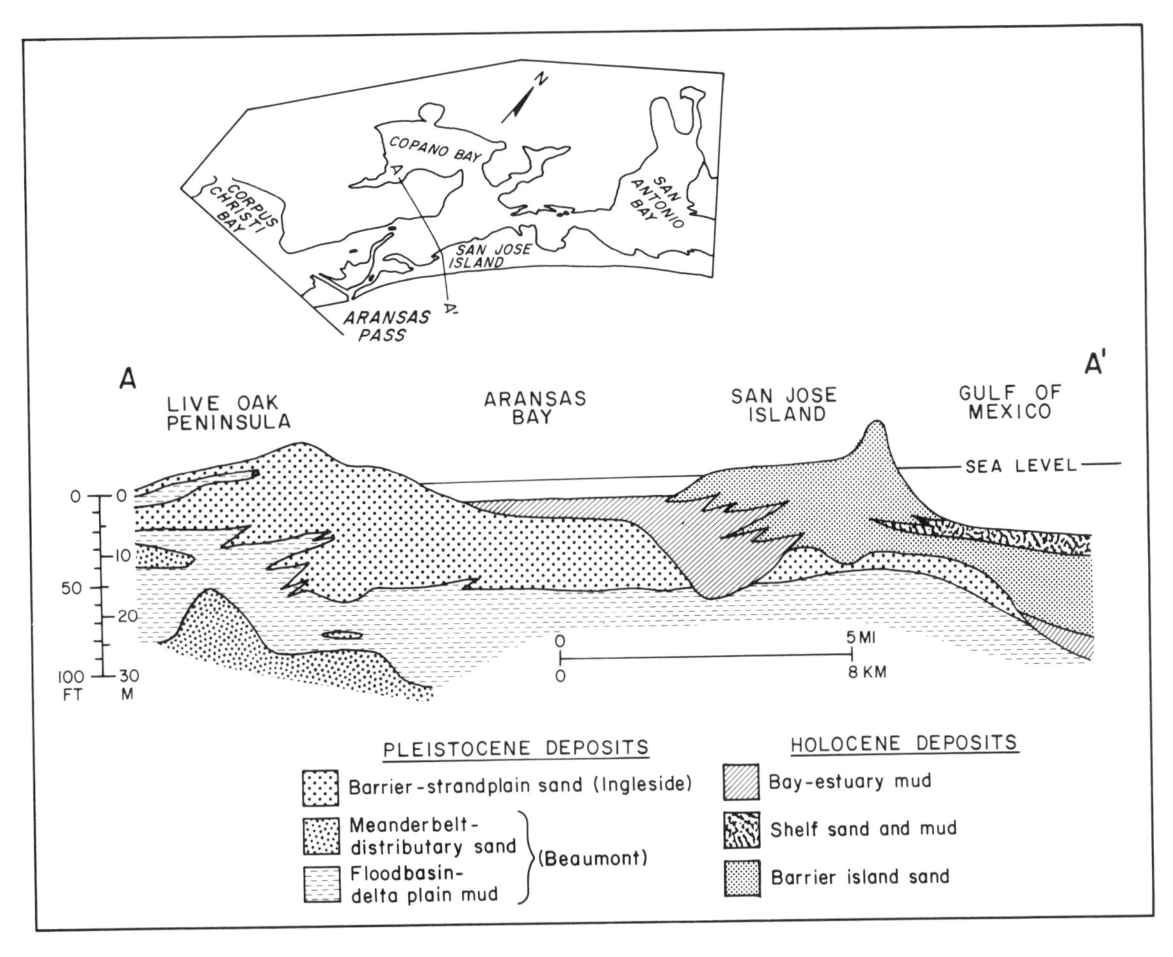

Figure 5. Late Quaternary depositional sequences of the middle Texas coast composed of aggradational and progradational barrier-lagoon sediments. General location shown in Figure 2. Subsurface data compiled from Shepard and others (1960). Winker (1979) and Morton and McGowen (1980).

valley wall. Both lines of evidence suggest a long history of marsh development associated with nearshore adjustment to the still-stand in sea level.

About 1.8 mi (3 km) from the visitors' center, the eroding shore of San Antonio Bay has exposed nearly a meter of Ingleside sandy marine facies. The outcrop consists of grass-covered muddy sand with layers of concentrated shell hash. Abundant specimens of the small surf clam *Donax* are good evidence that these sediments were deposited in a beach environment. The surface elevation of the barrier-strandplain (~25 ft; 8 m) suggests that sea level was near this height when the Ingleside sand was deposited.

At Dagger Point, surface elevations are about 50 ft (15 m) above sea level due to relict eolian dunes. The dunes and barrier core are truncated at the bay shore forming a steep sandy bluff partly stabilized by trees. These dunes are undoubtedly products of late Pleistocene or Holocene eolian activity unrelated to shore-line processes. About 2 mi (3.2 km) from Dagger Point, the observation tower provides a magnificent overview of the Pleistocene barrier, Holocene marsh, and modern bay systems. Thick oak mottes cover the barrier core, and salt-tolerant grasses delineate modern marshes fringing the bay where wave energy is low.

Bay margin sediments along western San Antonio Bay are composed of sand and shell eroded from high bluffs of Pleistocene interdistributary mud (north and west of the refuge) and low bluffs of Ingleside sand (within the refuge) representing marine reworked deposits of the delta front. Bluff elevations gradually decrease gulfward from about 25 ft (7.5 m) to sea level. Fresh-water inflow to the bay provides the nutrients and salinity balance necessary to maintain high biological productivity. Extensive oyster reefs and a great diversity of molluscan species explain why shell detritus constitutes a high percentage of bay center sediments. Before the mid-1970s, when dredging operations were economical, San Antonio Bay produced large quantities of shell aggregate that was washed and later used for construction material and for lime in manufacturing cement.

A late Pleistocene lagoon. Lagoons associated with the Ingleside barrier are either obscured by wind-blown sand, filled by younger deltaic sediments, or eroded by modern drainage systems. An exception is the exposure near the type locality at Ingleside (Fig. 2). West of Aransas Pass (97°13′30″W, 27°55′45″N, Aransas Pass Quadrangle), Texas 35 crosses the lagoon surface between FM 1069 and a roadside park, a distance of about 2.9 mi (4.6 km). Here the lagoon is identified on the basis of subtle changes in surface elevation, soil composition, and indigenous vegetation. The slightly depressed lagoon surface occurs between the Beaumont delta plain and the Ingleside back-barrier ramp. The lower elevation and clay substrate retain water, causing a noticeable difference in flora and land use.

Farmers cultivate the delta plain but not the muddy lagoon sediments, which are poorly drained and covered with clumps of dense grass (coastal sacahuista) and mesquite trees. Traveling toward Aransas Pass, one can observe that surface elevations increase, soils become sandier, and the vegetation grades into species that prefer loose, well-drained soils. This broad zone represents the transition between lagoon margin and backbarrier environments that is comparable to extant washover fans and wind-tidal flats. The barrier crest is about 33 ft (10 m) above sea level and densely covered with live oaks.

North Padre Island, a Holocene aggradational barrier. Padre Island National Seashore (97°17′30″W, 27°27′30″N, South Bird Island Quadrangle) provides access to a high-profile barrier that resembles other barriers along the central Texas coast. It can be reached by Park Road 22 southeast of Corpus Christi. Two roads, about 1.2 mi (2 km) and 2.2 mi (3.5 km) respectively from the park entrance, can be used to observe barrier subenvironments along a transect between Laguna Madre and the Gulf of Mexico. The first road goes to the gulf beach; the second road goes to South Bird Island boat basin on Laguna Madre.

Padre Island (Fig. 2) is one of the longest continuous barriers in the world. It is 3 to 5 mi (5 to 8 km) wide and encompasses large areas having elevations greater than 10 ft (3 m). The foredunes, which are tall, continuous, and well vegetated, prevent storm overwash except where the dune ridge has been breached by blowouts. Beginning at the Gulf shoreline, a transect across Padre Island encounters beach, foredunes with or without blowouts, vegetated barrier flat, deflation flats, active and stable back-island dunes, wind-tidal flats, and marine grassflats.

The beach is low, wide (250 ft; 75 m), and either horizontal or slopes gently toward the gulf. Incipient dunes, locally called coppice mounds, occupy the backbeach seaward of the foredune ridge. The foredunes commonly display erosional escarpments caused by storm waves. They are stabilized by salt-tolerant grasses except where gaps of barren sand (blowouts) are eroded by wind. Formerly active dunes that were subsequently stabilized occur landward of the foredune ridge. This hummocky topography grades toward the lagoon into a grass-covered ramp or barrier flat. In contrast to other barrier flats, this flat exhibits an unusual pattern of concentric deflation troughs and ridges that record the depth of deflation when extensive dune fields migrated across the island and into Laguna Madre. During the past few decades, most dunes have stabilized because of higher rainfall and reduced grazing, but some dune fields are still actively migrating at 30 to 80 ft/yr (10 to 25 m/yr). Broad, featureless, barren sand flats form the back-island environment where back-island dunes are absent. These wind-tidal flats are composed of alternating sand and mud layers and are covered by algal mats that grow when the flats are flooded but dry up and blow away when the flats are exposed (Fisk, 1959).

The barrier core of North Padre Island is predominantly sand with minor amounts of shell or clay. Vertical stacking of barrier and lagoon facies (Fisk, 1959) indicates that Padre Island grew mainly by aggradation. The Holocene sequence is about 30 to 40 ft (10 to 12 m) thick and overlies Pleistocene barrier sands that are stratigraphically equivalent to the Ingleside strandplain. Where subsurface lithology is essentially the same, only a change in color from reduced gray sand to oxidized tan or brown sand marks the Holocene–Pleistocene unconformity (Morton and McGowen, 1980).

SITE 100. GEOLOGY OF THE LOWER TEXAS COAST

South of Corpus Christi, U.S. 77 (Fig. 3) provides access to depositional environments that are different from those seen in other parts of the coastal plain. These differences arise partly because the climate is arid and partly because the late Quaternary sedimentologic and tectonic history of south Texas was different from other regions. Some of the important features of this region include a hypersaline bay, eolian sand sheet, large Holocene fluvial–deltaic system, restricted lagoon, and transgressive barrier island.

Baffin Bay. FM 771 links U.S. 77 with Riviera Beach located on the western shore of Baffin Bay (97°37′30″W, 21°15′00″N, Riviera Beach Quadrangle). The bay and its branching tributaries (Fig. 3) are drowned stream valleys that were partly filled during the Holocene rise in sea level (Fisk, 1959; Behrens, 1963). It owes its origin to Los Olmos Creek and San Fernando Creek, which are small ephemeral streams lacking significant baseflow because rainfall is low. After prolonged periods of limited freshwater inflow and high evaporation, the bay becomes hypersaline. At times, exceptionally high upland runoff, such as hurricane flooding, can flush the bay converting it to a freshwater system. Few faunal assemblages can tolerate such extreme fluctuations in salinity; consequently, species diversity is extremely low in Baffin Bay. The fauna are mostly concentrated around the bay margin whereas the deeper bay centers are nearly barren (Shepard and others, 1960).

Sediments accumulating in Baffin Bay are mostly dark gray organic-rich mud similar to fine-grain deposits of other coastal bays. The mud is partly derived from stream discharge, but mostly comes from wave erosion of low Pleistocene bluffs forming much of the bay shoreline (Fig. 3). Anomalous concentrations of beachrock and locally precipitated carbonates occur within this terrigenous clastic province. The beachrock is composed of aragonite-cemented shell hash deposited as storm berms on the northwestern shore of Laguna Madre (Shepard and others, 1960). Other skeletal carbonate deposits are large serpulid reefs composed of calcareous tubes secreted by worms. The reefs grew under normal marine conditions when the bay was connected to the Gulf, but reef-building worms died after Padre Island restricted tidal exchange, causing hypersaline conditions.

The warm carbonate-saturated water combines with locally generated waves to produce oolites and coated grains in Baffin Bay where the shallow water is constantly agitated. This nonskeletal carbonate deposition is restricted to the northern bay shores facing predominant southeast winds that cross the longest fetch and thus generate the greatest wave energy. Oolite rims and grain coatings are composed of aragonite that nucleates around quartz grains and shell fragments.

Minor evaporite deposits are also associated with sediments of Baffin Bay and Laguna Madre. Salt pans occur on broad sand flats that are periodically flooded by bay water; subsequent evaporation leaves a thin crust of halite crystals concentrated by cyclic repetition of the processes. Sandy gypsum roses are another chemical deposit found in lagoon sediments; they were precipitated below the water table from highly saline groundwater (Fisk, 1959). Numerous roses were dredged from Laguna Madre during construction of the Gulf Intercoastal Waterway.

Eolian sand sheet. The hummocky, sandy terrane traversed by U.S.-77 south of Sarita (97°47′30″W, 27°11′00″N, Sarita Quadrangle) reflects the combined influences of persistent southeast winds, sandy substrates, and low rainfall (Price, 1958; Brown and others, 1977). Topography of the area is mainly controlled by wind deflation, dune migration, and fluctuations in groundwater level. Extensive dune fields are reactivated and expanded during droughts when the water table is lowered and deflation removes sand down to the moist layers; these active dunes are barren or sparsely covered by grass. During wet periods, dunes are stabilized by higher internal moisture supporting dense vegetation and allowing only shallow deflation of barren areas. The oldest dune complexes are stabilized by live oak mottes like those growing on the Ingleside sand of the central Texas coast.

The broad dune fields are composed of well-sorted fine-grained sand. Some sand is locally derived from older dunes and subjacent Pleistocene meanderbelt, barrier, and delta-front deposits, and some sand is blown across central Padre Island and Laguna Madre (Fig. 3) near 27°N latitude (Fisk, 1959). This latitude has marked the average position of converging littoral currents since the sea reached its present level. Thus sand influx for dune building has been abundant in this area for several thousand years. Overgrazing by herds of cattle, coupled with severe droughts in the 1930s and 1950s, may have contributed to dune migration.

Price (1958) described the banner dune complexes that characterize the south Texas eolian sand sheet. The v-shaped complexes originate at spot blowouts and widen downwind. They are laterally restricted by marginal lag ridges that stand above the smaller dunes. Between the ridges, transverse dunes migrate to the northwest by downwind deposition and upwind erosion that normally intersects the water table. Wind deflation during severe droughts and subsequent elevation of the water table during wet periods creates interdune ponds and small lakes that are lined by freshwater wetland vegetation. Other interdune swales and depressions are sites of calcareous mud accumulation. Such fine-grained, low-permeability interdune deposits act as barriers to fluid flow in hydrocarbon reservoirs of eolian sandstones.

Holocene Rio Grande delta system. The Laguna Atascosa National Wildlife Refuge (Fig. 3) in far south Texas (97°20′00″W, 26°20′00″N, La Coma Quadrangle) is about 17 mi (27 km) east of U.S. 77 on FM 510. There, traces of the ancestral Rio Grande are well preserved as oxbow lakes. These meander cutoffs and abandoned river segments, locally called resacas (Price, 1958), are evidence of avulsion and other changes in channel position related to alterations in magnitude or location of fluvial discharge. The sinuous inactive channels and curved lakes are mostly filled with mud washed in as suspended sediment when the lower alluvial plain floods.

Vertically stacked, upward-fining sequences characterize fluvial sediments deposited by the Rio Grande since 7,000 years ago (Fulton, 1975; Fig. 6). These aggradational sand and mud deposits are 60 ft (18 m) thick, even though individual channels of the modern Rio Grande system are only 26 to 33 ft (8 to 10 m) thick. Dimensions of the modern Rio Grande are relatively small because they represent channel size for the delta plain rather than for the alluvial plain. Furthermore, these younger channels are smaller than older channels (Fig. 6) because thin but areally extensive delta-plain sediments were deposited over older, thicker, and more stable alluvial-channel sediments.

In the Rio Grande Embayment, the Holocene–Pleistocene contact coincides with the base of the youngest progradational sequence (Fig. 6). Overlying Holocene fluvial-deltaic sediments are subdivided into four lithofacies comprising a thick offlapping sequence. Brown to gray clays of prodelta origin compose the basal lithofacies, which is 10 to 36 ft (3 to 11 m) thick. This unit grades upward into thin gray sands interbedded with sediments similar to the prodelta muds. These alternating thin beds of sand and mud (second lithofacies) comprise the delta-front deposits that are up to 25 ft (7.5 m) thick. Sediments above the distributary-mouth bar are relatively thick (36 ft; 11 m) and composed either of brown to gray sandy clay and organic-rich clay or gray fine sand and sandy silt. These two lithofacies, which respectively originated in delta-plain and fluvial-channel environments, account for most of the Holocene Rio Grande fluvial–deltaic system.

Holocene sediments deposited by the Rio Grande are 5 ft (1.5 m) to nearly 100 ft (30 m) thick (Fulton, 1975; Fig. 6). Delta thickness generally increases away from the lateral contact with Pleistocene sediments and toward the extant river mouth where subsidence was greatest and the entrenched valley was deepest. Aggradation of the upper delta plain and fluvial channels accompanied valley filling and delta progradation across the inner shelf

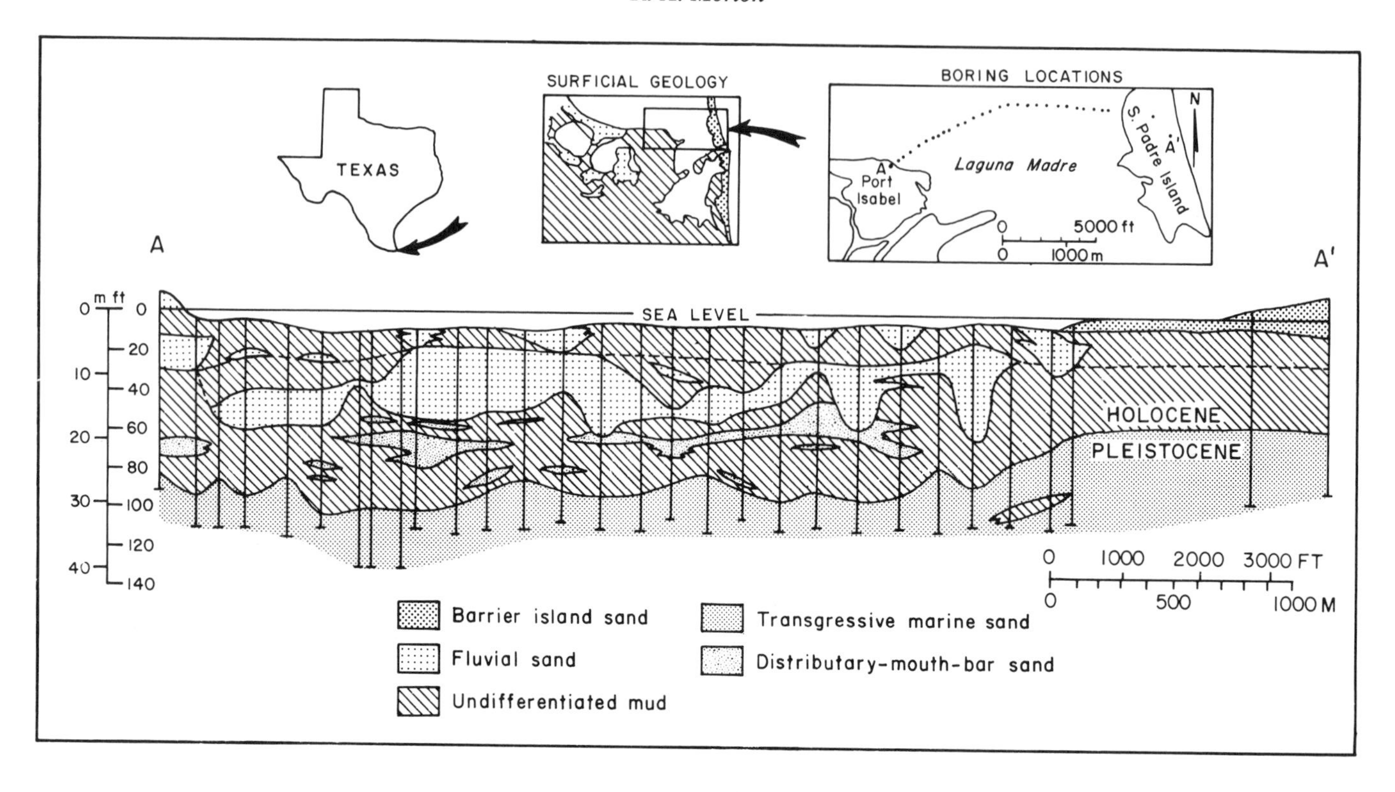

Figure 6. Late Quaternary depositional sequences of the lower Texas coast composed of fluvial–deltaic and transgressive barrier-island sediments. General location shown in Figure 3.

about 4,000 to 7,000 yr B.P., when the rate of sea-level rise diminished (Morton and McGowen, 1980).

Morphological and sedimentological evidence suggests that maximum extent of the Holocene Rio Grande delta was about 15 to 25 mi (25 to 40 km) seaward of its current position (Price, 1958). However, reductions in sediment load and discharge associated with climatic changes caused subsequent delta abandonment and headland retreat that is manifested as shoreline erosion along the south Texas coast. Delta abandonment was accompanied by subsidence, which in turn promoted the formation of Laguna Madre and the adjacent wind-tidal flats about 5,000 yr B.P. (Shepard and others, 1960).

Clay dunes are unique features of the Rio Grande delta plain where it subsided and formed broad wind-tidal flats (Price, 1958). Sediment for the clay dunes is locally derived, as wind blowing across the flats erodes and transports sand and silt-size particles of salty clay. Repeated dessication and deflation of the mudflats yields a large volume of airborne clay pellets. The clay dunes, which form near the coast, nucleate and accrete on topographic highs such as levees of the Holocene Rio Grande. In the Laguna Atascosa Refuge, clay dunes form the western boundary of Laguna Madre. The dunes are stable landforms attaining heights up to 26 ft (8 m) and are oriented both oblique and parallel to the lagoon shore. Most clay dunes have oblong shapes with internal concentric ridges, but some are shaped like the head of a hawk (Brown and others, 1980). The various orientations and shapes suggest that underlying topographic control is just as important as wind direction in determining dune growth.

Laguna Madre, a restricted lagoon. The Queen Isabella Causeway (97°12′30″W, 26°04′30″N, Port Isabel Quadrangle) crosses Laguna Madre between Port Isabel and South Padre Island (Figs. 3 and 6). The lagoon is a shallow elongate water body that formed about 5,000 yr B.P. as the rising sea inundated the flat coastal plain and Padre Island grew from the abundant marine-reworked sand on the inner shelf. Surface sediments of the lagoon are composed mostly of muddy sand and sandy mud with variable amounts of shell (Shepard and others, 1960). Sandy sediments filling the lagoon are supplied by storm washover and eolian transport across the barrier and adjacent wind-tidal flat; muddy sediments are generally derived from erosion of nearby Pleistocene low bluffs or deposition of suspended sediment transported by astronomical tides and wind-driven currents.

Tidal currents were probably important transporters of sediment shortly after the lagoon formed when tidal inlets were more numerous; presently astronomical tides are extremely low in the lagoon (about 12 in; 30 cm) because the long, continuous barrier greatly restricts tidal exchange with the Gulf. Water circulation chiefly depends on wind-driven currents and reversals in wind that periodically move water in opposite directions. Southeasterly winds drive water to the north, increasing flood currents through

Brazos Santiago Pass and raising water levels along the mainland shores; in contrast, northerly winds push water south, accelerating ebb currents through Brazos Santiago Pass and flooding the broad sand flats of Padre Island. Net water movement, independent of the wind tides, is into the lagoon whenever evaporation is excessive.

In some parts of the lagoon, a sandy bottom reduces turbidity, allowing penetration of sunlight and growth of thick stands of marine grasses. The grassflats are important nursery grounds for fish, shrimp, and crabs that constitute most of the marine fisheries industry. Common species of seagrass (algae) are *Halodule wrightii, Ruppia maritima,* and *Thalassia testudinum.* Total organic carbon in lagoon sediment is up to 8% by weight, depending inversely on grain size. Such high concentrations are attributable to the extremely high biological productivity of these waters.

Holocene sediments beneath the lagoon are 10 to 20 ft (3 to 6 m) thick (Shepard and others, 1960) with thickness being controlled by erosional relief of the Pleistocene surface and age of the lagoon. As a result, lagoon sediments progressively thin to the south where they onlap the Holocene Rio Grande delta plain.

South Padre Island, a transgressive barrier island. Comparing maps and aerial photographs of South Padre Island (97°10′00″W, 26°05′00″N, Port Isabel Quadrangle) clearly shows that the barrier is migrating as the gulf shoreline erodes and washover sediments are deposited in the lagoon (Morton and McGowen, 1980). The southern 5 mi (8 km) of the island, which is extensively developed, has been relatively stable since 1930 when the newly constructed jetties at Brazos Santiago Pass (Fig. 3) began trapping sand eroded from the ebb tidal delta and deposited by longshore currents. The island between the developed area and Mansfield Ship Channel to the north (accessible by 4-wheel drive vehicles) is still migrating landward.

Like other transgressive barriers, South Padre Island is low (<10 ft; 3 m), narrow (~1,000 ft; 300 m), and frequently flooded by storm surge. Along Park Road 100 north of the densely developed area, the fore-island dunes are generally low (10 to 15 ft; 3 to 4.5 m), sparsely vegetated, and discontinuous because washover channels create wide breaks in the dunes. The closely spaced washover channels are repeatedly occupied by high-velocity currents flowing across the island that damage or destroy the road. Washover fans composing most of the back-island environment coalesce to become broad, barren sand flats that are seasonally submerged by wind-tides blown from the adjacent lagoon. While the eastern margin of the lagoon is being filled by washover deposition, the western margin is encroaching on the mainland as a result of delta-plain subsidence and wave erosion of the shore (Brown and others, 1980).

South Padre Island began migrating several thousand years ago when fluvial sediment supply diminished, and the Holocene Rio Grande Delta began retreating. As a result of this transgression, the island consists of a thin sand wedge retreating over the subsiding delta plain (Fig. 6). The barrier wedge is thickest (16 ft; 5 m) near the shoreline, but it grades landward into muddy sand of the lagoon and seaward into sandy mud of the inner shelf.

Little sand is currently available for barrier maintenance, and that volume is probably decreasing as some sand is transported seaward during storms and deposited below normal wave base. Because of this deficit in sediment supply, transgressive barriers supporting high-density recreational communities such as South Padre, will be the first to be adversely affected by the relative sea-level rise (Morton and others, 1983).

REFERENCES CITED

Barton, D. C., 1930, Deltaic coastal plain of southeastern Texas: Geological Society of America Bulletin, v. 41, p. 359–382.

Behrens, E. W., 1963, Buried Pleistocene river valleys in Aransas and Baffin bays, Texas: University of Texas, Austin, Marine Science Institute Publication v. 9, p. 7–18.

Bernard, H. A., Major, C. F., Jr., Parrott, B. S., and LeBlanc, R. J., Sr., 1970, Recent sediments of southeastern Texas; A field guide to the Brazos alluvial and deltaic plains and the Galveston barrier island complex: University of Texas, Austin, Bureau of Economic Geology Guidebook 11, 132 p.

Brown, L. F., Jr., McGowen, J. H., Evans, T. J., Groat, C. G., and Fisher, W. L., 1977, Environmental geologic atlas of the Texas coastal zone, Kingsville area: University of Texas, Austin, Bureau of Economic Geology, 131 p.

Brown, L. F., Jr., Brewton, J. L., Evans, T. J., McGowen, J. H., White, W. A., Groat, C. G., and Fisher, W. L., 1980, Environmental geologic atlas of the Texas coastal zone, Brownsville–Harlingen area: University of Texas, Austin, Bureau of Economic Geology, 140 p.

Doering, J., 1935, Post-Fleming surface formations of coastal southeast Texas and south Louisiana: American Association of Petroleum Geologists Bulletin,v. 19, p. 651–688.

Donaldson, A. C., Martin, R. H., and Kanes, W. H., 1970, Holocene Guadalupe Delta of Texas Gulf Coast, *in* Morgan, J. P., ed., Deltaic sedimentation modern and ancient: Society of Economic Paleontologists and Mineralogists Special Publication 15, p. 107–137.

Fisher, W. L., Brown, L. F., McGowen, J. H., and Groat, C. G., 1972, Environmental geologic atlas of the Texas coastal zone, Galveston–Houston area: University of Texas, Austin, Bureau of Economic Geology, 91 p.

Fisk, H. N., 1959, Padre Island and the Laguna Madre flats, coastal south Texas: Louisiana State University, Baton Rouge, Second Coastal Geography Conference, p. 130–151.

Fulton, K. J., 1975, Subsurface stratigraphy, depositional environments, and aspects of reservoir continuity–Rio Grande Delta, Texas [Ph.D. thesis]: Cincinnati, Ohio, University of Cincinnati, 330 p.

Gould, H. R., and McFarlan, E., Jr., 1959, Geologic history of the Chenier Plain, southwestern Louisiana: Gulf Coast Association of Geological Societies Transactions, v. 9, p. 261–270.

Halbouty, M. T., 1936, Geology and geophysics showing caprock and salt overhang of High Island dome Galveston County, Texas, *in* Barton, D. C., and Sawtelle, G., eds., Gulf coast oil fields: Tulsa, American Association of Petroleum Geologists, p. 909–960.

McGowen, J. H., Proctor, C. V., Jr., Brown, L. F., Jr., Evans, T. J., Fisher, W. L., and Groat, C. G., 1976, Environmental geologic atlas of the Texas coastal zone, Port Lavaca area: University of Texas, Austin, Bureau of Economic Geology, 91 p.

Morton, R. A., and McGowen, J. H., 1980, Modern depositional environments of the Texas coast: University of Texas, Austin, Bureau of Economic Geology Guidebook 20, 167 p.

Morton, R. A., and Nummedal, D., 1982, Regional geology of the northwestern gulf coastal plain, *in* Sedimentary processes and environments along the Louisiana–Texas coast: Field trip guidebook, Geological Society of America annual meeting, p. 3–25.

Morton, R. A., Pilkey, O. H., Jr., Pilkey, O. H., Sr., and Neal, W. J., 1983, Living with the Texas shore: Durham, North Carolina, Duke University Press, 190 p.

Price, W. A., 1933, Role of diastrophism in topography of Corpus Christi area, South Texas: American Association of Petroleum Geologists Bulletin, v. 17, p. 907–962.

——, 1958, Sedimentology and Quaternary geomorphology of south Texas: Gulf Coast Association of Geological Societies Transactions, v. 8, p. 41–75.

Shepard, F. P., Phleger, F. B., and Van Andel, T. J., eds., 1960, Recent sediments, northwest Gulf of Mexico: Tulsa, Oklahoma, American Association of Petroleum Geologists, 394 p.

Suter, J. R., and Berryhill, H. L., Jr., 1985, Late Quaternary shelf-margin deltas, northwest Gulf of Mexico: American Association of Petroleum Geologists Bulletin, v. 69, p. 77–91.

Swanson, R. L., and Thurlow, C. I., 1973, Recent subsidence rates along the Texas and Louisiana coast as determined from tide measurements: Journal of Geophysical Research, v. 38, p. 55–94.

Winker, C. D., 1979, Late Pleistocene fluvial–detlaic deposition, Texas coastal plain and shelf [M.A. thesis]: Austin, University of Texas, 187 p.

Index

[Italic page numbers indicate major references]

Abilene anticline, Kansas, 11, 14, 29, 33
Acme Dolomite Member, Blaine Formation, Texas, 315
Ada Formation, Oklahoma, 160
Ada Valley anticline, Arkansas, 265
Adair County, Oklahoma, 88, 89, 90
Adams Branch Limestone, Texas, 350
Adetognathus sp., 48, 50, 52, 235, 237
Adiantum Capillus-Veneris, 155
Adobe Canyon Formation, Texas, 407, 409
Aguja Formation, Texas, 419, 431
Alamo Creek Member, Chisos Formation, Texas, 431, 433
Albany Group, Texas, 309, 310
Alder Gap, Arkansas, 245
Alfalfa County, Oklahoma, 135
algae
  Kansas, 25, 32, 39, 41, 64
  Oklahoma, 87, 89, 97, 142, 143, 153, 155, 185
  Texas, 339, 397, 457
  See also specific genus and species
algal reefs, Texas, *339*
Allamore Formation, Texas, 390
Allison, Arkansas, *215*
Alma Sandstone, Oklahoma, 145
Alsate Formation, Texas, 411
Altona Dolomite, Oklahoma, 83
Amarillo Uplift, Texas, 305
Ammodiscus sp., 33
amplexicaulis, 175
Anacacho Limestone Formation, Texas, *441*
Anacacho Mountains, Texas, 441
Anadarko Basin, Oklahoma, 79, 80, 83, 101, 122, 124, 127, 129, 136, 189
Anchicodium sp., 39
Anchignathodus, 48, 50
Anguila Fault Zone, Texas, 423
Anthracoceras discus, 235
Antlers Formation
  Arkansas, 282
  Texas, 323
Aphrodina lamarensis, 57
Appalachians, 152, 172, 176, 273
Arapaho Point, Oklahoma, 122
Arbuckle Anticline, Oklahoma, 178, 179, 181, 183
Arbuckle Group, Oklahoma, *93*, 99, 118, 123, 125, 127, 129, 130, 153, 155, 157, 163, 165, 177, 178, 181, 183, 185
Arbuckle Lake, Oklahoma, 177, 182
Arbuckle Mountains, Oklahoma, 122, *153*, *159*, *171*, *177*, *183*, 309
Arbuckle uplift, Oklahoma, 127, 147, 152, 178, 189
Archaeolithophyllum, 87, 143
Archaeoscyphia, 97
Archer City Formation, Texas, 311
Archer County, Texas, 310, 311
Arctodus simus, 295
Ardmore Basin, Oklahoma, 127, 178, *189*
Ardmore cyclothem, Kansas, 75
Arenicolites, 68, 102
Argentine Limestone, Kansas, 50
Arkadelphia, Arkansas, *277*
Arkansas Embayment, 41
Arkansas Novaculite Formation, Arkansas, 256, 258, 261, 263, *267*
Arkansas River, 85, 86, 87, 88, 89, *135*, 232, 249
Arkansas Valley, 211, 243, 254
Arkansas Valley Province, *249*
Arkoma Basin
  Arkansas, 231, *249*
  Arkansas-Oklahoma border, *91*
  Oklahoma, 139, *145*, 150
Ash Spring Basalt Member, Chisos Formation, Texas, 438
Ashley County, Arkansas, 265
Aspermont Dolomite Member, Blaine Formation, Texas, 315
Atchison County, Kansas, 5, 8
Atchison Formation, Kansas, *5*
Athens Plateau, Ouachita Mountains, 273
Atoka County, Oklahoma, 171
Atoka Formation
  Arkansas, 231, *239*, 241, *243*, 249, 251
  Oklahoma, 86, 87, 90, *145*, 150
Austin Chalk, Texas, 329, 332, 333, 441
Austin Group, Texas, 332, 441, 442
Aviculopecten, 30
Axinolobus sp., 88, 89

Bachelor Member, St. Joe Limestone, Arkansas, 204
Backbone Ridge, Texas, 353, 354
Bader Limestone, Kansas, 37
badlands, Kansas. *See* Borchers Badlands, Kansas
Baffin Bay, Texas, 454, 455
Bahamian Platform, 359
Bainbridge Limestone, Missouri, 224
Balcones Fault Zone, Texas, 329, 332
Baldwin coal bed, Arkansas, 89
Ballard Formation, Kansas, 69, 71, 72, 73
Bally Mountain, Oklahoma, *93*, 99, 101, 127
Barillos dome, Texas, 409
barite deposits, Oklahoma, 99, 100
Barneston Limestone, Kansas, *25*
Barnett Formation, Texas, 356, 357, 358
Barrel Springs Formation, Texas, 409
Basin and Range Province, *395*, 417, 421, 424, 435, 436, 437
Batesville Formation, Arkansas, 208
Bathysiphon sp., 33
Baylor County, Texas, 310, 312
Bear Mountain, Texas, 367
Beattie Cyclothem, *35*
Beattie Limestone
  Kansas, *35*
  Nebraska, 35
  Oklahoma, 35
Beaumont Formation, Texas, 445, 447, 452
Beaver Dam, Arkansas, *203*
Beaverburk Limestone, Texas, 311
Beavers Bend State Park, Oklahoma, *195*
Beavers Bend Tuff, Oklahoma, 198
Bee Cave Marl Member, Walnut Formation, Texas, 373, 376
Bell Canyon Formation, Texas, 396, 397, 398
Benevides Formation, Texas, 402, 404, 405, 406
Benton Uplift, Arkansas, 267
bentonite, Kansas, 1, 2, 57, 58, 59, 60
Berino Formation, Texas, 392, 393
Bethany Falls Limstone, Kansas, 50, 52
Bevier coal bed, Kansas, 78
bicornis, 174, 175
Big Bend National Park, Texas, *417*, *423*, *429*, *435*
Big Bluff, Arkansas, 208
Big Branch Gneiss, Texas, 362, 364, 365
Big Cedar–Kiamichi Mountain section, Ouachita Mountains, Oklahoma, *149*
Big Fork Chert, Oklahoma, 172, 175, 176
Big House Chalk, Texas, 442
Big Salt Plain, Oklahoma, 137
Big Yellow Member, Canoe Formation, Texas, 431
Bigfork Chert
  Arkansas, 256, 258, 263, 264
  Oklahoma, 172, 175, 176, 196, 200, 201
bioturbation
  Arkansas, 246, 250
  Kansas, 1, 19, 31, 32, 58
  Oklahoma, 97, 174
Birvanella, 339
Bishop Cap Formation, Texas, 392
Bison sp., 290, 295, 297
bivalves. *See specific genus and species*
Black Band Dolomite Member, Scenic Drive Formation, Texas, 392
Black Knob Ridge, Oklahoma 175, 176, 200
Black Prairies, Texas, 332

Black Shoals, Texas, 380
Blackwater Draw Formation, Texas, 294, 302, 303
Blaine County, Oklahoma, 83
Blaine Formation
  Oklahoma, 83
  Texas, 311, 314, 315
Blanco Basin, Texas, 303
Blanco Canyon, Texas, *299*
Blanco Formation, Texas, 294, 302
Blaylock Sandstone, Oklahoma, 195, 196, 197, 199, 200
Bliss Sandstone, Texas, 392
Bloyd Formation
  Arkansas, 231
  Oklahoma, 85
Bloyd Shale Formation, Arkansas, 85, 87, 88, 89, 208
Blue Beaver Creek, Oklahoma, 122
Blue Creek anticline, Oklahoma, 129
Blue Creek Canyon, Oklahoma, 123, *127*, 143
Blue Creek Canyon Fault, Oklahoma, 129, 130, 131, 132
Blue Creek horst, Oklahoma, 129, 131
Blue Creek Syncline, Oklahoma, 129, 130
Blue Mountain Dam, Arkansas, *243*
Blue Rapids Shale, Kansas, 29, 30
Blue Springs Shale Member, Matfield Shale, Kansas, 18, 24, 25, 26
Bluff Limestone, Texas, 403, 404, 405, 406
Bois d'Arc Formation, Oklahoma 183
Bolivar Peninsula, Texas, 449, 450
Boone County, Arkansas, 223
Boone Formation, Arkansas, *203*, 208, 210, *211*, 215, 220
Boquillas Limestone, Texas, 402, 424, 426, 428
Boracho Formation, Texas 402
Borchers Ash, Kansas, 70, 72, 73
Borchers Badlands, Kansas, *69*
Boston Mountains, Arkansas, 207, 211, 214, 215, 231, 239, 247
Bostrychoceras sp., 441, 444
Bowie Group, Texas, 309, 311
brachiopods
  Arkansas, 222, 240
  Kansas, 25, 32, 37, 39, 45
  Oklahoma, 87, 97, 174, 175, 185
Bradley County, Arkansas, 265
Brady, Texas, 352
Braggs Member, Sausbee Formation, Oklahoma, 86, 88, 90
Branneroceras branneri, 88
Brassfield Formation
  Arkansas, 211, 213, 221, 222, 223
  Kentucky, 223
  Missouri, 223
  Tennessee, 223
Bravo Dome, Texas, 305
Brazos Delta, Texas, 451
Brazos River, Texas, 293, 300, 310, 321, 378, 380, 445
Brazos River Formation, Texas, 317, 319, 320
Brazos River Valley, Texas, 377
Brentwood Limestone Member, Bloyd Shale, Arkansas, 85, 88, 89, 208
Brewer Bend Member, Sausbee Formation, Oklahoma, 85, 86, 87, 88, 89
Brewster County, Texas, 423, *429*
Bridwell Formation, Texas, 299, 301, 302
Briscoe County, Texas, 287
Broken Bow Uplift, Oklahoma, *195*
Brokeoff Mountains, New Mexico, 400
Bromide Formation, Oklahoma 163, 166, *172*, 179, 183, 184, 185, 187
Bromide Hill, Oklahoma, 159, 160, 163
Brown County, Texas, 343
Brownwood Shale, Texas, 350
Brushy Canyon Sandstone Formation, Texas, 398
bryozoans
  Kansas, 25, 37, 39
  Oklahoma, 87, 174, 175
Buck Creek Sandstone Member, Grandstone Creek Formation, Texas, 317
Buckhorn Caldera, Texas, 407, 409
Buda Limestone Formation, Texas, 331, 402, 405, 425, 426, 428
Buffalo National River, Arkansas, *207*
Buffalo River Valley, Arkansas, *211*, 212
Bull Creek Limestone, Texas, 376
Bunker Hill section, Kansas, *57*
Burnet County, Texas, 353
Burro Mesa fault zone, Texas, 429
Burro Mesa Member, South Rim Formation, Texas, 431, 432, 433, 439
Butler County, Kansas, 35
Butterly Dolomite, Oklahoma, 183, 185

Caballaro Formation, Texas, 393
Caballos Novaculite, 411, 413, 414, 418
Cabaniss Formation, Kansas, 75, 77
Cache Creek, Oklahoma, 122, 123
Cache Granite, Oklahoma, 119, 122
Caddell Formation, Texas, 381
Caddo County, Oklahoma, 103
Caddo Gap, Arkansas, *267*
Caddo Mountains, Arkansas, 269, 270, 271
Caddo River, Arkansas, 269, 271, 272
Calamites, 268
Camaraocrinus, 188
Cambrian
  Oklahoma, 93, 97, 99, 112, 119, 120, 121, 123, *127*, 147, 166, 177, 183
  Texas, *339*, 351
Camelops hesternus, 295
Camp San Saba, Texas, 339, 340
Cane Hill Member, Hale Formation, Arkansas, 89, 208, 210, 238
Caney Shale, Oklahoma, 184. *See also* Delaware Creek Shale
Canoe Formation, Texas, 431, 435
Canutillo Formation, Texas, 392, 393
Canville Limestone, Kansas, 51, 54, 55
Canyon Group, Texas, 317, 344, 350
Cap Mountain Limestone Member, Riley Formation, Texas, 352, 353
Cape Sable, Florida, 33
Capitan Limestone, Texas, 397, 398
Caprock Escarpments, Texas, 302, 305, 306, 307
Carboniferous, Oklahoma, 145, 149, 150, 196, 201
Carla Member, Vanoss Formation, Oklahoma, 159, 163
Carlile Shale, Kansas, 2, 57, 58
Carlsbad Caverns, New Mexico, 396
Carlsbad Formation, Texas, 396
Carlton Rhyolite Group, Oklahoma, *93*, 101, 111, *117*, 127, 129, 131, 132
Carroll County, Arkansas, 203
Carter County, Oklahoma, 171, 183, 190, 193
Casey Sandstone, Arkansas, 233
Cason Shale Formation, Arkansas, 208, 212, 215, 219, 222, 223
Cass Fault-Monocline System, Arkansas, 231
Castile Formation, Texas, 396
Castner Marble, Texas, 388, 391, 392
Catahoula Formation, Texas, *383*
Cecil Sprio Sandstone, Arkansas, 233, 234
Cedar Bluffs Till, Nebraska, 5
Cedar Creek, Arkansas, 252
Cedar Falls, Arkansas, 252
Cedar Park Limestone Member, Walnut Formation, Texas, 373, 376
Cedarton Shale, Texas, 350
Cement-Chickasha Anticline, Oklahoma, *103*
Central Lowland, 122
Central Mineral Region, Texas, 351. *See also* Llano Uplift
Central Texas Platform, 329
Ceratopea, 155
Cerro Castellan, Texas, *429*
Cerro Toledo Ash, 73
Chaetetes milleporaceous, 393
Chanute Shale, Kansas, 48
Chapman Ranch Fault, Oklahoma, 181, 185
Chappell Limestone, Texas, 356, 357
Chase Group, Kansas, 17, 25, 26
Chattanooga Shale, Arkansas, 203
chenier plain, Texas, 448
Cherokee County, Oklahoma, 86, 88
Cherokee County, Kansas, *75*
Cherokee County, Texas, 335
Cherokee Creek, Texas, 344
Cherokee cyclothem, Kansas, 75
Cherokee Group, Kansas, 75, 78
Cherry Bend area, Arkansas, *231*

Cherry Canyon Sandstone, Texas, 398
Chickachoc Chert Member, Wapanucka Formation, Oklahoma, 139, 141, 142
Chickasaw Creek Formation, Oklahoma, 150
Chihuahua, Mexico, *423*
Chihuahua Trough, Texas, 401, 402, 428
Childress County, Texas, 311
Childress Dolomite, Texas, 311, 315
Childress Gypsum, Texas, 311, 315
Chimneyhill Formation, Oklahoma, 223
Chinati Mountains, Texas, 431
Chisos Formation, Texas, 431, 432, 433, 435
Chisos Mountains, Texas, 429, 431, 432
Chisum Quarry Member, McCully Formation, Oklahoma, 87, 88
Choctaw Fault, Oklahoma, 145, 148, 150
Chondrites, 1
Christmas Mountains, Texas, 436, 439
Chromatiaceae, 99
Cimarron Evaporites, Oklahoma, 136
Cimarron River, Oklahoma, 137
Cisco Group, Texas, 309
Citronelle Formation, Arkansas, 226
Claiborne Group
  Arkansas, 225, 228, 229, 265, *283*
  Texas, *335*, 377, 380, 381
Clarita Limestone, Oklahoma, 184, 187
Clear Fork Group, Texas, 310, 312, 314
Cleveland County, Arkansas, 265
Click Formation, Texas, 364
Clifty Sandstone Formation, Arkansas, 203, 204
climatic changes
  Kansas, 19, 23, 37
  Oklahoma, 154, 177, 183
  Texas, 297, 303, 308, 446, 447
Cloud Chief Formation, Oklahoma, 104
Coahuila Platform, Texas, 401
Coal Creek Serpentinite, Texas, 362, 365
coal
  Arkansas, 77, 92
  Kansas, 75, 77, 78
  Oklahoma, 77, 92
  See also specific coal beds
Cochrane Limestone, Oklahoma, 184, 187
Codell Sandstone Member, Carlile Shale, Kansas, 2
Colbert Rhyolite, Oklahoma, 154, 177, 183
Colchester coal, Kansas, 78
Cold Springs Breccia, Oklahoma, 112, 113, 114, 115
Collingnoniceras woollgari, 59
Collings Ranch Conglomerate, Oklahoma, 159, 163, 179, 180, 184, 185, 187
Collings Ranch Graben, Oklahoma, 187
Colorado River, 445
Colorado River basin, Texas, 351
Colorado River Valley, Texas, *343*
Comanche County, Oklahoma, 79, 118
Comanche Peak Limestone Formation, Texas, 327, 376
Compton Member, St. Joe Limestone, Arkansas, 204
Concho Platform, Texas, 344
Conosticthus, 250
Conway County, Arkansas, 250, 251
Cook Creek Formation, Oklahoma, 97, 130, 132, 153, 155, 157, 181, 183, 185
Cook Mountain Formation, 335
Cooks Formation, Texas, 392
Cookson, Oklahoma, 88, 89
corals, Oklahoma, 87, 89, 142
Corbin Ranch Submember, Pooleville Member, Bromide Formation, Oklahoma, 172, 173, 174
Corbula, 376
Coronado Hills Conglomerate, Texas, 391
Cossatot, Arkansas, 269, 270
Coteau des Prairies, Minnesota, 9
Cotter Formation, Arkansas, 203, 204
Cottle County, Texas, 311
Cottonwood Limestone Member, Beattie Limestone, Kansas, 35, 37, 39, 41
Council Grove Group, Kansas, 11, 33, 35
Cow Creek Limestone Formation, Texas, 373, 374, 375, 376
Cox Sandstone Formation, Texas, 402, 403, 404, 405
Crassostrea, 67
Cravenocercas sp., 235
Cretaceous
  Arkansas, 281
  Kansas, 1, 2, 8, *57*, 61, 64, *67*, 70
  Oklahoma, 177
  Texas, 323, *401*, 431
Cribratina texana, 425
Criner Hills uplift, Oklahoma, 127, 189
Crockett Formation, Texas, 380, 381
Crooked Creek, Kansas, 69, 70
Crooked Creek fault, Kansas, 69
Crooked Creek Formation, Kansas, 69, 70, 71, 72, 73
Crosbyton Member, Couch Formation, Texas, 301
Cross County, Arkansas, 228, 229
Cross Mountain, Oklahoma, 122
Crouse Limestone, Kansas, *29*
Croweburg coal bed
  Kansas, 75, 77, 78
  Missouri, 77
  Oklahoma, 77
Croweburg cyclothem, Kansas, 75, 77
Crowley's Ridge, Arkansas, 225
Cuchillo Formation, Texas, 402
Culberson County, Texas, 395
Cuneiphycus, 87
cyclothems, Kansas, 17, 18, 25, 26, 35, *43*, 75. *See also specific cyclothems*

Dagger Flat Formation, Texas, 411, 414
Dakota Formation, Kansas, 58, *61*, 67, 68
Dakota ice lobe, 9
Davis Mountains, Texas, 402, *407*, 411
DeGray Lake, Arkansas, *273*
Del Carmen Formation, Texas, 426, 427
Del Carmen Mountains, Texas, 429
Del Rio Shale Formation, Texas, 402, 405, 425, 426
Delaho Formation, Texas, 431, 433, 437
Delaho Sub-block, Texas, 429, 431, 432
Delaware Basin, Texas, 395, 396
Delaware Creek Shale, Oklahoma, 184
Delaware Mountain Group, Texas, 397, 398, 400
Delepinoceras, 235
Dennis cyclothem, Kansas, 44, 50, 51, 52, 54
Denver Basin, 60
DeQueen Formation, Arkansas, *281*
Derbyia sp., 38
Desha Basin, Arkansas, 225, 265
Devil Ridge Mountains, Texas, 403
Devil's Graveyard Formation, Texas, 431
Devils Backbone
  Kansas, 63
  Oklahoma, 197
Devonian
  Arkansas, 203, 204, 207, 212
  Oklahoma, 183
  Texas, 357, 392
Diablo Platform, Texas, 401, 402, 404
Dictyoconus sp., 376, 402
Didymodon tophaceus, 154
Didymograptus, 185
Dimmit County, Texas, 443
Dimple Formation, Texas, 411, 414, 415
Dirocerithium, 337
Dirvanella, 339
Dobbs Valley Member, Mingus Formation, Texas, 317, 318, 319
Dockum Basin, Texas, 305
Dockum Formation, Texas, 290, 305, 306, 307
Dog Canyon, Texas, 420, 421
Dog Creek Shale, Oklahoma, 83, 315
Donax, 453
Donezella, 143
Doniphan County, Kansas, 5, 8
Douvilleiceras sp., 376, 406
Drew County, Arkansas, 265
Dufrenoya justinae, 374
Duncan Member, San Angelo Formation, Texas, 310, 314

Dunveganoceras pondi, 59
Dutch Creek Mountain, Arkansas, 247
Dye Shale Member, Bloyd Shale, Arkansas, 87, 88, 89, 90

Eagle Ford Group, Texas, 329, 331, 332
Eagle Ford Prairie, Texas, 332
Eagle Mountains
  Oklahoma, 117, 122
  Texas, 403
earthquakes
  Oklahoma, 80
  Texas, 437
Easly Creek Shale, Kansas, 29, 30
East Mountain Shale, Texas, 320
East Texas Basin, 323, 328, *329*, 402
East Timbered Hills, Oklahoma, 183
Eastern Lowlands, 122, 227
Eastland County, Texas, 323
Echinocorys texana, 444
Edwards Group, Texas, 402
Edwards Limestone Formation, Texas, 327, 328
Edwards Plateau, Texas, 351, 361, 367, 411
Eiss Limestone Member, Bader Limestone, Kansas, 37
El Capitan, Texas, 396, 398
El Paso Group, Texas, 392
Elk Creek, Oklahoma, 88
Elk Mountain, Oklahoma, 117, 122
Ellenburger Group, Texas, 353, 354, 356, 357, 375
Ellsworth County, Kansas, 67
Elm Creek, Texas, 442
Enchanted Rock batholith, Texas, 369
Enchanted Rock dome, Texas, 366, *369*
Engonoceras belviderense, 67
Eocene
  Arkansas, 225, 259, *265*, 283
  Texas, 335
Epiphyton, 339
Equus sp., 225
Erath County, Texas, 318
Ernst Member, Boquillas Formation, Texas, 425, 426, 428
Escondido Formation, Texas, 441
Eskridge Shale, Kansas, 37, 41
Eumorphoceras sp., 235
Eureka Quartzite, 170
Evansville Mountain, Arkansas, 85, 88, 89, 90
evaporites, Roman Nose State Park, Oklahoma, *83*
Everton Dolomite Formation
  Arkansas, 207, 208, 210, 212, 220
  Missouri, 170
Exogyra ponderosa, 442, 443, 444

Fairport Chalk Member, Carlisle Shale, Kansas, 57, 58, 59
Falls Creek, Oklahoma, 153
Farley Limestone, Kansas, 50
faults
  Arkansas, 241, 247
  Kansas, 7, 11
  Oklahoma, *79*, 103, 119, 122, 124, 129, 130, 131, 133, *145*, *177*, 185
  Texas, *411*
  See also specific faults
Fayetteville Shale Formation
  Arkansas, 208
  Oklahoma, 88
Fayettevillea, 235
Fernvale Limestone Formation
  Arkansas, *208*, 212, 213, 215, 216, 219, 220, 221, 222
  Oklahoma 173, 174
Finlay Limestone Formation, Texas, 402, 403, 404, 406
Flemming coal bed, Kansas, 78
Flint Hills, Kansas, 11, 18
Florena Shale Member, Beattie Limestone, Kansas, 35, 37, 41
Florence Limestone Member, Barnestone Limestone, Kansas, 25, 26
Florida Bay, 39
Florida Mountains Formation, Texas, 392
Flowerpot Member, San Angelo formation, Texas, 310
Floyd County, Texas, 287
Follets Island, Texas, 451
foraminifera
  Arkansas, 235
  Kansas, 1, 29, 33, 60
  Texas, 402, 404, 425
  See also specific genus and species
Fort Hays Limestone Member, Niobrara Chalk, Kansas, *1*
Fort Peña Formation, Texas, 411
Fort Sill Limestone Formation, Oklahoma, 95, 97, 99, 119, 122, 130, 132, 183, 185
Fort Sill North Anticline, Oklahoma, 123, 125
Fort Sill South Anticline, Oklahoma, 113
Fort Worth Basin, Texas, 309, 344, 358
fossils
  Oklahoma, 83, 87, 88, 97, 102, 119, 139, 146, 152, 174, 175, 183, 187
  Kansas, 1, 26, 30, 37, 46, 63, 67, 69, 70, 72
  Texas, 289, 302, 319, 332, 337, 357, 377, 444
Fourche La Fave, Arkansas, 249
Fourche Mountain, Arkansas, 259. *See also* Granite Mountain
Franklin Mountains
  New Mexico, 387
  Texas, *387*
Frazier Canyon Formation, Texas, 409
Fredericksburg Group, Texas, 323, 327, 423
Fremont cliffs, Nebraska, 9
French Lake Dam, Oklahoma 116, 117, 118
Frisbie Limestone, Kansas, 50
Frontal Fault Zone, Oklahoma, 79, 122, 124, 129
Frontal Ouachita Mountain Province, 249
Frost Bluff, Texas, 378
Fusselman Canyon, Texas, 387
Fusselman Dolomite Formation, Texas, 392, 393

Gaither Mountain Sandstone, Arkansas, 208, 210
Galesburg Shale
  Kansas, 50, 54
  Oklahoma, 54
Galveston Bay, Texas, 450
Game Refuge Formation, Oklahoma, 151
Gaptank Formation, Texas, 411
Garber Sandstone, Oklahoma, 123
Garland County, Arkansas, 259
Garza County, Texas, 307
Geary County, Kansas, 15, 17, 25, 26, 30, 31
Geisina, 33
Georgetown Limestone Formation, Texas, 327, 328, 402
Gillespie County, Texas, 364, *369*
Girvanella, 339
glaciation
  Kansas, 5, 72, 73
  Texas, 445
Glass Mountains, Texas, 411
Glen Mountains Layered Complex (GMLC), Oklahoma, *112*
Glen Rose Limestone Formation, Texas, 323, 324, 327, 373, 374, 402, 419, 426, 427
Globovalvulina sp., 33
Goat Mountain, Texas, 432, 433
Goat Mountain Member, Chisos Formation, Texas, 431, 433
Gober Chalk, Texas, 441
Goddard Formation, Oklahoma, 184, 190
Goen Limestone, Texas, 319
Golf Course Formation, Oklahoma, *189*
Gomez Ash-Flow Tuff, Texas, 407, 409
Gondolella, 48, 50, 52
Gorman Falls, Texas, 355
Gorman Formation, Texas, 355, 356
Grady County, Oklahoma 103, 104
Graneros Shale, Kansas, 57, 58, 59, 61
Granite Mountain
  Arkansas, *259*
  Texas, 365
Grant County, Arkansas, 265
Grayson Marl Formation, Texas, 328, 331
Great Plains, 300, 301, *395*
Great Salt Plains, Oklahoma, *135*
Green County Arkansas, 227
Greenhorn cyclothem, Kansas, 2, 58
Greenhorn Limestone, Kansas, 57, 58, 59
Greenleaf Lake Limestone Member,

McCully Formation, Oklahoma, 87, 88, 89
Greenwood County, Kansas, 35
Greenwood Shoal, Kansas, 39, 41
Grimes County, Texas, 382
Grindstone Creek Formation, Texas, 317
Grove Church Shale, Arkansas, 237
Gryphaea, 332
Guadalupe Delta, Texas, 451, 452
Guadalupe Mountains, Texas, *395*
Guadalupe River, Texas, 451, 452
Guadalupe River basin, Texas, 351
Gulf Basin, Texas, 446
Gulf Coast, Texas, *377*, 445
Gulf of Gaeta, Italy, 146
Gulfian Section, Texas, *329*
Gyrolithes, 381

Hale Formation, Arkansas, 88, 89, 90, 208, 210, 231
Hall County, Texas, 287
Halodule wrightii, 457
Hamburg Klippe, Appalachians, 172, 176
Hamilton County, Texas, 327
Hammett Shale Formation, Texas, 373, 374, 375
Hansel Sand Formation, Texas, 373, 374
Haragan Marlstone Formation, Oklahoma, 182, 184, 187
Hardeman County, Texas, 313, 315
Harding Formation, Colorado, 170
Hartland Shale Member, Greenhorn Limestone, Kansas, 58, 59
Hartshorne Coal, Arkansas, 241
Hartshorne Quarry, Oklahoma, 143
Hartshorne Sandstone
Arkansas, 241, 243, 245, 246, 247, 249, 251
Arkansas-Oklahoma Border, *91*
Hatton Tuff, Oklahoma, 198
Haymond Formation, Texas, 411, 414
Haymond thrust fault, Texas, 414
Hays County, Texas, 425
Hells Half Acre thrust fault, Texas, 413
Helminthopsis, 152
Helms Formation, Texas, 392, 393
Hemmed-In Hollows, Arkansas, 209
Hennessy Shales Group, Oklahoma, 113, 122, 123, 136
Henryhouse Formation, Oklahoma, 182, 184, 187, 224
Hickory Creek Shale, Kansas, 52
Hickory Sandstone Member, Riley Formation, Texas, 351, 352, 353, 366
High Island, Texas, 449
High Plains province, 63, 64, 69, 293, 299, 301, *305*
Highland Quarry, Arkansas, *281*
Hill County, Texas, 323, 328, 329, 331, 333
Hinesville Limestone Member, Batesville Formation, Arkansas, 208
Hog Mountain Sandstone Member, Mineral Wells formation, Texas, 320
Hollis Basin, Oklahoma, 127
Holmes Creek, Texas, 290
Holmesina septentrionale, 295
Holmsville Shale, Kansas, 26
Holocene
Oklahoma, 70, 119, 123
Texas, 293
Honey Creek, Oklahoma, 153, 154
Honey Creek Limestone Formation, Oklahoma, 94, 99, 102, 127, 130, 132, 133, 183
Honey Creek Valley, Oklahoma, 153
Honey Formation, Texas, 364
Honeycut Formation, Texas, 354, 355, 356, 357
Hood County, Texas, 324
Hoover Point, Texas, 353, 354
Horsehead Lake, Arkansas, *241*
Horseshoe Mountain, Arkansas, 247
Hosston Sand Formation, Texas, 374, 402
Hot Spring County, Arkansas, 259
Hot Springs Mountain, Arkansas, 263
Hot Springs, Arkansas, *263*
Hot Springs National Park, Arkansas, 263
Hot Springs Sandstone Member, Stanley Shale, Arkansas, 263
Houy Formation, Texas, 357
Howard County, Arkansas, 281
Huckleberry Ridge Ash, Kansas, 70, 72
Hudspeth County, Texas, 395
Hueco Canyon Formation, Texas, 392
Hueco Group, Texas, 392
Huelster Formation, Texas, 407, 409
Huhatanka kiowana, 67
Hunton arch, Oklahoma, 178, 181, 182
Hunton Group, Oklahoma, 127, 179, 182, 184, 187
Hushpuckney Shale, Kansas, 50, 52

Idiognathodus sp., 48, 50, 52, 193
Idiognathoides noduliferous, 88
Idioprioniodus, 48, 50, 52
Illa coal, Indiana, 77
Illinois Basin, 152, 273
Imo Formation, Arkansas, 235, 237
Independence County, Arkansas, 212, 221, 222, 223
Indian Creek, Arkansas, 208
Indio Mountains, Texas, 403
Ingleside barrier, Texas, 452
Inks Lake, Texas, 362
Inoceramus sp., 1, 59, 67, 332
Iola cyclothem, Kansas, 48, 50, 54, 55
Irving sycline, Kansas, 11, 29, 31, 33
Ischadites, 185
Island Creek Shale, Kansas, 50
Ives Breccia, Texas, 357
Izard County, Arkansas, 222

Jackfork Group, Oklahoma, 149, 150, 151, 152
Jackson County, Kansas, 29
Jackson Group
Arkansas, 225, 230, *265*
Texas, 377, 381, 385
Jasper County, Texas, 383
Javelina Formation, Texas, 431, 433
Jefferson County, Arkansas, 265
Jenkins Sandstone, Arkansas, 233
Jetmore Chalk Member, Greenhorn Limestone, Kansas, 58, 59
Jim Bluff graben, Arkansas, 212, 215, 217, 219
Johns Valley Shale Formation
Arkansas, 254, 273, 277
Oklahoma, 145, 150
Johnson County, Arkansas, 241
Johnson County, Kansas, 43
Johnston County, Oklahoma, *165*
Joins Formation, Oklahoma, 165, 166, 169, 183, 185
Jolliff Limestone Member, Golf Course Formation, Oklahoma, 193
Jose Formation, Texas, 392

Kaimichi Shale, Texas, 402, 405
Kanapolis Reservoir, Kansas, 67, 68
Kansas glaciation, 5, 9
Kansas River valley, Kansas, 9
Kansas Shelf Lagoon, 39
Kansas Till, 5, 6
Keel Limestone Member, Hunton Group, Oklahoma, 188
Keel Oolite, Oklahoma, 184
Kessler Limestone Member, Bloyd Formation, Arkansas, 85, 88, 89, 90
Ketch Creek, Oklahoma, 133
Keys Valley Marl Member, Walnut Formation, Texas, 376
Kiamichi Mountain, Oklahoma, *149*
Kimbell anticline, Oklahoma, 129, 131
Kimbell Mountain, Oklahoma, 131
Kimmswick Limestone, Arkansas, 212, 215, 219, 221, 223
Kincaid Formation, Texas, 378, 380
Kindblade Formation, Oklahoma, 97, 133, 155, 180, 181, 183, 185
King County, Texas, 311
King Mountain, Oklahoma, 109
Kingsdown Formation, Kansas, 72
Kinney County, Texas, 441, 442
Kinney Limestone, 18, 19, 21, 23
Kiowa County, Kansas, 67
Kiowa County, Oklahoma, 93, 99
Kiowa Formation, Kansas, 67
Konickophyllum, 142

La Tuna Formation, Texas, 392, 393
Labette County, Kansas, 44
Ladore Shale, Kansas, 50
Lafayette Gravel, Arkansas, 226
Lafferty Limestone, Arkansas, 213, 221, 222, 223, 224
Laguna Madre, Texas, 445, 454, 455, 456
Lake Ardmore Sandstone Member, Springer Formation, Oklahoma,

184, 190, 192, 194
Lake Ellsworth, Oklahoma, 123
Lake Elmer Thomas, Oklahoma, 122
Lake Lawtonka, Oklahoma, 122
Lake Pinto Sandstone Member, Mineral Wells Formation, Texas, 320
Lake Somerville, Texas, 382
Lake Tom Steed, Oklahoma, 112, 122
Lake Travis, Texas, 373
Lake Valley Group, Texas, 393
Lamar Limestone Member, Bell Canyon Formation, Texas, 396, 397
Lampass Cut Plain, Texas, 351
Lane Shale, Kansas, 50
Lane-Bonner Springs Shale, Kansas, 52
Lanoria Quartzite, Texas, 388, 391
Laramide deformation, 412, 417, 420, 423, 431, 435, 436, 440
Las Cruces Formation, Texas, 392, 393
Las Vigas Formation, Texas, 402, 404
Laurel Limestone, Tennessee, 223
Laverne Formation, Kansas, 70
Lee Creek, Arkansas, *239*
LeFlore County, Oklahoma, 149
Limestone County, Texas, 329, 333
Limestone Hills, Oklahoma, 121, 123
Lincoln County, Arkansas, 265
Lincoln County, Kansas, 61
Lincoln Limestone Member, Greenhorn Limestone, Kansas, 58
Lingos Formation, Texas, *287*
Linoproductus nodosus, 88
Lion Mountain Sandstone Member, Riley Formation, Texas, 353
Littig Member, Kincaid Formation, Texas, 378
Little Medicine Creek, Oklahoma, 122, 124
Little Red River, Texas, 290
Little Rock, Arkansas, *255*
Llano County, Texas, 351, 352, 366, *369*
Llano Estacado, Texas, 299, 300, 305
Llano Group, Texas, 361, 362
Llano River, Texas, 364, 365
Llano Uplift, Texas, 305, 344, 345, 348, 351, 361
Loma Plata Limestone Formation, Texas, 402, 405, 406
Longhorn Cavern, Texas, 354, 355
Lophoctenium, 152
Lost Creek Gneiss, Texas, 362, 363
Lost Valley, Arkansas, 210
Love Hollow Quarry, Arkansas, 221, 222
Loveland Silt, Arkansas, 226, 227, 230
Lower Kansas Till, Kansas, 5, 8, 9
Lower Kittanning coal, Appalachians, 77
Lowtonka graben, Oklahoma, 129, 130
Lubbock Lake, Texas, 293, 295, 297
Lugert Granite, Oklahoma, 110, 111
Lutetia, 337
Lynn Mountain Formation, Oklahoma, 145
Lynn Mountain Syncline, Oklahoma, 150

McAlester Formation, Arkansas, 243, 246, 252, 253
McCredie Formation, Missouri, 9
McCulloch County, Texas, 339, 352
McCully Formation, Oklahoma, *86*
McCutcheon fault, Texas, 409
McKelligon Canyon, Texas, 392
McKelligon Formation, Texas, 392
McKenzie Hill Limestone Formation, Oklahoma, 97, 119, 129, 132, 154, 155, 181, 183, 185
McKittrick Canyon, Texas, 397
McLish Formation, Oklahoma, 166, 179, 183, 185, 187
Madison County, Kentucky, 222
Magazine Mountain, Arkansas, 243
Magdelena Group, Texas, 392
Magnet Cove, Arkansas, *259*
Mammites nodosoides, 59
Mammuthus columbi, 295
Manhattan, Kansas, 29
Manning Formation, Texas, 382
Marathon Basin, Texas, 411, 418, 420
Marathon region, Texas, 175, 176
Marathon Uplift, Texas, 305, 402, *411*
Maravillas Chert Formation, Texas, 175, 176, 411, 413, 414, 418
Marble Falls Limestone Formation, Texas, 343, 344, 345, 347, 348, 358, 359, 366, 375
Marietta Basin, Oklahoma, 189
Markham Mill Formation, Oklahoma, 150
Marlow Formation, Oklahoma, 103
Marnet Cove, Arkansas, 259
Mason County, Texas, 351, 353
Maxon Sandstone, Texas, 402
Mazarn Basin, Arkansas, 263, *267*
Meade County, Kansas, *69*
Meade Formation, Kansas, 70
Mecca Quarry Shale Member, Linton Formation, Indiana, 77
Medicine Creek, Oklahoma, 122, 125
Medicine Park Granite, Oklahoma, 119
Medina County, Texas, 443
Meers Fault, Oklahoma, *79*, 119, 122, 125, 129
Meers Quartzite, Oklahoma, 119, 120, 121, 124, 125
Meers Valley, Oklahoma, 80, 97, 122, 123, 131
Menabites sp., 441
Merriam Limestone, Kansas, 52
Mesa de Anguiila, Texas, 424, 429, 431
Mesozoic, Oklahoma, 196
Mexia Fault Zone, Texas, 333
Mexia Member, Wills Point Formation, Texas, 380
Mictomys, 72
Midcontinent rift system, Kansas, 11, 15, 44
Midcontinent shelf, Kansas, 44, 47, 48
Middle Creek Limestone, Kansas, 50
Midland Basin, Texas, 305, 309, 310, 344
Midway Group
  Arkansas, 259
  Texas, 329, 333, 378
Milam Chalk, Texas, 441, 442, 444
Milankovitch orbital variations, 3, 48, 57, 59
Mill Creek fault, Oklahoma, 178
Mill Creek syncline, Oklahoma, 178
Miller County, Arkansas, 267, 285
Mills County, Texas, 343
Mineral coal bed, Kansas, 78
Mineral Wells Formation, Texas, 317, 320, 323
Mingus Shale Formation, Texas, 317, 318, 319
Minnesota ice lobe, 9
Miocene, Kansas, 63, 69, 70, 73
Mississippi River Embayment, 225, 226
Mississippian
  Arkansas, 203, 207, 208, 211, 214, 220, 235, 263
  Oklahoma, 86, 113, 121, 150, 151, 154, 183, 189, 196
  Texas, 357, 358, 392, 418
Missler Member, Ballard Formation, Kansas, 71
Missouri Mountain Shale Formation
  Arkansas, 258, 263, 271
  Oklahoma, 197
Missouri River valley, 9
Mitchell County, Texas, 306
Moccasin Creek gypsum member, Cloud Chief Formation, Oklahoma, 104, 105
Monopleura sp., 443
Montague County, Texas, 311
Montgomery County, Arkansas, 267
Montoya Group, Texas, 392
Moore Hollow Group, Texas, 351, 353, 354, 355
Morgan Creek Limestone Member, Wilberns Formation, Texas, 339, 353, 354, 355, 363
Morrill Limestone Member, Beattie Limestone, Kansas, 35, 37, 38, 41
Morris County, Kansas, 15
Moseley Limestone, Texas, 381
Mounds Member, McKelligon Formation, Texas, 392
Mt. Blanco, Texas, 294, 302
Mt. Enterprise Fault System, Texas, 337
Mt. Lincoln, Oklahoma, 117, 122
Mount Locke Formation, Texas, 409
Mt. Lugert, Oklahoma, 109
Mt. McKinley, Oklahoma, 122
Mount Magazine, Arkansas, 243
Mount Magazine syncline, Arkansas, 246

Mt. Marcy, Oklahoma, 117
Mt. Roosevelt, Oklahoma, 124
Mount Scott, Oklahoma, 119, 121, 122
Mount Scott Granite, Oklahoma, *116*
Mt. Sheridan, Oklahoma, 122, 124
Mt. Sheridan Gabbro Member, Roosevelt Gabbros, Oklahoma, 119, 121, 123, 124, 125
Mt. Sherman, Oklahoma, 122
Mount Wall, Oklahoma, 122
Mountain Fork Creek, Arkansas, 232
Mountain Fork River, Oklahoma, 195
Mountain Lake Member, Bromide Formation, Oklahoma, 185, 187
Mountain View Fault, Oklahoma, 122, 129
Moyers Formation, Oklahoma, 150
Mulberry fault system, Arkansas, 241
Mulberry River, Arkansas, 232
Mule Ear Springs Tuff Member, Chisos Formation, Texas, 433, 435
Muncie Creek Shale, Kansas, 48, 50, 52
Mundy Breccia, Texas, 388, 391
Murray County, Oklahoma, 183
Muskogee County, Oklahoma, 85, 86
Mytiloides sp., 59

Nameless Canyon Member, Scenic Drive Formation, Texas, 392
Navajoe Mountain Group, Oklahoma, 121
Navarro Group, Texas, 329, 333
Nebraska Shelf, 39
Nebraskan glacier, 5
Nebraskan Till, 9
Nemaha anticline, Kansas, 11, 14, 29, 33
Nemaha Ridge, Oklahoma, 124
Neochonetes granulifera, 37
Neognathodus sp., 88, 193
Neonereites, 152
Nereites, 152
Nescatunga Gypsum, Oklahoma, 83, 84
Newton County, Arkansas, 207, 208, 217
Newton Sandstone Member, Everton Formation, Arkansas, 207, 208, 210
Neylandville Marl, Texas, 333
Nickerson Till, Nebraska, 5, 9
Niger Province, Nigeria, 387, 392
Ninnescah Formation, Kansas, 67
Niobrara Chalk, Kansas, *1*, 64
Niobrara cyclothem, Kansas, 2
North American craton, 171
North Canadian River, 84
North Padre Island, Texas, 454
North Wichita Member, Blaine Formation, Texas, 315
Northview Mamber, St. Joe Limestone, Arkansas, 204
Nucula, 67, 337
Nuculana, 67
Nuia, 339

Oatman Creek, Texas, 361
Oatman Creek Granite, Texas, 362, 367
Octavia Fault, Oklahoma, 150
Ogallala Formation
  Colorado, 63
  Kansas, *63*, 69, 70, 73
  Nebraska, 63
  Oklahoma, 63
  Texas, 63, 290, *299*, 305, 307, 308
  Wyoming, 63
Oil Creek Formation, Oklahoma, *165*, 179, 183, 185
Oketo Shale Member, Barneston Limestone, Kansas, 25
Oklahoma City Field, Oklahoma, 183
Oldham County, Texas, 305
Omaha, Nebraska, 51
Ophiomorpha, 335
Orbitolina, 404, 406
Ordovician
  Arkansas, 203, 204, 207, 208, 210, 211, *215*, 263
  Oklahoma, 123, *127*, *153*, 165, 166, 169, *171*, 179, 183, 185, 196, 198, 200
  Texas, 357, 392
Osage Plains Physiographic Province, Kansas, 2
Osagia, 37, 39
Ostrea beloiti, 57
Ouachita County, Arkansas, 284
Ouachita frontal thrust belt, Oklahoma, *145*, 189
Ouachita Geosyncline, Oklahoma, *171*
Ouachita Mountain Province, 249
Ouachita Mountains, *139*, *149*, *171*, 177, 195, 196, 243, 247, 249, 254, 255, 263, *267*, 273, 360
Ouachita orogen, 91, 122, 145, 210, 211, 243, 271, 411
Overbook Member, Springs Formation, Oklahoma, 190, 192, 193
Ozan Formation, Texas, 333
Ozark Dome, Arkansas, 203, 211, 215, 247
Ozark Plateaus, Arkansas, *211*, 215, 222
Ozark uplift, 91, 147, 152

Packsaddle Mountain, Texas, 353
Packsaddle Schist, Texas, 352, 361, 369, 371
Pacydiscus sp., 441
Padre Island, Texas, 454, 455, 456, 457
Paisano Volcano, Texas, 407, 410
Paleocene, Arkansas, 259
Paleozoic
  Arkansas, 203, *207*, 214, *249*, 255, 259
  Oklahoma, 79, *93*, 127, 163, 169, 172, 177, 178, *183*, 200
  Texas, 308, *351*, *387*, 412, 413, 418, 419
Pallaea atropurpurea, 155

Palo Duro Basin, Texas, 311
Palo Duro Canyon, Texas, 305, 307
Palo Pinto County, Texas, 317, 318, 321
Palo Pinto Limestone Formation, Texas, 317, 321
Paluxy Sand, Texas, 402
Panther Seep Formation, Texas, 392, 393
Paola Limestone, Kansas, 48, 50
Paradox anticline, Oklahoma, 129, 131
Parapachydiscus sp., 441
Parker County, Texas, 317, 318
Pastoria Sand Member, White Bluff Formation, Arkansas, 266
Patterson Sandstone, Arkansas, 233
Pease River Group, Texas, 310, 314, 315
Pecan Bayou, Texas, 346
Pecan Gap Chalk, Texas, 333, 441, 444
Pecos River, Texas, 301
Pedernales River, Texas, 374
Pedernal Uplift, Texas, 399
Pellaea atropurpurea, 155
Pen Formation, Texas, 419
Peña Mountains, Texas, 432
Pennsylvanian
  Arkansas, 235, 243, 247, 249, 254, 267
  Kansas, 26, 28, 36, 43, 44, 54
  Oklahoma, 79, 97, 113, 122, 123, 124, *127*, *139*, 147, 150, 151, *159*, 177, 179, 184, 189
  Texas, *317*, *343*, 351, 358, 392
Peoria Loess, Arkansas, 226, 227, 229
Percha Shale, Texas, 392, 393
Permian
  Kansas, 11, 17, *25*, 29, *35*, 70
  Oklahoma, 79, 80, 83, 94, 97, 101, 103, 104, 110, 113, 116, 121, 122, 124, 127, 135
  Texas, 307, 308, *309*, *395*
Permian Basin, Oklahoma, 83, 135
Permophorus sp., 30, 33, 83
Perry County, Arkansas, 251, 253
Perry Mountain, Arkansas, 249, 253
Persimmon Gap, Big Bend National Park, Texas, *417*
Petalaxis, 87, 89
Petit Jean Mountain, Arkansas, 249, 251, 252
Petrolia Formation, Texas, 311
Peyton Creek, Arkansas, 235
Pfeifer Shale Member, Greenhorn Limestone, Kansas, 59
Phillips Creek, Oklahoma, 194
Philonotis calcares, 154
Phycosiphon, 152
Pierson Member, St. Joe Limestone, Arkansas, 204
Pike County, Arkansas, 267, 281
Pine Canyon Caldera, Texas, 431, 432, 439
Pinto Canyon, Texas, 406
Pisgah Member, Kincaid Formation, Texas, 378, 380

Pistoll Range Member, McKelligan Formation, Texas, 392
Pitkin Limestone
Arkansas, 214, 235
Oklahoma, 86
Pittsburg County, Oklahoma, *139*
Pitymys sp., 72, 73
Plains Border province, 69
Planolites, 1, 62, 166
Plattin Limestone, Arkansas, 208, 209, 212, *215*, 221
Plattsburg cyclotherm, Kansas, 44, 52, 54
Pleistocene
Arkansas, 205, 226, 228
Kansas, 5, 9, *69*
Oklahoma, 116, *153*
Texas, *293*, *429*
Plicochonetes arkansanus, 88
Pliocene
Arkansas, 226, 228
Kansas, 63, *69*
Texas, *293*, 307
Point Peak Member, Wilberns Formation, Texas, 339, 340, 354, 355, 356
Polk Bayou Limestone, Arkansas, 223. *See also* Fernvale Limestone
Polk Creek Shale
Arkansas, 256, 258, 263
Oklahoma, 175, 199, 200, 201
Ponca region, Arkansas, *207*
Pontotoc County, Oklahoma, 171
Pooleville Member, Bromide Formation, Oklahoma, 172, 174, 175, 179, 185, 187
Portals River, Texas, 301
Post Oak Conglomerate, Oklahoma, 80, 94, 97, 116, 117, *119*, 127, 129, 132, 133
Potato Hills, Oklahoma, 200
Poteau syncline, Arkansas, 247
Pottawatomie County, Kansas, 11
Powell Dolomote Formation, Arkansas, 203, 204, *207*, 212
Prairie Grove Member, Hale Formation, Arkansas, 88, 89, 90, 208, 210
Prairie Mountain Formation, Oklahoma, 150
Pratt Hill, Oklahoma, 122
Precambrian
Oklahoma, 177, 183
Texas, 351, *361*, *387*
Presidio County, Texas, 423
Presidio Formation, Texas, 406
Primrose Member, Golf Course Formation, Oklahoma, 190, 193, 194
Proplacenticeras sp., 441
Pseduorbitoides israelskyi, 442
Pseudhipparion gratum, 70
Pteria salinensis, 67
Pulaski County, Arkansas, 259
Pulchrilamina sp., 97, 392
Pycnodonte mutabilis, 444

Quanah Granite, Oklahoma, 110, 116, 117, 118, 119, 122
Quanah Mountain, Oklahoma, 117, 122
Quanah Parker Lake, Oklahoma, 117
Quartermaster Group, Texas, 287, 290, 305, 307, 311
Quartz Mountain, Oklahoma, 109, 111, 117, 118
Quartz Mountain State Park, Oklahoma, *109*
Quaternary
Kansas, 15
Oklahoma, 122, 135, 136
Texas, 287, 290, 293, 431, *445*
Queen City Formation, Texas, *335*
Quindaro Shale, Kansas, 50
Quitman Mountains, Texas, 403

Rader Limestone Member, Bell Canyon Formation, Texas, 396, 398
Raggedy Mountains, Oklahoma, 112, 122
Raggedy Mountains Gabbro Group, Oklahoma, 121
Rancheria Formation, Texas, 392, 393
Randall County, Texas, 307
Rastellum quadriplicata, 67
Rattlesnake Bluff, Oklahoma, 197, 198
Rattlesnake Mountains, Texas, 432, 437
Raytown Limestone, Kansas, 48, 50
Reagan fault, Oklahoma, 178, 179
Reagan Sandstone Formation, Oklahoma, 94, 99, 101, 102, 131, 132, 133, 177, 183
Receptaculites, 219
red beds
Oklahoma, *83*, 103, 105, 185
Texas, 302, 308, 310, 311
Red Bluff Granite Complex, Texas, 388, 391
Red Hill, Oklahoma, 129, 131
Red Mountain Gneiss, Texas, 361, 365
Red River, Oklahoma, 123
Redfield Formation, Arkansas, 265, 266
Reef Escarpment, Texas, 397, 398
Reformatory Granite, Oklahoma, 110, 111
Reklaw Formation, Texas, 335
Renalcis, 339
Rexroad Foramtion, Kansas, 69, 70, 73
Riley County, Kansas, 15, 17, 26, 29, 35
Riley Formation, Texas, 351, 353
Riley Mountains, Texas, 353, 356
Ring Top Mountain, Oklahoma, 131
Rio Grande, 404, 423, 445, 455
Rio Grande delta, Texas, 456, 457
Rio Grande Embayment, Texas, 335, 383, 455
Rio Grande Rift, 399, 423
Roaring Springs fault, Texas, 354
Rock Creek
Oklahoma, 159
Texas, 328
Rockdale Formation, Texas, 380
Rocktown channel sandstone body, Kansas, 61
Rocky Mountains, 69, 70, 71, 73, 300, 301
Rod Club Member, Springer Formation, Oklahoma, 190, 192, 193
Rolling Plains, Texas, *287*
Roman Nose State Park, Oklahoma, 83
Rooks County, Kansas, 1
Roosevelt Gabbros, Oklahoma, 113, 121, 123, 125
Rose Creek Mountain, Arkansas, 252
Ross Creek fault, Arkansas, 249, 254
Rough Ridge Formation, Texas, 364
Round Mountain
Arkansas, 252
Texas, 400
Roxana Silt, Arkansas, 226, 227, 229, 230
Royer Dolomite Formation, Oklahoma, 93, 95, 97, 181, 183, 185
Running Water Draw, New Mexico, 301
Ruppia maritima, 457
Rush Springs Formation, Oklahoma, 103, 104
Russell County, Kansas, 57

Sabine Uplift, Texas, 335
Saddle Mountain, Oklahoma, 123
Saddle Mountain Creek, Oklahoma, 99
St. Francis County, Arkansas, 230, 265
St. Joe Limestone Member, Boone Formation, Arkansas, 203, 204, 208, 210, 212, 220
St. Peter Sandstone
Arkansas, 170, 211, 212, *215*
Minnesota, 170
Salenia, 376, 406
Saline County, Arkansas, 259
Salt Basin, Texas, 396, 398, 399, 400
salt beds, Kansas, 69
salt domes, Texas, 449
Salt Fork of Arkansas River, Oklahoma, *135*
salt plains, Oklahoma, 135
San Angelo, Texas, 352
San Angelo Formation, Texas, 310, 314
San Antonio Bay, Texas, 452, 453
San Marcos Arch, Texas, 335, 383
San Marcos/Concho Arch, Texas, 373
San Martine Member, Boracho Formation, Texas, 402
San Saba County, Texas, 343, 357, 358
San Saba Member, Wilberns Formation, Texas, 339, 354, 356
San Saba River, Texas, 357
San Vincente Member, Boquillas

Formation, Texas, 425
Sanborn Group, Kansas, 72, 73
Sandy Creek, Texas, 369, 371
Sandy Formation, Texas, 364
Santa Elena Canyon, Texas, *423*, 432
Santa Elena Formation, Texas, 426, 427
Santiago Mountains, Texas, 417, 418, 420, 421, 429
Sausbee Formation, Oklahoma, 85
Savanna Sandstone Formation, Arkansas, 243, 246, 247
Scalarituba, 152
Scenic Drive Formation, Texas, 392
Schizodus sp., 315
Schroyer Limestone, Kansas, 18, 19, 21, 23
Schultztown coal, Kentucky, 77
Schwagerina sp., 37
Sciponoceras gracile, 58, 59
Scolicia, 313
Scott County, Kansas, 63
sea-level changes, 3, 19, 45, 47, 445
Searcy County, Arkansas, 216, 222, 235
Seco Creek, Texas, 444
seismic events, Kansas, 11
Septimyalina, 30
Sequoyah County, Oklahoma, 86
Sevier County, Arkansas, 281
Sevier orogeny, 3
Shafter Formation, Texas, 406
Shimer Gypsum, Oklahoma, 83, 84
Short Mountain, Arkansas, 247
Shovel Mountain, Texas, 375
Sheep Pasture Formation, Texas, 409
Sierra de Juarez, Mexico, 387
Sierra Diablo, Mexico, 424,429
Sierra Diablo Mountains, Texas, 399
Sierra Ponce, Mexico, 424, 429, 431
Sierra Quemada caldera, Texas, 438, 439
Sierra Rica caldera, Texas, 436
Sierrite Formation, Texas, 392
Sigillaria, 253
Signal Mountain Limestone Formation, Oklahoma, 95, 97, 99, 119, 130, 131, 133, 183, 185
Silurian
  Arkansas, 212, 212, 215, 221, 223
  Oklahoma, 200
  Texas, 392
Simpson Creek, Texas, 357
Simpson Group, Oklahoma, 127, 165, 166, 170, 183, 185
Simsboro Member, Rockdale Formation, Texas, 380
Sioux Quartzite, Minnesota, 9
Sisymbrium nasturtium-aquaticum, 155
Sixmile Granite, Texas, 361, 362
Skolithos, 102, 133, 166
Sleeping Lion Formation, Texas, 409
Slick Hills Range, Oklahoma, *93*, 99, 118, 121, 123, 125, *127*
Smithwick Shale, Texas, 343, 344, 348, 357, 359, 360, 375
Smoky Hill Chalk Member, Niobrara Chalk, Kansas, 2
Smoky River, Kansas, 67
Smuggler's Pass Formation, Texas, 392
Sneed Creek, Arkansas, 209
Somervell County, Texas, 325, 426
Sotol Vista, Texas, *429*
South Padre Island, Texas, 445, 457
South Rim Formation, Texas, 431, 432, 435
Southern High Plains, Texas, *293*, *299*, 308
Southern Oklahoma Aulacogen, 121, 127, 129, 155, 157, 159, 169, *171*, 177, 189
Spanish Oak Creek, Texas, 375
Sparta Formation, Texas, 335, 337, 381
Spillway fault system, Kansas, *11*
Spiro Sandstone
  Arkansas, 233
  Oklahoma, 139
Spring Hill Limestone, Kansas, 52, 54
Springer Formation, Oklahoma, *189*
Springfield Plateau, Arkansas, 211, 214, 222
Staendebach Member, Tanyard Formation, Texas, 356
Stanley Group, Oklahoma, 149, 150, 151
Stanley Shale Formation
  Arkansas, 263, 264, *267*, 273
  Oklahoma, *195*, 264
Star Mountain Formation, Texas, 407, 409, 410
Stark Shale
  Iowa, 51
  Kansas, 51, 54
  Oklahoma, 51
Stearns Shale, Kansas, 37, 38
Sterling County, Texas, 305
Stigmaria, 268
Stockton, Kansas, *1*
Stone City Bluff, Texas, 380, 381
Stone County, Arkansas, *215*, 223
Stonewall County, Texas, 314, 315
Strawn Group, Texas, 317, 343, 345, 346, 348, 359, 360
Stumbling Bear shear zone, Oklahoma, 129, 133
Sue Peaks Formation, Texas, 426, 427
Sulphur fault, Oklahoma, 178
Sulphur Springs Hollow, Arkansas, 235
Sunken Block, Texas, 429, 432
Swan Peak Formation, Oklahoma, 170
Swope cyclothem, Kansas, 44, 50, 52
Sycamore Creek anticline, Oklahoma, 179
Sycamore Formation, Texas, 373
Sycamore Limestone, Oklahoma, 154, 183, 184, 187
Sycamore Sandstone, Texas, 402
Sylamore Sandstone Member, Chattanooga Shale, Arkansas, 204, 212, 220
Sylvan Shale, Oklahoma, 127, 183, 185, 187
Synaptomys, 72, 73
Syringopora, 321

T-R units, Kansas, *19*
Taconic Allochthon, Appalachians, 172, 176
Taloga Formation, Kansas, 70
Tansill Formation, Texas, 398
Tanyard Formation, Texas, 354, 355, 356
Taylor Black Prairie, Texas, 333
Taylor Group, Texas, 329, 333, 441
Taylor Marl, Texas, 332, 444
Tecovas Formation, Texas, 305, 307
tectonism
  Kansas, 3
  Oklahoma, 129, 160
  Texas, *407*, *435*, *446*
Tehuacana Member, Kincaid Formation, Texas, 380
Telephone Canyon Formation, Texas, 426, 427
Tenmile Creek Formation
  Arkansas, 267
  Oklahoma, 150, 264, 267
Teredolithus, 62
Terlingua Creek, Texas, 424
Terlingua Fault, Texas, 427, 429
Terlingua Group, Texas, 423
Terlingua Uplift, Texas, 432
Tertiary
  Arkansas, 225
  Oklahoma, 122
  Texas, 307, *377*, *429*, *435*
Tesnus Formation, Texas, 411, 413, 414, 415, 418
Tetradium, 219
Tetrataxis sp., 33
Texas coastal plain, *445*
Texas Lineament, 423, 429, 435
Texigryphaea sp., 67, 327
Thalassia sp., 39, 457
Thalassinoides, 1, 313, 381
Thatcher Creek Member, Cool Creek Formation, Oklahoma, 130, 132, 155
Thorp Springs Member, Glen Rose Formation, Texas, 324, 327
Threadgill Member, Tanyard Formation, Texas, 356
Threemile Limestone, Kansas, 18
Thunderbird Group, Texas, 391
Thurber coal, Texas, 319
Ti Valley thrust fault, Oklahoma, 145, 147, 148, 150
Timber Mountain, Texas, 407
Timbered Hills Group, Oklahoma, 93, 94, 99, 123, 131, 183
Tishmingo Granite, Oklahoma, 178, 179, 183
Tokio Formation, Arkansas, 281
Tom Greene County, Texas, 352
Tom Myers Park Formation, Texas, 391
Toucasia sp., 443
Town Mountain Granite, Texas, 362, 365

Trace Creek Shale Member, Atoka Fm.
Arkansas, 233
Oklahoma, 90
Trans-Pecos Texas, 383, *401*, 407, *423*, *429*, 435
Trap Mountains, Arkansas, 263
Travis County, Texas, 373
Triassic, Texas, *305*
Trichichunus, 1
Trigonia, 376
Trinity Group, Texas, 323, 423, 426
Trinity Shelf, Texas, *323*
Triplesia alata, 222
Triticites, 28
Troy Granite, Oklahoma, 183
Trujillo Formation, Texas, 305, 307
Tubiphytes, 397
Tule Mountain Member, Chisos Formation, Texas, 433
Tulip Creek, Oklahoma, 184
Tulip Creek Formation, Oklahoma, 166, 183, 185, 187
Turner Falls, Oklahoma, *153*
Turritella, 67, 337
Tuttle Creek reservoir, Kansas, *11*
Twin Mountains Formation, Texas, 323, 324
Tyler Basin, Texas, 335
Type-B Ash, Kansas, 70, 72, 73

Upham Formation, Texas, 392
Upper Borchers Ash, Kansas, 73
Upper Glen Rose Member, Glen Rose Formation, Texas, 327
Upper Kanses Till, Kansas, 5, 7, 8, 9
Upson Formation, Texas, 442
uranium district, Texas, 383
Uvalde County, Texas, 441, 442, 443
Uvalde Salient, Texas, 443, 444

Val Verde County, Texas, 425
Valley Spring Gneiss, Texas, 362
Van Buren County, Arkansas, 235
Vanhem Formation, Kansas, 72
Vanoss Formation, Oklahoma, 159
Vascoceras birchbyi, 59
Vaucheria sp., 153
Venericardia, 337
Verdigris Limestone
Iowa, 77
Kansas, 75, 77
Oklahoma, 77
Victorio Hills Formation, Texas, 392
Vilas Shale, Kansas, 54
Village Bend Limestone Member, Mineral Wells Formation, Texas, 320
Vinton Canyon, Texas, 392, 393
Viola Group, Oklahoma, 127, 172, 174, 175, 183, 184, 185, 187
Viola Limestone Formation, Oklahoma, 163, 173, 174, 176, 179, 181
Viola Springs Formation, Oklahoma, 174. *See also* Viola Limestone Formation
Vokesula, 337
volcanic ash, Kansas, 69, 70, 72, 73
volcanic rocks, Oklahoma, 93
volcanism, Texas, *383*, *407*, 429, 431, 435, 443

Waggoner Ranch Formation, Texas, 311
Wallace County, Kansas, 63
Walnut Clay Formation, Texas, 327, 373, 376
Wapanucka Formation, Oklahoma, *139*
Washington County, Arkansas, 85, 88, 89, 90
Washington Group, Texas, 323
Washita Group, Texas, 67, 327, 329, 423
Washita Valley fault, Oklahoma, 179, 181
Wasp Spring Member, South Rim Formation, Texas, 431, 433
Watonga Dolomite, Oklahoma, 84
Webbers Falls, Oklahoma, 85
Weches Formation, Texas, 335
Wedge section, Texas, 354, 355
Weir-Pittsburg coal bed, Kansas, 78
Welge Sandstone Member, Wilberns Formation, Texas, 353, 355
Welling Formation, Oklahoma, 174. *See also* Fernvale Limestone Formation
Wellington Formation, Oklahoma, 106
Wesley Formation, Oklahoma, 150
West Atchison drift sections, Kansas, *5*
West Cache Creek, Oklahoma, 117, 118
West Mountain, Arkansas, 263
West Spring Creek Formation, Oklahoma, 93, 165, 183, 185
West Texas High Plains, 305
Western Interior Sea, 2, 60, 61, 67
Western Lowlands, 227
White Bluff, Arkansas, *265*
White Bluff Formation, Arkansas, 265
White Clay Creek, Kansas, 5
White River, Arkansas, 215
White River, Texas, 299, 300, 301
White Rock Cuesta, Texas, 332
Whitebrest coal bed, Iowa, 77
Whitehorse Group
Oklahoma, 103
Texas, 310, 311, 315
Whitestone Lentil Member, Walnut Formation, Texas, 373, 376
Whitsett Formation, Texas, 385
Wichita Frontal Fault. *See* Frontal Fault Zone, Oklahoma
Wichita Granite Group
Oklahoma, 111, 113, 114, 115
Texas, 309, 310
Wichita Mountains, Oklahoma, *109*, 118 121, 122, 124, 309, 383
Wichita uplift, Oklahoma, 79, 127, 129, 189
Wichita Valley fault system, Oklahoma, 164
Wilberns Formation, Texas, *339*, 353, 354, 355, 356, 363
Wilcox Group
Arkansas, 225, 227, 259
Texas, 377, 380
Wilcox Hills, Texas, 400
Wildhorse Mountain Formation, Oklahoma, 150
wildlife, Texas, 452
Wiles Limestone, Texas, 321
Williams Peak, Oklahoma, 109
Willow Creek Member, Blaine Formation, Texas, 315
Wills Point Formation, Texas, 380
Wilson County, Kansas, 44
Winchell Limestone Formation, Texas, 317, 321, 350
Windingstairs Fault, Oklahoma, 150
Winnipeg Formation, Great Plains, 170
Winterset Limestone
Iowa, 54
Kansas, 52, 54
Nebraska, 54
Oklahoma, 54
Wise County, Texas, 324
Witts Spring Formation, Arkansas, 235, 238
Wolf Mountain Shale, Texas, 317, 321
Wolfe City Escarpment, Texas, 333
Wolfe City Sand, Texas, 333
Womble Shale Formation
Arkansas, 256
Oklahoma, 172, 175, 176, 195, 196, 197, 200, 201
Woodbine Group, Texas, 331
Woodbine Sands, 329, 331
Woodford Shale, Oklahoma, 182, 183, 184, 187
Woods Hollow Formation, Texas, 175, 176, 411, 413, 414
Woolsley Member, Bloyd Shale, Arkansas, 88, 89
Wreford Megacyclothem, 28
Wyandotte cyclothem, Kansas, 50, 54
Wymore Shale, Kansas, 18
Wynne Limestone, Texas, 321

Yates Formation, Texas, 398
Yegua Formation, Texas, 335, 381
Yellowhouse Draw, Texas, 293, 294
Yucca Formation, Texas, 402, 405, 406

Zavala County, Texas, 443
Zeandale Dome, Kansas, 29, 30
Zigzag Mountains, Arkansas, 263
Zodletone Mountain, Oklahoma, 95, *99*

Typeset by WESType Publishing Services, Inc., Boulder, Colorado
Printed in U.S.A. by Malloy Lithographing, Inc., Ann Arbor, Michigan

Hawaiian Islands
130°
125°
120°
115°
110°
105°
100°